SOME FUNDAMENTAL PHYSICAL CONSTANTS

Constant	Symbol	Value
speed of light in a vacuum	c	3.00×10^8 m/s
gravitational constant	G	6.67×10^{-11} m³/kg·s²
Avogadro's number	N_A	6.02×10^{23} mol⁻¹
universal gas constant	R	8.31 J/mol·K
Boltzmann's constant	k	1.38×10^{-23} J/K
elementary charge	e	1.60×10^{-19} C
permittivity of free space	ϵ_0	8.85×10^{-12} C²/N·m²
	$1/(4\pi\epsilon_0)$	8.99×10^9 kg·m³·s⁻²·C⁻²
permeability of free space	μ_0	$4\pi \times 10^{-7}$ T·m/A
electron mass	m_e	9.11×10^{-31} kg
proton mass	m_p	1.67×10^{-27} kg
neutron mass	m_n	1.67×10^{-27} kg
Planck's constant	h	6.63×10^{-34} J·s
$h/2\pi$	$\hbar$	1.05×10^{-34} J·s
		$= 6.58 \times 10^{-22}$ MeV·s
	$\hbar c$	197 MeV·fm
electron charge-to-mass ratio	$-e/m_e$	-1.76×10^{11} C/kg
proton-electron mass ratio	m_p/m_e	1840
molar volume of ideal gas at STP		22400 cm³/mol
Bohr magneton	μ_B	9.27×10^{-24} J/T
magnetic flux quantum	$\Phi_0 = h/2e$	2.07×10^{-15} Wb
Bohr radius	a_0	0.529×10^{-10} m
Rydberg constant	R_∞	1.10×10^7 m⁻¹

SOME ASTRONOMICAL CONSTANTS

Constant	Symbol	Value
standard gravity at Earth's surface	g	9.80665 m/s²
equatorial radius of Earth	R_e	6.374×10^6 m
mass of Earth	M_e	5.976×10^{24} kg
mass of Moon		7.350×10^{22} kg $= 0.0123\, M_e$
mean radius of Moon's orbit around Earth		3.844×10^8 m
mass of Sun	$M_\odot$	1.989×10^{30} kg
mean radius of Earth's orbit around Sun	AU	1.496×10^{11} m
period of Earth's orbit around Sun	yr	3.156×10^7 s
diameter of our galaxy		7.5×10^{20} m
mass of our galaxy		2.7×10^{41} kg $= (1.4 \times 10^{11})\, M_\odot$
Hubble parameter	H	2.1×10^{-18} s⁻¹

SOME SI BASE UNITS

Physical Quantity	Name of Unit	Symbol
length	meter	m
mass	kilogram	kg
time	second	s
electric current	ampere	A
thermodynamic temperature	kelvin	K
amount of substance	mole	mol

SOME SI DERIVED UNITS

Physical Quantity	Name of Unit	Symbol	SI Unit
frequency	hertz	Hz	s^{-1}
energy	joule	J	$kg \cdot m^2/s^2$
force	newton	N	$kg \cdot m/s^2$
pressure	pascal	Pa	$kg/m \cdot s^2$
power	watt	W	$kg \cdot m^2/s^3$
electric charge	coulomb	C	$A \cdot s$
electric potential	volt	V	$kg \cdot m^2/A \cdot s^3$
electric resistance	ohm	Ω	$kg \cdot m^2/A^2 \cdot s^3$
capacitance	farad	F	$A^2 \cdot s^4/kg \cdot m^2$
inductance	henry	H	$kg \cdot m^2/A^2 \cdot s^2$
magnetic flux	weber	Wb	$kg \cdot m^2/A \cdot s^2$
magnetic flux density	tesla	T	$kg/A \cdot s^2$

SI UNITS OF SOME OTHER PHYSICAL QUANTITIES

Physical Quantity	SI Unit
speed	m/s
acceleration	m/s^2
angular speed	rad/s
angular acceleration	rad/s^2
torque	$kg \cdot m^2/s^2$, or $N \cdot m$
heat flow	J, or $kg \cdot m^2/s^2$, or $N \cdot m$
entropy	J/K, or $kg \cdot m^2/K \cdot s^2$, or $N \cdot m/K$
thermal conductivity	$W/m \cdot K$

SOME CONVERSIONS OF NON-SI UNITS TO SI UNITS

Energy:

1 electron-volt (eV) = 1.6022×10^{-19} J
1 erg = 10^{-7} J
1 British thermal unit (BTU) = 1055 J
1 calorie (cal) = 4.186 J
1 kilowatt-hour (kWh) = 3.6×10^6 J

Mass:

1 gram (g) = 10^{-3} kg
1 atomic mass unit (u) = 931.5 MeV/c^2
$\phantom{1 \text{atomic mass unit (u)}} = 1.661 \times 10^{-27}$ kg
1 MeV/c^2 = 1.783×10^{-30} kg

Force:

1 dyne = 10^{-5} N
1 pound (lb or #) = 4.448 N

Length:

1 centimeter (cm) = 10^{-2} m
1 kilometer (km) = 10^3 m
1 fermi = 10^{-15} m
1 Angstrom (Å) = 10^{-10} m
1 inch (in or ") = 0.0254 m
1 foot (ft) = 0.3048 m
1 mile (mi) = 1609.3 m
1 astronomical unit (AU) = 1.496×10^{11} m
1 light-year (ly) = 9.46×10^{15} m
1 parsec (ps) = 3.09×10^{16} m

Angle:

1 degree (°) = 1.745×10^{-2} rad
1 min (') = 2.909×10^{-4} rad
1 second (") = 4.848×10^{-6} rad

Volume:

1 liter (L) = 10^{-3} m^3

Power:

1 kilowatt (kW) = 10^3 W
1 horsepower (hp) = 745.7 W

Pressure:

1 bar = 10^5 Pa
1 atmosphere (atm) = 1.013×10^5 Pa
1 pound per square inch (lb/in^2) = 6.895×10^3 Pa

Time:

1 year (yr) = 3.156×10^7 s
1 day (d) = 8.640×10^4 s
1 hour (h) = 3600 s
1 minute (min) = 60 s

Speed:

1 mile per hour (mi/h) = 0.447 m/s

Magnetic field:

1 gauss = 10^{-4} T

Volume II

PHYSICS

FOR SCIENTISTS AND ENGINEERS

PHYSICS
Volume II
FOR SCIENTISTS AND ENGINEERS
THIRD EDITION

Paul M. Fishbane
University of Virginia

Stephen G. Gasiorowicz
University of Minnesota

Stephen T. Thornton
University of Virginia

Upper Saddle River, New Jersey 07458

Library of Congress Cataloging-in-Publication Data
Fishbane, Paul M.
    Physics for scientists and engineers / Paul M. Fishbane, Stephen G. Gasiorowicz,
Stephen T. Thornton.—3rd ed.
      p. cm.
    Includes indexes.
    ISBN 0-13-142094-1—ISBN 0-13-141883-1
    ISBN 0-13-141881-5—ISBN 0-13-141882-3—ISBN 0-13-035299-3
    1. Physics. I. Gasiorowicz, Stephen. II. Thornton, Stephen T. III. Title.

QC23.2.F58   2005
530—dc22

                        2003058209

Senior Editor: Erik Fahlgren
Editor in Chief, Science: John Challice
Development Editor: Catherine Flack
Executive Project Manager: Ann Heath
Editor in Chief, Development: Carol Trueheart
Executive Managing Editor: Kathleen Schiaparelli
Assistant Managing Editor: Beth Sweeten
Vice President ESM Production and Manufacturing: David W. Riccardi
Manufacturing Manager: Trudy Pisciotti
Manufacturing Buyer: Alan Fischer
Creative Director: Carole Anson
Art Director: John Christiana
Director of Creative Services: Paul Belfanti
Managing Editor of AV Management and Production: Patty Burns
AV Art Editor: Abigail Bass
Artwork: Imagineering
Media Editor: Paul Draper
Assistant Managing Editor, Science Media: Nicole Bush
Assistant Managing Editor, Science Supplements: Becca Richter
Executive Marketing Manager: Mark Pfaltzgraff
Associate Editor: Christian Botting
Editorial Assistant: Andrew Sobel
Interior Design: Dina Curro
Cover Design: Kiwi Design
Cover Photo: Antonio Mo/Getty Images, Inc. -Taxi
Photo Research: Yvonne Gerin
Photo Editor: Nancy Seise
Production Services/Composition: Preparé Inc.

 © 2005, 1996, 1993 Pearson Education Inc.
Pearson Prentice Hall
Pearson Education, Inc.
Upper Saddle River, New Jersey 07458

All rights reserved. No part of this book may be
reproduced, in any form or by any means,
without permission in writing from the publisher.

Pearson Prentice Hall ® is a trademark of Pearson Education, Inc.

Printed in the United States of America
10 9 8 7 6 5 4 3 2 1

ISBN 0-13-141881-5

Pearson Education Ltd., *London*
Pearson Education Australia Pty. *Limited, Sydney*
Pearson Education *Singapore Pte. Ltd.*
Pearson Education North Asia Ltd., *Hong Kong*
Pearson Education Canada, Ltd., *Toronto*
Pearson Educación de Mexico, *S.A. de C.V*
Pearson Education—*Japan, Tokyo*
Pearson Education *Malaysia, Pte. Ltd.*

Dedication

To our students, the most important element in the making of this book

Brief Contents

1 Tooling Up 1
2 Straight-Line Motion 28
3 Motion in Two and Three Dimensions 60
4 Newton's Laws 87
5 Applications of Newton's Laws 119
6 Work and Kinetic Energy 151
7 Potential Energy and Conservation of Energy 183
8 Linear Momentum, Collisions, and the Center of Mass 209
9 Rotations of Rigid Bodies 246
10 More on Angular Momentum and Torque 280
11 Statics 313
12 Gravitation 338
13 Oscillatory Motion 366
14 Waves 397
15 Superposition and Interference of Waves 435
16 Properties of Fluids 462
17 Temperature and Ideal Gases 491
18 Heat Flow and the First Law of Thermodynamics 515
19 The Molecular Basis of Thermal Physics 548
20 The Second Law of Thermodynamics ... 573
21 Electric Charge 609
22 Electric Field 633
23 Gauss' Law 661
24 Electric Potential 683
25 Capacitors and Dielectrics 714
26 Currents in Materials 738

27 Direct-Current Circuits 766
28 The Effects of Magnetic Fields 791
29 The Production and Properties of Magnetic Fields 820
30 Faraday's Law 847
31 Magnetism and Matter 873
32 Inductance and Circuit Oscillations 893
33 Alternating Currents 917
34 Maxwell's Equations and Electromagnetic Waves 943
35 Light 974
36 Mirrors and Lenses and Their Uses 998
37 Interference 1029
38 Diffraction 1050
39 Special Relativity
40 Quantum Physics
41 Atomic and Molecular Structure
42 Quantum Effects in Large Systems of Fermions and Bosons
43 Quantum Engineering
44 Nuclear Physics
45 Particles and Cosmology

Appendices A-1
Answers to "What Do You Think?" Questions Q-1
Answers to Odd-Numbered Understanding the Concepts Questions C-1
Answers to Odd-Numbered Problems O-1
Photo Credits P-1
Index I-1

Contents

	PREFACE	xvii

21 ELECTRIC CHARGE 609

21-1 Charge—a Property of Matter 609
21-2 Charge Is Conserved and Quantized 616
21-3 Coulomb's Law 617
21-4 Forces Involving Multiple Charges 621
Summary, 626   Understanding the Concepts, 627
Problems, 628

22 ELECTRIC FIELD 633

22-1 Electric Field 633
22-2 Electric Field Lines 639
22-3 The Field of a Continuous Distribution 642
22-4 Motion of a Charge in a Field 648
22-5 The Electric Dipole in an External Electric Field 650
Summary, 653   Understanding the Concepts, 654
Problems, 655

23 GAUSS' LAW 661

23-1 What Does Gauss' Law Do? 661
23-2 Gauss' Law 665
23-3 Using Gauss' Law to Determine Electric Fields 668
23-4 Conductors and Electric Fields 672
23-5 Are Gauss' and Coulomb's Laws Correct? 675
Summary, 677   Understanding the Concepts, 678
Problems, 678

24 ELECTRIC POTENTIAL 683

24-1 Electric Potential Energy 684
24-2 Electric Potential 686
24-3 Equipotentials 691
24-4 Determining Fields from Potentials 694
24-5 The Potentials of Charge Distributions 696
24-6 Potentials and Fields Near Conductors 700
24-7 Electric Potentials in Technology 703
Summary, 707   Understanding the Concepts, 708
Problems, 708

x | Contents

25 CAPACITORS AND DIELECTRICS 714

25-1 Capacitance 714
25-2 Energy in Capacitors 718
25-3 Energy in Electric Fields 720
25-4 Capacitors in Parallel and in Series 721
25-5 Dielectrics 723
25-6 The Microscopic Description of Dielectrics 729

Summary, 731   Understanding the Concepts, 732
Problems, 733

26 CURRENTS IN MATERIALS 738

26-1 Electric Current 739
26-2 Currents in Materials 742
26-3 Resistance 745
26-4 Resistances in Series and in Parallel 749
*26-5 Free-Electron Model of Resistivity 750
*26-6 Materials and Conductivity 752
26-7 Electric Power 755

Summary, 758   Understanding the Concepts, 759
Problems, 760

27 DIRECT-CURRENT CIRCUITS 766

27-1 EMF 766
27-2 Kirchhoff's Loop Rule 770
27-3 Kirchhoff's Junction Rule 772

27-4 Measuring Instruments 776
27-5 *RC* Circuits 779

Summary, 783   Understanding the Concepts, 783
Problems, 784

28 THE EFFECTS OF MAGNETIC FIELDS 791

28-1 Magnets and Magnetic Fields 792
28-2 Magnetic Force on an Electric Charge 793
28-3 Consequences of the Magnetic Force on a Charge 796
28-4 Magnetic Forces on Currents 804
28-5 Magnetic Force on Current Loops 807
28-6 The Hall Effect 810

Summary, 811   Understanding the Concepts, 812
Problems, 813

29 THE PRODUCTION AND PROPERTIES OF MAGNETIC FIELDS 820

29-1 Ampère's Law 820
29-2 Gauss' Law for Magnetism 826
29-3 Solenoids 828
29-4 The Biot–Savart Law 832
29-5 The Maxwell Displacement Current 838

Summary, 840   Understanding the Concepts, 841
Problems, 842

30 FARADAY'S LAW 847

- 30-1 Faraday's Discovery and the Law of Induction 848
- 30-2 Motional EMF 855
- 30-3 Forces and Energy in Motional EMF 858
- 30-4 Time-Varying Magnetic Fields 861
- 30-5 Generators 863
- *30-6 The Frame Dependence of Fields 864

Summary, 865   Understanding the Concepts, 865
Problems, 866

31 MAGNETISM AND MATTER 873

- 31-1 The Magnetic Properties of Bulk Matter 874
- 31-2 Atoms as Magnets 878
- 31-3 Ferromagnetism 881
- *31-4 Diamagnetism and Paramagnetism 884
- *31-5 Magnetism and Superconductivity 886
- *31-6 Nuclear Magnetic Resonance 886

Summary, 888   Understanding the Concepts, 889
Problems, 889

32 INDUCTANCE AND CIRCUIT OSCILLATIONS 893

- 32-1 Inductance and Inductors 893
- 32-2 Energy in Inductors 898
- 32-3 Energy in Magnetic Fields 900
- 32-4 Time Dependence in RL Circuits 901
- 32-5 Oscillations in LC Circuits 903
- 32-6 Damped Oscillations in RLC Circuits 905
- 32-7 Energy in LC and RLC Circuits 908

Summary, 910   Understanding the Concepts, 911
Problems, 911

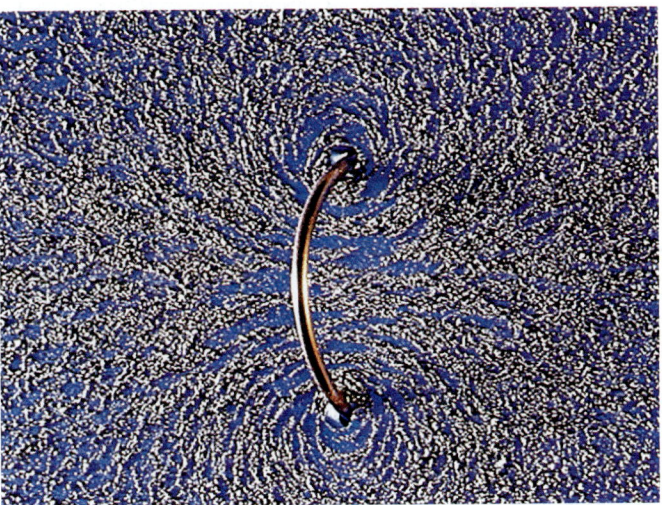

33 ALTERNATING CURRENTS 917

- 33-1 Transformers 917
- 33-2 Single Elements in AC Circuits 920
- 33-3 AC in Series RLC Circuits 926
- 33-4 Power in AC Circuits 929
- 33-5 Some Applications 931

Summary, 935   Understanding the Concepts, 936
Problems, 937

34 MAXWELL'S EQUATIONS AND ELECTROMAGNETIC WAVES 943

- 34-1 Maxwell's Equations 944
- 34-2 Electromagnetic Waves 945
- 34-3 Energy and Momentum Flow 954
- 34-4 Dipole Radiation 957
- 34-5 Polarization 959
- *34-6 Electromagnetic Radiation as Particles 964

Summary, 965   Understanding the Concepts, 968
Problems, 968

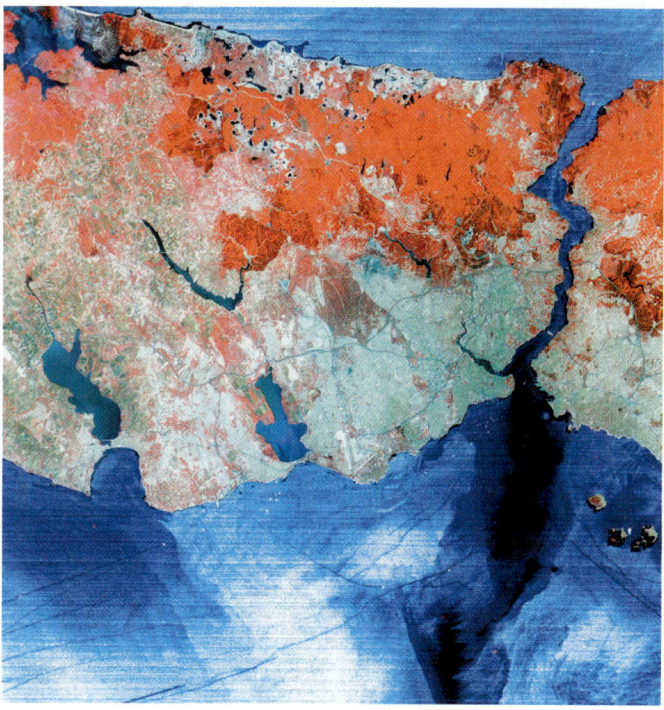

35 LIGHT 974

35-1 The Speed of Light 975
35-2 When Can Light Waves Be Treated as Rays? 977
35-3 Reflection and Refraction 979
*35-4 Fermat's Principle 985
35-5 Dispersion 988

Summary, 991   Understanding the Concepts, 992   Problems, 993

36 MIRRORS AND LENSES AND THEIR USES 998

36-1 Images and Mirrors 998
36-2 Spherical Mirrors 1001
36-3 Refraction at Spherical Surfaces 1009
36-4 Thin Lenses 1014
36-5 Optical Instruments 1018
*36-6 Aberration 1022

Summary, 1023   Understanding the Concepts, 1024   Problems, 1025

37 INTERFERENCE 1029

37-1 Young's Double-Slit Experiment 1029
37-2 Intensity in the Double-Slit Experiment 1034
37-3 Interference from Reflection 1037
*37-4 Interferometers 1042

Summary, 1044   Understanding the Concepts, 1044   Problems, 1045

38 DIFFRACTION 1050

38-1 The Diffraction of Light 1050
38-2 Diffraction Gratings 1052
38-3 Single-Slit Diffraction 1056
38-4 Resolution of Optical Instruments 1059
38-5 Slit Width and Grating Patterns 1062
*38-6 X-Ray Diffraction 1062
*38-7 Holography 1065

Summary, 1067   Understanding the Concepts, 1068   Problems, 1069

APPENDIX I
THE SYSTÈME INTERNATIONALE (SI) OF UNITS A-1

APPENDIX II
SOME FUNDAMENTAL PHYSICAL CONSTANTS A-2

APPENDIX III
OTHER PHYSICAL QUANTITIES A-3

APPENDIX IV
MATHEMATICS A-5

APPENDIX V
PERIODIC TABLE OF THE ELEMENTS A-8

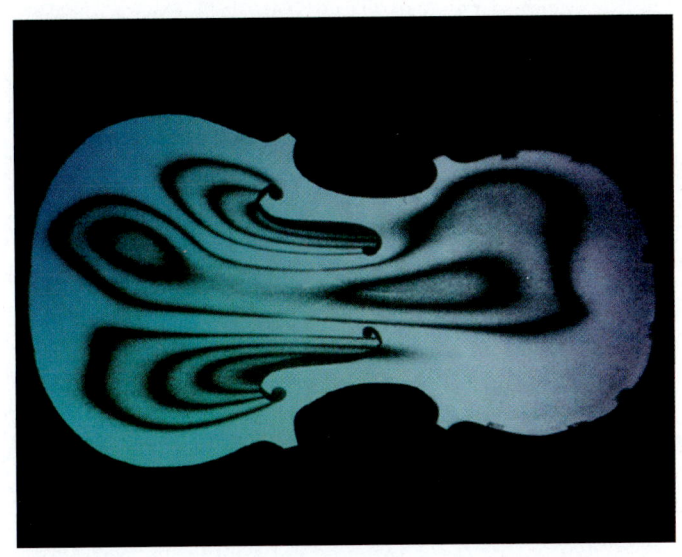

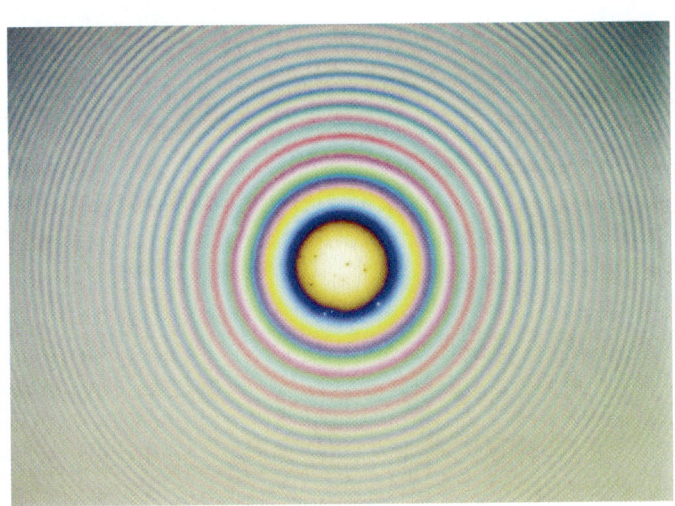

APPENDIX VI
SIGNIFICANT DATES IN THE DEVELOPMENT OF PHYSICS A-9

ANSWERS TO "WHAT DO YOU THINK?" QUESTIONS Q-1

ANSWERS TO ODD-NUMBERED UNDERSTANDING THE CONCEPTS QUESTIONS C-1

ANSWERS TO ODD-NUMBERED PROBLEMS O-1

PHOTO CREDITS P-1

INDEX I-1

Applications in the Text

Note: This list includes applied topics that receive significant discussion in the chapter text, Think About This sections, Conceptual Examples and worked Examples. Many other applied topics are introduced briefly or appear in the end-of-chapter Understanding the Concepts Questions and Problems. Topics that appear in a Think About This section appear in blue.

Chapter 21
If we accumulate a charge, where does it eventually go? 612
How many electrons are contained in one pure gold coin? 615
How do smoke detectors work? 616
Do quarks carry charge that is less than the electron charge? 617
Are there coulomb forces between home appliances? 620

Chapter 22
How does the field concept help? 636
When do electric dipoles appear in nature? 639
What is the total electric charge on the Earth? 647
How do we know that charges come in minimum size units? 648

Chapter 23
Could we use Gauss' law to find the field of a finite line of charge? 670
How can we shield a region from all electric fields? 676

Chapter 24
In what realm of physics is the electron-volt a natural unit? 690
How do lightning rods work? 703
Xerography 705

Chapter 25
Coaxial cable 717
Calculating Earth's capacitance 717
How does a stud finder work? 725
How large can the voltage across a capacitor be? 726
How are capacitors with very large capacitances constructed? 729

Chapter 26
Why does the light appear the moment you throw the light switch? 744
Can nonohmic material be useful in circuits? 746
Nichrome wire in a toaster 756

Chapter 27
Can capacitors replace batteries in laptop computers? 781

Chapter 28
Detecting charged particles in a bubble chamber 798
How does Earth's magnetic field create the northern lights? 803
How do electric motors work? 809
Does the Hall effect have technological use? 811

Chapter 29
Can you measure a current without entering the circuit? 824
What causes Earth's magnetic field? 832

Chapter 30
How does magnetic levitation work for superconductors? 860
The alternating-current generator 863

Chapter 31
Why aren't most materials magnetic? 881
What is magnetic resonance imaging? 888

Chapter 32
The chiming doorbell 898
How would you measure inductance? 903
Are there technological applications for current oscillations? 905

Chapter 33
Why is electric power transmitted along high-voltage lines? 920
Receiving FM radio station broadcasts 931
How does a car produce the electricity it needs? 932

Chapter 34
How do radio waves travel around the globe? 953
Solar energy on Earth's surface 955
The intensity of a lightbulb 958
Light and polarization in a transparency projector 962
When photons get reflected by a mirror, do they drop in frequency? 965

Chapter 35
What have we learned from corner reflectors on the Moon? 980

"Stealth" fishing 984
How does light propagate within optical fibers? 984
Why is the sky blue while clouds are white? 990

Chapter 36

Are ray-tracing techniques really useful? 1008
How are modern telescopes constructed? 1022

Chapter 37

Improving your radio's signal 1036

How would you check the flatness of a glass surface? 1040
Why do some eyeglasses reflect light while others don't? 1042

Chapter 38

What is the significance of diffraction gratings? 1052
Why do CDs glimmer with bright colors when tilted? 1056
Assessing the Hubble Space telescope 1061
What are the uses of holography? 1067

Preface

This text is designed for a calculus-based physics course at the beginning university and college level. It is written with the expectation that students either have taken or are currently taking a beginning course in calculus. Students taking a physics course based on this book should leave with a solid conceptual understanding of the fundamental physical laws, how these laws can be applied to solve many problems, and how physics is relevant to the world around them.

"Understanding" encompasses our three overriding goals for this book:

Doing Things Correctly. First, we want a book that is fundamentally *right*. This is sometimes taken to mean "rigorous," but we want to emphasize that we do not feel "rigorous" is a synonym for "difficult." Rather, we associate "right" with showing all the evidence, with using the evidence correctly to support the point being made, and with making the point in a way that will allow the student to say, "I see where that comes from." We try to avoid the phrase "It can be shown that . . . " and the attitude that "It is true because we say it is."

Conceptual Emphasis. Second, we want students to understand the material at as deep a conceptual level as possible. We are aware that there is a large gap between being able to get the right answer to a physics problem and comprehending the physical concepts that lie behind it. We want students to be able to answer the *why* as well as the *how*. The student who has a conceptual understanding can not only do problems for which he or she has models, but can also approach a new problem with confidence. Many of the changes to the third edition are designed to address this issue, including the addition of Conceptual Examples, a "What Do You Think" step at the end of most Examples, and a substantial increase in the number of end-of-chapter qualitative questions.

Modern Physics Integrated. Third, without sacrificing the essential aspects of classical physics, we have included modern notions throughout the book. Classical topics have lost none of their importance and must form the basis of any first course. However, what is traditionally called "modern" physics—the topics centered on relativity and quantum physics—began about a century ago. It hardly seems possible to ignore these topics in view of their importance for today's technology and, more critically, for understanding today's world. Many of the ideas of modern physics are not mathematically difficult. However, they can be nonintuitive, and we think it is important that students begin to develop intuition about this material as early as possible. Although much of this material appears in optional sections (marked with an asterisk), in many cases it is intertwined with the classical material. The uncertainty principle and its role in both classical and quantum physics, information on atomic structure and spectra, information on band structure or on blackbody radiation, and the nature and role of fundamental forces are a few of the topics that are included in this way. We conclude the text in what has become the traditional way, with chapters on modern physics. We think the preparation we laid down for this material in earlier chapters will make it more easily assimilated.

A few words about mathematics: The idea of getting it *right* applies to mathematical derivation as well as to qualitative aspects. Our approach is to introduce the mathematics that students need to know the first time they need to know it, in the context of the physics being presented. We try to make that material self-contained so that the student can understand the material without having to go elsewhere for mathematical help. In this way, the mathematics appears in progressive degrees of difficulty. We believe that this approach fosters better understanding and less reliance on formula memorization. We also feel that the ability to make quantitative estimates is one of the most important skills that a scientist or engineer can have. We have made the development of that ability an important part of our approach, both in the text and in the problems. Finally, as in real-world problems, we vary the number of significant figures in examples and in problems. We feel students should maintain an awareness of significant figures and not end up thinking all problems involve the same number of significant figures.

The Third Edition

With the help of reviewers and users of the second edition, we have made a thorough review of content and organization, with some material moved both within and among chapters to enhance the logical progression and structure of the material. We have rewritten much of the material with clarity in mind.

Organizational Changes.

Changes to the third edition include:

- Redistribution of the material in the 2nd-edition chapter "Properties of Solids"—for example, the material on heat conduction in solids now appears in the appropriate chapter on thermal physics, while material on stress and strain now appears in the chapter on statics.

- Redistribution of the material on waves between Chapters 14 and 15 to create a more logical division between single waves, and the superposition of several.

- Consolidation of some material that we feel does not affect the basic understanding of the subject—for example, both the "physical optics" and "magnetism in matter" chapters are more compact.

Conceptual Examples. We have added a new type of Example, designed to help the student think about the material in a way that emphasizes conceptual understanding of the content. These may have some modest algebraic content, such as a simple estimate or reasoning involving inequalities. There are two or more of these per chapter.

CONCEPTUAL EXAMPLE 17–4 We learn by experience that we can loosen a metal lid that is stuck on a glass jar by pouring hot water over the lid. Why does this work?

Answer As the lid's temperature rises, it expands. Movement occurs where the glass and metal are stuck and the lid releases (Fig. 17–11). In fact, you could dip the entire system (lid and glass container) in hot water, and the different expansions of the metal and the glass will lead to the same result. Note that the same difference in the thermal expansion is used in making a good seal: lids are placed on jars when the contents are hot.

▶ **FIGURE 17–11** By pouring hot water over this lid, you can take advantage of thermal expansion to free it from the jar.

Think About This. The primary purpose of these sections is to pose and answer questions about a new idea or the application of the material discussed. When writing these sections, we ask the kinds of questions a good student might be asking on his or her own, and which the majority of students will find intriguing.

Worked Examples. We have introduced a new structure into the Worked Examples to serve as a model and to build problem-solving skills. The goal of this new structure is to emphasize visualization and the identification of the knowns, unknowns, and concepts to be used. We avoid using a stiff and uncompromising framework, but most examples are broken down into a series of steps:

- *Setting It Up.* In most cases this step begins with a sketch and, for mechanics problems, the preparation of a free-body diagram. The figures accompanying most Examples are in a student sketch style that the student can realistically be expected to emulate. To reinforce this first step, all the figures in the end-of-chapter Problems are also drawn in this style. Because a sketch represents a first step in problem solving, the sketches in the end-of-chapter problems sometimes provide a crucial hint to the student. This step also includes some reasoning on how to determine what is being asked when it may not be completely obvious.

- *Strategy.* Here the concepts used to solve the problem are outlined and applied to the situation; it is where we "talk it through." This is the heart of the solution, the part where most students will succeed or fail.

- *Working It Out.* The strategy outlined in the preceding part is carried through in a series of well-defined steps. This part should be straightforward if the previous steps have been done properly.

- *What Do You Think?* The final step in solving a problem is to confirm that the answer makes sense. This section of the solution reinforces the example by asking the student a thought-provoking conceptual question associated with the problem just solved. These questions should confirm that the student has understood the concepts or send them looking for checks to the answers. Answers to "What Do You Think?" questions are provided at the end of the book.

Vectors. These are now represented with an arrow over the letter rather than in boldface (for example, $\vec{F} = m\vec{a}$) to be more consistent with the way professors write them in lecture and the way students write them in homework and exams.

Questions. The end-of-chapter material includes qualitative questions under the heading "Understanding the Concepts." We have increased the number of these conceptual questions by nearly 50 percent.

THINK ABOUT THIS...
HOW DOES AN AIR BAG PROTECT YOU IN A CRASH?

Large forces imply large accelerations. A car accident or a fall from a great height may be deadly because of the rapid deceleration, the result of large forces that your body may not be equipped to withstand. For protection it is necessary to find a way to bring you to a stop by providing a smaller deceleration over a larger time. Air bags in automobiles work on this principle; when a collision stops a car very suddenly, a passenger would suffer a very sudden deceleration in a subsequent collision with the steering wheel or the windshield. This is mitigated by the very rapid release of an air bag, which is deep enough and "soft" enough to allow the passenger to slow down over a longer period of time. Firefighters similarly use large elastic safety nets to catch people who have to jump from burning buildings. When the deceleration is for fun, the same principle applies. Bungee cords are made of a very elastic material, and there are no bungee *chains*, which would have the unfortunate effect of stopping you "on a dime." Still another application is provided by airplane ejection seats, which in the past were powered by explosives beneath the seat. The rapid acceleration of these mechanisms often led to serious damage to the pilot. Today ejection seats are powered by small rockets that can supply a smaller acceleration over a longer period of time, rather than a large acceleration over a very short period of time and hence a safer ejection (Fig. 4–12). One other example comes to mind: You may have seen drawings in which Superman catches Lois Lane just before she hits the ground. That very action would imply a rapid deceleration that would be just as bad for Lois as hitting the ground. Superman would do better not to wait for the last instant and instead slow Lois down over a longer period of time.

▲ FIGURE 4–12 Test ejection of a pilot from an AMX jet fighter. The jet plumes below the seat are due to the ejection rocket.

EXAMPLE 4–3 You need to deliver a box of bowling balls to a bowling alley. The balls will be placed in a box that is initially at rest but that you want to push into the bowling alley. The box itself has a mass that is very small compared to even one bowling ball. You start with one ball in the box, exert a given force of given strength upon the box for a time period Δt, and at the end of that time the box moves at a speed of 3.2 m/s. You then repeat the procedure with more bowling balls in the box; you exert the same amount of force on the box for the same period of time (Δt) and find the box to have a final speed of 0.4 m/s. How many balls are in the box now?

Setting It Up The two cases are shown in Figs. 4–11a and b at the particular time, after an interval Δt, when the speeds are v_1 and v_2, respectively. You know that an identical force of constant magnitude F acts on two different masses m_1 and m_2 for identical time periods Δt, where $m_2 = nm_1$. Here m_1 is the mass of one bowling ball and n is the number of balls in the box, which is the quantity we want to find. The resulting speeds after time Δt are v_1 and v_2, respectively, and are given.

Strategy In the two cases described the box containing the bowling ball(s) is subject to the same force. Moreover, we can ignore the mass of the box. Using Newton's second law, we can find the accelerations a_1 and a_2 during the period Δt when the force operates. These accelerations are

$$a_1 = \frac{F}{m_1} \quad \text{and} \quad a_2 = \frac{F}{m_2} = \frac{F}{nm_1}.$$

Although we do not know the numerical values of the two accelerations, we do know the speeds v_1 and v_2 after a fixed period of acceleration. Further, we learned in Chapter 2 that an object that starts at rest and undergoes a fixed acceleration $\vec{a}$ for a given period of time Δt has the velocity $\Delta \vec{v} = \vec{v} = \vec{a} \, \Delta t$. In our one-dimensional case, then, we have

$$v_1 = a_1 \, \Delta t = \frac{F \, \Delta t}{m_1} \quad \text{and} \quad v_2 = a_2 \, \Delta t = \frac{F \, \Delta t}{nm_1}.$$

We now have enough information to solve for the unknown, n.

Working It Out We can solve for the ratio F/m in terms of v_1 and Δt and substitute it into the equation for v_2, which we can then solve for n. Alternatively, we can simply take the ratio of the two speeds:

$$\frac{v_1}{v_2} = \frac{(F \, \Delta t/m_1)}{(F \, \Delta t/nm_1)} = n.$$

Numerical substitution gives $n = (3.2 \text{ m/s})/(0.4 \text{ m/s}) = 8$ bowling balls.

What Do You Think? Suppose the mass of the box is not negligible. What would be the effect?

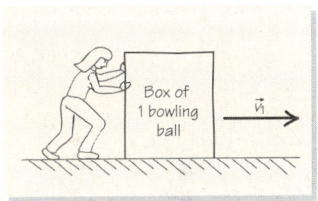

(a)

(b)

▲ FIGURE 4–11 Delivering bowling balls to an alley. In (a) the box has one ball, and in (b) there is an unknown number of balls.

Problems. Approximately 40 percent of the end-of-chapter Problems are new or revised. We have separated the Problems into two categories. The first group of Problems is organized by section. The second group, called General Problems, resembles the situations that are met in real-life science and engineering. They also help to develop the student's appreciation of the links that exist throughout physics as well as how to approach problems for which the clues may be more obscure. All Problems are labeled I, II, or III. Level I Problems are the least difficult. These Problems develop student recognition of particular physics concepts and build confidence. Level II Problems typically have more than one step and require an increased understanding of the material; the General Problems carry this requirement a step further. Level III Problems are especially challenging, in some cases demanding significant synthesis of concepts in the text. The gradations in problem range and difficulty allow you to tailor the Problems you assign to the capabilities of your class and to the subjects that interest you the most.

Versions of the Text

The third edition is available in two hardback versions, Extended with Modern Physics (Chapters 1–45) and Standard (Chapters 1–40), as well as in three softcover volumes: Volume I (Chapters 1–20), Volume II (Chapters 21–38), and Volume III (Chapters 39–45).

Supplements

Instructor Supplements

Instructor's Resource Center on CD-ROM (0-13-039150-6) This CD-ROM set includes virtually every electronic asset you'll need in and out of the classroom. Though you can navigate freely through the CDs to find the resources you want, the included software allows you to browse and search through the catalog of assets. The CD-ROMs are organized by chapter and include all text illustrations and tables from *Physics for Scientists and Engineers, Third Edition*, in JPEG, Microsoft PowerPoint, and Adobe PDF formats. Instructors can preview, sequence, and play back images, as well as perform keyword searches, add lecture notes, and incorporate their own digital resources. The IRC/CDs also contain TestGen, a powerful dual-platform, fully networkable software program for creating tests ranging from short quizzes to long exams. Questions from the third edition *Test Item File*, including algorithmic versions, are supplied, and professors can use the Question Editor to modify existing questions or create new questions. The CD-ROMs also contain additional Powerpoint Presentations, electronic versions of the *Test Item File*, the *Instructor's Solutions Manual*, the *Instructor's Resource Manual*, and the end-of-chapter Understanding the Concepts questions and problems from *Physics for Scientists and Engineers, Third Edition*.

Instructor's Solutions Manual (Vol. I: 0-13-039157-3; Vol. II: 0-13-144741-6) Authored by *Jerry Shi (Pasadena City College)*, the ISM contains detailed, worked solutions to every problem from the text, as well as answers to the "Understanding the Concepts" questions.

Instructor's Resource Manual (0-13-141738-X) By *Prabha Ramakrishnan (North Carolina State University)*. This IRM contains lecture outlines, notes, demonstration suggestions, and other teaching resources.

Test Item File (0-13-039158-1) This test bank contains approximately 3000 multiple choice, short answer, and true/false questions, many conceptual in nature. All are referenced to the corresponding text section and ranked by level of difficulty.

Transparency Pack (0-13-039166-2) Includes approximately 400 full-color transparencies of images from the text.

PH GradeAssist Instructor's Quick Start Guide (0-13-141740-1) This guide will help adopting instructors register for and use PH GradeAssist, Prentice Hall's own online homework system. It also contains the access code necessary to create their accounts and access the course material.

Peer Instruction: A User's Manual (0-13-565441-6)
Eric Mazur (Harvard University)
A manual with ready-to-use resources for an innovative new approach to teaching introductory physics, developed by a well-known physicist and leader in physics education.

Just-in-Time Teaching: Blending Active Learning with Web Technology (0-13-085034-9)
Gregor Novak (Air Force Academy), Andrew Gavrin (IUPUI), Wolfgang Christian (Davidson College), and Evelyn Patterson (Air Force Academy)
In this resource book for educators, the four authors fully discuss just what *Just-in-Time Teaching* is. *Just-in-Time Teaching (JiTT)* is a teaching and learning methodology designed to engage students by using feedback from pre-class Web assignments to adjust classroom lessons. This allows students to receive rapid response to the specific questions and problems they are having instead of more generic lectures that may or may not address what students actually need help with. Many teachers have found that this process also encourages students to take more control of the learning process and become active and interested learners.

Physlets®: Teaching Physics with Interactive Curricular Material (0-13-029341-5)
Wolfgang Christian and Mario Belloni (Davidson College)
This manual/CD package shows physics instructors—both Web novices and Java-savvy programmers alike—how to author their own interactive curricular material using Physlets—Java applets written for physics pedagogy that can be embedded directly into HTML documents and that can interact with the user. It demonstrates the use of Physlets in conjunction with JavaScript to deliver a wide variety of Web-based interactive physics activities. It also provides examples of Physlets created for classroom demonstrations, traditional and Just-in-Time Teaching homework problems, pre- and post-laboratory exercises, and Interactive Engagement activities. More than just a technical how-to book, the manual gives instructors some ideas about the new possibilities that Physlets offer and is designed to make the transition to using Physlets quick and easy.

Student Supplements

Student Study Guide with Selected Solutions
(Vol. I: 0-13-100070-5; Vol. II: 0-13-146500-7)
David Reid (Eastern Michigan University)
The print study guide provides the following for each chapter:

- Objectives
- Chapter Review with Examples and integrated quizzes
- Reference Tools & Resources (equation summaries, important tips, and tools)
- Practice Problems by *Carl Adler (East Carolina University)*
- Selected Solutions for several end-of-chapter problems

PH GradeAssist Student Quick Start Guide (0-13-141741-X) This guide contains information on how to register and use PH GradeAssist. It also contains the access code necessary for students to create their accounts and access the course.

Ranking Task Exercises in Physics: Student Edition (0-13-144851-X)
Thomas L. O'Kuma (Lee College), David P. Maloney (Indiana University–Purdue University at Fort Wayne), and Curtis J. Hieggelke (Joliet Junior College)
Ranking Task Exercises are a unique resource for physics instructors who are looking for tools to incorporate more conceptual analysis in their course. This supplement contains 218 Ranking Task Exercises that cover all classical physics topics. Ranking Tasks are an innovative type of conceptual exercise that ask students to make comparative judgments about a set of variations on a particular physical situation.

Physlet® Physics (0-13-101969-4)
Wolfgang Christian and Mario Belloni (Davidson College)
This text and CD-ROM package provides ready-to-run interactive Physlet-based curricular material for both teachers and students. Physlets, award-winning Java applets written by Christian and Belloni, have been widely adopted by the physics teaching community. This book provides the first class-tested collection of ready-to-run Physlet-based material that is easy to assign (like an end-of-chapter problem of a textbook) and easy to use (material is on a CD and in the book). Neither a Web server nor an Internet connection is required.

Interactive Physics Workbook, Second Edition (0-13-067108-8)
Cindy Schwarz (Vassar College), John Ertel (Naval Academy), MSC Software
This interactive workbook and hybrid CD-ROM package is designed to help students visualize and work with specific physics problems through simulations created with Interactive Physics files. Forty problems of varying degrees of difficulty require students to make predictions, change variables, run, and visualize motion on the computer. The workbook/study guide provides instructions, physics review, hints, and questions. The accompanying hybrid CD-ROM contains everything students need to run the simulations.

Tutorials in Introductory Physics and Homework Package (0-13-097069-7)
Lillian C. McDermott, Peter S. Shaffer, and the Physics Education Group (all of the University of Washington)
This landmark book presents a series of physics tutorials designed by a leading physics education research group. Emphasizing the development of concepts and scientific reasoning skills, the tutorials focus on the specific conceptual and reasoning difficulties that students tend to encounter. The tutorials cover a range of topics in Mechanics, Electricity and Magnetism, and Waves & Optics.

The Portable TA: A Physics Problem-Solving Guide, Second Edition (Vol. I: 0-13-231713-3; Vol. II: 0-13-231721-4)
Andrew Elby
This two-volume set contains a collection of problems with carefully detailed strategies and solutions that provide students with additional problem-solving techniques.

MCAT Physics Study Guide (0-13-627951-1)
Joseph Boone (California Polytechnic State University-San Luis Obispo)
Since most MCAT questions require more thought and reasoning than simply plugging numbers into an equation, this study guide is designed to refresh students' memories about the topics they've covered in class. Additional review, practice problems, and review questions are included.

Science on the Internet: A Student's Guide (0-13-028253-7)
Andrew Stull and Harry Nickla
This useful resource gives clear steps to access Prentice Hall's regularly updated science resources, along with an overview of general World Wide Web navigation strategies. Available free for students when packaged with the text.

Media Resources
Course Management Options
Course management systems offer you a robust architecture for communicating with your students and letting them communicate with each other, allowing you and your students to post course-related documents, managing your roster and gradebook, and providing on-line assessment. For schools with a local WebCT or Blackboard license, we offer a complete downloadable content cartridge that will give you a rapid start to your on-line course materials. Adapt and customize our materials to your needs, using

the tools of these systems. For instructors without the benefit of a local course management system, we also offer OneKey, Pearson Education's own nationally hosted course management system, powered by Blackboard. OneKey combines the power of a full-featured course management system with a quick, easy-to-use interface.

The content cartridge for *Physics for Scientists and Engineers, Third Edition*, includes all the material from the Companion Website, selected resources from the Instructor's Resource Center on CD-ROMs, and all of the questions from the TestGenerator test bank, plus additional materials designed specifically to work in concert with innovative teaching methods. The latter include several activities for *Just-in-Time Teaching*, by *Gregor Novak* and *Andrew Gavrin*; and conceptual, quantitative, and MCAT practice problem sets for on-line assessment.

On-line Assessment Options

You may not need the full capabilities of a course management system (such as posting documents or managing a bulletin board) but may prefer to use an on-line assessment system. On-line assessment provides students instantaneous feedback and repeated practice, and it offers instructors relief from hours of grading and managing a gradebook. On-line assessment systems feature algorithmically-generated questions and quizzes, allow you to create and modify assignments and question pools, and provide tools for analyzing your gradebook entries. Prentice Hall has a powerful new entry in the on-line assessment space: PH GradeAssist. Ask your PH representative about getting a PHGA demonstration and an Instructor's Quick Start Guide. In addition, Prentice Hall partners with other systems, such as WebAssign, to provide an ample selection of book-specific questions to include in your assignments.

PH GradeAssist

This nationally hosted system includes assessment banks associated with Just-in-Time Teaching materials, and other conceptual and quantitative questions. In addition, most of the even-numbered end-of-chapter Problems and questions for *Physics for Scientists and Engineers, Third Edition*, have been converted for use in PHGA, and the majority of them have an algorithmically-generated variant. You select which questions to assign, and you may edit them or create new questions. You also control various important parameters, such as how much questions are worth and when a student can take a quiz.

WebAssign

WebAssign's nationally hosted homework delivery service harnesses the power of the Internet and puts it to work for you by collecting and grading homework. You can create assignments from a database of end-of-chapter questions and problems from the third edition of *Physics for Scientists and Engineers*, or write and customize your own. You have complete control over the homework your students receive, including due date, content, feedback, and question formats.

Companion Website (*http://physics.prenhall.com/fishbane*)

The Companion Website is a quick, interactive resource that allows students to check their understanding with practice quizzes and to explore the material of each chapter further by mining the World Wide Web.

- Reference Tools and Resources by *David Reid (Eastern Michigan University)*
- Practice Questions by *Carl Adler (East Carolina University)*
- Algorithmically generated Practice Problems by *Carl Adler*
- On-line Destinations (links to related sites) by *Carl Adler*
- Applications, with links to related Internet sites, by *Gregor Novak* and *Andrew Gavrin (Indiana University–Purdue University, Indianapolis)*

All quiz modules are scored by the computer; results can be automatically e-mailed to the student's professor or teaching assistant.

Acknowledgments

Three authors alone cannot accomplish a project of this magnitude. We are grateful to the many people who have contributed to making the third edition a better text. A special thanks goes to Professor Jerry Shi (Pasadena City College). In addition to writing the *Instructor's Solutions Manual*, he was an invaluable resource in checking and refining the problem sets and answers and coordinating the independent critique and feedback of the problems and answers from Jenny Quan (Pasadena City College) and David Curott (emeritus, University of North Alabama).

We would like to acknowledge and thank all the instructors who provided valuable feedback for the third edition:

H. Biritz
Georgia Institute of Technology

Tim Bolton
Kansas State University

David Boness
Seattle University

Brian Borovsky
Grinnell College

Chih-Yung Chien
Johns Hopkins University

Krishna M. Chowdary
Bucknell University

Stephane Coutu
Pennsylvania State University

J. William Dawicke
Milwaukee School of Engineering

Michael Dennin
University of California–Irvine

Kathryn Dimiduk
University of New Mexico

N. John DiNardo
Drexel University

J. Finkelstein
San Jose State University

Francis M. Gasparini
SUNY University at Buffalo

Michael Grady
SUNY College at Fredonia

Benjamin Grinstein
University of California–San Diego

John B. Gruber
San Jose State University

Paul Haines
Boston College

Randy Harris
University of California–Davis

Charles A. Hughes
University of Central Oklahoma

Kevin Kimberlin
Bradley University

Amitabh Lath
Rutgers University

Ronald S. MacTaylor
Salem State College

Daniel Marlow
Princeton University

William W. McNairy
Duke University

M. Howard Miles
Washington State University (emeritus)

John Milsom
University of Arizona

Gary A. Morris
Rice University

Elena Murino Di Ventra
Virginia Polytechnic Institute and State University

Michael G. Nichols
Creighton University

Michael B. Ottinger
Missouri Western State College

Kevin T. Pitts
University of Illinois

Andrea Raspini
SUNY College at Fredonia

Dennis Rioux
University of Wisconsin-Oshkosh

Joseph Rothberg
University of Washington

Andrew Scherbakov
Georgia Institute of Technology

Bartlett M. Sheinberg
Houston Community College

Marllin L. Simon
Auburn University

Ross L. Spencer
Brigham Young University

Mark W. Sprague
East Carolina University

Michael G. Strauss
University of Oklahoma

Laszlo Takacs
University of Maryland–Baltimore County

Michael J. Tammaro
University of Rhode Island

Anatoli Vankov
Rochester Institute of Technology

T.S. Venkataraman
Drexel University

Walter D. Wales
University of Pennsylvania

Jeffrey L. Wragg
College of Charleston

Scott A. Yost
Baylor University

R. K. P. Zia
Virginia Polytechnic Institute and State University

We would also like to acknowledge reviewers of the previous editions:

Maris Abolins
Michigan State University

V. K. Agarwal
Moorhead State University

Ricardo Alarcon
Arizona State University

Bradley Antanaitis
Lafayette College

Thomas Armstrong
University of Kansas

Philip S. Baringer
University of Kansas

John E. Bartelt
Vanderbilt University

William Bassichis
Texas A & M University

Benjamin F. Bayman
University of Minnesota

Carl Bender
Washington University

Hans-Uno Bengtsson
University of California, Los Angeles

Robert Bowden
Virginia Polytechnic Institute and State University

Bennet Brabson
Indiana University

Michael Browne
University of Idaho

Timothy Burns
Leeward Community College

Alice Chance
Western Connecticut State University

Edward Chang
University of Massachusetts

Robert Clark
Texas A & M University

Albert Claus
Loyola University

Robert Coakley
University of Southern Maine

Lucien Cremaldi
University of Mississippi

W. Lawrence Croft
Mississippi State University

Chris L. Davis
University of Louisville

Robin Davis
University of Kansas

Jack Denson
Mississippi State University

James Dicello
Clarkson University

P. E. Eastman
University of Waterloo

Robert J. Endorf
University of Cincinnati

Arnold Feldman
University of Hawaii at Manoa

A. L. Ford
Texas A & M University

Gabor Forgacs
Clarkson University

William Fickinger
Case Western Reserve University

Rex Gandy
Auburn University

Alexander B. Gardner
Howard University

Simon George
California State University, Long Beach

James Gerhart
University of Washington

Robert E. Gibbs
Eastern Washington University

Wallace L. Glab
Texas Tech University

James R. Goff
Pima Community College

Alan I. Goldman
Iowa State University

Phillip Gutierrez
University of Oklahoma

Frank Hagelberg
SUNY–Albany

Robert F. Harder
George Fox College

Bruce Harmon
Iowa State University

Warren W. Hein
South Dakota State University

Joseph Hemsky
Wright State University

Jerome Hosken
City College of San Francisco

Joey Houston
Michigan State University

Francis L. Howell
University of North Dakota

Alvin Jenkins
North Carolina State University

Karen Johnston
North Carolina State University

Evan W. Jones
Sierra College

Garth Jones
University of British Columbia

Leonard Kahn
University of Rhode Island

Alain E. Kaloyeros
SUNY–Albany

Charles Kaufman
University of Rhode Island

Robert J. Kearney
University of Idaho

Thomas Keil
Worcester Polytechnic University

Carl Kocher
Oregon State University

Arthur Z. Kovacs
Rochester Institute of Technology

Claude Laird
University of Kansas

Vance Gordon Lind
Utah State University

A. Eugene Livingston
University of Notre Dame

B. A. Logan
University of Ottawa

Karl Ludwig
Boston University

Robert Marande
Pennsylvania State University

David Markowitz
University of Connecticut

Erwin Marquit
University of Minnesota

Marvin L. Marshak
University of Minnesota

Charles R. McKenzie
Salisbury State University

Norman McNeal
Sauk Valley Community College

Forrest Meiere
Indiana University–Purdue University

Roy Middleton
University of Pennsylvania

Irvin A. Miller
Drexel University

George Miner
University of Dayton

Thomas Muller
University of California, Los Angeles

Richard Murphy
University of Missouri–Kansas City

Lorenzo Narducci
Drexel University

Peter Nemethy
New York University

David Ober
Ball State University

Gottlieb S. Oehrlein
SUNY–Albany

Jay Orear
Cornell University

Micheal J. O'Shea
Kansas State University

Dan Overcash
Auburn University

Patrick Papin
San Diego State University

Kwangjai Park
University of Oregon

Robert A. Pelcovits
Brown University

R. J. Peterson
University of Colorado

Frank Pinski
University of Cincinnati

Lawrence Pinsky
University of Houston

Stephen Pinsky
Ohio State University

Richard Plano
Rutgers University

Hans Plendl
Florida State University

Shafigur Rahman
Allegheny College

Don D. Reeder
University of Wisconsin–Madison

Peter Riley
University of Texas, Austin

John Lewis Robinson
University of British Columbia

L. David Roper
Virginia Polytechnic Institute and State University

Ernest Rost
University of Colorado

Richard Roth
Eastern Michigan University

Carl Rotter
West Virginia University

Mendel Sachs
SUNY–Buffalo

Francesca Sammarruca
University of Idaho

Charles Scherr
University of Texas

Eric Sheldon
University of Lowell

Charles Shirkey
Bowling Green State University

Robert Simpson
University of New Hampshire

James Smith
University of Illinois

J. C. Sprott
University of Wisconsin

Malcolm Steuer
University of Georgia

Thor Stromberg
New Mexico State University

William G. Sturrus
Youngstown State University

Richard E. Swanson
Sandhills Community College

Leo H. Takahashi
Pennsylvania State University

Smio Tani
Marquette University

Robert Tribble
Texas A & M University

Rod Varley
Hunter College

Gianfranco Vidali
Syracuse University

John Wahr
University of Colorado

William Walker
University of California, Santa Barbara

Fa-chung Wang
Prairie View A & M University

Gail S. Welsh
Salisbury State University

George Williams
University of Utah

Finally, we would like to thank the publishing team at Prentice Hall who have helped us to carry this project through. Our editor, Erik Fahlgren, who directed the project, has been a constant source of ideas, encouragement, and material help. He has been a most constructive listener and has played the devil's advocate very well when it was necessary. Catherine Flack, our development editor, reminded us that this book is meant for students first and that a failure to communicate at the appropriate level is the worst kind of failure. She too has been willing to listen, and her role as the surrogate student has been immensely helpful. Her help has been crucial to the project. Thanks to Christian Botting, who coordinated the supplements program, and Paul Draper, who managed the media, including the Web-based material. Each of them have been crucial. Special thanks to Andrew Sobel, who worked behind the scenes and handled a massive string of entangled cords like Horowitz playing Chopin. It would be difficult to see how this work could have been completed without his unerring attention to detail. Last but not least, the production of a book such as this one is an enormous task demanding the most careful attention. We want to thank John Christiana for creating a design that is appealing and economical. We also want to thank Beth Sweeten and Fran Daniele for the elaborate management job that has led this book to press. To all the individuals listed above, and to the many others at Prentice Hall who have worked to make this book a success, we extend our heartfelt thanks.

The cumulative and accelerating nature of science and technology make it more imperative than ever that our emerging scientists and engineers understand how few and how solid are the pillars of the enterprise. From this view, the distinctions between "science" and "engineering" and between "classical physics" and "modern physics" melt. We want this book to make evident the pillars of physics as well as the highly interconnected structure that has been erected on those pillars.

Paul Fishbane
pmf2r@virginia.edu

Stephen Gasiorowicz
gasior@umn.edu

Stephen Thornton
stt@virginia.edu

About the Authors

Paul M. Fishbane

Paul Fishbane has been teaching undergraduate courses at the University of Virginia, where he is Professor of Physics, for more than 30 years. In addition to this text, he is the co-author of a modern physics text also published by Prentice Hall. He received his doctoral degree from Princeton University in 1967 and has published over 100 papers in his field, theoretical high-energy physics. He has held visiting appointments at the State University of New York at Stony Brook, Los Alamos Scientific Lab, CERN Laboratory in Switzerland, Amsterdam's NIKHEF laboratory, France's Institut de Physique Nucleaire, l'Université Paris-Sud, and l'École Polytechnique. He has been active for many years at the Aspen Center for Physics, where current issues in physics are discussed with an international group of participants. Among his many other interests, we'll mention in alphabetical order antiques, biking, the kitchen, and music. The rest of his time is spent trying to keep up with his family, including his grandchildren, Ruby and Ivy.

Stephen G. Gasiorowicz

Stephen Gasiorowicz was born in Poland and received his Ph.D. in physics at the University of California, Los Angeles, in 1952. After spending eight years at the Lawrence Radiation Laboratory in Berkeley, California, he joined the faculty of the University of Minnesota, where he is now Professor Emeritus. His field of research was theoretical high-energy physics. As a visiting professor, he has spent extended periods of time at the Niels Bohr Institute, NORDITA in Copenhagen, the Max Planck Institute for Physics and Astrophysics in Munich, DESY in Hamburg, Fermilab in Batavia, and the Universities of Marseille and Tokyo. He has been a frequent visitor to and officer of the Aspen Center of Physics. He has written books on elementary particle physics and quantum physics, and co-authored a modern physics text also published by Prentice Hall. He enjoys biking, canoeing, and skiing, and he is constantly amazed watching his grandchildren Hannah, Becca, Kyle, and Eliza grow up.

Stephen T. Thornton

Stephen Thornton completed his doctoral research at Oak Ridge National Laboratory while completing his Ph.D. at the University of Tennessee. He joined the faculty at the University of Virginia in 1968 and became the first Director of its Institute of Nuclear and Particle Physics in 1984. He has held a Max Planck fellowship and two Fulbright fellowships and has performed nuclear physics experiments at accelerators throughout the United States and Europe. His recent interests include teaching and developing distance-learning courses in physics and physical science for K–12 teachers, and he helped establish the Center for Science Education at the University of Virginia. He has had over 120 research papers published and co-authored three textbooks. He has two grown sons and is married to former NASA astronaut Dr. Kathryn Thornton, with whom he has three daughters. His interests include keeping up with his children, snow skiing, scuba diving, and traveling.

Volume II

PHYSICS

FOR SCIENTISTS AND ENGINEERS

◀ Investigations of the nature of lightning were important in the understanding of electrical phenomena. Here, Benjamin Franklin is represented performing a kite experiment, whose description he published between 1751 and 1753. The painting is inaccurate in that Franklin flew the kite before lightning struck, he did not stand on the open, and a key was attached to the string, which was then charged by the clouds. In fact, whether Franklin actually performed the experiment is not known.

Chapter 21

Electric Charge

Electricity, together with magnetism, governs virtually all that we see of the physical world. Electromagnetic forces control the structure of atoms and all materials. Light and other electromagnetic waves are pervasive. The understanding of these forces is one of the great success stories of science. In this chapter, we introduce electric charge, a property of atomic constituents, and we discuss the fundamental law of the interaction of two charges at rest, Coulomb's law. This force law is as fundamental as the universal law of gravitation. The interaction between charges has the same space dependence as gravitation, but the force described by Coulomb's law can be either attractive or repulsive.

21-1 Charge—a Property of Matter

Over the last century, we have become more and more dependent on electricity in our everyday lives, and most people are aware that electric charges exist. The experimental evidence for electric charges and the understanding of charge developed over a long time period. We begin with a very brief sketch of this development.

A Brief History of the Study of Electricity and Magnetism

The word *electricity* has its roots in the Greek word for "amber" (*electrum*), and the first written mention of the curious effects of rubbed amber dates from the fifth century B.C. It was not until the 1700s that the critical discovery that electric forces can be either repulsive or attractive was made. The idea developed that a quantity (which we now call electric charge) is associated with electric forces. Among the many important names associated with these discoveries are Stephen Gray, Charles Dufay, and Benjamin Franklin.

Benjamin Franklin is best known for his exploitation of the existing idea that electrical phenomena were associated with a kind of fluid contained in matter. He surmised that repulsion and attraction were the result of an excess or deficiency of the fluid. Although we now know that view is not quite right, implicit in Franklin's model is what we can recognize as the phenomenon of the conservation of charge. For example, if the "fluid" were to flow out of an object, it would leave behind a deficiency. Franklin introduced the terms "positive" and "negative" for the two types of charge and also set the standard sign convention in which the electron, the actual particle that moves in conductors, has negative charge. Franklin was known for his spectacular (and dangerous) experiments with lightning, which he recognized as an electrical effect. Franklin and his friend Joseph Priestley, as well as Henry Cavendish, are linked with the discovery that *the fundamental force between electric charges is proportional to the inverse square of the distance between them*. This law was confirmed more directly by John Robison and then by Charles Coulomb in the mid- and late eighteenth centuries, respectively. This inverse-square law is now known as **Coulomb's law**.

The nature of magnetism and its relation to electricity became clearer starting around 1820, primarily through the work of Hans Christian Oersted, André-Marie Ampère, and Michael Faraday. James Clerk Maxwell completed the unification of electricity and magnetism in the 1860s: Electricity and magnetism were aspects of a single fundamental set of phenomena, electromagnetism. This subject will occupy us for many chapters to come.

The Significance of Electric Forces

In this chapter we introduce a second basic force of nature. To the law of universal gravitation we add knowledge of the electrical interaction, as represented by the Coulomb force. Gravitational forces have an inverse-square dependence on the distance between interacting pointlike objects, and as we shall see, the same is true for the electric force. Both forces are proportional to the product of a characteristic attribute of the two objects—mass for the gravitational force and electric charge for the Coulomb force.

On the cosmic scale, gravity looms large. It is the force that keeps Earth rotating around the Sun, and the Moon rotating around Earth. The reasons why gravitation dominates electric forces on the astronomical scale are twofold. First, astronomical bodies have a great deal of mass. Second, astronomical bodies are almost exactly charge neutral, so the electric forces between them are relatively small. On anything less than an astronomical scale, however, the electric forces are normally much larger than the gravitational ones; apart from the direct effects of Earth's gravity, our everyday experience depends far more on the electric force than on the gravitational one.

As we shall see through study of the hydrogen atom, the electric force dominates the gravitational force on a microscopic scale. Even though a full explanation requires the inclusion of magnetic forces and quantum physics, we can now state that the electric force is responsible for

1. electrons binding to a positive nucleus, forming a stable atom;
2. atoms binding together into molecules;
3. atoms or molecules binding together into liquids and solids;
4. all chemical reactions; and
5. all biological processes.

The electric force is also responsible for such nonfundamental forces as friction and other contact forces. Electric energy fuels our homes, starts our cars, and runs our factories. We can say that electrical forces are dominant in the behavior of matter as we know it.

Matter and Electric Charge

In most of our discussions to this point, we have characterized bulk matter—and the atoms that make up matter—by a single attribute: mass. When we probe the structure of atoms more deeply, we find that atoms are made up of electrons and nuclei, which are characterized by an additional attribute, **electric charge** (usually labeled q). Electric charges exert (electric) forces on one another that are proportional to the product of their charges, just as masses exert gravitational forces on one another that are proportional to the product of their masses. Charges come in two types, termed positive and negative, and have the fundamental property that opposite charges attract and like charges repel. Electrons and nuclei carry opposite charges, and as we will describe in more detail later, it is the attractive electrical forces between them that hold the atom together. The set of phenomena associated with the forces between stationary charges form the subject of **electrostatics**.

A correct description of atoms requires quantum mechanics. But we can give a qualitative classical picture—the remarks about charge in this picture carry over to a quantum-mechanical picture. An atom's electrons (labeled by e) each carry the same unit of negative charge, $q_{electron} = -e$. (Note that the same symbol e stands for the electron and a charge quantity.) The electrons are bound in shell-like regions around the much heavier nucleus, which consists of electrically neutral neutrons (labeled n) and positively charged protons (labeled p). The proton charge is equal in magnitude but opposite in sign to that of the electron, $q_{proton} = +e$. When the number of protons in an atom equals the number of electrons, the atom is **electrically neutral**—it has no net charge.

Chemical elements differ in the number of protons in the nuclei of their atoms. With no ionization, the number of protons is the same as the number of electrons surrounding the nucleus. Chemical properties are closely linked to the behavior of the atom's electrons. The electrons that are, on average, closest to the nucleus are more strongly bound to the nucleus and are more difficult to dislodge from the atom and less available to interact with other atoms. The outermost electrons—farther away from the nucleus—are attracted less strongly and are more easily dislodged. The ease with which the outermost electrons can be removed from the atom, as well as how strongly an atom attracts other electrons, greatly influences the properties of the elements and how they interact chemically with one another. An atom that has lost one or more electrons and is thus left with a positive charge is called a *positive ion*, and a *negative ion* is an atom that has gained extra electrons. Bulk matter is typically electrically neutral or very nearly so. Even if there are chemical phenomena that produce positive and negative ions within the matter, these will appear in equal numbers to produce a neutral bulk.

Although bulk matter is overall neutral, it will have electrical properties that are associated with the outer electrons of its atoms. If the outer electrons of atoms in bulk matter are *weakly bound* to their respective nuclei, they behave as though they are almost free, and can move through the material almost unimpeded. Such materials—metals for the most part—are said to be good **conductors** of electricity; the term refers to how easily charge can move through the material. A certain class of materials—**superconductors**—have electrons that, in effect, move with *no* inhibition when the material is made sufficiently cold. The electrons of most nonmetallic solids do not travel easily; such solids, including rubber, glass, and plastics, are **insulators**. Silicon, germanium, and a large number of synthetic combinations of materials are substances that we can make into insulators or conductors by controlling either their electrical properties or their temperature. Such substances are called **semiconductors**, and they play an important role in electronic devices—transistors and so forth.

We can transfer charges back and forth between different materials by rubbing them together, allowing electrons to move from one material to the other. For example, when we rub a Teflon rod with fur, electrons are transferred from the fur to the Teflon rod (see Fig. 21–1a, b). The fur then has an excess positive charge: It has lost some electrons. The Teflon rod to which the electrons have been carried will have an excess of electrons and is now negatively charged. When charge is carried from one object to another in this way, the objects are said to be **charged by friction**. Both objects will be charged. We can similarly transfer electric charge by rubbing a glass rod on silk. The glass acquires a positive charge, because it transfers electrons to the silk, which in turn acquires an equal but opposite (negative) charge (Fig. 21–1c).

(a)

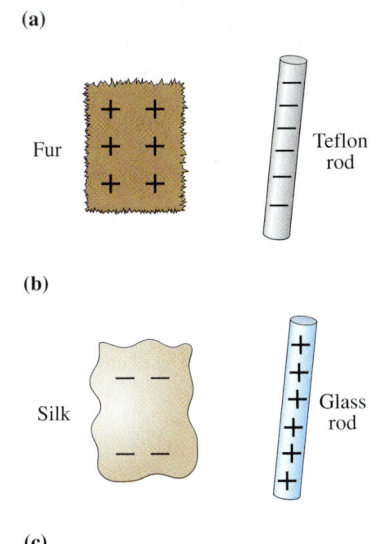

(b)

(c)

▲ **FIGURE 21–1** When Teflon is rubbed against fur (a, b) and glass is rubbed against silk (c), electric charge is transferred.

THINK ABOUT THIS...

IF WE ACCUMULATE A CHARGE, WHERE DOES IT EVENTUALLY GO?

When we walk across a carpet on a dry winter day, we may slowly accumulate a charge through the rubbing of our shoes on the carpet fibers. We specify that the air must be dry because if it is not, then this charge is easily transferred to water molecules in the air and carried off. And your shoes should be good insulators, because otherwise the charge will travel back to the carpet. Without these possibilities, you may later touch a radiator or another person. Your accumulated charge is then shared with the second object, sometimes in sudden and shocking fashion. (The passage of the charge may even be visible as a spark, an interesting subject on its own that we'll discuss in Chapter 25.) But just what is the object with which you are sharing your charge if you touch the radiator? The answer is that the metal pipes of the radiator may form a very long path, ultimately connecting you to Earth as a whole. If the carpet itself touches Earth through a sequence of conductors, you got your charge from Earth, and you are simply returning that charge to Earth. Even if the carpet is not well connected with Earth, by sharing your accumulated charge with Earth as a whole, your portion of it will be very small—Earth is large! By touching an object—here, it is you—to Earth, we say that the object is **grounded**. When a negatively charged object is grounded, electrons flow from the object to the ground, leaving the object (very nearly) neutral. If, instead, an object has an excess positive charge, then electrons flow from the ground and neutralize the object. The electric shock associated with a carpet is small, but there is potential for more damaging electric shocks around the home from the electricity supply. We therefore tend to ground any objects that might become highly charged, so that any excess charge will travel into Earth and not to you (Fig. 21–2).

▲ **FIGURE 21–2** The large copper rod that is being pounded into the ground will serve as the electrical ground for the household electrical service box.

Evidence That Charges Are of Two Types

A small mass will react to forces more visibly than a large mass, so we use small (initially neutral) masses to study the effects of forces between charges. We can do this by using small balls made of cork and coated with a conducting paint that allows charge to move around easily on the surface. Such a ball is hung by a thin insulating thread (Fig. 21–3a). If we touch a negatively charged Teflon rod to a cork ball, the ball is immediately repelled by the rod (Fig. 21–3b). If we touch the negatively charged Teflon rod to *two* suspended (neutral) cork balls, the balls will then strongly repel each other (Fig. 21–3c). Similar behavior occurs between two cork balls that have been touched by a positively charged glass rod, as they will both have acquired a positive charge. However, if we touch the Teflon rod to one cork ball and the glass rod to one cork ball, the oppositely charged balls attract each other (Fig. 21–3d).

▶ **FIGURE 21–3** (a) An insulated cork ball covered with a thin layer of conducting paint can indicate the presence of small electric charges. (b) A negatively charged Teflon rod approaches the neutral coated cork ball, which is initially attracted to the rod. After the rod touches the ball, the ball becomes charged and strongly repels the charged rod. (c) If we touch two initially neutral cork balls with a negatively charged Teflon rod, the two balls repel each other: Like charges repel. (d) If we touch one initially neutral cork ball with a negatively charged Teflon rod and a similar ball with a positively charged glass rod, the two balls attract each other: Unlike charges attract.

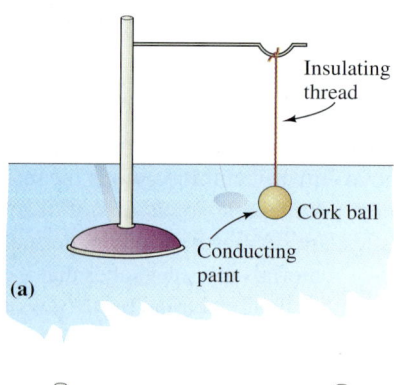
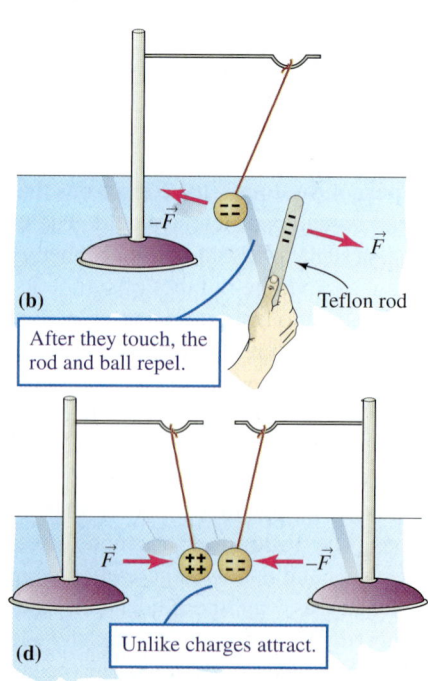
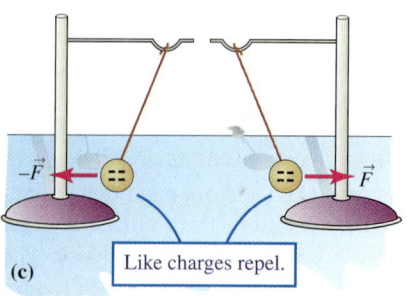

We conclude from these experiments that the electric charges on the Teflon and glass rods are different, and that

Like charges repel, and unlike charges attract.

For example, when the Teflon rod touches the cork ball, some of the rod's negative charge is transferred to the ball. Now both the ball and the rod have a negative charge. The ball, which has been charged by conduction, immediately jumps away from the rod. Our other observations are similarly explained by the rule that like charges repel and unlike charges attract.

Another, more subtle, effect is also present. Before the negatively charged Teflon rod actually touches the neutral cork ball, the ball is *attracted* to the rod, not repelled by it. How can we explain this initial attraction? Because we have coated the cork ball with conducting paint, there are mobile electrons on the surface of the ball. When the negatively charged Teflon rod comes near, the mobile electrons are repelled and move to the far side of the cork ball (Fig. 21–4). That leaves an equal amount of excess positive charge on the area of the ball near the rod. If we additionally infer that the electric force is stronger when the charges are closer, those positive charges are attracted to the rod more strongly than the negative charges on the other side of the ball are repelled. In other words, when the positive charges on the cork ball are closer than the ball's negative charges to the Teflon rod, the *net* force is attractive. We call this phenomenon, in which charges within an object are redistributed due to the presence of external charge, **charge polarization**. Figure 21–5 illustrates an 18th-century demonstration of charge polarization. The fact that electrical forces between interacting charges weaken with distance—the inference that we had to make to explain the initial attraction of the ball and rod—is of great importance, and we shall return to it.

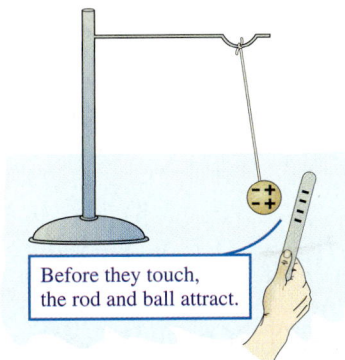

▲ **FIGURE 21–4** The neutral cork ball is initially attracted to the charged Teflon rod because some electrons on the ball move to the far side due to the repulsive force from the rod. The positive charges on the ball are on average closer to the rod, so the attractive force on them due to the rod is greater than the repulsive force on the redistributed electrons.

◀ **FIGURE 21–5** An eighteenth-century experiment on static electricity by Stephen Gray. The boy, suspended in air, carries a net charge (positive, let's say). As a result, charge polarization is induced in electrically neutral bits of paper near him. Negative charges on each paper bit tend to move toward the boy, leaving positive charges on the part of the paper bit farthest from the boy. The result is a net attractive force between the boy and each bit of paper.

Charge by Induction

An experiment closely related to the charge polarization experiments explains how initially neutral conductors can obtain a *charge by induction*, or an *induced charge*. Consider two neutral metal spheres, each standing on an insulated post and in side-by-side contact, so that they form a single conductor (Fig. 21–6a). If we bring a negatively charged Teflon rod close to one sphere, mobile electrons in the spheres move to the opposite side of the far sphere, leaving opposite charges on the two spheres (Fig. 21–6b). The spheres have a total charge of zero, but one is positive and the other negative. While the Teflon rod is still near, we separate the two spheres, leaving them oppositely charged (Fig. 21–6c). If we now remove the Teflon rod, the charges induced by the rod will remain on the two metal spheres (Fig. 21–6d). The spheres have been **charged by induction**. These charges can be transferred to two coated cork balls, by bringing a ball in contact with each sphere. The cork balls attract, demonstrating that the charges are opposite in sign. Note that only conductors can be charged by induction.

614 | Electric Charge

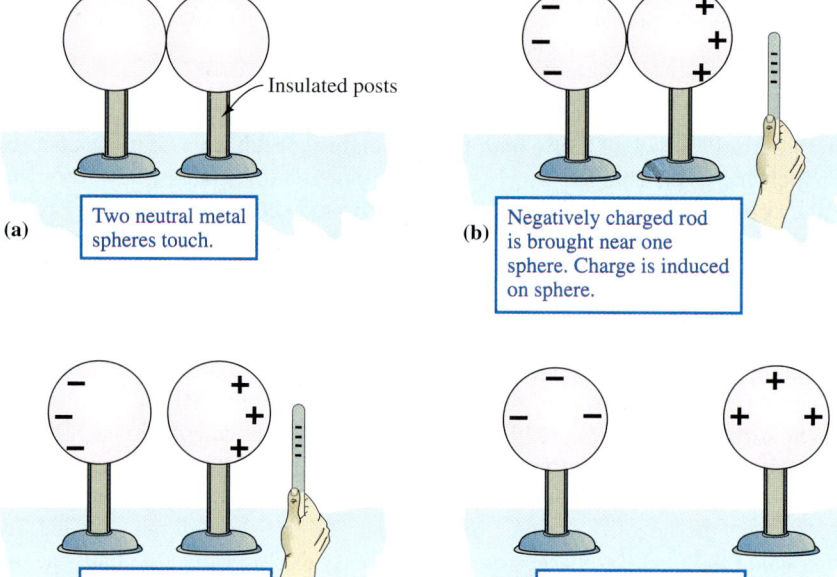

▶ **FIGURE 21–6** (a) Two neutral metal spheres on insulated posts touch. (b) A negatively charged Teflon rod polarizes the metal spheres. (c) If the metal spheres are separated while the Teflon rod is nearby, the spheres are charged oppositely. (d) When the Teflon rod is removed, the two metal spheres are still charged oppositely. Note that the total charge of the two spheres remains zero throughout.

Units of Charge

While the amount of charge on an electron is fixed, the numerical value of this amount of charge depends on how the scale for charge is defined. The SI unit of charge is called the **coulomb** (C). We can define the value of the coulomb by, say, specifying the magnitude of the force between two objects, each carrying 1 C of charge, and separated by a distance of 1 m. We'll assume that a procedure of this type has been carried out [see below Eq. (21–7)].

The magnitude of the charge on the electron—the smallest charge that can be isolated in nature—has been measured to high precision. An approximation sufficient for our purposes is

$$e \cong 1.60 \times 10^{-19} \text{ C}. \quad (21\text{–}1)$$

The mass and charge of the neutron, proton, and electron are given in Table 21–1 and also in the tables behind the front cover.

TABLE 21–1 • Mass and Charge of Atomic Constituents

	Mass (kg)	Charge (C)
Neutron, n	1.675×10^{-27}	0
Proton, p	1.673×10^{-27}	1.602×10^{-19}
Electron, e^-	9.11×10^{-31}	-1.602×10^{-19}

EXAMPLE 21–1 A glass rod rubbed with silk has a charge of +110 nC (110×10^{-9} C). By how many electrons is the rod deficient?

Setting It Up We have made an unspoken (and reasonable) assumption that the rod has started off electrically neutral. Thus the rod acquires a positive charge through either an excess of positive charge or a deficiency in negative charge. In thinking in terms of a deficiency, we are saying that the positive charge has come from transferring electrons from the rod.

Strategy Because we know that each electron has a charge of magnitude e, we can find the number of transferred electrons by dividing the remaining charge by e.

Working It Out We have

$$\text{transferred electrons} = \frac{\text{net charge}}{\text{charge magnitude of each electron}}$$

$$= \frac{110 \times 10^{-9} \text{ C}}{1.6 \times 10^{-19} \text{ C/electron}}$$

$$= 6.9 \times 10^{11} \text{ electrons.}$$

What Do You Think? Could we detect the number of missing electrons by the change in mass of the charged rod? You can find the mass of the electron in Table 21–1. Answers to **What Do You Think?** questions are given in the back of the book.

EXAMPLE 21–2 The largest American Eagle gold coin has a mass of 28.4 g. The atomic number of gold—the number of protons in the nucleus of an atom of gold—is 79, and thus the number of electrons in a neutral gold atom is also 79. The atomic mass of gold is 197, which means that 1 mol of gold has a mass $m_{Au} = 197$ g. How many electrons are contained in one pure gold coin? What is the total negative charge contained in the coin?

Setting It Up The only piece of information not specifically contained here is the fact that a mole of any element contains Avogadro's number, $N_A = 6.02 \times 10^{23}$, of atoms.

Strategy Avogadro's number N_A is the number of atoms in 1 mol of gold; the number of atoms in a given mass m of the gold is the ratio $(m)/(m_{Au})$ times N_A. We multiply this number by 79 electrons/gold atom to find the total number of electrons in the coin. Finally, we multiply the resulting number by the charge of one electron to find the total negative charge.

Working It Out The number of gold atoms in a mass of $m = 28.4$ g is

$$\frac{m}{m_{Au}} N_A = \frac{(28.4 \text{ g})}{(197 \text{ g/mol})}(6.02 \times 10^{23} \text{ atoms/mol})$$
$$= 8.68 \times 10^{22} \text{ atoms}.$$

The total number of electrons is then

number of electrons = (79 electrons/atom)(8.68 × 10²² atoms)
$$= 6.86 \times 10^{24} \text{ electrons}.$$

The total charge of these electrons is

total electron charge = (number of electrons)(charge per electron)
$$= (6.85 \times 10^{24} \text{ electrons})(-1.60 \times 10^{-19} \text{ C/atom}) = -1.1 \times 10^{6} \text{ C}.$$

What Do You Think? (a) If every electron were removed from the coin, what would be the charge left behind? (b) If one electron were removed from each gold atom, what would be the charge of the coin?

The Electroscope

The *electroscope* is a device used to detect excess free charge (Fig. 21–7a). There are two ways to use the electroscope. First, when charge is directly transferred to the electroscope by touching the metal ball at the top of the electroscope with a charged rod, the gold leaf inside separates from the vertical metal stem. This follows because the charge that is transferred to the electroscope is distributed throughout it, including to the leaf and stem; these each carry a charge of the same sign and hence repel one another, as in Fig. 21–7b. The leaf moves away from the stem until the vertical component of the electrostatic repulsion is balanced by the force of gravity on the leaf. Addition of more charge moves the leaf still more, and the angle made by the leaf is a measure of the amount of charge involved.

We also find that if we bring a charged rod near the metal ball at the top of the uncharged electroscope, the gold leaf still separates. What is happening? The overall charge on the electroscope is zero, and since the rod never actually touches the electroscope, the overall charge remains at zero. However, when the charged rod (positively charged, say) is brought close to the metal ball at the top of the electroscope, a negative charge is induced on the ball, as electrons are attracted by the positive charge on the rod. The positive ions collect on the leaf and stem, leaving them both positively charged. Again the leaf and stem repel (Fig. 21–7c). (An example of how this approach can be made more quantitative is given in Problem 10.)

▼ **FIGURE 21–7** (a) An *electroscope*, a device that detects the presence of charge. (b) When free charge is added to the metal conductor, the gold leaf and the vertical metal stem repel, causing the gold leaf to move away from the stem. (c) When a charged object is brought close to the metal ball at the top, a charge is induced on the ball, leaving a charge of the opposite sign on the stem and leaf. They again repel.

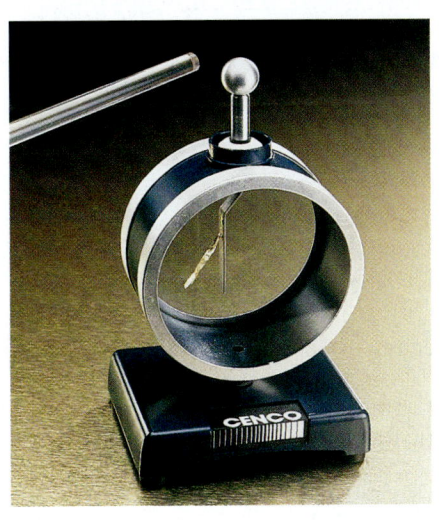

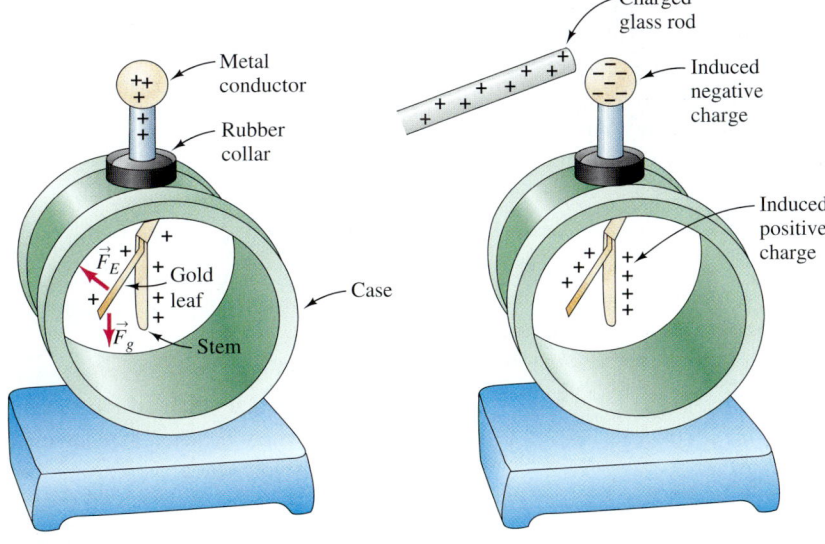

(a) (b) (c)

THINK ABOUT THIS...

HOW DO SMOKE DETECTORS WORK?

The smoke detectors common in many homes depend on the separation of charges. They contain a small radioactive source centered within a cylinder that is open at one end. The source steadily emits alpha (α) particles (nuclear constituents) with enough energy to knock electrons from air molecules with which the α particles collide. The outer cylinder contains a negative charge, and the positive ions produced when the air molecules are broken up are attracted to it. The rate at which the ions arrive at the outer cylinder is measured electronically. When there is a house fire, large organic molecules such as those contained in bacon fat or pinewood resin enter the cylinder. These large molecules are an easy target for the α particles, and as electrons are more loosely bound in these molecules, they are also more likely to be removed from the molecule in the collision. The increased number of ions arriving at the cylinder sets off the alarm. The only disadvantage to this system is that any large molecule in the air—paint solvents or ordinary cooking smoke, for example—will set off the alarm. Many detectors have on/off switches for such cases.

21–2 Charge Is Conserved and Quantized

The simple experiments described in Section 21–1 strongly suggest that *charge is conserved*. Further experiments show that the **conservation of charge** is a fundamental physical law: *Net* charge is the same before and after any interaction. Moreover, charge conservation is *local*. That means that if we have a big box, the charge is conserved not only in the box as a whole but in any subvolume, down to as small as we can measure with precision. If there is less charge in a subvolume than before, then that amount of charge will have crossed the boundary on its way out.

Evidence of Charge Conservation

The reactions of subatomic particles such as nuclei or their constituents allows us to test charge conservation at a fundamental level. Let's look at some of these reactions. One of the reactions between atomic nuclei that takes place in a nuclear reactor is[†]

$$n + {}^{235}_{92}U \rightarrow {}^{143}_{56}Ba + {}^{90}_{36}Kr + 3n + \text{energy}.$$

Here, the total number of protons (92) is the same on both "sides" of the reaction.

Even when the number of electrons or protons changes during a reaction, the total charge remains unchanged. Thus, another reaction that can take place in a nucleus is *electron capture*,

$$e^- + p \rightarrow n + \nu,$$

where ν stands for a neutral particle called the *neutrino*. (The neutrino, unlike the neutron, has a mass much smaller than even the electron mass.) In this reaction, the numbers of both protons and electrons change, but charge is still conserved.

Other particles, called *positrons*, are identical to electrons except for the *sign* of the charge, and are denoted by e^+. In the reaction (γ is the symbol for the photon, a package of electromagnetic radiation)

$$\gamma + p \rightarrow p + e^+ + e^-,$$

an electron is produced, but then only in partnership with a positron, whose charge has exactly the same magnitude (Fig. 21–8). In fact, in observed reactions involving the so-called elementary particles, *no one has ever seen a single case of net charge appearing or disappearing*.

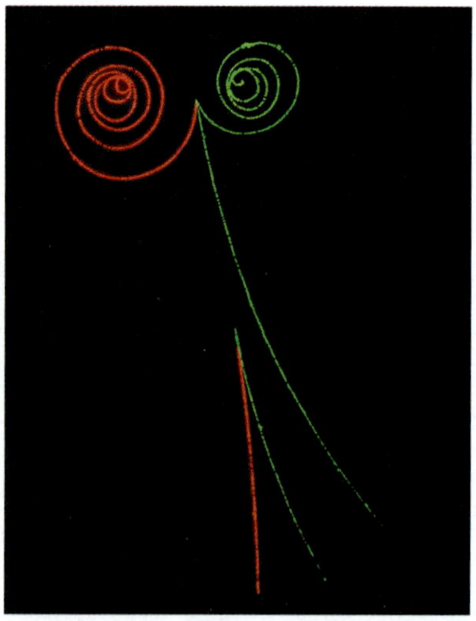

Is it possible for a little of the charge on an electron or a proton to wear off, like paint? Again, all the evidence points to the fact that the electron and the proton charges are the same, no matter where or when they are measured. In looking at quasars (distant and powerful sources of light), we are looking at matter that existed billions of years ago (it has

▲ **FIGURE 21–8** Production of an electron–positron pair. The event took place in a magnetic field, and the electron and positron spiral in opposite directions in this field as they lose energy.

[†]The superscript on the element symbol is the atomic mass, the sum of the numbers of protons and neutrons in one atom; the subscript is the number of protons.

taken that long for the light to reach Earth). Observations of the color of the light that quasars emit suggest that, to a very high accuracy, the properties of their atoms are identical to the properties of atoms here on Earth. This implies that the charge of electrons and protons are not only identical but have remained constant over billions of years.

CONCEPTUAL EXAMPLE 21–3 A new theory makes the unusual proposition that the equality of the magnitude of the electron and proton charge was not always true, so that perhaps a billion years ago they differed by one part in a billion. What sort of arguments could you advance to check whether this is true or not?

Answer We have already mentioned a couple of effects that a change in magnitude of both the electron and proton charges would make: differences in the color (spectrum) of the light emitted by atoms, for example. Such effects follow when, even though changed, the electron and proton each carry charge of *identical* magnitude. Here we are asked what would happen if the magnitudes were not the same, and that has even more dramatic consequences. If the relative numbers of electrons and protons were unchanged, it would mean that the entire universe would have a net charge. Indeed, each atom would also have a net charge, all of the same sign, and unless the difference in charge were extremely tiny, the repulsive force between atoms would overwhelm the attractive gravitational forces that allow matter to clump into galaxies and stars. Bulk matter would not form in the way that we know it, at least not on the time scale of the universe.

Charge Quantization

We have already indicated that charges appear to be organized in discrete amounts. The magnitude of this minimum amount of charge is that of one electron. Greater charges are always multiples of these values. The facts that, within experimental accuracy, charge occurs in integral multiples of the electron charge, known as **charge quantization**, and that charges are never observed with values smaller than the electron charge were first established in 1909 through the pioneering experiments of Robert Millikan.

In summary, we can say that

Charge is conserved absolutely

and that

Free charge is quantized in positive or negative integral multiples of e.

THINK ABOUT THIS...
DO QUARKS CARRY CHARGE THAT IS LESS THAN THE ELECTRON CHARGE?

In 1964, Murray Gell-Mann and George Zweig proposed that protons and neutrons are composed of even more fundamental particles, called **quarks**, whose charges are either $2e/3$ or $-e/3$. Thus the proton, say, contains three quarks whose total charge adds up to $+e$: two charge $(+2e/3)$ quarks and one charge $(-e/3)$ quark. There is strong experimental evidence that quarks really do make up particles such as protons, but strangely enough, and for reasons that are only partially understood, quarks cannot be isolated—they cannot be removed and separated from a proton in the way that a hydrogen atom can be separated into an electron and a proton. Despite many searches, quarks or any other freely moving object carrying fractional electron charges have never been observed. Most physicists now believe that only combinations of quarks possessing a net charge that is an integer multiple of e can ever be isolated and independently observed. We refer to any charge that can be isolated as **free charge**.

21–3 Coulomb's Law

Encouraged by Benjamin Franklin, Joseph Priestley concluded in the mid-eighteenth century from Franklin's and his own experiments that the electric force between two charged objects varies as the inverse square of the distance between the objects. Priestley made this deduction after he observed that there is no charge on the inside surface of a closed or nearly closed metal vessel—all the charge is on the outside surface—and that the force on a charged object placed inside such a vessel is zero. This is like the phenomenon we discussed in Chapter 12: There is no gravitational force on an object inside a uniform spherical shell of matter. As we argued in Chapter 12, this result is a direct consequence of the $1/r^2$ nature of the force law. By analogy with gravitation, Priestley argued that the electric force responsible for his observations must have a $1/r^2$ dependence.

618 | Electric Charge

In 1785, Charles Coulomb directly determined the force law for electrostatics. He performed the relevant experiments with a torsion balance similar to the one Henry Cavendish would use in 1798 to measure the gravitational constant, G (Fig. 21–9). The role played by massive balls in the Cavendish experiment (see p. 342) is here played by charged ones. Coulomb showed that the electrostatic force is central—directed on the line between the charges—and varies as

$$F \propto \frac{1}{r^2}, \tag{21-2}$$

where r is the distance between the centers of the charge sources. By changing the charge on the balls, Coulomb inferred that the force is proportional to the product of the charges q_1 and q_2 on the balls:

$$F \propto q_1 q_2. \tag{21-3}$$

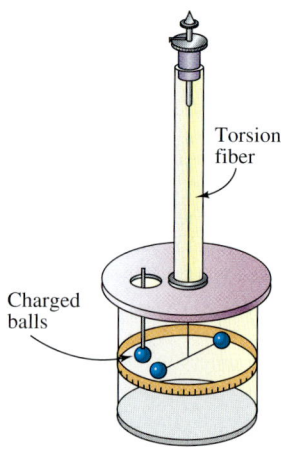

▲ **FIGURE 21–9** Coulomb's torsion balance, used to verify the inverse-square form of the force between electric charges.

To demonstrate the results of Eq. (21–3), we can ground one cork ball, neutralizing it, and charge another identical ball, giving it net (unknown) charge q. After we touch the two balls together, they each have a charge of $q/2$. Then we measure the force between these two balls. Next, we ground one ball again to neutralize it, and touch the balls together once more. Thus, each has a charge of $q/4$, and we measure the force between them to have decreased by a factor of 4 for the same amount of separation. This set of results is consistent with Eq. (21–3): In the first case, $F \propto (q/2)(q/2) = q^2/4$; in the second, $F \propto (q/4)(q/4) = q^2/16$.

Combining Eqs. (21–2) and (21–3) gives us a first view of Coulomb's law, the electrostatic force law. The magnitude of the force is

$$F = \frac{k|q_1 q_2|}{r^2}, \tag{21-4}$$

where k is a proportionality constant. The force is attractive when the charges have opposite sign and repulsive when they have the same sign. Moreover, the force obeys superposition: The force on a given charge from a collection of charges is the sum of the forces due to each charge in the collection.

The constant k plays the same role that the constant G plays in Newton's law of universal gravitation. The magnitude of k depends on the units used for charge—here we use a system of units that is consistent with the SI. It is then possible to define the coulomb by assigning a value to k:

$$k = \frac{1}{4\pi\varepsilon_0}, \tag{21-5}$$

where ε_0 is known as the *permittivity of free space*. (We shall see later that the value of ε_0 follows directly from the defined value of the speed of light, so in this sense ε_0 is itself defined.) To four significant figures, the permittivity is

$$\varepsilon_0 \cong 8.854 \times 10^{-12} \, \text{C}^2/\text{N} \cdot \text{m}^2. \tag{21-6}$$

The value of k (to four significant figures) follows from Eqs. (21–5) and (21–6):

$$k = 8.988 \times 10^9 \, \text{N} \cdot \text{m}^2/\text{C}^2. \tag{21-7}$$

Now that we have assigned a value to k, we can tentatively define the coulomb. From Eqs. (21–4) and (21–7), we say that

> **When the force between two identical charges separated by 1 m is equal to the numerical value of k in newtons (8.988×10^9 N), these charges are each 1 C.**

The definitive definition of the coulomb will come after we have discussed the forces between electric currents in Chapter 29.

CONCEPTUAL EXAMPLE 21–4 An electrically neutral object can be divided into a piece with charge $+q$ and another piece with charge $-q$. If these pieces are widely separated, they attract one another. Is there any way to break a neutral object into two pieces that repel each other?

Answer If an electrically neutral object is divided into two charged pieces, charge conservation demands that the two pieces are oppositely charged with charges of equal magnitude. If these two pieces are separated by distances large compared to their size, then they will look like point charges to one another and so attract. However, things can be different when the pieces are separated by relatively short distances, because then by distributing the charges properly within the pieces, we can use the distance dependence of Coulomb's law to produce a net repulsion. For example, imagine that we have one large and one small piece—the small piece with charge of $+q$ and the large piece with overall charge $-q$. However, we have distributed the charge in the large piece as in Fig. 21–10. According to Coulomb's law, the effect on a given charge of a nearby charge is much bigger than the effect of distant charge. The effect of having the positive charge concentrated at the side of the large piece that is closest to the positively charged small piece means that the overall force between the pieces is repulsive—the two pieces repel at short distances, even though at large distances they attract. While this scenario might be difficult actually to prepare, it reminds us that the distance dependence of the force between charges, with its potentially very large $1/r^2$ factor, can have important effects.

What Do You Think? Consider the larger piece with its charge arranged as in Fig. 21–10. If charge were free to flow within this piece, is this the way the charge would arrange itself?

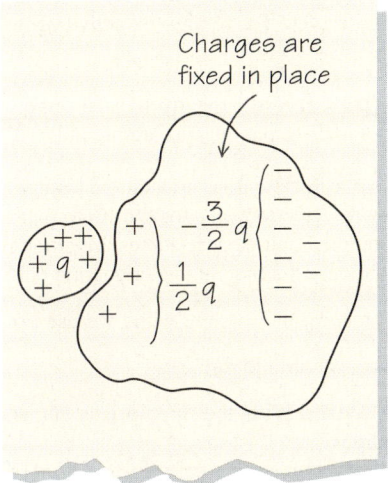

▲ **FIGURE 21–10** Two oppositely charged objects can repel at short distances if it is possible to rearrange the charge within one of them. Here the net charge $-q$ in the right-hand object has been shifted so that there is a positive piece very close to the positively charged object on the left. The magnitude of Coulomb forces falls off rapidly with distance, so that this arrangement could produce repulsion over short distances.

The electric force between point charges, the **Coulomb force**, has a direction and is described by a vector. We write Coulomb's law as

$$\vec{F}_{12} = \frac{1}{4\pi\varepsilon_0}\left(\frac{q_1 q_2}{r_{12}^2}\right)\hat{r}_{12}, \quad (21\text{–}8)$$

COULOMB'S LAW

where $\vec{F}_{12}$ is the force exerted on point charge q_1 due to point charge q_2 when they are separated by a distance r_{12}. Newton's third law tells us that the force exerted on point charge q_2 due to point charge q_1 is then $\vec{F}_{21} = -\vec{F}_{12}$. The unit vector $\hat{r}_{12}$ is directed from q_2 to q_1 along the line between the two charges (Fig. 21–11). Note that if q_1 and q_2 have opposite signs, Eq. (21–8) indicates that the force is attractive, along $-\hat{r}_{12}$. But rather than remembering the subscripts on $\vec{F}$ and the unit vector $\hat{r}$, it is easier to remember that like charges repel and unlike charges attract.

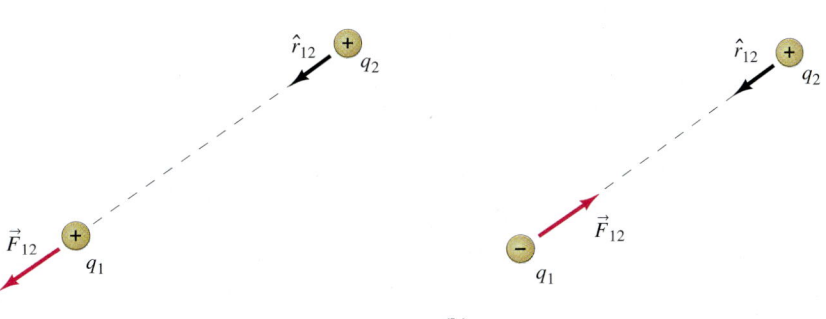

◀ **FIGURE 21–11** $\vec{F}_{12}$ is the force on q_1 due to q_2. The force is in the direction (a) $\hat{r}_{12}$ for like charges, and (b) $-\hat{r}_{12}$ for opposite charges.

EXAMPLE 21–5 Compare the electric force and the gravitational force between the single proton and single electron in a hydrogen atom. Assume a purely classical model of the hydrogen atom, in which the electron moves in a circular orbit around the proton, which is at the atom's center. The radius of a hydrogen atom is about 5×10^{-11} m.

Setting It Up Label as r the given distance between the proton and electron in a hydrogen atom (the atomic radius).

Strategy We calculate the gravitational force, obtaining the masses (m_e for the electron, m_p for the proton) from Table 21–1. Both the gravitational and electric forces are attractive in this case, and we calculate only the magnitudes. Equation (12–4) gives us the gravitational force, $F_g = \dfrac{Gm_e m_p}{r^2}$, while we use Eq. (21–8) to find the electric force, magnitude F_E.

Working It Out With the appropriate numerical values for masses and charges,

$$F_g = \dfrac{(6.67 \times 10^{-11}\,\text{N}\cdot\text{m}^2/\text{kg}^2)(9.11 \times 10^{-31}\,\text{kg})(1.67 \times 10^{-27}\,\text{kg})}{(5 \times 10^{-11}\,\text{m})^2}$$
$$= 4 \times 10^{-47}\,\text{N}.$$

$$F_E = \dfrac{(9 \times 10^9\,\text{N}\cdot\text{m}^2/\text{C}^2)(1.6 \times 10^{-19}\,\text{C})^2}{(5 \times 10^{-11}\,\text{m})^2} = 9 \times 10^{-8}\,\text{N}.$$

$$\dfrac{F_E}{F_g} = \dfrac{9 \times 10^{-8}\,\text{N}}{4 \times 10^{-47}\,\text{N}} \cong 2 \times 10^{39}.$$

This calculation of the ratio could more easily have been carried out directly because it is independent of r — the common factor $1/r^2$ would cancel in the ratio. We have shown that on the atomic scale, the electric force is much greater than the gravitational force and that we are justified in ignoring gravitation at the atomic level.

What Do You Think? If the force due to gravitation is so much smaller than that of the electric force, why do we even notice it?

EXAMPLE 21–6 Two small cork balls are both charged to 40 nC and placed 4.0 cm apart. What is the magnitude of the electric force between them? Each cork ball has a mass of 0.46 g. Compare the magnitude of the electric force between them to the weight of one of the balls. What will happen if the balls are arranged vertically within a tube, with one ball placed 4 cm above the other?

Setting It Up Denote each charge by Q, each mass by m, and the separation of the balls by d. Figure 21–12 shows the cork balls (in the arrangement appropriate for the last question).

Strategy We use Coulomb's law, Eq. (21–8), to find the electric force between the balls, magnitude F_E, and then use $W = mg$ to find the weight.

Working It Out The electric force has magnitude

$$F_E = \dfrac{kQ^2}{d^2} = \dfrac{(9 \times 10^9\,\text{N}\cdot\text{m}^2/\text{C}^2)(40 \times 10^{-9}\,\text{C})^2}{(4.0 \times 10^{-2}\,\text{m})^2}$$
$$= 9.0 \times 10^{-3}\,\text{N}.$$

The weight of each cork ball is

$$W = mg = (0.46 \times 10^{-3}\,\text{kg})(9.8\,\text{m/s}^2) = 4.5 \times 10^{-3}\,\text{N}.$$

We conclude that if the balls start in a vertical orientation as in Fig. 21–12, separated by 4 cm, the (repulsive) electric force would lift the upper cork ball. The electric force would balance the gravitational force only if the balls started out a little farther apart—you could easily calculate that distance.

What Do You Think? What would have happened if the experiment with the vertical placing were carried out as above, but each ball had twice the mass?

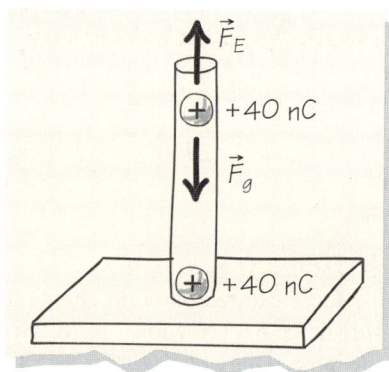

▲ **FIGURE 21–12** A small object can be suspended in space when equal but opposite gravitational and electric forces act on it.

THINK ABOUT THIS...
ARE THERE COULOMB FORCES BETWEEN HOME APPLIANCES?

A charge of 1 C is huge: Two such charges one meter apart exert a force of 9×10^9 N on each other. (By comparison, the force on you due to Earth's gravity is on the order of just 500 N.) In a typical household appliance, 1 C or more may move through the appliance every second, so why doesn't this produce enormous forces between appliances or the wires through which the charge moves? The answer is that the electrons move against a compensating background of positive charge, the stationary ions that are left behind when the electrons move away from their "parent" atoms in a conductor. Wires as well as the elements of appliances through which the electrons move are actually electrically neutral. Any Coulomb force is negligible. As we will see in later chapters, a different set of effects are associated with moving charges, even in an electrically neutral system, and these effects are the relevant ones for the operation of an appliance or device.

21–4 Forces Involving Multiple Charges

What happens if multiple charges are present? Experiment shows that the **principle of superposition** applies: The force on any one charge due to a collection of other charges is the vector sum of the forces due to each individual charge. In this respect, the Coulomb force is again like the gravitational force. The superposition principle allows us to find the force due to a set of charges on another charge or, for that matter, on another set of charges.

As an example of how superposition is applied, consider four charges, numbered 1, 2, 3, 4 (Fig. 21–13). The total force on, say, charge q_2 is the *vector sum* of the forces due to the other individual charges, q_1, q_3, and q_4:

$$\vec{F}_{2,\text{total}} = \vec{F}_{21} + \vec{F}_{23} + \vec{F}_{24}. \qquad (21\text{–}9)$$

If there are N charges—$q_1, q_2, \ldots, q_N$—all acting on a charge q, the total force $\vec{F}$ on charge q is the vector sum of the individual forces $\vec{F}_i$ on charge q due to charge q_i:

$$\vec{F} = \sum_{i=1}^{N} \vec{F}_i = \frac{q}{4\pi\varepsilon_0} \sum_{i=1}^{N} \frac{q_i}{r_i^2} \hat{r}_i \qquad (21\text{–}10)$$

The vector $\hat{r}_i$ is the unit vector from charge q_i to charge q. We have moved the common factor $q/4\pi\varepsilon_0$ out of the sum.

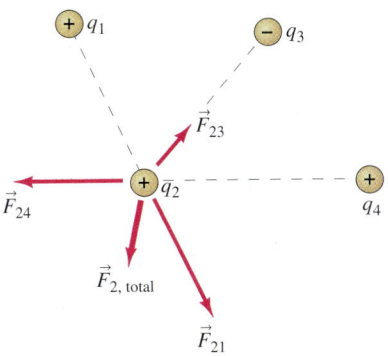

▲ **FIGURE 21–13** The superposition principle applies for multiple charges. The total force on charge q_2 is the vector sum of the individual forces on q_2 due to charges q_1, q_3, and q_4.

Problem-Solving Techniques

It is helpful to keep the following techniques in mind when looking at problems involving electric forces on a given charge in the presence of several other fixed charges or continuous distributions of charges:

1. Draw a clear diagram of the situation. Be sure to distinguish between the fixed external charges and the charges on which the forces must be found. The diagram should contain coordinate axes for reference.

2. Do not forget that the electric force that acts on a charge is a vector quantity; when many charges are present, the net force is a vector sum. In calculations, it is usually simplest to use unit vectors in a Cartesian coordinate system.

3. Search for symmetries in the distribution of charges that give rise to the electric force. When symmetries are present, the net force along certain directions will be zero. For example, if a point charge is midway between two identical charges, we know without performing any calculations that the net force on it will be zero.

EXAMPLE 21–7 Consider three point charges $q_1 = q_2 = 2.0$ nC and $q_3 = -3.0$ nC placed at the vertices of the triangle shown in Fig. 21–14. Find the net forces on q_1 and q_3, assuming that only Coulomb forces act.

Strategy The force on q_1, say, is due to the presence of charges q_2 and q_3. We find the vector forces on q_1 due to each of the charges q_2 and q_3 separately, then add them vectorially to find the net force on q_1. Coulomb's law depends on distance, and we will take these distances from Fig. 21–14. We can then do a similar calculation for the force on q_3.

Working It Out The force on q_1 is

$$\vec{F}_1 = \vec{F}_{12} + \vec{F}_{13} = \frac{q_1}{4\pi\varepsilon_0} \left[\left(\frac{q_2}{r_{12}^2}\right)\hat{r}_{12} + \left(\frac{q_3}{r_{13}^2}\right)\hat{r}_{13} \right].$$

From Fig. 21–14, we can deduce that $\hat{r}_{12} = -\hat{i}$ and $\hat{r}_{13} = -\hat{j}$. Thus

$$\vec{F}_1 = (9.0 \times 10^9 \text{ N} \cdot \text{m}^2/\text{C}^2)(2.0 \times 10^{-9} \text{ C})$$
$$\times \left[\frac{(2.0 \times 10^{-9} \text{ C})}{(2.0 \text{ m})^2}(-\hat{i}) + \frac{(-3.0 \times 10^{-9} \text{ C})}{(2.0 \text{ m})^2}(-\hat{j}) \right]$$
$$= (-9.0 \times 10^{-9} \text{ N})\hat{i} + (14 \times 10^{-9} \text{ N})\hat{j}.$$

The direction of force $\vec{F}_1$ is shown in Fig. 21–14.

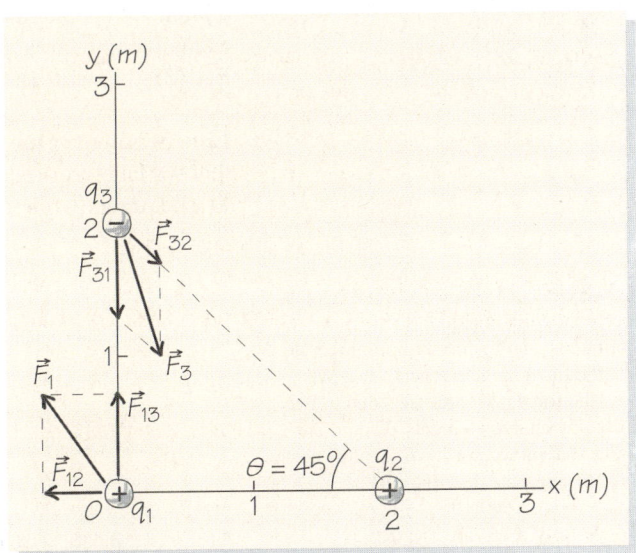

▲ **FIGURE 21–14** The positions of three point charges are indicated. Charges q_1 and q_2 are positive, while q_3 is negative. Forces $\vec{F}_{12}$ and $\vec{F}_{13}$ on charge q_1, and their resultant, $\vec{F}_1$, as well as forces $\vec{F}_{31}$ and $\vec{F}_{32}$ on charge q_3, and their resultant, $\vec{F}_3$, are drawn.

(continues on next page)

The force on q_3 is calculated in much the same way, with the unit vector $\hat{r}_{32}$, which points from q_2 to q_3, given by $(-\cos\theta)\hat{i} + (\sin\theta)\hat{j}$:

$$\vec{F}_3 = \vec{F}_{31} + \vec{F}_{32} = \frac{q_3}{4\pi\varepsilon_0}\left[\left(\frac{q_1}{r_{31}^2}\right)\hat{r}_{31} + \left(\frac{q_2}{r_{32}^2}\right)\hat{r}_{32}\right]$$

$$= (9.0 \times 10^9 \text{ N} \cdot \text{m}^2/\text{C}^2)(-3.0 \times 10^{-9} \text{ C})$$

$$\times \left[\frac{(2.0 \times 10^{-9}\text{ C})}{(2.0\text{ m})^2}\hat{j} + \frac{(2.0 \times 10^{-9}\text{ C})}{(2.0\text{ m})^2 + (2.0\text{ m})^2}((-\cos\theta)\hat{i} + (\sin\theta)\hat{j})\right].$$

The angle θ is 45°, or $\pi/4$ rad, so $\vec{F}_3$ becomes

$$\vec{F}_3 = (-14 \times 10^{-9}\text{ N})\hat{j} + (4.8 \times 10^{-9}\text{ N})\hat{i} - (4.8 \times 10^{-9}\text{ N})\hat{j}$$
$$= (4.8 \times 10^{-9}\text{ N})\hat{i} - (19 \times 10^{-9}\text{ N})\hat{j}.$$

What Do You Think? If q_1 and q_2 are fixed at the given locations, can we place q_3 at some point such that the net force on it is zero? If the point exists, is it a stable equilibrium point?

Continuous Distributions of Charges

The fact that charge is quantized will have no physical consequence when we deal with charges that are much larger than e. Such charges are composed of large numbers of electrons or protons. We can normally treat a large collection of point charges as a *continuous distribution* of charge. This is entirely analogous to thinking of a large object as a continuous distribution of mass, even though we know it is made of individual atoms. The techniques for treating continuous charge distributions will be very similar to the techniques we developed for treating continuous mass distributions.

We consider first the interaction of a point charge q with a large continuous charge distribution (Fig. 21–15). The force on q due to the tiny element of volume shown, which contains charge Δq and is a distance r from q, is

$$\Delta\vec{F} = \frac{q}{4\pi\varepsilon_0}\frac{\Delta q}{r^2}\hat{r}.$$

In turn, the net force on q is the sum over the forces due to the elements Δq:

$$\vec{F} = \sum \Delta\vec{F} = \sum \frac{q}{4\pi\varepsilon_0}\frac{\Delta q}{r^2}\hat{r} = \frac{q}{4\pi\varepsilon_0}\sum\frac{\Delta q}{r^2}\hat{r}. \quad (21\text{–}11)$$

We have to keep in mind that this is a vector sum.

At this point, we have not been very specific about how actually to do the sum. Just as for masses we found a more concrete way to proceed by working with a mass density, here it is useful to use the idea of a *charge density*. We separate the cases according to whether the continuous distribution of charge is distributed *along a line*, spread *over a plane*, or spread *throughout a volume*. In each case, we replace the finite charge Δq in a small length, area, or volume with the infinitesimal charge dq, and replace the sum in Eq. (21–11) with an integral. Let us look at each of these distributions in more detail:

Line Segment (One Dimension): If the charge distribution is distributed along a line that we label as the x-axis, we denote the *linear charge density* (charge/unit length) by $\lambda(x)$. (Actually the line of charge does not have to be a straight line; x really serves as a

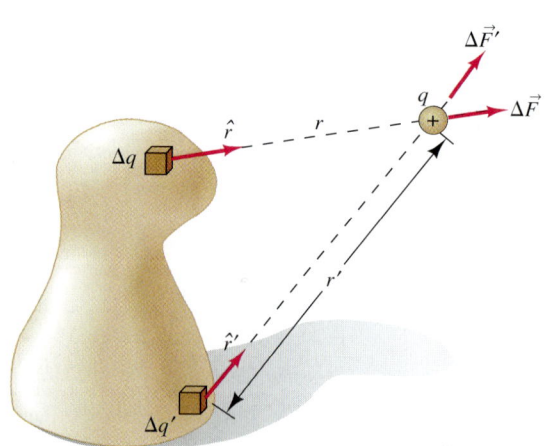

▶ **FIGURE 21–15** To find the total force on a point charge q due to a continuous charge distribution, integrate over the tiny charge elements Δq. We show the forces $\Delta\vec{F}$ and $\Delta\vec{F}'$ due to two of the tiny charge elements Δq and $\Delta q'$. Notice that the vector $\hat{r}$ will change as we move through the distribution.

way to label where you are along the distribution here.) The charge on an infinitesimal length dx of the line is

$$\text{for a charged line: } dq = \lambda(x)\, dx. \quad (21\text{-}12)$$

Note that λ can be a function of x; that is, the charge density can vary along the line. The force can be on the point charge q is then (Fig. 21–16)

$$\vec{F} = \frac{q}{4\pi\varepsilon_0} \int \hat{r}' \frac{\lambda(x)\, dx}{r'^2}. \quad (21\text{-}13)$$

Notice the meaning of the integration: We move along the line of charge, and each point along the line is a different distance r' and at a different direction $\hat{r}'$ from the charge q. Later we look at some examples to see how this works in practice. The line segment need not be straight. We must follow it wherever it goes in the integration of Eq. (21–13)—we will in the more general case refer to the integration as a line integral, as we did in our treatment of work (Chapter 6).

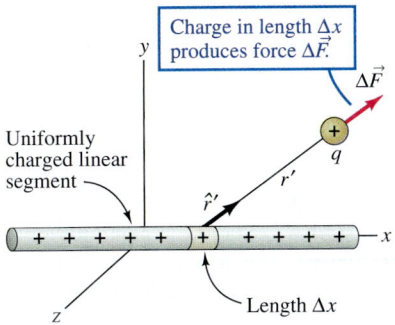

▲ **FIGURE 21–16** A one-dimensional charge distribution and the infinitesimal force on a point charge due to an infinitesimal piece of it.

Surface (Two Dimensions): Here the charge is distributed across a surface. We denote the *surface charge density* (charge/unit area) by σ. The density σ could be a constant (uniform charge density), or it could vary from point to point on the surface. The charge on an infinitesimal area dS of the surface is (Fig. 21–17)

$$\text{for a charged surface: } dq = \sigma\, dS. \quad (21\text{-}14)$$

The force on the point charge q is (Fig. 21–17)

$$\vec{F} = \frac{q}{4\pi\varepsilon_0} \int_{\text{surface}} \hat{r}' \frac{\sigma\, dS}{r'^2}. \quad (21\text{-}15)$$

Here, we are integrating over all the elements of the surface, as indicated by the subscript of the integral sign. In practice, such integrals can be done by working out one-dimensional integrals.

Volume (Three Dimensions): When the charge is distributed through a volume, we write the *volume charge density* of the distribution as $\rho(\vec{r}')$, which means that the infinitesimal charge dq contained in the infinitesimal volume dV is

$$dq = \rho\, dV. \quad (21\text{-}16)$$

In terms of the charge density of the continuous charge distribution, the *net* force due to the volume element shown in Fig. 21–15 is

$$\vec{F} = \frac{q}{4\pi\varepsilon_0} \int_{\text{volume}} \hat{r}' \frac{\rho\, dV}{r'^2}. \quad (21\text{-}17)$$

The integration is over the entire volume of the charge distribution, and that is why we have used the subscript. Again, such integrals often involve simpler one-dimensional integrals in practice.

In each of these cases, the argument of the charge distribution ρ is the vector displacement $\vec{r}'$, as it is the vector displacement from an element of the charge distribution to the point charge on which the force acts that is important. We may, however, have a *uniform* charge distribution, in which charge is distributed evenly throughout a region. In that case, the linear charge density λ is the total charge on the line divided by the length of the line, the surface charge density σ is the total charge on the surface divided by the area of the surface, and the volume charge density ρ is the total charge in the three-dimensional region divided by the volume of the region. All three quantities are constants that can be removed from the integral for the net force. Keep in mind that a uniform charge distribution is not possible with a conductor, within or on which charges are free to move.

The integrals that express the force may be simple to perform, particularly if there is some symmetry in the distribution. Without the symmetry, it may be hard to find an analytical answer for the integral, but numerical integration using a computer is always possible.

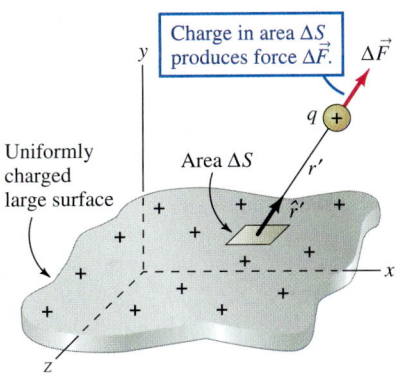

▲ **FIGURE 21–17** A two-dimensional charge distribution and the infinitesimal force on a point charge due to an infinitesimal piece of it.

CONCEPTUAL EXAMPLE 21–8 A point charge q is placed at the center of a uniformly charged ring (Fig. 21–18a). What is the net force on that point charge? Analyze this for both the point charge and the ring charge having the same sign and having different signs.

Answer Symmetry is very often a useful tool in dealing with forces due to charge distributions. In this case there is a good deal of symmetry. The ring is uniformly charged, and every point on the ring is equally distant from the point charge. There will always be equal and opposing forces due to the charged ring on either side of the charge q, as in Fig. 21–18b where we have identified areas of the ring that exert equal and opposing forces on point charge q. It does not matter whether q is positive or negative, because it will be repelled equally or attracted equally in all directions in the plane of the ring. The net force is zero, and no complex mathematical calculation is needed.

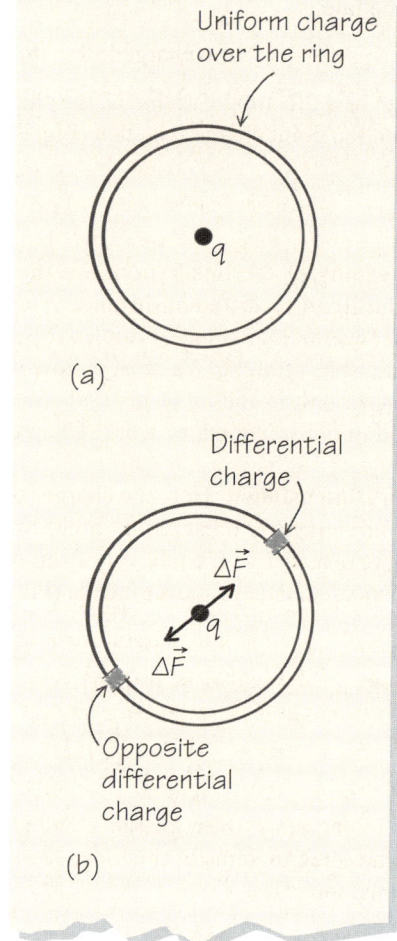

▶ **FIGURE 21–18** (a) The ring is uniformly charged. (b) Sections on opposite sides of the ring exert forces on the charge at the center that cancel.

EXAMPLE 21–9 Find the force on a point charge q_1 located on the axis of a uniformly charged ring of total charge Q. The radius of the ring is R, and q_1 is located a distance L from the center of the ring.

Setting It Up The geometry of the ring and point charge is shown in Fig. 21–19a, with appropriate labeling.

Strategy We want to find the force on q_1 due to a differential charge dq on the ring and then add the effects of all the parts of the ring. This is simplified by the recognition of the symmetry of this situation. Consider a small segment of the ring containing charge dq (Fig. 21–19a). All such segments are located a distance $r' = \sqrt{L^2 + R^2}$ from charge q_1, and the line to *any* segment on the ring makes the angle θ with the x-axis.

Next, look at the components of the force on q_1. Because every segment of the ring is the same distance r' from q_1, the *magnitude* of the infinitesimal force from each infinitesimal slice is the same. This is not true for the direction. The force from segment dq at the top of the ring ($z = 0$, $y = R$) is $d\vec{F}_{dq}$, and this force has components in the $+x$-direction and the $-y$-direction (Fig. 21–19b). The force from segment dq' at the bottom of the ring ($z = 0$, $y = -R$) is $d\vec{F}_{dq'}$, and this force has components in the $+x$-direction and the $+y$-direction. If the magnitude dq equals the magnitude dq', the y-components of the force will cancel each other while the x-components of the force will add. The y-components are the components perpendicular to the axis of the ring. This cancellation will hold for every perpendicular component of the force because we can always consider the charge elements in pairs. Thus we need compute only the component F_x by adding the (identical) infinitesimal components dF_x from each little element.

Working It Out The x-component from the element shown in Fig. 21–19a is

$$dF_x = \frac{q_1}{4\pi\varepsilon_0} \frac{dq}{r'^2} \cos\theta = \frac{q_1}{4\pi\varepsilon_0} \frac{\cos\theta}{(R^2 + L^2)} dq.$$

The net force has only an x-component and is the sum over the infinitesimal x-components:

$$F_x = \int dF_x = \int \frac{q_1}{4\pi\varepsilon_0} \frac{\cos\theta}{(R^2 + L^2)} dq.$$

At this point we see the symmetry come into play again: the coefficient of dq in the expression for dF_x is the same for every element and can therefore be placed outside the integral sign. Thus

$$F_x = \frac{q_1}{4\pi\varepsilon_0} \frac{\cos\theta}{(R^2 + L^2)} \int dq = \frac{q_1 Q}{4\pi\varepsilon_0} \frac{\cos\theta}{(R^2 + L^2)}.$$

We have used the fact that $\int dq = Q$, the entire charge. Finally, from trigonometry we find

$$\cos\theta = \frac{L}{\sqrt{R^2 + L^2}},$$

so

$$F_x = \frac{q_1 Q}{4\pi\varepsilon_0} \frac{L}{(R^2 + L^2)^{3/2}}. \quad (21\text{-}18)$$

A check is always desirable, and we can immediately find one: When the point charge q_1 is very far from the ring, the ring should appear as a distant point of total charge Q, and the force should take on the Coulomb form $q_1 Q/(4\pi\varepsilon_0 L^2)$; this is indeed the limit of Eq. (21-18) when $L \gg R$.

What Do You Think? What is the force on q_1 when $L = 0$? (This is a second check on the result.)

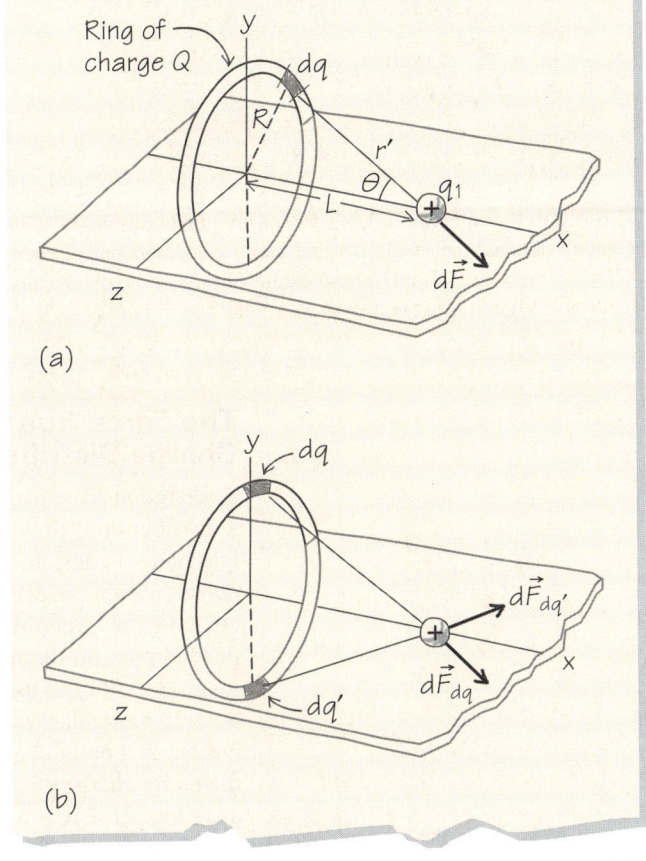

▶ **FIGURE 21-19** (a) The force on a point charge q_1 due to a ring with total charge Q. First we find the force between the point charge and a tiny ring segment with charge dq. (b) Only the x-component of the force needs to be determined, because the y- and z-components will cancel due to symmetry.

EXAMPLE 21-10 A straight rod of length L is aligned along the x-axis, with the ends at $x = \pm L/2$. The total charge on the rod is zero but the charge density is not; it is given by $\lambda(x) = 2\lambda_0 x/L$ (positive to the right of the origin, negative to the left). Find the force on a charge q located at a point $x = R$ on the x-axis, to the right of the right-hand end of the rod.

Setting It Up The geometry of the rod and of the charge on it is shown in Fig. 21-20a. The point charge for which we want to find the acting force is also indicated, at $x = R > L/2$.

Strategy We start by dividing the rod into elements that can be treated as points as far as figuring the force they exert on charge q. Thus we consider a thin slice of the rod with charge dQ located at point x, with thickness dx; this slice will have charge $dQ = \lambda(x)\,dx$. This is drawn in Fig. 21-20b. We find the force between each element of this type and q and then add these forces in the form of an integral. Although the symmetry in this case is not very marked, the integral will turn out to be an elementary one.

Working It Out With $dQ = \lambda\,dx = \dfrac{2\lambda_0}{L} x\,dx$, the infinitesimal force exerted by this charge on charge q is given by

$$d\vec{F} = \frac{q}{4\pi\varepsilon_0} \frac{2\lambda_0}{L} x\,dx\, \frac{1}{(R-x)^2}\,\hat{i}.$$

This force is aligned with the x-direction, as the unit vector indicates. The summation over the force due to the different slices to give the total force will also be aligned this way. This summation takes the form

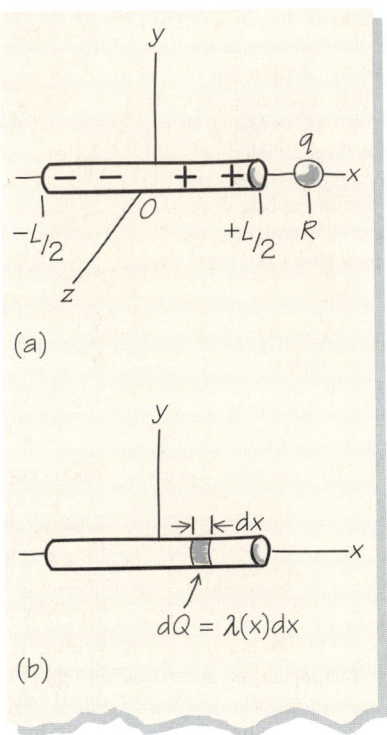

▲ **FIGURE 21-20** (a) A nonuniform charge density. (b) Isolating a section of the charge on the rod.

(continues on next page)

$$\vec{F} = \int d\vec{F} = \int_{-L/2}^{L/2} \frac{q}{4\pi\varepsilon_0} \frac{2\lambda_0}{L} \frac{x\,dx}{(R-x)^2} \hat{i}$$

$$= \frac{2q\lambda_0}{4\pi\varepsilon_0 L} \hat{i} \int_{-L/2}^{L/2} \frac{x\,dx}{(R-x)^2}$$

$$= \frac{2q\lambda_0}{4\pi\varepsilon_0 L} \hat{i} \left\{ \ln\left[\frac{R-(L/2)}{R+(L/2)}\right] + R\left[\frac{1}{R-(L/2)} - \frac{1}{R+(L/2)}\right] \right\}.$$

The result of the integration has come from a table of integrals, and we have evaluated the result at upper and lower limits and subtracted. These limits, $\pm L/2$, reflect the extent of the charge distribution. In this case, charge is present from $x = -L/2$ to $x = L/2$.

If we drew a numerical plot of the factor in curly brackets in the expression for the force, we would see that this force is positive—if the charge q is positive it acts to push it to the right. This is sensible because the right side of the rod is positively charged, and the point charge is closer to the right side of the rod. The space variation of the coulomb force makes the repulsion from the closer part of the rod more important than the attraction from the farther part of the rod.

What Do You Think? Assuming the charge q is positive, what is the force on it when it is placed very close to the right end of the rod?

The Force Due to a Spherically Symmetric Charge Distribution

A charge distribution that is *spherically symmetric* is often quite easy to analyze. Such a distribution is in the form of a sphere centered at, say, point P, with the charge density having a constant value at a given (radial) distance from P. (Notice that the charge density could nevertheless vary with the radial distance from P.) This case was discussed extensively in Chapter 12 for the gravitational force. The results for gravity depended only on the fact that the force due to each bit of the charge distribution varies inversely with the distance from it squared, and so we can use the Chapter 12 results here. In particular, we can say that the force of the spherically symmetric charge distribution on a point charge q outside the distribution (Fig. 21–21a) is the same as though the entire charge of the distribution were concentrated at P (Fig. 21–21b). Moreover, if as in Fig. 21–21c the point charge q is inside any part of the distribution, then the force on q due to the part of the distribution that lies outside q is zero (Fig. 21–21d).

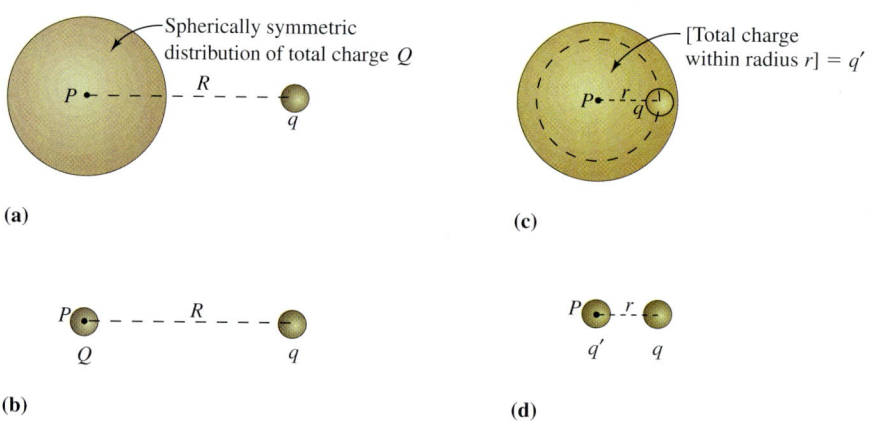

▶ **FIGURE 21–21** (a) A spherically symmetric charge distribution of total charge Q is centered on the point P. The force on a point charge q outside the distribution a distance R from P is the same as (b) the force it would experience if a point charge Q were located at P. (c) If q lies inside the distribution a distance r from P, and q' is the total charge that lies within a sphere of radius r centered on P, then it experiences the same force it would have (d) if there were a point charge q' at P.

Summary

Electric charge occurs in two forms, which we label as positive and negative charge. Charges of the same sign repel each other, and charges of unlike sign attract each other. In SI units, charge is measured in coulombs.

Much of the behavior of materials under the influence of electric forces is characterized by the ease with which electrons are dislodged from their constituent atoms and molecules and move through the material. Metals are normally good conductors of electric charge, whereas most nonmetals are not and are called insulators.

The basic electric charge is that of the electron. The electron has a charge of $-e$, and the proton has a charge of $+e$, with $e = 1.602 \times 10^{-19}$ C. Electric charge in matter is quantized in multiples of e. Charge is conserved in all interactions, meaning that the net charge before an interaction is the same as the net charge after the interaction.

The electric (Coulomb) force exerted by the point charge q_2 on the point charge q_1, when these are separated by a distance r_{12} is given by Coulomb's law:

$$\vec{F}_{12} = \frac{1}{4\pi\varepsilon_0}\left(\frac{q_1 q_2}{r_{12}^2}\right)\hat{r}_{12}, \tag{21-8}$$

where the factor $1/(4\pi\varepsilon_0)$ sets the units of charge.

The principle of superposition applies when multiple charges are present. The Coulomb forces on a point charge q due to all other charges add together vectorially. For continuous charge distributions, this addition takes the form of an integration, and the force of such a distribution on q depends on the charge distribution. For charges distributed on a line, over a surface, or through a volume, the force on the point charge q due to the distribution is, respectively,

$$\vec{F} = \frac{q}{4\pi\varepsilon_0}\int \hat{r}' \frac{\lambda(x)\,dx}{r'^2}, \tag{21-13}$$

$$\vec{F} = \frac{q}{4\pi\varepsilon_0}\int_{\text{surface}} \hat{r}' \frac{\sigma\,dS}{r'^2}, \tag{21-15}$$

and

$$\vec{F} = \frac{q}{4\pi\varepsilon_0}\int_{\text{volume}} \hat{r}' \frac{\rho\,dV}{r'^2}. \tag{21-17}$$

Here λ, σ, and ρ are the one-, two-, and three-dimensional charge densities, respectively.

On all but the astronomical scale, electric forces tend to be much stronger than gravitational forces. The electric force is responsible for making atoms, molecules, solids, and liquids stable, and all chemical reactions and biological processes are a result of electrical interactions.

Understanding the Concepts

1. Two identical positive charges are placed on a table and fixed there. Find all the places on the table where the net force on a test charge due to the two charges is zero.
2. Particles of opposite charge attract with an inverse-square law. Are there analogues of Kepler's laws for a system composed of such a pair, and what are they?
3. A balloon rubbed on a sweater and placed on a wall will often stay on the wall for a while. Explain how this happens.
4. When you walk across a carpet, you often pick up enough electric charge to cause a spark when you touch a doorknob. In climates that are dry in winter, this phenomenon is much more common in the winter than in the summer. Why?
5. Two metallic spheres on insulating stands are placed on an air-track. The mass of one sphere is five times larger than the other, and the charges are both positive in the ratio 3:1. The two objects are held at rest and then let go. What determines how far the two objects each move in a short time interval? How would you find the ratio of the distances that they travel in that interval?
6. By using the apparatus discussed in Section 21–1, how could you determine what charge you accumulate by walking across a wool rug?
7. Atoms consist of negatively charged electrons bound to the positively charged nucleus by the Coulomb force. The electrons are rearranged when two different chemicals are brought together. Would you expect the electrons that are closer to the nucleus or the ones that are farther from the nucleus to be more involved in chemical reactions?
8. Neutrons and protons are believed to be made of two types of charged particles called quarks, having charge $-1/3\,e$ and $2/3\,e$, as mentioned in Section 21–2. List the possible combinations of only three quarks that make up neutrons and protons.
9. When we unpack boxes, we often find that the "peanuts" used for cushioning stick to our hands, and it is difficult to shake them off. Why?
10. Some materials lose electrons easily by rubbing, so why are many of the objects around us not charged at all times?
11. *Earnshaw's theorem* states that a point charge cannot be in stable equilibrium while purely electrostatic forces act on the point charge. Consider a ring that is uniformly positively charged, with a positive charge at the center. It appears that the center charge suffers an identical repulsive force from every direction. How can the theorem be true?
12. How is the existence of a battery, which sends negative charges out of one of its contact points, consistent with the conservation of charge?
13. You have a cork ball with a charge of -4.8×10^{-19} C and three uncharged cork balls. Can you devise a method of touching cork balls together in sequence that will give a charge of -0.8×10^{-19} C to one of the balls?
14. You are given objects with two different charges. Can you determine whether the charges on these objects attract or repel? Can you determine with no further information whether the charges on the objects are positive or negative?
15. We spoke of generating a spark on a winter's day when we touch a conducting line to Earth and become grounded. Automobile tires are such good insulators that a car body is not connected to Earth by a conductor. How do you explain the spark that occurs when you touch a car door after you have rubbed the car upholstery?
16. Suppose that the electric charge of a fundamental particle such as an electron depends on the speed v of the particle, so that $e = e_0[1 + (\kappa v^2/c^2)]$, where e_0 is the particle's "rest charge," c is the speed of light, and κ is some tiny number. Discuss ways in which you might measure κ. Is there any experimental reason why κ must be small, if not zero?
17. Does the modification of the electric charge proposed in Question 16 necessarily violate the principle that it should not be possible to detect the absolute velocity of an object by means of any experiment?

18. The color of the light emitted by quasars is evidence that the charge on electrons has not changed over billions of years. Is saying that the charge on electrons and protons is unchanged equivalent to the statement that charge is conserved?
19. Suppose that electrons had charge $-e$ and protons had charge $+e(1 + \delta)$, with δ very small. Would there necessarily be an additional repulsive $1/r^2$ force between the Moon and Earth, for example, that could overpower the gravitational attraction between these bodies?
20. Consider a uniform, spherical positive charge distribution. A negative charge is placed at the center. Discuss the net force on that point charge. Discuss what happens to the point charge if it is placed a bit off center.
21. Describe what happens to $\vec{F}_1$ and $\vec{F}_3$ of Example 21–7 if the charge q_2 is doubled.
22. How would we know if at Alpha Centauri, the nearest star system nearest to us, the electric force had a $1/r^3$ dependence rather than the $1/r^2$ dependence on Earth?

Problems

21–1 Charge—A Property of Matter

1. (I) A cork ball is charged to $+1$ nC. How many fewer electrons than protons does the ball have?
2. (I) A uranium atom has undergone a violent collision that has stripped off 21 of its electrons. What is the charge of the resulting atom? If a uranium nucleus contains 92 protons and 146 neutrons, what is the charge of the nucleus?
3. (I) What is the total charge of all the electrons in 1 g of CO_2?
4. (I) Three identical metallic spheres are connected by wires, and a charge Q is placed on one of them. The wires are then removed. One of the spheres is then connected by wire to the ground. That wire is then removed. This particular sphere is then connected by a wire to one of the other spheres. What is the charge on each of the spheres when the process is completed?
5. (I) If you could remove 1 electron in 10^{13} from the gold coin of Example 21–2, how many would you be removing per atom, on the average?
6. (I) How many protons are contained in the gold coin of Example 21–2?
7. (II) A cork ball that is covered with conducting paint and charged to -4×10^{-10} C is touched by an identical but uncharged cork ball; the balls then separate. This second cork ball is then touched by a third uncharged cork ball, and they separate. What is the charge of each ball at the end, and how many excess electrons does each ball have?
8. (II) A cork ball covered with conducting paint is charged to -1.04×10^{-13} C. You have three similar but uncharged cork balls. Describe a method by which to produce a cork ball with a charge of -0.13×10^{-13} C. Do you need all three extra balls? Explain.
9. (II) An aluminum ball of mass 0.1 g is given a negative charge of 1 μC. What is the fractional increase in the number of electrons the ball contains?
10. (II) Two cork balls of mass 0.2 g hang from the same support point by massless insulating threads of length 20 cm (Fig. 21–22). A total positive charge of 3.0×10^{-8} C is added to the system. Half this charge is taken up by each ball, and the balls spread apart to a new equilibrium position. (a) Draw a free-body diagram for each cork ball. (b) What is the tension in the threads before the charge is added, and what is it after? (c) What is the value of angle θ in the figure? This device is a type of electroscope, or *electrometer*, a meter that measures electric charge. Angle θ measures the amount of charge on the balls if we can be sure that the charge is divided between them equally. This constraint is circumvented when the electrometer is made of a single strip of conducting material draped at its midpoint over a hook; the charge is then distributed over the strip equally, and half the strip repels the other half.

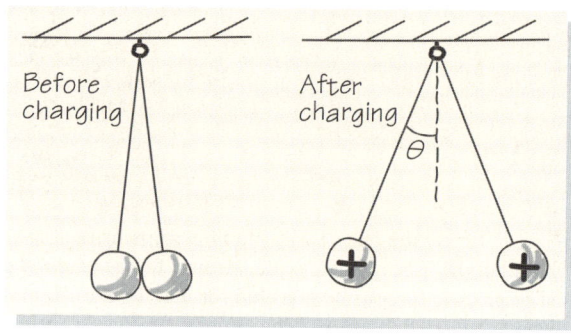

▲ **FIGURE 21–22** Problem 10.

11. (II) Silicon is the most abundant material on Earth's surface. (a) Assume that Earth is made of silicon (28 g/mol), and calculate the total number of negative charges contained within Earth. (b) When we neutralize a cork ball that has a charge of 1 μC by grounding it to Earth, what fractional change are we making in the total negative charge contained within Earth?

21–2 Charge Is Conserved and Quantized

12. (I) One possible result of the high-energy collision of two protons is the reaction $p + p \rightarrow X + p$. What is the electric charge of particle X?
13. (I) *Antiparticles* have the same mass as their counterpart particles but have an opposite charge. For example, the antiparticle of an electron, e^-, is the positron, e^+. Most antiparticles are denoted by a bar over the particle, so $\bar{p}$ is the antiparticle of the proton, and it has a charge of $-e$. Which of the following reactions satisfy the conservation of charge:
    (a) $p + \bar{p} \rightarrow e^+ + e^- + e^+ + e^- + 2n$;
    (b) $e^+ + e^- \rightarrow 2p + n + 2\gamma$,
    (c) $e^+ + e^- \rightarrow e^+ + e^- + p + \bar{p} + 2\gamma$;
    (d) $n + p \rightarrow e^- + p + \bar{p}$
14. (I) How much charge is contained in 6.5×10^{-4} g of electrons?
15. (II) The electric charge of an object is independent of the object's motion. Suppose that this were not true, but that the charge of a particle such as an electron or a proton that moves at speed v has the form $e = e_0[1 + (v^2/c^2)]$, where e_0 is the particle's charge when at rest and $c \cong 3 \times 10^8$ m/s is the speed of light. What would the net charge on a hydrogen atom be, assuming that the atom consists of a proton at rest and an electron orbiting the proton at average speed $v \cong (1/137)c$?

21–3 Coulomb's Law

16. (I) How far apart must two protons be for the Coulomb force on each other to be the same as the weight of one proton on Earth's surface?

17. (I) A proton is believed to consist of two "up" quarks of charge $+2/3\,e$ and one "down" quark of charge $-1/3\,e$. Assume that all three quarks are equidistant from each other at the distance of 1.5×10^{-15} m. What are the electrostatic forces between each pair of the three quarks?

18. (I) Two small balls, each of mass 16 g, are each charged with $+8.5$ nC. What distance apart must they be if the force on one of them has the same magnitude as the weight of that ball?

19. (I) Two identically charged sodium ions separated by 4.5×10^{-9} m have a force between them of 1.1×10^{-11} N. What is the charge of each ion, and how many electron charges does this represent?

20. (I) Two small cork balls have the same charge. When their centers are placed 2 cm apart, the force between them is observed to be 0.18 N. What is the cork balls' charge? Why do we have to assume that the size of the cork balls is small compared to 2 cm?

21. (I) Two tiny cork balls, both of mass 0.10 g, each have just one electron charge, $q = -1.6 \times 10^{-19}$ C. They are separated by 15 cm, which is much greater than their sizes. What is the ratio of the magnitudes of the Coulomb force between them to the gravitational force they exert on each other? Why is this result so different from that of Example 21–5?

22. (II) The experiment of Cavendish to determine the gravitational constant (see Chapter 12) relies on the measurement of a force of about 7×10^{-7} N between two masses separated by a distance of 0.1 m. One possible source of error is a small electric charge on the balls. Assuming the charges are equal, what is the magnitude of the largest allowed charge, if the force is to be measured to at least a 0.05 percent accuracy?

23. (II) Suppose that we were to measure a charge in some new unit, which we will call the esu, so defined that Coulomb's law reads, in magnitude, $F = q_1 q_2 / r^2$, and so that $F = 1$ dyne (10^{-5} N) when $q_1 = q_2 = 1$ esu and $r = 1$ cm. (a) How many esu are there in 1 C? (b) What is the charge of the electron in esu? (The esu is an actual unit, the *electrostatic unit*.)

24. (II) An electron and a proton attract each other with a $1/r^2$ electric force, just like the gravitational force. Suppose that an electron moves in a circular orbit about a proton. (a) If the period of the circular motion is 24 h, what is the radius of the orbit? (b) If the period is 4×10^{-16} s, as it is in a hydrogen atom, what is the radius of the orbit?

25. (II) A charge q is split into two parts, $q = q_1 + q_2$. In order to maximize the repulsive Coulomb force between q_1 and q_2, what fraction of the original charge q should q_1 and q_2 have?

26. (II) An alpha particle (a helium nucleus, composed of 2 protons and 2 neutrons) is directed onto a particular tungsten nucleus (^{184}W, with 74 protons and 110 neutrons). The alpha particle stops and turns around at a distance of 6.0×10^{-12} m from the tungsten nucleus (Fig. 21–23). Ignore the effects of electrons, and treat the alpha particle and tungsten nucleus as pointlike. What is the Coulomb force on the alpha particle at its closest approach to the nucleus?

27. (II) An electron orbits in uniform circular motion about a much heavier—and therefore nearly stationary—proton at a distance of 3×10^{-10} m. (a) What are the magnitude and direction of the Coulomb force exerted on the electron by the proton? (b) What is the speed of the electron in its circular orbit? (c) What is the frequency of the circular orbit? (d) Calculate the spring constant of a spring with an electron mass at its end and the frequency of part (c).

28. (II) Two pointlike objects are placed 8.75 cm apart and are given equal charge. The first object, of mass 31.3 g, has an initial acceleration of 1.93 m/s² toward the second object. (a) What is the mass of the second object if its initial acceleration toward the first is 5.36 m/s²? (b) What is the charge of each object?

29. (II) Two cork balls, each of mass 0.20 g, are hung by insulating threads 20.0 cm long from a common point. The cork balls are given an equal charge by a Teflon rod. The balls repel and deflect as shown in Fig. 21–24. What charge q was given to each cork ball? Assume uniform charge.

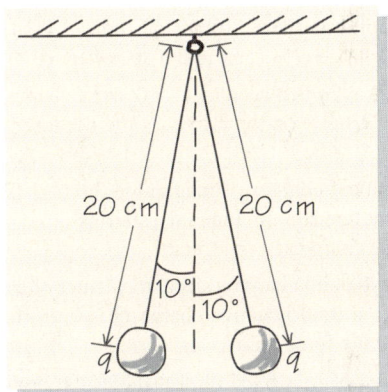

▲ **FIGURE 21–24** Problem 29.

30. (II) Astronomical data tell us that Earth's radius is 6.4×10^6 m, that its mass is 5.98×10^{24} kg, that the Moon's mass is 7.36×10^{22} kg, and that the mean Earth–Moon separation is 3.8×10^8 m. Suppose that, instead of being electrically neutral, as we believe, Earth and the Moon each have an excess positive charge of 8.5×10^{15} C. (a) What is the magnitude of the electrical repulsion between Earth and the Moon? (b) What is the ratio of this repulsive force to the attractive gravitational force? (c) If the charge on Earth were distributed uniformly throughout its volume, what would the excess charge density be, in coulombs per cubic meter (C/m³)? (d) Assume that the excess positive charge is due to excess protons, which have an electric charge of 1.6×10^{-19} C. Calculate the density of protons, in units of protons per cubic meter, that corresponds to the conditions in part (c). (e) Earth's mean density is 5.52×10^3 kg/m³, and a proton has mass 1.67×10^{-27} kg. Protons account for about half of Earth's mass. Compute the density of all protons in Earth, and compare this to your answer in part (d).

31. (II) Three unknown charges q_1, q_2, and q_3 exert forces on each other. When q_1 and q_2 are 15.0 cm apart (q_3 is absent), they attract each other with a force of 1.4×10^{-2} N. When q_2 and q_3 are 20.0 cm apart (q_1 is absent), they attract with a force of 3.8×10^{-2} N. When q_1 and q_3 are 10.0 cm apart (q_2 is absent), they repel each other with a force of 5.2×10^{-2} N. Find the magnitude and sign of each charge.

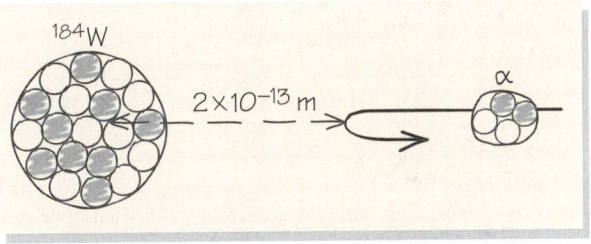

▲ **FIGURE 21–23** Problem 26.

32. (II) An electron has a mass of 0.9×10^{-30} kg and a charge of -1.6×10^{-19} C. Earth's mass is 6×10^{24} kg, and its radius is 6.4×10^6 m. Suppose that Earth has a net negative charge Q at its center. (a) How large would Q have to be for the charge repulsion on an electron to cancel the gravitational attraction at Earth's surface? (b) Suppose that this net charge is due to a discrepancy between the positive proton charge and the negative electron charge. Assume that half of Earth's mass is due to protons, each of which has a mass of 1.6×10^{-27} kg (the rest is neutrons, assumed to be neutral; electrons do not contribute much to the mass). What is the size of the charge discrepancy, compared to the electron charge?

33. (III) Use the similarity between Coulomb's law and the law of universal gravitation to calculate the distance of closest approach between a point charge of $+10^{-6}$ C, which starts at infinity with kinetic energy of 1 J, and a fixed-point charge of $+10^{-4}$ C. Assume that the moving-point charge is aimed straight at the fixed-point charge. [*Hint*: The similarity to gravity consists of using the notions of potential energy and energy conservation.]

21–4 Forces Involving Multiple Charges

34. (I) A charge $+4q$ is placed at $x = 0$, and a charge $+7q$ is placed at $x = 5$ units. Is there a point on the x-axis at which the net force on a charge Q is zero, and if so, where is it?

35. (I) A charge $-5q$ is placed at $x = 0$ and a charge $+3q$ is placed at $x = 10$. Where, on the x-axis, is the net force on a charge Q zero?

36. (I) Six identical charges of magnitude 3.6×10^{-8} C are placed on a straight line at 3-cm intervals, starting at $x = 0$. What is the force on the charge at $x = 15$ cm?

37. (I) A positive point charge q sits at the center of a circle of radius R on which a total negative charge Q is uniformly distributed. What is the net force on q?

38. (II) A charge of q is fixed on a plane at the origin $(0, 0)$ of an xy-coordinate system, and a charge $3q$ is fixed at $(3 \text{ cm}, -3 \text{ cm})$. Where must a charge of $-2q$ be placed at rest for it to be in equilibrium (that is, so that it remains at rest)? Is the equilibrium stable?

39. (II) What is the total force on each of the three quarks in Problem 17 due to the other two quarks?

40. (II) Three negative charges of magnitude $0.6 \, \mu$C are placed at the corners of an equilateral triangle of sides 18 cm. What is the net force on a charge of $1.5 \, \mu$C placed at the midpoint of one of the sides?

41. (II) Four positive charges $+q$ sit in a plane at the corners of a square whose sides have length d, as in Fig. 21–25. A negative charge, $-q$, is placed in the middle of the square. (a) What is the net force on the negative charge? (b) Is the equilibrium point at the center a stable equilibrium for motion of the negative charge in the plane of the square? (c) for motion of the negative charge perpendicular to the plane of the square?

42. (II) Calculate the force between two identical dipoles consisting of dumbbells with equal and opposite charges q and $-q$ at the end of a rigid rod of length $2d$. The dipoles are parallel, as shown in Fig. 21–26, and they are a distance x apart. Derive a first-order approximation for $d \ll x$. [*Hint*: Use $(1 + y)^n \cong 1 + ny + \cdots$ for $y \ll 1$.]

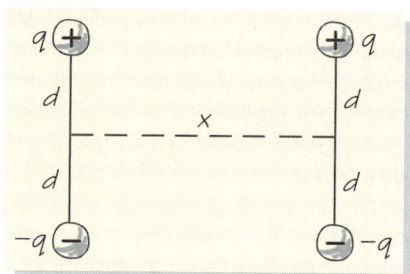

▲ **FIGURE 21–26** Problem 42.

43. (II) Charges q, $2q$, $-4q$, and $-2q$ (q is positive) occupy the four corners of a square of sides $2L$, centered at the origin of a coordinate system (Fig. 21–27). (a) What is the net force on charge q due to the other charges? (b) What is the force on a new charge Q placed at the origin?

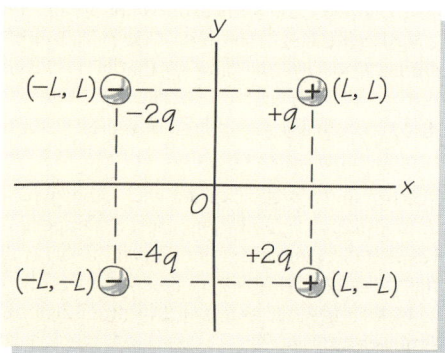

▲ **FIGURE 21–27** Problem 43.

44. (II) A charge Q is distributed uniformly along a rod of length $2L$, extending from $y = -L$ to $y = L$ (Fig. 21–28). A charge q is placed on the x-axis at $x = D$. (a) In what direction is the force on q, given that Q and q have the same sign? (b) What is the

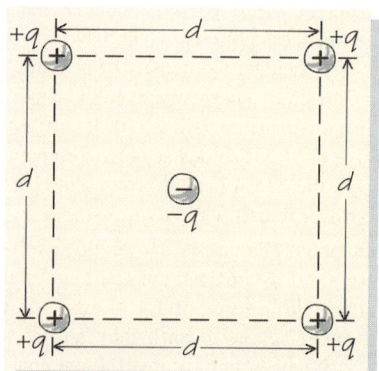

▲ **FIGURE 21–25** Problem 41.

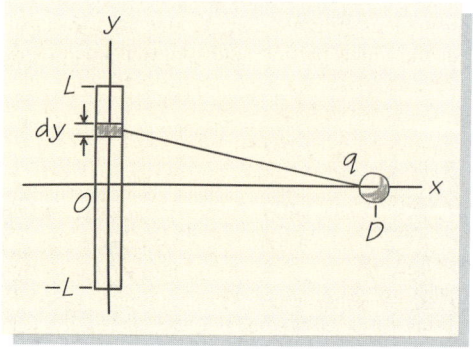

▲ **FIGURE 21–28** Problem 44.

charge on a segment of the rod of infinitesimal length dy? (c) What is the force vector on charge q due to the small segment dy? (d) Express an integral that describes the total force in the x-direction. (e) Compute the integral in order to find the total force in the x-direction.

45. (II) A charge is spread uniformly along the y-axis, stretching infinitely far in both the positive and negative directions. The charge density (charge per unit length) on the y-axis is λ. Find the force on a point charge q placed on the x-axis at $x = x_0$.

46. (II) A charge is spread uniformly along the y-axis from $y = 0$ to $y = +\infty$. The charge density on the y-axis is λ (Fig. 21–29). Find the force on a point charge q placed on the x-axis at $x = x_0$.

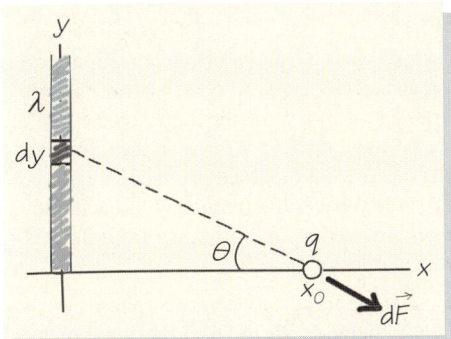

▲ **FIGURE 21–29** Problem 46.

47. (II) A long, thin rod of length L that contains a uniform distribution of charge Q points away from a point charge q. The nearest part of the rod is a distance d from the point charge. What is the electric force exerted on the charge q by the rod?

48. (II) Two uniformly charged rings of radii 25 cm and 40 cm respectively are placed parallel to each other, with a common axis. Each carries a charge of 2.2×10^{-4} C. and their centers are 100 cm apart. Where, along the common axis, should a charge q be placed so that the net force on it is zero?

49. (II) In the problem above, but now with the two rings having the same radius, the charge q is constrained to move only along the common axis. Is the position of equilibrium one of stable or unstable equilibrium? Does this depend on the sign of the charge q? [Hint: You may need the mathematical result that for $d \ll D$, $(D^2 + ad + bd^2)^n = D^{2n} + nadD^{2n-2} +$ terms that can be neglected.]

50. (II) A charge Q is distributed uniformly over a thin ring of radius R. The ring is oriented in the xy-plane, with its center at the origin. Find the force on a charge q located at the origin, and discuss the stability of its motion in the xy-plane. How does this compare with the case of a point charge placed at the center of a sphere whose surface is uniformly charged?

51. (II) Use the results of Example 21–9 to calculate the force on a positive point charge of magnitude $0.65\ \mu C$ located 5 cm above the center of a uniformly charged solid plate of radius 8 cm that carries a total positive charge of $1.6\ \mu C$. [Hint: Break the disk into concentric rings, use the results of Example 21–9 for each ring, and sum over the forces due to the rings.]

52. (II) Calculate the force exerted on a charge q by an infinite plane sheet with surface charge density (charge per unit area) σ. [Hint: Break up the plane into concentric rings centered below the charge, use the results of Example 21–9 for the force from each ring, then sum over the forces due to the rings.]

53. (II) Two rigid plates of equal size, made of different plastics, are rubbed against each other. This results in equal and opposite charges on the two plates. How large are these charges if it takes 0.1 N to separate the two plates? The area of each plate is $0.05\ m^2$, and the charge distribution may be assumed to be uniform. [Hint: Use the result of Problem 52.]

54. (II) Consider an infinite vertical sheet that carries a charge density of $+1.2 \times 10^{-6}\ C/m^2$. A cork ball of mass 8 g is suspended by a string 50 cm long at a distance of 55 cm from the charged sheet. What is the string orientation (a) if a charge $q = 0.8 \times 10^{-8}$ C is placed on the cork ball? (b) if instead a charge $q = -3 \times 10^{-8}$ C is placed on the ball?

55. (II) A total charge of $0.75\ \mu C$ is distributed uniformly over a thin, semicircular wire of radius 5.0 cm. What is the force on a charge of $0.30\ \mu C$ located at the center of the circle?

56. (II) A succession of $n + 1$ alternating positive and negative charges q are located along the x-axis at the points $x = 0$, $x = d$, $x = 2d, \ldots, x = nd$. An isolated charge Q is placed as shown in Fig. 21–30 at the point $x = D$ a very long distance away from the origin ($D \gg nd$). (a) Write a general expression for the electric force on charge Q. (b) Approximate your result, using the condition $D \gg nd$. Keep only leading and next-to-leading terms. [Hint: Use $(1 + x)^{-2} \cong 1 - 2x$ for $x \ll 1$.]

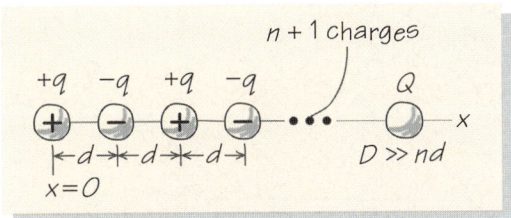

▲ **FIGURE 21–30** Problem 56.

57. (II) Charges $+q, -q, +q,$ and $-q$ are placed along the x-axis, at positions $x = 0, x = 1$ cm, $x = 2$ cm, and $x = 3$ cm, respectively. What is the force on a charge $Q = +3q$ placed at the point $(x, y) = (1.5\ \text{cm}, y_0)$, where y_0 is a variable?

58. (III) Consider a charge of $e = 1.6 \times 10^{-19}$ C distributed uniformly over a sphere of radius $R = 0.5 \times 10^{-10}$ m. Place a point particle of charge $-e\ (-1.6 \times 10^{-19}\ C)$ at the center of that sphere. Suppose that charge is displaced by a distance r (with $r < R$). Use the information given in Fig. 21–21 to show that the point charge will oscillate about the center of the sphere. Write down an expression for the frequency of oscillation in terms of R, the mass of the point charge m and the charge e. [Hint: If the acceleration for harmonic motion is given by $ma = -kr$, then the angular frequency of oscillation is $\omega = \sqrt{k/m}$.]

59. (III) What is the force per unit area between two infinite, uniformly charged plates with a surface charge density of $+10^{-5}\ C/m^2$ and $-10^{-5}\ C/m^2$, respectively, when the distance between the plates is 10 cm? What if the distance between the plates is doubled? [Hint: You may use the result of Problem 52.]

General Problems

60. (II) A cone of height h whose radius at the open end is R carries a total charge Q. Assuming that the charge is uniformly distributed over the surface, what is the charge density in C/m^2?

61. (II) Earth has a net charge of about 6×10^5 C. Assume the charge is evenly spread on the surface of the Earth. A cork ball of

10-g mass is hanging from a thin thread so the cork ball is 10 cm from the Earth's surface. What charge would the cork ball have to have to just barely rise up due to the electric force repulsion from the Earth?

62. (II) How much charge $+Q$ should be distributed uniformly over a square, horizontal plate of dimensions 60 cm × 60 cm if a ball of mass 1.5 g and charge 0.8 μC is to remain suspended 1 mm over the surface of the plate? Take gravity into account in this problem. How would your answer change if the ball were to be suspended 2 mm over the plate? *Qualitatively*, how would your answer change if the ball were to be suspended 1 m over the plate?

63. (II) A single charge $q_1 = +2 \times 10^{-8}$ C is fixed at the base of a plane that makes an angle θ with the horizontal direction. A small ball of mass $m = 0.5$ g and charge $+2 \times 10^{-8}$ C is placed in a smooth, frictionless groove in the plane that extends directly to the fixed charge (Fig. 21–31). It is allowed to move up and down until it finds a stable position $\ell = 8$ cm from the fixed charge. What is θ?

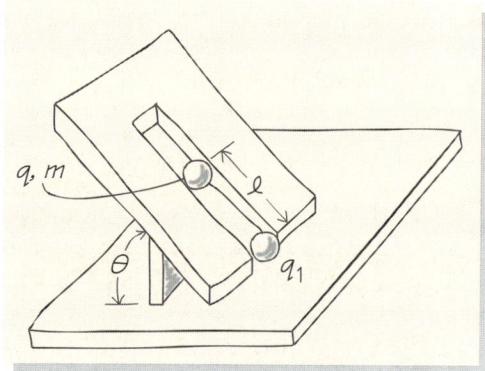

▲ **FIGURE 21–31**   Problem 63.

64. (II) The nucleus of an iron atom contains 26 protons within a sphere of radius 4×10^{-15} m. What is the Coulomb force between two protons at opposite sides of this nucleus? The answer to this problem illustrates that the force that holds the nucleus together against the Coulomb repulsion of its constituents must be strong indeed.

65. (II) An electron moves in a circular planetary orbit around a proton. (a) If the centripetal force is the attractive Coulomb force, what is the speed of the electron in terms of the charge e and the radius of the circular orbit? (b) What is the angular momentum, L, of the electron in the orbit? (c) Express the speed in terms of e and L. (d) Express the radius of the orbit in terms of e and L. (e) Express in terms of e and L the time it takes for the electron to go around the circle once. (f) Evaluate all these quantities, given that $L = 1.05 \times 10^{-34}$ kg·m²/s. This corresponds to a simplified version of the hydrogen atom.

66. (II) Suppose that the proton charge were slightly larger than the electron charge, so that $q_{proton} = (1 + \delta)e$ and $q_{electron} = -e$, where $0 < \delta \ll 1$. (a) Given that there are approximately 1.25×10^{57} protons (and electrons) in the Sun, and approximately 1.15×10^{44} protons and electrons in Earth, what is the upper limit on δ set by the fact that the resultant Earth–Sun electric repulsion cannot be large enough to cancel the attraction due to gravity? The mass of the Sun is approximately 2×10^{30} kg, that of Earth is approximately 6×10^{24} kg, and $G = 6.7 \times 10^{-11}$ N·m²/kg². Assume that the number of protons is equal to the number of electrons in the Sun as well as on Earth.

67. (II) Two fixed positive charges q are separated by a length ℓ. A third positive charge q of mass m is constrained to run on a line between the two fixed charges. (a) When the third charge is placed a distance x from the left-hand fixed charge, what is the net force on the third charge? Where is this force zero? In other words, where is the equilibrium point? (b) What is the net force as a function of the displacement of the third charge from the equilibrium point of part (a)? (c) For *small* values of the displacement from the equilibrium point, the third charge behaves as if a spring were acting on it. What is the value of the oscillation frequency?

68. (II) A positive charge q and a negative charge $-\alpha q$ ($\alpha > 1$) are fixed at a distance ℓ apart. Another positive charge q of mass m is constrained to move on the line connecting the two fixed charges. (a) Calculate the net force on the moving charge when it is at a distance x from the fixed positive charge. (b) Where is the force zero? (c) What is the frequency of oscillations if the moving charge is moved a small distance from its equilibrium position and then released, and if $\alpha = 40$?

69. (II) Show that the force between two spherically symmetric distributions of charge is identical to the force between two point charges that are located at the geometric center of each distribution and have the same total charge. [*Hint*: Use the fact that the force on a point charge due to distribution 1 is the same as if distribution 1 were concentrated at its center; then use similar reasoning for distribution 2, and then use Newton's third law.]

70. (III) Two rods, each of length $2L$, are placed parallel to one another a distance R apart. Each carries a total charge Q, distributed uniformly over the length of the rod. Write down an integral for the magnitude of the force between the rods, but do not evaluate it. Without working out any integrals, can you determine the force between the rods for $R \gg L$?

71. (III) Consider an infinite number of identical point charges q located at equally spaced points on the x-axis at the locations $x = na$ (n takes on integer values that range from $-\infty$ to $+\infty$) (Fig. 21–32). (a) Write an expression for the force on a charge Q, located at $x = 0$ and $y = R$, due to all the point charges q, and show the direction of the net force. (b) Take the limit of your result when the intercharge spacing $a \to 0$ and the charge $q \to 0$ such that $q/a = \lambda$ (a fixed charge density). Show that your expression can be written as an integral, and use dimensional analysis to determine the R-dependence of the force on charge Q.

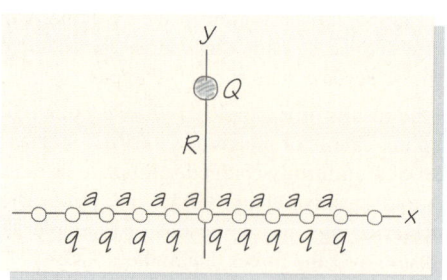

▲ **FIGURE 21–32**   Problem 71.

▸ An electric field forms between two charged objects, here two nails. The field is particularly strong in regions where the charged objects have sharp points, strong enough to allow charge to jump between the two objects. The charge ionizes atoms in the air, forming a series of sparks.

Electric Field

Charges exert forces on one another over large distances and across empty space. This idea of "action at a distance" poses both conceptual and technical difficulties. Action at a distance suggests that the object responsible for the force on a second object somehow reaches out, measures the distance to the second object, and then acts. Michael Faraday set out a better way to look at action at a distance: The first object influences the surrounding space by setting up a *field* around itself that is present whether there is a second object or not. When the second object is located at a given point, the field at this point acts on that object. This important idea can be developed quantitatively and, like any really good idea, leads to further ideas that go far beyond the original concept in utility and insight. In this chapter, we introduce and develop the concept of an electric field produced by static charges; here and in future chapters we will learn how it can be useful both as a practical tool and as a basis for a deeper understanding of electromagnetism.

▸ 22–1 Electric Field

It is useful to think of a distribution of charges as giving rise to an **electric field**, which acts on any other charge placed in that field. We can detect the electric field at any particular point by placing a small positive **test charge** q_0 at that location and seeing if it experiences a force. A test charge is only a probe: It does not *produce* the electric field that we are trying to measure; the field is due to other charges. The electric field $\vec{E}$ can

634 | Electric Field

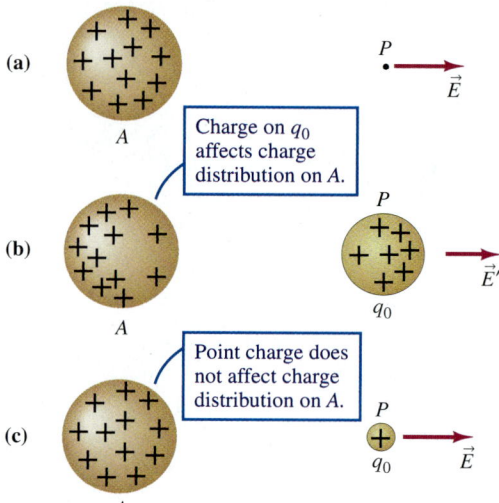

▲ **FIGURE 22–1** (a) An electric field exists at a point P due to charges on the sphere A. (b) A test charge q_0 that is too large will cause a redistribution of the charges on sphere A. A different electric field, $\vec{E}'$, is produced at P by sphere A, because the charges on A have been redistributed. (c) If we make the test charge q_0 small enough, it will hardly affect the charges on sphere A. The electric field produced by A at point P is now the same as in part (a). In each case the electric field is due to the charge on sphere A.

be defined by measuring the magnitude and direction of the electric force $\vec{F}$ acting on the test charge. The definition of the field is

$$\vec{E} \equiv \frac{\vec{F}}{q_0}. \tag{22-1}$$

We use a *small* test charge q_0 because a larger charge can affect the charges responsible for the electric field, perhaps making them move (Fig. 22–1b). This would affect the field itself. Thus, we more properly define the electric field by (Fig. 22–1c).

$$\vec{E} \equiv \lim_{q_0 \to 0} \frac{\vec{F}}{q_0} \tag{22-2}$$

DEFINITION OF ELECTRIC FIELD

Both the electric force and the electric field are vectors. We know a vector such as the electric field completely when we know both its magnitude and its direction *at every point in space*: $\vec{E} = \vec{E}(\vec{r})$.

From the definition in Eq. (22–2), the SI units of electric field are newtons per coulomb (N/C). Table 22–1 gives the magnitudes of the electric fields in various physical situations.

TABLE 22–1 • Values of Electric Fields (N/C)

Interplanetary space	10^{-3}–10^{-2}
Atmosphere at Earth's surface in clear weather	100–200
Value sufficient to cause electrical breakdown in dry air	3×10^6
Just outside large sphere of a Van de Graaff accelerator	10^6
In the Fermilab particle accelerator	1.2×10^7
In atoms within the radius of an electron orbit	10^9
In the electromagnetic radiation of the most intense laser	10^{12}
Outside a uranium nucleus, at a distance from the center of twice the nucleus's radius	5×10^{20}

The Electric Field of a Point Charge

The simplest example of an electric field is the field associated with a point charge, q_1. Consider two point charges, q_1 and q_0, located a distance r apart. The Coulomb force on q_0 due to q_1 is given in Eq. (21–8),

$$\text{for a point charge:} \quad \vec{F}_{01} = \frac{q_1 q_0}{4\pi\varepsilon_0 r^2} \hat{r}_{01}. \tag{22-3}$$

Here $\hat{r}_{01}$ is the unit vector pointing from q_1 to q_0. If q_0 is small, we use it as a test charge, and we can then use Eqs. (22–1) and (22–3) to find the electric field due to q_1:

$$\text{for a point charge:} \quad \vec{E}_1 = \frac{\vec{F}_{01}}{q_0} = \frac{q_1}{4\pi\varepsilon_0 r^2} \hat{r}_{01}. \tag{22-4}$$

The value of the test charge has canceled, so the limiting process in Eq. (22–2) introduces no complications. Equation (22–4) specifies that $\vec{E}_1$ is in the same direction as $\vec{F}_{01}$. Figure 22–2 shows the direction of $\vec{E}_1$ as determined by moving our test charge to various points a distance r away from q_1. This field is radial (Fig. 22–2a). At this point we drop the subscripts on $\hat{r}_{01}$ and q_1, and we use the radial unit vector $\hat{r}$ (measured from the point charge q) to specify completely the electric field $\vec{E}_{\text{pt}}$ due to q:

$$\vec{E}_{\text{pt}} = \frac{q}{4\pi\varepsilon_0 r^2} \hat{r}. \tag{22-5}$$

FIELD OF A POINT CHARGE

The electric field points *away* from a positive charge, as shown in Fig. 22–2b. When the charge is negative, the electric field has the same magnitude but is opposite in direction. The electric field due to a negative charge points *toward* that charge, as in Fig. 22–2c.

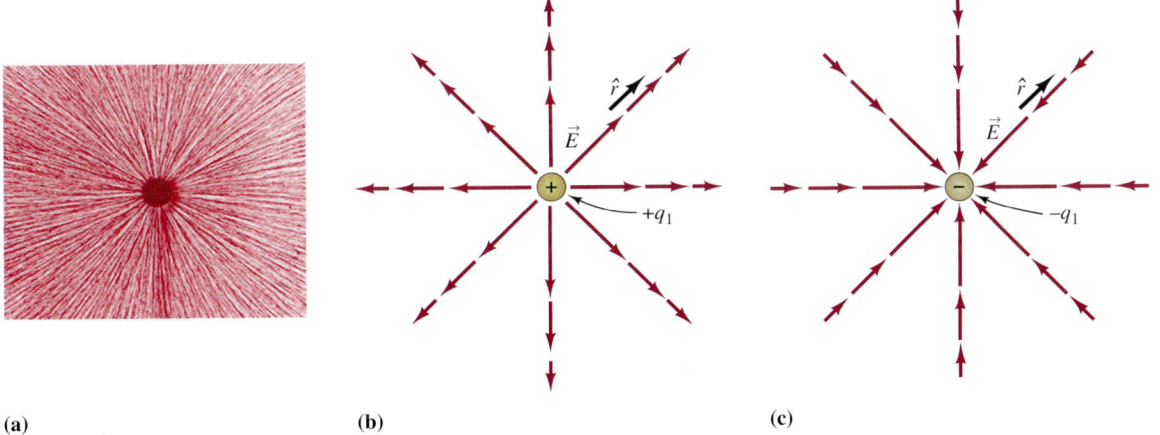

(a) (b) (c)

▲ **FIGURE 22–2** (a) Threads floating in oil become aligned with the electric field of this point charge. (b) The direction of the electric field $\vec{E}$ due to a point charge q_1 is radial. The charge is positive, and the field points away from it. (c) When the charge is negative the field points toward it.

CONCEPTUAL EXAMPLE 22–1 As you approach the charge in Fig. 22–2c, the magnitude of the electric field is large, and the arrows representing the field will become larger and larger, eventually overlapping each other and even passing through the (negative) charge. Does this pose a difficulty?

Answer This is a difficulty only for the artist. Remember that the arrows are just a visual representation, an aid in understanding. The tail of the arrow is placed on the position where you want to characterize the electric field. The magnitude of the electric field goes as $1/r^2$, so it can become quite large close to the point charge. But the length of the arrow has nothing to do with the scale of the drawing of charge positions.

The Usefulness of the Field Concept

Once we know the electric field $\vec{E}_{pt}$ produced by a point charge q, we can find the force on any point charge q' placed in that field by using Eq. (22–1); that is,

$$\vec{F} = q'\vec{E}_{pt}. \tag{22-6}$$

More important, *any* distribution of charges—not simply a point charge—produces an electric field throughout space. We use the subscript "ext" (for external) on the field to emphasize that this field is present *independent* of the charge q' on which the force acts, and once we know $\vec{E}_{ext}$, *the force on any point charge q' in the field* is the generalization of Eq. (22–6):

$$\vec{F} = q'\vec{E}_{ext}. \tag{22-7}$$

FORCE ON A POINT CHARGE IN AN EXTERNAL FIELD

The field concept has practical uses. We can calculate (or measure with a test charge) the field from a charge distribution once and for all, and then use Eq. (22–7) to find the effect of that distribution on any other charges within the field—we'll see numerous examples of this.

THINK ABOUT THIS...
HOW DOES THE FIELD CONCEPT HELP?

Why don't we deal with forces between charges instead of fields? In the chapter introduction, we mentioned the role the field plays in resolving the conceptual difficulties of action at a distance. We shall see in Chapter 24 that the field carries energy. To preserve the important idea of energy conservation, the field is a *necessary* concept. But the real power of the field concept appears when the field arises from *accelerating charges*. Even if these charges are limited to a small region (for example, within the arms of an antenna), the electromagnetic fields they produce spread through all of space at the speed of light. The supernova known as 1987A took place approximately 163,000 years ago; electric fields caused by the violent motion of many charges within and around the exploding star reached Earth on February 23, 1987. These traveling fields caused electrons in terrestrial antennas to move; this was the signal that supernova 1987A had occurred. This description of the process is easy to grasp; more, it is really the only practical way that we have to describe the process.

The notion of a field is useful in many areas of physical science. A mass distribution gives rise to a gravitational field analogous to the electric field of a charge distribution. We employ a velocity field in hydrodynamics; this field describes the velocity $\vec{v}$ at all points where fluid flow occurs, such as in the pipes of a city water system. In thermal physics, a temperature field describes the temperature at all points in a room. In this case, there is no directionality to the field; temperature forms what is called a scalar field instead of a vector field.

EXAMPLE 22–2 Find the electric field due to a point charge $+1.4\ \mu C$ at a distance of 0.10 m from the charge. What is the force on a second charge $-1.2\ \mu C$ that is placed 0.10 m from the first charge?

Setting It Up Figure 22–3a shows the electric field at point P due to a charge q. In Fig. 22–3b, we place a second charge q' at point P.

Strategy Equation (22–5) allows us to find the electric field $\vec{E}_{pt}$ at point P due to the point charge q. We then use Eq. (22–6) to find the force on the second point charge q' placed in this electric field.

Working It Out From Eq. (22–5):

$$\vec{E} = \frac{q}{4\pi\varepsilon_0 r^2}\hat{r} = \frac{(9.0 \times 10^9\ \text{N}\cdot\text{m}^2/\text{C}^2)(1.4 \times 10^{-6}\ \text{C})}{(0.10\ \text{m})^2}\hat{r}$$

$$= (1.3 \times 10^6\ \text{N/C})\hat{r}.$$

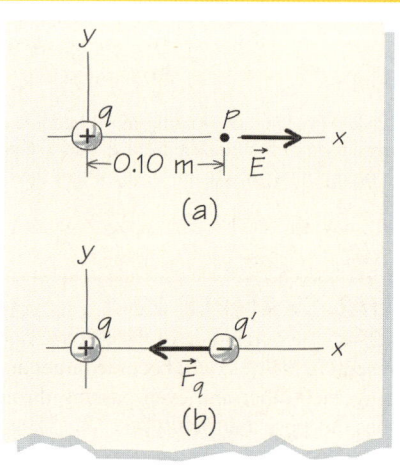

▲ FIGURE 22–3

The field is directed radially outward from the position of q (Fig. 22–3b). If q had been negative, the field would point radially inward rather than radially outward.

For the force, Eq. (22–6) gives

$$F = |q'|E = (1.2 \times 10^{-6} \text{ C})(1.3 \times 10^6 \text{ N/C}) = 1.6 \text{ N};$$
$$\vec{F} = (-1.6 \text{ N})\hat{i}.$$

Note that the direction of the force is in the opposite direction of the electric field at point P, that is, it is along $-\hat{i}$. The opposite charges attract (Fig. 22–3b). We could of course use Coulomb's law for the force between the two charges instead, Eq. (21–8).

What Do You Think? What would be the change in the electric field and force if q' were a positive charge instead of a negative one? Answers to **What Do You Think?** questions are given in the back of the book.

Superposition

In Section 21–4 we described how electric forces involving multiple charges superpose. The consequence of this is that if more than one point charge is responsible for producing a net electric field, that field is determined by the principle of superposition. The superposition principle states that the net electric force on an object is the vector sum of the forces due to individual point charges. Therefore *the net electric field is the vector sum of the fields of individual charges present*. The net force exerted on our test charge q_0 due to all the other charges in the region is

$$\vec{F}_{net} = \vec{F}_{01} + \vec{F}_{02} + \vec{F}_{03} + \cdots = \sum_i \vec{F}_{0i}. \quad (22\text{–}8)$$

Thus

$$\vec{E}_{net} = \frac{\vec{F}_{net}}{q_0} = \frac{\vec{F}_{01}}{q_0} + \frac{\vec{F}_{02}}{q_0} + \frac{\vec{F}_{03}}{q_0} + \cdots \quad (22\text{–}9)$$

$$= \vec{E}_1 + \vec{E}_2 + \vec{E}_3 + \cdots = \sum_i \vec{E}_i. \quad (22\text{–}10)$$

In Eq. (22–10), $\vec{E}_2$, for example, is the electric field due solely to the charge q_2 at the point in space where we have placed q_0. Using Eq. (22–5),

$$\vec{E}_{net} = \frac{1}{4\pi\varepsilon_0} \sum_i \frac{q_i}{r_i^2} \hat{r}_i. \quad (22\text{–}11)$$

FIELD OF A GROUP OF POINT CHARGES

In this equation, r_i is the distance from the ith charge q_i to the point at which the field is evaluated, and the unit vector $\hat{r}_i$ is directed from the position of the ith charge to the point where the field is evaluated.

EXAMPLE 22–3 Consider three charges placed on a line: $q_1 = +2\ \mu\text{C}$ at $x_1 = -2$ cm, $q_2 = +3\ \mu\text{C}$ at $x_2 = +4$ cm, and $q_3 = -2\ \mu\text{C}$ at $x_3 = +10$ cm. Find the electric field at point A, the origin of the coordinate system.

Setting It Up We sketch the configuration of charges in Fig. 22–4.

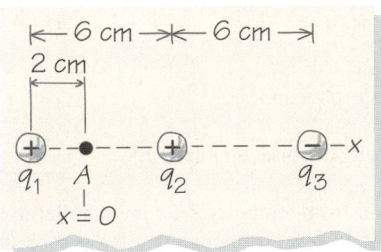

▲ **FIGURE 22–4** The electric field at point A is due to three charges. The distance x is measured from point A.

Strategy We solve this by a straightforward application of Eq. (22–11), the superposition of electric fields from point charges. Although it is generally important to remember that the required sum is vectorial, all the positions lie along a straight line in this case. The electric field at point A is

$$\vec{E}_A = \vec{E}_1 + \vec{E}_2 + \vec{E}_3,$$

where $\vec{E}_j$ ($j = 1, 2,$ or 3) is the field at point A due to charge q_j.

Working It Out Application of Eq. (22–11) gives

$$\vec{E}_A = \frac{1}{4\pi\varepsilon_0}\left[\frac{q_1}{x_1^2}(+\hat{i}) + \frac{q_2}{x_2^2}(-\hat{i}) + \frac{q_3}{x_3^2}(-\hat{i})\right]. \quad (22\text{–}12)$$

We must pay careful attention to signs. The unit vectors $\pm\hat{i}$ in parentheses indicate the direction of the unit vector $\hat{r}_j$, from the position of charge q_j to point A, but we must also include the signs of the individual charges. For example, the direction of $\vec{E}_3$ is $+\hat{i}$ because the negative sign of charge q_3 multiplied by $(-\hat{i})$ gives a direction $(+\hat{i})$. Putting the numbers into Eq. (22–12) gives

(continues on next page)

$$\vec{E}_A = (9 \times 10^9 \text{ N} \cdot \text{m}^2/\text{C}^2)\left[\frac{(2 \times 10^{-6} \text{ C})}{(0.02 \text{ m})^2}\hat{i} + \frac{(3 \times 10^{-6} \text{ C})}{(0.04 \text{ m})^2}(-\hat{i}) + \frac{(-2 \times 10^{-6} \text{ C})}{(0.10 \text{ m})^2}(-\hat{i})\right]$$
$$= (3 \times 10^7 \text{ N/C})\hat{i}.$$

The net electric field at point A is in the $+x$-direction, or toward the right.

What Do You Think? If point A were halfway between q_2 and q_3, in what direction would the electric field point?

Electric Dipoles and Their Electric Fields

The field of any single charge decreases as $1/r^2$. What happens if we have two charges, q_1 and q_2, spaced by a fixed distance L? Generally the field due to these charges will behave as $(q_1 + q_2)/r^2$ for $r \gg L$. Suppose now that the charges were equal and opposite. If these two charges were to sit precisely on top of one another, the two $1/r^2$ contributions to the field would exactly cancel to a zero electric field. But with the two charges separated by a finite distance, the field is not exactly zero, yet there is a partial cancellation and the field must fall off faster than $1/r^2$ for $r \gg L$. This arrangement—two charges, $+q$ and $-q$, of equal magnitude but opposite sign, that are separated by a fixed distance L—is an **electric dipole** (Fig. 22–5). We shall see that the field of the electric dipole decreases as $1/r^3$. This field depends only on the product qL, which is called the **electric dipole moment** of the neutral pair $(+q, -q)$; it is denoted by the letter p. We make $p = qL$ a vector through the displacement vector $\vec{L}$ directed from $-q$ to $+q$ (Fig. 22–5). The electric dipole moment $\vec{p}$ is then defined as

$$\vec{p} \equiv q\vec{L}. \tag{22-13}$$

ELECTRIC DIPOLE MOMENT DEFINED

The vector $\vec{p}$ points from the negative charge to the positive charge.

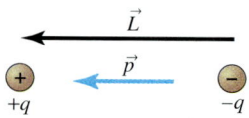

▲ **FIGURE 22–5** An electric dipole consists of equal but opposite charges separated by a distance L. The electric dipole moment $\vec{p}$ is directed from the negative charge to the positive charge.

EXAMPLE 22–4 Find the electric field of the electric dipole shown in Fig. 22–5 at a point P that lies along the perpendicular axis that bisects the line between the two charges and is at a large distance r ($r \gg L$) from each charge.

Setting It Up The sketch in Fig. 22–6 shows the point P, with xy-coordinates $(0, y)$.

Strategy We again use the superposition principle for point charges to find the net electric field $\vec{E} = \vec{E}_1 + \vec{E}_2$ at P, where the field $\vec{E}_1$ is due to the charge $+q$ and the field $\vec{E}_2$ is due to the charge $-q$. Since P is equidistant from both charges, the magnitudes of the two fields are the same, but $\vec{E}_1$ points away from $+q$, whereas $\vec{E}_2$ points toward $-q$. As usual, adding vectors means adding the components.

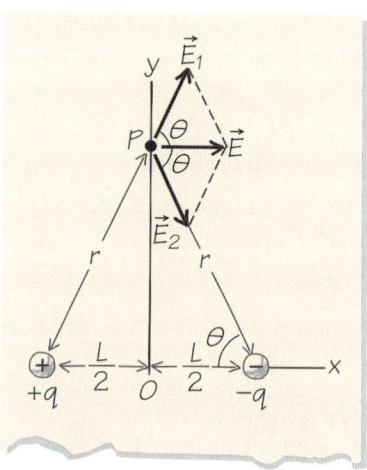

▲ **FIGURE 22–6** The net field at point P is parallel to the direction from $+q$ to $-q$.

Working It Out The y-components of $\vec{E}_1$ and $\vec{E}_2$ exactly cancel each other, and we are left with a net x-component toward the right that is twice the x-component of the field due to either charge:

$$\vec{E} = E_x\hat{i} = (E_{1x} + E_{2x})\hat{i} = 2E_{1x}\hat{i},$$

where

$$E_{1x} = \frac{q}{4\pi\varepsilon_0 r^2}\cos\theta.$$

From Fig. 22–6, we see that $\cos\theta$ is given by $\cos\theta = \frac{L/2}{r} = \frac{L}{2r}$. Thus the total electric field of the dipole along the perpendicular bisector is

$$\vec{E} = \left(\frac{2q}{4\pi\varepsilon_0 r^2}\right)\left(\frac{L}{2r}\right)\hat{i} = \frac{qL}{4\pi\varepsilon_0 r^3}\hat{i}. \tag{22-14}$$

We made no approximations here, and Eq. (22–14) is correct along the perpendicular bisector even when the distance from the charge pair is not large. The electric field decreases with r as $1/r^3$. This is the partial cancellation we spoke of earlier. Using Eq. (22–13) for the electric dipole moment, we can write Eq. (22–14) as

$$\vec{E} = -\frac{\vec{p}}{4\pi\varepsilon_0 r^3}. \tag{22-15}$$

If $r \gg L$, then $r \cong y$ and

along the bisecting axis: $\vec{E} \cong -\dfrac{\vec{p}}{4\pi\varepsilon_0 y^3}.$ (22–16)

The electric field from a dipole is *not* antiparallel to the dipole moment everywhere in space, although that is the case along the bisecting axis [Eqs. (22–15) and (22–16)].

What Do You Think? Is there any place in an electric dipole field that the value depends on only q or L and not on the product?

THINK ABOUT THIS...

WHEN DO ELECTRIC DIPOLES APPEAR IN NATURE?

Electric dipoles can be divided into two types. External fields (due to external charge distributions) frequently induce charge separations in electrically neutral molecules and materials, leading to an excess of positive (or negative) charge on one side (and the other) and hence to an **induced electric dipole moment** (Fig. 22–7). There are also examples in nature of charge configurations with **permanent electric dipole moments** (dipole moments that are not induced by external fields). Many molecules—water is an excellent example (Fig. 22–8)—have a structure with electrons distributed preferentially in certain regions, and so have permanent electric dipole moments. In cases such as common salt (NaCl) and hydrochloric acid (HCl), electrons cluster preferentially around one atom, giving that atom a negative charge. The other atom is left with a positive charge, so such molecules have a permanent electric dipole moment. The effects of electric dipole fields have great importance at the molecular level, where they can influence, for example, boiling and melting points. At this scale permanent dipole moments will typically dominate any induced dipole moments. For example, a water molecule has a permanent dipole moment $p \cong 6 \times 10^{-30}$ C·m, whereas a hydrogen atom in the rather strong field $E = 3 \times 10^6$ N/C acquires an induced dipole moment of $p \cong 3 \times 10^{-34}$ C·m.

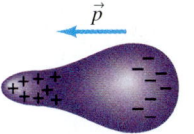 

| Induced electric dipole (polarized), total $q = 0$ | Nearby charge causing induced electric dipole |

▲ **FIGURE 22–7** A nearby charge can induce a polarized charge, and hence an electric dipole, on a neutral object.

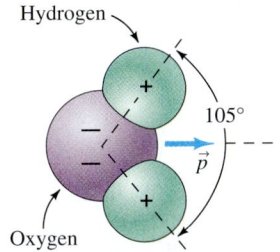

▲ **FIGURE 22–8** The water molecule, H_2O, is a permanent electric dipole. The electrons of both hydrogen atoms are shared with the oxygen atom, creating a strong electric bond that holds the molecule together (this is known as covalent bonding).

22–2 Electric Field Lines

We have already seen that we can map out the electric field by moving a test charge around in space. Unfortunately, the field is awkward to use in a visual sense. It is not easy to draw a vector at each point of a region of space that represents the magnitude and direction of the electric field at that point. The electric field can be more clearly visualized in terms of **electric field lines**. Their use was introduced by Michael Faraday around the middle of the nineteenth century, even before the concept of the electric field itself was clearly understood. Faraday used the phrase "lines of force." Electric field lines are continuous lines in space determined by the electric field, and in principle they fully describe the electric field. Keep in mind as we proceed that although the electric field itself has physical meaning, electric field lines are simply an aid to picturing the electric field and how a charge would react when placed in that field.

Electric field lines are determined according to two simple rules:

1. Electric field lines are drawn so that the tangent to the field line at each point specifies the direction of the electric field $\vec{E}$ at that point. This rule relates the *direction* of the electric field lines to the direction of the electric field. Note that we do not expect the electric field to change abruptly across any region of space that does not contain charge, so that the electric field lines in such regions are very nearly parallel to one another.

2. The *density* in space of electric field lines around a particular point is proportional to the strength of the electric field at that point (Fig. 22–9). What this means is that we take a small area oriented perpendicularly to the (nearly parallel) field lines and count the number of electric field lines that cross this small area; the line density is this number divided by the area. Thus the density is the number of lines *per unit area*. We'll explain below how we "count" the number of field lines.

Properties of Electric Field Lines

Let's draw the electric field lines of a positive point charge q. We know that the electric field points radially away from a positive charge at every location in space, so the lines are radial, pointing outward from the charge (Fig. 22–9). The field has the same magnitude at each point on a sphere centered on the charge, so the field lines are distributed uniformly about the charge. Finally, the magnitude of the electric field decreases with the distance r from the charge as $1/r^2$, so the density of lines falls off as $1/r^2$. And this is exactly how the density changes in Fig. 22–9 as long as we insist that electric field lines start (or stop, in the case of a negative charge) on the charge itself.

To establish this last point, suppose we draw N lines originating on our charge q (Fig. 22–9), and no lines appear (or disappear) as we move out. These lines radiate outward from the charge, and the number of lines that cut a sphere of radius R centered on the

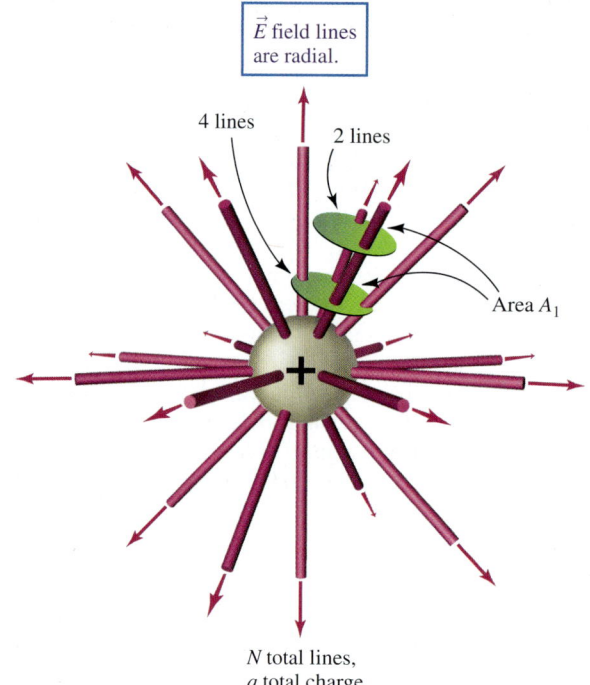

▶ **FIGURE 22–9** Representation of the radial electric field lines from a point charge. Fewer field lines pass through the same-size area farther from the charge. Note that field lines extend out to infinity.

charge is this same number N. Since the lines are evenly distributed over the sphere, the density of the lines is $N/(\text{sphere area}) = N/(4\pi R^2)$. Thus the density of the lines is proportional to $1/R^2$, and we have obtained the correct field strength dependence on R.

We can now relate the number of lines coming out of our charge to the magnitude of the charge: we take N to be any number we like! But once we have done so, the number of lines that leaves (or arrives at) any other charge is determined. For example, the number of field lines leaving a positive charge $q/2$ is $N/2$ and, more generally, the number of field lines leaving a positive charge q_i is $N_i = (q_i/q)N$. This will ensure that point 2 is satisfied: The field line density will be *proportional to* the strength of the electric field.

From the rules that determine field lines and the discussion above we can make a useful list of properties of field lines:

Property One: *Lines can start or terminate only on charges, never in empty space.* This property guarantees the connection between the strength of the field and the density of the electric field lines. We showed this above only for a single point charge, but with the help of the superposition principle, we can show that this is true in general.

Property Two: The electric field lines of a point charge go off to infinity, and by superposition this will be true for any *localized* collection of charges with a net charge—by "localized" we mean that the distribution is confined to a finite space. At distances that are large compared with the size of the charge distribution, the net charge of the distribution appears to be localized at a point, and electric field lines will be distributed evenly over a distant sphere centered on the distribution. If the net charge of a localized distribution is zero, then the field lines will curve back, not reaching to infinitely large distances.

Property Three: Electric field lines originate on, and run outward from, positive charges. They run toward, and terminate on, negative charges. This reflects the fact that electric charges are the sources of electric fields, which point away from positive charges and toward negative charges.

Property Four: *No two field lines ever cross, even when multiple charges are present.* They cannot cross because the electric field has a definite magnitude and direction at any point in space. If two or more electric field lines were to cross at some point, then the direction of the electric field at that point would be ambiguous.

Drawing Electric Field Lines

The properties above are a helpful guide for drawing the field lines associated with a given set of charges or a continuous charge distribution. Symmetry is another tool—and often a powerful one—for this process. A point charge looks the same when viewed from any direction. It has spherical symmetry, and the field lines follow the only direction that respects this symmetry—namely, they are radial. Similarly, if we are dealing with a long line of charge, there is a symmetry around the line, and the field lines must project radially outward from the line of charge, perpendicular to a cylinder that surrounds the line.

A Warning: We often draw field lines on a flat page, and since field lines are in three-dimensional space, the flat page presents certain limits. For example, you might draw Fig. 22–10 as the representation of the field lines for an isolated charge. *If you use such a drawing for determining the field strength, it should be done with care.* Figure 22–10 shows a circle of radius r centered on a positive charge. You cannot simply count the field lines that cross a particular circumference of this circle to find the field strength, as that would be equivalent to dividing by a length, not an area. If you did divide the (fixed) number of lines by the circumference $2\pi r$, you would find a field strength that falls as $1/r$, not the correct $1/r^2$. Nevertheless, planar drawings of electric field lines like Fig. 22–10 remain useful for visualizing the field and its effect on other charges.

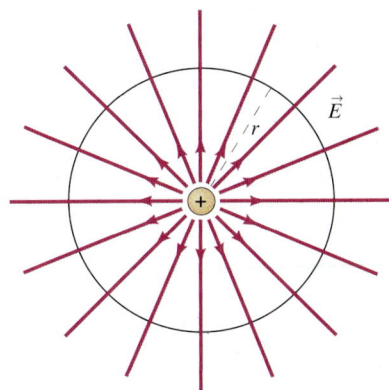

▲ **FIGURE 22–10** Electric field lines due to a point charge $+q$. Note the number of field lines that cross the circle (sphere) at radius r.

Some Examples

The easiest way to demonstrate the usefulness of electric field lines is to look at a couple of examples. Figures 22–11a and 22–11b show the electric field lines on a plane that passes through two positive charges of equal magnitude. The field lines all extend to infinity. The field lines that approach each other between the two positive charges appear to repel because two field lines cannot cross. If we were to place a positive test charge q' in Fig. 22–11b, the field lines at the charge's location would immediately show us the direction of the force on the charge (and likewise the acceleration).

Figures 22–12a and 22–12b depict the field lines of an electric dipole. The charges have equal magnitude, $\pm q$, so an equal number of field lines are attached to them, and every field line that originates on $+q$ terminates at $-q$. Near each charge the field lines are purely radial, but every line must eventually deviate from the radial direction in order to loop back to the other charge. Notice that the field lines in Fig. 22–12b are consistent with the field $\vec{E}$ determined in Example 22–4 (compare Figs. 22–12b and 22–6).

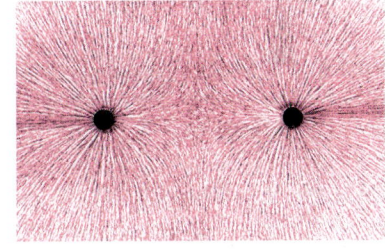

(a)

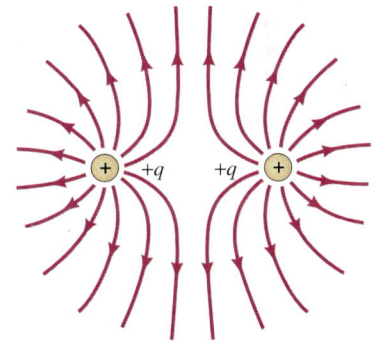

(b)

▲ **FIGURE 22–11** (a) The electric field lines due to two point charges $+q$, shown by threads in oil. (b) Schematic diagram of the field lines, which go off to infinity and appear to repel each other.

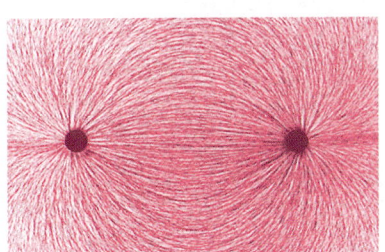

(a)

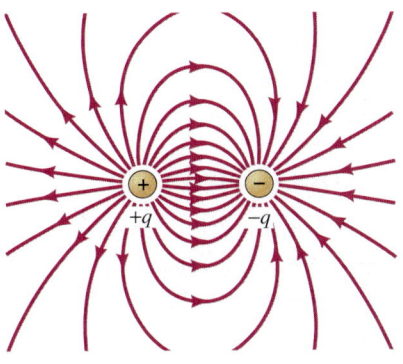

(b)

◀ **FIGURE 22–12** (a) The electric field lines due to point charges $+q$ and $-q$, a dipole, as indicated by threads in oil. (b) Schematic diagram of the field lines, all of which begin on $+q$ and end on $-q$; those that appear to head toward infinity do continue far from the charge, but eventually loop around and end on $-q$.

CONCEPTUAL EXAMPLE 22–5 The electric field lines of an isolated point charge go to infinity. How many field lines go to infinity for the electric dipole?

Answer Remember *Property Two:* Electric field lines always start on positive charges and terminate on negative charges. For an isolated point charge (say positive), the electric field lines go to infinity because there is no negative charge for them to end on. An electric dipole has one positive charge of a given magnitude and one negative charge of the same magnitude. Every line that starts on the positive charge will end on the negative charge, and so there will be none "left over" to go to infinity. We cannot see all the field lines doing this in Fig. 22–12, but if we made the figure large enough, we would see them all curve around and finally terminate on the negative charge!

EXAMPLE 22–6 Draw the electric field lines for a system of two charges, $+2q$ and $-q$, separated by a fixed distance.

Setting It Up We sketch a 2-dimensional representation of the field due to the two charges $+2q$ and $-q$ in Fig. 22–13a. Twice as many field lines leave the charge $+2q$ as end at the charge $-q$.

Strategy Arbitrarily close to each charge the field lines will be radial and uniformly spread over the area surrounding the charge. With twice as many lines coming from $+2q$ as go into charge $-q$, we can take half of the lines from the charge $+2q$ and connect them to the lines going into charge $-q$. The remaining lines go off to infinity.

Working It Out We choose, arbitrarily, to show 24 lines coming from $+2q$ so that 12 lines will go into $-q$ and 12 will go off to infinity. The final sketches are presented in Figs. 22–13b and 22–13c.

What Do You Think? What happens to the remaining 12 lines that emerge from $+2q$ that don't terminate on $-q$?

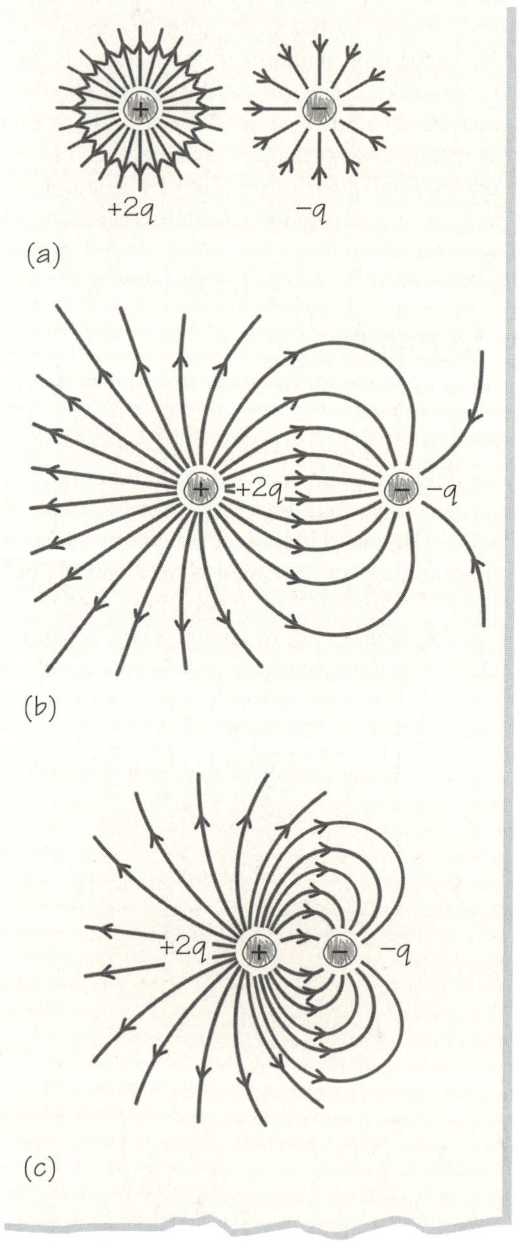

▶ **FIGURE 22–13** (a) The electric field lines close to the $+2q$ and $-q$ point charges are those of a point charge. (b) Half the electric field lines that emerge from $+2q$ end up on $-q$. (c) Far from the point charges, the electric field lines are those of a point charge $+q$. In this view you can begin to see the lines spread out to form a radial distribution.

22–3 The Field of a Continuous Distribution

We have thus far concentrated on electric fields due to point charges or collections of point charges. But *continuous* distributions of charge[†] also produce fields, and such distributions are very important in practice—for example, later we'll be interested in the

---

[†]Only point charges exist in nature; however, when we put many such charges together across a region, we will have a distribution that looks continuous to all but the sharpest measuring instrument.

field produced by charged capacitors. We will consider charges that are distributed *uniformly* throughout a region in space, whether a line, a surface, or a volume. We will also emphasize distributions where there is symmetry. For charge distributions that are not uniform or symmetrical, the problem of determining the resulting electric field can be more complex.

In considering forces on point charges due to charge distributions in Chapter 21, we have already set up a general framework for calculating electric fields due to line, surface, and volume distributions. Consider the calculation of the electric field at point P due to the charge distribution shown in Fig. 22–14. We divide the charge distribution into tiny elements, each of charge Δq. We first find the electric field $\Delta \vec{E}$ at point P that is due to a tiny charge element Δq, whose distance from P is r:

$$\Delta \vec{E} = \frac{\Delta q}{4\pi\varepsilon_0 r^2} \hat{r}. \qquad (22\text{–}17)$$

Here, $\hat{r}$ is the unit vector pointing away from the charge element. Superposition applies, and the total electric field at P is found by summing the infinitesimal fields $\Delta \vec{E}$:

$$\vec{E} = \sum \Delta \vec{E}. \qquad (22\text{–}18)$$

In the language of calculus, Eq. (22–17) becomes

$$d\vec{E} = \frac{dq}{4\pi\varepsilon_0 r^2} \hat{r}, \qquad (22\text{–}19)$$

whereas the summation that gives the total electric field becomes an integral over the entire charge distribution:

$$\vec{E} = \sum_{\lim \Delta q \to 0} \Delta \vec{E} = \int d\vec{E} = \frac{1}{4\pi\varepsilon_0} \int \hat{r} \frac{dq}{r^2}. \qquad (22\text{–}20)$$

The formal expression of Eq. (22–20) is identical to the one we found for the force on a point charge q due to a charge distribution, Eq. (21–11), divided by q. This is exactly how we define the electric field.

We can follow the development in Chapter 21 further to more specifically find the electric field due to a charge distribution along a line in terms of the linear charge density λ, the electric field due to a charge spread over a surface in terms of the surface charge density σ, and the electric field due to a charge distribution over a volume in terms of a volume charge density ρ (Fig. 22–15). In each case, the electric field is simply the expression for the force on the charge q divided by q; that is, Eqs. (21–13), (21–15), and (21–17) divided by q, respectively.

Constant Charge Densities

An important simplification for calculating the electric field due to a charge distribution occurs when the charge distribution, whether one-, two-, or three-dimensional, is *uniform*. This means that the charge densities are constants that can be removed from beneath the integrals expressing the electric field. Moreover, we can express the charge density in terms of the total charge Q contained in the distribution and the size of the distribution. In particular, if charge Q is distributed uniformly along a line segment of length L on a line, the linear charge density λ is

$$\lambda \equiv \frac{Q}{L}. \qquad (22\text{–}21)$$

Dividing Eq. (21–13) by the test charge q, we have

$$\vec{E} = \frac{\lambda}{4\pi\varepsilon_0} \int \hat{r} \frac{dx}{r^2}. \qquad (22\text{–}22)$$

Here, r is the distance to the point P at which the field is to be calculated, while $\hat{r}$ is a unit vector from the segment within the sum to point P. The integral is over a *one*-dimensional distribution.

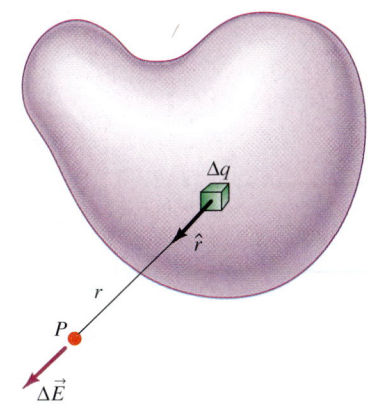

▲ **FIGURE 22–14** To find the electric field due to a continuous charge distribution, add all the electric fields $\Delta \vec{E}$ due to the charge elements Δq.

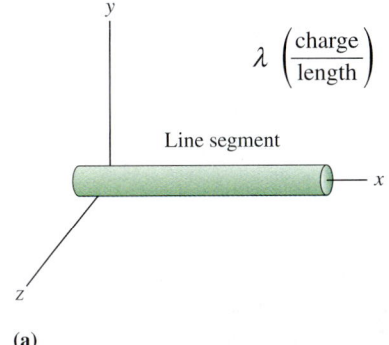

(a)

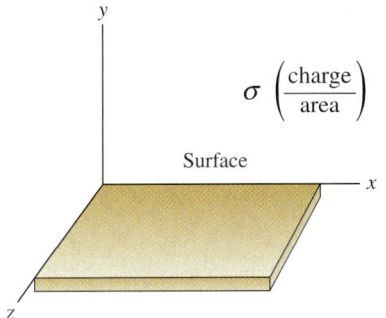

(b)

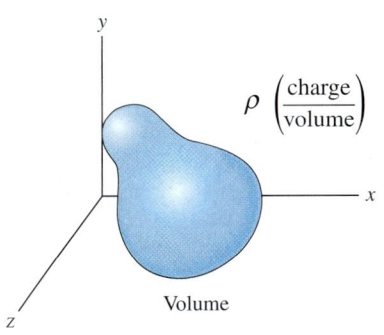

(c)

▲ **FIGURE 22–15** (a) One-dimensional, (b) two-dimensional, and (c) three-dimensional charge distribution. The charge density is labeled λ, σ, and ρ, respectively.

644 | Electric Field

For a charge Q distributed uniformly on a surface of area A, the surface charge density σ is

$$\sigma \equiv \frac{Q}{A}. \qquad (22\text{-}23)$$

Then

$$\vec{E} = \frac{\sigma}{4\pi\varepsilon_0} \int_{\text{surface}} \hat{r}\frac{dS}{r^2}. \qquad (22\text{-}24)$$

The integral in this case is over a *surface*.

Finally, if charge Q is distributed uniformly throughout a volume V, the volume charge density ρ is

$$\rho \equiv \frac{Q}{V} \qquad (22\text{-}25)$$

and the electric field is

$$\vec{E} = \frac{\rho}{4\pi\varepsilon_0} \int_{\text{volume}} \hat{r}\frac{dV}{r^2}. \qquad (22\text{-}26)$$

This time the integral is over a *volume*.

The three integrals in Eqs (22–22), (22–24), and (22–26) are formal, and to understand how they work in practice, it is best to look at some examples.

EXAMPLE 22–7 A straight insulating rod of length $2L$ carries a uniform linear charge density λ. Determine the electric field at point P, a distance R from the rod along the perpendicular bisector. First find the field in the limit that the rod is much longer than R ($L \gg R$). Then find it for a distance very far from the rod ($R \gg L$).

Setting It Up Figure 22–16a illustrates the situation. We include the origin of a coordinate system at the midpoint of the rod, which is aligned with the y-axis.

Strategy Equation (22–22) applies because the charge distribution is linear. We use Eq. (22–22) with dx replaced by dy (we are integrating along the rod) and with $\hat{r} = (\cos\theta)\hat{i} + (\sin\theta)\hat{j}$ (Fig. 22–16a) for the unit vector. Equation (22–22) becomes

$$\vec{E} = \frac{\lambda}{4\pi\varepsilon_0} \int_{-L}^{L} [(\cos\theta)\hat{i} + (\sin\theta)\hat{j}]\frac{dy}{r^2}. \qquad (22\text{-}27)$$

Before attempting the integration, we want to recognize any symmetry. The charge $dq = \lambda\,dy$ at a distance y *below* the x-axis gives rise to a field $d\vec{E}$ that is a mirror image of the field $d\vec{E}$ due to another charge dq' at a distance y *above* the axis (Fig. 22–16b). Thus we expect the net y-component of the field on the perpendicular bisector to vanish by symmetry. Here, we shall demonstrate this formally by performing the integration; normally, we would take advantage of the symmetry to reduce the mathematical calculation.

It is often true that the key to performing integrations such as that of Eq. (22–27) is to find the right variables. In this case, the simplest variable to use is the angle θ. To change from dy to $d\theta$, we recognize

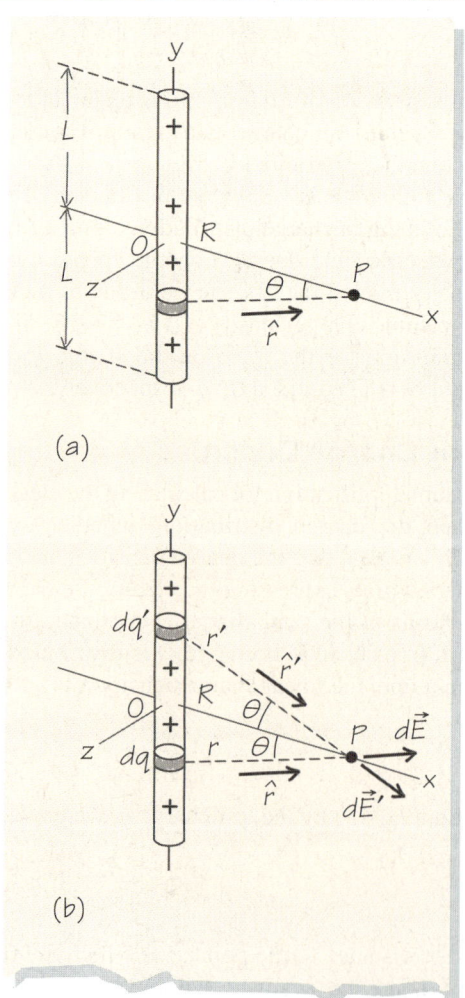

(a)

(b)

▶ **FIGURE 22–16** (a) Geometry of the situation described. (b) The y-components of electric field due to matching pieces above and below the origin cancel, whereas the x-components add.

that y, r, and $\hat{r}$ depend on θ. We must find the dependence of both y and r on θ. We have

$$\tan \theta = \frac{y}{R} \qquad (22\text{–}28)$$

and

$$\cos \theta = \frac{R}{r}. \qquad (22\text{–}29)$$

From Eq. (22–28), we obtain

$$dy = R\, d(\tan \theta) = R \sec^2 \theta\, d\theta = \frac{R}{\cos^2 \theta}\, d\theta.$$

With Eq. (22–29), the combination dy/r^2 that appears in Eq. (22–27) is

$$\frac{dy}{r^2} = \frac{1}{r^2} \frac{R}{\cos^2 \theta}\, d\theta = \frac{1}{r^2} \frac{R}{(R/r)^2}\, d\theta = \frac{1}{R}\, d\theta.$$

Now we turn to the integral itself. We use the algebra above; the factor $1/R$ is a constant and comes out of the integral, leaving

$$\vec{E} = \frac{\lambda}{4\pi\varepsilon_0 R} \int_{-\theta_0}^{\theta_0} [(\cos \theta)\hat{i} + (\sin \theta)\hat{j}]\, d\theta.$$

The limits $-\theta_0$ and θ_0 are the maximum values of θ, corresponding to the two ends of the line of charge.

Working It Out In this case our strategy has helped us a great deal, because the integral over $d\theta$ is simple. Let's consider each of the two terms in turn. When we integrate the second term ($\sin \theta$), we obtain a value proportional to $\cos \theta_0 - \cos(-\theta_0) = 0$. Therefore, the coefficient of $\hat{j}$, which is the y-component of the field, is zero, as we argued it must be by symmetry.

The coefficient of $\hat{i}$ is the x-component,

$$E_x = \frac{\lambda}{4\pi\varepsilon_0 R} \int_{-\theta_0}^{\theta_0} \cos \theta\, d\theta = \frac{\lambda}{4\pi\varepsilon_0 R} \sin \theta \Big|_{-\theta_0}^{\theta_0} = \frac{\lambda}{2\pi\varepsilon_0 R} \sin \theta_0. \qquad (22\text{–}30)$$

We can use $\sin \theta_0 = L/\sqrt{L^2 + R^2}$ if desired.

We next take the requested limits. For a rod with length $L \gg R$, $\sin \theta_0 \cong 1$, and the component E_x given by Eq. (22–30) becomes in this limit

$$\text{for } L \gg R: \quad E_x = \frac{\lambda}{2\pi\varepsilon_0 R}. \qquad (22\text{–}31)$$

Equation (22–31) gives the electric field for an almost infinitely long rod (or for a point very close to a finite rod). The direction of the field is perpendicular to the rod.

For the case of $R \gg L$, $\sin \theta_0 \cong L/R$, and in this limit Eq. (22–30) becomes

$$\text{for } R \gg L: \quad E_x = \frac{\lambda L}{2\pi\varepsilon_0 R^2} = \frac{Q}{4\pi\varepsilon_0 R^2}, \qquad (22\text{–}32)$$

where $Q = 2\lambda L$ is the total charge on the rod. In this case ($R \gg L$), we have obtained the point-charge result, because a rod of finite length looks like a point when it is viewed from large distances.

What Do You Think? The rod is a collection of point charges, and the field from a point charge falls off as $1/R^2$. So how can the field of the rod fall off as $1/R$, as in Eq. (22–31)?

EXAMPLE 22–8 Find the electric field at a distance L from an infinite plane sheet with a uniform surface charge density σ. (This situation relates to capacitors, which are important elements in electric circuits.)

Setting It Up We sketch the plane in Fig. 22–17a, where we place the sheet in the xz-plane. The point P where we want to find the field is on the y-axis.

Strategy Again, the main challenge is to find the simplest way to perform the integral of Eq. (22–24). For this we are going to use a technique that is useful in a variety of contexts. To integrate the effect of the entire plane, we break it up into pieces for each of which the field is easy to calculate. Here we break up the plane into a series of thin concentric circular regions centered around a point below P (Fig. 22–17a). These circles have a width Δr that is so small that the area of the circular regions is $2\pi r\, \Delta r$. To see why this is a good choice, take a look at Fig. 22–17b, which shows the contributions to the field at P from two segments on opposite sides of one of these concentric circles. You can see that their contributions to the field in the xz-plane cancel and that their contributions to the y-component of the field add. This is true all the way around the circle, so that we can first say that the field at P will have only a y-component. Moreover, the contribution dE_y to the y-component of the field of each of these segments is the same. Call the charge on a little segment of the ring dq. Then from Fig. 22–17b we have $dE_y = \dfrac{dq}{4\pi\varepsilon_0}\dfrac{1}{d^2} \cos \theta.$

Every factor in this expression is the same all the way around the ring, so that when we sum the contributions around the ring, the net field y-component ΔE_y from the ring is

$$\Delta E_y = \frac{\Delta Q}{4\pi\varepsilon_0 d^2} \cos \theta = \frac{\sigma(2\pi r\, \Delta r)}{4\pi\varepsilon_0 d^2} \cos \theta.$$

Here, ΔQ is the total charge on the ring, which in the second term is expressed as the surface-charge density times the area of the ring. Finally, we note that $\cos \theta = L/d$, so that

$$\Delta E_y = \frac{\sigma(2\pi r\, \Delta r)}{4\pi\varepsilon_0 d^2} \frac{L}{d} = \frac{\sigma L}{2\varepsilon_0} \frac{r\, \Delta r}{(r^2 + L^2)^{3/2}}.$$

(We have used the fact that $d = \sqrt{r^2 + L^2}$.) In thinking about the direction of $\Delta \vec{E}$, symmetry has been an important guide. We have now found the field from a single ring. To find the net field, we must now sum over the rings, and this is a straightforward integration.

Working It Out The net field is the sum over the fields from the set of rings of all different radii r from 0 to ∞; these rings cover the entire plane. This sum is a single integral over the radii r. We have

$$E_y = \sum \Delta E_y = \sum \frac{\sigma L}{2\varepsilon_0} \frac{r\, \Delta r}{(r^2 + L^2)^{3/2}} = \frac{\sigma L}{2\varepsilon_0} \int_0^\infty \frac{r\, dr}{(r^2 + L^2)^{3/2}}.$$

A table of integrals gives us

$$\int_0^\infty \frac{r\, dr}{(r^2 + L^2)^{3/2}} = \frac{-1}{(r^2 + L^2)^{1/2}} \bigg|_{r=0}^{r=\infty} = \frac{1}{L}.$$

(continues on next page)

Thus the field due to the entire plane has magnitude $\sigma/2\varepsilon_0$ and is oriented in the y-direction:

$$\vec{E} = \frac{\sigma}{2\varepsilon_0}\hat{j}, \qquad (22\text{--}33)$$

FIELD OF AN INFINITE UNIFORMLY CHARGED PLANE

where more generally the direction is perpendicular to the plane. The electric field is *constant* in both magnitude and direction: The field $\vec{E}$ does not even depend on how far the point is from the plane (Fig. 22–17c).

In reality, we cannot have planes of infinite extent. The result above holds for finite planes when the point where the field is specified is at a distance from the finite plane that is much less than the distance to the edge of the plane.

What Do You Think? The field does not depend on how far the point is from a plane, because (a) all charged planes have the same field; (b) the infinite, uniformly charged plane looks the same from any distance; (c) this is only an approximation, and very close to the plane the field is actually larger. Which of the above statements is true?

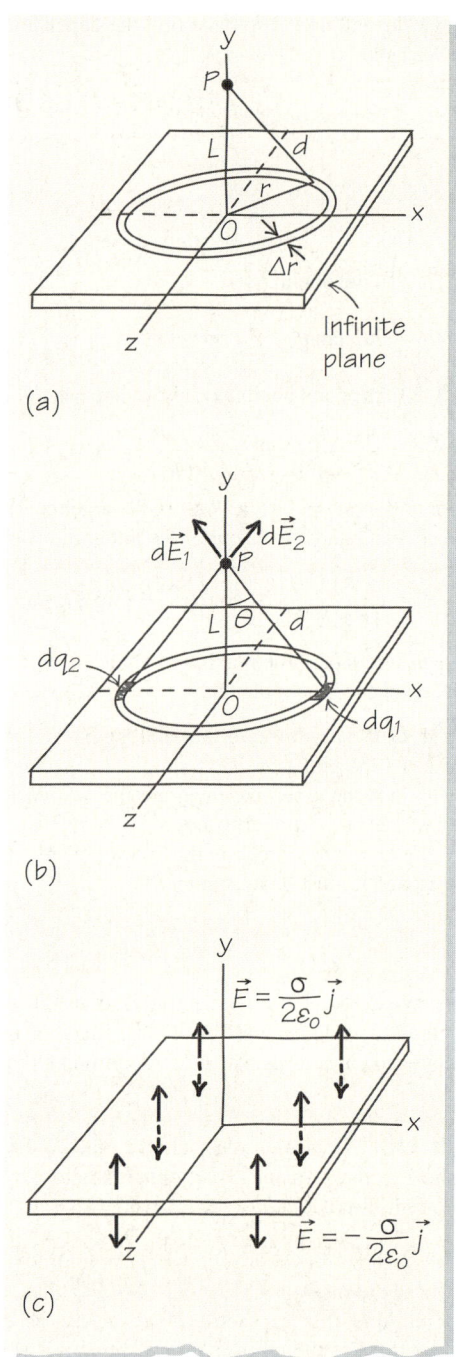

▶ **FIGURE 22–17** (a) To find the electric field at a point P a distance L above an infinitely charged plane, break up the plane into concentric circles about the axis passing through P. (b) The field due to one such circle has no horizontal component at P because the horizontal contributions from opposite points on the circle cancel; moreover, the vertical component is easy to find because every point of the circle is equidistant from P. (c) The net field after summation from all circles is uniform and perpendicular to the plane.

The preceding instance illustrates how a formal two-dimensional integration can be reduced to a single integration. There are often several ways to approach this type of problem. For instance, we could have divided our plane into narrow straight strips. The field due to a strip was found in Example 22–7, and we could have used that result. The effect of the entire plane is then found by summing over all the strips that make up the plane. Once again, the final integral would have been a single integral, with the integration variable being the differing distances of the strips from point P.

CONCEPTUAL EXAMPLE 22–9 Suppose that a very large sheet with uniform positive charge density is placed on an insulating stand parallel to Earth's (horizontal) surface. What will a charged cork ball placed a small distance above the plate do?

Answer We are told the sheet is very large and that the cork is only a small distance from the plate, so that to the cork the plate appears to be infinite and we can use the result of Example 22–6. We have a constant electric field, magnitude $E = \sigma/2\varepsilon_0$, that points upward where the cork is. Accordingly, a vertically oriented force of magnitude qE will act on the cork, where q is the charge on the cork. If q is positive, the cork will be repelled from the sheet, and if q is negative, the cork will be attracted to the sheet. For q negative, the cork will be accelerated toward the sheet. However, if q is positive, which way the cork moves depends on whether the electrical force is larger than or smaller than its weight.

EXAMPLE 22–10 Table 22–1 indicates there is an electric field near Earth's surface of about 150 N/C that points vertically down. Assume this field is constant around Earth and that it is due to charge evenly spread on Earth's surface. What is the total charge on the Earth?

Setting It Up We draw a diagram in Fig. 22–18.

Strategy Because the electric field points toward the Earth's center, the responsible charge is negative. To find the value of the charge, we can treat Earth's surface as a plane (see Question 24) and use Eq. (22–33) for the electric field. Given the field's magnitude, we can find the charge density σ, and because we know Earth's surface area, we can then find the total charge Q.

Working It Out From Eq. (22–33),

$$\sigma = 2\varepsilon_0 E = 2(8.85 \times 10^{-12}\, \text{C}^2/\text{N} \cdot \text{m}^2)(150\, \text{N/C})$$
$$= 2.7 \times 10^{-9}\, \text{C/m}^2.$$

The total charge Q is found by multiplying this charge density by Earth's surface area, $4\pi R^2$:

$$Q = \sigma 4\pi R^2 = (2.7 \times 10^{-9}\, \text{C/m}^2)(4\pi)(6.37 \times 10^6\, \text{m})^2$$
$$= 1.4 \times 10^6\, \text{C}.$$

While this may seem like a lot of charge, don't forget that it is spread out over Earth's entire surface. Washington, D.C., has an area of 68 square miles ($1.8 \times 10^8\, \text{m}^2$), which would mean that it contains only a half coulomb.

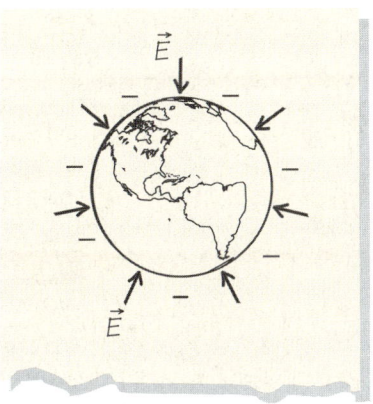

▲ **FIGURE 22–18**

What Do You Think? Suppose the Moon had a similar total charge on its surface. As a result, (a) the Moon and Earth would fly apart; (b) the Moon's surface charge density would be less than that of Earth's; (c) there would be very little effect on the Earth–Moon system; (d) the electric field at the Moon's surface would be the same as that of Earth's. Which of these statements are true?

The Electric Field Between Two Uniformly Charged Planes with Opposite Charge

Example 22–8 shows that the electric field for a positively charged plane of uniform surface charge density σ is uniform and perpendicularly directed away from the plane (Fig. 22–19a). If the plane were negatively charged, the field would be similar but would be directed *toward* the plane (Fig. 22–19b). What happens if we place the two planes, oppositely charged but with the same magnitude of charge density σ, parallel to each other? (This forms a capacitor.) As shown in Fig. 22–19c, the fields outside the

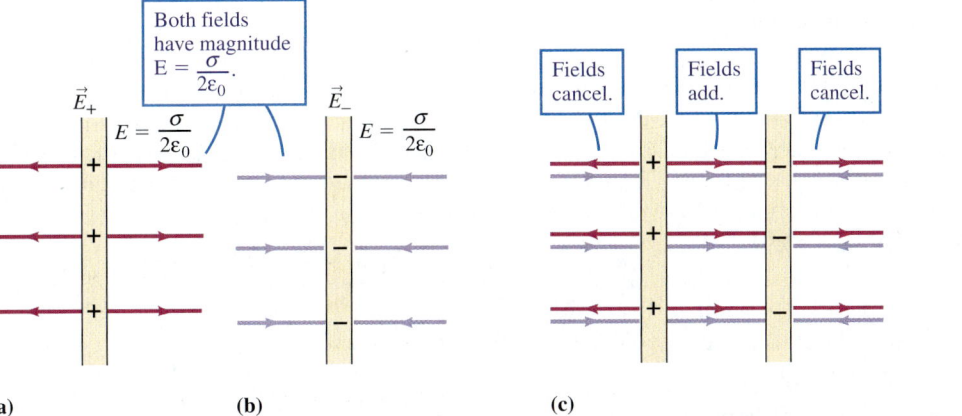

▲ **FIGURE 22–19** (a) The electric field due to a positively charged plane is directed away from the plane; (b) that due to a negatively charged plane is directed into the plane. (c) With two parallel planes carrying equal but opposite charge, the electric field cancels to zero outside the planes but is additive inside. (d) The field inside is σ/ε_0 and is directed from the positive plane to the negative plane.

parallel planes will exactly cancel each other, but the fields between the planes are additive. The resulting field is shown in Fig. 22–19d. For two parallel, oppositely charged planes, the electric field is zero everywhere except between the planes, where the field has magnitude

$$E = \frac{\sigma}{\varepsilon_0} \qquad (22\text{–}34)$$

ELECTRIC FIELD BETWEEN PARALLEL PLANES
OF OPPOSITE UNIFORM CHARGE DENSITY

and is directed from the positively to the negatively charged plane. (Remember that the direction of the electric field is always the direction of the force on our positive test charge q_0.)

> **THINK ABOUT THIS...**
> HOW DO WE KNOW THAT CHARGES COME IN MINIMUM SIZE UNITS?
>
> The existence of a minimum charge is an experimental result. The determination of this fact and of the value of the minimum charge was carried out by Robert Millikan at the beginning of the twentieth century. The *electron* had been discovered in 1897 by the British physicist J. J. Thomson—the method of discovery will be explained in a later chapter—but all that was known about the electron at that time was the ratio q/m, the charge-to-mass ratio. Millikan's experiment of 1910, the result of several years of preparation, involves spraying of very small oil drops (size 0.1–1.0 μm) into a cavity. These oil drops fall with a very small constant velocity under the influence of gravity, the buoyancy of air, and the air's viscosity. The drops were allowed to fall through a hole into a region between two parallel horizontal plates, oppositely charged, so that there was a known, constant electric field pointing downward. In formation some of the drops acquired small electric charges, and any drops that were negatively charged were subject to a force qE that would arrest their fall, or lead to a constant upward velocity. By studying a large number of such drops, Millikan determined that the charges on the drops came in multiples of a charge $|e| = 1.6 \times 10^{-19}$ C. This smallest charge is the charge of the electron. See Problem 66 for a quantitative treatment of Millikan's experiments.

22–4 Motion of a Charge in a Field

We have been concerned thus far with finding the electric field of a given collection of charges. Let's turn now to the force that charged particles will experience in an external electric field. Newton's second law becomes

$$\vec{F} = q\vec{E}_{\text{ext}} = m\vec{a}, \qquad (22\text{–}35)$$

where a particle of mass m and charge q has an acceleration $\vec{a}$ due to a given external electric field $\vec{E}_{\text{ext}}$. We then solve Newton's second law as usual. Example 22–11 demonstrates this.

EXAMPLE 22–11 Consider two oppositely charged parallel plates. The magnitude of the surface charge density on each plate has a constant value of $\sigma = 1.0 \times 10^{-6}$ C/m², and the plates are 1.0 cm apart. (a) If a proton is released from rest near the positively charged plate, with what speed will it strike the negatively charged plate? (b) What will the proton's transit time be?

Setting It Up We show a sketch of the two plates in Fig. 22–20, with the x-axis to the right and perpendicular to the plates. This is the direction toward which the proton will accelerate from rest.

Strategy Given the surface charge density, we can find the electric field and hence the force on the proton; the force will be constant. This then becomes a problem in one-dimensional constant acceleration kinematics in which we determine the proton's speed and transit time.

Working It Out Equation (22–34) gives us the electric field between the plates. In this case the field has only an x-component. From Eq. (22–35), the acceleration due to the electric field has only an x-component, magnitude a_x:

$$a_x = \frac{qE_x}{m} = \frac{q\sigma}{m\varepsilon_0} = \frac{(1.6 \times 10^{-19}\,\text{C})(1.0 \times 10^{-6}\,\text{C/m}^2)}{(1.67 \times 10^{-27}\,\text{kg})(8.85 \times 10^{-12}\,\text{C}^2/\text{N}\cdot\text{m}^2)}$$
$$= 1.1 \times 10^{13}\,\text{m/s}^2, \qquad (22\text{–}36)$$

where we use the known charge q and mass m of the proton.

(a) Now we turn to the kinematics, starting with the speed. From Section 2–5 we have $v^2 - v_0^2 = 2a_x x$. With the initial speed v_0 zero,

$$v^2 = 2a_x x. \qquad (22\text{–}37)$$

Equation (22–36) gives us a_x, while the distance traveled between the plates is $x = 1.0$ cm. Thus

$$v^2 = 2(1.1 \times 10^{13} \text{ m/s}^2)(1.0 \times 10^{-2} \text{ m}) = 2.2 \times 10^{11} \text{ m}^2/\text{s}^2;$$
$$v = 4.7 \times 10^5 \text{ m/s}.$$

(b) The transit time is the final speed divided by the acceleration:

$$t = \frac{v}{a_x} = \frac{4.7 \times 10^5 \text{ m/s}}{1.1 \times 10^{13} \text{ m/s}^2} = 4.3 \times 10^{-8} \text{ s}.$$

The plates accelerate protons and thus represent a charged-particle accelerator.

What Do You Think? What would happen to an electron released from the same position as the proton?

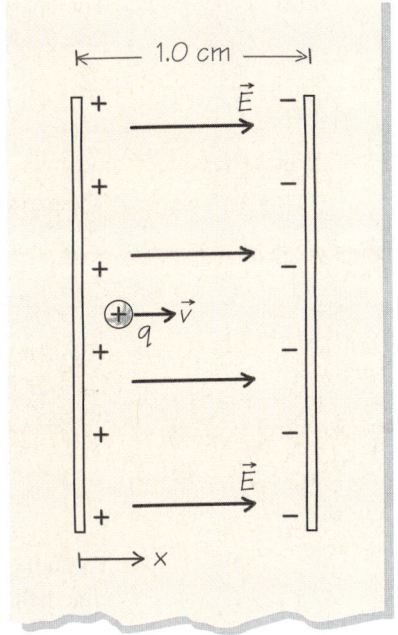

▶ **FIGURE 22–20** A charge $+q$ moving between parallel plates.

Deflection of Moving Charged Particles

Let's consider what happens when we inject a negatively charged particle (an electron, for example) into a region of uniform $\vec{E}$ between two plates (Fig. 22–21; we ignore edge effects). The particle has initial velocity $\vec{v}_0$ perpendicular to $\vec{E}$, and since it carries negative charge, it will be deflected upward in Fig. 22–21. From Eq. (22–35) the acceleration vector is

$$\vec{a} = a_x \hat{i} + a_y \hat{j} = \frac{qE}{m} \hat{j}. \qquad (22\text{–}38)$$

Note that the x-component of the acceleration is zero. Because the initial velocity is only in the x-direction ($\vec{v}_0 = v_0 \hat{i}$), the velocity vector becomes

$$\vec{v} = v_x \hat{i} + v_y \hat{j} = v_0 \hat{i} + \left(\frac{qE}{m} t\right) \hat{j}. \qquad (22\text{–}39)$$

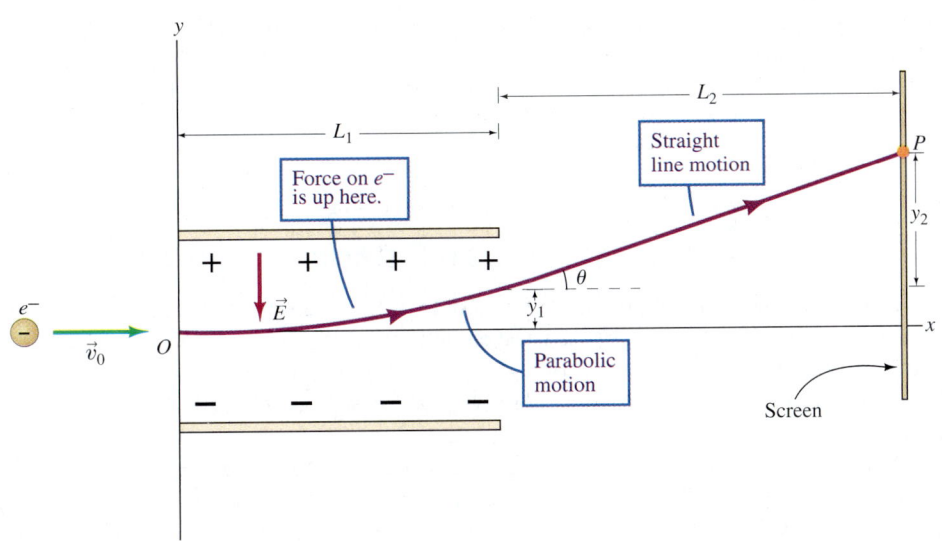

◀ **FIGURE 22–21** An electron passing between plates that deflect vertically.

The charged particle travels a horizontal length L_1 between the charged plates in the time T, determined by

$$x = v_0 T = L_1; \tag{22-40}$$

$$T = L_1/v_0. \tag{22-41}$$

The particle's deflection in the y-direction is then

$$y_1 = \frac{1}{2} a_y t^2 = \frac{1}{2} \frac{qE}{m} T^2 = \frac{1}{2} \frac{qE}{m} \frac{L_1^2}{v_0^2}. \tag{22-42}$$

The charged particle emerges from the plates at a position (x, y) given by Eqs. (22–40) and (22–42). The charged particle is then free from the influence of any force (ignoring gravity) and continues past the plates in a straight line at an angle θ from its initial direction:

$$\tan \theta = \frac{v_y}{v_x} = \frac{(qE/m)(L_1/v_0)}{v_0} = \frac{qEL_1}{mv_0^2}. \tag{22-43}$$

The setup described here allows control of a beam of charged particles and is the principle behind the picture tubes of televisions. In a television, there is also a second set of plates that controls deflection in the third direction, and a beam of electrons can then be directed anywhere on the screen.

EXAMPLE 22–12 An electron moving horizontally at a speed $v_0 = 3 \times 10^6$ m/s enters the region between two horizontally oriented plates of length $L_1 = 3$ cm. A fluorescent screen is located $L_2 = 12$ cm past these plates. Find the electron's total vertical deflection on the screen from its initial direction if the electric field between the plates points downward with a magnitude of $E = 10^3$ N/C.

Setting It Up Figure 22–21 again illustrates the situation, with the x-axis along the initial direction of the electron and the y-direction perpendicular. The electron will be deflected along the y-direction.

Strategy A constant electric field will lead to a constant acceleration, and we must apply the kinematic equations that describe constant acceleration. These equations are worked out above. Let us label as y_1 the vertical deflection of the electron between the plates, and as y_2 the additional vertical deflection of the electron after it has passed between the plates, where it continues in a straight path until it strikes the screen. We use Eq. (22–42) to find the deflection y_1 of the electron while it is between the plates. After the electron leaves the region between the plates, it travels at an angle to its original direction, given by Eq. (22–43) as $\tan \theta = qEL_1/mv_0^2$, and the additional deflection will be $y_2 = L_2 \tan \theta$. Finally, the total deflection is $y = y_1 + y_2$.

Working It Out We have from Eq. (22–42) $y_1 = \frac{1}{2} \frac{qE}{m} \frac{L_1^2}{v_0^2}$, and from Eq. (22–43) $y_2 = L_2 \tan \theta = \frac{qEL_1 L_2}{mv_0^2}$. The net deflection is then

$$y = y_1 + y_2 = \frac{1}{2} \frac{qEL_1^2}{mv_0^2} + \frac{qEL_1 L_2}{mv_0^2} = \frac{qEL_1}{mv_0^2} \left(\frac{1}{2} L_1 + L_2 \right).$$

Numerical evaluation with a minus sign for E ($\vec{E}$ points downward) gives

$$y = \frac{(-1.6 \times 10^{-19} \text{ C})(-10^3 \text{ N/C})(3 \times 10^{-2} \text{ m})}{(9.11 \times 10^{-31} \text{ kg})(3 \times 10^6 \text{ m/s})^2}$$

$$\times \left[\frac{1}{2}(3 \times 10^{-2} \text{ m}) + (12 \times 10^{-2} \text{ m}) \right]$$

$$= 8 \times 10^{-2} \text{ m}.$$

Of this 8 cm, the deflection $y_1 = 1$ cm, and the deflection $y_2 = 7$ cm.

What Do You Think? If the electric field between the plates doubles, the vertical deflection (a) is unchanged; (b) doubles; (c) is half of the calculated amount; (d) changes in a complicated way that would require a numerical reevaluation.

22–5 The Electric Dipole in an External Electric Field

In Section 22–1 we introduced the electric dipole, which has a total charge of zero but a positive and a negative center of charge separated by a distance L, and we remarked that many molecules—water or salt, for example—have permanent electric dipole moments. Let's discuss how permanent electric dipoles such as these molecules move in external electric fields.

Consider the permanent electric dipole discussed in Example 22–4. The electric dipole moment of the dipole has a vector character, and we assigned its direction as indicated in Fig. 22–7. The expression for the dipole moment $\vec{p}$ is given by Eq. (22–13).

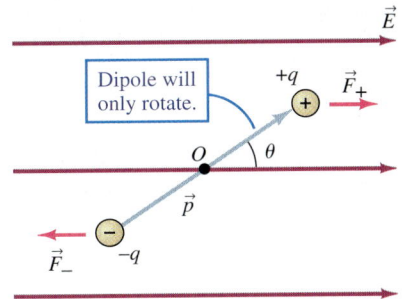

FIGURE 22–22 A dipole placed in a uniform external electric field experiences no net force but will experience a torque.

If we place the electric dipole in a uniform external field (Fig. 22–22), then the forces on $+q$ and $-q$ are, respectively,

$$\vec{F}_+ = q\vec{E},$$
$$\vec{F}_- = -q\vec{E} = -\vec{F}_+.$$

We notice that the two forces are equal and opposite and therefore cancel. *There is no net force on the dipole.*

There is, however, a torque that tends to rotate the dipole. To calculate the torque and the corresponding rotation, we must choose a reference point, and it is convenient to choose this point to be the midpoint of the dipole, which is located at point O in Fig. 22–22. The actual motion will be independent of the choice of reference point. The torque, τ, about a point due to a force that acts on another point that is a displacement $\vec{r}$ away is given by Eq. (10–6):

$$\vec{\tau} = \vec{r} \times \vec{F}, \tag{22–44}$$

where $\vec{r}$ is measured from point O. The resulting torque from the force on each charge is then clockwise, with magnitudes

$$\tau_+ = \left(\frac{L}{2}\right)qE\sin\theta, \quad \tau_- = \left(\frac{L}{2}\right)qE\sin\theta,$$

where the subscripts $+$ and $-$ refer to the charges. Because both τ_+ and τ_- are clockwise rotations, the total torque is also clockwise, with magnitude

$$\tau = \tau_+ + \tau_- = qLE\sin\theta. \tag{22–45}$$

We can represent this expression for the torque on a dipole as the vector product of $\vec{p}$ and $\vec{E}$:

$$\vec{\tau} = \vec{p} \times \vec{E}, \tag{22–46}$$

TORQUE ON A DIPOLE DUE TO A FIELD

which gives both the torque's magnitude ($pE\sin\theta$) and direction (into the page in Fig. 22–22).

The maximum torque ($\tau = pE$) occurs when $\vec{p}$ and $\vec{E}$ are perpendicular ($\theta = \pi/2$). The torque is zero when $\vec{p}$ and $\vec{E}$ are parallel ($\theta = 0$) or antiparallel ($\theta = \pi$). The torque tends to rotate the electric dipole until $\vec{p}$ is parallel to $\vec{E}$. The position $\theta = 0$ corresponds to a stable equilibrium, but the position $\theta = \pi$ is one of unstable equilibrium because a small deviation will cause the dipole to rotate toward $\theta = 0$.

Suppose the dipole is initially aligned along some angle $\theta \neq 0$ with respect to an electric field. What will the motion look like? This question is most easily answered when we look at the potential energy U of the dipole in the field (see the paragraph below). There we shall see that *for small angles, U is a constant $+$ a term proportional to θ^2*. This corresponds to the behavior of an ordinary pendulum, or a spring, and the motion is oscillatory about $\theta = 0$. As the dipole rotates toward $\theta = 0$, it gains kinetic energy and passes through $\theta = 0$ to the other side. The torque, however, then becomes counterclockwise, and the dipole slows down, stops, returns to $\theta = 0$, and passes through it again to the original side. Table 22–2 illustrates a time sequence for a rotating dipole in an electric field. Without a mechanism to dissipate the dipole's energy, the oscillations will continue.

TABLE 22–2 • An Electric Dipole Rotating in a Uniform Electric Field

Electric Field	Torque, τ	Angular Velocity, ω
Initial condition	Maximum, into page	Zero
	Decreasing, into page	Increasing, into page
$\theta = 0$	Zero	Maximum, into page
	Changed direction, out of page, increasing	Decreasing, into page
	Maximum, out of page	Zero
	Decreasing, out of page	Changed direction, increasing, out of page

The Energy of a Dipole in an External Electric Field

When a dipole rotates under the influence of an external electric field, the field does work on it. From this work we can find the potential energy of the dipole in the field. To find the work done by the field as the dipole rotates from an initial angle θ_0 to a final angle θ, we use the form $W = \int_{\theta_0}^{\theta} \vec{\tau} \cdot d\vec{\theta}$. We refer to Fig. 22–22 to see that for a counterclockwise rotation, the infinitesimal angle change is out of the page, while the torque $\vec{\tau} = \vec{p} \times \vec{E}$ is into the page. Thus the scalar product $\vec{\tau} \cdot d\vec{\theta} = -\tau\, d\theta = -(pE \sin \theta)\, d\theta$, and the work done by the field is

$$W = \int_{\theta_0}^{\theta} (-pE \sin \theta)\, d\theta = -pE(\cos \theta_0 - \cos \theta). \tag{22–47}$$

Now, as we know from Chapter 7, the potential energy $U(\theta)$ at point θ minus its value $U(\theta_0) = U_0$ at some initial point θ_0 is the *negative* of the work done by the associated force as the object moves from θ_0 to θ. Thus we have

$$U - U_0 = pE(\cos \theta_0 - \cos \theta). \tag{22–48}$$

We are free to choose the constant U_0, and we choose it such that $U_0 = 0$ at $\theta_0 = \pi/2$. Thus the potential energy at angle θ is given by

$$U = -pE \cos \theta. \tag{22–49}$$

Notice that Eq. (22–49) is consistent with our choice for the zero of the potential energy, because $U_0 = -pE \cos \theta_0$, which is zero when $\theta_0 = \pi/2$.

Equation (22–49) can be written more compactly by using the scalar dot product of $\vec{p}$ and $\vec{E}$:

$$U = -\vec{p} \cdot \vec{E}. \quad (22\text{–}50)$$

POTENTIAL ENERGY OF A DIPOLE IN A FIELD

Above, we discussed the stability of the equilibrium of the dipole in an external field. We can see directly from Eq. (22–50) that the orientation in which $\vec{p}$ is aligned with $\vec{E}$ is a point of stable equilibrium because U has a minimum there. In contrast, U has a maximum when $\vec{p}$ is antiparallel to $\vec{E}$, and therefore this is a point of unstable equilibrium.

CONCEPTUAL EXAMPLE 22–13 What is the motion, if any, of an electric dipole placed in a *nonuniform* external electric field?

Answer A possible setup of this situation is sketched in Fig. 22–23. We have used a cork ball that has a dipole moment induced by the field itself. If the field is nonuniform, the forces on the two charges in the dipole will generally differ in magnitude. Then, in addition to a torque, there will be a net force on the dipole. The resulting motion would be a combination of linear acceleration and rotation. The details of the motion depend on the particular electric field. While we have used the tip of a charged rod in this figure, we could equally well have placed a dipole in the region of a point charge, as long as the point charge field lines "spread" significantly over the size of the dipole.

The effect of a nonuniform electric field on an induced dipole explains the attraction of a neutral cork ball coated with conducting paint to a Teflon rod rubbed against fur (Fig. 22–23). The charged Teflon rod induces an electric dipole on the cork ball, which is in the nonuniform electric field of the rod. Another example of the action of a nonuniform electric field on an induced dipole is the attraction between small bits of paper and a comb that has just been charged by passing it through hair. The bits of paper have induced dipole moments and are attracted to the comb in its nonuniform field.

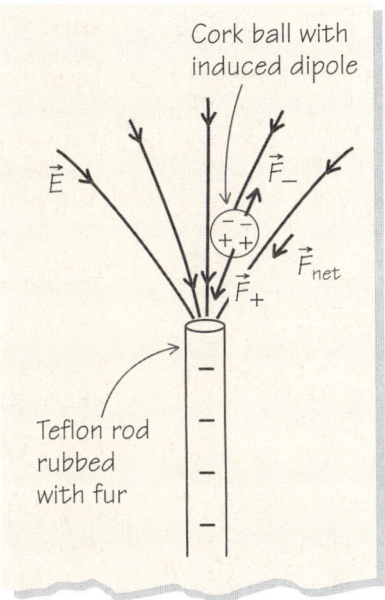

▲ **FIGURE 22–23** A dipole placed in a nonuniform external electric field can experience a net force. In this case the external electric field induces a dipole in the cork ball, which then experiences a force due to the electric field. This effect can be understood by using Coulomb's law.

Summary

Charge distributions set up electric fields in the space around them. The electric field vectors can be mapped out by moving a small, positive test charge q_0 around in this field. The field $\vec{E}$ is defined as the force $\vec{F}$ on this test charge, divided by q_0:

$$\vec{E} \equiv \lim_{q_0 \to 0} \frac{\vec{F}}{q_0} \quad (22\text{–}2)$$

The electric field has units of N/C (or V/m). The force on a point charge q' in a given external electric field is

$$\vec{F} = q'\vec{E}_{\text{ext}}. \quad (22\text{–}7)$$

Electric field lines aid in the visualization of the direction and magnitude of the electric field produced by various charge configurations. We can summarize the rules for drawing them and their properties by saying that

- Electric field lines are drawn so that the tangent to the field line specifies the direction of the electric field itself.
- The density in field lines per unit area is proportional to the magnitude of the electric field.
- Electric field lines start and end on electric charges, never in empty space. In particular, they start on positive charges and terminate on negative charges.
- Electric field lines never cross.

From Coulomb's law, the electric field due to a point charge q is

$$\vec{E} = \frac{q}{4\pi\varepsilon_0 r^2}\hat{r}. \tag{22-5}$$

Electric fields obey the principle of superposition:

$$\vec{E}_{\text{net}} = \vec{E}_1 + \vec{E}_2 + \vec{E}_3 + \cdots = \sum_i \vec{E}_i, \tag{22-10}$$

where $\vec{E}_i$ labels the field of the components (normally point charges) that make up a charge distribution.

In its simplest form, an electric dipole consists of a positive charge q separated by a distance L from a negative charge $-q$. Such a configuration, or any configuration that is electrically neutral but has an imbalance of positive and negative charge from one side to another, occurs often in nature and produces an electric field. This field decreases with distance r as $1/r^3$ and, for the simple dipole, is proportional to the magnitude of the electric dipole moment $\vec{p}$, given by

$$\vec{p} = q\vec{L}. \tag{22-13}$$

The direction of $\vec{L}$ (and of $\vec{p}$) is from the negative charge to the positive charge. This direction determines the angular dependence of the electric dipole field. In addition to producing an electric field, an electric dipole experiences a torque in a uniform external electric field:

$$\vec{\tau} = \vec{p} \times \vec{E}. \tag{22-46}$$

The dipole in the external field has a potential energy of

$$U = -\vec{p} \cdot \vec{E}. \tag{22-50}$$

The dipole is a configuration of two point charges, but we also deal with continuous charge distributions. The electric field due to a continuous charge distribution is

$$\vec{E} = \frac{1}{4\pi\varepsilon_0} \int \hat{r}\frac{dq}{r^2}. \tag{22-20}$$

Here, r is the distance of a charge element dq from the point where the field is measured. To use this result, it is necessary to know how dq varies throughout space.

The electric field due to an infinitely long wire is radial and perpendicular to the wire. A charged plane, infinite in area, with a charge per unit area of σ has an electric field that is uniform and directed perpendicular to the plane, with

$$E = \frac{\sigma}{2\varepsilon_0}. \tag{22-33}$$

In later chapters we'll have ample opportunity to apply the electric field concept.

Understanding the Concepts

1. In older movies, you may see gasoline trucks dragging metal chains along the road. Why?
2. Why can't two electric field lines cross?
3. We have introduced the concept of an electric field. Why might it be useful to introduce an analogous gravitational field? In what ways would this field be like an electric field, and in what ways would it be different?
4. An inflated rubber balloon is charged by rubbing it with fur. Explain what happens when that balloon is placed against (a) a metal wall; (b) an insulating wall.
5. Two charges of opposite polarities are close together. Five times as many field lines leave one charge as end up on the other charge. What is the ratio of charges?
6. Electric field lines originate from positive charges and terminate at negative charges, as exemplified by the field lines due to a dipole. Does this statement contradict the depiction of field lines due to a single positive point charge?
7. If in Example 22–3 there were a point (or points) on the line connecting the three charges at which the electric field is zero, where, qualitatively, would it (or they) be?
8. A pair of equal and opposite charges forms a dipole, and the electric field of a dipole is not zero. But if we were to look at a dipole from very far away, the two charges would appear to be on top of one another and to cancel; that is, we would see no charge, and hence we should see no field. How do you reconcile these statements?
9. In Example 22–10, we asked how many electrons the charge on Earth represents. How do we know that the charge responsible for Earth's field is a negative one?
10. Can the electric dipole induced on a spherical conducting ball cause the ball to rotate? How about the electric dipole induced on a long rod?
11. After you comb your hair, the comb can often attract small pieces of paper. The act of combing may induce an electric charge on the comb, but the combing does not itself affect the paper. What accounts for the attraction?
12. Is it possible for electric field lines to go all the way to infinity?
13. Explain how the water molecule, H_2O, acts as an electric dipole (see Fig. 22–8), given that there are *two* spatial regions (around the H atoms) with negative charge.

14. We called the arrangement in Example 22–11 an accelerator; how could we get the accelerated particle out in order to use it?
15. Explain why the electric-field-line technique would not be useful for a point charge if Coulomb's experiments had shown the electric force to decrease as $1/r$ or as $1/r^{2+\delta}$.
16. The internal motion of a liquid can be described by a velocity field, which is the velocity vector of the element of fluid at a given point. In what ways is this field like an electric field, and in what ways is it different?
17. Can you invent an arrangement of charges whose electric field would be radially directed into a point in some region of empty space? The correct answer has implications for the stability of charges placed in static electric fields.
18. Suppose that a small electric dipole ($+q$ and $-q$) is placed somewhere on a line that is perpendicular to, and bisects, a second (fixed) dipole ($+Q$ and $-Q$), as seen in Fig. 22–24. If the small dipole is free to pivot about its center, what will it do?
19. We have charges $+q$ at $x = \pm a$ and charge $-2q$ at $x = 0$. Is it useful to think about this configuration as two electric dipoles placed back-to-back? If so, the dipole moments for the two dipoles are opposite; is the distant field on the y-axis therefore exactly zero?
20. Consider a large number of identical dipoles centered in the xy-plane and pointing in the z-direction, distributed with uniform density. What is the electric field in the limit that the dipoles form a continuous distribution?
21. A large, flat, positively charged plate (of uniform charge density) is placed on the ground. A positively charged pellet, starting from rest, is released from above the plate. Ignore all air resistance. Qualitatively describe the motion of the pellet, according to the height from which it is dropped.
22. Suppose that a positively charged pellet is dropped from above onto the north pole of a large, positively charged sphere (of uniform charge density). Disregarding air resistance, and any small instabilities that would make the pellet move away from the vertical, describe the motion of the pellet.
23. Suppose in Example 22–3 q_1 were at $x_1 = -1$ mm rather than -1 cm. What useful approximation could you make to find the field at A?
24. In Example 22–10, we found Earth's charge by treating its surface as a plane rather than thinking of it as a sphere. Why is this a very good approximation?

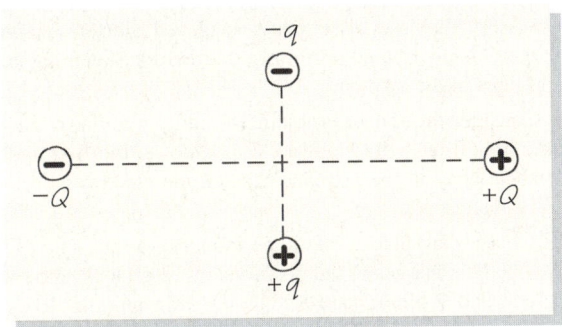

▲ **FIGURE 22–24** Question 18.

Problems

22–1 Electric Field

1. (I) A 5-μC charge is located at $(x, y) = (5 \text{ cm}, 0 \text{ cm})$. Determine the electric field at (2 cm, 4 cm).
2. (I) Calculate the electric field at the origin due to the following distribution of charges: $+q$ at $(x, y) = (a, a)$, $+q$ at $(-a, a)$, $-q$ at $(-a, -a)$, and $-q$ at $(a, -a)$ (Fig. 22–25).

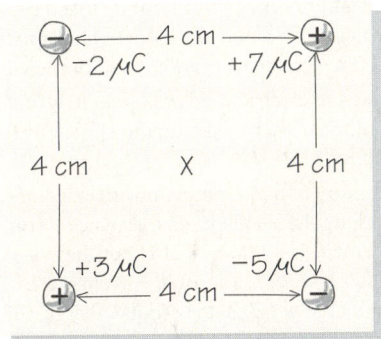

▲ **FIGURE 22–26** Problem 4.

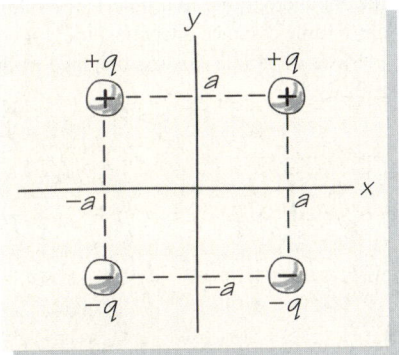

▲ **FIGURE 22–25** Problem 2.

3. (I) Find the electric field due to the nucleus of an atom of gold ($Z = 79$) at a point 1 nm from the nucleus. Treat the nucleus as a point. What is the force on an electron at that point?
4. (II) Charges of $+3\ \mu$C, $-5\ \mu$C, $+7\ \mu$C, and $-2\ \mu$C are located at the four corners of a square 4 cm on each side (Fig. 22–26). Calculate the electric field at the center of the square. [*Hint*: You may want to use a coordinate system with axes along the diagonals.]
5. (II) Five charges are located at five of the corners of a regular hexagon with sides of 10 cm, as shown on Fig. 22–27 (see next page). Find the electric field at the sixth corner of the hexagon.
6. (II) Charges $q_1 = 1.5\ \mu$C and $q_2 = -3.5\ \mu$C are a distance 22 cm apart. (a) Calculate the electric field of q_1 at the position of q_2; (b) the force acting on q_2, and (c) the total electric field at the midpoint between q_1 and q_2.
7. (II) A charge $-q$ is located at $y = -\ell/2$, and a second charge $+q$ is located at $y = +\ell/2$ (Fig. 22–28 see next page). (a) What is the electric field at the origin? (b) If the charge at $-\ell/2$ were instead $+q$, what would the electric field be at the origin? (c) For part (b), what would the electric field be in the entire xz-plane specified by $y = 0$?

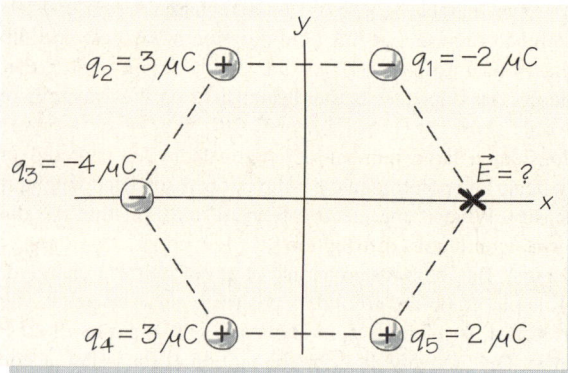

▲ FIGURE 22–27 Problem 5.

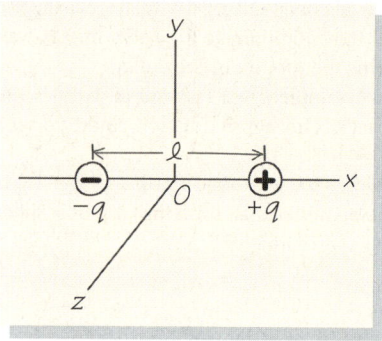

▲ FIGURE 22–28 Problem 7.

8. (II) Identical positive charges Q are placed at $x = a$ and $x = -a$, respectively. (a) What is the electric field at $x = 0$? (b) Suppose that a positive test charge q_0 is placed at $x = 0$. Will it be in stable or unstable equilibrium? [*Hint*: Assume that the test charge is displaced a distance δ in a direction perpendicular to the *x*-axis. What will the net force on the test charge be at the new location?]

9. (II) Calculate the electric field *along the axis* of a dipole at a distance r from the center of the dipole shown in Fig. 22–5. Work out the field for $r \gg L$.

10. (II) A succession of n alternating positive and negative charges q are placed along the *x*-axis, each a distance d from its neighbors. The arrangement is symmetrical about the *y*-axis, with the first $+q$ charge at $x = d/2$, the first $-q$ charge at $x = -d/2$, the second $-q$ charge at $3d/2$, the second charge $+q$ at $-3d/2$, and so forth (Fig. 22–29). What is the field at a distant point $y = Y$ (where $Y \gg nd$) on the *y*-axis?

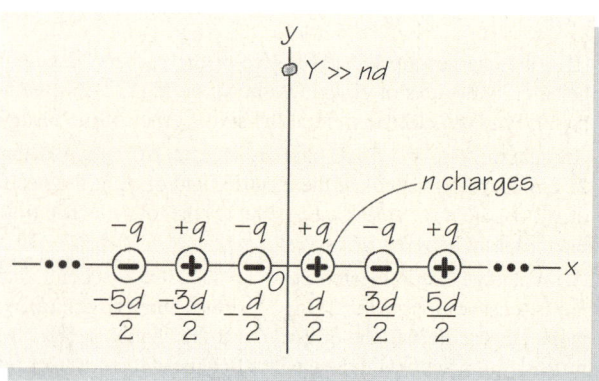

▲ FIGURE 22–29 Problem 10.

11. (III) Suppose that the positive test charge in Problem 8 is constrained to move along the *x*-axis only. Will $x = 0$ be a stable equilibrium position? If it is, then the test charge should oscillate about $x = 0$ for small enough displacements. If that were the case, what would the frequency of oscillation be for a test charge of mass m? [*Hint*: Assume that the charge is displaced to a point $x = \delta$, where $\delta \ll a$, and calculate the magnitude and the direction of the electric field there. Use the approximation $1/(a + \delta)^2 = (1/a^2) - (2\delta/a^3) + \cdots$, valid for $\delta \ll a$.]

22–2 Electric Field Lines

12. (I) Draw the electric field lines due to charges of $+0.6\ \mu\text{C}$ and $+1.8\ \mu\text{C}$ located 15 cm apart.

13. (I) A pair of parallel plates have equal and opposite uniform charge distributions. The field is represented by parallel lines drawn with a density N per m^2. The charge density on the plates is tripled. How should the density of electric field lines be changed?

14. (I) A very thin rod of length L is placed along the *x*-axis. A charge Q is distributed uniformly on the rod. Sketch the electric field lines in the *xy*-plane.

15. (I) Consider the rod in Problem 14, and put another rod of the same length but with charge $-Q$ parallel to the first and some distance away in the *xy*-plane. Sketch the electric field lines in the *xy*-plane.

16. (II) Charges are placed on the perimeter of a turntable of radius 0.15 m. The charges at the positions of 12 o'clock, 3 o'clock, 6 o'clock and 9 o'clock are 8, 8, -8, -8 in units of 10^{-5} C, respectively. Draw the field lines for this arrangement. What is the electric field at the center of the circle?

17. (II) Six equal charges of alternating opposite sign $(+, -, +, -, +, -)$ are aligned along the *x*-axis, each separated from its neighbor by a distance d. Draw the configurations of field lines.

18. (II) Consider charges q placed along the *x*-axis at $x = na$, with $n = 0, \pm 1, \pm 2, \pm 3, \ldots$. Sketch the electric field lines.

19. (II) The field lines due to an electric dipole $\vec{p}$ (dipole 1) are shown in Fig. 22–12b; by definition, the direction of $\vec{p}$ points from $-q$ to $+q$. Sketch the field lines for the combination of this dipole and (a) a dipole $-\vec{p}$ adjacent and parallel to dipole 1; (b) a dipole $\vec{p}$ adjacent and parallel to dipole 1; (c) a dipole $-\vec{p}$ on the axis of dipole 1 some distance away past the $+q$ charge; (d) a dipole $\vec{p}$ on the axis of dipole 1 some distance away past the $+q$ charge (Fig. 22–30).

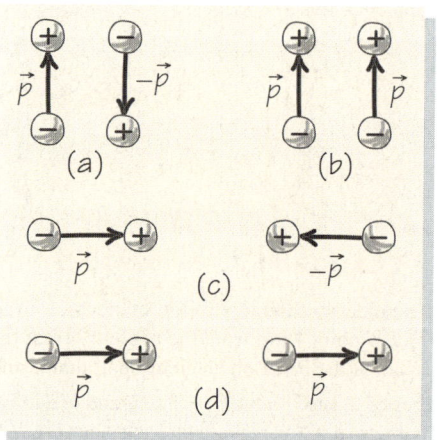

▲ FIGURE 22–30 Problem 19.

20. (II) Charges q, q, and $-q$ form an equilateral triangle whose sides each have length 12 cm, with the negative charge at the top. Sketch the electric field lines. What is the magnitude of the electric field 0.080 cm from and directly above $-q$; what is it at a distance of 35 m from and directly above $-q$?

22–3 The Field of a Continuous Distribution

21. (I) Calculate the electric field due to an infinitely long, thin, uniformly charged rod with a charge density of 0.3 μC/m at a distance 20 cm from the rod. Assume that the rod is aligned with the x-axis.

22. (I) A thin, uniformly charged rod with a total charge of 6 μC and length 25 cm is placed along the z-axis, centered at the origin. Find the electric field at $(x, y, z) = (6$ cm, 0 cm, 0 cm$)$ and $(0$ cm, 6 cm, 0 cm$)$.

23. (I) Sketch the electric field lines between a point charge Q and a uniformly charged, flat square of area L^2 and total charge $-Q$. The point charge is located a distance L above the center of the plane.

24. (I) Consider two infinite plane sheets of insulator with uniform surface charge densities σ_1 and σ_2, respectively. The two sheets are parallel to each other and a distance L apart. What is the force on a point charge Q placed midway between the sheets?

25. (II) A negative charge is distributed uniformly on a long cylindrical shell. Sketch the field lines both inside and outside the shell. Do not include the ends of the cylinder.

26. (II) Consider positive charges distributed uniformly with a charge density λ on a circle of radius R. (a) Use symmetry arguments to deduce the direction of the electric field at a point in the plane of the circle but outside the circle. (b) What is the magnitude of the electric field at a distance L along the axis of the circle for $L \gg R$?

27. (II) A rod with a uniform negative charge is bent into a semicircle. Make a rough sketch of the electric field lines in the plane of the rod.

28. (II) Two infinite plates with a uniform charge density of 1.2 μC/m^2 are placed along the yz-plane with one plate passing through $x = 2$ cm and the other through $x = -2$ cm. Determine the electric field at $(x, y, z) = $ (a) $(0$ cm, 0 cm, 0 cm$)$; (b) $(8$ cm, 0 cm, 0 cm$)$; (c) $(8$ cm, 1 cm, 2 cm$)$.

29. (II) Two large, flat, vertically oriented plates are parallel to each other, a distance d apart. Both have the same uniform positive charge density σ. What is the electric field in the space around and between them?

30. (II) The axis of a hollow tube of radius R and length L is aligned with the x-axis; the tube's left-hand edge is at $x = 0$, as shown in Fig. 22–31. It carries a total charge q distributed uniformly along its surface. By integrating the result for a field due to a hoop of charge along the axis of the hoop (see Example 22–8), find the electric field along the x-axis due to the tube as a function of x.

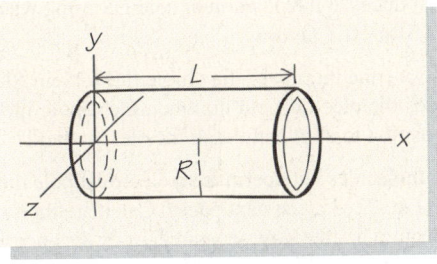

▲ FIGURE 22–31 Problem 30.

31. (II) A thin, circular disk of radius R is oriented in the xy-plane with its center at the origin. A charge Q on the disk is distributed uniformly over the surface. (a) Find the electric field due to the disk at the point $z = z_0$ along the z-axis. (b) Find the field in the limit $z_0 \to \infty$. (c) Find the field in the limit that $R \to \infty$. Are the limits of parts (b) and (c) the same?

32. (II) Consider a thin, uniformly charged rod 18 cm long that is bent into a semicircle. The total charge on the rod is 0.36 μC. What are the magnitude and direction of the electric field at the center of the semicircle?

33. (II) A rod 30 cm long is charged uniformly with a charge density of 15 μC/m. A charge of 3 μC is placed 30 cm from the midpoint of the rod along a line perpendicular to the rod. Calculate the electric field at a point halfway between the point charge and the center of the rod.

34. (II) Determine the electric field from the large plane in Example 22–8 by assuming the plane is made of a series of charged rods, using the results of Example 22–7.

35. (II) Consider a spherical shell of radius R uniformly charged with a total negative charge $-Q$. Starting at the surface of the shell going outward, there is a uniform distribution of positive charge in space such that the electric field at the radius $(R + h)$ vanishes, where $R \gg h$. What is the positive charge density? [Hint: With $R \gg h$ you can assume that $(R + r)^n \cong R^n + nrR^{n-1}$ for $0 \le r \le h$.]

36. (III) A total charge Q is distributed uniformly over a rod of length L. The rod is aligned on the x-axis, with one end at the origin and the other at the point $x = L$ (Fig. 22–32). Calculate the electric field at a point $(0, D)$, and compare this result with the field at the point $(L/2, D)$.

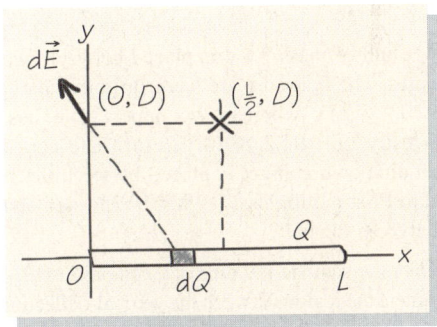

▲ FIGURE 22–32 Problem 36.

37. (III) Consider a point at a height z_0 directly above the midpoint of a square with sides of length $2L$. The (nonconducting) square carries a uniform charge density σ. (a) Use the method of Example 22–8 to write an integral for the electric field at z_0. (b) How does the integral simplify in the limit $L \to \infty$? (c) $z_0 \to 0$?

22–4 Motion of a Charge in a Field

38. (I) An infinite plate carries a uniform charge density $\sigma = 2.17 \times 10^{-6}$ C/m^2. A pellet of mass 0.555 g is placed at rest 0.175 m from the plate. The pellet carries a negative charge $q = -1.08 \times 10^{-6}$ C. What is its speed when it reaches the plate? Ignore all forces except the electrostatic attraction.

39. (I) A small object, mass 120 mg, is observed to undergo an acceleration of magnitude 4.6 m/s² when it is placed in a constant electric field of magnitude 850 N/C. What is the charge on the object?

40. (I) A sheet of uniform charge density 6.1×10^{-9} C/m² is placed at $z = 0$ in the xy-plane. An electron with zero initial velocity is placed at $z = 0.45$ m. What will be its velocity after 17.5 ns?

41. (I) A large, flat plate with unknown, uniform charge density σ is placed on a horizontal tabletop. A cork ball of mass 0.83 g, carrying a charge 8.5×10^{-7} C, is placed at rest above the plate and remains at rest. What is σ?

42. (I) An alpha particle approaches a gold atom head on, stops, and turns around at a distance of 10^{-11} m from the nucleus. What is the electric field due to the gold nucleus at this point? Ignore the effects of the gold atom's orbiting electrons. What is the acceleration of the alpha particle when it is stopped? An alpha particle is a helium nucleus, composed of two protons and two neutrons.

43. (II) Consider an infinite wire with uniform charge density λ along the z-axis. A negatively charged particle moves in a circle in the xy-plane centered on the wire. Calculate the particle's speed, and show that the speed is independent of the radius of the circle. Ignore all forces except those due to the wire.

44. (II) A negative charge $-q$ is restricted to move in a plane in which there is a continuous line of positive charge and a charge density λ. The negative charge, of mass m, can pass the line of positive charge freely. What is the equation of motion for the negative charge?

45. (II) A positive charge q can travel in a circular orbit about a negatively charged line with uniform charge density λ. Show that the period of the orbit is proportional to the radius of the orbit. Compare this to the dependence of the period of a circular orbit on the radius of the orbit for a point charge that interacts with another point charge.

46. (II) A cork ball of mass 5.6 g is placed between two large horizontal plates. The bottom plate has a uniform charge density of $+1.6 \times 10^{-6}$ C/m², whereas the upper plate has a uniform charge density of -0.22×10^{-6} C/m². The cork ball, which carries an unknown charge, is placed between the plates and is observed to float motionlessly. What are the sign and magnitude of the charge on the ball?

47. (II) Consider the cathode-ray tube of Example 22–12. This time an electron enters the region between the vertical-deflection plates with a total speed of $v_0 = 5.0 \times 10^6$ m/s. The direction is such that the velocity has a vertical component $v_{0y} = +2.0 \times 10^5$ m/s. Find the total vertical deflection of the electron when it reaches the screen.

48. (II) A beam of electrons is accelerated by passing it through a region between two large charged parallel plates (Fig. 22–33). Calculate the charge density on the plates if the electrons accelerate from 1.4×10^6 m/s to 3.0×10^7 m/s between the plates.

49. (II) A cork ball of mass 5 g, carrying a charge of $-2\ \mu$C, is suspended from a string 1 m long above a horizontal, uniformly charged plate of charge density 1 μC/m². The ball is displaced from the vertical by a small angle and allowed to swing. Show that the ball moves in simple harmonic motion, and calculate the angular frequency of that motion.

50. (III) A proton moves at speed $v = 5 \times 10^5$ m/s in the $+x$-direction and enters a certain region. An electric field in the region also is oriented in the $+x$-direction. The field's strength

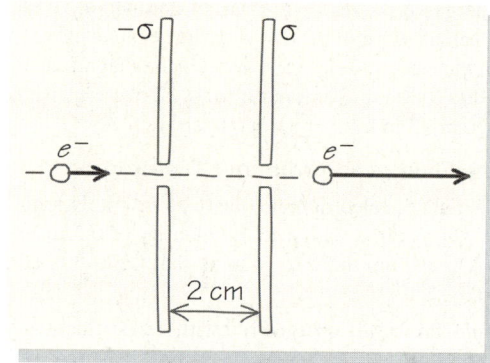

▲ FIGURE 22–33 Problem 48.

drops linearly with x: At the beginning of the region, $x = 0$ m, the field strength is 500 N/C; at $x = 3$ m, the field strength is zero. How much time does it take for the proton to traverse this region? [*Hint*: The equation of motion will be more familiar in terms of the variable $x' = x - 3$.]

22–5 The Electric Dipole in an External Electric Field

51. (I) An electric dipole consists of two opposite charges of magnitude 2 μC placed 10 cm apart (Fig. 22–34). The dipole is placed in a uniform electric field of 10 N/C along the x-axis, with the direction of $\vec{p}$ at an angle of $+45°$ from the x-axis in the xy-plane. Determine the torque on the dipole.

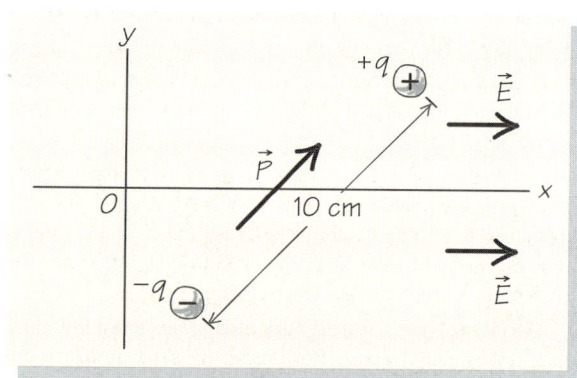

▲ FIGURE 22–34 Problem 51.

52. (I) The magnitude of the two opposite charges that form an electric dipole is increased by a factor of 5 while the separation between the charges is tripled. What is the change in magnitude of the torque on the dipole in a uniform electric field?

53. (I) A water molecule has a permanent electric dipole moment of magnitude 6×10^{-30} C·m. Estimate the size of the electric field it produces at the position of a neighboring water molecule, which is 3×10^{-9} m away.

54. (II) Describe the motion of the dipole in Problem 51. How much work does the electric field do when the dipole moves from its initial position to alignment with the electric field?

55. (II) Two molecules with permanent electric dipole moments $\vec{p}$ are aligned (Fig. 22–35, see next page). Calculate the force between the molecules if they are separated by a distance that is large compared with the dimension of the dipoles. [*Hint*: The relation $(1 + x)^{-2} \cong 1 - 2x + 3x^2 - \cdots$ for small x is useful.]

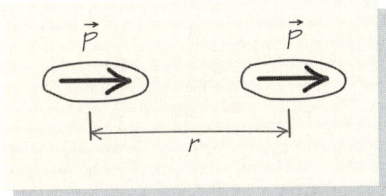

▲ FIGURE 22–35 Problem 55.

56. (II) A molecule of lithium fluoride (LiF) has a permanent dipole moment. The molecule is placed in a uniform electric field of strength 10^4 N/C, and the difference between the maximum and minimum potential energies of the molecule in this field is 4.4×10^{-25} J. What is the electric dipole moment of the LiF molecule?

General Problems

57. (II) A point charge $-q$ is fixed at the center of a hollow spherical conductor of charge $+q$. Draw the electric field lines both inside and outside the sphere.

58. (II) A point charge $+q$ is fixed at the center of a hollow spherical conductor also of charge $+q$. Draw the electric field lines both inside and outside the sphere.

59. (II) Draw the electric field lines for a point charge $+q$ near an infinitely long, positively charged wire.

60. (II) A cork ball of radius 1.2 cm and a charge of $+3.5$ nC is covered with conducting paint. What is the electric field strength just outside the surface? A nickel nucleus, with a radius of 5×10^{-15} m, has a positive charge of $28e$. What is the electric field strength just outside the surface of the nucleus?

61. (II) Two infinitely long, uniformly charged rods, with charge densities of λ and $-\lambda$, respectively, are lined up parallel to each other and separated by a distance R. What are the magnitude and direction of the electric field due to the two rods at points that lie (a) on a line joining the two rods, and (b) along a perpendicular bisector of that line? Draw a figure to show the configuration, and use symmetry.

62. (II) What is the force per unit length that one of the two rods in Problem 61 exerts on the other?

63. (II) Two uniformly charged infinite plates with charge densities -5 μC/m^2 and 3 μC/m^2 are placed at right angles, the first one along the xz-plane, the second along the yz-plane. A test particle of mass 1 g and charge 1×10^{-7} C is placed a distance of 1 m from both planes; that is, its initial position is $(x, y, z) = (1$ m$, 1$ m$, 0$ m$)$. What is the location of the test particle after a short time t (before it hits a plate)?

64. (II) Two infinite lines of charge density 5 μC/m are parallel to the z-axis. One line passes through $(x, y) = (0$ cm$, 1$ cm$)$; the other, through $(x, y) = (0$ cm$, -1$ cm$)$ (Fig. 22–36). Find (a) the electric field at the origin; (b) the force on a 0.5-μC charge at the origin; (c) the force on a 6-μC charge located at $(x, y, z) = (4$ cm$, -3$ cm$, 0$ cm$)$.

65. (II) A proton with kinetic energy of 2×10^6 eV is fired perpendicular to the face of a large metal plate that has a uniform surface charge density of $\sigma = 8.0 \times 10^{-6}$ C/m^2. (a) Calculate the magnitude and direction of the force on the proton. (b) How much work must the electric field do on the proton to bring it to rest? (c) From what distance should the proton be fired so that it stops right at the surface of the plate?

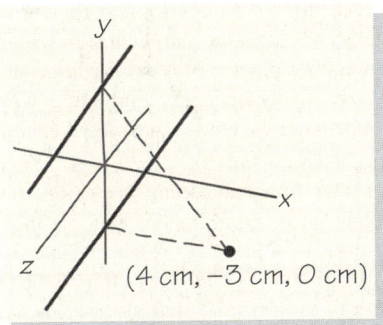

▲ FIGURE 22–36 Problem 64.

66. (II) The electric charge with the smallest magnitude that can be isolated is the charge on the electron or the proton. In 1909, Robert A. Millikan developed a classic method to measure this charge, known as the *oil drop experiment*. Millikan was able to place charges on tiny droplets of oil, which would fall at a given terminal velocity under the influence of gravity and air drag. By placing these droplets between parallel, horizontal charged plates, as in Fig. 22–37, the electric field between the plates produces a force on the charged droplet that is directed upward and can partly cancel the gravitational force. If the mass and size of the droplet are known, then, by seeing how fast droplets fall with and without the electric field, the charge can be measured.

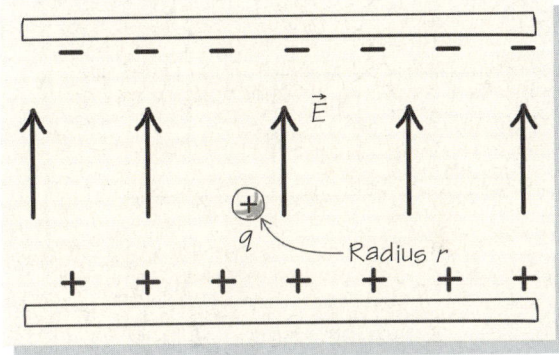

▲ FIGURE 22–37 Problem 66.

The drag force on a droplet of radius r that falls at a steady speed v through air is also directed upward and is given by Stokes's law, $F_{\text{drag}} = 6\pi\eta r v$, where η is the viscosity of air. (a) Show from Newton's second law that the terminal velocity v_0 of the *uncharged* drop is $v_0 = (2/9)r^2\rho g/\eta$, where ρ is the density of the oil and g is the acceleration due to gravity. (b) Suppose that the charge on the drop, q, is positive and that the field is directed vertically upward, as in the figure, so that the electric force points up. Show by using Newton's second law that the charge is given by

$$q = \frac{18\pi(v_0 - v_1)}{E}\sqrt{\frac{v_0\eta^3}{2\rho g}},$$

where v_1 is the terminal velocity when the electric field E is imposed. (c) Take the minimum charge as 1.6×10^{-19} C, the oil's density as 0.85 g/cm^3, and the radius of the droplet as 2.0×10^{-4} cm. The droplet has the minimum charge. Find the value of E that will hold the droplet stationary between the plates.

67. (II) We will learn in Chapter 23 that the electric field near a conductor *must be perpendicular to the conducting surface*.

Using this fact, draw the electric field lines for the following configurations: (a) a point charge $+q$ above an infinite, uncharged conducting plane; (b) a point charge $-q$ near an infinitely long, positively charged conducting wire; (c) a point charge $+q$ a distance $L/2$ above a charged conducting plane of area L^2 and charge $+q$.

68. (II) The field due to a line of uniform charge density λ varies with a radial distance r from the line as $1/r$. Suppose that a point charge q of mass m is placed at rest a distance R from the line, and that the force on the point charge due to the field of the line is attractive. Use dimensional analysis to calculate how the time it will take for the charge to drop to the charged line depends on λ, q, m, R, and ε_0.

69. (II) Consider two thin insulating rods of equal length (0.20 m) placed parallel to each other, and 0.18 m apart. Each of the rods has a charge 2.2×10^{-4} C and -2.2×10^{-4} C placed at opposite ends, with the positive charges on the two rods nearest to each other. What is the force that one of the rods experiences due to the presence of the other rod? What torque about an axis through its midpoint is experienced by one of the rods due to the presence of the other?

70. (II) An electron of speed 3×10^6 m/s enters a region of constant electric field at an angle of $40°$ as shown in Fig. 22–38. How far away from where the electron enters will it strike the bottom plate?

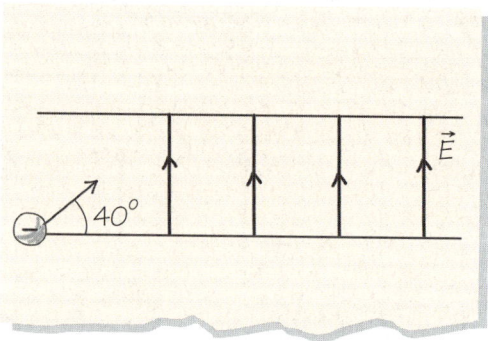

▲ **FIGURE 22–38** Problem 70.

71. (III) Consider the straight, nonuniformly charged rod of length L aligned along the x-axis, with the ends at $x = \pm L/2$, in Example 21–11. We showed there that the force on a charge q located at a point $x = R$ on the x-axis, to the right of the right-hand end of the rod, is

$$\vec{F} = \frac{q\lambda_0}{2\pi\varepsilon_0 L}\left\{\ln\left[\frac{R - (L/2)}{R + (L/2)}\right] + R\left[\frac{1}{R - (L/2)} - \frac{1}{R + (L/2)}\right]\right\}\hat{i}.$$

Show that for $R \gg L$, the force reduces to that of a dipole acting on q, $\vec{F} \cong (q\lambda_0 L^2/12\pi\varepsilon_0 R^3)\hat{i}$. What is the dipole moment? [*Hint*: Use the approximate forms $(1 - x)^{-1} \cong 1 + x + x^2 + \cdots$ and $\ln(1 + x) \cong x - (x^2/2) + (x^3/3) - \cdots$, both appropriate for $x \ll 1$.]

72. (III) The field of an electric dipole decreases as $1/r^3$ when the distance of a given point to the dipole, r, is much larger than the separation between the charges. The only way to arrange two charges with a total charge of zero is to form a dipole. There are, however, many ways to arrange four charges with a total charge of zero in a compact pattern. An arrangement with an electric field that behaves at great distances as $1/r^4$ is an *electric quadrupole*. (a) For four charges aligned with alternating sign (such as $(+ - - +)$ so that the combination acts like dipoles of opposite orientation) along an axis, show that the field on the axis perpendicular to the line of charges decreases as $1/r^4$, where r is much larger than any separation distance within the quadrupole. [*Hint*: Use the approximation $(r^2 + \delta^2)^{-3/2} \cong \frac{1}{r^3} - \frac{3\delta^2}{2r^5} + \cdots$ (good when $\delta \ll r$) for each of the four charges; do not forget the sign of the charge in determining the field.] (b) Sketch the field for the arrangement shown in Fig. 22–39, using the field-line technique.

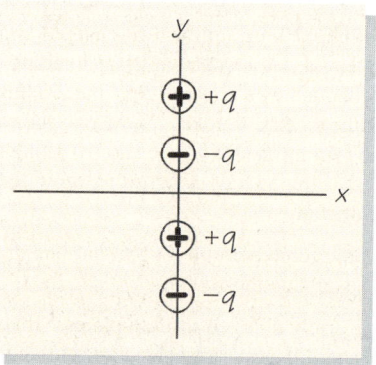

▲ **FIGURE 22–39** Problem 72.

Chapter 23

Gauss' Law

◀ Sparks from a high-voltage electrostatic generator at the Boston Museum of Science do not harm the operator who is sitting inside a grounded Faraday cage.

The concept of the electric field allows us to study the interaction between two charges as a two-step process: one involving the interaction of a charge with an electric field, the other involving the determination of the electric field due to electric charges. We saw in Chapter 22 that this division simplifies the problem of calculating interactions between charges. Gauss' law, the subject of this chapter and one of the fundamental laws of electromagnetism[†], gives us a new way to think about the electric field due to charge distributions. This law contains all the information in Coulomb's law and will allow us to go further. When there is symmetry in the charge distribution, Gauss' law is a powerful tool for the direct evaluation of the electric field. Gauss' law also gives us valuable insight into the behavior of conductors.

23–1 What Does Gauss' Law Do?

For a simple insight into Gauss' law, consider the electric field lines associated with a charge Q. We suppose that the charge gives rise to N electric field lines. (The choice is arbitrary, although once it is made we must use it as a normalizing factor—for example, a charge $2Q$ will give rise to $2N$ lines.) We can think of the field lines as a collection of N rigid wires that emerge radially from a point—the location of the charge—and extend to infinity. One way of finding the value of Q would be to imagine that we could somehow make a sphere of tissue appear at any fixed radius around the charge and count the holes

---

[†] Karl Friedrich Gauss, a great mathematician of the nineteenth century, worked on celestial mechanics, electromagnetism, optics, and the theory of errors.

662 | Gauss' Law

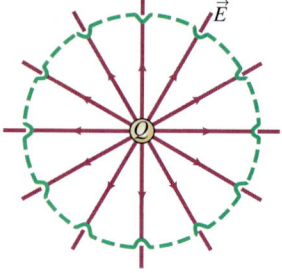

(a)

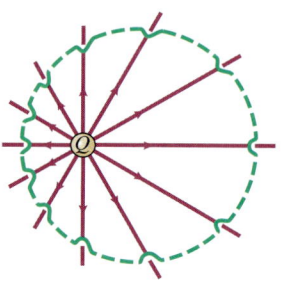

(b)

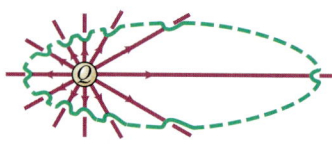

(c)

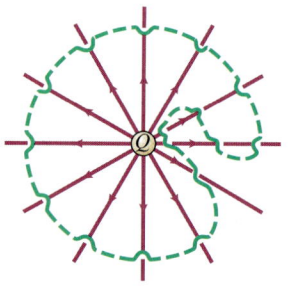

(d)

▲ **FIGURE 23–1** (a) Twelve wires representing field lines pass through this two-dimensional representation of a three-dimensional tissue forming a sphere with a charge Q at its center. (b) If the sphere is displaced the number of wires (field lines) passing through it remains 12, at least as long as the charge is somewhere within. (c) The tissue does not have to form a sphere; it can be another shape that still encloses the charge, and the count of wires that pass through it is the same. (d) If the surface enclosing the charge has folds, the counting looks as though it fails. But if one counts the number of times a line enters the surface from the outside, and assigns this number a minus sign, the algebraic total of the number of wires through the closed surface surrounding the charge is the same whatever the shape, including the case illustrated here, in which the surface has a fold.

that the wires make. There will be N holes in the tissue. Figure 23–1a represents this situation in two dimensions, not three, with the holes on the two-dimensional circle. (You should, however, be thinking about this in three dimensions.) For purposes of illustration, we have also chosen to draw 12 field lines in this figure, i.e. we have supposed that $N = 12$, but this has no fundamental significance. Suppose now that the charge were not at the center of the sphere of tissue, but somewhere else within it (Fig. 23–1b). The number of holes would still be the same, even if they are clustered more on the part of the surface that is closer to the charge. If the tissue still formed a closed surface but no longer made a sphere (Fig. 23–1c), the number of holes would again be the same. Now imagine a fold in the closed surface of tissue, as in Fig. 23–1d. This time there would be more holes in the surface, $12 + 4$ in Fig. 23–1d. But two of the four extra holes are caused by a wire going *into* the surface, and if we keep track of holes made by wires going *into* the surface and assign a negative count to such holes then in Fig. 23–1d, the number of holes would be $12 - 2 + 2 = 12$, the same as in all the other cases. The distinction between going out of the closed surface and into the surface will be an important one. To summarize, what we have seen is that we can determine the charge Q by counting the net number of "holes" made by the "field lines" through a closed surface of any shape that surrounds the charge, including a minus sign for field lines going into the surface.

Using the fact that we count lines going into the surface from the outside with a minus sign, then we see that if the charge inside our closed surface were a negative one, $-Q$ instead of $+Q$, then the electric field lines will go toward the negative charge, and the number of holes in our tissue would be -12, the sign of the number of holes correctly accounting for the charge's sign. Without a distinction about whether the field lines were ingoing or outgoing, we cannot tell the sign of Q.

The situation we have described is the essence of Gauss' law: We count "holes" created by "wires"—by this we mean the passage of field lines through an imaginary closed surface that surrounds any charge or set of charges, with a rule that field lines going out of the surface make positive holes, and field lines going into the surface make negative holes. The net number of field lines that pass through the closed surface is a measure of the charge enclosed. The charge Q does not have to be a point charge. It can be an extended charge, the only proviso being that the imaginary surface must enclose all of the charge. This is really all there is to Gauss' law, which as we'll see in the next section expresses a relation between the charge contained within a closed surface and an appropriate generalization to the number of "holes" made in the surface by the electric field.

Drawing field lines and literally counting the passage of the lines through a surface is not a practical possibility, and we now need to provide a mathematical expression for this process. We will do this by replacing the idea of holes made by rigid wires with a calculation of what is known as the *flux* of electric field, or, more simply, electric flux.

Electric Flux

We'll stay for a moment with the idea of electric field lines as represented by rigid wires, but this time we'll consider a sheet of tissue that is not closed. Rather than the field of a point charge, consider the electric field due to an infinite plane of uniformly distributed charge. As we showed in Example 22–8, the field lines are uniformly distributed (let's say with N lines per unit area of the plane) and are perpendicular to the infinite sheet of charge (aligned with the field *direction* $\hat{E}$, where the hat indicates that this is a unit vector that specifies the direction, but not the size, of the electric field). Suppose we now somehow place a flat piece of tissue, area A, above the plane and parallel to it. The wires are perpendicular to the piece of tissue, and we'll take the charge density on the plane such that there will be NA holes in the tissue (Fig. 23–2a). Now, if instead of the tissue aligning parallel to the charge plane, we could make it appear in a tilted position (Fig. 23–2b), there would be fewer holes—think of fewer field line crossings—for the simple reason that the *perpendicular* area presented to the wires by the square of tissue is smaller. We can in fact give a mathematical description to the idea that our surface makes an angle to the sheet of charge by defining the *normal* to the surface. More precisely, the normal vector $\hat{n}$ will be a unit vector *perpendicular* to the surface. In Fig. 23–2b we draw it in the direction away from the sheet of charge. Suppose now $\hat{n}$ and $\vec{E}$ make an angle θ (Fig. 23–2b). Then you can see from the figure that the area presented to the wires coming from the surface is the original area A multiplied by $\cos\theta$. A compact way of stating this is that the number of holes will be given by $NA\cos\theta = NA(\hat{n}\cdot\hat{E})$. When the tissue is aligned parallel to the plane, then $\hat{n}$ and $\vec{E}$ are parallel, and the expression for the number of holes—again, the number of field lines that cross the surface—reduces correctly to NA.

At this point, we can say that the number of field lines crossing the surface is proportional to the surface area, the orientation of the surface, and the field strength. The orientation factor is in the scalar product $\hat{n}\cdot\hat{E}$. The field strength factor is in the number of lines N, which you will recall is an arbitrary normalization of the field strength. We can take care of the factor N and remove any arbitrariness by replacing N with the magnitude of the field itself, E. Thus we replace the factor $(\hat{n}\cdot\hat{E})N$ with $\hat{n}\cdot\vec{E}$. Finally, the area A remains. But we have chosen in this discussion a large flat area—the flatness makes the normal the same. For a more general surface, there will be various curvatures, and the normal will vary from point to point. Thus, rather than A, we work with an area dA that is sufficiently tiny so that we can treat it as flat and there is no ambiguity in the value of $\vec{E}$ over it.

We have now found our equivalent to the number of holes through a piece of tissue of area dA: It is the **electric flux** $d\Phi$ through the area dA,

$$d\Phi \equiv (\hat{n}\cdot\vec{E})\, dA.$$

We have written this expression in differential form as a reminder that the area to which it refers is small; the field $\vec{E}$ is the field at the location of the surface element. Now just as the total number of holes in our tissue is the sum of the number of holes through the entire surface, we will want to define the electric flux Φ_S across a surface S as the sum over the fluxes through the elements that make it up:

$$\Phi_S = \int_S (\hat{n}\cdot\vec{E})\, dA. \tag{23–1}$$

Finally, we noted above that Gauss' law will concern *closed* surfaces. We indicate that the surface being summed over is closed with a circle on the integral sign,

$$\Phi = \oint_S \hat{n}\cdot\vec{E}\, dA. \tag{23–2}$$

There is another item to deal with: For an open surface, it is not clear from which side of the surface the normal $\hat{n}$ points. In Fig. 23–2b we chose to draw it in one of two possible directions, but this choice was in no way compelled by anything in the situation; for an open surface, the direction of $\hat{n}$ is inherently ambiguous. For a closed surface, we can decide this in a completely unambiguous way: The direction of $\hat{n}$ is defined as pointing toward the *outside* of the closed surface. With this choice, the flux through a closed surface containing a positive charge is positive because $\vec{E}$ points outward as $\hat{n}$ does, and $\hat{n}\cdot\vec{E}$ is positive. For a negative charge, $\vec{E}$ points inward and $\hat{n}\cdot\vec{E}$ is negative (Fig. 23–3).

We simplify the notation by combining the magnitude of the infinitesimal area dA with the associated normal direction, writing $\hat{n}\, dA = d\vec{A}$. Thus

$$\Phi = \oint_S \vec{E}\cdot d\vec{A}. \tag{23–3}$$

This, then, is the mathematical generalization of counting the number of field lines crossing a closed surface.

We have now established that when electric flux through a closed surface is not zero, there is a net charge within the surface. The surface, which can be picked at will just as long as it encloses the charge, is an imaginary construct generally known as a **Gaussian surface**. We are going to use such surfaces to help us calculate values for the electric field, and we shall see that surfaces that have a special symmetry in relation to the charge distribution are especially useful. For example, for spherically symmetric charge distributions, a concentric spherical Gaussian surface is particularly useful, and for charges uniformly distributed along a straight line, a Gaussian surface in the shape of a cylinder, with the line charge forming the axis, is useful.

We note in passing that the use of the term *electric flux* comes from an analogy with the flow of water through a surface, a concept we met in Chapter 16, and an analogy between electric flux and the flux of water can be made.

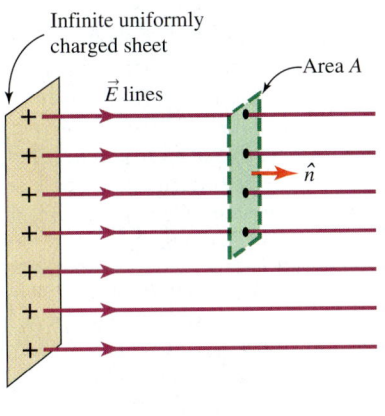

(a)

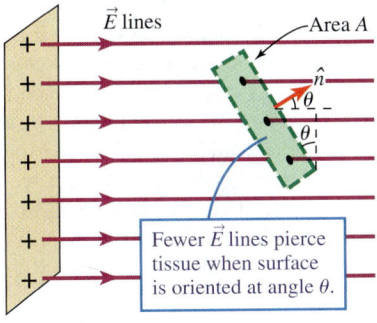

(b)

▲ **FIGURE 23–2** A plane of area A is placed in a uniform electric field. In (a) the plane is aligned perpendicular to the field. (b) The perpendicular to the plane makes an angle θ with respect to the field.

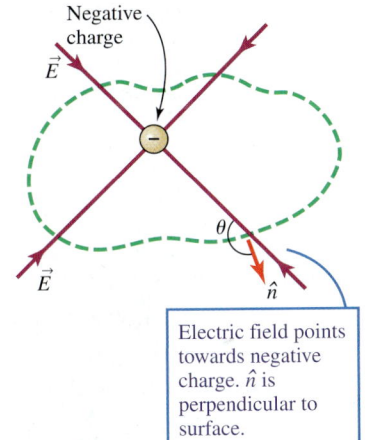

▲ **FIGURE 23–3** If the charge forming the field is negative, the angle between the perpendicular to the surface and the field is larger than 90°.

CONCEPTUAL EXAMPLE 23–1 Consider the combination of a positively charged ring and a plate with an equal and opposite charge, as in Fig. 23–4a, then think about a Gaussian surface that surrounds the ring but not the plate (Fig. 23–4b). Suppose the net flux through that surface is determined to be Φ_0. (a) What is the flux through the surface if the charge on the ring is tripled? (b) What is the flux through a Gaussian surface that encloses both the ring and the plate (Fig. 23–4c) before and after the tripling of the ring charge?

Answer (a) We have not yet made an explicit connection between the flux and the enclosed charge, but we do know that the number of field lines is *proportional* to the charge that gives rise to them and that the flux through a closed surface is also proportional to the number of field lines emerging from (or entering) the enclosed charge. From this we see that the tripling of the charge on the ring will result in the tripling of the flux through a Gaussian surface enclosing the ring only: If the initial flux due to the ring is Φ_0, the final flux is $3\Phi_0$.

(b) Initially the net charge of ring and plate is zero, so that none of the electric field lines go off to infinity. The total charge enclosed by the Gaussian surface is zero. Although field lines may cross a Gaussian surface on their way out, they return, and by our rule of counting "in" lines as yielding negative flux, the total flux is zero. When the ring charge is tripled without a change in the plate charge, the total flux through a Gaussian surface may be calculated as 3Φ due to the ring and $-\Phi$ due to the plate, so that the flux goes from an initial value of 0 to a final value of $2\Phi_0$.

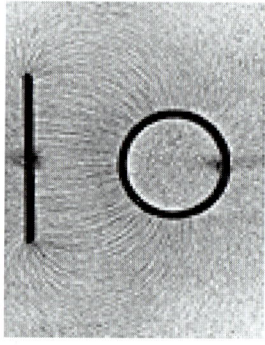

(a)

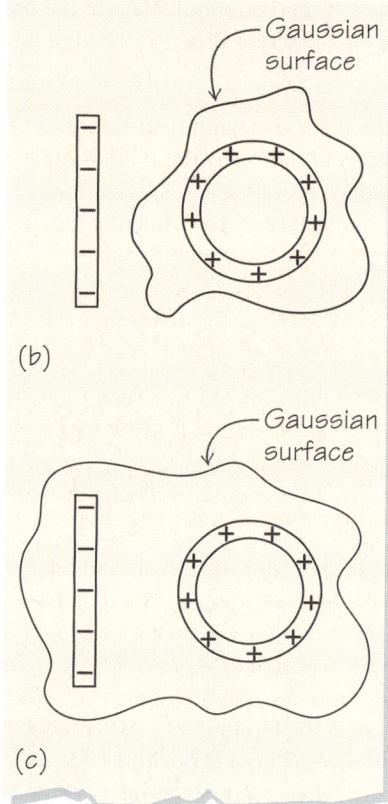

▶ **FIGURE 23–4** (a) The electric field lines due to a charged conducting cylinder placed close to an oppositely charged conducting plate, as shown by threads in a shallow dish of oil. (b) A Gaussian surface surrounds the cylinder. (c) A Gaussian surface surrounds both the cylinder and the plate.

Let us look at a calculation of electric flux through an *open* surface, where there is an ambiguity: Whereas for a closed surface there is a clear distinction between a normal to the surface $\hat{n}$ pointing out and pointing in, for an open surface this is subject to definition. The choice, once made, must be consistently preserved throughout the calculation. To take an example, we can look again at the large uniformly charged plane, with a flat square surface of area A oriented as in Fig. 23–2b. With the unit vector oriented as in that figure and a positive charge on the plane, the factor $\vec{E} \cdot \hat{n}$ in the flux is $+E \cos\theta$. This is constant over the surface, so that the flux is just $(E \cos\theta)A$. On the other hand, if we had initially chosen the normal in the opposite direction, as in Fig. 23–5, then the sign of the flux would be reversed. For an open surface such as this planar one, there is no way to decide what the "right" direction for the normal is. For a closed surface, the use of the word "outside" is unambiguous, as is the choice of direction for the normal.

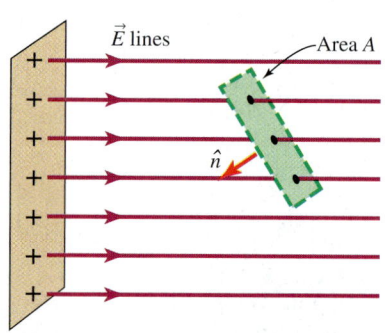

◀ **FIGURE 23–5** The perpendicular to an open Gaussian surface has an ambiguous orientation. Here we have placed it in the opposite direction to that chosen for Fig. 23–2b. This means the flux has an ambiguous sign for an open surface. This is not the case for a closed surface.

CONCEPTUAL EXAMPLE 23–2 You have a plane sheet of charge, infinite in extent, with a uniform positive charge density. You want to consider the flux through a flat, open surface. Are there circumstances in which the flux through the surface is zero?

Answer Yes. Aside from the possibility that the charge on the plane is zero, so that $E = 0$, the surface could be aligned so that the electric field is parallel to the surface. In other words, $\hat{n}$ is perpendicular to the direction of E, and therefore $\hat{n} \cdot \vec{E} = 0$.

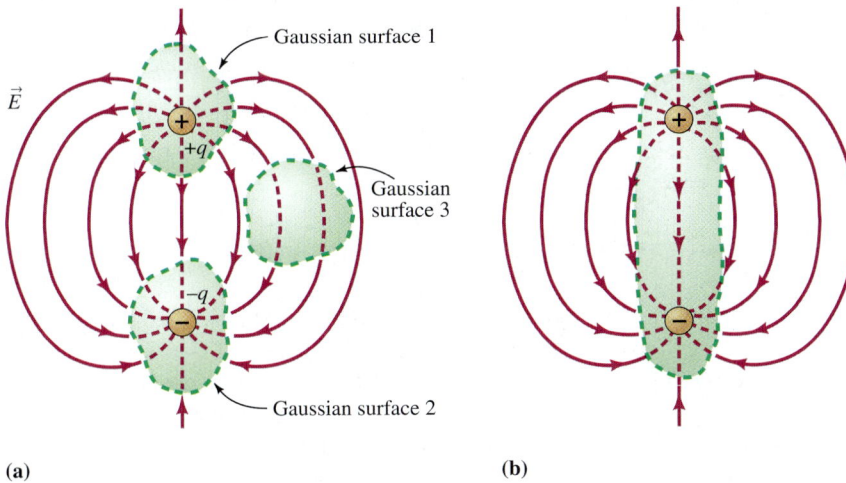

◀ **FIGURE 23–6** (a) Three Gaussian surfaces (shown dashed to remind you that they are imaginary) in the electric field of a dipole. For surface 1, which surrounds the $+q$ charge, the electric flux is positive; for surface 2, which surrounds the $-q$ charge, the flux is negative; and for surface 3, which surrounds no charge, the flux is zero. (b) A Gaussian surface surrounding both charges. The flux through this surface is proportional to the net charge and is therefore zero.

23–2 Gauss' Law

Gauss' law describes the relation between a charge and the electric flux through a closed surface—a Gaussian surface—that surrounds that charge. To start, we can most easily show that the flux through a Gaussian surface is zero if no *net* charge is enclosed by the surface by using a field-line argument. No matter how many field lines we employ, it will be true that if the enclosed charge adds to zero, then as many field lines go back into the surface as leave it. This follows by virtue of the fact that outgoing lines contribute a positive sign and ingoing lines a negative sign. For example, Fig. 23–6 shows the electric field due to a dipole as described in Chapter 22. Imagine a series of Gaussian surfaces of any convenient shape placed wherever we choose. For example (Fig. 23–6a), if we place an imaginary Gaussian surface (surface 1) around charge $+Q$ of a dipole, all the electric field lines exit the Gaussian surface, and the total electric flux is positive. If we place a second Gaussian surface (surface 2) around charge $-Q$, all the electric field lines enter the Gaussian surface, and the electric flux is negative. Any Gaussian surface, such as surface 3, that surrounds *neither* charge has no net electric flux through it because the same number of electric field lines enter and exit such a surface. If the Gaussian surface surrounds *both* charges (Fig. 23–6b), then the number of field lines that enter and exit the surface is again equal, and the total flux is zero. We conclude that

> **The electric flux through a closed surface that encloses no net charge is zero.**

To go further, we must find the relation between the electric flux Φ and the enclosed charge. We can do so by thinking about a point charge, then using superposition. Figure 23–7 shows an (imaginary) Gaussian sphere of radius R centered on a point charge q. The centered sphere is chosen because the electric field has constant magnitude at a fixed distance from a charge, and it will be easy to find the flux through the sphere. We use Eq. (23–3) to find the electric flux that passes through the Gaussian surface. The electric field due to a point charge q was found to be [Eq. (22–5)]

$$\vec{E} = \left(\frac{q}{4\pi\varepsilon_0 r^2}\right)\hat{r}.$$

The electric field points in the radial direction and is directed outward for positive q. Because the direction of the infinitesimal area $\hat{n} \cdot dA = d\vec{A}$ for a small area on the sphere also points outward in the radial direction, $\vec{E} \cdot d\vec{A} = E\, dA$. Because the electric

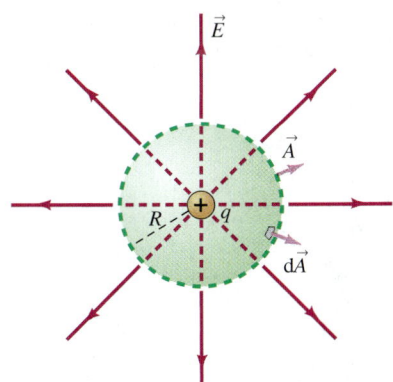

▲ **FIGURE 23–7** A simple choice for a Gaussian surface for a point charge q is a sphere of radius R.

field has the constant value $q/(4\pi\varepsilon_0 R^2)$ everywhere on our sphere, the infinitesimal electric flux through the infinitesimal area dA is

$$d\Phi = E\, dA = \frac{q}{4\pi\varepsilon_0 R^2}\, dA.$$

We can now move the field E (constant on the surface A) outside the integral that represents the total flux [Eq. (23–3)]:

$$\Phi = \oint \vec{E}\cdot d\vec{A} = \oint E\, dA = \oint \frac{q}{4\pi\varepsilon_0 R^2}\, dA = \frac{q}{4\pi\varepsilon_0 R^2}\oint dA.$$

The sum of the area elements dA over the closed surface is the total area of the closed surface, $A = 4\pi R^2$. Thus,

$$\Phi = \frac{q}{4\pi\varepsilon_0 R^2} A = \frac{q}{4\pi\varepsilon_0 R^2} 4\pi R^2 = \frac{q}{\varepsilon_0}. \tag{23–4}$$

This result establishes the relation between the flux and the enclosed charge for a particularly simple case: a point charge at the center of the spherical Gaussian surface. The relation is evidently independent of the radius of the sphere. More generally, we have argued that this relation is independent of the shape of the Gaussian surface and of the location of the charge inside it; this follows from a field-line argument. The number of field lines passing through a closed surface placed around the charge will be the same for any form and any location as long as it encloses the charge. Then it must also be true that the electric flux through any Gaussian surface around the charge is the same for a surface of any form and any location as long as it encloses the charge, and *that value is given by Eq. (23–4)*.

Using superposition, we can easily generalize Eq. (23–4) to the case of multiple point charges and then continuous charge distributions. Start with an assembly of point charges q_i that add up to a net charge Q. From the superposition principle, we know that the total electric field $\vec{E}$ is the sum of the fields $\vec{E}_i$, due to point charges q_i. The total flux Φ through a Gaussian surface S enclosing a net charge is the sum of the fluxes Φ_i due to the charges q_i:

$$\Phi = \sum_i \Phi_i = \oint_S \sum_i \vec{E}_i \cdot d\vec{A} = \frac{1}{\varepsilon_0}\sum_i q_i = \frac{Q}{\varepsilon_0}.$$

It is clear from the way we derived this result that it also holds for a continuous distribution of charge—the only difference is that the sum over point charges is replaced with an integral, but that integral remains the net charge Q.

The general statement of **Gauss' law** is therefore

$$\oint_S \vec{E}\cdot d\vec{A} = \frac{Q}{\varepsilon_0}. \tag{23–5}$$

GAUSS' LAW

The closed surface is *any* Gaussian surface that surrounds the *net* charge Q. The case in which the net charge is zero is included here—either because no charge whatsoever is enclosed by S or because there is an equal amount of positive and negative charge.

Coulomb's Law and Gauss' Law

In Chapter 22 we used Coulomb's law to determine the electric field of a point charge, and this in turn has led us to Gauss' law. Actually, the statement of Gauss' law is more general, and we may reverse the procedure to show that Gauss' law implies Coulomb's law. To do so, we center a Gaussian sphere on a point charge q (Fig. 23–7). The electric field $\vec{E}$ of the charge is assumed to be unknown. Gauss' law tells us only that the electric flux integrated over the surface of the sphere is q/ε_0. This is insufficient to determine the field, because the flux through any tiny surface element of the sphere depends on the value of the field in that region. We can, however, use a symmetry argument. There is no preferred direction for the field of a point charge. The only configuration of field around a charge that does not favor some particular direction is a radial field. The surface element $d\vec{A}$ of a Gaussian sphere is also radial. Let's assume that $\vec{E}$ is parallel to $d\vec{A}$ at all locations (the other option is antiparallel). It then follows that

$$\vec{E} \cdot d\vec{A} = E\, dA.$$

Moreover, symmetry—that is, the absence of a preferred direction—also implies that $\vec{E}$ will have the same magnitude E everywhere on the centered sphere. We can then remove E from the integral that expresses the total flux through the sphere:

$$\oint \vec{E} \cdot d\vec{A} = \oint E\, dA = E \oint dA = EA = E(4\pi r^2) = \frac{q}{\varepsilon_0},$$

where r is the radius of the Gaussian sphere. The last term in this equality is just Gauss' law. The equation can be solved for the magnitude of the electric field:

$$E = \frac{q}{4\pi\varepsilon_0 r^2}.$$

This result is consistent with Eq. (22–5). Because E is positive, we correctly chose the direction of $\vec{E}$ to be radially outward for a positive charge. The symmetry of the situation tells us only that the electric field must be radial: either outward or inward. Gauss' law determines the orientation of $\vec{E}$ to be radially outward. Coulomb's law follows directly from our result for the electric field: We put another charge, q', in the electric field and use the relation $\vec{F} = q'\vec{E}$.

Gauss' law does not require us to use any particular surface. This has practical importance, because the flux through one surface may be much easier to calculate than the flux through another.

CONCEPTUAL EXAMPLE 23–3 Find the electric flux through the Gaussian surfaces in Fig. 23–8: (a) a cube of sides L that surrounds the point charge q; (b) a sphere of radius R that surrounds the charge q, which is off center; (c) a sphere of radius b that surrounds the charges $-2q$ and $+q$.

Answer We are asking for quantitative answers, yet this is in fact a conceptual example; this illustrates the power of Gauss' law. Although the expression for the flux looks as if it involves a formidable integral, Gauss' law tells us that it is just equal to the net charge inside the surface, divided by ε_0. We can therefore immediately answer the questions.

(a) We use Gauss' law here, rather than doing a direct integration of the electric field over the cube. The total electric flux is simply q/ε_0.

(b) It does not matter that the Gaussian sphere is off center. The total electric flux is still q/ε_0.

(c) The total net charge Q enclosed by the Gaussian surface is $-2q + q = -q$, and the total electric flux through the Gaussian surface is $-q/\varepsilon_0$. We do not need to concern ourselves with the positions of the two charges within the cube.

What Do You Think? Suppose the cube in Fig. 23–8a contained three more point charges of magnitude $2q$, $-7q$ and $4q$. What would the flux through the sides of the cube be? *Answers to* **What Do You Think?** *questions are given in the back of the book.*

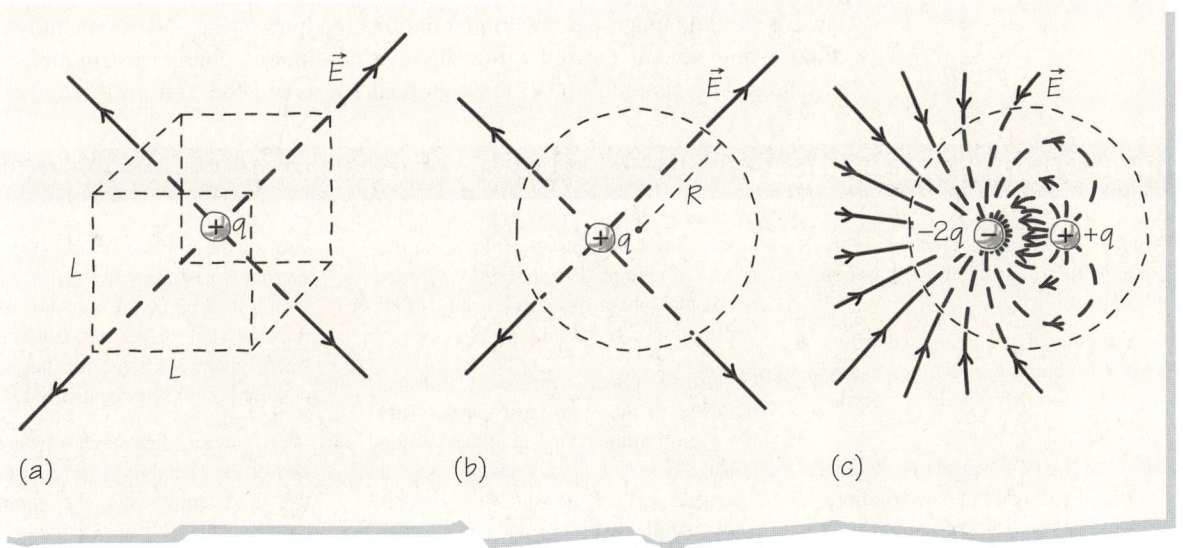

▲ FIGURE 23–8

EXAMPLE 23–4 Consider a point charge $q = 1$ mC placed at a corner of a cube of sides 10 cm. Determine the electric flux through each face of the cube.

Setting It Up The situation is sketched in Fig. 23–9a.

Strategy We spoke about the utility of symmetry in solving problems with the help of Gauss' law. Here we'll use the symmetry of the situation, which involves the sides joining at the corner at which the charge resides. You can see from Fig. 23–9a that for these sides, $\vec{E} \cdot \hat{n} = 0$, since the normal is perpendicular to the surfaces while the electric field goes off in a spherically symmetric pattern and lies in the sides. In other words, the electric field that originates at the charge is tangential to the surface of these three sides. This means there is no flux through these sides. The electric flux through each of the remaining three faces of the cube must be equal by symmetry. We'll refer to these sides with the label F.

To find the flux through each of the sides F, we can use a technique that puts the single charge in the middle of a larger cube. It takes seven other similarly placed cubes to surround the point charge q completely (Fig. 23–9b). The charge is at the dead center of the new, larger cube, so the flux through each of the six sides of the large cube will now have an electric flux of one sixth of the total flux. The large sides of the cube consist of four smaller squares, one of which is in fact one of the sides F, so given that the total structure is completely symmetric, the flux through a side F is one fourth of the flux through the large side.

Working It Out The total flux is q/ε_0, so that the flux through each of the sides of the large cube is $q/6\varepsilon_0$, and one quarter of that, $q/24\varepsilon_0$, goes through each of the far sides of the small cube. Numerical evaluation gives

$$\Phi_F = \frac{q}{24\varepsilon_0} = \frac{1 \times 10^{-3} \text{ C}}{24(8.85 \times 10^{-12} \text{ C}^2/\text{N} \cdot \text{m}^2)} = 5 \times 10^6 \text{ N} \cdot \text{m}^2/\text{C}.$$

What Do You Think? Could you use this technique to determine the flux through each of the sides if a charge q were placed at each corner of the original cube?

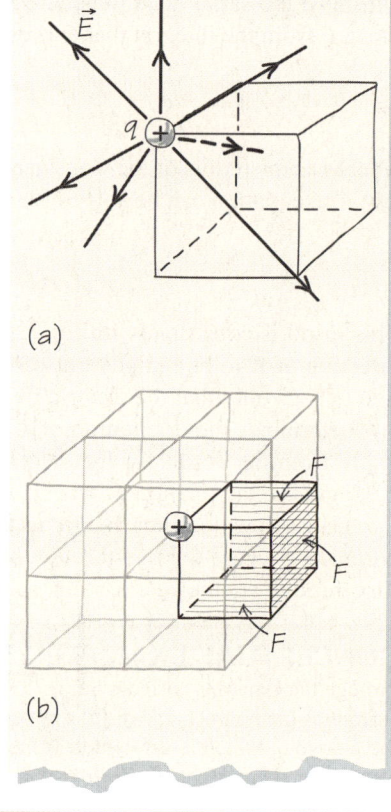

▶ **FIGURE 23–9**
(a) A charge q is placed at the corner of a cube.
(b) By surrounding the charge with a series of cubes such that the charge is at the center of a larger cube, we have created an arrangement sufficiently symmetric to be able to solve for desired flux values.

23–3 Using Gauss' Law to Determine Electric Fields

Gauss' law is a fundamental law in its own right. It is also a powerful tool for the determination of electric fields in situations where there is a high degree of symmetry. With enough symmetry the electric field will be constant over a simple surface, and this can then be removed from the integral that expresses the flux. We can then solve Gauss' law for the field magnitude, without complicated integrations. We shall study this technique using several examples that involve continuous charge distributions: the line of charge, the charged plane, the spherical shell, and the uniformly charged sphere.

Problem-Solving Techniques

To use Gauss' law to find electric fields given a charge distribution, the following steps are helpful:

1. Make a sketch of the charge distribution—it will help you recognize any appropriate symmetry. Add a coordinate system.

2. Identify any spatial symmetry of the charge distribution and the electric field it produces. For example, a uniform sphere of charge has spherical symmetry because it looks the same from all around any other sphere centered on the sphere of charge. The spherical symmetry of the sphere of charge implies that the field must be radial.

3. Choose a Gaussian surface that is matched to the symmetry. This is the most important step in determining electric fields by Gauss' law. Choose a Gaussian surface for which the field is either parallel to the surface ($d\Phi_E = 0$) or perpendicular to the surface ($d\Phi_E = E \, dA$) at various locations; further, choose the surface so that the field is constant over the part of the surface to which it is perpendicular. For example, the Gaussian surface best suited to a uniform sphere of charge is a larger sphere centered on the charge distribution.

4. With surfaces chosen as in step 3, it should be possible to remove the electric field from inside the integral that expresses the flux. Then Gauss' law becomes an algebraic expression for the magnitude of the field.

EXAMPLE 23–5 *Line of charge.* Determine the electric field due to an infinitely long, straight charged rod with positive, constant charge density λ.

Setting It Up Figure 23–10a illustrates the situation. We have oriented the rod along the z-axis.

Strategy To find the appropriate Gaussian surface, we want to see what symmetry tells us about the direction and magnitude of the electric field lines. These lines must leave the positively charged rod and, to be symmetric, the electric field lines must extend away from the rod radially in the xy-plane (Fig. 23–10b). The electric field lines cannot have a component along the z-direction (the direction along which the rod lies), because the symmetry prevents a choice between the $+z$- or $-z$-direction. Moreover, again by symmetry, the magnitude of the field must be the same on every point of a circle centered on the rod. Thus the field magnitude can depend only on the radial distance from the rod. The Gaussian surface that takes advantage of the symmetry is a closed cylinder whose axis coincides with the line charge. We choose it to be of radius r and height h (Fig. 23–10c). We apply Gauss' law for this surface and use the symmetry to extract the value of the field. Note that for application of Gauss' law we'll need to know that the total charge inside the cylinder is $q = \lambda h$, where λ is the charge per unit length.

Working It Out The flux through the cylinder is

$$\Phi = \int_{\text{top surface}} \vec{E} \cdot d\vec{A} + \int_{\text{bottom surface}} \vec{E} \cdot d\vec{A} + \int_{\text{side surface}} \vec{E} \cdot d\vec{A}.$$

The flux through the top and bottom surfaces is zero. That is because the electric field is parallel to these surfaces, so the surface element $d\vec{A}$ is perpendicular to the field. As for the cylindrical side surface, the electric field is perpendicular to that surface, so that $\vec{E} \cdot d\vec{A} = E\, dA$. This expression must be integrated over the curved surface of the cylinder to find the flux through the side of the cylinder. But we have chosen the cylindrical surface so that the electric field has constant magnitude over the surface. The field magnitude can therefore be removed from the integral. Thus

$$\Phi = \int_{\text{side surface}} \vec{E} \cdot d\vec{A} = E \int_{\text{side surface}} dA = E(2\pi rh).$$

Here we use the fact that the surface area of the curved part of the cylinder is the circumference $2\pi r$ multiplied by the height.

The enclosed charge is λh, so Gauss' law reads

$$E(2\pi rh) = \frac{\lambda h}{\varepsilon_0},$$

or

$$E = \frac{\lambda}{2\pi\varepsilon_0 r}. \tag{23–6}$$

The arbitrary height h has canceled. In SI units, the charge density is in coulombs per meter. We can thus check that $\varepsilon_0 E$ has the units of coulombs per square meter. Compare the ease with which we obtained Eq. (23–6) with the direct integration technique [Eq. (22–31)].

What Do You Think? What can you say about the external field due to an infinitely long cylinder whose surface is uniformly charged?

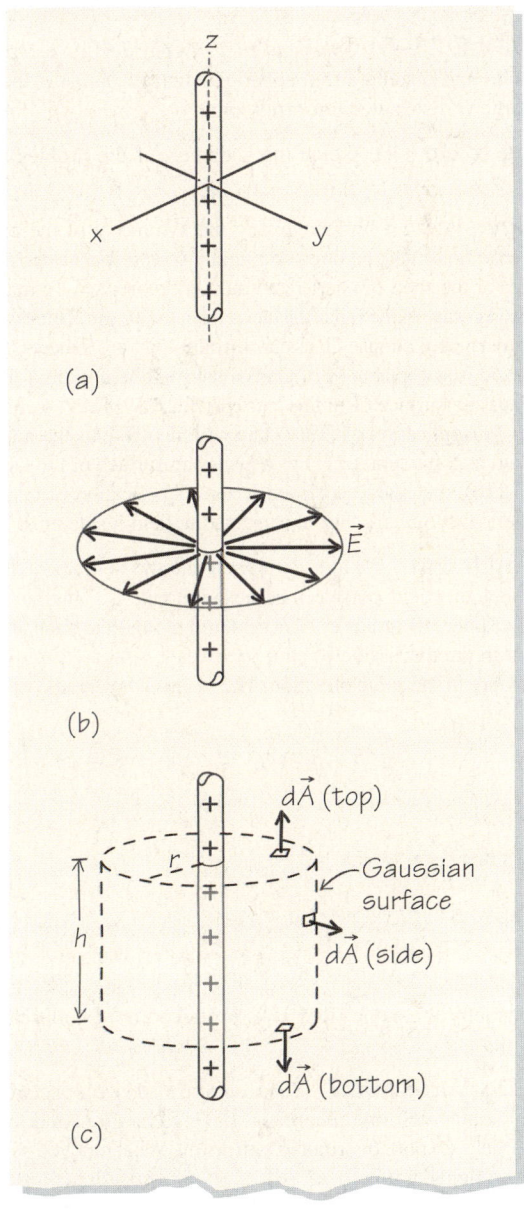

▲ **FIGURE 23–10** (a) A line of charge is oriented along the z-axis. (b) By symmetry, the direction of the electric field $\vec{E}$ is radial in the xy-plane. (c) The best Gaussian surface to use to determine the electric field of a line charge is a cylinder. The directions of the areas $d\vec{A}$ for the various surfaces of the cylinder are shown.

THINK ABOUT THIS...
COULD WE USE GAUSS' LAW TO FIND THE FIELD OF A FINITE LINE OF CHARGE?

Gauss' law holds for *any* distribution of charge, but for a line of charge of finite length the symmetry that allows us to determine the direction of $\vec{E}$ and remove it from the flux integration is not present. If the ends of the line are in view, they provide a guide to tell us where we are along the wire—for example, we can look to see that we are close to one end or the other. The symmetry along the wire is lost. This loss of symmetry has two consequences: First, the electric field will have a component along the wire; and second, the magnitude of the field will vary *along* the line. In fact, from far enough away, a finite charged line is indistinguishable from a point charge, and the electric field will point in a direction normal to a Gaussian sphere centered on the line charge whose radius is much larger than the line length. Gauss' law is always true but not always useful.

EXAMPLE 23–6 *Spherical shell.* Determine the electric field both inside and outside a thin spherical shell of radius R that has a total charge Q distributed uniformly on it.

Setting It Up We sketch the geometry of the problem in Fig. 23–11a.

Strategy We start by recognizing the symmetry of the problem: Any electric field must be directed radially with respect to an origin at the center of the shell of charge and must moreover have a magnitude that depends only on the radial distance from the origin. Once we know this symmetry and choose Gaussian surfaces that are spheres centered at the origin, the application of Gauss' law will help us find the field. For a Gaussian surface of radius $r > R$ (Fig. 23–11a), the charge enclosed is Q. Application of Gauss' law will then give us the magnitude of the field as a function of r. For a Gaussian surface of radius $r < R$ (Fig. 23–11b), the charge enclosed is zero. Application of Gauss' law with spherical symmetry then shows that the field inside must be zero.

Working It Out We already argued that the field inside the shell is zero. For the field outside the shell, we take a Gaussian surface forming a sphere of radius $r > R$ centered around the shell. Then the argument in the discussion of strategy gives $\vec{E} \cdot d\vec{A} = E\,dA$, because $\vec{E}$ and $d\vec{A}$ are in the same direction. Then Gauss' law reads

$$\frac{Q}{\varepsilon_0} = \oint_{\text{surface}} \vec{E} \cdot d\vec{A} = E \oint_{\text{surface}} dA = E 4\pi r^2. \qquad (23\text{--}7)$$

It follows that for $r \geq R$,

$$E = \frac{Q}{4\pi\varepsilon_0 r^2}. \qquad (23\text{--}8)$$

This is exactly the same field as is produced by a point charge at the origin.

What Do You Think? Suppose you had *two* concentric thin spherical shells, with the inner shell having charge Q and the outer shell charge $-Q$, both distributed uniformly. What can you say about the electric fields in this case?

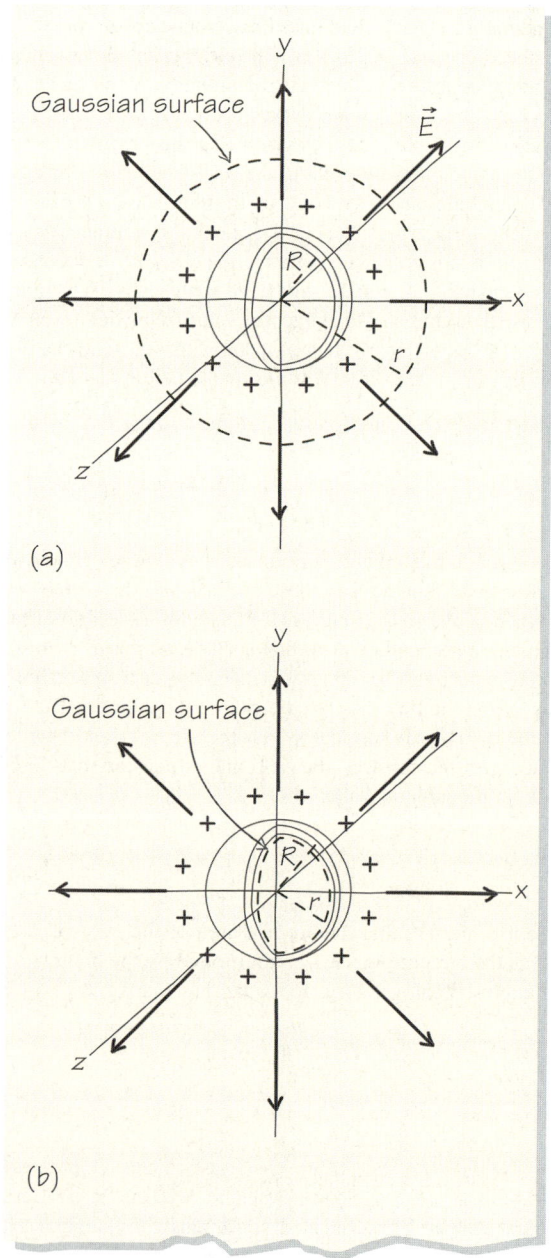

▶ **FIGURE 23–11** (a) The best Gaussian surface to determine the electric field outside a uniformly charged spherical shell is a sphere, because the symmetry is spherical. (b) The best Gaussian surface to determine the electric field inside a uniformly charged spherical shell is a sphere inside the shell.

EXAMPLE 23–7 *Solid sphere.* Find the electric field outside and inside a solid, nonconducting sphere of radius R that contains a total charge Q uniformly distributed throughout its volume.

Setting It Up The charge distribution is sketched in Fig. 23–12a.

Strategy We again make use of the symmetry of the problem and take for our Gaussian surface a concentric sphere of radius r. By symmetry, the electric field is radial and uniform over the surface of the sphere, so that we can again use Gauss' law,

$$\frac{Q_{enclosed}}{\varepsilon_0} = \oint_{sphere} \vec{E} \cdot d\vec{A} = E \oint_{sphere} dA = E4\pi r^2.$$

Here, E is the field at a distance r from the center. The quantity $Q_{enclosed}$ requires a little thought. In working it out, we note that if our Gaussian surface is outside the sphere (Fig. 23–12a), with $r > R$, the enclosed charge is the total charge. But if our Gaussian surface is inside the sphere (Fig. 23–12b), with $r < R$, we must *calculate* the charge included within the sphere of radius r, and this is done by using the fact that the charge density is given.

Working It Out We have for this case the general result

$$E = \frac{Q_{enclosed}}{4\pi\varepsilon_0 r^2}.$$

For the field *outside* the solid sphere, the charge enclosed by a Gaussian sphere at $r > R$ is just Q and, just as for the field for a point charge or for the region outside a spherical shell,

$$E = \frac{Q}{4\pi\varepsilon_0 r^2}. \quad (23\text{–}9)$$

Inside the solid sphere, however, the situation is different (Fig. 23–12b). The enclosed charge can be calculated in terms of the *uniform* charge density, which we denote by ρ:

$$Q_{enclosed} = (\text{volume of sphere}) \times \rho = \frac{4\pi r^3}{3}\rho.$$

The density is determined by the condition that the charge Q is uniformly distributed throughout a sphere of radius R. This means that

$$\rho = \frac{Q}{(\text{volume})} = \frac{Q}{(4\pi/3)R^3}.$$

Putting all this together, we get within the sphere

$$E = \frac{Q_{enclosed}}{4\pi\varepsilon_0 r^2} = \frac{1}{4\pi\varepsilon_0 r^2} \times \frac{4\pi r^3}{3}\rho = \frac{1}{4\pi\varepsilon_0 r^2} \times \frac{4\pi r^3}{3} \times \frac{Q}{(4\pi/3)R^3}$$

$$= \frac{Q}{4\pi\varepsilon_0} \frac{r}{R^3}. \quad (23\text{–}10)$$

The electric field due to a solid sphere has the radial dependence displayed in Fig. 23–12c. As symmetry demands, the field is zero at the center of the sphere. The field increases linearly with r up to the radius of the sphere and then decreases inversely as the square of r.

The fields in Eqs. (23–9) and (23–10) match at the point $r = R$, as the field is continuous.

What Do You Think? Suppose a point negative charge $-q$ is inserted at rest inside the uniformly charged sphere, a distance r from the center. How do you expect it to move?

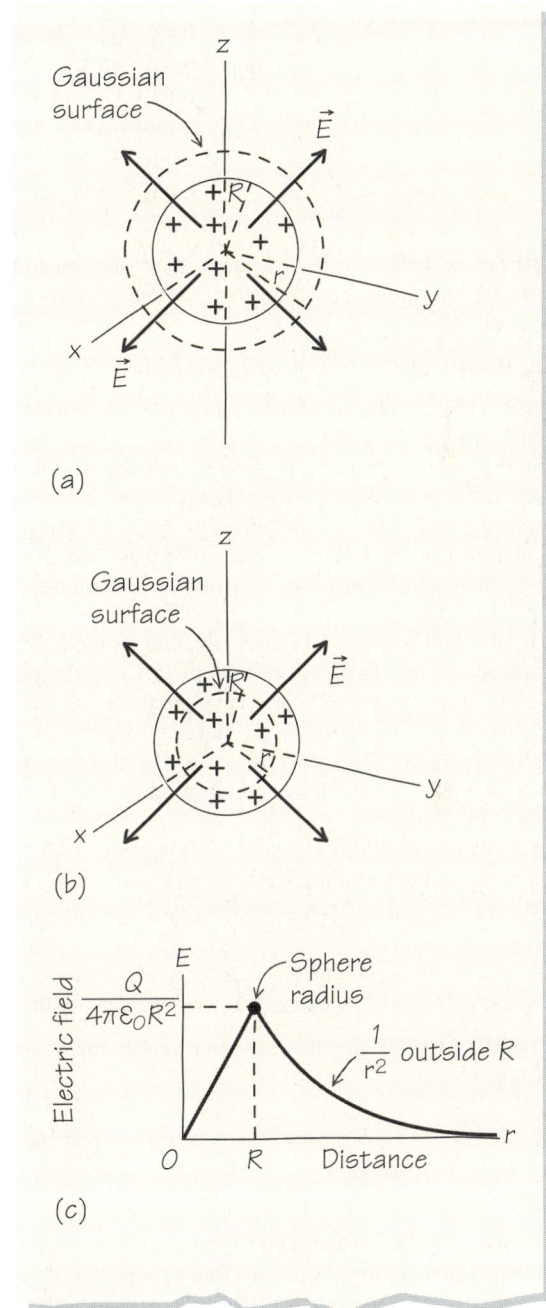

▲ **FIGURE 23–12** (a) The best Gaussian surface to determine the electric field outside a uniformly charged, nonconducting sphere is a concentric spherical surface. The symmetry is spherical. (b) The best Gaussian surface to determine the electric field inside a uniformly charged, nonconducting sphere is a Gaussian sphere inside the solid sphere. Only the charge inside the Gaussian sphere contributes to the electric field at r. (c) The electric field due to a uniformly charged, nonconducting sphere as a function of the distance from the center of the sphere.

From the two previous examples, we can draw the general conclusion that for a spherically symmetric charge distribution, the field at a radius r outside the charge distribution is that of a point charge at the center whose magnitude is the total charge within the sphere of radius r. We have seen that this is easy to prove by using Gauss' law. It holds not only for thin shells and solid spheres but indeed for *any* distribution of charge whose charge density varies only with the radius and is therefore spherically symmetric.

We noted in Chapter 12 that these same results hold for the force of gravity due to a spherical shell of matter. The mathematical problem is identical because the gravitational force has the same inverse-square form as the Coulomb force. We gave only the results in Chapter 12—without derivation—because the direct integration technique is fairly complicated. The Gauss' law derivation provided here is a very simple one. It is interesting to note that Newton delayed the publication of his theory of gravitation by some 20 years because of his lack of a simple proof of these results. If he had known Gauss' law, Newton would have saved a lot of time!

EXAMPLE 23–8 *Plane of charge.* Find the electric field outside an infinite, nonconducting plane of charge with uniform charge density σ.

Setting It Up We show a charged plane in Fig. 23–13.

Strategy The geometry here shares some aspects with that of the infinite line charge in Example 23–5. Whatever the sign of the charge on the plane, symmetry dictates that the field will be perpendicular to the plane. (If the plane is positively charged, as we'll assume here, then the electric field will point away from the plane.) Symmetry also dictates that the electric field has a magnitude that depends at most on the perpendicular distance from the plane. Because the electric field is perpendicular to the plane, a good choice for the Gaussian surface is any solid (such as a cylinder) that has its top and bottom (area A) parallel to the charged plane, that pierces the charged plane (Fig. 23–13). Every facet of this Gaussian surface is either parallel or perpendicular to the electric field. The differential areas $d\vec{A}$ for the top and bottom of the Gaussian surface also point away from the charged plane, so the product $\vec{E} \cdot d\vec{A}$ for the three surfaces is

for the top: $\vec{E} \cdot d\vec{A} = E\, dA$;
for the bottom: $\vec{E} \cdot d\vec{A} = E\, dA$;
for the side: $\vec{E} \cdot d\vec{A} = 0$.

The last equation follows because $d\vec{A}$ for the side points everywhere parallel to the plane, but $\vec{E}$ is everywhere perpendicular to the plane.

Working It Out With the reasoning above, Eq. (23–5) for Gauss' law reads

$$\frac{Q}{\varepsilon_0} = \oint \vec{E} \cdot d\vec{A} = \int_{\text{top surface}} \vec{E} \cdot d\vec{A} + \int_{\text{bottom surface}} \vec{E} \cdot d\vec{A} + \int_{\text{side}} \vec{E} \cdot d\vec{A}$$
$$= 2EA,$$

where we have used the fact that E is constant over the top and bottom area A of the Gaussian surface while the flux through the side is zero.

The total charge enclosed by the Gaussian surface is the charge on the plane within the surface. Because the charge density is σ and the area enclosed is A, we must have $Q = \sigma A$. The previous equation becomes

$$\frac{Q}{\varepsilon_0} = \frac{\sigma A}{\varepsilon_0} = 2EA,$$

from which we get the result

$$E = \frac{\sigma}{2\varepsilon_0}. \tag{23–11}$$

Equation (23–11) is the same result that we found with much more difficulty by direct integration in Chapter 22 [Eq. (22–33)]. Note that E is independent of the distance from the plane.

What Do You Think? Does your result tell you anything about the field outside a *finite* uniformly charged plane?

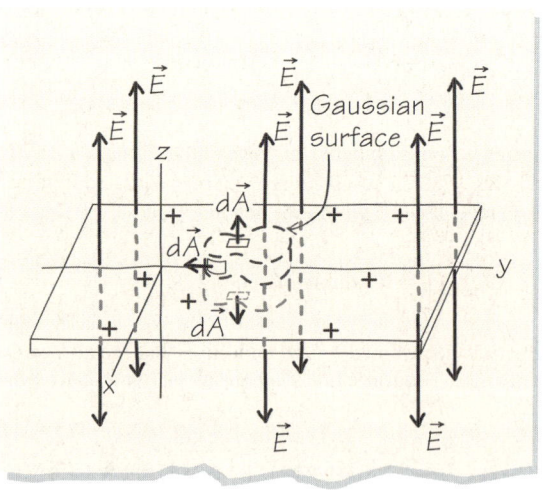

▲ **FIGURE 23–13** A convenient Gaussian surface for a uniformly charged infinite plane can be any shape whose sides are perpendicular to the plane and whose top and bottom are parallel to the plane.

23–4 Conductors and Electric Fields

A good conductor, such as silver, copper, or aluminum, has a large number of "free" electrons, which can move within the (electrically neutral) material. Any electric field that may appear inside the metal due to the presence of an external electric field will cause these electrons to move. In less than a microsecond, they rearrange themselves into a configuration that cancels the electric field inside the material. If any field whatsoever remained inside the material, it would cause the electrons of the conductor to move until they reached equilibrium. *Conductors in electrostatic equilibrium have no internal static electric field.*

The mechanism for the field cancellation within conductors (think of metals) is illustrated in Fig. 23–14. A metal is placed in a spatially constant and static external field that points to the right (Fig. 23–14a). Some electrons in the metal move to the left side of the conductor, which leaves a deficiency of electrons on the right side of the conductor. The arrangement of excess electrons on the left and a deficiency of electrons on the right forms a new, internal electric field that points to the left. This internal field will precisely cancel the external field, with the result that there is no net field within the conductor (Fig. 23–14b).

The fact that there are no static electric fields within conductors has implications for the behavior of conductors when charges are put on or near them, or when they are placed in external electric fields, and this behavior can be determined using Gauss' law. Let's consider what happens when an excess charge is added to a conductor. We might guess that because the components of this excess charge repel each other, and because they can move freely within the conductor, they move as far apart as possible. Figure 23–15 shows such a conductor as well as a Gaussian surface just inside the metal surface. If we apply Gauss' law to this surface, we find that because there is no field, there is no flux, and hence there is no net charge inside the metal. Where is the excess charge? *In electrostatic equilibrium, all excess charge is on the outside surface of a conductor.*

We can establish that the remark above is true even if our conductor has one or more cavities within it. Imagine such a cavity, filled with a nonconducting medium such as air or even a vacuum (Fig. 23–16a). Suppose that there is no excess charge within the cavity. Could charge accumulate on the surface of the cavity—an *interior* surface of the conductor? A Gaussian surface surrounding the cavity, but drawn within the conductor, has no electric flux through it because there is no static field within any conductor. Thus there is no net charge within that Gaussian surface. We have thereby shown that there can be no net charge on the interior surface of the conductor. *Any excess charge placed on a conductor, even if the conductor contains nonconducting cavities, moves to the outside surface of the conductor,* provided there is no charge within the nonconducting cavities.

We must modify our reasoning when there is charge within nonconducting cavities in the conductor. Suppose that such a cavity contains a charge $+Q$ (Fig. 23–16b). Again, draw a Gaussian surface within the metal to surround the cavity. Because there is no field inside the metal, the net charge enclosed must be zero. In this case, a charge of $-Q$ will be induced on the *inner* surface of the metal, that is, on the cavity surface. This induced negative charge keeps the electric field zero *inside the conductor.*

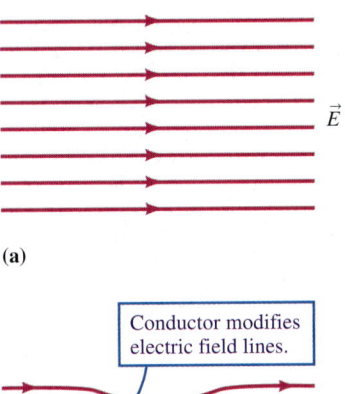

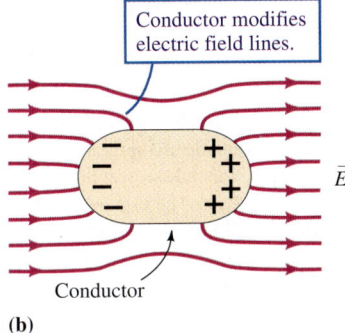

▲ **FIGURE 23–14** An uncharged conductor in an external electric field. (a) The electric field before a conductor is introduced. (b) Charges are induced on the surface of the conductor such that the electric field inside the conductor is zero. The induced charges modify the field outside the conductor, so the field no longer has its original form.

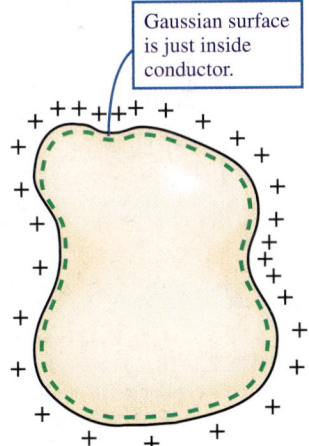

▲ **FIGURE 23–15** To find where excess charge placed on a conductor of arbitrary size and shape goes, choose a Gaussian surface just inside the surface. There is no field within the conductor, hence no flux through our surface, so all the excess charge is on the outside surface of the conductor.

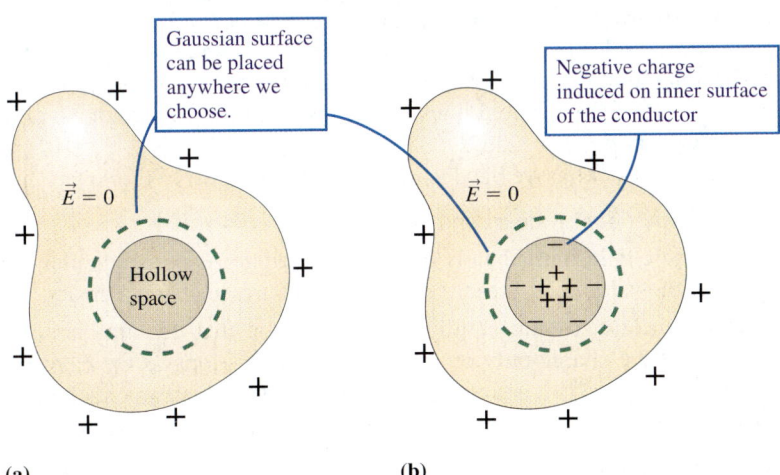

▲ **FIGURE 23–16** (a) A nonconducting cavity inside a conductor; the cavity contains no charge. Any net charge on the conductor must be on the outer surface of the conductor. (b) If we place a charge inside the hollow space, an induced charge will appear on the inside surface of the conductor, such that the electric field within the conducting material is zero. A Gaussian surface drawn just outside the cavity illustrates these results.

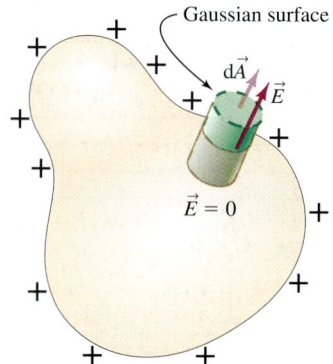

▲ **FIGURE 23–17** To find the electric field outside a conductor of arbitrary size, we choose a small right circular cylinder for the Gaussian surface. The only part of the cylinder through which there is a nonzero electric flux is the outside end of the cylinder.

Electrostatic Fields Near Conductors

We can draw two important conclusions about electrostatic fields around metals or other conductors from this discussion. First, *the electric field immediately outside a conductor must be perpendicular to the conductor's surface*. If there were a parallel component, then there would be a force on charges resting on the surface; the charges would react and move along the surface and we would not have the static situation we assumed. The charges would readjust themselves until there was no parallel component. Second, by using Gauss' law, we can find the value of this perpendicular electric field at a point near the surface in terms of the charge density at that point on the surface. Consider the conductor shown in Fig. 23–17, with a tiny Gaussian surface whose side is perpendicular and whose top is parallel to the conductor's surface. We take the Gaussian surface small enough so that the surface charge density σ is constant within it, even though σ may vary over the conductor. We refer to σ only at the point where the Gaussian surface is erected. The electric field is zero inside the metal surface and outside it is parallel to the side of the Gaussian surface. Thus, the only contribution to the flux comes from the top of the Gaussian surface. If the Gaussian surface is small enough, $\vec{E}$, which is perpendicular to the top surface, can be regarded as constant over it, and

$$\frac{Q}{\varepsilon_0} = \oint \vec{E} \cdot d\vec{A} = EA,$$

where A is the area of the top of the Gaussian surface. The total charge Q enclosed by the Gaussian surface is σA, so that in the previous equation we set $Q = \sigma A$. We see that the area cancels and we get, *just outside the surface*,

$$\vec{E} = \frac{\sigma}{\varepsilon_0}. \qquad (23\text{–}12)$$

From the way we found it, we see that this result holds only near the conductor's surface. The magnitude of the field will vary around the surface of the conductor as σ does, and the field direction will always be perpendicular to the conductor near its surface (Fig. 23–18). Whether this result is useful or not depends on whether we know the charge density. We can check our result by considering a conductor that is a sphere of radius R and total charge Q. In this case, symmetry demands that the charge is spread evenly over the surface, and

$$\sigma = \frac{Q}{\text{area}} = \frac{Q}{4\pi R^2}.$$

For the field just outside the sphere, Eq. (23–12) would then give $E = Q/4\pi\varepsilon_0 R^2$, which agrees with our earlier result, Eq. (23–9).

The field just outside a conductor ($E = \sigma/\varepsilon_0$) is twice as large as the field of a nonconducting charged plane with the same charge density ($E = \sigma/2\varepsilon_0$), Eq. (23–11). A simple way of understanding this is to think in terms of field lines. The charge on a surface of area dA, $\sigma\, dA$, gives rise to a certain number of field lines. For a nonconducting plane, the field lines divide equally between the two sides of the plane. For a thick conducting plane, there are no field lines on the inside of the conductor, so all the field lines must emerge on the open side.

We can summarize what we have learned about conductors as follows:

1. The electrostatic field inside a conductor is zero.
2. The electrostatic field immediately outside a conductor is perpendicular to the surface and has the value σ/ε_0, where σ is the local surface charge density.
3. A conductor in electrostatic equilibrium—even one that contains nonconducting cavities—can have charge only on its outer surface, as long as the cavities contain no net charge. If there is a net charge within the cavity, then an equal and opposite charge will be distributed on the surface of the conductor that surrounds the cavity.

We can add one more important result. Suppose we have a charge-free cavity in a metal. We know that there is no field within the metal and, moreover, no *net* charge on the inner surface of the metal surrounding the cavity. But even for nonsymmetric situations, it can be shown that as long as there is no charge within the cavity, *the electric field is zero everywhere within the cavity*.

We'll discuss fields around conductors further in Chapter 24.

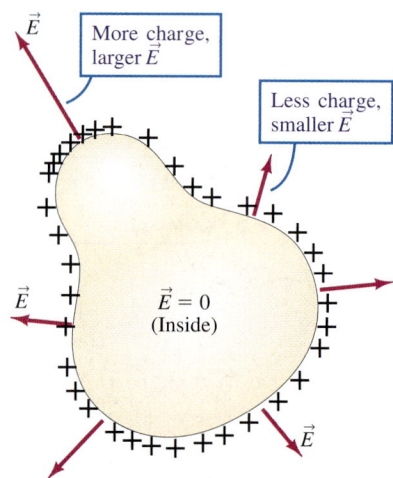

▲ **FIGURE 23–18** The electric field in and around a conductor in equilibrium. The electric field inside the conductor is zero, and just outside the conducting surface it must be perpendicular to the surface. The magnitude of the electric field varies according to the surface charge density σ, which may not be constant everywhere on the surface.

23-5 Are Gauss' and Coulomb's Laws Correct?

It is one of the characteristics of science to be eternally skeptical of yesterday's experiment. It is not so much that yesterday's experiment is wrong, although that certainly can happen; rather, a more accurate experiment can be done with more modern apparatus.

The equivalence of Gauss' and Coulomb's law makes it clear that testing one tests the other. Gauss' law is one of the cornerstones of our understanding of electricity and magnetism, and we must therefore ask just how well it is known and how it can be tested as precisely as possible. Many such tests rely on measurements of the $1/r^2$ behavior of Coulomb's law—you will recall that the equivalence of the two laws rests on this precise dependence. In this way, the errors implicit in measurements of Coulomb's law set limits on our knowledge of Gauss' law; these limits have been continually improved up to the present time. There are also more direct ways to look at Gauss' law, and we'll study a particularly sensitive technique for testing Gauss' law in this section.

Testing Gauss' Law with a Null Experiment

In 1773 Henry Cavendish, who you will recall from Chapter 12 measured the gravitational constant, made an early test of what later became known as Gauss' law. He placed one conducting sphere inside another and connected the two by a wire. After placing a charge on the apparatus, he disconnected the wire and looked for any charge that remained on the inner sphere. To the accuracy of his experiment, he found none. Cavendish's experiment now goes under the general name of the *Faraday "ice-pail" experiment*, after Michael Faraday, who presumably literally used an ice pail (as a conducting container) in a version of the experiment that he performed. This experiment is the basis for many of the modern high-precision tests of Gauss' law and—because Gauss' law is equivalent to Coulomb's law—of Coulomb's law.

For a simple version of the Faraday ice-pail experiment we require an electroscope (the free-charge detector introduced in Chapter 21). We also need a hollow metal container with a hole in the top, as in Fig. 23–19a, and a small metal ball on the end of an insulated rod that can be used to introduce charge to the inside of the container. The electroscope is attached to the outside of the container and thus indicates whether there is charge on the outside.

Next a positive charge, $+Q$, is placed on the small metal ball, and the ball is inserted through the small hole into the hollow container without touching it (Fig. 23–19b). Gauss' law states that there is no net charge *inside* the nearly closed metal container; therefore, a charge of $-Q$ is induced on the inside surface of the container. (The hole can be made smaller and smaller until its presence does not matter.) Because the metal container is neutral, a charge of $+Q$ must then be induced on its outside surface, and the electroscope indicates this charge. If the ball is moved around, there is no change whatsoever in the electroscope, consistent with Gauss' law—it makes no difference where in the cavity the charge is. The metal ball is subsequently touched to the interior of the hollow container (Fig. 23–19c). If Gauss' law is correct, the charge on the ball neutralizes the $-Q$ charge induced on the inside surface, leaving the $+Q$ charge on the outside surface. The electroscope indicates this result by not changing at all. When the metal ball is removed from the container, the container's outer surface remains charged (Fig. 23–19d). By touching the metal ball to another electroscope, we can verify that it carries no charge.

The description of this experiment shows why it is potentially so precise: If Gauss' law is correct, there is *no change* in the position of the gold leaf when the inner surface is touched. Equivalently, Cavendish's experiment tests for the *absence* of charge on the inner of two spheres. Experiments such as Coulomb's, the ones that try to measure departures from a $1/r^2$ fall-off of the force between two charges, look for small changes in comparison with larger effects. Such experiments are inherently less precise than experiments such as Cavendish's, which look for small changes in comparison with *no* effect. Experiments that test for small change versus no change are called **null experiments**. It is far easier to make a precise test of Gauss' law than of Coulomb's law because a null experiment can be done.

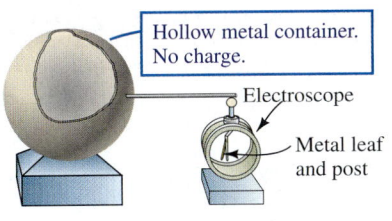
(a)

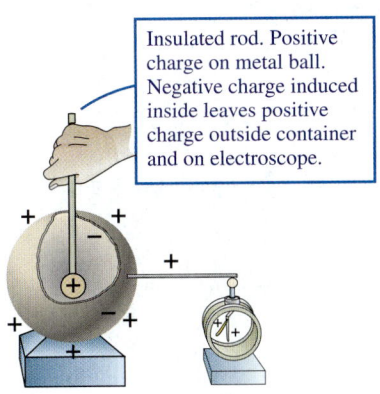
(b)

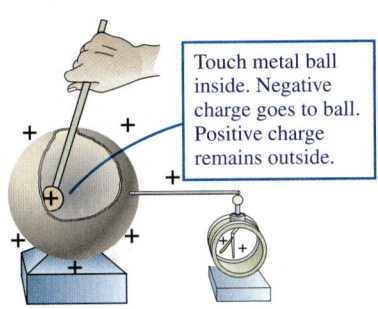
(c)

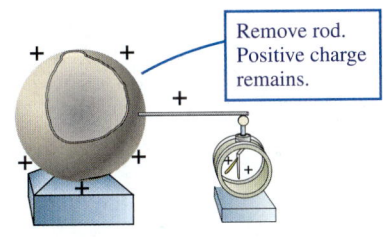
(d)

▲ **FIGURE 23–19** An electroscope is attached to the outside surface of a hollow conducting sphere to show the presence of charge. (a) No charge is present, and the gold leaf hangs down. (b) A charged metal ball on the end of an insulated rod is placed inside the sphere, and charge is induced. (c) If the metal ball touches the inside surface of the hollow conductor, all the charge passes to the outside surface. The electroscope's gold leaf indicates no change in the charge on the outside of the hollow conductor. (d) When the insulated metal ball is removed, the charge remains on the outside of the hollow conducting sphere, with no charge remaining on the metal ball.

THINK ABOUT THIS...
HOW CAN WE SHIELD A REGION FROM ALL ELECTRIC FIELDS?

Shielded regions or rooms in laboratories are often necessary so that electronic measurements are unaffected by outside electrical interference (Fig. 23–20). Such shielded rooms, known as *Faraday cages* and formed by enclosing a region with copper screens or sheets, rely on the fact that there are no electric fields within charge-free cavities. The enclosure is simply a cavity within a metal—in this case the conducting material is the copper that forms the screens. As long as there is no net charge inside of the enclosure, there is no electric field within it due to any external effects. If there were a net charge inside, charge would be induced on the inside of the copper screens, forcing the electric field in the copper to be zero, and there would be an electric field inside the enclosure.

Faraday cages occur beyond the laboratory. The interior of your car is a safe place in the event of nearby lightning, as any charge on the vehicle will go to the metal outer surface, but for the same reason, your car radio does not work as well when the car is located within the "cage" formed by a metal bridge.

◀ **FIGURE 23–20** By Gauss' law, there is no static electric field in an empty cavity in a metal. To the extent that this radio receives signals with only relatively slowly varying electric fields, it will not work very well within its cage. (*Copyright Jim Krider/Arizona State University.*)

*Coulomb's Law Holds over Small and Large Distances

Table 23–1 gives a summary of the accuracy to which Coulomb's law is known through experiments of the ice-pail type. It is characterized by expressing a *deviation* from Coulomb's law in the form

$$F \propto \frac{1}{r^{2 \pm \delta}}.$$

When $\delta = 0$, the inverse-square law is exact; the smaller the limit on δ, the closer the law is known to be an inverse-square law. Table 23–1 expresses what is known experimentally about the possible value of δ. The limits on δ that are placed by the most recent experiments are astonishing.

TABLE 23–1 • Experimental Measurements of Deviation from an Inverse-Square Force Law[†]: Force $\propto 1/r^{2 \pm \delta}$

Investigators	Date	Maximum δ
Robison	1769	0.06
Cavendish	1773	0.02
Coulomb	1785	0.10
Maxwell	1873	5×10^{-5}
Plimpton and Lawton	1936	2×10^{-9}
Williams, Faller, and Hill	1971	3×10^{-16}

[†]For more information on this subject, see A. S. Goldhaber and M. M. Nieto, "The Mass of the Photon," *Scientific American*, p. 86, May 1976.

This is not the end of the story, however. First, the experiments that we have listed in Table 23–1 test the laws only over a distance of about 1 m. Yet the laws of electrodynamics are supposed to hold in atomic systems and over galactic distances. Second, other evidence about the framework of the laws of physics suggests strongly that *a deviation from Coulomb's law of the form* $1/r^{2+\delta}$ *is not possible*. Instead, a way to characterize a deviation from Coulomb's law is with the *approximate* form

$$F \propto \frac{e^{-\mu r}}{r^2},$$

where e is the exponential constant $2.78\ldots$ and μ is a constant. If Coulomb's law is correct, the parameter $\mu = 0$. The exponential function decreases with r over a distance that depends on μ. The larger μ is, the faster the exponential decreases, and the larger the violation of Coulomb's law. Any violation is, we now know, more properly expressed by limits on μ. We can determine limits on μ, and hence tests of the accuracy of Coulomb's law, from the previously reported experiments. The experiment of Williams, Faller, and Hill, for example, implies that μ is smaller than 6×10^{-8} m^{-1}. These limits can be extended by observing the space dependence of Earth's magnetic field and also of Jupiter's magnetic field, as measured by the spacecraft *Pioneer 10*. Although we have not yet studied magnetism, we can say that the limits on μ found thereby are indeed those associated with Gauss' law. In addition to being direct, the planetary measurements give values of μ that are smaller by an order of magnitude or more than those given by the laboratory experiments; they have the further advantage of testing Gauss' law out to large distances.

Finally, how well do we know Gauss' law at short distances? The colors of light given off by excited hydrogen atoms are very sensitive indicators of the Coulomb force at distances on the atomic scale, about 10^{-10} m. The accuracy with which Gauss' (and therefore Coulomb's) law is known is comparable to the accuracy of the experiments of Plimpton and Lawton (see Table 23–1); that is, to about one part in 1 billion. Even down to nuclear distances—about 10^{-15} m—experiments indicate consistency with the basic theory that leads to Coulomb's law.

Summary

The electric flux due to the electric field $\vec{E}$ that intersects a surface S is

$$\Phi_S = \int_S (\hat{n} \cdot \vec{E})\, dA \tag{23–1}$$

where we must find the normal to the surface at each point to carry out the integration. Gauss' law relates the electric flux through a Gaussian surface—an imaginary *closed* surface—to the total charge enclosed by the surface, Q:

$$\oint_S \vec{E} \cdot d\vec{A} = \frac{Q}{\varepsilon_0}. \tag{23–5}$$

Gauss' law is equivalent to Coulomb's law for static situations; it is indeed one of the fundamental equations of electromagnetism.

Gauss' law is a powerful tool for determining electric fields due to charge distributions with a high degree of symmetry. It can be used to derive in simple fashion the electric fields due to a straight-line charge or due to a conducting plane. For a general spherically symmetric charge distribution centered at the origin of a coordinate system, Gauss' law gives a simple derivation of the field at a distance r from the origin. If q is the total charge contained within a Gaussian sphere of radius r, then the electric field at r is the same as that of the field of a point charge q at the origin, $E = q/(4\pi\varepsilon_0 r^2)$.

Using Gauss' laws, we learned that conductors have the following properties:

1. The electrostatic field inside a conductor is zero.
2. The electrostatic field immediately outside a conductor is perpendicular to the surface and has the value σ/ε_0, where σ is the local surface charge density (which is not necessarily constant).
3. If there are no nonconducting cavities containing charge, a conductor in electrostatic equilibrium can have charge only on its outer surface.

Gauss' law (and its equivalent, Coulomb's law) has been subjected to many experimental tests since the mid-eighteenth century. The inverse-square law dependence on distance has been verified to a precision that ranges from one part in 10^9 to one part in 10^{16} over distances between 10^{-10} m and 10^9 m.

Understanding the Concepts

1. A temperature field is defined when the temperature of every point of a region of space is specified. Is it possible to compute a flux associated with this field?
2. In the text we refer to the Faraday ice-pail experiment and discuss one version of it in detail. The discussion concerns a sphere with a hole cut in it, and we speak of the inside and outside of this open sphere (Fig. 23–19). Yet an open sphere does not have a clear inside and outside because, unlike a closed, hollow sphere, it can be deformed continuously to a plane. Why is it possible to talk of the inside and outside of an open sphere, and why does the open sphere behave like a closed, hollow sphere (a cavity) in the experiment?
3. Consider a surface enclosing no electric flux. Discuss whether the electric field is zero everywhere on the surface.
4. Use Gauss' law to show that electric field lines must be continuous and must originate from and end on charges.
5. Describe the way in which Gauss' law would fail if the field of a point charge were to decrease as $1/r$ rather than as $1/r^2$.
6. Suppose the charge in Example 23–4 were located in the middle of the face of a cube. What symmetric arrangement would you set up, if any, to calculate the flux through the remaining sides of the cube?
7. If a large, thin, flat plate is positively charged, the field extends in both directions from the plate and has a magnitude of $\sigma/2\varepsilon_0$. If a second plate of equal but opposite charge is placed parallel to the first plate, the field around the first plate extends only toward the second plate and has a magnitude of σ/ε_0, where σ is exactly the same as before. How do you reconcile this second case with Gauss' law?
8. Consider an electric field $\vec{E}$ that is zero at every point on a closed surface S. Does this mean that there are no charges within this surface? Give an example for which there are charges inside a surface while $\vec{E} = 0$ on the surface.
9. What is the force on a charge Q that is just inside a shell of uniformly distributed charge?
10. Analyze Gauss' law as it applies to the flow of fluids. How would you formulate it? Suppose that in a certain region there are no sources of fluid, only sinks. What would you learn from Gauss' law? How does the possibility of evaporation of a fluid affect Gauss' law? What would be its counterpart in the electric Gauss' law?
11. You are sitting in your spaceship, inside an interior cubicle made entirely of aluminum. Your arch enemy shoots a trillion coulombs of charge at you with his coulomb gun. Do you survive?
12. What would Gauss' law look like for the gravitational field, which is defined by force/unit mass of a test body?
13. Charge is distributed uniformly on a wire forming a circle that is surrounded by a torus (doughnut) for which the wire serves as an axis. Does the symmetry allow us to say anything about the electric field due to the charge on the circular wire?
14. A positive point charge and a negative point charge of equal magnitude are fixed on the surface of a conductor of arbitrary shape. What, if anything, can be said about the resulting electric field lines?
15. A region in space has a uniform electric field. What can we say about whether or not any charges are inside the region?
16. You have a charge at the origin of some coordinate system, and a hemispherical surface whose center of curvature is located at the origin. The charge is tripled. How does the flux through the hemispherical surface change?
17. To derive the electric field of an infinitely long line of charge, we used a Gaussian surface in the form of a right cylinder centered on the line. Why does the use of such a surface not allow us to find the field of a line of charge of *finite* length?
18. What would be the flux if both the charge and the length of the sides were doubled in Example 23–4?
19. Suppose that the electric field in some region is known to have only x- and y-components and that the components depend only on x and y, not on z. What can you deduce about the charge distribution that gives rise to this field?
20. You have a probe that measures the electric field at any point in space. For a region in which you know independently that the charge density is constant, how can you use the probe to measure that charge density?
21. Would the result of Example 23–6, which finds the field of a thin spherical shell of uniformly distributed charge, be the same if the distribution of charge weren't spherically symmetric?

Problems

23–1 What Does Gauss' Law Do?

1. (I) An infinitely large, nonconducting, thin plate carries a uniform charge density σ. (a) What is the electric flux through a circle of radius R placed parallel to the plane? (b) What is the flux through that circle if the plane of the circle is tilted at a 30° angle with respect to its original orientation (Fig. 23–21)?

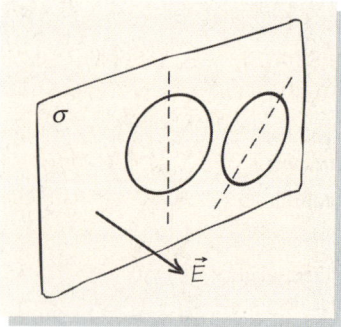

▲ **FIGURE 23–21** Problem 1.

2. (I) A region of space contains a constant electric field of magnitude 1325 N/C. A wire frame forming a square 0.27 m on a side is placed in the region, oriented so that the perpendicular to the plane of the square makes an angle of 48° with the field. What is the magnitude of the electric flux through the frame?

3. (I) The electric field due to an infinitely long, straight line of charge with uniform charge density λ points straight away from the line and has magnitude $E = \lambda/(2\pi\varepsilon_0 r)$, where r is the distance from the wire. Calculate the flux of this electric field through a right cylinder of height h and radius R, concentric with the charged line. Repeat the calculation for a cylinder of radius $2R$.

4. (I) The electric field in a certain region of space points in the z-direction and has magnitude $E = 5xz$, where x and z are measured from some origin. Calculate the flux of that field through a square perpendicular to the z-axis; the corners of the square are at $(x, y, z) = (-1, -1, 1)$, $(-1, 2, 1)$, $(2, 2, 1)$, and $(2, -1, 1)$. (All fields are measured in N/C, all distances in m.)

5. (I) An electric field has the components $E_x = 5x$, $E_y = -3y$, and $E_z = 4z$. Calculate the electric flux through the sides of a unit cube, whose corners are $(x, y, z) = (0, 0, 0)$, $(1, 0, 0)$, $(1, 1, 0)$, $(0, 1, 0)$, $(0, 0, 1)$, $(1, 0, 1)$, $(1, 1, 1)$, and $(0, 1, 1)$. (All fields are measured in N/C, all distances in m.)

6. (I) An electric field of 150 N/C points in the *x*-direction. A wire loop that is 4 cm^2 in area is placed so that its plane is perpendicular to the *x*-axis. (a) What is the electric flux through the loop? (b) If the loop is rotated about the *y*-axis so that the normal to the loop makes an angle of 25° with the *x*-axis, what is the flux through the loop now? (c) How does the flux change if the angle is increased to 335°?

7. (II) An electric field that is constant in direction is perpendicular to the plane of a circle of radius R. This electric field has a magnitude of $E_0(1 - r/R)$ at a distance r from the center of the circle. Calculate the electric flux through the plane of the circle.

8. (II) By direct calculation (that is, without using Gauss' law), find the flux of a constant electric field $\vec{E}$ through a hemispherical surface of radius R whose circular base is perpendicular to the direction of the field. Your result should be the same as the flux through the top surface of a cylinder whose circular base, of radius R, is oriented perpendicular to the field direction (Fig. 23–22). [*Hint*: The area of an infinitesimal strip at a latitude θ and a thickness $R\, d\theta$ is $2\pi R^2 \sin\theta\, d\theta$; θ varies from 0 at the North Pole to $\pi/2$ at the equator.]

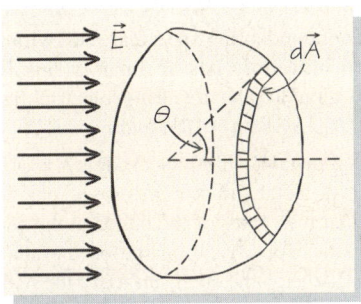

▲ **FIGURE 23–22** Problem 8.

9. (II) A point charge q is placed in the middle of a cylindrical surface of radius R, height $2h$. Find the electric flux through the surface by direct integration. [*Hint*: Use the angle θ in Fig. 23–23 as your variable of integration. Only the upper half of the cylinder is shown in the figure.]

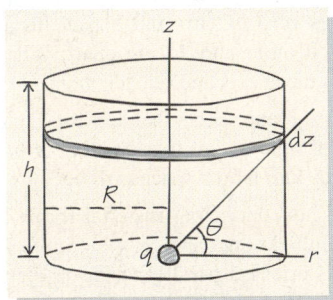

▲ **FIGURE 23–23** Problem 9.

10. (II) A charge q is placed just above the center of a horizontal circle of radius r, and a hemisphere of this radius is erected about the charge. Compute the electric flux through the closed surface that consists of the hemisphere and the planar circle (Fig. 23–24). Do not use Gauss' law.

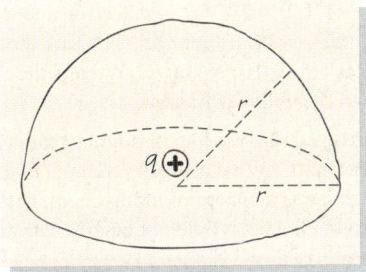

▲ **FIGURE 23–24** Problem 10.

11. (III) Consider an infinitesimal parallelepiped located at the point (x, y, z) with sides dx, dy, and dz along the *x*-, *y*-, and *z*-axes (Fig. 23–25). Show that the electric flux of the electric field given by $\vec{E} = E_x \hat{i} + E_y \hat{j} + E_z \hat{k}$ through the surface that bounds this volume is given by

$$\Phi = \left(\frac{\partial E_x}{\partial x} + \frac{\partial E_y}{\partial y} + \frac{\partial E_z}{\partial z} \right) dx\, dy\, dz.$$

The quantity in parentheses (the coefficient of $dx\, dy\, dz$) is called the *divergence* of the vector field $\vec{E}$.

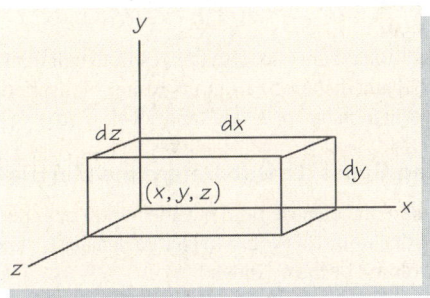

▲ **FIGURE 23–25** Problem 11.

23–2 Gauss' Law

12. (I) The flux through a closed surface surrounding a single charge is -5.7×10^{-5} N·m^2/C. What is the value of the charge?

13. (I) A charge of 10^{-3} C is distributed uniformly on the surface of a sphere of radius 1 cm. Calculate the total electric flux through a concentric sphere (a) just within the charged surface, and (b) just outside the charged surface.

14. (I) A 120-nC point charge is placed just inside the center of one face of an imaginary Gaussian cube. What is the flux that passes through all six faces of the cube?

15. (I) Consider the charge distribution shown in Fig. 23–26, where $q = 1\ \mu$C. Draw a spherical Gaussian surface centered on the origin with a radius of 5 cm. (a) What is the electric flux through the Gaussian sphere? (b) Do any of the electric field lines from the three charges pierce the Gaussian surface?

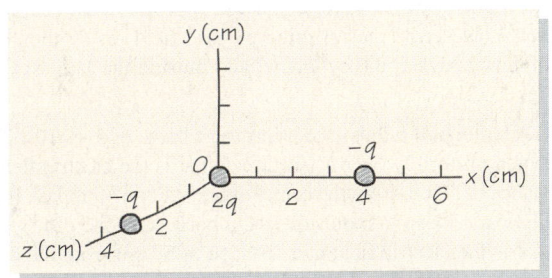

▲ **FIGURE 23–26** Problem 15.

16. (I) A charge of 1.2×10^{-4} C is placed inside a cylinder at the midpoint of the axis of the cylinder. The flux through one end of the cylinder is 4.5×10^6 N·m²/C. What is the flux through the curved part of the cylinder's surface?

17. (I) The net electric flux passing through a given closed surface is -4×10^2 N·m²/C. What charge is contained inside the surface if that surface is (a) a sphere of radius 3 cm, (b) a cube of sides 3 cm, (c) a right circular cylinder of height 3 cm and radius 1 cm?

18. (II) A 420-μC charge is placed at the center of a cube of sides 8 cm. Determine the electric flux through each of the sides.

19. (II) A given region has an electric field that is a sum of two contributions: a field due to a charge $q = 5 \times 10^{-8}$ C at the origin, plus a uniform field of strength $E_0 = 3000$ N/C in the $-x$-direction. Calculate the flux through each side of a cube with sides of length 20 cm that are parallel to the x-, y-, and z-directions; the cube is centered at the origin.

20. (II) The *gravitational field* $\vec{g}$ due to a point mass M may be obtained by analogy with the electric field by writing an expression for the gravitational force on a test mass, and dividing by the magnitude of the test mass, m. Show that Gauss' law for the gravitational field reads $\Phi = \oint_S \vec{g} \cdot d\vec{A} = -4\pi GM$, where G is the gravitational constant. Use this result to calculate the gravitational field at a distance r from the center of a sphere of radius R and uniform density for $r > R$ and for $r < R$.

23–3 Using Gauss' Law to Determine Electric Fields

21. (I) Calculate the electric field outside a long cylinder of finite radius R with a uniform (volume) charge density ρ spread throughout the volume of the cylinder.

22. (I) Use Gauss' law to show that the electric field outside a large, thin, nonconducting plate with uniform charge density σ is given by $E = \sigma/2\varepsilon_0$.

23. (I) Charge is distributed on a long, straight rod with uniform density $\lambda = 6.5 \times 10^{-8}$ C/m. Compare the magnitude of the field 1 cm from the rod to the field 1 cm from a point charge $q = 6.5 \times 10^{-8}$ C.

24. (II) An infinitely long cylinder of radius R carries a uniform (volume) charge density ρ. Calculate the field everywhere inside the cylinder.

25. (II) On a clear day in Nebraska, the electric field just above the ground is 110 N/C and points toward the ground. Our planet Earth is a reasonable conductor and contains no electric field. How much net charge is contained on the surface of a 60-acre corn field (1 acre ≅ 4000 m²)?

26. (II) Two long, thin cylindrical shells of radii r_1 and r_2, respectively, are oriented coaxially (one cylinder is centered inside the other). The cylinders carry equal and opposite linear charge densities λ. Describe the resulting electric field inside the smaller cylinder, between the cylinders, and outside the larger cylinder (Fig. 23–27).

27. (II) A balloon of radius 15 cm carries a charge of 5×10^{-7} C distributed uniformly over its surface. What is the electric field at a distance of 50 cm from the center of the balloon? Suppose that the balloon shrinks to a radius of 10 cm but loses none of its charge. What is the electric field at a distance of 50 cm from the center?

28. (II) A thin, cylindrical copper shell of diameter 6.0 cm has a thin metal wire of diameter 0.10 mm along its axis. The wire and the

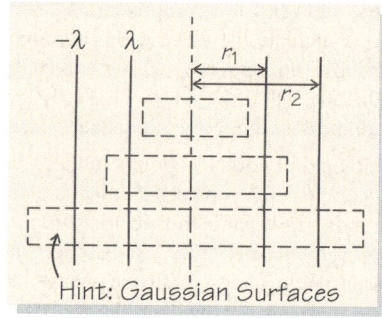

▲ **FIGURE 23–27** Problem 26.

shell carry equal and opposite charges of 8.5×10^{-9} C/cm, distributed uniformly. Calculate the electric field in the region between the wire and the cylinder, and the magnitude of the electric field at the surface of the wire and at the inner surface of the cylinder.

29. (II) A long, cylindrical shell of inner radius r_1 and outer radius r_2 carries a uniform volume charge density ρ. Find the electric field due to this distribution of charge everywhere in space.

30. (II) A Teflon rod of radius 4.0 mm and height 7.0 cm is being charged uniformly over its cylindrical surface. How much charge can the rod hold before the surrounding air breaks down electrically, which happens when the electric field in air is 2.0×10^6 N/C? Ignore the likelihood of breakdown at the sharp edges.

31. (II) A thick, nonconducting spherical shell with a total charge of Q distributed uniformly has an inner radius R_1 and an outer radius R_2. Calculate the resulting electric field in the three regions $r < R_1$, $R_1 < r < R_2$, and $r > R_2$.

32. (II) A spherical metal shell of inner radius R is isolated and carries no net charge. A metal ball of radius r with charge q is suspended inside it, so that the center of the ball is at the center of the shell. Write expressions for (a) the charge on the inner surface of the shell; (b) the charge on the outer surface of the shell, (c) the electric field at a distance d from the center, where $r < d < R$.

33. (II) Consider two infinite parallel charged plates with surface charge densities of σ_1 and σ_2 respectively. (a) What is the electric field in the three regions on the far sides of the plates and between the plates? (b) What are the fields if an infinite uncharged metallic plate is inserted between and parallel to the two charged plates? Is it enough to just use Gauss' law? What else is needed?

34. (II) Two infinite-plane nonconducting, thin sheets with uniform surface charges of $3 \mu C/m^2$ and $-1 \mu C/m^2$, respectively, are parallel to each other and 12 cm apart. What are the electric fields between the sheets and outside them?

35. (II) Two infinite-plane sheets that are just like those of Problem 34 are placed at right angles to each other. What are the fields in the four regions into which space is divided by the planes?

36. (II) A slab of nonconducting material forms an infinite plane. The slab has a thickness t and carries a uniform positive charge density ρ. It is oriented parallel to the xy-plane, with its upper surface at $z = t/2$ and its lower surface at $z = -t/2$. Use Gauss' law to find the electric field both above and below the surface, as well as at an arbitrary value of z in the interior of the slab.

37. (II) Consider a solid sphere of radius 3 cm that carries a negative charge of $2 \mu C$ distributed uniformly. The sphere is placed concentrically in a spherical shell of radius 8 cm that has a positive charge of $5 \mu C$ distributed uniformly over it. Calculate the electric field as a function of radius r for $0 < r < 15$ cm.

38. (III) Charge is distributed throughout a sphere with the charge density given by $\rho = \rho_0$ for $r < a$, $\rho = \rho_0(r - R)/(a - R)$ for $a < r < R$, and $\rho = 0$ for $R < r$ (Fig. 23–28). Calculate the flux through the spherical surfaces at $r = a$, $r = R$, and $r = 10R$, and calculate the corresponding electric fields at these radii.

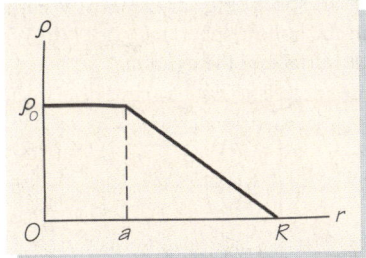

▲ FIGURE 23–28 Problem 38.

39. (III) Consider the charge distribution given in Problem 38. Plot the charge density, the flux through a concentric shell of radius r, and the electric field as a function of r. Use $R = 3a$.

23–4 Conductors and Electric Fields

40. (I) Two large, thin, metallic plates are placed parallel to each other, separated by 11 cm. The top plate carries a uniform charge density of 6.5 μC/m^2, while the bottom plate carries a uniform charge density of -4.8 μC/m^2. What is the electric field halfway between the plates?

41. (I) Two concentric metallic shells—conductors—have radii of R and $2R$, respectively. A charge q is placed on the inner shell, and a charge $-2q$ is placed on the outer shell. What are the electric fields in all of space due to the two shells?

42. (I) Two oppositely charged, parallel metal plates give rise to a field of 3×10^6 N/C between them. The plates are square and have dimensions 0.1 m $\times$ 0.1 m. How much charge must there be on each plate? Assume that the charge distribution and electric field are uniform, as if the plates were infinite in size. This will be a good approximation if the distance between the plates is much smaller than 0.1 m.

43. (I) Charge is placed on a large spherical surface. What is the maximum surface charge density that avoids electrical breakdown in air ($E_{max} = 3 \times 10^6$ N/C)?

44. (I) A metal sphere of radius 15 cm is concentrically surrounded by a thin spherical metal shell whose inner radius is 25 cm. The electric flux through a concentric spherical Gaussian surface at a radius of 40 cm is 1.6×10^7 N·m^2/C, and that through a concentric spherical Gaussian surface at a radius of 18 cm is 0.80×10^7 N·m^2/C. What is the ratio of the charges on the inner and outer spheres?

45. (I) What is the ratio of the charge densities on the inner and outer spheres in Problem 44?

46. (I) A solid copper cube is placed in a constant electric field that points in the +x-direction. The faces of the cube are parallel to the xy-, yz-, and xz-planes, and one corner is at the origin. Draw the field lines as they would be observed looking down on the cube toward the xy-plane. Show at least two electric field lines starting or stopping on each of the four sides of the cube perpendicular to the xy-plane.

47. (II) A metal sphere of radius a is surrounded by a metal shell of inner radius b and outer radius R. The flux through a spherical Gaussian surface located between a and b is Q/ε_0, and the flux through a spherical Gaussian surface just outside radius R is $2Q/\varepsilon_0$ (Fig. 23–29). What are the total charges on the inner sphere and on the shell? Where are the charges located, and what are the charge densities?

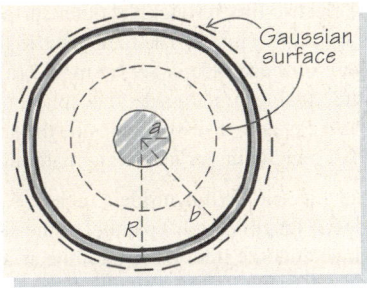

▲ FIGURE 23–29 Problem 47.

48. (II) The electric field near Earth's surface on a given day is 100 N/C, pointing radially inward. If this were true everywhere on the Earth's surface, what would the sign and magnitude of the total charge on Earth be? If Earth is treated as a conductor, where is the charge located? What is the charge density?

49. (II) A point charge q is placed a distance $L/2$ over the center of a conducting square plate of area L^2. (a) Draw the electric field lines on both sides of the plate, which has charge $-q$. (b) Repeat part (a) for a charge on the plate of $2q$.

50. (II) The center of a solid conducting sphere of radius 18 cm and charge 380 μC is placed 15 cm above and away from the center of a flat, horizontal conducting square plate of area 1000 cm^2 and charge 15 μC. Draw the electric field lines.

General Problems

51. (II) Consider a cube of sides a located at the origin (Fig. 23–30). Suppose that an electric field is present and given by $\vec{E} = bx^2\hat{\imath}$, where b is a constant. Calculate the flux through each side of the cube, and use this to find the charge within the cube.

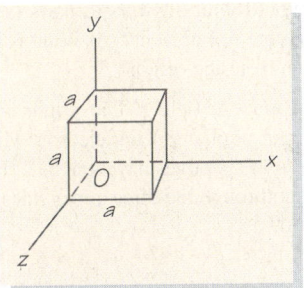

▲ FIGURE 23–30 Problem 51.

52. (II) Consider a solid sphere of radius R with a charge Q distributed uniformly. Suppose that a point charge q of mass m, with a sign opposite that of Q, is free to move within the solid sphere. Charge q is placed at rest on the surface of the solid sphere and released. Describe the subsequent motion. In particular, what is the period of the motion, and what is the total energy of the point charge? [*Hint*: Recall the properties of the motion for which the force varies linearly with the distance from a fixed point and is a restoring force.]

53. (II) Consider a point charge Q at the center of a Gaussian sphere of radius R. The sphere has a cap "sliced off," with the area of the cap forming a fraction of 0.067 of the total area of the sphere. (This corresponds to making the slice at a latitude of 60°, measured from the equatorial plane). The cap is replaced by a flat surface. (a) What is the total flux through the flat surface? (b) What is the magnitude of the electric field at the boundary of the flat surface?

54. (II) The total electric flux through the outer surface of a uniform solid sphere is 17.1 N·m²/C. The total electric flux through the top hemisphere of the sphere is 8.9 N·m²/C. (a) What can you say about the total charge inside the sphere? (b) What can you say about the charge distribution inside the sphere? (c) Is the sphere made of conducting or insulating material?

55. (II) A constant electric field is inside a tube of square cross section with sides of length L and is parallel to the sides of the tube. A plane surface cuts the interior of the tube at an angle θ (Fig. 23–31). Show by explicit calculation that the flux through this surface is independent of the angle θ. How would you show this without explicitly calculating the flux through the surface?

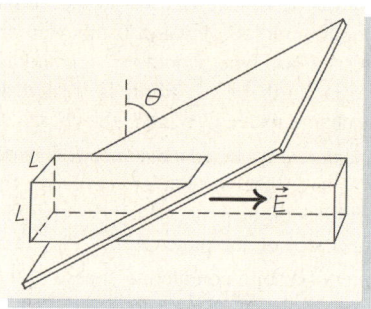

▲ **FIGURE 23–31** Problem 55.

56. (II) A conducting sphere of radius 25 cm is centered at the origin of a coordinate system, as is a surrounding conducting shell of radius 75 cm. The inner sphere has a charge density of 16 μC/m² over its surface, and the outer sphere has a uniform charge density half that large. (a) Find the electric field at a distance 50 cm from the origin; (b) at a distance 70 cm from the origin. (c) How would your answers to parts (a) and (b) change if the outer shell were not present? (d) What is the electric field at a distance 1.5 m from the origin?

57. (II) A constant electric field E that points in the $+z$-direction passes through an equilateral tetrahedron whose base is in the xy-plane and whose six edges have length L (Fig. 23–32). Calculate the total flux through the three upper sides of the tetrahedron.

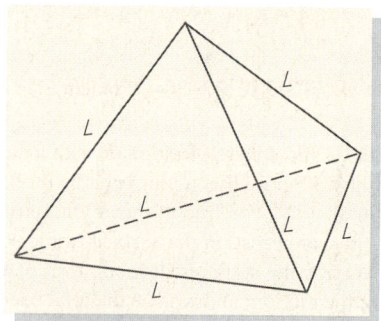

▲ **FIGURE 23–32** Problem 57.

58. (II) How should the charge density of a sphere of radius R vary with the distance from the center of the sphere to give a radial field of constant magnitude within the sphere? What happens at the origin, and why?

59. (II) A certain experiment requires an electric field that points symmetrically away from an axis and has a constant magnitude. Describe the charge distribution capable of creating such a field.

60. (II) A right solid conducting cylinder has a charge of -0.55 mC. Inside the cylinder a $+0.12$-mC charge rests at the center of a hollow spherical space (Fig. 23–33). (a) What is the charge on the surface of the hollow spherical space? (b) What is the charge on the outside surface of the cylinder?

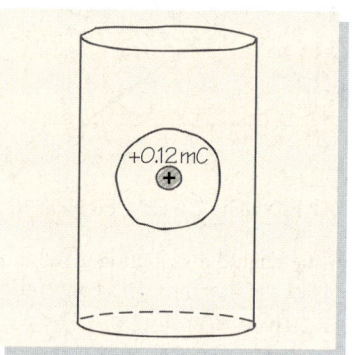

▲ **FIGURE 23–33** Problem 60.

61. (II) A conductor has a surface oriented in the yz-plane that marks the boundary of a region in which there is an electric field oriented in the $+x$-direction. The strength of this field increases linearly as x increases from $x = 0$ m to $x = 0.5$ m. At the beginning of the region, at $x = 0$ m, the field strength is 0; at $x = 0.5$ m, the field strength has increased to 3000 N/C. Describe the distribution in the x-direction of the charge that produces this field.

62. (III) A nonconducting sphere of radius R is charged uniformly with charge density ρ. Use Gauss' law to show that the electric field inside the sphere at a point P whose displacement vector from the sphere's center is $\vec{r}$ is given by $\vec{E} = (\rho/3\varepsilon_0)\vec{r}$. A small sphere centered at the point whose displacement from the origin is $\vec{a}$ is cut out of the sphere (Fig. 23–34). Use the superposition principle to calculate the electric field inside the cavity. [*Hint:* The cavity can be created by inserting in the original sphere a sphere of opposite charge density, $-\rho$, and radius b, centered at $\vec{a}$.]

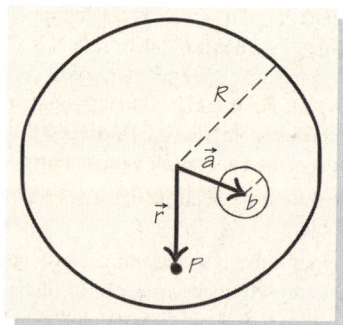

▲ **FIGURE 23–34** Problem 62.

63. (III) Use Gauss' law to show that a test charge in the electric field due to any given static charge distribution cannot be in stable equilibrium. [*Hint:* At an equilibrium point, the net electric field must be zero. What must the fields in the vicinity of that point be so that the equilibrium is stable?]

Chapter 24

◀ The discharge of lightning bolts provides an impressive demonstration that there is energy in electric fields. Lightning forms when a significant electric potential difference exists between Earth and the clouds or between different clouds.

Electric Potential

We know that there is a potential energy associated with conservative forces, that this potential energy is a term in the total energy of systems, and that the conservation of energy is a powerful tool for solving problems. The electric, or Coulomb, force is a conservative force. Thus we expect a collection of charges to have a potential energy, the subject of this chapter. Many of the results we develop here are similar to those for gravitation (Chapter 12) because the gravitational force and Coulomb's law have the same form.

Electric force concerns the interaction of a charge distribution and a second charge and is the product of the second charge and the electric field due to the distribution. Similarly, the electrical potential energy is the potential energy of the system made up of the charge distribution interacting with a second charge. And just as it is useful to remove the second charge from the electric force and deal with the electric field due to the distribution, it is useful to remove the second charge from the electrical potential energy and deal with the *electric potential*, or just potential, a property of the charge distribution alone.

There is one more important point to keep in mind. One of the advantages of using the potential energy rather than the force is that the potential energy is a scalar quantity, while force is a vector. If we know the potential energy, we may find the force by appropriate derivatives. Similarly, the potential is a scalar quantity, easier to handle in many respects than the electric field, which is a vector quantity, and if we know the electric potential of a charge distribution, we can derive the electric field due to that distribution from it.

24–1 Electric Potential Energy

The concept of a potential energy is extremely useful. For example, in thinking about gravity we learned that a mass m at a height h (much less than Earth's radius) above Earth's surface has a potential energy that can be written as $U(h) = mgh$. This helps us to determine the object's speed at any other height if we know its speed at one height. *Any* conservative force has a potential energy associated with it. This potential energy is a function of position, and it can be converted to kinetic energy in accordance with the conservation of mechanical energy. The total energy is $E = K + U$, where K is the kinetic energy. Conservation of energy means that the change in E is zero, so $\Delta E = 0 = \Delta K + \Delta U$, or $\Delta K = -\Delta U$. Any change in U will be matched by an equal but opposite change in K.

Let us now look at Coulomb's law. The electric force on charge q_0 due to charge q, separated by a distance r, is

$$\vec{F} = \frac{qq_0}{4\pi\varepsilon_0}\frac{1}{r^2}\hat{r}, \qquad (24\text{--}1)$$

where $\hat{r}$ is the unit vector that points radially outward from the position of q to the position of q_0. Recall that the gravitational force between a mass m_0 and a mass m separated by a distance r is

$$\vec{F} = -Gmm_0\frac{1}{r^2}\hat{r}. \qquad (24\text{--}2)$$

This is just the same form as Coulomb's law, although the gravitational force is always attractive, whereas the electric force is attractive or repulsive according to whether qq_0 is negative or positive. Each force is conservative, so a potential energy is associated with each force. This potential energy must of course take the same form for both cases.

Recall that only *changes* in potential energy have meaning. From Chapter 7, we know that we can express the change in potential energy of our system as the charge q_0 (or, in the case of gravitation, the mass m_0) moves from an initial point a at position $\vec{r}_a$ to a final point b at position $\vec{r}_b$ through a displacement (Fig. 24–1) as

$$\Delta U = U_f - U_i = U(\vec{r}_b) - U(\vec{r}_a) = -\int_{\vec{r}_a}^{\vec{r}_b} \vec{F} \cdot d\vec{s}. \qquad (24\text{--}3)$$

The integral in this expression is a line integral whose value is *independent of the path of integration* between points a and b—that is precisely why Eq. (24–3) involves only a difference in the value of the potential energy at the two end points.

Let's now evaluate the change in electric potential energy for the point charge q at the origin and the point charge q_0 when q_0 moves from point a to point b. We start with the simplest situation (Fig. 24–2a), in which point a is on the same radius as point b.

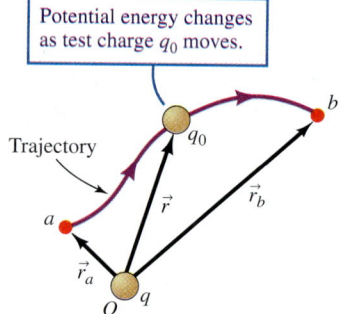

▲ **FIGURE 24–1** When a test charge q_0 moves from point a to point b in the presence of a charge q that is fixed in place, the potential energy of the system changes.

▶ **FIGURE 24–2** The change in potential energy of the system of two charges q and q_0 when the charge q_0 moves from point a to point b, in terms of a path-independent line integral.
(a) Charge q_0 moves along a radius.
(b) The two points are not along the same radius. The path is here taken to run radially outward to the radius of point b, and then follow the circumference at that radius.

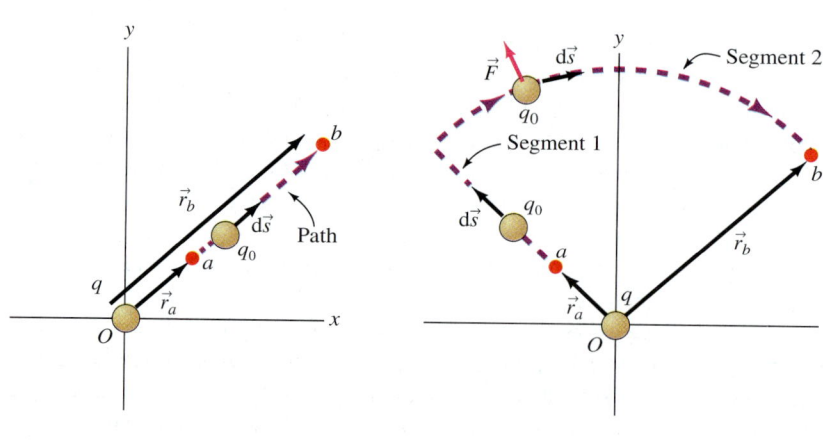

The path from a to b is along the dashed line shown in Fig. 24–2a, and as the Coulomb force points outward along this radial direction, we have for our path

$$\vec{F} \cdot d\vec{s} = F\, dr.$$

Then, from Eq. (24–3), the potential energy change when charge q_0 moves from a to b is

$$\Delta U = -\int_{r_a}^{r_b} F\, dr = -\int_{r_a}^{r_b} \frac{qq_0}{4\pi\varepsilon_0 r^2}\, dr$$

$$= -\frac{qq_0}{4\pi\varepsilon_0} \int_{r_a}^{r_b} \frac{dr}{r^2} = -\frac{qq_0}{4\pi\varepsilon_0}\left(\frac{-1}{r}\right)\bigg|_{r_a}^{r_b} = \frac{qq_0}{4\pi\varepsilon_0}\left(\frac{1}{r_b} - \frac{1}{r_a}\right). \quad (24\text{--}4)$$

What if charge q_0 moves between two points that do not lie on the same radius, as in Fig. 24–2b? In this case, we follow the dashed path shown. [Remember, the result of the integration in Eq. (24–3) is path independent.] For segment 1, which runs outward radially from a to a distance r_b from the origin, the result is identical to Eq. (24–4). For segment 2, which follows a circumference at a distance r_b from the origin, the integral is zero because the force is perpendicular to the path segment $d\vec{s}$ everywhere. The result for the change in potential energy is still given by Eq. (24–4).

Let's look more closely at what Eq. (24–4) tells us. Suppose first that the charges move closer together ($r_a > r_b$). If the charges repel (qq_0 is positive), the change in potential energy is positive. This is like moving a mass *up* a mountain. If the charges attract (qq_0 is negative), the system loses potential energy when the charges move closer together. This is like moving a mass *down* the mountain. As with any potential energy, electric potential energy can be converted into kinetic energy. If there are no additional forces acting, then like-sign charges slow down—or lose kinetic energy—when they move closer together. Similarly, charges of opposite sign speed up—or gain kinetic energy—when they move closer together. We draw a similar set of conclusions when the charges move farther apart ($r_a < r_b$). Charges that repel lose electric potential energy and, if there are no other forces, gain kinetic energy. Opposite charges (which attract) gain electric potential energy when they move farther apart and lose kinetic energy in the absence of other forces.

Equation (24–3) shows that the change in electric potential energy is given by the difference of two functions, $U(r_b)$ and $U(r_a)$. We can therefore choose the zero of the potential energy function to be at whatever value of r we like. It is convenient and natural to choose zero potential energy to be at infinity. We can do this if we let $r_a \to \infty$ and let r_b take on a general value r in Eq. (24–4):

$$\Delta U = U(r) - U(r_a)\bigg|_{r_a \to \infty} = \frac{qq_0}{4\pi\varepsilon_0}\frac{1}{r}.$$

We then say that the potential energy of a charge q_0 a distance r from charge q is the difference in potential energy between that point and infinity. When we reverse the roles of q and q_0, the potential energy of q at a distance r from q_0 is again $qq_0/4\pi\varepsilon_0 r$. We can then say that the **electric potential energy** $U(r)$ for a system of two point charges q and q_0 separated by a distance r is

$$U(r) = \frac{qq_0}{4\pi\varepsilon_0}\frac{1}{r}. \quad (24\text{--}5)$$

It is indeed true that $U(r) = 0$ in the limit $r \to \infty$. Thus the system has no potential energy when the two charges are infinitely far apart. Note that the potential energy of the two charges depends *only* on the distance r between them and on the magnitudes and signs of the charges. Equation (24–5) has the same form as Eq. (12–9), calculated in Chapter 12 for the gravitational potential energy.

CONCEPTUAL EXAMPLE 24–1 How much work is done by the electrical force when a point charge is brought from infinity to rest at a distance r from a fixed charge of the opposite sign? What is the meaning of the sign of your result?

Answer The work done is given by $W = \int_{\infty}^{r} \vec{F} \cdot d\vec{s}$, where $\vec{F}$ is the electrical force between the charges. The quantity is the negative of the corresponding change of electric potential energy, as in Eq. (24–3), namely $-[U(r) - U(\infty)]$. Since the zero of the potential energy is at infinity, this is just $-U(r)$, with U given by Eq. (24–5). For charges of opposite sign, $-U(r)$, and therefore the work done, is positive. This is a sensible result, because when the electrical force is the only force acting the work done by it is the change in kinetic energy, and that change would indeed be positive for charges of opposite sign, which attract one another.

EXAMPLE 24–2 One measure of the strength of the nuclear forces that hold the constituents of the nucleus together is provided by the comparative strength of the repulsive Coulomb force between the (positively charged) protons in the nucleus. Find the electrostatic potential energy between two protons that are separated by the average separation between the nuclear constituents, roughly 2×10^{-15} m.

Strategy This is a straightforward application of Eq. (24–5), an expression for the electrostatic potential energy. This expression assumes that the potential energy is zero at infinity.

Working It Out The proton charge is equal and opposite to that of the electron, $+e = 1.6 \times 10^{-19}$ C. Then

$$U = \frac{(e)^2}{4\pi\varepsilon_0 r} = \frac{(9 \times 10^9 \text{ N} \cdot \text{m}^2/\text{C}^2)(1.6 \times 10^{-19} \text{ C})^2}{(2 \times 10^{-15} \text{ m})} = 10^{-13} \text{ J}.$$

This is a typical energy value on the nuclear scale and is about 10^5 times larger than the energy of a proton and electron that make up a hydrogen atom (see Example 24–4). This energy can be converted into kinetic energy in nuclear reactions (see Chapter 44).

24–2 Electric Potential

A point charge q is the source of an electric field $\vec{E}$ that exists in the surrounding space. The electric field affects any charge q_0 introduced into that space through a force $\vec{F}$ on q_0 given by $\vec{F} = q_0 \vec{E}$. We saw in Section 24–1 that the introduction of a charge q_0 a distance r from q gives rise to the potential energy $U(r)$ of Eq. (24–5). If we write $U(r) = q_0 V(r)$, we can make a statement analogous to the statement about the electric field: The charge q is the source of an *electric potential* (or just *potential*) $V(r)$, within which any charge q_0 a distance r from q will have potential energy $U(r) = q_0 V(r)$. Strictly speaking q_0 should be a small test charge, so that its presence does not disturb charge q or any other charge distribution that gives rise to the electric potential. This description tells us how to define the **electric potential** due to a charge distribution:

$$V(\vec{r}) \equiv \lim_{q_0 \to 0} \frac{U(\vec{r})}{q_0}, \qquad (24\text{–}6)$$

where $U(\vec{r})$ is the potential energy of the test charge q_0 in the presence of the charge distribution. The potential $V(\vec{r})$ is independent of the test charge q_0, just as the electric field, defined by $\vec{E} = \vec{F}/q_0$, is independent of the test charge.

The Electric Potential of a Point Charge

Let's calculate the electric potential of the simplest possible system: one point charge. Consider a test charge q_0 separated by a distance r from a single point charge q. As Eq. (24–5) shows, the potential energy of the system is $U(r) = q_0 q / 4\pi\varepsilon_0 r$, and hence $U/q_0 = q/4\pi\varepsilon_0 r$. We have found *the electric potential of a point charge q at a distance r from the charge*:

$$V(r) = \frac{q}{4\pi\varepsilon_0 r}. \qquad (24\text{–}7)$$

In Eq. (24–7), we have assumed that zero potential energy is at infinity and, as a consequence, we have taken the electric potential due to a charge q to be zero at infinity. To emphasize this point, we might say that Eq. (24–7) is the potential of a single charge with respect to infinity.

As for potential energy, the only physically relevant feature of the potential is how it differs between two points. The *electric potential difference* due to the charge q between the points a and b at locations $\vec{r}_a$ and $\vec{r}_b$ is given by (Fig. 24–3):

$$\Delta V = V_b - V_a = \frac{U_b - U_a}{q_0} = \frac{q}{4\pi\varepsilon_0}\left(\frac{1}{r_b} - \frac{1}{r_a}\right). \qquad (24\text{–}8)$$

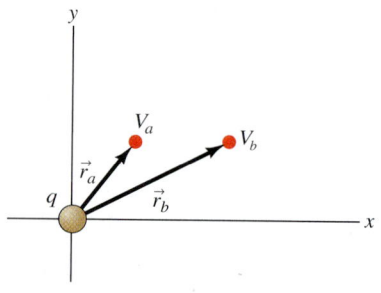

▲ **FIGURE 24–3** The potential at two different locations in space. Only the potential *difference* has physical meaning.

Here, we have abbreviated V as a function of r_a, or $V(r_a)$, as V_a, and so forth.

We can obtain another formulation of the electric potential difference by using Eqs. (24–3) and (24–8) and substituting $\vec{F} = q_0\vec{E}$:

$$\Delta V = \frac{U_b - U_a}{q_0} = -\int_{r_a}^{r_b} \vec{E} \cdot d\vec{s}. \qquad (24\text{–}9)$$

Here, the electric potential difference is expressed as a path-independent integral over an electric field. The electric field in Eq. (24–9) is not necessarily the electric field of a point charge. Equation (24–3) is the potential energy change when a test charge q_0 moves from point a to point b in the field of *any* charge distribution. Thus Eq. (24–9) is a general expression for the electric potential difference between two points. Any charge distribution produces an electric field, and an electric potential is associated with any charge distribution.

Recall (Section 7–1) that the change in the potential energy of a system is equal to the negative of the work done by the system in moving an object from point a to point b. Equivalently, $U_b - U_a$ is the work done by an external agent to move the object. By the parallel relation between force and field we can then interpret Eq. (24–9) to mean that

The electrical potential difference $V_b - V_a$ is the work per unit charge that must be done by an external agent to move a test charge from point a to point b without changing its kinetic energy.

If there is no external agent, then a change in potential, which corresponds to a change in potential energy of the test charge, must be accompanied by a corresponding change in the kinetic energy of the test charge.

With knowledge of the electric potential $V(\vec{r})$ due to a charge distribution and the magnitude of a test charge q_0 we immediately have the potential energy $U(\vec{r})$ of the system composed of the distribution and the test charge q_0 placed at the position $\vec{r}$:

$$U(\vec{r}) = q_0 V(\vec{r}). \qquad (24\text{–}10)$$

In the absence of other forces, this equation tells us that a positive test charge q_0 in the presence of an electric potential will move toward lower values of the potential because the potential energy decreases in that way. The charge speeds up as it moves to lower potentials.

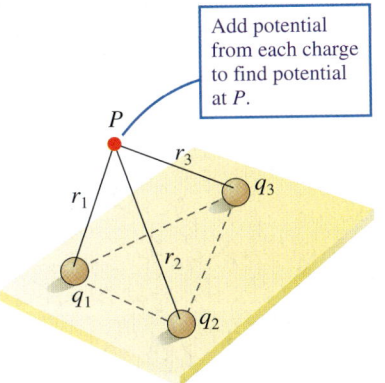

▲ **FIGURE 24–4** The superposition principle determines the potential at point P due to multiple charges. We simply add the potential due to each of the charges.

The Electric Potential of Charge Distributions

The electric field obeys the superposition principle. Therefore the electric potential of a system of charges can also be determined from the superposition principle. The superposition principle states that the electric field of a collection of charges is the sum of the electric fields of each charge. Thus *the electric potential at a point P due to n point charges* $q_1, q_2, \ldots, q_n$ (Fig. 24–4 shows three charges) at distances $r_1, r_2, \ldots, r_n$ from point P is

$$V_P = \frac{q_1}{4\pi\varepsilon_0 r_1} + \frac{q_2}{4\pi\varepsilon_0 r_2} + \cdots + \frac{q_n}{4\pi\varepsilon_0 r_n} = \frac{1}{4\pi\varepsilon_0}\sum_{i=1}^{n} \frac{q_i}{r_i}, \qquad (24\text{–}11)$$

where r_i is the distance from point charge q_i to point P. The electric potential due to a collection of charges is the *scalar* sum of the potentials due to single charges. This scalar sum is much easier to perform than the vector sum that expresses the electric field due to a collection of point charges, illustrating in another way the usefulness of the concept of potential.

The calculation of the electric potential due to a continuous charge distribution is also straightforward. We first find the electric potential dV at a point P due to a small charge dq that is part of an arbitrary charge distribution (Fig. 24–5). Because electric potential is a scalar quantity, the addition of all the tiny potentials dV is given by scalar integration. Thus *the potential due to a continuous charge distribution* takes the symbolic form

$$V = \int dV = \frac{1}{4\pi\varepsilon_0}\int \frac{dq}{r}. \qquad (24\text{–}12)$$

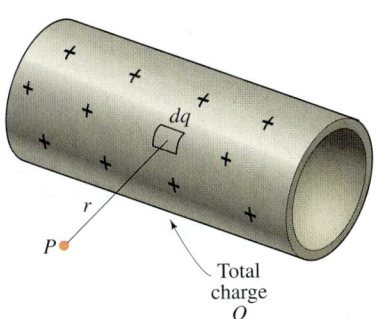

▲ **FIGURE 24–5** To find the potential at point P due to a continuous charge distribution, here a cylinder, integrate over the contribution from the differential charges dq as if each dq were a point charge.

The integration must be done over the entire charge distribution. In Section 24–5 we discuss techniques for the calculation of V in specific situations.

Units of Electric Potential

The dimension of electric potential is energy per charge; thus the SI unit is joules per coulomb (J/C). Because electric potential is used frequently, it has a separate name in the SI: the **volt(V)**:

$$1 \text{ V} \equiv 1 \text{ J/C}. \tag{24-13}$$

(Don't confuse the roman letter V that symbolizes the unit with the italic V that is the usual symbol for the function electric potential.) It is named after Alessandro Volta, who did research at the beginning of the nineteenth century on the nature of electric energy. Note that electric potential has the dimensions of electric field times length, so the dimensions of electric field must be the dimensions of potential divided by length (V/m):

$$1 \text{ N/C} = 1 \text{ V/m}. \tag{24-14}$$

The Potential Energy of a System of Charges

Equation (24–10) gives the potential energy $U(r) = q_0 V(r)$ of a test charge q_0 placed in the electric potential of a charge distribution. If the charge distribution is a collection of charges, then the electric potential V_P is given by Eq. (24–11), and the potential energy of the test charge is $U(r) = q_0 V_P$. But it would be incorrect to call this the potential energy of the entire system of charges $q_0, q_1, q_2, \ldots, q_n$, because the product $q_0 V_P$ just represents the work that needs to be done to bring charge q_0 in from infinity. It does *not* take into account the work that must be done to bring the charges $q_1, q_2, \ldots, q_n$ in from infinity. To calculate the potential energy of a collection of three charges, for example, we assemble them one by one. To bring the first charge, q_1, in to the point P_1 requires no work by the external agent if the kinetic energy of the charge is unchanged. To bring the second charge, q_2, in from infinity to the point P_2 does require work because of the potential due to q_1. For our two charges, the work the external agent must do to bring q_2 in from infinity—the potential energy—is given by

$$U_{12} = q_2 V_1 = \frac{q_1 q_2}{4\pi\varepsilon_0 r_{12}}, \tag{24-15}$$

where r_{12} is the distance between charges q_1 and q_2.

What happens if we bring a third charge, q_3, in from infinity? We must calculate the additional work done by an external force to bring q_3 in. This work is given by the product of q_3 and electric potentials V_1 and V_2 due to q_1 and q_2 in place. Thus, the additional contribution to the potential energy of the system is

$$U_{13} + U_{23} = \frac{q_1 q_3}{4\pi\varepsilon_0 r_{13}} + \frac{q_2 q_3}{4\pi_0 r_{23}}, \tag{24-16}$$

where r_{13} and r_{23} are the distances between q_3 and q_1, q_3 and q_2, respectively. The total potential energy U of the system is the sum of U_{12}, U_{13}, and U_{23}:

$$U = \frac{1}{4\pi\varepsilon_0}\left(\frac{q_1 q_2}{r_{12}} + \frac{q_1 q_3}{r_{13}} + \frac{q_2 q_3}{r_{23}}\right). \tag{24-17}$$

This can be generalized to any number of charges, and the resulting formula for the *electric potential energy of the system* is a simple generalization of Eq. (24–17):

$$U = \frac{1}{4\pi\varepsilon_0} \sum_{i<j} \frac{q_i q_j}{r_{ij}}, \tag{24-18}$$

where r_{ij} is the distance between the locations of the charges q_i and q_j. The sum over i and j includes all charge pairs in the system, and the inequality $i < j$ avoids the counting of pairs more than once. We can eliminate that restriction by writing the equivalent expression

$$U = \frac{1}{2} \sum_{\substack{i,j \\ i \neq j}} \frac{q_i q_j}{4\pi\varepsilon_0 r_{ij}}.$$

Now the sum is unrestricted, except that we omit the case $i = j$, which is not in the original sum, Eq. (24–18). Thus we can rewrite Eq. (24–18) as

$$U = \frac{1}{2} q_1 \sum \frac{q_j}{4\pi\varepsilon_0 r_{1j}} + \frac{1}{2} q_2 \sum \frac{q_j}{4\pi\varepsilon_0 r_{2j}} + \frac{1}{2} q_3 \sum \frac{q_j}{4\pi\varepsilon_0 r_{3j}} + \cdots \quad (24\text{–}19)$$

$$= \frac{1}{2} q_1 V_1 + \frac{1}{2} q_2 V_2 + \frac{1}{2} q_3 V_3 + \cdots,$$

where V_1 is the electric potential due to all the other charges at the location of charge q_1, and so on. It should be stressed that the potential energy of q_1 in a given potential V_1 is still $q_1 V_1$; this means that $q_1 V_1$ can be converted into the kinetic energy of the particle that carries charge q_1. This potential energy must be distinguished from the potential energy of the *entire* charge configuration, Eq. (24–18) or (24–19). The potential energy of the entire charge configuration is the energy that would be made available if *all* the charges that appear in the problem were to move to infinity.

In Examples 24–3 and 24–4, we illustrate calculation techniques for the electric potential energy and the electric potential when two or more point charges are involved.

EXAMPLE 24–3 In an experiment to investigate the effects of electricity, Benjamin Franklin could well have placed two point charges, $q_1 = 2.0\ \mu\text{C}$ and $q_2 = -4.0\ \mu\text{C}$, at some distance apart (points P_1 and P_2, respectively) (Fig. 24–6). (a) Find the electric potential at points a and b due to these two point charges. (b) Find the potential difference between points b and a. (c) How much energy would Franklin have had to supply to bring a third charge, of magnitude $3.0\ \mu\text{C}$, in from infinity to point b?

Setting It Up The only quantity that requires new labeling here is the third charge [part (c)], which we shall call q_3.

Strategy For part (a) we can use Eq. (24–11) to determine the electric potential. We'll require various distances; for example, to find the potential at point a, we'll need the distance from point a to points P_1 and P_2. These distances are a matter of geometry. Part (b) requires taking the difference of the potentials we found in part (a). Finally for part (c), we note that the work that must be done to bring q_3 in from infinity to point b is equal to the change in the potential energy of the system. But we know from part (a) the electric potential of the original system of two charges at point b, so we use Eq. (24–10), $U_b = q_0 V_b$, to find the potential energy of the new charge at point b.

Working It Out (a) The distance from point a to point P_1 is $r_{1a} = 2$ m, and the distance from point a to point P_2 is $r_{2a} = \sqrt{(2.0\ \text{m})^2 + (3.0\ \text{m})^2} = 3.6$ m (note that we work with two significant figures here). The electric potential V_a at point a is then

$$V_a = \frac{1}{4\pi\varepsilon_0} \left(\frac{q_1}{r_{1a}} + \frac{q_2}{r_{2a}} \right)$$

$$= (9.0 \times 10^9\ \text{N} \cdot \text{m}^2/\text{C}^2) \left(\frac{2.0 \times 10^{-6}\ \text{C}}{2.0\ \text{m}} + \frac{-4.0 \times 10^{-6}\ \text{C}}{3.6\ \text{m}} \right)$$

$$= 1.0\ \text{kV}.$$

The SI units of potential are volts. A check is always useful—in this case, the unit combination is $\text{N} \cdot \text{m}/\text{C} = \text{J}/\text{C} = \text{V}$.

For the electric potential at point b, the distance from charge q_1 to b is $r_{1b} = 2.0$ m; similarly, $r_{2b} = 1.0$ m. Therefore, the potential V_b is

$$V_b = \frac{1}{4\pi\varepsilon_0} \left(\frac{q_1}{r_{1b}} + \frac{q_2}{r_{2b}} \right)$$

$$= (9.0 \times 10^9\ \text{N} \cdot \text{m}^2/\text{C}^2) \left(\frac{2.0 \times 10^{-6}\ \text{C}}{2.0\ \text{m}} + \frac{-4.0 \times 10^{-6}\ \text{C}}{1.0\ \text{m}} \right)$$

$$= -27\ \text{kV}.$$

(b) The potential difference $V_b - V_a = -27\ \text{kV} - 1\ \text{kV} = -28\ \text{kV}$. Thus the electric potential is higher at point a than at point b.

(c) We have

$$U_b = q_3 V_b = (3.0\ \mu\text{C})(-28\ \text{kV})$$

$$= (3.0 \times 10^{-6}\ \text{C})(-28 \times 10^3\ \text{V}) = -8.4 \times 10^{-2}\ \text{J}.$$

The work done is in the SI units of joules.

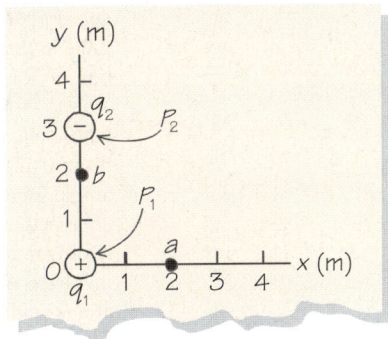

▲ **FIGURE 24–6**

What Do You Think? True or false: The sign of our result for the work that must be done in part (c) is wrong. *Answers to* **What Do You Think?** *questions are given in the back of the book.*

EXAMPLE 24–4 The classical model of a hydrogen atom in its normal, unexcited configuration has an electron that revolves around a proton at a distance of 5.3×10^{-11} m. What is the electric potential due to the proton at the position of the electron? Determine the electrostatic potential energy between the two particles. This energy is relevant to understanding the chemical activity of atoms.

Setting It Up We sketch this situation in Fig. 24–7.

Strategy The electric potential V_P due to the proton can be found by using Eq. (24–7), and we can subsequently find the electrostatic potential energy by using Eq. (24–15), $U = (-e)V_P$, where $-e$ is the electron charge.

Working It Out We have

$$V_P = \frac{+e}{4\pi\varepsilon_0 r} = \frac{(9.0 \times 10^9 \text{ N} \cdot \text{m}^2/\text{C}^2)(1.6 \times 10^{-19} \text{ C})}{5.3 \times 10^{-11} \text{ m}} = 27 \text{ V}.$$

In turn,

$$U = (-e)V_P = (-1.6 \times 10^{-19} \text{ C})(27 \text{ V})$$
$$= -4.3 \times 10^{-18} \text{ J}. \quad (24\text{–}20)$$

What Do You Think? What does the sign of U tell you about whether or not you must supply positive energy to separate the electron from the proton?

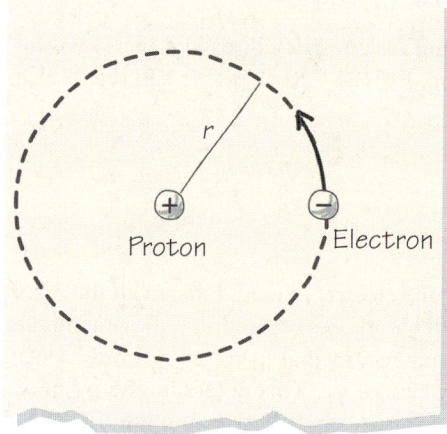

▲ **FIGURE 24–7** A simplistic representation of an electron that orbits a proton in the hydrogen atom.

The Electron-Volt

We have seen that the energy of a system is determined by multiplying charge by electrostatic potential. Because the charge on an electron is such a frequently used quantity, a useful unit of energy is that of the charge magnitude of an electron (or proton) times 1 V. We call this unit of energy an **electron-volt** (eV). An electron-volt is simply the energy an electron gains when it is accelerated though a potential difference of one volt. The electron-volt is not an SI unit. The relation between the electron-volt and the SI unit joule is

$$1 \text{ eV} = (1.6 \times 10^{-19} \text{ C})(1 \text{ V}) = 1.6 \times 10^{-19} \text{ J}.$$

The electron-volt is especially valuable for calculations in atomic, nuclear, and particle physics.

> **THINK ABOUT THIS...**
> **IN WHAT REALM OF PHYSICS IS THE ELECTRON-VOLT A NATURAL UNIT?**
>
> As the name indicates, this is a unit of energy appropriate to an object with a charge comparable to that of an electron, moving in a potential of the magnitude of volts. It is in the atomic domain that we deal with particles whose charges are those of a single electron or proton, of the order of 10^{-19} C. We saw in Example 24–4 that the electrostatic potential energy between the proton and electron of the hydrogen atom is -27 eV. Chemical reactions involve the rearrangement of electrons among atoms, and in accordance with the result of Example 24–4, involve energies of the order of fractions of electron-Volts. This can translate into temperatures that are involved in chemical reactions, because energy can be expressed that way via the relation energy $= k_B T$, where k_B is Boltzmann's constant. As an example, consider the temperature at which paper burns, 451°F. This corresponds to 506K, and therefore to an energy of $E = (1.38 \times 10^{-23} \text{ J/K})(506\text{K})/(1.6 \times 10^{-19} \text{ J/eV}) = 4.4 \times 10^{-2}$ eV. This number is much smaller than the ionization energy of hydrogen—the energy that must be supplied to remove the electron entirely—because rearrangements of electrons cost much less energy than stripping them off entirely.
>
> In nuclear reactions the forces are much stronger than electrical ones, and distances are much shorter than typical atomic dimensions, and the natural unit is a million electron-volts, the MeV. For example, a typical "ionization energy" for nuclei—the energy required to remove a constituent particle from a nucleus—is 8 MeV, while rearrangement energies are in the hundreds of kilo-electron volts (keV).

EXAMPLE 24–5 Calculate the electric potential due to an electric dipole whose dipole moment has magnitude p at an arbitrary point Q. Work in the limit in which the distance of point Q from the dipole is much larger than the dipole length.

Setting It Up We draw the dipole in Fig. 24–8, which also contains some necessary geometrical and label information. In particular, the charges are $\pm q$, and they are separated by a distance $\ell = p/q$. The distances to the point P from the charges $\pm q$ are r and $r + \Delta r$, respectively.

Strategy A dipole consists of two pointlike charges, so Eq. (24–11)— superposition—determines the potential, which will be zero at infinity. We can see from the figure that point Q is specified in part by the angle between $\vec{p}$ and the line between the $-q$ charge and Q, and as the figure shows, Δr can be expressed in terms of that angle and the separation between the charges in the limit in which $r \gg \Delta r$, a limit that is equivalent to the limit that the distance to the dipole is much greater than the linear size of the dipole.

Working It Out Equation (24–11) gives

$$V = \frac{q}{4\pi\varepsilon_0 r} + \frac{-q}{4\pi\varepsilon_0(r + \Delta r)} = \frac{q}{4\pi\varepsilon_0}\left(\frac{1}{r} - \frac{1}{r + \Delta r}\right)$$

$$= \frac{q}{4\pi\varepsilon_0}\frac{(r + \Delta r) - r}{r(r + \Delta r)} = \frac{q}{4\pi\varepsilon_0}\frac{\Delta r}{r(r + \Delta r)}. \quad (24\text{–}21)$$

From the figure we see that in the large r limit the dotted line is normal to the longer line, so that

$$\Delta r = \ell \cos\theta = \frac{p\cos\theta}{q}. \quad (24\text{–}22)$$

When this result is substituted into Eq. (24–21), we find

$$V = \frac{p\cos\theta}{4\pi\varepsilon_0}\left[\frac{1}{r(r + \ell\cos\theta)}\right] = \frac{pq\cos\theta}{4\pi\varepsilon_0}\left[\frac{1}{r(qr + p\cos\theta)}\right], \quad (24\text{–}23)$$

where in the last step we have multiplied and divided by q.

It is a mathematical exercise to calculate the electric dipole potential of Eq. (24–23) in the large r limit. While Eq. (24–23) is rather complicated, it takes a simple approximate form far from the dipole, when $r \gg \ell$. The easiest place to make an approximation is in Eq. (24–21), where we can use the fact that $\Delta r \ll r$. The numerator has one power of Δr, and if we want a calculation that is correct to first order in the small quantity Δr, we can drop the Δr in the denominator. From that point on we make our substitutions and end up with

$$\text{for } r \gg \ell: V = \frac{p\cos\theta}{4\pi\varepsilon_0 r^2}. \quad (24\text{–}24)$$

[Another way of getting the same result is to note that the condition $r \gg \ell$ is equivalent to $qr \gg q\ell = p$, so that we can drop the second term in the denominator of Eq. (24–23).] Because of the large r approximation, we can now measure θ from anywhere between the two charges of the dipole. Note that the potential of the dipole for distant points decreases as $1/r^2$, as compared to the $1/r$ dependence for a point charge, just as the electric field of the dipole falls off as $1/r^3$ compared to the $1/r^2$ dependence of the point charge.

This example is of some importance because, as we mentioned in Section 22–1, the dipole charge distribution occurs repeatedly in nature.

What Do You Think? Our picture of the hydrogen atom in Example 24–4 consists of a charge $+e$ and $-e$ a certain distance apart. Does this neutral atom give rise to a dipole field, and if not, why not?

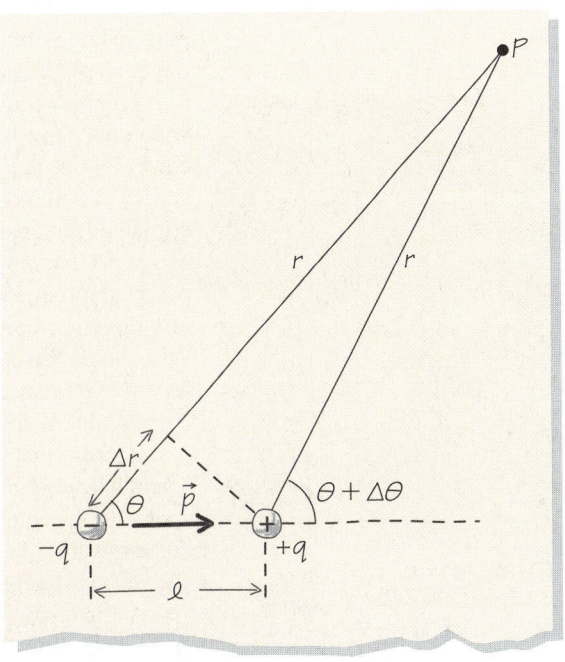

▲ **FIGURE 24–8** Using geometry to find the potential at a point P for an electric dipole. The dipole moment $p = q\ell$.

24–3 Equipotentials

Regions for which the electric potential of a charge distribution has constant values are called **equipotentials**. They are particularly interesting and worth investigating—we shall see, for example, that the surfaces of conductors form equipotentials. Suppose that a system of charges produces a certain potential. The positions in space that have the same electric potential form surfaces in three dimensions and lines in two dimensions. We say that the places where the potential has a constant value form **equipotential surfaces** in three dimensions or **equipotential lines** in two dimensions. As an example, consider the equipotential surfaces formed by a point charge. The electric potential is proportional to $1/r$ and has a constant value at any fixed radial distance from the charge.

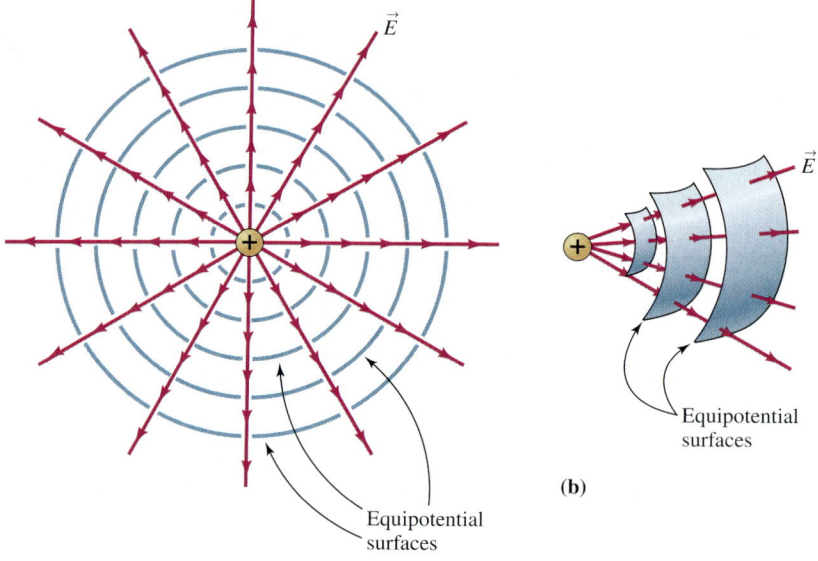

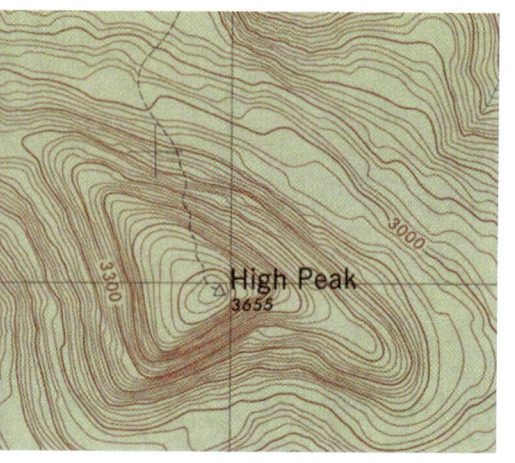

► **FIGURE 24–9** (a) The equipotential surfaces for a point charge in a two-dimensional representation. (b) The full three-dimensional representation; the equipotential surfaces are spheres centered on the charge.

▲ **FIGURE 24–10** The contour lines on topographic maps are lines of constant elevation. These are also lines of constant gravitational potential energy. The force of gravity has no component *along* contour lines, only perpendicular to them.

Therefore, a sphere centered on the charge forms an equipotential surface (Fig. 24–9). Any sphere with a different radius centered on the charge forms a different equipotential because the potential varies only with the radius of the sphere.

Equipotentials are analogous to contour lines on a topographic map—lines for which the elevation from sea level is constant (Fig. 24–10). The lines on such maps are equipotential lines for the potential energy of local gravity. The gravitational potential energy of a mass depends only on the mass's elevation, so the gravitational potential energy does not change when a mass moves along a contour line. Consequently, the force of gravity has no component along contour lines. Gravity only has a component perpendicular to a contour line; a ball that starts on a particular contour line will accelerate in a direction perpendicular to the line, or what we would call straight down the hill. What holds for contour lines holds for any equipotential surface or line, and any conservative force acts in a direction perpendicular to the equipotential because it can have no component along the equipotential.

Because the potential has exactly the same value along an equipotential, so does the potential energy of a test charge. No work is done when the test charge moves at constant speed on an equipotential surface or line. The equipotentials for the point charge in Fig. 24–9 are spheres centered on the charge, and a test charge can move freely about any one such spherical surface without work being done by the electric field.

As no work is done by the electric force when a test charge moves on an equipotential, we can understand why the electric field does not have a component along an equipotential surface. If it did, then that component of the electric field would do work to move a charge on the equipotential surface, which is not possible. Thus

the electric field is everywhere perpendicular to the equipotential surface.

Furthermore, because all the charge on a conductor in equilibrium resides on the surface, a potential difference between two points on the surface would be quickly equalized by a flow of free charge, so

the surface of a conductor is an equipotential.

The same reasoning shows that in static equilibrium the entire conductor will be at that same electric potential—there is no field within the conductor, so it takes no work to move a charge through it.

Electric Field Lines from Equipotentials and Vice Versa

The fact that the electric field and the equipotentials are everywhere perpendicular to each other is helpful in finding equipotential surfaces if the field is known, and in finding the electric fields if the equipotentials are known. We illustrate this process for some charge

24–3 Equipotentials

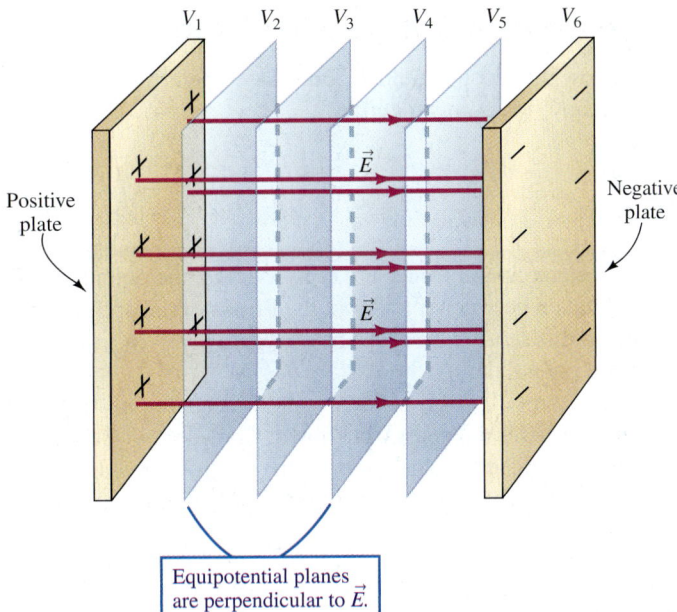

◀ **FIGURE 24–11** The electric field lines (burgundy) and the equipotentials (blue) for two oppositely charged parallel plates.

configurations for which the fields are known. We have already seen how the equipotential surfaces of a point charge are perpendicular to the (radial) electric field. For another example, this time starting from known field lines, consider two oppositely charged plates, for which the field lines are as in Fig. 24–11. The equipotential surfaces are perpendicular to these lines, and hence form planes parallel to the charged plates. If the charged plates are conductors, then they must also be equipotential surfaces. Thus we have a series of n equipotential planes $V_1, V_2, V_3, \ldots, V_n$ between and including the two charged plates.

CONCEPTUAL EXAMPLE 24–6 Sketch the equipotentials for an electric dipole given the field lines shown in Fig. 24–12a.

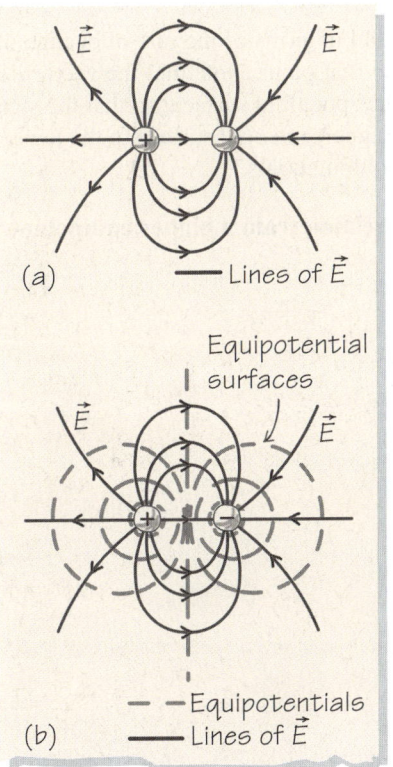

◀ **FIGURE 24–12** (a) The electric field lines for an electric dipole. (b) The equipotentials are drawn as dashed lines.

Answer Starting from Fig. 22–12a, we draw our equipotential surfaces everywhere perpendicular to the electric field lines, arriving at the dashed lines shown in Fig. 24–12b. These lines represent the locations in the plane of the page where the equipotential surfaces cut the page. You can visualize them in three dimensions as a series of nested but not concentric closed surfaces—each is a sort of flattened sphere—that encircles each charge. By sketching in this way we can visualize what the equipotential surfaces are like, even without using the algebraic form of the electric dipole potential derived in Example 24–5. A computer-generated drawing of the dipole potential is shown in Fig. 24–13.

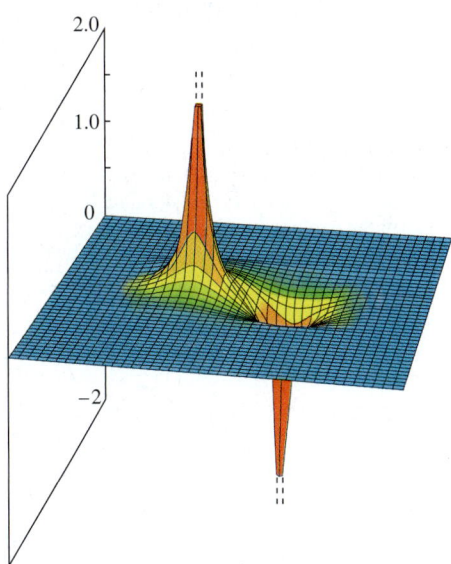

▶ **FIGURE 24–13** The dipole potential. Height represents the potential here.

24-4 Determining Fields from Potentials

If we know the electric field $\vec{E}$, Eq. (24–9) determines the potential difference $V_b - V_a$ between any two points a and b:

$$V_b - V_a = \int_{r_a}^{r_b} dV = -\int_{r_a}^{r_b} \vec{E} \cdot d\vec{s}.$$

Because electrostatic forces are conservative, the potential difference is independent of the path taken between a and b in the line integral, and we can choose a path for convenience. In this section we'll see how we can invert this relation and find the electric field given the potential.

Finding the field given the potential is analogous to finding the force between objects if their potential energy is known. Consider two equipotentials, labeled by b and a, that are very close together, so that the electric field can be considered constant in the region between them. In that region Eq. (24–9) takes the form

$$V_b - V_a = dV = -\vec{E} \cdot d\vec{s}, \tag{24–25}$$

where $d\vec{s}$ is an infinitesimal displacement vector pointing from equipotential a to equipotential b. (Actually, this relationship is true for any $d\vec{s}$ connecting the equipotentials, even a large displacement, as long as the electric field can be treated as constant along the displacement.)

It is simplest to take our infinitesimal path as in Fig. 24–14, with $d\vec{s}$ perpendicular to the two equipotential surfaces. As the electric field also points in that direction, Eq. (24–25) reads

$$dV = -E\,ds.$$

Equivalently,

$$E = -\frac{dV}{ds}. \tag{24–26}$$

This equation gives the magnitude of the electric field in terms of the rate of change of V in a direction perpendicular to the equipotential at that point. Note that the vector $d\vec{s}$ pointed from the lower-potential surface to the higher-potential surface, so that the sign in Eq. (24–26) shows that the field points from the higher-potential surface to the lower-potential surface. In summary, for closely spaced equipotentials,

the electric field points along the shortest direction from a higher equipotential to a lower one.

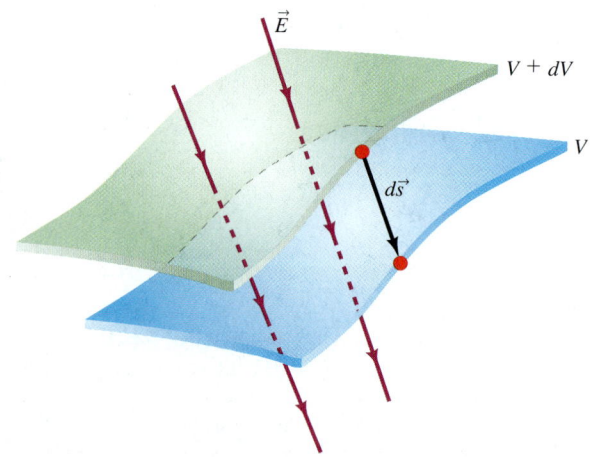

▶ **FIGURE 24–14** Two equipotentials differ by dV. The displacement $d\vec{s}$ along the direction of $\vec{E}$ between the equipotentials is perpendicular to them.

How the Potential Determines the Field in Cartesian Coordinates

For a different point of view, suppose that an arbitrary displacement vector $d\vec{s}$ is decomposed into Cartesian coordinates:

$$d\vec{s} = dx\,\hat{i} + dy\,\hat{j} + dz\,\hat{k}.$$

Here $\hat{i}$, $\hat{j}$, and $\hat{k}$ are the unit vectors in the x-, y- and z-directions, respectively. Then the scalar product in Eq. (24–25) takes the form

$$dV = -\vec{E} \cdot d\vec{s} = -E_x\,dx - E_y\,dy - E_z\,dz, \qquad (24\text{--}27)$$

where we have separated the field $\vec{E}$ into its Cartesian components. In general, the potential depends on all three space coordinates, $V = V(x, y, z)$. The change in V in going from an initial position $\vec{r} = x\hat{i} + y\hat{j} + z\hat{k}$ to a new position $\vec{r} + d\vec{s} = (x + dx)\hat{i} + (y + dy)\hat{j} + (z + dz)\hat{k}$ is

$$dV = \frac{\partial V}{\partial x}dx + \frac{\partial V}{\partial y}dy + \frac{\partial V}{\partial z}dz. \qquad (24\text{--}28)$$

Note the use of the partial derivatives here; this is necessary because V depends on all three Cartesian coordinates. Recall that partial derivatives are simple to use: The partial derivative with respect to x means that y and z are held fixed while the ordinary derivative with respect to x is taken. To illustrate, if $V = xz^2$, then $\partial V/\partial x = z^2$, $\partial V/\partial y = 0$, and $\partial V/\partial z = 2zx$. We can equate the coefficients of dx, dy, and dz in Eqs. (24–27) and (24–28):

$$E_x = -\frac{\partial V}{\partial x}, \quad E_y = -\frac{\partial V}{\partial y}, \quad E_z = -\frac{\partial V}{\partial z}.$$

Equivalently, *the electric field vector is given in terms of derivatives of the electric potential* by

$$\vec{E} = -\hat{i}\frac{\partial V}{\partial x} - \hat{j}\frac{\partial V}{\partial y} - \hat{k}\frac{\partial V}{\partial z}. \qquad (24\text{--}29)$$

THE FIELD IN TERMS OF THE POTENTIAL

Equation (24–29) gives the Cartesian components of the electric field in terms of the potential. We have found a way to express a particular vector, the electric field, in terms of the derivatives of a scalar, the electric potential. It can be simpler to calculate the electric field of a charge distribution by first finding the potential and then differentiating, rather than doing a vector integration to get the field.

EXAMPLE 24–7 Use the electric potential of a point charge q to find its electric field.

Setting It Up Specifying the electric field means finding both its magnitude and its direction.

Strategy From the discussion above, we know that the direction of the electric field is perpendicular to the equipotential surfaces, pointing from a region of higher to lower potential. Here the potential is a function only of the radial distance from the charge, $V = q/4\pi\varepsilon_0 r$. The equipotential surfaces are therefore spheres at a constant distance from the charge. This means that the electric field lines point in the radial direction, and they point from higher potential (smaller r) to lower potential (larger r), that is, outward. We can find the magnitude of the field by using Eq. (24–26).

Working It Out We know already that the direction is the *outward radial direction*, and the magnitude is

$$E = -\frac{dV}{dr} = -\frac{q}{4\pi\varepsilon_0}\frac{d}{dr}\left(\frac{1}{r}\right) = \frac{q}{4\pi\varepsilon_0 r^2}.$$

Although we already know this answer from previous discussion, the technique is useful in contexts where we do not already know the answer!

24–5 The Potentials of Charge Distributions

We rarely deal with the electric field and potential of a single point charge. More often, we have collections of charges spread over regions of space, as when charges spread over the surface of a metal, or when the field of a complicated ionic molecule determines its chemical or biological behavior. We must therefore be able to find the potentials of continuous charge distributions. These charge distributions may not be simple ones, and we must develop strategies for calculating the corresponding electric potentials. In this section, we first summarize the underlying techniques, then illustrate them with a series of examples.

The qualitative shapes of equipotential surfaces due to a charge distribution are most easily found by graphical techniques. For quantitative calculations, we have learned two different ways to determine the electric potential of a charge distribution:

1. If the electric field is known, then Eq. (24–9) can be used to determine the potential:

$$\Delta V = V_b - V_a = -\int_{r_a}^{r_b} \vec{E} \cdot d\vec{s}.$$

2. If the electric field is not known, we may calculate the potential directly by using one of various forms:

for one point charge, Eq. (24–7): $\quad V = \dfrac{q}{4\pi\varepsilon_0 r};$

for many point charges, Eq. (24–11): $\quad V = \dfrac{1}{4\pi\varepsilon_0} \sum_i \dfrac{q_i}{r_i};$

for a continuous charge distribution, Eq. (24–12): $\quad V = \dfrac{1}{4\pi\varepsilon_0} \int \dfrac{dq}{r}.$

In a direct calculation of electric potential, we must decide the location of zero potential. In fact, the convention that zero potential is at infinity is already implicit in Eqs. (24–7), (24–11), and (24–12), and is almost always the most convenient choice for a charge distribution that does not extend all the way to infinity. If a potential *difference* is calculated directly, no decision need be made about the zero level.

Examples

Let's first look at the relation between electric field and potential for two parallel conducting plates (a *parallel-plate capacitor*, a circuit device to be revisited in Chapter 25), each brought to different potentials (Fig. 24–15a). We suppose that the plates are close enough together or large enough that we can ignore the distortions of the field near the edges, as shown in Fig. 24–15b. In Fig. 24–15a, the left-hand plate is at a lower potential than the right-hand plate. The electric field between parallel plates is known to be constant, and it runs from regions of higher potential to lower potential—from right to left in this case. In Eq. (24–9), we take the path to be a straight line from the left to the right plate, so that $\vec{E}$ is antiparallel to $d\vec{s}$. We will let the direction from left to right define the x-axis, with the left plate at $x = 0$. Then the potential difference between the plates is given in terms of the field E between the plates and the separation ℓ of the plates by

$$\Delta V = V_{\text{right}} - V_{\text{left}} = -\int_{\text{left}}^{\text{right}} \vec{E} \cdot d\vec{s} = +E \int_0^\ell dx = E\ell,$$

and the electric field between parallel plates has magnitude

$$E = \frac{\Delta V}{\ell}. \tag{24–30}$$

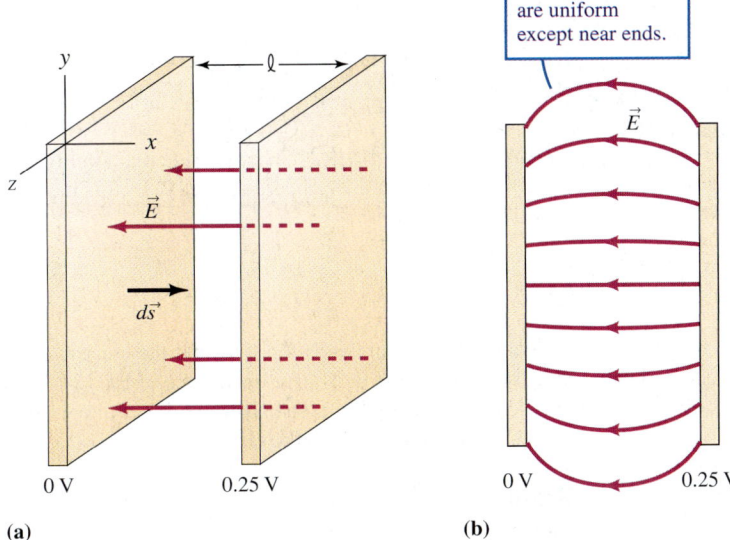

▲ **FIGURE 24–15** (a) Two parallel plates, viewed from the side, with a potential difference between them. A differential displacement $d\vec{s}$ is indicated. (b) The fields do show distortions at the edges for a real (not infinitely large) pair of plates.

Equation (24–30) and the preceding discussion yield an important practical result:

> **The magnitude of the constant electric field between parallel conducting plates is the potential difference between the plates divided by the distance between the plates. It points from the higher to the lower potential plate.**

This result gives the electric field between two parallel-plane conducting plates whose potential difference is ΔV. We can also use it to find the equipotential surfaces associated with *any* constant field. These surfaces are planes perpendicular to the field, and the potential difference for a plane a distance ℓ from a reference equipotential changes *linearly* with ℓ, $\Delta V = E\ell$. Note the sign: The potential decreases along the direction in which the electric field points.

EXAMPLE 24–8 Two parallel metal plates carrying equal and opposite charges have area $A = 225$ cm² and are separated by 0.50 cm. There is a potential difference of 0.25 V between them. Find the numerical value of the electric field. What is the charge density and total charge on each plate? Draw the equipotential surfaces at 0.10 V and 0.20 V.

Setting It Up Figure 24–15a labels axes and places the plates within them. We call ΔV the given potential difference between the plates and ℓ the known separation.

Strategy Equation (24–30) directly applies here and allows us to calculate the magnitude of the electric field. The field points from the higher to the lower potential plate. Given the electric field, we can calculate the charge density, since $E = \sigma/\varepsilon_0$, and given the area, we can calculate the total charge on each plate.

Working It Out
$$E = \frac{\Delta V}{\ell} = \frac{0.25 \text{ V}}{0.0050 \text{ m}} = 50 \text{ V/m},$$

and it points from right to left in Fig. 24–16. Since the electric field between the plates is σ/ε_0,

$$\sigma = E\varepsilon_0 = (50 \text{ V/m})(8.85 \times 10^{-12} \text{ C}^2/\text{N} \cdot \text{m}^2)$$
$$= 4.4 \times 10^{-10} \text{ C/m}^2.$$

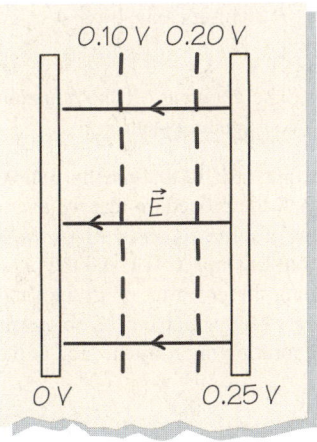

▲ **FIGURE 24–16** The electric field for parallel plates points from right to left when the potential of the left-hand plate is lower than that of the right-hand plate.

(continues on next page)

The answer is in coulombs per square meter—SI units. Given the area of the plates, the magnitude of the total charge on each plate is

$$Q = \sigma A = (4.4 \times 10^{-10} \text{ C/m}^2)(225 \text{ cm}^2)(10^{-4} \text{ m}^2/\text{cm}^2)$$
$$= 1.0 \times 10^{-11} \text{ C}.$$

The electric field is constant between the parallel plates. Thus, from Eq. (24–30), we find that at a distance d from the left plate the potential differs by an amount $\Delta V = Ed$ from its value at the left plate, $d = 0$. The equipotential surface for 0.10 V is then

$$d = \frac{\Delta V}{E} = \frac{0.10 \text{ V}}{50 \text{ V/m}} = 0.20 \text{ cm}$$

from the left plate. For 0.20 V, we determine a distance of 0.40 cm (Fig. 24–15b).

What Do You Think? The left-side plate in Fig. 24–15 is at a lower potential than the right-side plate. Does this mean that it has a negative charge on it?

EXAMPLE 24–9 *The Charged Ring.* Find the electric potential due to a uniformly charged ring of radius R and total charge Q at a point P on the axis of the ring.

Setting It Up The charge distribution is shown in Fig. 24–17, together with point P and a distance labelled r.

Strategy Symmetry plays an important role in the solution, with each point on the ring the same distance $r = \sqrt{R^2 + x^2}$ from the point P. The potential due to an infinitesimal section of the ring that carries charge dq is given by $dV = \dfrac{dq}{4\pi\varepsilon_0}\dfrac{1}{\sqrt{R^2 + x^2}}$. An integration of this yields the answer—the integration is simple because of the fact that each point on the ring is the same distance from P.

Working It Out The integral expressing the potential is

$$V = \int dV = \int \frac{dq}{4\pi\varepsilon_0} \frac{1}{\sqrt{R^2 + x^2}} = \frac{1}{4\pi\varepsilon_0} \frac{1}{\sqrt{R^2 + x^2}} \int dq$$
$$= \frac{Q}{4\pi\varepsilon_0} \frac{1}{\sqrt{R^2 + x^2}}. \tag{24-31}$$

In the second to last step, we have removed the constant distance r from the integral, leaving an integral over dq, which gives the total charge Q.

We could use our result for V to find the electric field along the axis by applying the derivative operations of Eq. (24–29) (see Problem 43). This method is easier than the direct integration technique presented in Chapter 22 for the electric field.

What Do You Think? What is the potential at the center of the ring? Interpret your answer.

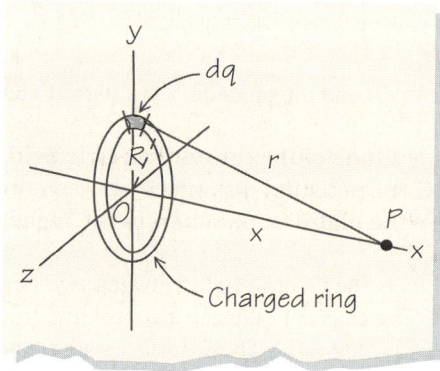

▲ **FIGURE 24–17** Geometry to find the potential at a point P on the axis of a charged ring of radius R by using a differential charge dq.

EXAMPLE 24–10 *The Charged Disk.* Find the electric field due to a thin, flat, uniformly charged disk of radius R and total charge Q at a point P along its axis, by first calculating the electric potential at this point.

Setting It Up The geometry of the situation is shown in Fig. 24–18, including a coordinate system.

Strategy The first task is to find the potential. Figure 24–18 contains in a nutshell the procedure that we can use: We divide the disk up into a series of thin concentric rings. We know the potential for a single ring from Example 24–9, and the potential of the disk is obtained by summing the potential of all the rings. We first require the charge on a single ring, which can be obtained by calculating the charge density and multiplying it by the area of the strip.

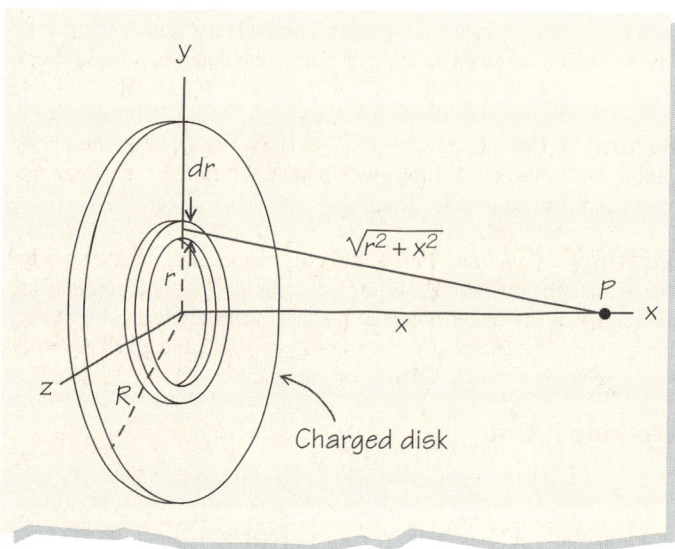

▶ **FIGURE 24–18** Geometry to find the potential at a point P on the axis of a charged disk of radius R. The potential due to the ring of radius r and width dr is first found and then integrated.

For the second part of the problem we can find the field given the potential by taking the appropriate derivative of the potential. Symmetry is helpful here: You can see that along the axis the electric field can only have an x-component. The reason is that a component in a direction perpendicular to the x-axis coming from any point on any of the concentric rings will have an equal and opposite component coming from the point on the ring at the opposite side of that ring.

Working It Out The charge density is $\sigma = Q/\pi R^2$. The area of a ring with inner and outer radius r and $r + dr$ respectively is $dA = \pi(r + dr)^2 - \pi r^2 \cong 2\pi r\, dr$ for infinitesimal dr and the charge on the ring is then $\sigma(2\pi r\, dr)$. From Example 24–9, the potential due to the ring on the axis is

$$dV = \frac{\sigma(2\pi r\, dr)}{4\pi\varepsilon_0} \frac{1}{\sqrt{r^2 + x^2}}$$

The potential for the whole disk is obtained by summing the potentials from all the rings, i.e. integrating this result from $r = 0$ to $r = R$:

$$V = \int dV = \int_0^R \frac{\sigma}{4\pi\varepsilon_0} \frac{1}{\sqrt{r^2 + x^2}} (2\pi r\, dr)$$

$$= \frac{\sigma}{2\varepsilon_0} \int_0^R \frac{r\, dr}{\sqrt{r^2 + x^2}}$$

$$= \frac{\sigma}{2\varepsilon_0} \sqrt{r^2 + x^2}\bigg|_0^R = \frac{\sigma}{2\varepsilon_0}\left(\sqrt{R^2 + x^2} - x\right)$$

$$= \frac{Q}{2\pi\varepsilon_0 R^2}\left[\sqrt{R^2 + x^2} - x\right]. \qquad (24\text{-}32)$$

In turn, the only component of the electric field is

$$E_x = -\frac{\partial V}{\partial x} = -\frac{Q}{2\pi\varepsilon_0 R^2}\left(\frac{x}{\sqrt{R^2 + x^2}} - 1\right).$$

What Do You Think? The result for the x-component of the field for $x \gg R$ is (a) $\sigma x/(2\varepsilon_0)$; (b) $\sigma/(2\varepsilon_0 x)$; (c) $Q/(4\pi\varepsilon_0 x)$; (d) $(Q/4\pi\varepsilon_0 x^2)$?

EXAMPLE 24–11 *The Charged Line.* Find the electric potential as a function of the radial distance R from an infinite charged line of uniform charge density λ.

Strategy We found the field for this charge configuration earlier [Eq. (22–32)], and we can here use Eq. (24–9) to find the potential from the electric field. Since the electric field has only a radial component, we integrate along a radial direction in Eq. (24–9), that is, $d\vec{s} = d\vec{r}$ in Fig. 24–19a.

Working It Out Equation (24–9) becomes

$$\Delta V = -\int E_r\, dr = -\frac{\lambda}{2\pi\varepsilon_0} \int \frac{dr}{r}.$$

The potential difference depends on the end points of the integration. Let zero potential be at $r = a$, so that

$$\Delta V = V_R - V_a \equiv V = -\frac{\lambda}{2\pi\varepsilon_0} \int_a^R \frac{dr}{r} = -\frac{\lambda}{2\pi\varepsilon_0} \ln r\bigg|_a^R,$$

or

$$V = -\frac{\lambda}{2\pi\varepsilon_0} \ln \frac{R}{a}. \qquad (24\text{-}33)$$

Note that it is not possible to set zero potential at infinity in this case because the logarithm is infinite at $a = \infty$. Physically, this is because the line itself reaches to infinity. We graph the potential of Eq. (24–33) in Fig. 24–19b, assuming that the charge of the line is positive.

What Do You Think? It is not possible to do this calculation directly by putting the charge along the z-axis, then calculating the potential by integrating the potential $\lambda\, dz/\left(4\pi\varepsilon_0\sqrt{R^2 + z^2}\right)$ due to a charge $\lambda\, dz$ a distance $\sqrt{R^2 + z^2}$ away from the point P. Why?

▶ **FIGURE 24–19** (a) An infinite charged line has a radial electric field. To find the potential at the point P, we consider a displacement $d\vec{s}$ in the direction of the electric field $\vec{E}$ and use the known expression for the electric field. (b) The resulting potential, defined to be zero at $r = a$, goes to positive infinity at $r = 0$ and continues to negative infinity for large r.

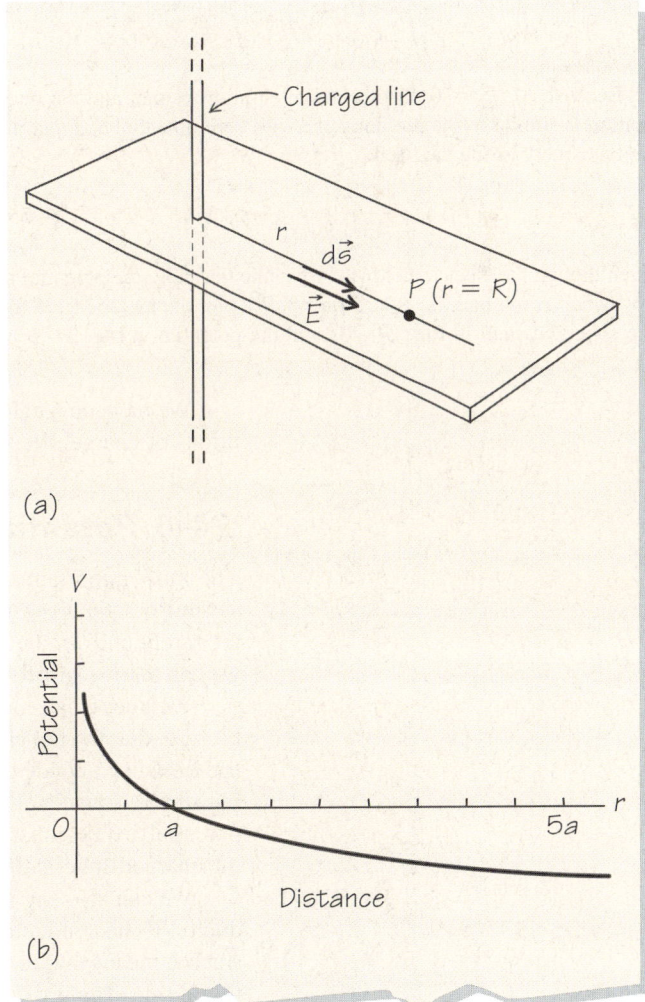

EXAMPLE 24–12 *The Charged Spherical Shell.* Find the potential for a uniformly charged spherical shell of total charge Q and radius R at positions both inside and outside the shell. Set zero potential at infinity.

Strategy As in the previous example, we can best work backwards from the electric field to the potential. We already know the electric field of the spherical shell from Example 23–4 and can use Eq. (24–9) to determine the electric potential, or more precisely a potential difference. But as zero potential is set at infinity, the potential $V(r)$ is the potential difference $\Delta V = V(r) - V(\infty)$. From Example 23–6, we have

outside a spherical shell, $r > R$: $\quad E = \dfrac{Q}{4\pi\varepsilon_0 r^2}$;

inside a spherical shell, $r < R$: $\quad E = 0$.

We must integrate these to find the potential.

Working It Out The electric field is purely radial, so by choosing a radial path from ∞ to a radial position r, we have in Eq. (24–9) $\vec{E} \cdot d\vec{s} = E\, dr$. Then for a point *outside* the spherical shell,

$$V(r) = V(r) - V(\infty) = -\int_\infty^r E\, dr = -\frac{Q}{4\pi\varepsilon_0}\int_\infty^r \frac{dr}{r^2}$$

$$= \frac{Q}{4\pi\varepsilon_0}\left(\frac{1}{r} - \frac{1}{\infty}\right) = \frac{Q}{4\pi\varepsilon_0 r} \qquad (24\text{–}34)$$

If r is *inside* the shell, we have

$$V(r) = -\int_\infty^R E_{\text{outside}}\, dr - \int_R^r E_{\text{inside}}\, dr.$$

Because $E_{\text{inside}} = 0$, the second integral drops out, and the integration is similar to the previous one with zero potential again at infinity, namely inside the shell,

$$V(r) = \frac{Q}{4\pi\varepsilon_0 R} = \text{a constant.} \qquad (24\text{–}35)$$

Even though the electric field is zero inside the shell, the potential is not (when zero potential is at infinity). We plot the electric field for the spherical shell in Fig. 24–20a and the potential in Fig. 24–20b.

What Do You Think? How is the fact that the field is zero inside the shell consistent with a potential that is not zero?

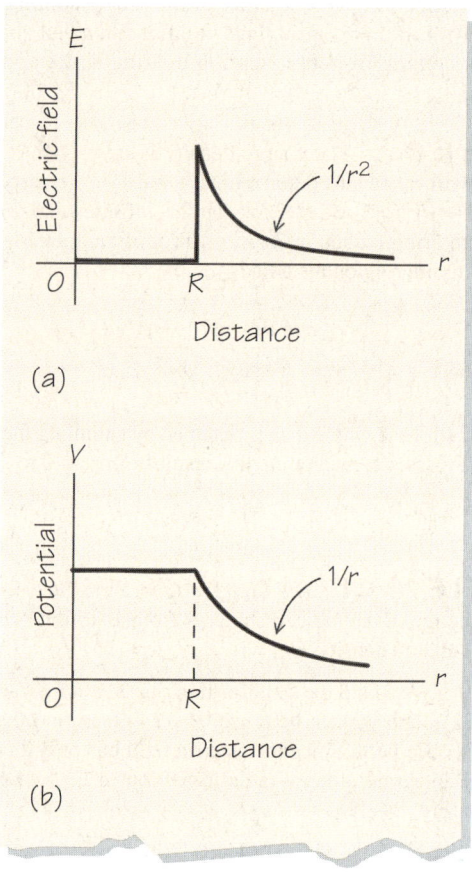

▲ **FIGURE 24–20** (a) The electric field and (b) electric potential for a spherical shell of radius R. Even though the electric field is zero inside the shell, the potential has a constant value equal to that on the shell's surface.

We have now calculated the electric field and potential for several different distributions of charge. We summarize these results in Table 24–1.

24–6 Potentials and Fields Near Conductors

The most important cases of continuous charge distributions are those on metals; for several practical examples you can glance ahead to Section 24–7. These distributions are rarely uniform because charges are free to move on and within metals. Despite this, we can learn a surprising amount about the electric potentials near metals.

We know that, in electrostatics, the electric field inside a conducting material must be zero, that the net charge on a conductor lies on the outside surface, and that the electric field just outside that surface must be normal to the surface. Moreover, the conducting surface must itself be an equipotential and, as there is a zero electric field inside, the potential inside must have the same value as it has at the surface. Example 24–12, the charged spherical shell, provides an illustration of these features.

We can also say something about properties of the electric potential outside a conductor. If the conductor is charged, the fact that the electric field is perpendicular to the surface means that the equipotentials near the surface will be parallel to the surface, and this is true even if the conductor is uncharged. Consider, for example, the constant electric field shown in Fig. 24–21a due to two parallel plates (not shown). If we place an uncharged conductor of arbitrary size in this electric field, the field around the conductor

TABLE 24–1 • Electric Fields and Potentials for Various Charge Configurations

Charge Configuration	Magnitude of Electric Field	Electric Potential	Location of Zero Potential
Point charge	$\dfrac{q}{4\pi\varepsilon_0 r^2}$	$\dfrac{q}{4\pi\varepsilon_0 r}$	∞
Infinite line of uniform charge density λ	$\dfrac{\lambda}{2\pi\varepsilon_0 r}$	$-\dfrac{\lambda}{2\pi\varepsilon_0} \ln \dfrac{r}{a}$	$r = a$
Parallel, oppositely charged plates of uniform charge density σ, separation d	$\dfrac{\sigma}{\varepsilon_0}$	$\Delta V = -Ed = -\dfrac{\sigma d}{\varepsilon_0}$	anywhere
Charged disk of radius R, along axis at distance x	$\dfrac{Q}{2\pi\varepsilon_0 R^2}\left(\dfrac{\sqrt{R^2 + x^2} - x}{\sqrt{R^2 + x^2}}\right)$	$\dfrac{Q}{2\pi\varepsilon_0 R^2}\left(\sqrt{R^2 + x^2} - x\right)$	∞
Charged spherical shell of radius R	$r \geq R: \dfrac{Q}{4\pi\varepsilon_0 r^2}$ $r < R: 0$	$r \geq R: \dfrac{Q}{4\pi\varepsilon_0 r}$ $r \leq R: \dfrac{Q}{4\pi\varepsilon_0 R}$	∞ ∞
Electric dipole	Along bisecting axis only, far away: $\dfrac{p}{4\pi\varepsilon_0 r^3}$	Everywhere, far away: $\dfrac{p\cos\theta}{4\pi\varepsilon_0 r^2}$	∞
Charged ring of radius R, along axis	$\dfrac{Qx}{4\pi\varepsilon_0(R^2 + x^2)^{3/2}}$	$\dfrac{Q}{4\pi\varepsilon_0\sqrt{R^2 + x^2}}$	∞
Uniformly charged nonconducting solid sphere of radius R	$r \geq R: \dfrac{Q}{4\pi\varepsilon_0 r^2}$ $r < R: \dfrac{Qr}{4\pi\varepsilon_0 R^3}$	$r \geq R: \dfrac{Q}{4\pi\varepsilon_0 r}$ $r < R: \dfrac{Q}{8\pi\varepsilon_0}\left(3 - \dfrac{r^2}{R^2}\right)$	∞ ∞

will be greatly modified (Fig. 24–21b). Charge will be induced on the outside of the conductor in equilibrium, thereby forcing the electric field to be normal to the conducting surface and, again, *the equipotential surfaces near a conductor of arbitrary shape are parallel to the conductor's surface.*

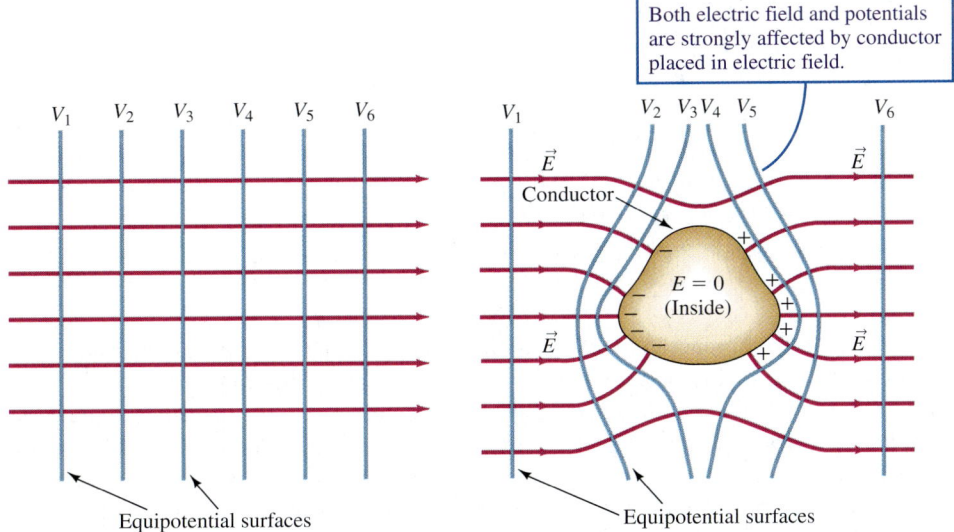

◀ **FIGURE 24–21** (a) A uniform electric field before an uncharged conductor is placed in the field. (b) Afterward, the electric field is changed dramatically, with no electric field inside the conductor. Induced charges, which make the electric field inside the conductor zero, appear on the outside surface of the conductor. These charges affect the electric field outside the conductor.

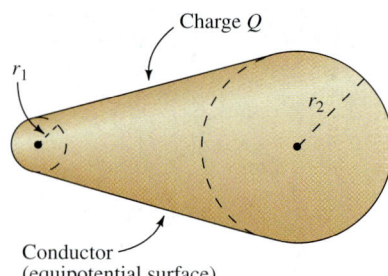

FIGURE 24–22 A conductor of irregular shape can be modeled by spheres of radii r_1 and r_2 at its ends.

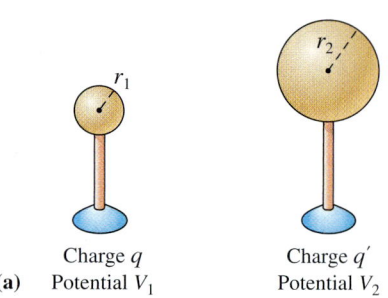

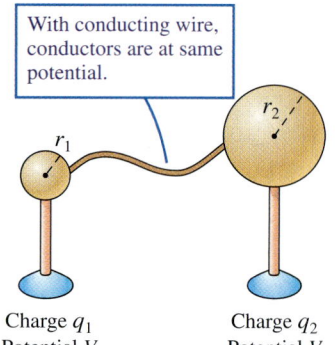

FIGURE 24–23 (a) Two conductors initially are at different potentials. (b) If the conductors are connected by a wire, charge must flow to make the potential equal everywhere throughout the connected conductors.

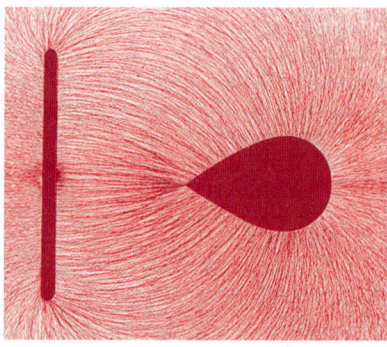

FIGURE 24–24 The electric field near small radii of curvature can be quite large, as seen for the point of this charged object.

There is a last important result to consider: Charge density on a conductor will vary if the surface has an irregular shape. In the next subsection we shall see how the concept of potential allows us to say more about the charge density and hence the fields near irregularly shaped charged conductors.

The Role of Sharp Points on Conducting Surfaces

Many real conductors have irregular shapes and the shape of a conductor affects the electric field in its vicinity. Consider the irregular conductor shown in Fig. 24–22. To simplify its analysis, we can think of the two ends shown as approximated by sections of spheres with radii r_1 and r_2, respectively. The left region is more sharply curved than the right region, so $r_1 < r_2$. We model this conductor with a two-step process; for the first step, we consider the field and potential for the two spherical conductors shown in Fig. 24–23a. The sizes of these spheres match the two ends of our irregular conductor (Fig. 24–22). The charges q and q' are placed on the two spheres, and the electric potentials at the spheres are, respectively,

$$V_1 = \frac{q}{4\pi\varepsilon_0 r_1} \quad \text{and} \quad V_2 = \frac{q'}{4\pi\varepsilon_0 r_2}.$$

If we now connect the two spheres by a long conducting wire (Fig. 24–23b), charge will flow between the two spheres, and the entire system will come to the same potential. This potential is

$$V = \frac{q_1}{4\pi\varepsilon_0 r_1} = \frac{q_2}{4\pi\varepsilon_0 r_2},$$

where q_1 and q_2 are the equilibrium charges on the two spheres. Since the entire connected object is at a single potential, these charges and radii must be related by

$$\frac{q_1}{r_1} = \frac{q_2}{r_2}.$$

In this discussion we did not use the fact that the spheres are connected by a thin wire, only that they are connected; this is the information that ensures that the entire system is at a single potential. Therefore this calculation will apply to our irregular conductor (Fig. 24–22). In particular, the connected spheres form a model for the relative size of the electric fields at the two ends of the conductor.

Let us now relate what we have done to the surface charge density σ. For a sphere of radius r this quantity is the charge q on the sphere divided by the surface area: $\sigma = q/(\pi r^2)$, or $q = \sigma\pi r^2$. For our two connected spheres, therefore, the equation $q_1/r_1 = q_2/r_2$ is

$$\frac{\sigma_1 \pi r_1^2}{r_1} = \frac{\sigma_2 \pi r_2^2}{r_2};$$

$$\sigma_1 r_1 = \sigma_2 r_2. \quad (24\text{–}36)$$

The electric field E_i just outside each conducting sphere has magnitude σ_i/ε_0, so that we can replace σ_i by $\varepsilon_0 E_i$. Thus $\varepsilon_0 E_1 r_1 = \varepsilon_0 E_2 r_2$, or

$$\frac{E_1}{E_2} = \frac{r_2}{r_1}. \quad (24\text{–}37)$$

The electric fields are inversely related to the radii, as are the surface charge densities. We can say that at a region of the charged conductor's surface that can be inscribed with a sphere of a small radius—a region of small radius of curvature—*the surface charge density and the corresponding electric field are larger than at a region of large radius of curvature.*

The effect just described is important if a conductor has a sharp point, because a sharp point is a region with a very small radius of curvature; the sharper the point, the more intense the field (Fig. 24–24). Even if the conductor is at a low electric potential, there will be intense fields near sharp points.

THINK ABOUT THIS...

HOW DO LIGHTNING RODS WORK?

Observations show that there is an electric potential over much of the land and seas of Earth that increases with height at a rate of about 100 V/m; that is, there is an electric field that points downward and has a nearly constant strength of about 100 V/m. (To be more exact, the magnitude of the field decreases slowly till it is almost zero at about 50–100 km above the surface of Earth.) Depending on weather conditions, the electric field can be very much larger—a lightning strike is preceded by the presence of large fields. Air is a fairly good insulator; what, then, conducts electricity in air? Molecules of nitrogen and oxygen are occasionally *ionized*—electrons are removed from them—by collisions with incoming *cosmic rays*, energetic particles that come from beyond the solar system. The resulting ratio of ions (positive and negative) to molecules that are not ionized is about 1 in 10^{16}, and it is these charged particles that conduct electricity in the atmosphere.

We know that there are large electric fields near the pointed parts of a charged conductor, and ions of the opposite charge are attracted to this part of the conductor while ions of the same charge are repelled. The positive and negative ions may be strongly accelerated, and ionize other neutral molecules in their path. In this way the conductivity of the air near the pointed conductor increases measurably, and a flow of charge acts to cancel the field. On a small scale this gives rise to sparks (see Fig. 24–25); a more massive charge transfer produces a *corona discharge*. Sailors long ago saw the glow of corona discharge at the pointed tops of their masts and spars and dubbed the phenomenon *St. Elmo's fire* (see Fig. 24–26). More generally we speak of *dielectric breakdown*.

In electric storms clouds acquire a large charge by mechanisms that are not entirely understood,[†] and this gives rise to high fields extending over large regions. An ionization cascade can develop from small seeds of ionized material in the atmosphere and result in massive discharges from clouds to Earth or from cloud to cloud. *Lightning* (see the chapter-opening photo) is an example of such a large-scale discharge. When lightning strikes, the passage of electric charge heats the air along its path. This causes a rapid expansion of the air, and surrounding air rushes back into the partial vacuum, creating the sound that we term *thunder*.

If there is a pointed, grounded conductor (lightning rod) beneath the highly charged cloud, then the strong field at the point acts to lead the charged particles to the rod. More important than providing a convenient path for a lightning stroke, the steady discharge from the rod will serve to decrease the potential difference between the rod and the clouds and discourage the formation of a catastrophic discharge in the vicinity.

▲ **FIGURE 24–25** The electric field at the tips of these fork tines is large enough to cause a discharge from them into the air in the form of many small sparks.

▲ **FIGURE 24–26** St. Elmo's fire at the end of the masts.

[†]A detailed discussion of electricity in the atmosphere may be found in Chapter 9, Vol. II of *Lectures on Physics*, by R. P. Feynman.

24–7 Electric Potentials in Technology

Electrostatics represents only a small portion of our study of electromagnetism, but it has important applications. We briefly mention a few of them here to indicate how the understanding of basic principles can be put to good practical use.

The Van de Graaff Accelerator

If we place a charge anywhere in a conductor, the charge will move to the outside surface, and the field inside the conductor will be zero. Robert Van de Graaff took advantage of this concept in 1931 to build an *accelerator*: an apparatus that produces highly energetic charged particles. Such particles are useful for microscopic probes of matter and as cancer treatments. Van de Graaff used a device similar in concept to the apparatus shown schematically in Fig. 24–27. An insulated belt (or chain) continuously brings charge to the inside of a hollow conductor, which then moves to the outside surface of the conductor. The electric potential on the spherical conducting surface increases as charge flows to its surface ($V = q/4\pi\varepsilon_0 R$). An *ion source* produces charged atoms whose sign is such as to be repelled from the region of high potential and thus accelerated. Such devices are called **Van de Graaff accelerators** or **Van de Graaff generators** (Fig. 24–27), and the beams they or other accelerators produce play an important role in modern technology—for example, such beams are used to make microcircuits.

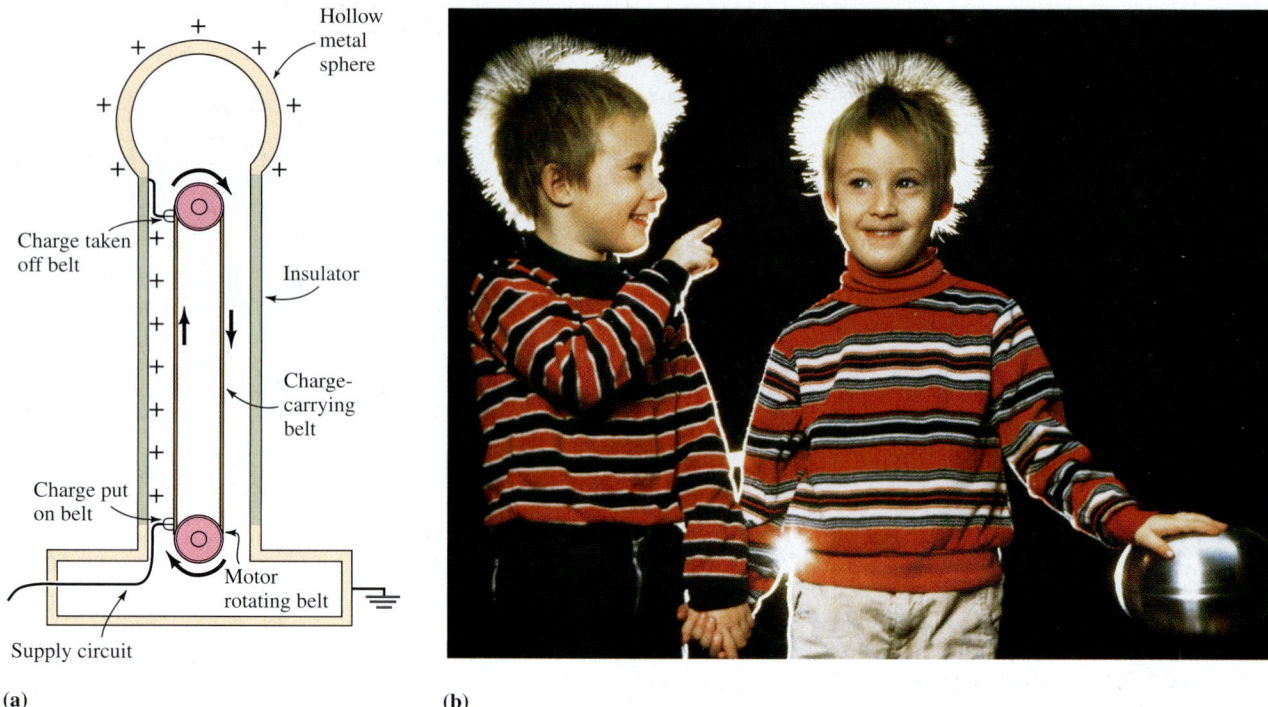

▲ **FIGURE 24–27** (a) Schematic diagram of a simple Van de Graaff accelerator. Charge is sprayed on the rotating belt at the bottom and taken off at the top. The charge goes to the outside surface of the conductor, and the potential continues to build up to high values. The symbol in the bottom right indicates that the base of the accelerator has been grounded. (b) The children touching this Van de Graaff generator are brought to a high electric potential. The individual hairs behave like the leaves of an electroscope.

The Field-Ion Microscope

The phenomenon that large electric fields occur at sharp points on a conductor is carried to its extreme in **field-ion microscopy**. The high electric fields involved in this technique allow us to produce images of individual atoms in the crystalline structure of the sharp point, or tip, of a metal. A fine tip of the crystalline material is prepared, commonly by dipping a mechanically formed tip in a substance that dissolves atoms off the end of the tip. These tips can be as small as 200 nm across, depending on the particular metal and the crystal being prepared. On the 200-nm scale, such a tip looks smooth, but it is still very rough at the atomic level. The tip is then introduced into a vacuum, and a large positive potential of several kilovolts is applied (Fig. 24–28). The end of the tip is thereby smoothed off even further as protruding groups of atoms of

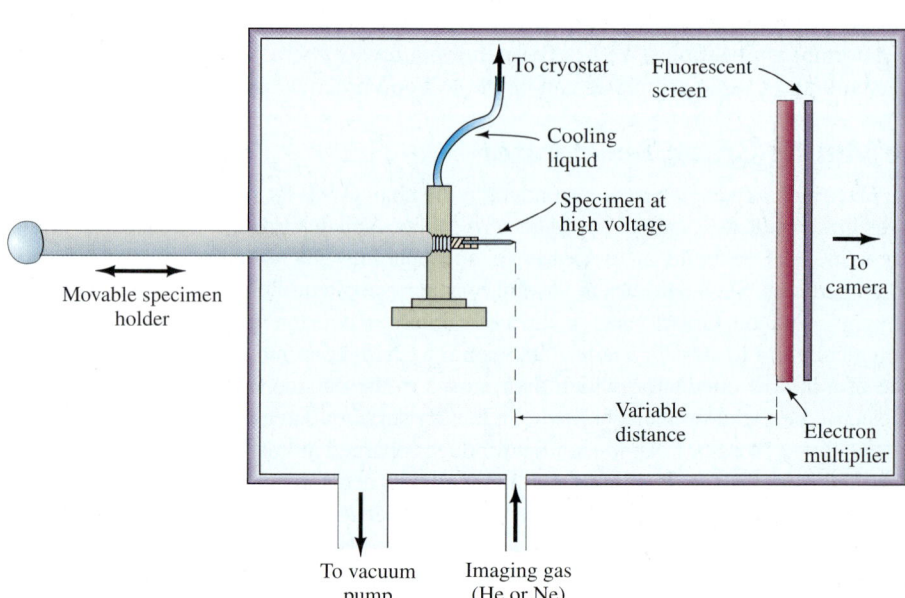

▶ **FIGURE 24–28** Schematic diagram of a field-ion microscope. The cryostat maintains a steady, low temperature in the chamber.

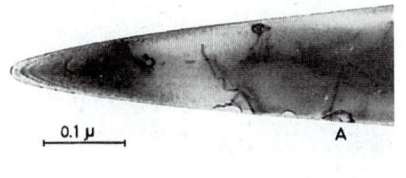
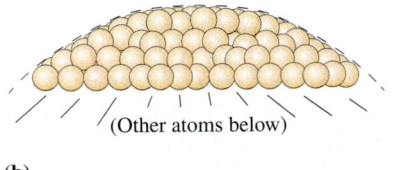

◀ **FIGURE 24–29** (a) Magnified iron tip. (b) The spheres represent individual atoms in a field-ion tip.

the metal itself are driven off in the form of positive ions by the large fields. This smoothing process leaves a tip like those in Figs. 24–29a and 24–29b. Oranges stacked in layers into a semipyramidal shape provide a familiar example of the possible structure of such a tip.

In the next stage, a dilute gas such as helium or neon—called the *imaging* gas—is introduced into the chamber that contains the tip. The positive potential on the tip is increased, again to several kilovolts, until the gas atoms just begin to be ionized. This happens only where the field is largest: right above individual atoms of the tip, where the field is strong enough to ionize the gas. The gas ions are then driven away from the tip, following the electric field lines out from the tip atoms to a grounded screen, where the impinging gas ions leave a visible trace. The image formed corresponds to the position of the individual atoms of the tip, which thus become visible in a picture such as Fig. 24–30. Field-ion microscopy is useful for observing crystal structures and the effects of impurities and defects in crystals. Even some noncrystalline materials can be investigated this way. Learning about the structure of materials is certainly one of the important steps in our ability to use them in applications.

▲ **FIGURE 24–30** This field-ion micrograph of a platinum crystal allows the atomic structure of the crystal to be determined from the many overlapping geometrical patterns. The magnification is 200,000.

Xerography

Photocopying machines take advantage of electrostatics in several steps of **xerography**, or photoreproduction. The process is illustrated in Figure 24–31. It begins with a positively charged plate coated with photoconducting material (a photoconductor is a good conductor in light but not in the dark, because light releases conducting electrons within it) such as selenium (Fig. 24–31a). Light reflected from the original to be copied passes through a lens onto the charged plate, where the dark areas remain charged, but charges flow to the plate underneath at the areas that receive light (Fig. 24–31b). The resulting image of the dark areas is represented by the remaining charges. Negatively charged toner (a black powder) is added to the positively charged plate, leaving the original dark areas with black toner on the plate (Fig. 24–31c). In the next step, paper that has also been positively charged is placed over the plate and attracts the black, negatively charged toner (Fig. 24–31d). Heat is used to fuse the toner (and thus the image) to the paper.

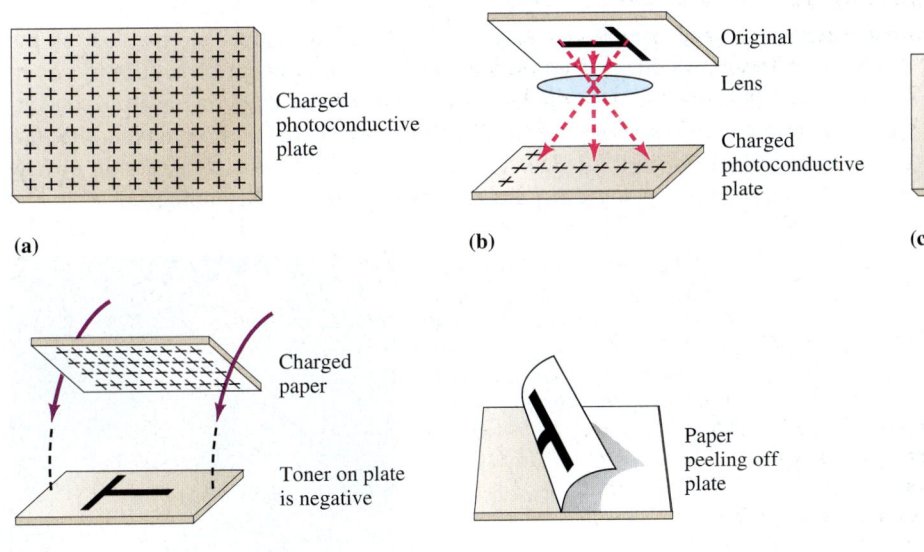

◀ **FIGURE 24–31** Schematic diagram of xerography. (a) A positively charged photoconductive plate. (b) Light from the white areas on the original neutralizes the positive charges on the plate. (c) The negatively charged toner is attracted to the positive charge. (d) The positively charged paper picks up the toner. Heat seals the toner on the paper, (e) which is then peeled off the plate.

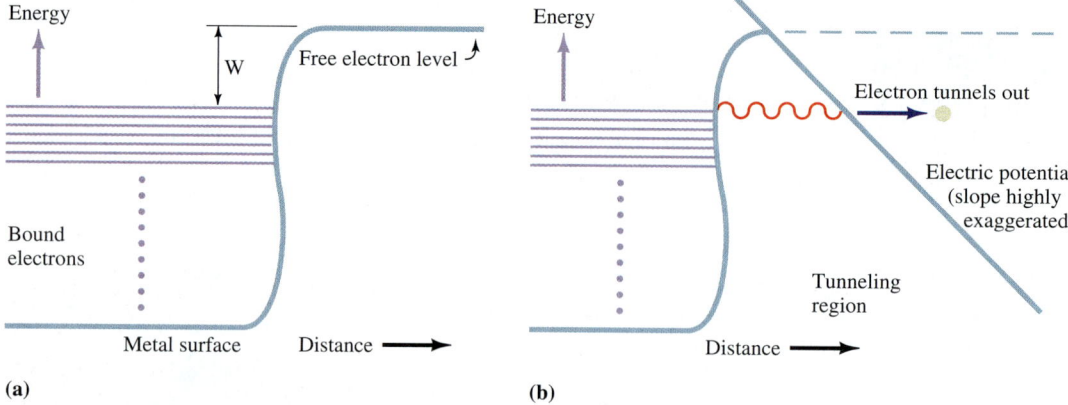

▲ **FIGURE 24–32** (a) Electrons in a metal take on a range of energies up to a highest level. This level is still negative compared to the energy of a free electron. It takes an energy W—typically 4 to 5 eV—to sufficiently raise the electron energy to allow the electron to leave the metal. The potential energy curve is drawn in blue. (b) When an external potential is applied to the metal, that potential plus the existing potential yields an effective potential barrier. Electrons can tunnel through this barrier.

Electric Potentials and Quantum Engineering

Electrons are trapped within a metal by electrostatic attractions to their parent ions. These forces can be represented in classical physics by a potential barrier that the electrons cannot cross. Figure 24–32a is an energy diagram of the potential energy of an electron as well as the electron's (constant) total energy. Classically, the electron cannot enter the region where its total energy is less than its potential energy.

A positively charged object brought near the metal surface pulls on the electrons. The positively charged object has an electric potential with respect to the metal, and an electron has a potential energy due to the external object (Fig. 24–32b). When the potential energy due to the external object is added to the original potential energy that holds the electron within the metal, the barrier is, in effect, reduced. Even if the external potential is too weak to lower the maximum potential energy below the electron's total energy and allow the electrons to escape classically, penetration through a barrier (*barrier tunneling*) is possible in quantum physics (Chapter 40). The tunneling is enhanced when the barrier is lowered. By tunneling, the electrons can emerge on the other side of the barrier and escape the electrostatic attraction of the parent ion, though many of them are absorbed on the way. This effect is utilized in the *scanning tunneling microscope* (Fig. 24–33). A weak positive potential is placed on an ultrafine tungsten needle. The needle scans the surface of a sample and provides the necessary potential to help electrons escape the sample by making it easier for them to penetrate the barrier. These electrons are attracted to the needle and form a flow of charge (*electric current*) through it, whose magnitude depends on the distance between the needle and the surface. This effect is used in two ways.

▼ **FIGURE 24–33** Schematic diagram of a scanning tunneling microscope. The fine-tipped needle in the scanning head comes to within 1 nm of the sample; this distance is the tunnel gap. The tunneling current across this gap holds the gap distance constant as the tip scans the sample surface and thereby provides a map of that surface. The base voltage leads to the tunneling current, which can be used to form the driving voltage. The driving voltage moves the tip by means of piezoelectricity, a phenomenon in which a voltage applied to crystals distorts them in a predictable way.

1. A feedback mechanism that constantly repositions the needle can be set up so that the current is constant. The distance between the needle tip and the sample's surface is therefore constant. The repositioning can be measured, and the topography of the surface is thereby mapped (Fig. 24–34).

2. The potential on the needle can exert a slight pull on whole atoms. Just as the nonuniform field of a charged comb induces a dipole moment of neutral pieces of paper and attracts them, the needle tip induces a dipole moment on the atoms and attracts them out of the sample material. In this way, atoms can be moved *one at a time* to new positions (Fig. 24–35). This effect may allow the construction of new molecules and ultrasmall logic circuits (switching circuits in computers).

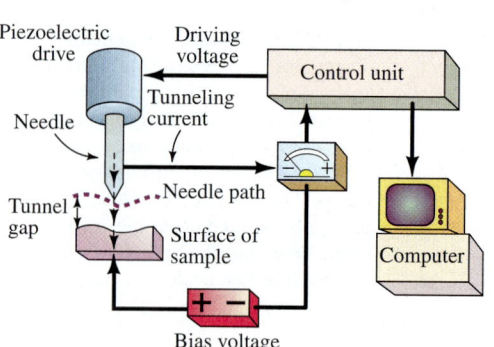

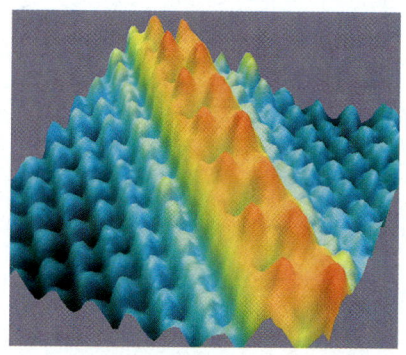

▶ **FIGURE 24–34** This is a scanning tunneling microscope (STM) image of a chain of cesium atoms (reddish yellow) on a gallium arsenide surface (blue). Each bump in this image represents a single atom.

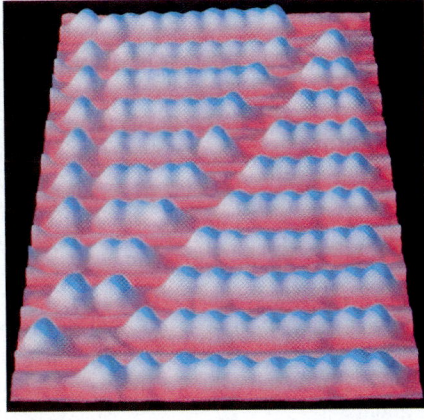

FIGURE 24–35 Individual atoms, whose size is less than a nanometer, have been moved one at a time to form the pattern shown. The vertical scale of the figure is enhanced by a factor of ten for a more dramatic effect.

The combination of electrostatics and quantum mechanics (through the use of the wavelike behavior of electrons) is an important tool in what is aptly called *quantum engineering*.

Summary

The Coulomb force is conservative, so a potential energy—electric potential energy—is associated with it. If a test charge q_0 moves from point a to point b in the presence of a point charge q at the origin, the change in potential energy is given by

$$\Delta U = \frac{qq_0}{4\pi\varepsilon_0}\left(\frac{1}{r_b} - \frac{1}{r_a}\right). \tag{24-4}$$

The electric potential difference due to any charge distribution between points a and b is defined as the change in potential energy divided by the magnitude of a test charge q_0:

$$\Delta V = \frac{U_b - U_a}{q_0} = -\int_{r_a}^{r_b} \vec{E} \cdot d\vec{s}. \tag{24-9}$$

Here $\vec{E}$ is the electric field due to the charge distribution. The integral in Eq. (24–9) is independent of the path between the end points. The potential difference $V_b - V_a$ is the work done per unit charge by an external agent in moving a test charge from point a to point b with no change in kinetic energy. The potential is independent of the test charge.

The electric potential can be determined by the following methods, in addition to graphical methods:

1. If the electric field is known, then Eq. (24–9) may be used.
2. If the electric field is not known, it is generally easier to calculate the potential directly by using one of these forms:

for one point charge:
$$V(r) = \frac{q}{4\pi\varepsilon_0 r}; \tag{24-7}$$

for many point charges:
$$V_P = \frac{1}{4\pi\varepsilon_0}\sum_{i=1}^{n}\frac{q_i}{r_i}; \tag{24-11}$$

for a continuous charge distribution:
$$V = \frac{1}{4\pi\varepsilon_0}\int\frac{dq}{r}. \tag{24-12}$$

In each of these cases, zero potential is chosen to be at infinity.

The SI unit of electric potential is the volt (V); 1 V = 1 J/C. A useful unit of energy for atomic and subatomic systems is the electron-volt (eV); 1 eV = 1.6×10^{-19} J.

If the potential is known, then the electric field can be determined in terms of derivatives of the potential:

$$\vec{E} = -\hat{i}\frac{\partial V}{\partial x} - \hat{j}\frac{\partial V}{\partial y} - \hat{k}\frac{\partial V}{\partial z}. \tag{24-29}$$

The electric field between two parallel plates is constant and is given by the potential difference divided by the distance between the plates:

$$E = \frac{\Delta V}{\ell}. \tag{24-30}$$

The electric field and potential for several charge configurations are given in Table 24–1.

Electric Potential

Equipotential surfaces are surfaces at a fixed potential. The electric field is perpendicular to equipotentials. The surfaces of conductors form equipotentials, and the potentials inside conductors in equilibrium are everywhere the same as the potential on the surface. Electric fields just outside conductors are inversely proportional to the radius of curvature, so there are high electric fields near sharp points on conductors even if the conductors are at low potentials.

Understanding the Concepts

1. How many joules are in 1 V·C?
2. An infinite plane is uniformly charged with positive charge of density σ. How would you use a known negative test charge to measure σ?
3. How would you create an electric field inside the hollow space of a spherical metal shell that is constant in magnitude and direction in a small region of the interior space?
4. In good weather, the electric field in the lower atmosphere is approximately 100 V/m, pointing downward. What happens when a 3-m metal rod is planted in the ground?
5. When an electric field moves a charge by doing work on it, what is the source of the energy to do the work? Where did this energy come from originally?
6. In describing the potential difference as the work per unit charge to move a test charge, we added the phrase "without changing its kinetic energy" (see the boldface statement on p. 687). Why is this important?
7. For a dipole $\theta = 0$ is a point on the axis, while $\theta = \pi/2$ is a point on the perpendicular bisector of the line between the two charges. Without looking at the equations, and given that one is at the same distance from the dipole center, at which angle is the field larger, and why?
8. Are the children in Fig. 24–27 grounded?
9. Will a conductor always be an equipotential? If not, under what circumstances will that occur?
10. Using Eq. (24–29), explain why changing the location of zero potential does not affect the value of the electric field.
11. A small Van de Graaff generator can be used as a lecture demonstration device. If a person touches the dome, his or her hair stands up (see Fig. 24–27b). Explain why. Why should the person stand on an insulated mat during this demonstration?
12. Earth is typically defined to be at zero potential with respect to infinity. Does this mean that Earth can have no net charge? If Earth does have a net charge, can it still be at zero potential?
13. If we know the electric potential at a certain point, do we also know the electric field? What can we know about the electric field if we know the electric potential at two points arbitrarily close to one another?
14. Examine Fig. 24–24. In what regions do the field lines seem to be denser? In what regions do you expect dense lines, and why?
15. Is the electric potential energy of a system of point charges independent of the order in which the system is assembled?
16. Why are there so many curved surfaces on the Van de Graaff generator?
17. How do we really know that electric forces are conservative?
18. In the potential associated with a point charge, we chose zero potential to be infinitely far from the charge. What would change in our predictions about electric charges if we had chosen the potential to be zero at $r = 10^{-10}$ m from the charge?
19. If we start with point charges, for each of which zero potential is at infinity, is it possible for a superposition of charges to have zero potential other than at infinity?
20. The potential of a configuration of point charges is zero at certain points. Does this mean that the force on a test charge is zero at these points?
21. Is it possible to arrange charges so that the potential is zero over a small but finite region?
22. Consider Fig. 24–10. What is the easiest way off High Peak? Think in particular about whether or not your path is perpendicular or not to the contour lines, and how the spacing between the contours affect your choice. Describe the relevance of your reasoning to a possible set of equipotentials for a two-dimensional distribution of charge.

Problems

24–1 Electric Potential Energy

1. (I) Two protons are separated by a nuclear diameter of 5×10^{-15} m. What is their mutual electrostatic energy?
2. (I) What is the electrical potential energy between the nucleus of a uranium atom (92 protons) and a single electron located 3×10^{-12} m from the nucleus?
3. (I) A charge of 7.0×10^{-7} C is fixed at the origin of a coordinate system. A charge of 3.0×10^{-6} C is placed on a raisin of mass 0.30 g. The raisin is then brought from far away to a point 20 cm from the origin. What is the electric potential energy of the system?
4. (I) Suppose that the raisin of Problem 3 is released from rest from its position 20 cm from the origin. If no other forces act on the raisin, where will it move? What will its final kinetic energy be?
5. (I) A 3-μC charge is brought in from infinity and fixed at the origin of a coordinate system. (a) How much work is done? (b) A second charge, of 5 μC, is brought in from infinity and placed 10 cm away from the first charge. How much work does the electric field of the first charge do when the second charge is brought in? (c) How much work does the external agent do to bring the second charge in if that charge moves with unchanging kinetic energy?
6. (I) Charges $q_1 = 2.0 \times 10^{-5}$ C and $q_2 = -9.0 \times 10^{-5}$ C are placed at rest 0.50 mm apart. How much work must be done by an outside agent to move these charges slowly and steadily until they are 0.05 mm apart?
7. (II) A positive charge of magnitude 3.0×10^{-6} C is placed 5.0 cm above the origin of a coordinate system, and a negative charge of the same magnitude is placed 5.0 cm below the origin, both on the z-axis. What is the potential energy of a positive charge of magnitude 0.20×10^{-6} C placed at the position $(x, y, z) = (30$ cm, 0 cm, 50 cm)? at (30 cm, 0 cm, 0 cm)?
8. (II) Repeat the calculation of Problem 7 for the case that (a) both charges on the z-axis are positive and the third charge is negative; (b) the signs and magnitudes of all charges are the same.
9. (II) A charge of 1.5 μC is placed at the point $x = 12, y = 25, z = 0$ (all distances given in centimeters). Calculate the work done in bringing a charge of -3 μC from $x = 12, y = 60, z = 50$ to the point $x = 12, y = 50, z = -25$, assuming that the charge is moved at a steady speed.
10. (II) Use potential energy arguments to show that charges of the same sign cannot form a system with a closed circular orbit.

24–2 Electric Potential

11. (I) Two equal charges of -4 mC are placed along the y-axis at -3 mm and 4 mm, respectively. Where is the electric potential zero?

12. (I) Two charges are placed along the x-axis; 3 μC at 14 cm, and -4 μC at 15 cm. Find those points along the x-axis where the potential is zero.

13. (I) The electrostatic potential of a single unknown charge 2.5 mm from that charge is 0.12 V. (The value of the potential at infinity is zero.) What is the value of the charge?

14. (I) A proton moves from point A to point B under the sole influence of an electric field, losing speed as it does so from $v_A = 8 \times 10^5$ m/s to $v_B = 5 \times 10^4$ m/s. What is the potential difference between the two points?

15. (I) An external force steadily moves a point charge of $+3.0 \times 10^{-7}$ C from a negatively charged to a positively charged plate. The plates are large and parallel, and the negatively charged plate is at a potential of $+3.0$ kV, whereas the positively charged plate is at a potential of $+17$ kV. How much work does the external force do?

16. (I) Consider two very long coaxial cylinders that carry opposite charges. The interior cylinder, negatively charged, is at a potential of $+16$ kV, whereas the exterior cylinder, positively charged, is at a potential of $+27$ kV. An external force steadily moves a point charge of -3×10^{-8} C from the negatively to the positively charged cylinder (Fig. 24–36). How much work does the external force do?

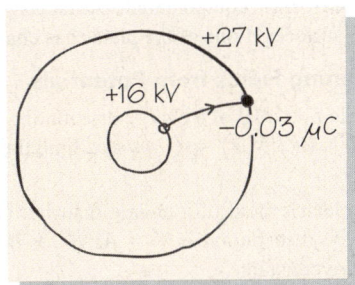

▲ **FIGURE 24–36** Problem 16.

17. (II) Three charges are at rest on the z-axis, $q_1 = 2$ mC at $z = 0$ m, $q_2 = 0.5$ mC at $z = 1$ m, and $q_3 = -1.5$ mC at $z = -0.5$ m. What is the potential energy of this system?

18. (II) Charges $+q$, $-q$, $+q$, and $-q$ are placed on successive corners of a square in the xy-plane. Plot all the locations in the xy-plane where the potential is zero.

19. (II) Consider two charges of $24 \times 10^{-2} \mu$C and $-10 \times 10^{-2} \mu$C, respectively, at opposite ends of a diameter of a circle of radius 25 cm (Fig. 24–37). (a) What is the potential on a point of the circle that is 30 cm from the positive charge? (b) How much work is required to bring a charge of -0.2 μC from infinity to that point on the circle?

20. (II) The origin of a coordinate system is at the intersection point of the perpendicular bisectors of the sides of an equilateral triangle of sides 3 cm. Calculate the potential at the origin due to three identical charges of 0.5 μC placed at the corners of the triangle.

21. (II) Consider a square of sides 14 cm. Charges are placed on the corners of the square as follows: 2 μC at (0 cm, 0 cm); -3 μC at (0 cm, 14 cm); 5 μC at (14 cm, 14 cm); $+3$ μC at (14 cm, 0 cm). What is the potential at the point (30 cm, 30 cm)?

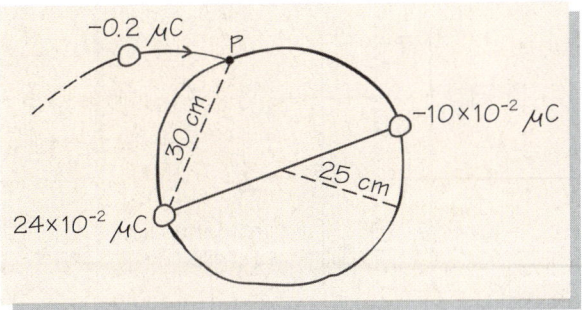

▲ **FIGURE 24–37** Problem 19.

22. (II) A 5-μC charge is fixed at $(x, y) = (15$ mm, 20 mm$)$, a -3-μC charge is fixed at $(15$ mm, 30 mm$)$, and a -2-μC charge is fixed at $(25$ mm, 20 mm$)$ (Fig. 24–38). What is the potential energy of the system? Does the order in which the charges are brought in from infinity matter?

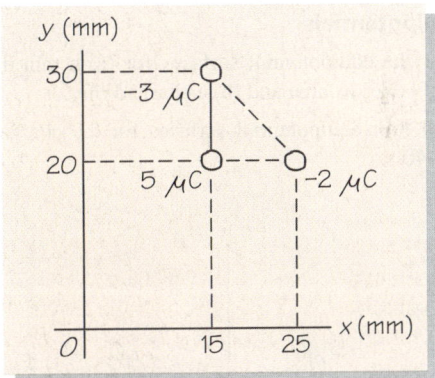

▲ **FIGURE 24–38** Problems 22 and 23.

23. (II) Calculate the electric potential at the origin due to the three charges considered in Problem 22.

24. (II) Two parallel conducting plates are brought to a potential difference of 600 V, and a small pellet of mass 2 mg carrying a charge of 3×10^{-7} C accelerates from rest at one plate. With what speed will it reach the other plate?

25. (II) Charges of $+12$ μC and -20 μC are placed along the y-axis at positions $y = +5.0$ cm and $y = -9.0$ cm respectively. (a) What is the potential at $y = 0$ and $x = 12.0$ cm? (b) What is the potential at $y = 0$, $x = 0$? (c) What is the electric field at that point?

26. (II) Two positive charges of magnitude 2.5 μC and 7.5 μC are placed a distance of 0.80 m apart. What is the electric potential along a line joining them? Where along that line will the electric field vanish? If a test charge is placed at that point, will it be in stable or unstable equilibrium as far as its motion along the line is concerned? Does your answer depend on the sign of the test charge?

27. (II) Figure 24–39 shows the cross section of a very large insulating slab that is uniformly charged to a charge density of 10^{-5} C/m^3. The thickness of the slab is 2 cm. (a) Determine the electric field of the charge on the slab at points A, B, and C. (b) Calculate the potential at points B and C, assuming that it is zero at A. (c) Plot the electric field and the potential as a function of distance from the center of the slab.

28. (II) A charge Q is distributed uniformly over the surface of a spherical shell of radius R. How much work is required to move these charges to a shell with half the radius? The charges are again distributed uniformly.

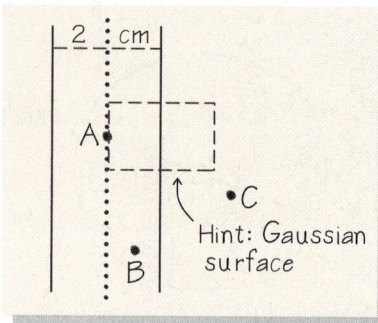

FIGURE 24–39 Problem 27.

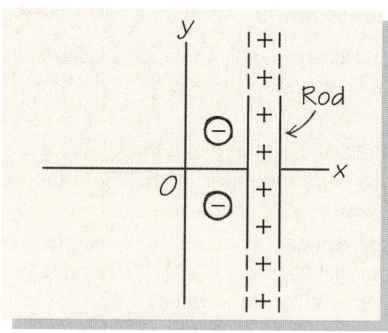

FIGURE 24–42 Problem 33.

29. (III) Calculate the potential inside and outside a sphere of radius R and charge Q, in which the charge is distributed uniformly throughout. [*Hint*: The additive constant for the potential inside the charged sphere must be chosen so that the two potentials, inside and outside, agree at $r = R$.]

24–3 Equipotentials

30. (I) Draw the equipotential surfaces for (a) a thin disk charged uniformly over its area and (b) a charged ring.

31. (I) Draw four equipotential surfaces for the charges shown in Fig. 24–40.

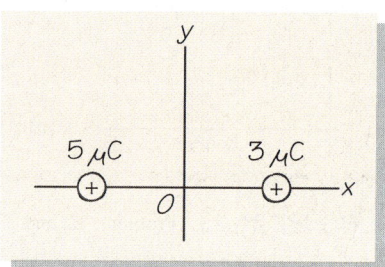

FIGURE 24–40 Problem 31.

32. (I) Sketch the equipotential surfaces for the charges shown in Fig. 24–41. Assume that the rod is an insulator.

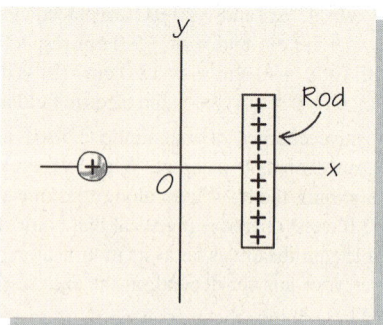

FIGURE 24–41 Problem 32.

33. (I) Sketch the electric fields and the equipotentials for the charge distribution shown in Fig. 24–42. Assume that the rod (of infinite length) is an insulator.

34. (I) Three metallic spheres are placed on the corners of an equilateral triangle. The radii of the spheres are 1/4 of the side of the triangle. All the spheres are at the same potential V_0. Sketch the equipotentials for this system.

35. (II) A uniformly charged metal rod is placed parallel to an infinite, uncharged metal plate. Sketch the equipotentials in a plane perpendicular to the plate and to the rod, and in a plane perpendicular to the plate but parallel to the rod.

36. (II) Sketch the equipotentials in the xy-plane due to an infinite number of identical point charges q that lie on a line and are separated by a distance a, so that the coordinates of the point charges are $x_n = na$ and $y_n = 0$, where $n = 0, \pm1, \pm2, \pm3, \ldots$.

37. (II) Two charges of equal magnitude but opposite sign are separated by a distance L. Sketch the equipotentials. What equipotential surfaces will have a potential of zero when the separate potentials for the two charges are chosen to be zero at infinity?

38. (II) Two infinite plates, each charged uniformly with charge density σ, are placed at right angles to each other and are almost touching. What are the equipotential surfaces? What are the equipotential surfaces if one of the plates has charge density $-\sigma$?

24–4 Determining Fields from Potentials

39. (I) The electric potential of a charge distribution within some region of space is $V(x, y, z) = Q/4\pi\varepsilon_0 x$. Find the electric field in this region.

40. (I) Find the electric field of a charge distribution if the electric potential of the distribution is $V = Ax^2y^2 + Byz^2 + C$, where A, B, and C are constants.

41. (I) In a certain region of space, the electric potential due to a charge distribution varies only with x, changing according to $V = a_0 + a_1x$ where $a_0 = 12.7$ V, $a_1 = -6.68$ V/m, and x is in meters. Find the electric field, magnitude, and direction in this region.

42. (II) Starting from the solution in Example 24–9 of the potential due to a uniformly charged ring, use the derivative operations in Eq. (24–29) to find the electric field along the axis of the ring.

43. (II) Find the electric field far away along the bisecting axis of an electric dipole from the potential given in Eq. (24–24).

44. (II) Consider charge distributed in an infinitely long cylinder of radius R whose axis forms the z-axis. The charge distribution depends only on the distance r from the z-axis. The potential is given for $r < R$ by $V(r) = (Q/2\pi\varepsilon_0)[A(r/R) + B(r/R)^2 + C]$, where A, B, and C are constants. What is the electric field within the rod? What is the value of C if the potential is defined to be zero on the cylinder's surface?

45. (II) The potential $V(r)$ of a spherically symmetric charge distribution is given by $V(r) = (Q/4\pi\varepsilon_0 R)[-2 + 3(r/R)^2]$ for $r < R$ and $V(r) = Q/4\pi\varepsilon_0 r$ for $r > R$ (Fig. 24–43, see next page). Calculate the electric field.

46. (II) Use the results of Problem 45 and Gauss' law applied to Gaussian surfaces at various radii to calculate the charge distribution that gives rise to the potential given in that problem.

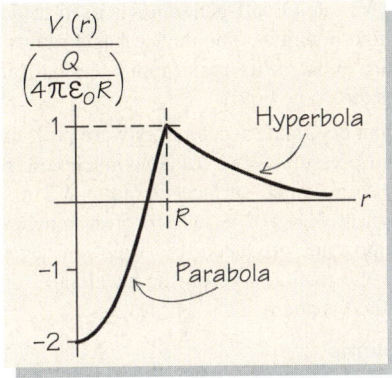

▲ **FIGURE 24–43** Problem 45.

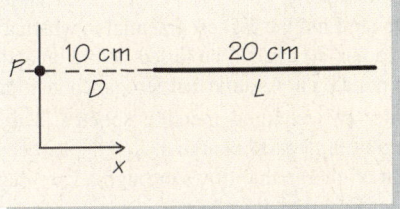

▲ **FIGURE 24–44** Problem 53.

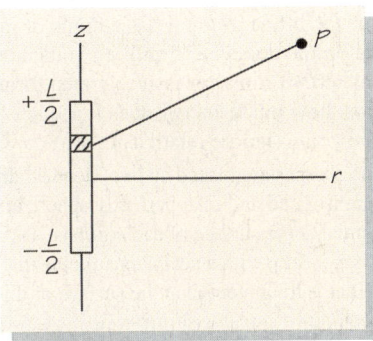

▲ **FIGURE 24–45** Problem 54.

47. (III) The potential in the xy-plane due to a certain charge distribution is given by

$$V(x, y) = \frac{Q}{4\pi\varepsilon_0 L} \times \left[\arctan\left(\frac{y}{x - a_0}\right) - 2\arctan\left(\frac{y}{x}\right) + \arctan\left(\frac{y}{x + a_0}\right) \right],$$

where L and a_0 are constant lengths. Show that the electric field at distances $x \gg a_0$, $y \gg a_0$ is proportional to a_0^2, and find its dependence on x and y. Express your answer in terms of r, the distance to the origin, and θ, the angle that the line from the origin to the point (x, y) makes with the x-axis.

24–5 The Potentials of Charge Distributions

48. (I) Two large, metal, parallel plates have a potential difference of 200 V, and the electric field between them has magnitude 7×10^3 V/m. What is the separation distance between the plates?

49. (I) The voltage along the axis of a uniformly charged ring of radius 10 cm is 5 V at a point 15 cm from the center of the ring. How much charge is on the ring?

50. (I) In fair weather, there is a constant electric field near Earth's surface whose magnitude is roughly 100 V/m, directed downward. (a) Find the potential associated with this field. (b) What is the most convenient point to choose for zero potential? (c) How does the potential energy of a test charge near Earth compare in form with the potential energy of gravity? (d) How much negative charge would have to be placed on a person of mass 50 kg to have the electric force balance the force of gravity?

51. (II) Find the potential as a function of the perpendicular distance R from an infinite line of uniform charge density by using Gauss' law and Eq. (24–9).

52. (II) Charges are distributed with uniform charge density λ along a semicircle of radius R, centered at the origin of a coordinate system. What is the potential at the origin?

53. (II) A rod that is 20 cm long is given a uniformly distributed charge of 2 μC (Fig. 24–44). Calculate the potential at a point P, which is a distance of 10 cm from the end of the rod, assuming that $V = 0$ at infinity.

54. (III) Find an expression for the electric potential at all points due to a rod of length L and uniform charge density λ, using Eq. (24–12). The rod is oriented along the z-axis, with its center at the origin (Fig. 24–45). Show that at distances much greater than L from the rod, the potential reduces to that of a point charge $Q = \lambda L$ at the origin.

55. (III) A charge $3q_0$ is placed on the x-axis at the point $x = x_0$ (where x_0 is positive), and a second charge, $-q_0$, is placed on the x-axis at the point $x = -x_0/2$. (a) What is the potential on the x-axis of this distribution of charges? Assume that zero potential for a point charge is at infinity. (b) Show that your result for part (a) can be approximated for large x by a term proportional to $1/x$, plus a term proportional to $1/x^2$, plus higher powers of $1/x$. (c) Show that the expansion of part (b) is that of a point charge at the origin, plus an electric dipole oriented along the x-axis and centered at the origin, plus other terms. Find the strength of the point charge as well as the dipole moment of the electric dipole. (d) How large must x be so that the approximation of a point charge plus a dipole comes within 1 percent of the exact answer? [*Hint*: Use the approximation $(1 + z)^k = 1 + kz + \frac{1}{2}k(k - 1)z^2 + \cdots$ good for $z \leq 1$.]

24–6 Potentials and Fields Near Conductors

56. (I) A thin disk of radius 2.8 cm carries a total charge of 6.0×10^{-8} C spread evenly over its surface. What is the minimum work required to bring a charge $q = 3.2 \times 10^{-7}$ C at rest from infinity to a distance of 8.8 cm from the disk along its axis?

57. (I) A thin ring of radius 24 cm carries a uniformly distributed charge of 3.5×10^{-7} C. A negative charge $q = -8.5 \times 10^{-8}$ C is placed on the axis of the ring 28 cm from the plane of the ring (Fig. 24–46). How much work must an external agent do to move the charge slowly and steadily to a distance 85 cm away, also on the axis?

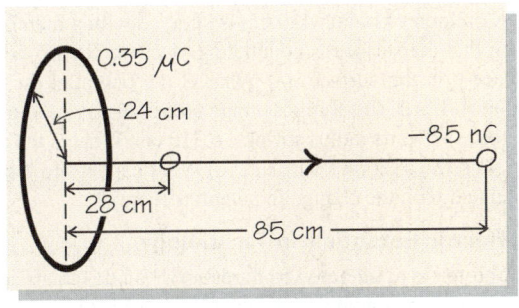

▲ **FIGURE 24–46** Problem 57.

58. (I) An electric field of 2.8×10^6 V/m is sufficiently large to cause sparking in air. Find the highest potential to which a spherical conductor of radius 30 cm can be raised before breakdown occurs in the air surrounding it. Assume that zero potential is taken at infinity.

59. (I) Consider two charged metallic spheres. The spheres have radii r_1 and r_2, and carry charge q_1 and q_2, respectively. What is the amount of charge that flows through a wire that is brought in and connected to the two spheres?

60. (II) A spark plug has a gap of 2 mm. What must be the potential difference across the gap in order for it to ionize a bridge across the gap? Assume air fills the gap.

61. (II) A Van de Graaff accelerator has a dome of radius 0.61 m. If the potential on the dome is 5.5 million volts, how much charge has accumulated? If a proton is accelerated through this potential to ground, how much energy does it acquire? Calculate how fast it will be going (ignore relativity).

62. (II) The same charges are placed on two identical drops of mercury. The drops are isolated and take perfectly spherical shapes, and the electric potential at the surface of each drop is 70 V. The drops coalesce into a larger drop with a net charge double that of either smaller charge. What is the potential at the surface of this larger charge?

63. (II) Two conducting spheres of different sizes are connected by a thin conducting wire. The radius of the larger sphere is three times that of the smaller sphere. If a total charge Q is placed on this apparatus, what fraction of Q sits on each sphere?

64. (II) Concentric metal shells, perfect conductors, have radii R and $1.5R$, respectively. A charge q is placed on the inner shell, and a charge $-3q$ is placed on the outer shell. (a) What are the electric fields in all space due to the two shells? (b) What is the potential difference between the two shells? (c) If a thin, perfectly conducting wire now joins the two shells, how does the charge redistribute itself?

65. (II) A cloud is made up of raindrops, each carrying a positive charge of 16×10^{-19} C. There are 1.2×10^{10} raindrops per cubic meter. When the electric field at the surface of the cloud builds up to 3.2×10^6 V/m there will be electrical breakdown and lightning will be seen. What will be the radius of the cloud, assuming it is spherical, when this occurs?

66. (II) Two metallic spheres of radii 0.05 m and 0.08 m respectively are placed far away from each other. They are then connected by a thin wire and a charge of 40 μC is placed on the system. Assuming that we can neglect the charge on the wire, how will the charge be distributed between the two spheres? What is the potential of the system?

67. (II) Two spherical conductors of radii 20 mm and 100 mm are connected by a thin wire and carry charges q_1 and q_2, respectively. If the wire is cut and the centers of the spheres are 250 mm apart, there is a repulsive force of 3.5 N between them. Use this information to calculate (a) q_1 and q_2 and (b) the electric fields at the surfaces of the conductors when they are connected by the wire.

68. (II) A balloon of radius 430 cm is sprayed with a metallic coating so that the surface is conducting. A charge of 1.5×10^{-5} C is placed on the surface. (a) What is the potential on the balloon's surface? (b) Suppose that some air is let out of the balloon, so that its radius shrinks to 310 cm. What is the new potential on the balloon's surface? (c) What happens to the energy associated with the change in potential energy?

24–7 Electric Potentials in Technology

69. (I) A proton is accelerated from rest in a Van de Graaff accelerator through a potential of 5.5×10^6 V. (a) What energy does the proton have—in electron-volts and joules? (b) What is the proton's final speed?

70. (I) A small Van de Graaff generator is used to demonstrate the effects of high potential. The device has a radius of 41 cm and stands in air. What is its maximum potential, and how much charge does the dome hold?

71. (II) Early Van de Graaff accelerators were built to operate in air without high-pressure gases. (a) How much voltage could an accelerator with a domed surface of radius 1.3 m produce if air breaks down at 3×10^6 V/m? (b) How much kinetic energy could the protons produced by such an accelerator have? (c) What is the total charge on the accelerator dome when the maximum field is attained?

General Problems

72. (I) We have a high-voltage power supply capable of producing 15,000 V, and we want to ionize the air molecules between parallel plates. What plate separation will give us electrical breakdown?

73. (I) The potential at a point x due to a thin, flat, uniformly charged disk of radius R and total charge Q at a point on the disk's axis a distance x away from the disk is given in Eq. (24–32). Consider two such identical disks, one in the yz-plane at $x = -a$, and another in the yz-plane at $x = +a$, both centered on the x-axis (Fig. 24–47). What is the potential at an arbitrary point on the x-axis between the two disks?

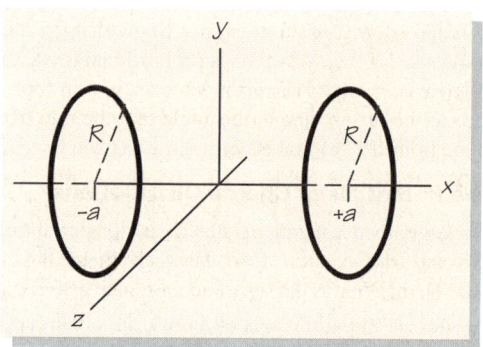

▲ FIGURE 24–47 Problem 73.

74. (I) A metallic ring of radius R, carrying charge Q, has an associated electric potential at a distance x from the center of the ring along its axis given in Table 24–1. What is the potential at some point x due to two rings, one carrying charge Q and the other $-Q$, both in the yz-plane, centered on the x-axis, one located at $x = a$, the other at $x = -a$?

75. (II) What is the electric field at a point x on the axis due to the two charged rings described in the previous problem for $x \gg a$? [Hint: Use the approximation $(x^2 + 2ax + b^2)^{-1/2} \cong x^{-1}(1 - a/x + (3a^2 - b^2)/2x^2)$.]

76. (II) Write an expression for the total energy of two point charges—one positive and of magnitude Q, fixed at the origin; the other negative and of magnitude q and mass m, located at a point a distance r from the origin. Suppose the charge q, instead of being stationary, moves in a circular orbit of radius r around the charge Q. Assuming that the Coulomb attraction is responsible for the centripetal acceleration, calculate the energy. Why is the angular velocity of this motion constant?

77. (II) A nonconducting sphere of radius R carries a charge $+Q$ distributed uniformly throughout its volume. What is the potential energy of a point charge $-q$ a distance r ($r < R$) from the center of the sphere? Show that if there is a hole drilled through the sphere so that the point charge can move through it, then the point charge oscillates as though it were attached to a spring. Find the effective spring constant.

78. (II) Three electrons are located along the x-axis at positions $-6\,\mu m$, $0\,\mu m$, and $6\,\mu m$, respectively. How much energy was required to move each of the electrons in turn from infinity? Does the order in which they were moved matter?

79. (II) A salt crystal consists of an array of positive Na and negative Cl ions, both carrying an elementary charge of magnitude e. Assume that a small "seed" crystal consists of four ions, forming a square of side 0.25 nm (Fig. 24–48). Find the electric force acting on one of the sodium atoms due to the other atoms of the seed and the work needed to remove this ion from the seed. Give your result in electron volts.

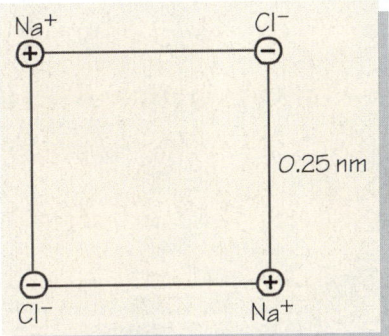

▲ **FIGURE 24–48** Problem 79.

80. (II) Calculate the potential at the point $P(x, y)$ due to the dipole in Fig. 24–49, which consists of a charge $+q$ placed at $(0, a)$ and a charge $-q$ placed at $(0, -a)$, and use this potential to calculate the electric field at point P.

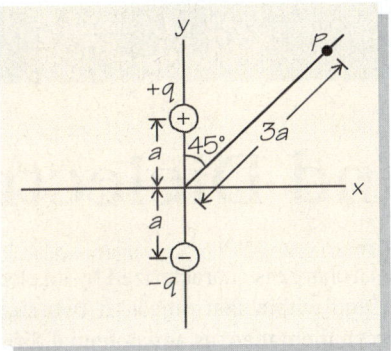

▲ **FIGURE 24–49** Problem 80.

81. (II) Find the electric field along the axis of a uniformly charged ring of radius R and total charge Q by taking the appropriate derivatives of the potential found in Example 24–9. Set up the problem by using the direct integration techniques presented in Chapter 23. Compare the difficulties of the two ways of calculating the electric field.

82. (II) A positron (charge $+e$ and same mass as that of the electron) approaches a proton (charge $+e$) head-on. As a result of the repulsion, the positron turns around a distance $r_0 = 6.5 \times 10^{-10}$ m from the proton. What is the kinetic energy of the positron when it is very far from the proton? You may assume that the proton motion can be neglected.

83. (II) An electron is moving in the field of a helium nucleus (atomic number $Z = 2$). What is the change in the electron's potential energy when it moves from a circular orbit of radius 3×10^{-10} m to one of radius 2×10^{-10} m? What is the change in kinetic energy? in the total energy of the electron? Energy conservation is not violated in this process because it can be carried away by radiation that is emitted during the change of orbits.

84. (II) Four charges $+Q$, $-Q$, $+Q$ and $-Q$ are placed at corners of a square of side L. What is the electric potential magnitude at a point P which lies on one of the diagonals of the square at a distance R from the center of the square, with $R \gg L$. [Hint: Use the binomial expansion $(R^2 \pm x^2)^n = R^{2n} \pm nR^{2n-2}x^2 + \frac{n(n-1)}{2}R^{2n-4}x^4 \pm \cdots$, good for $x < R$.] The field's dependence on R is characteristic of an *electric quadrupole*.

85. (II) Two concentric metal shells of radii r_1 and r_2 respectively carry charge q_1 and q_2 respectively. Assuming $r_1 < r_2$, what is the potential in the range $0 \le r \le \infty$?

86. (II) An electric dipole fixed in space consists of a charge $+q$ at the point $x = -0.2$ m and a charge $-q$ at the point $x = +0.2$ m, where $q = 5\,\mu C$. A test charge $q_0 = 3\,\mu C$ is steadily moved from the point $x = +0.6$ m to the point $x = -0.4$ m by following a semicircular path of radius 0.5 m that takes the test charge through the y-axis (Fig. 24–50). How much work is required to move the test charge?

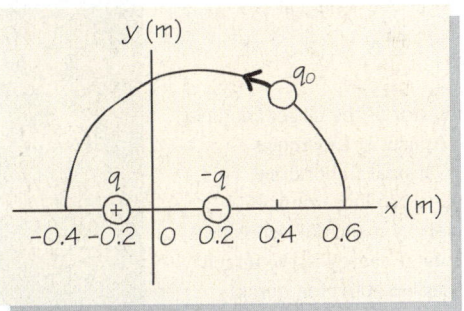

▲ **FIGURE 24–50** Problem 86.

87. (II) Two identical cork balls of charge $2.0\,\mu C$ are suspended from the same point by thin threads 0.80 m long. (a) Calculate the mass of the cork balls if the threads each make a 30° angle with the vertical. (b) Calculate the potential energy of the system of two balls due to the presence of charges and to the presence of gravity as a function of the angle θ the threads make with the vertical. Choose zero gravitational potential energy to correspond to $\theta = 0$.

88. (II) A large, square plane with sides of length L, parallel to the yz-plane and located at x_1, has charge density σ_1. A similar plane, located at x_2, has charge density σ_2. How much work must be done to bring the second plane to within a distance a of the first one? Neglect end effects; that is, calculate the fields as though the planes were infinite.

89. (III) An infinitely long cylinder of radius R is filled with uniform charge density ρ. Calculate the potential inside and outside the cylinder.

90. (III) The inner radius of a spherical dielectric shell is 16 cm, and the outer radius is 45 cm. The shell carries a charge of 5.0×10^{-6} C, distributed uniformly. Sketch the shape of the potential for all values of r, the distance from the center of the shell, and evaluate it at the center and at the inner and outer radii.

91. (III) A solid sphere of radius R has uniform charge density ρ. Calculate the total potential energy by calculating the energy required to bring a spherical shell of thickness dr and charge density ρ from infinity to a distance r from the sphere's center in the potential due to a uniformly charged sphere of radius r.

Chapter 25

▶ The interior of the target chamber of the NOVA laser at Lawrence Livermore National Laboratory. This project aims to produce controlled nuclear fusion by means of depositing large amounts of energy at the target; the energy can be delivered quickly because it is stored in large capacitor banks.

Capacitors and Dielectrics

Any conducting object that carries a charge is characterized by an electric potential that is constant everywhere on and within that object. If two such conductors have a potential difference between them then, as any potential difference is able to accelerate charges, the system effectively stores energy. A capacitor is a device that can maintain a potential difference, storing energy by storing charge. The relation between the amount of charge a capacitor stores and the potential difference it maintains depends on the geometry of the capacitor. The storage of charge is also affected by the presence of insulating (nonconducting) material—dielectric material. In this chapter, we study the role capacitors play in electrical circuits, where they are indispensable elements, how they can be used directly as energy storage devices, as well as how geometry and dielectric materials affect the properties of capacitors. We also extend our fundamental knowledge about the behavior of matter by thinking about the microscopic structure of dielectrics and how that microscopic behavior shows itself at the everyday level.

25–1 Capacitance

A pair of conductors, whether separated by empty space or by an insulating material, forms the simplest type of **capacitor**. Capacitors store separated equal and opposite charges. In their most common and useful form, capacitors are made from two conductors with charge $+Q$ on one conductor and charge $-Q$ on the other. In Chapter 24 we saw that the potential difference[†] between the conductors is linearly dependent on this

---

[†] We will also use the terms "potential," "voltage drop," or "voltage" rather than "potential difference."

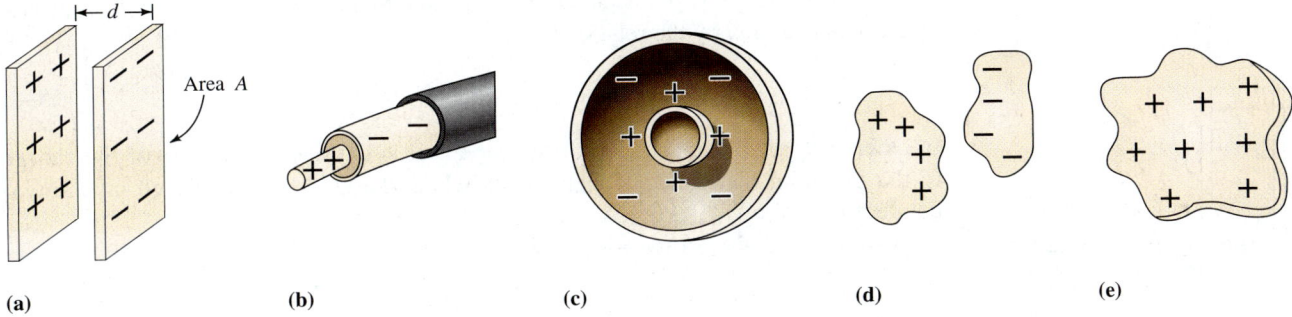

▲ **FIGURE 25–1** Various kinds of capacitors: (a) parallel-plate; (b) coaxial cable; (c) spherical (two hollow conducting spheres); (d) conductors of arbitrary size and shape; (e) isolated conductor infinitely far from second conductor.

charge; that is, $V \propto Q$. Thus, if we double the charge, we double the potential difference between the two conductors, so that the ratio of Q to V between two conductors is constant. We call this constant ratio Q/V the **capacitance** C. It depends on the shape and arrangement (the geometry) of the two conductors of the capacitor, and on the material between the conductors. Figures 25–1a to 25–1e illustrate different configurations of conductors that can act as capacitors.

CONCEPTUAL EXAMPLE 25–1 In the case of Fig. 25–1e, there is a charge of only one sign. Where is the charge of the other sign?

Answer An isolated conductor has a capacitance because when charge is placed on it, the charge must be brought in from infinity. Remember that charge must be conserved, so that the second piece of the capacitor is, in effect, at infinity. The potential difference in this case is the difference between the potential at infinity, normally taken to be zero, and the potential on the isolated conductor itself. Accordingly, an isolated conductor held high above Earth forms a capacitor. Even Earth itself is a capacitor—we'll calculate its capacitance below.

Why are capacitors important? Capacitors of differing capacitance allow us to hold different amounts of charge for a given potential difference or to maintain different potential differences for a given amount of charge. Capacitors thus control the storage and delivery of charge. Since it requires work to separate charges, capacitors store energy as well as charge. This energy is quite visible in certain circumstances, particularly through the often rapid release of the energy. A lightning strike is the spectacular discharge of a large capacitor formed by the system of a cloud and Earth. In its simplest form, a camera photoflash contains a capacitor that stores energy and then discharges it when the flash is fired. The slow but smooth delivery of energy when capacitors are coupled with other circuit elements represents another type of application. Emergency backup systems for computers and electrical power distribution systems use capacitors in this way. Lastly, because capacitors play a crucial role in controlling the time dependence of the electric charges that move through circuits, almost any device with an electronic circuit contains capacitors. We regularly use capacitors both large and small. Devices for the possible generation of fusion energy employ gigantic capacitors whose energy content rivals that of the largest lightning strike. In the realm of microscopic circuits—nanotechnology, or quantum engineering—capacitors are used that may consist of small collections of atoms.

Consider a capacitor formed by placing equal but opposite charges, $+Q$ and $-Q$, on two conductors. This is easily done by touching the two conductors with wire leads attached to the $+$ and $-$ terminals of a battery (Fig. 25–2). The battery will give the two pieces a potential difference, and the amount of charge that accumulates depends on the shape of the conductors and on their relative positions. We call Q the *charge on the capacitor* even though the *net* charge on the oppositely charged pair of elements is zero.

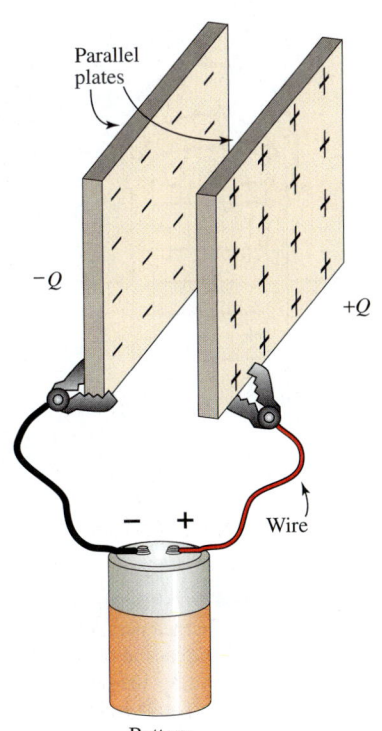

▶ **FIGURE 25–2** We can construct a capacitor by connecting two wires to two conductors such as the plates shown here, and then attaching one wire to the positive terminal of a battery and the other wire to the negative terminal of the battery. This will place charge $+Q$ on one conductor and charge $-Q$ on another conductor.

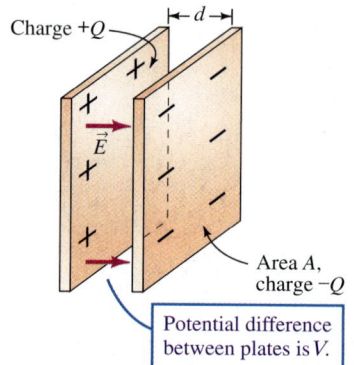

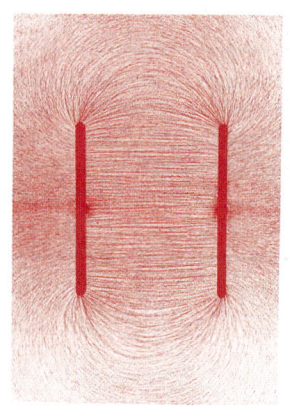

(a)

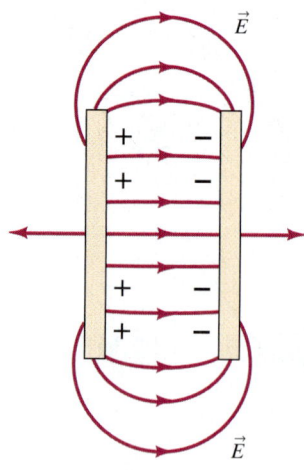

(b)

▲ FIGURE 25–3 Two parallel plates with equal and opposite charges make up the most basic capacitor.

▲ FIGURE 25–4 (a) The electric field lines due to a charged parallel-plate capacitor, shown by threads in oil. (b) The electric field lines between the two conducting plates of a parallel-plate capacitor.

With the charge on our capacitor proportional to the potential difference, its capacitance, C, is defined by the relation

$$Q = CV. \tag{25–1}$$

In other words, the capacitance of the capacitor is defined as the ratio of the charge to the potential difference V that results when charges $\pm Q$ are placed on the two conductors:

$$C \equiv \frac{Q}{V}. \tag{25–2}$$

DEFINITION OF CAPACITANCE

When there is charge on a capacitor, we say it is *charged*. When a capacitor *discharges*—for example, when it is used to fire a flashbulb—it can deliver its stored energy rapidly as the charges on the plates flow off.

C is always taken to be positive; that is, Eq. (25–2) should contain magnitudes only. The unit of capacitance is coulombs per volt (C/V), but capacitance occurs so frequently that it has been given its own SI unit, the **farad** (F), in honor of Michael Faraday:

$$1\,\text{F} \equiv 1\,\text{C/V}. \tag{25–3}$$

In practice, the farad is inconveniently large, and for practical use units of μF, nF, and pF ("puffs") are more common.

Calculating Capacitance

The capacitance of a capacitor can be calculated easily if the geometry is simple. (Keep in mind that even if we can't calculate a capacitance, we can always measure it.) The most basic capacitor consists of two parallel conducting plates of area A, separated by a distance d, with charges $+Q$ and $-Q$, respectively, distributed uniformly over the plates (Fig. 25–3). If the dimensions of the plates are large compared with d, then the electric field between the plates is to a very good approximation constant. Neglecting (small) edge effects, we found previously that the field between the plates has magnitude $E = \sigma/\varepsilon_0$, where σ is the charge density Q/A and ε_0 is the permittivity of free space. The potential difference between the plates is $V = Ed$, and we can combine these results to find that

$$V = Ed = \frac{\sigma}{\varepsilon_0} d = \frac{Q}{A} \frac{d}{\varepsilon_0} = Q \frac{d}{\varepsilon_0 A}.$$

Thus $Q/V = \varepsilon_0 A/d$, and the capacitance $C \equiv Q/V$ of a parallel-plate capacitor is

$$C = \frac{Q}{V} = \frac{\varepsilon_0 A}{d}. \tag{25–4}$$

The fields near the edge, the *fringe fields*, are not uniform (Fig. 25–4). Their effect is small if the linear dimensions of the plates are much larger than the separation. Fringe fields do not affect the linear relationship between the charge and the potential, but do modify the simple form of Eq. (25–4). We will use the approximation of Eq. (25–4) throughout this chapter in reference to parallel-plate capacitors.

Equation (25–4) gives us a second commonly used unit for the permittivity of free space, ε_0, namely farads per meter (F/m):

$$\varepsilon_0 = 8.85 \times 10^{-12}\,\text{C}^2/\text{N} \cdot \text{m}^2 = 8.85 \times 10^{-12}\,\text{F/m} = 8.85\,\text{pF/m}.$$

Either unit is consistent with the SI.

EXAMPLE 25–2 (a) Calculate the capacitance C of parallel plates of area $A = 100$ cm^2 separated by a distance $d = 1.0$ cm. (b) Find the area of a parallel-plate capacitor with plate separation of 1.0 cm and a capacitance of 1.0 F.

Setting It Up The setup is similar to that of Fig. 25–3.

Strategy For part (a) we use Eq. (25–4) to determine the unknown capacitance. For part (b) we can again use this equation, but this time we solve it for A.

Working It Out (a) Substituting the known values into Eq. (25–4) gives

$$C = \frac{\varepsilon_0 A}{d} = \frac{(8.85 \text{ pF/m})(1.0 \times 10^{-2} \text{ m}^2)}{1.0 \times 10^{-2} \text{ m}} = 8.9 \text{ pF}.$$

The plate area is rather large, yet the capacitance is only 8.9 pF.

(b) We again use Eq. (25–4), this time solving for A:

$$A = \frac{dC}{\varepsilon_0} = \frac{(1.0 \times 10^{-2} \text{ m})(1.0 \text{ F})}{8.85 \times 10^{-12} \text{ F/m}} = 0.11 \times 10^{10} \text{ m}^2.$$

This represents a square with sides of length 0.33×10^5 m $= 33$ km! Most common practical capacitors have capacitances much smaller than 1 F.

What Do You Think? Suppose the charge on each plate doubles. By how much does C change? Answers to **What Do You Think?** questions are given in the back of the book.

EXAMPLE 25–3 Consider a system made of a solid conducting cylinder of radius a surrounded by a thin coaxial conducting tube, or sheath, of radius b. Find the capacitance per unit length of this system, assuming that there is a vacuum between the central wire and the sheath. This is a model of a *coaxial cable*, a wire used for the high-speed transmission of signals. In a real coaxial cable, there would be some type of material, not a vacuum, between the central wire and the sheath, but for simplicity we assume here that there is a supporting material that has the electrical properties of a vacuum.

Setting It Up The system is sketched in Fig. 25–5. (Keep in mind that the material that mechanically supports the outer cylinder has the electrical properties of the vacuum.) We want to find the potential difference for a given charge. For an infinitely long system, however, we can specify only the charge *per unit length*, so we must similarly calculate the capacitance per unit length.

Strategy The capacitance is the ratio of the charge to the potential difference, and the capacitance per unit length is the ratio of the charge per unit length to the potential difference. Therefore we set a charge per unit length $+\lambda$ on the outer (and $-\lambda$ on the inner) conductor of the cable and calculate the resulting potential between the conductors. This calculation was already done in Example 24–11. There, we saw how Gauss' law gives the electric field, and hence the potential, outside a wire of finite radius. That example applies directly to the potential in the space *between* the conductors of our system.

Working It Out From Example 24–11, we have

$$V_b - V_a = \frac{\lambda}{2\pi\varepsilon_0} \ln \frac{b}{a}.$$

Dividing the charge per unit length λ by this potential difference gives the capacitance per unit length, a quantity we denote by c:

$$c = \frac{\lambda}{V_b - V_a} = \frac{2\pi\varepsilon_0}{\ln(b/a)} \quad (25\text{–}5)$$

Equivalently, the capacitance of a length L of the cable is L times this value.

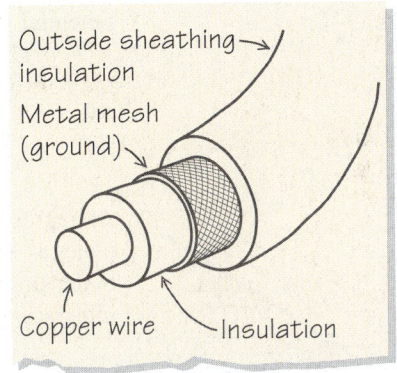

▲ **FIGURE 25–5** The coaxial cable has an inner solid wire of radius a and a cylindrical metal sheath of radius b. An insulator is placed between the two conductors.

EXAMPLE 25–4 *The Isolated Sphere.* What is the capacitance of an isolated conducting sphere of radius R? Calculate Earth's capacitance.

Strategy We place a charge Q on the sphere and find its potential V, assuming the potential at infinity is zero; because the "second piece" of the capacitor is at infinity, V is indeed the appropriate potential difference. The ratio Q/V is, by definition, the capacitance.

Working It Out We found in Example 24–12 that, for the potential at infinity set at zero, the potential of a conducting sphere is

$$V = \frac{Q}{4\pi\varepsilon_0 R}.$$

The capacitance of the isolated sphere is Q divided by V:

$$\text{for an isolated sphere: } C = \frac{Q}{V} = 4\pi\varepsilon_0 R. \quad (25\text{–}6)$$

Earth's capacitance is determined by setting R equal to Earth's radius, $R_E = 6.38 \times 10^6$ m. Therefore

$$C = 4\pi\varepsilon_0 R = 4\pi(8.85 \times 10^{-12} \text{ F/m})(6.38 \times 10^6 \text{ m})$$
$$= 7.10 \times 10^{-4} \text{ F}.$$

It is interesting to compare this number to the value of the largest capacitors in use, of the order of 1 F.

What Do You Think? Perform a dimensional analysis to confirm the dependence of the capacitance on the radius.

The technique for determining the capacitance is always the same. We assume that the conductors have a charge $\pm Q$; then we find the potential difference V between the conductors due to this charge. The ratio Q/V gives the capacitance.

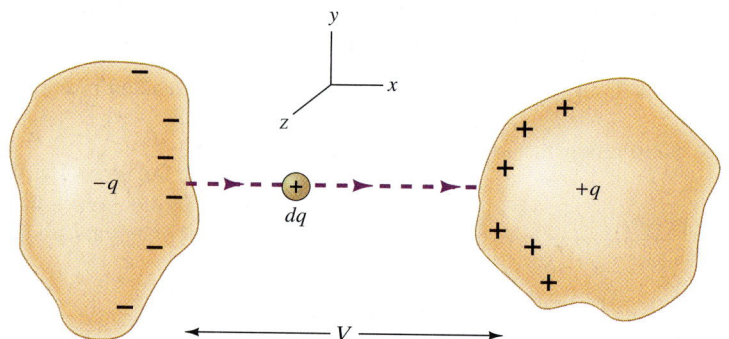

▶ FIGURE 25–6 Work is done when an infinitesimal charge dq is moved between the two conductors of a capacitor already charged to a potential difference V.

25-2 Energy in Capacitors

There is an electric field between the two conductors of a charged capacitor, and this field can accelerate a test charge. Thus, a charged capacitor is capable of doing work and must contain energy. The energy contained in a charged capacitor is in fact the work required to charge it, and we can calculate this work to find the energy the capacitor contains. Start by taking a positive charge dq from one neutral conductor and moving it to the other neutral conductor. The second conductor then has charge $+dq$, and the first conductor has charge $-dq$. If we do this for a time, the conductors will be charged to a potential difference V with charge q (Fig. 25–6), and V is given in terms of q by $V = q/C$. To move an *additional* charge dq from the negatively charged conductor to the positively charged one, we must then do work

$$dW = V\,dq = \frac{q}{C}\,dq.$$

As represented in Fig. 25–7, the infinitesimal work done to place a charge dq on the plates is zero when the plates are uncharged and increases as the charge on the plates builds up to $\pm Q$. The total work done as the charge increases from 0 to Q is obtained by integrating the above expression from $q = 0$ to $q = Q$:

$$W = \int dW = \int_0^Q \frac{q}{C}\,dq = \frac{1}{C}\int_0^Q q\,dq = \frac{Q^2}{2C}. \tag{25–7}$$

This is a general result for all capacitors.

Once charged, the stored energy in the capacitor can do work. We can think of the stored energy as a potential energy. This potential energy can be used to move a test charge placed between the conductors or to cause a flashbulb to flash.

To summarize, the potential energy of a charged capacitor is

$$U = \frac{Q^2}{2C}. \tag{25–8}$$

ENERGY IN A CAPACITOR

Because $Q = CV$, this result is equivalent to

$$U = \frac{CV^2}{2}. \tag{25–9}$$

ENERGY IN A CAPACITOR

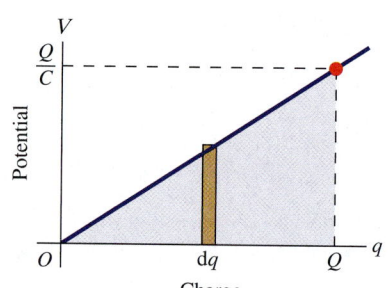

▲ FIGURE 25–7 The amount of work $dW = V\,dq$ that is done to move a charge element dq from one part of a capacitor to the other increases as the charge on the conductors—and hence the potential V across them—builds up. The total work is found by adding the work elements together. This work is the area under a curve of potential difference versus charge on the plates: the integral of $V\,dq$.

The first form is used when the charge is known; the second when the potential is known. Another equivalent form that is useful when the charge and voltage are both known is

$$U = \frac{QV}{2}. \qquad (25\text{--}10)$$

ENERGY IN A CAPACITOR

As we know from Chapter 24, the electric potential energy associated with the movement of a charge Q through a fixed potential V is $U = QV$. This expression differs by a factor of 2 from the capacitor energy $QV/2$, Eq. (25–10). The reason for this difference is that, as a capacitor is charged, the potential increases from 0 to V, and in effect the average potential as the charging takes place is $V/2$.

Batteries vs. Capacitors

A *battery* is a chemical device for the storage of energy. Whereas the potential of a capacitor decreases as the capacitor delivers its charge, a battery ideally maintains a fixed potential between two points (terminals) as it delivers charge. If we want to charge a capacitor to a certain potential, we can use a battery because it can hold the desired potential even as it delivers charge to the capacitor.

EXAMPLE 25–5 A 12-V car battery is used to charge a 100-μF capacitor. (a) How much energy is stored in the capacitor? (b) Compare this energy with the energy stored in the battery itself, if the battery is capable of delivering a total charge of $Q = 3.6 \times 10^5$ C at the given voltage. (This is the charge that can be delivered by a battery rated at 100 ampere-hours; the ampere-hour is a standard unit for charge.)

Setting It Up By a "12-V battery" we mean a battery with a 12-V potential difference across its leads. In the charging process the battery brings the potential difference across the capacitor plates to that same value.

Strategy (a) To find the energy in the capacitor given the capacitance and the potential difference across it, we apply Eq. (25–9). (b) The electric potential energy associated with the movement of a charge Q through a fixed potential V is $U = QV$. If the battery is capable of moving a given amount of charge through a given potential, as described here, then the battery must contain at least this amount of energy.

Working It Out (a) We have

$$U = \frac{CV^2}{2} = \frac{(100 \times 10^{-6}\text{ F})(12\text{ V})^2}{2} = 7.2 \times 10^{-3}\text{ J}.$$

The answer is in joules—we have used SI units throughout.

(b) The potential energy in the battery is (at least)

$$U = QV = (3.6 \times 10^5\text{ C})(12\text{ V}) = 4.3 \times 10^6\text{ J}.$$

The battery contains a factor of 6×10^8 more energy than is stored in the capacitor!

What Do You Think? Why is it possible to store so much more energy in the battery?

The car battery of Example 25–5 has a potential energy of about 10^6 J. This is far more energy than that of a typical practical capacitor of 100 μF charged to a moderate potential. A typical capacitor would require a potential of about 10^{10} V to store the amount of energy in a car battery. The largest available commercial capacitors have capacitance on the order of 1 F, and it would require a potential of about 1000 V for a 1-F capacitor to contain as much energy as that contained in the car battery. However, high voltages across capacitors present their own difficulties, and large capacitors can be charged to a potential of only a few volts. A battery's energy is stored in chemical bonds rather than in the macroscopic separation of charge. Batteries provide a practical way to store large amounts of energy for long periods, but they are not a practical way to quickly deliver electrical energy when it is needed. (By electrical energy, we essentially mean the movement of charge. Chemical energy can in other situations be delivered quickly, as in chemical explosives.) Conversely, whereas a capacitor cannot store as much energy as a battery, it can deliver electrical energy much more quickly. This is achieved if a charged capacitor is "shorted," meaning that the two plates are connected by a conducting wire, as in Fig. 25–8.

720 | Capacitors and Dielectrics

▶ **FIGURE 25–8** (a) Energy is stored in a capacitor by charging it. (b) Shorting the capacitor allows the charges on the capacitor to recombine and results in a rapid release of the stored energy, which makes a spectacular display of light and heat.

(a)　　　　　　　　　　　(b)

25–3 Energy in Electric Fields

The electric field is of fundamental physical significance. In this section, we develop the concept that the energy of a capacitor is contained in the electric field itself.

We start by relating the expression for the energy in a capacitor to the strength of the electric field in that capacitor. The parallel-plate capacitor is convenient for this purpose because both the capacitance and the field are known (Fig. 25–4). Equation (25–4) gives the capacitance for this case, $C = \varepsilon_0 A/d$, where A is the plate area and d is their separation. The field has constant strength E, and the potential difference between the plates is $V = Ed$. Thus Eq. (25–9) gives the energy

$$U = \frac{CV^2}{2} = \frac{\varepsilon_0 A}{2d}(Ed)^2 = \left[\frac{\varepsilon_0 E^2}{2}\right](Ad). \quad (25\text{–}11)$$

We have written Eq. (25–11) so that the *volume* of the space between the plates, Ad, stands out. This is the volume that contains the electric field and, because the field is constant, the coefficient of the volume (the quantity in square brackets) in Eq. (25–11) is the **energy density**, u, or energy per unit volume:

$$u \equiv \frac{U}{\text{volume}} = \frac{\varepsilon_0 E^2}{2}. \quad (25\text{–}12)$$

ENERGY DENSITY IN ELECTRIC FIELD

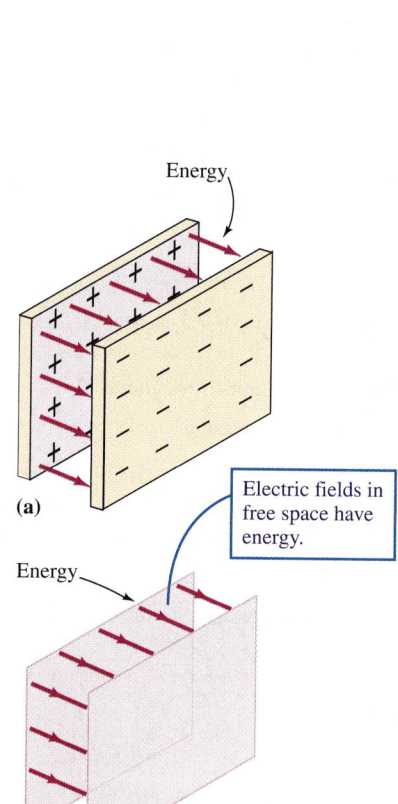

▲ **FIGURE 25–9** Electric fields have energy whether they are (a) inside a capacitor or (b) in free space.

This result suggests that the energy of a capacitor is located where the associated electric field exists, in the space between the plates (Fig. 25–9a). Now imagine that the plates aren't there but the field remains (Fig. 25–9b). Equation (25–12) is in fact *a general expression for the local energy density in free space, for a constant or variable electric field*. Wherever there is an electric field—even one that varies throughout space—the energy density, or energy per unit volume, at a particular location in space is found by squaring the electric field there and multiplying by $\varepsilon_0/2$. This makes sense if we use the reasoning that if a physical configuration can do work, it must contain energy—after all, if we place a test charge in a region of space where there is an electric field, it will accelerate, gaining energy in the process. In Section 25–5 we shall see that we must modify the coefficient $\varepsilon_0/2$ when dielectrics are present, but the energy density is always *proportional* to the square of the field.

CONCEPTUAL EXAMPLE 25–6 The plates of an isolated parallel-plate capacitor with a fixed charge on the plates are pulled slightly apart. Does the energy density change in the region between the plates? Does the overall energy in the field increase and, if so, why?

Answer We have argued that the energy density u depends on the electric field, but the field magnitude is σ/ε_0, and as long as the plate separation is small compared to their size, this doesn't depend on the distance between the plates. The energy density is constant. However, the volume between the plates has increased, so that the total energy in the field increases. This makes sense because the two oppositely charged plates attract, and it takes positive work to separate them further. We have added energy to the system by moving them apart, and this shows up in the greater total energy in the field.

EXAMPLE 25–7 An isolated spherical conductor of radius R carries a total charge Q. (a) Determine the energy density at each point in space as a function of the distance r from the sphere's center. (b) Use this energy density to compute the system's total energy. (c) Compare this total energy to the work done in charging the sphere.

Strategy (a) Equation (25–12) expresses the energy density in terms of the electric field. Equation (22–5) gives the electric field at a radius r outside the charged sphere. This field is radial and has magnitude $E = Q/(4\pi\varepsilon_0 r^2)$. Inside the conducting sphere, the field is zero.

(b) We must integrate the energy density over all space to find the total energy. As we have spherical symmetry, we can reduce the three-dimensional integral over all space to a one-dimensional integral over the radial distance by considering the volume as a series of concentric spherical shells. Because the energy density is zero inside the sphere, we need integrate only over radii larger than R.

(c) To calculate the work required to charge the sphere by bringing charge in from infinity, suppose that the sphere already has charge q and is at a potential of $V = q/(4\pi\varepsilon_0 R)$. The additional work to bring charge dq in is $dW = V\,dq$, and this must be integrated from $q = 0$ to $q = Q$.

Working It Out (a) Given our value for the electric field, the energy density at distance r is

$$u = \frac{\varepsilon_0 E^2}{2} = \frac{Q^2}{32\pi^2 \varepsilon_0 r^4}$$

outside the sphere ($r > R$) and zero inside ($r < R$).

(b) The volume dV of a thin shell of thickness dr is the product of dr and the surface area of the shell, $4\pi r^2$. Then the volume element $dV = 4\pi r^2\,dr$, and the system's total energy is

$$U = \int_{\text{volume}} u\,dV = \frac{Q^2}{32\pi^2 \varepsilon_0} \int_R^\infty \frac{4\pi r^2}{r^4}\,dr$$
$$= \frac{Q^2}{8\pi\varepsilon_0} \int_R^\infty \frac{dr}{r^2} = -\frac{Q^2}{8\pi\varepsilon_0} \frac{1}{r}\Big|_R^\infty = \frac{Q^2}{8\pi\varepsilon_0 R}.$$

(c) We integrate $V\,dq$ from $q = 0$ to Q to determine the total work done. This gives

$$W = \int dW = \int V\,dq$$
$$= \int_0^Q \frac{q}{4\pi\varepsilon_0 R}\,dq = \frac{1}{4\pi\varepsilon_0 R} \int_0^Q q\,dq = \frac{Q^2}{8\pi\varepsilon_0 R}.$$

The work done to bring the charge in from infinity is indeed equal to the total energy of the system found in part (b). This is a good check on the validity of our calculations.

What Do You Think? Our results indicate that the energy density increases dramatically for smaller R. Does this make sense?

25–4 Capacitors in Parallel and in Series Circuits

Here we begin our discussion of electric circuits, in which the physics of electric and magnetic fields have direct application to a myriad of electrical devices that we use each day. We have already mentioned two circuit elements: capacitors and batteries. We use the universal symbols shown in Fig. 25–10 for batteries and capacitors in circuit diagrams, which are schematic drawings that illustrate the connection of various circuit elements. The lines connecting circuit elements in these diagrams are assumed to be perfectly conducting wires, which means that they form equipotentials. We can use circuit diagrams to single out some interesting capacitor combinations. In a **parallel** combination, capacitors are connected as in Fig. 25–11a, with equal voltage drops across the capacitors; in a **series** combination, capacitors are connected as in Fig. 25–11b, with equal charges on the capacitors. We can simplify the analysis of circuits by finding the capacitance of the single *equivalent capacitor* that can replace these combinations without changing the potential across them.

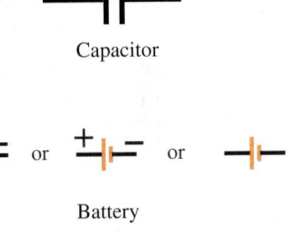

▲ **FIGURE 25–10** The symbols used to indicate (a) capacitors and (b) batteries in electric circuits.

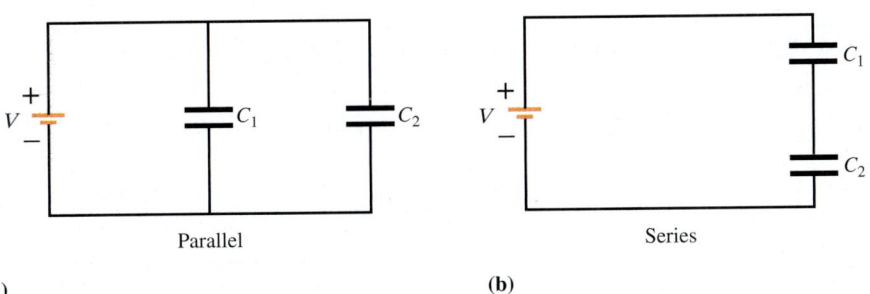

▲ **FIGURE 25–11** (a) Circuit with two capacitors connected in parallel. (b) Circuit with two capacitors connected in series.

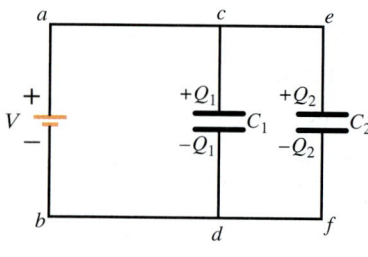

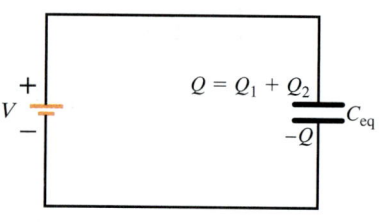

▲ **FIGURE 25–12** (a) A battery is used to place the same voltage V across the two capacitors connected in parallel. The charge on each capacitor depends on their individual capacitances. (b) The two capacitors can be replaced by an equivalent capacitor C_{eq}.

Parallel Connection

The battery in Fig. 25–12a maintains a potential V across the points a and b. Because the connecting wires are perfect conductors, the line ace is an equipotential, as is the line bdf. The potential difference from points c to d and from e to f must therefore also be V. *Capacitors connected in parallel have the same potential between their conductors.*

What is the equivalent capacitance, C_{eq}, of the single capacitor—defined by Fig. 25–12b—that could replace the parallel combination of capacitors C_1 and C_2 in such a way that the behavior of the circuit outside the parallel combination is unchanged? The equivalent capacitor will have the same potential difference V and total charge Q as the parallel combination of C_1 and C_2 and it is this condition that determines C_{eq}. The charges on C_1 and C_2 are related to the voltage across each capacitor:

$$Q_1 = C_1 V \quad \text{and} \quad Q_2 = C_2 V.$$

The charge for the equivalent capacitor is $Q = C_{eq}V$, and the total charge produced by the battery for the circuit in Fig. 25–12a is $Q_1 + Q_2$, so

$$Q = Q_1 + Q_2 = C_1 V + C_2 V = (C_1 + C_2)V = C_{eq}V.$$

The last equality shows that

$$C_{eq} = C_1 + C_2. \tag{25-13}$$

With n capacitors connected in parallel, we can similarly show that the equivalent capacitance is

$$C_{eq} = C_1 + C_2 + \cdots + C_n. \tag{25-14}$$

When capacitors are arranged in parallel, the total capacitance is larger than any of the individual capacitances.

Series Connection

We can also determine the single capacitor equivalent to capacitors connected in series (Fig. 25–13a). Because the battery maintains a fixed potential V, charge $+Q$ appears at c and charge $-Q$ at f. The positive charge at c induces a charge $-Q'$ at d, and because the isolated piece of metal enclosed by the dashed line in Fig. 25–13a is neutral, $+Q'$ also appears at e. We now show that $Q' = Q$ by drawing a Gaussian surface like that shown in Fig. 25–13b. Because there is no field within the metal plates and because there is no electric flux through the short side portions, the flux through the surface is zero, the net charge enclosed must be zero, and $Q' = Q$. Capacitors C_1 and C_2 thus have identical charges Q. *Capacitors connected in series have identical charges.*

The single equivalent capacitor C_{eq} in the circuit of Fig. 25–13c must carry the identical charges $+Q$ and $-Q$ and have the same potential difference V across it as exists between c and f in the circuit of Fig. 25–13a. Capacitor C_1 has potential $V_1 = Q/C_1$; similarly, capacitor C_2 has potential $V_2 = Q/C_2$. The total potential is $V = V_1 + V_2$, so we have

$$V = V_1 + V_2 = \frac{Q}{C_1} + \frac{Q}{C_2} = Q\left(\frac{1}{C_1} + \frac{1}{C_2}\right) = \frac{Q}{C_{eq}}.$$

The last equality shows that the value of the equivalent capacitance is determined by

$$\frac{1}{C_{eq}} = \frac{1}{C_1} + \frac{1}{C_2}; \tag{25-15}$$

$$C_{eq} = \frac{C_1 C_2}{C_1 + C_2}. \tag{25-16}$$

With n capacitors connected in series, the equivalent capacitance is given by

$$\frac{1}{C_{eq}} = \frac{1}{C_1} + \frac{1}{C_2} + \cdots + \frac{1}{C_n}. \tag{25-17}$$

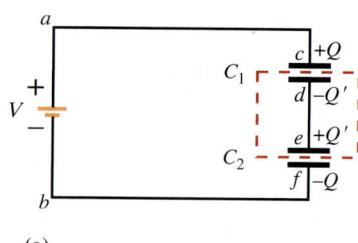

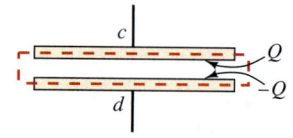

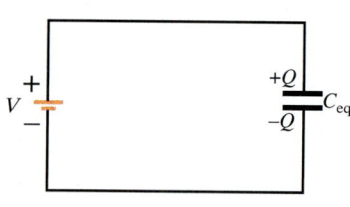

▲ **FIGURE 25–13** (a) Two capacitors connected in series must have identical charges $\pm Q$ but can have different voltages. The net charge within the dashed region is zero. (b) By drawing a Gaussian surface we establish that $Q' = Q$. (c) The two capacitors can be replaced by an equivalent capacitor C_{eq}.

When capacitors are arranged in series, the total capacitance is given by the reciprocal of the expression in Eq. (25–17), and this capacitance is less than any of the individual capacitances.

CONCEPTUAL EXAMPLE 25–8 Suppose you add an additional capacitor in series to a group of capacitors already in series. Does the equivalent capacitance increase or decrease?

Answer According to Eq. (25–17), if we add another capacitor in series, the value of $1/C_{eq}$ increases, so that C_{eq} decreases.

EXAMPLE 25–9 Determine the equivalent capacitance for the capacitors in the circuit shown in Fig. 25–14a.

Strategy In this circuit, the capacitors are combined both in series and in parallel, and we want to find the single equivalent capacitance for the whole combination. Our first step is to combine the two capacitors in parallel into one capacitor of value C'_{eq} (Fig. 25–14b); we then combine the three remaining capacitors in series into one final equivalent capacitor (Fig. 25–14c).

Working It Out We use Eq. (25–14) to combine the parallel capacitors in Fig. 25–14a:

$$C'_{eq} = 10\ \mu F + 6\ \mu F = 16\ \mu F.$$

Having reduced the capacitor arrangement to the one shown in Fig. 25–14b, we combine the series capacitors by using Eq. (25–17), as in Fig. 25–14c:

$$\frac{1}{C_{eq}} = \frac{1}{5\ \mu F} + \frac{1}{16\ \mu F} + \frac{1}{2\ \mu F} = \frac{61}{80}(\mu F)^{-1};$$

$$C_{eq} = 1.3\ \mu F.$$

What Do You Think? If the 2 μF capacitor is 1 μF instead, is the equivalent capacitance smaller, larger, or the same?

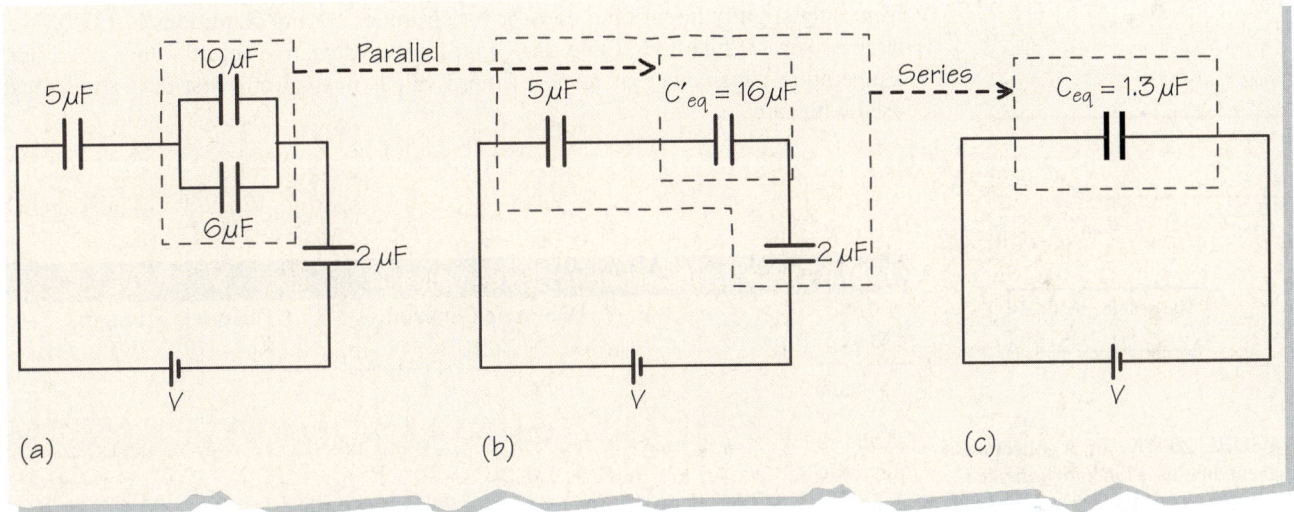

▲ **FIGURE 25–14** (a) Four capacitors of this electric circuit can be combined into one. (b) The two parallel capacitors are combined, giving three capacitors in series, (c) which are then combined into one *equivalent capacitor* C_{eq}.

25–5 Dielectrics

Many materials, including paper, plastics, and glass, do not conduct electricity easily; we referred to them earlier as *insulators*. Nevertheless, they modify the external electric fields in which they are placed. We call these materials **dielectrics**. We shall see that a dielectric placed in a capacitor allows more charge on the capacitor for a given voltage. A solid dielectric placed between a capacitor's two conductors also lends strength and mechanical stability to the capacitor. Finally, a dielectric can reduce the possibility of charge leakage and of sparking across the plates of a capacitor.

In this section, we take an experimental point of view of dielectrics, and in the next section we explore the microscopic origin of this behavior. Before we describe the basic experimental facts, however, it is worthwhile to take a peek ahead and describe in a conceptual way the underlying mechanism. Electrons are not free to move in dielectrics as they do in conductors, but that does not mean there is no response to an applied field. Dielectrics contain molecules that can behave as dipoles, either permanent ones or induced ones, and we can illustrate the possible effects with a collection of permanent dipoles. When such a collection starts completely unaligned (Fig. 25–15a) and is exposed to an external field, the dipoles align with the field as in Fig. 25–15b. But once there is a set of aligned dipoles, the set of dipoles can make its own electric field, which as in Fig. 25–15c opposes the external field. The net field within the material is thereby decreased. It is helpful to keep this qualitative picture in mind as we turn to the experimental facts.

Michael Faraday is generally given credit for having performed the first experiments, in the early nineteenth century, that showed that the capacitance increases when insulating materials are placed between the two conductors of a capacitor. If C_0 is the capacitance of a given capacitor in a vacuum (or in air), then the capacitance C of the same capacitor with dielectric between its conductors is larger than C_0 by a factor called the **dielectric constant**, κ. We have

$$C = \kappa C_0; \qquad (25\text{-}18)$$

CAPACITANCE MODIFICATION IN DIELECTRICS

$$\kappa \equiv \frac{C}{C_0}. \qquad (25\text{-}19)$$

The dielectric constant κ, which is larger than unity for all materials, depends on the material as well as on external conditions such as temperature. The value of κ can run from only slightly larger than one—κ for air under normal conditions is 1.0005—to as large as several hundred. Table 25–1 gives a representative set of values of κ, but the temperature dependence of many of these values is so strong that the values must be used with care.

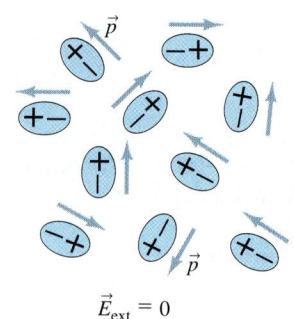

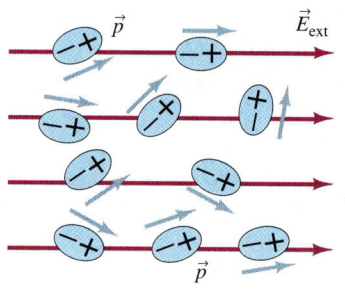

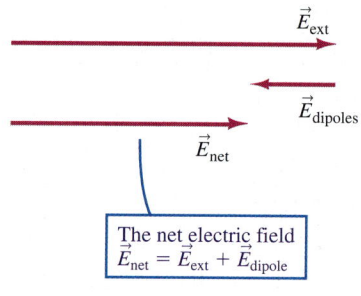

▲ **FIGURE 25–15** (a) A collection of permanent dipoles is randomly oriented. (b) In the presence of an external electric field, the dipoles tend to align themselves along the external field. (c) The aligned dipoles produce their own field $\vec{E}_{dipole}$ that tends to cancel the external field $\vec{E}_{ext}$ that aligned them, producing a net field $\vec{E}_{net}$.

TABLE 25–1 • Dielectric Properties of Materials†

Material	Dielectric Constant, κ	Dielectric Strength, E_{max} (10^6 V/m)
Vacuum	1.0	
Air	1.00054	3
Paraffin	2.0–2.5	10
Teflon	2.1	60
Polystyrene	2.5	24
Lucite	2.8	20
Mylar	3.1	
Plexiglas	3.4	40
Nylon	3.5	14
Paper	3.7	16
Fused quartz	3.75–4.1	
Pyrex	4–6	14
Bakelite	4.9	24
Neoprene rubber	6.7	12
Silicon	12	
Germanium	16	
Water	80	
Strontium titanate	332	8

†Values for some materials depend strongly on temperature and the frequency of oscillating fields.

You may have seen a device that makes use of the modifications due to dielectrics, an electric stud finder (Fig. 25–16). A unit with a metal plate acting as a capacitor in a circuit slides along a wall. Unless the plate is over a stud, there is air behind the wall, but when a stud is encountered, the capacitance changes because wood is a dielectric, and this is detectable in the circuit.

Experimental Evidence for the Behavior of Dielectrics

To simulate the experiment that led Faraday to his conclusions about dielectrics, let's first use a battery to charge a parallel-plate capacitor in air to a potential V_0 and a charge $Q_0 = C_0 V_0$ (Fig. 25–17a). (By "air" we mean an approximate vacuum—air itself has a dielectric constant, but it is very close to unity.) Here, the subscripts refer to the quantity in air—for example, C_0 is the capacitance when there is air between the plates. We disconnect the battery and measure the voltage, or potential difference (Fig. 25–17b). We then slide a dielectric, such as Plexiglas, between the plates (Fig. 25–17c). *The voltage is reduced.* The charge is unchanged—Plexiglas is an insulator—and the reduction factor is what we called above the dielectric constant:

$$V = \frac{V_0}{\kappa}.$$

To see that κ is indeed the dielectric constant defined above, note that the charge did not change, so we can interpret the reduction of the voltage as due to a change in the capacitance from the original value C_0 to a new value C:

$$C = \frac{Q_0}{V} = \frac{Q_0}{V_0/\kappa} = \kappa \frac{Q_0}{V_0} = \kappa C_0.$$

This verifies Eq. (25–18).

In another experiment we leave the battery connected to the capacitor after it is charged in air (Fig. 25–18a). After we insert the Plexiglas the potential remains the battery voltage V_0 (Fig. 25–18b). We observe, however, that *the charge on the conducting plates increases* by a factor of κ ($Q = \kappa Q_0$). Our experimental result remains in agreement with Eq. (25–18):

$$C = \frac{Q}{V_0} = \frac{\kappa Q_0}{V_0} = \kappa C_0.$$

Let's reinterpret these results in terms of permittivity. If we take a parallel-plate capacitor, for which $C_0 = \varepsilon_0 A/d$, we have

$$C = \kappa C_0 = \frac{\kappa \varepsilon_0 A}{d}. \tag{25-20}$$

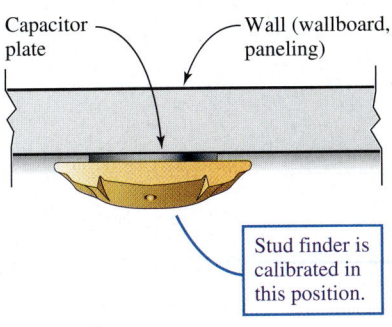

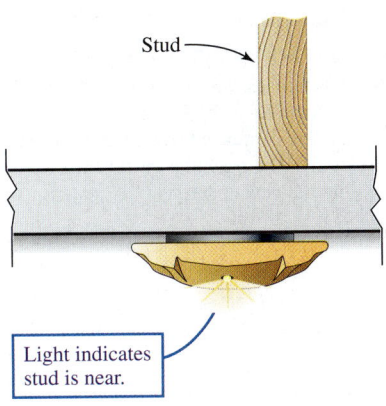

▲ **FIGURE 25–16** A schematic diagram of a stud finder, as seen from above the wall looking down. In (a) the stud finder is calibrated with no dielectric material (as for example a building stud) behind the wall. (b) The capacitance changes when the stud finder approaches a stud.

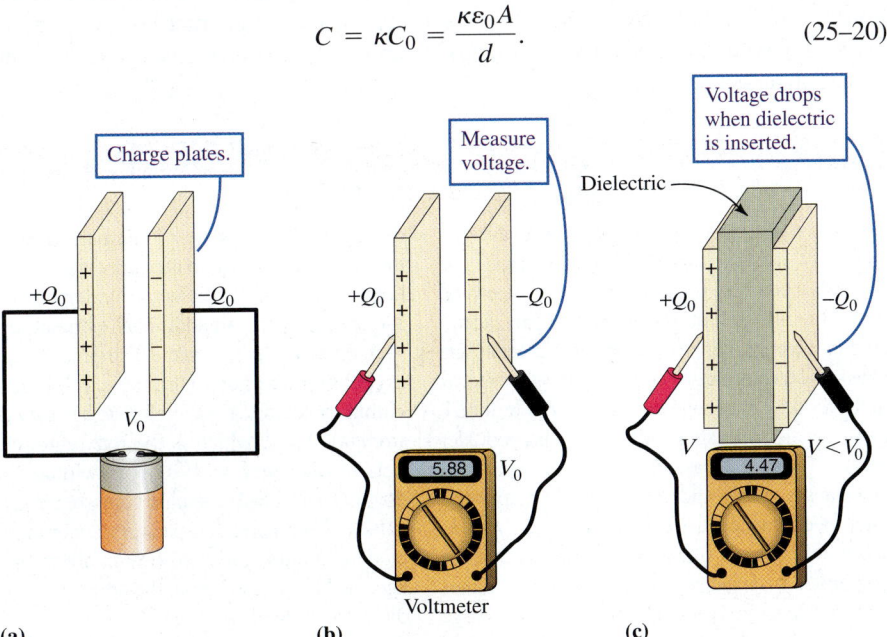

◀ **FIGURE 25–17** (a) A battery charges a capacitor to charge Q_0 and potential V_0. (b) If we take the battery away and measure the voltage with a voltmeter, we measure V_0 for the voltage. (c) If a dielectric is inserted into the capacitor, the voltage drops to $V < V_0$.

726 | Capacitors and Dielectrics

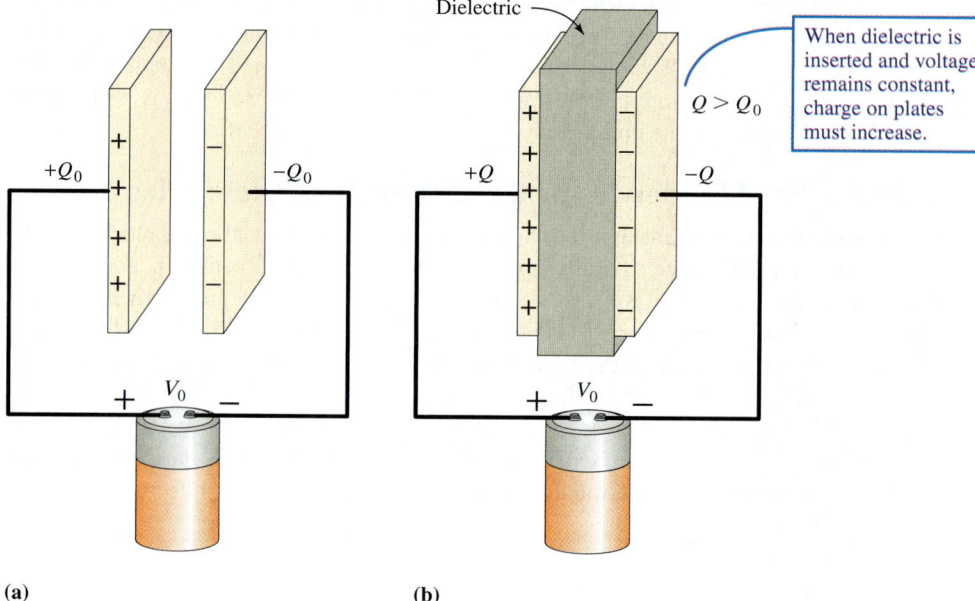

▶ FIGURE 25–18 (a) Again, a battery charges the capacitor to charge Q_0 and potential V_0. (b) This time we leave the battery connected when we insert the dielectric. The potential must remain at V_0, but the new charge is $Q > Q_0$.

If the charge is held fixed, the new voltage is

$$V = \frac{V_0}{\kappa} = \frac{Q_0}{\kappa C_0} = \frac{\sigma_0 A}{\kappa C_0} = \frac{\sigma_0 d}{\kappa \varepsilon_0}. \qquad (25\text{–}21)$$

The new electric field between the plates is reduced in magnitude to

$$E = \frac{V}{d} = \frac{\sigma_0}{\kappa \varepsilon_0} = \frac{E_0}{\kappa}, \qquad (25\text{–}22)$$

where $E < E_0$. Note that in each of Eqs. (25–20) through (25–22) we can *replace the permittivity of free space, ε_0, by a new permittivity, ε*, which depends on the dielectric used and on external conditions:

$$\varepsilon = \kappa \varepsilon_0. \qquad (25\text{–}23)$$

PERMITTIVITY MODIFICATION IN DIELECTRICS

Although we have shown this simple rule only for the parallel-plate case, the substitution of ε for ε_0 when a dielectric is involved applies to all geometries and to all equations in which the permittivity appears, such as the expressions for field strength, potential, and energy density.

THINK ABOUT THIS...
HOW LARGE CAN THE VOLTAGE ACROSS A CAPACITOR BE?

Above we described a lightning bolt as the discharge of a capacitor. The voltage across the "plates" becomes so large that the electrical forces literally pull atoms in the air apart, accelerating charges across the space between the plates. As these charges collide with other atoms in the air they may have enough energy to ionize these atoms, releasing still more charged elements. There is a cascade of charges that will end on the plates and discharge the capacitor. This phenomenon is called *dielectric breakdown*. It is evident from our description that the properties of whatever is between the plates plays a role in breakdown; moreover, a discharge can carry enough energy to damage irrevocably this material and hence the capacitor itself. Thus how easily a given dielectric undergoes dielectric breakdown is an additional element, beyond such elements as mechanical strength, cost, etc., in deciding whether it is suitable to fill the space within capacitors. Each dielectric has a *dielectric strength*, E_{max}, which is the maximum electric field a dielectric will support without breakdown. Table 25–1 contains some representative values. Commercial capacitors accordingly include a limitation to a maximum allowable voltage. We'll say more about the interesting topic of dielectric breakdown later.

EXAMPLE 25–10 A parallel-plate capacitor has plate area 20.0 cm² and a plate separation 4.0 mm. (a) Find the capacitance in air and the maximum voltage and charge the capacitor can hold. (b) A Teflon sheet is slid between the plates, filling the entire volume. Find the new capacitance and maximum charge. (c) Before the insertion of the Teflon, the plates are set to a voltage of 24 V by a battery that is then disconnected. What are the energies in the capacitor before and after the Teflon is inserted?

Setting It Up We have a standard parallel-plate capacitor with known plate area A and separation d. We denote the unknown capacitance as C for air and C' for Teflon, according to what fills the space between the plates.

Strategy For parts (a) and (b), this is a matter of applying Eq. (25–4), but with ε_0 replaced by ε in part (b). We can look up the value of the dielectric constant of Teflon in Table 25–1 ($\kappa = 2.1$), and this gives us $\varepsilon = \kappa\varepsilon_0$; equivalently, $C' = \kappa C$. As for the maximum voltages, V_{max} and V'_{max} for air and Teflon respectively, that will be determined by the dielectric strengths of air and Teflon, also found in Table 25–1. Finally, the maximum charge that can be put on the plates is the capacitance multiplied by the maximum voltage.

For part (c), we fix the voltage across the capacitor in the two cases and use Eq. (25–9) to determine the energy given this voltage and the known capacitances.

Working It Out (a) For air we have

$$C = \frac{\varepsilon_0 A}{d} = \frac{(8.85 \times 10^{-12} \text{ F/m})(2.00 \times 10^{-3} \text{ m}^2)}{(4.0 \times 10^{-3} \text{ m})}$$
$$= 4.4 \times 10^{-12} \text{ F} = 4.4 \text{ pF},$$

while for Teflon,

$$C' = \kappa C = (2.1)(4.4 \text{ pF}) = 9.2 \text{ pF}.$$

(b) From Table 25–1, the dielectric strength of air and Teflon are 3×10^6 V/m and 6.0×10^7 V/m respectively, so for air

$$V_{max} = E_{max}d = (3 \times 10^6 \text{ V/m})(4.0 \times 10^{-3} \text{ m})$$
$$\cong 10^4 \text{ V} = 10 \text{ kV},$$

and

$$Q_{max} = CV_{max} = (4.4 \times 10^{-12} \text{ F})(10^4 \text{ V}) \cong 4 \times 10^{-8} \text{ C}.$$

For Teflon

$$V'_{max} = E'_{max} d$$
$$= (6.0 \times 10^7 \text{ V/m})(4.0 \times 10^{-3} \text{ m}) = 2.4 \times 10^5 \text{ V}$$

and

$$Q'_{max} = C'V'_{max} = (9.2 \times 10^{-12} \text{ F})(2.4 \times 10^5 \text{ V}) = 2.2 \text{ }\mu\text{C}.$$

Both the maximum voltage and maximum charge are greatly increased after the Teflon is inserted.

(c) Before the Teflon is inserted, the energy is

$$U = \frac{CV^2}{2} = \frac{(4.4 \times 10^{-12} \text{ F})(24 \text{ V})^2}{2} = 1.3 \times 10^{-9} \text{ J}.$$

After the Teflon is inserted, C increases by the factor κ, whereas V decreases by the same factor. The product CV^2 thus *decreases* by a factor of κ:

$$U' = \frac{C'V'^2}{2} = \frac{(\kappa C)(V/\kappa)^2}{2}$$
$$= \frac{U}{\kappa} = \frac{1.3 \times 10^{-9} \text{ J}}{2.1} = 6.2 \times 10^{-10} \text{ J}.$$

What Do You Think? Did the capacitor do work as the Teflon was inserted? If so, what was its sign?

Example 25–10 shows that when the material in the space between the plates of a capacitor is replaced by a dielectric of higher dielectric constant, the energy decreases. As we implied in the "What do you think?" question for that example, the capacitor therefore does positive work as the new dielectric is inserted. In turn, this means that there must be a force that pulls the dielectric into the space between the plates, and with sensitive instruments, the tug on the dielectric can be measured.

EXAMPLE 25–11 Suppose that the Teflon sheet inserted between the capacitor plates in Example 25–10 is only 2.0 mm thick and fills just half the volume. The capacitor is isolated, and both before and after the Teflon is inserted, the capacitor carries a charge of $Q = 1.0$ nC. Find the electric field everywhere inside the capacitor, and find the new capacitance.

Setting It Up We show the new configuration in Fig. 25–19. Although the Teflon sheet is on the right in Fig. 25–19, this choice is immaterial.

Strategy We previously found the electric field between the plates in air to be $E_0 = \sigma/\varepsilon_0$. Where Teflon is present rather than air, ε_0 is replaced by $\varepsilon = \kappa\varepsilon_0$, or equivalently the electric field is $E_{Tef} = E_0/\kappa$ [Eq. (25–22)]. In this case, we will have two successive values of electric field, and by integrating the field all the way across the space, we find the voltage between the plates. Given the known charge, we can then determine the capacitance from Eq. (25–4).

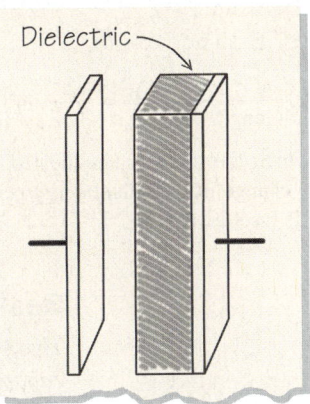

▲ **FIGURE 25–19** A dielectric inserted in a parallel-plate capacitor fills only half the volume.

(continues on next page)

Working It Out The electric field strengths in air and Teflon are

$$E_0 = \frac{\sigma}{\varepsilon_0} = \frac{Q}{\varepsilon_0 A} = \frac{1.0 \times 10^{-9}\,\text{C}}{(8.85 \times 10^{-12}\,\text{F/m})(2.0 \times 10^{-3}\,\text{m}^2)}$$
$$\cong 5.6 \times 10^4\,\text{V/m},$$

$$E_{\text{Tef}} = \frac{E_0}{\kappa} = \frac{5.6 \times 10^4\,\text{V/m}}{2.1} \cong 2.7 \times 10^4\,\text{V/m}.$$

The direction of these fields is from the positively charged plate to the negatively charged one. To find the total voltage drop across the plates (which we need to find the capacitance), we integrate the electric field over the distance between the plates:

$$V = \int_0^d E\,dx = \int_0^{2\,\text{mm}} E_0\,dx + \int_{2\,\text{mm}}^{4\,\text{mm}} E_{\text{Tef}}\,dx$$

$$= E_0 \int_0^{2\,\text{mm}} dx + E_{\text{Tef}} \int_{2\,\text{mm}}^{4\,\text{mm}} dx = E_0(2\,\text{mm}) + E_{\text{Tef}}(2\,\text{mm})$$

$$= (5.6 \times 10^4\,\text{V/m})(2 \times 10^{-3}\,\text{m})$$
$$+ (2.7 \times 10^4\,\text{V/m})(2 \times 10^{-3}\,\text{m}) \cong 170\,\text{V}.$$

This calculation makes it clear that it does not matter whether the Teflon sheet is on one side or the other, or even in the middle.

Finally, the capacitance is, by definition, $C = Q/V$:

$$C = \frac{1.0 \times 10^{-9}\,\text{C}}{170\,\text{V}} = 5.9\,\text{pF}.$$

This value is intermediate to the capacitances of the system empty or filled with Teflon. This capacitor is equivalent to two capacitors in series—one empty and of width $d/2$; the other filled with Teflon and of width $d/2$.

What Do You Think? If we used Bakelite instead of Teflon, would the new capacitance be smaller, larger, or unchanged?

EXAMPLE 25–12 Consider the system shown in Example 25–3—what we'll refer to here as a coaxial cable. (We ignore the necessary mechanical presence of a support for the structure.) With equal but opposite charges per unit length λ on the two elements of the cable, the latter acts as a capacitor. A plug of insulating material of dielectric constant κ is inserted between the wire and the sheath to a depth x from the end. What is the change in potential energy of the charged cable? What electric force, if any, acts on the material as it is inserted? Assume that the charge density on the cable is unchanged as the material is inserted.

Setting It Up We show the coaxial cable in Fig. 25–20 with the radii of the central wire and outer tube a and b, respectively.

Strategy In the presence of the dielectric the charge on the elements is unchanged, while the field and hence the potential drop across the elements will change. In Example 25–3 we found the potential difference between the conductors in the coaxial cable to be $V_b - V_a = V = \dfrac{\lambda}{2\pi\varepsilon_0} \ln \dfrac{b}{a}$. By using a charge per unit length λ in place of a total charge Q, Eq. (25–10) will give us an energy per unit length, and we will want to see if this changes as the dielectric is inserted—we can anticipate from previous examples that it will. A change in the energy will imply a force, and its sign will in turn depend on the sign of the energy change: A decrease in the energy in the cable as the dielectric is inserted means a force pulling the material in. More precisely, if U is the energy in the cable as a function of the insertion distance x, the force will be $-dU/dx$.

Working It Out From Eq. (25–10),

$$\frac{U}{\text{unit length}} = \frac{1}{2}\lambda V = \frac{\lambda^2}{4\pi\varepsilon_0} \ln \frac{b}{a}.$$

In the region in which the dielectric is located, ε_0 is replaced by $\varepsilon = \kappa\varepsilon_0$. Thus the change in potential energy per unit length when the dielectric plug is inserted is

$$\frac{\Delta U}{\text{unit length}} = \frac{U_{\text{dielectric}}}{\text{unit length}} - \frac{U_{\text{air}}}{\text{unit length}} = \frac{\lambda^2}{4\pi}\left(\frac{1}{\varepsilon} - \frac{1}{\varepsilon_0}\right) \ln \frac{b}{a}.$$

Thus if the plug penetrates to a depth x, the total change in potential energy is

$$\Delta U = \frac{x\lambda^2}{4\pi}\left(\frac{1}{\varepsilon} - \frac{1}{\varepsilon_0}\right) \ln \frac{b}{a} = \left(\frac{1}{\kappa} - 1\right)\frac{x\lambda^2}{4\pi\varepsilon_0} \ln \frac{b}{a}.$$

Because $\kappa > 1$, $\Delta U < 0$. The energy decreases when the plug is inserted, so we expect the plug to be pulled into the space between the conductors. The force exerted on the dielectric as it moves into the cable is

$$F = -\frac{dU}{dx} = \left(1 - \frac{1}{\kappa}\right)\frac{\lambda^2}{4\pi\varepsilon_0} \ln \frac{b}{a}.$$

F is positive because $\kappa > 1$. The plug is indeed pulled into position.

What Do You Think? The force on the plug is independent of the depth to which it is inserted. Why?

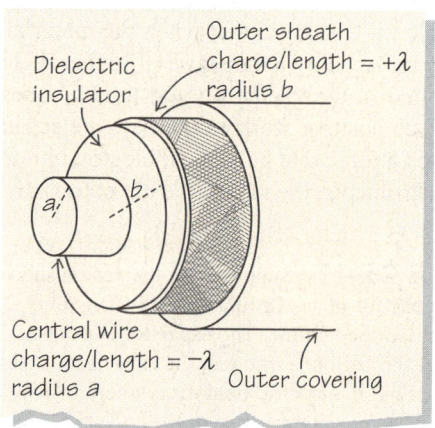

▲ **FIGURE 25–20** A coaxial cable.

Real Capacitors

Two types of capacitors comprise the bulk of modern capacitors. In *multilayer ceramic capacitors*, metal sheets separated by ceramic insulators with dielectric constants as high as 20,000 are folded into a compact form. The dielectric constant is so high that capacitances on the order of millifarads can be reached. *Electrolytic capacitors* can achieve capacitances of roughly the same size in even smaller volumes. In this case, the dielectric—a nonconducting metal oxide—is deposited in a thin layer on a sheet of metal. The second conductor is a conducting paste or liquid that adheres well to the metal oxide. The dielectric layer between the conductors can be made quite thin—as thin as 10^{-8} m. Moreover, by etching the metal before the dielectric layer is deposited, a series of sharp valleys is created in the metal, greatly increasing its surface area. If we recall that the capacitance of parallel plates is inversely proportional to the distance between the plates and proportional to the area of the plates, we see that electrolytic capacitors can achieve large capacitances.

THINK ABOUT THIS...
HOW ARE CAPACITORS WITH VERY LARGE CAPACITANCES CONSTRUCTED?

Capacitors with large capacitance are potentially important in many applications, and this has encouraged extensive work. The technology now exists for making capacitors with capacitances on the order of farads, but such capacitors cannot be built with a couple of flat plates! With a suitable dielectric, a 6-V potential difference across a gap of as little as 0.25 μm is possible without breakdown of the dielectric material. A 1-F capacitor for this situation must therefore have an area $A = (0.25 \times 10^{-6} \text{ m})(1 \text{ F})/(8.85 \times 10^{-12} \text{ F} \cdot \text{m}^{-1}) = 2.8 \times 10^4$ m^2; this corresponds to a square more than 100 m on a side, twice as large as a football field. One way to make capacitors with capacitances this large is to make one of the "plates" from a conducting powder which is compressed only enough to join the grains but not so much as to close the pores. The result is a piece of material with a huge surface area. A dielectric is deposited to make a very thin layer on the compacted powder, and the entire piece is then immersed in a conducting liquid that forms the second "plate." While such capacitors can hold a lot of charge, the paths that current follows through them is long and, in fact, the capacitor cannot be discharged very rapidly.

25–6 The Microscopic Description of Dielectrics

We briefly argued at the start of the previous section that the presence of dipoles within dielectrics explains their macroscopic behavior in electric fields. In this section we shall expand on these comments. We refer to molecules with permanent electric dipole moments, such as H$_2$O, as *polar* molecules. In the absence of an external electric field, the directions of the dipole moments of polar molecules in a material are randomly distributed (Fig. 25–21a). However, when the material is placed in an external electric field, as

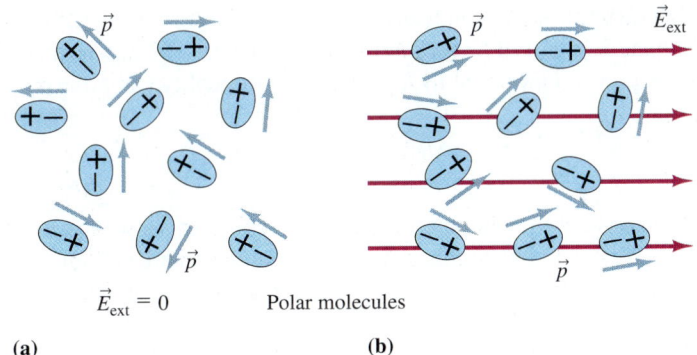

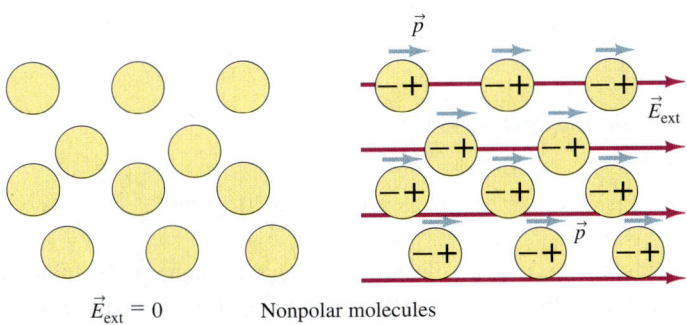

◀ **FIGURE 25–21** (a) In the absence of an external electric field, the dipole moments of polar molecules, here denoted $\vec{p}$, are randomly oriented. (b) In the presence of an external electric field, the dipole moments align themselves with the field. (c) In the absence of an external electric field, nonpolar molecules have no dipole moment. (d) In the presence of an external electric field, nonpolar molecules obtain an induced dipole moment aligned with the electric field.

in Fig. 25–21b, the dipoles are subject to a torque and tend to align themselves with the field. Thermal agitation disturbs the individual alignments, leaving only an average alignment which will be more pronounced for stronger electric fields and lower temperatures. The average degree of alignment grows *linearly* with the external electric field over a wide range of values of the field.

Nonpolar molecules are those without a permanent dipole moment. In the absence of an external electric field, their charge distributions are symmetric; that is, no particular direction is apparent (Fig. 25–21c). As we discussed in Section 22–5, when these molecules are placed in an external electric field, their charge components separate, and they acquire an induced dipole moment that is fully aligned with the field (Fig. 25–21d). The magnitude of this dipole moment increases as the external field increases, and again there is an important range of values of the external electric field for which the dipole moment grows *linearly* with the field.

The effect of placing an insulating slab made of either polar or nonpolar molecules in a charged capacitor is shown in Fig. 25–22a. As opposed to a conductor, in an insulator there is no free charge to move through the material. However, the dipole moments, either permanent or induced, become aligned with the electric field. The inside of the dielectric remains electrically neutral, but the charge distribution is *polarized*, and *induced charge* appears on the two outside surfaces of the slab. We denote the induced surface charge density as σ_{ind}. The two dielectric surfaces have equal but opposite induced charge densities. The induced surface charge density is proportional either to the degree to which the permanent dipole moments of polar molecules are aligned, or, for a material consisting of nonpolar molecules, to the magnitude of the induced dipole moments. Both of these are linear in the external field over a large range of values of the external field, so there is an important range of values of the external field for which σ_{ind} *is proportional to the external field*.

By assuming that the induced charge is proportional to the external field, we can distinguish the electric fields that appear in dielectrics. There are three fields: The external field $\vec{E}_0$ is present whether the dielectric is present or not. An *induced electric field* $\vec{E}_{ind}$ is produced by the induced surface charge on the dielectric, and this is proportional to the external field. Because of the way the induced charge forms, $\vec{E}_{ind}$ is opposite to $\vec{E}_0$. Finally, the net electric field $\vec{E}$ inside the dielectric is

$$\vec{E} = \vec{E}_0 + \vec{E}_{ind}. \tag{25-24}$$

The direction of $\vec{E}$ is indicated in Fig. 25–22b. From Eq. (25–24), we can see that the resultant field $\vec{E}$ will be proportional to $\vec{E}_0$ because if $\vec{E}_{ind}$ (which we know is proportional to σ_{ind}) is proportional to $\vec{E}_0$. This proportionality can be expressed with the constant κ:

$$\vec{E} = \frac{\vec{E}_0}{\kappa}.$$

The constant κ is, in fact, the dielectric constant, as we can see from Eq. (25–22).

We can translate this in terms of charge densities. Suppose that the original electric field $\vec{E}_0$ is produced by a surface charge density σ on the capacitor plates. We refer to σ as the density of *free charge*. We know that the magnitude of $\vec{E}_0$ is $E_0 = \sigma/\varepsilon_0$ and that $\vec{E}_{ind}$ has magnitude $E_{ind} = \sigma_{ind}/\varepsilon_0$ and points to the left if $\vec{E}_0$ points to the right. From Eq. (25–22), we have $E = \sigma/\kappa\varepsilon_0$, so Eq. (25–24) becomes

$$\frac{\sigma}{\kappa\varepsilon_0} = \frac{\sigma}{\varepsilon_0} - \frac{\sigma_{ind}}{\varepsilon_0}. \tag{25-25}$$

We cancel ε_0 and solve for σ_{ind}:

$$\sigma_{ind} = \sigma\left(1 - \frac{1}{\kappa}\right). \tag{25-26}$$

Because $\kappa > 1$ for all dielectrics, the induced charge is always less than the free charge. This is evident in our microscopic model; if E_{ind} were to exceed E_0, we would actually reverse the field in the material.

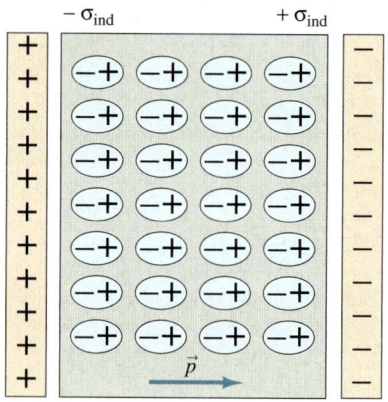

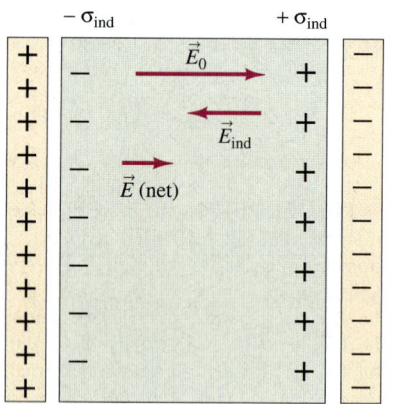

▲ **FIGURE 25–22** (a) When a dielectric is inserted in a capacitor, induced charges appear on the surface. (b) The induced charges cause an induced electric field $\vec{E}_{ind}$ opposite to the external electric field $\vec{E}_0$ caused by the free charge on the capacitor plates. The net effect is a reduced electric field $\vec{E} = \vec{E}_0 + \vec{E}_{ind}$ within the dielectric.

Gauss' Law and Dielectrics

How does the fact that ε_0 is replaced by ε in a dielectric affect Gauss' law? For the Gaussian surface drawn in Fig. 25–23, Gauss' law gives

$$\int_{\text{closed surface}} \vec{E} \cdot d\vec{A} = \frac{Q_{\text{encl}}}{\varepsilon_0} = \frac{Q - Q_{\text{ind}}}{\varepsilon_0}, \qquad (25\text{--}27)$$

where Q is the free charge enclosed, the charge that sets up the original field $\vec{E}_0$. The enclosed charge is the free charge minus the induced charge. But from Eq. (25–25),

$$\frac{Q}{\kappa} = Q - Q_{\text{ind}},$$

and this gives us an alternative form of Gauss' law when dielectrics are present, namely

$$\int_{\text{closed surface}} \vec{E} \cdot d\vec{A} = \frac{Q}{\kappa \varepsilon_0} = \frac{Q}{\varepsilon}. \qquad (25\text{--}28)$$

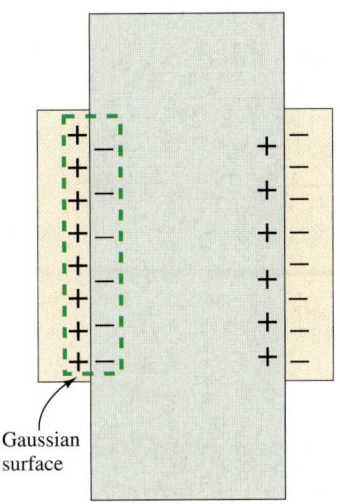

▲ **FIGURE 25–23** A dielectric fills the entire volume of a capacitor. A Gaussian surface surrounding the interface region between the dielectric and each plate surrounds both free and induced charges. The total charge enclosed by the Gaussian surface when Gauss' law is applied includes both charges.

The constant ε_0 is replaced by $\varepsilon = \kappa \varepsilon_0$ provided Q is the free charge. This is a general result even if Fig. 25–23 refers to a parallel-plate capacitor. If the dielectric is not uniform throughout, κ will not have the same value throughout. In this case, κ should be brought under the integral, and Gauss' law reads

$$\int_{\text{closed surface}} \kappa \vec{E} \cdot d\vec{A} = \frac{Q}{\varepsilon_0}. \qquad (25\text{--}29)$$

Consequences of the Microscopic Model of Dielectrics

Our model of induced charges explains the experimental behavior of the dielectrics we have described to this point. This model is also the basis for a variety of testable experimental predictions:

1. There are two classes of dielectrics: those made of nonpolar molecules and those made of polar molecules. The dipole moments of induced dipoles are generally much smaller than those of permanent dipoles, so the value of κ for nonpolar dielectrics should be much closer to one than that for polar dielectrics. Water is an example of a polar dielectric, and you can see from Table 25–1 that its dielectric constant κ is large.

2. Thermal effects that disrupt the alignment are less strong, and polar dielectrics should line up more easily—have larger values of κ—at lower temperatures. Kinetic theory (see Chapter 19) shows that the dielectric constant takes the more precise form

$$\kappa = 1 + \frac{\text{a constant}}{T}. \qquad (25\text{--}30)$$

This temperature dependence, which is *Curie's law*, holds rather well. The mechanism by which nonpolar dielectrics react to external fields suggests that they should not obey Curie's law, and this is indeed the case.

3. The polarization of crystalline solids with a permanent dipole moment can change if the planes of their lattice structure are stressed by being twisted or pressed. Under such stress, the internal electric fields change, and the changing fields produce an electrical signal. This phenomenon, known as *piezoelectricity*, is the principle behind the operation of some microphones and strain gauges.

Summary

Capacitors are devices for storing electric charge and energy and typically consist of two conductors with equal and opposite charges of magnitude Q and potential difference V. Capacitance is defined as

$$C \equiv \frac{Q}{V}. \qquad (25\text{--}2)$$

The capacitance of a parallel-plate capacitor in air is given by

$$C = \frac{\varepsilon_0 A}{d}. \tag{25-4}$$

where A is the plate area and d is the plate separation. The SI unit of capacitance is the farad: $1\,\text{F} = 1\,\text{C/V}$.

The potential energy of a capacitor can be written as

$$U = \frac{Q^2}{2C} = \frac{CV^2}{2} = \frac{QV}{2}. \tag{25-8, 25-9, 25-10}$$

The energy density, or energy per unit volume, of an electric field is

$$u = \frac{\varepsilon_0 E^2}{2}. \tag{25-12}$$

Capacitors connected in parallel can be replaced by an equivalent capacitor with capacitance

$$C_{eq} = C_1 + C_2 + \cdots + C_n. \tag{25-14}$$

Capacitors connected in series can be replaced by an equivalent capacitor according to the relation

$$\frac{1}{C_{eq}} = \frac{1}{C_1} + \frac{1}{C_2} + \cdots + \frac{1}{C_n}. \tag{25-17}$$

Dielectrics are insulators with a characteristic property called the dielectric constant, κ, $\kappa > 1$. When a dielectric fills the space between the two conducting plates of a capacitor, the value of the capacitance is increased:

$$C = \kappa C_0, \tag{25-18}$$

where C_0 is the capacitance of the capacitor with a vacuum (or air) between its conductors. Our previous results can be modified for the presence of dielectrics by replacing the permittivity of free space, ε_0, by the permittivity ε given by

$$\varepsilon = \kappa \varepsilon_0. \tag{25-23}$$

Each insulator also has a characteristic property called the dielectric strength, which gives the approximate maximum electric field that the insulating material can withstand before it breaks down and ionizes.

The behavior of capacitors can be understood by considering the molecular structure of matter. Polar and nonpolar molecules of a dielectric become aligned with the external electric field, reducing the effects of that field. An alternative form of Gauss' law when dielectrics are present in a capacitor is

$$\int_{\text{closed surface}} \vec{E} \cdot d\vec{A} = \frac{Q}{\varepsilon}, \tag{25-28}$$

where Q is the free charge.

Understanding the Concepts

1. There are two common ways to write SI units of permittivity. Does the fact that there is more than one way present a problem?
2. You have two parallel plates, a battery, a voltmeter, and a piece of unknown plastic. Devise a method to determine the dielectric constant of the plastic.
3. What argument can you give to show that the electric field of a parallel-plate capacitor cannot drop abruptly to zero as we pass outside the region between the plates? Recall the fact that the voltage drop around any closed path must be zero.
4. What is the meaning of a capacitor with zero capacitance?
5. If the radius of the inner wire of the coaxial cable in Example 25–3 approaches zero, the capacitance per unit length of the coaxial cable also approaches zero. What is the physical significance of that?
6. It is not possible to break up every combination of capacitors into a sequence of parallel and series capacitors. Find an example of a combination that cannot be decomposed in this way.
7. We say that a capacitor has a charge Q. But is that what we really mean? Is the total charge on a capacitor actually Q?
8. From our discussion of the physical nature of dielectrics, can you imagine a physical system in which the dielectric constant is less than one?
9. The plates of a charged parallel-plate capacitor are disconnected from the charging battery and are pushed together. What happens to the potential difference, the capacitance, and the stored energy?
10. The plates of a parallel-plate capacitor, still connected to a battery with potential difference V, are pushed together. What happens to the charge on the plates, the capacitance, and the stored energy?
11. You are given a thin metal sheet of area A. You can make it into a spherical shell, roll it into two concentric cylinders, or cut it to make a parallel-plate capacitor. Which arrangement would give the largest possible capacitance?
12. What happens if you short out (connect with a conductor) the two plates of a large, charged capacitor? Could this be dangerous?
13. For finite parallel plates there is a fringe field (see Fig. 25–4). What effect would you expect this phenomenon to have on the capacitance of a parallel-plate capacitor?

14. Is it possible for a pair of nonconductors carrying equal but opposite charges to act as a capacitor? In what ways would such an arrangement differ from, or be similar to, the capacitors treated in this chapter?
15. Consider Example 25–5. Do you think you could charge more than one 12-V capacitor with a car battery? What about a thousand?
16. Why is it a good idea to short out (connect with a conductor) the plates of a large capacitor when the capacitor is not in use?
17. Would you expect the term "dielectric strength" to have meaning for a vacuum?
18. Air, particularly on humid days, can cause charge leakage. Why, then, can capacitors with air between their plates hold charge in a way that is useful for circuits?

Problems

25–1 Capacitance

1. (I) (a) What is the capacitance of two square metal plates, each 50 cm² in area, separated by 1 mm? (b) What is the radius of a conducting sphere with the same capacitance?
2. (I) A coaxial cable has an inner wire of radius 0.8 cm and an outer sheath of radius 1.2 cm. What is the capacitance of a kilometer of the cable?
3. (I) At different times, a 4-μF capacitor has a charge of (a) 4 μC, (b) 10 μC, and (c) 1 mC. What is the voltage across the capacitor in each case?
4. (I) How much charge can be stored on the plates of a 1-μF capacitor if the plates are attached to a battery that can give a potential difference of (a) 2 V? (b) 12 V?
5. (I) You must design a capacitor to store 2×10^{-6} C of charge, but you have only a 3-kV power supply and two metal plates of area 250 cm² each. What limits do you put on the separation between the plates?
6. (I) What is the capacitance of a piece of coaxial cable 180 cm long for which the radius of the inner conductor is 1.0 cm and the radius of the outer conducting sheath is 1.5 cm?
7. (II) Calculate the capacitance of two concentric spherical conductors of radii r and R, respectively. Discuss the limits of (a) finite r, $R \to \infty$; (b) $(R - r) \ll r$.
8. (II) Two concentric conducting spheres have radii of 3.0 cm and 15 cm, respectively, and an equal but opposite charge of 1.4×10^{-7} C. What is the potential difference between them? [*Hint*: Use the results of Problem 7.]
9. (II) A parallel-plate capacitor of area 0.040 m² carries a charge $q = 4.0 \times 10^{-8}$ C. The potential across the plates increases with time t according to the equation $V = 50.0 \text{ mV} + (0.10 \text{ mV/s})t$, as a result of a time-dependent increase of the separation between the plates. Find the function of time that describes the separation.
10. (II) A parallel-plate capacitor has square plates 6 cm on a side, separated by 0.3 mm. The capacitor is charged to 3 V, then disconnected from the charging power supply. What is the charge density on the plates? The total charge on each plate?
11. (II) The capacitance of a variable capacitor used in a radio varies from 0.2 μF to 0.01 μF. The capacitor is charged to a potential difference of 300 V at maximum capacitance and then isolated. At minimum capacitance, what is the voltage?

25–2 Energy in Capacitors

12. (I) A thundercloud has a charge of 900 C and a potential of 90 MV with respect to the ground 1 km below it. (a) What is the capacitance of the system? (b) How much energy is stored in the thundercloud system?
13. (I) A capacitor holds a charge of 0.068 C at a potential of 2900 V. How much energy was required to charge the capacitor?
14. (I) A capacitor in a computer with a capacitance of 0.7 pF has 2 V across it. How much energy is contained within the electric field of the capacitor?
15. (I) A fully charged flash attachment for a camera has electrical energy of 27 J. The potential across it is 300 V. What is the capacitance of the flash attachment?
16. (I) How much energy is stored on a metal sphere of radius 35 cm when a charge of 3.0×10^{-5} C is placed on it?
17. (II) A coaxial cable with an inner wire of diameter 3 mm and an outer sheath wire of diameter 8 mm has a potential of 1 kV between the wires. (a) What is the capacitance of 10 m of the cable? (b) How much energy is stored in the 10-m piece of cable? In a 1-km piece?
18. (II) Two concentric conducting spheres of radii 4.0 cm and 12.0 cm, respectively, are given equal but opposite charges of 6.0×10^{-8} C. How much energy is stored in the system?
19. (II) A capacitor consists of two parallel plates, each of area A. It is charged using a battery of potential V_0, which is then disconnected. (a) How much does the energy of the capacitor change if the separation of the plates is changed from d_0 to d_1? (b) How much work is done by the external force used to move the plates? (c) Suppose that the plates of the capacitor remain connected to the battery as they are moved. How much does the energy stored in the capacitor change under these conditions? (d) Is this change related to the work done by the force moving the plates?

25–3 Energy in Electric Fields

20. (I) The electric field in a large thunderstorm is 125,000 V/m. How much energy is contained in 1 m³? In 1 km³?
21. (I) The electric field due to an infinitely long, uniformly charged wire is calculated in Example 23–5. What is the electrical energy density in the space around the wire as a function of the distance from the wire?
22. (I) A Van de Graaff accelerator with a spherical dome of radius 0.75 m has a potential of 20,000 V in air. Assume that the accelerator is, in effect, a charged sphere. How much energy is stored in its electric field?
23. (I) Approximately how much energy is stored in a cube of sides 5 cm that is 1.0 m from a point charge of magnitude 5×10^{-4} C?
24. (I) The energy density in the space between the plates of a parallel-plate capacitor is 10^{-6} J/m³. What is the voltage between the plates if the separation of the plates is 1 cm?
25. (II) An isolated metal sphere of radius 18 cm is at potential 8300 V. What is the charge on the sphere? What is the energy density of the electric field outside the sphere? Integrate this to obtain the total energy in the electric field.
26. (II) A metal sphere of diameter 25 cm carries a charge of 6.0×10^{-7} C. How much energy is contained in a spherical region of radius 50 cm that is concentric with the sphere?
27. (II) A Geiger–Muller tube is a device used to detect ionizing particles (radioactive products). It is a cylindrical capacitor with the outer metal cylinder at zero potential and the central wire at about 500 V (Fig. 25–24). (a) Calculate the capacitance of the tube if its length is 15 cm, the radius of the outer cylinder is

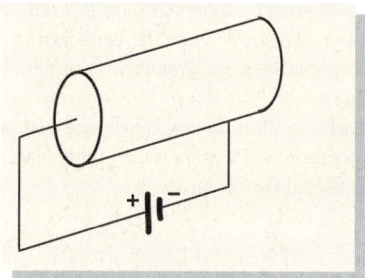

▲ FIGURE 25–24 Problem 27.

2 cm, and that of the central wire is 0.02 cm. (b) When an ionizing particle enters the detector, it creates free electrons and ions, the gas breaks down, and the capacitor discharges. How much energy is needed to recharge the Geiger–Muller tube?

28. (III) A nonconducting sphere of radius 0.070 m carries a uniformly distributed charge of 7.3×10^{-6} C. How much energy is contained in a spherical region of radius 25 cm that is concentric with the sphere?

29. (III) The plates of a parallel-plate capacitor are 400 cm² in area and 0.5 cm apart. The potential difference between the plates is 1500 V. (a) What is the field between the plates? (b) the charge on each plate? (c) the force exerted by the field on one of the plates? (d) Suppose that the plates are pulled apart so that the separation increases by 20 percent. What is the change in the stored energy? Is this consistent with the answer to part (c)? [If not, you probably answered (c) incorrectly.]

30. (III) Assume that an electron consists of a sphere of radius R with its charge distributed uniformly on the surface. (a) What is the electric field outside of the radius R? (b) What is the total electrostatic energy stored in the electric field? (c) Assume that all the energy of part (b) is solely responsible for the rest energy of the electron. (Rest energy is the energy associated with an object's mass, according to the theory of special relativity, even if the object is at rest. It takes the form mc^2, where in this case m is the electron's mass, 0.9×10^{-30} kg, and c is the speed of light, 3×10^8 m/s.) What must the radius R of the electron be?

25–4 Capacitors in Parallel and in Series

31. (I) When two capacitors are connected in parallel, the resulting combination has capacitance 6.5 μF. When the same two capacitors are connected in series, the resulting combination has capacitance 1.4 μF. What are the capacitances of the two capacitors?

32. (II) Find the capacitance of the parallel-plate system in Fig. 25–25. Can this system be represented by two pairs of parallel plates of half the total area connected in series or in parallel?

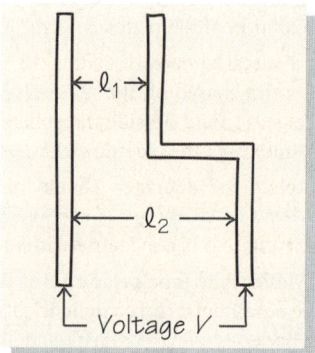

▲ FIGURE 25–25 Problem 32.

33. (II) Find the equivalent capacitance of the circuit shown in Fig. 25–26.

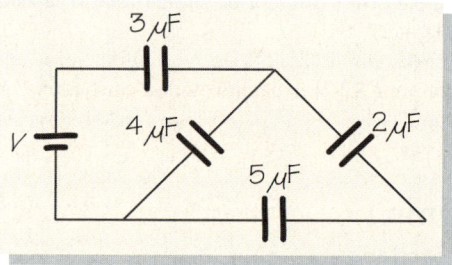

▲ FIGURE 25–26 Problem 33.

34. (II) Two large, thin metal plates of area A and thickness d, carrying charges Q and $-Q$, respectively, are placed a distance D apart (Fig. 25–27). Suppose that an uncharged, thin metal plate of the same area and thickness is placed between them, such that the distance between the uncharged plate and the positively charged plate is x. What is the capacitance of the combined system as a function of x?

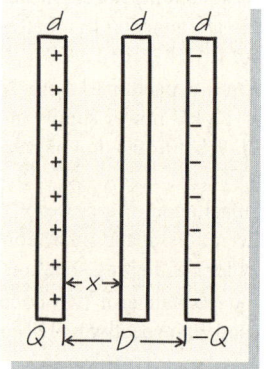

▲ FIGURE 25–27 Problem 34.

35. (II) What is the capacitance of the two concentric, spherical conductors of radii 3.0 mm and 12 mm, respectively, connected as shown in Fig. 25–28a? Suppose that the conductors are connected as shown in Fig. 25–28b. What is the capacitance now?

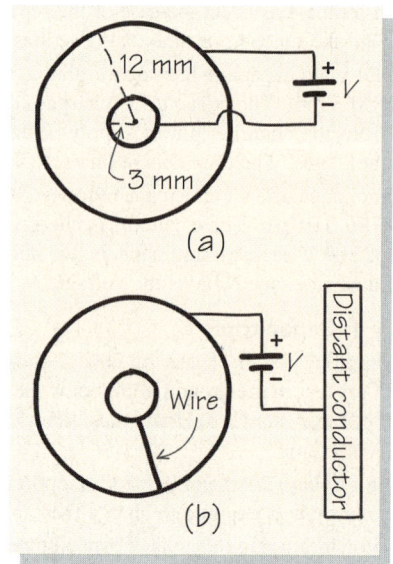

▲ FIGURE 25–28 Problem 35.

36. (II) Find the equivalent capacitance of the circuit shown in Fig. 25–29. The capacitance of each capacitor is 18 μF.

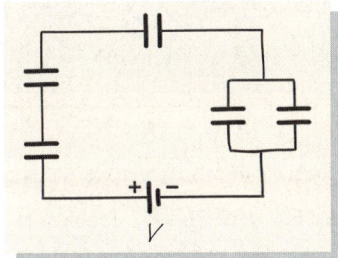

▲ **FIGURE 25–29** Problem 36.

37. (II) (a) Find the equivalent capacitance of the combination of capacitors shown in Fig. 25–30. (b) Assume that the potential difference between b and a is 300 V and find the charge on each of the capacitors in the figure.

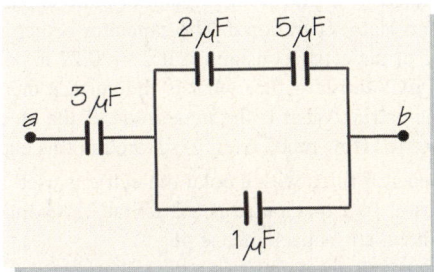

▲ **FIGURE 25–30** Problem 37.

38. (II) Figure 25–31 illustrates a set of five capacitors connected together across the points a and b. What is the value of a single capacitor that could replace this system and collect the same total charge for a given voltage drop V_{ab}?

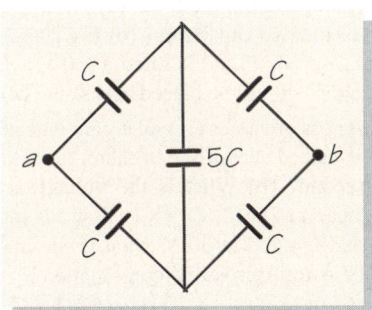

▲ **FIGURE 25–31** Problem 38.

39. (II) Figure 25–32 shows a network of identical capacitors. What is the equivalent capacitance between the points a and b? What is it between a and c? Between b and d?

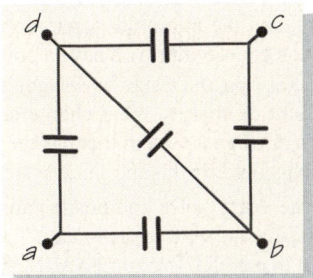

▲ **FIGURE 25–32** Problem 39.

40. (II) Capacitor C_1 has a capacitance of 175 μF; capacitor C_2 has a capacitance of 18 μF. A charge of $q = 4\ \mu$C is placed on C_1, whereas C_2 is brought to a potential difference between its plates of 3 V. (a) What is the total energy stored in the two capacitors? (b) The negatively charged plate of C_1 is connected to the positively charged plate of C_2. What will change in the system, if anything? Neglect the fringe fields at the ends of the capacitors.

41. (II) Consider the capacitors of Problem 40, with C_2 modified so that it holds a charge of 25 μC at a potential difference of 5 V between the plates. (a) How is the capacitance of C_2 modified? (b) What is the charge on the capacitor equivalent to the whole system when the negatively charged plate of C_1 is connected to the positively charged plate of C_2?

42. (II) You have four capacitors whose capacitances are 2 μF, 3 μF, 4 μF, and 5 μF, respectively. Describe a circuit with an equivalent capacitance smaller than the 5-μF capacitor by 0.032 μF.

25–5 Dielectrics

43. (I) Consider a parallel-plate capacitor in which the space between the plates is filled with Teflon. With the charge held fixed, the Teflon is replaced by Plexiglas. If the voltage across the capacitor was 600 V in the first case, what is it after the change?

44. (I) You have a piece of plastic whose dielectric constant you want to measure with two parallel plates, a 4-V battery, and a voltmeter. You charge the plates with the battery and then disconnect them. After you slide the plastic into the full volume between the plates, the voltmeter indicates a voltage drop from 4 V to 3.6 V. What is the dielectric constant?

45. (I) A 12-V automobile battery can store 4×10^6 J of energy. Find the area of a parallel-plate capacitor that can store the same amount of energy, if the separation between the plates is 1 mm and a dielectric with dielectric constant $\kappa = 3$ is between the plates.

46. (II) Repeat Problem 17 for polystyrene placed between the wires of the coaxial cable.

47. (II) A homemade capacitor is constructed out of two sheets of aluminum foil 20 cm by 15 cm in size. The foil is placed on either side of a single page of a telephone book. One hundred pages of the telephone book are 7.7 mm thick. (a) If the paper has $\kappa = 2.9$, what would the capacitance be? (b) The capacitance was measured to be 4.6 nF. What would you estimate for the actual value of the dielectric constant? (c) Make a plot of the capacitance for 1, 2, 4, 10, and 20 telephone pages between the foils. What does this curve look like? (Such low capacitances can be measured with a typical undergraduate physics laboratory multimeter. However, one does have to take account of the capacitance of the multimeter leads themselves.)

48. (II) Electrolytic capacitors involve a very thin coating of a dielectric between two conducting plates that are very close together. Let's imagine such a capacitor with plates of area 2 cm^2, but an insulating layer (and hence a separation) only 1 atom thick. If an atom is approximately 0.1 nm in diameter and has a dielectric constant of 10, what capacitance could we attain? Such a capacitor would have to operate at a relatively low voltage.

49. (II) Calculate the change in capacitance of an isolated sphere that becomes embedded in a dielectric with dielectric constant κ. If the capacitance change is due to a charge induced on the surface of the dielectric, what is the ratio of the induced charge density to that of the original surface charge density?

50. (II) Two large, parallel metal plates have a potential difference of 12 kV, and the electric field between them has a magnitude of 0.90×10^6 V/m. A material with a dielectric constant of 1.5 is inserted between the plates, with the plate separation adjusted so that the capacitance is unchanged. Calculate the new plate separation.

51. (II) A parallel-plate capacitor carrying charge q_0 is modified by the insertion of a dielectric with $\kappa = 1.8$ between the plates. As a consequence, the energy stored in the capacitor triples. What will the charge be after the dielectric is inserted?

52. (II) A coaxial cable has an inside wire of radius 3.5 mm and an outside metal sheath of radius 5.0 mm. The intermediate region is filled with a material of dielectric constant 2.2. (a) What is the capacitance of such a cable 100 m long? (b) If the potential difference between the inner and outer conductors is 500 V, what is the charge on the inner conductor, and how much energy is stored in 100 m of cable?

53. (II) A dielectric slab of thickness d and dielectric constant κ is inserted in the middle of a parallel-plate capacitor of plate separation D. What is the new capacitance of the capacitor, given that the area of each plate is A?

54. (II) A parallel-plate capacitor of dimensions 20 cm × 28 cm and separation distance 1.6 cm contains a dielectric slab of thickness 0.6 cm and dielectric constant 1.8. The potential difference between the plates is 600 V (Fig. 25–33). What are the electric fields in the empty space and inside the dielectric?

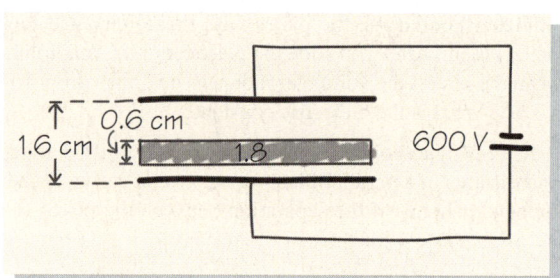

▲ FIGURE 25–33 Problem 54.

55. (II) A parallel-plate capacitor of area 10 cm² and plate separation 5 mm holds how much free charge if the voltage between its plates is 300 V, and the following materials are inserted between its plates: air, paper, neoprene, Bakelite, and strontium titanate? (Use Table 25–1.)

56. (II) Two parallel plates of area 0.80 cm² with Plexiglas inserted between them break down when a voltage of 6 kV is applied to the plates. How much charge will the plates hold when the Plexiglas is removed? (Use Table 25–1.)

57. (II) A capacitor consists of two concentric spherical shells of radii r_1 and r_2, respectively. Calculate the capacitance if the space between the shells is filled with a dielectric of dielectric constant κ. If the capacitor starts out with air between the shells and carries a charge Q, and if the space is then filled with the dielectric, what is the change in energy?

58. (II) A parallel-plate capacitor has area $L \times L$ and separation $D \ll L$. One-half the space between the plates is filled with a dielectric for which $\kappa = \kappa_0$, and the other half with a dielectric for which $\kappa = \kappa_1$ (Fig. 25–34). Find the capacitance of this capacitor.

59. (II) A capacitor consists of 10 plates attached alternately to a positive and negative terminal. The plates are 6.0 cm × 8.0 cm in size and are 1.2 mm apart. What is the capacitance? Suppose that the region between the plates is filled with material of dielectric constant 2.8. What will the capacitance be?

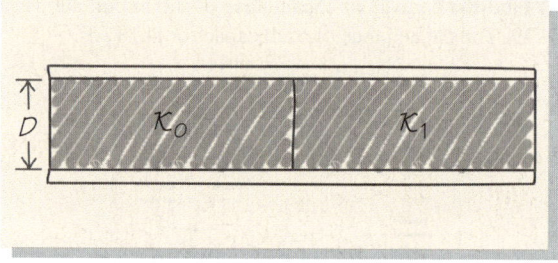

▲ FIGURE 25–34 Problem 58.

25–6 The Microscopic Description of Dielectrics

60. (I) By measuring the capacitance and voltage of a capacitor containing a dielectric with dielectric constant 4.5, the free charge on the capacitor is measured to be 18 μC. What is the induced charge?

61. (II) Use Gauss's law and Eq. (25–24) to show, from Fig. 25–23, that $E_{ind} = \sigma_{ind}/\varepsilon_0$.

62. (II) A charge Q is placed on a parallel-plate capacitor of area $L \times L$ and plate separation d. The capacitor is then filled with a dielectric of dielectric constant κ. If $L = 0.22$ m, $d = 1.8$ mm, $Q = 0.3$ μC, and $\kappa = 3.5$, what is the surface charge induced on the dielectric? What is the magnitude of the electric field in the dielectric? How much energy is stored in this capacitor?

63. (II) A capacitor filled with a polar dielectric is used as a temperature sensor. Its capacitance is 3.2 μF at 23°C and 2.65 μF at 87°C. What is the capacitance at 48° C?

General Problems

64. (I) An uncharged metal plate is inserted midway between the plates of a parallel-plate capacitor carrying charges Q and $-Q$ on the plates. Will the plate be sucked in or will it have to be pushed in? Give a simple explanation for your result.

65. (I) Two capacitors of values C_1 and C_2 are placed in series, and a voltage V is placed across the total. (a) What is the total voltage across each of the two capacitors? (b) If $C_1 = 2$ nF and is rated at 10 V, and $C_2 = 3$ nF and is rated at 30 V, what is the maximum voltage that should be placed across the combination?

66. (I) Two capacitors of values C_1 and C_2 are placed in parallel, and a voltage V is placed across one of them. (a) What is the voltage across the second? (b) What is the voltage across the single equivalent capacitor? (c) If $C_1 = 2$ nF and is rated at 10 V, and $C_2 = 3$ nF and is rated at 30 V, what is the maximum voltage that should be placed across the combination?

67. (II) *Estimate* how much charge you pick up when you walk across a carpet on a dry winter day. [*Hint*: View yourself as a good conductor, spherical in shape, and notice how close your hand has to come to a doorknob before the inevitable spark occurs. Use Table 25–1.]

68. (II) You have 300 cm² of aluminum plate (which you can cut into pieces) and a 500-cm² sheet of 5-mm-thick Bakelite (which you can also cut). Neither material can be sliced into thinner sheets or rolled, and the minimum separation between any aluminum plates you cut is 5 mm. You have a power supply of a single voltage, 1200 V. (a) Design a system that will hold the maximum amount of charge. What charge and energy can this system hold? (b) Design a system that has the maximum electric field, and find this field. Is this the same system as part (a)?

69. (II) Calculate the energy of a composite capacitor that consists of N identical capacitors of capacitance C_1 that are connected (a) in series; (b) in parallel. In parts (a) and (b), the total potential difference across the composite capacitor is V. (c) Assume that the total charge is Q, and repeat the calculation.

70. (II) A capacitor consists of two flat metal plates of area 0.28 m² and plate separation of $d = 1.5$ cm. A flat metal plate of the same area and of thickness 0.5 cm is inserted midway between the plates of the capacitor, leaving two spaces of thickness 0.5 cm each. (a) Find the new capacitance. (b) If the original capacitor has charge Q, what is the surface charge density induced on the intermediate plate? (c) Suppose that the original charge on the external plates remains the same. How does the energy of the new system compare to the energy of the system without the inserted plate? (d) Compare the capacitor with the metal inserted to the same capacitor with a dielectric of the same dimensions inserted.

71. (II) A parallel-plate capacitor has an area of $L \times L$ and a plate separation of $D \ll L$. It is filled with a nonuniform dielectric whose dielectric constant varies linearly across the capacitor (Fig. 25–35). At $x = 0$, $\kappa = \kappa_0$, and at $x = L$, $\kappa = \kappa_1$. We can express κ as a function of x: $\kappa = \kappa_0 + [(\kappa_1 - \kappa_0)x/L]$. Treat the capacitor plates as broken into a set of capacitors connected in parallel with plates that are strips of width dx, and calculate the capacitance.

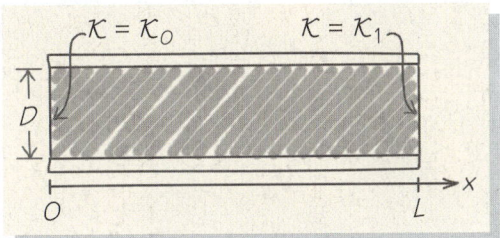

▲ **FIGURE 25–35** Problem 71.

72. (II) A thunderstorm is a fairly complicated phenomenon in terms of the distribution of charges, but we can estimate that there is a voltage drop of as much as 10^8 V between Earth and the bottom of a thundercloud, and the charges involved may run into the hundreds of coulombs. Estimate the capacitance of the Earth–cloud system and the energy contained in the space between the cloud and Earth.

73. (II) Consider the arrangement of the four initially uncharged capacitors shown in Fig. 25–36. Capacitors A, B, C, and D have capacitances $5.4\,\mu\text{F}$, $4.3\,\mu\text{F}$, $3.2\,\mu\text{F}$, and $2.1\,\mu\text{F}$, respectively. Suppose that a battery applies a potential difference of 3000 V across the circuit, which is then disconnected from the battery. What is the potential difference across each capacitor?

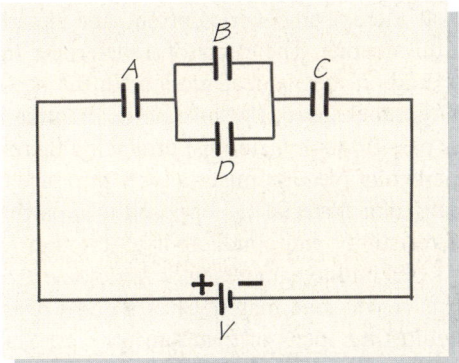

▲ **FIGURE 25–36** Problem 73.

74. (II) Consider a parallel-plate capacitor of plate area 0.40 m² and plate separation 3.0 mm. (a) Assume that the maximum electric field strength (before breakdown) in air is 2.7×10^6 V/m. What are the capacitance and the charge stored at the maximum voltage? (b) Suppose that the capacitor is immersed in oil of dielectric constant $\kappa = 6.0$, and the maximum charge that can be stored is a factor of 10 larger than that without the oil. What is the maximum field strength the oil can maintain?

75. (II) A parallel-plate capacitor has a capacitance of $3.0\,\mu\text{F}$. The plates are charged to 1500 V. What is the energy stored in the capacitor? How much work is required to insert a dielectric of $\kappa = 2.8$ between the plates? Assume that the capacitor is disconnected from the voltage source before the dielectric is inserted.

76. (II) A dielectric of dielectric constant κ is inserted a distance x into a parallel-plate capacitor with square plates of area A and plate separation d. What is the capacitance as a function of x? Calculate the amount of energy stored in the capacitor for a potential difference V.

77. (II) A parallel-plate capacitor has an area of $L \times L$ and a plate separation of $D \ll L$. It is filled with a nonuniform dielectric whose dielectric constant varies linearly from one plate to another (Fig. 25–37). At the bottom plate, the dielectric constant is κ_0; at the upper plate, it is κ_1. If y is the distance measured up from the bottom plate to the top plate, then $\kappa = \kappa_0 + [(\kappa_1 - \kappa_0)y/D]$. Treat the capacitor as a set of capacitors connected in series, and calculate the capacitance.

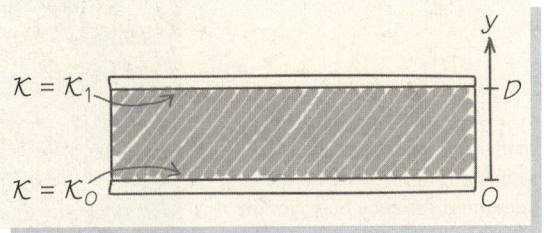

▲ **FIGURE 25–37** Problem 77.

78. (II) Three capacitors of strengths $2\,\mu\text{F}$, $4\,\mu\text{F}$, and $9\,\mu\text{F}$, respectively, can be connected in various ways between two points. What arrangement gives the smallest equivalent capacitance, and what arrangement gives the largest equivalent capacitance?

79. (II) Show that when capacitors are arranged in series, the total capacitance is less than any of the individual capacitances.

80. (III) Two identical capacitors of capacitance C are connected in series across a total potential V. A dielectric slab of dielectric constant κ can fill one of the two capacitors and is slowly inserted into that capacitor (Fig. 25–38). Compute the changes in the total electric energy of the two capacitors, in the charge on each capacitor, and in the potential drop across each capacitor. Account for any energy change by a corresponding change in energy in some other part of the system.

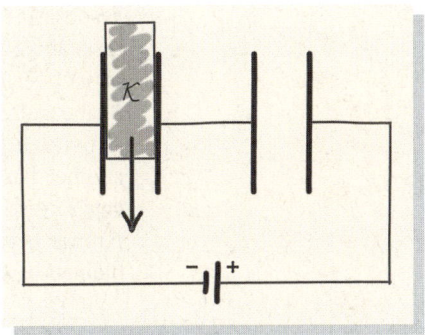

▲ **FIGURE 25–38** Problem 80.

Chapter 26

▶ An electric arc longer than one meter jumps through air due to the large potential difference between the two electrodes. The current carried in such an arc can be quite large.

Currents in Materials

In the last few chapters we have been discussing electrostatics, but we also encountered some examples of charge movement, such as the charging and the rapid discharge of a capacitor in Chapter 25, and of the motion of electrons as an electron beam in Section 22–4. In fact the motion of charge within conducting materials—for example, within the wiring, appliances, and other circuits in your home—is everywhere around us. When charges move, we say there is an electric current, and electric currents within conductors form the topic of this chapter. The effect of a material's structure on electric current is like the effect of drag on mechanical motion, in the sense that the charges move on average at constant terminal speed. Because of the draglike forces, we must expend energy to make charges pass through materials, producing thermal energy. To describe these phenomena for a particular piece of material, we introduce the idea of resistance; to describe them for a particular material (as opposed to a particular piece of it), we introduce the concepts of resistivity and conductivity. Although we can describe a classical model that is an aid to the intuitive understanding of some aspects of resistivity, a fundamental understanding of why one material has a given resistivity and another has a different resistivity requires the ideas of quantum physics. Materials are categorized according to how easy it is for a current to move through them, and the difference between conductors, insulators, semiconductors, and superconductors also is explained by quantum physics.

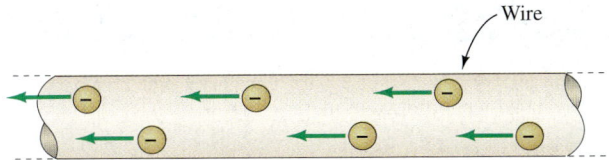

◀ **FIGURE 26–1** Charges move in a cross section of wire, meaning that a current is flowing.

26–1 Electric Current

Electric current (or just **current**) is defined as the total charge that passes through a given cross-sectional area per unit time. Current can be composed of moving negative charges such as electrons or positive charges such as protons; it may occur within an overall-neutral material such as a conductor, or it may occur as a charged beam such as the electron beam of a television tube. Here, we concentrate on the general notions of current—whether that current describes the motion of charges within free space or within conducting materials. In Fig. 26–1, we have drawn the charge that passes through a conducting wire. Recall that *charge is conserved* (see Chapter 21), meaning that unless charge accumulates within a region, as much charge enters that region as leaves it.

A simple picture may be helpful to you in thinking about electric current within the metals that conduct electric current easily. In this picture, electrons under the influence of the potential that makes them move are like the balls that pass through a pinball machine impelled by gravity. The parent ions that form the lattice structure of the material are like the pins of the pinball machine, and multiple collisions with these fixed ions introduce a random element in the motion of the electrons, slowing them down, giving them a terminal velocity on average. While this picture "explains" many features of electric current, don't take it too seriously; we'll see later in the chapter that it is quite incomplete!

If ΔQ is the amount of charge passing through the cross sectional area of a wire in a time interval Δt, then the *average current*, I_{av}, through the wire is defined as

$$I_{av} \equiv \frac{\Delta Q}{\Delta t}. \tag{26-1}$$

If the current changes with time, we define the *instantaneous* current, I, by taking the limit $\Delta t \to 0$, so that the current is the instantaneous rate at which charge passes through an area:

$$I \equiv \frac{dQ}{dt}. \tag{26-2}$$

DEFINITION OF ELECTRIC CURRENT

The unit of current is the coulomb per second; this unit is also called the *ampere* (A, or, frequently, "amp"), after André Marie Ampère, who performed pioneering work in electricity and magnetism early in the nineteenth century. While the ampere will be defined more precisely in Chapter 28, that definition is equivalent to the simple relation

$$1 \text{ A} \equiv 1 \text{ C/s}.$$

The ampere is a fairly large unit, and it is also convenient to express current in milliamps (mA, 10^{-3} A); microamps (μA, 10^{-6} A); or even nanoamps (nA, 10^{-9} A). You might note that since an electron has a charge magnitude 1.6×10^{-19} C, some 0.6×10^{19} electrons pass each second in a 1 A current.

Currents occur over a wide range of values (Table 26–1). When the current has no time dependence we refer to a direct current (DC). Currents that have a harmonic time dependence, alternating currents (AC), are commonplace; we shall study them in more detail in Chapter 33. The values in Table 26–1 that refer to AC represent the average magnitude of the oscillating current.

TABLE 26–1 • Values of Various Currents

Situation	Current (A)
Advanced-technology computer chips	10^{-12} to 10^{-6}
Electron beam of a TV set	10^{-3}
Minimum current that is dangerous when it passes through the human body (AC or DC)	10^{-2} to 10^{-1}
Proton beam of the Fermilab accelerator	0.3
Flashlight bulb	0.3
Household lightbulb (AC)	1
Automobile starter	200
Peak current in a lightning strike	10^4
Maximum current carried by a superconducting niobium wire of 1 cm² cross section	10^7

The Direction of Current

Current is a scalar quantity, as both charge and time are scalars, but it has a sign associated with it. It is useful to indicate the sign of the current by a directional arrow. Figure 26–2 depicts the direction of current flow within a conductor. By historical convention we associate the direction of current flow with the direction of "flow" of positive charges—even though it is actually the negative charges that move in conducting materials such as metals. In a conductor the positive charges—the atomic ions left behind by the electrons—are fixed in an ordered crystal lattice. In an ionized gas or a chemical solution, the charges that actually move and create the current may be positive or negative. This arbitrary convention for current direction causes no real problem, because a flow of positive charge to the right and a flow of the same amount of negative charge to the left represent the same current. By simply measuring the current, it is *not possible* to determine the sign of the charges that move (the *charge carriers*). By convention:

> **The direction of the current is the direction in which positive charge carriers would move, even if the actual charge carriers are negative charges moving in the opposite direction.**

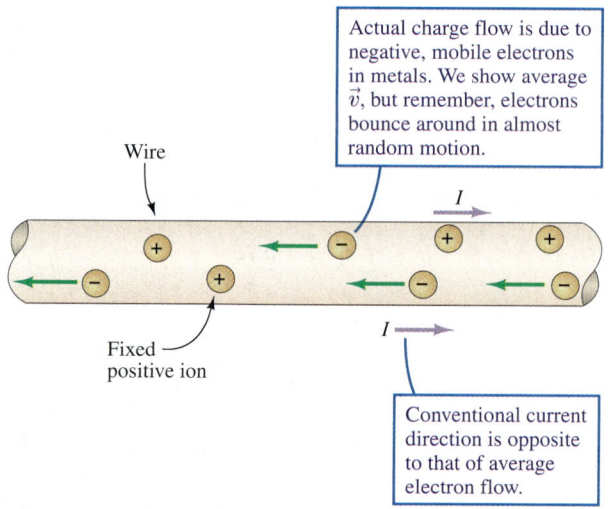

▲ FIGURE 26–2 By convention, the current direction is the direction in which positive charge effectively moves, even though in conductors it is the negative charge, electrons, that actually moves.

EXAMPLE 26–1 An accelerator used for medical research emits protons at the rate of 2.0×10^{13} protons per second. What is the current carried by this beam of protons?

Setting It Up Let us call N the number of protons emitted per unit time. We will need to use the charge per proton, $e = 1.6 \times 10^{-19}$ C.

Strategy The current I is defined as the rate at which charge passes, i.e. the charge per unit time. Here that is simply given by $I = Ne$.

Working It Out

$$I = Ne = (2.0 \times 10^{13} \text{ protons/s})(1.6 \times 10^{-19} \text{ C/proton})$$
$$= 3.2 \times 10^{-6} \text{ A}.$$

What Do You Think? In what way does this current differ from a current in a wire described above? *Answers to* **What Do You Think?** *questions are given in the back of the book.*

CONCEPTUAL EXAMPLE 26–2 A colliding beam experiment at the accelerator at the Fermi National Laboratory involves a beam of protons moving in the positive x-direction, and a beam of antiprotons moving in the opposite direction. Is there a net current here, and what is its sign? (For our purpose it is enough to know that antiprotons are particles with the same mass as protons, but with a charge of equal magnitude and opposite sign.)

Answer Suppose that the number of protons (charge e) that pass a given point each second is N, and the number of antiprotons (charge $-e$) per second is N'. Then the current associated with the protons is Ne in the positive x direction. The current of antiprotons is $N'(-e)$ in the *negative* x direction, and this is equivalent to a current of $N'e$ in the positive direction. Thus the total current is $(N + N')e$ in the positive x-direction. As far as current-carrying capacity is concerned, the antiprotons are equivalent to electrons.

What Do You Think? Consider also the fact that an antiproton and a proton can collide and annihilate each other, with the result that both disappear. Suppose that $N = N'$ and that all the antiprotons annihilate all the protons in some region where they can collide. Will this violate current conservation?

Current Density

Sometimes we must deal with the *details* of charge motion, not just an overall movement of charge, as for example within nonuniform materials, or even gases and plasmas. In that case we work with **current density**, $\vec{J}$, which is the rate of charge flow per unit area through an infinitesimal area. To define the current density, we must take into account the *local* magnitude and direction of the charge flow. Unlike current, which is a scalar, current density is a *vector*, with units of amperes per square meter. The direction of $\vec{J}$ is defined to be the direction of the net flow of positive charges at that particular infinitesimal element of area—since the area is tiny, it forms a plane and its direction is unambiguous.

What is the relation between current density and current? We determine this relation in a wire by dividing the finite area A through which the charge flows into infinitesimal areas $d\vec{A}$ (Fig. 26–3). This procedure is analogous to one we followed in treating fluid flow (in Chapter 16) or electric flux (in Chapter 23). The differential current dI flowing through $d\vec{A}$ is

$$dI = \vec{J} \cdot d\vec{A} = J \, dA \cos \theta, \quad (26\text{–}3)$$

where θ is the angle between $\vec{J}$ and the area element $d\vec{A}$. From Eq. (26–3), we see that dI is a maximum when $\vec{J}$ and $d\vec{A}$ are parallel and dI is zero when $\vec{J}$ is perpendicular to $d\vec{A}$. The total current passing through the area A is a sum over the differential currents dI:

$$I = \int_{\text{surface } A} \vec{J} \cdot d\vec{A}. \quad (26\text{–}4)$$

If the current density is uniform across a wire, the current through the wire is just the product of the wire's cross-sectional area with the magnitude of the current density.

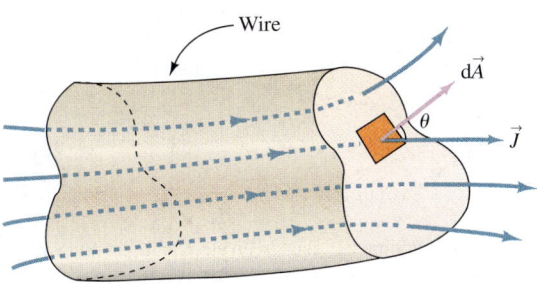

◀ **FIGURE 26–3** The area of a finite wire is divided up into differential areas $d\vec{A}$ with the current density, $\vec{J}$, whose direction is the direction of the local current, defined at every point. The direction of $d\vec{A}$ is normal to the differential area.

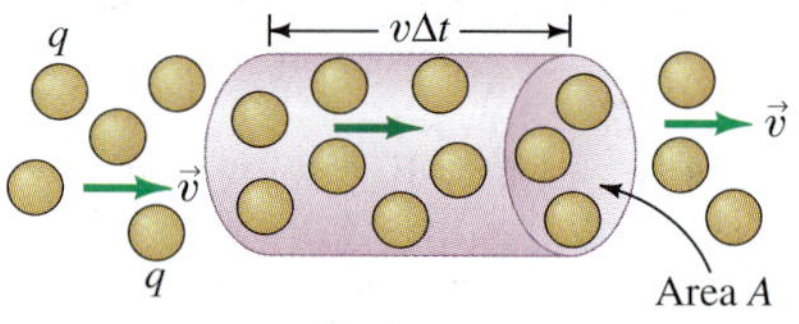

FIGURE 26–4 A collection of particles (each with charge q) with number density n_q all move to the right with velocity $\vec{v}$. The total charge passing through an area A in time Δt is $\Delta Q = n_q q v A \, \Delta t$.

Current Density of Moving Charges

Let us now connect current density to the microscopic motion of individual charges. Consider a collection of moving charges each of charge q, as in Fig. 26–4. In some small region, the number of charged particles per unit volume—the *number density*—is n_q. Suppose also that these particles all move with velocity $\vec{v}$. Then in a time interval Δt, the amount of charge passing through a given area A perpendicular to $\vec{v}$ is ΔQ, the charge contained in the volume $A(v\,\Delta t)$ swept out by the moving charges. We have

$$\Delta Q = \left(\frac{\text{charge}}{\text{volume}}\right)(\text{volume}) = (n_q q)(Av\,\Delta t) = n_q q v A\,\Delta t, \qquad (26\text{–}5)$$

where we have used the fact that the charge per unit volume is the number density of the charge carriers times the charge per particle. Thus the current is given by

$$I = \frac{\Delta Q}{\Delta t} = n_q q v A. \qquad (26\text{–}6)$$

Finally, the current density is I divided by A in the limit of *small A*, or $J = I/A$. The direction of $\vec{J}$ is specified by the direction of $\vec{v}$ with a sign determined by the sign of q:

$$\vec{J} = n_q q \vec{v}. \qquad (26\text{–}7)$$

26–2 Currents in Materials

We have referred to *conductors* as materials through which charge moves easily, *insulators* as materials through which charge does not move easily, *semiconductors* as materials intermediate to conductors and insulators, and *superconductors* as materials that under certain circumstances—in particular, at sufficiently low temperatures—carry current with no opposition whatsoever. How materials carry charge is of central interest for technology and of fundamental scientific importance. In Section 26–1 we mentioned a simple guiding picture (to be developed further in Section 26–5) in which electrons within conductors behave as though they were free to move, except for the presence of the stationary ions that form the crystal lattice of the metal, which form a set of obstacles for the electron motion. In this scenario electrons, constantly moving because the metal is at a finite temperature, undergo frequent collisions with the positive ions that form the crystal lattice of the metal whether the field is present or not (Fig. 26–5). When there is no field, the electrons do not, *on average*, move in any particular direction, and there is no macroscopic current. The motion of the electrons is random, like the motion of air molecules. In the presence of an electric field $\vec{E}$ each "free" electron experiences a force $\vec{F} = -e\vec{E}$ and thus accelerates between collisions in a direction opposite to the electric field.[†] In this case there is a *net* movement of electrons in

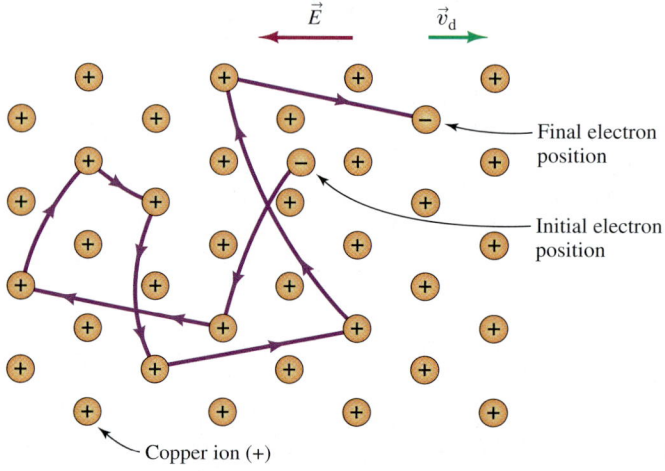

▶ **FIGURE 26–5** In one picture of current flow in a metal, an electron collides frequently with the ions and impurities and scatters randomly from the ions, here indicated with circles enclosing plus signs. When an electric field is present, the electron picks up a component of velocity opposite the field. The electron's path in the electric field is slightly parabolic. The extra component of velocity present due to the electric field is, on average, much less than the average speed of the electrons when no field is present.

[†] In electrostatics, metals contain no electric fields. But when we deal with currents, we are no longer in electrostatics: Charges are moving continuously.

the direction of the electric force they experience. The collisions, in effect, give rise to a drag force on the flow of electrons. As with the fall of a parachute, drag acts to settle the motion to a steady flow in the direction of the force. The electrons move on average with a constant terminal velocity called the **drift velocity**, $\vec{v}_d$—this is the velocity indicated in Figs. 26–4 and 26–5. Equation (26–6) gives the relation between drift speed and current. Here the charge carriers are the electrons, so that n_q equals n_e, the density of free electrons in the metal, and Eq. (26–6) then gives

$$I = n_e q v_d A, \qquad (26\text{–}8)$$

where A is the cross-sectional area of a metal wire.

We can solve Eq. (26–8) to find the drift speed in terms of the current:

$$v_d = \frac{I}{n_e q A}. \qquad (26\text{–}9)$$

Equations (26–8) and (26–9) show the relations between current and drift speed. Remember that the direction of the electron's drift velocity is opposite to the defined direction of the current density because of the positive charge-carrier convention.

In Fig. 26–6, we show the relationships among the external electric field $\vec{E}$, the current I, the current density $\vec{J}$, and the electron's drift velocity $\vec{v}_d$. For the case of the wire, $J = I/A$, and from Eq. (26–9), we have

$$\vec{J} = n_e q \vec{v}_d. \qquad (26\text{–}10)$$

Note the formal confirmation that $\vec{J}$ is opposite to the direction of $\vec{v}_d$: A negative sign is introduced when we recognize that $q = -e$.

◀ **FIGURE 26–6** Electrons drift in the direction opposite that of the current, I, current density, $\vec{J}$, and electric field, $\vec{E}$.

EXAMPLE 26–3 Estimate the drift speed v_d for electrons in a copper wire of diameter $d = 1.0$ mm that carries a current $I = 100$ mA. Copper has about one free electron per atom available to carry charge and has a mass density of 8.92 g/cm^3 and a molecular weight of 63.5 g/mol.

Setting It Up We refer to Fig. 26–6 for a pictorial representation.

Strategy The drift speed can be calculated from Eq. (26–10) if the current density and electron density are known. The current density involves the total current and the area of the wire, both obtainable from the data given. We have for the magnitude of the current density $J = \dfrac{I}{A} = \dfrac{I}{\pi r^2} = \dfrac{I}{\pi (d/2)^2}$, where d is the diameter of the wire and I is the total current. To calculate the free electron density, we make use of the fact that there is one free electron per atom, so that the free electron density equals the density of copper atoms. Here we proceed as follows:

(No. of atoms)/m^3 = [(No. of atoms)/kg][mass density in kg/m^3]

= [(No. of atoms/mole)/(No. of kg/mole)][mass density in kg/m^3].

In the last step, the number of atoms per mole is Avogadro's number N_A. To get the number of kg in 1 mole, we use the fact that the molecular weight of Cu is 63.5 g/mol = 63.5 × 10^{-3} kg/mol, and the mass density is 8.92 g/cm^3 = 8.92 × 10^3 kg/m^3. With all this information in hand we can then use Eq. (26–10) to find v_d.

Working It Out With the number of free electrons per atom denoted by n (here $n = 1$), we have

$$n_e = n N_A \frac{1}{M} \rho_{\text{Cu}}$$

$$= (1 \text{ electron/atom})(6.02 \times 10^{23} \text{ atoms/mol})$$

$$\times \frac{1}{63.5 \times 10^{-3} \text{ kg/mol}} (8.92 \times 10^3 \text{ kg/m}^3)$$

$$= 8.46 \times 10^{28} \text{ electrons/m}^3.$$

A calculation of the current density gives

$$J = \frac{I}{\pi (d/2)^2} = \frac{4(100 \times 10^{-3} \text{ A})}{\pi (1.0 \times 10^{-3} \text{ m})^2} = 1.3 \times 10^5 \text{ A/m}^2,$$

from which we calculate

$$v_d = \frac{J}{n_e q} = \frac{1.3 \times 10^5 \text{ C/s} \cdot \text{m}^2}{(8.5 \times 10^{28} \text{ electrons/m}^3)(1.6 \times 10^{-19} \text{ C/electron})}$$

$$= 9.6 \times 10^{-6} \text{ m/s}.$$

Does this seem like a small value to you? See the "Think about this . . ." box below.

What Do You Think? If the current is increased by a factor of 4 in this problem, by how much would the radius of the wire have to change to keep the current density constant?

THINK ABOUT THIS...
WHY DOES THE LIGHT APPEAR THE MOMENT YOU THROW THE LIGHT SWITCH?

In Example 26–3, we saw that the drift speed in a realistic wire carrying a realistic current is only a tiny fraction of a centimeter per second. Certainly you don't have to wait for hours for the electrons to drift to the lightbulb you want to put to use. In fact, household circuits carry alternating current (the subject of Chapter 33), in which there is no net flow of electrons in one direction or another! So what happens when we switch on a light in our house? When the switch is thrown, the electric field that influences the electrons to move in the wire is set up throughout the wire at speeds approaching the speed of light. The free electrons are spread throughout the wire, and they are each affected by the electric field almost simultaneously. They start moving at once in response to this field—those nearest the switch as well as those nearest the electrical appliance. A similar effect occurs in fluid flow, and we can compare an electric wire with free electrons to a garden hose full of water. If you want to move a sprinkler while you water the lawn, you turn the water off, move the sprinkler, and then turn the water back on. Because the hose is already full of water, the sprinkler starts immediately: The force of the water at the faucet end is quickly transmitted all along the hose, and the water at the sprinkler end of the hose flows from the sprinkler almost the moment the faucet is opened. Similarly, the electrons that move within the filament of the lightbulb are already present when the switch is thrown. When this happens the electric field and hence the current that it drives, are set up almost instantaneously.

Current and the Conservation of Charge

How does the conservation of charge affect currents in materials? The current in a wire describes the flow of charge along the wire. For steady flow (constant current), the rate at which charge enters a section of the current-carrying wire in a given time must equal the rate at which charge leaves the section: There is no accumulation of charge in any part of the wire, and net charge is neither created nor annihilated. This is just the statement that *current is conserved*.

A consequence of this principle is that if the cross-sectional area of the wire changes, the *current density* must change accordingly so that the product JA is unchanged (Fig. 26–7a). Equation (26–10) in turn implies that *for a fixed current*, the drift speed, like the current density J, is inversely proportional to the area of the wire (Fig. 26–7b). The conservation of current will continue to act as a constraint if the wire

▶ **FIGURE 26–7** (a) A steady current is the same in all parts of a wire, even if the area of the wire varies. This means that the current density and drift speed will vary with the area: Both are larger when the wire cross section is smaller. (b) More precisely, the drift speed and current density are inversely proportional to the wire area.

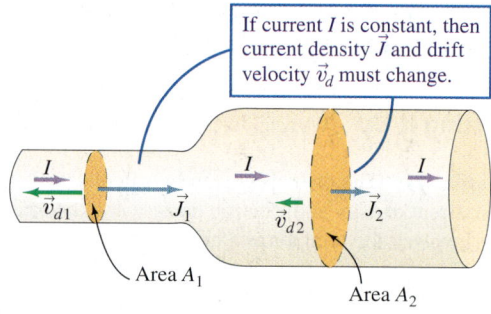

(a)

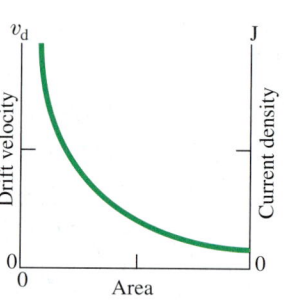

(b)

splits into two or more wires, even if, as we will learn in Section 26–4, the currents in the individual wires are determined by the properties of these wires. The situation is analogous to the flow of water along a pipe. As we learned in Chapter 16, for water in steady flow the conservation of mass implies the conservation of flux of the water. The rate at which water enters a pipe is equal to the rate at which water leaves. If the pipe narrows, the speed of the flow increases, while if the pipe widens, the speed of the flow decreases. If the pipe splits, the rate of flow in each section adjusts itself in a way that is consistent with conservation of the total flow rate.

26–3 Resistance

We have seen that a current flows when an electric field is applied to a conductor. We can equally consider the flow to be due to a potential difference V, with the current flowing from a higher to lower potential. The amount of current that flows through a material for a given potential difference across that material depends on the material's properties and its geometry.

The **electrical resistance** of a piece of material is a measure of how easily charge flows within that material. The resistance R is defined to be the ratio of the voltage (potential difference) across the material to the current that flows through it:

$$R \equiv \frac{V}{I}. \tag{26-11}$$

DEFINITION OF RESISTANCE

If a material has a low resistance, then more current will flow for a given voltage than it will for a higher resistance. The units of resistance are volts per ampere, but a separate SI unit called the **ohm** (Ω) has been defined as the resistance through which a current of 1 A flows when a potential difference of 1 V is applied:

$$1 \, \Omega \equiv 1 \, \text{V/A}.$$

Units of kilohms (kΩ or $10^3 \, \Omega$) or megohms (MΩ or $10^6 \, \Omega$) are also commonly used.

Georg Simon Ohm was the first to study the resistance of different materials systematically. In 1826, he published his experimental result that, for many materials including most metals, *the resistance is constant over a wide range of potential differences*. This statement is called *Ohm's law*. It is not, properly speaking, a law, but rather an empirical statement about the behavior of materials. When the resistance of a material is constant over a range of potential differences, we say that the material is *ohmic*. We shall continue the traditional practice of referring to this linear relation between voltage and current for these materials as a "law" and writing it as

$$V = IR, \tag{26-12}$$

OHM'S LAW

where *R is independent of V*. Figure 26–8 illustrates the consequence of the independence of V and R.

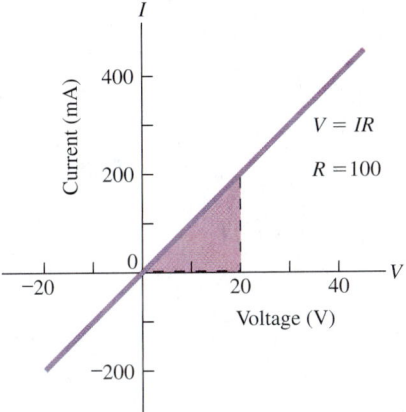

▲ **FIGURE 26–8** For ohmic materials, Ohm's law states that the ratio V/I is a constant.

Resistors

A **resistor** is a piece of ohmic material with a specific value of resistance. It forms a part of many circuits; indeed, since all circuit elements through which current flows have some resistance, it is ubiquitous in circuits. Common resistors have resistances that range from fractions of an ohm to millions of ohms (Fig. 26–9). A resistor of given resistance R with a given potential V between its terminals allows the flow of a current $I = V/R$. Resistors are represented in circuit diagrams by zigzag lines, 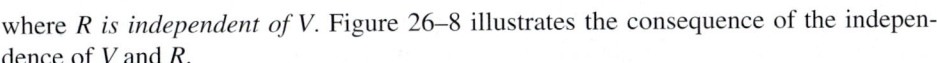. They are connected to each other and to other elements such as capacitors or batteries by conducting wires that are generally assumed to have negligible resistance.

▲ **FIGURE 26–9** Resistors are color coded to indicate the value of their resistance.

THINK ABOUT THIS...
CAN NONOHMIC MATERIAL BE USEFUL IN CIRCUITS?

There are many *nonohmic materials*: materials for which the voltage and current do not obey the linear relation of Ohm's law. In fact, if you look closely enough, there are no perfectly ohmic materials, so this is partly just a question of degree. But some materials have conduction properties that differ spectacularly from Ohm's law (see Section 26–6). Moreover, one can tailor these conduction properties through precise control of the relative proportions of different atoms. Such materials and devices play an enormous role in computing and other electronic applications. Figure 26–10a shows current-versus-voltage curves (ideal and typical) for a **diode**. A diode is a device that transmits current easily when the voltage difference has one sign, but prevents charge flow (that is, it has a very high resistance) when the voltage difference has the other sign. The curve is very far from the linear curve demanded by Ohm's law. Diodes are used in many electric devices (Fig. 26–10b). They may be used to charge a battery while preventing it from discharging, or, more interestingly, they prevent the "backflow" in AC and therefore are important in devices that convert AC to DC.

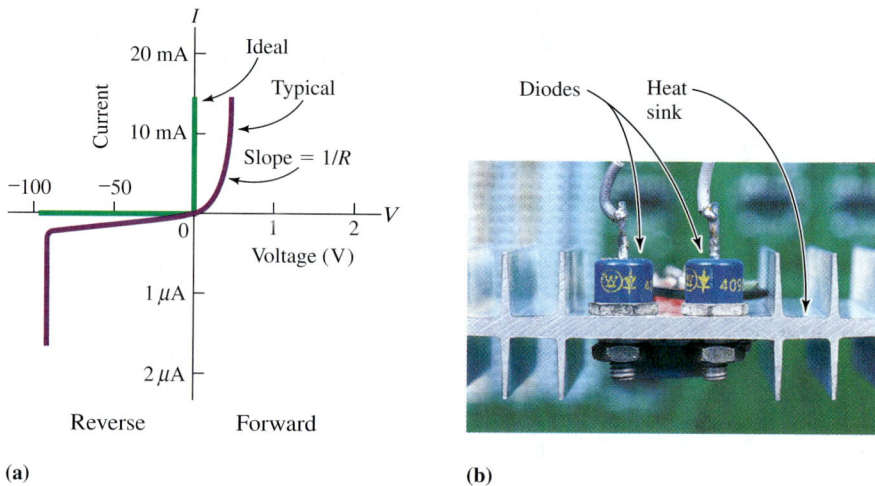

▲ **FIGURE 26–10** (a) Plot of current versus voltage for an ideal diode (green) and a typical real diode (purple). Note that, for the ideal diode, there is no current when the voltage is negative: The diode allows current to flow in only one direction. (b) Two (blue) diodes used to convert AC voltage to DC are bolted to an aluminum plate to dissipate thermal energy (heat sink).

Resistivity and Conductivity

The resistance of a conducting wire of a given material depends both on the shape (length and cross-sectional area) of the wire, and on the intrinsic properties of the material. Our guiding picture for resistance ascribes the difference between the acceleration of a charge in free space and the steady flow in a conductor to the effect of collisions with the ions within the material. These provide a retarding force that opposes the electric field E. This retarding force will clearly vary from material to material. Since the effect of the collisions is a local one—that is, it does not depend on the precise shape or size of the piece of material—we expect that it is $J = I/A$ that will depend on E and the properties of the material. If we define the intrinsic (geometry- and size-independent) **resistivity** ρ of a material by E/J, we see that this can be expressed in terms of I/A and the voltage V, which is related to the electric field by $E = V/L$, where L is the length of the wire. We write

$$\rho \equiv \frac{E}{J}. \qquad (26\text{–}13)$$

DEFINITION OF RESISTIVITY

Using $E = V/L$ and $J = I/A$, and recalling $R = V/I$, Eq. (26–13) reads $\rho = R(A/L)$, or

$$R = \rho \frac{L}{A}. \qquad (26\text{–}14)$$

The reciprocal of the resistivity is the **conductivity**, σ.

$$\sigma \equiv \frac{1}{\rho}. \qquad (26\text{–}15)$$

DEFINITION OF CONDUCTIVITY

Typical values of conductivity are listed in Table 26–2. The resistivities and conductivities of the materials shown in Table 26–2 vary over many orders of magnitude. The conductivity of a metal such as aluminum is a factor of 10^{21} higher than that of a good insulator such as Teflon. It is worthwhile noting here that while our guiding model—acceleration of free electrons with drag due to collisions—may provide us with a concrete picture to work with, it is not remotely adequate to explain the huge range of values for conductivity observed in nature. Only quantum mechanics can do that, and we shall discuss this a little in Section 26–6.

In the basic relation that defined the resistivity in Eq. (26–13), both the electric field and the current density are vectors, and a better version of this relation is

$$\vec{E} = \rho \vec{J}. \qquad (26\text{–}16)$$

RELATION BETWEEN FIELD AND CURRENT DENSITY

TABLE 26–2 • Resistivities, Conductivities, and Temperature Coefficients (at 20°C)

Material	Resistivity, ρ ($\Omega \cdot m$)	Conductivity, σ $(\Omega \cdot m)^{-1}$	Temperature Coefficient, α $(°C)^{-1}$
Conductors			
Elements			
Aluminum	2.82×10^{-8}	3.55×10^{7}	0.0039
Silver	1.59×10^{-8}	6.29×10^{7}	0.0038
Copper	1.72×10^{-8}	5.81×10^{7}	0.0039
Iron	10.0×10^{-8}	1.0×10^{7}	0.0050
Tungsten	5.6×10^{-8}	1.8×10^{7}	0.0045
Platinum	10.6×10^{-8}	1.0×10^{7}	0.0039
Alloys			
Nichrome	100×10^{-8}	0.1×10^{7}	0.0004
Manganin	44×10^{-8}	0.23×10^{7}	0.00001
Brass	7×10^{-8}	1.4×10^{7}	0.002
Semiconductors			
Carbon (graphite)	3.5×10^{-5}	2.9×10^{4}	-0.0005
Germanium (pure)	0.46	2.2	-0.048
Silicon (pure)	640	1.6×10^{-3}	-0.075
Insulators			
Glass	10^{10} to 10^{14}	10^{-14} to 10^{-10}	
Neoprene rubber	10^{9}	10^{-9}	
Teflon	10^{14}	10^{-14}	

EXAMPLE 26–4
Determine the current density, resistance, and electric field for the copper wire of Example 26–3 if the wire is 10 m long.

Setting It Up In addition to the quantities given in the problem statement, we will want to use data for copper in Table 26–2.

Strategy Given the current and the wire's cross section, the current density is $J = I/A = I/(\pi r^2) = 4I/(\pi d^2)$. We are also given the length of the wire, so that the resistance can be calculated—$R = \rho L/A$—and we can look up $\rho = 1.72 \times 10^{-8}\, \Omega \cdot \text{m}$ for copper. Finally we find the desired field using $E = \rho J$.

Working It Out We calculate in order

$$J = \frac{4I}{\pi d^2} = \frac{4(100 \times 10^{-3}\,\text{A})}{\pi(1.0 \times 10^{-3}\,\text{m})^2} = 1.3 \times 10^5\,\text{A/m}^2.$$

Next we have

$$R = \rho \frac{L}{A} = \frac{(1.72 \times 10^{-8}\,\Omega \cdot \text{m})(10\,\text{m})}{\pi(0.50 \times 10^{-3}\,\text{m})^2} = 0.22\,\Omega.$$

Finally

$$E = \rho J = (1.72 \times 10^{-8}\,\Omega \cdot \text{m})(1.3 \times 10^5\,\text{A/m}^2)$$
$$= 2.2 \times 10^{-3}\,\text{V/m}.$$

What Do You Think? If the electric field we calculated above acts within an iron wire of the same size as the copper wire in this example, how will the current differ, if at all?

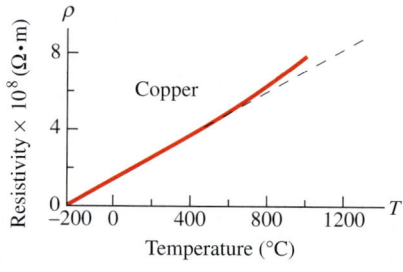

▲ **FIGURE 26–11** The resistivity of copper as a function of temperature.

The Temperature Dependence of Resistivity

Resistivities of some materials, copper for example, have a strong temperature dependence (Fig. 26–11). We can represent the temperature dependence with a linear approximation that is sufficiently accurate for most purposes:

$$\rho \cong \rho_0[1 + \alpha(T - T_0)]. \qquad (26\text{–}18)$$

The parameter α is the *temperature coefficient of resistivity*, and ρ_0 is the resistivity at the reference temperature T_0, normally 20°C. Values of ρ, σ, and α are given in Table 26–2 for $T = 20$°C. Resistivities for most metals increase with temperature, as Fig. 26–11 shows for copper, and we'll discuss this further in the next section.

EXAMPLE 26–5
Find the resistance of a coil of platinum wire of diameter 0.5 mm and length 20 m at 20°C and at 1000°C.

Setting It Up The numbers for resistivity as well as for temperature dependence can be found in Table 26–2.

Strategy Calculation of resistance in terms of the dimensions of the wire and of the resistivity involves a simple application of Eq. (26–14). We are given the diameter d of the wire, and the area is then $A = \pi(d/2)^2$. The resistance is then given in terms of the resistivity. The latter has a well-defined temperature dependence, given by Eq. (26–18). If we take that into account, then

$$R = \rho \frac{L}{A} = \rho \frac{4L}{\pi d^2} = \rho_0[1 + \alpha(T - T_0)]\frac{4L}{\pi d^2}.$$

Here ρ_0 for platinum is $10.6 \times 10^{-8}\,\Omega \cdot \text{m}$, and the coefficient $\alpha = 3.9 \times 10^{-3}\,°\text{C}^{-1}$. All we need to do is to evaluate the above formula for two different values of T.

Working It Out For $T = T_0 = 20$°C,

$$R = \rho \frac{L}{A} = (10.6 \times 10^{-8}\,\Omega \cdot \text{m})\frac{4 \times 20\,\text{m}}{\pi(0.5 \times 10^{-3}\,\text{m})^2} = 11\,\Omega.$$

For $T = 1000$°C, we multiply the above result by the factor $[1 + \alpha(T - T_0)]$ to get

$$R = (11\,\Omega)[1 + (3.9 \times 10^{-3}\,°\text{C}^{-1})(1000°\text{C} - 20°\text{C})] = 53\,\Omega.$$

This large spread in values suggests that, depending on the materials used for resistors, the operation or design of a circuit must account for temperature dependence. This is particularly true since energy is dissipated within resistors in the form of thermal energy (see Section 26–7).

What Do You Think? The melting temperature of platinum is over 1700°C, while that of copper is a little over 1000°C. Consider their temperature coefficient of resistivity and decide which would be more useful for a lightbulb filament. (Real filaments are made of tungsten.)

26–4 Resistances in Series and in Parallel

In the previous chapter we studied capacitors, and we saw that a set of capacitors connected in a combination within an electric circuit could sometimes be replaced by a single "equivalent" capacitor, i.e. one that could replace the combination and leave the relevant circuit characteristics external to the combination unchanged. Resistors can be treated similarly, and in this section we develop the rules that tell us how to find equivalent resistors. As for capacitors, so long as any particular configuration of resistors can be reduced to a sequence of series and parallel connections, it is possible to find a single equivalent resistance. This can be helpful in any number of contexts, starting with the most direct: simplifying a circuit that has been "overdesigned."

Resistors in Series

Current conservation allows us to calculate the effective resistance when a number of resistors are connected in series. Figure 26–12a shows a wire connecting two points A and B, with a resistor of resistance R in between. If the current flowing through the resistor R is I, then the potential difference between the points A and B is $V_{AB} = IR$. Suppose we now put two resistors along the same wire—we have connected them in series—and take a point C that lies on the wire between the two resistors (Fig. 26–12b). Current conservation demands that the current I is the same all along the wire. If the first resistor has resistance R_1, then the potential difference between A and C must be $V_{AC} = IR_1$. If the second resistor between C and B has resistance R_2, then the potential difference between these points is $V_{CB} = IR_2$. Let us now put a box around the two resistors in series, as in Fig. 26–12c. We see that the potential difference between the points A and B is given by $V_{AB} = V_{AC} + V_{CB}$. This sum, however, is $V_{AB} = IR_1 + IR_2 = I(R_1 + R_2)$. We may now ask for the resistance R_{eq} equivalent to the two resistors in series, by which we mean that $V_{AB} = IR_{eq}$. It follows that

$$R_{eq} = \frac{V_{AB}}{I} = R_1 + R_2. \qquad (26\text{--}19)$$

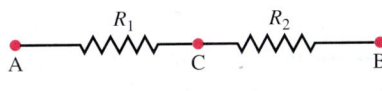
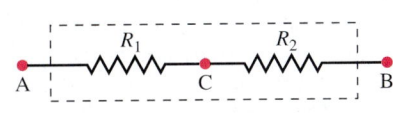

▲ **FIGURE 26–12** (a) A single resistor between points A and B on a wire, with potential difference V_{AB} between the points. (b) Two resistors in series between points A and B. (c) Two resistors in series act as a single equivalent resistor.

A repeat of the exercise with three resistors R_1, R_2, R_3 between the points A and B, with intermediate breaks at the points C and D, as in Fig. 26–13, shows that the equivalent resistance is

$$R_{eq} = \frac{V_{AB}}{I} = \frac{V_{AC} + V_{CD} + V_{DA}}{I} = R_1 + R_2 + R_3. \qquad (26\text{--}20)$$

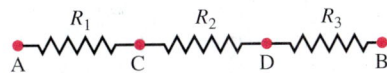

▲ **FIGURE 26–13** Three resistors in series, acting as a single equivalent resistor.

This is easily generalized to n resistors in series:

$$R_{eq} = R_1 + R_2 + R_3 + \cdots + R_n. \qquad (26\text{--}21)$$

Resistors in Parallel

Like capacitors, resistors can be placed in parallel, as in Fig. 26–14. A current I flows into a junction point A, from which n wires sprout, carrying resistors of resistance $R_1, R_2, \ldots R_n$. These wires come together again at the junction point B, out of which the original (*conserved*) current I flows. The potential difference between the points A and B is given as V_{AB}, and *this is the same as the potential difference across any of the resistors*. The current flowing out of the junction point A breaks up into parts that flow through the different resistors. How big these partial currents are is determined by the resistors that they flow through. Thus $I_1 = V_{AB}/R_1$, $I_2 = V_{AB}/R_2$, and so on. The equivalent resistance is given by rewriting the conserved current I as a sum of the partial currents, all of which is written in terms of the potential difference V_{AB} and the individual resistances. In other words, the equivalent resistance R_{eq} is defined by V_{AB}/I, so that $I = V_{AB}/R_{eq}$. Current conservation then implies that

$$I = \frac{V_{AB}}{R_{eq}} = I_1 + I_2 + \cdots + I_n = \frac{V_{AB}}{R_1} + \frac{V_{AB}}{R_2} + \cdots + \frac{V_{AB}}{R_n}$$
$$= V_{AB}\left(\frac{1}{R_1} + \frac{1}{R_2} + \cdots + \frac{1}{R_n}\right).$$

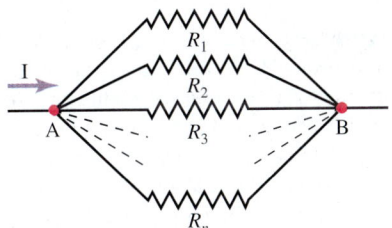

▲ **FIGURE 26–14** An arrangement of n resistors in parallel, carrying current from A to B, with potential difference V_{AB} from A to B.

Dividing out V_{AB} yields the relation that holds for resistors **in parallel**,

$$\frac{1}{R_{eq}} = \frac{1}{R_1} + \frac{1}{R_2} + \cdots + \frac{1}{R_n}. \tag{26-22}$$

EXAMPLE 26-6 Consider a current $I = 3.0$ A flowing as shown into the combination of resistors in Fig. 26-15. (a) Calculate the potential difference between the points A and B. (b) What is the potential difference across the 4 Ω resistor?

Setting It Up We remark here only that the desired potential differences are best labeled as V_{AB} for part (a), and for part (b), V_{CB} across the 4 Ω resistor.

Strategy The current entering at point A divides up into two branches, so that we have a situation with resistances in parallel. In each branch we have successive resistances, so there are also resistances in series. We can apply our rules to find equivalent resistances: first the equivalent resistance in each branch, then an overall equivalent resistance for the parallel combination. By finding the overall equivalent resistance R_{eq} we can find the potential difference between A and B, namely $V_{AB} = IR_{eq}$. This gives us the answer to (a). For part (b), we note that with the known V_{AB} and the equivalent resistance in each branch, we can find the current in each branch. The potential difference V_{CB} is then the current in the left branch times the resistance 4 Ω.

Working It Out The equivalent resistance in the left branch (the ACB path), which contains two resistors in series, is $R_{ACB} = 8\,\Omega + 4\,\Omega = 12\,\Omega$. The equivalent resistance in the right branch is similarly $R_{ADB} = 6\,\Omega + 2\,\Omega = 8\,\Omega$. Finally we want the overall equivalent resistance,

$$1/R_{eq} = 1/R_{ACB} + 1/R_{ADB} = 1/(12\,\Omega) + 1/(8\,\Omega) = 5/(24\,\Omega),$$

or $R_{eq} = (24\,\Omega)/5 = 4.8\,\Omega$. With these values we can proceed to the potential differences.

(a) We have immediately

$$V_{AB} = R_{eq}I = (4.8\,\Omega) \times (3\text{ A}) = 14\text{ V}.$$

(b) The current through the branch ACB is $I_{ACB} = V_{AB}/R_{ACB} = (14\text{ V})/(12\,\Omega) = 1.2$ A. In turn,

$$V_{CB} = I_{ACB} \times (4\,\Omega) = (1.2\text{ A}) \times (4\,\Omega) = 4.8\text{ V}.$$

As a check we can calculate the potential difference between A and C, $V_{AC} = (1.2\text{ A}) \times (8\,\Omega) = 9.6$ V, which when added to the 4.8 V yields the total potential difference of 14 V.

What Do You Think? In what way would the calculation be simpler if we were to interchange the 8 Ω and 6 Ω resistors?

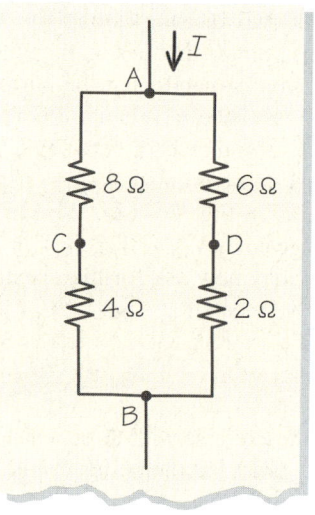

▲ **FIGURE 26-15**

CONCEPTUAL EXAMPLE 26-7 Equation (26-14) tells us that the resistance of a wire is proportional to its length L and inversely proportional to its cross-sectional area A. Is this consistent with the results obtained within this section?

Answer The linear dependence on the length is just an application of the equivalence rule for resistors in series. If we cut a uniform wire of length L into n equal segments, each of resistance r, then the rule for the calculation of the resistance equivalent to n identical resistors in series gives $R = nr$. Since r is the resistance per segment, and n segments give the total length of the wire, we find that R is proportional to the length L. The $1/A$ dependence of the resistance is a direct application of the equivalence rule for resistors in parallel. We may view a wire of total cross-section A as a total of N wires in parallel, each of cross-sectional area A/N. Suppose each of these N wires has a resistance r. Then the rule for calculating the equivalent resistances yields $1/R_{eq} = N(1/r)$, or $R_{eq} = r/N$. Now according to Eq. (26-14), $r = \rho \frac{L}{(A/N)} = N\rho \frac{L}{A}$. This gives us $R_{eq} = r/N = \rho L/A$, a consistent result. (Another way to see this is to note that R should be independent of N—of how many pieces you cut the original resistor into—and the only way that can happen is for r to be *inversely proportional* to the area A/N; this leaves R_{eq} inversely proportional to the area A.)

*26-5 Free-Electron Model of Resistivity

A more fundamental understanding of resistivity requires quantum mechanics. Nevertheless, there is a simple classical model of electrons and resistivity that is consistent with Ohm's law. It was first proposed in 1900 by Paul Drude and is known as the **free-electron model**, or the **Drude model**. Although the model has fundamental deficiencies, its study is worthwhile for two reasons: First, the model allows us to focus on the physics of resistivity. Second, the model illustrates how model-building in the physical sciences proceeds, and how we can judge the success or failure of a model.

26–5 Free-Electron Model of Resistivity

We start with the idea that solids contain "free" electrons, which can move within the material and carry charge. The density of free electrons n_e depends on the material, and this is the factor responsible for the differences among conductors, insulators, and semiconductors (which we shall discuss in Section 26–6). In metals, the number of loosely attached electrons per atom (these are the electrons that behave as though they were free) lies on average in the range 1.0 to 1.3, although it can be as large as 3.5 (for aluminum).

The model postulates that free electrons form a "gas" of independent particles at temperature T. The electrons move erratically due to collisions with the atoms or ions that form the crystal lattice when there is no electric field, but when there is an applied field, they are accelerated by it in between the collisions. The result is an on-average movement that forms the current. The effect of the collisions, in other words, is to produce a drag force that results in a constant drift velocity, producing the steady current. As we saw in Chapter 5, one of the simplest drag forces is proportional to the electrons' speed. If we assume this form, then Newton's second law for the component of electron motion that is parallel to the applied field is

$$ma = -eE - (\text{a constant})v,$$

where m is the mass of an electron. The constant must have dimensions of mass/time, and we write it as m/τ, where τ is a quantity with dimensions of time. From the kinetic theory of Chapter 19 one can establish that τ is the *collision time*, the average time between successive collisions of an electron with the lattice ions. The acceleration drops to zero when the speed of the electrons reaches the drift speed, v_d, meaning that $ma = -eE - (m/\tau)v_d = 0$, or

$$v_d = -\frac{eE\tau}{m}. \qquad (26\text{–}23)$$

The minus sign indicates that the direction of the drift velocity is opposite to that of the electric field, as must be the case with electrons in motion. When this expression is inserted into Eq. (26–10) for the current density, we find

$$J = n_e(-e)v_d = \frac{n_e e^2 \tau}{m} E. \qquad (26\text{–}24)$$

Comparison with Eq. (26–17) yields

$$\sigma = \frac{n_e e^2 \tau}{m}, \quad \text{or} \quad \rho = \frac{1}{\sigma} = \frac{m}{n_e e^2 \tau} \qquad (26\text{–}25)$$

for the conductivity σ and the resistivity ρ. The quantities e and m are independent of the type of material. The average time between collisions may be expressed in terms of the *mean free path* λ and the average speed v_{av} of the electrons in the free-electron "gas" by using Eq. (19–51), $\tau = \lambda/v_{av}$.

For electric fields not so large that the material itself is disrupted, none of the quantities in Eq. (26–25) depend on E, and thus the resistivity (or conductivity) is constant over a wide range of applied electric fields. This was the basis of Drude's (and independently Hendrik Lorentz's) claim to understand Ohm's law at an atomic level, dating from 1900.

EXAMPLE 26–8 What is the free-electron model's prediction for the collision time of current-carrying electrons in copper, given that the resistivity of copper is $1.7 \times 10^{-8} \ \Omega \cdot \text{m}$? You may use the parameters of Example 26–2.

Setting It Up The collision time τ is related to the resistivity ρ given above by Eq. (26–25). The other parameters in the equation are the known charge $e = 1.6 \times 10^{-19}$ C, the known electron mass $m = 0.91 \times 10^{-30}$ kg, and the electron density $n_e = 8.5 \times 10^{28}$ electrons/m^3, a number calculated in Example 26–3.

Strategy A rearrangement of Eq. (26–25) gives us the desired quantity, $\tau = \dfrac{m}{n_e e^2 \rho}$.

Working It Out We have

$$\tau = \frac{0.91 \times 10^{-30} \text{ kg}}{(8.5 \times 10^{28} \text{ electrons/m}^3)(1.6 \times 10^{-19} \text{ C})^2 (1.7 \times 10^{-8} \ \Omega \cdot \text{m})}$$

$$= 2.5 \times 10^{-14} \text{ s}.$$

The Failure of the Free-Electron Model

The free-electron model assumes electrons moving among a lattice of ions. A closer look at the predictions of this model reveals significant discrepancies with experiment. In particular:

- The measured average speed of electrons in metals is *more than a factor of 10* higher than the model predicts for copper at room temperature.
- We know from Chapter 19 [see Eq. (19–33)] that the rms speed of the particles of a gas is proportional to $\sqrt{T}$, where T is the temperature. However, experimental values for the mean speed of conduction electrons are essentially independent of temperature.
- The mean free path—the average distance an electron travels between collisions [see Eq. (19–15)]—should be independent of temperature according to the model. Experimentally, this quantity is much larger than expected and has a $1/T$ dependence.

Although the free-electron model is *qualitatively* correct in many aspects, it cannot be taken too literally, because it is a classical model in what turns out to be the domain of quantum mechanics. A correct model of electrical conduction *requires* the use of quantum mechanics. Indeed, quantum physics comes in at several levels, the major ones as follows:

- Conduction electrons do not act as a classical gas of noninteracting electrons; rather, they obey a velocity-distribution law based on quantum physics.
- Quantum physics requires us to treat electrons as though they were waves scattering from the lattice structure of the material. In particular, this treatment leads to a surprise: Quantum physics predicts that *there would be no resistance to electron flow in a fixed, perfectly ordered crystal with no impurities*. Finite conductivities result from departures from the perfect lattice structure. These departures occur in two ways: impurities are present in real materials, and the positions of lattice atoms vibrate at finite temperatures due to thermal excitation. At high temperatures, resistivity to electron flow is caused primarily by thermal vibrations. At low temperatures, resistivity is due to electrons being scattered by impurities.

There is ample evidence that the quantum physics ideas are correct. Indeed, *all* the properties described are correctly explained with these ideas. We'll discuss them in more detail in the following optional section and then later in Chapter 43.

*26–6 Materials and Conductivity

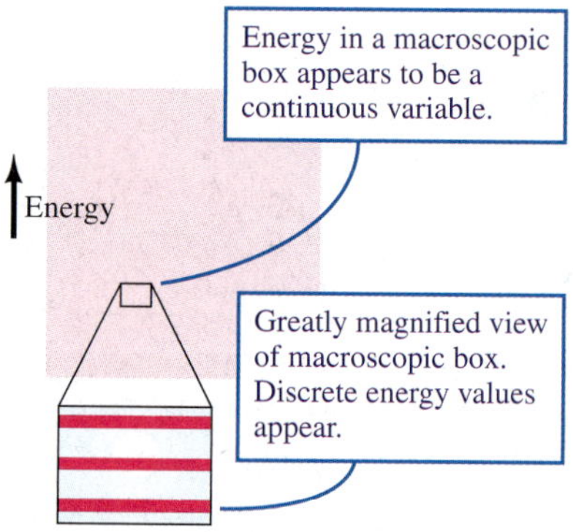

Materials differ in their ability to conduct electricity over an enormous range. A good conductor might have a resistivity of $10^{-8}\ \Omega \cdot m$; a good insulator, about $10^{14}\ \Omega \cdot m$. The resistivity of semiconductors ranges from 10^3 to $10^{-5}\ \Omega \cdot m$ and depends sensitively on temperature. Superconductors have no measurable resistance at all below a so-called critical temperature, a quantity that is specific to the material. We argued in the previous section that a proper quantitative explanation of the resistivity of all materials requires quantum physics. In this section, we employ a minimal amount of the ideas that explain quantum physics to describe the critical properties that distinguish conductors, insulators, semiconductors, and superconductors.

In classical physics, the energy of a "free" electron within a metal can take on any value; we say that the energy values form a *continuum*. In contrast, a quantum description of electrons confined to the interior of a metal but otherwise free shows that the possible energy values of such electrons are *quantized*; that is, the possible energies have discrete values. In other words, an electron cannot have *any* energy value, much as the frequencies of standing waves on a string cannot have any value, just a set of discrete values. In a sample of material whose size is large compared with atomic sizes (10^{-10} m), these energy values are so close together that they appear to be continuous, just as the separate dots in a newspaper photograph are not distinguishable from a large distance. Figure 26–16, an energy diagram, illustrates the allowed energy levels. It is important to keep in mind that this diagram illustrates only the *possible* energy levels.

▲ **FIGURE 26–16** Energy diagram that shows the possible energy levels of an electron in a solid. (It takes no account of the crystalline structure formed by the parent ions of the electrons, whose presence has a strong effect on the possible energy levels—see Fig. 26–17). Classical physics predicts a continuum of possible energies, but quantum mechanics shows that the possible levels are actually discrete but so closely spaced that they are hard to distinguish.

We do not necessarily have electrons in each energy level. Some states are empty, while others have electrons in them and are said to be *occupied*.

When a set of atoms forming a regular background lattice is added to the picture, the possible energy values of the electrons are modified still further. The allowed energies of an electron are still discrete, but instead of a tiny separation between neighboring levels, there are *energy gaps*, which are large regions of energy forbidden to the electron. The regions where the energy levels are close together are called *allowed bands* of energy levels (Fig. 26–17). The gaps are quite sizable on the scale of atomic physics—of the magnitude of electron-volts.

According to quantum physics, *there are at most two electrons in any one energy level*. This property, proposed by Wolfgang Pauli in 1925 and called the *Pauli exclusion principle*, has no counterpart in classical physics, and it plays a crucial role in determining the properties of materials. Let's consider a solid with many "free" electrons. In an equilibrium state of that material, at least at low temperatures, these electrons fill the lowest energy levels available in the allowed bands—up to two in each level. When all the electrons are placed in the lowest possible energy states, we have two possible situations. In the first, the highest level to be filled is some intermediate level within a band; in the second, the electrons fill one or more bands completely.

Suppose that we now add some energy to the free electrons—by imposing an electric field, for example. The electrons in the lower energy levels cannot accept that energy, because they cannot move into a higher energy level that already has its quota of two electrons. The only electrons that can accept energy are those that lie in the top levels, and then only if there are nearby unoccupied levels into which they can move. This will be quite easily done with a small input of energy if the topmost occupied levels are in a partially filled band, but difficult if the topmost occupied levels fill out a band. Materials with a partially filled band are *conductors*. When the electrons in the highest occupied energy levels move freely into the empty energy levels immediately above, a current is produced. The electrons that jump from a lower level to a higher level are said to be *excited*. The energy-band structure for conductors is shown in Fig. 26–18a. *Conductors are characterized by having a highest-energy band with levels only partly occupied.*

If the highest-energy electrons of a material fill a band completely, then a small electric field will not give these electrons enough energy to jump the large energy gap to the bottom of the next (empty) band. We then have an *insulator* (Fig. 26–18b). An example of a good insulator is diamond (a form of carbon), whose energy gap is 6 eV.

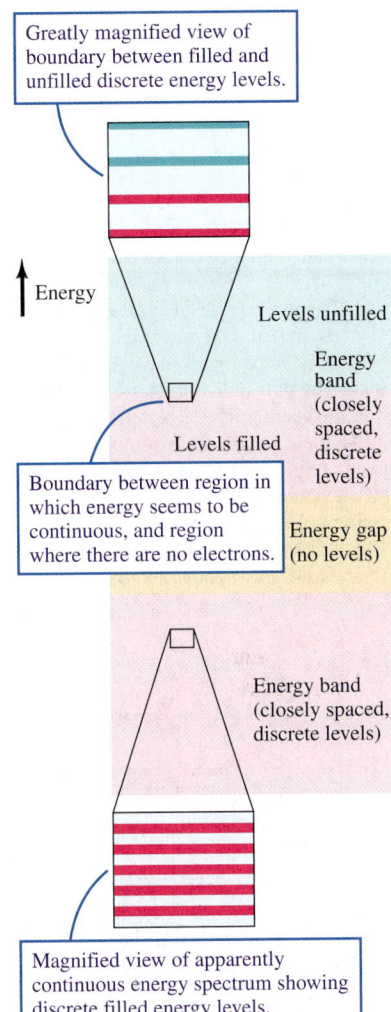

▲ **FIGURE 26–17** Energy diagram that shows the possible energy levels of an electron within a material made of a regular lattice of atoms. In contrast to the possibilities of Fig. 26–16, the electron energies are restricted to lie within allowed bands, and there is a large energy gap where no electrons are allowed. Even within the allowed bands, the possible electron energies are closely spaced discrete levels, as the magnified view shows. In the pink regions, the electron energy levels are filled; in the green regions, electron levels are present but are unfilled.

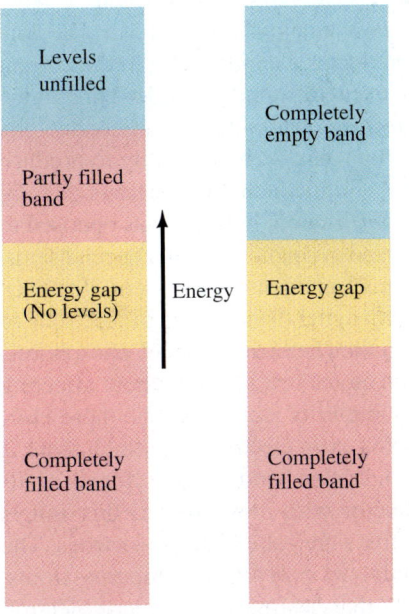

◀ **FIGURE 26–18** (a) Conductors have electrons in partly filled bands, whereas (b) insulators have an energy gap between a completely filled band and the next completely empty band. The pink and blue regions indicate where the allowed energy levels are filled and unfilled, respectively. Within each of the allowed bands, the possible energy levels form a set of closely spaced discrete levels.

754 | Currents in Materials

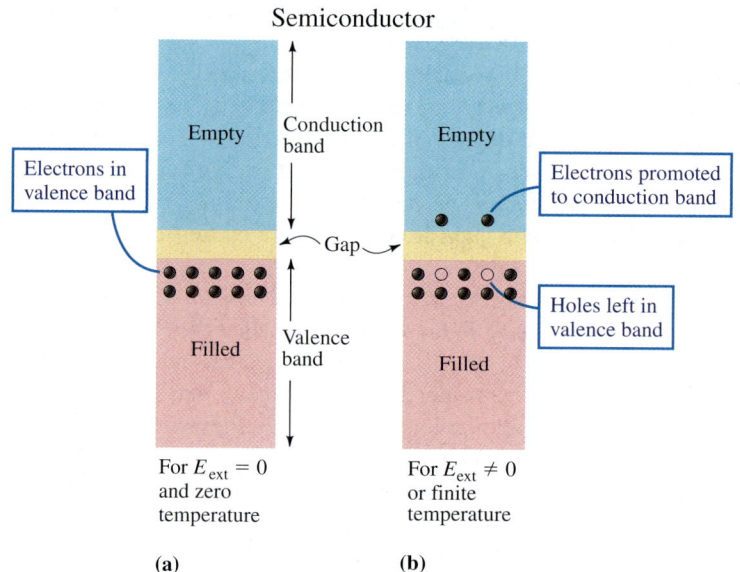

▶ **FIGURE 26–19** (a) For zero temperature and no external electric field, semiconductors have only a small energy gap between a completely filled band (the valence band) and the next highest, completely empty band (the conduction band). The dots indicate electrons, here all in the valence band. (b) A modest electric field $\vec{E}_{ext}$ or finite temperatures are enough to give some of the electrons sufficient energy to jump the energy gap, leaving holes (open circles) in the valence band and conduction electrons in the previously empty conduction band.

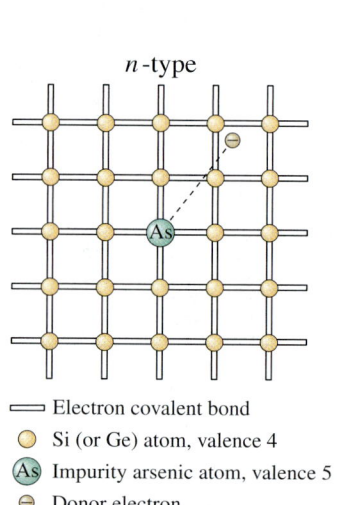

(a)

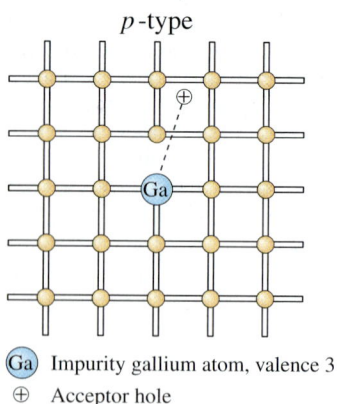

(b)

▲ **FIGURE 26–20** An (a) n-type and a (b) p-type semiconductor are created by doping the original lattice with atoms that have, respectively, more and less valence electrons than the atoms of the original lattice have.

In semiconductors, the highest-energy electrons fill a band (the *valence band*) at $T = 0$, as in insulators. However, semiconductors have only a small energy gap between that band and the next, the *conduction band* (Fig. 26–19a). Silicon and germanium have energy gaps of 1.1 eV and 0.7 eV, respectively, and are semiconductors. Because the energy gap is so small, increasing the temperature or applying a modest electric field will allow some electrons to jump the gap, and electric current will flow (Fig. 26–19b). A semiconductor with an electric field too small to boost conducting electrons to the next energy band will act as an insulator; once the field increases above the strength necessary to allow electrons to "jump the gap," the material acts as a conductor. In other words, there is a minimum electric field under the influence of which a material changes from an insulator to a conductor. In addition, an increase in temperature will give a fraction of the electrons enough thermal energy to jump the gap in semiconductors and thus *lowers* the resistivity. For a conductor, a rise in temperature *increases* the resistivity because the atoms, which are obstacles to electron flow, vibrate more vigorously.

When an electron in the valence band of a semiconductor crosses the energy gap and conducts electricity, it leaves behind what is known as a **hole**. Other electrons in the valence band near the top of the stack of energy levels can move into this hole, leaving behind their own holes, into which still other electrons can move, and so forth. The hole behaves as a positive charge and, since it can move, forms a contribution to the current on its own as a positive charge carrier. An electron excited from the valence band to the conduction band is thus doubly effective at conducting electricity in semiconductors.

One of the major advances in materials technology has been our ability to produce new semiconductors, those with tailored properties. Semiconductor materials that are compounds, such as gallium arsenide, are called *hybrid* semiconductors, as opposed to *intrinsic* elemental semiconductors, such as silicon and germanium. Other special semiconductors are made by introducing impurities, small amounts of different elements, into the lattice. These impurities have the effect of changing the numbers of electrons or holes available to conduct. For example, an atom in the chemical group of phosphorus, arsenic, and antimony can replace one of the silicon atoms in a lattice without affecting the lattice itself too much. However, each of these impurity atoms has one more electron in its valence level than does a silicon atom; this extra electron, for which there is no room in the valence band, takes a place in the conduction band and therefore is available to carry current (Fig. 26–20a). A semiconductor with impurities of this sort is called an *n*-type semiconductor, and the extra electrons are called *donor* electrons. The semiconducting material, silicon in this case, is said to be *doped* by the impurity atoms.

We can alternatively add atoms of an element that has one less valence electron than silicon, such as boron, aluminum, or gallium. Figure 26–20b shows that in this case we are one electron short of what is needed to form the atomic-level bond that holds the lattice together. This electron must be provided by the electrons of the valence band of the lattice material, and so holes are created in this band. These holes act as positive charge carriers. The impurity atoms are called *acceptors*, and a semiconductor with impurities of this sort is called a *p-type semiconductor*.

Many electronic devices, such as the diode mentioned in Section 26–3, depend heavily on the properties of semiconductors. Probably the best known and most widely used of these devices are transistors, which can amplify electronic signals, and form the basic building blocks of logic devices.

Superconductors

In 1911, H. Kammerlingh Onnes found that mercury abruptly loses *all* of its resistance—at least as far as he could tell—at a *critical temperature* T_c of 4.1K (Fig. 26–21a). This state of zero resistance persists at temperatures below T_c. A material that exhibits zero resistance at some critical temperature is called a *superconductor*. An experiment on a superconducting ring in which a current had been induced showed that there was no observable decrease in the current after a full year. From the measurements in this experiment, it was possible to deduce that if there were any resistive decrease of the current, it had to occur over a period of at least 10^9 years!

The prospect of having an *electric current that lasts forever* is an enticing one. It implies, among other things, the cheap transmission of electricity. The phenomenon of superconductivity cannot be understood as an extension of ordinary conductivity. The abruptness with which resistance disappears completely suggests that an ordinary conductor makes a transition to a totally different state of matter at T_c, much as liquid water turns into a crystal (ice) at 273K. In 1957, John Bardeen, Leon Cooper, and Robert Schrieffer satisfactorily explained the *superconducting phase* with quantum physics in what is now known as the BCS theory.

Until 1986, the materials with the highest-known values of T_c became superconducting at 23K. Helium is liquid at such temperatures and is thus used for cooling superconductors. However, liquid helium is a relatively expensive medium, as is the equipment that keeps it cold. That limited the uses of superconductors to scientific applications, such as magnets for particle accelerators (see Chapter 28) or to important situations such as nuclear magnetic resonance imaging machines in hospitals (Fig. 26–21b). In 1986, however, K. Alex Muller and J. George Bednorz discovered a new class of materials for which T_c is much higher; superconductors are now known that have a T_c above 120K ($-153°C$). This discovery has great technological implications because such materials can be cooled relatively cheaply with nitrogen (which is liquid at 77K). Research in both the basic physics behind these materials and their technological applications is of active interest.

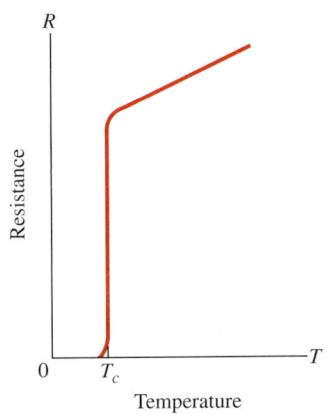

(a)

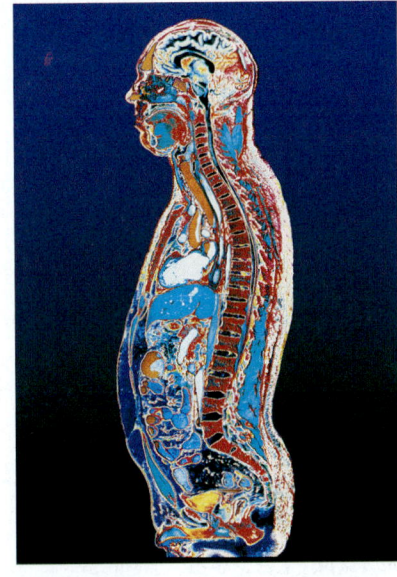

(b)

▲ **FIGURE 26–21** (a) For superconductors the resistance drops to zero at the critical temperature T_c. (b) An MRI image of a human made using a superconducting magnet.

26–7 Electric Power

Electric energy is sent to our homes and workplaces and composes much of the energy used in our society. Efficient delivery of this energy is of paramount importance, and in this section, we shall look at the ways in which resistance affects the delivery of electric energy.

We have compared electrical resistance to mechanical drag. When there is drag in mechanical motion, mechanical energy is converted to thermal energy. The second law of thermodynamics (Chapter 20) shows that some of this thermal energy is irretrievably lost in the sense that it cannot all be converted to mechanical work. Similarly, just as mechanical friction generates heat, the passage of a current through a resistor generates heat, and in this way some electric energy is lost due to resistance. While sometimes we want to use the thermal energy, as in the heating element of an electric stove, it cannot all be converted to useful mechanical work, and in the transmission of electricity the thermal energy created is lost energy.

To calculate the energy lost per unit time (the power lost) when a charge moves in a material, consider a small charge dq that moves through a potential difference V. The change in the potential energy of the charge (dU) is equal to the work done (dW)

by the electric force due to the potential difference, and is given by $dU = V\,dq$. It follows that the power, the rate at which energy is expended by the force that pushes the charge, is

$$P = \frac{dW}{dt} = V\frac{dq}{dt}. \tag{26–26}$$

Because the current $I = dq/dt$, the electric power lost, which is the power that must be delivered to move I through the potential V, is

$$P = VI. \tag{26–27}$$

POWER LOST IN RESISTANCE

This result is a general one, independent of the type of material—in particular, whether the material is ohmic or nonohmic—and of the nature of the charge movement. Power has SI units of watts (W), with $1\text{ W} = 1\text{ J/s}$. By using Eq. (26–27), we have another unit for power:

$$1\text{ W} = 1\text{ V}\cdot\text{A}. \tag{26–28}$$

For ohmic materials, $V = IR$, where R is a constant. Thus the power expenditure for ohmic materials is

$$P = VI = V\left(\frac{V}{R}\right) = \frac{V^2}{R}. \tag{26–29}$$

Equivalently, we can use $V = IR$ in Eq. (26–29) to find that

$$P = I^2 R. \tag{26–30}$$

Whether we use Eq. (26–29) or Eq. (26–30) depends on what is known in a particular application. The power lost (rate of energy loss) in a resistor appears in the form of thermal energy and is variously called *ohmic heating, Joule heating,* and $I^2 R$ loss.

EXAMPLE 26–9 Nichrome is an alloy of nickel, chromium, and iron often used as a heating element in electrical devices. A nichrome wire (1.0 m in length) is crisscrossed along the interior of a toaster (Fig. 26–22) that can carry a maximum current of 8 A when there is a 120-V potential difference from one end of this wire to the other[†]. If the resistivity of nichrome is $1.0 \times 10^{-6}\ \Omega\cdot\text{m}$, what is the radius of the wire? What power does the toaster use?

Setting It Up We call the given current I, the given voltage V, the given wire length L, and the given resistivity $\rho = 1.0 \times 10^{-6}\ \Omega\cdot\text{m}$. The desired wire radius will be labeled r. The current and resistance are known, so the power can be calculated.

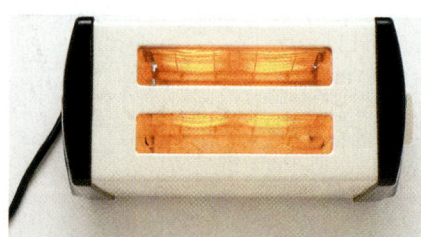

▲ FIGURE 26–22 The glowing nichrome wires of a toaster.

Strategy With the help of Eq. (26–14) we can calculate the cross-sectional area A once the resistance is determined: $A = \rho L/R$, and given A the wire radius follows from $A = \pi r^2$. Thus the first step is to calculate the resistance using $R = V/I$. The power can be calculated using $P = V^2/R$.

Working It Out We have

$$R = \frac{V}{I} = \frac{120\text{ V}}{8\text{ A}} = 15\ \Omega.$$

It follows from $A = \rho L/R$ that

$$r = \sqrt{\frac{A}{\pi}} = \sqrt{\frac{\rho L}{\pi R}} = \sqrt{\frac{(1.0 \times 10^{-6}\ \Omega\cdot\text{m})(1.0\text{ m})}{\pi(15\ \Omega)}}$$

$$= 0.15 \times 10^{-3}\text{ m}.$$

Finally, the power is

$$P = \frac{V^2}{R} = \frac{(120\text{ V})^2}{15\ \Omega} = 960\text{ W}.$$

What Do You Think? Suppose the toaster is accidentally connected into a 240-V outlet. With the voltage doubled, by how much will the power consumed change? (Assume that the toaster doesn't burn out, and don't try this!)

[†]Real household electricity involves an oscillating voltage difference and an oscillating (or alternating) current. Ignore these effects here and in Example 26–10.

26–7 Electric Power

We described the commercial units of electric energy in Section 6–5: The energy unit in the electric power industry is the kilowatt-hour (kWh), 3.6×10^6 J. When electric energy is delivered to a home, the energy delivered per unit time is also called the electric power.

EXAMPLE 26–10 A 100-W bulb is left on in an outdoor storage room to keep paint from freezing. The 100-W rating refers to the power dissipated in the bulb's filament, which is a resistor. If electricity costs 8 cents/kWh, about how much does it cost to burn the lightbulb for three months during winter?

Setting It Up The given power of 100 W expresses an energy consumed per second. We know the total time that the bulb burns, as well as the cost per kWh of energy.

Strategy The number of joules of energy used over three months is found by converting three months to seconds, then multiplying the power, 100 W, by the number of seconds the bulb is on. Since the cost rate is given in dollars per kWh, we'll want to convert the energy units to kWh.

Working It Out The number of seconds the bulb is lit is

$$(3 \text{ months})(30 \text{ days/month})(24 \text{ h/day})(3600 \text{ s/h}) = 7.8 \times 10^6 \text{ s}.$$

The energy used is

$$(100 \text{ W})(7.8 \times 10^6 \text{ s}) = 7.8 \times 10^8 \text{ J}$$
$$= (7.8 \times 10^8 \text{ J}) \times [1 \text{ kWh}/(3.6 \times 10^6 \text{ J})] = 220 \text{ kWh}.$$

From this we find the cost, $(8 \times 10^{-2} \text{ \$/kWh}) \times (220 \text{ kWh}) = \$17.60$.

Although the primary purpose of a lightbulb is to produce light, most of the electric energy it dissipates is converted into heat, not light.

What Do You Think? Assuming the voltage is fixed, will it be more or less expensive to have *two* of these bulbs in (a) series, (b) parallel?

EXAMPLE 26–11 As we described in Eq. (17–17), a piece of metal such as a wire heated to a temperature T (in units of degrees kelvin) radiates energy in the form of electromagnetic waves, with the power emitted per unit surface area given by Stefan's formula, $P_{\text{Stefan}} = \sigma T^4$, where the constant σ is roughly 5.7×10^{-8} W/(K$^4 \cdot$m^2). A bus-stop shelter is heated by a 1-m-long metallic coil that is 1 mm in diameter. The temperature of the coil is 2000K. If the resistivity of the coil material at that temperature is given by $5 \times 10^{-7} \, \Omega \cdot$m, estimate the current that must flow through the wire to maintain its temperature. (We are assuming that at this temperature the fraction of the power dissipated due to convection of the air around the coils is small.)

Setting It Up In addition to knowing the temperature T, we are given the diameter d of the wire and its length L. These will suffice to find the power emitted in radiation, and since we are given the resistivity ρ, we can match this power emission to a resistive power loss in terms of the unknown current.

Strategy The first step is to use Stefan's formula to calculate the power P radiated by the wire, given that the temperature is $T = 2000$K. This will give for the total radiated power $P = P_{\text{Stefan}} \times A$, where A is the surface area of the wire. This power must come from resistive loss in the wire, because that is what heats the wire to high temperatures. Therefore we calculate the resistance R of the wire in terms of the resistivity of the material, using Eq. (26–14) and the wire dimensions, and match P to the resistive loss RI^2, a relation that allows us to solve for the desired current I.

Working It Out With the wire's surface area $A = \pi d \times L$, the emitted power is

$$P = \sigma T^4 \times (\pi d \times L) = [5.7 \times 10^{-8} \text{ W/(K}^4 \cdot \text{m}^2)] \times$$
$$(2 \times 10^3 \text{ K})^4 \times \pi \times (10^{-3} \text{ m}) \times (1.0 \text{ m}) = 3 \times 10^3 \text{ W}.$$

The resistance of the wire is

$$R = \rho \frac{L}{A_{\text{cross section}}} = \rho \frac{L}{\pi (d/2)^2}$$
$$= (5 \times 10^{-7} \, \Omega \cdot \text{m}) \frac{1.0 \text{ m}}{\pi (0.5 \times 10^{-3} \text{ m})^2} = 0.6 \, \Omega.$$

Finally we use $P = RI^2$, with P the radiated power, so that

$$I = \sqrt{\frac{P}{R}} = \sqrt{\frac{3 \times 10^3 \text{ W}}{0.6 \, \Omega}} = 70 \text{ A}.$$

What Do You Think? Suppose you would like to keep the material and the temperature of the radiating coil the same but at the same time reduce the current. Could this be done by changing the diameter d of the wire, and if so, how?

Resistors used in circuits are characterized not only by their resistance, but also by a power rating. This power rating states the maximum power that the resistor can dissipate without being damaged due to overheating. The power rating is measured in watts. According to Eq. (26–30), which states that the power dissipated in a resistor is $P = I^2 R$, we can deduce the maximum allowed current from the power rating. One class of relatively inexpensive resistors, so-called carbon film resistors, is limited to about 2 W; a second more expensive type known as wire-wound resistors have a power rating up to 50 W.

Summary

Electric current is the rate at which charge passes. The instantaneous current is given by

$$I \equiv \frac{dQ}{dt}. \tag{26-2}$$

The unit of current is the ampere (A), 1 C/s. Currents in wires are depicted as though the positive charges are moving, but it is actually electrons (negative charge) that are mobile.

The current density $\vec{J}$ is a vector quantity representing the current that passes through an area per unit time. The current is related to the current density by

$$I = \int_{\text{surface } A} \vec{J} \cdot d\vec{A}. \tag{26-4}$$

The free-electron model of conduction is useful as a qualitative description of current in a solid. The average, or drift, speed of the electrons that pass through the material is

$$v_d = \frac{I}{n_e q A}, \tag{26-9}$$

where n_e is the density of free electrons and q is the charge of an electron.

Electrical resistance R is the ratio of voltage to current

$$R \equiv \frac{V}{I}. \tag{26-11}$$

Many conducting metals show a linear relationship between voltage and current. The resistance is then constant over a wide range of voltages. This relation is called Ohm's law, $V = IR$.

Resistivity ρ is the quantity that distinguishes the part of the resistance that is intrinsic to each particular type of material. For wires of length L and cross-sectional area A, we have

$$R = \rho \frac{L}{A}. \tag{26-14}$$

The inverse of the resistivity is the conductivity σ which expresses how well a type of material conducts current:

$$\sigma \equiv \frac{1}{\rho}. \tag{26-15}$$

Both ρ and σ depend on temperature.

The electric field and current density are related by

$$\vec{E} = \rho \vec{J} \tag{26-16}$$

and by

$$\vec{J} = \sigma \vec{E}. \tag{26-17}$$

When a number of resistors of resistance $R_1, R_2, \ldots R_n$ are placed in series, then they act together as a single resistor of equivalent resistance

$$R_{\text{eq}} = R_1 + R_2 + R_3 + \cdots + R_n. \tag{26-21}$$

When a number of resistors of resistance $R_1, R_2, \ldots R_n$ are placed in parallel, then they act together as a single resistor of equivalent resistance given by

$$\frac{1}{R_{\text{eq}}} = \frac{1}{R_1} + \frac{1}{R_2} + \cdots + \frac{1}{R_n}. \tag{26-22}$$

The free-electron model, which provides a classical model for Ohm's law and the mechanism of electrical conduction, cannot fully account for the observed behavior of conductors, and the correct explanation of electrical conduction in metals is dependent on how electrons fill allowed energy bands and on the energy gap between these bands. Materials that conduct current easily have electrons in partially filled bands. Semiconducting materials, such as silicon and germanium, can be doped by impurity atoms to increase the density of charge carriers. The explanation of superconductivity requires both quantum mechanics and the presence of a new phase of matter in which electrons collectively transport electric current.

When a current moves through a potential difference, electric power P is dissipated (or produced), given by

$$P = VI. \tag{26-27}$$

For resistive materials, the power is also given by

$$P = \frac{V^2}{R} = I^2 R. \qquad (26\text{--}29,\ 26\text{--}30)$$

Understanding the Concepts

1. Consider the electron beam in a cathode-ray tube. The velocity of the electrons in the beam changes as the electrons are accelerated. Is the current the same everywhere in the beam?
2. How does the free-electron model for electrical resistance account for power dissipation? Does our microscopic picture agree with the voltage/current result?
3. The same current passes through two similar wires of unequal areas. Which wire will get hotter, and why?
4. The same current passes through two wires of the same area. One of the wires is made of aluminum, whereas the other is made of brass. Which wire will get hotter, and why?
5. What factors determine the differences in drift velocity of electrons in wires if the dimensions and current are the same?
6. Knowing what you now know about parallel and series resistance, why does the potential drop linearly along the length of a resistor that is a uniform cylinder?
7. If the movement of charges in a wire is similar to the flow of water in a hose, why, when a new hose is hooked up to a faucet, do we have to wait for a while until the water comes out, but when we hook a new wire up to a circuit, we do not have to wait for charge to come out the other end when the switch is turned on?
8. According to the discussion of Section 26–5, the resistivity in the free-electron model should vary with the square root of the temperature and thus should be zero at $T = 0$. Is this reasonable? How would you interpret this result?
9. We know that the resistivity of a metal is temperature dependent, and so therefore is the resistance of a wire. In Chapter 17, we saw that the dimensions of a piece of metal—such as a wire—change when the wire is heated. Does this provide an additional reason to change the resistance of a wire as it undergoes Joule heating? Would you expect the effect to be large?
10. When you throw a switch and charge flows in a household wire, does the wire become charged?
11. Suppose that we orient a wire between the plates of a charged capacitor so that there is an electric field along the cross section of the wire. Will the resistance of the wire change because all the charge-carrying electrons crowd to one side of the wire, thus effectively reducing the wire's cross section?
12. Gauss' law states that free charge in a conductor moves to the surface of the conductor. Does this mean that the current flowing through a wire is actually on the wire's surface?
13. The resistivity of most metals is on the order of $10^{-8}\ \Omega \cdot m$. Discuss why this might be so in terms of the result given by Eq. (26–25).
14. Is it possible to break up any combination of resistors into a sequence of parallel and series resistances? If not, give an example of a combination that cannot be so decomposed.
15. What is likely to happen when a current is so large that the power dissipation in a resistor through which the charge flows exceeds the resistor's power rating? What mechanism is responsible for such a disaster scenario?
16. What considerations would you have to take into account in designing a lightbulb filament with trade-off between the length of filament and the diameter of the filament?
17. The resistance of a wire is proportional to the length of the wire. Think of a wire as consisting of a series of shorter wires placed together in series. Can you use this information to predict the effective resistance of two different resistors of resistance R_1 and R_2 when they are placed in series in a circuit?
18. In Fig. 26–7a the drift velocity is smaller in area A_2 than in A_1, yet the current is the same. How is that possible?
19. Consider three identical bulbs between points A and B as shown in Fig. 26–23. The potential difference between A and B is fixed. The switch between C and bulb 3 may be open or closed. Will bulb 2 be brighter when the switch is open, or if it is closed? Will bulb 1 be brighter when the switch is open or if it is closed?

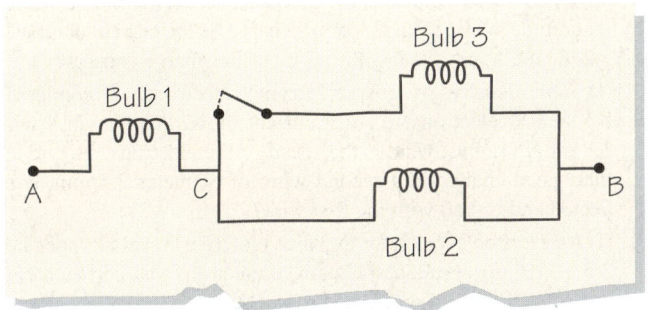

▲ **FIGURE 26–23** Question 19.

20. A copper rod of a given length and diameter has resistance R. Suppose the same rod is drawn into a wire with diameter 1/10 of what it was before. By what factor will the resistance change? Will it be larger or smaller?
21. Suppose all the resistors in Fig. 26–14 have the same resistance and the potential difference V_{AB} is held fixed. If the number of resistors is changed from n to $n + 1$, will the total current flowing between A and B increase or decrease?
22. Three resistors of resistance 1 Ω, 2 Ω and 4 Ω are placed in a network shown in Fig. 26–24. Can you decide in which places the resistors are to be put so as to minimize the overall resistance between A and B? What arrangement would maximize the overall resistance?

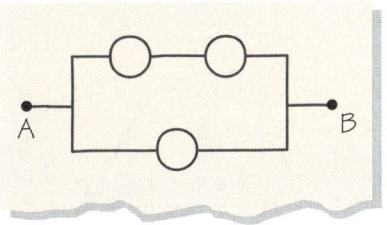

▲ **FIGURE 26–24** Question 22.

23. In high wattage lightbulbs, the filament is typically coiled. Why?

Problems

26–1 Electric Current

1. (I) A wire of diameter 2.2 mm carries a current of 0.46 A. What is the average current density? How much charge crosses a fixed point in the wire per second?

2. (I) There is a 1.2-A current in a wire of cross-sectional area 4.2×10^{-5} m^2. The drift speed of the electrons that carry the current is 0.32×10^{-5} m/s. Find the density of current-carrying electrons.

3. (I) A jumper cable used to start a car carries a current of 100 A and has a cross-sectional area of 36 mm^2 and a length of 2 m. The free-electron density in the cable is 8.5×10^{28} electrons/m^3. How long does it take a free electron to pass from one end of the cable to the other?

4. (I) Three straight wires of area 0.02 mm^2, 0.2 mm^2, and 2 mm^2, respectively, are aligned along the x-axis. They carry current densities along the x-axis of magnitude 3×10^5 A/m^2, 13×10^4 A/m^2, and 15×10^4 A/m^2, respectively. Find the current in each wire.

5. (I) A wire of radius 1.6 mm carries a current of 0.092 A. How many electrons cross a fixed point in the wire in 1 s?

6. (I) Charge carriers in a semiconductor have a number density $n_q = 3.5 \times 10^{24}$ carriers/m^3. Each carrier has a charge whose magnitude is that of an electron's charge. If the current density is 7.2×10^2 A/m^2, what is the speed of the charge carriers?

7. (I) The density of charge-carrying electrons in copper is 8.5×10^{28} electrons/m^3. If a current of 1.2 A flows in a wire 1.8 mm in radius, what is the speed of the electrons? How does that speed change in a second wire, of diameter 2.4 mm, connected end-to-end with the first wire?

8. (I) An electron accelerator in which electrons travel at a speed of 3.5×10^7 m/s produces a beam of electrons that carries a current of 6.1 mA. The effective area occupied by the beam is 0.50 cm^2. What is the density of electrons in the beam? Ignore all relativistic effects.

9. (II) In the National Synchrotron Light Source x-ray device at Brookhaven National Laboratory, there is an electron beam with an average current of 200 mA. The electrons have a kinetic energy of 2.5 GeV and a speed extremely close to the speed of light. How many electrons pass a given point in the accelerator per hour? How many electrons are contained in a 1-m length of the beam? Ignore all relativistic effects (a poor approximation, in this case).

10. (II) A cube of material is placed with one corner at the origin of a coordinate system; its sides, 1 cm long, are parallel to the three axes. The current density is $A\hat{i} + B\hat{j} + C\hat{k}$ throughout the cube. The units of A, B, and C are mA/cm^2. What are the currents along the x-axis, y-axis, and z-axis?

11. (II) The current density in a cylindrical wire of radius R is $J = J_0(1 - r^2/R^2)$, parallel to the axis of the wire (Fig. 26–25). Calculate the total current across a section perpendicular to the axis.

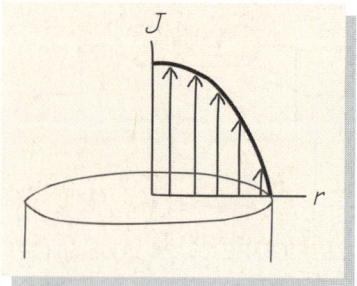

▲ FIGURE 26–25 Problem 11.

12. (II) In a plasma containing equal densities n of electrons and (positive) ions, the ions move to the right. Their speed is a factor of 1.5×10^{-3} smaller than the speed with which the electrons move to the left. What is the (net) current density? Give its direction and magnitude.

13. (II) An aqueous solution contains 0.1 mol/L of NaCl. The NaCl is dissolved in the form of Na$^+$ and Cl$^-$ ions. Calculate the velocities of the Na$^+$ and Cl$^-$ ions, respectively, if there is a measured total current density of 40 A/m^2. Assume that the velocity of the Cl$^-$ ions is about 50 percent greater than that of the Na$^+$ ions.

26–2 Currents in Materials

14. (I) Calculate the drift speed of electrons in the conduction cables of an automobile starter cable, which is made of copper and has a diameter of 4 mm, if you suppose that the cable carries 100 A. How would this speed change if the diameter of the wire were doubled? [Hint: Useful data are contained in Example 26–2.]

15. (I) A single charged elementary particle ($q = 1.6 \times 10^{-19}$ C) travels with a speed very close to that of light in a circular accelerator of diameter 5 km. What is the current represented by the particle, taking into account multiple traversals of the charge past a given point?

16. (I) How many particles like that described in Problem 15 must be present at a given time to give rise to a current of 42 mA?

17. (II) An aluminum wire of area 50 mm^2 placed along the x-axis passes 10,000 C in 1 h. Assume that there is one free electron for each aluminum atom. Determine the current, current density, and drift speed. The mass density of aluminum is 2.7 g/cm^3.

18. (II) Gold has one electron per atom available to carry charge. Given that the mass density of gold is 19.3×10^3 kg/m^3 and that its molecular weight is 197 g/mol, calculate the drift speed of the electrons in a gold wire that carries 0.3 A and has a circular cross section 0.5 mm in radius.

19. (II) Two parallel metal wires of diameter 0.2 cm and a charge-carrier density $n_e = 7 \times 10^{22}$ electrons/cm^3 carry a current of 3 A each. The wires join and then split into three identical but separate wires, each with a radius one-half that of the original wire (Fig. 26–26). All the wires are made of the same material. What are the drift speeds in both the larger and smaller wires? Can you explain the difference in speeds in terms of the speeds of water flow in pipes?

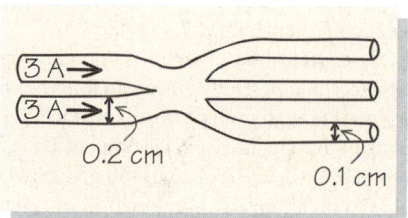

▲ FIGURE 26–26 Problem 19.

20. (II) The charge carriers in a certain wire of circular cross section and radius R have a drift speed down the wire that is not constant across the wire. Instead, the drift speed rises linearly from zero at the circumference ($r = R$) to v_0 at the center ($r = 0$). Compare the total current carried by this wire with the current carried by a wire of the same radius, same density of charge carriers, and a constant drift speed of $v_0/2$.

21. (II) A thin copper wire carrying a current I is welded to the center of a circular copper plate capping a copper tube (Fig. 26–27). The radius of the tube is R, the thickness of its wall and the top plate is d, and $d \ll R$. What is the current density in the tube and in the top plate?

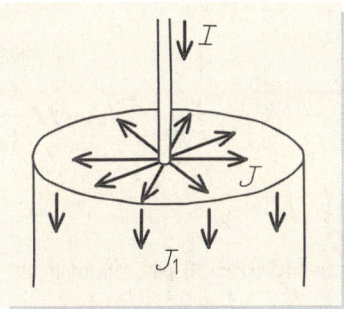

▲ **FIGURE 26–27** Problem 21.

22. (III) Charges q move longitudinally down a rod of circular cross section and radius R. The density of the charge carriers n decreases as a function of the radial distance r from the center of the rod according to $n = n_0 - n'r$. The speed v of the charge carriers varies with r according to $v = v_0 - v'r^2$, where $n_0, n', v_0,$ and v' are constants. Calculate the current that passes through the rod.

26–3 Resistance

23. (I) You have two solid cylinders of the same material. Piece 2 has half the length and half the diameter of piece 1. What is the ratio of the resistances of the two pieces?

24. (I) The conductivity of silver is 1.5 times that of gold. What is the ratio of the diameter of a silver wire to that of a gold wire of the same length if both wires are designed to have the same resistance?

25. (I) An underground wire made of aluminum is 528 m long and has an area of 0.12 cm^2. (a) What is its resistance? (b) What is the radius of a copper wire of the same length and resistance?

26. (I) An old house is wired with AWG #18 copper wire, which has a diameter of 0.0403 in. (a) What is the wire's resistance per 100 ft? (b) One circuit consists of only one wire behind walls and has a resistance of 7.5 Ω. How long is this wire?

27. (I) The resistivity of copper is $1.72 \times 10^{-8} \, \Omega \cdot \text{m}$. What is the resistance of a section of gauge #10 wire (diameter 0.2588 cm) that is 10 m long?

28. (I) A carbon rod used in a welding machine is 5.0 mm in diameter and 20.0 cm in length. What is its resistance and how much current will pass through it if the welding machine puts a voltage of 380 V across it?

29. (I) Cables used to jump an automobile can get hot if used for more than a few seconds. Calculate the resistance at 20°C of a 2-m-long copper cable of cross-sectional area 36 mm^2. By how much does the resistance increase as the temperature rises from 20°C to 100°C?

30. (I) In the text, we refer to a power line with a total resistance of 10 Ω. Suppose that the power line is made of copper with a resistivity of $1.72 \times 10^{-8} \, \Omega \cdot \text{m}$, and is 175 km long. What is the radius of the wire?

31. (I) How long would a tungsten wire have to be if it is to be used in a toaster at 120 V and the current to be carried is 15 A? The cross-sectional area of the wire is $0.20 \times 10^{-6} \text{ m}^2$.

32. (I) The electron current densities of copper and aluminum are $9 \times 10^{28} \text{ m}^{-3}$ and $18 \times 10^{28} \text{ m}^{-3}$, respectively. A wire of copper is joined to a wire of aluminum of the same diameter. If the same current is passing through the wires, compare the electron drift velocities and the resistances per unit length.

33. (II) A current passes through a tungsten wire in an appliance. If you assume that there is a fixed potential drop from one end of the wire to the other, what is the fractional change in the power consumed as the temperature of the tungsten wire changes from 800°C to 1200°C? Ignore any effects due to the change in the wire's length by thermal expansion.

34. (II) An electrician tests for a short circuit by putting a potential difference of 1.5 V across two neighboring parallel wires that would be independent of each other if there were no short. A current of 0.14 A then flows in the wires. The wires consist of material with a resistivity of $1.7 \times 10^{-8} \, \Omega \cdot \text{m}$ and have a diameter of 0.24 mm (Fig. 26–28). Given that the short effectively makes the wires act like a single wire, how far away is the short?

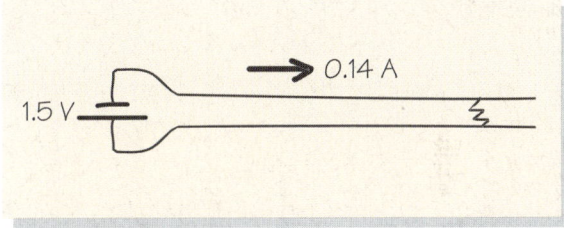

▲ **FIGURE 26–28** Problem 34.

35. (II) A nichrome wire of diameter 0.5 mm and length 50 cm is connected to a 50-V battery. What current passes through the wire at room temperature (25°C) and after the wire heats up to 400°C?

36. (II) The change in the resistance of a thin platinum wire can be used to measure temperature. Suppose that a constant 6.0 mA current passes through a platinum wire and that the potential drop measured at room temperature is 8.5 mV. What is the temperature of the wire when the potential drop is 8.7 mV?

37. (II) An aluminum wire of length L and a copper wire of length $5L$ have precisely the same resistance. Given that the resistivity of aluminum and copper are $2.8 \times 10^{-8} \, \Omega \cdot \text{m}$ and $1.7 \times 10^{-8} \, \Omega \cdot \text{m}$, respectively, what is the ratio of the radii of the two wires?

38. (II) You have a 100-m-long wire of area 0.5 mm^2 with a thin coating of insulation, but you cannot identify the type of material that makes up the wire. You have a 12.0-V battery and a device to measure current. When the battery is placed across the two ends of the wire, you measure a current of 1.07 A. What is the wire material? (Use Table 26–2.)

39. (II) A coil used to produce a magnetic field is made of copper wire of area 1.5 mm^2 wound many times around a spool of diameter 20 cm. The resistance of the wire is 1.35 Ω. We must know the number of turns of wire to know the magnetic field. How many turns of wire are there on the spool?

40. (II) You wish to double a current that flows through a wire of fixed length, but you can increase the voltage that drives the current by only a factor of 1.8. You have other wires made of the same material but of different radii. What is the smallest factor by which the radius of a replacement wire should differ from the radius of the original wire?

41. (II) Aluminum has a density of $2.7 \times 10^3 \text{ kg/m}^3$. What is the resistance of an aluminum wire 0.12 cm in diameter and 80 m long? What is the mass of the wire? What is the mass of a copper wire, of density $8.9 \times 10^3 \text{ kg/m}^3$, with the same length and same total resistance?

762 | Currents in Materials

42. (II) What are the length and the radius of a copper wire (of circular cross section) whose resistance is 2 Ω and whose mass is 1.5 kg?

43. (II) How much silver (of density 10.5×10^3 kg/m^3) would be needed to make a wire 1 km long, with a resistance of 5 Ω?

44. (II) A copper pipe has an inside diameter of 2.75 cm and an outside diameter of 2.45 cm. What length of copper pipe will have a resistance of 3.5 Ω?

45. (II) A copper resistor has the shape of a cylindrical shell. What is the resistance of this resistor if its length is 1 m, its inner radius is 0.1 cm, and its outer radius is 0.2 cm? What is the radius of a solid wire of circular cross section with the same length and the same resistance? Compare the masses of the two resistors.

46. (II) A *zener diode*, named for Clarence Zener, has the *I–V* curve shown in Fig. 26–29. Sketch the resistance of the diode versus both current and voltage. What is special about the critical voltage V_c?

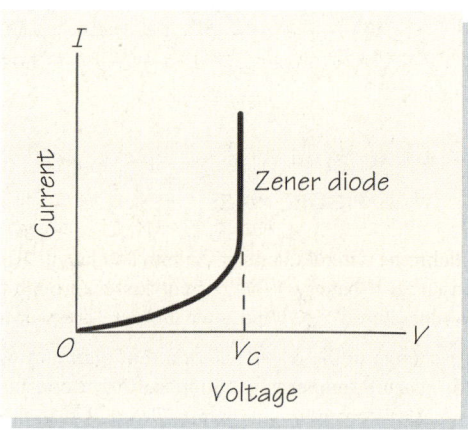

▲ **FIGURE 26–29** Problem 46.

26–4 Resistances in Series and in Parallel

47. (I) We have three identical lightbulbs available to us. Two of them are connected in series to a battery. What is the relative brightness of the two bulbs? If the third bulb is connected in parallel with the second of the above bulbs, what are the relative brightnesses of the three bulbs?

48. (I) Consider the circuits shown in Fig. 26–30. The lightbulbs in all circuits are identical and the batteries are the same in the two circuits. Before working with these circuits, you are asked to make some predictions: (a) What is the brightness of the bulbs in circuit II relative to each other and to the bulb in circuit I? (b) If one of the bulbs is removed in circuit II, how will the brightness of the other bulb be affected? Does it matter which bulb is removed?

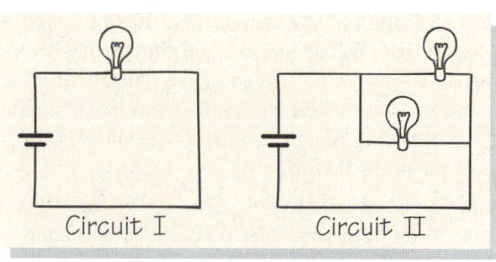

▲ **FIGURE 26–30** Problem 48.

49. (I) Consider the circuit in Fig. 26–31 in which there are two resistors in series. The resistance of one of them, x, is unknown. If the resistor R is 10 Ω, the voltage drop across x is 8 V. If R is 5 Ω, the voltage drop across x is 12 V. What is the resistance of x, and what is the total potential drop across the two resistors in the case that R is 10 Ω?

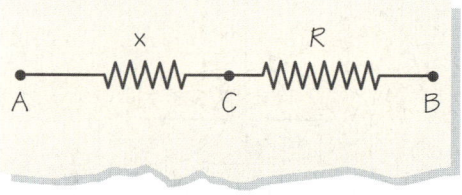

▲ **FIGURE 26–31** Problem 49.

50. (I) Suppose we have two resistors in parallel. Their values are 20 Ω and 12 Ω respectively. What resistance placed in parallel with these will make the total effective resistance equal to 4 Ω?

51. (I) Five resistors of 18 Ω each are connected in series. If the potential difference between the ends of this set of resistors is 16 V, what current flows through the resistors? What is the power expended in the circuit?

52. (I) Two 60-Ω resistors are placed in series across two terminals whose potential difference is 120 V. What is the total power dissipated?

53. (I) Two resistors are placed in parallel. One of the resistors has twice the resistance of the other; the lesser of the two resistances is 150 Ω. What is the resistance of the parallel combination?

54. (I) Find the equivalent resistance of the circuit shown in Fig. 26–32.

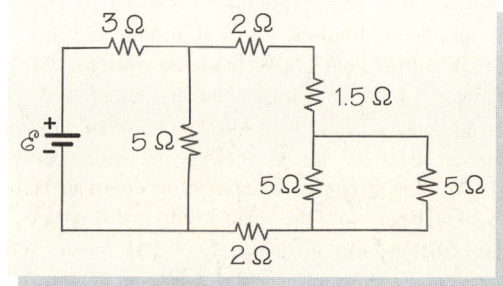

▲ **FIGURE 26–32** Problem 54.

55. (II) Consider the combination of resistors shown in Fig. 26–33. Calculate the current in each resistor, given that $V_{AB} = 16$ V.

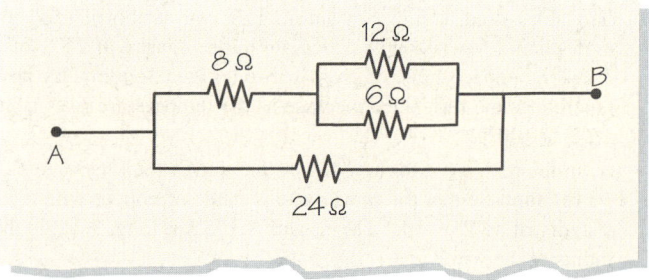

▲ **FIGURE 26–33** Problem 55.

56. (II) Take the resistor combination in Fig. 26–33. What is the power dissipated in each of the resistors?

57. (II) Consider the network of resistors shown in Fig. 26–34. If a total current of 20 A flows through the network from A to B, what is the current in each of the resistors, and what is the voltage across each of the resistors?

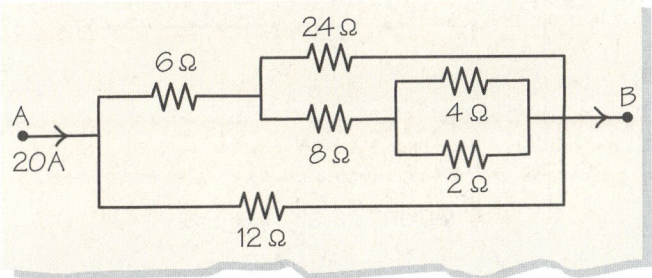

▲ **FIGURE 26–34** Problem 57.

58. (II) Consider the circuit shown in Fig. 26–35. What must the value of the resistance x of the unknown resistor be so that the total equivalent resistance of the network is also x?

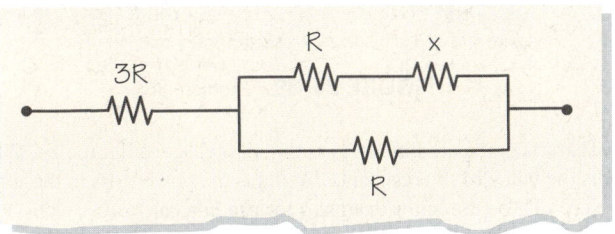

▲ **FIGURE 26–35** Problem 58.

59. (II) Points a and b are connected by the system of resistors shown in Fig. 26–36. A battery of 12 V and negligible internal resistance is connected across points a and b. (a) What is the equivalent resistance between points a and b? (b) The potential difference across the 75-Ω resistor? (c) The current flowing through the 33-Ω resistor?

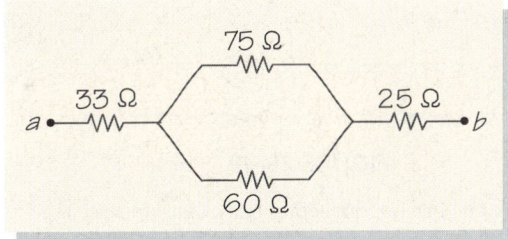

▲ **FIGURE 26–36** Problem 59.

*26–5 Free-Electron Model of Resistivity

60. (I) Using the average time between collisions as calculated in Example 26–5, determine the drift speed of charge carriers for a material in which the electric field is 2.0×10^{-3} V/m.

61. (I) Assuming the collision time from Example 26–8 and an average speed of 2.7×10^{6} m/s, estimate the mean free path for an electron in copper.

62. (I) Recall Eq. (19–46), which relates the collision cross section to the mean free path of a particle. Use that result together with the results of Problem 52 to estimate the collision cross section of an electron with an ion in a copper lattice.

63. (II) In Problem 20, we described a wire of radius R within which the drift speed of charge carriers varies with the distance from the center of the wire as $v_d = v_0[1 - (r/R)]$. Supposing that this wire is made of ohmic material, describe how the resistivity must vary with r to produce this drift-speed profile.

*26–6 Materials and Conductivity

64. (II) If you treat electrons as a gas of independent particles, at what temperatures would an average electron have sufficient energy to cross the energy gap for silicon (1.1 eV), germanium (0.7 ev), and carbon (6 eV)?

26–7 Electric Power†

65. (I) What is the resistance of a 65-W headlight used on a 12-V car battery?

66. (I) What is the maximum voltage that can be applied to a 1000-Ω resistor rated at 1.5 W?

67. (I) Your little sister leaves a 100-W lightbulb burning for an unnecessary hour. Assuming that electric power costs 10 cents per kilowatt-hour, what is the cost of her inaction?

68. (I) A graduate student in engineering has a collection of 100-Ω resistors with different power ratings of 1/8, 1/4, 1/2, 1, and 2 W. What is the maximum current that the student should use in each resistor?

69. (I) What is the maximum allowable current for (a) a 160-Ω, 5-W resistor? (b) A 2.5-kΩ, 3-W resistor?

70. (I) An electrostatic accelerator has a maximum attainable voltage of 8×10^6 V. If a current of 100 μA is produced by accelerating charges in this potential difference, what is the nominal power required to operate the accelerator? Assume that this power can be converted to the accelerator voltage with 100 percent efficiency.

71. (I) Consider a resistor of resistance R. If the maximum allowed power dissipation is P, what is the maximum allowed operating voltage?

72. (I) An electric heater draws a current of 10 A from a 120-V circuit. What is the cost per hour of operating the heater if electrical energy costs 7 cents per kWh?

73. (I) A 2-m-long copper cable with cross-sectional area 36 mm² draws 100 A when it is used to jump start an automobile. It takes 20 s to start the engine. Given the resistance is 9×10^{-4} Ω, calculate how much energy is dissipated in the cable during this operation.

74. (II) A heater uses nichrome wiring ($\rho = 10^{-6}$ $\Omega \cdot$m) and generates 1250 W when connected across a 110-V line. How long must the wire be if its cross-sectional area is 0.2×10^{-6} m²?

75. (II) Consider the terminals of a 12-V battery connected by a copper wire. How long must the wire be if its cross-sectional area is 8×10^{-6} m² and if the power dissipated is 0.8 kW?

76. (II) Buildings have circuit breakers, devices that switch the current off when it exceeds a critical value, to protect the electrical system from damage. One circuit for a building's lights has a 15-A breaker. (a) What is the maximum power that can be delivered by a 110-V line to this circuit? (b) How many lightbulbs, each requiring 75 W, can this circuit handle?

---

†When we refer to household electricity applications, assume in each case that their currents and voltage differences are of the simple, constant type discussed in this chapter.

77. (II) A 6-V battery is connected to two metal wires dipped in a pot of water (Fig. 26–37). A current of 50 mA flows for 18 h. How much energy is taken out of the battery during that time?

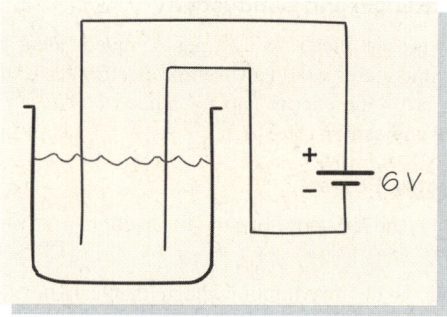

▲ FIGURE 26–37 Problem 77.

78. (II) A 500-W electric heater is designed to operate on a line of 115 V. As the result of a brownout (a partial interruption of electrical power) the line voltage drops to 105 V. Assuming that the heating unit has a fixed resistance, what is the power of the heater now?

79. (II) Consider the bus stop in Example 26–11. If it were necessary to reduce the current in the heater by a given factor, one could proceed by decreasing the temperature T of the wire or the wire's diameter (see "What do you think?" for that example). What changes would be preferable and why? Ignore the change of the resistivity with temperature—this is typically a smaller effect than the others that you will deal with here.

General Problems

80. (I) The power dissipated in a resistor through which a current I passes is P_0. What is the power dissipated if the same current passes through three such resistors connected in series?

81. (II) An electric hot plate is used to boil water. The current drawn by the hot plate is 4 A. Calculate the voltage and resistance, based on a rough estimate of how long it takes to make a pot of tea.

82. (II) One month's electricity bill for an apartment is $25.33, and the cost of electricity is 8 cents/kWh. All appliances used in this apartment work at 120 V. How many electrons passed through the apartment's electrical meter that month?

83. (II) A wire of resistance r is drawn—pulled like taffy—to double its length. Assuming a constant voltage and a fixed volume, by how much does the power dissipation change?

84. (II) A Van de Graaff accelerator delivers 4-MeV protons at a current of 5 μA through a target onto a piece of tungsten that serves to stop the proton beam. (a) How many protons stop in the tungsten in 1 h? (b) How much energy is delivered to the tungsten in 1 h? (c) What is the power of the proton beam?

85. (II) A piece of brass is machined into a long, tapering cylinder. Its radius is expressed by $r = r_0 + \alpha x$, where α is a constant and x is measured from the narrow end of the tapering cylinder and runs from 0 to L (Fig. 26–38). Find an expression for the resistance of this piece.

86. (II) The voltage at an electrical outlet is a constant 120 V. You have 10 identical lightbulbs whose maximum power consumption is 5 W (Fig. 26–39). (a) What is the resistance in each bulb if the power consumption is 50 W when the bulbs are connected in series? (b) If the 10 bulbs are connected in parallel, an additional resistor is needed so that the bulbs do not burn out. The resistor is connected in series with the total set of lightbulbs. What is the value of the resistance? What is the power loss in the resistor? (These are bulbs designed for use in a car.)

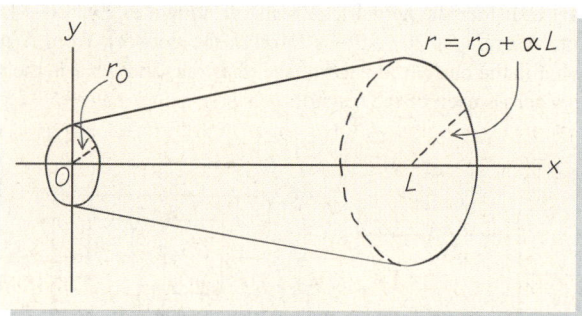

▲ FIGURE 26–38 Problem 85.

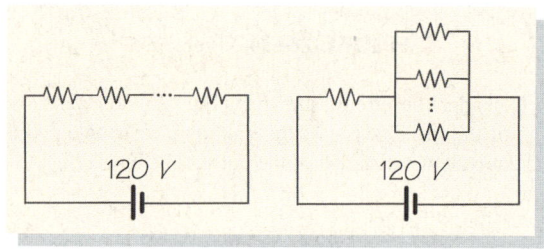

▲ FIGURE 26–39 Problem 86.

87. (II) Consider the circuit shown in Fig. 26–40. Calculate the current and the power dissipated in the 4-Ω resistor as a function of the unknown resistance R_x.

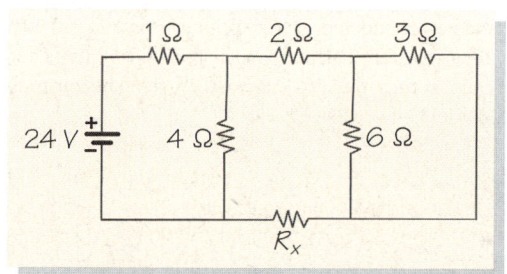

▲ FIGURE 26–40 Problem 87.

88. (II) A bus bar (a conducting bar meant to carry a good deal of current) made of copper, of resistivity 1.72×10^{-8} $\Omega \cdot$m, is meant to carry 100 A over a distance of 0.25 m at a temperature of 300°C. What is the minimum cross-section of the bus bar if no more than 0.2 W of power is to be dissipated?

89. (II) A generator delivers 75 A at a voltage of 12 V. What power does the generator deliver? How long would it take to raise the temperature of 10^{-3} m^3 of water by 7.5°C? How long would it take to boil away 0.5 L of water, starting at 25°C?

90. (II) Figure 26–10 shows the I–V curve of a typical semiconductor diode. Use the data from the figure to sketch the power dissipated in the diode as a function of the current. What is the power dissipated in the ideal diode, as shown in the figure?

91. (II) A potential is set up from one end of a copper wire to the other; as a result, current flows. The copper is thermally isolated to some extent. As the charge flows, the wire heats up, causing the resistivity to increase. Suppose that, during a short time period, the temperature of the wire as a function of time t is given by $T = T_0 + kt^2$. (a) Describe the current in the wire during this period. (b) What is the power dissipated by the wire as a function of time? (c) From the change with time of the dissipated power, will the wire continue to heat up and, perhaps, melt?

92. (II) The density of charge-carrying electrons in copper is 8.5×10^{28} electrons/m^3, its resistivity is 1.7×10^{-8} $\Omega \cdot$m, and the drift speed in a copper wire is 1.2×10^{-5} m/s. The wire has a diameter of 1 mm and a length of 3 m. At what rate must thermal energy be carried off by a cooling medium if the wire is to maintain its temperature?

93. (II) A single layer of 200 turns of closely spaced wire of radius $r_1 = 0.6$ mm is wound in a coil of diameter $D_1 = 5$ cm (Fig. 26–41). A second coil of the same length but of diameter $D_2 = 8$ cm is composed of a single layer of closely spaced wire of radius $r_2 = 0.4$ mm. The wires are made of the same material. Find the ratio of the resistances of the two coils.

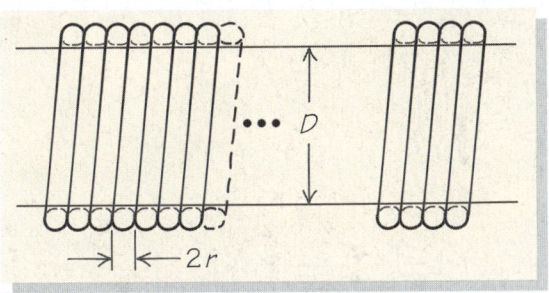

▲ FIGURE 26–41 Problem 93.

94. (III) A thin wire of length L and cross-sectional area A oriented in the x-direction is made of an ohmic material whose resistivity varies along the wire according to the empirical law $\rho = \rho_0 e^{-x/L}$. (a) Describe how the field within the wire varies with position if the end at $x = 0$ is at a potential V_0 greater than the end at $x = L$. (b) How does the potential vary as you move along the wire? (c) What is the total resistance of the wire?

95. (III) If all the energy lost from Joule heating stays in a wire, and the temperature increases as a result, the resistivity will increase according to Eq. (26–18). The current will therefore change as a function of time, the Joule heating will change, and so forth. If the wire material has a constant heat capacity, the rate of energy loss in the wire will be proportional to the rate of change of temperature. Assuming that the potential stays constant, set up a differential equation that describes the rate of temperature change. If this equation is solved, how can the current be found as a function of time?

Chapter 27

▶ This experimental aircraft runs on sunlight. The solar cells act as batteries, with energy from the Sun, and drive the electric circuits that make the propellors turn.

Direct-Current Circuits

We have seen how charges move through potential differences under the influence of electric fields, and how resistors control current and capacitors store charge. When resistors, capacitors, and batteries are connected together by conducting wires, they form electric circuits. We can understand the flow of currents in circuits by applying just two simple physical principles: the conservation of current and the conservation of energy. In this chapter, we learn to apply these principles systematically to the analysis of circuits. Our earlier work on capacitors or resistors connected in parallel and in series will be useful in this analysis. We also discuss some of the instruments available to measure and monitor the current and voltage of electric circuits. The flow of energy to and from circuit elements is an important theme that leads us to the concept of time-varying currents and voltages.

▶ 27–1 EMF

The sources of energy that cause charges to move in electric circuits have historically been called sources of **electromotive force**. They are actually sources of energy, not of force, and because the word force is misleading here, we use the abbreviation **emf** instead. Briefly, a source of emf is a device that *maintains* a potential difference across a conductor, thereby allowing a current to run continuously through the conductor. (Contrast this to the potential difference supplied by, say, a capacitor. The charged capacitor does indeed have a potential difference between its plates, but the current running under its influence is composed essentially of the charge on the plates running off, and as it runs off, the potential difference quickly drops to zero.) When we think of sources of emf, we often think of batteries. A battery can be thought of as a device that expends chemical energy to pump charges, just as a water pump expends mechanical energy to pump water uphill to a tank with a higher gravitational potential energy. There is a wide variety of sources of electric energy beyond batteries. A battery converts chemical energy into an emf; a solar cell converts the energy of sunlight into an emf; a thermocouple produces an emf as a result of a difference in temperature; a large commercial electric power plant may burn coal, gas, or nuclear fuel, or use falling water, to drive a generator that produces an emf (Fig. 27–1).

In this chapter, we use the term "battery" to refer to any source of emf. We shall restrict ourselves to batteries for which the emf is *constant* with time. Up to Section 27–5, we focus on phenomena such as current flows or potential differences that are similarly constant in time. We refer to this as *equilibrium*, or *steady-state*, behavior, and use the terms **direct-current**, or **DC**, behavior as a label for it.

Circuits

Sources of emf supply energy to circuits. But just what is a circuit? When batteries, resistors, capacitors, or other circuit elements (some of which will be introduced later) are connected by wires (ideally of negligible resistance), they form a **circuit**. For example, when a switch is closed—meaning that a gap in a wire is closed—and a battery establishes current through the filament of the lightbulb of a flashlight, a circuit has been formed. Figure 27–2 illustrates this simple circuit with the conventional symbols for resistors, ideal wires (wires with no resistance), and batteries; the lightbulb is a simple resistor.

Circuit analysis typically requires us to relate the currents and potential differences within the circuit. For example, we may want to know the potential drop across a capacitor or the current that passes through a resistor when there is a particular emf in the circuit.

The Meaning of Emf

The fact that the force involved in moving charges under the influence of an emf is a conservative one provides us with a powerful method for the analysis of circuits. Consider the circuit of Fig. 27–2, consisting of a battery and a single resistor known as the **load** resistance. The battery has a potential difference across its terminals called the **terminal voltage**. Current flows away from the battery terminal at the higher potential—the one marked positive. How much current flows depends on the load resistance (or, in more complicated circuits, on the characteristics of the components of the rest of the circuit). Because current is defined as moving in a direction opposite to that of electrons, it is helpful to imagine positive charges flowing to the negative terminal (the terminal at the lower potential), equivalent to electrons flowing to the positive terminal, which is in fact what occurs.

Inside a chemical battery, a chemical process carries the positive charges back to the positive terminal. It is this process—the internal pumping action of the battery—that gives a precise definition of the emf. Suppose that it takes work dW to move a charge dq from the negative to the positive terminal. Then the emf of the battery is defined to be

$$\mathcal{E} \equiv \frac{dW}{dq}. \tag{27–1}$$

EMF DEFINED

The SI unit of emf is the volt, one joule per coulomb. The word *voltage* is sometimes used loosely to describe the emf $\mathcal{E}$, but voltage more properly refers to the potential difference or terminal voltage across the emf terminals, which as we shall see below may be different from $\mathcal{E}$.

When a battery is connected to a circuit such as that in Fig. 27–2, it sets charges into motion, driving a current from the positive (higher-potential) terminal around the circuit to the negative (lower-potential) terminal, and we say that the battery *discharges*. In discharging, the battery is expending its chemical energy. If a current is driven from the negative to the positive terminal, a process that can be accomplished in conjunction with a battery of larger emf, the battery of smaller emf is said to be *charging*; this process replaces its spent chemical energy.

(a)

(b)

(c)

▲ **FIGURE 27–1** Various sources of emf that produce electrical energy include (a) solar panels (photovoltaic), (b) burning fuel, here nuclear, can heat water, produce steam, and turn turbines, (c) batteries, which produce their emf through chemical means.

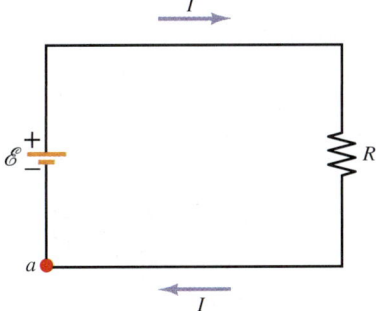

▲ **FIGURE 27–2** A simple circuit with a source of emf, $\mathcal{E}$, and a resistor, R.

Let us now turn to the analysis of our circuit. Start by assuming that the potential difference across the battery terminals in Fig. 27–2 is in fact the emf $\mathcal{E}$. (We'll refine this in the next subsection.) A current will flow around the circuit, and to find this current we can use the fact that the electric potential is associated with a conservative force. If the force is conservative, then the net work done by the force in sending a charge around a closed loop is zero. In turn, the total potential drop involved in any round trip that starts from any point on a closed loop must be zero. Make such a round trip that starts at point a of Fig. 27–2 and follow the current around the circuit. There is no change in potential as we pass through the ideal (resistanceless) wire. In crossing the battery from the negative to the positive terminal, the potential *increases* by $\mathcal{E}$. When we cross the ohmic resistance, the potential *decreases* by an amount IR [see Eq. (26–12)]. The potential drop implies a decrease in the potential energy of the charges. This potential energy is converted into thermal energy in the resistor. The net potential change as we travel once around the circuit is zero, so

$$\mathcal{E} - IR = 0. \tag{27-2}$$

This equation determines the current, I:

$$I = \frac{\mathcal{E}}{R}. \tag{27-3}$$

Internal Resistance

Above we assumed that the voltage across a battery's lead is identical to the emf. We can imagine the existence of an ideal emf of, say, 9 V that will always have a potential difference of 9 V between its terminals. But any real source of emf will entail some energy loss as charge moves through it. You may know, for example, that a car battery heats up noticeably when it discharges, a result of resistive heating. This is true for all batteries. The reasons for this are in general rather complex, but at the very least the charge passing through a battery driving a circuit has to pass through the material of the battery itself, and this material will have some ordinary resistivity. Thus a real battery contains an *internal resistance r* in addition to maintaining an emf. This resistance is sometimes shown separately from the emf (Fig. 27–3). If we calculate the net potential change around the circuit as before, we find that Eq. (27–2) becomes

$$\mathcal{E} - Ir - IR = 0. \tag{27-4}$$

Because of internal resistance, the potential difference across the battery terminals is no longer just $\mathcal{E}$; it is given instead by

$$\text{with internal resistance: } V = \mathcal{E} - Ir. \tag{27-5}$$

This potential difference is a function of the current. Depending on the direction of current flow, the voltage across the terminals of a battery can be greater or less than the battery's emf. A second modification that results from internal resistance is that the current depends on it. From Eq. (27–4), the current in our circuit is

$$I = \frac{\mathcal{E}}{r + R}. \tag{27-6}$$

Compare this to Eq. (27–3).

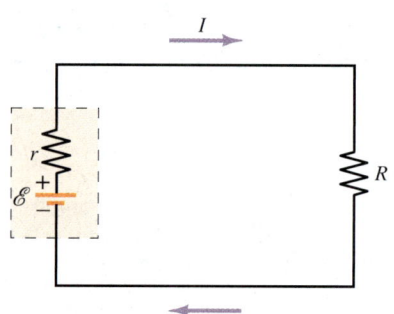

▲ **FIGURE 27–3** A source of emf also contains an internal resistance r, depicted in circuit diagrams as the emf symbol with a resistor in series close by.

CONCEPTUAL EXAMPLE 27–1 An engineer has inadvertently placed a battery into a circuit (Fig. 27–3) that has a much larger internal resistance r than the resistance R outside the battery. As a result, which of the following, if any, is true: (a) The current through the circuit will be much less than the emf of the battery; (b) the voltage drop across the terminal will be much greater than it would have been if the internal resistance were less; (c) the current through the circuit doesn't depend very much on what R is.

Answer (a) is a nonsensical statement. The current and the emf may be related, but they are not comparable quantities. (b) is false—the voltage drop is in fact much less than it would have been if the internal resistance were small. In effect, the internal resistance "uses up" some of the voltage drop that the battery could otherwise supply for use within the circuit. (c) is true—the current is determined by the sum of the internal and external resistance, as in Eq. (27–6), and this is insensitive to R if $r \gg R$.

It is worth looking at (b) a little more closely, and in a quantitative way. The voltage drop across the terminals is given by Eq. (27–5), and if we insert the current from Eq. (27–6), we see the voltage drop is

$$V = \mathcal{E} - Ir = \mathcal{E} - [\mathcal{E}/(r+R)]r$$
$$= \mathcal{E}[r + R - r]/(r + R) = \mathcal{E}R/(r+R) \cong \mathcal{E}R/r,$$

and this is a small fraction of the emf.

We can see from Eq. (27–5) that for small internal resistance the potential across the terminals is approximately the same as the emf. This is generally true if there is no current at all, so a reading of the potential across the terminals is also a reading of the emf when the external, or load, resistance in the circuit is very large. It is desirable that the internal resistance be "small," but we need to remember that we mean small in comparison with external resistances, which vary depending on the application. The typical internal resistance of a car battery is less than $0.01\ \Omega$, but it may be as large as $0.1\ \Omega$ for a flashlight battery. In many situations, the internal resistance is so small relative to the resistances within the circuit that it can be ignored in electric-circuit analyses. Ordinary batteries run down with age not because their emf decreases, but because their internal resistance increases, which means that the current they can supply decreases.

EXAMPLE 27–2 One of two different resistors with respective resistances $R_1 = 5.00\ \Omega$ and $R_2 = 10.0\ \Omega$ can be placed into the circuit shown in Fig. 27–3. The emf $\mathscr{E}$ and internal resistance r of the battery are unknown. When only R_1 is inserted, the current is $I_1 = 0.291$ A; when only R_2 is inserted, the current is $I_2 = 0.147$ A. Find $\mathscr{E}$ and r.

Strategy The total resistance for the two cases under consideration are $R_1 + r$ and $R_2 + r$, and in both cases, when these resistances are multiplied by the appropriate currents I_1 and I_2, respectively, they yield the emf $\mathscr{E}$. We will then have two linear equations that we can solve for the two unknowns r and $\mathscr{E}$.

Working It Out The fact that the net potential drop around the circuit is zero for the two cases yields the two equations

$$\mathscr{E} - I_1 r - I_1 R_1 = 0,$$
$$\mathscr{E} - I_2 r - I_2 R_2 = 0.$$

Multiplying the first equation by I_2 and the second by I_1 and then subtracting to solve for $\mathscr{E}$ we find that

$$\mathscr{E} = \frac{I_1 I_2}{I_1 - I_2}(R_2 - R_1).$$

If we insert this into either of the first two equations, we can solve for r:

$$r = \frac{I_2 R_2 - I_1 R_1}{I_1 - I_2}.$$

Numerical evaluation gives

$$\mathscr{E} = \frac{(0.291\ \text{A})(0.147\ \text{A})}{0.291\ \text{A} - 0.147\ \text{A}}(10.0\ \Omega - 5.00\ \Omega) = 1.49\ \text{V}$$

and

$$r = \frac{(0.147\ \text{A})(10.0\ \Omega) - (0.291\ \text{A})(5.00\ \Omega)}{0.291\ \text{A} - 0.147\ \text{A}} = 0.104\ \Omega.$$

What Do You Think? Suppose R_2 had been $1.0 \times 10^5\ \Omega$, known to 2 significant figures. Could you have found r? Answers to *What Do You Think?* questions are given in the back of the book.

Electric Power and Batteries

A source of emf (or electric energy) is also a source of *electric power*. The power is the rate at which the source delivers energy. From Eq. (26–30), the power of the source is the potential drop across the source times the current that passes through it. For a source of emf $\mathscr{E}$ we have

$$P = I\mathscr{E}. \qquad (27\text{–}7)$$

Let's see how this works for the circuit of Fig. 27–3. Equation (27–6) tells us that $\mathscr{E}$ is given by $I(r + R)$, where I is the current in the circuit and r and R are the internal and load resistances, respectively. If we use this relation for $\mathscr{E}$ in Eq. (27–7), we find that

$$P = I^2 R + I^2 r. \qquad (27\text{–}8)$$

Energy conservation implies that we had to find this result. The electric power of the source of emf is balanced by the sum of the power dissipated in both the internal and load resistances.

EXAMPLE 27–3 (a) A battery of emf $\mathscr{E}$ and negligible internal resistance r acts in the circuit of Fig. 27–3 with a resistor R. What is the power delivered in terms of $\mathscr{E}$ and R? (b) Suppose r is not negligible. What is the power delivered in terms of $\mathscr{E}$, R and r?

Setting It Up We want the result for power not in terms of I and $\mathscr{E}$, or in terms of I and resistance, as in the text discussion above, but in terms of $\mathscr{E}$ and resistance.

Strategy By using the expression $P = \mathscr{E}I$ and expressing the current in terms of the emf and the resistances, we will find the power in terms of the emf and the resistances.

Working It Out For case (a) $I = \mathscr{E}/R$, while for case (b) $I = \mathscr{E}/(R + r)$. We substitute these into the expression for power $P = I\mathscr{E}$: For case (a)

$$P = I\mathscr{E} = \frac{\mathscr{E}}{R}\mathscr{E} = \frac{\mathscr{E}^2}{R},$$

while for case (b)

$$P = I\mathscr{E} = \frac{\mathscr{E}}{R + r}\mathscr{E} = \frac{\mathscr{E}^2}{R + r}.$$

What Do You Think? The value of r determines the maximum power P_{max} a battery of a given emf can deliver in this circuit. What is P_{max}?

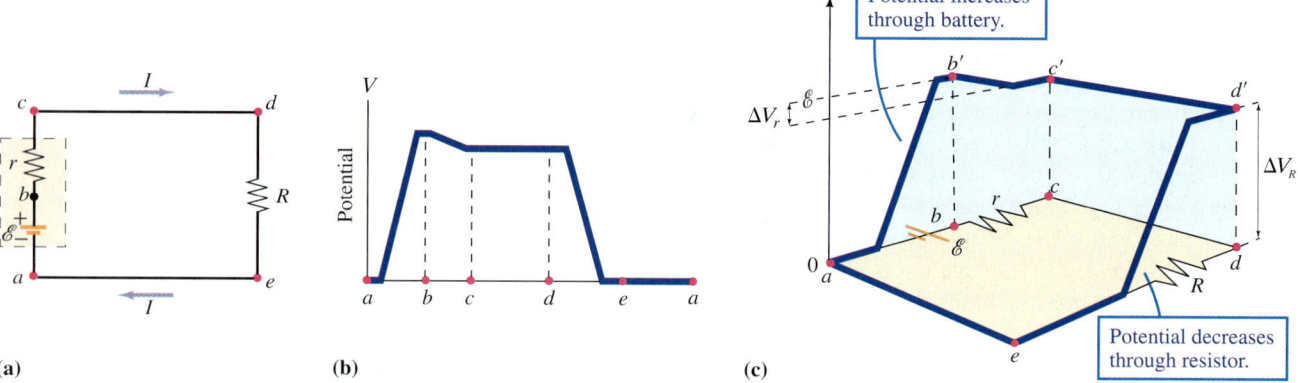

▲ **FIGURE 27–4** (a) A single-loop circuit showing the emf ($\mathcal{E}$), internal resistance (r), and load resistance (R). (b) The potential differences for the points labeled in part (a). (c) A three-dimensional view of part (b).

27–2 Kirchhoff's Loop Rule

A **single-loop circuit** is a circuit with a single path for the current. The simple circuit discussed in Section 27–1 (Fig. 27–3) is an example of such a circuit. Let's go around the loop and examine the potential change at every step. First we'll redraw the circuit in Fig. 27–4a and mark the points a through e. The potential is graphed in Fig. 27–4b, c, where we follow the circuit along the current direction, with zero potential chosen arbitrarily at point a. We draw the internal resistance, r, in Fig. 27–4a as though it were separate from the emf; in Fig. 27–4b, c, we draw the rise of the emf as gradual. There is no potential change in an ideal conducting wire. There is some very small resistance in a real wire, but it can usually be ignored, and we do so here.

As we observed earlier, the net potential change in traversing the complete circuit is zero. This is an expression of the conservation of energy: For this particular circuit the charge that moves round the loop gains energy from the emf and loses the same amount of energy in the resistors. If we had followed the circuit in the opposite direction—against the current—the changes would all be of the opposite sign, but the end result would remain: The potential change in a complete circuit is zero. In the context of circuits, this simple law is given a special name, after the nineteenth-century physicist Gustav Kirchhoff, **Kirchhoff's loop rule**:

The sum of the potential changes around a closed path is zero,

$$\sum_{\text{closed path}} \Delta V = 0. \quad (27\text{–}9)$$

KIRCHHOFF'S LOOP RULE

The loop rule is applicable to any closed path in any electric circuit. When a circuit is laid out in a diagram, many closed paths may be possible, and as we shall see in several of the examples that follow, the loop rule applied to these loops is an important tool for finding the desired circuit parameters, which is the aim of any circuit analysis. In Fig. 27–4a, there is only one closed loop.

In applying the loop rule, we must have knowledge of the potential differences across various parts of a circuit. It is therefore useful to summarize what we have learned here and in previous chapters about the potential changes ΔV across individual circuit elements—batteries, resistors, and capacitors—as a set of rules. Figure 27–5 is a summary figure for these rules. The quantity ΔV is the change in potential when moving from an initial point to a final point; for example, the potential change in going from point a to point b is $\Delta V = V_{ab} = V_b - V_a$.

FIGURE 27–5 Rules for potential differences across various circuit elements.

1. In going from the negative to the positive terminal of a battery with emf $\mathcal{E}$, the potential change is positive, $\Delta V = +\mathcal{E}$. In going from the positive to the negative terminal, the potential change is negative, $\Delta V = -\mathcal{E}$. This rule ignores the presence of an internal resistance, which we treat as a separate resistance in series with the battery. The actual potential difference across the terminals is the voltage change $\mathcal{E}$ plus a term associated with the internal resistance that depends on the current (see point 2, below).

2. In moving across a resistance R *along* the direction of the current, I, the potential change is negative, $\Delta V = -IR$. The sign is opposite, $\Delta V = +IR$, in moving *against* the direction of the current.

3. In moving from the negatively to the positively charged plate of a capacitor of capacitance C and charge Q, the potential change is positive, $\Delta V = +Q/C$. The potential change is negative, $\Delta V = -Q/C$, when we move from the positively charged plate to the negatively charged plate. It is worth recalling here that no actual current flows across a capacitor.

The simplest application of the loop rule involves several resistors in series in a single loop. As an example, consider the loop shown in Fig. 27–6a. Here we have an emf and three resistors. (The internal resistance of the source of emf will be ignored.) We start at point a and move toward the battery in the (clockwise) direction of the assumed current. (We could just as well follow the circuit in the opposite direction; each term would change sign, but this is irrelevant if the sum of voltage changes is zero.) Kirchhoff's loop rule, Eq. (27–9), gives

$$\sum \Delta V = \mathcal{E} - IR_1 - IR_2 - IR_3 = 0.$$

The solution for the current is

$$I = \frac{\mathcal{E}}{R_1 + R_2 + R_3}. \quad (27\text{–}10)$$

Of course, we know from Section 26–4 that we can replace resistors in series with an equivalent resistor R_{eq} given by the sum of the individual resistances, Eq. (26–21). This is illustrated in the circuit of Fig. 27–6b. There the current is $I = \mathcal{E}/R_{eq}$, and comparison of this result with Eq. (27–10) confirms Eq. (26–21).

In a single-loop circuit the internal resistance is simply included by adding it to the load resistance, since it always occurs *in series* with it. It is important in multi-loop circuits to place the battery's internal resistance adjacent to the battery.

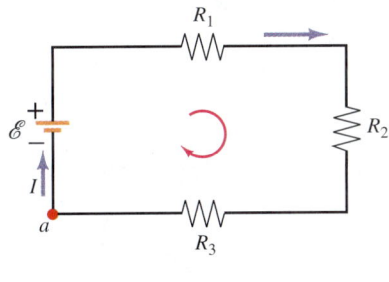

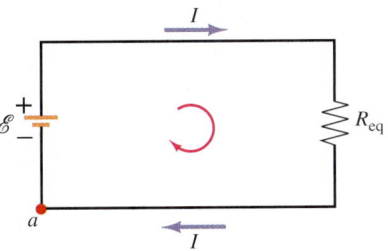

▲ **FIGURE 27–6** (a) The circuit with three resistors connected in series is equivalent to (b) the circuit with one resistor of value R_{eq}.

EXAMPLE 27–4 Find the current for the two-battery circuit shown in Figs. 27–7a and b. The values of the emfs and resistances are $\mathcal{E}_1 = 12$ V, $r_1 = 0.4\ \Omega$, $R_3 = 3\ \Omega$, $\mathcal{E}_2 = 6$ V, $r_2 = 0.1\ \Omega$, and $R_4 = 10\ \Omega$.

Setting It Up We have indicated the direction of the current in the figure; this assumption is necessary because the voltage change across resistors depends on the choice of direction. The solution to the problem will give us the actual current direction; if I is positive, our assumed direction was correct, and if I is negative, the current flows the other way.

Strategy We can regard the loop rule as an equation for the current, and we know all the other quantities that appear in the loop rule, including the internal resistances associated with the two emfs. These emfs are oriented in opposite directions and so will appear in the loop rule with opposite signs.

Working It Out Proceeding counterclockwise from point a, the loop rule reads

$$\sum \Delta V = +Ir_1 - \mathcal{E}_1 + IR_4 + \mathcal{E}_2 + Ir_2 + IR_3 = 0,$$

and we can solve this equation for the current. The result is

$$I = \frac{\mathcal{E}_1 - \mathcal{E}_2}{r_1 + R_4 + r_2 + R_3}.$$

Numerically,

$$I = \frac{12\ \text{V} - 6\ \text{V}}{0.4\ \Omega + 10\ \Omega + 0.1\ \Omega + 3\ \Omega} = \frac{6\ \text{V}}{13.5\ \Omega} \cong 0.4\ \text{A}.$$

Note that the only time you might see two opposing emfs in a single-loop circuit is when one battery is charging another.

What Do You Think? What would the consequence of having drawn the current in the opposite direction have been?

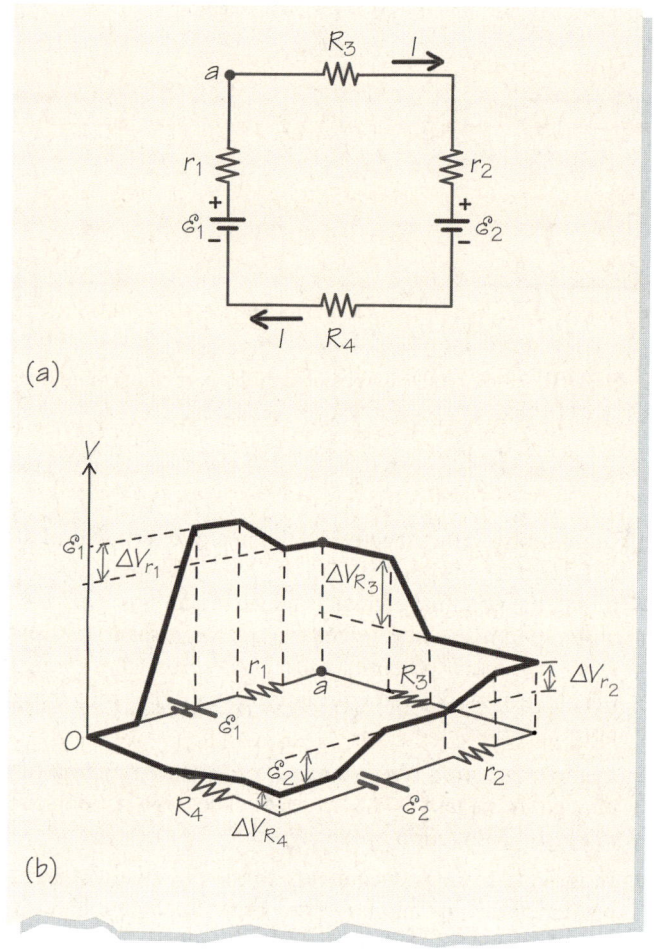

▲ FIGURE 27–7

27-3 Kirchhoff's Junction Rule

Few circuits are as simple as the one-loop circuits discussed so far. Examples of more complex circuits—multi-loop circuits—are shown in Figs. 27–8a and 27–8b. These examples have four- and three-line junctions. *Current conservation*, an experimental fact whose meaning and consequences were discussed in some detail in Chapter 26, is an additional tool that will allow a complete analysis of multi-loop circuits. We can summarize current conservation with **Kirchhoff's junction rule**. The rule states that the sum of the currents that enter a junction equals the sum of the currents that leave the junction. If we view currents that leave a junction as *negative*, and currents that enter the junction as *positive*, then we may rephrase this as

The algebraic sum of the currents that enter a junction equals zero,

$$\sum I_{\text{in}} = 0. \tag{27–11}$$

KIRCHHOFF'S JUNCTION RULE

This equation applies for every junction. Application of Kirchhoff's two rules gives us further information for currents, voltage differences, etc., in multi-loop circuits.

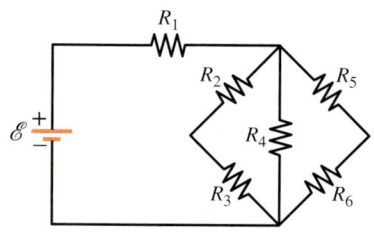

(a)

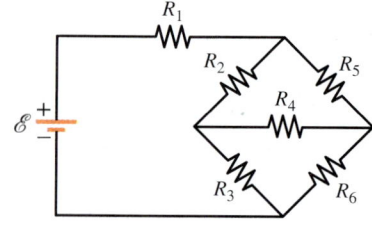

(b)

▲ FIGURE 27–8 Two multi-loop circuits.

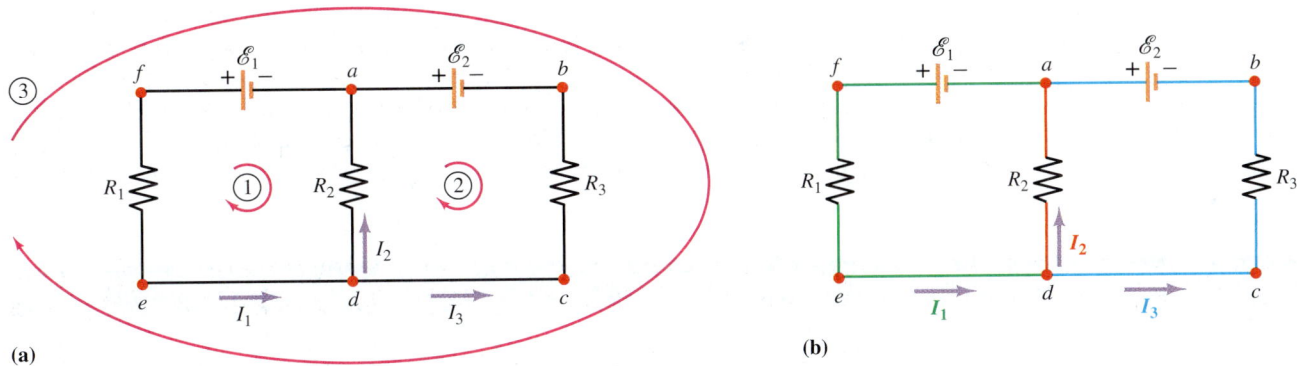

▲ **FIGURE 27–9** (a) A multi-loop circuit with two independent loops and two junctions. (b) The three legs of the circuit are shown in different colors.

Solving for the Behavior of Multi-Loop Circuits

Let's take a look at how the junction rule and loop rule help us solve for the unknown values in multi-loop circuits. Consider the circuit in Fig. 27–9a. Current flows between every junction. Thus we have currents I_1, I_2, and I_3 in the three separate *legs*, or *branches afed* (green), *da* (red), and *dcba* (blue), respectively (Fig. 27–9b). (A leg is any line of the circuit that starts and finishes on a junction but does not itself include a junction. For example, the three possible lines that connect *a* and *d* in Fig. 27–9b are all legs.) We can arbitrarily choose the direction of the three currents, and as we shall see, the algebraic equations for the currents will determine the actual direction for us: If these equations give a negative value for any of the currents, then that current actually travels in the opposite direction than the one we postulated. The junction rule can be applied at junctions *a* and *d* to obtain

$$\text{for junction } a: \quad -I_1 + I_2 + I_3 = 0; \tag{27-12a}$$

$$\text{for junction } d: \quad I_1 - I_2 - I_3 = 0. \tag{27-12b}$$

Notice that currents drawn as leaving the junction have minus signs, while currents that are drawn as arrows entering the junction have plus signs. Notice also that these two equations are not independent of each other; in fact, they are identical! There is a general rule—we won't prove it—that if there are N' junctions in a circuit, then there are $N' - 1$ independent junction rules—the meaning of an independent set of equations is that one equation cannot be derived as a combination of any others. But it is perhaps not so important to remember this rule; it is always possible to write an equation for each junction, then check whether or not they are independent.

How do we use the loop rule for this circuit? We must start by counting the number of *independent* loops. (While the counting is a bit more complicated than for the junctions, it is worthwhile looking at it in more detail here.) The rule for counting the number of such loops is as follows: For planar circuits (circuits that can be drawn in a plane with no two wires crossing over each other), the number of independent loops is just the number of enclosed areas through which you could poke a pencil. According to this idea, the circuit in Fig. 27–9a has two *independent* loops. But we can draw three possible loops, those indicated by the circled integers 1, 2, and 3 of Fig. 27–9a, where 3 is around the perimeter of the circuit. Only two of these three loops can be independent, and we can choose *any* two of the three possible loops as our independent loops. Before we choose, though, let's apply the loop rule for all three loops and verify that only two are independent. Point *a* is part of each of the three loops, so for convenience we'll begin at point *a* and apply the loop rule of Section 27–2. The internal resistances for the two emfs are neglected. We traverse each loop in the direction of the loop arrow:

$$\text{for loop 1}: \quad +I_2 R_2 + I_1 R_1 - \mathcal{E}_1 = 0; \tag{27-13a}$$

$$\text{for loop 2}: \quad -\mathcal{E}_2 + I_3 R_3 - I_2 R_2 = 0; \tag{27-13b}$$

$$\text{for loop 3}: \quad -\mathcal{E}_2 + I_3 R_3 + I_1 R_1 - \mathcal{E}_1 = 0. \tag{27-13c}$$

Only two of these three loop-rule equations are independent; the sum of the first two produces the third.

If we now take one equation from Eqs. (27–12), the junction rule, and the first two equations from Eqs. (27–13), the loop rule, *we have a total of three equations to solve for the three unknowns* I_1, I_2, and I_3. If we insert values for resistances and emfs, we can solve the three linear equations for the currents. This example illustrates features common to many circuit problems. We can summarize the important features with a set of problem-solving techniques.

Problem-Solving Techniques

In problems associated with multi-loop circuits, we must find unknown circuit parameters (such as resistance or current) when other parameters are given. To solve these problems, the following procedure may be helpful.

1. Draw a diagram with sources of emf, resistors, capacitors, and so forth clearly labeled. List the known and unknown parameters.

2. Assign a separate current for each leg of the circuit, and indicate that current on the diagram. The direction of current flow may not be immediately obvious; any direction can be assumed for the current, and the final algebraic solution will determine the correct direction. If the solution for a current turns out to be negative, the actual direction of current is opposite to your initial guess.

3. Apply the junction rule for the currents at each junction. Currents that you have chosen to draw as incoming have plus signs; those you have chosen to draw as outgoing have minus signs. If the circuit has N' junctions, then $N' - 1$ of the equations relating currents at the junctions will be independent.

4. Identify the number of independent loops N by counting the number of different ways that a pencil can poke through the circuit—a simple procedure for planar circuits. Indicate N loops on the diagram (for example, the loops labeled 1 and 2 on Fig. 27–9a).

5. Apply the loop rule to each of these loops.

6. Check to see that the number of linear equations from steps 3 and 5 matches the number of unknowns.

7. Solve these equations for the unknowns—whether they are currents or other parameters of the circuit. It is usually best to solve these equations *algebraically* and substitute numerical values later. Any checks, in the form of special cases or simple limits, can easily be made this way.

EXAMPLE 27–5 Find the currents for the circuit of Fig. 27–10, given that $\mathcal{E}_1 = 6.00$ V, $\mathcal{E}_2 = 12.0$ V, $R_1 = 100\ \Omega$, $R_2 = 10.0\ \Omega$, and $R_3 = 80.0\ \Omega$.

Strategy We follow the Problem-Solving Techniques (p. 774). The circuit diagram is given, and all parameters are labeled on Fig. 27–10. In this case, the unknowns are I_1, I_2, and I_3. For the application of the junction rules, we note that there are only two junctions, and hence only one of them is independent. For the loop rules, we choose loop 1 and loop 2 as the two independent ones. In fact, we have already written the junction and loop equations for this circuit: Eqs. (27–12) and (27–13), respectively. Accordingly, our single independent junction equation will be Eq. (27–12b), and our two independent loop equations will be Eqs. (27–13a) and (27–13b). We have three equations to solve for the three unknowns.

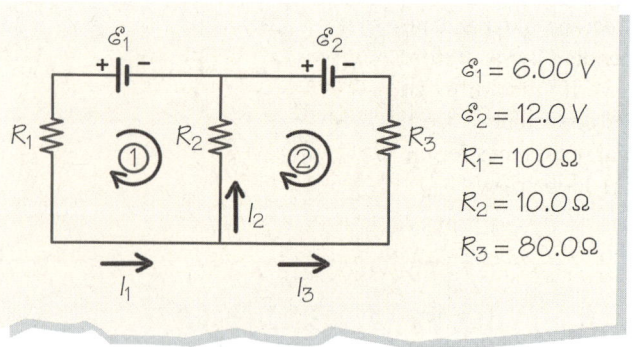

▲ **FIGURE 27–10**

$\mathcal{E}_1 = 6.00$ V
$\mathcal{E}_2 = 12.0$ V
$R_1 = 100\ \Omega$
$R_2 = 10.0\ \Omega$
$R_3 = 80.0\ \Omega$

Working It Out Equation (27–12b) gives $I_1 = I_2 + I_3$. We substitute this into the loop equations:

$$I_2 R_2 + (I_2 + I_3)R_1 - \mathcal{E}_1 = 0; \tag{27–14a}$$

$$-\mathcal{E}_2 + I_3 R_3 - I_2 R_2 = 0. \tag{27–14b}$$

Solve Eq. (27–14b) for I_3:

$$I_3 = \frac{I_2 R_2 + \mathcal{E}_2}{R_3}. \tag{27–15}$$

When this result is inserted into Eq. (27–14a), we find an equation for I_2, namely $I_2(R_1 + R_2) + (I_2 R_2 + \mathcal{E}_2)\dfrac{R_1}{R_3} - \mathcal{E}_1 = 0$, and this has solution

$$I_2 = \frac{R_3 \mathcal{E}_1 - R_1 \mathcal{E}_2}{R_1 R_2 + R_1 R_3 + R_2 R_3}. \tag{27–16}$$

If we substitute this result into Eq. (27–15), we get an expression for I_3:

$$I_3 = \frac{R_2(\mathcal{E}_1 + \mathcal{E}_2) + R_1 \mathcal{E}_2}{R_1 R_2 + R_1 R_3 + R_2 R_3}. \tag{27–17}$$

Finally, I_1 is the sum of I_2 and I_3:

$$I_1 = \frac{R_2(\mathcal{E}_1 + \mathcal{E}_2) + R_3 \mathcal{E}_1}{R_1 R_2 + R_1 R_3 + R_2 R_3}. \tag{27–18}$$

Equations (27–16), (27–17), and (27–18) constitute the desired algebraic solution. Numerically, we can most simply start with the denominator combination, $(100\ \Omega)(10.0\ \Omega) + (100\ \Omega)(80.0\ \Omega) + (10.0\ \Omega)(80.0\ \Omega) = 9.80 \times 10^3\ \Omega^2$. In turn,

$$I_1 = \frac{(10.0\ \Omega)(6.00\ \text{V} + 12.0\ \text{V}) + (80.0\ \Omega)(6.00\ \text{V})}{9.80 \times 10^3\ \Omega^2}$$
$$= 67.3\ \text{mA}$$

$$I_2 = \frac{(80.0\ \Omega)(6.00\ \text{V}) - (100\ \Omega)(12.0\ \text{V})}{9.80 \times 10^3\ \Omega^2} = -73.5\ \text{mA}$$

$$I_3 = I_1 - I_2 = 140.8\ \text{mA}.$$

Note that I_2 is negative, meaning that it flows in a direction opposite to the one we assumed in Fig. 27–10.

What Do You Think? What do you expect to happen if the resistance R_2 is much larger than any other resistance in the problem? Verify your expectation by using the expressions above for the currents.

EXAMPLE 27–6 Assume that the emf and resistors are known for the circuit in Fig. 27–8b. Express the linear equations that can be solved to find all currents.

Setting It Up Figure 27–11 identifies our choice of junctions and loops to which the Kirchhoff rules are to be applied.

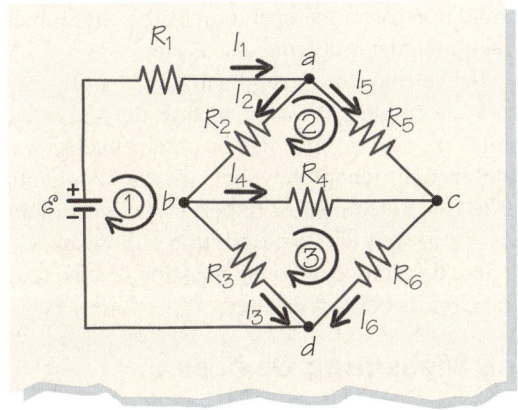

▲ **FIGURE 27–11** The circuit of Fig. 27–8b is labeled with loops and junctions to be solved by circuit analysis.

Strategy As we see from the figure, there are six currents, so we need six linear equations. Three independent loops in the circuit are drawn in Fig. 27–11. There are four junctions, hence three independent junction rule equations—one of the four junction rule equations will not be independent of the others. Thus we choose any three junctions to apply the junction rules, giving three equations. The three loop equations provide the necessary remaining three equations.

Working It Out The six equations are

for junction a: $\quad I_1 - I_2 - I_5 = 0;$

for junction b: $\quad I_2 - I_3 - I_4 = 0;$

for junction c: $\quad I_4 + I_5 - I_6 = 0;$

for loop 1: $\quad \mathcal{E} - I_1 R_1 - I_2 R_2 - I_3 R_3 = 0;$

for loop 2: $\quad -I_5 R_5 + I_4 R_4 + I_2 R_2 = 0;$

for loop 3: $\quad -I_6 R_6 + I_3 R_3 - I_4 R_4 = 0.$

With sufficient patience, we can solve these six equations for the six unknown currents. This, however, is not required in this example.

What Do You Think? Suppose loop 1 were chosen so that it goes all the way around (battery–a–c–d) rather than (battery–a–b–d), as chosen in the example. Could you still solve for the unknowns?

EXAMPLE 27–7 Find the steady-state currents I_1 and I_2 in the circuit drawn in Fig. 27–12. Also find the resistance of resistor R_3 that will give a steady-state current $I_3 = 50$ mA. Finally, determine the potential drop across the capacitor. The values of the known elements are $\mathcal{E} = 6$ V, $R_1 = 100\ \Omega$, $R_2 = 80\ \Omega$, and $C = 2\ \mu$F.

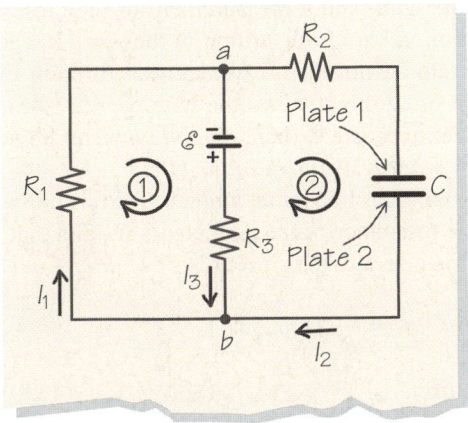

▲ **FIGURE 27–12** Notice that, in steady-state operation, the capacitor acts as an open switch despite the fact that it has voltage across its plates.

Strategy A capacitor acts as an open switch for steady-state current flow. Nevertheless, there is a voltage drop across the capacitor, because charge can build up on the capacitor plates under the impetus of the battery as the steady state is achieved. (Once steady state is reached, there is no further buildup of charge.) We must accordingly include the capacitor in application of the loop rule, and we cannot simply ignore, for example, the loop (a–capacitor–b–a). There will therefore be two loop equations and one junction equation—with two junctions, there is always only one independent equation. With three equations, we will be able to solve for the three unknowns I_1, R_3, and the potential drop across the capacitor.

Working It Out We first write down the loop equations, remembering that the currents are constant in time, so that no current passes through the leg containing the capacitor, and $I_2 = 0$. This simplifies the independent junction (b) equation which reads

$$I_1 = I_2 + I_3 = I_3.$$

For the loop equations: loop 1 gives

$$\mathcal{E} - I_3 R_3 - I_1 R_1 = 0,$$

from which we obtain the resistance

$$R_3 = \frac{\mathcal{E} - I_1 R_1}{I_3} = \frac{\mathcal{E} - I_3 R_1}{I_3} = \frac{\mathcal{E}}{I_3} - R_1.$$

(continues on next page)

776 | Direct-Current Circuits

Finally, the equation for loop 2 reads $-I_2R_2 + V_C + I_3R_3 - \mathcal{E} = 0$, where V_C is the potential difference across the capacitor, and with $I_2 = 0$ we get

$$V_C = \mathcal{E} - I_3R_3.$$

The numerical results are

$$R_3 = \frac{6\text{ V}}{50\text{ mA}} - 100\text{ }\Omega = \frac{6\text{ V}}{50 \times 10^{-3}\text{ A}} - 100\text{ }\Omega = 20\text{ }\Omega$$

and

$$V_C = (6\text{ V}) - (50 \times 10^{-3}\text{ A})(20\text{ }\Omega) = 5\text{ V}.$$

What Do You Think? (a) Could the equation for R_3 lead to a negative value for this resistance under certain circumstances? (b) Which of the two capacitor plates is at the higher potential?

27–4 Measuring Instruments

We have referred to the currents and voltages in various circuit elements, but we have not yet explained how these quantities are measured. A variety of instruments exists for this purpose: **Ammeters** measure current, **voltmeters** measure voltage, and **ohmmeters** measure resistance. These devices are often combined into one instrument called a **multimeter** (Fig. 27–13). We shall focus our attention on ammeters and voltmeters. Whatever the detailed operation of such instruments, a general principle must be respected: The measuring device should not distort the operation of the circuit being measured. We shall emphasize this principle in the following discussion.

For most applications, analog devices have been supplanted by digital ones, which are usually less expensive and more accurate. However, analog devices (dials)—the gasoline gauge of your automobile, or the moving hands of an analog watch, for example—remain superior for the recognition of how the measured quantities are changing. It will be simplest for us to illustrate the most important features of ammeters and voltmeters with analog devices. We'll start with a description of how analog ammeters and voltmeters can be constructed, then move on to the question of how they can be used in ways that do not disturb the circuits being measured.

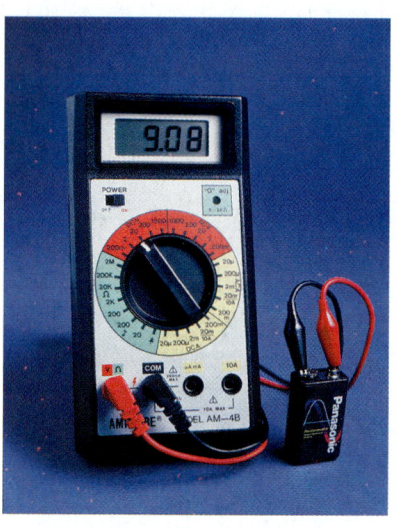

▲ **FIGURE 27–13** A multimeter can be used as a voltmeter, ohmmeter, or ammeter.

How to Construct Analog Measuring Devices

Analog ammeters and voltmeters typically utilize a *galvanometer*, a device that relies on magnetic effects. The galvanometer, indicated by a circled "G" in circuit diagrams, consists of a coil of wire that rotates in the magnetic field produced whenever a current passes through the wire. (We'll explore this when we study the effects of magnetic fields, in Chapter 28.) A needle connected to the coil is deflected by an amount proportional to the current that passes through the coil. The position of the needle, properly calibrated, is the analog measurement.

Ammeters: An ammeter measures the current in a circuit wire by direct insertion into the wire—in other words, it is placed in series in the circuit segment through which the current to be measured passes. That means that the current flowing through the ammeter is the current flowing through the wire, and a measurement of the current passing through the ammeter itself is a measurement of the current in the wire. Recall now the general principle that an ammeter should not disturb the current through the circuit. From the fact that the ammeter is inserted directly into the wire we see that ideally the ammeter should have zero resistance. In practice, then, *a good ammeter should have a resistance that is small compared to other resistances in the circuit*.

Figure 27–14a shows a circuit being probed by an ammeter, indicated by a circled "A" in Fig. 27–14b. Suppose that the resistance of the ammeter is R_A and that the series resistance is R. Before the ammeter is inserted, the current is

$$I = \frac{\mathcal{E}}{R}.$$

After the ammeter is inserted, the current is

$$I = \frac{\mathcal{E}}{R + R_A}. \tag{27-19}$$

The current will be the same with or without the ammeter attached only if $R_A \ll R$.

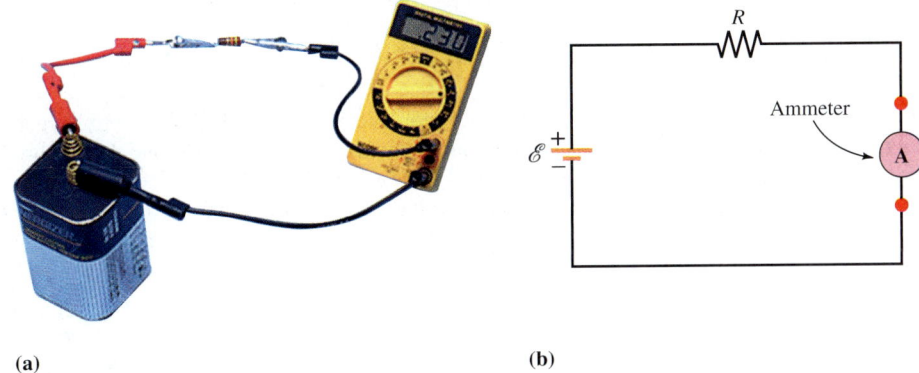

(a)   (b)

◀ **FIGURE 27–14** (a) An ammeter is placed in series where the current is to be measured. The resistance of an ammeter should be small so as not to change the current. (b) Schematic diagram of the circuit.

Voltmeters: Voltmeters measure potential differences across circuit elements. They do so by being placed in parallel with those elements, so that the potential drop across the voltmeter is the same as across the circuit element. Figure 27–15a shows a voltmeter—indicated by the circled "V" in Fig. 27–15b—used to measure the potential across a resistor. A voltmeter should not disturb the potential difference being measured, and so *a good voltmeter should have a large resistance*. Why is this? If the internal resistance of the voltmeter is R_V, the combination of voltmeter and resistance in Fig. 27–15b forms a parallel resistance circuit with equivalent resistance

$$\frac{1}{R_{eq}} = \frac{1}{R_V} + \frac{1}{R}.$$

When $R_V \gg R$, then $R_{eq} \cong R$, and none of the parameters of the original circuit are affected.

A galvanometer connected in series with a resistor of large resistance can serve as a voltmeter. If the current I passing through the galvanometer is measured (and that is just what the needle deflection measures), then the potential drop across the voltmeter is approximately IR_V. For example, take a galvanometer that can measure a maximum current of 100 μA. This is equivalent to a measurement of 10 V if the internal resistance is set at

$$R_V = \frac{10 \text{ V}}{100 \times 10^{-6} \text{ A}} = 10^5 \text{ } \Omega.$$

An analog voltmeter is then a galvanometer with a series resistor of resistance R_V (Fig. 27–16). If there are potential differences within a circuit of 1000 V, then the 100 μA current would be exceeded and the resistor in our example would not be adequate. With a 10^5-Ω resistor the full-scale reading is 10 V. To measure a full-scale voltage of 1000 V with 100 μA, we must take $R_V = 10^7$ Ω.

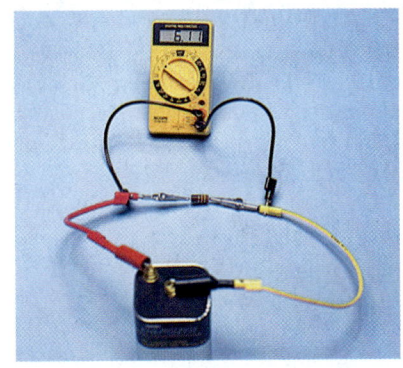

(a)

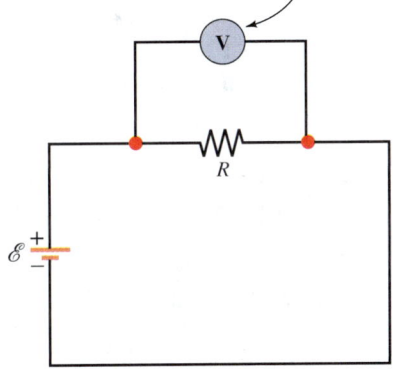

(b)

▲ **FIGURE 27–15** (a) A voltmeter is placed in parallel across the circuit element whose potential drop is to be measured. The resistance of a voltmeter should be large so as not to change the circuit. (b) Schematic diagram of the circuit.

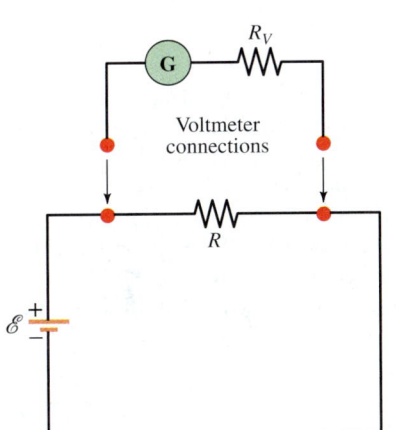

◀ **FIGURE 27–16** A galvanometer with a large series resistance can serve as a voltmeter.

CONCEPTUAL EXAMPLE 27–8
A voltmeter with a large resistance R_V is used as in Fig. 27–15. Will it be a good voltmeter for all circuits?

Answer The point of the apparatus is that the current that flows through the voltmeter be small compared to the current in the circuit. This means that R_V must be large compared to the parallel equivalent resistance. In this way, the total current, which is

$$I = \mathcal{E}\left(\frac{1}{R_V} + \frac{1}{R_{eq}}\right) = I_{voltmeter} + I_{circuit},$$

will be approximately equal to $I_{circuit} = \mathcal{E}/R_{eq}$. The word "large" in the question statement is not a very precise one, and the important point is whether it is large when compared with the corresponding quantity, here R_{eq}, for the circuit. When someone uses a word like "large," a suitable response is "large compared to what?"

EXAMPLE 27–9
A voltmeter with an internal resistance of $10^5 \, \Omega$ is used to measure the voltage across resistor R_1 in the circuit of Fig. 27–17. To learn the error in measurement caused by the voltmeter itself compare the potential drop with and without the voltmeter for $\mathcal{E} = 6$ V, $R_1 = 10 \, k\Omega$, and $R_2 = 5 \, k\Omega$.

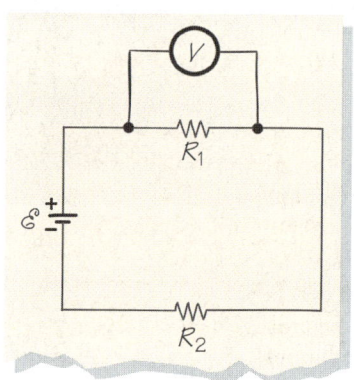

▲ FIGURE 27–17

Setting It Up As Fig. 27–17 makes clear, we are asked to compare the voltage across the resistance R_1 without and with the voltmeter placed in parallel with R_1.

Strategy In the case where the voltmeter is not present, we deal with R_1 and R_2 in series, in a single loop with a load resistance $R_1 + R_2$. In the case where the voltmeter is present, we can proceed either by applying the Kirchhoff rules to two loops or, more directly, by treating the voltmeter as a resistor parallel to R_1 and replacing the combination of these parallel resistors with a single equivalent resistance in a single-loop problem. The second alternative is simpler.

Working It Out With the voltmeter absent and a load resistance of $R_1 + R_2$, the current flowing is

$$I = \frac{\mathcal{E}}{R_1 + R_2}.$$

From this we calculate the voltage across the resistor R_1 to be

$$V_1 = IR_1$$

$$= \frac{\mathcal{E}R_1}{R_1 + R_2}.$$

When the voltmeter is connected, the resistance R_1 is replaced by the equivalent resistance R_1^* given by

$$\frac{1}{R_1^*} = \frac{1}{R_1} + \frac{1}{R_V} \quad \text{or} \quad R_1^* = \frac{R_1 R_V}{R_1 + R_V}.$$

The current is now

$$I^* = \frac{\mathcal{E}}{R_1^* + R_2}$$

and the voltage drop is just the voltage drop across the effective resistance R_1^*,

$$V_1^* = R_1^* I^* = \frac{\mathcal{E} R_1^*}{R_1^* + R_2}.$$

Numerically,

$$V_1 = IR_1 = \frac{6 \text{ V}}{10 \, k\Omega + 5 \, k\Omega}(10^4 \, \Omega) = 4 \text{ V}.$$

With the voltmeter we have

$$R_1^* = \frac{R_1 R_V}{R_1 + R_V}$$

$$= \frac{(10^4 \, \Omega)(10^5 \, \Omega)}{(10^5 \, \Omega)(1 + 0.1)} = 0.9 \times 10^4 \, \Omega$$

and therefore

$$V_1^* = R_1^* I^* = \frac{E R_1^*}{R_1^* + R_2}$$

$$= \frac{(6 \text{ V})(0.9 \times 10^4 \, \Omega)}{(0.9 \times 10^4 \, \Omega) + (0.5 \times 10^4 \, \Omega)} = 3.9 \text{ V}.$$

The difference between 4 V and 3.9 V is equivalent to a 3 percent error due to the presence of the voltmeter. The larger the voltmeter resistance, the smaller the error introduced. Whether this is a tolerable error depends on the requirements.

What Do You Think? A good voltmeter will have an internal resistance of 10 MΩ, a factor of 100 larger than the resistance used in this example. Would such a device give a significantly more accurate reading?

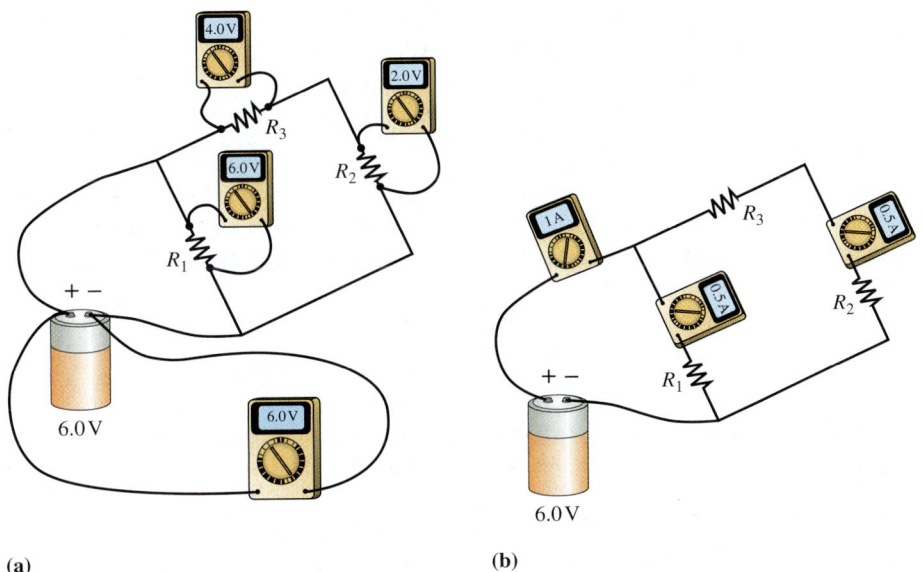

◀ **FIGURE 27–18** (a) Measuring voltage across circuit elements. (b) Measuring current through circuit elements.

In Fig. 27–18, we summarize the use of voltmeters and ammeters in a typical circuit.

27–5 RC Circuits

RC circuits are circuits that contain both resistors and capacitors. Unlike the circuits we have considered so far that have a steady current flowing through them, the currents and potentials of *RC* circuits can exhibit time-varying behavior when we introduce another element, a switch (Fig. 27–19). Even for circuits containing steady sources of emf, we introduce time dependence in a circuit every time we open or close the switch. *RC* circuits with switches have time-dependent effects that are useful for the control of motors, machinery, or computers.

We first observed the effect of a fully charged capacitor in an electric circuit in steady-state operation in Example 27–7. Now we want to examine the more complex, transient behavior that occurs when a capacitor is being charged and discharged. Consider the circuit shown in Fig. 27–19, with an initially uncharged capacitor. When the switch is closed (to position *a*) at $t = 0$, current begins to flow from the positive terminal of the battery, and positive charge begins to collect on plate 1 of the capacitor, while an equal amount of negative charge collects on plate 2. Current flows everywhere in the circuit *except* through the plates of the capacitor. Immediately after the switch is closed, the current has its maximum value, but the charge that builds up on the capacitor plates opposes further charge flow, and the current decreases. When the potential across the capacitor plates equals the emf and equilibrium is reached, the current is zero. This occurs when the charge on the capacitor plates, Q_0, is such that $\mathcal{E} = Q_0/C$.

After equilibrium has been reached and the current has become zero, we change the switch to position *b*, effectively taking the battery out of the circuit. The circuit now consists only of the charged capacitor and the resistor, and it is not in equilibrium. Current flows through the circuit from plate 1 of the capacitor to plate 2. The rate of flow is limited by the resistor. At first the current is high, but it decreases as the capacitor discharges through the resistor. Eventually the capacitor discharges completely, and the current again falls to zero when equilibrium is reached.

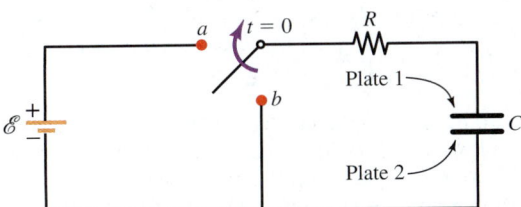

◀ **FIGURE 27–19** A circuit used to charge and discharge a capacitor through a resistor. When the switch is closed at *a*, the capacitor is charged by the source of emf, whereas the capacitor discharges through *R* when the switch is thrown to *b*.

We first apply Kirchhoff's loop rule to the circuit of Fig. 27–19 for the switch at position a, when the capacitor is being charged. The loop rule gives

$$\mathcal{E} - IR - \frac{Q}{C} = 0. \tag{27-20}$$

In this equation, neither the current nor the charge on the capacitor is constant while the capacitor charges. Because $I = dQ/dt$ (conservation of charge ensures that the rate at which current flows through the wire is the rate at which charge builds up on the capacitor), we can rewrite Eq. (27–20) as

$$\mathcal{E} - R\frac{dQ}{dt} - \frac{Q}{C} = 0. \tag{27-21}$$

The single variable in this equation is the charge Q. The differential equation (27–21) is straightforward to solve; let's omit the mathematical complexities and present its solution:

$$Q = C\mathcal{E}(1 - e^{-t/RC}). \tag{27-22}$$

By differentiating Eq. (27–22) with respect to time and substituting into Eq. (27–21), we can see that it satisfies Eq. (27–21) (see Problem 54). More important, does it agree physically with what we expect? According to Eq. (27–22), the charge on the capacitor is zero at $t = 0$ and builds smoothly to $C\mathcal{E}$ at large times, in agreement with our earlier discussion.

We can find the current in the circuit by differentiating Eq. (27–22) with respect to time:

$$I = \frac{dQ}{dt} = C\mathcal{E}\left(\frac{1}{RC}e^{-t/RC}\right) = \frac{\mathcal{E}}{R}e^{-t/RC}. \tag{27-23}$$

The sign of the current is positive, so we chose the correct current direction (clockwise). The maximum value of the current is $\mathcal{E}/R$, at $t = 0$, and the current is zero at $t = \infty$, which also agrees with our earlier discussion. Just after the switch is closed, the potential drop across the resistor is $\mathcal{E} = IR$, with no potential drop across the uncharged capacitor. As the capacitor charges, the current drops *exponentially* to zero.

Equations (27–22) and (27–23) show that the time dependence of both charge and current is determined by the product $\tau \equiv RC$; τ is called the **time constant**. It has units of time; with R and C in SI units, RC will be in seconds. The time constant determines how fast a capacitor charges and discharges. The smaller the value of RC, the more quickly the exponentials in the equations for Q and I fall; similarly, the larger the value of RC, the more slowly the exponentials change. Figure 27–20a and c show the current

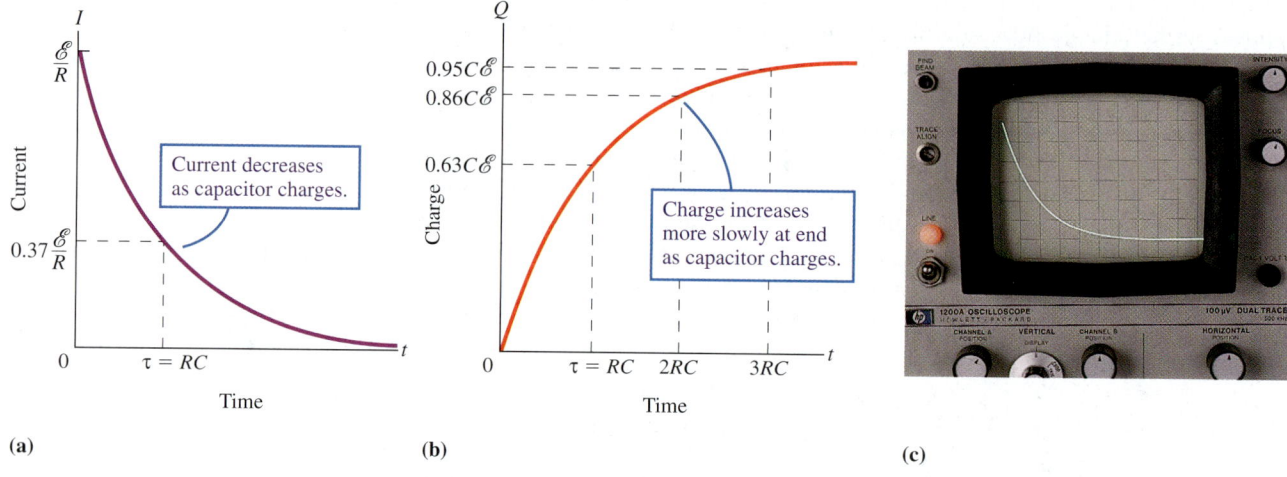

▲ **FIGURE 27-20** The time response of (a) the current I and (b) the charge Q stored in a capacitor as the capacitor is charged. The characteristic time response of the exponential behavior is RC. The value 0.37 in the graph of current is the factor e^{-1}; the value 0.63 in the graph of charge is the factor $(1 - e^{-1})$. (c) This oscilloscope screen shows the exponential current drop on a charging capacitor.

in the circuit, and Fig. 27–20b shows the charge on the capacitor as a function of time while the capacitor is being charged. After a time RC, the current has dropped to $e^{-1} \cong 0.37$ times its original value. After this same amount of time, the capacitor is $(1 - e^{-1}) \cong 63$ percent fully charged. It is 86 percent charged at time $2RC$ and 95 percent charged at time $3RC$.

Let's return to the circuit of Fig. 27–19. Suppose that the switch has been in position a for a long time, the capacitor is fully charged, and there is no current. At time $t = 0$, we throw the switch to position b. Only the discharging capacitor and the resistor are now in the circuit (Fig. 27–21). The positive charge is on plate 1, and we assume as before that the current is clockwise. The loop rule now gives

$$-IR - \frac{Q}{C} = 0. \quad (27\text{–}24)$$

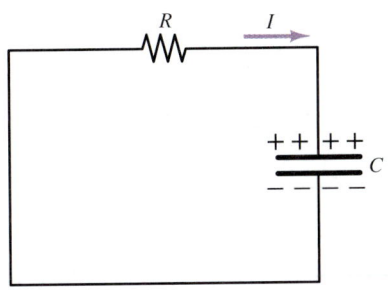

▲ **FIGURE 27–21** The circuit of Fig. 27–19 after the switch has been thrown to position b.

Using $I = dQ/dt$, we have

$$R\frac{dQ}{dt} + \frac{Q}{C} = 0. \quad (27\text{–}25)$$

This differential equation is solved by the function

$$Q = Q_0 e^{-t/RC}, \quad (27\text{–}26)$$

where Q_0 is the initial charge on the capacitor when the switch is changed, $Q_0 = C\mathcal{E}$. Equation (27–26) may be substituted into Eq. (27–25) to verify that it is a solution (see Problem 55). The charge on the capacitor decreases exponentially with the time constant RC, and after a long time, there will be no charge on the capacitor.

We find the current by differentiating Eq. (27–26):

$$I = \frac{dQ}{dt} = -\frac{Q_0}{RC} e^{-t/RC}. \quad (27\text{–}27)$$

The current in this case is negative, indicating that the actual current is counterclockwise, opposite in direction to the current we assumed when we drew the diagram. It is again a maximum at $t = 0$, when the magnitude of the current is $Q_0/RC = \mathcal{E}/R$. After a long time, the current is again zero.

The behavior of the charge and current for the capacitor that discharges through a resistor is qualitatively what we expected from our earlier discussion. The magnitude of the current for this case is just as shown in Fig. 27–20a for the charging capacitor. The charge on the capacitor is plotted as a function of time in Fig. 27–22. Again, the factor 0.37 is e^{-1}.

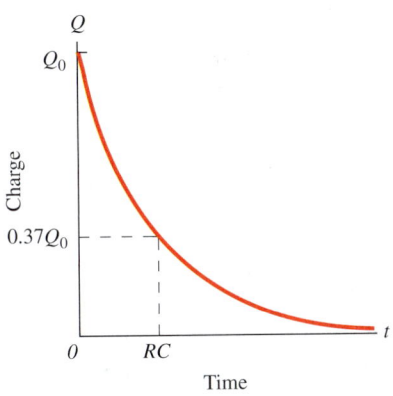

▲ **FIGURE 27–22** The capacitor charge as the capacitor of Fig. 27–21 discharges through the resistor as a function of time. The characteristic time response of the exponential behavior is again RC.

THINK ABOUT THIS...
CAN CAPACITORS REPLACE BATTERIES IN LAPTOP COMPUTERS?

Rechargeable batteries in laptop computers work only for a few hours, and recharging takes a long time. Could a capacitor replace the battery? The computer requires a steady current of around 2 μA but can work with a range of voltage from 6 V to a minimum voltage of 2.5 V (the lower limit is determined by the necessity to maintain memory function). If the charged capacitor is placed in series with a resistance, the current produced by the capacitor as it discharges will drop as the voltage drop V_C across the capacitor decreases. A constant current I calls for a non-ohmic device for which I is independent of the voltage drop V across it over some range of V. (The device that achieves this is a *zener diode*.) When such a device is placed in series with a capacitor, then the relation $Q = CV$ implies that the current obeys the equation $I = dQ/dt = C \, dV/dt$. Since the device produces constant I, V must drop linearly with time. If $I = 2 \, \mu$A how long will it take before the potential difference across the capacitor changes from 6 V to 2.5 V? From the relation $I = C \, dV/dt$ with constant I we calculate $\Delta t = C \times$ (change in V)/I = $C(3.5 \text{ V})/(2 \times 10^{-6} \text{ A})$. With a capacitor of $C = 1$ F this time scale is 1.8×10^6 s $\cong$ 20 days. Whether this is a practical option depends on the existence of capacitors with capacitance of farads—a very large capacitance—yet compact enough to be used in portable devices, and as we already discussed in Chapter 26 this technology is being actively pursued.

Energy in RC Circuits

Let's now examine the role of energy and its conservation during the charging of a capacitor. From the definition of potential, the amount of work done by the battery emf during the charging process is $\mathcal{E}$ times the total charge processed by the battery. This charge is the final charge $C\mathcal{E}$ on the capacitor plates after a long period of time. Thus, the work done by the battery, W_{bat}, is

$$W_{bat} = \mathcal{E}(C\mathcal{E}) = C\mathcal{E}^2. \qquad (27\text{–}28)$$

How do we account for the energy that matches this work? In part, the energy is stored in the capacitor. We know from Eq. (25–9) that the total energy stored by a capacitor is $CV^2/2$. The voltage V in this case is $\mathcal{E}$, so the energy stored by the capacitor, E_{cap}, is

$$E_{cap} = \tfrac{1}{2}C\mathcal{E}^2. \qquad (27\text{–}29)$$

Where has the other half of the work done by the battery gone? The only other circuit element is the resistor, and the other half of the work has gone into Joule heating of that resistor. From Eq. (26–27), we know that the power loss in the resistor is $P = I^2R$. We can integrate the power over time to find the energy loss in the resistor, E_{res}, using the current from Eq. (27–23)

$$\begin{aligned}E_{res} &= \int_0^\infty I^2 R\, dt = \frac{\mathcal{E}^2}{R}\int_0^\infty e^{-2t/RC}\, dt \\ &= \frac{\mathcal{E}^2}{R}\left(-\frac{RC}{2}e^{-2t/RC}\right)\bigg|_0^\infty = \frac{C\mathcal{E}^2}{2}.\end{aligned} \qquad (27\text{–}30)$$

The thermal energy loss in the resistor accounts for the other half of the work done by the battery. This 50-percent split of energy between the resistor and the capacitor is *independent* of $\mathcal{E}$. For the case of the discharging capacitor, all the energy stored in the capacitor dissipates as heat in the resistor. A spectacular demonstration of the rapid release of the energy contained within a capacitor was shown in Fig. 25–8.

EXAMPLE 27–10 The charging circuit shown in Fig. 27–23 (with a switch thrown to position a at $t = 0$) has a 12 V emf, a 100 Ω resistance, and a 100 µF capacitance. (a) Find the time constant, the final charge on the capacitor, and the work done by the battery. (b) How long does it take for the capacitor to be charged to 99.9 percent of its final charge?

Setting It Up We label the given emf, resistance, and capacitance $\mathcal{E}$, R, and C, respectively.

Strategy (a) Knowing both R and C, the time constant is $\tau = RC$. The final charge on the capacitor is $Q_f = C\mathcal{E}$ (as we already know from Chapter 25), and the work done by the battery is $W_f = C\mathcal{E}^2$. As for part (b), this is a matter of solving the expression $Q = Q_f(1 - e^{-t/\tau})$ for t knowing that $Q = 0.999 Q_f$.

Working It Out (a) We need only plug numerical values into the appropriate expressions:

$$\tau = RC = (100\ \Omega)(10.0 \times 10^{-6}\ \text{F})$$
$$= 1.00 \times 10^{-3}\ \text{s} = 1.00\ \text{ms},$$
$$Q_f = C\mathcal{E} = (10.0 \times 10^{-6}\ \text{F})(12\ \text{V}) = 1.2 \times 10^{-4}\ \text{C},$$
$$W_{bat} = C\mathcal{E}^2 = (10.0 \times 10^{-6}\ \text{F})(12\ \text{V})^2 = 1.4 \times 10^{-3}\ \text{J}.$$

(b) With $Q = 0.999 Q_f$, we want to solve Eq. (27–22) for t. The factor Q_f cancels, leaving $0.999 = (1 - e^{-t/\tau})$, or

$$e^{-t/\tau} = 1 - 0.999 = 0.001.$$

Taking the natural logarithm of both sides gives

$$-\frac{t}{\tau} = -6.91.$$

The time to reach 99.9 percent of its final charge is thus

$$t = 6.91\tau = (6.91)(1.00\ \text{ms}) = 6.91\ \text{ms}.$$

What Do You Think? If it takes 6.91 ms to charge the capacitor to 99.9% of its final voltage, how long do you estimate it will take to reach 99.99%?

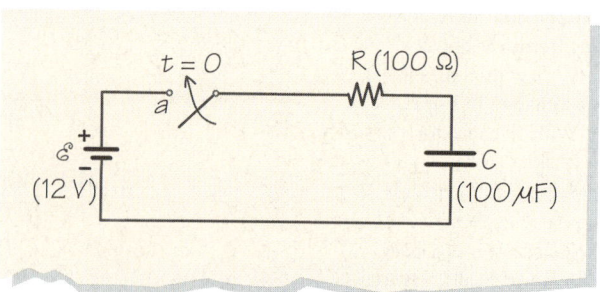

▲ FIGURE 27–23

Summary

Sources of emf (electromotive force) $\mathcal{E}$, such as chemical batteries, are sources of electric energy. The emf is defined by the amount of work it can do to move charge:

$$\mathcal{E} \equiv \frac{dW}{dq}. \tag{27-1}$$

Batteries cause charges to move in circuits. The simplest circuits to analyze are direct-current circuits, in which no circuit parameters change with time.

Analysis of single- or multi-loop circuits is accomplished by the use of Kirchhoff's two rules. Kirchhoff's loop rule states that the sum of the potential changes around a closed path of a circuit is zero:

$$\sum_{\text{closed path}} \Delta V = 0. \tag{27-9}$$

We can specify the potential change across batteries, resistors, and capacitors. Any source of emf has an internal resistance r that may be large enough to require consideration.

Kirchhoff's junction rule is "circuit language" for the conservation of electric current. It states that the sum of the currents that enter a junction equals the sum of the currents that leave the junction. If we interpret an outgoing current as an ingoing current with a minus sign in front of it, then the junction rule takes the simple form:

$$\sum I_{\text{in}} = 0. \tag{27-11}$$

Ammeters, voltmeters, and ohmmeters measure current, voltage, and resistance, respectively. Ammeters must have a small internal resistance so that they do not affect the circuit leg in which the current is measured. Conversely, voltmeters need a large internal resistance because they are used in parallel with the circuit element being measured, and they too should not affect the circuit being measured.

A circuit exhibits time-varying behavior when a capacitor is being charged with a source of emf or when the capacitor is being discharged. For a simple circuit with an emf $\mathcal{E}$, a resistor R, and a capacitor C, an initially uncharged capacitor has charge

$$Q = C\mathcal{E}(1 - e^{-t/RC}) \tag{27-22}$$

and current

$$I = \frac{\mathcal{E}}{R} e^{-t/RC}. \tag{27-23}$$

When the source of emf is disconnected from the circuit and the capacitor is allowed to discharge through the resistor, the charge decreases exponentially:

$$Q = Q_0 e^{-t/RC}. \tag{27-26}$$

The time constant RC determines the time dependence of exponential increase or decrease of the charge and the current during the charging and discharging of a capacitor.

Understanding the Concepts

1. Why is it dangerous to be in a bathtub when an electrical appliance is standing on the edge of the tub?
2. One voltmeter measures the voltage of a flashlight battery as 0.9 V, whereas another measures 1.5 V. What might cause this difference?
3. Show that it is irrelevant whether, in drawing a circuit diagram, the internal resistance of a source of emf is placed before or after the emf itself.
4. What sense does it make to draw a circuit diagram with resistanceless wires when real wires always have some resistance?
5. By taking a special combination of batteries of constant emf, resistors, and capacitors, is it possible to construct a circuit in which the emf around a closed loop is not zero?
6. You measure a potential difference of 3 V across the terminals of a 3-V battery with some age to it and proclaim it to be just fine. Are you right?
7. A flash unit on a camera discharges by means of an RC circuit. Find out (say, from a local camera store) what the R and C values are, and obtain the RC time.
8. You have an RC circuit and three identical capacitors available to you. How should you arrange them to make the discharge time as short as possible?
9. It would appear superficially that, in Section 27–5, if we had chosen the current in the wrong direction, we would have ended up with the equation $IR - (Q/C) = 0$ rather than Eq. (27–7) to describe the current in an RC circuit. This would be disastrous

because the solution of our new equation would be a rising exponential, $e^{+t/RC}$, rather than a falling exponential, and this expression would grow without limit. What is wrong with this reasoning?
10. When you reverse the polarities of all batteries in a circuit, the magnitudes of all currents stay the same. Why?
11. Given two identical sources of emf and two identical light bulbs, how would you arrange these elements so that both bulbs glow with maximum brightness?
12. Suppose that you connect the terminals of two batteries of different emfs + to + and − to −. What do you expect to happen?
13. What happens if the emf of battery 2 in Example 27–4 is reversed?
14. It might appear that the only effect of the internal resistance of a battery in a circuit is to change the battery's emf from $\mathcal{E}$ to $\mathcal{E}' = \mathcal{E} - Ir$, where $\mathcal{E}'$ acts as though there were no internal resistance. Could this be true? If not, why not?
15. Show that the circuit in Fig. 27–8(a) can be reduced to a one-loop circuit. Is this also true of the circuit in Fig. 27–8(b)?
16. Two reckless teenagers hang by both hands on a wire that can be connected to a constant voltage source. One teenager hangs from a position in which there is a resistor between his hands; the other one does not. The resistance of the resistor is much larger than that of the wire, which is very small. When the wire is connected to the battery, how do the teenagers experience that fact?
17. What happens to the current I_1 of Example 27–7 if the resistance R_3 is increased?
18. To start a car in cold weather, one sometimes first turns the lights on for a while. How could this help?

Problems

27–1 EMF

1. (I) A 12-V car battery is rated at 80 A, meaning that it will send 80 A through a wire connected to its terminals. What is its internal resistance?

2. (I) The *Magellan* spacecraft that studied Venus in 1990 used two solar panels capable of producing 1200 W. If the solar array was capable of producing a total of 40 A, what was the terminal voltage of the device?

3. (I) A battery with an emf of 3.00 V sends a current of 1.99 A when it is connected in series with a 1.50-Ω resistor. What is the internal resistance of the battery?

4. (I) Nickel–cadmium batteries used in space flight are rated at 30 A·h (they can put out 30 A for 1 hr) and 30 V. How much energy do the batteries contain?

5. (I) A defibrillator used by emergency medical staff to restart an accident victim's heart has an internal resistance of 20 Ω, and a power supply that produces an average 5000 V over a short period of time. If the body resistance between the two electrodes is 230 Ω, how much current passes through the body to restart the heart?

6. (II) A flashlight battery with an internal resistance of 0.06 Ω produces a 170-mA current through a 15-Ω resistor. What is the emf of the battery? What is the terminal voltage of the battery in this usage?

7. (II) A certain automobile battery has an emf of 12 V. When it produces a current of 100 A, the terminal voltage reads 9.0 V. Calculate the internal resistance of the battery. What is the power dissipated in the battery when it produces this current?

8. (II) The resistance of the starter of a car, including the cables, is 0.11 Ω. On a frigid winter morning, cranking the engine for 10 s reduces the terminal voltage of the car's 12-V battery to 8 V. The car does not start. How much heat (in kilocalories) is produced in the battery during this time? Does it improve your chances of starting the car on a second try?

9. (II) The internal resistance of a battery whose emf is 12 V varies with the current according to the equation $r = (\alpha + \beta I)$, where $\alpha = 0.15\ \Omega$ and $\beta = 0.018\ \Omega/A$. Find the terminal voltages and the power dissipated in the battery when $I = 1.0$ A and $I = 10$ A.

27–2 Kirchhoff's Loop Rule

10. (I) Two bulbs in series are connected across an emf of 110 V. The normal operating ratings of the bulbs are given as (2.5 V, 0.5 A) for one bulb and as (110 V, 10 W) for the other. Calculate how much power will actually be developed in each lamp.

11. (I) For what value of R_2 in the circuit of Fig. 27–24 is the voltage across the points a and b zero? For what value is the current in the circuit zero?

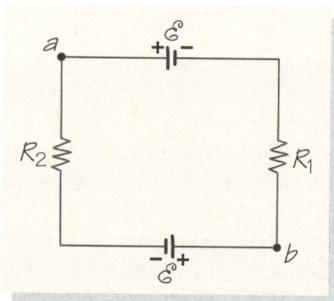

▲ FIGURE 27–24 Problem 11.

12. (II) A generator (a "battery" that uses mechanical rather than chemical energy) of emf 110 V and internal resistance 0.50 Ω is used to charge a series of 20 batteries, each with emf 2.2 V and internal resistance 0.06 Ω. A series resistor is used to limit the charging current. (a) What is the terminal voltage of the generator? (b) the terminal voltage of the bank of batteries? (c) What series resistance must be included to allow a charging current of 15 A? (d) What is the power dissipated in all the resistors?

13. (II) A portion of a larger circuit is shown in Fig. 27–25. The potential drop between the pairs of points labeled in the figure is $V_{ba} = 2$ V, $V_{cb} = 3.5$ V, $V_{cd} = 2$ V, and $V_{df} = -0.5$ V. Find the potential differences V_{gf}, V_{ag}, and V_{ca}.

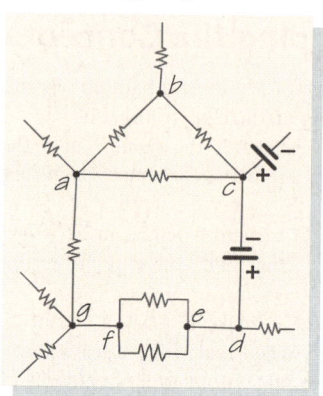

▲ FIGURE 27–25 Problem 13.

14. (II) A flashlight consists of two 1.5-V batteries connected in series to a bulb with resistance 10 Ω. (a) What is the power delivered to the bulb? (b) Batteries run down when they acquire an (internal) resistance. How large is the additional resistance if the power delivered to the bulb has decreased by one-third of its initial value?

27–3 Kirchhoff's Junction Rule

15. (I) In Fig. 27–26, the currents $I_1 = 2$ A, $I_2 = 0.5$ A, $I_3 = -3$ A, $I_4 = -0.5 I_6$, and $I_5 = -I_6$. Find the unknown currents I_4, I_5, and I_6.

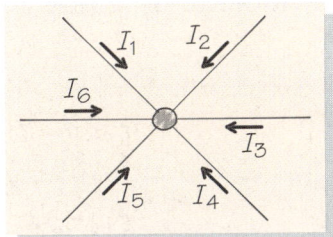

▲ **FIGURE 27–26** Problem 15.

16. (I) Five identical lightbulbs are placed in a circuit as shown in Fig. 27–27. What is the brightness of bulb #3 relative to bulb #1? Even though this problem can be worked without them, use the Kirchhoff rules.

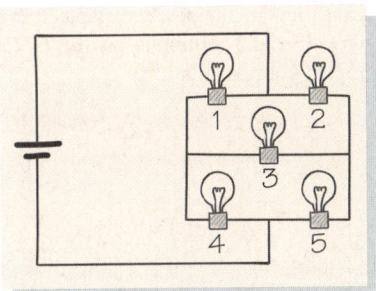

▲ **FIGURE 27–27** Problem 16.

17. (II) Consider the part of a circuit shown in Fig. 27–28. What is the total resistance if (a) the switch is open, and (b) the switch is closed?

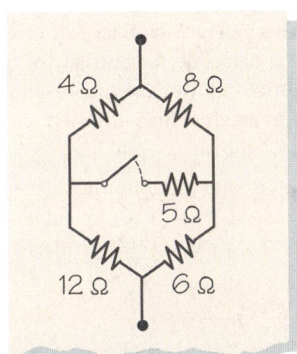

▲ **FIGURE 27–28** Problem 17.

18. (II) Consider the circuit in Fig. 27–29. Calculate the current through the 16 Ω resistor, and the voltage across ab.

19. (II) Find the current that passes through each of the resistors in the circuit shown in Fig. 27–30.

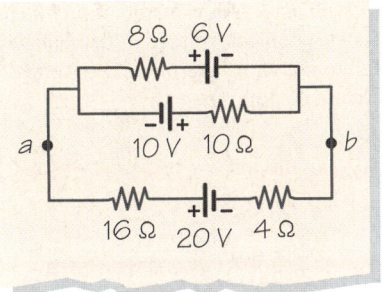

▲ **FIGURE 27–29** Problem 18.

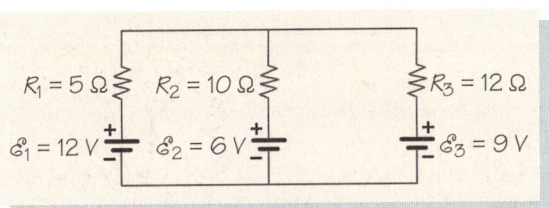

▲ **FIGURE 27–30** Problem 19.

20. (II) Find the current that passes through the 4-Ω resistor in the circuit shown in Fig. 27–31.

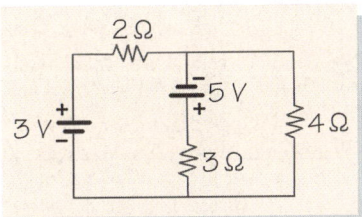

▲ **FIGURE 27–31** Problem 20.

21. (II) Three resistors connected in parallel have resistances of 250 Ω, 420 Ω, and 510 Ω, respectively. The total current passing through the set is 0.020 A. What is the potential difference across the set, and what are the currents in each of the resistors?

22. (II) Can the resistors of the circuit in Fig. 27–32 be reduced to a single equivalent circuit by application of the rules for circuits with connections in parallel and in series? Solve for the currents through the three resistors.

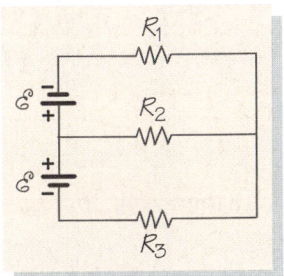

▲ **FIGURE 27–32** Problem 22.

23. (II) N identical batteries, with emf $\mathcal{E}$ and internal resistance r, are connected in parallel across a resistance R. Obtain the value for the current, and compare its value with that obtained if the batteries are connected in series.

786 | Direct-Current Circuits

24. (II) Figure 27–33 shows an example of a *voltage divider*, a device that allows a reduced voltage to be obtained. Calculate the potential difference across the line CD in terms of the potential difference across the line AB.

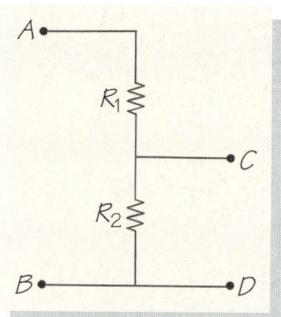

▲ **FIGURE 27–33** Problem 24.

25. (II) The circuit shown in Fig. 27–34 is an example of a loaded voltage-divider circuit (see Problem 24). By varying the values of R_1 and R_2, different values of V_L can be obtained; R_L represents the load. Let $R_1 = R_2 = 3.3$ kΩ and $\mathcal{E} = 10$ V. For load resistances of 20 kΩ, 200 kΩ, and 2 MΩ, how much different is V_L than 5 V?

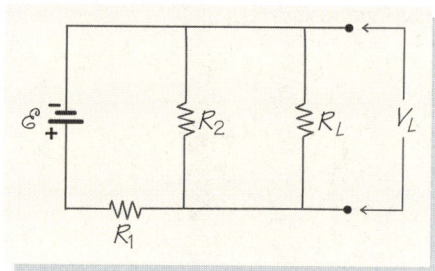

▲ **FIGURE 27–34** Problem 25.

26. (II) Consider the circuit shown in Fig. 27–35. If $\mathcal{E} = 2.8$ V, and if all three resistors have identical resistances, find the resistance that ensures that the current I_3 will be 55 mA. Even though this problem can be worked without them, use the Kirchhoff rules.

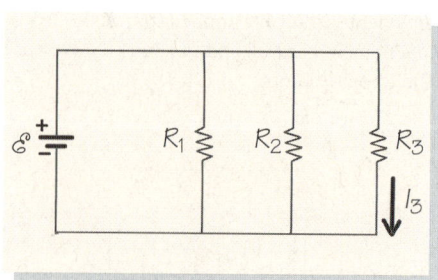

▲ **FIGURE 27–35** Problem 26.

27. (II) How many independent junctions are there in the circuit shown in Fig. 27–36? To verify your answer, solve for all currents.

28. (II) Replace the network of resistors in Fig. 27–37 by a single equivalent resistor. Can the combination of resistors be reduced to a single resistor by successive application of the rules for parallel and series resistors?

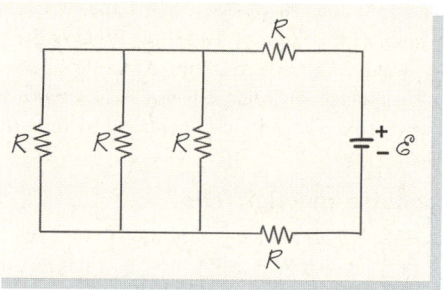

▲ **FIGURE 27–36** Problem 27.

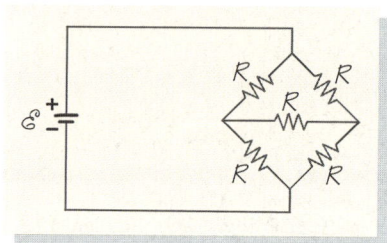

▲ **FIGURE 27–37** Problem 28.

29. (II) Two batteries with emf $\mathcal{E}_1 = 6$ V and $\mathcal{E}_2 = 9$ V, respectively, are connected to resistors with the resistances as marked in Fig. 27–38. (a) Calculate the power dissipated in the 50-Ω resistor. (b) Assume that the terminals on the 6-V battery are reversed, and repeat your calculation.

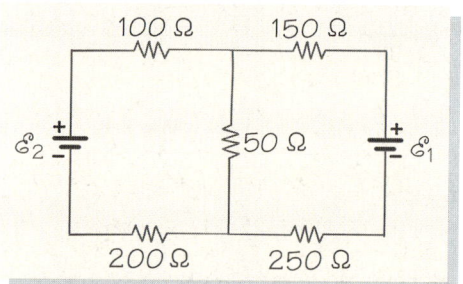

▲ **FIGURE 27–38** Problem 29.

30. (II) Given 20 batteries, each with an emf of 12 V and an internal resistance of 2 Ω, calculate the current for the following cases: (a) batteries in series, with an external load of 80 Ω; (b) batteries in parallel, with an external load of 20 Ω; (c) 4 parallel rows of 5 batteries in series, with an external load of 40 Ω.

31. (II) Consider the circuit shown in Fig. 27–39. Calculate the current through the 50-Ω resistor (a) by calculating the equivalent resistance for the circuit, and (b) by using Kirchhoff's rules.

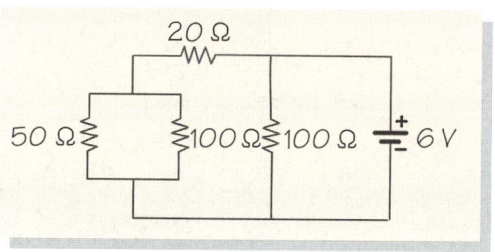

▲ **FIGURE 27–39** Problem 31.

32. (II) The known elements of the circuit in Fig. 27–40 are indicated. Find the value of R_3 that will give a current I_3 of 0.1 A with the indicated sign. Is there a value of R_3 that will give a current I_3 of the same magnitude but of opposite sign? If so, what is it?

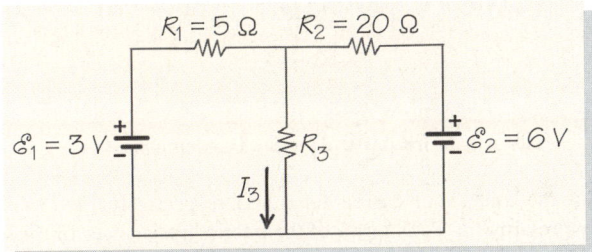

▲ FIGURE 27–40 Problem 32.

33. (II) Two batteries are connected in parallel as in Fig. 27–41 and supply current to a load resistor of 5 Ω. One of the batteries is freshly charged, with an emf of $\mathcal{E}_1 = 12$ V and an internal resistance of $r_1 = 0.1$ Ω. The other one is almost dead, with an emf of $\mathcal{E}_2 = 10$ V and an internal resistance of $r_2 = 10$ Ω. What is the current through the load resistor? How much of this current is supplied by each of the batteries?

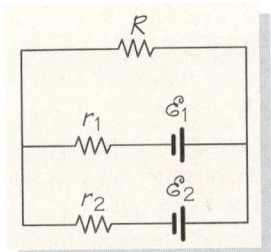

▲ FIGURE 27–41 Problem 33.

34. (III) Consider a tetrahedron whose sides consist of identical wires, each with resistance 1 Ω (Fig. 27–42). Suppose that this arrangement is attached at two of its corners to a generator with potential 4 V. What is the power dissipated in each of the wires?

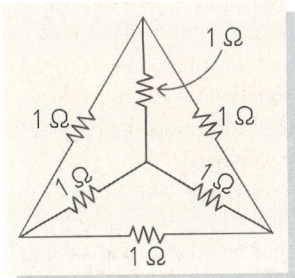

▲ FIGURE 27–42 Problem 34.

35. (III) A cube consisting of identical wires, each of resistance R, is put across a line with voltage V (Fig. 27–43). What is the equivalent resistance of the cube? What is the current in each of the wires?

36. (III) Figure 27–44 shows a ladder of resistors with n rungs. (a) Find the equivalent resistance between points P_1 and P_2 for $n = 1$; (b) for $n = 2$; (c) for $n = 3$; (d) for the limit $n \to \infty$. [Hint: Write an expression for R_n (the equivalent resistance of a ladder of n rungs) in terms of R_{n-1} (the equivalent resistance of a ladder of $n - 1$ rungs) and R, and use that equation in the limit $n \to \infty$.]

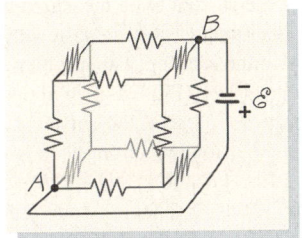

▲ FIGURE 27–43 Problem 35.

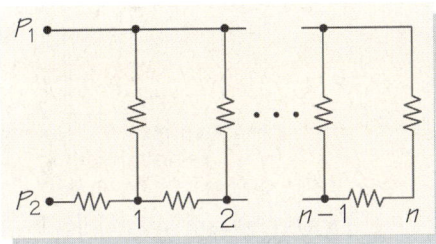

▲ FIGURE 27–44 Problem 36.

27–4 Measuring Instruments

37. (I) Look at Example 27–9 again, with emf $= 6$ V, $R_1 = 1400$ Ω, and $R_2 = 10$ kΩ. Calculate the voltage across R_1 for the two cases of when the voltmeter's internal resistance is 200 kΩ and 10 MΩ.

38. (I) A voltmeter with an internal resistance of 60 kΩ measures the voltage of a D-cell flashlight battery (of nominal voltage 1.5 V) as 1.45 V. What is the internal resistance of the battery?

39. (I) Currents produced with a 12-V source of emf and a range of resistances from 10 Ω to 1000 Ω are to be measured to an accuracy of at worst 0.1 percent with an ammeter. How small must the resistance of the ammeter be?

40. (I) A voltmeter is to be used to measure the voltage across a range of resistances from 5 Ω to 5000 Ω. What is the minimum value of the internal resistance of the voltmeter such that a measurement can be carried out to 0.1 percent accuracy?

41. (I) A certain voltmeter has an internal resistance of 10^5 Ω. The voltmeter is used to measure the potential drop across resistors of (a) 10 Ω, (b) 10^5 Ω, and (c) 100 MΩ. In each case, what is the equivalent resistance of the voltmeter and the resistor across which it is placed? (The voltmeter is a suitable one if this equivalent resistance is as close as possible to the resistance of the resistor across which the potential drop is measured.)

42. (II) An ammeter that can measure a maximum current of 5 mA has an internal resistance of 1.8×10^{-4} Ω. What series resistance will convert it to a 0-to-3-V voltmeter?

43. (II) A battery with emf $\mathcal{E}$ and internal resistance r is connected across a variable resistance. A voltmeter is also connected across the resistance. When the resistance R is set to 20 Ω then the voltmeter reads 23 V; if R is set to 5 Ω, the voltmeter reads 16 V. What will it read when R is set to 50 Ω?

44. (II) A galvanometer whose resistance is 20 Ω will go through its full deflection when a current of 2×10^{-4} A flows through it. What would you do to have its full deflection occur for a current of 2×10^{-3} A? What would you have to do to make it into a voltmeter so that the full deflection corresponds to 0.2 V?

45. (II) Suppose that the current to be measured by an ammeter is so large that a galvanometer deflected by the current would be pinned at its maximum reading. This problem can be resolved by the use of a *shunt resistor* (Fig. 27–45). Show that with the shunt resistor (resistance R_s) present, the current I is given in terms of a reduced current I_G flowing through the galvanometer by the formula $I = I_G[1 + (R_G/R_s)]$, where R_G is the resistance of the galvanometer. Thus, a reading of the reduced current I_G allows us to determine the current I.

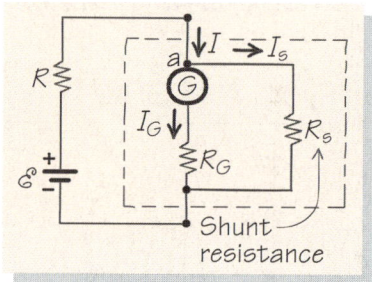

▲ **FIGURE 27–45** Problem 45.

46. (II) The output of the voltage-divider network shown in Fig. 27–46 is to be measured with two voltmeters of internal resistances $500 \text{ k}\Omega$ and $100 \text{ M}\Omega$, respectively. What voltage will each indicate?

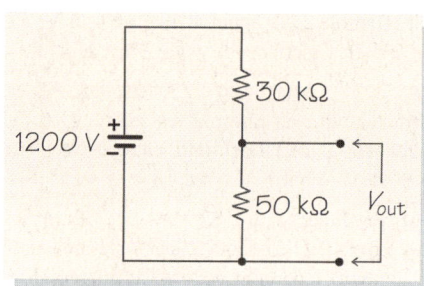

▲ **FIGURE 27–46** Problem 46.

47. (II) A *Wheatstone bridge* is a device that measures resistances. In the circuit shown in Fig. 27–47, R is an unknown resistance. The resistances R_1, R_2, and R_3 are variable. A galvanometer, G, can be used to determine when the potential difference between B and C is zero, given that the battery is connected between A and D. The variable resistances are varied until there is no current in the galvanometer when the circuit is closed at the switch S. Obtain an expression for R in terms of R_1, R_2, and R_3.

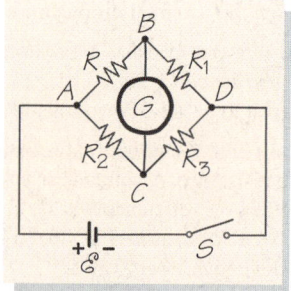

▲ **FIGURE 27–47** Problem 47.

48. (II) The circuit shown in Fig. 27–48 is used to measure the resistance R_x. Draw the circuit including internal resistances. V and I

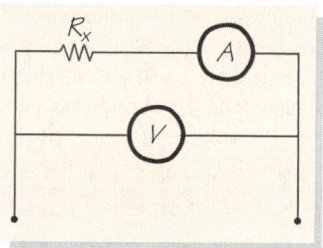

▲ **FIGURE 27–48** Problem 48.

are the voltage and current measured, respectively. Find an exact expression for R_x in terms of the internal resistances of the voltmeter and ammeter. Under what conditions is $R_x = V/I$?

49. (II) Repeat Problem 48 for the circuit shown in Fig. 27–49.

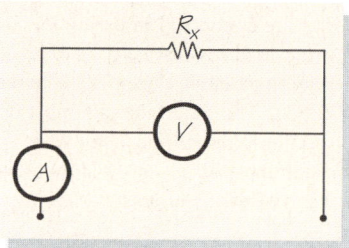

▲ **FIGURE 27–49** Problem 49.

27–5 RC Circuits

50. (I) A flashbulb in an RC circuit discharges with a time constant of 5×10^{-4} s. If the capacitor has a capacitance of $16 \, \mu\text{F}$, what is the resistance in the RC circuit?

51. (I) A flashbulb mechanism operating through an RC circuit has a capacitor charged with a time constant of 2.0 s. If the resistance in the RC circuit is $10^5 \, \Omega$, what is the capacitance of the charging mechanism?

52. (I) Show that the product RC has units of seconds. Find the time constants for the following values of R and C: $5 \text{ M}\Omega$, $30 \, \mu\text{F}$; $8 \text{ k}\Omega$, $3 \, \mu\text{F}$; $20 \, \Omega$, 50 pF.

53. (I) The flash attachment of a camera has a capacitance of $600 \, \mu\text{F}$. The flash time—the characteristic time for the discharge of the capacitor—is $1/500$ s. What is the resistance in the RC circuit?

54. (II) Show by direct substitution that Eq. (27–22) is a solution for the differential equation (27–21).

55. (II) Show by direct substitution that Eq. (27–26) is a solution for the differential equation (27–25).

56. (II) A resistor of resistance $3 \text{ M}\Omega$ and a capacitor of capacitance $350 \, \mu\text{F}$ are connected in series to a 50-V power supply. Calculate (a) the time constant and (b) the current at a time when the charge on the capacitor has acquired 90 percent of its maximum value.

57. (II) Calculate the current in the battery as a function of time for the circuit shown in Fig. 27–50 if the switch S is closed at time $t = 0$.

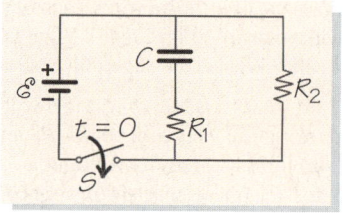

▲ **FIGURE 27–50** Problem 57.

58. (II) The circuit of Fig. 27–19 has $\mathcal{E} = 200$ V, $R = 350$ kΩ, and $C = 20$ μF. What are the voltage across the resistor and the charge on the capacitor 4 s after the switch is closed to position a?

59. (II) You have two capacitors of capacitance 5 μF and three resistors, one of resistance 250 Ω and the remaining two of resistance 300 Ω. Find the connection between these elements that will make a circuit whose time constant is 1 ms.

60. (II) Show that the time constant of a parallel-plate capacitor filled with a dielectric with a finite resistivity is independent of the area and separation of the plates.

61. (II) Polycarbonate, a so-called polar polymer, is a material with a dielectric constant $k = 3.2$. It has a resistivity $\rho = 2 \times 10^{14}$ $\Omega \cdot$m. Suppose that it is used to fill the space in a parallel-plate capacitor of area 0.03 m^2 and plate separation 0.50 mm. A charge of $Q = 2$ μC is placed on the plates of the isolated capacitor. How long does it take for 70 percent of the charge to leak away?

62. (III) A capacitor C_1 is charged to Q_0 and connected to an uncharged capacitor C_2 via the resistor R (Fig. 27–51). Find the charge on each capacitor as a function of time, assuming that the switch is closed at $t = 0$.

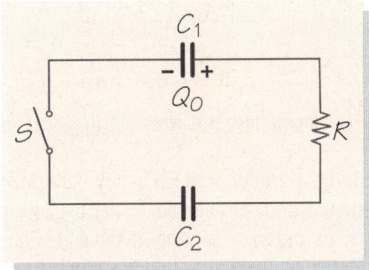

▲ FIGURE 27–51 Problem 62.

General Problems

63. (I) Imagine that a household circuit uses direct current. Compare the current drawn from the main supply at 120 V if three household appliances of resistances 50 Ω, 60 Ω, and 20 Ω, respectively, are connected in parallel. (The appliances would not work in series, because each appliance requires a voltage drop of 120 V. Actual household circuits use alternating currents; see Chapter 33.)

64. (II) Consider the circuit shown in Fig. 27–52, in which a 12-V battery is used to charge a 6-V battery. The resistance in the circuit is 20 Ω. Calculate (a) the current in the circuit; (b) the rate at which the energy of the smaller battery increases; (c) the total rate of energy dissipation in the resistor.

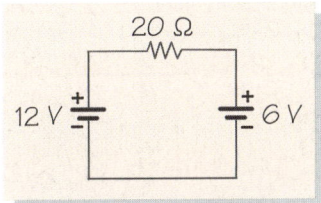

▲ FIGURE 27–52 Problem 64.

65. (II) A student has a wide range of resistors, all rated for 5 W. How can a student combine identical resistors to obtain an effective resistance of 100 Ω rated for 30 W?

66. (II) If a battery of fixed emf and internal resistance r is connected to an external resistor of resistance R, show that the maximum power delivered to the external resistor occurs when $R = r$.

67. (II) Automobile batteries typically have 12.6 V output and an internal resistance of 0.05 Ω. They have a rating of ampere-hours, which demonstrates how long they can produce suitable currents for 20 hours. A good battery has a rating of 75 A $\cdot$ hr. The battery is charged at the rate of 2.5 A for 10 hours. (a) How much current is the battery capable of providing? (b) What is the voltage across the battery terminal during recharging? (c) What total electrical energy is stored in the battery after recharging?

68. (II) Suppose you have n identical batteries, each with emf $\mathcal{E}$ and internal resistance r. The cells may be placed in series, or in parallel across a load resistance R. Which arrangement should you use to maximize the current through the external resistor? Does your answer depend on how large r is?

69. (II) The battery considered in Problem 9 is connected to a load resistor R. Calculate and plot the current in the circuit as a function of R in the range $R = 0$ (short circuit) to $R = 5$ Ω. Plot the ratio of the power delivered to the load to that dissipated in the internal resistance of the battery.

70. (II) When separately connected across a line with voltage V, two resistors generate power P_1 and P_2, respectively. What is the power generated when the two resistors are connected in series? In parallel?

71. (II) A student picks up an electric heater at a yard sale. She discovers later that, even at the lowest setting, it delivers too much power to be used in her small room. She determines that the heater has two heating elements, one delivering 1 kW, the other 2 kW. The highest setting turns on both heaters simultaneously. How can she connect the elements to get a new setting lower than the lowest setting, and what is the power output for the new setting?

72. (II) A resistor R forms a single loop with an arrangement of two batteries of emf $\mathcal{E}$ and internal resistance r. The batteries are arranged (a) in series and (b) in parallel. Find the current through the resistance in both cases. Which arrangement gives the larger current for large R? for small R?

73. (II) An 800-W kitchen mixer, a 600-W vacuum cleaner, and a chandelier with 10 60-W bulbs are all plugged into the same outlet in a 120-V circuit. A *fuse* acts as a switch that opens if the current exceeds 15 A. How much current does each device draw? What is the minimum number of screwed-in bulbs that will blow the circuit? Do not worry about the oscillations of the current and voltage in real household circuits, but assume that all currents and voltages are DC.

74. (II) Consider three resistors of 30 Ω each. Each resistor can dissipate 2 W at most. What are the four possible distinct ways of arranging all three resistors, and calculate the maximum power that can be dissipated in each of the ways.

75. (II) Find the currents in each leg of the circuit shown in Fig. 27–53.

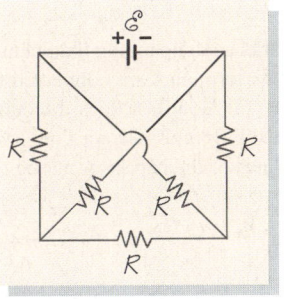

▲ FIGURE 27–53 Problem 75.

790 | Direct-Current Circuits

76. (II) By using Table 26–2, compare the current density, electric field strength, and power loss in two cylindrical wires of the same length and same radius, one made of aluminum and the other made of copper. The wires are connected (a) in series and (b) in parallel. (c) If wires of a given length and radius were constructed from all the materials listed in Table 26–2, and if the same current were passed through each, which wire would have the largest current density, the weakest electric field, and the least power loss? Assume throughout that the temperature dependence is unimportant.

77. (II) A simple *potentiometer* circuit used to measure unknown voltages accurately is shown in Fig. 27–54. Here V_s is the known source voltage, V_x is the unknown voltage, and the resistor is a variable one from which the values R_1 and R_2 can be read from the position of the pointer. These resistances are varied until the current in the ammeter is zero. Show that the unknown voltage then has the value $V_x = V_s R_2/(R_1 + R_2)$.

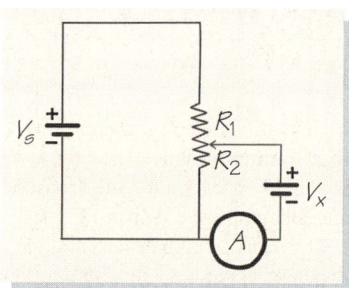

▲ **FIGURE 27–54** Problem 77.

78. (II) Two resistors and two capacitors are connected in series to a battery as shown in Fig. 27–55. Calculate the potential at B relative to that at A: (a) shortly after the closing of the switch and (b) a long time after the closing of the switch. (c) How fast does the circuit reach a steady state? (Give a time scale.)

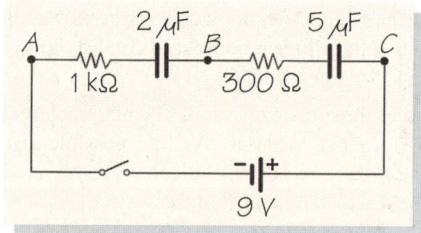

▲ **FIGURE 27–55** Problem 78.

79. (II) To avoid sparks accompanying the opening of a high-current circuit breaker, its terminals are connected to a large capacitor, as in Fig. 27–56. (a) How fast does the current decrease in the circuit shown in the figure? (Give a time scale.) (b) What is the charge on the plates of the capacitor a long time after the switch is opened?

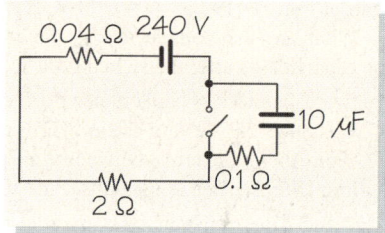

▲ **FIGURE 27–56** Problem 79.

80. (II) The circuit shown in Fig. 27–57 has been established for a long time. (a) What is the charge on the capacitor? Indicate which plate carries the positive charge and which one carries the negative charge. (b) Calculate the current flowing through the 35-Ω resistor.

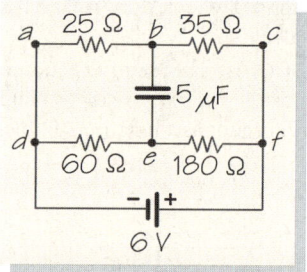

▲ **FIGURE 27–57** Problem 80.

81. (II) A single-loop circuit contains a battery of emf V_0 and negligible internal resistance and, connected in series with the battery, two circular plates of radius r separated by a distance $d \ll r$. The space between the plates is filled with a material of conductivity σ and dielectric constant 1. (a) What is the electric field between the two plates? (b) What is the current flowing in the circuit?

82. (III) Consider the infinite network of resistors shown in Fig. 27–58a. Calculate the resistance R^* of the network by noting that with an infinite set of resistors, adding one more rung to the ladder does not change the resistance. Thus the network may be broken up as shown in Fig. 27–58b.

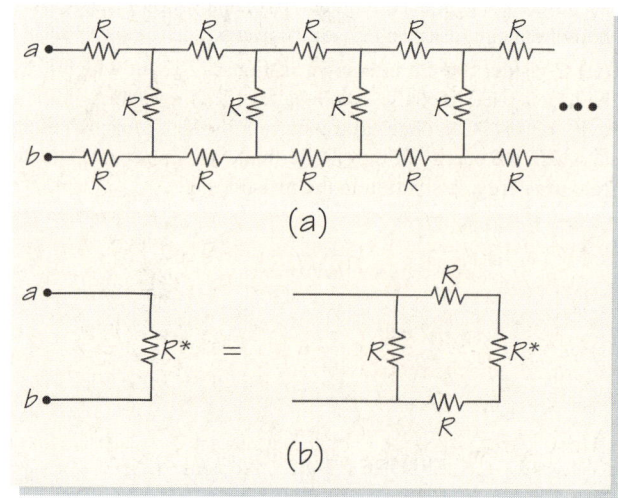

▲ **FIGURE 27–58** Problem 82.

Chapter 28

The Effects of Magnetic Fields

◀ The aurora is a phenomenon that depends on the existence of Earth's magnetic field. When the stream of charged particles known as the solar wind enters Earth's atmosphere under the influence of Earth's magnetic field, its interaction with the atoms of the upper atmosphere produce excitations that lead to the emission of light: the aurora.

We are all familiar with magnets. Sailors have used navigational compasses made from treated iron or lodestone—the natural mineral magnetite—for at least 800 years, and possibly far longer. The behavior of compass needles was systematically studied around the year 1600 by William Gilbert, an English physician. He correctly suggested that the action of a compass is a manifestation of magnetic forces. These forces arise because Earth is itself a giant lodestone, or magnet. Magnetism was not associated with electricity until 1820, when André Ampère used his own experiments and those of Hans Christian Oersted to show that magnetic effects arise when electric charges move. In the 1820s, Michael Faraday uncovered another connection between electricity and magnetism, but it was James Clerk Maxwell who, in the late 1860s, made the ultimate synthesis of electricity and magnetism. We now know that electrical and magnetic phenomena are both aspects of the interactions of electrically charged objects. The synthesis of electricity and magnetism is described entirely by Maxwell's equations, and our understanding of light and other electromagnetic waves rests on Maxwell's great achievement. Maxwell's equations, together with Newton's work, the ideas of thermodynamics, and Einstein's special theory of relativity, summarize virtually all of classical physics. In Chapters 28 to 34 we shall study magnetic phenomena, their connection to electrical phenomena, their practical applications, and other remarkable consequences of Maxwell's equations.

In this chapter we describe the laws of magnetic forces through experiments involving magnets and electric currents. We'll also learn how magnetic forces are associated with magnetic fields, just as electric forces are associated with electric fields. We will here concentrate on the effects of magnetic fields on test objects, leaving the description of how magnetic fields are generated for the next chapter.

28–1 Magnets and Magnetic Fields

When two bar **magnets** are brought close to each other, the forces between them—**magnetic forces**—become evident. These forces are of a type we have not yet encountered. In some positions they attract each other, while in other positions they repel (Fig. 28–1); in still other positions, they exert torques on each other. These forces suggest that bar magnets have an orientation, or an axis. We arbitrarily label the end of a magnet that is attracted to a point very near Earth's geographic South Pole as the *south pole*, S, while the other end of the magnet is called the *north pole*, N. If we experiment with two bar magnets labeled in this way, we find that the N end of one attracts the S end of the other, whereas the two N ends repel each other, as do the two S ends, as Fig. 28–1 shows. (The fact that the south pole of a bar magnet is attracted toward Earth's South Pole means that Earth's South Pole actually behaves like the north pole of a magnet! Similarly, Earth's North Pole behaves like the south pole of a magnet. These labelings are of course historical artifacts.) Based on our experience with charges, we might be tempted to conclude that a bar magnet contains "magnetic charges" (or *magnetic monopoles*) at each end and, further, that we could somehow extract them. A little bit of experimentation suggests that this is not possible; when you break a bar magnet in two, you end up with two bar magnets, not two separated magnetic charges. (Magnetic monopoles remain, nevertheless, the subject of modern searches.)

Iron and a few other materials (all known as *ferromagnetic* materials) have a particular property: If we place a piece of iron, for example, near a lodestone (a "natural" magnet), the piece of iron also becomes a magnet. Later we'll discuss the reasons why this occurs. For now we note that we can use this property to turn tiny shavings of iron (iron filings) into tiny magnets that can be used as test probes of magnetic forces.

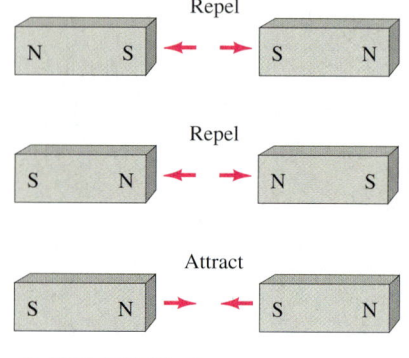

▲ **FIGURE 28–1** Bar magnets exert magnetic forces on each other.

Magnetic Fields

If we scatter iron filings on a sheet of plastic above a bar magnet, as in Fig. 28–2, the magnetic forces acting on them line up the iron filings in certain directions, which are dependent on their position in relation to the magnet. Moreover, the filings clump more densely in certain regions, such as near the poles. For different magnets, the scattered filings have different densities and alignments. Figure 28–2 shows the distribution of iron filings around a straight bar magnet, a horseshoe magnet, and—surprise!—a wire that carries current. This last example is a clue that magnetic forces are associated with moving charges as well as with magnets. The magnetic force acts at a distance, just like gravitational and electric forces. Just as an object with mass sets up a gravitational field and a charged object sets up an electric field, a magnet, a moving charge, or an electric current sets up a **magnetic field** throughout space. We denote this field with the symbol $\vec{B}$. The bar magnet, horseshoe magnet, and current-carrying wire each set up a characteristic magnetic field. Magnetic forces have a marked directional character, and the magnetic field, like the electric field, is a vector.

How do we find the magnetic field due to a magnet or a current? Just as an electric field can be mapped with a small test charge, the response of iron filings to the presence of a magnet or a current can be used to map out the magnetic field. The magnetic field $\vec{B}$ is oriented along the alignment direction of the filings, and the magnitude of $\vec{B}$ is proportional to the density of the filings. (Actually, the density-magnitude connection is only an approximation, which we will refine later.) We can picture the magnetic field by using magnetic field lines—continuous lines that run parallel to the direction of the field at every point and whose density (the number of lines per unit area) is proportional to the strength of the field, in analogy to electric field lines representing electric fields. As an example of this mapping process, take a look at Fig. 28–2a

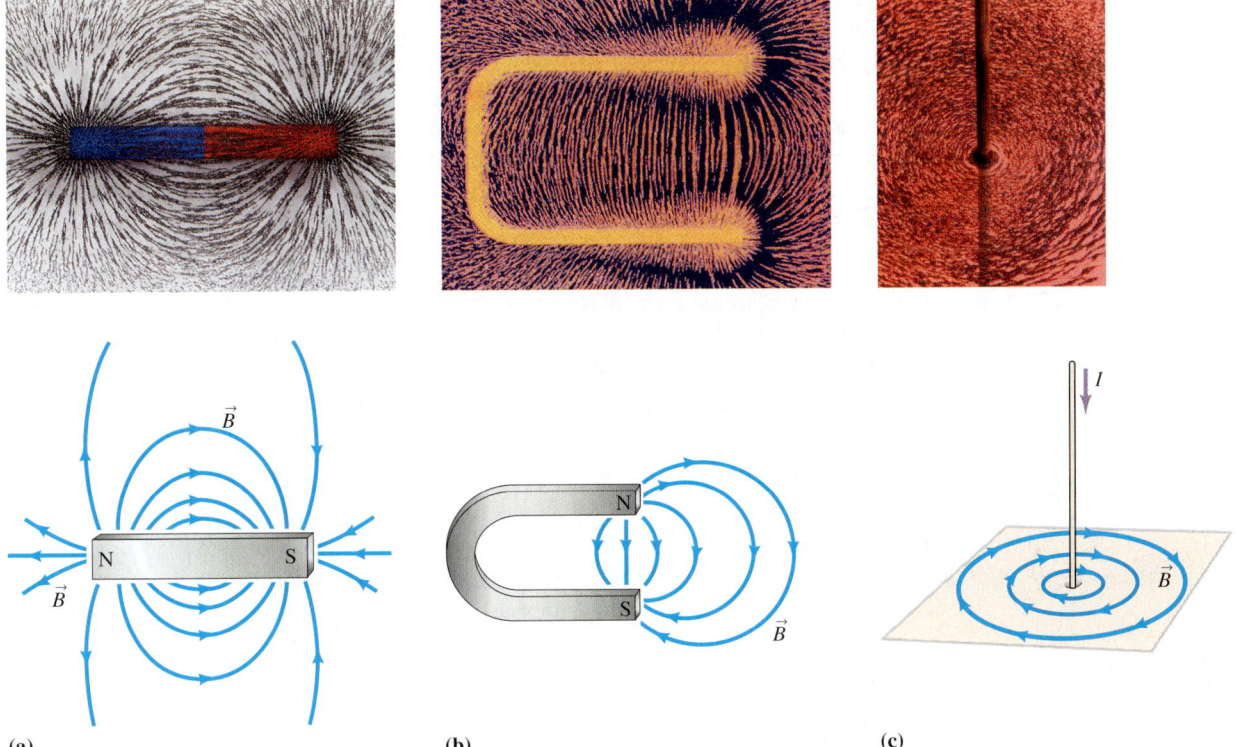

▲ **FIGURE 28–2** Iron filings map the magnetic field for (a) a straight bar magnet, (b) a horseshoe magnet, and (c) a current-carrying wire.

and b: The iron filings align themselves between the poles of magnets, and, therefore, the magnetic field lines associated with the magnet run from pole to pole. We take the direction of the magnetic field of a magnet to run *from the N pole to the S pole*, just as we assign the electric field to run from positive electric charges to negative charges. Notice, however, that the magnetic field around a current-carrying wire has no magnetic pole—it has no starting or ending point. Once we have mapped out a magnetic field in this way, or by the technique of Fig. 28–3, we can further investigate its effects and find the force laws associated with magnetism.

28–2 Magnetic Force on an Electric Charge

Experiments show that compass needles or iron filings are not the only objects to experience forces in the presence of magnetic fields. Moving charged particles also experience forces due to magnetic fields. The effect of a magnetic field on electric charge is most easily studied by using a bar magnet to deflect the electron beam of an oscilloscope. When a bar magnet is placed in different orientations near the beam, the beam deflects in various ways. The deflection allows us to measure the magnetic forces on the beam.

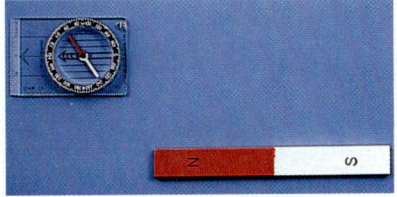
(a)

(b)

◀ **FIGURE 28–3** A compass needle is sensitive to a magnetic field and can be used to map the field due to a magnet.

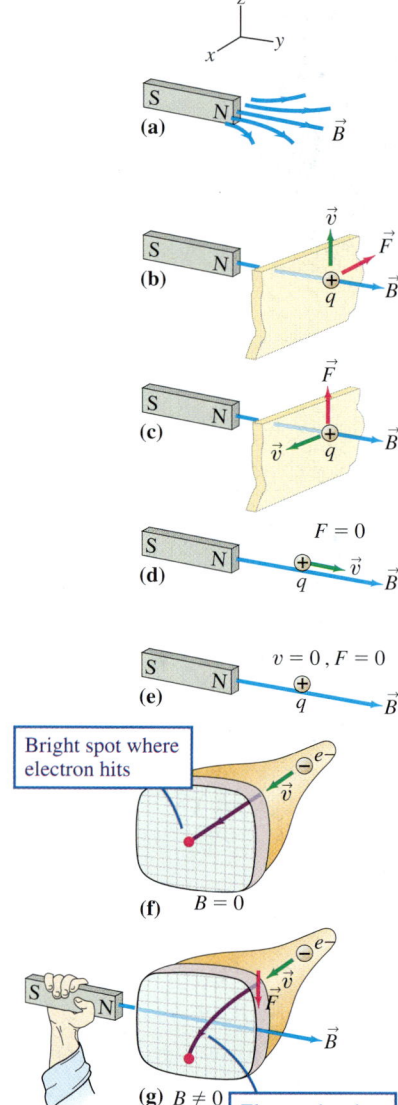

▲ **FIGURE 28–4** (a) The magnetic field of a bar magnet will be used to influence a moving charge. (b)–(e) Experiments on a positive charge moving in a magnetic field. (f) An oscilloscope can be used to measure the effects of a magnetic field on moving negative charges, here an electron in a cathode-ray tube. (g) In the orientation shown the electron is deflected down, whereas for the positive charge in part (c) the force was upward.

Consider a bar magnet with its N pole oriented so that the magnetic field is in the $+y$-direction (Fig. 28–4a). The magnitude of the magnetic field at the position of a moving electric test charge q can be varied by altering the distance of the magnet from the charge. The results of a series of experiments are shown in Fig. 28–4, with the deflected particle chosen to have a *positive* charge. (If the actual experiment is done with an oscilloscope, keep in mind that the test charge, a moving electron, has a negative charge.) We observe the following (below the list, we summarize these observations in a different way):

1. If q moves at speed v in the $+z$-direction and the magnetic field points in the positive y-direction, then q is deflected in the $-x$-direction (Fig. 28–4b). Furthermore, the larger v is, the stronger is the force $\vec{F}$. Detailed measurements show that the magnitude of $\vec{F}$ due to the magnetic field is proportional to v.
2. If q moves in the $+x$-direction, $\vec{F}$ is in the $+z$-direction, again proportional to v (Fig. 28–4c).
3. If q moves in the y-direction ($+$ or $-$), there is no change in the charge's direction or speed; that is, there is no force (Fig. 28–4d).
4. If q moves at speed v in an arbitrary direction, $\vec{F}$ is proportional to the velocity component perpendicular to the magnetic field, $v_\perp$, and perpendicular to the directions of both $\vec{v}$ and $\vec{B}$. This result summarizes points 1 through 3. In particular, if the charge is at rest, so that $v = 0$, there is no force (Fig. 28–4e).
5. $\vec{F}$ is proportional to the magnitude of $\vec{B}$.
6. $\vec{F}$ is proportional to the sign and magnitude of q (Figs. 28–4f and 28–4g).

One important feature of this collection of results is the dependence on $\vec{v}$. In results 1 and 2, the initial velocity is purely in the z- or x-directions and is therefore perpendicular to $\vec{B}$. In result 4, there is no $\vec{v}$ and also no force. A stationary charge in a magnetic field experiences no force. Moreover, Figure 28–4 indicates that the force is always perpendicular to both $\vec{v}$ and $\vec{B}$.

To summarize:

The magnetic force $\vec{F}$ on a moving charge is proportional in magnitude to q, $v_\perp$, and B, where $\vec{v}_\perp$ is the velocity component perpendicular to the field, while the direction of $\vec{F}$ is perpendicular to both $\vec{B}$ and $\vec{v}$ and depends on the sign of q.

A direction perpendicular to both $\vec{v}$ and $\vec{B}$ can be represented by the *vector product*, or *cross product*, which was discussed extensively in Chapter 10 in connection with torques and rotational motion (see the problem-solving techniques box, p. 284). Recall that a vector $\vec{c} = \vec{a} \times \vec{b}$ (the vector product of $\vec{a}$ and $\vec{b}$) has magnitude $ab \sin \theta$, where θ is the angle between vectors $\vec{a}$ and $\vec{b}$ and is always taken to be less than $180°$ (Fig. 28–5). Vector $\vec{c}$ is perpendicular to both $\vec{a}$ and $\vec{b}$, in a direction determined by the right-hand rule. Thus our experiments have determined that the magnetic force on a test charge q moving with velocity $\vec{v}$ in a magnetic field $\vec{B}$ is given by

$$\vec{F} = q\vec{v} \times \vec{B}. \quad (28\text{–}1)$$

MAGNETIC FORCE LAW

This important result is the **magnetic force law**. If θ is the angle between vectors $\vec{v}$ and $\vec{B}$, the magnitude of $\vec{F}$ is given by

$$F = qvB \sin \theta = qv_\perp B. \quad (28\text{–}2)$$

Figure 28–6 shows how $\vec{F}$ is perpendicular to the plane formed by $\vec{v}$ and $\vec{B}$ according to a right-hand rule. Recall that the vector product of two *parallel* vectors is zero; this describes the fact that there is no magnetic force on a charge that moves along the axis of a bar magnet and no magnetic force associated with the component of $\vec{v}$ parallel to $\vec{B}$.

Equation (28–1) shows that the dimensions of the magnetic field are quite different from those of the electric field. The SI unit of magnetic field is called the *tesla* (T), in honor of Nikola Tesla, who made important contributions to the technology of electrical energy generation. In terms of previously defined SI units,

$$1 \text{ T} = 1 \frac{\text{kg}}{\text{C} \cdot \text{s}}. \tag{28-3}$$

Another (non-SI) unit for the magnetic field that is in common use is the *Gauss* (G); 10^4 G = 1 T.

Table 28–1 contains some representative values of magnetic fields.

TABLE 28–1 • Some Magnetic Fields

Location or Source	Magnitude (T)
Interstellar space	10^{-10}
Near Earth's surface	5×10^{-5}
Refrigerator magnet for notes	10^{-2}
Bar magnet near poles	10^{-2}–10^{-1}
Near surface of Sun	10^{-2}
Large scientific magnets	2–4
Largest steady-state magnet	30
Largest pulsed field in laboratory	500–1000
Near surface of pulsar	10^8
Near surface of atomic nucleus	10^{12}

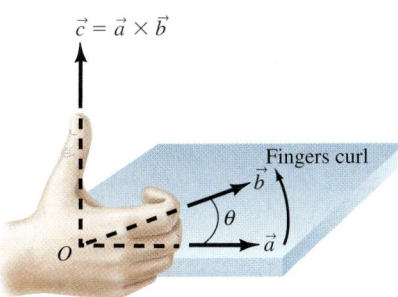

▲ **FIGURE 28–5** The vector product $\vec{c} = \vec{a} \times \vec{b}$ and the right-hand rule that determines the direction of the vector product. See Chapter 10, Fig. 10–5.

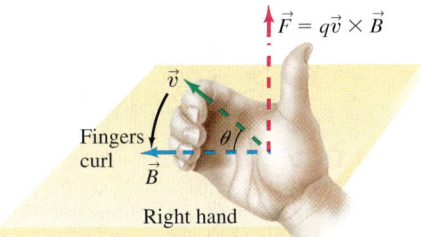

▲ **FIGURE 28–6** The right-hand rule for the magnetic force law $\vec{F} = q\vec{v} \times \vec{B}$.

CONCEPTUAL EXAMPLE 28–1 Consider the situation depicted in Fig. 28–4f, with the electron beam showing a spot on the oscilloscope screen. Suppose we now bring a bar magnet down from above to the top of the oscilloscope, with the S pole of the magnet closest to the beam. As we bring the magnet down slowly, how will the spot move on the screen?

Answer We see from Fig. 28–2a that our action effectively brings in an increasing B field that points up in the vertical direction. An application of the right-hand rule shows that the direction of the cross-product of the velocity and the field points in the left-side direction as seen when facing the screen. Note, however, that the electrons in the beam are *negatively* charged, so that the force points toward the right side of the screen, as seen from the front. The electrons are therefore accelerated to the right and will therefore strike the screen to the right of the original spot (always as seen from the front). As the magnet is slowly brought down from above, the field strength at the electron location increases, and the deflection will also increase, so the spot will be seen to move to the right.

EXAMPLE 28–2 The undisturbed electron beam of an oscilloscope moves along the x-direction. The north pole of a bar magnet approaches the cathode-ray tube from the right, as seen facing the screen, and deflects the beam. The magnitude of the magnetic field from the magnet is 0.050 T in the vicinity of the beam, and the speed of the electrons in the beam is 2.0×10^5 m/s. What is the magnitude of the magnetic force on the electrons? What is the direction of this force; that is, which way is the beam deflected?

Setting It Up We draw the situation in Fig. 28–7 and note that we are given both the magnitude of the velocity v and the magnitude of the magnetic field B, as well as their directions. The electron charge $-e$ is known.

Strategy We can obtain the direction of the force by using the right-hand rule, but to get some experience with the notion of vector products, we here set up a coordinate system and use unit vectors along the x-, y- and z-axes to work out the direction of the force. It is convenient to choose the direction of the beam as the positive x-axis. We also choose the horizontal line pointing to the right on the screen to be the y-axis, so that the conventional (right-hand) choice of z-axis is upward. We then simply apply the force law.

Working It Out We have $\vec{v} = v\hat{i}$ and $\vec{B} = -B\hat{j}$, and with the help of the right-hand rule for unit vectors, $\hat{i} \times \hat{j} = \hat{k}$, we find

$$\vec{F} = (-e)(v)(-B)\hat{k} = (-1.6 \times 10^{-19} \text{ C})(2.0 \times 10^5 \text{ m/s})(-0.050 \text{ T})\hat{k}$$
$$= (1.6 \times 10^{-15} \text{ N})\hat{k}.$$

The force is upward, and this is the direction in which the spot moves.

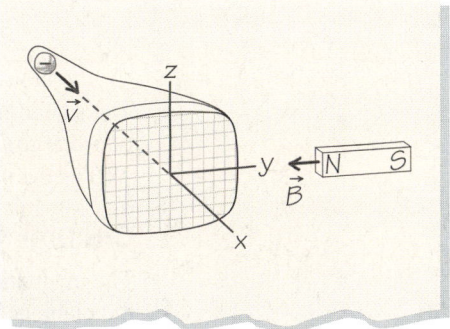

▲ **FIGURE 28–7** The magnetic field points away from the north pole of the bar magnet and is therefore oriented in the $-y$-direction. The spot at which the beam reaches the screen is deflected in the $+z$-direction.

(continues on next page)

What Do You Think? Given the force, we can easily calculate the acceleration. Is the deflection of the spot as simple to calculate as the drop of a rock thrown in an initially horizontal direction and subject to gravity? *Answers to* **What Do You Think?** *questions are given in the back of the book.*

The Lorentz Force

Further experimentation shows that charges react independently to electric and magnetic fields. Thus if an electric field is present in addition to a magnetic field, it produces an additional force $\vec{F} = q\vec{E}$ on a charge. The net force on a charged particle in an electric and a magnetic field is then

$$\vec{F} = q(\vec{E} + \vec{v} \times \vec{B}). \qquad (28\text{--}4)$$

LORENTZ FORCE LAW

This equation is known as the **Lorentz force law**, named after the late-nineteenth-century physicist Hendrik A. Lorentz, who influenced the development of many areas of classical physics.

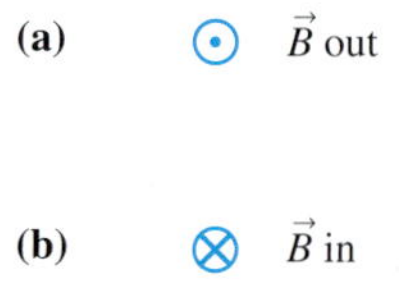

A Notation for Vectors Perpendicular to the Page

Because the three-dimensional aspect of magnetic forces is so important, it is useful to have a notation for vectors oriented perpendicular to the page. Figures 28–8a and 28–8b show a vector coming out of and going into the page, respectively. We shall often use this convention when we illustrate magnetic fields.

▲ **FIGURE 28–8** Conventions to indicate that a vector is (a) out of the page or (b) into the page. A useful mnemonic is to think of the circle with a dot as the point of the arrow coming toward you, and the circle with an x as the feathered tail of the arrow moving away from you.

28–3 Consequences of the Magnetic Force on a Charge

Magnetic forces on charged particles have important implications that range from medical imaging devices to complicated phenomena in astrophysics and plasma physics. In this chapter, we'll keep things simple and assume that the magnetic fields are constant and do not change over time. When magnetic fields are independent of time, we are dealing with **magnetostatics**.

Energy of a Charged Particle in a Static Magnetic Field

Magnetic forces have a special property: *A charged particle moving in a static magnetic field has a constant kinetic energy.* One way to see this is to note that the force on a charge is perpendicular to its velocity, and hence to its infinitesimal displacement. The work done by the force is then zero, and by the work-energy theorem, the kinetic energy is unchanging. *A static magnetic field does no work on a charge.* Another way to see it is to calculate the rate of change of the kinetic energy. We have

$$\frac{d}{dt}\left(\frac{1}{2}m\vec{v}^2\right) = \frac{1}{2}m\frac{d}{dt}(v_x^2 + v_y^2 + v_z^2) = m\left(v_x\frac{dv_x}{dt} + v_y\frac{dv_y}{dt} + v_z\frac{dv_z}{dt}\right)$$

$$= m\vec{v} \cdot \frac{d\vec{v}}{dt} = m\vec{v} \cdot \vec{a}.$$

The scalar product on the right is zero, since magnetic forces, and hence the acceleration due to those forces, always point in a direction perpendicular to the velocity (as well as the magnetic field).

Circular Motion in a Constant Magnetic Field

The magnetic force law—Eq. (28–1)—states that only the component of the velocity in the plane perpendicular to $\vec{B}$ contributes to the expression for the force. *The component of the velocity of a charged particle parallel to the magnetic field is not affected by the field*, and it is therefore unchanging in the absence of any other forces. In addition, Eq. (28–1) states that *the force on the charge, and hence the charge's acceleration, is perpendicular to $\vec{B}$ and thus acts only in the plane perpendicular to $\vec{B}$.*

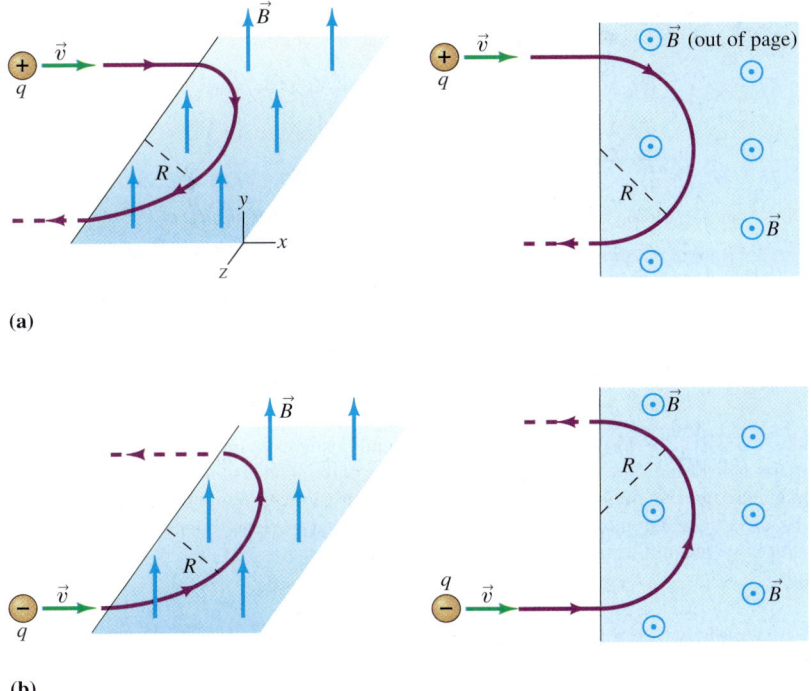

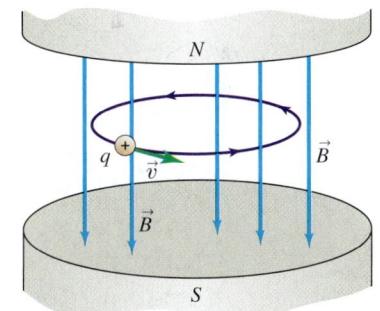

◀ **FIGURE 28–9** (a) A positively charged particle moves perpendicularly to a constant magnetic field, $\vec{B}$, shown in both an oblique perspective and a view from above. The charged particle traces a circular path in the plane perpendicular to $\vec{B}$, which is directed out of the plane of the page. (b) The direction of the curvature is opposite for a negatively charged particle.

To explore the consequences of these observations more closely, consider a magnetic field $\vec{B}$ that is uniform in some region of space and a test charge q that enters this region with a velocity $\vec{v}$ perpendicular to the field (Fig. 28–9). What is the consequent motion of the charge? According to Eq. (28–1), the force will be perpendicular to $\vec{v}$ and have magnitude $F = qvB$. We saw in Chapter 3 that when the acceleration (and hence the force) is constant in magnitude and perpendicular to the velocity, there is *circular motion at constant speed*. A charged particle moving perpendicularly to a constant, spatially uniform magnetic field will move in a circle (Fig. 28–9a). The magnitude of the acceleration for circular motion is $a = v^2/R$, where R is the radius of the particle's circular path; the direction of the acceleration is toward the center of the circular path. By Newton's second law, the force must have magnitude $ma = mv^2/R$. In our case, the force responsible for the acceleration has magnitude F, and Newton's second law ($F = ma$) becomes (Fig. 28–10)

$$F = qvB = \frac{mv^2}{R}.$$

We solve for R:

$$R = \frac{mv}{qB}. \tag{28-5}$$

R is proportional to the product of m and v—that is, to the momentum of the moving particle, $p = mv$—and is inversely proportional to the magnitudes of the charge q and of the field, $\vec{B}$. Equation (28–5) holds *only* when the velocity is perpendicular to $\vec{B}$. Whether the motion is clockwise or counterclockwise depends on the sign of the charge, according to the right-hand rule. Figure 28–9b depicts the motion for a test charge of opposite sign to the test charge in Fig. 28–9a. The larger B is, the larger is the magnetic force, and the "tighter" is the curved path, which corresponds to a smaller radius of curvature (the radius of the segment of a circle along which the charge moves at a given moment). The smaller the magnetic field, the smaller the force, and the larger is R. If the magnetic field varies in strength from place to place, then so will the radius of curvature of the path (Fig. 28–11).

The circular motion has a period $T = 2\pi R/v$, or, from Eq. (28–5),

$$T = \frac{2\pi}{v}\frac{mv}{qB} = \frac{2\pi m}{qB}. \tag{28-6}$$

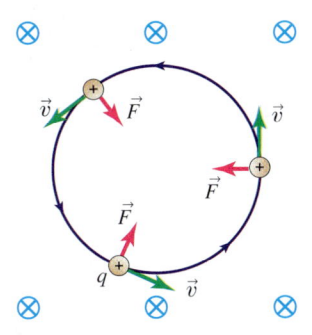

▲ **FIGURE 28–10** Forces on a charged particle that is moving perpendicularly to a uniform magnetic field.

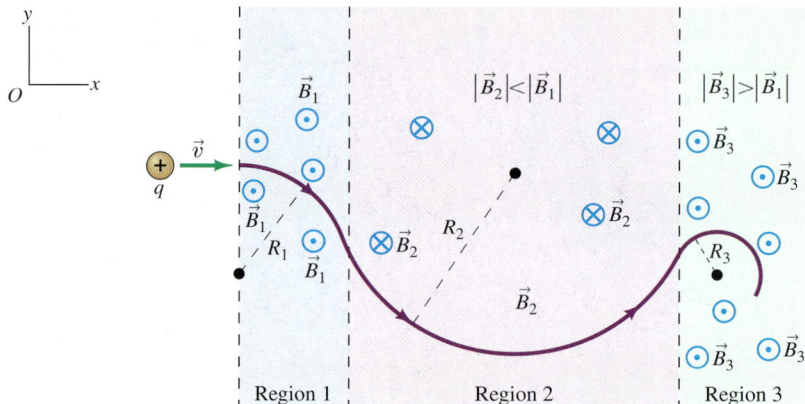

▲ **FIGURE 28–11** A particle of mass m and positive charge q moves with an initial velocity $\vec{v}$ in the x-direction. The curvature of the particle's path indicates the strength and direction of the magnetic field. In region 1, the magnetic field $\vec{B}_1$ is of medium magnitude and oriented in the $+z$-direction. In region 2, $\vec{B}_2$ is small in magnitude and oriented in the $-z$-direction. In region 3, $\vec{B}_3$ is of large magnitude and oriented in the $+z$-direction.

Equivalently, the frequency $f = 1/T$ is

$$f = \frac{qB}{2\pi m}. \qquad (28\text{–}7)$$

CYCLOTRON FREQUENCY

This frequency is called the **cyclotron frequency**. Notice that the period and frequency are *independent* of the speed. A slow particle traces out a tight circle in the same time that a fast particle traces out a large circle. The constancy of the cyclotron frequency is a guiding principle of a device called the *cyclotron* (see Problem 31).

Equation (28–5), which specifies the radius of the circular path of a charged particle, has found application in many particle-detection devices—the *bubble chamber* illustrates the principle well. When charged particles produced in high-energy collisions speed through liquid hydrogen, they leave tracks that consist of very tiny bubbles, like a jet leaving a vapor trail in the atmosphere (Fig. 28–12). The momentum of these particles can be obtained by measuring the radius of curvature of their tracks when an external magnetic field is imposed. As we know from Chapter 8, information about momentum is helpful in deciphering collisions—in this case, the collisions of subatomic particles. (Although today bubble chambers are rarely used in such experiments, the more modern detectors that have replaced them rely on the same principles for the measurement of momentum.)

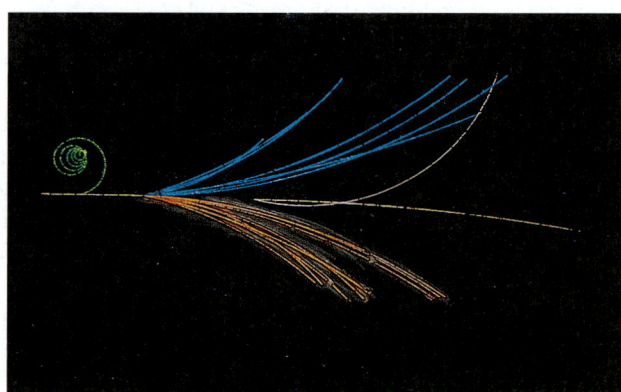

▲ **FIGURE 28–12** The tracks left by charged particles moving through a bubble chamber in a magnetic field. The colors are computer generated.

28–3 Consequences of the Magnetic Force on a Charge

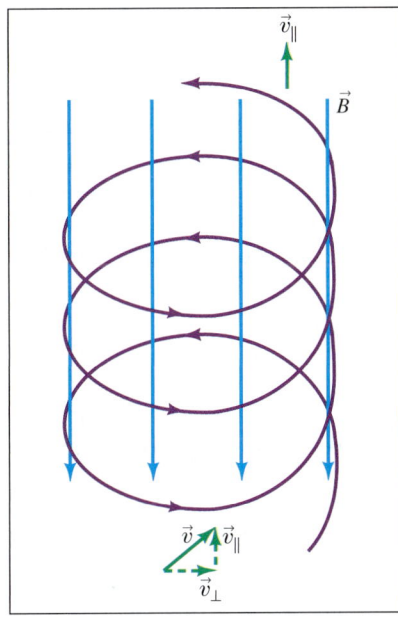

(a)

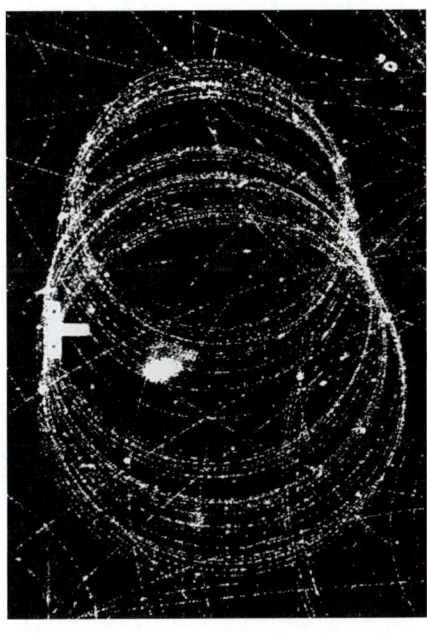

(b)

◄ **FIGURE 28–13** (a) A charged particle follows a helical path in a region where the magnetic field is constant. (b) An electron in a cloud chamber produced this 10-m-long spiral track. The electron's path begins at the bottom. The helix becomes more tightly wound about halfway up because the electron loses energy by radiation while moving in the helical path.

Finally, when there is a component $v_\parallel$ of velocity $\vec{v}$ that lies along $\vec{B}$, that component of the velocity does not change. The particle advances along $v_\parallel$ while it moves in a circle in the plane perpendicular to $v_\parallel$ (Fig. 28–13a). The resulting trajectory forms a *spiral* (or *helix*), with its axis along $\vec{B}$ (Fig. 28–13b). The circular motion in the plane perpendicular to $\vec{B}$ has a radius (the "radius of curvature") given by

$$R = \frac{mv_\perp}{qB}. \tag{28-8}$$

EXAMPLE 28–3 A particle of unknown charge q and unknown mass m moves at speed $v = 4.8 \times 10^6$ m/s in the $+x$-direction into a region of constant magnetic field. The field has magnitude $B = 0.50$ T and is oriented in the $+y$-direction. The particle is deflected in the $+z$-direction and traces out a segment of a circle of radius $R = 0.10$ m. What is the sign of the particle's charge, and what is the ratio q/m?

Setting It Up The situation described is drawn in Fig. 28–14.

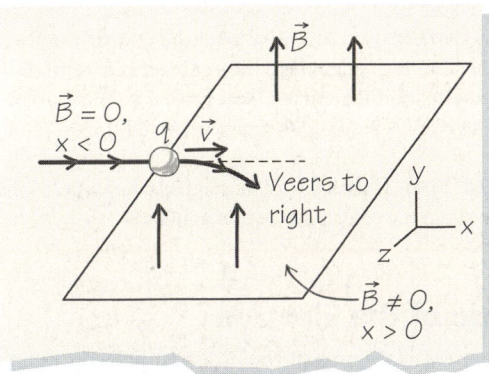

▲ **FIGURE 28–14** If the speed of a particle of unknown charge and mass that moves in a region of constant magnetic field is known, then a measurement of the radius of curvature of the particle's path gives the charge-to-mass ratio of the particle.

Strategy Given the velocity and the magnetic field, we can find the force. The acceleration will be in the same direction as the vector product $\vec{v} \times \vec{B}$ when the charge is positive, and opposite to the vector product if the charge is negative. The information on the deflection of the particle thereby tells us about the charge's sign. For the ratio q/m, we use the equation of motion, which links this ratio to the radius and the field magnitude if the field is constant, as in Eqs. (28–8) or (28–5).

Working It Out The standard rules for vector products give $\vec{v} \times \vec{B} = vB\hat{i} \times \hat{j} = vB\hat{k}$. For a positive charge this would lead to a deflection in the $+z$-direction, and since the observed deflection is in that direction, the *charge is positive*.

To calculate q/m, we use Eq. (28–5), which gives

$$\left|\frac{q}{m}\right| = \frac{v}{BR} = \frac{4.8 \times 10^6 \text{ m/s}}{(0.50 \text{ T})(0.10 \text{ m})} = 9.6 \times 10^7 \text{ C/kg}.$$

It is possible that this particle is a proton. We can see why if we assume that the unknown charge is that of a proton, $q = 1.6 \times 10^{-19}$ C. Then $m = q/(9.6 \times 10^7 \text{ C/kg}) = (1.6 \times 10^{-19} \text{ C})/(9.6 \times 10^7 \text{ C/kg}) = 1.7 \times 10^{-27}$ kg, which is just the mass of a proton. Note, however, that the experiment described in this example can measure only the charge-to-mass ratio, not the charge or the mass alone.

What Do You Think? A deuteron is one type of hydrogen nucleus. It has the charge of a proton and (approximately) the mass of two protons. What fraction of the proton's speed would a deuteron have if it followed exactly the same path as the protons in the example above?

Velocity Selectors

Beams of charged particles moving with a precisely known velocity are required in a wide range of applications, including in the manufacture of computer chips. A particular arrangement of electric and magnetic fields makes a **velocity selector**, which passes only particles of a specific velocity out of a beam of identical charged particles with a variety of velocities. Consider a region with uniform, mutually perpendicular $\vec{E}$ and $\vec{B}$ fields (Fig. 28–15). (Such fields are said to be *crossed*.) A particle of mass m, charge q (positive), and velocity $\vec{v}$ is directed perpendicularly to both $\vec{E}$ and $\vec{B}$ when it enters this region. We will now show that there is a certain value of v for which the particle traverses the region undeflected. At a speed other than v, the same particle *is* deflected; thus, in a beam of particles with a variety of speeds, only those particles with a certain speed pass through undeflected.

Both $\vec{E}$ and $\vec{B}$ fields are present, so we must use the Lorentz force law [Eq. (28–4)] and compute the contributions to the force from both fields. Referring to Fig. 28–15, the electric force is

$$\vec{F}_E = qE\hat{k}.$$

By the right-hand rule, the magnetic force $q\vec{v} \times \vec{B}$ is

$$\vec{F}_B = -qvB\hat{k}.$$

The electric and magnetic forces will cancel if their magnitudes are equal, because they point in opposite directions; in this case, the particle will travel undeflected. This cancellation occurs for $qvB = qE$, so the speed of a charged particle that passes through the crossed fields undeflected is

$$v = \frac{E}{B}. \tag{28-9}$$

If we reverse the sign of charge q, the electric force points in the $-z$-direction while the magnetic force points in the $+z$-direction; the forces still cancel when v is given by Eq. (28–9).

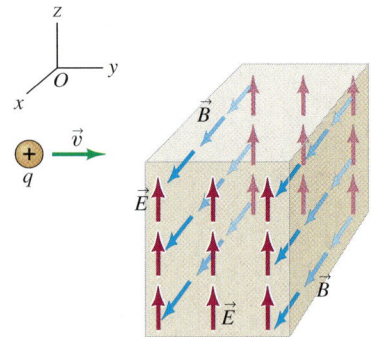

▲ **FIGURE 28–15** A charged particle enters a region of crossed electric and magnetic fields. If the speed of the particle is $v = E/B$, the particle will cross the region undeflected.

EXAMPLE 28–4 A hole is cut in the wall of a container of a plasma (matter that is composed of positively charged ions and electrons acting as a kind of gas), creating a beam of particles of various charges. It is necessary to choose particles from the beam that have a speed of 3.2×10^6 m/s in order to carry out some materials testing. The engineer designing the speed selector uses crossed $\vec{E}$ and $\vec{B}$ fields, and the $\vec{B}$ field comes from a magnet with field strength $B = 6.5 \times 10^{-4}$ T. What must the engineer choose as the magnitude of the electric field? Will all the particles so chosen have the same momentum and energy?

Strategy With B and v given, there is enough information to calculate the magnitude of the electric field required, and we can use the relation $v = E/B$. Since that relation is independent of the charge q and the mass m of the particle, the particles selected by the apparatus are only characterized by the velocity v. The momentum and the energy of a particle depends not only on its velocity but on the mass of the particle, and this is not given, so we have no information on the momentum and energy. Thus, for example, electrons will be chosen in the same way as positive ions of a variety of charges. Both momentum and energy will take on a variety of values. This fact would make the particles so chosen not so useful to test materials, because their effect on materials depends more on their momentum and energy than on their speed.

Working It Out The relation in Eq. (28–9) leads to

$$E = vB = (3.2 \times 10^6 \text{ m/s})(6.5 \times 10^{-4} \text{ T}) = 2.1 \times 10^3 \text{ V/m}.$$

The magnetic field used here is smaller than those usually associated with permanent magnets; unless the engineer is careful, Earth's magnetic field will spoil the design (see Table 28–1). A larger magnetic field, however, will require a larger electric field.

What Do You Think? What happens to particles that move a little faster than the selected v in this apparatus?

The Charge-to-Mass Ratio of the Electron

In 1897 Sir Joseph John Thomson, the discoverer of the electron, performed a series of wide-ranging experiments whose results were crucial in the development of our understanding of the electrical nature of matter. A velocity selector was an important component of his experiment to measure the charge-to-mass ratio of the electron (Fig. 28–16). He first accelerated electrons in an electric field—not the electric field of the velocity selector—by passing them through an electric potential V. The work thereby done on

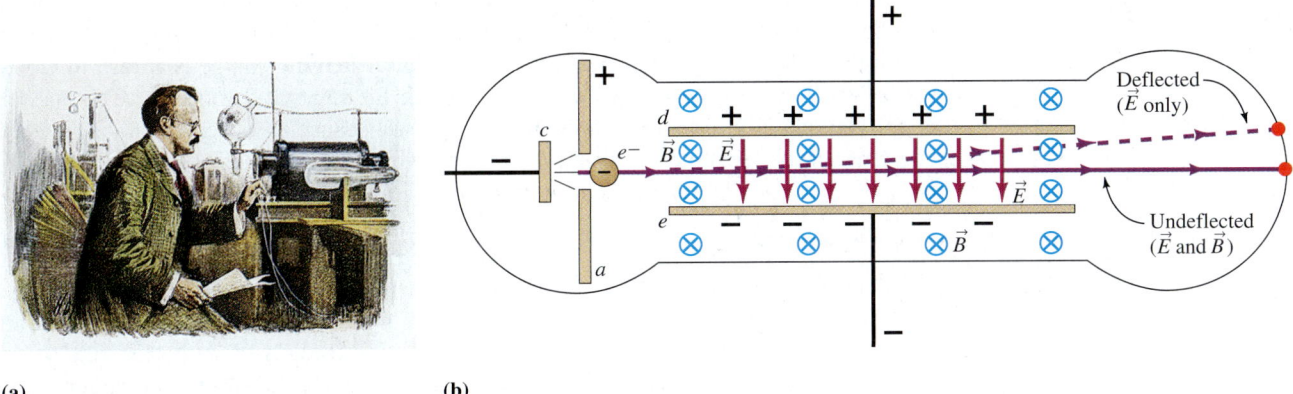

▲ **FIGURE 28–16** (a) J. J. Thomson at work in his laboratory. (b) Schematic diagram of Thomson's apparatus for measuring the charge-to-mass ratio of the electron. The magnetic field is directed into the plane of the page.

the electrons is qV. Assuming they started from rest, the electrons gained a speed v determined by $mv^2/2 = qV$, so

$$v = \sqrt{\frac{2qV}{m}}. \qquad (28\text{–}10)$$

The electrons accelerated in this way continued into a region of crossed electric and magnetic fields. Thomson adjusted the magnitudes of these fields until the electrons passed through the apparatus undeflected. When we combine Eqs. (28–9) and (28–10), we find that

$$v = \frac{E}{B} = \sqrt{\frac{2qV}{m}}.$$

When both sides of this equation are squared, we can solve for q/m:

$$\frac{q}{m} = \frac{E^2}{2VB^2}. \qquad (28\text{–}11)$$

The electrons in Fig. 28–16b are accelerated between plates c and a. They travel on to the region between plates d and e where the magnetic field was adjusted until electrons were undeflected. The electric field and the required magnetic field may be used in Eq. (28–11) to obtain q/m. More refined experiments based on Thomson's scheme have led to a value for q/m of 1.759×10^{11} C/kg.

*The Mixing of Electric and Magnetic Fields

CONCEPTUAL EXAMPLE 28–5 Is there an arrangement of non-zero $\vec{E}$ and $\vec{B}$ fields that will leave a particle at rest if it starts at rest?

Answer This question has a simple answer: When a particle is at rest, so that $v = 0$, there is no magnetic force on the particle. In the presence of a non-zero electric field, there will be an electric force and therefore an acceleration. The particle cannot remain at rest unless the electric field is zero.

The Conceptual Example above is more interesting than it might appear. In Section 4–2 we described an overriding law of physics, what we referred to as the relativity principle: Observers in different inertial frames—frames that move at constant velocity with respect to one another—cannot decide by any experiment which of them is at rest and which is moving. In particular, observers in different inertial frames all see the same forces. But above we constructed a velocity selector, in which a combination of electric and magnetic fields produces no net force on a moving charge. The charge moves without acceleration in the fields of the velocity selector, yet this would not be the case if it started at rest in the presence of these same fields—rather, as the Conceptual Example indicates, it would accelerate. This is a way of distinguishing the frame in which the charge moves uniformly from a frame in which the charge is at rest, in violation of the relativity principle. Must we sacrifice the relativity principle? The answer is no. The relativity principle holds, and it does so because an observer moving with a constant velocity relative to a frame in which we have given constant electric and magnetic fields $\vec{E}$ and $\vec{B}$ will see *different* electric and magnetic fields $\vec{E}'$ and $\vec{B}'$. The transformation is such that an observer who moves with the charge, and thereby sees it at rest, sees *no electric field at all*, that is, the *primed* electric field vanishes. This guarantees that the particle remains at rest.

We can go farther by looking at the Lorentz force law, Eq. (28–4). In a frame in which a charged particle has velocity $\vec{v}$ (frame 1), the force is given by Eq. (28–4). In the frame moving with velocity $\vec{u}$ relative to the previous frame (frame 2), the force has the same form as Eq. (28–4), but with the changes $\vec{E} \to \vec{E}'$; $\vec{B} \to \vec{B}'$; $\vec{v} \to \vec{v} - \vec{u}$. Since according to the relativity principle the forces are the same in the two inertial frames, we must have

$$\vec{E} + \vec{v} \times \vec{B} = \vec{E}' + (\vec{v} - \vec{u}) \times \vec{B}'.$$

Since this holds whatever $\vec{v}$ is, we conclude that

$$\vec{B}' = \vec{B}, \quad \vec{E}' = \vec{E} + \vec{u} \times \vec{B}.$$

For our velocity selector, the combination $\vec{E} + \vec{v} \times \vec{B}$ is zero. We now see that if $\vec{u}$ is chosen to equal $\vec{v}$, so frame 2 is the frame in which the particle is at rest, then the electrical force on the object, $q\vec{E}'$, is also zero. This shows that the Lorentz force is not merely the sum of two forces, electrical and magnetic, that are independent of each other, but that it involves both the electric and the magnetic field intrinsically intertwined. The relativity principle inevitably leads to a close relation between electric and magnetic forces.

We conclude this brief discussion with a caveat that the results we have described are only approximate. When speeds are comparable to the speed of light, the discussion needs refinement. We'll discuss this further in Chapter 39.

Magnetic Fields in Outer Space

Magnetic fields exist in outer space. Throughout our galaxy, the magnetic field strength is in the range of 10^{-10} T (although the fields also have a great deal of structure). Charged particles (*cosmic rays*) are generated and accelerated by various stellar processes. If their momentum is less than a certain critical value p_c, they drift in gigantic circles within the galaxy due to the magnetic forces on them.[†] Cosmic rays with a momentum greater than p_c move on a circle with a radius of curvature greater than the galaxy's radius, and they therefore escape the galaxy. To estimate p_c for a cosmic ray whose charge has the magnitude e of the electron charge, we use the observation that the radius of the galaxy is about 5×10^{21} m. From Eq. (28–5), the critical momentum has magnitude

$$p_c = eBR = (1.6 \times 10^{-19} \text{ C})(10^{-10} \text{ T})(5 \times 10^{21} \text{ m}) = 8 \times 10^{-8} \text{ kg} \cdot \text{m/s}.$$

For a particle such as an electron or a proton, this momentum is enormously large. For comparison, an electron in the beam of a television picture tube typically has a momentum of 10^{-22} kg·m/s, while protons in the huge Fermilab proton accelerator attain momenta of 5×10^{-16} kg·m/s. Because cosmic rays with a momentum greater than p_c leave the galaxy, we should expect to detect more cosmic rays that strike Earth with momenta lower than p_c than with momenta greater than p_c. Experimental observations of particles arriving from outer space help us to estimate the value of the interstellar magnetic field.

[†]We use momentum rather than speed because a calculation of the speed here gives a critical speed greater than the speed of light. This indicates that special relativity is necessary, and special relativity shows that momentum should be used here.

THINK ABOUT THIS...
HOW DOES EARTH'S MAGNETIC FIELD CREATE THE NORTHERN LIGHTS?

Earth has a magnetic field like that of a huge bar magnet, directed from the geographic South Pole to the geographic North Pole. As Fig. 28–17 shows, the lines get denser near the poles. (The field extends down into the atmosphere, but ions within the atmosphere tend to be affected more by the presence of the atmosphere than by Earth's magnetic field.) The field interacts with charged particles, one evidence of which is the aurora, also known in the northern hemisphere as the northern lights. Charged elementary particles reach Earth from two sources. One is the *solar wind*, which consists of electrons and positive ions ejected from the sun. Most of these particles are of relatively low energy, and in the vicinity of Earth they are trapped by the magnetic field, moving in spiral paths along Earth's magnetic field lines. The helical orbits of particles around the field lines get flattened as the lines approach each other near the pole, and ultimately the particles are turned around. The mechanism for the reversal is known as a **magnetic mirror** (the process has to do with magnetic fields of varying strength; we'll look at the dynamics of this situation in Chapter 29.). In this way, the trapped charged particles, mainly electrons, bounce back and forth between the poles creating the *outer Van Allen radiation belt*. Particles also reach Earth as **cosmic rays**, consisting of very energetic particles coming from distant space. Some of these lose energy by collisions with atmospheric particles, and if they are charged, they get trapped along magnetic field lines closer to Earth than the outer Van Allen belt. Electrons and protons originating with cosmic rays have spiral paths wide enough that they touch the atmosphere and, because of collisions with atmospheric molecules, don't stay around very long. The neutral particles, *neutrons*, which are produced in the high-energy collisions of the cosmic rays with the atmosphere, will eventually decay into protons, electrons, and neutrinos. The protons that come from some of these neutron decays are captured by the magnetic field lines lying relatively close to Earth, and they also accumulate slowly to form the stable *inner Van Allen radiation belt*. The outer belt averages about 15,000 km from the surface of Earth, the inner one about 3000 km, although both approach Earth more closely in the polar regions.

An aurora occurs (see photo on page 791) when a sufficient number of charged particles from the belts (or directly from the solar wind) enter the atmosphere and excite the molecules in the air through collisions. When the excited atoms decay back to their normal configuration, they emit radiation with wavelength (i.e. color) that is characteristic of the state of excitation (see Chapter 41). The auroras are more noticeable near the poles, where the magnetic field lines, and therefore the Van Allen belts, dip toward Earth.

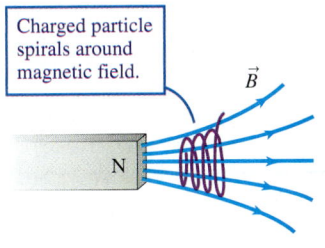

(a)

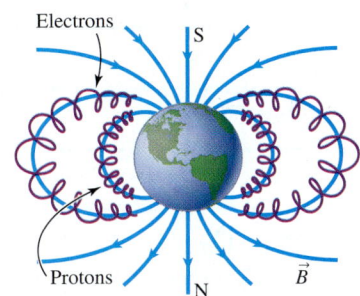

(b)

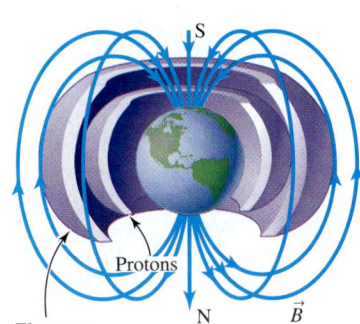

(c)

▶ **FIGURE 28–17** (a) Charged particles spiraling around the varying magnetic field near the pole of a magnet. The pitch, or inclination, of the helix decreases to zero as particles near the poles, where the field lines are denser, and the particles reverse the direction of their helical path. (b) When this is applied to particles spiraling around Earth's magnetic field, we see that the particles are trapped, bouncing back and forth between the polar regions. The trapped particles form the two Van Allen belts, one for electrons and one for protons. (c) A three-dimensional view of the magnetic field lines and of the belts.

EXAMPLE 28–6 Assume that a proton of speed 1.5×10^7 m/s approaches Earth at an angle of 40° to Earth's magnetic field lines and is captured in the lower Van Allen belt (at a mean altitude of 3000 km) without a change in speed. If the mean strength of the field at this altitude is 10^{-5} T, find the cyclotron frequency and the radius of curvature for the circular part of the proton's helical motion.

Setting It Up Figure 28–18 shows the path of the proton. The curvature is the radius of the circular part of the motion. Given the angle of approach of 40° and the speed, we know the speed perpendicular to the magnetic field. We also know the field strength B and therefore have all the data necessary for the calculation.

Strategy The cyclotron frequency, Eq. (28–7), depends only on B and on the proton mass $m_p = 1.67 \times 10^{-27}$ kg and charge $e = +1.6 \times 10^{-19}$ C. For the calculation of the curvature of the proton's motion, we note that the proton does not change speed when it starts its helical motion along the magnetic field line, whose B value is given. We can use Eq. (28–8). Given a speed v, the component of velocity that is perpendicular to the direction of B is $v \sin 40°$, and it is this quantity that appears in Eq. (28–8).

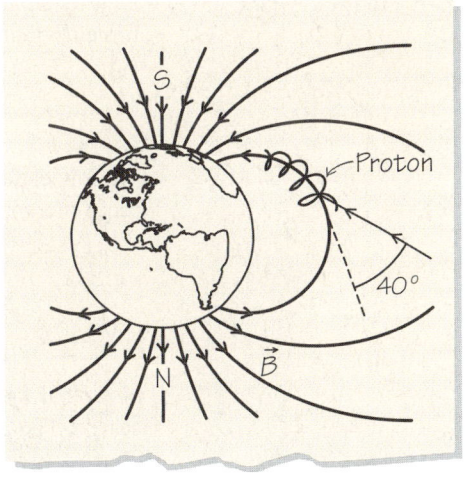

▲ **FIGURE 28–18** A proton approaches and passes into Earth's magnetic field.

(continues on next page)

Working It Out The cyclotron frequency is

$$f = \frac{qB}{2\pi m} = \frac{(1.6 \times 10^{-19} \text{ C})(10^{-5} \text{ T})}{2\pi (1.67 \times 10^{-27} \text{ kg})} = 150 \text{ Hz}.$$

The proton's velocity component perpendicular to the magnetic field is $v_\perp = v \sin 40° = (1.5 \times 10^7 \text{ m/s})(0.64) = 9.6 \times 10^6 \text{ m/s}$. Using Eq. (28–8), we then find that

$$R = \frac{mv_\perp}{qB} = \frac{(1.67 \times 10^{-27} \text{ kg})(9.6 \times 10^6 \text{ m/s})}{(1.60 \times 10^{-19} \text{ C})(10^{-5} \text{ T})} = 10^4 \text{ m} = 10 \text{ km}$$

The radius of curvature is much less than the altitude of the Van Allen belt. Since it is collisions with the atmosphere that tend to be most effective at removing particles from the belt, this proton is likely to stay within the belt for a while.

28–4 Magnetic Forces on Currents

In the previous sections, we have seen that moving charges in a magnetic field may experience a force due to the field. Because electric currents in wires consist of moving charges, we can expect that a magnetic field will exert a force on the charges in a current-carrying wire, and thus on the wire itself (Fig. 28–19), and experiment shows this is the case.

A wire contains moving charges throughout its length, and a magnetic field may vary significantly along its length. The total force on a current-carrying wire is the vector sum of the magnetic forces on all of the moving charges within it. To find the total force, we first determine the force on a small segment of a current-carrying wire, and subsequently sum (integrate) the infinitesimal force on each segment.

Magnetic Forces on Infinitesimal Wires with Currents

Let's denote the small segment of a thin current-carrying wire by $d\vec{\ell}$. It has both an infinitesimal magnitude $d\ell$ and a direction along the instantaneous current carried by the wire at the location of the segment. If the moving charge dq contained in the segment has velocity $\vec{v}$ along the wire (Fig. 28–19), its displacement $d\vec{\ell}$ in time dt is $d\vec{\ell} = \vec{v} \, dt$ so

$$\vec{v} = \frac{d\vec{\ell}}{dt}. \tag{28–12}$$

Because the current, I, is dq/dt by definition, the amount of moving charge within the segment is

$$dq = I \, dt. \tag{28–13}$$

Note that the magnetic field will be uniform over the length of the segment if the segment is small enough. With Eqs. (28–12) and (28–13) we can calculate the magnetic force $d\vec{F}$ that acts on our charge element dq and hence on the wire element:

$$d\vec{F} = dq(\vec{v} \times \vec{B}) = (I \, dt)\left(\frac{d\vec{\ell}}{dt} \times \vec{B}\right).$$

We cancel the factor dt to find the infinitesimal force on a wire element $d\vec{\ell}$ carrying current I in a magnetic field $\vec{B}$:

$$d\vec{F} = I \, d\vec{\ell} \times \vec{B}. \tag{28–14}$$

MAGNETIC FORCE ON WIRE SEGMENT

Note that the current is the same everywhere along the wire because current is conserved. The magnitude of the magnetic force $d\vec{F}$ is given by

$$dF = I(d\ell)B \sin \theta, \tag{28–15}$$

where θ is the angle between the direction of the wire segment (the current's direction) and the direction of the magnetic field.

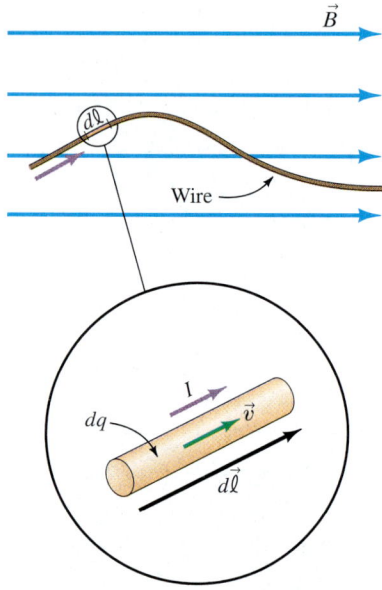

▲ **FIGURE 28–19** A wire carrying a current in a magnetic field. We isolate a segment $d\ell$ of the wire that contains the moving charge dq.

CONCEPTUAL EXAMPLE 28–7
What is the direction of the force on a current-carrying wire (a) perpendicular to and (b) parallel to the direction of a constant magnetic field?

Answer We sketch the three different placements of a segment of current-carrying wire in a uniform magnetic field that points in the $+x$-direction in Fig. 28–20. In each case, the direction of the force on the wire segment is given by the right-hand rule and Eq. (28–14). In Fig. 28–20a, $d\vec{\ell}$ points in the $+y$-direction, so that the force points in the $-z$-direction. In Fig. 28–20b, $d\vec{\ell}$ points in the $+z$-direction, so the force points in the $+y$-direction. In Fig. 28–20c, $d\vec{\ell}$ points in the $+x$-direction, parallel to $\vec{B}$, and since the vector product of two parallel vectors is zero, there is no force.

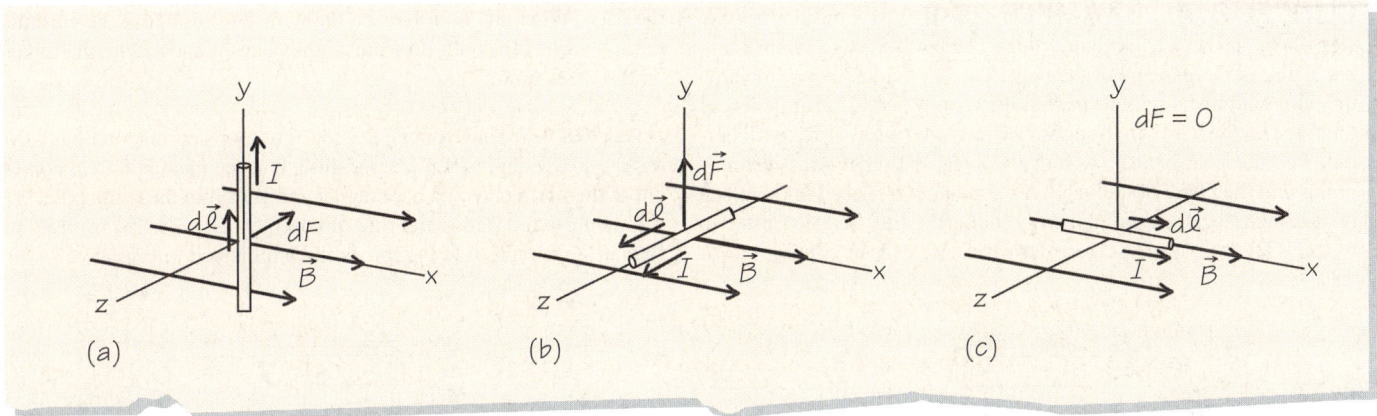

▲ **FIGURE 28–20** A wire segment, aligned with the (a) y-axis, (b) z-axis, and (c) x-axis, in a magnetic field aligned along the $+x$-direction, with the infinitesimal forces $d\vec{F}$ that act on the segment indicated.

Magnetic Forces on Finite Wires with Currents

The *net* force $\vec{F}$ on a *finite* section of wire is the vectorial sum of the forces on the various infinitesimal segments that make up the wire. We find the net force by integrating $d\vec{F}$ [Eq. (28–14)] over the total length of the wire. Because I is the same everywhere along the wire, the summation of Eq. (28–14) has the form

$$\vec{F}_B = I \int (d\vec{\ell} \times \vec{B}). \qquad (28\text{–}16)$$

Whether the integral is easy to perform or not depends on the particular situation. A straight wire within a constant magnetic field represents an important special case. Let's suppose first that the wire, with length L and current I, is oriented perpendicular to the field (Fig. 28–21a). To perform the integration of Eq. (28–16), notice that each segment $d\vec{\ell}$ of the wire points in the y-direction, $d\vec{\ell} = d\ell\,\hat{j}$, and $\vec{B}$ is constant at each segment, $\vec{B} = B\hat{i}$. The infinitesimal force on each segment is therefore identical:

$$d\vec{F} = I[(d\ell\,\hat{j}) \times (B\hat{i})] = I(d\ell)B(\hat{j} \times \hat{i}) = I(d\ell)B(-\hat{k}),$$

which is a vector that points in the $-z$-direction. We can easily check the direction with the right-hand rule. The net force $\vec{F}$ is an integral over $d\vec{F}$:

$$\vec{F} = \int d\vec{F} = I \int d\ell B(-\hat{k}) = IB(-\hat{k}) \int_0^L d\ell = -IBL\hat{k}. \qquad (28\text{–}17)$$

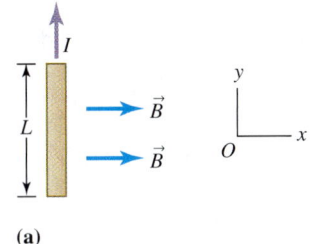

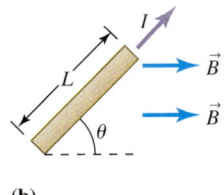

▲ **FIGURE 28–21** (a) A wire segment is oriented perpendicularly to a magnetic field. (b) The same wire segment at an angle to the field.

The net force points in the $-z$-direction (here, into the paper), and its magnitude is IBL.

Next, suppose that the wire makes an angle θ to the field (Fig. 28–21b). The only change we must make is in the expression for the magnitude of the infinitesimal force element: There is an additional factor $\sin\theta$ in dF [Eq. (28–15)]. The direction of the force remains in the $-z$-direction by the right-hand rule. Because θ is constant, it does not enter into the integration, which thus remains as in Eq. (28–17). The result is identical to Eq. (28–17) with an additional factor of $\sin\theta$:

$$\vec{F} = -IBL\sin\theta\,\hat{k}. \qquad (28\text{–}18)$$

Equations (28–17) and (28–18) are useful and important results. They can be combined into one vectorial equation that gives the force on a thin, straight wire of length L in a magnetic field, namely,

$$\vec{F} = I\vec{L} \times \vec{B}, \quad (28\text{–}19)$$

where the vector $\vec{L}$ is oriented along the wire in the direction of the current.

CONCEPTUAL EXAMPLE 28–8 (a) Consider a wire carrying a current I coming out of the plane of the page. A bar magnet is placed in the plane of the page. The bar magnet points towards the wire with the N pole nearest to the wire. What is the direction of the force on the wire? (b) Suppose we have two wires—one with current I coming out of the plane of the page, the other with current I going into the plane of the page. A bar magnet is placed in the plane of the page, located symmetrically along the line between the two wires, with the N pole nearest the wire with current coming out of the page. What are the forces on the wires? (c) Suppose the currents in the wires of (b) both flow out of the page. What will be the forces on the wires now?

Answer We have drawn the configurations corresponding to the different cases in Fig. 28–22, recalling that the magnetic field comes out of the north pole of a bar magnet and goes into the south pole. We can then use the right-hand rule imposed by Eq. (28–19) in straightforward fashion to find the forces, each indicated in Fig. 28–23.

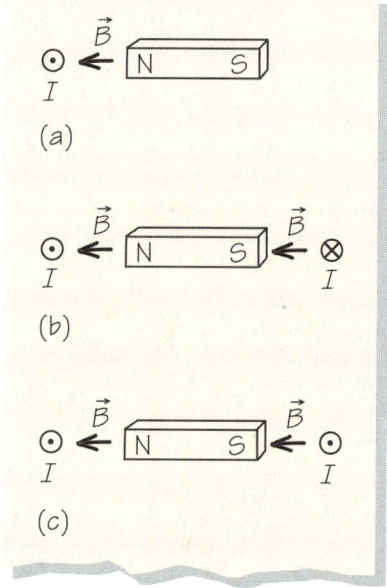

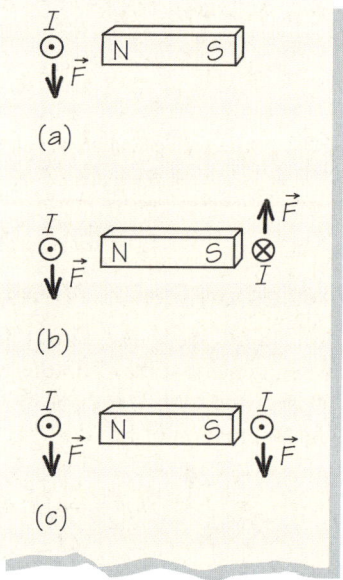

▲ **FIGURE 28–22** Magnet positions for each of the configurations described.

▲ **FIGURE 28–23** Forces for each of the configurations described.

EXAMPLE 28–9 A 12-cm-long straight segment of wire carrying a current of 7.2 A is maneuvered entirely within a region known to contain a constant magnetic field until the force on the wire has a maximum magnitude of 0.37 N. Find the magnitude of the magnetic field.

Strategy Figure 28–24 shows the orientation of a wire relative to the direction of the magnetic field in the plane which contains both the field and the direction of the wire. Since there is no force on the component of the wire parallel to the field, the maximum force occurs when the wire and the magnetic field are perpendicular to each other. In the relation $F = IBL \sin\theta$ [Eq. (28–18)] that expresses the magnitude of the force on the wire, the maximum occurs for $\theta = \pi/2$ (90°). The magnetic field can be calculated from this relation.

Working It Out For $\theta = \pi/2$, we have $F = ILB$, and hence

$$B = \frac{F}{IL} = \frac{0.37 \text{ N}}{(7.2 \text{ A})(12 \times 10^{-2} \text{ m})} = 0.43 \text{ T}.$$

The measurement of forces on wires represents an important tool for the accurate measurement of magnetic fields.

What Do You Think? What is the direction of the magnetic force on the wire?

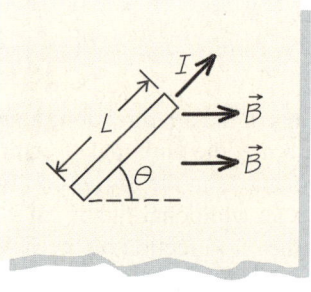

▲ **FIGURE 28–24**

The expression for the magnetic force on an isolated moving charge has led us directly to the expression for the magnetic force on a current-carrying wire. Historically, the order of discovery was just the reverse: Oersted, François Arago, and Ampère, who early in the nineteenth century performed the first quantitative experiments on magnetic forces, observed those forces on current-carrying wires. Their results then led to an understanding of magnetic forces on moving charges.

28–5 Magnetic Force on Current Loops

Magnetic fields exert forces on all kinds of current-carrying wires, including those of closed loops. As we shall see, a uniform magnetic field actually exerts only a torque on a current loop. This phenomenon provides the torque that runs direct-current electric motors and the galvanometer (the device cited in Chapter 27 for use in ammeters and voltmeters).

Figure 28–25a shows a stiff rectangular loop of wire carrying current I in the presence of a constant magnetic field along the $+x$-direction. The rectangular wire loop has sides of length a and b (denoted 1, 2, 3, and 4 in Fig. 28–25), and can be thought of as a series of straight wire segments. Figure 28–25b is a side view of the apparatus along the $+y$-direction. A perpendicular to the plane of the loop (the direction of the thumb when the fingers of the right hand follow the current direction) makes an angle ψ with the magnetic field. Fig. 28–25b also shows the angle θ between the direction of the magnetic field and the direction of the current for leg 1. We can calculate the force on each leg of the loop by using Eq. (28–19), and we find:

$$F_1 = IaB \sin\theta, \text{ in the } -y\text{-direction;} \qquad (28\text{–}20\text{a})$$

$$F_2 = IaB, \text{ in the } -z\text{-direction;} \qquad (28\text{–}20\text{b})$$

$$F_3 = IaB \sin\theta, \text{ in the } +y\text{-direction;} \qquad (28\text{–}20\text{c})$$

$$F_4 = IaB, \text{ in the } +z\text{-direction.} \qquad (28\text{–}20\text{d})$$

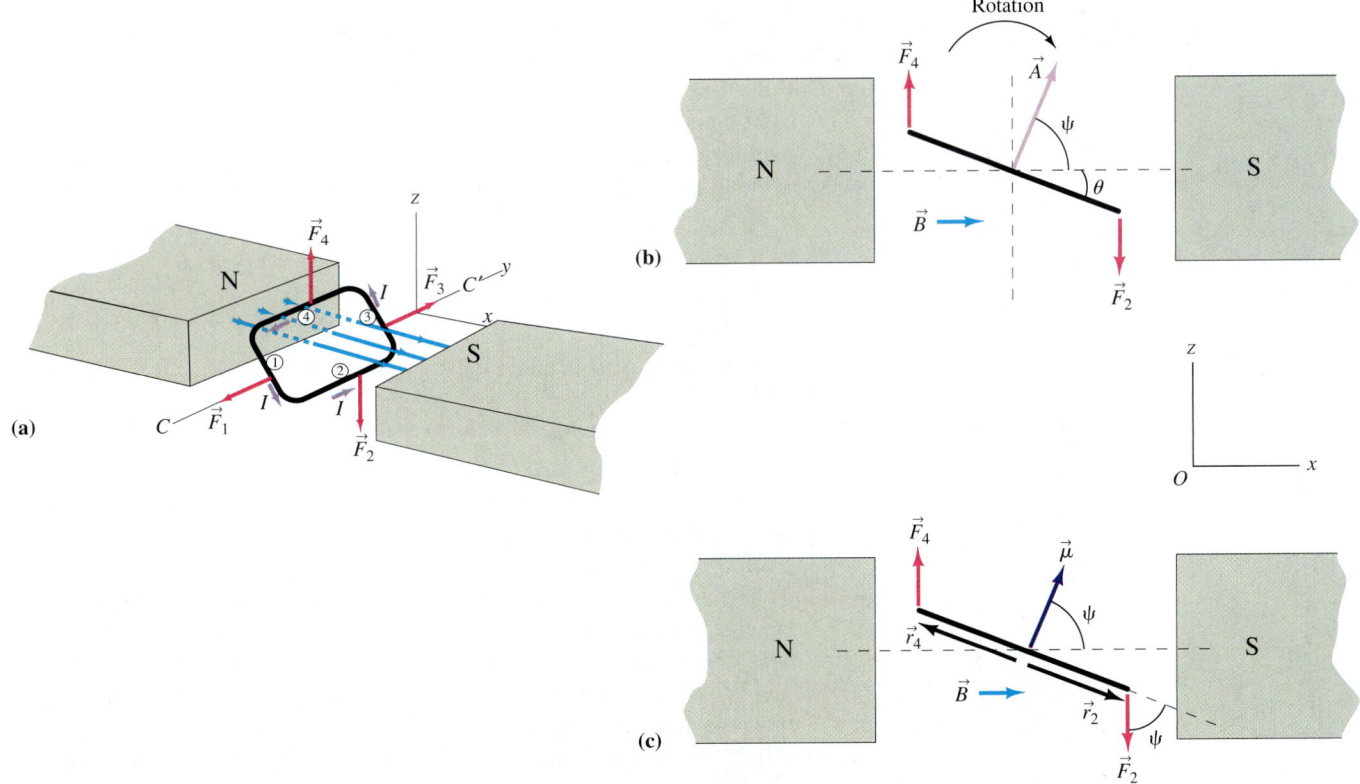

▲ **FIGURE 28–25** (a) A stiff, rectangular loop of wire is placed in a constant magnetic field. (b) A side view, looking along the y-axis, of the loop. (c) Geometry of the loop that allows us to calculate the torque on the loop. The torque tends to align the vector $\vec{\mu}$ with the magnetic field, $\vec{B}$.

These forces are indicated in Figs. 28–25a and 28–25b. Forces $\vec{F}_1$ and $\vec{F}_3$ are equal and opposite, as are forces $\vec{F}_2$ and $\vec{F}_4$, so *there is no net force on a current loop in a uniform magnetic field.* However, there is an important difference between these two sets of forces: $\vec{F}_1$ and $\vec{F}_3$ act along the same axis (CC' in Fig. 28–25a) and exert no torque on the loop. As Fig. 28–25b shows, $\vec{F}_2$ and $\vec{F}_4$ act along different axes and therefore produce a torque that causes the wire loop to rotate clockwise in the magnetic field. When the wire has rotated into the yz-plane (when $\theta = 90°$ in Figure 28–25b), $\vec{F}_2$ and $\vec{F}_4$ act along the same axis, and there is no torque. When the loop is in the xy-plane ($\theta = 0°$), the torque is a maximum. Finally, when θ changes sign, so does the torque, and the loop will tend to rotate counterclockwise.

We can find the torque about the central axis CC' in Fig. 28–25a by using the results of Chapter 10. From Eq. (10–6), the net torque about this axis is

$$\vec{\tau} = (\vec{r}_2 \times \vec{F}_2) + (\vec{r}_4 \times \vec{F}_4),$$

where $\vec{r}_2$ and $\vec{r}_4$ are the perpendicular vectors from axis CC' to legs 2 and 4, respectively (Fig. 28–25c). Both $\vec{r}_2$ and $\vec{r}_4$ have magnitude $a/2$. Figure 28–25c shows that ψ is the angle between $\vec{r}_2$ and $\vec{F}_2$ and between $\vec{r}_4$ and $\vec{F}_4$. The torque has magnitude

$$\tau = r_2 F_2 \sin \psi + r_4 F_4 \sin \psi = (a/2)(IbB) \sin \psi + (a/2)(IbB) \sin \psi$$
$$= IabB \sin \psi. \qquad (28\text{--}21)$$

Here, we have used the values for F_2 and F_4 given by Eq. (28–20). According to the right-hand rule, both terms in the equation for τ point in the $+y$-direction, so the net torque is in this direction.

The torque on a current loop in a magnetic field as given by Eq. (28–21) can be summarized and generalized. For a rectangular loop, the factor ab is the area A of the current loop. We can generalize to *any* planar loop of area A, whatever its shape, by the calculus technique of decomposing a planar loop of any shape into tiny rectangles and then treating each small rectangle in the way we have discussed above. We can neatly handle the vectorial nature of the torque by defining a vector $\vec{\mu}$ perpendicular to the plane of the loop. There are two possibilities for the direction of a vector perpendicular to any plane—which do we choose for $\vec{\mu}$? We choose the direction of $\vec{\mu}$ with a right-hand rule: Curl the fingers of the right hand in the direction of the current around the loop, and the right thumb gives the direction of $\vec{\mu}$ (Fig. 28–26). Try it for Fig. 28–25c, where we have indicated $\vec{\mu}$. The angle between $\vec{\mu}$ and $\vec{B}$ is ψ. We have thus shown that the direction and magnitude of the *torque on a current loop* are given by

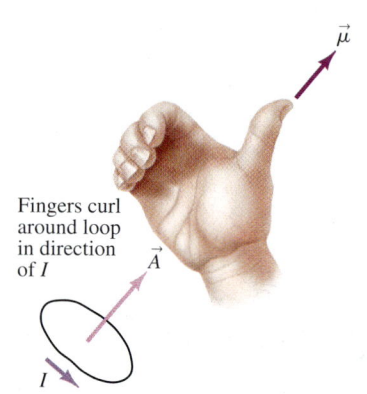

Fingers curl around loop in direction of I

▲ **FIGURE 28–26** A right-hand rule indicates the direction of the magnetic dipole moment, $\vec{\mu}$, of a current loop.

$$\vec{\tau} = \vec{\mu} \times \vec{B}, \qquad (28\text{--}22)$$

TORQUE ON A CURRENT LOOP

provided that the magnitude of $\vec{\mu}$ is taken to be

$$\mu = IA. \qquad (28\text{--}23)$$

From these equations, the magnitude of the torque is $\tau = \mu B \sin \psi$, exactly as in Eq. (28–21), and the direction of the torque is the $+y$-direction. A torque in this direction acts to align $\vec{\mu}$ and $\vec{B}$. It is generally true that the torque tends to rotate a current loop or coil in such a way that $\vec{\mu}$ and $\vec{B}$ become aligned.

The torque on a current loop in a uniform magnetic field is completely analogous to the torque on an electric dipole (a pair of equal and opposite electric charges) in a uniform electric field, discussed in Chapter 21. The electric dipole was described by the *electric dipole moment* $\vec{p}$, a vector aligned with the two charges and equal in magnitude to the charge times the distance of charge separation. In terms of the response to an external magnetic field, the current loop is in all measurable respects a **magnetic dipole**. We therefore call $\vec{\mu}$, which plays a role analogous to $\vec{p}$, the **magnetic dipole moment** of the loop. One of the measurable properties—the defining property—of a dipole, magnetic or electric, is the characteristic field that the dipole itself *produces*. We deal with the calculation of the magnetic dipole field in Chapter 29.

CONCEPTUAL EXAMPLE 28–10 Suppose instead of one turn of wire we take N turns of wire, each surrounding the same plane area, thus creating a *coil*. What is the magnetic moment of the coil?

Answer Each turn of the coil experiences the forces we have described, and the torque is multiplied by N. This factor is included with the other factors intrinsic to the coil, so the magnetic dipole moment will now be multiplied by the factor N:

$$\mu = INA. \tag{28-24}$$

MAGNETIC DIPOLE MOMENT OF A COIL

We have seen another system that aligns itself with external magnetic fields: the iron filings we used to make our preliminary definition of magnetic fields. These iron filings behave like little bar magnets that are rotated by magnetic fields. Bar magnets react to fields just like current loops do. As we shall see in Chapter 29 when we find the magnetic fields *produced* by magnetic dipoles, bar magnets are themselves magnetic dipoles. We can explain this by looking at the microscopic level, where we find that metals contain the quantum-mechanical equivalent of circulating currents. We shall study this behavior in more detail in Chapter 31.

Galvanometers

In Chapter 27 we encountered the *galvanometer*—a device that measures currents. The principle behind a galvanometer is the fact that a magnetic field exerts a torque on a current loop. For example, we can attach a spring to a loop to balance the torque due to a known magnetic field, and the amount the spring stretches is a measure of the torque on the loop and hence of the current that passes through it.

THINK ABOUT THIS...
HOW DO ELECTRIC MOTORS WORK?

In one fundamental type of electric motor, we have a permanent magnet to supply a magnetic field and an emf to supply a current. The current is run through a loop that is placed in the magnetic field. With the apparatus shown in Fig. 28–25a, as soon as the wire loop rotates past the position in which it is aligned with the field due to the poles, the torque on it changes sign and becomes counterclockwise. In fact, the torque changes direction when ψ goes through 0° or 180°. This device oscillates and would not make a motor, which requires a continuous turning motion. However, if we can make the *current* switch directions every time the loop passes $\psi = 0$ or π (180°), then the torque will continuously produce a clockwise rotation. A device that does just this is known as a *split-ring commutator*, a device whose mechanism is visible in Fig. 28–27. With this device, the loop accelerates its rotation, always in the same direction, under the influence of a torque whose sign does not change, and an *electric motor* has been created. Once we have this arrangement, and a continuously turning loop, it is no great step to imagine tranferring this motion to, for example, a set of wheels. But that is another story more properly treated in a chapter on mechanics.

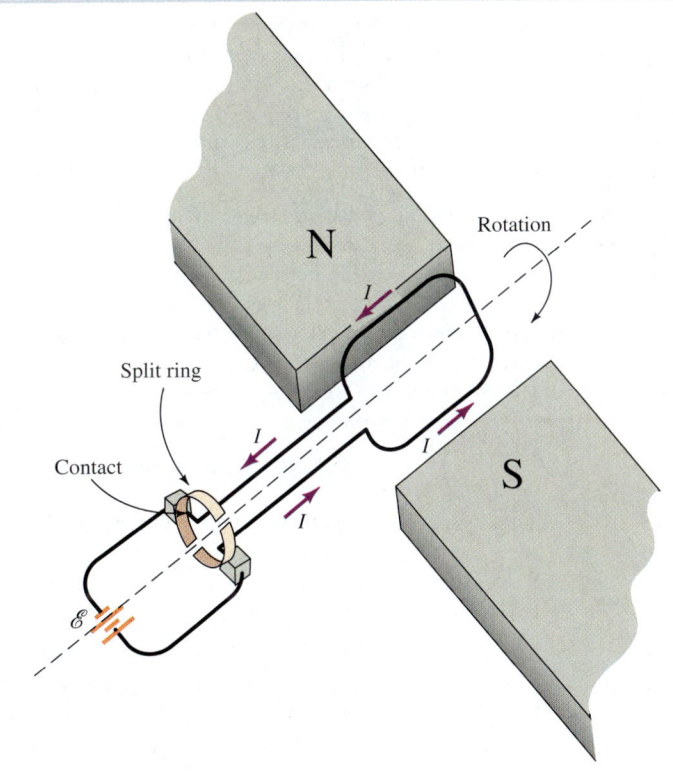

▲ **FIGURE 28–27** A current-carrying loop aligned in a magnetic field is fitted with a split-ring commutator, as shown in a schematic diagram. The torque on the loop serves to turn the loop and makes a motor.

Energy and the Torque on Loops

When a magnetic field rotates a current loop, the field does work. For a constant field, the only variable in the work is the angle of rotation, ψ. We know from Chapter 7 that we can use the concept of potential energy when the force (or the torque) depends only on position, as here. Accordingly we associate a potential energy $U(\psi)$ with a loop in a magnetic field, where ψ is the angle between $\vec{\mu}$ and $\vec{B}$. As always, only *changes* in the potential energy have physical consequences. The change in potential energy in rotating the coil from some initial angle ψ to a final angle of $\pi/2$ (90°) is given by the negative of the work done by the magnetic field in moving the coil through these angles:

$$U(\psi) - U(\pi/2) = -\int_{\psi}^{90°} \tau \, d\psi' = -\int_{\psi}^{90°} \mu B \sin \psi' \, d\psi'$$

$$= -\mu B \int_{\psi}^{90°} \sin \psi' \, d\psi' = \mu B \cos(\pi/2) - \mu B \cos \psi.$$

The cosine of $\pi/2$ is zero, so

$$U(\psi) - U(\pi/2) = -\mu B \cos \psi. \tag{28-25}$$

We choose the location of $U = 0$ for convenience, and it is customary to choose U to be zero at $\psi = \pi/2$; that is, when $\vec{\mu}$ is perpendicular to $\vec{B}$. Setting the term $U(\pi/2)$ in Eq. (28–25) to zero gives *the potential energy of a current loop with a given magnetic dipole moment $\vec{\mu}$ in a constant magnetic field $\vec{B}$*:

$$U(\psi) = -\vec{\mu} \cdot \vec{B}. \tag{28-26}$$

POTENTIAL ENERGY OF A MAGNETIC DIPOLE

The potential energy has a *minimum* when $\vec{\mu}$ is aligned along $\vec{B}$ (that is, when $\psi = 0$). Thus the orientation in which $\vec{\mu}$ is aligned with $\vec{B}$ is a stable equilibrium point. This agrees with our earlier result that the torque tends to rotate the loop to line up $\vec{\mu}$ and $\vec{B}$.

CONCEPTUAL EXAMPLE 28–11 A current loop in a constant magnetic field is initially aligned so that $\vec{\mu}$ points in a direction slightly different from that of $\vec{B}$, and the loop is then released. There is no mechanism such as friction for energy loss—that is, no damping. What is the subsequent motion of the loop?

Answer At the (stable) equilibrium point, $\vec{\mu}$ is aligned with the magnetic field. A small deviation is measured by an angle ψ. The subsequent motion will be determined by the potential energy expressed in terms of ψ for small ψ. This potential has a minimum around the equilibrium, and we can express it for small deviations from the minimum. More precisely, the potential energy is $U(\psi) = -\mu B \cos \psi$, with a minimum $-\mu B$ for $\psi = 0$, and it rises as ψ moves away from the minimum value, independent of the sign of ψ. Any such curve varies as $C\psi^2/2$ for ψ close enough to $\psi = 0$, where C is a constant that depends on the potential energy. This, however, is just the potential energy for a harmonic oscillator, and we may therefore expect that the motion will be simple harmonic, of the general form $\psi = \psi_m \sin(\omega t + \delta)$, where ω involves the coefficient C as well as the rotational inertia of the loop (see Problem 64). In Chapter 13 we saw that this is a general property of motion around stable equilibrium points.

What Do You Think? What is the behavior of an undamped compass needle slightly displaced from the direction of Earth's magnetic field?

28–6 The Hall Effect

The direction of a current does *not* itself determine the sign of the charge carriers in that current because a current to the right can be produced by the movement either of positive charges to the right or of negative charges to the left. The Hall effect allows us to find this sign, that is, whether a current is formed of moving electrons or of moving positive charges such as holes in semiconductors (see Section 26–6).

The Hall effect results from the fact that charges moving along a wire in a magnetic field experience a force whose sign depends on their charge. Consider a metal strip of length L along which a current flows. The strip is placed in a uniform magnetic field that is perpendicular to the strip, as in Fig. 28–28. As a result, there will be a potential difference between points a and b whose cause and qualitative nature we now describe. Equation (28–19) gives the total force on the strip, $\vec{F} = I\vec{L} \times \vec{B}$. This force, by the right-hand rule, is directed in the $-x$-direction—it acts to the *left* in Fig. 28–28. By using the equivalent force law $\vec{F} = q\vec{v} \times \vec{B}$, we can show that the force on the charge carriers acts to the left, *whatever the sign of the charge carriers*. If the moving charges are positive (ions) and the current flows in the $+y$-direction, then the velocity of the charges is also in the $+y$-direction. According to the right-hand rule, $\vec{F} = q\vec{v} \times \vec{B}$ is then directed toward point a in Fig. 28–28. If the moving charges are negative (electrons), however, then the velocity of the charges is in the $-y$-direction. The vector product $\vec{v} \times \vec{B}$ is directed to the right, but q is negative, and $q\vec{v} \times \vec{B}$ again points to the left, moving these negative charges toward a. Either way, there is a buildup of the charge carriers at the left side of the strip. This buildup cannot continue indefinitely: Once enough charge carriers have moved to the left, they will supply a repulsive Coulomb force against the movement of other charge carriers that is large enough to stop the process. An equilibrium is established in which an electric potential is set up between points a and b that prevents further leftward drift of charge carriers. Charges then move up the strip as they would if there were no magnetic field. (The charge separation has led to an electric field between points a and b. We have crossed $\vec{E}$ and $\vec{B}$ fields, with the value of the electric field such that the charges travel undeflected up the strip just as in the velocity selector discussed in Section 28–3.)

The **Hall effect** is that *the sign of the potential difference between points a and b determines the sign of the charge carriers*. If the charge carriers are negative, negative charges build up on the left side of the metal strip, and point a is at a lower potential than point b. Conversely, if the carriers are positive, positive charges build up on the left side of the strip, and point a is at a higher potential than point b. The first measurement of the sign of this *Hall potential* was performed by the American physicist Edwin H. Hall in 1879. His measurement proved that the carriers of current in metals are negatively charged.

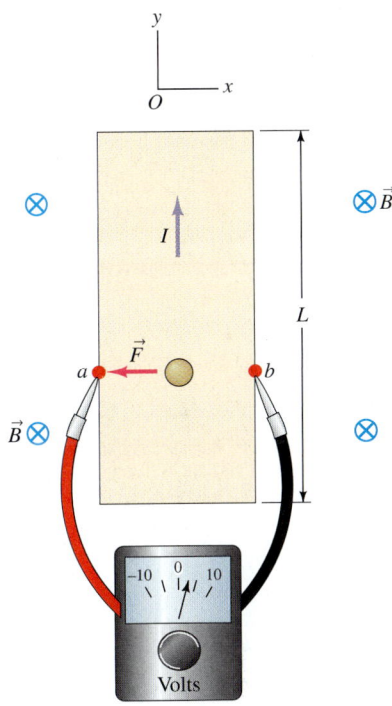

▲ **FIGURE 28–28** A conducting strip perpendicular to a constant magnetic field develops a potential called the Hall voltage between points a and b when the strip carries a current I.

THINK ABOUT THIS...
DOES THE HALL EFFECT HAVE TECHNOLOGICAL USE?

Our discussion of the Hall effect shows that the potential across the strip is proportional to the strength of the magnetic field. Potential changes are easy to measure accurately, and in a Hall strip they signal changes in the magnetic field. In particular, they can be used to sense the change in distance of a magnet from the Hall strip. This makes the Hall effect useful in the detection of the motion of magnets, even slow motion.

An example is in the computer keyboard. Constant and heavy use would wear out the keyboard rather rapidly if the use involved establishing direct contact between two solid pieces. The Hall effect is used to signal the application of force to a key on a keyboard without this type of contact. A small, permanent magnet is attached to the bottom of each key on the keyboard, and underneath it there is a Hall probe. Modern probes consist of a thin layer of a conductor deposited on a rigid substrate. Because of the spatial dependence of the magnetic field, the change in the position of the magnet relative to the conducting layer can change the magnetic field at the probe by an order of magnitude. Such a change leads to changes in a Hall potential that is then communicated to the computer, signaling that a given key has been pressed. In an electronic piano this application can be carried further. The rate of change in the Hall potential signals the speed at which a key is depressed, so that the keyboard can signal whether the key has been pressed rapidly or gently, and produce a loud or soft sound in response.

EXAMPLE 28–12 Figure 28–28 shows a Hall strip whose width is 1.0 cm in a magnetic field of 2.0 T with a voltage of magnitude 7.2 μV across the strip. What is the speed v of the charge carriers in the strip?

Strategy The velocity of the charge carriers is that of charges passing through a crossed electric and magnetic field. We are given the magnetic field B and can find the electric field in terms of the strip width d and the voltage V across the strip. The charge flow is that of a velocity selector, with $v = E/B$. The electric field is determined from the potential difference and the strip width to be $E = V/d$.

Working It Out Combining the equations above, the velocity is
$$v = \frac{V}{Bd} = \frac{7.2 \times 10^{-6} \text{ V}}{(2.0 \text{ T})(1.0 \times 10^{-2} \text{ m})} = 3.6 \times 10^{-4} \text{ m/s}.$$

This is a measurement of an electron's speed in a current, what we earlier called the drift speed.

What Do You Think? The width of the strip is doubled. Which of the following is true: the voltage (a) is doubled, (b) is halved, (c) remains the same?

Summary

Magnets, moving electric charges, and electric currents all experience magnetic forces. These forces can be described in terms of a magnetic field, $\vec{B}$, whose spatial dependence can be mapped out with iron filings or by observing its effect on a moving electric test charge or a test current element. In terms of this field, the magnetic force on an electric charge q depends on the charge's velocity according to the magnetic force law,

$$\vec{F} = q\vec{v} \times \vec{B}. \qquad (28\text{–}1)$$

The SI unit of magnetic field is the tesla, T: 1 T = 1 kg/(C·s). When both magnetic and electric fields are present, the Lorentz force law holds:

$$\vec{F} = q(\vec{E} + \vec{v} \times \vec{B}). \quad (28\text{-}4)$$

In a static magnetic field, the component of a charged particle's velocity parallel to the field is unaffected by that field. The magnitude of the force on the particle due to the field is proportional to the component of the velocity perpendicular to the field, and the direction of the force is perpendicular to this component of the velocity and to the field itself. It follows that the kinetic energy of a charged particle in a magnetic field is unchanging. When the field is constant, a charged particle traveling perpendicular to the field moves in a circle of radius

$$R = \frac{mv}{qB}. \quad (28\text{-}5)$$

The frequency of the particle's circular motion is the cyclotron frequency,

$$f = \frac{qB}{2\pi m}, \quad (28\text{-}7)$$

which is independent of the particle's velocity. The general path followed by a moving charge is a spiral around the magnetic field lines.

When a charged particle has a particular velocity perpendicular to constant crossed electric and magnetic fields, the electric and magnetic forces cancel, and the particle passes through the fields undeflected. The magnitude of this special velocity is

$$v = \frac{E}{B}. \quad (28\text{-}9)$$

With the help of a velocity selector, an apparatus based in part on this phenomenon, the charge-to-mass ratio of the electron can be measured.

The infinitesimal magnetic force on an infinitesimal length of thin wire $d\ell$ that carries a current I in the presence of a constant magnetic field is

$$d\vec{F} = I\, d\vec{\ell} \times \vec{B}. \quad (28\text{-}14)$$

To find the net force on a wire of finite length in a magnetic field, Eq. (28–14) is integrated. For example, the force on a straight wire of length L in a uniform magnetic field is given by Eq. (28–19), $\vec{F} = I\vec{L} \times \vec{B}$. Another important example is a wire that carries a current and is formed into a loop (or coil) of N turns; the area of the face of the loop is A. When it is placed in a constant magnetic field, such a loop experiences a torque

$$\vec{\tau} = \vec{\mu} \times \vec{B}. \quad (28\text{-}22)$$

The loop reacts as a magnetic dipole with magnetic dipole moment $\vec{\mu}$. For a coil of N turns, $\vec{\mu}$ has magnitude

$$\mu = INA \quad (28\text{-}24)$$

and direction perpendicular to the face of the coil, oriented by a right-hand rule on the current. The torque tends to rotate the loop so that $\vec{\mu}$ and $\vec{B}$ become aligned. The potential energy of the loop in a constant magnetic field can be expressed as

$$U(\psi) = -\vec{\mu} \cdot \vec{B}. \quad (28\text{-}26)$$

The Hall effect exploits the equivalence between the force on a moving charge and the force on a current-carrying wire. This effect proves that the current carriers in metals are negatively charged.

Understanding the Concepts

1. A wire carrying a current is electrically neutral, yet a magnetic field acts on it. Why?
2. Explain how you might define and measure a magnetic field if magnetic monopoles existed.
3. An electron beam in an oscilloscope is deflected to the right on the screen. Could this be caused by an electric field *or* by a magnetic field? Explain how you could distinguish these possibilities.
4. An electron beam makes a spot in the center of the screen of a cathode-ray tube. A bar magnet is brought in from the left side (as seen from the front of the tube), with the S pole nearest to the beam. Which way will the spot move? Suppose that the N end of the bar magnet is brought near the beam from above. Which way will the spot move?
5. Much of the description of magnetic forces depends on the use of a right-hand rule. Does the magnetic force depend fundamentally on the fact that we have chosen the right rather than the left hand?
6. Currents flow in two parallel wires. Wire 1 carries 2 A to the right, and wire 2 carries 1 A to the left. (a) What is the direction of the magnetic field near wire 2 due to wire 1? (b) What is the direction of the magnetic force on wire 2?
7. If you have just used a velocity selector for electrons and you wish to use it to choose positrons with the same speed, do you have to change any settings on the selector? Positrons are like electrons, but positively charged.

8. Induced charges give rise to electric forces even between electrically neutral objects. How do we know that the forces between bar magnets are not induced electric forces?
9. The aurora is a manifestation of Earth's magnetic field. Why do these lights occur frequently near Earth's poles and only rarely elsewhere?
10. Imagine that an electrically neutral wire carrying a current moves in the presence of an external magnetic field. Do you expect that there will be an additional force on the wire due to the movement?
11. You have a fixed length of wire and want to use it to make a magnetic dipole with the largest possible magnetic dipole moment. Into what shape should you wind it? Are you better off making a single loop or N loops?
12. A small bar magnet forms a magnetic dipole; a current-carrying wire in the shape of a small loop also forms a magnetic dipole. If that is the case, the current loop should give rise to a magnetic field. Use this analogy to sketch the magnetic field lines that would be generated by such a current loop.
13. Consider two small circular current loops. Suppose the two loops are placed flat on a table close to each other (but not touching) and the two currents both flow in a counterclockwise direction. Will the two loops attract or repel? What happens if the directions of the currents are opposite?
14. A proton coming from outer space reaches Earth and is now traveling across the magnetic north pole parallel to Earth's surface. Will it be deflected (a) up, (b) down, (c) to the left, (d) to the right, or (e) not at all?
15. Consider two small circular current loops. Suppose one loop is placed above the other (but not touching), with their areas oriented similarly. If the currents flowing in the two loops are both in the same direction, will the loops attract or repel?
16. Is it possible for an electron to move in a straight line through a magnetic field? If so, how?
17. Suppose that the coil of a direct-current electric motor consists of many turns rather than one turn of wire that carries a current I. Does the coil rotate faster than a single loop would? Does the split-ring commutator still work?
18. You have a large pail of water, a bar magnet with its N and S ends unmarked, a straight pin, and a cork. How could you make a compass? One of the things you need to know to construct this compass is how to distinguish north from south; you are allowed to watch the Sun to help with this part of the question.
19. Do magnetic north poles repel positive electric charges?
20. A classmate tells you that 1 T is 1 N/A·m. Is your classmate correct?
21. True or false? The magnetic force on an electron moving in Earth's magnetic field is zero only if the electron (a) is at the N and S poles, (b) is at the equator, (c) is moving slowly, (d) is moving along the magnetic field lines, or (e) is moving radially away from Earth.

Problems

28–1 Magnets and Magnetic Fields

1. (II) Sketch the magnetic fields for the arrangements of bar magnets shown in Fig. 28–29.

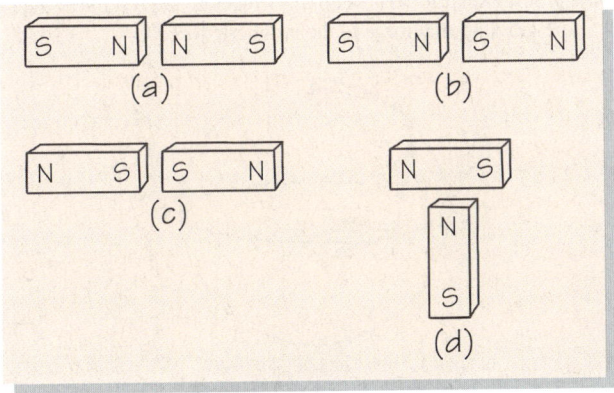

▲ **FIGURE 28–29** Problem 1.

2. (II) Consider the magnetic field generated by a current-carrying wire, as depicted in Fig. 28–2. Assuming that a reversal in the direction of the current also reverses the direction of the magnetic field, sketch the magnetic field due to two wires that are parallel to each other, and whose currents flow in the same direction. You may assume that the magnetic fields add vectorially, just like electric fields. Repeat the sketch for the situation in which the currents flow in opposite directions. Your sketches should show the field lines as seen by someone looking along the wires.

28–2 Magnetic Force on an Electric Charge

3. (I) An electron moving in the $-y$-direction enters a region of constant magnetic field and is observed to deflect to the $-x$-direction. What is the direction of the magnetic field?

4. (I) A proton with velocity $\vec{v} = (1.7 \times 10^6 \text{ m/s})\hat{i} + (0.8 \times 10^6 \text{ m/s})\hat{j} - (4.5 \times 10^5 \text{ m/s})\hat{k}$ moves through a magnetic field $\vec{B} = (0.70 \text{ T})\hat{i} - (0.50 \text{ T})\hat{j} + (0.10 \text{ T})\hat{k}$. Calculate the force on the proton.

5. (I) A proton of energy 100 keV moving in the $+x$-direction enters a region of uniform magnetic field perpendicular to the x-axis. Upon entry into that region, the proton experiences an acceleration of 3×10^{12} m/s^2 in the $+y$-direction. What are the magnitude and direction of the magnetic field?

6. (I) A pith ball charged to $+1$ μC falls vertically at the equator. At this location, the magnitude of Earth's magnetic field is 0.5 Gauss (0.5×10^{-4} T) and the field points to the north. When the ball reaches a speed of 5 cm/s, what is the magnetic force (magnitude and direction) on it?

7. (I) A proton moving with speed v enters a narrow (1.0 cm wide) region of magnetic field perpendicular to the direction of the proton's motion. As a result, the proton acquires a small component of speed perpendicular to its original direction of motion. This speed is much less than the original speed, and it is measured to be 3.3×10^5 m/s. What is the strength of the magnetic field?

8. (II) A cork ball carrying charge q has a mass of 0.4 g and is set in straight-line motion perpendicular to a uniform magnetic field of 0.007 T. What is the value of q if its direction of motion changes by 0.05° in 3.0 s?

9. (II) (a) A rapidly moving charged particle of charge e, mass m, and speed v passes through a region of magnetic field $\vec{B}$, which points in a direction perpendicular to the motion. The particle spends a time interval Δt in the region. Estimate the angle θ through which it will be deflected during Δt, assuming that θ is small. (b) The particle is a proton, with $m = 1.7 \times 10^{-27}$ kg, and $e = 1.6 \times 10^{-19}$ C, and the speed is 1.4×10^7 m/s. The size of the magnetized region is 0.1 m across. How large must B be to give rise to a deflection of 0.1 rad?

10. (II) In an oversimplified model of Earth's magnetic field, the field is parallel to the rotation axis, has a constant magnitude of 10^{-4} T up to a height of 100 km, and then quickly drops to zero. A cosmic-ray particle with charge 1.6×10^{-19} C and mass 9.5×10^{-26} kg moves at a speed of 10^8 m/s directly toward the equatorial region from above. (a) In what direction is the particle deflected? (b) Estimate how much it will be deflected from the point of impact it would have if it were uncharged. (In fact, this is not a realistic example. Cosmic-ray particles as massive as this have greater charge.)

11. (II) Electrons travel at a speed of 6.0×10^7 m/s in a television tube. The electrons are affected by Earth's magnetic field. The tube, which is 0.40 m long, is located at a region where the magnetic field has a vertical component of 18 μT and a horizontal component of 24 μT (Fig. 28–30). If the initial direction of the electron beam is in the same direction as the horizontal component of Earth's magnetic field, in which direction, and by how much, is the electron beam deflected?

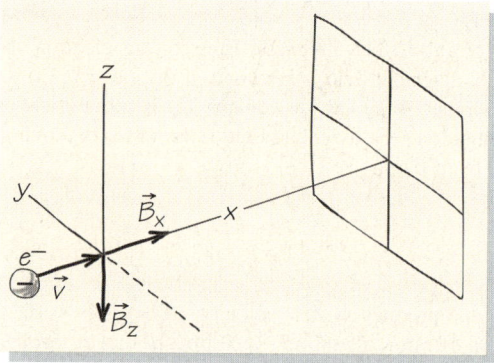

▲ **FIGURE 28–30** Problem 11.

28–3 Consequences of the Magnetic Force on a Charge

12. (I) A proton is sent into a region of constant magnetic field, oriented perpendicular to the proton's path. There the proton travels at a speed of 3×10^6 m/s in a circular path of radius 240 cm. What is the magnitude of the magnetic field?

13. (I) (a) Suppose that electrons from an electron gun with a voltage of 1600 V are injected into a region of constant magnetic field perpendicular to the electrons' velocity. What magnetic field will give the electrons a radius of curvature of 6 cm? (b) A magnetic field of what magnitude is necessary to give an alpha particle (charge $q = 2e$ and mass $m_\alpha = 7360$ times the mass of an electron) with a kinetic energy of 1200 eV a path with radius of curvature of 20 cm?

14. (I) Show that the radius of curvature of a proton moving at a velocity of 25 km/s in a magnetic field of 10^{-10} T is small compared to interplanetary distances. The protons therefore spiral around interplanetary magnetic field lines; we say that the protons are "tied to the magnetic field lines" in cosmic magnetic fields.

15. (I) The magnetic field at the surface of a neutron star has magnitude 3×10^7 T. What is the radius of the circular orbit of an electron that moves there at 0.1 percent of the speed of light? What is the magnitude of the magnetic force on the electron?

16. (I) With what frequency will deuterons, which have the same charge as protons but twice the mass, circulate in a cyclotron with a magnetic field of 0.65 T?

17. (I) If we want to triple the cyclotron frequency associated with a proton accelerator from an initial value of 6.1 MHz, what quantity must we change, and from what initial value to what final value?

18. (I) In a certain region, the average radius of curvature of the trajectory of electrons trapped in the Van Allen belt is 300 m and the average electron energy is 100 keV. What is the value of Earth's magnetic field in this region?

19. (I) Electrons of speed 10^6 m/s and protons of speed 10^4 m/s perpendicularly enter a region of constant magnetic field 10^{-5} T above Earth. What are the radii of their orbits? Why is the proton's radius greater? If the proton's speed were the same as that of the electron (10^6 m/s), what would be the radius of its orbit?

20. (II) An accelerator designer envisages circulating protons moving in a ring of radius 17 km by means of magnetic fields of magnitude 7.0 T. What is the magnitude of the momentum of a proton that moves in this way? For protons with this momentum, the energy is given to excellent accuracy by the formula $E = pc$, where c is the speed of light, about 3×10^8 m/s. Calculate the energy of the proton in megaelectron-volts; 1 MeV = 1.6×10^{-13} J.

21. (II) Assume that the electrons in a television picture tube have an energy of 10 keV and move perpendicularly to Earth's magnetic field (see Table 28–1). (a) Calculate the final velocity (vector) of an electron when it hits the screen if the horizontal distance the electron travels is 40 cm. (b) What is the deflection (distance) of the electron perpendicular to its original direction?

22. (II) Earth acts as a giant magnet whose field lines are like those of a bar magnet, running from the magnetic north pole to the magnetic south pole. The magnetic field at the equator is approximately constant, of magnitude 5×10^{-5} T, and runs from the geographic South Pole to the geographic North Pole (Fig. 28–31). If we ignore air resistance and the gravitational force, a charged object could orbit Earth at the equator as a result of the magnetic force if it has just the right velocity. Suppose that such an object has a charge of -1 mC and a mass of 1.0 g. (a) What would its velocity have to be for it to travel in such an orbit? (b) Suppose that the gravitational force acts on this object as well. What is the ratio of the gravitational force to the magnetic force?

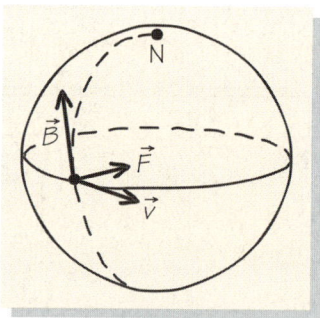

▲ **FIGURE 28–31** Problem 22.

23. (II) In Section 28–3, we calculated the critical momentum for an electron to stay within the galaxy. (a) Given that the energy of a high-energy particle is related to its momentum by $E = pc$ (see Problem 20), what is the energy of an electron with the critical momentum? What are the critical momentum and energy for (b) an alpha particle (charge $2e$ and mass four times the mass of a proton) and (c) an ion of uranium, with charge e and mass 240 times the proton mass?

24. (II) A proton moves horizontally perpendicular to a constant magnetic field oriented so as to deflect the proton instantaneously upward. The magnitude of the field is 0.7×10^{-2} T. What is the speed of the proton so that the magnetic force just cancels the gravitational force on the proton, leaving it in horizontal flight? This problem illustrates how very weak the gravitational force is compared to electromagnetic forces.

25. (II) An electron moves at a speed $v = 3.0 \times 10^5$ m/s in a region of constant magnetic field of magnitude 0.12 T. The direction of the electron when it enters this region is at 40° to the field, and the electron follows a helical path. When you look along the direction of the magnetic field, the path is a projected circle. How far has the electron traveled along the direction of $\vec{B}$ when one projected circle has been completed?

26. (II) An electron enters a bubble chamber that contains a constant magnetic field of strength 0.035 T and follows a helical path. The spacing between the turns of the path is 8.5 mm, as is the radius of the circular part of the path. Find the components of the velocity parallel and perpendicular to the field.

27. (II) A vacuum tube contains two axial cylinders (Fig. 28–32). The potential difference between these cylinders is 500 V. Some electrons are released from the inner cylinder and are accelerated by the electric field toward the outer cylinder; a small current thereby travels through the tube. Suppose that a uniform magnetic field is set up parallel to the axis of the tube. This curves the trajectories of the electrons, and at a critical magnetic field, the electrons will no longer reach the outer cylinder, and the current ceases to flow. What is the kinetic energy of the electrons that hit the outer cylinder at a magnetic field just below the critical value?

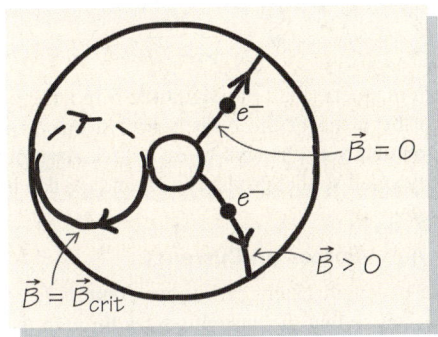

▲ FIGURE 28–32 Problem 27.

28. (II) A proton and an alpha particle, which has twice the charge and four times the mass of the proton, are each accelerated through the same potential difference and enter a region of constant magnetic field perpendicular to their paths. (a) What is the ratio of the radii of their orbits? (b) What is the ratio of the frequencies of their orbits?

29. (II) An electron is injected at $t = 0$ s with velocity $\vec{v}_0 = (2 \times 10^6 \text{ m/s})\hat{i}$ into a region with parallel electric and magnetic fields $\vec{E} = (1500 \text{ V/m})\hat{j}$ and $\vec{B} = (-0.2 \text{ T})\hat{j}$, respectively. Calculate the subsequent motion.

30. (II) You want to be able to tune a velocity selector such that you have the capacity to select electrons that have been accelerated from rest by a potential that runs from 1500 V to 15,000 V. If the magnetic field, B, is fixed at 0.40 T, what range of electric field strengths must be available? If the electric field strength were fixed at 15 V/cm, what range of magnetic field strengths must be available?

31. (II) Figure 28–33 is a schematic diagram of a cyclotron. A charged particle starts out at the central point and, for a given magnetic field perpendicular to the plane of motion, follows a circular path. The cyclotron takes advantage of the fact that the time for the particle to execute a half-circle is independent of the particle's velocity. An alternating voltage is applied across the gap between the two "dees" (the semicircular regions), so that when the particle crosses the gap, the voltage acts to accelerate it. When the particle gets to the gap again after having completed a half-circle, the voltage has changed sign, and the particle is once again accelerated. The frequency of the oscillating voltage must match the cyclotron frequency. In this way, the particle is always accelerated, completing ever bigger circles in the same time, until the beam is extracted at the maximum radius. (a) If the magnetic field has strength 1.0 T and the circulating particle is a proton, $q = +e$ and $m = 1.7 \times 10^{-27}$ kg, what is the cyclotron frequency? (b) What is the maximum velocity of the proton for a maximum radius of 50 cm? (c) the corresponding maximum kinetic energy? (d) If the maximum voltage across the gap is 50 kV, how many full circles does the proton make before it reaches its maximum energy? (e) How much time does the proton spend in the accelerator?

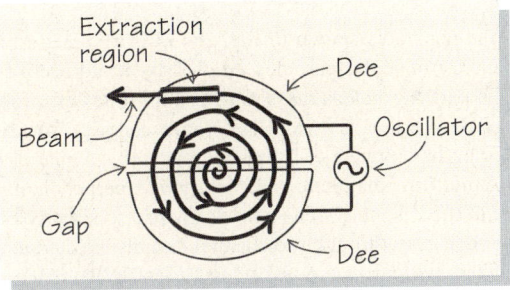

▲ FIGURE 28–33 Problem 31.

32. (II) A cyclotron used for accelerating protons has a magnetic field of magnitude 1.7 T. The circular region in which the magnetic field exists has a radius of 40 cm. (a) What is the cyclotron frequency? (b) What is the largest kinetic energy that a proton accelerated in this machine can have? (c) Repeat parts (a) and (b) for a doubly ionized helium nucleus, $^4\text{He}^{++}$, with four times the mass of a proton and twice the charge.

33. (II) The particle accelerator at Fermilab, the Fermi National Accelerator Laboratory in Batavia, Illinois, can accelerate protons to relativistic speeds. The accelerator is circular and holds the protons in circular paths by increasing the strength of a magnetic field perpendicular to this path as the protons' momentum increases. (The momentum increases because the protons pass repeatedly through regions of electric potential.) The radius of the main Fermilab accelerator is 6.2 km, and the magnets are capable of maintaining magnetic field strengths between 1 T and 4.5 T. Given that the magnitude of a proton's electric charge is 1.6×10^{-19} C, what range of momenta can be accommodated in this accelerator? Because protons of such momenta are highly relativistic, their energies are given by the approximate relativistic formula $E = pc$. What range of energies can be reached at Fermilab? What would the speed of a baseball, mass 0.15 kg, be if it had the energy of the most energetic protons at Fermilab? (For the baseball, use the normal nonrelativistic formulas that relate energy and speed.)

34. (II) A proton, with charge $q_p = +e$ and mass m_p, is accelerated through an electric potential V. The proton then enters a region of constant magnetic field $\vec{B}$ oriented perpendicular to its path. In this region, the proton's path is circular with radius of curvature R_p. Another particle with the same charge as the proton but with mass m_x follows under the same conditions. Its radius of curvature in the magnetic field, R_x, is 1.4 times as large as R_p. What is the ratio of m_x to m_p?

The device we have described is a type of *mass spectrometer*, which can be used to identify a material by the masses of that material's constituent molecules (Fig. 28–34, see next page). Sometimes, instead of a simple electrostatic potential as in our example, a velocity selector of crossed $\vec{E}$ and $\vec{B}$ fields is used to select particles of a given speed.

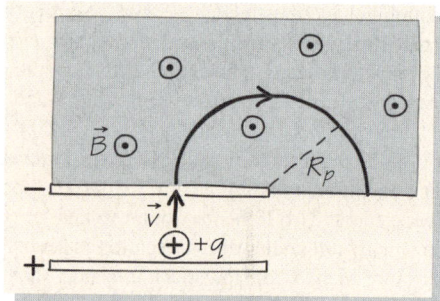

▲ **FIGURE 28–34** Problem 34.

35. (II) The apparatus shown in Fig. 28–35 is designed to measure the energy of alpha particles emitted by a radioactive source. (Alpha particles have a mass roughly four times the proton mass and a charge that is twice the proton charge.) The source is placed at the entrance of a channel that forms a quarter of a circle. A uniform magnetic field is applied perpendicular to the plane of the channel. Alpha particles with a specific velocity will make their way through the channel and be detected at the exit. All others will strike the walls and be lost. What is the range of values of B necessary to analyze alpha particles whose energies range up to 6 MeV?

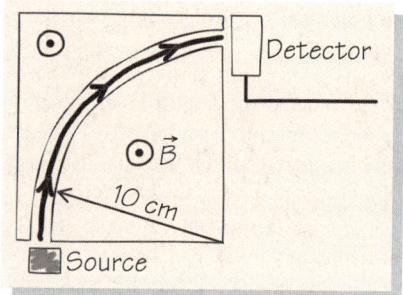

▲ **FIGURE 28–35** Problem 35.

36. (II) A 150-MeV proton, moving in the x-direction, enters a region in which there is a magnetic field. The proton experiences an acceleration of 7.0×10^{12} m/s^2 in the y-direction. What can you say about the magnetic field?

37. (II) A 2.3-keV electron is moving horizontally and passes perpendicular to the Earth's magnetic field at a location where the field magnitude is 0.52 Gauss. What is the magnetic force on the electron? What is the ratio of this force to the gravitational force on the electron? What energy electron would be needed to make the magnetic and gravitational forces equal?

38. (II) The 88-inch (diameter) cyclotron was put into operation at the Lawrence Berkeley Laboratory in 1962 and for 40 years produced significant scientific results. It was able to produce protons up to 55 MeV energy, and heavier ions of mass A up to 5 MeV/A. (a) What was the maximum magnetic field of the magnet? (b) If a ^{86}Kr ion with charge $+19e$ is accelerated in this cyclotron, what maximum energy can be expected?

39. (II) Constant electric and magnetic fields are perpendicular to each other, the electric field pointing from left to right on the page, the magnetic field from the bottom to the top of the page. A proton with charge q and speed v moves into the region of the fields in the direction out of the page. (a) What are the directions of the magnetic and electric forces? (b) What is the magnitude of the resulting force on the proton? (c) When could this force ever be zero?

40. (III) You have an apparatus that can form an electric field of 2000 N/C and a magnetic field of 0.3 T. You want to build a velocity selector to select electrons of speed 2×10^4 m/s. (a) Draw the orientation of your apparatus, showing $\vec{E}$, $\vec{B}$, and $\vec{v}$. (b) What are the minimum and maximum values of v that you can select? [*Hint:* Set the apparatus up so that $\vec{v}$ and $\vec{B}$ are not perpendicular to each other (Fig. 28–36).]

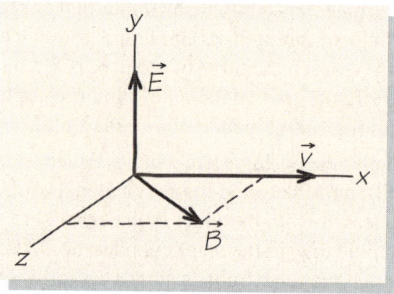

▲ **FIGURE 28–36** Problem 40.

41. (III) A particle of mass m and charge $-q$ moves in a circular orbit of radius R about a fixed charge Q. The angular frequency for the orbit is given by

$$\omega_0^2 = \frac{qQ}{4\pi\varepsilon_0 mR^3}.$$

A uniform magnetic field of magnitude B in a direction perpendicular to the plane of the orbit is turned on. As a result, the angular frequency is changed to $\omega_0 + d\omega$. Assuming that B is sufficiently small so that products of B and $d\omega$ can be neglected, calculate $d\omega$.

28–4 Magnetic Forces on Currents

42. (I) A wire of length 12 cm carries a current of 1.3 A. When the wire is placed so that the current passes through the wire along the x-direction, the force on the wire due to an external magnetic field B is measured to be $-(0.02 \text{ N})\hat{j}$. When the wire is moved so the current through the wire moves along the y-direction, the force is measured to be $(0.02 \text{ N})\hat{i}$. What is the magnetic field?

43. (I) A straight wire segment is placed in a region known to contain a constant magnetic field of unknown strength. A current of 6 A runs through the wire, which can be turned in various directions until the force per unit length acting on it takes on a maximum value of 0.18 N/m. What is the value of the magnetic field?

44. (I) A long wire carries a current of 6.5 A. A bar magnet is brought near the wire so that the charge carriers, of speed 1.7×10^{-4} cm/s, experience a magnetic field of 0.50 T perpendicular to their direction of motion. Calculate the force (a) on each moving charge carrier (electron) and (b) on a 1-m length of the wire.

45. (I) A thin, straight wire carries a current of 10 mA and makes an angle of 60° with a constant magnetic field of magnitude 10^{-6} T. The portion of the wire in this field has a length of 10 cm. Calculate the force, both direction and magnitude, on this segment of the wire.

46. (I) The length of a vertical lightning conductor from roof to ground is 20 m. A lightning stroke leads to a current of 10^4 A flowing through the conductor. Given that Earth's magnetic field is horizontal and of magnitude 0.5×10^{-4} T at the location of the building, what is the force on the conductor during the period the current flows?

47. (I) A straight wire is placed in a uniform magnetic field of magnitude 0.010 T. The direction of the field makes an angle of 30° with that of the wire, which carries a current of 10 A. What is the force on a 1.0-m segment of the wire?

48. (II) A current I flows through a circular wire loop of radius R that lies in the xy-plane (Fig. 28–37). Consider a constant magnetic field of magnitude B that points in the x-direction. Calculate the force on an element of the loop formed by an angle $d\theta$, located at an angle θ from the $+x$-axis.

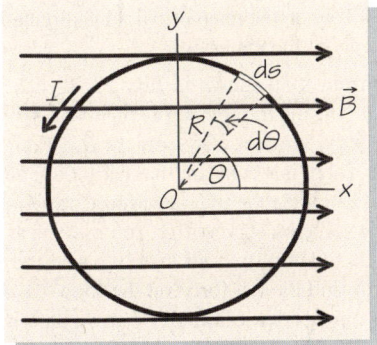

▲ **FIGURE 28–37** Problem 48.

49. (II) A wire of length L is suspended from two springs of spring constant k attached to a current source (Fig. 28–38). A magnetic field B, in a horizontal direction perpendicular to the wire (out of the page), is turned on; a current I then flows in the wire, which moves to a new equilibrium position. Which way will the wire move, and by how much?

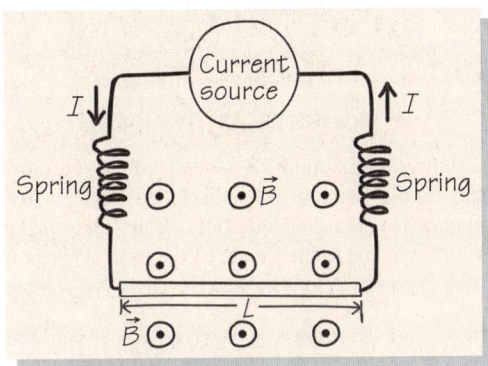

▲ **FIGURE 28–38** Problem 49.

50. (II) In a physics lecture demonstration, a thick copper wire of length 0.8 m and mass 70 g is attached to two thin wires and suspended so that it is horizontal (Fig. 28–39). A 0.03-T magnetic field pointing in the downward direction is turned on. What angle will the supporting wires make with the vertical if a current of 0.5 A flows through the wire?

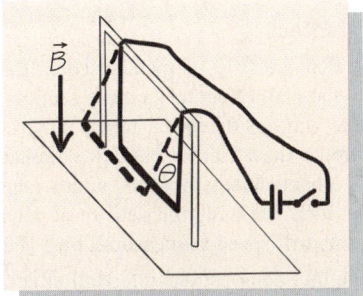

▲ **FIGURE 28–39** Problem 50.

51. (II) Figure 28–40 shows a possible device for measuring magnetic fields. A loop carrying a current I is dipped into a region of

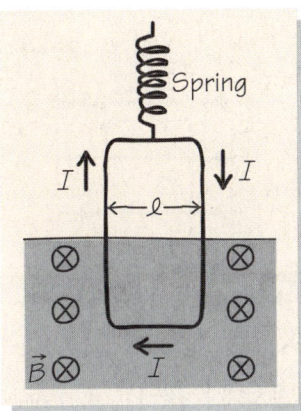

▲ **FIGURE 28–40** Problem 51.

magnetic field. The loop is suspended from a spring of spring constant k that stretches if the magnetic field points in a certain direction. Here the loop has width $\ell = 1.2$ cm, $I = 100$ mA, the spring stretches 0.6 cm, $k = 5 \times 10^{-2}$ N/m, and the magnetic field is uniform. What is the magnitude of the field? How could such a device be used, or modified, to measure fields that are not uniform?

28–5 Magnetic Force on Current Loops

52. (I) A wire coil of area 20 cm² with 180 turns experiences a maximum torque of 2×10^{-2} N·m when placed in a magnetic field of 0.3 T. What is the current through the coil?

53. (I) A rectangular wire loop of height 5 cm and width 3 cm consists of 60 turns and carries a current of 1.2 A. What are the magnitude and direction of the magnetic dipole moment? If a uniform magnetic field of 0.5 T is applied to the loop, and the field's direction makes an angle of 26° with the normal to the current loop, what is the torque (magnitude and direction) that acts on the loop?

54. (I) A circular coil of diameter 2.5 cm, consisting of 1500 turns of wire, carries a current of 50 mA. How much work must be done to flip the coil through 180° when it is placed in a uniform magnetic field of 0.75 T? The field makes an initial angle of 50° with the direction of the coil's dipole moment.

55. (I) A wire forms a circular coil of N turns and radius R and carries a current I. The coil's magnetic dipole moment is initially aligned with a fixed external magnetic field, $\vec{B}$. How much work must be done by an external torque to rotate the coil through an angle θ?

56. (I) A current loop of area 3.0 cm², carrying a current of 5.0 A, is placed in a uniform magnetic field of 0.25 T such that the normal to the loop is perpendicular to the direction of the magnetic field. There is a torque, and the loop changes direction. Because of friction in the bearings, it settles to the minimum energy orientation. How much energy was dissipated in the process?

57. (I) An atom can have a magnetic dipole moment of 10^{-23} J/T. Such an atom is placed in a magnetic field of 10 T. What is the range of potential energies involved?

58. (II) A wire carrying a current I splits into two channels of resistance R_1 and R_2, respectively, forming a circuit. The wire enters the space between the two poles of a magnet with a uniform magnetic field that runs from one pole piece to the other (Fig. 28–41, see next page). The circuit forms a loop; the field lies in the plane of the loop. What is the torque on the circuit about the wire axis, given that the wires are a distance d apart and that the length of the split is L?

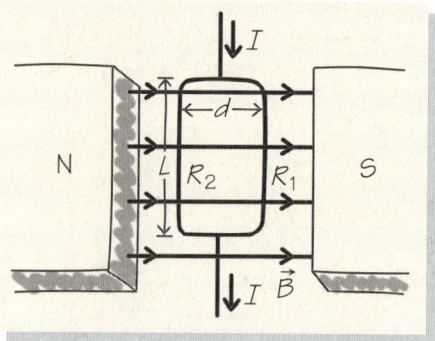

▲ **FIGURE 28–41** Problem 58.

59. (II) A circular wire coil of area 6 cm² has 50 turns. When the coil is placed in a magnetic field of 0.2 T, the maximum torque is 3×10^{-5} N·m. (a) What is the current in the coil? (b) What work is required to rotate the coil 180° in the magnetic field? Does the work depend on the initial angle?

60. (II) An electric motor consists of a current-carrying wire loop in a constant magnetic field $\vec{B}$ (Fig. 28–42). The field produces a torque that tends to rotate the loop so that the loop's magnetic dipole moment, $\vec{\mu}$, and $\vec{B}$ become aligned. When that happens, a split-ring commutator reverses the current direction, so that $\vec{\mu}$ changes its orientation by 180°, and the torque acts to continue the rotation. Suppose that $\vec{\mu}$ and $\vec{B}$ start out almost antiparallel. Plot the magnitude of the torque as a function of the angle between $\vec{\mu}$ and $\vec{B}$, as this angle runs from $-\pi$ to 0. At 0° the commutator reverses the current. Plot the torque through another half turn. What is the average value of the torque through a full turn if the current in the motor is 6.2 A, the magnitude of $\vec{B}$ is 0.45 T, and the area of the loop is 54 cm²?

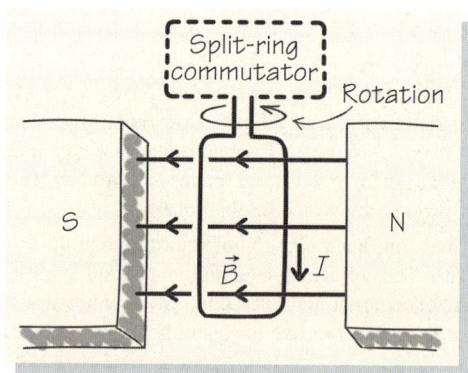

▲ **FIGURE 28–42** Problem 60.

61. (II) An electron, of charge $q = -1.6 \times 10^{-19}$ C, has a "size" of about 3×10^{-15} m, called its classical radius. The magnetic dipole moment of the electron is roughly 10^{-23} A·m². (a) Suppose that this magnetic moment were due to the entire charge q orbiting at the classical radius. What would the speed of the charge be to generate this magnetic moment? (b) Suppose that the electron's magnetic moment were perpendicular to a magnetic field of magnitude 1 T. What is the torque on the electron?

62. (II) (a) Calculate the magnetic dipole moment of a single atom, based on the following model: One electron travels at speed 2.2×10^{6} m/s in a circular orbit of diameter 10^{-10} m. (b) The individual atomic magnetic dipoles of magnetic materials (such as iron) are preferentially lined up to point in the same direction. If a fraction f of the dipoles are so aligned along the long axis (with the rest oriented randomly so that their magnetic dipole moments add vectorially to zero), what is the net magnetic dipole moment of a piece of such material 1 cm² in area and 10 cm long? (The material may be viewed as an array of cubes, each of which contains one atom and is 10^{-10} m on a side.) (c) What is the torque experienced by the piece of material in part (b) in a field of 10^{-3} T when the magnetic field is directed at right angles to the long axis of the material?

63. (III) The current loop shown in Fig. 28–43 lies in the xy-plane and consists of a straight segment α of length $2R$ in the x-direction and a semicircular segment β, which has a radius of curvature R. There is a constant magnetic field of strength B into the page. (a) Compute the magnetic force on segment α. (b) Find the magnetic force on segment β. You may wish to use symmetry arguments to simplify your task. (c) Add the results of parts (a) and (b) to find the net force on the loop. (d) How could you generalize your results to a loop of any shape in the xy-plane?

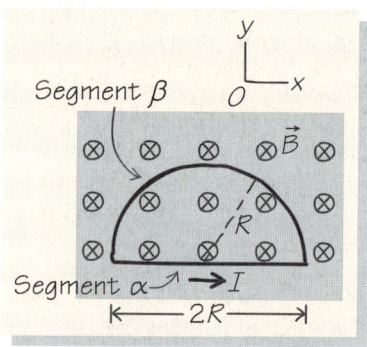

▲ **FIGURE 28–43** Problem 63.

64. (II) We showed in Example 28–11 that, when a current loop with magnetic dipole moment $\vec{\mu}$ is displaced slightly from perfect alignment of $\vec{\mu}$ and a magnetic field $\vec{B}$, the rotational motion of the current loop due to the torque of the field is harmonic. Given the usual expression for the kinetic energy associated with a rotation of angle ψ, namely $K = \frac{1}{2}I_M\left(\frac{d\psi}{dt}\right)^2$, and the approximate expression $\cos\psi \cong 1 - \psi^2/2$, calculate the angular frequency of the harmonic motion.

65. (III) A coil carrying current $I = 50$ mA has a moment of inertia $I_M = 7.5 \times 10^{-7}$ kg·m² about a rotational axis and an area of 6.0×10^{-4} m². The coil is placed in a magnetic field of magnitude 0.6 T, displaced 5° from alignment between its magnetic dipole moment, μ, and the field, and released from rest. Describe the subsequent motion. What is the maximum angular speed of the coil in that motion?

28–6 The Hall Effect

66. (II) Suppose that the strip of metal used in the apparatus that demonstrates the Hall effect has a cross section of width w and depth d_0. (The width is the space across which the Hall voltage ΔV is measured.) Show that the density n of charge carriers with charge e is independent of the width and is given by $n = IB/(d_0 e \Delta V)$. Knowing the density of carriers, find an expression for the drift speed as measured by a Hall apparatus.

67. (II) The probe that demonstrates the Hall effect is used to measure the density of charge carriers in an unknown sample of metal. A sample of the material 1.5 mm thick is placed in a magnetic field of 1.2 T. When a current of 1.8 A passes through the material, a Hall voltage of 6.2 μV is measured. What is the density of charge carriers?

68. (II) A Hall-effect probe can be used to measure the magnitude of a magnetic field. A researcher has lost the instruction booklet and forgotten the calibration procedure. However, when she places the Hall probe inside a known magnetic field of 7500 G, she measures a Hall voltage of 165 mV. What is the field of a magnet with a Hall voltage of 390 mV?

General Problems

69. (II) The wire coil of a galvanometer has an area of 2 cm^2 and 500 turns. The coil is placed in a magnetic field of magnitude 0.18 T and oriented so that its plane is initially parallel to the field. The restoring torque of the galvanometer spring is proportional to the angular deflection, with a proportionality constant of 10^{-8} N·m/° (see Example 28–9). What current corresponds to a deflection of 70°?

70. (II) The masses of atomic ions of known charge can be precisely measured by finding the time an atom takes to complete a circular trajectory in a known magnetic field. With a magnetic field of magnitude 3.0 T and an apparatus capable of measuring times to an accuracy of 10^{-9} s, how accurately can the mass of an ion with charge $+e$ be measured in 1 rev? If the mass is to be measured to an accuracy of 5×10^{-31} kg, how many revolutions must be measured?

71. (II) When an electron orbits a proton, the smallest circular orbit is one with a radius of about 0.5×10^{-10} m, the Bohr radius. The proton's electric field must have what magnitude to make the electron follow this orbit? Compare the magnitude of the magnetic field that would be required to make an electron move in a circle of the same radius at the speed it would have if it were orbiting a single proton.

72. (II) A massive charge Q is fixed at the origin of a coordinate system. A magnetic field $\vec{B}$ points in the $+z$-direction. A light particle of charge q and mass m moves in a circular orbit of radius r about the origin. For what value of $\vec{B}$ (as a function of r) is such motion possible if Q and q have the same sign and if the angular momentum of the motion is a fixed constant L?

73. (II) Consider a parallel-plate capacitor with charge density $\pm 8.0 \times 10^{-7}$ C/m^2 on the two plates and an electric field that points in the $+z$-direction. What magnetic field is necessary to provide a velocity selector for 60-keV deuterons that move in the $+y$-direction? A deuteron has a mass of 3.2×10^{-27} kg and a charge of 1.6×10^{-19} C; 1 keV = 1.6×10^{-16} J.

74. (II) A narrow beam of particles of mass m and charge q travels in free space at speeds between v_1 and v_2. It enters a region of length L with a constant magnetic field that points perpendicular to the beam direction and parallel to the boundary between the field-free region and the region with the field. There, it follows a circular path with radius of curvature R until it exits that region. Show that the beam widens when it emerges from the region with the field, and calculate the spread in terms of a range of angles.

75. (II) An electron moving in the xy-plane is subject to forces due to a constant magnetic field $\vec{B}$ that points in the $+z$-direction. Assuming that the electron loses 10 percent of its energy after 20 turns, as a consequence of frictional forces, what will the fractional change in the radius of the orbit be after 20 turns?

76. (II) For the motion described in Problem 75, (a) what will the fractional angular-momentum change be during the 20 turns? (b) What is the torque exerted by the frictional forces in terms of the initial kinetic energy?

77. (II) Electrons are injected into a region with a constant magnetic field $\vec{B}$ by an electron gun with known voltage V. The electrons move in a plane perpendicular to $\vec{B}$ and follow an arc of radius R. Determine the charge-to-mass ratio e/m for the electrons in terms of the given parameters.

78. (II) Particles with mass $M_A = A(1.6 \times 10^{-27}$ kg$)$ and charge $q = 1.6 \times 10^{-19}$ C are accelerated by a potential difference of 6.0×10^4 V and directed perpendicularly into a region of uniform magnetic field of strength 2.2 T. The region with the field is 35 cm deep. Calculate the angular deflection of the particles, θ, as a function of A.

79. (II) N electrons move at speed v in a circular orbit of radius R. (a) What is the angular momentum of the system of electrons? (b) The magnetic dipole moment associated with the current loop? (c) The ratio of the quantities in parts (a) and (b)?

80. (II) A rectangular wire loop of width a and height b is connected to a current source that, when turned on, gives rise to a current I in the wire. The loop is suspended in a uniform magnetic field $\vec{B}$ that points in a vertical direction (Fig. 28–44), and it would hang vertically if there were no current. We assume that the wire is massless, but two masses m are suspended at the lower corners. What is the angle θ at which the loop is in equilibrium? Calculate this in two ways: by using torques, and by expressing the potential energy as a function of θ and minimizing it. What happens if the direction of the current is reversed?

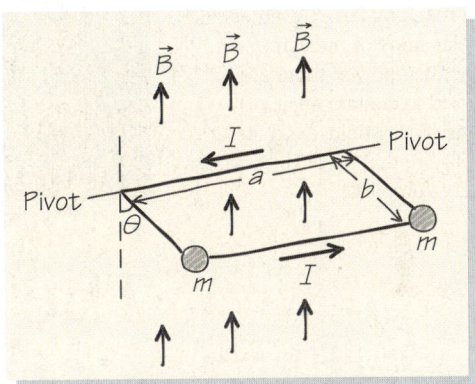

▲ FIGURE 28–44 Problem 80.

81. (III) A particle of charge q and mass m is subject to gravity acting downward on the page, and a magnetic field of magnitude B acting in a direction perpendicular to the page, pointing into the page. We assume that the initial velocity of the particle is zero. (a) Show that the motion will only be in the plane of the page; (b) if there are many such particles, and their mutual Coulomb repulsion is negligible, what is the direction of the average current? (c) Check your conclusions by writing out the form of Newton's second law for the three components of the motion.

82. (III) Suppose that an experimental apparatus can have both electric and magnetic fields constant in magnitude and direction. In this apparatus a proton moving at a speed of 5.0×10^4 cm/s in the $+z$-direction does not accelerate, whereas a proton moving at a speed of 8.0×10^4 cm/s with no x-component at an angle of 42° with respect to the z-axis experiences an initial acceleration of magnitude 3.5×10^8 m/s^2 in the $-x$-direction. A proton moving in the xy-plane has a circular orbit. Find the values of $\vec{E}$ and $\vec{B}$ in the apparatus.

Chapter 29

▶ In the tokamak, an experimental device for the study of nuclear fusion–generated power, magnetic fields are used to contain a gas of positive ions. These fields, and the windings of wire that produce them, are topologically complex.

The Production and Properties of Magnetic Fields

We have seen how magnetic fields influence moving charges and current-carrying wires. But where do the magnetic fields themselves come from? We have seen that an electric charge creates an electric field, and that an electric field exerts a force on a charge. Similarly, a *moving* charge or a current *creates* a magnetic field, just as we know that a magnetic field exerts a force on a *moving* charge or a current. In other words, currents and moving charges exert magnetic forces on each other. In this chapter, we will describe and explore the ways in which magnetic fields are produced, and learn the properties of these fields, some quite different from those of electric fields. We shall also continue to explore the close relation between electric and magnetic fields.

▶ 29–1 Ampère's Law

Over the winter of 1819–1820, Hans Christian Oersted discovered that electric currents influence compass needles (Fig. 29–1). Until this discovery, there was only a suspicion of a connection between electricity and magnetism. Oersted, as well as André-Marie Ampère, soon showed that *current-carrying wires exert forces on each other*. Because such wires are everywhere electrically neutral, these forces are not electric. As we saw in the previous chapter, a current-carrying wire aligns iron filings on a plane perpendicular to the wire in a circular pattern (Figure 28–2c). This suggests that a current-carrying wire creates the magnetic field.

The Magnetic Field of a Straight Wire

Figure 29–2 shows a set of experiments that tell us much about the nature of the magnetic fields produced by a current-carrying wire. To start with, in Figs. 29–2a and 29–2b two parallel wires with current flowing in the same direction are attracted to each other, while in Figs. 29–2c and 29–2d we see that if the currents flow in opposite directions, the parallel wires repel. If we interpret the force between the two current-carrying wires in Fig. 29–2 as a magnetic force on one wire due to the magnetic field produced by the other and perform some measurements, we can determine the magnetic field of a straight wire. In Fig. 29–3a we redraw the situation with wire segments that are long, $L \gg d$, where L is the wire length and d their separation. In this limit all other segments of the wires are distant and can be ignored, because the forces between the wires weaken considerably as the separation between the wires increases. Only the relatively close long, parallel segments come into the picture. In Fig. 29–3a, the force on wire 2, which is directed to the left, is due to the magnetic field of wire 1. Equation (28–19) describes the force on a segment of current-carrying wire. Using this equation, the force on wire 2 is of the form

$$\vec{F}_2 = I_2 \vec{L}_2 \times \vec{B}_1, \qquad (29\text{–}1)$$

provided only that the field $\vec{B}_1$ due to wire 1 is the same all along wire 2, an assumption that is justified for arbitrarily long wires. Here the vector $\vec{L}_2$, of magnitude L, is oriented along the direction of I_2.

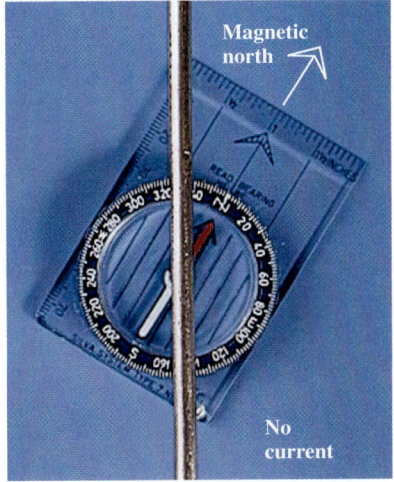

(a)

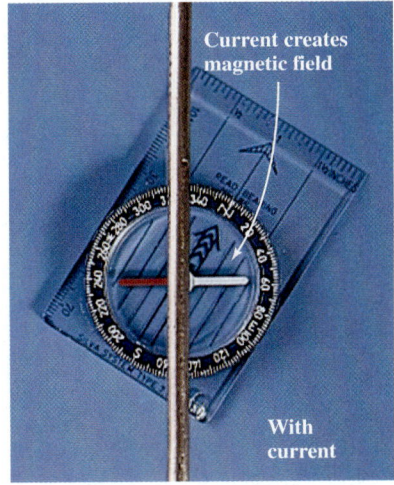

(b)

▲ **FIGURE 29–1** (a) The compass needle continues to point north when there is no current in the wire, but (b) when current flows through the wire from the bottom to the top of the photo, the needle reacts to the magnetic field produced by the current.

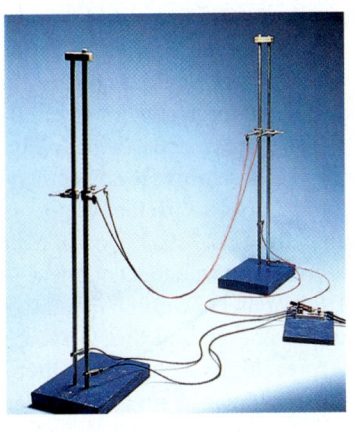

(a)

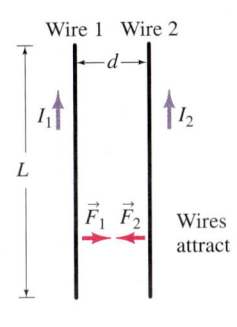

(b)

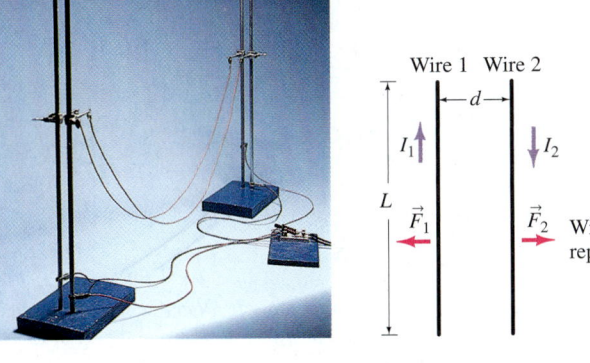

(c)   (d)

◀ **FIGURE 29–2** Two parallel wires that carry currents exert forces on one another; these forces are larger when the wires are closer. (a) and (b) The currents are parallel, and the forces are attractive. (c) and (d) The currents are antiparallel, and the forces are repulsive.

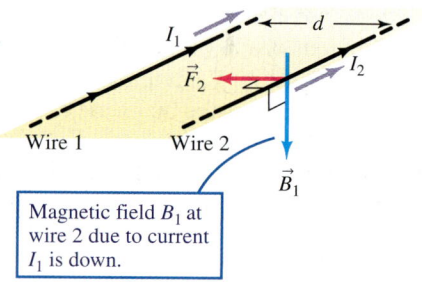

▶ **FIGURE 29–3** Determining the direction of the magnetic field due to wire 1. Currents I_1 and I_2 are parallel to each other. (a) According to the right-hand rule, $\vec{B}_1$ due to wire 1 is directed down when wire 2 is to the right of wire 1. (b) $\vec{B}_1$ due to wire 1 is directed up when wire 2 is to the left of wire 1.

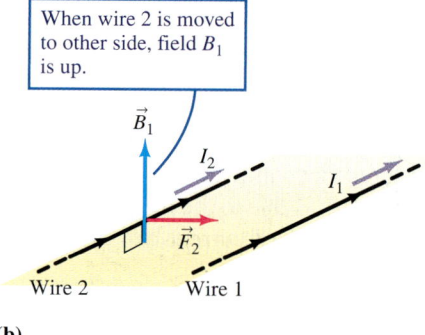

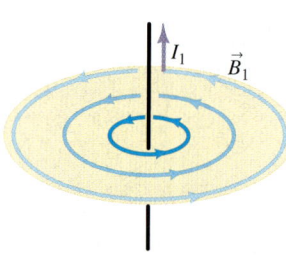

(a)

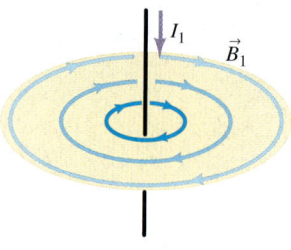

(b)

▲ **FIGURE 29–4** (a) The magnetic field due to wire 1, $\vec{B}_1$, traces out a circle around the wire in the direction shown. (b) If the current in wire 1 were reversed, the orientation of $\vec{B}_1$ would change.

If we move wire 2 around wire 1 in such a way that the wires remain parallel and at the same separation (moving from the position in Fig. 29–3a to that in Fig. 29–3b), we find that the force remains attractive and does not change in magnitude. This observation and Eq. (29–1) are consistent with the interpretation that *the magnetic field $\vec{B}_1$ due to wire 1 follows a circle around wire 1.* Application of a right-hand rule in Eq. (29–1) shows that $\vec{B}_1$ must be directed down when wire 2 is in its original position to the right of wire 1 (Fig. 29–3a). If wire 2 is moved to the left of wire 1, however, field $\vec{B}_1$ at wire 2 will be directed up because the two wires continue to attract each other (Fig. 29–3b). By using this argument for other positions, we find that the magnetic field lines due to wire 1 make circles about wire 1 (Fig. 29–4a; see also Fig. 28–2c). Equation (29–1) shows that force $\vec{F}_2$ is insensitive to any component of $\vec{B}_1$ that is *parallel* to the wires, because the vector product of two parallel vectors is zero—there may be a component of $\vec{B}_1$ parallel to the wire, but due to Eq. 29–1, we can't tell from the experiments that we have so far discussed. We have to resort to another argument—see Conceptual Example 29–1—to rule out a component of $\vec{B}_1$ parallel to wire 1.

If we reverse the current in wire 1 as we did in Fig. 29–2b, we find that the force on wire 2 is also reversed—the wires repel—and we interpret this as a change in direction of the magnetic field around wire 1 (Fig. 29–4b). The field lines again form circles around the wire, but in the opposite direction.

CONCEPTUAL EXAMPLE 29–1 In an experiment, a wire (wire 2) is wrapped in a circle around a straight wire (wire 1), as in Fig. 29–5. Measurements show that when constant currents run through them, there are no forces between them, and it is claimed that this shows that wire 1 produces no component of magnetic field parallel to itself. Is this correct?

Answer Yes. Symmetry tells us that any component of $\vec{B}_1$ oriented parallel to wire 1 will be the same all the way around wire 1, so that every segment of the circle formed by wire 2 will be in the presence of the same field $\vec{B}_1$. But then an application of the right-hand rule shows that a component of $\vec{B}_1$ parallel to wire 1 would cause a force to be exerted on wire 2. As Eq. (29–1) indicates, such forces would tend to expand or contract the circle traced by wire 2, and this effect would be measurable. Wire 1 does not produce a component of the magnetic field parallel to itself.

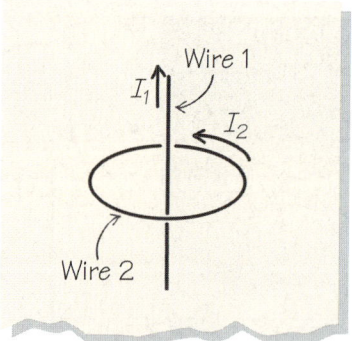

▲ **FIGURE 29–5** If wire 2 traced a circle around wire 1, it would react to any components of $\vec{B}_1$ due to wire 1 that are parallel to wire 1. No such forces are found.

We can summarize the results above by saying that the direction of the magnetic field produced by a straight wire is determined by a right-hand rule (Fig. 29–6):

If the thumb of the right hand is oriented along the direction of current flow in a wire, the fingers curl in the direction of the magnetic field.

We have found the direction of the magnetic field produced by the current. How do we find its magnitude? We do this by measuring the magnitude of the force between the wires, and such measurements show that the magnitude of the force between two parallel, straight segments of wire is

$$F = \frac{C I_1 I_2 L}{d},$$

where I_1 and I_2 are the currents in wires 1 and 2, respectively, d is the separation between the wire segments, and L is their length. The proportionality constant C depends on how we define the units of current. Conversely, if we use a *defined* proportionality constant, the force between two current-carrying wires determines the units of current. This latter alternative is the one used in the SI: C is defined according to

$$F = \frac{\mu_0 I_1 I_2 L}{2\pi d}, \tag{29-2}$$

where the constant μ_0, called the **permeability of free space**, is

$$\mu_0 \equiv 4\pi \times 10^{-7} \text{ T} \cdot \text{m/A}. \tag{29-3}$$

With this definition of C (that is, of μ_0) 1 A is defined as the current that travels in two long, parallel wires of length L that are 1 m apart, such that the attractive force between them is $(2 \times 10^{-7} \text{ N/m}) L$. Is this result consistent with previous definitions of the current? In Chapter 21, we defined the coulomb as the charge on two pointlike objects such that there is a certain force between them, and we provisionally defined 1 A as 1 C/s in Chapter 26. The definition of the coulomb in terms of a force between charges depends on another constant, ε_0, in exactly the same way that the definition of the ampere depends on μ_0. Thus, for our relations to be consistent, ε_0 must be a measured constant. If both μ_0 and ε_0 are known, the same must be true for their product, which is given by

$$\mu_0 \varepsilon_0 \equiv c^{-2} = (2.99792458 \times 10^8 \text{ m/s})^{-2}. \tag{29-4}$$

The constant c is precisely the speed of light! We shall see in Chapter 34 why this is so.

Comparison of Eqs. (29–1) and (29–2) shows that a long, straight wire that carries a current I gives rise to a magnetic field whose magnitude at a distance r from the wire is

$$B = \frac{\mu_0 I}{2\pi r}. \tag{29-5}$$

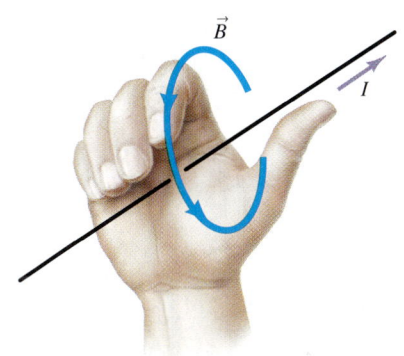

▲ **FIGURE 29–6** A right-hand rule determines the direction of the magnetic field around a current-carrying wire.

Ampère's Law

We can find a more universal form for the magnetic field produced by a current by expressing Eq. (29–5) in a different form, one that relates the magnetic field along a closed path to the electric current that the closed path encloses. This relation is what is known as Ampère's law. We'll see that we can use it to find the magnetic field in symmetric situations much as we use Gauss' law to find the electric field in symmetric situations.

To arrive at Ampère's law, imagine a line integral over the magnetic field $\vec{B}$ that follows a circular path of radius r all the way around a long wire, as in Fig. 29–7. The integration path, labeled C, thus follows the direction of $\vec{B}$. The path is broken into infinitesimal distance elements $d\vec{s}$ that are parallel to the magnetic field, so $\vec{B} \cdot d\vec{s} = B \, ds$. Finally, B is a constant when the distance r from the wire is constant and we can remove it from the integral:

$$\oint \vec{B} \cdot d\vec{s} = B \oint ds = B(2\pi r). \tag{29-6}$$

(The sign $\oint$ indicates that the path of the line integral is closed, going all the way around the circle.) The factor $2\pi r$ is the length of the path, the circumference of the circle of radius r. If we use Eq. (29–5), we find that

$$\oint \vec{B} \cdot d\vec{s} = \frac{\mu_0 I}{2\pi r} 2\pi r = \mu_0 I. \tag{29-7}$$

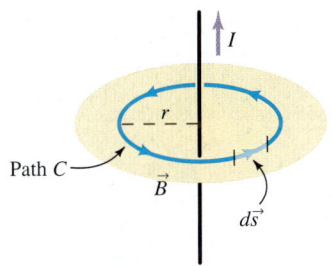

▲ **FIGURE 29–7** Path C circles a current-carrying wire at a constant distance r from the wire and follows the direction of the magnetic field, $\vec{B}$, around the wire.

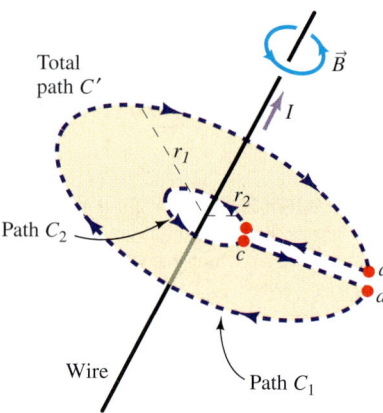

▲ **FIGURE 29–8** This path does not form a closed path around the wire. Path C' consists of a clockwise circle C_1 of radius r_1, a leg from a to b that moves inward to a distance r_2 from the wire, a counterclockwise circle C_2 of radius r_2, and a leg from c to d that moves outward to r_1.

Equation (29–7) includes a right-hand-rule convention in which path C must be in the direction of the fingers of the right hand when the thumb is oriented along I. Equation (29–7) is the first step in the development of Ampère's law.

A current "pierces" the closed path C described above (or more properly pierces a surface whose edge is formed by the path C). Now let us consider a similar integral, but this time over a closed path *not* pierced by a current, for example path C' shown in Fig. 29–8. We thus wish to compute $\oint_{C'} \vec{B} \cdot d\vec{s}$. This time we shall see that the integral vanishes. We first break the path C' into the segment from a to b, the nearly full circle C_2, the segment from c to d, and the nearly full circle C_1. The total contribution of the two paths from a to b and c to d is zero, because $\vec{B}$ is perpendicular to the path there. Thus in the limit that the "nearly" full circles get arbitrarily close to closing,

$$\oint_{C'} \vec{B} \cdot d\vec{s} = \int_{C_1} \vec{B} \cdot d\vec{s} + \int_{C_2} \vec{B} \cdot d\vec{s} = -B_1(2\pi r_1) + B_2(2\pi r_2). \quad (29\text{–}8)$$

Here B_1 is the magnitude of the magnetic field at a distance r_1 from the wire, and B_2 is the magnitude of the field at a distance r_2. The first term is negative because $\vec{B}$ is oriented opposite to the path direction on segment C_1. From Eq. (29–5) we see that the two terms on the right of Eq. (29–8) cancel:

$$\oint_{C'} \vec{B} \cdot d\vec{s} = -\frac{\mu_0 I}{2\pi r_1}(2\pi r_1) + \frac{\mu_0 I}{2\pi r_2}(2\pi r_2) = -\mu_0 I + \mu_0 I = 0. \quad (29\text{–}9)$$

The difference between the integral over path C and that over path C' arises from the fact that path C encloses current I, whereas path C' encloses no current. This is the first step toward the following generalization. Let the quantity I_{enclosed} be the total current enclosed by *any closed path*. Then

$$\oint \vec{B} \cdot d\vec{s} = \mu_0 I_{\text{enclosed}}, \quad (29\text{–}10)$$

AMPÈRE'S LAW

where the integral is taken around that closed path. Equation (29–10), which was formulated by Ampère in the 1820s during his extensive work on magnetism, is known as **Ampère's law**. The direction of the loop integral must be specified: If the fingers of the right hand curl in the same sense as the integral path, the thumb points in the direction a positive current takes in passing through the loop. The total current can include both positive and negative contributions, and *the path does not have to be circular, just closed*.

Ampère's law includes an experimental result that is worth pointing out: *The magnetic fields produced by different currents add, or superpose*, just as the electric fields of different charges add according to the superposition principle.

> **THINK ABOUT THIS...**
>
> CAN YOU MEASURE A CURRENT WITHOUT ENTERING THE CIRCUIT?
>
> It is not always easy to insert an ammeter directly into a circuit wire, for example a wire from your car's battery. If we want to measure the current carried by one or more wires in a bundle of many wires without physically entering the circuit, we can use a device known as a clip-on ammeter, shown in Fig. 29–9. It works by measuring the magnetic field that the current in the wire produces; Ampère's law then gives the current. In its simplest form the magnetic field is measured with a permanent dipole magnet pointer mounted on a center point bearing much like a magnetic compass. When a current is present in the wire, the magnetic field generated deflects the pointer. The deflection angle depends on the torque induced by the magnetic field—and hence the current. In more sophisticated versions of the clip-on ammeter, the magnetic field sensor is a Hall effect sensor (see Chapter 28). Clip-on ammeters can be found in automobile parts stores; mechanics use them to measure the current through a starter motor.

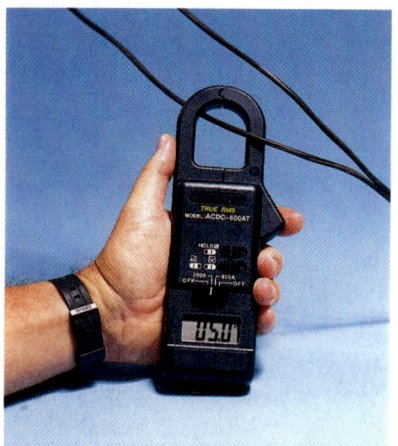

▲ **FIGURE 29–9** This clip-on ammeter uses Faraday's law to measure the current through the wire it surrounds, here 5.0 A. The jaws of the ammeter can be opened to allow its positioning.

Using Ampère's Law to Find the Magnetic Field

If there is some symmetry that suggests that the integral over a particular path is simple, then Ampère's law [Eq. (29–10)] can be used to *find* the magnetic field, in analogy with the way we use Gauss' law to find electric fields. In the case of Gauss' law, the integral is taken over a closed surface, and $\vec{E}$ is related to the electric charge enclosed. In the case of Ampère's law, the integral taken is along a closed path, and $\vec{B}$ is related to the electric current enclosed by the path.

EXAMPLE 29–2 The current I within a wire that has a circular cross section of radius R is known to be distributed uniformly over that cross section. (Real currents in real wires would show some variance from this.) What is the magnetic field as a function of the distance r from the wire's axis outside the wire, and what is it within the wire?

Setting It Up We show current moving through the wire in Fig. 29–10. The wire has cylindrical symmetry—it looks the same as we move around it at a fixed distance.

Strategy The cylindrical symmetry means that any magnetic field will not vary with the angle around the wire; the field is a function only of the radial distance r from the central axis. We can apply Ampère's law—Eq. (29–10)—for a circular path of radius r centered on the middle of the wire, and use the fact that $\vec{B}$ will be the same all along this path. This will allow us to pull the magnitude of the magnetic field from the integration of Ampère's law. In turn, we can then use information about the current enclosed by the path to solve for the field as a function of r. The amount of current enclosed depends on whether the path lies outside or inside the wire. Figure 29–10a shows a path *outside* the wire that will determine the field outside, and Fig. 29–10b shows a path *inside* the wire that determines the field inside.

Working It Out By the right-hand rule, $\vec{B}$ is oriented in the direction of the path, so $\vec{B} \cdot d\vec{s} = B\,ds$. The magnetic field magnitude is constant over the chosen path and thus comes out of the integral, leaving just the circumference of the path. If the circular path is outside the wire, the current enclosed is the *total* current carried by the wire. Thus Ampère's law becomes

$$\oint \vec{B} \cdot d\vec{s} = \oint B\,ds = B \oint ds = B(2\pi r) = \mu_0 I.$$

We can solve for B to find that

$$B = \frac{\mu_0 I}{2\pi r},$$

the same result we found for a thin wire. Note that the magnetic field outside the wire is independent of the size of the wire, just as the electric field outside a spherically symmetric charge distribution is independent of the size of the distribution.

We continue to use symmetry to find the field inside the wire, but this time we take our circular path *inside* the wire (Fig. 29–10b). The current enclosed by the path is I times the ratio of the area of the circle of radius r to the area of the wire:

$$I_{enclosed} = I\frac{\pi r^2}{\pi R^2}.$$

As before, Ampère's law gives

$$\oint \vec{B} \cdot d\vec{s} = B(2\pi r) = \mu_0 I\left(\frac{\pi r^2}{\pi R^2}\right).$$

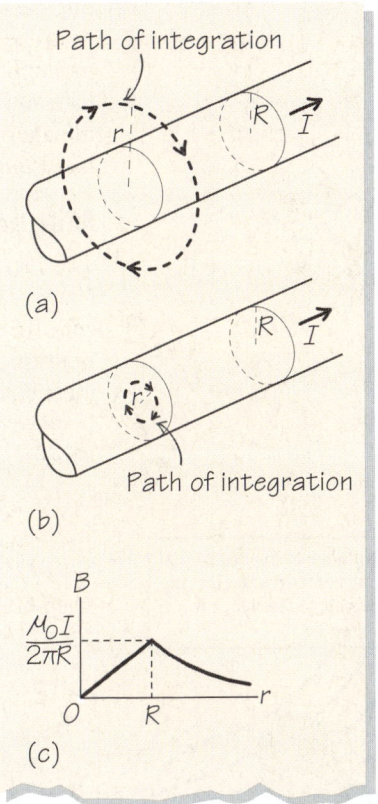

▲ **FIGURE 29–10** (a) A circular path of radius r is used to determine the magnetic field outside a wire that carries a current I. (b) A similar path inside the wire. (c) The magnitude of the magnetic field versus r.

If we solve for B, we find that

$$B = \frac{\mu_0 I}{2\pi R^2}r.$$

By analogy with Gauss' law for electricity, any current outside a circle of radius r makes no contribution to the net magnetic field at radius r. Inside the wire, the magnetic field decreases linearly to zero as r approaches zero. As a check, we see that the results for outside and inside the wire agree at $r = R$. Figure 29–10c is a graph of the magnitude of the magnetic field.

What Do You Think? Would this method give us the same magnitude and direction for the magnetic field if the current were reversed? Answers to **What Do You Think?** questions are given in the back of the book.

29–2 Gauss' Law for Magnetism

We have already studied Gauss' law for electricity. Is there a Gauss' law for magnetism similar to that for electricity? In this section, we shall see in what sense such a law holds.

There is an important difference between electricity and magnetism that determines the form of Gauss' law for magnetism: Despite much experimental effort, *magnetic charges* (**monopoles**) *have never been observed*. A magnetic monopole would be a source of magnetic field analogous to electric charges as sources of the electric field. Finding magnetic monopoles would be the equivalent of being able to isolate the N or S poles of a bar magnet. The bar magnet's magnetic field looks from the outside like the electric field of an electric dipole, and we can of course separate the plus and minus charges that form the electric dipole. But when we try to separate the N and S poles of a bar magnet by cutting the magnet in two, we end up with two smaller bar magnets.

The consequence of these observations for the magnetic field is that if magnetic monopoles analogous to electric charges existed, then magnetic field lines would originate and terminate on magnetic monopoles, just as electric field lines originate and terminate on electric charges. But since there are no magnetic monopoles, *magnetic field lines, unlike electric field lines, must form closed curves.*

Magnetic Flux and Gauss' Law for Magnetism

The fact that there are no magnetic charges means that a relation similar in form to Gauss' law for electricity holds for magnetism, but with the electric charge replaced by zero. In other words, Gauss' law for magnetism is equivalent to the statement that if field lines are continuous, then the number of field lines entering any closed surface must be the same as the number leaving.

We come to the explicit statement of the law by defining the **magnetic flux** Φ_B for a magnetic field $\vec{B}$ over a surface S, open or closed, by

$$\Phi_B(S) \equiv \int_{\text{surface } S} \vec{B} \cdot d\vec{A}. \tag{29-11}$$

Then **Gauss' law for magnetism** is

$$\text{for a closed surface: } \Phi_B = \int_{\text{closed surface}} \vec{B} \cdot d\vec{A} = 0. \tag{29-12}$$

GAUSS' LAW FOR MAGNETISM

As for the electric flux, infinitesimal surface elements $d\vec{A}$ are perpendicular to the surface and, for a closed surface, are oriented outward. We can easily see why this holds by thinking in terms of field lines and noting that the field lines for a magnetic field do not terminate. The magnetic flux is a measure of the number of field lines that go out of a surface minus the number of lines that go in. But since magnetic field lines are continuous, the number of magnetic field lines that enter a closed surface minus the number that leave the surface is zero (Fig. 29–11). Any magnetic field line entering a closed surface must leave it somewhere because there are no magnetic monopoles inside on which magnetic field lines can begin or end.

The SI unit for magnetic flux is the unit of magnetic field times area, that is, tesla times square meters $(T \cdot m^2)$. This unit occurs often enough to be given its own name in SI, the **weber** (Wb), after Wilhelm Eduard Weber:

$$1 \text{ Wb} \equiv 1 \text{ T} \cdot m^2. \tag{29-13}$$

The Field Lines of a Bar Magnet

When we drew the magnetic field lines for a bar magnet in Chapter 28, it may have seemed natural to think of the field lines as starting on the north pole and ending on the south pole. In light of what we have just learned, we now realize that magnetic field lines never start or stop—they are *continuous*. Therefore, we must reconsider our view of the

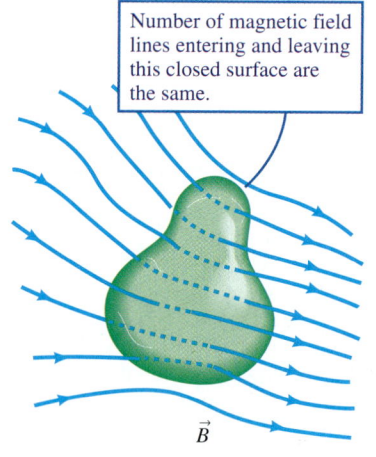

▲ **FIGURE 29–11** Magnetic field lines are everywhere parallel to the magnetic field; their density measures the field's strength. There are no magnetic charges, so magnetic field lines do not end, as there is nothing for them to originate or end upon, and the magnetic flux through a closed surface is zero. This is Gauss' law for magnetism.

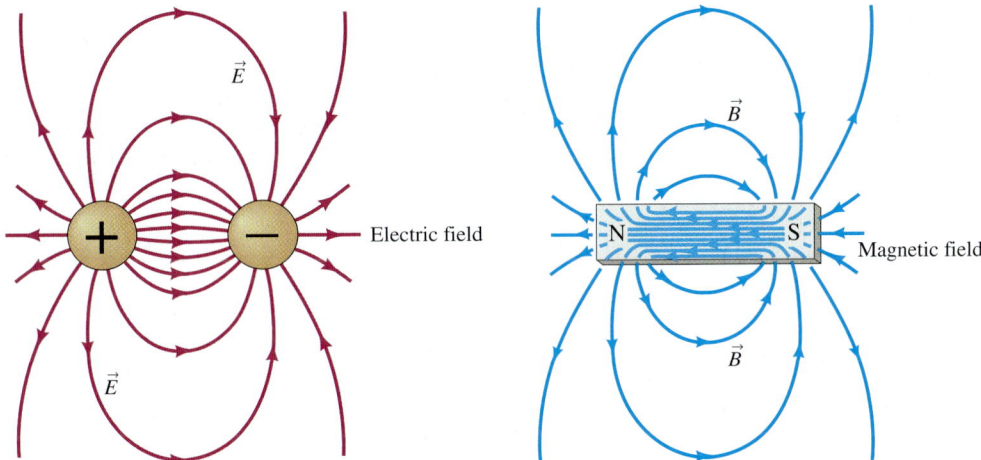

▲ **FIGURE 29–12** Although electric field lines outside an electric dipole resemble magnetic field lines outside a bar magnet, the fields within are quite different. The electric field lines begin and end on the electric charges, whereas the magnetic field lines are continuous.

magnetic field lines for a bar magnet. In fact, the field lines do not start or stop at the poles *but pass through the bar magnet.* This remark underlines a critical difference between the electric field of an electric dipole (equal and opposite charges) and the magnetic field of a bar magnet, both shown in Fig. 29–12. "Outside" the ends, the field lines have the same form. However, while the field lines for the electric dipole begin and end on the charges, the field lines that run from the north pole to the south pole outside the magnet *return within the magnet to form closed loops*. This view is consistent with Gauss' law for magnetism, Eq. (29–12), because for any closed surface that can be drawn in and around a bar magnet, the same number of field lines enter the surface as leave it.

Using Gauss' Law to Find Magnetic Fields

Gauss' law for magnetism is useful for limiting the forms a magnetic field may take. As an example, let's use Gauss' law for magnetism to show that the magnetic field around a straight current-carrying wire can have no radial component, as we concluded in Section 29–1. We need a suitable closed (imaginary) surface to construct about the wire in order to exploit any symmetries, and for a straight wire cylindrical symmetry is appropriate. Our closed surface will therefore be a cylinder of radius R and length L whose central axis lies on the wire (Fig. 29–13). The wire looks the same from any point on the surface of the cylinder, so the magnetic field cannot depend on the angle around the axis of the cylinder. Thus, if there were a radial component B_r at some fixed radial distance, it would have to be the same all around the wire.

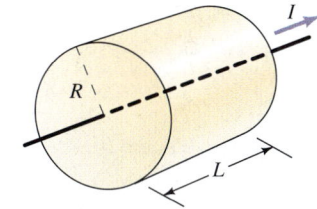

▲ **FIGURE 29–13** A Gaussian surface that exploits the symmetry of a long, straight wire.

To find the net magnetic flux through the closed cylinder, Φ_B, we must consider contributions from its ends and sides. Only a longitudinal component of the field (along the wire) contributes to the flux at the ends. But the contribution from one end must cancel the contribution from the other end—if the longitudinal component enters the surface at one end, it must leave the surface at the other end. The net flux through the ends is therefore zero.

The contribution to the magnetic flux from the sides is due to the radial component of the field, B_r. We have for the net flux

$$\Phi_B = \underbrace{\int \vec{B} \cdot d\vec{A}}_{\text{end}} + \underbrace{\int \vec{B} \cdot d\vec{A}}_{\text{side}} = \underbrace{\int \vec{B} \cdot d\vec{A}}_{\text{side}} = B_r(2\pi RL),$$

where $2\pi RL$ is the area of the cylinder's sides. This must equal zero by Gauss' law and, because the area of the cylinder sides is not zero, B_r must be zero. We have shown by Gauss' law that there can be no radial component of the magnetic field.

We shall see in later chapters that the magnetic flux plays an important role in other fundamental laws of electromagnetism, and it is important to know how to calculate the magnetic flux for both closed and open surfaces. Example 29–3 is an exercise of this type.

EXAMPLE 29–3 The region between the poles of an electromagnet† contains a constant magnetic field, $B = 0.0030$ T, oriented in the $+x$-direction. A square wire loop of sides $L = 1.0$ cm is oriented at a 30° angle to the field (Fig. 29–14a). Find the magnetic flux through the loop.

Setting It Up In order to find the flux through a given surface, we must first specify an area element $d\vec{A}$ of the surface and find its scalar product with the magnetic field, $\vec{B} \cdot d\vec{A}$. We draw a view from the side in Figure 29–14b that includes the orientation of the surface element $d\vec{A}$. This surface element is perpendicular to the plane of the wire loop and makes an angle $\theta = 60°$ with $\vec{B}$.

Strategy Having established the orientation of the area element, we calculate $\vec{B} \cdot d\vec{A}$, then integrate over the whole surface of the loop to find the magnetic flux.

Working It Out The scalar product $\vec{B} \cdot d\vec{A} = B \cos\theta \, dA$. Because B and θ are constants, they can be removed from the integral for the magnetic flux [Eq. (29–11)]:

$$\Phi_B(S) = \int_{\text{surface } S} \vec{B} \cdot d\vec{A} = \int_{\text{surface } S} B \cos\theta \, dA$$

$$= B \cos\theta \int_{\text{surface } S} dA = BL^2 \cos\theta.$$

The numerical value is

$$\Phi_B(S) = (0.0030 \text{ T})(1.0 \times 10^{-2} \text{ m})^2 (\cos 60°)$$
$$= 1.5 \times 10^{-7} \text{ Wb}.$$

What Do You Think? For what orientation of the loop will the flux be a maximum?

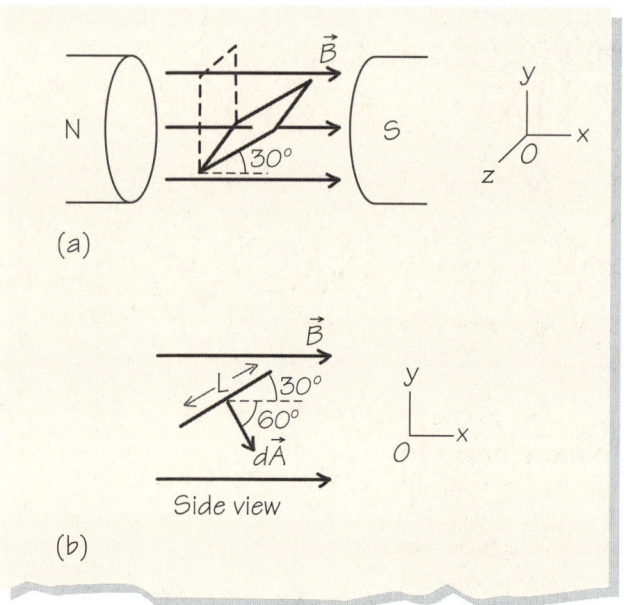

▲ **FIGURE 29–14** (a) A tabletop electromagnet for which $\vec{B}$ is oriented in the $+x$-direction. (b) The surface element $d\vec{A}$ of the square wire loop is oriented perpendicular to the surface.

†An electromagnet is a magnet whose field is produced by appropriately circulating electric currents. It is typically set up so that there are two parallel flat faces separated by a space, with a roughly constant magnetic field running from one face (the N pole) to the other (the S pole)—in the next section, we'll describe how this might be done. It thus resembles a horseshoe magnet whose tips have been formed into the two flat faces.

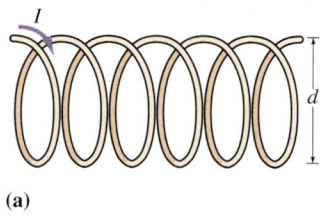

(a)

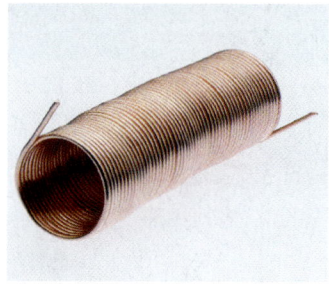

(b)

▲ **FIGURE 29–15** (a) An ideal solenoid is an infinitely long cylinder made from a uniformly wound coil carrying current I (view shown is exaggerated). (b) A real solenoid.

29–3 Solenoids

A parallel-plate capacitor produces a uniform, constant electric field between its plates. The equivalent device for magnetic fields, a **solenoid**, is a length of wire coiled uniformly into a long cylinder, ideally infinitely long (Fig. 29–15). A solenoid generates a constant magnetic field in the interior of the cylinder it forms just as a parallel-plate capacitor sets up a constant electric field in the space between its plates. In Chapter 32 we'll see that, like capacitors, solenoids play an important role in circuits. We mention one use as a mechanical switch: An iron piece near an end of a solenoid moves in response to a magnetic force that appears when a current starts to run through the solenoid.

Let's sketch what the magnetic field of a solenoid might look like. Consider a solenoid whose cylinder diameter is d carrying current I, with the wires wound so that there are n turns per unit length, where the length is measured along the axis of the solenoid. In Fig. 29–15a we have exaggerated the spacing between the wires, which normally are tightly wound. Figure 29–16a is a cross-sectional view of several loops of the solenoid, which are again spaced more widely than in reality. Very near the wire, the magnetic field lines form circles around the wire because the field approximates that of a single straight wire. Figure 29–16a shows that *between* adjacent turns of the wire where two segments of wire are near one another, the fields tend to cancel. *Within* the solenoid, the fields from adjacent turns add together to form a large component that points to the right along the axis of the solenoid. *Outside* the solenoid the fields from the bottom of the wire loop tend to cancel the fields from the top of the wire loop, so that as the number of loops becomes large, the field outside is small (Fig. 29–16b).

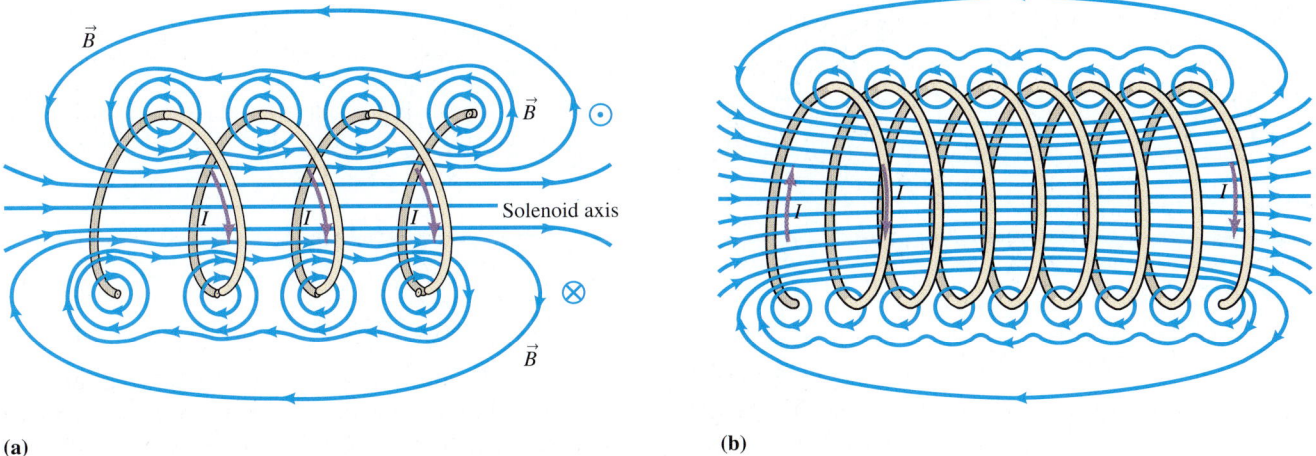

▲ **FIGURE 29–16** (a) One three-turn section of a solenoid, showing the superposed magnetic fields (view shown is exaggerated). (b) As the winding density increases, the magnetic field takes on a simpler form.

To summarize, the fields from the different loops of the coil reinforce inside the cylinder to create a net magnetic field that is parallel to the cylinder axis (Fig. 29–17a) and whose direction is determined by a right-hand rule: If the fingers curl around the solenoid in the direction of the current, the thumb shows the direction of the magnetic field (Fig. 29–17b). Outside the cylinder, the field points primarily in the opposite direction and is much weaker. A quantitative analysis shows that *even though the magnetic field is not exactly zero outside a real solenoid, to a good approximation the field there is insignificant.*

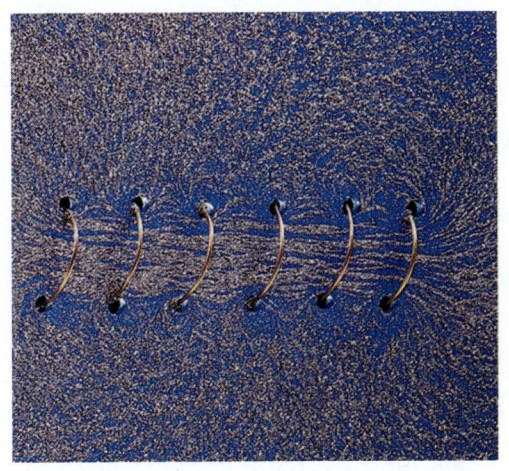

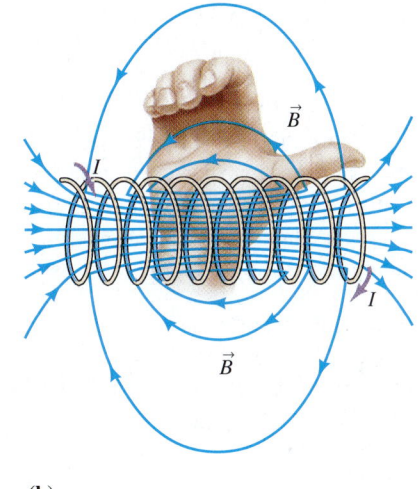

◀ **FIGURE 29–17** (a) Magnetic field lines of a solenoid, as shown by iron filings that align with the field. (b) A right-hand rule gives the direction of the magnetic field within a solenoid.

(a)   (b)

Using Ampère's Law to Find the Magnetic Field in a Solenoid

Now that we understand qualitatively that a long solenoid has a large magnetic field inside—parallel to the solenoid axis—and a weak field outside, we can apply Ampère's law to calculate quantitatively the magnetic field inside an ideal solenoid—one so long that the field outside can be taken to be zero. Figure 29–18 shows a solenoid that carries a current I, and an *imaginary* closed area, bounded by a loop consisting of four legs in a rectangle of length ℓ and height w, on which to apply Ampère's law. The wire of the solenoid passes N times from above through the imaginary loop. The path about the imaginary loop is taken to be clockwise, so the net current into the imaginary loop, NI, is positive by the right-hand rule. We now calculate the line integral on the left-hand side of Eq. (29–10). There is only a very small contribution from leg 2 (point b to point c), because the field outside is insignificant. There is no contribution from leg 1 (point a

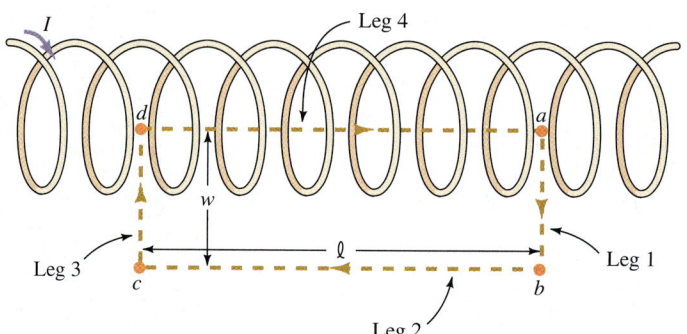

▶ **FIGURE 29–18** An imaginary rectangular loop is drawn half inside and half outside a solenoid. This loop provides a path for the application of Ampère's law.

to point *b*) or from leg 3 (point *c* to point *d*) for two reasons. First, the field outside is insignificant, and the field inside is parallel to the cylinder axis and hence perpendicular to the path. Second, any contributions from these two legs would cancel each other because they are in opposite directions. From point *d* to point *a* (leg 4), the field is parallel to the path. Along this portion of the path, the field has a constant unknown value *B*. The contribution to the integral is $B\ell$, and thus Ampère's law gives

$$\oint \vec{B} \cdot d\vec{s} = B\ell = \mu_0 I_{\text{enclosed}} = \mu_0 NI. \quad (29\text{–}14)$$

We can eliminate the explicit dependence on ℓ in Eq. (29–14) by noting the total number of turns $N = n\ell$, where n is the *turn density* of the solenoid, the number of turns per unit length. We have

$$B\ell = \mu_0 n\ell I,$$

and the *interior magnetic field of a long solenoid* has magnitude

$$B = \mu_0 nI. \quad (29\text{–}15)$$

MAGNETIC FIELD WITHIN A SOLENOID

Note that Eq. (29–15) contains no reference to the distance from the axis on the inside of the loop. Our derivation is completely independent of how close the imaginary path in Fig. 29–18 comes to the solenoid axis, and any choice of this distance would give the same field. *The magnetic field inside a long solenoid, not too close to the ends, is uniform.* The magnetic field depends linearly on the current.

EXAMPLE 29–4 A solenoid consists of wires—each of diameter $d = 0.6$ mm—that can carry a maximum current of $I = 0.03$ A; the wires are tightly wound in a single layer. What is the maximum magnitude of the field inside the solenoid?

Strategy Equation (29–15) gives us the magnitude of the unknown magnetic field inside the solenoid in terms of the current I and the turn density, n. We are given I, but we must calculate n. If the wire has diameter d, and the wires are tightly wound in a single layer, then we have one turn every length d, and $n = 1/d$.

Working It Out We have

$$B = \mu_0 nI = \frac{\mu_0 I}{d} = \frac{(4\pi \times 10^{-7}\ \text{T}\cdot\text{m/A})(0.03\ \text{A})}{0.6 \times 10^{-3}\ \text{m}}$$
$$= 0.6 \times 10^{-4}\ \text{T}.$$

This is comparable in magnitude to Earth's field.

What Do You Think? What would happen to the magnetic field if we double the number of turns of wire by wrapping another row on top of the existing one?

CONCEPTUAL EXAMPLE 29–5 Assume that the maximum current a wire can carry in the previous example is proportional to the area of the wire, and that the wire's diameter is the only variable under consideration. The winding will continue to be in a single layer. How should we change the diameter of the wire to double the magnetic field inside?

Answer If B is to double, then the product nI must double according to Eq. (29–15). Let's see how these factors depend on the diameter d. Because the solenoid is tightly wound, the turn density $n \propto 1/d$. The maximum current is proportional to the wire area, which in turn is proportional to the wire diameter squared. Thus $I \propto d^2$. Combining our results, $nI \propto (1/d)d^2 \propto d$. If the field is to be doubled, the diameter of the wire used in the solenoid should also be doubled.

We have used Ampère's law to calculate the magnetic field inside a long, cylindrical solenoid, and we can also apply the law in the same way to noncylindrical geometry—say, a solenoid with square cross section—with exactly the same results. Equation (29–15) holds *even if the winding does not form a cylindrical tube*. We require only that the solenoid be long and that the cross-sectional area be constant.

Our results for the ideal solenoid hold rather well for a solenoid of finite length. Figure 29–19 shows the magnetic field lines, numerically calculated, of a solenoid in a plane that cuts through the center of the solenoid, whose length is four times its diameter.

The exterior field of the solenoid of finite length illustrated in Fig. 29–19 looks just like the magnetic field of a bar magnet, Fig. 28–2a. Does this mean that the field of a bar magnet has the same physical origin as that of a solenoid? As Ampère himself suggested, the answer is yes. A bar magnet is made of the equivalent of aligned current loops of atomic size. In fact, the interior field of a bar magnet is also the same as the interior field of a solenoid. The origin of magnetism in matter is discussed further in Chapter 31.

A Toroidal Solenoid

A real solenoid has finite length, and therefore its magnetic field departs from its constant value as one approaches either end. These end effects can be eliminated by making the solenoid into a doughnut shape, or *torus* (Fig. 29–20). This shape does, however, introduce some variation in the magnetic field within the solenoid. To look at this variation, consider a toroidal solenoid whose coil radius is r_0 and whose overall radius—the distance from the center to the circular axial line—is R. Symmetry implies that the magnetic field within the coil must be parallel to the cylinder walls. The same arguments that we gave for the straight solenoid imply that this field is oriented in the direction of the thumb when the fingers of the right hand are curled in the direction of the current. Ampère's law can then be used to find the magnitude of $\vec{B}$. Take a circular path for Ampère's law that lies within the coil a distance R' from the center (Fig. 29–20). The magnitude of the field is the same all along the path, so

$$\oint \vec{B} \cdot d\vec{s} = B(2\pi R') = \mu_0 NI.$$

Here, N is the total number of loops in the coil. We solve to find the field magnitude at a distance R',

$$B = \frac{\mu_0 NI}{2\pi R'}. \quad (29\text{--}16)$$

Although there are no end effects, the field does depend on R'—it is not constant across the cross-sectional area of the torus. But if the coil's radius r_0 is much less than the overall radius R of the torus, the possible values of R', from $R - r_0$ to $R + r_0$, do not vary much, and the magnetic field within the torus will not vary very much either.

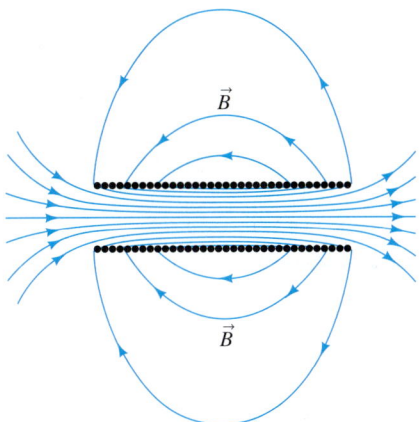

▲ **FIGURE 29–19** The magnetic field of a solenoid of finite length. (After E. M. Purcell, *Berkeley Physics Course: Electricity and Magnetism*, McGraw-Hill, 1990, p. 229.)

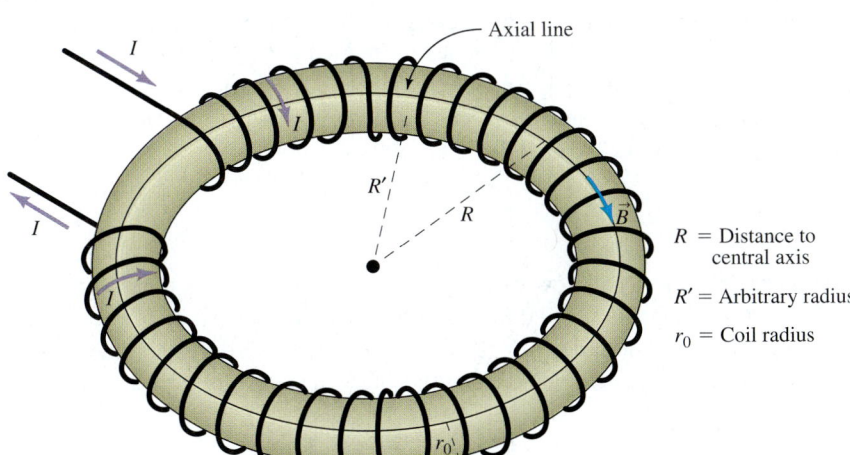

◀ **FIGURE 29–20** A torus wrapped with a wire that carries a current I has a magnetic field inside, which we can calculate by using Ampère's law. The overall radius of the torus is R, whereas the radius of the coil is r_0. The distance R' is not equal to R.

R = Distance to central axis

R' = Arbitrary radius

r_0 = Coil radius

> **THINK ABOUT THIS...**
> **WHAT CAUSES EARTH'S MAGNETIC FIELD?**
>
> In the regions outside Earth itself, Earth's field is that of a magnetic dipole, which is the form of the magnetic field due to a bar magnet or to a current loop or coil. The field's origin is in Earth's high-temperature core, which is mainly composed of iron, with a solid inner core and a fluid outer core. In fact, Earth's core cannot be a gigantic bar magnet, which is formed by the alignment of microscopic magnetic dipoles in substances like solid iron (Section 31–3), because when a bar magnet is too hot, the alignment is spoiled by the high temperature, and Earth's core is far too hot to maintain such an alignment. It is rather associated with the convective motion of the fluid iron, as well as with the rotation of the solid core at a slightly greater speed than Earth's surface. These motions effectively carry current that result in magnetic field lines with the same form as that of a bar magnet or current loop. The behavior of this system is too complex to yield an analytic solution, but it has been successfully modeled. The best mathematical model, which requires large supercomputers to run, has some surprising features. This model confirms the speed of rotation of Earth's solid core. It also explains the hitherto puzzling fact that Earth's magnetic field reverses rather suddenly every hundred thousand years or so (although the reversal is not entirely regular). These reversals are obvious in the geological rock record, because molten rock solidifies with a record of Earth's magnetic field frozen within it. (The Sun reverses its magnetic field every 11 years, the difference with Earth being that the Sun does not have a solid core.)

29–4 The Biot–Savart Law

Ampère's law is a general one, but its usefulness as a tool for calculating magnetic fields depends on the symmetry of the system of currents that create the magnetic field. Here we find a direct expression known as the *Biot–Savart law* for the magnetic field produced by a current, one that can be applied even when there is no symmetry. There is a simple analogue to this procedure in electrostatics. When there is symmetry in a charge distribution, Gauss' law provides a powerful tool for finding the electric field. When there is no symmetry, we can always find the net electric field by using the superposition of the electric fields of point charges (as determined by Coulomb's law). Similarly, the Biot–Savart law gives us the magnetic field due to an infinitesimal distribution of current segments. We then use the superposition principle to determine the magnetic field of a finite arrangement of currents.

Let's start with a result we already know. The magnitude of the magnetic field at a radial distance r from a long, straight wire that carries a current I is

$$B = \frac{\mu_0 I}{2\pi r},$$

Eq. (29–5). The field lines form circles around the wire with the direction given by the right-hand rule. We expect this field to be the sum of the contributions of all the infinitesimal current elements $I\,d\ell$ that make up the wire. To find the form of the individual contributions, note that the $1/r$ dependence of the magnetic field resembles the $1/r$ dependence of the electric field due to a long, charged rod of constant linear charge density λ, as given in Eq. (22–31):

$$E = \frac{1}{2\pi\varepsilon_0}\frac{\lambda}{r}.$$

It will be helpful to recall how we arrived at this result and then to look for a similar procedure for the magnetic field due to a long wire. The electric field result was obtained by integrating the component of the electric field perpendicular to the wire due to the charge in an infinitesimal length $d\ell$ of the charged rod, from Eq. (22–27):

$$dE_\perp = \frac{1}{4\pi\varepsilon_0}\frac{\lambda\,d\ell}{r^2}\cos\phi.$$

Here, the particular element of the charged rod is a distance $r = \sqrt{L^2 + d^2}$ from the point where the field is measured, and ϕ is as shown in Fig. 29–21a. The factor $\dfrac{\lambda\, d\ell}{4\pi\varepsilon_0 r^2}$ is just the electric field strength for a point charge $dq = \lambda\, d\ell$. The second factor, $\cos\phi$, is present only because we are looking at the perpendicular component. If we compare Eqs. (22–31) and (29–5), we see that they have the same form, but with $1/\varepsilon_0$ replaced by μ_0 and λ replaced by I. This suggests that the individual contribution to the magnetic field of a length $d\ell$ of a wire that carries a current I can similarly be found by taking the contribution to the electric field from an element of the charged wire, Eq. (22–27), and replacing $1/\varepsilon_0$ with μ_0 and λ with I:

$$dB = \frac{\mu_0}{4\pi}\frac{I\, d\ell}{r^2}\cos\phi. \qquad (29\text{–}17)$$

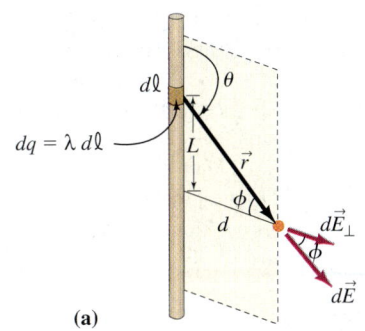

There is, however, a difference in the vectorial aspect of the electric and magnetic fields. The electric field is directed radially away from the charge element, whereas the magnetic field lines form circles. The direction of $\vec{B}$ is shown in Fig. 29–21b for point P and follows the right-hand rule. The direction of the current is indicated by making the infinitesimal length $d\ell$ a vector $\vec{d\ell}$ whose direction is along I. We see that $\vec{B}$ is perpendicular to both $\vec{d\ell}$ and $\vec{r}$. You should recall that there is a direct way to produce a vector that is perpendicular to two other vectors, and that is by using the vector product (see Chapter 10). Using trigonometry, we can see that if $\theta = \phi + 90°$ (as in Fig. 29–21b), then $\cos\phi = \cos(\theta - 90°) = \cos\theta\cos 90° + \sin\theta\sin 90° = \sin\theta$. Equation (29–17) becomes

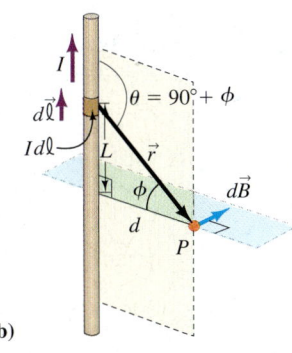

$$dB = \frac{\mu_0}{4\pi}\frac{I\, d\ell\, \sin\theta}{r^2} = \frac{\mu_0}{4\pi}I\frac{r\sin\theta}{r^3}d\ell \qquad (29\text{–}18)$$

in the direction perpendicular to $\vec{d\ell}$ and $\vec{r}$. The magnetic field $\vec{dB}$ produced by a segment of wire $d\ell$ that carries a current I at a displacement $\vec{r}$ from the segment is:

$$\vec{dB} = \frac{\mu_0}{4\pi}\frac{I\, \vec{d\ell}\times\vec{r}}{r^3}. \qquad (29\text{–}19)$$

THE BIOT–SAVART LAW

▲ **FIGURE 29–21** (a) A rod that carries a net charge density λ can be broken up into segments that contribute an electric field at any point. Integration over the contributions from the segments gives the net electric field. (b) A wire carrying current I similarly has a magnetic field that can be calculated by integrating the contributions of segments of the wire. For a segment of length $d\ell$, the vector $\vec{d\ell}$ is oriented along the segment in the direction of I.

The vector product $\vec{d\ell}\times\vec{r}$ has a magnitude $d\ell\, r\sin\theta$, and for the situation in Fig. 29–21 the direction is as drawn there. Note that there is no ambiguity as to the vector $\vec{d\ell}$; if the segment is short enough, it may be treated as a straight line, and that has a definite direction. Equation (29–19) is the **Biot–Savart law**, named after the two physicists who first formulated it, Jean-Baptiste Biot and Félix Savart. This law is analogous to Coulomb's law in electricity. It even has the same overall distance dependence of $1/r^2$—note that the magnitude r appears in the numerator of Eq. (29–19). The angular factors are quite different, however.

Using the Biot–Savart Law

We can use the Biot–Savart law to find the magnetic field due to nonsymmetric current distributions, in just the same way that Coulomb's law is used to find the electric field. The net magnetic field is found by integrating over $\vec{dB}$:

$$\vec{B} = \int d\vec{B} = \frac{\mu_0}{4\pi}\int\frac{I\, \vec{d\ell}\times\vec{r}}{r^3}. \qquad (29\text{–}20)$$

Evaluation of this expression depends on the details of the currents. For most cases Eq. (29–20) is too complicated to give an analytic expression. In that case it is always possible to make a numerical evaluation of the integral. Sophisticated computer programs exist that allow one to design the form of the windings of wire needed to produce a desired field. Such programs superpose infinitesimal field contributions $\vec{dB}$, each given by the Biot–Savart law. Finally it is of course always possible to measure the magnetic fields due to nonsymmetric current distributions experimentally.

834 | The Production and Properties of Magnetic Fields

EXAMPLE 29–6 A straight segment of wire of length L carries a current I. Use the Biot–Savart law to find the magnetic field due to the wire segment in the plane perpendicular to the wire and passing through the midpoint of the wire segment.

Setting It Up Orient the wire along the x-axis, as in Fig. 29–22, with its center positioned at the origin. We then wish to find the field in the yz-plane.

Strategy Since the wire looks the same from anywhere on any circle in the yz-plane that is centered on the wire, we can pick any point on such a circle. Here we choose to find the field at a distance D from the wire along the y-axis with $z = 0$. The right-hand rule shows that the quantity $d\vec{\ell} \times \vec{r}$ points out of the plane where $d\vec{\ell}$ and $\vec{r}$ are as shown in the figure. The field forms circles around the wire, as before.

Working It Out From Eq. (29–18), the magnetic field at the chosen point due to the segment $d\vec{\ell}$ has magnitude

$$dB = \frac{\mu_0 I}{4\pi} dx \frac{\sin(\pi - \theta)}{r^2} = \frac{\mu_0 I}{4\pi} dx \frac{\sin \theta}{r^2} = \frac{\mu_0 I}{4\pi} dx \frac{\cos \phi}{r^2},$$

where we have replaced $d\ell$ by dx and used the angle ϕ defined in Fig. 29–22. To find the net magnetic field, we sum over the contributions of segments from $x = -L/2$ to $x = L/2$:

$$B = \frac{\mu_0 I}{4\pi} \int_{-L/2}^{L/2} \frac{\cos \phi}{r^2} dx.$$

Both ϕ and r depend on x. The integral is computed most simply if we use trigonometric variables; we therefore change variables from x to ϕ. We have by simple geometry $x/D = \tan \phi$. Hence

$$dx = D \, d(\tan \phi) = D(\sec^2 \phi) \, d\phi = \frac{D}{\cos^2 \phi} d\phi.$$

In addition, $r = D/\cos \phi$, so the combination that appears in the integral becomes

$$\frac{\cos \phi}{r^2} dx = \cos \phi \frac{1}{(D/\cos \phi)^2} \frac{D}{\cos^2 \phi} d\phi = \frac{1}{D} \cos \phi \, d\phi.$$

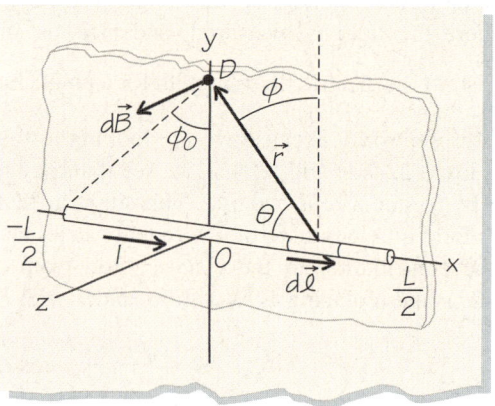

▲ **FIGURE 29–22** A straight segment of wire of length L, carrying a current I, is oriented on the x-axis and centered at the origin.

Thus

$$B = \frac{\mu_0 I}{4\pi D} \int_{-\phi_0}^{+\phi_0} \cos \phi \, d\phi,$$

where $\pm \phi_0$ is the limit of integration—the largest values taken on by ϕ. The integral of the cosine is the sine, and so

$$B = \frac{\mu_0 I}{4\pi D} [\sin \phi_0 - \sin(-\phi_0)] = \frac{\mu_0 I}{2\pi D} \sin \phi_0.$$

We can reexpress this result in terms of L and D by using the geometrical relation

$$\sin \phi_0 = \frac{L/2}{\sqrt{(L/2)^2 + D^2}}$$

to find that

$$B = \frac{\mu_0 I}{4\pi} \frac{L}{D\sqrt{(L/2)^2 + D^2}}. \quad (29\text{--}21)$$

This form shows that the field depends not only on the distance D from the wire but also on the relative magnitudes of D and L.

What Do You Think? Find the magnetic field when $L \gg D$. Compare your result with Eq. (29–5), which is the field of an infinitely long wire.

What is the numerical significance of the fact that the wire of Example 29–6 has finite length? Suppose that the current is 1 A and that the wire segment is 10 cm long. Then, 1 cm from the wire, Eq. (29–21) gives a field of 1.96×10^{-5} T, whereas the magnetic field of the infinitely long wire carrying the same current [Eq. (29–5)] is 2.00×10^{-5} T. In this case, we are a distance from the wire of one-tenth the length of the wire and the field is only 2 percent less than the infinite length result. But let's now move away from the wire so that we are one wire's length away: $L = 0.1$ m, and $D = 0.1$ m. In this case, the field from the wire has a value some 2.2 times smaller than the field of the wire of infinite length.

EXAMPLE 29–7 A wire forms a circular loop of radius $R = 12$ cm. A current $I = 8.0$ A flows counterclockwise in the wire. Find the magnetic field at the center.

Setting It Up We show the wire loop and an element of the wire $d\vec{\ell}$ in Fig. 29–23 along with an appropriate coordinate system.

Strategy There is no path along which the magnetic field is constant, so we cannot use Ampère's law. We must use the Biot–Savart law and integrate over the contributions $d\vec{B}$ of the different elements of the wire to find the unknown total field $\vec{B}$. Vector $\vec{r}$ runs from the current element $d\vec{\ell}$ to the center of the circle, the point where we want to find $\vec{B}$. The quantity $d\vec{\ell} \times \vec{r}$ for the current element shown in Fig. 29–23 is directed along the *x*-axis, and this will be true for all the current elements that make up the loop. The net magnetic field at the center is thus directed along the *x*-axis.

Working It Out The magnitude of the field due to the element shown is [Eq. (29–18)]

$$dB = \frac{\mu_0}{4\pi} \frac{I\, d\ell}{R^2}.$$

There is no sine factor here because $d\vec{\ell}$ is perpendicular to $\vec{r}$. The integral of $d\vec{\ell}$ around the circle is the circumference $2\pi R$, so the net field at the center has magnitude

$$B = \int dB = \int \frac{\mu_0}{4\pi} \frac{I\, d\ell}{R^2} = \frac{\mu_0 I}{4\pi} \frac{1}{R^2} \int d\ell = \frac{\mu_0 I}{4\pi} \frac{2\pi R}{R^2} = \frac{\mu_0 I}{2R}. \quad (29\text{–}22)$$

The numerical value is

$$B = \frac{(4\pi \times 10^{-7}\, \text{T}\cdot\text{m/A})(8.0\, \text{A})}{2(0.12\, \text{m})} = 4.2 \times 10^{-5}\, \text{T}.$$

This is roughly the size of Earth's magnetic field at Earth's surface.

What Do You Think? Is the orientation of the magnetic field we have calculated consistent with a right-hand rule?

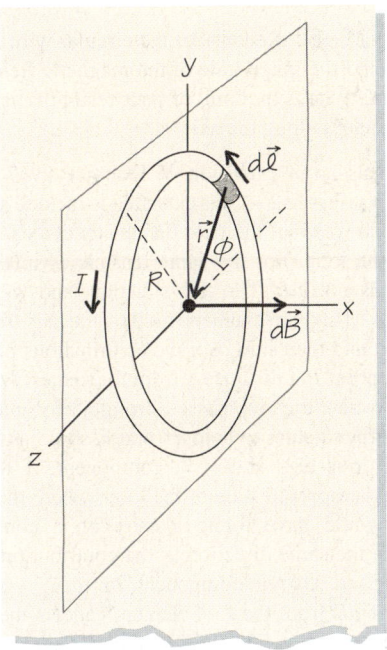

▲ **FIGURE 29–23** An integration is required to find the net magnetic field at the center of the loop.

Problem-Solving Techniques

In the examples of this chapter and Chapter 28, we have studied two aspects of problems on static magnetic fields: We may need to find the magnetic fields produced by a given time-independent set of currents, or we may need to find the magnetic forces on currents or on moving charges. Two sets of key laws contain all that is generally necessary to approach such problems, and you should understand the symbols in these formulas and what the laws mean. First, we have the laws that determine the magnetic field due to currents, which can be written in the two forms

Ampère's law: $\oint \vec{B}\cdot d\vec{s} = \mu_0 I_{\text{enclosed}};$
  (29–10)

Biot–Savart law: $d\vec{B} = \frac{\mu_0}{4\pi} \frac{I\, d\vec{\ell} \times \vec{r}}{r^3}.$
  (29–19)

Second, we have the laws that express the force on a moving charge or a current due to a given magnetic field, namely,

$\vec{F}_B = q\vec{v} \times \vec{B},$ (28–1)

$d\vec{F} = I\, d\vec{\ell} \times \vec{B}.$ (28–14)

Each set of laws involves a right-hand rule. For Ampère's law, if the thumb of the right hand follows the current, the fingers curl in the direction of the integration path. For the force laws and the Biot–Savart law, the right-hand rule for a vector product applies.

Based on these laws, we can suggest a list of habits to develop when solving static magnetic field problems. Many of these are the very same habits that are useful for solving *any* problem in physics.

1. Draw a figure that indicates the physical situation with the quantities known; include directions if appropriate.
2. Write down what is known and what is to be determined. Are you dealing with moving charges or with currents?
3. What physical principles connect the unknown quantities to the known ones?
4. If the problem concerns a force, do you have sufficient information to determine the force directly from the force laws? If not, you may need to compute a magnetic field or integrate an infinitesimal force.
5. In a situation with enough symmetry (for example, for long, straight wires), we can use Ampère's law to calculate a magnetic field. When it is applicable, Ampère's law will usually give the answer more easily than the Biot–Savart law.
6. If the system is not sufficiently symmetric to use Ampère's law, the Biot–Savart law is always available. In using it, be sure that the infinitesimal element $d\vec{\ell}$ and the position vector $\vec{r}$ are identified properly. A partial symmetry may rule out one or more directions for the magnetic field. If only one direction is indicated by symmetry, then the other components will cancel in the calculation of the integral over $d\vec{\ell}$, and you need only integrate for the component desired.
7. Superposition can be a useful tool; the field due to a complex system of currents can sometimes be found by adding the fields due to pieces of the current. For example, the field due to a wire that has an angle bend in it can be found by adding the fields due to the two straight segments.
8. It is always useful to check dimensions and units.
9. Substitute numbers only at the last stage. Any checks you can find of limits or special cases are always helpful.

EXAMPLE 29–8 Reconsider the circular wire loop of Example 29–7. (a) Find the magnitude of the magnetic field all along the axis of the loop. What is the limit of your result (b) at the center and (c) at large distances along the axis?

Strategy This is an extension of Example 29–7 to include all points on the axis. Figure 29–24 is extended to include such points; the axis along which we want the field defines the *x*-axis, with $x = 0$ at the center of the loop. For the same reasons as in Example 29–7, Ampère's law is not useful in this situation, and we must use the Biot–Savart law to find the magnetic field on the axis from a particular current element and then sum over the contributions of the elements.

Consider a point *P* a distance *x* from the center. We have chosen an element $d\vec{\ell}$ where the loop passes through the $+y$-axis. Its contribution $d\vec{B}$ is perpendicular to both $d\vec{\ell}$ and $\vec{r}$, so it has *both* a component along the loop axis and a $+y$-component. If we had chosen an element on the opposite side of the loop, where the loop cuts the $-y$-axis, we would have found a $d\vec{B}$ with a component along the loop axis in the same direction as the contribution from the first loop element, and also a component in the $-y$-direction. The *y*-component of $d\vec{B}$ from the first element cancels the *y*-component from the second element. This will be true for all pairs of elements around the loop, so we must calculate only the component of $d\vec{B}$ along the loop axis, which is the *x*-component. This understanding of the role played by symmetry is important, and a moment or two spent looking for such symmetries is time well spent.

From Fig. 29–24, we see that the vectors $d\vec{\ell}$ and $\vec{r}$ are perpendicular, so the cross product $d\vec{\ell} \times \vec{r} = (d\ell)(r)$. Then

$$dB_x = \frac{\mu_0}{4\pi} \frac{I \, d\ell}{r^2} \cos \gamma.$$

We sum these elements to find the net field. Notice that the *x*-dependence of the field is contained both in the distance *r* and in the angle γ.

Parts (b) and (c) simply require substitutions and limits of the more general result.

Working It Out

(a) From Fig. 29–24 we can use trigonometry for the following simplifications: $d\ell = R \, d\phi$, $\cos \gamma = \dfrac{R}{\sqrt{R^2 + x^2}}$, and $r = \sqrt{R^2 + x^2}$, so that

$$dB_x = \frac{\mu_0}{4\pi} \frac{IR^2 \, d\phi}{(R^2 + x^2)^{3/2}}.$$

The net field is then

$$B_x = \int dB_x = \frac{\mu_0 IR^2}{4\pi (R^2 + x^2)^{3/2}} \int_0^{2\pi} d\phi = \frac{\mu_0 IR^2}{2(R^2 + x^2)^{3/2}}. \tag{29-23}$$

(b) At the center of the loop, $x = 0$ in Eq. (29–23). The result of substituting $x = 0$ correctly gives Eq. (29–22).

(c) At large distances, $x \gg R$, the axial magnetic field in Eq. (29–23) reduces to

$$B = \frac{\mu_0 I}{2} \frac{R^2}{x^3} = \frac{\mu_0}{2\pi} \frac{I\pi R^2}{x^3}. \tag{29-24}$$

What Do You Think? Does the $1/r^3$ (i.e. $1/x^3$ in this case) behavior of the magnetic field remind you of the electric field of (a) a point charge or (b) an electric dipole?

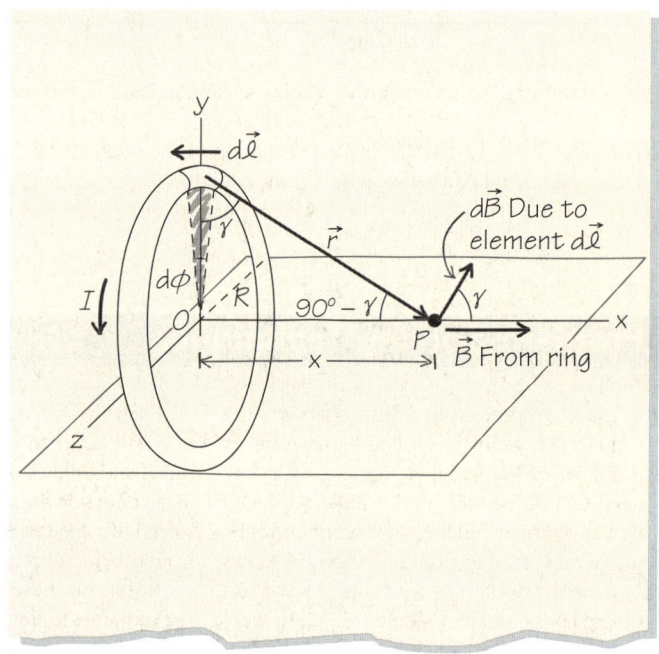

▲ **FIGURE 29–24** The net magnetic field along the loop axis due to a current loop carrying a current *I* is oriented along the axis, according to a right-hand rule. The contribution from an infinitesimal element $d\vec{\ell}$ has components along its axis and in other directions as well.

Magnetic Dipoles

Calculation of the magnetic field over all space due to the current-carrying loop of Examples 29–7 and 29–8 shows that the magnetic field lines shown in Fig. 29–25 are just like the electric field lines of the electric dipole (Fig. 22–12), apart from the region between the two electric charges of the electric dipole. In fact, it is the form of the field, not how that field is produced, that labels the magnetic field of the current loop as a *magnetic dipole field*. The loop forms a magnetic dipole whose strength is characterized by a magnetic dipole moment, μ. The magnetic dipole moment μ plays the same role for the magnetic dipole that the electric dipole moment *p* plays for the electric dipole. (Don't confuse μ with the permeability of free space, μ_0!) By looking at Eq. (29–24), we see that a sensible definition is

for a circular current loop: $\mu \equiv I\pi R^2$. (29–25)

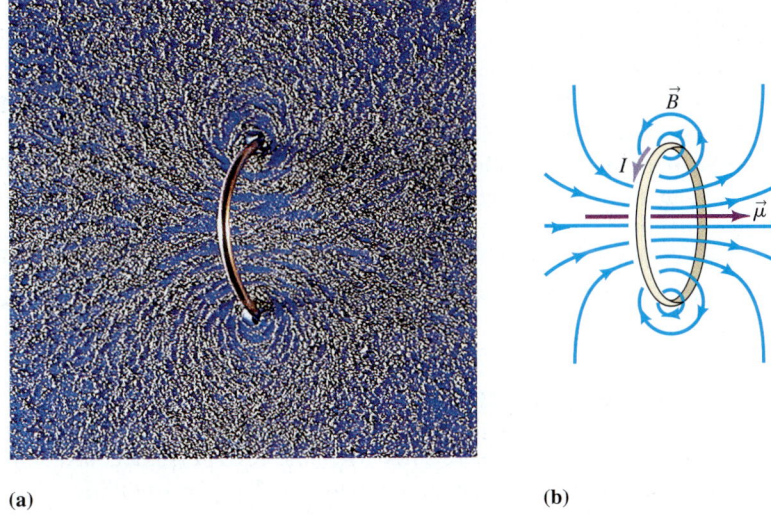

(a) (b)

► FIGURE 29–25 (a) Magnetic field lines for a circular loop of current, as shown by iron filings. (b) The field for such a loop is a magnetic dipole field.

From Eq. (29–24) we know that at a distance x far from the loop the magnitude of the magnetic field along the axis is

$$\text{along the axis: } B = \frac{\mu_0}{2\pi}\frac{\mu}{x^3}. \tag{29–26}$$

In fact, for *any* closed loop of area A that carries a current I, there is a similar result: The magnetic field decreases as $1/r^3$ far from the loop. The strength of the field is proportional to a dipole moment μ that is equal to the product of the current in the loop and the area of the loop:

$$\mu = IA. \tag{29–27}$$

The magnetic dipole moment, like the electric dipole moment, forms a vector $\vec{\mu}$: For any plane loop, $\vec{\mu}$ is perpendicular to the plane of the loop according to a right-hand rule—curl the fingers of the right hand around the direction of current flow in the loop, and the thumb points in the direction of $\vec{\mu}$. The vectorial aspect of the magnetic dipole moment ensures that the magnetic field of the current loop has an appropriate directionality.

One feature of this discussion is quite striking. The electric dipoles that we know about are formed by a pair of equal and opposite electric charges, while the magnetic dipole above is formed by a current loop. Magnetic charges do not exist; the magnetic dipole *cannot* be formed from a pair of equal but opposite magnetic charges, rather it is formed from a closed loop of current. The magnetic field of a current loop (Fig. 29–26a) has the same form as the

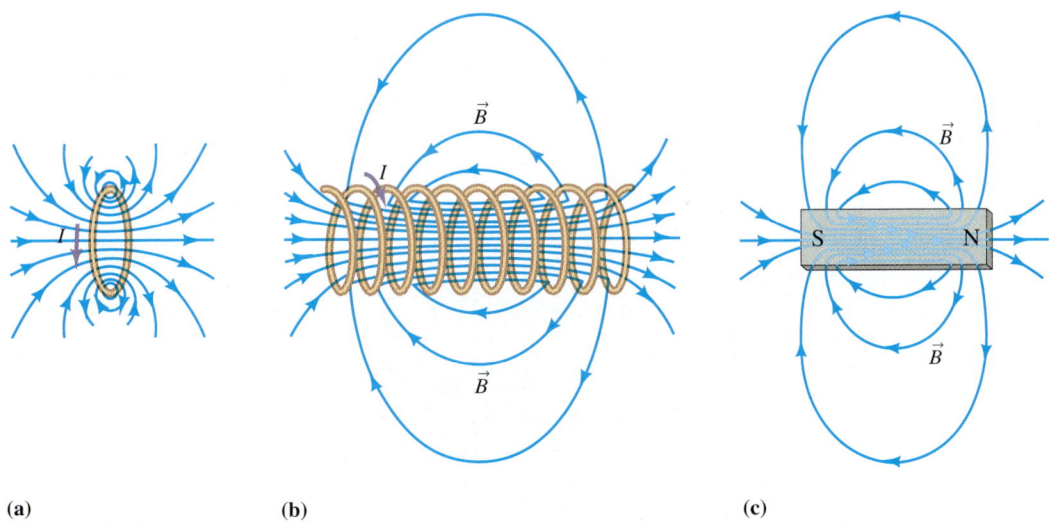

(a) (b) (c)

▲ FIGURE 29–26 (a) The magnetic field lines of a circular loop of current have the same form as the magnetic field lines of (b) a solenoid and of (c) a bar magnet.

fields of a solenoid and of a bar magnet (Figs. 29–26b, c). A solenoid is of course just a particular form of a current loop. The bar magnet is another story, and to understand it fully requires a deeper understanding of materials at the atomic level. But we can say here that the materials that form magnets behave at a microscopic level as if they were composed of microscopic current loops—see Chapter 31 for more on this subject.

In Section 28–5 we studied the *response* of a current loop to an external magnetic field. We saw that the magnetic dipole responds to an external field by rotating, and the torque on the loop is proportional to the magnetic dipole moment. In more detail, the field $\vec{B}_{ext}$ exerts a torque $\vec{\tau} = \vec{\mu} \times \vec{B}_{ext}$ on the loop that tends to line up the vector $\vec{\mu}$ with $\vec{B}_{ext}$. As Eqs. (29–27) and (28–22) show, both the magnetic field of the loop and the reaction of the loop to an external magnetic field are given in terms of the dipole moment.

29–5 The Maxwell Displacement Current

There is a logical flaw in Ampère's law when the current is not constant. In 1865, James Clerk Maxwell modified the law to remove this flaw. This modification was crucial to the completely unified theory of electricity and magnetism that will be discussed in Chapter 34.

Ampère's law is applied with an integration over some closed path. The right-hand side of Ampère's law, Eq. (29–10), contains what we called the current enclosed by a path. By "the current enclosed by a path," we mean the rate of charge flow through a surface whose boundary is the closed path. Such a surface can be chosen in many different ways (Fig. 29–27), but *when the current is continuous, the current that crosses any one of these surfaces must be the same as the current that crosses any other*. For constant current, the freedom to choose a surface therefore presents no problem. There is a situation, however, in which the freedom to choose the surface presents a difficulty, and this situation arises, for example, when the current deposits charge on the plates of a capacitor. Figure 29–28 shows two surfaces with the same loop as their boundary. A current I crosses surface 1 in the positive sense, while no current crosses surface 2, as the charge collects and remains at the capacitor plate. Ampère's law is ambiguous in this situation, because either surface is allowed in its expression.

Maxwell noted that even if no current passes through surface 2, there is a distinguishing feature for this surface: *There is a changing electric flux through it*. As the charge builds up on the plates of the capacitor, the resulting electric field between the plates also builds up, and hence so does the electric flux in the space between the plates.

(a)

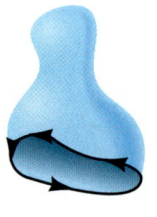

(b)

(c)

▲ **FIGURE 29–27** A closed path defines an infinite number of surfaces.

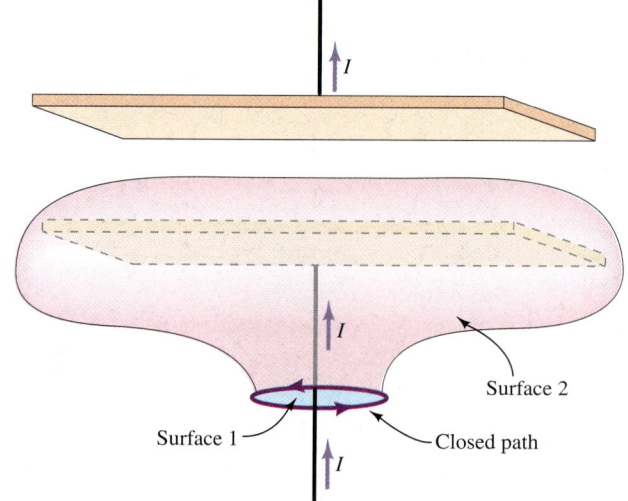

▲ **FIGURE 29–28** Two surfaces bounded by the same closed path. A current passes through surface 1 but not through surface 2.

Suppose that we have a capacitor with large, planar plates of area A (Fig. 29–29). As the current I flows, it deposits a charge $+q$ on one of the plates. We can then use the results of Chapters 22 and 25: The field $\vec{E}$ of such a capacitor is uniform between the plates and small outside the region between the plates. The field points from the plate with positive charge (where the charge accumulates) to the plate with negative charge. The electric flux associated with a surface that passes between the plates, Φ_E, is then just EA. The magnitude of the electric field is given by $E = (1/\varepsilon_0)(q/A)$. Because $\Phi_E = EA$ in this case, this result is equivalent to

$$\varepsilon_0 \Phi_E = q. \tag{29–28}$$

If we take a time derivative, we find the relation

$$\varepsilon_0 \frac{d\Phi_E}{dt} = \frac{dq}{dt} = I. \tag{29–29}$$

Equation (29–29) implies that whatever the value of the current that passes through the wire that leads to the capacitor, that current equals the quantity $\varepsilon_0 \, d\Phi_E/dt$ between the plates. Therefore, if we replace I in Ampère's law by the *sum* of the two terms in Eq. (29–29),

$$I + \varepsilon_0 \frac{d\Phi_E}{dt}.$$

Ampère's law would be satisfied for *any* surface we could draw for the path of Fig. 29–28. For surface 1, only the term I in this sum applies; for surface 2, only the changing flux term applies. The second term, $\varepsilon_0 \, d\Phi_E/dt$, is written in a way that does not refer explicitly to the plane geometry. Indeed, Maxwell was able to show that if the sum of these two terms is used, any surface gives the same answer in Ampère's law. Maxwell called the changing flux term the **displacement current**, I_d:

$$I_d \equiv \varepsilon_0 \frac{d\Phi_E}{dt}. \tag{29–30}$$

DISPLACEMENT CURRENT

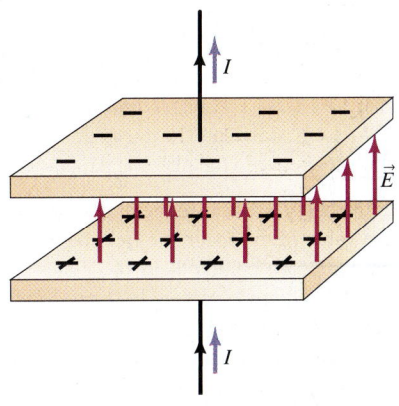

▲ **FIGURE 29–29** The electric field between two parallel plates that carry charges $+q$ and $-q$, respectively, is uniform inside and small outside the region between the plates.

The term "current" is used here because the displacement current has the same dimensions as, and appears in the place of, the current composed of moving charge. Note that the displacement current is present only when there are changing electric fields.

Maxwell's insight that changing electric flux gives rise to a magnetic field, and his quantification of the magnetic field that results, allows Ampere's law to be extended to cover all situations, including those in which there is varying current. Even though the current is not continuous when capacitors are present, *the sum of the ordinary current and the displacement current is continuous.*

Maxwell's generalized form of Ampère's law is accordingly

$$\oint \vec{B} \cdot d\vec{s} = \mu_0(I + I_d) = \mu_0 I + \mu_0 \varepsilon_0 \frac{d\Phi_E}{dt}. \tag{29–31}$$

GENERALIZED AMPÈRE'S LAW

Here, the sum $I + I_d$ is calculated with reference to *any* surface that spans the closed path defining the line integral of the magnetic field. We have concluded from this generalization that changing electric flux produces magnetic field just as moving charges do.

That a changing electric flux produces a magnetic field has great importance for electromagnetic waves, as we shall see in Chapter 34. It may also be of importance when a very large capacitor discharges very quickly, but as Example 29–9 shows, it may otherwise have very little practical effect.

EXAMPLE 29–9 The planar circular plates of a capacitor are being charged. At a given moment, the charge is being built up at the rate of 1 C/s. The plates have a radius $R = 0.1$ m and a separation $d = 1$ cm. Calculate the magnetic field due to the displacement current midway between the plates at a radius equal to half the plate radius.

Strategy Because the plates are circular, symmetry requires that the value of the magnetic field be the same everywhere on path C, a circular path centered on the plates' axis and of radius $R/2$ (Fig. 29–30). Then we can apply the generalized Ampère's law using this path, extracting the magnetic field magnitude B from the integration. The problem statement gives us enough information to find the rate of change of the electric field across the plates and hence the rate of change of electric flux through our path. We can then solve for B.

Working It Out The line integral in Ampère's law is taken in the sense drawn. Because $\vec{B}$ has constant magnitude on this path and points along the path, we can remove it from the integral. The remaining integral is the circumference of the path, $2\pi(R/2)$:

$$\oint \vec{B} \cdot d\vec{s} = B \oint ds = B\left(2\pi\frac{R}{2}\right).$$

To calculate the displacement current, we note that, in terms of the charge q on the plates, the electric field, which is uniform across the region between the plates, has magnitude

$$E = \frac{1}{\varepsilon_0}\frac{q}{\pi R^2}.$$

We must now calculate the electric flux through the area bounded by path C (and not the *total* electric flux in the capacitor). The flux through path C is E times the area $\pi(R/2)^2$:

$$\Phi_E = \frac{1}{\varepsilon_0}\frac{q}{\pi R^2}\pi\left(\frac{R}{2}\right)^2 = \frac{1}{4\varepsilon_0}q.$$

Thus the displacement current is

$$I_d = \varepsilon_0 \frac{d\Phi_E}{dt} = \varepsilon_0\left(\frac{1}{4\varepsilon_0}\frac{dq}{dt}\right) = \frac{1}{4}\frac{dq}{dt}.$$

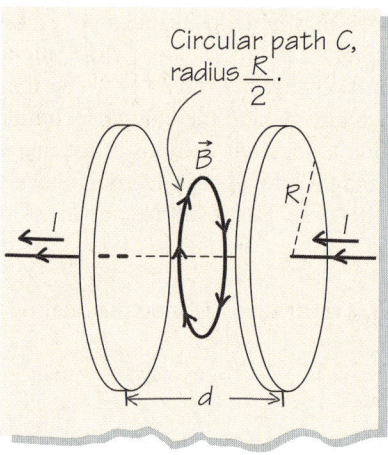

▲ **FIGURE 29–30** Symmetry requires that the value of the magnetic field be the same everywhere on path C.

We can find the magnitude of the magnetic field $\vec{B}$ by using Eq. (29–31):

$$B\left(2\pi\frac{R}{2}\right) = \mu_0 I_d = \frac{\mu_0}{4}\frac{dq}{dt};$$

then

$$B = \frac{\mu_0}{4\pi}\frac{1}{R}\frac{dq}{dt}.$$

Numerically, $dq/dt = 1$ C/s $= 1$ A and $R = 0.1$ m, so

$$B = \frac{4\pi \times 10^{-7}\,\text{N/A}^2}{4\pi}\frac{1}{0.1\,\text{m}}(1\,\text{A}) = 10^{-6}\,\text{T}.$$

This is a small field; for comparison, recall that Earth's magnetic field at Earth's surface is around 10^{-4} T.

What Do You Think? What happens if the distance d between the plates is halved?

Summary

Magnetic fields are produced by electric currents. The magnetic field lines about a long, straight wire that carries a constant current form circles around the wire in the plane perpendicular to the wire. The direction of the field lines in these circles is determined when the thumb of the right hand points along the direction of the current flow: The fingers then curl in the direction of the magnetic field. The magnitude of the field at a radial distance r from the wire is

$$B = \frac{\mu_0 I}{2\pi r}. \tag{29–5}$$

The defined constant $\mu_0 = 4\pi \times 10^{-7}$ T·m/A is the permeability of free space.

The magnetic fields produced by unchanging currents obey Ampère's law:

$$\oint \vec{B} \cdot d\vec{s} = \mu_0 I_{\text{enclosed}}. \tag{29–10}$$

Here, the line integral follows any closed path through which the current I_{enclosed} passes. A second law obeyed by the magnetic field results from the absence of magnetic equivalents to the electric charge. Because there are no magnetic charges on which magnetic field lines begin or end, magnetic field lines must close on themselves. This fact is expressed by Gauss' law for magnetism:

$$\text{for a closed surface: } \Phi_B = \int_{\text{closed surface}} \vec{B} \cdot d\vec{A} = 0. \tag{29–12}$$

This law states that the magnetic flux, Φ_B, through any closed surface is zero; equivalently, the number of magnetic field lines that enter a closed surface is the same as the number of lines that leave the surface.

Ampère's law is an important practical tool for determining magnetic fields when there is enough symmetry to allow a path choice in which the integral simplifies, as in the determination of the interior field of a long solenoid. A solenoid is a wire wound uniformly into a coil to form a tube. When current flows, a magnetic field is produced within the tube that has a constant magnitude and is aligned with the tube axis. The magnitude of the interior field is

$$B = \mu_0 n I, \qquad (29\text{--}15)$$

where n is the number of windings of wire per unit length of the solenoid. Because its interior magnetic field is constant, a solenoid is to magnetism what a capacitor is to electricity.

When there is not enough symmetry to allow Ampère's law to be used to determine the magnetic field produced by a given configuration of currents, the Biot–Savart law can be used instead. According to this law, the magnetic field $d\vec{B}$ produced by a segment of wire $d\vec{\ell}$ that carries a current I at a displacement $\vec{r}$ from the segment is given by

$$d\vec{B} = \frac{\mu_0}{4\pi} \frac{I\, d\vec{\ell} \times \vec{r}}{r^3}. \qquad (29\text{--}19)$$

The magnetic field from an infinitesimal segment can be integrated to find the net magnetic field due to a finite segment of wire.

Application of the Biot–Savart law shows that the magnetic field due to a ring of current, or the exterior field of a solenoid of finite length, is a magnetic dipole field. The form of this magnetic dipole field is the same as that of a bar magnet or, equivalently, has the same form as the exterior electric field produced by an electric dipole. The current loop forms a magnetic dipole, characterized by a magnetic dipole moment μ which is aligned perpendicular to the surface of the loop according to a right-hand rule. Its magnitude is

$$\mu = IA, \qquad (29\text{--}27)$$

where A is the loop area.

If currents are not constant in time, as when wires are interrupted by the presence of charging capacitor plates, then one surface that spans a closed path might not cross the wire that another surface might cross, and the concept of the current enclosed by a path becomes ambiguous. This ambiguity is remedied by Maxwell's modification of Ampère's law to

$$\oint \vec{B} \cdot d\vec{s} = \mu_0(I + I_d) = \mu_0 I + \mu_0 \varepsilon_0 \frac{d\Phi_E}{dt}. \qquad (29\text{--}31)$$

The quantity I_d, proportional to the rate of change of electric flux, is known as the Maxwell displacement current. The surface through which the sum of I and I_d passes is any surface that spans the closed integration path. We conclude from this modification that a changing electric flux creates a magnetic field.

Understanding the Concepts

1. Suppose you move a compass needle near a straight wire that carries a current. Describe how the compass needle reacts when the compass is moved slowly in a circle centered on the wire and perpendicular to it.
2. In the definition of the ampere, does the length of the two parallel wires matter?
3. A wire connected to a battery is placed in the yoke of a tabletop electromagnet when a switch is open. When the switch is closed, the wire may take a big jump upward or downward, according to which side of the battery terminals the wires are attached. Why?
4. Suppose the torus of Figure 29–20 is replaced by a tube that has an almost rectangular cross section and whose shape is irregular (e.g., an extended ellipse). Current-carrying wire is wrapped closely around the tube. What can you say about the magnetic field?
5. In the definition of the ampere, must we worry about the Coulomb forces between the charges in the two wires?
6. Does the statement of the generalized Ampère's law, including displacement current, imply that without the displacement current charge would *not* be conserved?
7. Why is the Biot–Savart law written in differential form? Explain why it cannot be written as in Eq. (29–19) but without the differential signs.
8. Consider the solenoid, length L, of Example 29–4. You increase the number of windings, always using the same wire, so that the magnetic field is doubled. How will the resistance of the coil change?
9. Why is it preferable to define current in terms of the force between two long, parallel wires rather than in terms of the rate at which charge passes a point?

842 | The Production and Properties of Magnetic Fields

10. Is it possible to arrange a set of electric currents and produce a magnetic field that, at large distances from the apparatus responsible, is everywhere directed radially away from the apparatus? Feel free to choose your apparatus, and give either a proof that it is impossible or a description of the apparatus.
11. Suppose that the space between the plates of the capacitor discussed in Section 29–5 is not empty but is filled with a dielectric. How would the treatment in that section, and the determination of the displacement current, change?
12. Suppose that magnetic charges were discovered. What are some practical consequences?
13. In Example 29–8 we looked at the field along the axis of a loop. Suppose you had two identical loops parallel to one another and aligned along the same axis. Consider the point P that is on the axis and midway between the loops. What, qualitatively, is the field at P if the current travels in the same direction in both loops? If the current travels in opposite directions?
14. What are the SI units of the ratio E/B, where E is an electric field and B is a magnetic field?
15. When two bar magnets are placed side by side, they will (a) attract or (b) repel if the adjacent poles are (a) opposite or (b) the same. If you draw magnetic field lines for the combination of two magnets in both cases, the net magnetic field between the magnets will tend to (a) cancel or (b) be doubled. What conclusions can you draw?
16. A current passes through a helical coil. Which of the following statements is true? (a) The helical coil will tend to get shorter; (b) the helical coil will tend to get longer; (c) the coil will get charged; (d) no magnetic field will get produced.

Problems

29–1 Ampère's Law

1. (I) Sketch the magnetic field lines due to two current-carrying wires that are parallel to each other for the cases that (a) the currents move parallel to each other, and (b) the currents move in opposite directions.
2. (I) Consider an array of parallel, current-carrying wires arranged so that they all lie in a plane and are separated by equal distances. Sketch the magnetic field due to this array. Assume that all the currents point in the same direction.
3. (I) What is the force per meter between two long, parallel wires, each carrying 50,000 A but in opposite directions, if the two wires are 30 cm apart? Such currents are normal in the electrolytic production of aluminum.
4. (I) A long, straight wire carries a current of 17 A. What is the magnetic field at a distance of 35 cm from the wire?
5. (I) A lightning conductor carries a current of 5×10^4 A for a short period. During that time, what is the magnitude of the magnetic force per unit length exerted on a parallel wire 4 m away in which a current of 100 μA flows?
6. (I) Two 12-cm-long parallel wires in a handheld calculator are 1.2 cm apart. The currents are parallel, running in the same direction, and have values 35 μA and 8 μA, respectively. What is the force between the wires due to the currents?
7. (I) Find the dimensions of μ_0 and ε_0, and use your expressions to show that the product $\mu_0\varepsilon_0$ has dimensions of $(1/\text{speed})^2$. Find the value of that speed in SI units.
8. (I) A coaxial cable consists of a central wire that carries current I to the right and a tube centered on the central wire that carries the same current to the left. Find the magnetic field outside the cable.
9. (II) (a) In a thick, straight wire carrying a current that is uniform through its cross section, where is the magnetic field the greatest? (b) If the radius of the wire is R and the current is I, what is the value of the maximum magnetic field? (c) What is the minimum magnetic field, and where does this occur? Consider regions both inside and outside the wire. (d) Plot the magnetic field as a function of the distance from the center of the wire.
10. (II) The current-carrying capacity of superconducting wires is limited by the fact that superconductivity breaks down if a large magnetic field is present. Estimate the largest current that can be transported via a NbTi wire, 0.8 mm in diameter, if the critical (breakdown) magnetic field is 10 T. [*Hint*: What is the wire's own magnetic field?]
11. (II) Plot the curves of constant magnetic field in the xy-plane for values B_0, $2B_0$, $3B_0$, and $4B_0$ of the field about a straight wire that carries current along the z-axis. B_0 is some field value that you can choose. These curves are the intersections with the xy-plane of the surfaces of constant field.
12. (II) A very thin, infinitely long metal sheet lies in the xy-plane, between $x = -w$ and $x = w$. A current of density h A/m flows in the $+y$-direction (Fig. 29–31). What are the magnitude and direction of the magnetic field at a distance $z \ll w$ above and below the sheet? Neglect end effects.

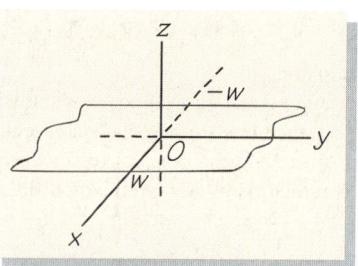

▲ **FIGURE 29–31** Problem 12.

13. (II) Current is carried from a battery to a device by a copper "ribbon" 1 in wide and 1/32 in thick. What is the magnetic field over the surface of the ribbon, if the current it carries is 120 A?
14. (II) Consider two parallel metal sheets, such as the sheet of Problem 12, with currents flowing in opposite directions. What are the magnetic fields between and outside the sheets? What is the situation when the currents are parallel rather than anti-parallel?
15. (II) Consider a wire that passes through the origin along the z-axis and carries a current I (Fig. 29–32a). (a) Calculate the x- and

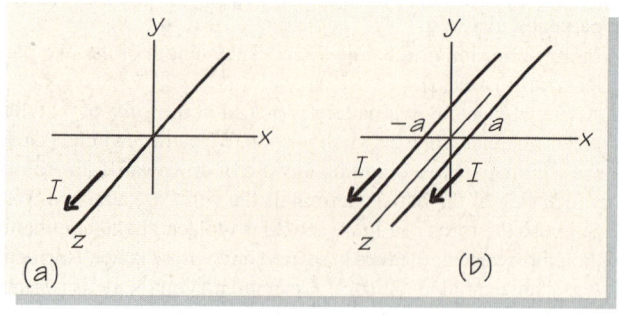

▲ **FIGURE 29–32** Problem 15.

y-components of the magnetic field at a point whose coordinates are $(x, y, 0)$. (b) Use this result to obtain the magnetic field due to two wires that are parallel to the *z*-axis, cross the *xy*-plane at $(a, 0)$ and $(-a, 0)$, and carry current *I* in the $+z$-direction (Fig. 29-32b). (c) What are the fields when the currents are in opposite directions?

16. (II) A uniform current with current density J A/m flows parallel to the *z*-axis on a cylindrical metal sheath, where the radius of the cylinder is R. What is the magnetic field outside the sheath? inside the sheath?

17. (II) Current flows up the inner cylinder of a coaxial cable and returns on the outside cylinder. The radius of the inner cylinder is 0.1 cm, and the radius of the thin outer cylindrical shell is 0.5 cm. Calculate the magnetic field on the cylindrical surface midway between the inner and outer surfaces, given that the current is 10 A. Ignore end effects.

18. (II) Two long, parallel wires carrying a current I in the same direction each have a mass density λ. The wires are initially a distance D apart and are then released. Write a differential equation for the distance between the wires that describes the relative motion of the wires. Ignore all forces other than the magnetic force.

19. (III) An electron beam contains electrons that move along the $+x$-axis at $0.020c$. The beam enters a region of length 1.0 m and runs parallel to a wire that carries a 0.20-A current in the $+x$-direction. The beam is 10.0 cm from the wire. (a) Specify the direction in which the beam is deflected, if at all. (b) Find the deflection of the beam as it passes through the 1-m region by calculating the impulse it receives during its brief passage through that region. (c) After it has passed the wire, does the beam have the same energy it had when it entered the region that contains the wire?

29-2 Gauss' Law for Magnetism

20. (I) Figure 29-33 shows the magnetic field lines that emerge from one pole of a bar magnet; these lines resemble the electric field lines that emerge from one end of an electric dipole. Sketch the magnetic field lines in the region of the pole inside the magnet. For comparison, also sketch the electric field lines in the central region of an electric dipole. What would the magnetic lines look like if the N and S poles of a magnet represented magnetic monopoles, which would be point sources of magnetic field?

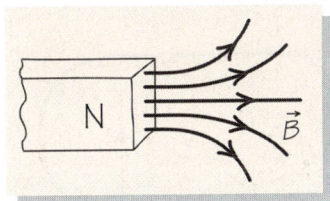

▲ **FIGURE 29-33** Problem 20.

21. (II) A long, current-carrying wire is oriented vertically; next to it is drawn a square whose area lies in the same plane as the wire (Fig. 29-34). Using the distances indicated, find the magnetic flux through the square.

22. (II) Using Gauss' law for magnetism, show that a magnetic field with only an *x*-component must be constant as *x* varies.

23. (II) Show that Gauss' law is satisfied for the magnetic field due to a straight wire that carries a current I in the $+z$-direction for a volume that represents a portion of a cylindrical shell of height h, extending from a radius r to a radius R and formed by an angle θ (Fig. 29-35).

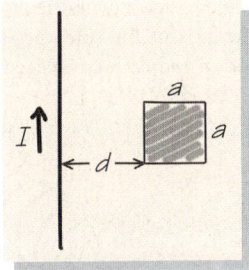

▲ **FIGURE 29-34** Problem 21.

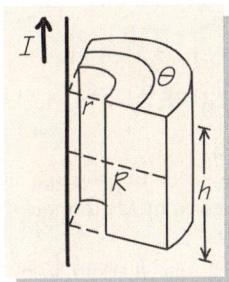

▲ **FIGURE 29-35** Problem 23.

24. (III) Apply Gauss' law for magnetism to a parallelepiped of dimensions a, b, and c, one of whose corners is located at point (x, y, z), as in Fig. 29-36. Assume that the dimensions in the *x*-, *y*-, and *z*-directions (a, b, and c) are small enough so that $B(x + a, y, z) = B(x, y, z) + a\, \partial B/\partial x$, and so on. Show that Gauss' law leads to the condition $(\partial B_x/\partial x) + (\partial B_y/\partial y) + (\partial B_z/\partial z) = 0$ in this limit.

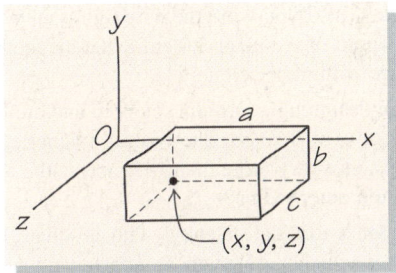

▲ **FIGURE 29-36** Problem 24.

29-3 Solenoids

25. (I) A solenoid of diameter 5 cm has a length of 25 cm and 320 turns of wire. What is the magnetic field at the center of the solenoid when the current in the coil is 3 A?

26. (I) A long, superconducting solenoid is wound with fine niobium–tin wire so that there are 16×10^4 turns/m. If a power supply produces 12 A, what is the magnetic field inside the solenoid?

27. (I) You are told that a toroidal solenoid, carrying a current of 0.36 A through 2500 turns, produces a magnetic field of 5.1×10^{-4} T at its central axis. What can you conclude about the radius of the circle made by that axis?

28. (I) A wire is wound around a torus with outer radius of 32 cm and inner radius of 30 cm. There are 6400 turns in all and the wire carries a current of 2.2 A. What is the range of the magnetic field inside the torus? What percentage change is there from the center to the outside?

29. (I) The magnetic field inside a cylindrical solenoid of area 4 cm² is 0.15 T along the axis of the solenoid. What is the magnetic flux through a disk of radius 3 cm placed perpendicular to the solenoid axis (Fig. 29–37)?

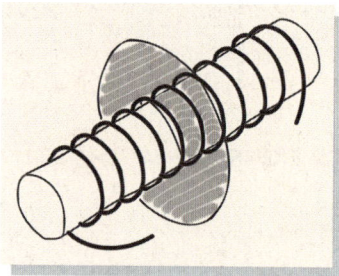

▲ **FIGURE 29–37** Problem 29.

30. (I) Show that the magnetic flux through an ideal cylindrical solenoid of radius R is given by the formula $\Phi_B = \mu_0 n I \pi R^2$, where n is the turn density.

31. (I) What is the current that produces a magnetic field of 0.40×10^{-4} T in the middle of a solenoid that is 1.6 m long and that has 800 turns?

32. (II) A toroidal solenoid consists of a cylinder bent into a circle. The axis of the cylinder is 60 cm from the center of the torus. The cross-sectional area of the cylinder is 70 cm², and the coil has 1200 turns. Given that the current in the coil is 25 A, what is the magnetic field inside the solenoid, along the axis of the cylinder?

33. (II) You have 20 m of #16 AWG copper wire (0.051 inch diameter) that can carry a maximum current of 22 A. (a) If you form the wire into a large circle and pass the maximum current through it, what magnetic field would you produce at the center of the circle? (b) If you wind the wire tightly in a single layer to form a solenoid of diameter 3.0 cm, what magnetic field would you produce within the solenoid?

34. (II) You are designing a toroidal solenoid and are constrained to an inner radius of 0.76 m. The central magnetic field must be 3.0 T and cannot vary more than 15% across the toroid. What is the maximum outer radius?

35. (II) Consider a toroidal solenoid with a square cross section, each side of which has length L. The inner wall of the torus forms a cylinder of radius R. The torus is wound evenly with N loops of wire, and a current I flows through the wire. What is the total magnetic flux through the torus?

36. (II) A toroidal solenoid—similar to the one considered in Problem 35—has a square cross section of side length 1.1 cm and an inner radius of $R = 15$ cm. It is wound with 500 turns of 0.25-mm-diameter copper wire. The wire is connected to a 1.5-V battery with negligible internal resistance. (a) Calculate the largest and smallest magnetic field across the cross section of the toroid. (b) Calculate the magnetic flux through the torus. (c) Do you need to cool the solenoid?

29–4 The Biot–Savart Law

37. (I) Two long wires are placed along the y- and z-axes, respectively. They carry the same current I in the positive directions. Calculate the magnetic field along the x-axis.

38. (I) A single loop of wire forms a rectangle whose sides have lengths 17 mm and 150 mm. The wire carries a current of 36 mA. What is the magnetic dipole moment of the loop?

39. (II) An infinitely long L-shaped wire is placed so that a current I flows in along the y-axis toward the origin, then out from the origin along the x-axis. What is the magnetic field at a point on the z-axis at a height H above the origin?

40. (II) Consider a straight segment of wire of length L that carries a current I. Use the Biot–Savart law to find the magnetic field along the axis of the wire, beyond the wire itself, due to this segment.

41. (II) A differential length $d\vec{\ell}$ of wire carrying a current of 2 A is positioned at the origin of a coordinate system and points in the $+x$-direction. Find the magnetic field due to this wire segment at the following (x, y, z) positions, given in centimeters: (a) $(0, 0, 3)$, (b) $(0, 6, 0)$, (c) $(3, 0, 0)$, and (d) $(6, 0, 6)$. Give both the magnitude and direction of the magnetic field.

42. (II) Consider a thin dielectric ring 7 cm in diameter that rotates around a stem perpendicular to the plane of the ring and through its center at the rate of 120 rev/s. Assume that the ring is charged uniformly and carries a total charge of 8×10^{-7} C. What is the magnetic field produced at the center of the ring by the rotating charge?

43. (II) Repeat the calculation of Problem 42 for a solid disk 5 cm in diameter, with the same total charge.

44. (II) A wire carries a current of 10 A, starting from $x = -\infty$. The wire is laid along the negative x-axis to the point $x = -15$ cm. The wire then follows in the positive y direction for 10 cm, then continues parallel to the x-axis to the point $(x = 15$ cm, $y = 10$ cm$)$, then drops back in a straight line to the x-axis, and continues along the positive axis. What is the magnetic field at the origin?

45. (II) Calculate the magnetic field at the center of a rectangular wire, with sides a and b respectively, if the current flowing through the wire is I. What happens when $a = b$? What happens when $b \gg a$?

46. (II) A current loop consists of a square with sides of length L. A current I circulates counterclockwise around the loop. Find the direction and magnitude of the magnetic field at the center of the square. Compare this to the field at the center of a circular loop of diameter L that carries the same current.

47. (II) Consider the wire shown in Fig. 29–38. Calculate the magnetic field at point P, the center of the half-circle of radius R around which the wire turns, as a function of R and the current I carried by the wire.

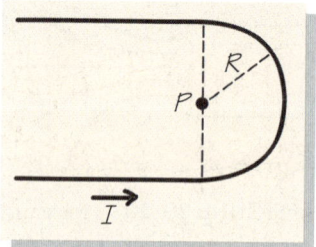

▲ **FIGURE 29–38** Problem 47.

48. (II) A very long wire is aligned along the $+x$- and $+y$-axes, making a right angle at the origin. A current I travels in the $-y$-direction and continues in the $+x$-direction. What is the magnetic field at the point (x, y), where both x and y are positive?

49. (II) Consider the wire shown in Fig. 29–39, see next page, with the inner and outer radii of the semicircle given as 5 cm and 8 cm, respectively. Given that the current in the wire is 12 A, what is the magnetic field at point P, the center of the semicircles?

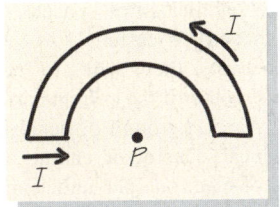

▲ FIGURE 29–39 Problem 49.

50. (II) Calculate the magnetic field at the center of a wire square that consists of 80 loops and has sides of length 5 cm and carries a current of 0.7 A.

51. (II) A charge q moves at instantaneous speed v when it crosses the axis of a ring of current with a magnetic dipole moment μ. At that instant, q is located a distance d from the center of the ring in the direction of the dipole moment vector and is moving perpendicular to the axis (Fig. 29–40). What is the resulting instantaneous motion of the charge? Find the instantaneous radius of curvature of its motion.

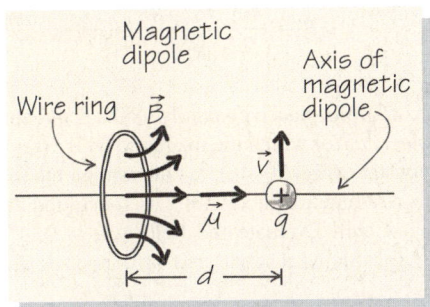

▲ FIGURE 29–40 Problem 51.

52. (II) A circular current loop of radius R produces a magnetic field. At what distance along the axis of the loop does the field have magnitude 0.5 times the magnitude at the center of the loop? At what distance is the magnitude of the field reduced to $1/100$ the value at the center? Give your answer in units of R.

53. (II) Find the magnetic field at point P in Fig. 29–41 if a current of 8 A flows in the infinitely long wire; the radius R of the semicircle is 1.2 cm.

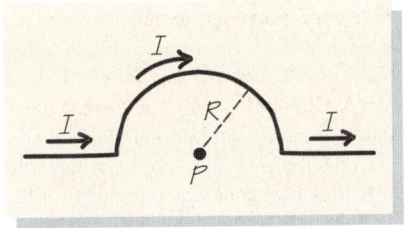

▲ FIGURE 29–41 Problem 53.

54. (III) A segment of wire forms a straight line of length L and carries a current I. Find the magnetic field due to the wire segment in the plane perpendicular to it and passing through one end.

55. (III) By integration, find the magnetic dipole moment of a spherical shell of radius R that carries a total charge Q, distributed uniformly, if the shell rotates with angular velocity ω oriented along the z-axis.

56. (III) Consider a long, thin-walled metal pipe that carries a total current I distributed evenly along the walls of the pipe. A simple application of Ampère's law indicates that the magnetic field inside the pipe is zero. Show by a simple geometric argument that the same result follows from the Biot–Savart law.

29–5 The Maxwell Displacement Current

57. (I) Consider the RC circuit shown in Fig. 29–42. Switch S is closed at time $t = 0$. Calculate the displacement current in the capacitor as a function of time.

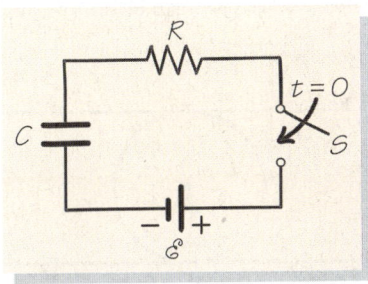

▲ FIGURE 29–42 Problems 57, 58.

58. (I) Consider the RC circuit shown in Fig. 29–42; this time the switch is closed. At some time, the switch is opened. What is the displacement current?

59. (I) A parallel-plate capacitor is being charged at a rate of $I = 0.2$ A. The plates have an area of 0.25 m^2 and are separated by 1.0 cm (Fig. 29–43). What is the value of $\int \vec{B} \cdot d\vec{\ell}$ for a closed path midway between the plates and covering an area of 5.0×10^{-2} m^2?

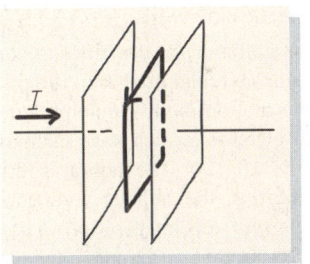

▲ FIGURE 29–43 Problem 59.

60. (II) A 15-μA current starts flowing in a circuit with a 3.5 μF capacitor of area 40 cm^2 at $t = 0$ s. (a) How fast is the voltage across the capacitor plates changing at $t = 0$ s? (b) Use the result of (a) to calculate *explicitly* $d\Phi_E/dt$ and the displacement current at $t = 0$ s.

61. (II) An alternating voltage of the form $V = V_0 \cos(\omega t)$ is connected across a capacitor C. What is the displacement current in the capacitor?

62. (II) A voltage of the form $V = V_0 \cos(\omega t)$, with $\omega = 2 \times 10^4$ rad/s and $V_0 = 0.1$ V, is applied across the plates of a 5-nF capacitor; the plates are 1.5 cm apart. (a) What is the maximum rate of change in electric field between the plates? (b) The maximum value of current leading to the capacitor?

63. (II) A conducting sphere of radius R initially has a uniform surface-charge density σ_0. Beginning at $t = 0$, this charge is drained off over a period t_0 such that $\sigma = \sigma_0[1 - (t/t_0)]$. Find the displacement current at the surface of the sphere as a function of time. Compare the displacement current to the current carried off by the wire.

General Problems

64. (I) Three wires lie in a plane, placed parallel to one another and equally spaced by 0.20 m. If the wires are oriented up and down on the page, the right-hand wire carries 100 A to the top, the middle wire 300 A to the bottom, and the left-hand wire carries 200 A to the top. What are the forces per unit length on each of the wires?

65. (II) Two wires shown in Fig. 29–44 have identical currents flowing. Use the superposition principle and symmetry to obtain as much information as you can about the magnetic field at point P.

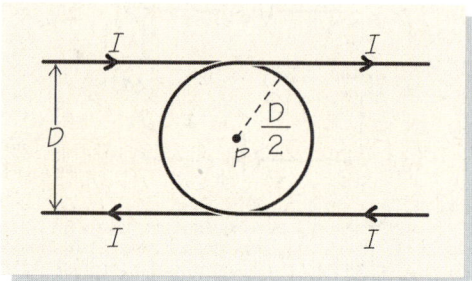

▲ **FIGURE 29–44** Problem 65.

66. (II) Calculate the force per unit area between two metal sheets that carry identical currents in the same direction. The sheets carry a current of linear density h A/m as in Problem 12.

67. (II) Equal but opposite currents I travel in the inner and outer wires of a coaxial cable. As a function of the distance from the central axis, find the magnetic field (a) inside the inner wire; (b) in the region between the wires; (c) in the outer (tubular) wire; (d) outside the outer wire.

68. (II) A hydrogen atom may be described as consisting of an electron that moves in a circular orbit around a proton. The force that gives rise to the motion is the Coulomb attraction between the proton and the electron, which have charges $\pm e$, respectively, where $e = 1.6 \times 10^{-19}$ C. The motion is further constrained by the requirement that the angular momentum has the value $nh/2\pi$, where n is an integer and $h = 6.63 \times 10^{-34}$ J·s, Planck's constant. Calculate the magnitude and direction of the magnetic field at the location of the proton. What is the magnetic moment of the current loop?

69. (II) Find the force between the long, straight wire and the rectangular wire loop shown in Fig. 29–45 for currents $I_1 = 10$ A and $I_2 = 5$ A.

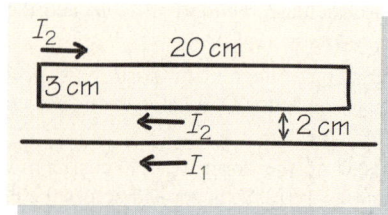

▲ **FIGURE 29–45** Problem 69.

70. (II) The mechanical integrity of solenoids may present a problem that has to be anticipated in technical design. To illustrate the forces that can be present, calculate the force between two neighboring turns of a superconducting solenoid. The radius of the solenoid is 3 cm, the diameter of the wire is 0.8 mm, and the current in the solenoid is 150 A.

71. (II) Consider two parallel wires spaced a distance $d = 1$ cm apart, which each carry a current $I = 1$ A. (a) Compare the magnetic force between these wires to the electric force they would exert on each other if the current carriers (electrons) were not neutralized by a background of positive charges. Use 10^{21} per cm as the linear density of charge carriers in the wire. (b) What excess of electrons per unit length over the positive background would make the electric force equal the magnetic force between the wires? (c) What fraction of the total number of charge carriers is the excess calculated in part (b)?

72. (II) A long wire carries a current I_1. A segment of a second wire, which carries a current I_2, is oriented radially away from the first wire. The segment has length L, and its closest end is a distance d from the first wire. Calculate the torque, direction and magnitude, on the wire segment about the axis defined by the long wire.

73. (II) In Example 29–8, we found the magnetic field due to a circular wire of radius R, carrying a current I, at a point a distance x away from the center of the ring but along the axis to be

$$B = \frac{\mu_0 I}{2} \frac{R^2}{(R^2 + x^2)^{3/2}}.$$

A pair of such coils placed coaxially a distance R apart makes up a *Helmholtz coil*, for which the magnetic field everywhere inside is fairly constant (Fig. 29–46). (a) Determine the magnetic field on the axis as a function of x, with $x = 0$ marking the location of the left-hand coil. Evaluate the field at $x = 0$, $x = R/4$, and $x = R/2$. (b) Show that $dB_x/dx = 0$ and $d^2B_x/dx^2 = 0$ at $x = R/2$.

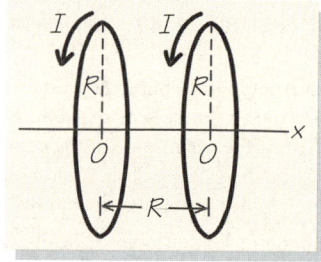

▲ **FIGURE 29–46** Problem 73.

74. (II) A sensitive experiment has to be performed in zero magnetic field. To achieve this, a Helmholtz coil of radius $R = 50$ cm with 50 turns is used to compensate for Earth's magnetic field of 5×10^{-5} T. (A Helmholtz coil is a current loop that sets up a magnetic field to cancel an external field, such as that due to Earth.) (a) What should be the current in the coil? (b) *Estimate* the residual field if the sensitive component of the equipment is confined to a thin cylindrical volume 10 cm long.

75. (III) A demonstration apparatus consists of a large glass bulb containing a small electron gun. The bulb is filled with rarified inert gas, which makes the trajectory of the electron visible. By placing the equipment in the magnetic field of a Helmholtz coil, the experiment described in Example 28–3 can be performed, and the ratio e/m for an electron can be determined. Design the equipment and select appropriate parameters for this demonstration.

76. (III) A certain electric current distribution produces a magnetic field of the form $\vec{B} = \beta(y\hat{i} - x\hat{j})$ near the origin of a coordinate system. Find the current distribution responsible.

◀ Michael Faraday delivering one of his famous public lectures in 1856. These lectures earned him great popular success.

Chapter 30

Faraday's Law

Our treatment of the magnetic field, its sources, and its effects on moving charges and currents brought in elements of the intimate connection between electricity and magnetism. In this chapter, a new feature of this connection is introduced in the form of a new physical law: Faraday's law. This law describes how changes in magnetic fields produce electric fields. Faraday's law has far-reaching technological applications. It lies behind our entire system of electrical power generation and plays a role in most of the electronic devices we use.

Before we look further at Faraday's law, we should acknowledge Faraday the man. Apprenticed as a bookbinder at age 13, he was inspired at age 22 by a series of lectures at the Royal Institution in London to become a scientist. In spite of the fact that he had little mathematical training, in 1813 he became an assistant to Sir Humphry Davy, an already famous scientist. Davy soon recognized Faraday's qualities, and Faraday gradually became more independent in his experimental inquiries. After Oersted discovered in 1819 that an electrical current in a wire deflected a magnetic compass, considerable excitement developed around the possible connections between electrical currents and magnetism. In particular, the inverse question presented itself: Could magnetic fields produce an electric current? Faraday pursued this sort of question for several years—an entry dated 1822 in his notebooks sets the goal "Convert magnetism into electricity." Faraday exhibited all the qualities of a great scientist in the discovery of what we now know as Faraday's law, which states that *changing magnetic fields generate electric fields*. We will study the details of his discovery and some of its ramifications in this chapter.

848 | Faraday's Law

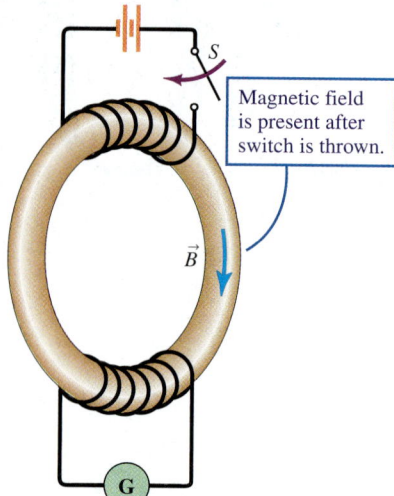

▲ **FIGURE 30–1** Faraday's ring. A changing magnetic flux in the iron ring induces a current in the galvanometer coil at the bottom; the changing flux is due to the opening or closing of a switch connected to the battery of the coil on the top.

▼ **FIGURE 30–2** (a) A figure from one of Faraday's lectures. The movement described in the text creates a changing magnetic flux and induces a current even when no battery is present. Adapted from Faraday's book *Experimental Researches in Electricity* in 1839; the labeling is our own. (b), (c) A more modern version of Faraday's experiment. There is a current produced in the coil only when the bar magnet moves toward (or away from) the coil.

30–1 Faraday's Discovery and the Law of Induction

A great experimentalist such as Michael Faraday recognizes the significance of an odd or unexpected measurement. He or she realizes that a small effect is not always experimental error, and pursues the effect systematically, checks its reality, and considers its ramifications from as many points of view as possible. In 1831, Faraday carried out the experiment shown schematically in Fig. 30–1. The battery sends a current through the coil on the top side of an iron ring, which acts as a solenoid. The galvanometer is used to indicate any current in the coil on the bottom side of the ring. The only unfamiliar element is the iron ring, which does two things: It carries the magnetic field set up by the upper coil within the torus, and hence through the bottom coil (recall that magnetic field lines are closed), and—as we shall discuss in Chapter 31—it *magnifies* the size of the field set up in the upper coil. We say that the iron ring *links* the two coils.

To Faraday's disappointment, he observed no effect on the galvanometer when a *steady* current passed through the upper coil, but Faraday's intuition served him well when he noticed a very small twitch of the galvanometer *when the switch that controlled the flow of current in the upper coil was opened or closed*. Within several days of his observations of this small effect, he completed a series of experiments that revealed essentially all the aspects of magnetic induction.

One of the first things Faraday did was to eliminate the possibility that the battery was itself important to the effect. In a second experiment, illustrated in Fig. 30–2a, two bar magnets make a V shape. A large magnetic flux passes through an iron rod when it touches the ends of the two magnets as shown. This rod is surrounded by a coil attached to a galvanometer. The galvanometer deflects—indicating a current in the coil—both when the rod is brought into contact with the ends of the two bar magnets and when the rod is pulled away. A more modern version of this experiment is illustrated in Fig. 30–2b, c; when a bar magnet remains stationary in the presence of a coil, there is no effect (Fig. 30–2b), but when the magnet moves toward the coil, a current is observed (Fig. 30–2c). The critical observation is that it is the *change* in the magnetic flux through the coil that leads to the creation of the current. The general result, which we will describe in detail below, is that the *change* of the magnetic flux through any surface bounded by a closed line causes an emf around that line. If this closed line is a circuit, then this emf can induce a current in the circuit. Faraday referred to the emf as an **induced emf** and called the current produced by a changing magnetic flux an **induced current**; he called the general phenomenon **magnetic induction**.

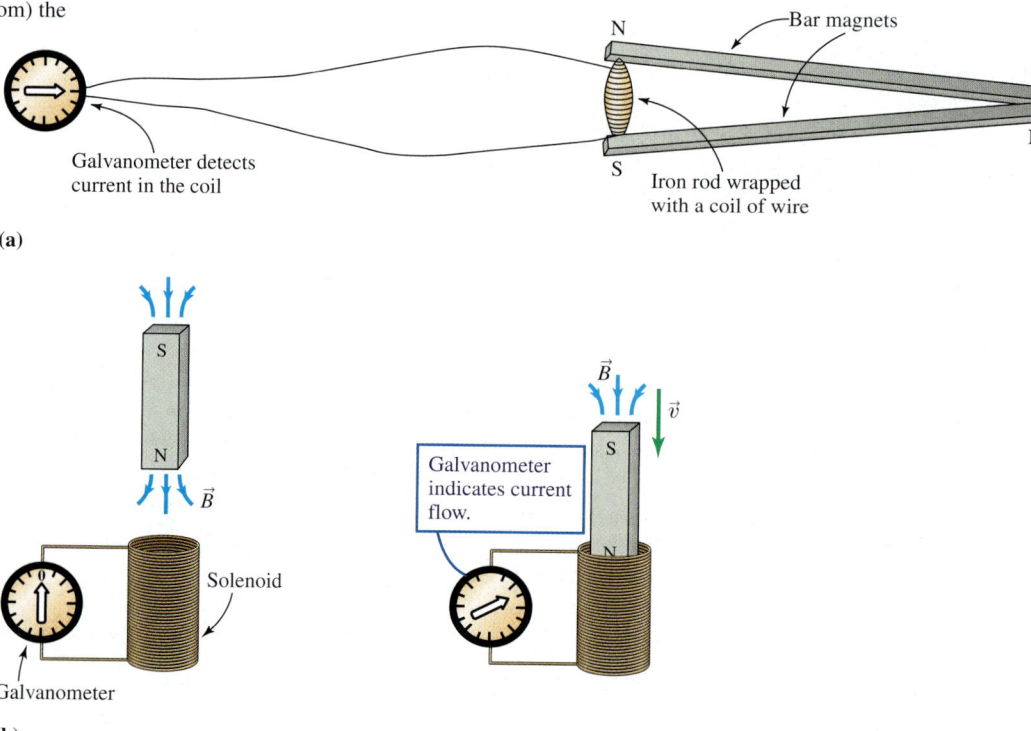

Faraday's discovery was greeted enthusiastically. With it, the possibility of converting mechanical energy to electrical energy became a reality, and electricity generation worldwide is based on Faraday's results. Faraday's discovery could be said to have had a greater effect on the material welfare of humans than any other discovery before or since.

Faraday's Law of Magnetic Induction

We can clarify and add detail to Faraday's law of magnetic induction through a series of experiments, shown in Fig. 30–3.

1. Figure 30–3a shows a bar magnet in the vicinity of a wire loop. If we move the bar magnet toward or away from the wire loop, we observe a current in the wire—an induced current. In Section 29–2 we discussed the magnetic flux, which is associated with the number of magnetic field lines passing through a surface. We have such a flux here, and when the magnet moves as described, the flux through the loop made by the wire changes. Current flows *only* when the magnetic flux through the loop changes. A faster movement—a more rapidly changing flux—results in a larger current than does a slower movement. If the magnet's motion is reversed, then the current reverses.

2. Figure 30–3b shows that as a switch closes in the circuit of a first loop, an induced current momentarily appears in the second loop. Current flow in the first loop produces a magnetic field. When the switch closes, this magnetic field changes as the current in the first loop builds up from zero to its steady value. In turn, this results in a changing magnetic flux through the second loop, and a changing flux through the second loop creates an emf and hence an induced current within it.

3. In Figure 30–3c, there is a constant current in the first (upper) loop, and therefore the magnetic field it generates is constant. But if we move the second loop up or down, the flux through that loop due to the field from the first loop changes. We find that once again an induced current flows in the second loop while it is in motion within the magnetic field of the first loop, and the size of the induced current depends on how fast we move the second loop up or down.

4. Figure 30–3d shows an experiment similar to the third, except that in this case the magnetic flux through the second loop due to the magnetic field of the first loop changes because we change the orientation of the two loops. Again a current flows in the second loop.

5. A last experiment is revealing. We can take the lower coil in Fig. 30–3c and simply squeeze it, thereby changing its area. The flux through it accordingly changes, and once again we observe an induced current, and the size of the current depends on how fast we change the area of the second loop.

▼ **FIGURE 30–3** Ways to make a magnetic flux through a loop change and thereby induce a current I_{ind}. (a) The distance between a wire loop and a bar magnet changes, and a current is induced in the loop. (b) A switch is closed to start a current in one loop, and a current is induced in a second, nearby loop. (c) The distance between a current-carrying loop and a second loop changes, and a current is induced in the second loop. (d) A second loop rotates in the presence of the current-carrying loop, and a current is induced in the second loop.

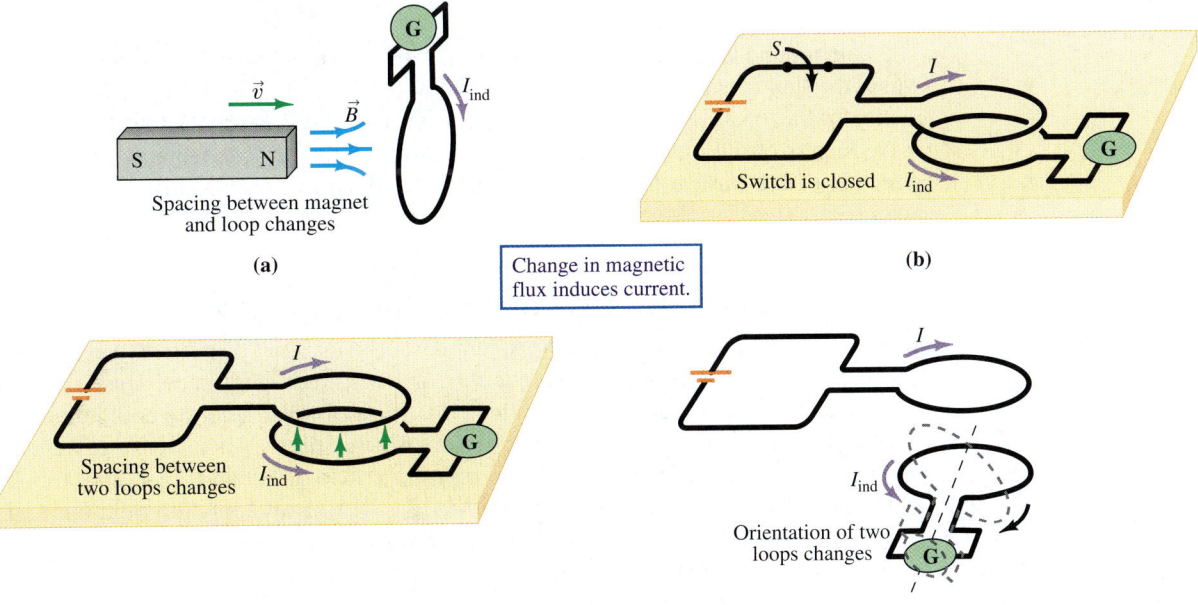

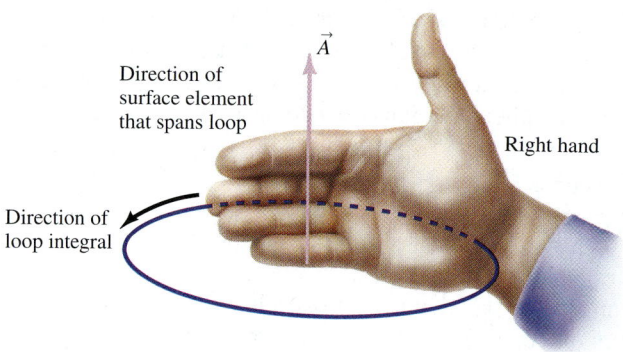

► **FIGURE 30–4** When the direction around a loop is given, the orientation of the surface that spans the loop is specified by a right-hand rule. Here we show a single vector $\vec{A}$ for the entire surface, which is flat. For a curved surface, the directions of infinitesimal areas $d\vec{A}$ vary from point to point.

The common feature to all these experiments is that a changing magnetic flux through a loop induces an emf around that loop—if the loop is formed by a wire, then the induced emf leads to an induced current. How fast the flux through the loop changes determines how large the emf is. Faraday's law summarizes these observations, and also the sign: The negative of the *time rate of change* of the magnetic flux through a surface, Φ_B, equals an emf around the closed loop that bounds the surface. The negative sign dictates the direction of the induced emf, as we will discuss in the next subsection. We know from Chapter 28 that an emf $\mathcal{E}$ is the line integral of an electric field. In this case we are interested in the line integral around a *closed* loop:

$$\mathcal{E} = \oint \vec{E} \cdot d\vec{s}. \tag{30–1}$$

The precise statement of **Faraday's law of induction** is

$$\mathcal{E} = \oint \vec{E} \cdot d\vec{s} = -\frac{d\Phi_B}{dt}. \tag{30–2}$$

FARADAY'S LAW OF MAGNETIC INDUCTION

Here Φ_B is the magnetic flux through the surface S that spans the loop, a quantity defined in Chapter 29:

$$\Phi_B = \int_{\text{surface } S} \vec{B} \cdot d\vec{A}. \tag{29–11}$$

The loop around which the emf is defined, Eq. (30–1), must bound the surface through which the flux is calculated, and the orientation of that surface is determined by the direction of the loop integral and a right-hand rule. This right-hand rule works as follows: If the fingers of the right hand curl in the direction of the loop, the thumb indicates the direction of the surface for calculating the flux—the direction the surface element $d\vec{A}$ takes (Fig. 30–4).

Lenz's Law and the Direction of Induced Current

The minus sign in Eq. (30–2) is critical and deserves a special discussion. Let's look at a loop of wire and suppose the magnetic flux *increases* through it in the sense shown in Fig. 30–5a. (This could happen in a number of ways; for example, we could thrust the N pole of a magnet toward the loop.) When we say that the flux is increasing, we mean that its time derivative is positive, so the right-hand side of Faraday's law [Eq. (30–2)] is negative. The induced emf is therefore *negative*. When we apply the right-hand rule to Fig. 30–5a, we see that the positive direction is counterclockwise; thus the negative sign means that the induced emf is *clockwise* in that figure. The resulting induced current will similarly be clockwise.

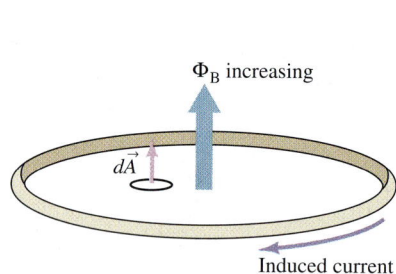

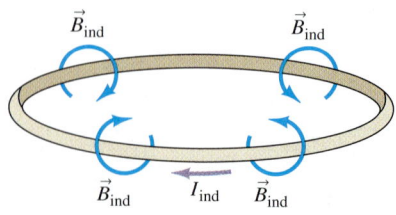

▲ **FIGURE 30–5** (a) A magnetic flux change induces a current. (b) The induced current produces its own magnetic field; that field tends to oppose the flux change that induced the current.

We know that currents produce magnetic fields, and the induced current is no exception. By using the right-hand rule, we can see that the magnetic field produced by the induced current is directed down through the loop (Fig. 30–5b). The direction of this field is such that it tends to *decrease* the magnetic flux through the loop. Because the original flux change that induced the current in the first place was positive, we conclude that the induced current has acted to oppose the flux change that caused it. Further analysis shows that the induced current always opposes the *change* in flux and therefore tends to keep the flux from changing. This way of thinking about Faraday's law is due to Heinrich Emil Lenz, and it is called **Lenz's law**:

> Induced currents produce magnetic fields that tend to oppose the flux changes that induce those currents.

Lenz's law is useful in determining the *direction* of an induced current.

CONCEPTUAL EXAMPLE 30–1 The north pole of a bar magnet is thrust toward the face of a fixed metal ring (Fig. 30–6). Use Lenz's law to determine the direction of any induced current in the ring.

Answer As the north pole of the magnet approaches the ring, the magnetic field lines near the ring become denser. Hence the magnetic flux through the surface of the ring, which is perpendicular to the magnet, *increases*. Lenz's law states that the induced current will *oppose* the change of magnetic flux that passes through the ring. The induced current must therefore produce a field that serves to decrease the magnetic flux (and magnetic field) through the ring. This induced magnetic field will be directed to the left. If we use the right-hand rule, the current in the ring that will produce this field is oriented clockwise as seen by an observer looking toward the north pole of the magnet.

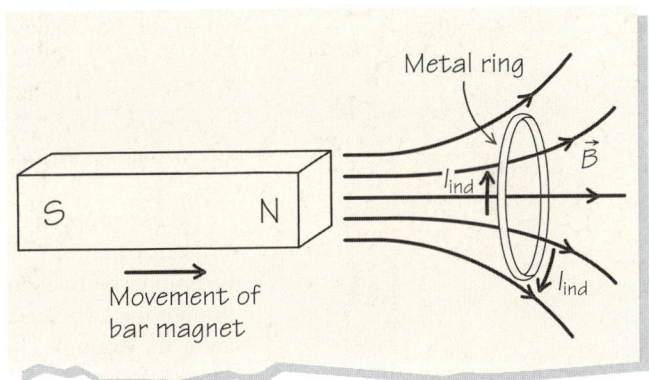

▲ **FIGURE 30–6** When the bar magnet approaches the ring, a current is induced in the ring.

EXAMPLE 30–2 A constant magnetic field has only a y-component B_0 in a large region for which $x < 0$, and is zero for $x > 0$ (Fig. 30–7). A square metal loop with sides of length L is oriented in the xz-plane and pulled through the field in the $+x$-direction with steady velocity $\vec{v} = v\hat{\imath}$. The total resistance of the loop is R. Find the magnitude and direction of any induced current in the wire as a function of time, assuming that the front edge of the square crosses the line $x = 0$ at $t = 0$. Evaluate your result for $B_0 = 1.0$ T, $L = 0.10$ m, $R = 0.065$ Ω, and $v = 10.0$ cm/s.

Strategy Starting at $t = 0$, the wire loop passes out of the region where there is magnetic field ($x < 0$). As it does so, the magnetic flux upward through the loop decreases, and an emf is induced in the loop. We find the magnetic flux through the planar surface of the loop and in particular its time dependence. We can then use Faraday's law, Eq. (30–2), to determine the emf and subsequently the induced current in the wire loop. The loop integral as seen from above is arbitrarily chosen to be counterclockwise. With this orientation for the surface element, the infinitesimal surface elements $d\vec{A}$ that make up the integral are oriented upward, in the same direction as $\vec{B}$, and $\vec{B} \cdot d\vec{A} = B_0 \, dA$. The constant B_0 can then be removed from the integral.

Working It Out For $t < 0$, the flux through the loop has a constant value,

$$\Phi_B = \int_{\text{surface } S} \vec{B} \cdot d\vec{A} = B_0 \int_{\text{surface } S} dA = B_0 L^2,$$

so there is no induced emf and no current.

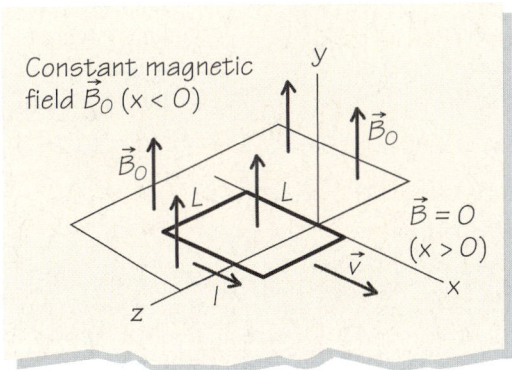

▲ **FIGURE 30–7**

In the time period $t = 0$ to $t = L/v$, the loop is in the process of leaving the region of the magnetic field, and the magnetic flux through it changes. The portion of the loop in the field runs from $x = 0$ to the back of the loop, which is at $x = -(L - vt)$. In other words, only an area $(L - vt)L$ remains in the field, and

$$\Phi_B = B_0(L - vt)L.$$

This flux is not constant, and for this time period

$$\frac{d\Phi_B}{dt} = -B_0 vL.$$

(continues on next page)

The emf counterclockwise around the loop equals the *negative* of this value: $\mathcal{E} = +B_0vL$. A counterclockwise current is induced in the loop during the time interval $t = 0$ to $t = L/v$:

$$I = \frac{\mathcal{E}}{R} = \frac{B_0vL}{R}. \quad (30\text{--}3)$$

The numerical value of this induced current is

$$I = \frac{(1.0\text{ T})(10.0 \times 10^{-2}\text{ m/s})(0.10\text{ m})}{0.065\ \Omega} = 0.15\text{ A}.$$

Finally, for $t > L/v$, the loop has moved out of the region of constant field, so the flux takes on a constant value (zero), and there is neither an induced emf nor a current.

What Do You Think? What happens if the loop is pushed back into the region of magnetic field? (a) nothing, (b) the induced current is in the same direction as before, (c) the induced current is in the opposite direction as before. *Answers to* **What Do You Think?** *questions are given in the back of the book.*

What Sets the Surface Used in Faraday's Law?

In our statement of Faraday's law, we did not specify the surface formed by the loop to which the law makes reference. We shall now show that *any surface bounded by the given loop is suitable because the flux is the same through any such surface*. Let's recall two features of magnetic flux. *All magnetic field lines are continuous; they neither begin nor end on "charges"*; indeed, there are no magnetic charges. Also, *the magnetic flux through a surface is proportional to the net number of field lines that pass through a surface*.

These properties of magnetic field lines have some consequences: Consider two surfaces, S_1 and S_2, bounded by a loop (Fig. 30–8a). Because the lines are continuous, the number of lines that pass through the two surfaces must be the same; hence the flux through the two surfaces must be the same. Some lines may not pass through all surfaces, for example, line 1 in Fig. 30–8b does not pass through S_1, but if a line passes into some third surface S_3 and not into surface S_1, then the line must also pass out of surface S_3 and therefore does not contribute to the net flux through S_3.

From this analysis we can conclude that *the magnetic flux through one surface bounded by a closed loop is the same as the magnetic flux through any other surface bounded by the same loop. Both surfaces must be oriented by the right-hand rule*. This is a very helpful result because it shows that, within the constraint of a given bounding loop, we can choose any surface we like to calculate the magnetic flux. A good problem-solving technique is to find a surface over which the flux is easily calculated when it is necessary to compute the flux, as it is in Faraday's law. Example 30–3 illustrates this technique.

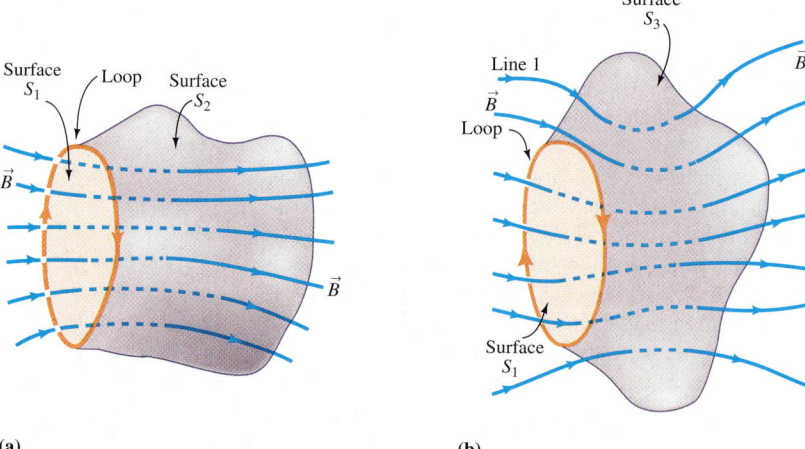

▲ **FIGURE 30–8** (a) The same net number of magnetic field lines pass through any two surfaces S_1 and S_2 bounded by a closed loop. (b) Surfaces S_1 and S_3 are both bounded by a loop. If magnetic field line 1 passes through surface S_3 but not through surface S_1, it must pass through surface S_3 again and hence does not contribute to the net magnetic flux through surface S_3.

EXAMPLE 30–3 Suppose that a certain region has a constant magnetic field of magnitude B_0. Find the magnetic flux upward through a hemisphere of radius R whose base is perpendicular to the field.

Setting It Up In Fig. 30–9 we sketch the situation, including making a choice of axes: We choose the field in the y-direction, $\vec{B} = B_0\hat{j}$, so that the base of the hemisphere forms a circle on the xz-plane.

Strategy We can find the flux through the hemisphere by finding the flux through any other surface bounded by the hemisphere's bounding perimeter, namely the circle of radius R in the xz-plane. The simplest surface that spans this perimeter is the planar disk in the xz-plane. For the disk, $\vec{B}$ is parallel to $d\vec{A}$.

Working It Out Mathematically, the equality of the flux through the hemisphere and through the planar disk can be written as

$$\Phi_{B\text{ hemisphere}} = \int_{\text{planar disk}} \vec{B} \cdot d\vec{A} = B_0 \int_{\text{planar disk}} dA = B_0 \pi R^2.$$

Although the direct calculation of the flux through the hemisphere would be complicated, the calculation becomes trivial when the flat surface is used.

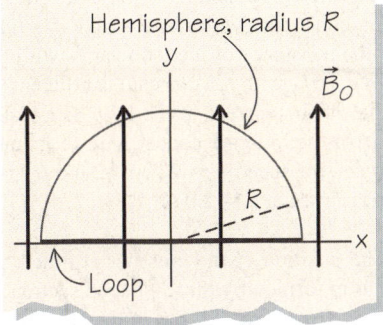

▲ **FIGURE 30–9**

CONCEPTUAL EXAMPLE 30–4 A loop of wire and a straight wire each lie on a tabletop (Fig. 30–10a). The straight wire, which is connected to a battery and carries a current in the direction shown, is moved toward the loop. Is a current induced in the loop? If so, what is its direction?

Answer First, does the movement of the straight wire lead to an induced emf in the loop? In Fig. 30–10b, we have drawn some of the magnetic field lines for the constant current in the straight wire. This field is stronger near the wire, so as the wire moves toward the loop, there is an increase in the flux down through the loop into the table. Because there is a changing flux through the loop, there is indeed an induced emf around it. What is its direction (which will also be the direction of the induced current)? We can use Lenz's law to answer a question about the sign of the induced emf. If we look down on the table, the current is induced in the counterclockwise direction, and the magnetic field associated with this current will come up through the center of the loop, creating a flux that opposes the flux change from the movement of the straight wire (Fig. 30–10c). Thus the direction of the induced current is counterclockwise.

What Do You Think? Will there still be a current induced in the loop if instead of the straight wire with the battery moving toward the loop, the loop moved toward the straight wire?

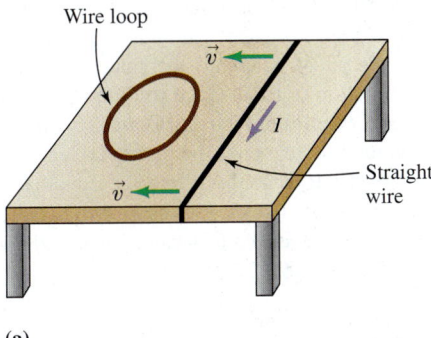

(a)

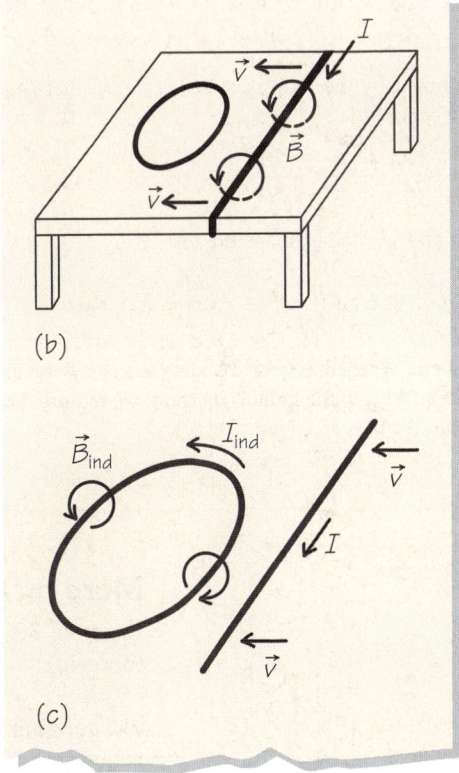

(b)

(c)

▶ **FIGURE 30–10** (a), (b) The field associated with the current in the wire. (c) The current induced in the loop produces a magnetic field of its own.

EXAMPLE 30–5 A closed loop is constructed of a fixed wire shaped as a square-ended U (Fig. 30–11a) and a conducting crossbar that is free to move in the x-direction and makes electrical contact with the U-shaped section. The loop lies in the xz-plane with the square base of the U-shaped section at $x = 0$. A constant magnetic field is oriented in the y-direction, $\vec{B} = B_0 \hat{j}$. (a) The movable crossbar is pulled at a constant speed v to the right, starting from $x = 0$ when $t = 0$. If the resistance of the loop varies with the total length L according to $R = \alpha L$, with α a constant coefficient, what is the direction and value of the current in the loop as a function of time? (b) All segments of the loop are copper wire of radius 0.25 cm, the length $D = 7.0$ cm, the speed $v = 28$ cm/s, and $B_0 = 0.18$ T. What is the current magnitude at $t = 2.0$ s?

Strategy The position of the crossbar at time t is $x = vt$, and the area of the loop formed by the U-shaped section and the crossbar increases. In the presence of the magnetic field, and with the area of the loop oriented upward, parallel to the magnetic field, that means that the flux through the loop also increases, and Faraday's law of induction applies. Lenz's law states that the magnetic field due to the induced current must be directed down in order to decrease the flux (Fig. 30–11b), so the induced current is in the clockwise direction as seen from above. To calculate the magnitude of the induced emf, we must calculate the magnetic flux Φ_B through the loop as a function of time. Once we know the induced emf, we can find the induced current by using Ohm's law. In this case the resistance of the loop increases as the perimeter length increases, and that must be taken into account to find the time dependence of the current. Finally, for part (b) we use the known resistivity of copper to find the coefficient α in the resistance, then find the value of the current at a particular time.

Working It Out (a) The flux through the loop is

$$\Phi_B = \int_{\text{surface}} \vec{B} \cdot d\vec{A} = \int_{\text{surface}} B_0 \, dA = B_0 \int_{\text{surface}} dA = B_0 A.$$

The area formed by the loop is $A = (vt)D$, so that $\Phi_B = B_0 A = B_0 Dvt$, and hence

$$\frac{d\Phi_B}{dt} = B_0 Dv.$$

We now apply Faraday's law, Eq. (30–2):

$$\oint \vec{E} \cdot d\vec{s} = \mathcal{E} = -\frac{d\Phi_B}{dt} = -B_0 Dv.$$

The induced emf is negative, corresponding to a clockwise induced emf.

To relate the emf to the induced current, we require the resistance R of the loop. We have

$$R = \alpha L = \alpha(2D + 2vt) = 2\alpha(D + vt).$$

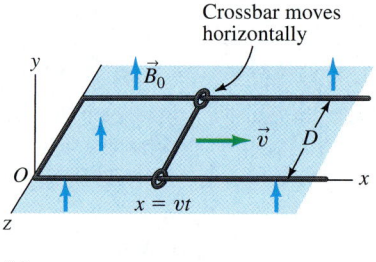

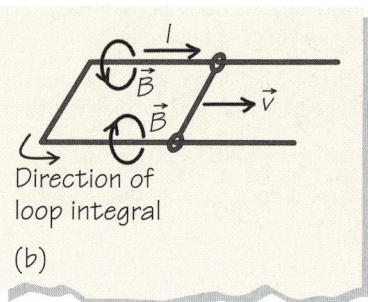

▲ **FIGURE 30–11** (a), (b) Lenz's law shows that the current induced in the circuit that contains the moving crossbar will be in the direction shown.

The induced current is therefore

$$I_{\text{ind}} = \frac{\mathcal{E}}{R} = -\frac{B_0 Dv}{2\alpha(D + vt)}.$$

Try checking the dimensions of this current.

(b) To find α, note that the resistivity of copper is $\rho = 1.72 \times 10^{-8} \, \Omega \cdot \text{m}$ and that, according to Eq. (27–14), the resistance of a wire of length L and cross-section A is $R = \rho(L/A)$. By comparing the form $R = \alpha L$, we see that

$$\alpha = \frac{\rho}{A} = \frac{1.72 \times 10^{-8} \, \Omega \cdot \text{m}}{\pi(0.25 \times 10^{-2} \, \text{m})^2} = 8.8 \times 10^{-4} \, \Omega/\text{m}.$$

At this point, all quantities in the equation for the current are known and, at $t = 2$ s, the magnitude of the induced current is

$$I_{\text{ind}} = \frac{B_0 Dv}{2\alpha(D + vt)}$$

$$= \frac{(0.18 \, \text{T})(0.070 \, \text{m})(0.28 \, \text{m/s})}{2(8.8 \times 10^{-4} \, \Omega/\text{m})[(0.070 \, \text{m}) + (0.28 \, \text{m/s})(2.0 \, \text{s})]}$$

$$= 3.2 \, \text{A}.$$

What Do You Think? Copper has a low resistivity. Nichrome has a much higher one. If nichrome wire is used throughout, the induced current will (a) increase, (b) decrease, (c) stay the same, (d) go to zero.

More on Magnetic Induction

Example 30–5 illustrates some important features of Faraday's law, among them the following:

A Changing Flux Does Not Necessarily Mean a Changing Magnetic Field: The magnetic flux can change not only because the magnetic field changes with time but also because the area of the loop through which the flux is calculated may change with time.

In Example 30–5, the magnetic field is constant in some region, yet there is nevertheless an induced emf. In the next section we see some immediate applications of this fact.

Induced Electric Fields Are Nonconservative: We note that induced fields differ fundamentally from the electric fields we have previously encountered. In our earlier work, electric fields were always associated with *conservative* forces. The work done by those fields in moving a charge around a closed loop is always zero:

$$\text{conservative: } \mathcal{E} = \oint \vec{E} \cdot d\vec{s} = 0,$$

and this fact is what allows us to describe such electric forces with a potential. This is precisely what is *not* true for the fields that result from Faraday's law; the emf about a closed loop is specified by the changing flux:

$$\text{nonconservative: } \mathcal{E} = \oint \vec{E} \cdot d\vec{s} = -\frac{d\Phi_B}{dt}.$$

The induced electric field cannot be described by a potential that is a function of space.

30–2 Motional EMF

We have now developed the idea that an emf can be induced in a conductor that moves in a magnetic field, as in Example 30–5. We call this emf a **motional emf**. Figure 30–12a illustrates this in the case of a conducting rod (length L) moving with constant speed v in a constant magnetic field of magnitude B; in Fig. 30–12a, the axes are chosen so that the position of the rod along the x-axis is $x = vt$. The rod is not part of a circuit, and there is no current flowing through it, but there is an effect on the charge carriers within the rod: There is an accumulation of positive charges at the near end of the rod and negative charges at the far end. The charge accumulation leads to an emf that cancels the motional emf and allows an equilibrium to be established. This effect is described alternatively by Faraday's law or by the Lorentz force law, and here we want to see how that occurs.

Let us first look at Faraday's law applied to this situation. The rod itself does not form a loop, so you may wonder how Faraday's law applies here. But Faraday's law applies to *any* loop; we can think of the rod as forming one leg of a loop that is part real conductor and part imaginary, as indicated by the rod and the fixed dashed lines that complete the loop in Fig. 30–12a. The loop formed by the imaginary line and the rod is situated in the xz-plane with its area elements $d\vec{A}$ oriented in the $+y$-direction. Its area

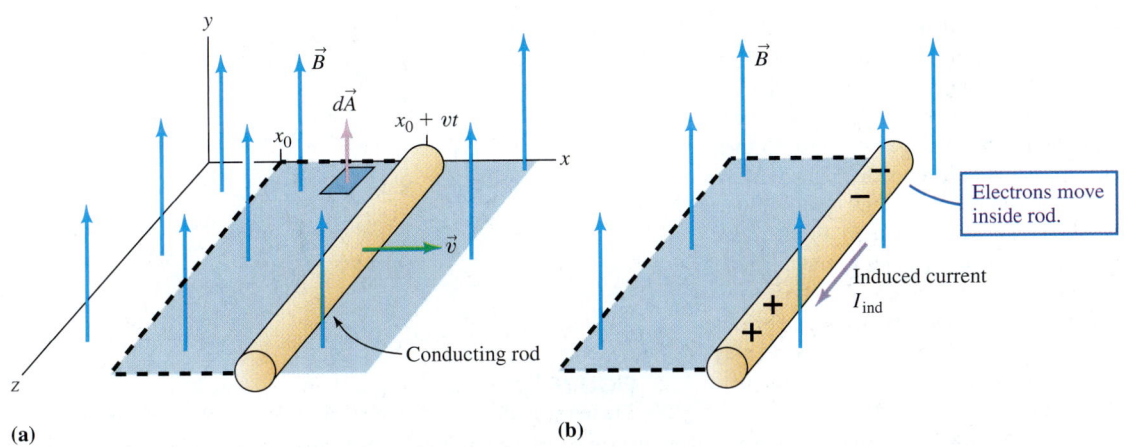

▲ **FIGURE 30–12** (a) A conducting rod of length L moves with constant speed v in the $+x$-direction through a constant magnetic field directed in the y-direction. The rod is oriented in the z-direction. The dashed lines represent the imaginary closure of a loop, to which we apply Faraday's law. (b) The induced emf in the rod can be computed by Faraday's law.

is $L(vt - x_0)$. The flux through the closed loop is therefore

$$\Phi_B = BL(vt - x_0),$$

and the rate of change of this flux is

$$\frac{d\Phi_B}{dt} = BLv. \tag{30–4}$$

The emf taken in the counterclockwise direction looking down—the direction that must be taken if we orient the loop area in the $+y$-direction—is then

$$\mathcal{E} = -BLv. \tag{30–5}$$

$\mathcal{E}$ is the *motional emf*. If the dashed line were conducting wire, this emf would drive a current clockwise (as seen from above) around the circuit. Because there is no actual closed circuit, there is movement of charge, and therefore current flow, only until sufficient positive charges accumulate toward the near end of the rod to set up an emf that cancels the motional emf (Fig. 30–12b). At that point, charge carriers no longer move within the rod. Note that the charges move in such a way that the potential will be higher at the near end of the bar.

The effect of the motional emf—the fact that positive charge moves to accumulate at the near end—is simple to understand in terms of the Lorentz force law. Each charge carrier in the rod is moving in a magnetic field and therefore feels a force that equals $q\vec{v} \times \vec{B}$. This force acts in the $+z$-direction for positive charges, just as the motional emf argument in the paragraph above indicates. The force per unit charge, or the electric field that produces this force, has magnitude vB. Because this electric field is constant along the entire length of the rod, the potential difference from one end of the rod to the other is $\Delta V = EL = BvL$. This potential is just the emf that cancels the motional emf, Eq. (30–5). *We conclude that we can view the effect of motional emf as due either to the Lorentz force law or to Faraday's law of induction.*

CONCEPTUAL EXAMPLE 30–6 In Example 30–2, we looked at how Faraday's law describes induced currents in a square loop of wire as it is pulled through a region of constant magnetic field into a region of no magnetic field. Will a Lorentz force analysis lead to the same results? Assume at $t = 0$ that the loop of wire is entirely within the magnetic field region and is about to start to leave it.

Answer We redraw the situation in Figure 30–13. According to Faraday's law, there is no current in the loop for as long as the loop is entirely within the constant field. A Lorentz force analysis leads to the same result: The force on positive charges in leg a is directed downward, as is the force on the positive charges in leg c, so these forces oppose each other within the loop and no charges will move around the loop. A similar conclusion holds for legs b and d. Once leg a passes out of the field region (as shown in Fig. 30–13), however, there is a net force due to the charges in leg c, and this tends to push the positive charges in the counterclockwise direction, as we deduced from Faraday's law in Example 30–2. Also, the emf will be B_0vL, just as we found in Example 30–2. Finally, there are no magnetic forces once the loop has entirely left the region of the magnetic field, and hence no current, just as Faraday's law states.

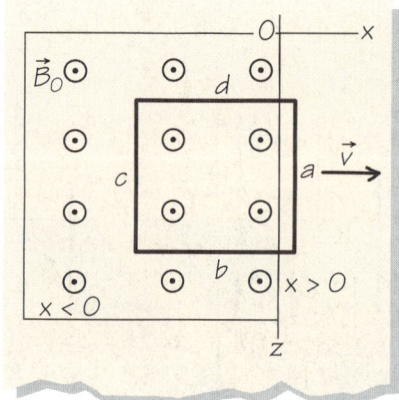

▲ **FIGURE 30–13** An analysis of the moving loop of Example 30–2 in terms of the Lorentz force law.

EXAMPLE 30–7 A rod of length L rotates counterclockwise with constant angular velocity ω about one end in a constant magnetic field of magnitude B_0 oriented perpendicular to the rod's plane of rotation. Find the motional emf in the rod by applying Faraday's law of induction.

Setting It Up We illustrate this situation in Fig. 30–14a, which also includes specification of a coordinate system.

Strategy This is a case similar to the one described at the start of this section. The rod itself does not make a loop, but we construct one through the dashed line shown in Fig. 30–14b. We can then apply Faraday's law of induction by calculating the time dependence of the flux through the loop. It is simplest to orient the area elements of the loop parallel to the magnetic field. Then the flux through the loop is just the constant field times the area of the loop. We shall first find the area of our imaginary loop. This will allow us to find the magnetic flux and subsequently to find its change with time that will give us the emf.

Working It Out With the angle $\theta = \omega t$ specified in the figure, the loop area is

$$\frac{1}{2}\theta L^2 = \frac{1}{2}\omega t L^2.$$

Since the area is oriented upward, the loop integral is followed counterclockwise as seen from above. The magnetic flux through the loop is B_0 times the area; the rate of change of the flux is

$$\frac{d\Phi_B}{dt} = \frac{d}{dt}\left(\frac{1}{2}B_0 \omega t L^2\right) = \frac{1}{2}B_0 \omega L^2.$$

The motional emf is then

$$\mathcal{E} = -\frac{1}{2}B_0 \omega L^2.$$

The sign corresponds to a clockwise emf—that is, the emf drives positive charges radially out from the origin.

What Do You Think? Does the application of the Lorentz force law imply the same direction of charge flow?

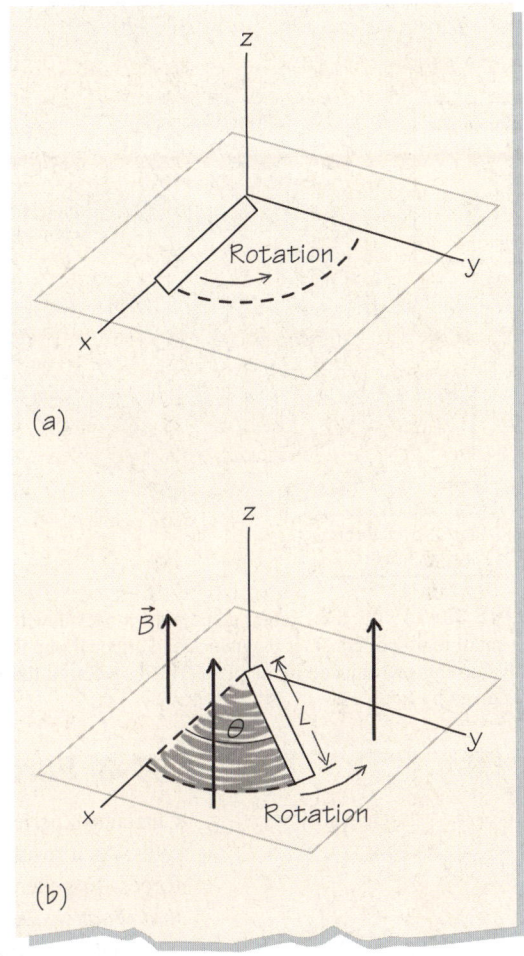

▲ **FIGURE 30–14** (b) The dashed lines represent the imaginary closure of a loop, to which we apply Faraday's law.

Eddy Currents

We have talked about induced emfs in the motion of wires and rods. But in many applications large pieces of metal move in a magnetic field, and there the effects of induced emfs can be significant, either desirable or undesirable. In large pieces the induced currents are spread through the material, and we refer to them as **eddy currents**. Figure 30–15 shows the eddy currents set up in a flat, vertically oriented plate moving downward through a region with a horizontal magnetic field.

Eddy currents are dissipated in Joule heating through the resistivity of the metal plate. The eddy currents or the related Joule heating can be a significant advantage in certain applications; eddy currents in a piece of metal that moves through a magnetic field act as brakes. Brakes of this type have practical applications that range from large electric motors to sensitive measuring devices within which oscillations or other motion would be a disadvantage. When eddy currents are undesirable, they can be reduced by eliminating paths for the current flow. This is done either by cutting slots in the metal plate (Fig. 30–16) or by laminating the metal with an insulator.

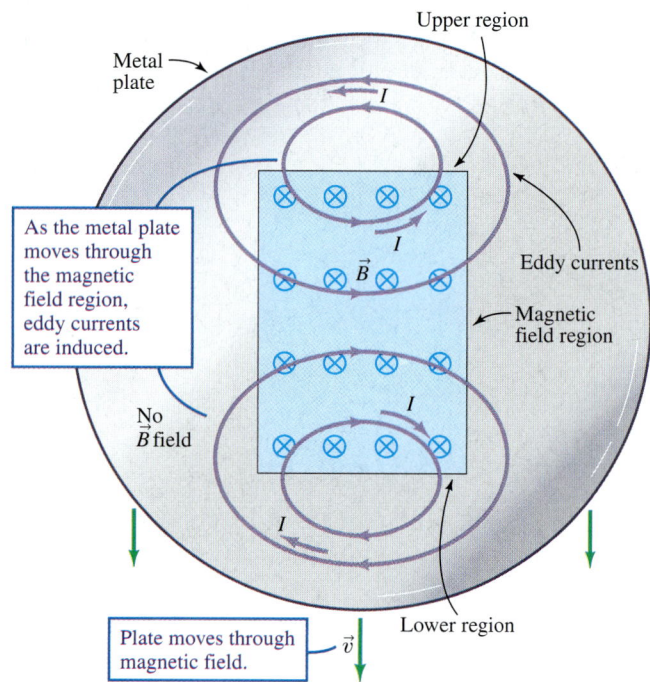

▲ **FIGURE 30–15** As the circular metal plate moves down through a small region of constant magnetic field directed into the plate, eddy currents are induced in the plate. The direction of these currents is given by Lenz's law.

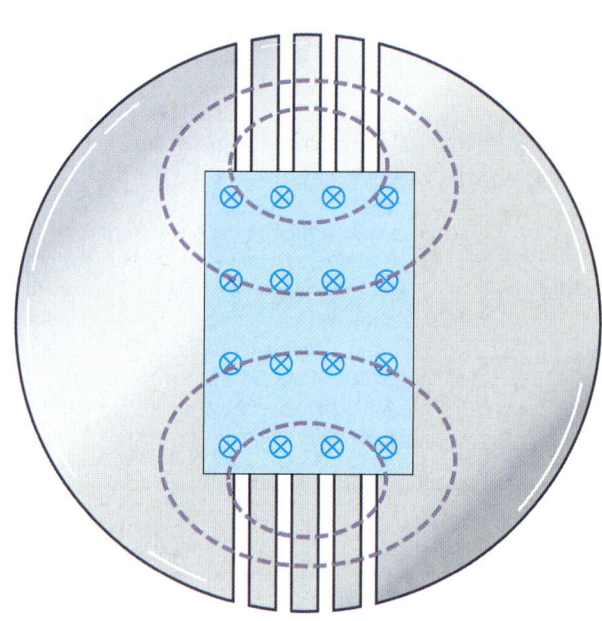

▲ **FIGURE 30–16** To inhibit the development of eddy currents in the moving metal plate, slots can be cut in the plate.

30–3 Forces and Energy in Motional EMF

Currents experience forces in magnetic fields, and this is also the case for any induced currents. As a result, wires or other materials in which currents are induced will experience forces. In general, *the magnetic force on an induced current always inhibits the motion that produces the motional emf.* As we'll now argue, this is a consequence of Lenz's law.

In the moving loop of Example 30–2 (Fig. 30–17), an induced current I appears as the loop leaves the region of magnetic field. The force on the wire is given by Eq. (28–16),

$$\vec{F}_B = I \int d\vec{\ell} \times \vec{B}, \tag{30-6}$$

where $d\vec{\ell}$ describes an element of the wire and $\vec{B}$ is the magnetic field at that wire element. Let's once again draw the loop—this time including the forces on each leg (Fig. 30–17). There is no force on those legs or portions of legs that are out of the magnetic field. Using the right-hand rule, the force on the portion of leg b that is located in the field is directed down, and it is canceled by the force on the portion of leg d located in the field. The only contribution to the net force comes from leg c. Here, application of Eq. (30–6) gives a force

$$F_c = ILB_0 \text{ to the left.} \tag{30-7}$$

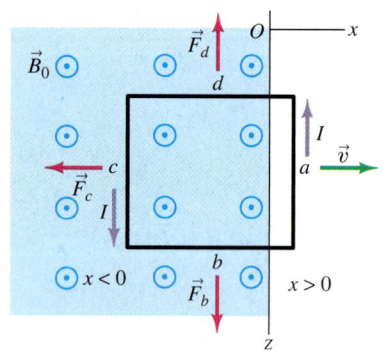

▲ **FIGURE 30–17** The magnetic forces on the loop of Example 30–2 act to slow down the loop.

Recall from Example 30–2, Eq. (30–3), that the magnitude of the current is $I = B_0 vL/R$, where R is the total resistance of the loop and v is the speed with which the loop moves. The force on the loop is thus

$$F_c = \frac{vL^2 B_0^2}{R}. \tag{30-8}$$

This force acts to slow down the loop—it is the typical drag force of a viscous medium.

Because the magnetic force tends to slow the motion, the loop can move at constant velocity only if there is an external force that cancels the magnetic drag. We can imagine that we pull the loop in the direction of the velocity in Fig. 30–17. In this case, we are the external agent supplying a force $\vec{F}$ to the right. The magnitude of this force is F_c, Eq. (30–8).

By applying the external force, we are doing work, or equivalently we are supplying energy at a certain rate. In other words, there is an expenditure of power given by

$$P = \vec{F} \cdot \vec{v} = F_c v = \frac{v^2 L^2 B_0^2}{R}. \qquad (30\text{-}9)$$

CONCEPTUAL EXAMPLE 30-8 The expenditure of power by an external force in a system whose kinetic energy does not change means that there is an energy dissipation in the system. Where is the energy lost in the moving loop of Example 30-2?

Answer There is one place where we lose energy in electrical phenomena, and that is in current flow through a resistance: Joule heating. And we do indeed generate a current in this situation, the induced current which is running through a system with a resistance—the wire. Therefore we have a mechanism to account for the power supplied by the external force. An analytic calculation of the rate of Joule heating should match Eq. (30-9), and we show this below.

Let us check that the power lost in the resistive heating matches the rate at which energy must be supplied to the loop to maintain its motion at a constant speed. The power loss (or energy loss per unit time) in the resistance of the wire is

$$P = I^2 R = \left(\frac{B_0 v L}{R}\right)^2 R = \frac{v^2 L^2 B_0^2}{R}. \qquad (30\text{-}10)$$

Equations (30-9) and (30-10) are the same. *The power loss due to the current flow through the resistor is matched by the power required to keep the loop moving.* The principle of conservation of energy suggests that this result was a foregone conclusion.

The magnetic force on the loop [Eq. (30-8)] is proportional to the speed at which the loop moves and hence is a typical viscous drag (see Chapter 16). This is generally true for the forces on induced currents due to motional emf because their origin is the Lorentz force, which is also proportional to the speed.

EXAMPLE 30-9 A square loop of wire has sides of length 5.0 cm. This loop falls at speed v under the influence of gravity through a region with a constant magnetic field of magnitude 15 T (only magnets built for scientific purposes can attain such a high field), into a region with no magnetic field (Fig. 30-18). The loop is constrained to remain vertically oriented, and the field is horizontal—perpendicular to the loop and into the page. The total resistance of the loop is 1.0 Ω, and its mass is 150 g. (a) Find the terminal speed of the loop as it passes the boundary between the two fields. (b) Calculate the total energy lost to Joule heating in the loop during this period. As an approximation, assume that the loop moves at its terminal speed when it enters the magnetic field region and that this speed remains constant during the loop's passage.

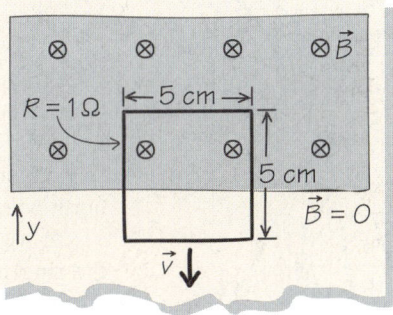

▲ FIGURE 30-18

Strategy (a) There will be a drag force for the situation shown in Fig. 30-18 only when the loop of wire is either partway into or partway out of the region with magnetic field. If this is the case, we can use Eq. (30-8) for the magnetic force on the loop, which will be upward. The magnetic force, $F = vL^2 B_0^2/R$, will be equal in magnitude to the force of gravity mg when the loop acceleration is zero and the loop reaches its terminal speed v_t.

Working It Out We set $v_t L^2 B_0^2 / R = mg$ and solve to obtain the terminal speed v_t,

$$v_t = \frac{mgR}{L^2 B_0^2}. \qquad (30\text{-}11)$$

Numerically,

$$v_t = \frac{(0.15 \text{ kg})(9.8 \text{ m/s}^2)(1.0 \text{ }\Omega)}{(5.0 \times 10^{-2} \text{ m})^2 (15 \text{ T})^2} = 2.6 \text{ m/s}.$$

Strategy (b) As long as the loop moves at its terminal speed, the total energy lost to Joule heating will be a constant given by the product of Joule power and the time the loop spends in the transition region. Given that the loop is moving at v_t, the time spent in the transition region is $t = L/v_t$, where $L = 0.050$ m is the length of a side.

Working It Out The Joule power is, according to Eq. (30-10),

$$P = I^2 R = \frac{v_t^2 L^2 B_0^2}{R}.$$

Thus the energy loss is

$$\Delta E = Pt = \frac{v_t^2 L^2 B_0^2}{R} \frac{L}{v_t} = \frac{v_t L^3 B_0^2}{R}.$$

(continues on next page)

Substituting for v_t from Eq. (30–11), we find that

$$\Delta E = \frac{mgR}{L^2 B_0^2} \frac{L^3 B_0^2}{R} = mgL.$$

This is just the change in the loop's gravitational potential energy. Numerically

$$\Delta E = (0.15 \text{ kg})(9.8 \text{ m/s}^2)(0.050 \text{ m}) = 0.074 \text{ J}.$$

What Do You Think? The calculated energy change is just the change in the gravitational potential energy of the loop as it falls through the transition region. Why?

THINK ABOUT THIS...
HOW DOES MAGNETIC LEVITATION WORK FOR SUPERCONDUCTORS?

As we have remarked above, magnetic drag can act to slow the fall of a metal plate in a magnetic field. We can carry this to an interesting extreme by supposing that the metal plate is a superconductor, a material with no resistance and with the property that the magnetic field does not penetrate its interior. (We'll discuss superconductors and their magnetic properties further in Chapter 31.) If we drop such a piece of material into the field region, the piece is more than slowed down by its entry into the field; it is repelled and bounces back up. This bounce is perfectly elastic because there is no Joule energy loss in a material with no resistance. In this extreme thinking about drag, we are led to a system without energy losses; we have an ideal form of magnetic levitation (Fig. 30–19).

A related application is in magnetically levitated trains, with the levitation used both to hold the train off the surface and guide it along its track (Fig. 30–20). This is useful because there is no rolling friction in the operation of such a train.

(a)

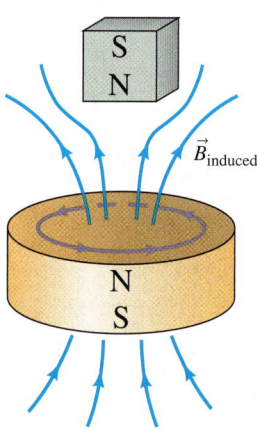

(b)

▲ **FIGURE 30–19** (a) A levitated magnet. A bar magnet moves toward the superconducting material, inducing persistent currents in the superconductor. (b) The magnetic forces between the superconductor and the magnet are repulsive and sufficiently strong to support the magnet's weight.

Forces and Lenz's Law

Lenz's law gives us a second way to think about the forces on induced currents. The magnetic field of a current loop or solenoid is the same as that of a bar magnet. We have already seen that both magnets and current loops produce magnetic dipole fields. Suppose that the north pole of a bar magnet moves toward a conducting ring that initially has no current. The magnetic flux through the ring increases, and an emf is induced in the ring, which causes a current to flow (Fig. 30–21a). The induced current, in turn, produces a magnetic field whose flux tends to cancel the increase of flux due to the moving bar magnet, because the direction of this induced magnetic field is opposed to the field of the bar magnet. We can think of the induced magnetic field as the magnetic field of a second bar magnet, with a field as shown in Fig. 30–21b. The situation is one of two north poles meeting, and two north poles repel each other.

If we now pull the original bar magnet away from the ring, the induced current in the ring points in the opposite direction (Fig. 30–21c). The induced magnetic field changes direction, and the ring's magnetic field is like that of a bar magnet, with its south pole adjacent to the north pole of the original bar magnet (Fig. 30–21d). We know that opposite poles attract each other, so the magnet pulls the ring along with it.

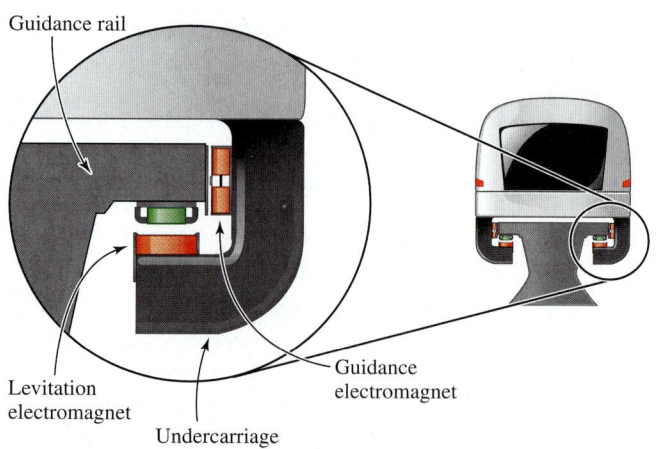

▶ **FIGURE 30–20** Magnetically levitated trains use magnetic induction to hold the train above the travel track as well as to guide it within the track. Even the propulsion system is contained in the time-dependent fields of the magnets. Friction is reduced to a minimum.

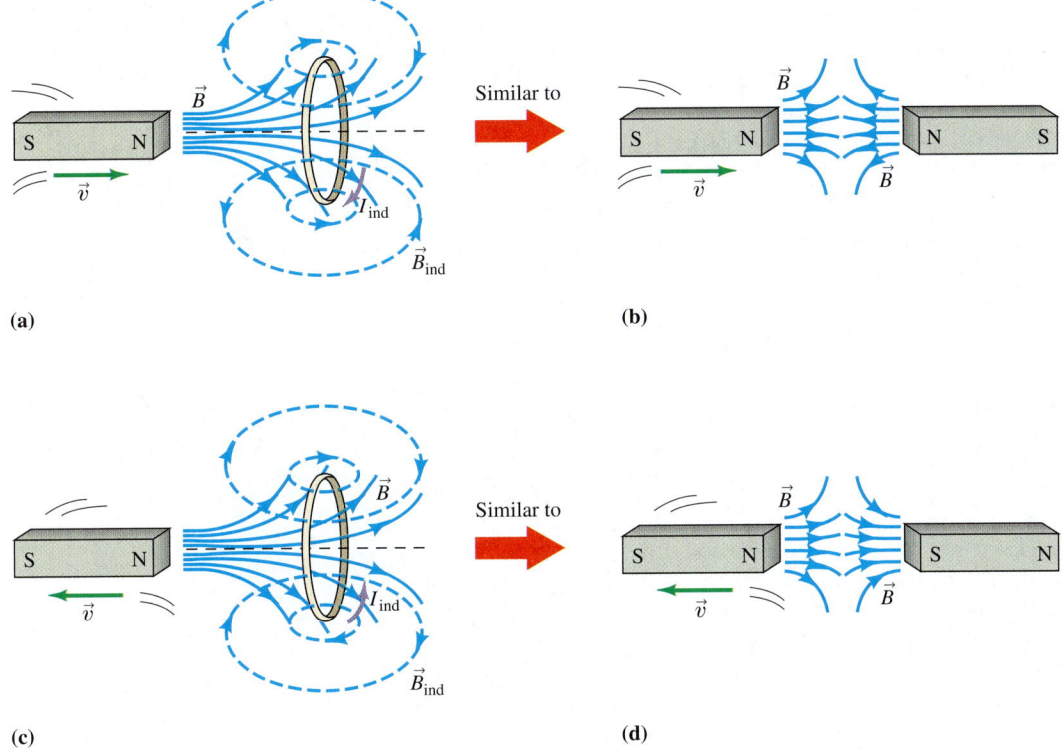

▲ **FIGURE 30–21** (a) The magnetic flux through a conducting ring is increasing because the north pole of a bar magnet moves toward it. The ring is repelled. (b) The direction of the induced current in the ring gives the ring the field of a bar magnet with its north pole to the left, and two north poles repel. (c) The bar magnet is pulled away. In this case, a current is induced in the ring in the opposite direction. (d) The magnetic field of the ring is then that of a bar magnet with its south pole to the left, and it is attracted to the receding bar magnet.

30–4 Time-Varying Magnetic Fields

The magnetic flux through a loop can change in a variety of ways.

1. The loop can move or rotate in the presence of a magnetic field that is not changing with time.
2. The source of the magnetic field can move, as when a bar magnet moves.
3. The source of the magnetic field and hence the field itself can have explicit time dependence, such as when the current through a solenoid is made to change.

We have seen that for case (1) it is possible to analyze the situation using the Lorentz force law. This is not true for cases (2) and (3). Yet Faraday's law doesn't care whether it is the loop that has moved or the field that has changed. *Motional* emf can be interpreted in terms of the Lorentz force law, but a time-varying magnetic field and the emf induced by it is a truly new aspect of Faraday's law. In this section, we look more closely at this situation.

A New Way to Make an Electric Field

When a magnetic field changes with time, then an electric field is induced in space that satisfies Eq. (30–2). *If* there is sufficient symmetry in a situation, then it is possible to calculate the induced electric field in a way similar to the way in which we used Ampère's law to determine a magnetic field. This is most easily illustrated with an example.

EXAMPLE 30–10 The two circular pole faces of an electromagnet, both of radius $R = 0.5$ m, are oriented horizontally with the north pole underneath. The electromagnet produces a field that is uniform throughout the volume between the faces. The field is increased linearly from 0.1 T to 1.1 T over a period of 10 s. Describe the electric field that results in the region between the poles.

Setting It Up We show the electromagnet with its pole faces in Fig. 30–22a. The magnetic field $\vec{B}$ is uniform across the entire face, oriented upward. Its magnitude varies linearly with time according to $B = B_0 + \alpha t$, with both B_0 and α given.

Strategy The unknown electric field is induced by the changing magnetic flux. There is cylindrical symmetry between the pole faces, so the induced electric field can vary only with the distance r from the central axis of the pole faces; it cannot vary with the angle around this axis. By Faraday's law, the rate of change of magnetic flux through a horizontal disk of radius r centered on the axis of the pole faces will determine the integral of the electric field along the circle forming the edge of the disk. Symmetry then allows us to argue that this integral is the (constant) field strength E times the circumference $2\pi r$, and that will determine E. The induced electric field is oriented along this circumference—Faraday's law will tell us whether the field lines run clockwise or counterclockwise.

Working It Out The magnetic flux through the pole face is

$$\Phi_B = \pi r^2 B.$$

Using the given time dependence of B, we have

$$\frac{d\Phi_B}{dt} = \pi r^2 \frac{dB}{dt} = \pi r^2 \frac{d}{dt}(B_0 + \alpha t) = \pi r^2 \alpha.$$

Faraday's law, Eq. (30–2), then gives

$$\oint \vec{E} \cdot d\vec{s} = -\frac{d\Phi_B}{dt} = -\pi r^2 \alpha. \quad (30\text{–}12)$$

The direction of the loop integration is counterclockwise as we look down the north pole because the flux is oriented upward (Fig. 30–22b). With the minus sign in Eq. (30–12), the induced electric field lines run clockwise. Application of the symmetry argument gives

$$\oint \vec{E} \cdot d\vec{s} = E 2\pi r = -\pi r^2 \alpha$$

or

$$E = \frac{\pi r^2 \alpha}{2\pi r} = \frac{1}{2}\alpha r.$$

The magnitude of the induced electric field increases as we go out from the center.

Numerically, the coefficient α is found by knowing that B increases from 0.1 T to 1.1 T in 10 s, so the rate of increase is $\alpha = (1.1 \text{ T} - 0.1 \text{ T})/10 \text{ s} = 0.1 \text{ T/s}$. Then

$$E = \tfrac{1}{2}(0.1 \text{ T/s})r = (0.05 \text{ T/s})r,$$

which increases from 0 N/C at $r = 0$ m to a maximum of $E_{max} = (0.05 \text{ T/s})(0.5 \text{ m}) = 2.5 \times 10^{-2}$ N/C at $r = 0.5$ m.

What Do You Think? What is the induced electric field at distances far outside the poles?

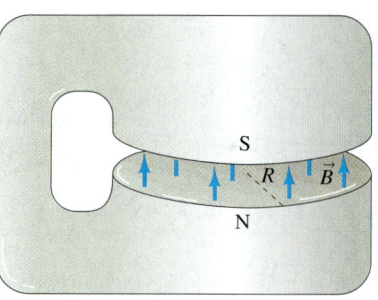

(a)

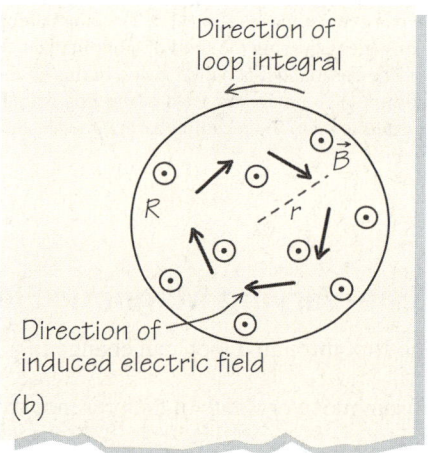

(b)

▲ **FIGURE 30–22** (a) The current windings that produce the field are not shown. (b) A view straight down at the north pole. A changing magnetic field between the pole faces of an electromagnet induces an electric field. Symmetry allows us to specify the electric field, not just its integral around an arbitrary loop.

Is a Magnetic Field Present Where a Current Is Induced?

A changing magnetic flux induces electric fields even where the magnetic field itself is very small. This aspect of Faraday's law is not very intuitive and deserves special attention. Suppose that charged particles such as protons are moving, but they are so far away from the region of the pole faces of an electromagnet that the magnetic field in their location is very small. Would the protons accelerate when the magnetic field is changed? Is an electric field induced even in regions with no magnetic field whatsoever, as in the space outside a toroidal solenoid, or as in the situation described in Example 30–10, "What Do You Think?" Experiment confirms that the answers to our questions are affirmative. According to Faraday's law, all that counts is the change of the flux through the loop in question, regardless of the loop's location.

30–5 Generators

The generation of electric energy in our society is based largely on the Faraday induction law. The mechanical energy of the rotating blades of a steam turbine turned by steam or by rushing water is converted to the energy of moving charges—electric current—with the **alternating-current (AC) generator**.

Imagine a coil of N turns of wire that makes a circle of area A. The coil is placed in a constant magnetic field, $\vec{B}$, and rotated at angular speed ω around an axis perpendicular to the field (Fig. 30–23a). The ends of the wire that make up the coil are brought to the exterior through some sort of sliding contact with a fixed wire. As the coil rotates in the magnetic field, the magnetic flux through it changes, and an emf is induced. Figure 30–23b, which is a side view of the coil, shows that the magnetic flux through the loop is $\Phi_B = \vec{B} \cdot \vec{A} = BA \cos\theta$. If we imagine starting the rotation at $t = 0$, so that $\theta = \omega t$, the time derivative of the magnetic flux is

$$\frac{d\Phi_B}{dt} = BA \frac{d}{dt} \cos \omega t = -BA\omega \sin \omega t. \tag{30–13}$$

There are N turns of wire, and the total emf induced across the two ends of the coil is

$$\mathcal{E} = -N\frac{d\Phi_B}{dt} = NBA\omega \sin \omega t. \tag{30–14}$$

This arrangement makes up our generator, denoted in circuits by a circle enclosing a wavy line (Fig. 30–24).

If the wire of the generator coil is connected as a series element of a circuit with a resistance R, then a current is generated in the circuit:

$$I = \frac{\mathcal{E}}{R} = \frac{NAB\omega}{R} \sin \omega t. \tag{30–15}$$

This *alternating current* oscillates in sign and has a maximum magnitude of $NAB\omega/R$.

The power P delivered to this circuit is the product of the emf and the current:

$$P = \mathcal{E}I = INAB\omega \sin \omega t. \tag{30–16}$$

The mechanical force that rotates the loop is the source of this power. We know that a loop that carries a current forms a magnetic dipole; we also know that a magnetic dipole experiences a torque that tends to align it with the direction of the magnetic field (Section 28–5). Thus the force that rotates the coil must do work against this torque. Let's compute the rate at which this work is done. The torque on a dipole of magnetic dipole moment $\vec{\mu}$ in a field $\vec{B}$ has magnitude

$$\tau = |\vec{\mu} \times \vec{B}| = \mu B \sin \theta,$$

where we refer to Fig. 30–23b for θ and recall that the magnetic dipole moment is perpendicular to the current loop. The mechanical power P_{mech}, or work per unit time, that must be expended by the force that rotates the loop against this torque is

$$P_{\text{mech}} = \tau\omega = \mu B \omega \sin \theta.$$

The magnetic dipole moment of a current loop with N turns is INA [Eq. (28–24)]. Thus

$$P_{\text{mech}} = INAB\omega \sin \theta. \tag{30–17}$$

This result is the same as that given by Eq. (30–16). As expected, the electric power is accounted for entirely by the mechanical power expended.

The explicit time dependence of the power is found by taking the product of the current, Eq. (30–15), and the emf, Eq. (30–14) or, more simply, by evaluating the product V^2/R. We then find

$$P = \frac{(NAB\omega)^2}{R} \sin^2 \omega t. \tag{30–18}$$

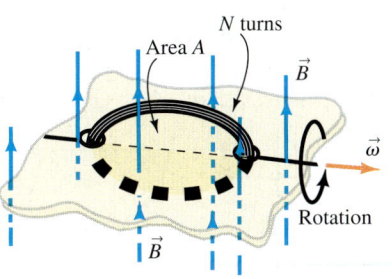

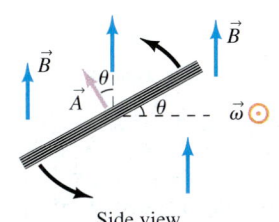

▲ **FIGURE 30–23** (a) An external force rotates a coil with angular velocity $\vec{\omega}$ in a magnetic field. An emf with sinusoidal time dependence of angular frequency ω is induced in the coil. (b) Side view of the process, looking at $\vec{\omega}$.

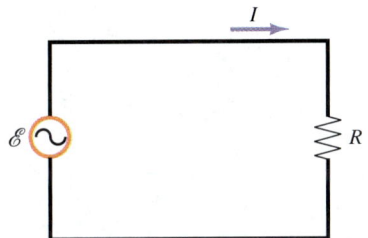

▲ **FIGURE 30–24** When the sinusoidally varying emf that results from rotating a coil in a constant magnetic field (as in Fig. 30–22a) is part of a circuit, a sinusoidal current with the same angular frequency ω results.

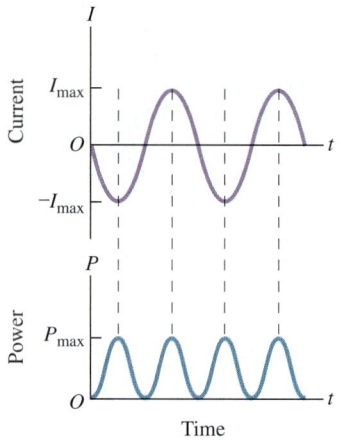

▲ **FIGURE 30–25** The power dissipated in the circuit of Fig. 30–24 is always positive, unlike the current, which alternates in sign. The maximum current is $I_{max} = NAB\omega/R$, while the maximum power is $P_{max} = I_{max}^2 R$.

This quantity is always positive, as opposed to the emf or the current, both of which alternate in sign. The distinction is illustrated in Fig. 30–25, which shows a plot of the current and the power in the circuit of Fig. 30–24 as a function of time.

In practical situations it is important to consider the *time average* of the power. To do so, we note that the sine-squared function, which oscillates between 0 and 1, averages over one period to 1/2. The average power dissipated in this circuit is then

$$P_{av} = \frac{1}{2}\frac{(NAB\omega)^2}{R} = \frac{1}{2}\frac{V_{max}^2}{R}. \tag{30-19}$$

Electric power is generated by using mechanical energy to produce electric current; the current is transmitted as an alternating current (AC)—see Chapter 33. If we run a generator in reverse, we can convert electric energy into mechanical energy, and this is the basis of an *electric motor*. As we described in Chapter 28, the proper operation of a motor requires a split-ring commutator or something similar to periodically reverse the torque on a loop such that the torque is always in the same direction as the loop turns.

*30–6 The Frame Dependence of Fields

In Section 28–3 we discussed the fact that in order to maintain the relativity principle, electric and magnetic fields had to "mix," meaning that observers in different inertial frames see different combinations of electric and magnetic fields. You should recall that according to the relativity principle, there should be no way to tell which of two inertial frames—frames moving at constant velocity with respect to one another—is moving and which is not. The laws of motion should be the same in all inertial frames, so they do not provide a way to tell which frame is moving. Thus an observer O and an observer O', who is moving with constant velocity $\vec{u}$ relative to observer O, should express the laws of motion in exactly the same way. In particular, Newton's second law,

$$\frac{d\vec{p}}{dt} = \vec{F},$$

with $\vec{p} = m\vec{v}$, must look the same to observer O, who sees a particle move with velocity $\vec{v}$, and to observer O', who sees the same particle move with velocity $\vec{v}' = \vec{v} - \vec{u}$. This will be the case if $\vec{u}$ is constant (as it must be for inertial frames) and if the force on the particle is the same to both observers.

If the particle is charged and subject to a Lorentz force, there is an apparent difficulty because the Lorentz force depends on the velocity of the particle. The difficulty is that since the force describes the motion of the object on which the force acts, observation of the motion would represent a way to decide which of the observers is moving, and this is a violation of the principle of relativity. The problem would only be an apparent one if both observers saw the same Lorentz force. We applied this argument in Chapter 28 with the velocity selector and suggested that the problem is resolved if the fields seen by our two observers take the form

$$\vec{B}' = \vec{B}, \quad \vec{E}' = \vec{E} + \vec{u} \times \vec{B}. \tag{30-20}$$

Equation (30–20) is the correct solution to the problem of reconciling the forces when speeds are not large compared to the speed of light. *Under the transformation from one inertial frame to another, electric and magnetic fields get mixed up (transform among themselves) in a very special way, as specified in Eq. (30–20).*

It is interesting to think about how stationary and moving observers view the source and effect of motional emf. Suppose that an observer O is in a reference frame where there is a magnetic field but no electric field. If a conducting rod moves through this field, there is a motional emf. Its source, according to observer O, is the magnetic force on the conducting electrons in the rod (Section 30–3). Observer O', moving with the rod, sees the rod at rest and also sees the same constant magnetic field seen by observer O, from Eq. (30–20). Therefore observer O' sees no magnetic force. However, the same transformation equation shows that observer O' also sees an electric field $\vec{E}'$, and this electric field has a magnitude that makes the conducting electrons in the rod accelerate in just the way that observer O sees them accelerate. Each observer attributes the observed effects to different combinations of fields.

The question of the consistency of physical law for observers moving with respect to one another is a very important one; it led Albert Einstein to formulate the theory of special relativity, the subject of Chapter 39.

Summary

When the magnetic flux Φ_B through an open surface changes with time, an emf $\mathcal{E}$ is induced around the line that bounds the surface. Faraday's law states the relation:

$$\mathcal{E} = \oint \vec{E} \cdot d\vec{s} = -\frac{d\Phi_B}{dt}. \tag{30–2}$$

Here, the line integral is over the closed bounding line that forms the edge of the surface. When the loop is a physical object capable of carrying current, such as a loop of wire, then the induced emf results in an induced current. In this case, the minus sign in Faraday's law can be interpreted in more physical terms as Lenz's law: Induced currents produce magnetic fields that tend to cancel the flux changes that induce them.

The flux change to which Faraday's law refers can occur either because the magnetic field changes with time or because the area or orientation of the surface through which the flux is calculated changes with time. In the latter case, the induced emf is called a motional emf, and it can be derived directly from application of the Lorentz force law. Application of Lenz's law to motional emfs shows that the induced current must lead to forces that inhibit the motion of the object in which the emf is induced. The power loss due to resistive flow of induced currents is matched by the power required to keep the conductor moving. When an induced emf occurs because magnetic fields change with time, Faraday's law predicts a new type of electric field, one that cannot be described with a conventional potential and that is therefore nonconservative.

When a changing magnetic flux passes through a conducting solid, Faraday's law manifests itself by the induction of eddy currents in the material.

The AC generator, the foundation of electrical power generation, is an application of Faraday's law. When a coil rotates in a magnetic field, an emf is induced in the coil. The mechanical energy of the rotation is thus transformed into electric energy in the form of a current in circuits connected to the coil.

Understanding the Concepts

1. A spherical surface is placed in a changing magnetic field. Will there be an induced electric field along the equator?
2. Must there be a real conducting loop in a region with a changing magnetic flux in order for an electric field to be induced?
3. When the end of a magnet is brought to a stationary loop of wire, which of the following are true: (a) the magnetic flux through the loop decreases, (b) Lenz's law does not apply, (c) a voltage but no current are induced in the wire, (d) the induced voltage depends on the resistance in the wire, (e) the induced current depends on the resistance in the wire.
4. Electric leads (the wires that run from one part of the apparatus used for experiment to another) for sensitive experiments are almost never separated but are close together or even twisted around one another. Explain why this might be done.
5. Can the magnetic field change over some region without a change in the magnetic flux through a surface in the region? If so, give as many examples as you can.
6. A bar magnet is dropped vertically through a loop of wire, with the north pole crossing the loop first. Looking down at the loop from above, the induced current will be (a) clockwise and then counterclockwise, (b) counterclockwise and then clockwise, (c) zero, (d) only clockwise, or (e) only counterclockwise.
7. Each part of Fig. 30–26 shows a current being induced in a conducting loop by a changing magnetic flux through the loop. In each case, is the direction of the induced current correct as

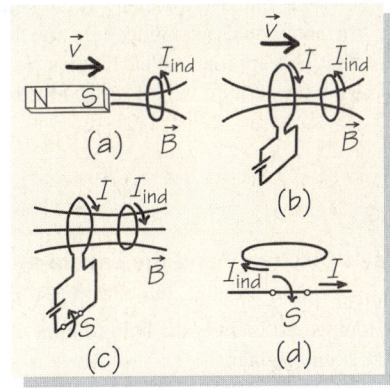

▲ FIGURE 30–26 Question 7.

shown? (a) A magnet approaches a loop; (b) a current-carrying conducting loop approaches a loop at rest; (c) a switch is closed in the first loop, causing a current to flow; (d) a switch is closed in the straight wire, causing a current to flow.

8. A conducting hoop is rolled in a straight line at a constant speed in an east–west direction in the northern hemisphere through Earth's magnetic field. Will a current flow in the hoop? If so, in what direction will it circulate?

9. A rectangular loop is moving across a uniform magnetic field such that the induced emf is zero. Can you tell how the loop is oriented?
10. In Example 30–2 the induced current jumped from zero to a constant value and then dropped to zero again. Sketch, roughly, what the time-dependence of the induced current would look like if the loop were circular.
11. A sheet of metal is placed between the pole pieces of a permanent magnet, perpendicular to the direction of the field lines. Does it take positive work to pull the sheet of metal out? If so, why?
12. What would happen if you dropped a strong cylindrical magnet down inside a long vertical copper tube? Down an aluminum tube? Down a plastic tube?
13. When a bar magnet is moved toward a current loop, a current is induced in the loop. How, if at all, will physically measurable quantities change if the loop is moved toward the magnet rather than vice versa?
14. If a flat metal plate hung by a cord and oriented parallel to the pole faces of a magnet—the faces are in a vertical plane, parallel to each other—moves as a pendulum bob through the pole faces, it slows down. If the magnet is sufficiently strong, the plate comes to rest. Why? How could this phenomenon be prevented?
15. The Tennessee Valley Authority constructed several dams on the Tennessee River to produce electricity. How is water used to produce electricity?
16. When a gas is heated sufficiently, its atoms ionize into electrons and positive ions. This type of material is called a plasma. Plasma flow may be viewed as a superposition of equal and opposite currents. If the plasma is forced to flow in a channel perpendicular to a magnetic field, an electric potential builds up across the channel. The device based on this phenomenon is the *magnetohydrodynamic (MHD) generator*. Given the magnetic field strength and the potential, we can calculate the velocity of the plasma. The charge and density of the charge carriers in the plasma do not enter into the result. Why not? [*Hint*: Review the Hall effect.]
17. What happens when a bar magnet is dropped down a long, vertical copper tube?
18. In Example 30–4 we attached a battery to the straight wire and asked what current would be induced in the loop when the straight wire and loop approached each other. Would there be a current induced in the straight wire if instead the battery were on the loop?
19. In a demonstration, an aluminum ring is placed around a projection of an iron core wound with a wire and connected to a battery (Fig. 30–27). The ring jumps when the circuit is closed. Why? What happens if a gap is cut in the ring?

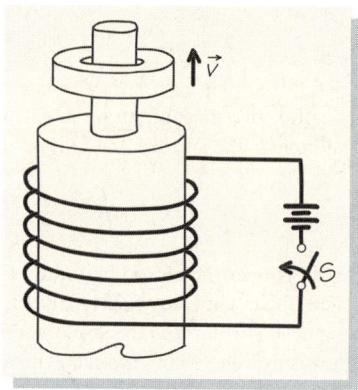

▲ **FIGURE 30–27** Question 19.

20. A cylindrical piece of iron is inserted inside a solenoid to increase the magnetic field. A voltage varying harmonically with time is placed across the solenoid leads. A copper ring is slipped down over the solenoid, so that the solenoid passes through the ring, and is held there. (a) Explain why the copper ring becomes hot even though nothing touches it. (b) What is the source of the thermal energy? (c) Explain how energy is conserved in this case.
21. In Question 20, the solenoid axis is vertical. It is possible to find a particular ring of copper that, when slipped over the solenoid and placed horizontally, remains suspended in space around the solenoid. (a) Why does this work? (b) What are the criteria for selecting the particular piece of copper?
22. Discuss how a bicycle light generator utilizing the Faraday effect might work.
23. At many traffic lights cars drive over a wire embedded in the pavement, and the traffic light soon changes from red to green. The wire forms a loop of about 1.5 m on a side. How might such a system work?
24. Describe some of the interesting things you might see if the sign of the flux change term in Faraday's law were positive.

Problems

30–1 Faraday's Discovery and the Law of Induction

1. (I) You are given a horseshoe magnet, a coil of wire, and a flashlight bulb. How would you get the bulb to light up? How would you make the light brighter?
2. (I) Explain in words why the induced currents are as shown in each case in Figure 30–3. Describe the induced magnetic field.
3. (I) A loop of wire of area 12 cm^2 is placed between the pole pieces of an electromagnet, at right angles to the direction of the magnetic field lines. What is the emf generated around the loop if the magnetic field is changed at a uniform rate from 1.5 T to 2.0 T in 5.7 s? Assume that the magnetic field is uniform across the area of the loop.
4. (I) Suppose that the wire in Problem 3 has a resistance of 7.7 Ω. How much power will be lost to ohmic heating while the magnetic field increases?
5. (I) A magnetic field that changes with time but is uniform in space is directed along the *x*-axis. A conducting ring of diameter 7 cm and resistance 1.5×10^{-3} Ω is placed in the *yz*-plane. If the current in the ring is 2 A, how fast is the magnetic field changing?
6. (II) The magnetic field in a region is uniform. It varies with time as shown in Fig. 30–28 (see next page). Plot the current through a ring that has an area of 14 cm^2 and a resistance of 0.02 Ω, and whose plane is perpendicular to the magnetic field.
7. (II) A square wire loop of dimensions $L \times L$ oriented in the *xz*-plane enters a region where the magnetic field is first oriented in the $+y$- and then in the $-y$-direction (Fig. 30–29, see next page). The width of each region is *L*. The loop moves at speed *v* in the $+x$-direction. Find the emf, sign and magnitude, induced in the loop as it enters and passes through the regions with the magnetic field.

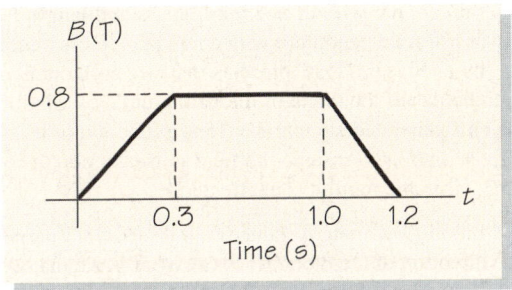

▲ **FIGURE 30–28** Problem 6.

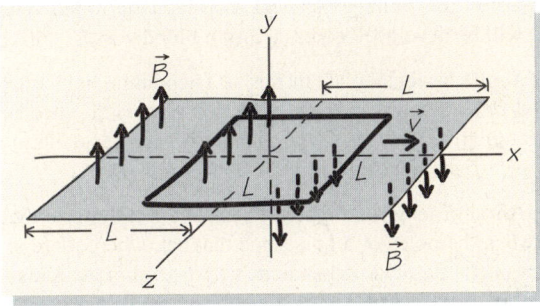

▲ **FIGURE 30–29** Problem 7.

8. (II) A long, straight wire oriented in the z-direction carries a current of 120 mA. A square loop with sides of length 2.5 cm is in the xz-plane with its nearest edge 15 cm from the wire (Fig. 30–30). In a time of 0.05 s, the square loop moves uniformly 1.0 cm closer to the wire. What is the emf induced in the loop while it is moving? Ignore the variation in the wire's magnetic field *across* the loop.

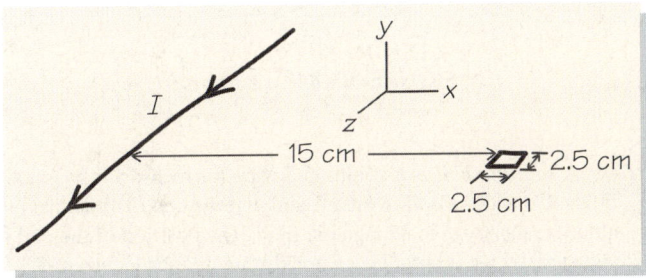

▲ **FIGURE 30–30** Problem 8.

9. (II) What is the peak emf produced by a 100-turn square coil 8.0 cm on each side, rotating on a diagonal axis with a frequency of 15 Hz in a magnetic field of 0.30 T perpendicular to the axis?

10. (II) A coil with 450 turns, a radius of 2.5 cm, and a resistance of 12 Ω is rotating about a diameter in a uniform magnetic field of 0.35 T. How fast must it rotate to produce a maximum current of 3.0 A in the coil?

11. (II) There is a constant magnetic field $\vec{B} = B_0(\hat{i} + \hat{j} + \hat{k})$ in the region $x > 0, y > 0, z > 0$. A square loop of dimensions $L \times L$ whose sides are parallel to the x- and y-axes moves with constant velocity $\vec{v} = v_0(\hat{i} + \hat{j})$ in the xy-plane such that its center moves along the line $x = y$. Calculate the emf induced in the loop, given that its leading corner passes the origin at the time $t = 0$.

12. (II) A vertical loop rotates with angular velocity $\vec{\omega}$ as shown in Fig. 30–31. At time $t = 0$, it is aligned perpendicular to a constant magnetic field oriented in the x-direction. Use Lenz's law to find the direction of the emf induced in the loop at $t = 0$, $t = T/4, t = T/2$, and $t = 3T/4$, where T is the rotation period of the loop.

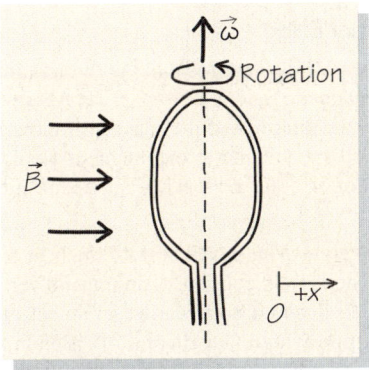

▲ **FIGURE 30–31** Problem 12.

13. (II) A closed loop is constructed of a fixed wire shaped as a squared-off U and a crossbar free to move in the x-direction, all in the xz-plane. The square base of the U-shaped segment is at $x = 0$. A magnetic field oriented in the y-direction varies with x according to $\vec{B} = Cx\hat{j}$; it is zero at $x = 0$. The situation and the relevant dimensions are as in Fig. 30–32. Suppose that the movable crossbar is pulled at a constant speed v to the right, starting at $x = 0$ when $t = 0$. Its position at any time is $x = vt$. If the resistance of the loop varies with the total length L according to $R = \alpha L$, what is the current in the loop as a function of time? Compare your answer with Example 30–5, and explain any differences.

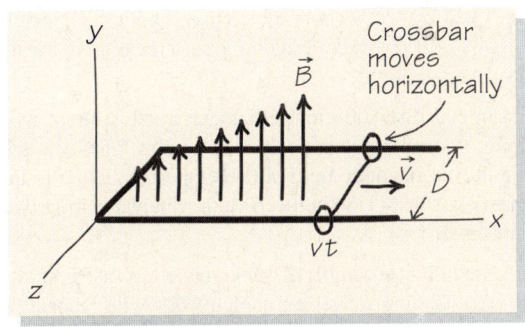

▲ **FIGURE 30–32** Problem 13.

14. (II) Suppose the magnetic field in Example 30–5 is a constant field oriented in the z-direction, $\vec{B} = B_0\hat{k}$. Find the induced current as a function of time.

15. (II) Suppose the magnetic field in Example 30–5 varies linearly with z and is oriented in the y-direction, $\vec{B} = Cz\hat{j}$. Find the induced current as a function of time.

16. (II) A metal ring is constructed so as to expand or contract freely. In a region with a constant magnetic field $\vec{B}_0$ oriented perpendicular to it, the ring expands, with its radius growing linearly with time as $r = r_0(1 + \alpha t)$. As the ring expands and grows thinner, its resistance *per unit length* changes according to the empirical rule $R = R_0(1 + \beta t)$. Find the current induced in the ring as a function of time. Specify the direction as well as the magnitude of the current.

17. (II) A circular loop of area A rotates with angular frequency ω about its vertical diameter. The rotating loop is placed in a horizontal constant magnetic field, B. What is the emf induced in the loop?

18. (II) Work Example 30–3 by direct computation of the magnetic flux through the hemispherical surface.

30–2 Motional EMF

19. (I) The spacecraft *Voyager I* is moving through interstellar space, where the magnetic field is 2×10^{-10} T. Assume that *Voyager I* has an antenna 5 m long. If the spacecraft moves so that the antenna rod is perpendicular to the magnetic field when *Voyager I* has a speed of 8×10^3 m/s, what is the emf induced across the antenna?

20. (I) A 747 is flying due north at 940 km/h in a location where Earth's magnetic field consists of an upward vertical component of 2.8×10^{-5} T and a northward component of 2.5×10^{-5} T. If the wingtip-to-wingtip length of a 747 is 35 m, find the emf induced across the wings. If the airplane were flying due east instead of due north, how would your answer change?

21. (I) A metal rod is pulled through a magnetic field perpendicular to it with a velocity perpendicular to both the rod and the magnetic field as in Fig. 30–12a. The rod has length 0.25 m, its speed is 1.7 cm/s, and the magnetic field has magnitude 0.069 T. What is the magnitude of the potential difference, if any, from one end of the rod to the other?

22. (I) A metal rod 6.0 cm long falls to the ground from a height of 6.0 m. It stays horizontal and oriented in an E–W direction throughout the fall. If we assume that, in the region where the metal falls, Earth's magnetic field is 1.5×10^{-4} T and points in the N–S direction, then at what rate does the potential difference between the ends of the rods increase?

23. (I) A rod 10 cm long lying in the xy-plane pivots with angular speed 100 rad/s counterclockwise about the origin (see Fig. 30–14). If the measured emf across the rod is 100 mV, what is the magnetic field?

24. (II) A metal disk 7.0 cm in diameter rotates about its axis of symmetry at an angular speed of 150 rad/s. The disk is situated in a uniform magnetic field of 0.10 T perpendicular to the plane of the disk. What is the induced voltage between the axis and the rim of the disk?

25. (II) A metal bar of length 0.7 m is moved to the right at a speed of 5 m/s. The bar makes an angle of 60° with respect to its direction of motion. It is passing through a region of uniform magnetic field of magnitude 5×10^{-3} T oriented perpendicular to the plane swept out by the bar (out of the page in Fig. 30–33). What is the potential difference between the two ends of the bar as it moves through the magnetic field?

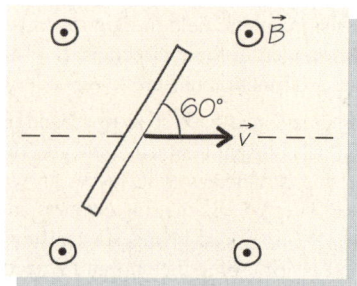

▲ **FIGURE 30–33** Problem 25.

26. (II) A thin copper wire of negligible mass and length L is oscillating as a simple pendulum with a mass bob m. A uniform, constant magnetic field is present that is horizontal, oriented perpendicular to the plane of the pendulum's oscillations. What is the emf generated along the wire as a function of its angle θ of small oscillation? (Assume the field is weak enough so that the simple harmonic motion is unaffected.)

27. (II) A blood-flow meter produces a 0.030 T field perpendicular to the direction of the blood flow. (a) What is the magnetic force on an ion of charge e flowing with the blood at 0.25 m/s? (b) Positive and negative ions in the blood are forced in opposite directions, producing an electric field and a voltage across the blood vessel. The ion separation stops when the electric force caused by this voltage balances the magnetic force. What voltage will be developed across a 3-mm blood vessel?

28. (II) A circular metal plate moves as a pendulum bob between the poles of a tabletop electromagnet. The plate is oriented so that it is parallel to the faces of the magnet. Describe qualitatively the eddy currents induced in the plate as it moves.

29. (II) A rod of length L moves at constant speed v into the region between the poles of a horseshoe magnet, where there is a constant magnetic field perpendicular to the rod in a circular region (Fig. 30–34). $L = 2R$, the radius of the circular region. What is the emf induced in the rod as a function of time?

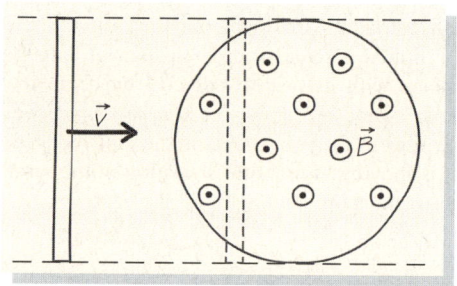

▲ **FIGURE 30–34** Problem 29.

30. (II) A *rotating coil* is a common device for measuring magnetic fields. Consider a coil of area A and N turns that is rotated at angular frequency ω in a magnetic field. The position of the coil is adjusted so as to produce a maximum induced current I_{max}, which can be measured by using an appropriate ammeter. R is the total resistance of the coil circuit. Find the relationship between the unknown magnetic field and I_{max}.

31. (II) If the rotating coil of Problem 30 is used with the split-ring commutator described in Chapter 29, DC current can be measured with a sensitive galvanometer. (a) Sketch the current as a function of time for several periods, where the period T is given by $2\pi/\omega$. (b) Calculate the average of this rectified current. (c) Find the relationship between the unknown magnetic field and the average measured DC current. [*Hint:* The average of an oscillating function with period T is $(1/T) \int_0^T f(t)\, dt$.]

32. (II) A long, straight wire carries a current of 650 mA. A thin metal rod 45 cm long is oriented perpendicular to the wire and moves with a speed of 2.2 m/s in a direction parallel to the wire. What are the size and direction of the emf induced in the rod if the nearest point of the rod is 3.5 cm away from the wire, and if the rod moves in a direction opposite to the current?

30–3 Forces and Energy in Motional EMF

33. (I) A loop of metal, total resistance $R = 25\,\Omega$, moves with speed 35 m/s through a region of magnetic field such that at time $t = 0$ the rate of change of magnetic flux through the loop is given by $17\,\text{T}\cdot\text{m}^2/\text{s}$. The magnetic field in the vicinity of the loop has instantaneous value 0.16 T at $t = 0$. Assume the shape of the loop is such that the resulting net force on the loop is due entirely to a straight section of the loop, 12 cm in length, that is perpendicular both to the magnetic field and to the direction of the loop's motion. What is the drag force on the loop at $t = 0$, and what is the instantaneous power at $t = 0$ expended by the force that must be used to keep the loop moving with constant velocity?

34. (II) A square wire loop of dimensions $L \times L$ lies in a plane perpendicular to a constant magnetic field. The field exists only in a certain region, with a sharp boundary (Fig. 30–35). The sides of the loop make a 45° angle with this boundary, and an external force moves the loop at a speed v out of the region of constant field. How much power must be supplied by the external force as a function of time?

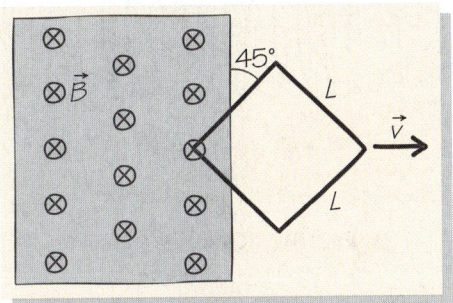

▲ FIGURE 30–35 Problem 34.

35. (II) A conducting bar slides frictionlessly on two parallel horizontal rails 30 cm apart. The bar and rails form a closed circuit with a resistor of resistance $0.050\,\Omega$, assumed to be constant throughout the motion. The circuit is placed in a uniform vertical magnetic field of 0.28 T perpendicular to the circuit's plane. The bar is pulled at a constant speed of 60 cm/s along the rails. (a) What is the magnitude of the force required to pull the bar? (b) What is the rate of Joule heating in the resistor?

36. (II) A circuit with a moveable cross wire is indicated in Fig. 30–36. The resistance is $2.5\,\Omega$, the battery voltage is 5 V, the moveable wire length is 12 cm, and there is a uniform, constant magnetic field into the plane of the circuit (the page) with magnitude 0.3 T. (a) What is the current if the moveable wire is locked into position? (b) What is the force on the moveable wire at the moment it is released from the locked position? (c) If there is an opposing force of 0.02 N on the wire, how fast will it move as a function of the time after release?

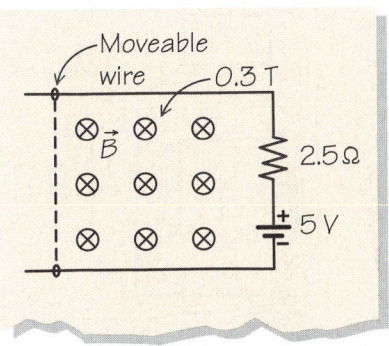

▲ FIGURE 30–36 Problem 36.

37. (II) A horizontal wire of mass m is free to slide on vertical rails a distance L apart. These are connected at the bottom with a resistance R between them. There is a constant, uniform magnetic field B in a horizontal direction, normal to the area enclosed by the rails and sliding wire. Use the fact that the power generated by gravity is equal to the power dissipated by ohmic heating, to show that the wire ultimately moves with a constant velocity, and obtain an expression for this velocity in terms of mg, B, R and L.

38. (II) A long, straight wire carries a constant current I_0. A square loop with sides of length L and two sides parallel to the wire is pulled away at uniform speed v in a direction perpendicular to the wire. The nearest side of the loop is initially a distance D from the wire; the resistance of the loop is R. (a) Calculate the force necessary to pull the loop. (b) At what rate is work being done by the force? (c) How does your answer to part (b) compare with the Joule heating in the loop?

39. (II) What happens if the initial speed of the loop in Example 30–9 is (a) less than v_t, and (b) greater than v_t?

40. (II) When eddy currents are induced in a piece of metal moving through a magnetic field, drag forces that are proportional to the velocity of the metal act on it. Consider a thin metal disk rotating in a plane between the poles of a magnet. Show that the equilibrium angular velocity of the disk is proportional to the torque on the disk. Note that in household electricity meters, this arrangement allows us to correlate the number of turns of the disk with the power consumption.

30–4 Time-Varying Magnetic Fields

41. (I) Consider a length of wire looped back on itself in a magnetic field B. The shape of the loop is not given, but it lies in a plane. The wire has negligible resistance, but there is a resistor R at one end. A cross bar, which rests on opposite legs of the loop, is pulled along so that the flux enclosed by the wire and crossbar varies with time as $\Phi_B(t)$. What is the instantaneous power needed to pull the crossbar?

42. (I) A long, straight wire carries a current $I = I_0 \cos(220\pi t)$, where t is time. Two sides of a fixed rectangular loop are 9.0 cm long and are parallel to the wire; the other sides are 0.80 cm long. The nearest long side is 2.0 cm from the wire. What is I_0 if the maximum emf induced in the loop is $1.3\,\mu\text{V}$? (Ignore the small variation of the magnetic field across the loop.)

43. (I) A loop of wire is placed between the poles of a large electromagnet. The loop is oriented so that the vector that characterizes the orientation of its planar surface runs from one pole to the other. When the magnet is turned on, its magnetic field builds up according to the formula $B = B_0(1 - e^{-at})$. We may assume that the magnetic field has the same value all across the surface of its poles and runs from one pole to another. (This formula gives $B = 0$ at $t = 0$ and $B = B_0$ at $t = \infty$.) What is the magnitude of the emf around the loop?

44. (I) A metal ring of diameter 3.0 cm is left between the pole tips of an electromagnet, such that the plane of the loop is perpendicular to the magnetic field. The electromagnet is turned on and reaches its full magnetic field of 2.7 T in 50 ms at a linear rate. If the ring has resistance $0.15\,\Omega$, how much current passes through the ring during the 50 ms?

45. (II) A solenoid of radius r wound with n turns per unit length carries a current given by $I = I_0 \cos(\omega t)$, where t is the time. What are the magnitude and direction of the induced electric field just outside the solenoid?

870 | Faraday's Law

46. (II) A wire of 5.5 Ω resistance and 20 m length is wound into a coil of radius 4.3 cm. A magnetic field of 2.4 T is perpendicular to the plane of the coil. (a) If the current that forms the magnetic field is switched to the opposite direction, what is the total charge that passes through the wire coil? Assume the current switching takes place in 85 ms. (b) What is the average current and emf induced in the coil over that period?

47. (II) A single-turn circular loop of radius 0.08 m and resistance 6 Ω is coaxial with a solenoid of length 90 cm and a radius of 0.04 m, with 1800 turns. A variable resistor, as indicated in Fig. 30–37, is varied in such a way that the solenoid current falls linearly from 5 A to 1.2 A in 0.3 sec. (a) What is the induced current in the loop? (b) In what direction does the induced current flow?

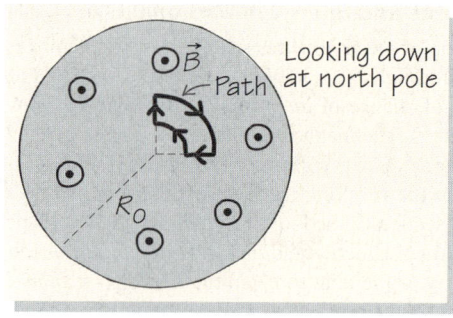

▲ **FIGURE 30–39** Problem 49.

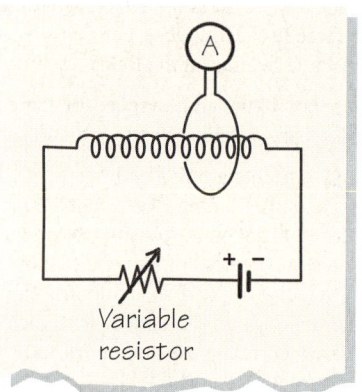

▲ **FIGURE 30–37** Problem 47.

48. (II) A very long cylindrical solenoid of radius r made from n turns of wire per unit length carries a current with the time dependence $I = I_0 e^{-t/t_0}$. Coaxial with and surrounding the solenoid are two turns of wire that make a circular loop slightly larger than the circular cross section of the solenoid (Fig. 30–38). The loop with two turns is far from the ends of the solenoid and has a resistance R. Find the current in the loop with two turns, I', as a function of time.

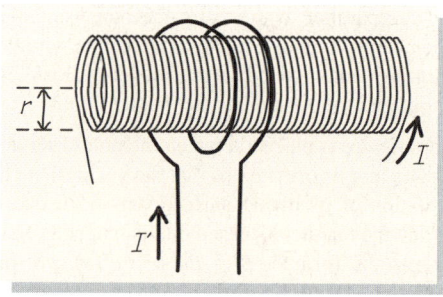

▲ **FIGURE 30–38** Problem 48.

49. (II) The uniform magnetic field of the electromagnet of Example 30–10, with circular pole faces of radius $R_0 = 0.08$ m, decreases linearly from 1.5 T to 0.7 T in 25 ms. What is the emf induced around the path drawn in Fig. 30–39 that consists of quarter arcs at radial distances $R_0/4$ and $R_0/2$, connected by radial lines? The path is clockwise.

50. (II) A long solenoid of radius R and n turns per unit length carries an alternating current $I = I_0 \sin(\omega t)$ (Fig. 30–40). What are the electric fields induced within the solenoid at a distance $R/2$ and outside the solenoid at a distance $2R$? [*Hint*: Apply Faraday's law to the two paths shown, and use symmetry.]

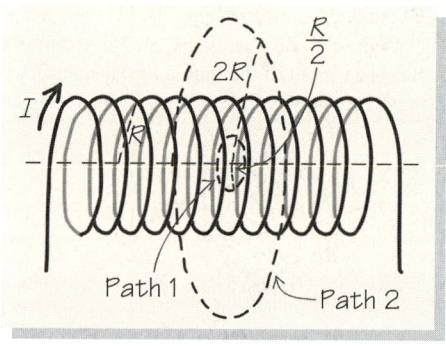

▲ **FIGURE 30–40** Problem 50.

30–5 Generators

51. (I) A coil of area 6.0 cm² with 180 turns of wire is connected to a resistor of resistance 3 Ω. It is rotated by hand at a frequency of 0.6 rev/s in a magnetic field of 0.40 T. (a) What is the maximum amount of current produced? (b) the average power produced?

52. (II) You have 18 m of wire, a constant magnetic field of 0.45 T, and a device that can rotate a coil at a fixed frequency of 300 Hz. What size circular coil will produce an AC emf of maximum voltage 120 V?

53. (II) The headlight of a bicycle is powered by a small generator that is driven by a wheel of the bicycle. The generator contains two coils fixed at the sides of the generator and connected in series with appropriate polarity (Fig. 30–41). Each coil consists of 70 turns and has an area of 8 cm². A small permanent magnet is rotated in front of the coils, so that the magnitude of the magnetic field in the coils varies between 0.1 T and zero. At what speed of the bicycle will the maximum emf be 6.4 V, given that the radius of the friction wheel is 1 cm?

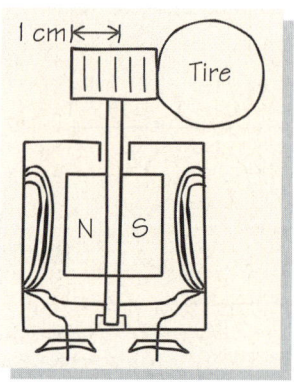

▲ **FIGURE 30–41** Problem 53.

54. (II) A bicycle wheel of radius $R = 33$ cm rotates at angular speed 53 rad/s in a plane perpendicular to a constant magnetic field of magnitude 0.55 T. What is the emf generated between the center of the wheel and its rim? When one end of a wire is attached to the center and the other end to a circular track in contact with the rim, a direct current is generated in the wire. Such a device is called a *homopolar generator*.

*30–6 The Frame Dependence of Fields

55. (II) Suppose that observer O sees an electric field $\vec{E} = E\hat{i}$ and a magnetic field $\vec{B} = B\hat{k}$. In what direction and at what (constant) speed u should a second observer move so as to see no electric field whatsoever? Use the nonrelativistic relation Eq. (30–20). If $E = 10^3$ V/m, for what range of values of B is the nonrelativistic approximation appropriate?

General Problems

56. (II) A 120-cm-long wire of square cross section with a mass of 65 g and a resistance of 2.0 Ω slides without friction down parallel conducting rails of negligible resistance (Fig. 30–42). The rails are connected to each other at the bottom by a resistanceless rail parallel to the wire so that the wire and rail form a closed rectangular conducting loop. The plane of the rails makes an angle of 15° with the horizontal, and a uniform vertical magnetic field of 0.68 T, pointing upward, exists throughout the region. What is the steady speed of the wire?

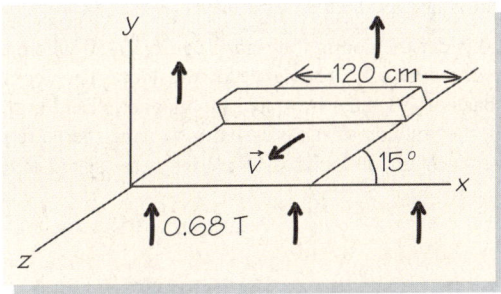

▲ **FIGURE 30–42** Problem 56.

57. (II) A straight wire carries a current $I = 150$ A near a rod that moves across two conducting wires (Fig. 30–43). The resistor has $R = 0.20$ Ω, and the rod moves at speed 45 cm/s. (a) What is the emf induced in the rod? (b) What is the current in the circuit? (c) How much work is done to move the rod 100 cm to the right? What force does this work?

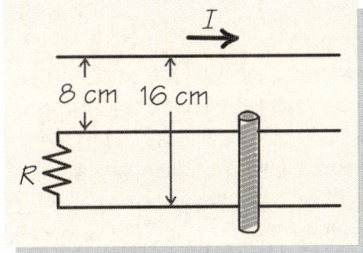

▲ **FIGURE 30–43** Problem 57.

58. (II) A conducting crossbar bracketing two vertical conducting wires slides down the wires. The wires are connected with a resistor R to form a closed circuit (Fig. 30–44a). (a) If there is a horizontal magnetic field B perpendicular to the plane of the loop, how fast does the bar fall after the initial accelerating period? (b) A battery is added to the circuit (Fig. 30–44b). What polarity and emf of the battery are needed to lift the bar with the same velocity?

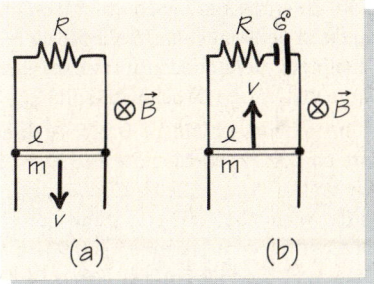

▲ **FIGURE 30–44** Problem 58.

59. (II) A large, circular coil of N turns and radius R carries a steady current I and is rotated at a constant angular speed ω about a horizontal diameter. At the center of this coil is a small, fixed, horizontal circular ring of radius r. (a) What is the emf induced in the small ring? (b) What is the angle between the plane of the coil and that of the ring when this emf is a maximum?

60. (II) Consider a 9-V battery attached to two conducting, frictionless rails 0.1 m apart. There is a magnetic field $\vec{B}$ of magnitude 0.5 T perpendicular to the rails, and a conducting bar can slide over the rails perpendicular to them as well as to the field (Fig. 30–45). The bar is placed on the rails, starts from rest, and accelerates. (a) What is the direction of its motion? (b) the direction of the emf induced? (c) Given that the total resistance of the closed circuit is 3 Ω, calculate the current in the bar when its speed is 3 m/s.

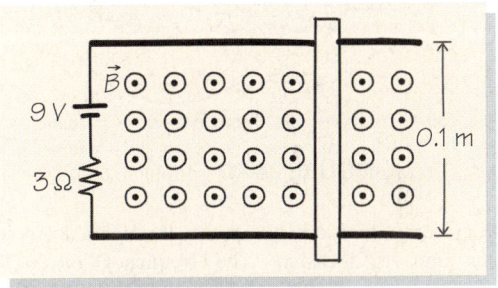

▲ **FIGURE 30–45** Problem 60.

61. (II) A wire carrying a current I is oriented in a horizontal direction. To its side, a wire loop is oriented so that it and the straight wire lie in the same horizontal plane. The straight wire is moved toward the loop. If a current is induced in the loop, what is its direction, and what is the direction of the force on the loop?

62. (II) If the plasma in a magnetohydrodynamic generator (see Question 16) is forced to flow in a channel perpendicular to a magnetic field, an electric potential builds up between points a and b, which are 1 m apart (Fig. 30–46). If the magnetic field has a strength 2.5 T, what must the speed of the plasma be in order that the potential be 1000 V?

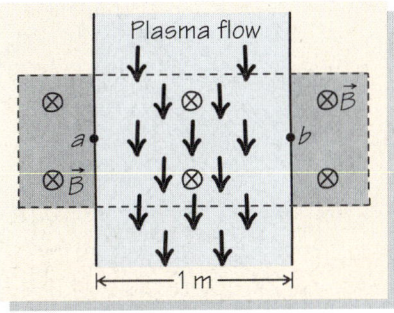

▲ **FIGURE 30–46** Problem 62.

63. (II) A coil with 200 turns, a diameter of 8.0 cm, and a resistance of 5.6 Ω is placed perpendicular to a uniform magnetic field of 1.4 T. The magnetic field suddenly reverses direction. What is the total charge that passes through the coil?

64. (II) A constant magnetic field of 0.5 T is directed along the x-axis. A wire coil of 200 turns and area 12 cm² is placed in the yz-plane. The coil of wire, called a *flip coil*, is then turned over (in other words, rotated by 180°). (a) If the total charge that passes through the coil when it is flipped is 0.007 C, what is the resistance of the coil circuit? (b) The same flip coil is used to measure an unknown magnetic field. The coil is flipped in several directions until it attains its maximum charge of 0.02 C, when the coil is flipped with its face in the xy-plane. What is the magnitude of the magnetic field? (c) What is the direction of the magnetic field in part (b)?

65. (II) A circular ring of area 100 cm² is connected to a 15-μF capacitor. The circuit has a resistance of 2 Ω. A uniform time-dependent magnetic field of magnitude $B = (0.03 \text{ T/s})t$ is perpendicular to the ring (Fig. 30–47). Calculate the current in the ring and the charge on the capacitor. Give the direction of the current and the polarity of the charge.

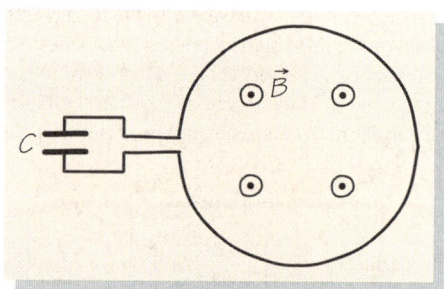

▲ **FIGURE 30–47**   Problem 65.

66. (II) A current $I = I_0 \cos(\omega t)$ passes through a solenoid of area 10 cm² and 10^5 turns/m. The frequency is 60 Hz, and $I_0 = 10$ A. A small coil—a sense coil—is used to sense the changing flux. This sense coil has an area of 20 cm² with 10 turns and is placed across the solenoid so that the face of the coil is perpendicular to the solenoid axis; the two coils are concentric. (a) What is the emf induced in the sense coil? (b) If the resistance of the sense coil circuit is 5 Ω, what is the current?

67. (II) A wire that is bent into a semicircle is rotated with angular velocity ω about its diameter, as shown in Fig. 30–48. The bent wire and its supports are placed in a uniform magnetic field perpendicular to the plane of the supports. What is the emf induced in the circuit shown? If the resistance of the closed loop is R, what is the average power dissipated?

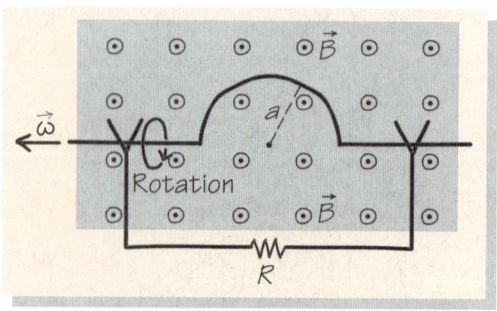

▲ **FIGURE 30–48**   Problem 67.

68. (II) A pendulum consists of a metal bar suspended from two thin wires attached to a fixed conducting bar (Fig. 30–49). The resistance of this closed circuit is 0.030 Ω. The pendulum is placed in a vertical magnetic field of magnitude 0.12 T. The pendulum is displaced by a small angle from the equilibrium position and allowed to oscillate. What is the ratio of the power dissipated to the energy of the oscillator?

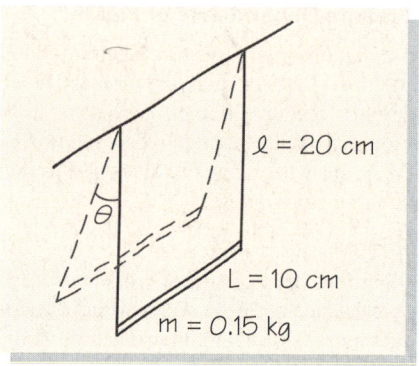

▲ **FIGURE 30–49**   Problem 68.

69. (II) A train travels in the northerly direction at 60 mph on tracks that are 1.5 m apart. In the region, Earth's magnetic field is 0.7×10^{-4} T making an angle of 30° with the vertical. What is the emf generated by the train?

70. (II) Consider the configuration in Figure 30–50 where the resistor R dominates the resistance of the loop. The length of the crossbar is L, its mass is m, and the magnetic field is B. At time $t = 0$, the velocity of the crossbar is v_0, and there are no external forces acting on the crossbar. What is the speed of the crossbar as a function of time?

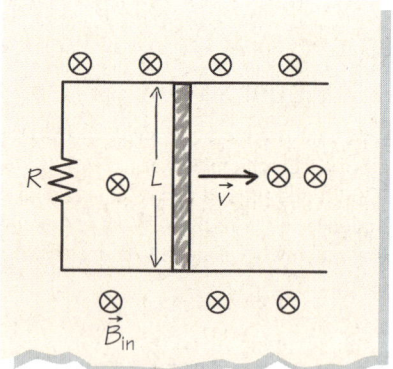

▲ **FIGURE 30–50**   Problem 70.

71. (III) An electron follows a circular path of radius $R = 1$ m while traveling in a plane perpendicular to a spatially constant magnetic field of magnitude 10^{-6} T. As viewed along the magnetic field lines, the electron follows a counterclockwise path. (a) What is the speed of the electron? (b) Assuming that the motion of the electron is nonrelativistic, what is the energy E of the electron? (c) The magnitude of the magnetic field is reduced smoothly by a certain percentage during an interval Δt. Show that the fractional energy change of the electron, $\Delta E/E$, is independent of the radius of the electron's orbit as well as of the electron's speed. (This effect is the basis for low-energy operation of a particle accelerator called the *betatron*.) (d) If the magnetic field is reduced in time $\Delta t = 5$ s by 10 percent, estimate $\Delta E/E$.

Chapter 31

◀ As part of the manufacturing process for high-temperature superconductor wire, technicians at American Superconductor operate capstans that pull the wire through a metal die, reducing its diameter and increasing its length. The successful production of high-capacity wire made of a high-temperature superconductor would revolutionize many aspects of our power distribution system.

Magnetism and Matter

While we have made frequent use of bar magnets in the previous chapters, we have not asked questions as to why the materials that make up such magnets behave the way they do. Is there something in such materials that is like the electric currents that produce magnetic fields according to the laws we have developed? How do we explain the fact that magnets pick up needles but not pieces of paper? Why is it possible to make a slab of iron act like a bar magnet at some times but not at other times? Why can't the same be said for a piece of aluminum? What explains the magnetic properties of superconductors? In other words, what lies behind the magnetic properties of matter? The answers to these types of questions lie in the microscopic structure of matter, all the way down to the subatomic level. We utilize the magnetic properties of material when we construct computer disk drives, electric motors, generators, transformers, particle accelerators, and medical scanners, so that understanding the magnetic properties at a fundamental level is essential. When we considered the electric properties of materials, we saw that the dielectric properties of materials depend on the polarizability of atoms and molecules. In the same way, the magnetic properties of materials depend on the magnetic properties of atoms and molecules and their constituents. This chapter is concerned with the origins of these properties.

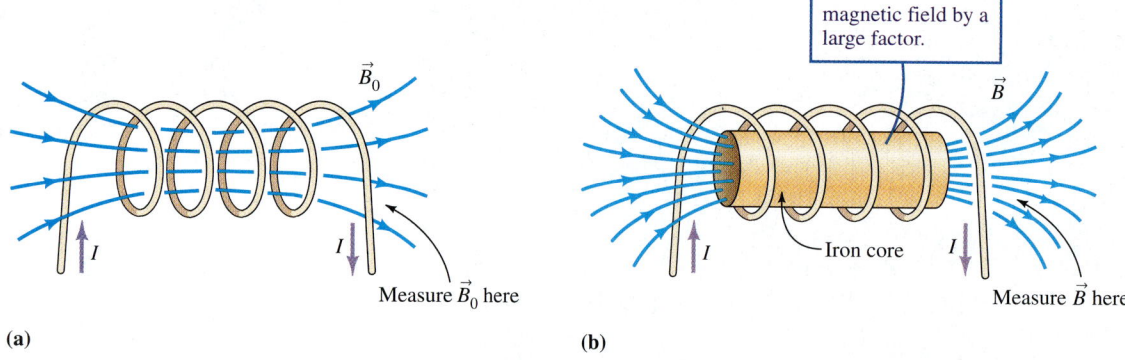

▲ **FIGURE 31–1** (a) A solenoid that carries a current has a magnetic field. (b) The cylindrical volume of the solenoid is filled with an iron core.

31–1 The Magnetic Properties of Bulk Matter

We begin with a description of the different kinds of behavior that are observed on an everyday scale. We can explore the magnetic properties of various materials using a solenoid. We measure the magnetic field $\vec{B}_0$ produced by, and near one end of, a current-carrying solenoid (Fig. 31–1a) and ask how the presence of various materials affect the magnetic field. (Throughout this chapter $\vec{B}_0$ represents any field, not necessarily constant, that is present before we introduce materials.) If we insert a wooden or copper core into the solenoid and repeat the measurement, it would take a very sensitive instrument to reveal that the field has been changed by one part in 10^6. But if we place an iron core in the solenoid (Fig. 31–1b), the measured field is increased by a factor many times that of the original field of the hollow solenoid. If we now remove it, the iron core will act like a bar magnet, even if it showed no such properties before insertion. The solenoid somehow magnetizes the iron core; the wooden or copper cores show no such effect. Magnetic properties depend on the material!

The magnetic behavior of bulk materials—gases, liquids, and solids—is characterized by the **magnetization**, $\vec{M}$, which we define as the *magnetic dipole moment per unit volume* of material. The field outside a material with a net magnetization is that of a magnetic dipole, the same as that of a bar magnet. A magnetic moment has dimensions of current times area [see, for example, Eq. (29–27)], so magnetization has dimensions of current times area divided by volume, or current per length. The SI units of $\vec{M}$ are amperes per meter (A/m). Our aim here is to find a connection between an applied field $\vec{B}_0$ and the magnetization.

The net magnetic field of a solenoid S_1 with a core of some material in it is the vector sum of contributions from the original magnetic field of the solenoid, $\vec{B}_0$, and the contribution of the magnetization of the core. The core can be viewed as a second solenoid S_2 as any magnetic dipole has a magnetic field with the same general form as that of a solenoid. Suppose that S_2 can be treated as a solenoid of area A, length L, and N loops that carry current I. From Eq. (29–15), the magnetic field of S_2 is

$$B_2 = \frac{\mu_0 NI}{L} = \frac{\mu_0 NIA}{LA} = \mu_0 \frac{m_2}{V},$$

where[†] m_2 is the magnetic moment of S_2, $m_2 = NIA$, and μ_0 is the permeability of free space. The factor m_2/V is the magnetic moment per unit volume—the magnetization size $|\vec{M}|$. Thus *the contribution of the core to the total magnetic field is $\mu_0 \vec{M}$*, and the total field of the solenoid with the core present is

$$\vec{B} = \vec{B}_0 + \mu_0 \vec{M}. \tag{31-1}$$

[†]To avoid confusion with the magnetic permeability μ to be introduced shortly, we use the notation m for the magnetic dipole moment throughout this chapter. Don't confuse it with a mass!

Equation (31–1) applies whatever the source of the original field $\vec{B}_0$ in which the bulk material is placed, and it need not be the field due to a solenoid. This situation is similar to that of dielectrics, where an electric field within a material has contributions from an original electric field due to free charges and from an internal, induced distribution of charge. We saw in Chapter 25 that we can separate the effect of these types of charges through the introduction of a dielectric constant and a generalized permittivity. Similarly, we want to separate the effects of the internal magnetization, which has contributions from what must be the equivalent of internal currents, from the magnetic fields due to ordinary currents. We refer to the latter as *free*, or *real*, currents. The effects of free currents are isolated by defining the **magnetic intensity** $\vec{H}$, a quantity for which the effect of the magnetization of the material is subtracted out:

$$\vec{H} \equiv \frac{\vec{B}}{\mu_0} - \vec{M}. \tag{31-2}$$

Notice that the dimensions of $\vec{H}$ are those of $\vec{M}$, *not* of $\vec{B}$. By replacing $\vec{B}$ in Eq. (31–2) with $\vec{B}_0 + \mu_0 \vec{M}$ according to Eq. (31–1), we find

$$\vec{B}_0 = \mu_0 \vec{H}. \tag{31-3}$$

Equation (31–3) shows that *the magnetic intensity measures the magnetic field due to free currents*. Another form for the relation among $\vec{B}$, $\vec{H}$, and $\vec{M}$ is found by combining Eq. (31–1) and (31–3):

$$\vec{B} = \mu_0 \vec{H} + \mu_0 \vec{M}. \tag{31-4}$$

Most materials have no magnetization unless they are in the presence of an external magnetic field $\vec{B}_0$ that induces magnetization. The important exceptions to this are iron and some other materials that we will refer to below as ferromagnetic—these are the materials that make permanent magnets. We know by experiment that over a large range of conditions, *the size of the magnetization of nonferromagnetic materials varies linearly with the external magnetic field*. For these materials, $\vec{M}$ depends linearly on $\vec{B}_0$, and hence on $\vec{H}$. The **magnetic susceptibility** χ_m is defined as the coefficient of the linear relation between the magnetization and the magnetic intensity:

$$\vec{M} \equiv \chi_m \vec{H}. \tag{31-5}$$

The direction of the magnetization is more complicated: The original field $\vec{B}_0$ and the field $\mu_0 \vec{M}$ due to the magnetization are parallel for one class of materials but antiparallel for a second class. If the susceptibility of a material is positive, its magnetization is aligned along the external field; if the susceptibility is negative, the magnetization is aligned opposite to the external field. Because $\vec{M}$ and $\vec{H}$ have the same dimensions, χ_m is dimensionless. Table 31–1 gives a range of susceptibilities found in nature. Examination of the table reveals the three general classes of materials: those with small negative susceptibility (**diamagnetic materials**), those with small positive susceptibility (**paramagnetic materials**), and iron and its cousins (**ferromagnetic materials**), which have a large positive susceptibility and can have a magnetization even with no external field present. The nature and properties of these classes of materials is the major topic of this chapter.

With our definition of magnetic susceptibility χ_m, we can express the relation between the magnetic field in a material and the magnetic intensity. From Eq. (31–1),

$$\vec{B} = \vec{B}_0 + \mu_0 \vec{M} = \mu_0 \vec{H} + \mu_0 \chi_m \vec{H} = \mu_0 (1 + \chi_m) \vec{H}. \tag{31-6}$$

We define the coefficient of $\vec{H}$ in this equation as the **permeability**, μ, of the material:

$$\mu = \mu_0 (1 + \chi_m). \tag{31-7}$$

TABLE 31–1 • Some Magnetic Susceptibilities (at 20°C unless indicated otherwise)

Material	Susceptibility, χ_m
Diamagnetic	
Water	-13×10^{-6}
Copper	-5.5×10^{-6}
Silver	-2.0×10^{-5}
Carbon (diamond form)	-5.9×10^{-6}
Bismuth	-2.4×10^{-4}
Paramagnetic	
Sodium	1.6×10^{-6}
Cupric oxide (CuO)	2.6×10^{-4}
Aluminum	1.7×10^{-5}
Liquid oxygen (90 K)	8.7×10^{-3}
Oxygen gas	3.5×10^{-3}
Ferromagnetic	
Iron (annealed)	5.5×10^3
Permalloy (55% Fe, 45% Ni)	2.5×10^4
Mu-metal (77% Ni, 16% Fe, 5% Cu, 2% Cr)	1×10^5

The relation between the total magnetic field in a material and the magnetic intensity, which is a measure of the effect of free currents, is then

$$\vec{B} = \mu \vec{H}. \tag{31-8}$$

Just as when materials are present, the electric permittivity ε replaces the permittivity of free space ε_0 in expressions for electric fields that contain only the free charge, so too when materials are present, μ replaces μ_0 in expressions for magnetic fields that contain only the free current. Equation (31–3) shows that when there is a magnetic field in a vacuum, that field is related to the intensity by a relation like that of Eq. (31–8), but with μ_0 appearing in the place of μ; thus μ_0 is the permeability of the vacuum. From Table 31–1 we can see that μ is very close to μ_0 for nonferromagnetic materials.

The various quantities we have defined are all useful in characterizing the bulk magnetic behavior of materials. In Table 31–2, we summarize these quantities and their relations.

TABLE 31–2 • Magnetic Bulk Properties and Their Relations

Symbol	Property
$\vec{B}_0$	Applied magnetic field, produced independently of type of material by a nearby magnet or currents
$\vec{H}$	Magnetic intensity, proportional to the applied magnetic field
$\vec{M}$	Magnetization, the magnetic dipole moment per unit volume of a material
$\vec{B}$	Net magnetic field, the sum of the applied magnetic field and a term proportional to the magnetization
μ_0	Permeability of free space
χ_m	Magnetic susceptibility of a material
μ	Permeability of a material, $\mu = \mu_0(1 + \chi_m)$.

Some Relations	$\vec{B}_0$	$\vec{H}$	$\vec{M}$	$\vec{B}$
$\vec{H} =$	$\vec{B}_0/\mu_0$	—	$\vec{M}/\chi_m$	$\vec{B}/\mu$
$\vec{M} =$	$\chi_m \vec{B}_0/\mu_0$	$\chi_m \vec{H}$	—	Not used
$\vec{B} =$	$(1 + \chi_m)\vec{B}_0$	$\mu_0(1 + \chi_m)\vec{H}$	$\dfrac{\mu_0(1 + \chi_m)}{\chi_m}\vec{M}$	—

EXAMPLE 31–1 A straight solenoid of diameter 5 cm and length 25 cm is wrapped with 200 turns of wire that carries a current of 5 A. The solenoid is filled with iron, magnetic susceptibility $\chi_m = 5.5 \times 10^3$. Find (a) the magnetic intensity within the solenoid and (b) the magnetic field within the solenoid. (c) By what factor is the magnetic field changed due to the presence of the material?

Setting It Up We assign labels of the known and unknown quantities: The current is I, the number of turns of wire per unit length is n, the magnetic intensity has magnitude H, and the magnetic field has magnitude B.

Strategy The first step is to calculate the magnetic intensity $H = B_0/\mu_0$, where B_0 is the magnetic field associated with the free currents in the solenoid. This is given by $B_0 = \mu_0 nI$, so that $H = nI$. We can then calculate $B = \mu H$, using, for example, Eqs. (31–8) and (31–7), and hence the factor by which the field changes due to the material in the solenoid. The diameter of the solenoid does not enter here, except to confirm that the solenoid is long compared to its width, so that our formulas for ideal solenoids apply.

Working It Out (a) With $n = (200 \text{ turns})/(25 \text{ cm}) = 800$ turns/m, and $I = 5$ A,

$$H = nI = (800 \text{ turns/m})(5 \text{ A}) = 4000 \text{ A/m}.$$

(b) The magnetic field includes the effect of the field due to the iron that fills the solenoid. We have

$$B = \mu H = \mu_0(1 + \chi_m)H \cong \mu_0 \chi_m H$$
$$= (4\pi \times 10^{-7} \text{ T}\cdot\text{m/A})(5.5 \times 10^3)(4000 \text{ A/m}) = 28 \text{ T}.$$

(c) The factor by which the field changes is

$$\frac{B}{B_0} = 1 + \chi_m \cong \chi_m.$$

For a ferromagnetic material such as iron, this is a large factor, and very substantial enhancements of magnetic fields occur using these kinds of materials.

What Do You Think? Do you expect B inside to be (a) larger, (b) smaller, or (c) unchanged if copper were used in the core? *Answers to* **What Do You Think?** *questions are given in the back of the book.*

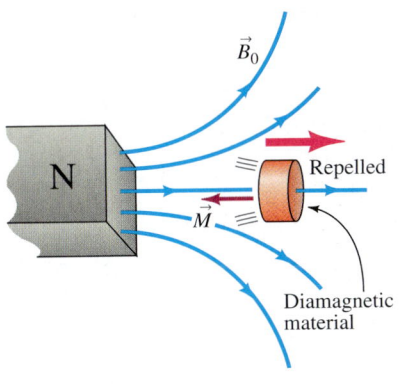

▲ **FIGURE 31–2** Diamagnetic substances are repelled by one pole of a nearby bar magnet.

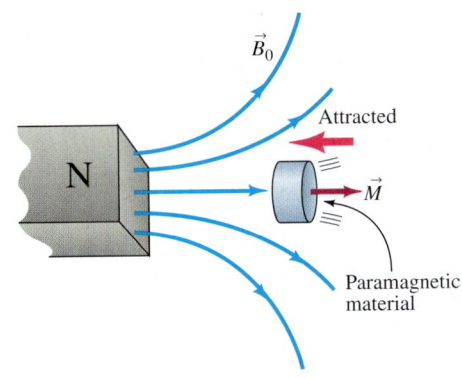

▲ **FIGURE 31–3** Paramagnetic substances are attracted to one pole of a nearby bar magnet.

Let us summarize the properties of the three broad classes of materials that Table 31–1 reveals. The class of materials that are most often used for their magnetic properties is composed of *ferromagnetic* materials, those with large positive susceptibilities. These substances can have magnetization without the presence of an external field, forming permanent magnets.

Diamagnetic substances are those with very small negative susceptibilities. In such materials, the magnetization direction is *opposite* to the direction of the inducing field. The magnetic field inside such materials is *reduced* a little from its value outside the material. If a diamagnetic material is placed near the north pole of a magnet, the magnetization produces a field that points toward that pole (Fig. 31–2). The diamagnetic material acts as though it has a north pole adjacent to the external north pole: The diamagnetic material is *repelled* by the magnet. The behavior of diamagnetic substances is similar to that of dielectrics, for which polarization effects tend to cancel the electric field associated with free charges.

Paramagnetic substances are those with small positive susceptibilities. The magnetization points in the same direction as the field of an external magnet (Fig. 31–3), as though the paramagnetic substance has a south pole oriented toward the magnet's north pole: The piece of paramagnetic material is *attracted* to the magnet.

Table 31–3 gives the forces acting on some samples near a sizable (3.0 T) electromagnet. Only the relative scale of the numbers matters in the table, since we are providing no details about the geometry of the magnet. The force is expressed in units of the weight of each sample. The minus sign reflects repulsion (diamagnetism); the plus sign reflects attraction (paramagnetism and ferromagnetism). Note the differences in the sizes of these forces for the different classes of materials.

We'll discuss the underlying atomic origins of ferromagnetism, diamagnetism, and paramagnetism in Sections 31–3 and 31–4.

TABLE 31–3 • Magnetic Forces on Materials near a Large Electromagnet

Material	Material Class	Force (in units of sample weight)
Copper (pure)	Diamagnetic	-1.3×10^{-3}
Lead	Diamagnetic	-19×10^{-3}
Graphite	Diamagnetic	-56×10^{-3}
Sodium	Paramagnetic	$+10.2 \times 10^{-3}$
Copper chloride	Paramagnetic	$+143 \times 10^{-3}$
Iron	Ferromagnetic	$+20$
Magnetite	Ferromagnetic	$+61$

CONCEPTUAL EXAMPLE 31–2 Suppose a small cylinder of a paramagnetic substance and a small cylinder of a diamagnetic substance are placed between the (parallel) pole pieces of a strong magnet, so that they are aligned along the direction of the field lines. Will they attract or repel each other?

Answer We visualize the situation in Fig. 31–4. In this way, the magnetic field lines point upward. We now consider two possible situations, one in which the two cylinders are placed one above the other (Fig. 31–4a) and another in which the two cylinders are side by side (Fig. 31–4b). For case a, we place the paramagnetic cylinder just below the S pole piece and the diamagnetic cylinder below the paramagnetic cylinder. (You can verify that the order does not matter.) The external magnetic field induces magnetism in the paramagnetic substance so that the cylinder acts like a very weak magnet lined up in such a way that the field points up. Thus the top of the paramagnetic cylinder acts like a N pole of a magnet, and the bottom like the S pole of a magnet. The situation with a diamagnetic cylinder is just the opposite. The top of the diamagnetic cylinder acts like the S pole, and the bottom like a N pole. Thus the two S poles of the materials are adjacent, and they repel each other. Figure 31–4b shows that when the two cylinders are placed side by side, they attract.

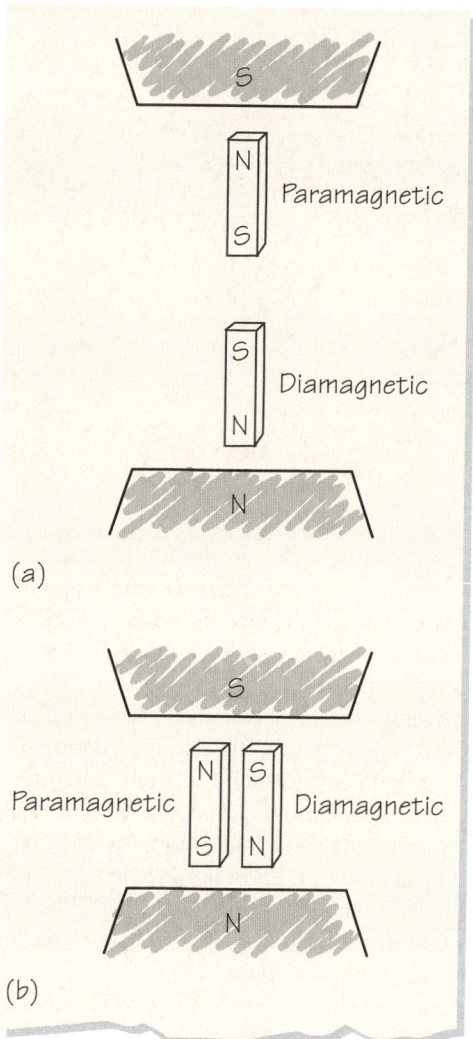

▶ **FIGURE 31–4** In (a) the samples are one above the other; in (b) they are side by side.

31–2 Atoms as Magnets

The magnetic field of a solenoid is the same as that of a bar magnet. How can two such apparently dissimilar systems have the same magnetic field? The answer lies in the magnetic properties of atoms. To understand these properties qualitatively, we start with a classical planetary model for atoms, in which electrons orbit a nucleus, and add some necessary quantum mechanical features. The orbiting electrons form ring currents like those of a solenoid, and the atoms then act as magnetic dipoles, complete with magnetic dipole moments. In most situations, the magnetic moments of a large assembly of atoms point in random directions, so that the net magnetic moment of a macroscopic material adds to zero. The exception occurs in ferromagnetic materials, in which forces between atoms act to line up atomic magnetic moments with their neighbors, giving rise to a net magnetization.

The magnetic properties of individual atoms are affected by a (quantum-mechanical) tendency of their electrons to pair off such that the magnetic moment due to an individual electron's current points in the opposite direction from that of its partner. The net result is that a pair of electrons has no magnetic moment. In a classical picture of the atom as a collection of electrons orbiting about a nucleus, this effect would be interpreted as the cancellation of moments due to clockwise-moving electrons with the moments due to counterclockwise-moving electrons. This suggests that in many cases, atoms with an even number of electrons will have no magnetic dipole moment, whereas only a last unpaired electron makes a difference in atoms with an odd number of electrons.

The Magnetic Dipole Moment of Atoms

Consider the simplest possible *classical* planetary model of an atom, in which a single electron of charge $-e$ moves at speed v in a circular orbit of radius r around a heavy nucleus—as in Fig. 31–5a. The orbital period is $T = 2\pi r/v$. Because electric current is charge per unit time, the current around the nearly stationary nucleus is

$$I = \frac{-e}{T} = -\frac{ev}{2\pi r}. \quad (31\text{-}9)$$

The minus sign indicates that the current is in a direction opposite to that of the motion of the electron. The current loop has an area of πr^2; following Eq. (29–27), the magnitude of the *orbital magnetic dipole moment* is therefore given by

$$m_{\text{orbital}} = |I|\pi r^2 = \frac{ev}{2\pi r}\pi r^2 = \frac{1}{2}evr. \quad (31\text{-}10)$$

The direction of the magnetic moment vector $\vec{m}$ is determined by a right-hand rule (Fig. 31–5b).

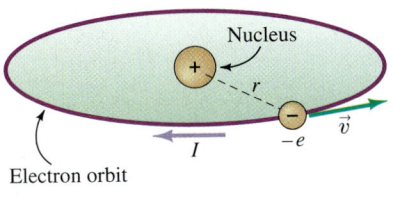
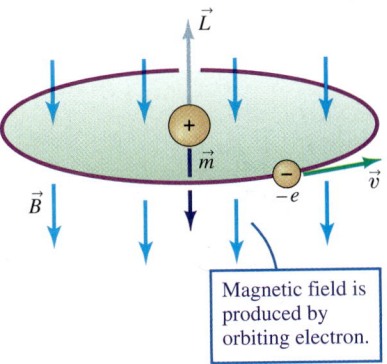

▶ **FIGURE 31–5** (a) An electron in a circular orbit around a nucleus. Over numerous orbits, the effect is the same as a continuous current ring about the nucleus. (b) The circulating electron forms a magnetic dipole with magnetic moment $\vec{m}$ ($\vec{m}_{\text{orbital}}$) oriented downward, opposite to the angular momentum $\vec{L}$.

EXAMPLE 31–3 Estimate an atomic orbital magnetic moment by taking the radius of the electron orbit to be of roughly atomic size, 10^{-10} m, and the kinetic energy to be a typical atomic energy of 1 eV. Compare this to the magnetic moment of a macroscopic loop of area 1 cm^2 that carries a current of 1 mA.

Strategy To use Eq. (31–10) for the magnetic moment, we must first calculate the speed. With the kinetic energy $K = m_e v^2/2$, where m_e is the electron mass, we get $v = \sqrt{2K/m_e}$. Thus

$$m_{\text{orbital}} = \frac{1}{2}er\sqrt{\frac{2K}{m_e}}.$$

The moment of the specified current loop is IA, where A is the area of the loop.

Working It Out If we convert to SI units, our energy estimate of 1 eV is 1.6×10^{-19} J. Thus

$$v = \sqrt{\frac{2(1.6 \times 10^{-19}\text{ J})}{m_e}} \cong \sqrt{\frac{3.2 \times 10^{-19}\text{ J}}{10^{-30}\text{ kg}}} \cong 6 \times 10^5 \text{ m/s},$$

where we have used (to the same accuracy as the orbit radius) $m_e \cong 10^{-30}$ kg. Equation (31–10) then gives the magnetic moment

$$m_{\text{orbital}} = \tfrac{1}{2}(1.6 \times 10^{-19}\text{ C})(6 \times 10^5 \text{ m/s})(10^{-10}\text{ m})$$
$$\cong 5 \times 10^{-24} \text{ A} \cdot \text{m}^2.$$

For comparison, the magnetic moment of the macroscopic loop is

$$m = (10^{-3}\text{ A})(10^{-4}\text{ m}^2) = 10^{-7} \text{ A} \cdot \text{m}^2,$$

which is some 2×10^{16} times larger than the magnetic moment of the single atom.

What Do You Think? We learned in mechanics that in planetary motion, both the light particle (the electron here) and the heavy particle (here, the positively charged nucleus) rotate about the center of mass of the system. Do we need to worry about this complication?

In Example 31–3, we can think of a magnetic dipole moment as due to a circulating charge, with the magnetic moment proportional to the angular momentum of the circulating charge. It is useful to express magnetic moments in terms of angular momentum because angular momentum is a fundamental physical quantity. For an electron in an atom, the orbital magnetic dipole moment [Eq. (31–10)] has the magnitude

$$m_{\text{orbital}} = \frac{1}{2}evr = \frac{e}{2m_e}m_e vr = \frac{e}{2m_e}L, \quad (31\text{-}11)$$

where m_e is the electron's mass and $L = m_e vr$ is the angular momentum of the electron in its circular orbit. If we include the vectorial properties of both the angular momentum and the magnetic moment, then Eq. (31–11) becomes

$$\vec{m}_{\text{orbital}} \equiv g_L \vec{L}. \quad (31\text{-}12)$$

The coefficient g_L connecting the magnetic moment and the angular momentum is known as the **gyromagnetic ratio**. For the orbital motion, we have just seen that

$$g_L = -\frac{e}{2m_e}. \quad (31\text{-}13)$$

The minus sign is present because $\vec{m}_{\text{orbital}}$ and $\vec{L}$ point in opposite directions (Fig. 31–5b).

At this point, we make a necessary leap to a nonclassical atom. According to classical mechanics, the magnitude of the orbital angular momentum depends on the size of the orbit and the velocity of the electron, but in fact atoms cannot be described in this way. Quantum mechanics shows that orbital angular momentum cannot take on a continuous range of values; rather, the magnitude of the orbital angular momentum is limited to the values $\ell\hbar$, where $\hbar = 1.05 \times 10^{-34}$ J·s is a constant that appears in all aspects of atomic physics and ℓ is a positive integer (or zero).[†] When we use this restriction in Eq. (31–12), we find that this also places restrictions on the possible values of the magnetic moment:

$$m_{\text{orbital}} = \left(\frac{e\hbar}{2m_e}\right)\ell \equiv m_B \ell. \tag{31-14}$$

The quantity m_B is the **Bohr magneton**, after Niels Bohr, one of the founders of quantum mechanics. Its numerical value is

$$m_B = \frac{e}{2m_e}\hbar = 9.27 \times 10^{-24} \text{ A} \cdot \text{m}^2. \tag{31-15}$$

This is a value that is characteristic of all magnetic moments on the atomic scale.

In addition to a magnetic moment associated with orbital motion, experiments show that electrons also carry an *internal* magnetic moment that cannot be identified with any real current. *The electron itself behaves as a tiny magnet!* This is another quantum phenomenon, one with no analog in classical physics. As a consequence, we must add the contribution of the electron's *intrinsic magnetic moment*, $m_{\text{intrinsic}}$, to the orbital magnetic moment. The value of $m_{\text{intrinsic}}$ turns out to be approximately m_B to an accuracy of 0.1%.

Bulk Effects Are Due to the Alignment of Atomic Magnetic Dipoles

The magnetic field of an individual atom is tiny compared to the magnetic field of a bar magnet. But there are so many atoms that if all the atomic magnetic moments were perfectly aligned in a material, we would have a truly huge effect. As Example 31–4 shows, the alignment of the atomic magnetic moments need be only very slight to produce noticeable bulk effects.

EXAMPLE 31–4 Consider 1 mol of atoms with individual magnetic moments $m_0 = 10^{-23}$ A·m². Assume that the magnetic moments can point only in the $+z$- and $-z$-directions, with a fraction f pointing "up" and a fraction $1 - f$ pointing "down." What value of f gives the same magnetic moment as a 1-cm² wire loop that carries a current of 10 mA?

Strategy The total magnetic moment is the vectorial sum of individual magnetic moments $m_0 = 10^{-23}$ A·m². In this case the vectors that represent the individual moments take only the two directions *up* and *down*. Thus with altogether N_A atoms in a mole, where N_A is Avogadro's number 6×10^{23}, the total magnetic moment is the algebraic sum of fN_A moments with a plus sign (pointing up) and $(1 - f)N_A$ identical moments with a minus sign (pointing down). This must match the magnetic moment IA of the wire loop, where A is the area of the loop and I is the current running through it. We can then solve this equation for f.

Working It Out The relation above reads

$$m_{\text{tot}} = N_A f m_0 - N_A(1-f)m_0 = N_A m_0 (2f - 1) = IA.$$

Solving this relation for f, we find

$$f = \frac{1}{2}\left(1 + \frac{IA}{N_A m_0}\right).$$

Numerical evaluation gives

$$f = \frac{1}{2} + \frac{(10^{-2} \text{ A})(10^{-4} \text{ m}^2)}{2(6 \times 10^{23} \text{ atoms})(10^{-23} \text{ A} \cdot \text{m}^2/\text{atom})}$$

$$= \frac{1}{2} + (8 \times 10^{-8}).$$

In other words, a departure from complete randomness ($f = \frac{1}{2}$) of only one part in 10 million leads to macroscopic effects.

What Do You Think? Suppose we place the atoms between the poles of a magnet. The magnetic field will to some degree line up the magnetic moments so that they tend to point in the direction of the magnetic field. Why don't we expect f to be 1 in this case?

[†]We would have to refine this further to be in accord with the full predictions of quantum mechanics; for now this is a sufficiently good description of the situation.

CONCEPTUAL EXAMPLE 31–5 Take a bar magnet 30 cm long and cut it into two 15- cm pieces. On the basis of our understanding of the magnetism of materials, would you predict that the magnetization of the two pieces is (a) larger, (b) smaller, or (c) comparable to the magnetization of the original magnet?

Answer If we use the simple model of a magnet treated in Example 31–4, we see that net magnetism is a consequence of a small imbalance between atomic magnetic moments pointing *up* and *down*. There is no reason to believe that this imbalance is in any way distributed inhomogeneously throughout the sample, so that the fraction f discussed in Example 31–4 should be the same across different parts of the sample. Since the magnetization is the magnetic moment per unit volume, the size of any pieces is not relevant—magnetization is an intensive quantity. We thus expect the magnetization of the two pieces to be the same as that of the single larger piece from which they came.

THINK ABOUT THIS...
WHY AREN'T MOST MATERIALS MAGNETIC?

We saw in Example 31–4 that only a small deviation from randomness leads to significant bulk effects provided that individual atoms or molecules have *some* magnetic moment (as indeed they do). Some random fluctuation of this size might then produce macroscopic effects. But in fact the "small" deviation in the example is not so small. In statistics, $\sqrt{N}$ is a typical fluctuation from the mean when N objects or events are involved. Thus in a sample of 10^{24} atoms, statistical fluctuations away from an average magnetic moment of zero will lead, on average, to an excess of only 10^{12} atoms pointing, say, up. This translates to a fraction $f = 10^{12}/10^{24} = 10^{-12}$ of atoms pointing up, and this corresponds to a net magnetic moment much smaller than the one we saw in Example 31–4. It is only when there are forces between atoms that cause neighboring atoms to line up with each other that we form permanent magnets.

The Connection Between Microscopic and Macroscopic Quantities

We have seen that a piece of material will have significant magnetization if the directions of the magnetic dipole moments of its many component atoms or molecules are not completely random. In that case, the vector sum of the atomic magnetic moments will not be zero. If we divide the vector sum of the magnetic moments by the number of atoms, we get an *average* magnetic moment $\vec{m}_0$ per constituent (atom or molecule). The magnetization is then

$$\vec{M} = n\vec{m}_0, \tag{31–16}$$

where n is the number of constituents per unit volume. Once we have determined $\vec{M}$, we can determine the other bulk magnetic properties from the discussion in Section 31–1.

31–3 Ferromagnetism

Ferromagnetic materials, which include the elements iron, cobalt, nickel, gadolinium, and dysprosium, together with their alloys, can have large permanent magnetizations, and the direction and size of the magnetization can be set by an external magnetic field. These materials and their properties come into play every time you use computer memory or a credit card, as well as throughout the electrical transmission network.

In ferromagnetic materials, the intrinsic magnetic dipole moments of the electrons in atoms align themselves to some degree and lead to large magnetic effects. The explanation for the alignment of the intrinsic magnetic moments, which was suggested in 1928 by Werner Heisenberg, is purely quantum mechanical—there is no classical mechanism that results in a sufficiently strong alignment. As a consequence of the quantum-mechanical exclusion principle (see Chapter 26), electrons with parallel *intrinsic* magnetic moments arrange themselves in ferromagnetic materials in orbits that tend to maximize the distance between them. This reduces the potential energy of Coulomb repulsion between them and makes a state with parallel magnetic moments a state of lower energy. As a result, there is a preference based on energy for the intrinsic magnetic moments of electrons to line up parallel with one another. This argument shows that the forces that tend to keep the magnetic moments parallel are electrostatic forces rather than the much weaker magnetic forces, and this explains why ferromagnetic materials must be heated to hundreds of degrees Kelvin before they become demagnetized.

882 | Magnetism and Matter

▶ **FIGURE 31–6** Photomicrograph of magnetic domains in a sample of iron with 3 percent silicon. A strong net magnetic field is associated with each domain. Domains with different orientations appear in different colors.

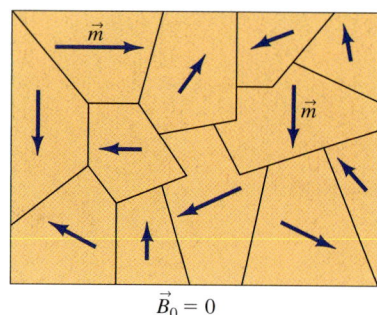

(a)

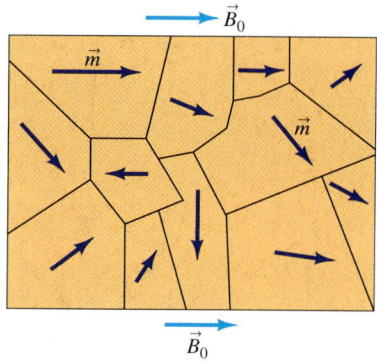

(b)

◀ **FIGURE 31–7** (a) Domain formation in ferromagnetic materials in the absence of an external magnetic field. The arrows indicate the magnetic moments of individual domains. (b) The presence of an external magnetic field influences the domains, making some larger and realigning others.

The intrinsic magnetic moments of the unpaired electrons of different atoms—for example, each iron atom has two such electrons—do not ordinarily become aligned *throughout* a piece of ferromagnetic material. Rather, the alignment takes place between adjacent atoms in regions called *magnetic domains*, which may contain 10^{17} to 10^{21} atoms and occupy a volume on the order of 10^{-12} to 10^{-8} m³ (volumes from 0.1 mm to 1 mm on a side). The magnetic field within these domains is quite large, but a piece of ferromagnetic material may be made of thousands of such domains, and because each domain will have its magnetization aligned differently, the magnetization of the entire material will average to zero. Figure 31–6, a photograph of iron, clearly shows the boundaries (*domain walls*) between domains. Figure 31–7a is a schematic diagram of the domains showing their individual magnetic moments.

Given the description above, how does a piece of iron gain and maintain a significant magnetization? An external magnetic field $\vec{B}_0$ provides the mechanism that can align the magnetizations of different domains. This field acts in two ways. First, the size of domains with their magnetic moments already aligned with $\vec{B}_0$ may enlarge at the expense of neighboring domains. Second, the magnetic moments of some of the domains may rotate to the direction of $\vec{B}_0$ through an overall realignment of their constituents (Fig. 31–7b). Both of these mechanisms work because a state with a magnetic moment aligned along $\vec{B}_0$ is a state of lower energy.

The process we have described is a little like that of a marching band whose members face in random directions. The band leader orders them to face the same direction but fails to say *which* direction. Influenced by immediate neighbors, small groups of the band may align themselves together, but these groups may not be aligned with each other. If the band leader adds an instruction as to a precise direction, the groups will realign along that direction and there will be uniformity across the entire band.

When a magnet made of ferromagnetic material—we'll refer to this more simply as a ferromagnet—is heated, the increased movement of the atoms leads to a randomization of their orientation and thus to a decrease in the alignment. At the *Curie temperature*, T_c (after Pierre Curie), the randomization is complete, and the material loses its magnetization—it is no longer a ferromagnet. The value of T_c varies from material to material; in iron, $T_c = 1043$K, in gadolinium, $T_c = 292$K. Now we drop the temperature again. As T drops below T_c, ferromagnetism appears through a phase change, just as water forms the ordered lattice we know as ice below 273K. The ordering occurs in domains, as described by our marching-band analogy. Thus when the material cools below T_c, it will not act as a permanent magnet, because the domains are randomly oriented. Water supplies another physical example of this behavior. When a lake surface freezes, the freezing takes place in domains rather than as one huge ice crystal.

EXAMPLE 31–6 Estimate the maximum possible magnetization in a piece of iron.

Setting It Up To solve this problem, we require some knowledge about iron: We may know the intrinsic magnetic moment of a single electron, but we also need the knowledge that there are *two* unpaired electrons per iron atom, and both will contribute to the total magnetic dipole moment. In addition, to find the number of electrons per unit volume we also need the number of iron atoms per cubic meter; this is information that can come from the mass density of iron and the atomic weight of iron.

Strategy The maximum possible magnetization corresponds to having *all* the magnetic moments of the contributing electrons lined up—only one domain, with perfect alignment within the domain. The magnetization is then the number of contributing electron magnetic moments in a unit volume, times the individual electron magnetic moment $m_B = 9.3 \times 10^{-24}$ A·m² [Eq. (31–15)]. The number of iron atoms in a cubic meter of iron is

$$n_{Fe} = \frac{N_A \text{ atoms}}{1 \text{ mol}} \frac{1 \text{ mol}}{\text{mass of iron in 1 mol}} \times \frac{\text{mass of iron}}{\text{per unit volume}};$$

the number of unpaired electrons per atom is twice this value.

Working It Out The mass of iron in 1 mol is 56 g, and the mass density of iron—the third factor on the right in the expression for n_{Fe}—is 7.8 g per cm³ = 7.8×10^6 g per m³. Using $N_A = 6.02 \times 10^{23}$, we then find $n_{Fe} = 0.84 \times 10^{29}$ atoms per m³, hence n = the number of unpaired electrons per m³ = 1.7×10^{29}. This has to be multiplied by the magnetic moment per unpaired electron to get the maximum magnetization:

$$M_{max} = nm_B = (1.7 \times 10^{29} \text{ electrons/m}^3)$$
$$(9.3 \times 10^{-24} \text{ A·m}^2 \text{ per electron}) = 1.6 \times 10^6 \text{ A/m}.$$

Compare this value to the magnetic dipole moment of the macroscopic loop in Example 31–3—the value found here is quite large by comparison.

What Do You Think? An experimental value of M_{max} for iron in a particular device is 1.7×10^4 A/m. What does this say about our calculation or about the piece of iron?

Hysteresis

The relation between the magnetic field B and the magnetic intensity H is more complicated in ferromagnets than in other materials. In order to measure the relation between B and H in a ferromagnetic material, that material is first demagnetized by heating. It is then cooled, shaped into a ring, and wound with a wire that carries a current I. This experimental arrangement is called a *Rowland ring* (Fig. 31–8). Without the ferromagnetic material, the magnetic field inside the ring, or toroidal solenoid, has the nearly constant value

$$B_0 = \mu_0 H = \mu_0 nI, \quad (31\text{--}17)$$

provided that the torus is "thin," like a bicycle tire. Here n is the number of windings per unit length. As we know, when the ferromagnetic material is inserted into the torus, the magnetic field increases tremendously to a new value B. We measure B by using a sense coil outside the torus (Fig. 31–8). The sense coil measures an induced emf proportional to the time rate of change of the magnetic field. As we raise the current in the toroidal coil at a given rate, we know the magnetic intensity $H = nI$, and the sense coil measures B. Figure 31–9 shows one example of a measured relation between H and B; a plot of H versus B is called a *magnetization curve*. Knowing H and B, we can determine the magnetization from the relation $B = \mu_0 H + \mu_0 M$.

For ferromagnetic materials, we observe a magnetization curve like that shown in Fig. 31–9; this curve is known as a **hysteresis loop**, and it indicates the phenomenon of **hysteresis**. The presence of hysteresis demonstrates an irreversibility in the magnetization process. When the current I in the solenoid is slightly changed and then changed back again, the original magnetization is generally not recovered. For example, if we start on curve c in Fig. 31–9 at a value of 10^{-4} T for $\mu_0 H$, B in the ferromagnetic material is negative. If $\mu_0 H$ is increased to 3×10^{-4} T and then brought back to 10^{-4} T, B is now positive, following curve b. Hysteresis results from the fact that the magnetic domains do not return to their original zero-external-field status when the current decreases. They "remember" the rise in field and do not automatically revert to their original alignments.

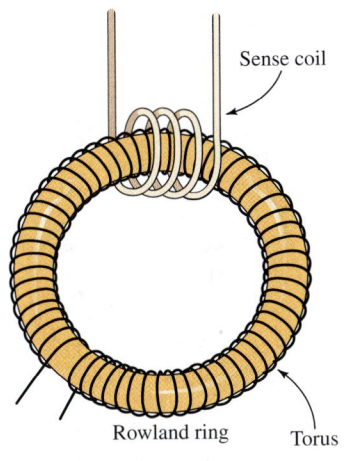

▲ **FIGURE 31–8** A Rowland ring is a wound core (solenoid) of material that may be used to measure the relation between $\vec{B}$ and $\vec{H}$ of that material. A sense coil measures changes in $\vec{B}$.

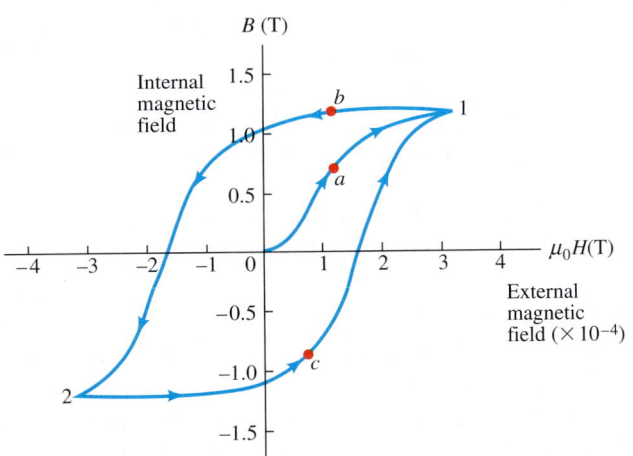

◀ **FIGURE 31–9** A magnetization curve illustrates the phenomenon of hysteresis in ferromagnetic materials. The material starts at the origin with zero magnetization. When a magnetic intensity H is applied, the material responds by becoming magnetic and is magnetic even when H is again zero.

884 | Magnetism and Matter

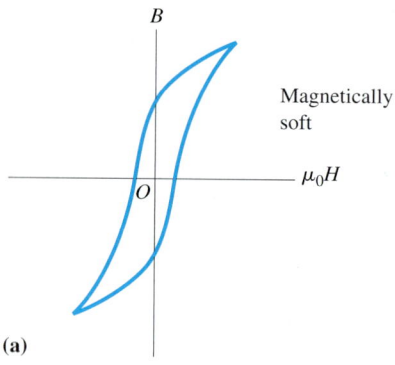

(a)

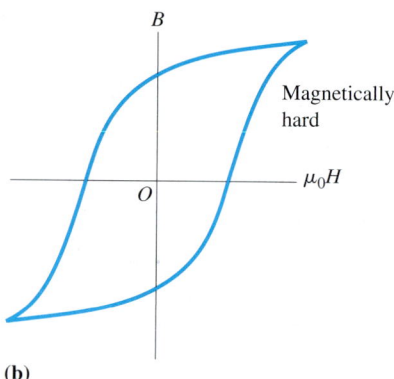

(b)

▲ **FIGURE 31–10** Hysteresis loops for materials that are (a) magnetically soft and (b) magnetically hard.

▲ **FIGURE 31–11** Closeup view of the magnetic structure of recording media, here a hard disk. These consist of a series of magnetized regions laid out along grooves. Here the grooves are some 3 nm wide, and the storage density on this still-experimental medium would be 1.7 Gbit/in^2.

Some materials have narrow hysteresis loops, meaning that the alignment of the domains follows the external field rather closely (Fig. 31–10a). This is the type of curve that materials considered *magnetically soft*, such as properly prepared iron, will follow. Such materials are often used in transformer cores. (Transformers are devices that transform AC currents—or voltages—from one value to another.) Other materials have broad hysteresis loops, meaning that their domains respond only to large external fields (Fig. 31–10b). Such materials, including alloys of iron with carbon or tungsten, are said to be *magnetically hard*. They are difficult to magnetize but, once magnetized, they retain much of the magnetization and make good permanent magnets because they are equally difficult to demagnetize. Magnetically hard materials are especially important for making magnetic tapes or memory disks because such materials are stable against changes due to nearby magnetic fields (Fig. 31–11). Which materials are magnetically hard and which are magnetically soft is a complex question whose understanding demands a detailed understanding of atomic and solid state structure.

*31–4 Diamagnetism and Paramagnetism

In this section we'll examine the origin of the phenomena of *diamagnetism* and *paramagnetism*, both generally small but nevertheless significant manifestations of the way materials respond to external magnetic fields.

Diamagnetism

Diamagnetic materials respond to an external magnetic field $\vec{B}_0$ by reducing it slightly; the induced field is *opposite* to $\vec{B}_0$. The resulting net magnetic field is *less* than $\vec{B}_0$. Diamagnetism, as distinguished from ferromagnetism and paramagnetism, occurs as a primary effect in materials whose atoms have no permanent magnetic dipole moments, either orbital or intrinsic, and is also a secondary effect in any material. A classical model helps us to understand the phenomenon qualitatively. Let's consider two electrons with identical orbits, except that the motion is counterclockwise in one and clockwise in the other (Fig. 31–12a). With no external magnetic field, the orbital magnetic moments of the two electrons cancel ($\vec{m}_1 + \vec{m}_2 = 0$), and there is no magnetization.

Now suppose that an applied magnetic field $\vec{B}_0$ perpendicular to the orbits of the electrons is turned on (Fig. 31–12b). For the electron on the left of the figure, the flux through its orbit increases as the external field increases; by Lenz's law, the electron responds to counter the increasing flux. Accordingly, the negatively charged electron speeds up, increasing its angular momentum as well as the magnitude of its orbital magnetic moment $\vec{m}_1$, which points downward. When the external field levels off, angular momentum conservation ensures that the new value of $\vec{m}_1$ persists. Similarly, the electron on the right slows down to oppose the increase in flux through its orbit, so the magnitude of its magnetic moment $\vec{m}_2$, which points upward, is reduced. The result is that $\vec{m}_1 + \vec{m}_2$ now has a net value that points downward, and a magnetic field is produced that opposes the increasing external field. This is the origin of the negative magnetic susceptibility. The classical model must be revised for a proper quantum mechanical treatment of the atom because, in quantum mechanics, the angular momentum is *quantized*, and cannot change only slightly when the field changes slightly. Nevertheless, a correct treatment of a large collection of atoms reproduces the effect of the classical discussion.

When it is applied quantitatively, the simple model just described leads to reasonable estimates for the size of diamagnetic effects. The model correctly implies that *diamagnetism is present in all materials*, although it is masked for materials whose atoms have permanent magnetic moments.

Paramagnetism

Paramagnetic materials respond to an external field by reinforcing the field slightly; the induced field is in the same direction as the external field. These materials contain molecules with permanent magnetic dipole moments due to the intrinsic magnetic moments of unpaired electrons. In the absence of an external magnetic field, these dipoles are

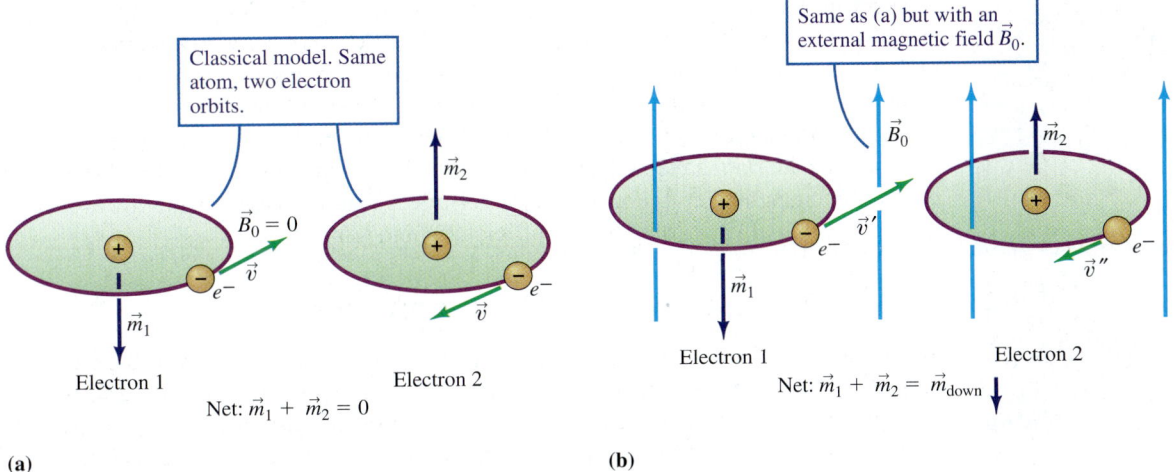

▲ **FIGURE 31–12** (a) A two-electron atom with a net orbital magnetic moment of zero in the absence of an applied magnetic field. (b) In the presence of an applied magnetic field, the orbital magnetic moment associated with each electron is changed, and there is a net magnetic moment that points down, opposite to the external field.

randomly oriented due to thermal motion, and the net magnetization of the materials is zero. Recall from Chapter 28 that the energy of a magnetic dipole moment $\vec{m}$ in a magnetic field $\vec{B}$ is, by Eq. (28–26), $U = -\vec{m} \cdot \vec{B}$. The lowest energy occurs when $\vec{m}$ and $\vec{B}$ are *parallel*. Thus an external magnetic field $\vec{B}$ tends to align the atomic magnetic moments along $\vec{B}$ and produces a positive magnetic susceptibility.

Two effects determine the extent to which the permanent magnetic dipoles become aligned. The first is the external field, which by the energy argument above encourages alignment, and the second is the thermal motion, which randomizes the alignment. The relative importance of these two factors is measured by the relative size of the magnetic energy factor mB and the thermal energy factor kT, where T is temperature. If T is so large that $kT \gg mB$, the average alignment over a large number of electrons will be weak. Conversely, if T is so low that $kT \ll mB$, the average alignment will be strong. For intermediate temperatures and fields, the average alignment is proportional to the ratio of these energies, i.e. to $(mB)/(kT)$. At room temperature, the intrinsic magnetic moments of most paramagnetic materials are only very slightly aligned, but as we already saw in Example 31–4, large bulk effects can come from very small alignments. In 1895, Pierre Curie observed the linear relation that we now call *Curie's law*:

$$\vec{M} = C\frac{\vec{B}}{T}, \tag{31-18}$$

where C is *Curie's constant*, a quantity that varies from one material to another. This law, which holds best for small values of mB/kT, is often expressed in terms of magnetic susceptibility, defined according to Eq. (31–5) as $\vec{M} = \chi_m \vec{H}$. If we anticipate that the susceptibility will be small, as it is for paramagnetic materials, then we can replace $\vec{B}$ in Eq. (31–18) by $\mu_0 \vec{H}$:

$$\vec{M} = C\frac{\mu_0 \vec{H}}{T},$$

or

$$\chi_m = \frac{\mu_0 C}{T}. \tag{31-19}$$

The value of C depends on the material. The susceptibility is positive, which is characteristic of paramagnetism.

The temperature dependence in Eq. (31–18) is the same as that of the analogous phenomenon for dielectrics [see Eq. (25–30)]. There, too, this dependence is called *Curie's law*. We expect the law to fail at sufficiently low temperatures and/or large fields.

▲ **FIGURE 31–13** Oxygen is paramagnetic and is therefore attracted by the poles of a magnet. Here, liquid oxygen poured between two poles is held in place by the forces between it and the permanent magnet.

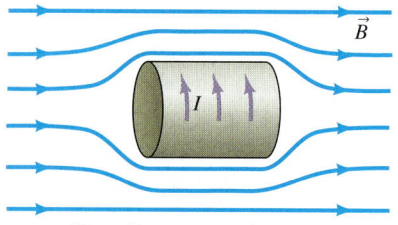

Type I superconductor

▲ **FIGURE 31–14** A Type I superconductor expels magnetic field from its interior by acting as a perfect diamagnet: Surface currents that just cancel the applied field inside are established.

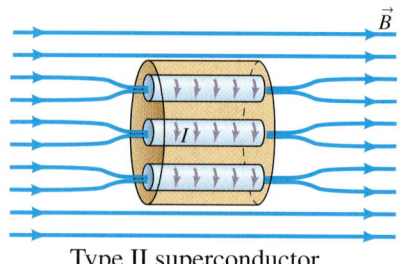

Type II superconductor

▲ **FIGURE 31–15** In Type II superconductors, the magnetic field is confined to filamentary structures. Inside the filaments, the material is not in its superconducting phase.

If the intrinsic magnetic moments are aligned perfectly with the field, they cannot be further aligned to produce still higher magnetization. This *saturation* phenomenon is observed. A more quantitative version of our arguments can be used to predict—successfully—the size of the paramagnetic susceptibility.

Unlike diamagnetism, paramagnetism is not a universal phenomenon, because relatively few materials have molecules with unpaired electrons (Fig. 31–13). When it is present, paramagnetism is normally a larger effect than diamagnetism. However, diamagnetism will always dominate at sufficiently high temperatures.

*31–5 Magnetism and Superconductivity

Superconductors have magnetic properties that are just as extraordinary as their electric properties. The same collective quantum physical mechanism that makes the resistance of a superconductor exactly zero *also makes the magnetic field inside zero*. A superconductor acts as a perfect diamagnet in the sense that currents are induced that precisely cancel any magnetic field inside. Alternatively, we say that the magnetic field lines are *expelled* from the superconductor—a phenomenon known as the *Meissner effect*. In fact, superconductors come in two types according to just how the field is expelled: In Type I superconductors, the field is expelled entirely (Fig. 31–14). In Type II superconductors, the field is isolated in nonsuperconducting filamentary structures within the material (Fig. 31–15). Currents circulate on the surfaces of these filaments, shielding the rest of the material from the magnetic field within the filaments. Quantum physics sets a minimum amount of magnetic flux within each filament, and the flux found within any given filament is an integer multiple of this minimum amount. This is yet another example of how physical quantities are quantized in a quantum-mechanical world.

The expulsion of magnetic fields from the interior of a superconductor, whether Type I or II, translates into a statement about its magnetic susceptibility. Equation (31–6) expresses the internal field in terms of the magnetic intensity $\vec{H}$, and this internal field must be zero:

$$\vec{B} = \mu_0(1 + \chi_m)\vec{H} = 0.$$

The magnetic intensity is due to the free currents and is not zero. Thus $1 + \chi_m = 0$, or

$$\chi_m = -1.$$

If there is no magnetic field inside a superconductor, then there can be no currents inside. To see this, imagine that some internal region of a superconductor carries current. Then we can draw a loop around this region and apply Ampère's law. If there were a current through the loop, there would be a magnetic field, but there is no such field. We conclude that *all current carried by superconductors must be carried on their surfaces* (any boundary between superconducting and nonsuperconducting phases). For example, current can be carried on the walls of the filaments in Fig. 31–15.

In the presence of a large enough magnetic field (a *critical field*), a material in a superconducting phase jumps back to the normal (nonsuperconducting) phase even at a fixed temperature, a phenomenon that can cause serious practical problems. For example, many large electromagnets are made from superconductors because they do not undergo Joule heating in spite of the large currents in them. However, the large magnetic field may itself destroy the superconductivity, and one is left with a lot of current in nonsuperconducting wires and a potential meltdown from Joule heating. To overcome this problem, Type II superconductors are generally a better choice for superconducting wires in the electromagnets, because they channel the magnetic field into filaments, providing a way to make superconductors with much higher critical fields.

*31–6 Nuclear Magnetic Resonance

Atomic nuclei consist of protons and neutrons, far more massive than electrons. Because it is so massive, the orbital motion of the nucleus itself about the atomic center of mass is negligible, and the small size of the nucleus allows us to ignore orbital motions of the protons and neutrons within the nucleus. However, protons and neutrons, like electrons, have intrinsic magnetic dipole moments. These magnetic moments are some 2000 times smaller than that of the electron because protons and neutrons are some 2000 times more massive than electrons [see Eq. (31–15)]. They arise from an intrinsic angular momentum called the *spin* and labeled $\vec{S}$. Unlike the orbital angular momentum, which according to quantum mechanics can occur only in integer multiples of $\hbar$ (Planck's constant, h, divided by 2π), the electron or proton spin has the value $\frac{1}{2}\hbar$. If the proton magnetic moment is labeled $\vec{m}_p$, then as for the orbital angular momentum [see Eq. (31–12)], there is a gyromagnetic ratio g_p that gives the relation between the spin and the magnetic moment,

$$\vec{m}_p \equiv g_p \vec{S}. \qquad (31\text{–}20)$$

What is the effect of an external magnetic field $\vec{B}$ on the proton? Consider the torque on it due to such a field, given by $\vec{\tau} = \vec{m}_p \times \vec{B}$. Now, the torque on the proton is the rate of change of its internal angular momentum, $d\vec{S}/dt$, so with $\vec{S} = \vec{m}_p/g_p$ we have

$$\frac{1}{g_p}\frac{d\vec{m}_p}{dt} = \vec{m}_p \times \vec{B}. \qquad (31\text{–}21)$$

This is an equation that gives the rate of change of the magnetic moment of a proton in a magnetic field. As the analysis of this equation in Chapter 10 showed, the magnitude of $\vec{m}_p$ cannot change, but its direction can, and Eq. (31–21) describes the *precessional motion* of $\vec{m}_p$ about the direction of $\vec{B}$ (Fig. 31–16). This precession, called *Larmor precession*, is analogous to the precession of a spinning top under the influence of gravity (see Section 10–8). In a classical picture of Larmor precession, the direction of $\vec{m}_p$ traces a cone around the direction of $\vec{B}$, as in Fig. 31–16. However, quantum mechanics tells us that the analog of only two such cones are allowed: one with spin "up" and one with spin "down." In Problem 41, we find the angular speed of classical precession,

$$\omega_0 = g_p B. \qquad (31\text{–}22)$$

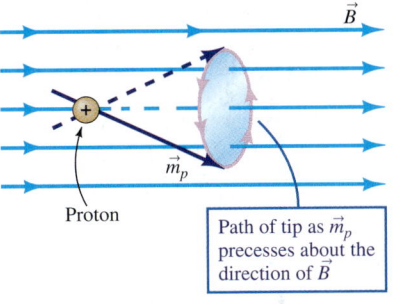

▲ **FIGURE 31–16** The magnetic moment of a proton, $\vec{m}_p$, precesses about the direction of an external magnetic field with an angular speed ω_0.

Finally, let's recall one other feature of this motion: There is a potential energy associated with a magnetic moment $\vec{m}$ in an external field, given by $U = -\vec{m} \cdot \vec{B}$, and the motion with the "up" cone has lower energy than does the motion with the "down" cone.

Now consider exposing the atom to an oscillating magnetic field. (Such a field occurs in electromagnetic waves; see Chapter 34.) This field would normally not have much effect, but when the angular frequency ω of the oscillating field *exactly matches* the angular frequency ω_0 of the precession (a condition known as *resonance* and described in Section 13–8), the direction of the magnetic moment can flip. In this case, the oscillating magnetic field either supplies *just the precise amount of energy*—$2\vec{m}_p \cdot \vec{B}$—necessary to flip the spin of the proton from up to down, or absorbs exactly this amount of energy to flip the spin from down to up.

This effect is called **nuclear magnetic resonance** (**NMR**). When the spin flips due to a transfer of energy between the oscillating field and the proton, there is a detectable signal. Thus, by tuning the frequency of the oscillating magnetic field, we can measure the frequency $\omega_0 = g_p B$ with very high precision.

If the external magnetic field is known, the NMR method may be used to measure the gyromagnetic ratio, g_p. It is in this way that the gyromagnetic ratios for protons and neutrons are known to contain the coefficients 2.79 and -1.91, respectively, multiplying an expression like Eq. (31–13) (but with the proton or neutron mass in place of the electron mass). These coefficients suggest that the internal structures of protons and neutrons are more complicated than that of electrons. NMR measurements can also be extended to nuclei, where they are used to study nuclear structure and the forces that give rise to it. For example, the deuterium nucleus—which may be described as a bound state of a proton and neutron structured in such a way that the intrinsic spins, and therefore the intrinsic magnetic moments, are parallel to each other—is expected to have a magnetic moment that is the sum of the moments of the proton and the neutron. NMR measurements give a slightly smaller result than this sum. It is possible to conclude from this that the deuteron is slightly cigar-shaped, and this gives us information about the forces that bind it.

THINK ABOUT THIS...
WHAT IS MAGNETIC RESONANCE IMAGING?

NMR has important applications in the study of materials and in medical diagnostics, where the procedure is called *magnetic resonance imaging* (MRI). (The word "nuclear" was dropped due to patients' fears that nuclear radiation was being used. In fact, whatever it is called, it is an especially safe procedure.) One knows the value g_p associated with the nucleus of hydrogen, and by placing a complex combination of materials such as a human body within a field whose values vary with position, one can tell by setting up the conditions for spin flip and hence absorption of energy from the field just where there are large numbers of hydrogen atoms—only at locations where the B-field has just the right value will the spin flip occur. In this way MRI locates concentrations of hydrogen atoms in patients. Fat, which has a high concentration of hydrogen, can be distinguished from muscle, which has a much lower hydrogen concentration, and tumors can be distinguished from nerve tissue. Bones have little hydrogen and are hardly seen at all (Fig. 31–17). Before MRI it was difficult to image soft tissues, which don't show in X-rays.

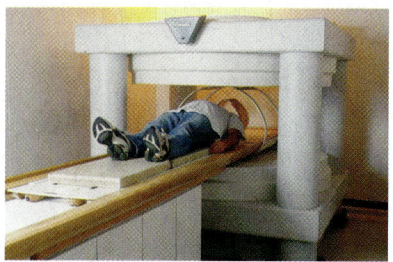

(a)

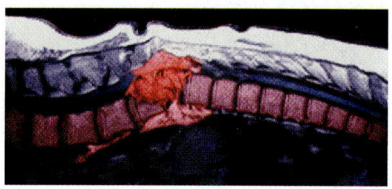

(b)

▲ **FIGURE 31–17** (a) An MRI diagnostic machine, as used in a medical facility. (b) The spinal tumor visible on this MRI image would not be as evident in a conventional X-ray.

Summary

The magnetic properties of bulk matter are summarized in the magnetization, $\vec{M}$, the magnetic dipole moment per unit volume. In the presence of an external **magnetic field** $\vec{B}_0$, there is a total magnetic field in a material given by

$$\vec{B} = \vec{B}_0 + \mu_0 \vec{M}. \tag{31-1}$$

The effect of free (real) currents (as opposed to the induced atomic effects) is contained in the magnetic intensity, $\vec{H} = \vec{B}_0/\mu_0$:

$$\vec{H} \equiv \frac{\vec{B}}{\mu_0} - \vec{M}. \tag{31-2}$$

The magnetic susceptibility χ_m describes the response of a material to a magnetic field of external origin:

$$\vec{M} \equiv \chi_m \vec{H}. \tag{31-5}$$

The net magnetic field is given in terms of χ_m by

$$\vec{B} = \mu_0(1 + \chi_m)\vec{H} = \mu\vec{H}. \tag{31-6, 31-8}$$

Here, μ is the permeability of the material:

$$\mu = \mu_0(1 + \chi_m). \tag{31-7}$$

Magnetism in matter is due ultimately to the magnetism of its atomic constituents and particularly to the unpaired electrons of atoms. An orbiting electron produces an atomic orbital magnetic moment

$$\vec{m}_{\text{orbital}} = g_L \vec{L}, \tag{31-12}$$

where g_L is the gyromagnetic ratio. Quantum mechanics implies that these magnetic moments take the value

$$m_{\text{orbital}} = \left(\frac{e\hbar}{2m_e}\right)\ell \equiv m_B \ell, \tag{31-14}$$

where the factor m_B is the Bohr magneton and ℓ is an integer. In addition, electrons have intrinsic magnetic moments equal in magnitude to m_B. Even a very slight alignment of atomic magnetic moments leads to large magnetic effects in bulk matter.

Ferromagnetic materials have large permeabilities. The atomic dipole moments are lined up in small regions called domains due to forces of quantum mechanical origin. The imposition of an external field leads to the dipole moments of the domains lining up together and produces permanent magnets. The fact that a ferromagnetic material "remembers" the orientation of the external field that magnetizes it leads to the phenomenon of hysteresis, in which the magnetization curve depends on how the magnetization was produced.

Diamagnetic materials have small negative susceptibilities that are ultimately due to Faraday's law. Diamagnetism is always present but may be masked by paramagnetic or ferromagnetic effects. Paramagnetic materials have small positive susceptibilities due to the intrinsic magnetic moments of unpaired electrons, which find it energetically favorable to line up with an external field. Paramagnetism is strongly temperature dependent. As well as having no measurable resistance, superconductors also have some remarkable magnetic properties; namely, they completely expel magnetic field from their superconducting regions.

In nuclear magnetic resonance (NMR), the intrinsic magnetic moments of nuclei and nuclear constituents precess about an applied magnetic field. This precession is detected by the response of a material to the imposition of an electromagnetic wave of just the right frequency—a frequency that is characteristic of the material.

Understanding the Concepts

1. When we calculated the magnetic dipole moment associated with orbital motion, why was it reasonable to think of the nucleus as stationary and the electron as circulating around it?
2. When an electron orbits the nucleus in a planetary model, the system forms an electric dipole. Why does this electric dipole not produce a measurable electric dipole field around the atom?
3. Devise an experiment that would determine whether a certain material was diamagnetic or paramagnetic.
4. Under what circumstances will Gauss' law for the magnetic field also hold for the magnetic intensity?
5. In a Rowland ring measurement of the magnetic field inside a piece of magnetic material, is it helpful to wrap the sense coil around the material many times?
6. Why would speaker coils and computer disks use magnetically hard material, while transformers and computer read/write heads use magnetically soft material?
7. Does iron exhibit diamagnetic properties? How could you determine them?
8. Aluminum is separated in junk yards by using large magnets. How is this possible?
9. Should the magnetic latch on a refrigerator door be made from magnetically hard or soft material?
10. Why should computer floppy disks not be made from magnetically soft material?
11. Why does diamagnetism dominate over paramagnetism in most materials at sufficiently high temperatures?
12. Explain how a permanent bar magnet attracts an unmagnetized iron needle.
13. Will a paramagnetic sample be attracted or repelled from a region of increasing magnetic field?
14. You are given two identical iron rods—one magnetized, the other not. How can you determine which is the magnet, without using a third magnet (for example, Earth)?
15. Suppose that an electron in a circular orbit around a nucleus is placed in an external magnetic field. Will the angular momentum of the electron change if the field is aligned perpendicular to the plane of motion? Parallel to the plane of motion?
16. Is it possible to arrange for a classical current loop to have a magnetic moment but no angular momentum? Assume first that you have both positive and negative charge carriers to work with, and then that you have only negative ones.
17. What is the value of $\vec{H}$ in an isolated permanent magnet?
18. It takes an external field to establish a macroscopic magnetization inside a permanent magnet cooled below its Curie temperature. What could have done this for lodestones, which are permanent magnets found in nature?
19. In a uniform magnetic field, a magnetic dipole experiences no net force, only a torque. How do two bar magnets repel or attract each other?
20. Suppose you are given the magnetic moment due to a single charge orbiting in a circle about a center. Describe qualitatively how you would calculate the magnetic moment of (a) a uniformly charged ring rotating with angular momentum perpendicular to its plane; (b) a uniformly charged disk rotating with angular momentum perpendicular to its plane; and (c) a uniformly charged sphere rotating about an axis through its center.

Problems

31–1 The Magnetic Properties of Bulk Matter

1. (I) A cylindrical rod of palladium (magnetic susceptibility $\chi_m = 8 \times 10^{-4}$), of radius 1 cm and length 5 cm, is placed in and aligned with a uniform magnetic field of 1.0 T. What is the magnetic dipole moment of the rod?

2. (I) A thin, toroidal coil of total length 34 cm is wound with 1600 turns of wire. A current of 0.62 A flows through the wire. What is the magnitude of $\vec{B}$ inside the torus if the core consists of a ferromagnetic material of magnetic susceptibility $\chi_m = 2.8 \times 10^3$? What is the magnitude of $\vec{H}$?

3. (I) The coil of a solenoid wound with a turn density of 3400 turns/m is filled with a material of unknown magnetic susceptibility χ_m. When the wire carries 0.450 A, the magnetic field within is 1.907×10^{-4} T. What is χ_m?

4. (I) A solenoid magnet wound with a turn density of 1300 turns/m, with permalloy inserted inside the windings, has a magnetic field of 3.5 T inside. How much current flows in the windings?

5. (I) A permalloy magnet is 5 cm in diameter, 30 cm long, and has magnetic intensity $\vec{H} = 30$ A/m at its pole. How many turns/m must an empty solenoid of the same dimensions have to give rise to the same intensity if it carries a current of 6 A?

6. (I) What is the magnetic moment of the equivalent solenoid in Problem 5?

7. (I) Earth's magnetic moment is about 10^{23} A·m^2. If the core, which is responsible for the magnetic moment, is about 20% of Earth's volume, what would be the core's magnetization? Assume the core is spherical.

8. (I) The core of a solenoid wound with 400 turns/m is filled with some material. When the current is $I = 3.1$ A, the magnetic field inside the core is measured to be 0.20 T. What is the magnetic susceptibility of the material placed in the core?

9. (I) A toroidal solenoid has a mean outer radius of 8.7 cm with 487 turns of wire. How much current is required to produce a magnetic field of 0.40 T if the solenoid is filled with permalloy?

10. (I) A toroid of core radius of 1.5 cm and toroidal radius of 20 cm is wound with silver wire. There are 1200 turns of wire and a current of 8 A flows through the wire. What is the magnetization per turn of wire?

11. (II) In a vacuum, a solenoid with a current I has a magnetic field B_0. (a) If silver is placed inside the solenoid, what is the change in the magnetic field? (b) What happens if cupric oxide is placed inside the solenoid?

12. (II) A 0.80-cm^3 cube of copper is placed between the poles of a magnet with a magnetic field of 0.48 T. What is the induced magnetization in the copper?

13. (II) A long solenoid filled with ferromagnetic material of permeability $\mu = 1320\mu_0$ is wound with wire so that there are 15 turns per cm. What current must flow through the wire to produce a magnetic field of 0.16 T within the solenoid?

31–2 Atoms as Magnets

14. (I) Suppose that 1 mol of atoms in a material have individual magnetic moments of 1.8×10^{-23} A·m^2. In the absence of any alignment, the magnetic moments form an *average* angle of 90° with some external axis. By how much does the average angle differ from 90° if the material has the same magnetic moment as a 1-cm^2 loop of wire that carries a current of 0.3 A? (The magnetic moment of the loop is aligned with the external axis.) Assume that the components of the atomic magnetic moments add algebraically.

15. (I) The atomic number of iron is 26, its atomic weight is 55.8, and its density is 7.87 g/cm^3. (a) How many electrons are there in 10 cm^3 of iron? (b) Suppose that each electron has the magnetic moment estimated in Example 31–3 (5×10^{-24} A·m^2), and that the magnetic moments "up" and "down" make up the fractions $\frac{1}{2}(1 + 2 \times 10^{-7})$ and $\frac{1}{2}(1 - 2 \times 10^{-7})$, respectively. What is the magnetization of the iron?

16. (II) Consider an electron in a circular orbit around a single proton (a hydrogen nucleus) whose total energy is -13.5 eV. Find the value of the orbital magnetic moment.

17. (II) Consider a ring of radius R rotating about the axis through its center and perpendicular to the plane of the ring with an angular velocity ω. It carries a uniform charge, with charge density λ C/m. What is the magnetic moment of the ring?

18. (II) A disk of radius R rotates with angular velocity ω about an axis through its center, perpendicular to the surface of the disk. The disk carries a uniform charge, with charge density σ C/m^2. What is the magnetic moment of the disk?

19. (II) The electron has a *classical radius* given by $r_0 = e^2/4\pi\varepsilon_0 m_e c^2 = 2.8 \times 10^{-15}$ m. This quantity is suggested by dimensional analysis: The particular combination of classical quantities is the only one that can be formed with dimensions of length. Use Eq. (31–11), with $m_{orbital}$ equal to the Bohr magneton, m_B, to show that any charge at the distance of the classical radius will be moving faster than the speed of light. (Assume that all the charge is concentrated at a belt of radius r_0.) Treatment of the magnetic moment of an electron as a classical quantity leads to trouble!

20. (II) The current I in a circular loop of radius R is due to the flow of free electrons in the loop. Show that the gyromagnetic ratio of the loop is independent of I, R, and the density of atoms.

31–3 Ferromagnetism

21. (I) A torus is wound with 800 turns/m of wire. A current of 5 A runs through the wire. If the core of the torus is iron, the internal magnetic field in the core is 1.8 T. What is the magnetization? What is the value of μ/μ_0 for the iron core?

22. (I) An electromagnet with a ferromagnetic core, $\chi_m = 15,000$, produces a maximum magnetic field of 0.36 T. What is the maximum current carried by the coil if the turn density of the coils is 14 turns/cm?

23. (I) A long, tightly wound solenoid contains a magnetic field of magnitude $B = 2.4 \times 10^{-3}$ T. An iron core, with susceptibility χ_m, is inserted so that it fills the space inside. What is the new value of B?

24. (II) We may view the iron core in the previous problem as equivalent to another solenoid that is concentric with the outer one and essentially coincident with it. How could you view the situation if the iron core were quite a bit shorter than the length of the solenoid? If the radius of the iron core were smaller than that of the solenoid? Draw magnetic field lines for these cases.

25. (II) A current of 0.5 A flows through a solenoid with 400 turns/m. An iron bar, with $\mu/\mu_0 = 640$, is placed along the solenoid axis (Fig. 31–18). (a) What is the magnetic field inside the iron bar? (b) Outside the iron bar, but still within the solenoid?

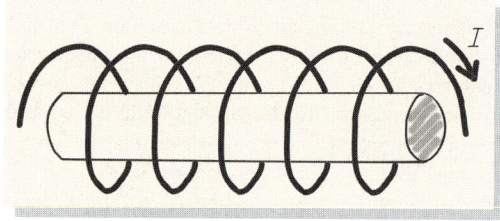

▲ **FIGURE 31–18** Problem 25.

26. (II) A disk-shaped permanent magnet has a thickness of 3 mm and a diameter of 1.5 cm. It is magnetized perpendicular to the plane of the disk, with the magnetic field on the axis near its north pole of magnitude 0.05 T. What is the current carried by a 80-turn coil of the same dimensions that gives this same value of the magnetic field on the axis (Fig. 31–19)?

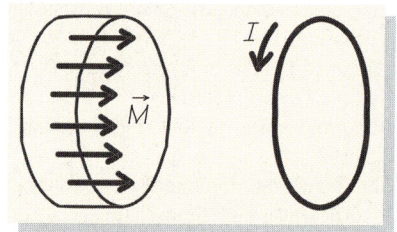

▲ **FIGURE 31–19** Problem 26.

27. (II) A current of 0.16 A is carried by a 50-turn coil 5.5 cm in diameter and 1.0 cm in length. Suppose a piece of iron with susceptibility 4.8×10^3 is placed inside the coil. What is the magnetic field inside the iron? What is the magnetic intensity there?

28. (II) A torus with a central radius of 25 cm and a tube radius of 2.0 cm is filled with iron of permeability $2800\mu_0$ (Fig. 31–20). There are 1200 turns around the torus. How much current must flow in the winding coil to produce a magnetic field of 1.5 T inside the torus? Treat the torus as having a constant magnetic field equal to the field at the central radius of the torus.

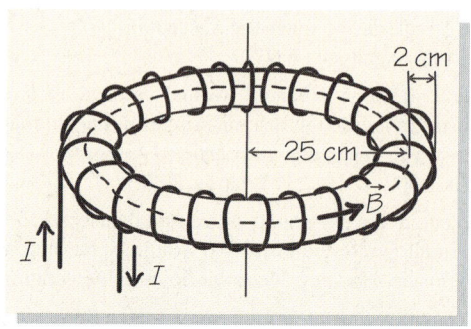

▲ **FIGURE 31–20** Problem 28.

29. (II) Iron (susceptibility = 6000) is used as the core of a transformer (see Chapter 33 for a discussion of this device). If large distortions of current are to be avoided, the transformer must be designed in such a way that the proportionality between $\vec{B}$ and $\vec{H}$ applies even for the strongest field. Estimate the maximum allowable value of H, knowing that the magnetic moment of individual iron atoms is $2.2\ m_B$.

30. (II) A Rowland ring measures the charge Q that passes through a sense coil by integrating the current in the sense coil over time (see Fig. 31–8). Both the sense coil and primary coil are wrapped tightly around a material. These coils have an area A. The number of turns in the sense coil is N. The emf induced in the sense coil by a changing magnetic flux in the material (due to a switch that passes current through the primary coil when the switch closes) is $\mathcal{E}$, and the sense coil has resistance R. Obtain a relation between the change in magnetic field, ΔB, and the charge Q. [Hint: Remember that $I = dQ/dt$.]

31. (II) A sense coil with a resistance of $0.1\ \Omega$ is wrapped tightly in 40 turns around a magnetic material of area $0.02\ m^2$. When a switch is closed in the primary coil, a charge of 5 mC flows through the sense coil (Fig. 31–21). If the magnetic field was initially zero, what is the new magnetic field in the material? (See Problem 30.)

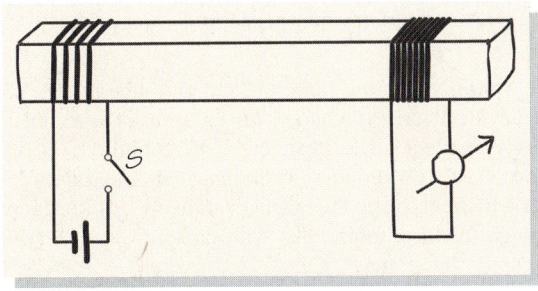

▲ **FIGURE 31–21** Problem 31.

*31–4 Diamagnetism and Paramagnetism

32. (III) An electron under the influence of some central force moves at speed v_i in a counterclockwise circular orbit of radius R. A uniform magnetic field $\vec{B}$ perpendicular to the plane of the orbit is turned on (Fig. 31–22). Suppose that the magnitude of the field changes at a given rate dB/dt. (a) What are the magnitude and direction of the electric field induced at the radius of the electron orbit? (b) The tangential force on the electron due to the induced electric field increases the electron's speed. Find the value of dv/dt. (c) Assuming that the initial orbital speed was v_i, find the final speed v_f as the magnitude of the magnetic field steadily increases from zero to a final value B_f by integrating dv/dt with respect to time. (d) Using your result for the change in speed, find the change in orbital angular momentum. (e) Use Eq. (31–11) to relate a change in the orbital magnetic moment to the change in the angular momentum.

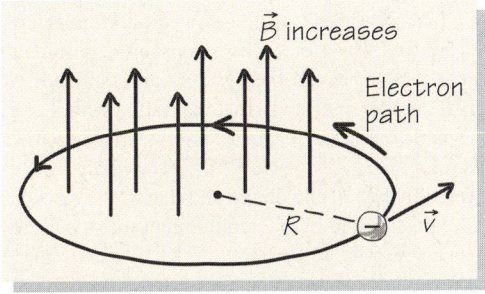

▲ **FIGURE 31–22** Problems 32 and 33.

33. (III) Refer to Fig. 31–22, but this time assume that the electron circulates clockwise rather than counterclockwise at speed v_i. By applying the same sequence of steps, show that the change in the magnetic moment of the electron's orbit is opposite the direction of change in the external field, just as in the case in which the electron circulates counterclockwise.

34. (III) Refer to Problems 32 and 33. Suppose that there are now two electrons moving at speed v_i in circular orbits of radius R, one clockwise and one counterclockwise. (a) What is the net orbital magnetic moment when the external field is zero? (b) After the external field has reached $\vec{B}_f$? (c) Show that the magnetic susceptibility for this system is $\chi_m = -(\mu_0 e^2 R^2/4m_e)\rho_e$, where ρ_e is the electron density.

35. (III) Using the techniques of Problems 32 through 34, estimate the magnetic susceptibility of copper, which has 29 electrons per atom. Assume that all the electrons move in orbits of the same radius, and that 14 move clockwise while 15 move counterclockwise. You will need to calculate the number density of electrons in copper.

36. (II) The temperature of a sample of $FeCl_3$ (the ferric ions have an intrinsic magnetic moment) inside a magnetic field is held constant as the field is increased. Sketch the induced magnetic moment as the magnetic field is increased.

37. (II) A long, straight conducting wire is embedded within an insulating paramagnetic material of magnetic susceptibility 2.6×10^{-4} at 300K and carries a current of 10 mA. Find the value of the magnetic intensity as a function of the distance from the wire, as well as the magnetic field. What is the change in the magnetic field when the temperature is lowered to 86K?

*31–6 Nuclear Magnetic Resonance

38. (I) Find the magnetic moment of the neutron, given that its gyromagnetic ratio is $-3.82e/2m_n$.

39. (II) Assume that it is possible to align perfectly the magnetic moments of protons in 1 mol of hydrogen gas at standard temperature and pressure. What are the magnetization and magnetic field inside the gas?

40. (II) In ^{17}O (oxygen with 17 nucleons in its nucleus) the nuclear magnetic moment is $-9.54 \times 10^{-27}\ A\cdot m^2$. The atomic electrons are lined up in such a way that they make no contribution to the magnetic moment of the atom. Suppose it were possible to align the oxygen atoms such that 50.05 percent pointed in one direction and 49.95 percent pointed in the opposite direction. What would be the magnetization of 1 mol of ^{17}O gas under those conditions at standard temperature and pressure?

41. (II) Express the equation $d\vec{m}/dt = g_p \vec{m} \times \vec{B}$ relevant to NMR in component form for the case that $\vec{B} = B\hat{k}$ and $\vec{m} = m_x\hat{i} + m_y\hat{j} + m_z\hat{k}$. (a) Show that m_z is a constant; (b) that $m_x^2 + m_y^2 + m_z^2$ is a constant; and (c) that $m_x = m_1\cos(\omega t)$ and $m_y = -m_1\sin(\omega t)$ satisfy the equation of motion, where ω is the *angular frequency of precession*.

42. (II) Archaeological objects are often located by detecting their minute influence on Earth's magnetic field. The equipment used for such measurements is the *proton magnetometer*, which measures the intensity of the magnetic field by measuring the angular frequency of protons in that field. Determine the angular frequency of protons in Earth's field at a typical location ($B = 80\ \mu T$) and the change in frequency caused by a change $\Delta B = 12$ nT.

43. (II) Calculate the frequency of precession (see Problem 42) for a proton's magnetic moment in a field of 10^{-1} T. This frequency is in the so-called rf (radio-frequency) range.

44. (II) Gauss' law for magnetism,

$$\int_{\text{closed surface}} \vec{B} \cdot d\vec{A} = 0,$$

remains unchanged when materials are present because materials do not give rise to magnetic monopoles. Use this law to show that the magnetic field $\vec{B}$ does not change at the interface of two materials if the interface is perpendicular to the direction of the field. [*Hint:* Recall the method used to derive the electric field due to a surface distribution of charges.]

45. (II) If magnetic matter is present, Ampere's law changes to

$$\oint (\vec{B}/\mu) \cdot d\vec{s} = \oint \vec{H} \cdot d\vec{s} = I_{\text{enclosed}}.$$

Use this equation to find the magnetic field in a narrow gap cut through the core of a toroidal coil (Fig. 31–23). (The purpose of such an arrangement is to allow samples to be placed within the gap.) The average radius of the torus is 30 cm, $\mu = 1200\mu_0$ for the core material, and the coil consists of 500 turns carrying a current of 8 A. Calculate B for gap widths of 1 mm and 3 cm.

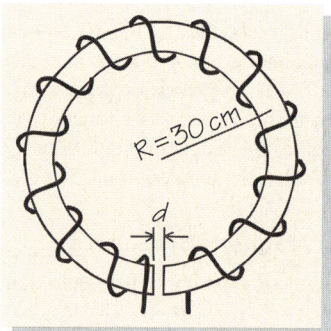

▲ **FIGURE 31–23** Problem 45.

General Problems

46. (II) Estimate the diamagnetic susceptibility of the diamond form of carbon, using the formula derived in Problem 34. Take the density of diamond to be 3.5 g/cm³; the atomic weight, 12; the atomic number, 6; and the atomic radius, 0.75×10^{-10} m. Assume that all the electrons circulate at this radius. Your estimate should be rather good. Compare this result to that found in Table 31–1.

47. (II) Large magnets typically consist of wound toruses of ferromagnetic material. There is a gap in the torus that forms a space between pole faces. Show that the magnetic field across the pole faces is the same as the magnetic field inside the ferromagnet by applying Gauss' law for magnetism to a closed surface partly in and partly out of one of the pole faces. (Gauss' law for magnetism holds independent of the presence of materials. Its validity rests on the fact that there are no magnetic monopoles, and materials introduce no such animals.)

48. (II) Two parallel conducting strips are each 3.0×10^{-3} m thick and 2.5 cm wide and are separated by a distance of 1.0 cm. The space between the strips is filled with a ferromagnetic material whose permeability is $650\mu_0$. Each strip carries a uniform current of 2.0 A, in opposite directions. Find the value of the magnetic field and magnetic intensity in the space between the strips.

49. (II) Cobalt has atomic mass number $A = 59$, and mass density of 8.7×10^3 kg/m³. If each atom has a magnetic dipole moment of the order of 1.7 Bohr magnetons, what is the maximum magnetic moment of a cubic centimeter of Co?

50. (II) The magnetic moment of a uniformly charged sphere of radius a and total charge Q rotating with angular velocity ω is $m = Q\omega a^2/5$. A neutral spherical object has a net magnetic moment 0.45μ. If we were to model this by a sphere of charge $-Q/2$ of radius r, surrounded by a spherical shell of charge $+Q/2$ and thickness $a - r$, what is the ratio r/a?

51. (II) Earth's magnetic field is close to that of a dipole, and the strength of the field at the magnetic north pole is about 0.6×10^{-4} T. Calculate Earth's magnetic moment. If this magnetic moment is due to a magnetized iron core whose radius is half Earth's radius, what is the magnetization of the core (Fig. 31–24)? If the magnetic moment were due to a circulating belt of current at the radius of the core, what would be the magnitude of this current (Fig. 31–24)?

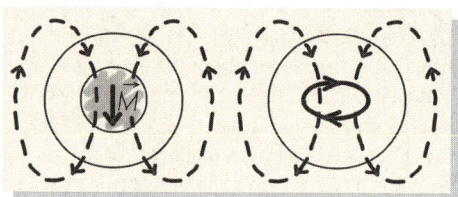

▲ **FIGURE 31–24** Problem 51.

52. (II) A torus of central radius 12 cm and tube radius 1.5 cm is filled with silver. It is wound with 450 turns of wire and carries a 0.80-A current. The magnetic susceptibility of silver is -2.4×10^{-5}. Determine (a) the magnetic intensity, $\vec{H}$; (b) the magnetic field, $\vec{B}$; (c) the magnetization, $\vec{M}$. (d) Repeat parts (a) through (c) for a torus filled with nickel instead of silver. The susceptibility of nickel is 95.

53. (II) In one of two simple classical models of the electron spin, the charge circulates at the classical electron radius r_0 (see Problem 19). In the other, the total electron charge is spread uniformly over a disk whose radius is the classical radius. Calculate the ratio of magnetic moments for the case in which the overall charge occurs entirely at the classical radius versus the case in which the charge is spread over the disk.

54. (II) The neutron has an internal spin $\vec{S}$ of magnitude $\hbar/2$ and a magnetic moment related to the spin by the gyromagnetic ratio g_p, as in Eq. (31–20). The gyromagnetic ratio is $g_p = -3.82(e/2m_n)$. Suppose that a neutron consists of a heavy, positively charged particle of mass M and magnetic moment $e\hbar/2M$, with a lighter, negatively charged particle of mass m but no intrinsic magnetic moment orbiting the heavier particle with orbital angular momentum $\hbar$. What would mass m have to be to explain the observed magnetic moment of the neutron? (For simplicity, ignore the motion of the heavier particle about the center of mass.)

55. (III) In our discussion of kinetic theory (Chapter 19), we noted that, according to Boltzmann, the number of systems with a given energy E in a collection of systems in equilibrium at temperature T is given by $Ce^{-E/kT}$. Here, C is a constant determined by the requirement that, when all the systems are summed, we find the same total number of systems that we started with. For a collection of N magnetic dipoles at rest in an external magnetic field, we have $N(T) = Ce^{-(-\vec{m}\cdot\vec{B})/kT} = Ce^{(mB\cos\theta)/kT}$, where θ is the angle between the direction of the dipole and that of the external magnetic field. C is determined by the requirement that the total number of systems N is $N = C\int_0^\pi 2\pi \sin\theta e^{(mB\cos\theta)/kT} d\theta$. (a) Calculate C. (b) Calculate the average value of $\cos\theta$. (c) Plot $\langle\cos\theta\rangle$ as a function of mB/kT.

Chapter 32

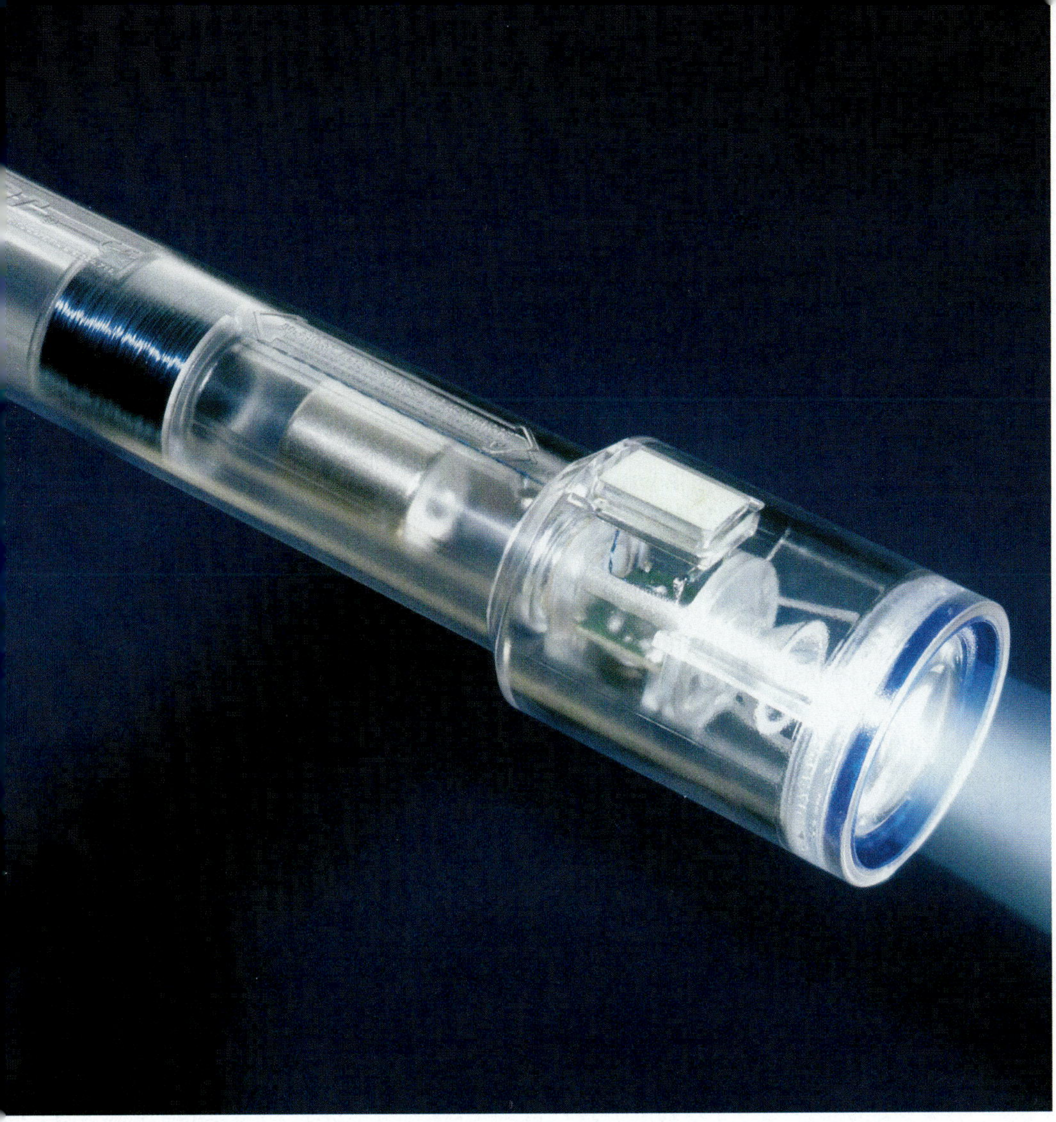

Inductance and Circuit Oscillations

◀ Mechanical motion charges this flashlight. When it is shaken the plug of magnetized material visible at the center moves through the coil and through Faraday's law generates an electric current. With the addition of a capacitor or small chargable battery this is a flashlight that will never run out on you.

We have already seen that we can store energy in the electric field of a capacitor. Energy can also be stored in a magnetic field, using *inductors* as the circuit elements. Their operation is based on Faraday's law, which describes the phenomena that occur when magnetic fields (or more generally magnetic fluxes) change. These phenomena occur frequently in circuits because so many applications of circuits, from computers to televisions to the functioning of the power grid, involve time dependence in currents and hence in magnetic fields. For this reason, inductors are crucial to the control of time dependence in circuits. We will see that electric circuits containing inductors, capacitors, and resistors are analogous to damped harmonic oscillators, and all the features of such mechanical systems are also seen in these circuits.

32–1 Inductance and Inductors

When you are at home vacuuming the rug, you may see a spark at the wall socket if you accidentally pull the plug from it. Why? This is just Faraday's law at work, holding up the currents that were already flowing. When a circuit contains a *changing* electric current, the magnetic field associated with that current also changes, changing the magnetic flux through the circuit. According to Faraday's law, that means that the circuit's *own* changing current will induce an additional emf in the circuit. Lenz' law tells us that this additional emf will tend to maintain the existing current.

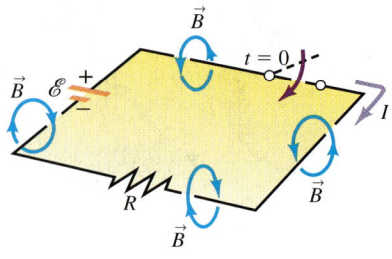

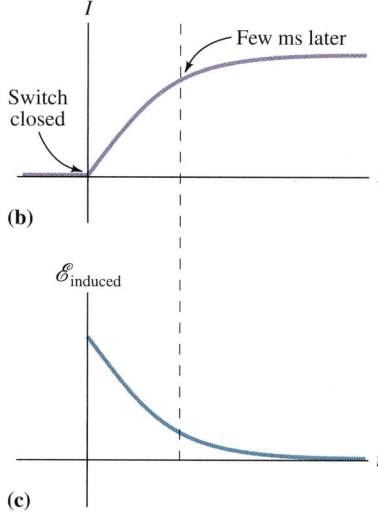

▲ FIGURE 32–1 (a) When the current in a circuit changes, the flux through the circuit also changes. (b) The current in the circuit increases as a function of time. (c) The induced emf is proportional to the derivative of the current according to Faraday's law.

For example, Fig. 32–1a shows a circuit with a switch that closes at $t = 0$. When as a result the current increases from zero (Fig. 32–1b), the magnetic field around the wire also increases. As the magnetic field grows, the magnetic flux through the area enclosed by the loop increases—in the figure, the flux is directed downward. According to Faraday's law, as formulated by Lenz, an emf is induced in the loop and opposes this increase in flux (Fig. 32–1c). The induced emf *opposes* the emf of the battery and slows down the flow of current. The principle illustrated here is a simple one: *Changing currents in circuits lead to effects that act to reduce the rate of change of those currents.*

If a second circuit loop is in the general vicinity of the first, there may be a changing magnetic flux through the second circuit due to the magnetic field produced by the first circuit, and a current will be induced in the second circuit. This in turn produces a changing flux that can affect the first circuit, and so forth. In this case, the two circuits are said to be **linked**.

When the first loop induces an emf in itself, we say that there is a **self-inductance**, or **inductance** for short. When the first loop induces a current or emf in a second loop, we say that there is a **mutual inductance** between the two loops. Faraday's law is the principle that lies behind both self-inductance and mutual inductance.

Self-Inductance: When a wire carries a current I, a magnetic field is set up whose strength is proportional to I. In turn, the magnetic flux appearing through a loop of that wire is also proportional to the current. The proportionality constant is defined to be the *inductance* L. L depends on the particular geometry of the loop around which the emf is induced as well as on the number of turns of the loop, and is defined by

$$\Phi_B = LI. \tag{32–1}$$

If the current and therefore the flux changes, then according to Faraday's law, the emf induced in this loop, $\mathcal{E}$, is the rate of change of the flux through the loop:

$$\mathcal{E} = -\frac{d\Phi_B}{dt} = -L\frac{dI}{dt}. \tag{32–2}$$

EMF BY SELF-INDUCTANCE

The minus sign in this equation is the manifestation of Lenz' law. We will see that the presence of this induced emf can have a marked effect on how charges flow through the circuit.

Mutual Inductance: Let's now consider the two adjacent circuits shown in Fig. 32–2. If a current I_1 flows in loop 1 and a current I_2 flows in loop 2, there is a magnetic flux $\Phi_B(1)$ through the area of loop 1 given by

$$\Phi_B(1) = L_1 I_1 + M_{12} I_2. \tag{32–3}$$

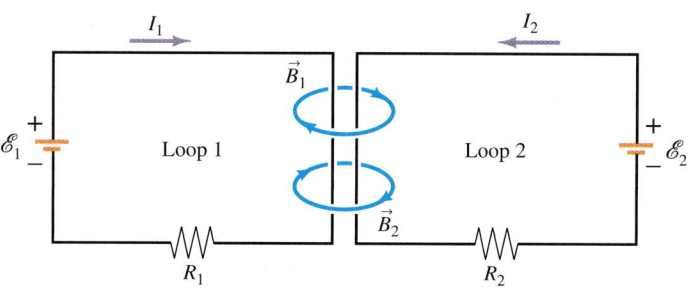

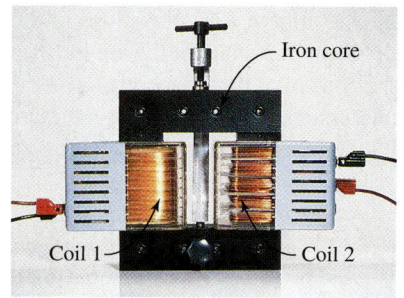

▲ FIGURE 32–2 (a) The flux through a circuit or circuit element may be due to its own current or to the current carried by an adjacent circuit or circuit element. (b) The two coils mounted on the iron core demonstrate mutual induction.

The first term is due to the current flowing in loop 1, and the constant of proportionality L_1 is the self-inductance of loop 1. The second term is due to the current flowing in loop 2 (Fig. 32–3a), and the constant M_{12} is what we define as the *mutual inductance* of loop 1 due to loop 2. Both L_1 and M_{12} are positive by definition. They depend *only* on the geometry of the loop and on the materials in its vicinity, but not on the currents themselves.

The term *mutual* implies a degree of symmetry between the two loops. The magnetic flux through loop 2 has a term proportional to its own current and also a term proportional to the current in loop 1 (Fig. 32–3b):

$$\Phi_B(2) = L_2 I_2 + M_{21} I_1. \tag{32-4}$$

The second term introduces what might appear to be a new constant, M_{21}, the mutual inductance of loop 2 due to loop 1. We will not provide the proof, which is not simple, but *the mutual inductances are equal*: $M_{12} = M_{21}$. It is customary to drop the subscripts and write $M = M_{12} = M_{21}$, the mutual inductance of two loops.

Faraday's law gives the emf induced in loop 2 due to the change in current of loop 1:

$$\mathcal{E}_{21} = -M \frac{dI_1}{dt}. \tag{32-5}$$

EMF BY MUTUAL INDUCTANCE

A similar expression gives the emf induced in loop 1 due to the current in loop 2.

The inductances L and M have SI units of magnetic flux divided by current, or webers per ampere (Wb/A). Inductance is given its own unit in the SI, the **henry** (H), named after Joseph Henry (Fig. 32–4):

$$1 \text{ H} = 1 \text{ Wb/A} = 1 \text{ T} \cdot \text{m}^2/\text{A}. \tag{32-6}$$

To give you an idea of how large a henry is, a cylindrical solenoid of area 10 cm², length 20 cm, and a winding density of 10 turns/cm has an inductance of 0.25 mH. An inductance of 1 H is large but not unrealizable; typical values of self- and mutual inductance range from μH to tens of mH. Self-inductance is usually the more important effect in circuits; the role of mutual inductance in linked circuits is most important in *transformers*. These are devices used to change the magnitude of time-varying voltages and are crucial elements of the electric-power distribution system. We shall encounter transformers again in Chapter 33.

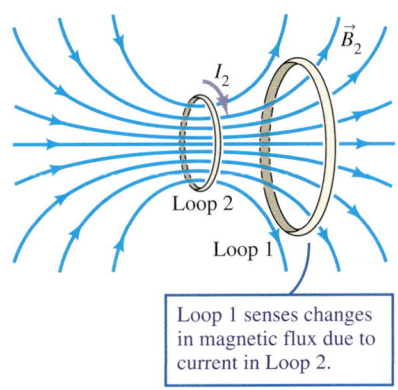

(a)

Loop 1 senses changes in magnetic flux due to current in Loop 2.

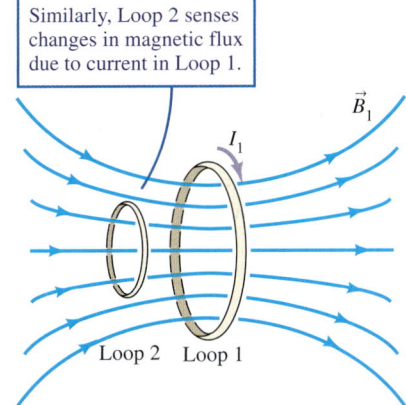

Similarly, Loop 2 senses changes in magnetic flux due to current in Loop 1.

(b)

▲ **FIGURE 32–3** Mutual inductance. (a) There is a magnetic flux through loop 1 due to the magnetic field from the current I_2 in loop 2. (b) There is a magnetic flux through loop 2 due to the current I_1 in loop 1.

◀ **FIGURE 32–4** Joseph Henry, as depicted in a stained-glass window in the First Presbyterian Church of Albany, New York, the site of Henry's baptism. Henry investigated many effects of induction at about the same time as did Faraday.

Elements within circuits with a significant self-inductance—we refer to them as **inductors**—provide sources of emf, which must be included when Kirchhoff's loop rule is used for potential changes around a circuit. Inductors, usually in the form of solenoids, are useful devices that are placed in circuits to control time dependence and to store energy. They join capacitors, resistors, and batteries as the basic elements of circuits. They are represented in circuit diagrams by the symbol ⌇⌇⌇⌇⌇. We add the following to our list of rules for application of Kirchhoff's loop rule to circuits:

> **In moving across an inductor of inductance L along (or *against*) the presumed direction of the current I, the potential change is $\Delta V = -L\,dI/dt$ (or $+L\,dI/dt$, respectively).**

The actual sign of the potential drop depends on the sign of the rate of change of the current and is determined when the circuit equations are solved.

Finding the Inductance

To use the loop rule with inductors present in a circuit, we must know the values of the self-inductance or mutual inductance. As for capacitance, there are only a few simple, but important, geometries for which inductance can easily be calculated. The most important of these is the ideal solenoid. Consider the ideal solenoid of Fig. 32–5, which has length ℓ and radius R. For $\ell \gg R$, the magnetic field within the solenoid is longitudinal and constant and is given by Eq. (29–15):

$$B = \mu_0 n I,$$

where μ_0 is the permeability of free space, n is the number of turns per unit length of solenoid, and I is the current the solenoid carries. The magnetic flux through one turn of the solenoid is the field B times the cross-sectional area A, $\Phi_B = BA = \mu_0 A n I$. The total magnetic flux is this value times the *total* number of turns $N = n\ell$:

$$\Phi_B = \mu_0 A n^2 \ell I. \tag{32-7}$$

By comparison with Eq. (32–1), the self-inductance is the coefficient of the current:

$$\text{for an ideal solenoid: } L = \mu_0 A \ell n^2. \tag{32-8}$$

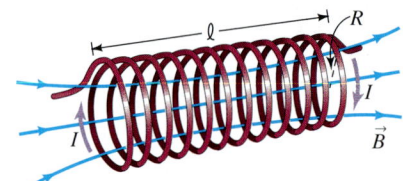

▲ **FIGURE 32–5** A solenoid of length ℓ, radius R, and turn density n carries a current I. Only the magnetic field $\vec{B}$ inside is shown; the field outside the solenoid is zero in the limit that the solenoid is infinitely long.

EXAMPLE 32–1 During a short time period, the current in a cylindrical coil of length 10 cm, radius 0.5 cm, and 1000 turns of wire in a single layer is increased at the steady rate of 10^3 A/s. Find the emf induced during this period.

Strategy We can treat the cylindrical coil as an ideal solenoid, as it is much longer than its coil width. The induced emf is given by Eq. (32–2); the inductance that appears in that equation is in turn given for a coil by Eq. (32–8). Hence this problem is simply a matter of inserting numbers into these equations.

Working It Out We start with the inductance. The turn density is $n = (1000 \text{ turns})/(0.1 \text{ m}) = 10^4$ turns/m. The area of the solenoid is given by $A = \pi r^2$, and

$$L = \mu_0 A \ell n^2$$
$$= (4\pi \times 10^{-7} \text{ T} \cdot \text{m/A})[\pi(0.005 \text{ m})^2](0.1 \text{ m})(10^4 \text{ m}^{-1})^2$$
$$= 10^{-3} \text{ H}.$$

As for the induced emf, the current change rate is $dI/dt = 10^3$ A/s, so if we follow the circuit in the direction of the current, we have

$$\mathcal{E} = -L\frac{dI}{dt} = -(10^{-3} \text{ H})(10^3 \text{ A/s}) = -1 \text{ V}.$$

The induced emf is negative.

What Do You Think? Which of the following will increase the induced emf the most? In each case the change mentioned is the *only* change. (a) Doubling the length of the solenoid, (b) doubling the rate at which the current changes, (c) doubling the area of the solenoid, (d) doubling the number of turns in the solenoid. Answers to **What Do You Think?** questions are given in the back of the book.

CONCEPTUAL EXAMPLE 32–2 A friend claims that he can increase the inductance L of a solenoid with given area and length and that must be constructed with a single layer of wire by using finer wire of the same material. Is he correct? What are the implications of his method of increasing L on the resistance R of the solenoid?

Answer The inductance is proportional to the square of the turn density n^2, and for a single layer $n \propto 1/d$, where d is the diameter of the wire. Thus

$$L \propto 1/d^2,$$

and decreasing d will certainly increase L. To see the implication of the use of finer wire for the resistance, we can recall from Chapter 26 that the resistance of a wire of length ℓ and area A made of a material of resistivity ρ is $R = (\ell/A)\rho$. In our case, with fixed solenoid length and area, each turn uses the same wire length, and the total number of turns N is proportional to the turn density $n \propto 1/d$. Thus the length of wire needed is proportional to $1/d$. The wire cross-section $A \propto d^2$, so $\ell/A \propto d^{-1}/d^2$, i.e.

$$R \propto 1/d^3.$$

R increases faster than L does as the wire diameter decreases. Whether this has important consequences depends on the circumstances, but the changed value of R would almost certainly have to be taken into account in determining the properties of the circuit.

As an example of a calculable mutual inductance, consider a solenoid (of length ℓ_1, radius R_1, winding density n_1, and current I_1) that contains within it a single loop of radius R_2 whose area $A_2 = \pi R_2^2$ is oriented perpendicular to the axis of the solenoid (Fig. 32–6a). The magnetic field of the solenoid is given by Eq. (30–15), $B = \mu_0 n_1 I_1$. The magnetic flux that passes through the single loop is

$$\Phi_B = BA_2 = \mu_0 A_2 n_1 I_1.$$

By definition, the mutual inductance, M, is the coefficient of I_1:

$$M = \mu_0 A_2 n_1. \qquad (32\text{–}9)$$

If the single loop is replaced with a second solenoid (Fig. 32–6b) with a total number of turns N_2, then the total flux that links that second solenoid contains a factor N_2, and M must be increased by this same factor:

$$M = \mu_0 A_2 n_1 N_2. \qquad (32\text{–}10)$$

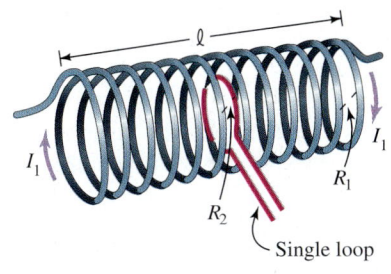

(a)

(b)

▶ **FIGURE 32–6** (a) The area of the single, small loop is oriented perpendicular to the axis of the solenoid. (b) Similar qualitative effects are associated with the replacement of the single loop (a solenoid with but a single turn) by a solenoid with multiple turns.

EXAMPLE 32–3 Consider the two solenoids in Fig. 32–6b and take their radii, and hence the areas, to have the same value. Compare the ratio of mutual inductance M to self-inductance L_1 of solenoid 1.

Setting It Up We denote the common area as A. We must assume values for the turn density n_1 and the length ℓ_1 of solenoid 1, as well as the total number of turns N_2 in solenoid 2.

Strategy This is a straightforward use of Eqs. (32–8) and (32–10).

Working It Out We have from Eqs. (32–8) and (32–10)

$$\frac{M}{L_1} = \frac{\mu_0 A n_1 N_2}{\mu_0 A n_1^2 \ell_1} = \frac{\mu_0 A n_1 N_2}{\mu_0 A n_1 N_1}.$$

In the last step, we have used the fact that the factor $n_1 \ell_1$ is the total number of turns N_1 in solenoid 1. Both the area A and the turn density n_1 cancel, and we are left with

$$\frac{M}{L_1} = \frac{N_2}{N_1}.$$

If solenoid 2 has many fewer windings than solenoid 1, then the effect of the mutual inductance on solenoid 1 is small compared to that of the self-inductance.

What Do You Think? (a) If the radius R_1 is doubled, what happens to L_1 and M? (b) If instead, the radius R_2 is doubled, what happens to L_1 and M?

The Effects of Magnetic Materials on Inductance

In Chapter 31, we considered modifications to a magnetic field due to a current in the presence of materials with magnetic properties. We saw that the presence of materials can modify, and in the case of ferromagnetic materials greatly magnify, the magnetic field present and hence the flux. This is made explicit in expressions that include the free (or real) current by replacing the permeability of free space μ_0 with the permeability of the material μ. The permeability is given by Eq. (31–7),

$$\mu = \mu_0(1 + \chi_m).$$

Here, χ_m is the magnetic susceptibility of the material, which is negative and small for diamagnets, positive and small for paramagnets, and positive and large for ferromagnets. If a solenoid is filled with a magnetic material, its self-inductance would change, with the replacement of μ_0 by μ in Eq. (32–8). For ferromagnetic materials, self-inductances can thereby be increased manyfold (Fig. 32–7). In Fig. 32–8, we illustrate the application of inductors that use the magnetic properties of materials in a doorbell.

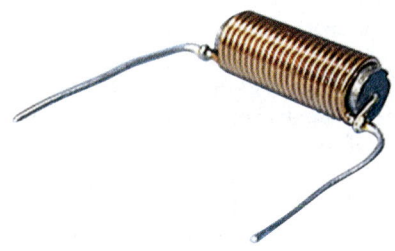

▲ **FIGURE 32–7** The inclusion of an iron core in this solenoid (an actual inductor) increases the magnetic field produced by a given coil current.

898 | Inductance and Circuit Oscillations

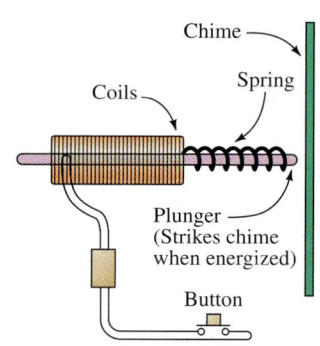

► **FIGURE 32–8** The chiming doorbell is a simple application of inductance. The touch of a doorbell button induces a current in the rods within the solenoid, which then are thrust from the solenoid according to Lenz's law and strike a chime.

EXAMPLE 32–4 Consider the single loop and the solenoid shown in Fig. 32–6. Suppose the loop carries a current I_2 that is a function of time. Find the emf in the solenoid induced by current I_2.

Strategy This is a case of mutual inductance—we want the effect within solenoid 1 of a changing current in loop 2. The single loop of radius R_2 carries the current I_2 in this case, which changes as dI_2/dt. To apply Faraday's law as in Eq. (32–5), we must find the magnetic flux Φ_B due to loop 2 through solenoid 1. This flux is the mutual inductance of the loop and solenoid times current I_2. A direct calculation of the flux might be complicated here, but we can fortunately use Eq. (32–9) for the inductance together with the "mutuality" of the inductance—the equation gives M_{21}, while what we want is M_{12}, but the two are equal.

Working It Out Using Eq. (32–9) for M gives

$$\Phi_B = MI_2 = \mu_0 \pi R_2^2 n_1 I_2.$$

The emf in the solenoid is then the negative time derivative of this flux:

$$\mathcal{E} = -\frac{d\Phi_B}{dt} = -\mu_0 \pi R_2^2 n_1 \frac{dI_2}{dt}.$$

What Do You Think? If a ferromagnetic material is placed inside the solenoidal coil (without the single loop), the self-inductance of the coil will (a) increase, (b) decrease, (c) stay the same.

CONCEPTUAL EXAMPLE 32–5 A hollow solenoid is carrying a steady current I when a cylindrical piece of iron is sent through the cylindrical space and out the other side. The cylinder of iron fits the interior of the solenoid and is a little shorter than the solenoid. Describe the emf induced in the circuit containing the solenoid and include a sketch. What is the effect on the current?

Answer As the iron passes into the solenoid (Fig. 32–9a), the magnetic field (and magnetic flux) inside the solenoid will suddenly increase, because the space is increasingly filled with a material with a permeability much larger than the permeability of the air, which is to a very good approximation that of the vacuum. According to Lenz's law, an emf will be generated within the solenoid that opposes this increase in flux. The sign of the emf corresponds to this "opposition," and we will call it negative, as shown in Fig. 32–9b. If the speed of the cylinder entering the space is held constant, then the rate of increase in flux will be constant, and the induced emf will be constant. The emf will induce a current $I_{induced}$ in the coil that will be opposite to the original current. Once the iron cylinder is entirely within the solenoid, there is no change in flux and no further emf; for this brief period the induced emf is zero. Finally, the cylinder exits the space, this time leading to a decreasing flux and an induced emf that tries to hold up the flux. This positive emf is also indicated in the figure; it results in a positive current. Just as quickly, the induced emf (and the induced current) returns to zero once the iron has left the solenoid.

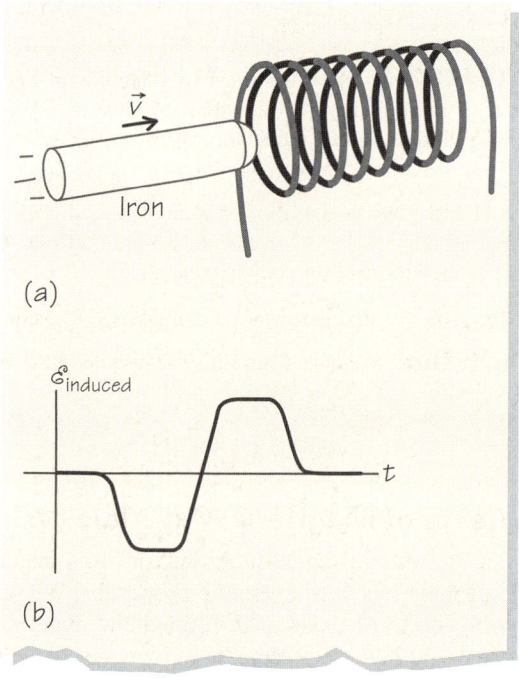

▲ **FIGURE 32–9** (a) A ferromagnetic cylinder moves into and through a coil. (b) The emf induced in the coil as a result of steady movement of the cylinder.

32–2 Energy in Inductors

Just as a capacitor is a device for storing energy in an electric field, an inductor is a device for storing energy in a magnetic field. Because any emf induced in the inductor opposes the change in current, work must be done by an external source, such as a battery, to cause a current to pass through an inductor. By the work–energy theorem, just how much work is done is a measure of the energy stored in the inductor. To calculate this energy, we proceed as we did for the capacitor in Chapter 25, and calculate the work that must be done by some external emf to pass a current through the inductor.

We derive the general expression for the rate dW/dt (the power) at which an external emf $\mathcal{E}_{\text{ext}}$ does work when a current I flows from Eqs. (26–26) and (26–27):

$$\frac{dW}{dt} = I\mathcal{E}_{\text{ext}}.$$

If we have only the external emf and an inductor, the external emf must be equal but opposite to the induced emf in the inductor, given by Eq. (32–2). Thus

$$\frac{dW}{dt} = +LI\frac{dI}{dt}. \tag{32–11}$$

If the current is increasing, the power is positive, meaning that the external source must do positive work in supplying energy to the inductor; the internal energy U_L in the inductor is increasing. If the current is decreasing, the power is negative, meaning that the external source takes energy from the inductor; the inductor's internal energy is decreasing. The net change ΔU_L in the total magnetic energy of the inductor as the current changes from a value I_1 to a value I_2 between the times t_1 and t_2 can be found by integrating the work done by the external source as the current changes. We integrate Eq. (32–11) for dW/dt from an initial time t_1 to a later time t_2:

$$\Delta U_L = \int_{t_1}^{t_2} \frac{dW}{dt} dt = \int_{t_1}^{t_2} LI\frac{dI}{dt} dt = L\int_{I_1}^{I_2} I\, dI$$
$$= \frac{1}{2}LI_2^2 - \frac{1}{2}LI_1^2. \tag{32–12}$$

In particular, if the inductor carries a current I, then the increase in energy as the current increases from zero up to I, which we refer to simply as the energy of the inductor, is

$$U_L = \frac{1}{2}LI^2. \tag{32–13}$$

ENERGY IN AN INDUCTOR

In applications to circuits (see Section 32–4 as well as later chapters), we'll see that it is not only the storage of energy that is interesting but the time dependence of that process.

Equation (32–13) should be compared to the expression for the energy U_C contained in a capacitor of capacitance C that carries charge Q from Eq. (25–8):

$$U_C = \frac{1}{2}\frac{Q^2}{C}.$$

EXAMPLE 32–6 A solenoid is designed to store $U_L = 0.10$ J of energy when it carries a current I of 450 mA. The solenoid has a cross-sectional area A of 5.0 cm² and a length ℓ of 0.20 m. How many turns of wire must the solenoid have?

Strategy We know the energy, which is a function of the inductance L and the given current; L is in turn a function of N, through Eq. (32–8). The only modification necessary in Eq. (32–8) is to write it in terms of the total number of turns N rather than the turn density n by using $N = n\ell$, where ℓ is the length of the solenoid. We can then solve for N in terms of L and hence in terms of the energy.

Working It Out With the change to Eq. (32–8), we have

$$L = \frac{\mu_0 A N^2}{\ell}.$$

The expression for the energy, Eq. (32–13), then reads

$$U_L = \frac{1}{2}\frac{\mu_0 A N^2}{\ell} I^2.$$

We solve this for N and insert numbers:

$$N = \frac{1}{I}\sqrt{\frac{2U_L \ell}{\mu_0 A}}$$
$$= \frac{1}{4.5 \times 10^{-1}\text{ A}}\sqrt{\frac{2(0.10\text{ J})(0.20\text{ m})}{(4\pi \times 10^{-7}\text{ N/A}^2)(5.0 \times 10^{-4}\text{ m}^2)}}$$
$$= 1.8 \times 10^4 \text{ turns}.$$

What Do You Think? What is the easiest way to increase the energy stored in a solenoid if its size is fixed and the wire (of which you have an abundant supply) can carry no additional current?

900 | Inductance and Circuit Oscillations

An inductor has an energy given by Eq. (32–13) even if the current is steady. We have, however, argued that the origin of the effects of inductance is Faraday's law, which involves changes in current. How can we reconcile these two facts? As we are going to describe in more detail in the next section, the energy of an inductor is in the magnetic field within it, just as the energy of a capacitor lies within the electric field between the plates. The energy of an inductor carrying steady current must arise from the original buildup of current—even if it occurred in the distant past. It is when the current changes that the magnetic field changes and, along with it, the energy in the magnetic field.

32–3 Energy in Magnetic Fields

In Chapter 25, we demonstrated that the electric energy associated with a capacitor is located in the electric field within the capacitor. Similarly, the energy of an inductor is located in its magnetic field. Just as the ideal parallel-plate capacitor was a means to learn about energy in the electric field, the ideal solenoid presents us with a tool to explore the energy density in a magnetic field.

The inductance of an ideal solenoid of area A and length ℓ is given by Eq. (32–8), so from the expression for the total energy of an inductor [Eq. (32–13)], we find

$$U_L = \frac{1}{2}LI^2 = \frac{1}{2}\mu_0 A \ell n^2 I^2. \tag{32–14}$$

We also know that the magnetic field in the solenoid is proportional to the current. Equation (29–15) gives the precise connection, $B = \mu_0 n I$. If we substitute for I in terms of B in Eq. (32–14), we obtain

$$U_L = \frac{1}{2}\frac{B^2}{\mu_0} A\ell. \tag{32–15}$$

The volume enclosed within the solenoid is $A\ell$. Because the magnetic field is uniform within the solenoid, we can identify the **energy density** u_B, the energy per unit volume of the magnetic field, as

$$u_B = \frac{1}{2}\frac{B^2}{\mu_0}. \tag{32–16}$$

ENERGY DENSITY IN A MAGNETIC FIELD

This result generalizes to the case of a nonuniform magnetic field, no matter how it is produced. It should be compared to our expression for the energy density of an electric field, Eq. (25–12):

$$u_E = \frac{1}{2}\varepsilon_0 E^2,$$

a result derived in a similar way. It is important to realize that *energy is located within the electric and magnetic fields themselves.*

When both magnetic and electric fields are present, the energy density is the sum of both magnetic and electric energy densities:

$$u = u_B + u_E = \frac{1}{2}\left(\frac{B^2}{\mu_0} + \varepsilon_0 E^2\right). \tag{32–17}$$

EXAMPLE 32–7 A large electromagnet produces a magnetic field of 1 T. Compare the energy density associated with this field to that of the largest electric field in air, about 10^6 V/m (beyond this value, there is breakdown).

Strategy Equations (32–16) and (25–12) express the energy density in magnetic and electric fields, respectively, and we can directly obtain the ratio of the magnetic and electric energy densities from them.

Working It Out We have

$$\frac{u_B}{u_E} = \frac{\frac{1}{2}\frac{B^2}{\mu_0}}{\frac{1}{2}\varepsilon_0 E^2} = \frac{1}{\mu_0 \varepsilon_0}\frac{B^2}{E^2}.$$

In our case, the magnitudes of both the magnetic and electric fields are given, and

$$\frac{u_B}{u_E} = \frac{1}{(4\pi \times 10^{-7}\,\text{N/A})(8.85 \times 10^{-12}\,\text{F/m})} \frac{(1\,\text{T})^2}{(10^6\,\text{V/m})^2}$$
$$= 9 \times 10^4.$$

Depending on the medium, electric breakdown properties may limit the size of electric fields that can be easily maintained and make magnetic fields more suitable for storing energy at high density. How rapidly that energy can be delivered back for other uses is another story, however, and capacitors remain the best way to deliver a lot of energy as fast as possible.

What Do You Think? What puts a limitation on the production of large magnetic fields?

32–4 Time Dependence in RL Circuits

The fact that the potential drop across an inductor depends on how rapidly the current passing through it changes leads to new time-dependent behavior in circuits with inductors. For example, if we attempt to stop a current already flowing—say, by opening a switch—then an emf is induced that attempts to keep the current flowing, possibly creating an arc across the switch. The simplest illustration of time-dependent behavior involves a circuit with a source of emf $\mathcal{E}$, a resistor of resistance R, and an inductor of inductance L (Fig. 32–10). We call such a circuit an **RL circuit**. A switch allows us to control the initial conditions. When we close the switch, the inductor acts to oppose the changing current, and as a result, the current cannot jump suddenly but must build up over time. If we apply the loop rule to the circuit in the direction of the pink arrow in Fig. 32–10, we can see this quantitatively. We have

$$\mathcal{E} - IR - L\frac{dI}{dt} = 0. \tag{32-18}$$

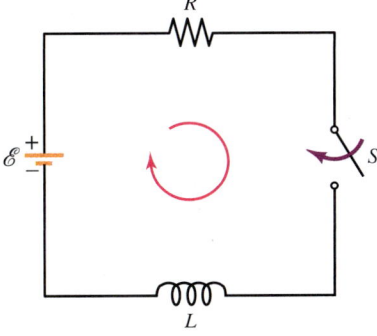

▲ **FIGURE 32–10** An RL circuit, including a source of constant emf.

To find the solution to this differential equation for I, compare Eq. (32–18) with Eq. (27–21), which comes from applying the loop rule to the RC circuit shown in Fig. 32–11:

$$\mathcal{E} - \frac{Q}{C} - R\frac{dQ}{dt} = 0. \tag{27-21}$$

Here, Q is the charge on the capacitor. The first differential equation determines the current in RL circuits and has exactly the same *form* as the equation for the charge in RC circuits. The similarities between the RC and RL circuits are summarized in Table 32–1, and the solution of Eq. (32–18) for the current in the circuit is the same as the solution of Eq. (27–21) for charge if we make the substitutions indicated in the table. In particular, if we replace RC with L/R, the current in the RL circuit will have the time dependence $\exp[-t/(L/R)]$. The RL circuit is said to have a *time constant L/R*. Note that, as for the RC circuit, the time dependence is transient. Large values of the time constant (large L and/or small R) mean that the transient behavior is slow to disappear, so that long times are required for the current to build up or decay; small values of the time constant (small L and/or large R) mean that the transient behavior quickly disappears, and current builds up or decays rapidly. Of course, the "large" and "small" times are relative terms.

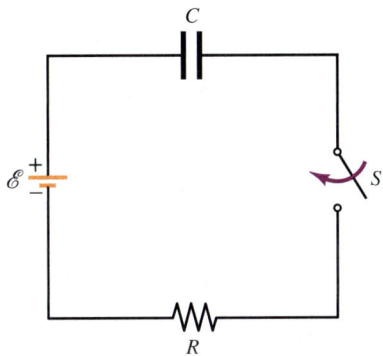

▲ **FIGURE 32–11** An RC circuit is analogous to the RL circuit of Fig. 32–10.

For complete solutions of Eq. (32–18) for the current, including the initial conditions, we can directly apply the results of Section 27–5 for the solution of the differential equation for the RC circuit. As in Section 27–5, it is important to understand

TABLE 32–1 • Analogy Between RC and RL Circuits

	RC Circuit Parameter	RL Circuit Parameter
Variable	Q	I
Coefficient of variable	$1/C$	R
Coefficient of $\frac{d}{dt}$ (variable)	R	L
Time constant	RC	L/R

902 | Inductance and Circuit Oscillations

physically how different initial conditions will affect the solution. As a guiding principle, keep in mind the following:

The current in an inductor never changes instantaneously, but after the current settles down to a constant value, the inductor plays no role in the circuit.

Finally, we should mention that real (as opposed to ideal) inductors may contain relatively long lengths of wire and always have some resistance. Depending on the circuit, this resistance can be significant.

Let's look again at time dependence in an *RL* circuit, this time being explicit about some boundary conditions. Suppose that the switch in Fig. 32–10 has been open for a very long time and is suddenly closed at $t = 0$. What happens? Since the inductor acts to prevent any sudden change in current, the current must ramp up smoothly from zero following the switch closing. We also know that after a very long time, the system will have stabilized, with any transient effects having disappeared. The inductor will play no role at these late times, and the circuit effectively consists of only an emf $\mathcal{E}$ and a resistance R (including any resistance in the inductor). The constant current will be $I = \mathcal{E}/R$. In between these times the current has gone from 0 at $t = 0$ to $\mathcal{E}/R$ at large times with a time constant L/R—in other words, the current must reach its asymptotic value exponentially over this time scale.

Without going through the formal steps (see Problem 47 to verify the solution to the differential equation), we can say that the function that fits these criteria is

$$I = \frac{\mathcal{E}}{R}[1 - e^{-t/(L/R)}] = \frac{\mathcal{E}}{R}(1 - e^{-Rt/L}). \quad (32\text{–}19)$$

Figure 32–12 is a graph of current versus time for this case. At $t = 0$, the exponential term is unity and $I = 0$, consistent with the principle that the current cannot change instantaneously. As $t \to \infty$, the transient exponential term drops out and I approaches $\mathcal{E}/R$—as though the inductor were not present at all.

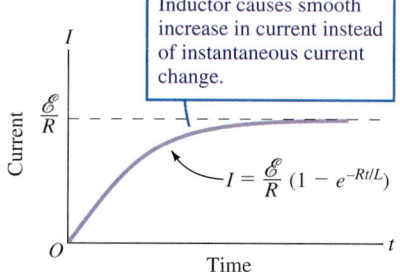

▲ **FIGURE 32–12** The switch of Fig. 32–10 is closed at $t = 0$, and the current rises from zero to a steady-state value only after a period of time determined by the ratio L/R.

EXAMPLE 32–8 Find the potential magnitude across the resistance and the inductor in the circuit of Fig. 32–10 as a function of time, assuming that the switch is closed at $t = 0$. Draw graphs of the potential across the resistor and the potential across the inductor as a function of time, starting at $t = 0$.

Strategy We have already found, in Eq. (32–19), the current as a function of time in this situation. This is sufficient to find the potentials across these elements, which in the case of the resistor is IR and in the case of the inductor is $L\, dI/dt$.

Working It Out The potential across the resistor is just Eq. (32–19) multiplied by R; we won't write that out again. For the magnitude of the potential across the inductor, we must evaluate

$$V_L = L\frac{dI}{dt} = L\frac{\mathcal{E}}{R}\left[-\frac{-R}{L}e^{-Rt/L}\right] = \mathcal{E}e^{-Rt/L}.$$

This starts at $t = 0$ with a maximum value of $\mathcal{E}$, then drops off exponentially to zero at long times; there is a potential across an ideal inductor only when the current through it is changing. In this case, it is changing fastest at $t = 0$ and not changing at all at large values of time.

We have plotted V_R and V_L in Figs. 32–13a and b.

What Do You Think? If L is arbitrarily large, the argument of the exponential in the expression for V_L is zero and the potential across the inductor never drops to zero. Is that correct?

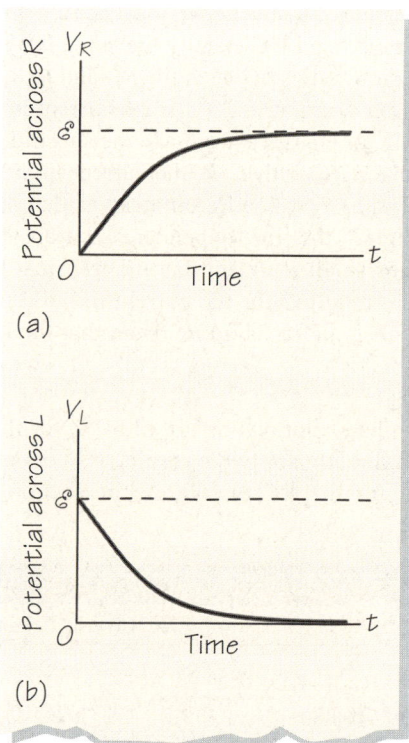

▲ **FIGURE 32–13** (a) The potential across the resistor in the *RL* circuit. (b) The potential across the inductor.

THINK ABOUT THIS...
HOW WOULD YOU MEASURE INDUCTANCE?

We have described the calculation of inductance for simple geometries—for example, a long coil—but in many cases the geometry is complicated, and it is not possible to calculate L. The time dependence we have described in this section and the section to come provides us with very simple tools to measure L directly. Here we have seen that the presence of an inductor in a circuit that additionally contains a resistor of known resistance R exhibits exponential time dependence with a time constant L/R. Measurement of a curve of rise time or of fall time allows us to extract L. In the next section we'll see that circuits with an inductor and a capacitor but no (or negligible) resistance exhibit oscillatory behavior whose frequency depends on L and C. Since it is possible to measure the time for N oscillations of such a circuit and to count N with precision, one can learn the value L of an unknown inductance very accurately this way. Finally, in the next chapter we shall learn and discuss still another way in which an unknown L can be determined, using a resonance phenomenon.

32–5 Oscillations in *LC* Circuits

A single-loop circuit with inductance and capacitance but negligible resistance exhibits a behavior we have not yet encountered in electric circuits but which played a central role in our study of mechanics—oscillatory behavior. Such a circuit is called an **LC circuit**, and it is analogous to a mass on the end of a spring—a familiar mechanical system. If we look at Fig. 32–14, we can see what the oscillatory behavior is and why it is plausible. Imagine that the capacitor is fully charged at an initial time. Then current will flow off the positively charged plate of the capacitor through the wire toward the negatively charged plate; without the inductor, the discharge through the wire would be rapid and direct. With the inductor present, the current is held back initially and grows smoothly from zero. Moreover, once the current reaches some maximum value, the inductor will induce an emf that tends to maintain this maximum current. This allows the discharge to "overshoot," with positive charge building up on the initially negatively charged plate. Conditions are then right for the process to repeat in the other direction.

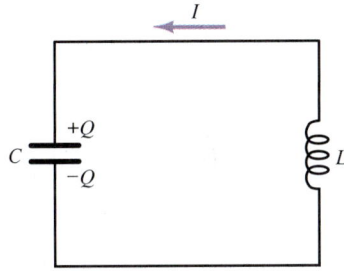

▲ **FIGURE 32–14** An *LC* circuit, consisting simply of an inductor and a capacitor in series.

At this point we need to establish not simply oscillatory behavior but harmonic behavior, and this must be done in a quantitative fashion. We do this by establishing a direct correspondence between the *LC* circuit and the mass on an ideal spring. We start with the loop rule for our *LC* circuit (Fig. 32–14). Following the clockwise direction, we have

$$\frac{Q}{C} + L\frac{dI}{dt} = 0, \qquad (32\text{–}20)$$

where Q is whatever charge is on the capacitor and I is the current in the circuit with direction indicated. The current I is dQ/dt, so this expression is actually a *second-order* differential equation for the charge:

$$\frac{Q}{C} + L\frac{d^2Q}{dt^2} = 0. \qquad (32\text{–}21)$$

At this point recall the equation of motion for a mass m on the end of an ideal spring of spring constant k, namely $-kx = ma$. The acceleration is the second derivative of displacement, so that this equation of motion is

$$kx + m\frac{d^2x}{dt^2} = 0. \qquad (32\text{–}22)$$

This equation for x has the same form as the equation for Q, and so the solutions for these variables have the same form. But we already know that the mechanical motion is harmonic—$x = A\sin\omega t + B\cos\omega t$—and so the same is true for the charge on the capacitor. We must change the variable names using the correspondences summarized in Table 32–2. In particular, the angular frequency of the mechanical system is $\omega = \sqrt{k/m}$, so from the table of equivalences the angular frequency of the *LC* circuit is

$$\omega = \frac{1}{\sqrt{LC}}, \qquad (32\text{–}23)$$

TABLE 32–2 • Analogy Between *LC* Circuit and Mass on a Spring

	Mass on Spring Parameter	LC Circuit Parameter
Variable	x	Q
Coefficient of variable	k	$(1/C)$
Coefficient of $\frac{d^2}{dt^2}$ (variable)	m	L
Natural frequency	$\sqrt{k/m}$	$1/\sqrt{LC}$

and the charge on the capacitor varies as

$$Q = Q_1 \sin(\omega t) + Q_2 \cos(\omega t). \quad (32\text{–}24)$$

Equivalently, this can be written as

$$Q = Q_0 \cos(\omega t + \phi). \quad (32\text{–}25)$$

The constants, either Q_1 and Q_2 or Q_0 and ϕ, are determined by the initial conditions. For example, if we know that the capacitor has a given total charge at time $t = 0$ and that the current at $t = 0$ is also zero (because a switch has been open until that time), then we can determine the two unknown constants.

Once we know the charge on the capacitor, we also know the current flowing through the inductor; we need only apply the relation $I = dQ/dt$. For example, Eq. (32–25) gives

$$I = \frac{d}{dt}[Q_0 \cos(\omega t + \phi)] = -Q_0 \omega \sin(\omega t + \phi). \quad (32\text{–}26)$$

Thus the current, like the charge, is oscillatory, and with the same angular frequency $\omega = 1/\sqrt{LC}$. Here we have rather directly used the equivalent of the equations of motion, but we shall see in Section 32–7 that it is also possible to understand the oscillations in terms of energy flow.

EXAMPLE 32–9 The capacitor ($C = 0.80\ \mu\text{F}$) in the circuit of Fig. 32–14 carries a charge in the form $Q = Q_0 \cos(\omega t)$. Find the voltage drop across the inductor ($L = 12\ \mu\text{H}$) and the period of oscillation of that voltage.

Strategy The current through the circuit is the time rate of change of the charge on the capacitor, so the voltage across the inductor is

$$V_L = -L\frac{dI}{dt} = -L\frac{d^2Q}{dt^2}.$$

The time dependence of Q is given, and the derivative can be carried out to find V_L. Since two derivatives of a cosine give the cosine back again, the time dependence of V_L, like that of the charge, will be $\cos \omega t$. The angular frequency ω is given for our circuit by Eq. (32–23), $\omega = 1/\sqrt{LC}$, and from this the period $T = 1/f = 2\pi/\omega = 2\pi\sqrt{LC}$ can also be evaluated.

Working It Out We have

$$\frac{d^2Q}{dt^2} = Q_0 \frac{d}{dt}\left(\frac{d}{dt} \cos \omega t\right)$$

$$= Q_0 \frac{d}{dt}[-\omega \sin \omega t] = -\omega^2 Q_0 \cos \omega t = -\omega^2 Q.$$

Thus

$$V_L = \omega^2 L Q = \omega^2 L Q_0 \cos \omega t = (Q_0/C) \cos \omega t,$$

where in the last step we used $\omega^2 = 1/LC$. The oscillation period of this voltage drop is given by

$$T = \frac{2\pi}{\omega} = 2\pi\sqrt{LC} = 2\pi\sqrt{(12 \times 10^{-6}\ \text{H})(0.80 \times 10^{-6}\ \text{F})}$$

$$= 1.9 \times 10^{-5}\ \text{s}.$$

Circuits such as this are useful where internal timing is necessary—here we have a clock with a 20 μs "tick," or equivalently a 0.05 MHz frequency. When you realize that FM radio is in the range of 100 MHz, then you can begin to see why circuits such as this have direct relevance to FM radio broadcasting; for example, decreasing the inductance by a very reasonable factor will lead to oscillations in the FM realm.

What Do You Think? Does the maximum voltage across the inductor depend more strongly on the inductance L or the capacitance C in this example?

THINK ABOUT THIS...
ARE THERE TECHNOLOGICAL APPLICATIONS FOR CURRENT OSCILLATIONS?

Every time you use your cell phone you are making use of the phenomenon. In what has become a very important application, wireless communication relies on circuits with natural frequencies of what are effectively *LC* circuits produced on chips with a so-called CMOS (Complementary Metal Oxide Semiconductor) technology. These circuits, which are microscopic and involve very little energy, are used for both the generation and reception of signals—the reception occurs through what we will learn are resonance phenomena analogous to those seen in mechanical systems.

Another class of application is in sound synthesizers. We have seen in Chapter 15 that the sounds we hear are a complex combination of different frequencies, and successful simulation of, say, a trumpet or a violin is possible only when the entire set of frequencies and their relative amplitudes approximates the spectrum of a real instrument. For this purpose *LC* circuits with enough of a variety of inductances and capacitances can generate fairly complex spectra.

32–6 Damped Oscillations in *RLC* Circuits

When resistance, inductance, and capacitance are all present in a single-loop circuit, as in Fig. 32–15, we have an **RLC circuit**. The addition of a resistance to the *LC* circuit gives us an element for which there is energy loss—energy dissipated in Joule heating—and as we shall see next, the resistance term in Kirchhoff's loop rule is analogous to a drag force, proportional to the speed of the mass, in a mechanical harmonic oscillator. Thus we would expect behavior characteristic of damped mechanical oscillations in *RLC* circuits. It is worth noting that since any inductor has some resistance, all *LC* circuits are really *RLC* circuits.

To analyze this case, we assume that a current is present in the circuit and apply the loop rule by following the direction of the current:

$$-L\frac{dI}{dt} - IR - \frac{Q}{C} = 0. \tag{32-27}$$

Here Q is the charge on the capacitor. Because $I = dQ/dt$, Eq. (32–27) can also be written as

$$L\frac{d^2Q}{dt^2} + R\frac{dQ}{dt} + \frac{Q}{C} = 0. \tag{32-28}$$

This equation is a differential equation for the charge Q; with the appropriate initial conditions, it determines the charge on the capacitor, including its time dependence. The current in the circuit is then found by differentiation of the charge.

Equation (32–28) has an analogue in mechanics problems that involve masses on springs in the presence of drag: To Newton's second law for a mass on an ideal spring [Eq. (32–22)], add a term for a drag force. The drag force $-bv$ is proportional to the velocity or first derivative of x. With this term Newton's second law for the damped harmonic oscillator becomes

$$m\frac{d^2x}{dt^2} + b\frac{dx}{dt} + kx = 0. \tag{32-29}$$

This equation [which is Eq. (13–45)] describes the motion of a mass at the end of a spring immersed in a fluid that gives rise to a drag force, and it is a physical system about which we have prior knowledge—it was the subject of Section 13–7. Intuition about this system and the mathematical equivalence of Eqs. (32–28) and (32–29), described in detail in Table 32–3, is immensely helpful in understanding the *RLC* circuit. We note in particular that the mechanical drag term $bv = b\,dx/dt$ corresponds to the circuit resistance term $RI = R\,dQ/dt$.

As you can review in Section 13–7, damping modulates the harmonic behavior of a mass on an ideal spring by changing the period slightly and imposing an envelope on the harmonic motion in the form of a falling exponential. The harmonic motion occurs with an exponentially decreasing amplitude. In the same way, the charge on the capacitor of an *RLC* circuit will display harmonic oscillatory time dependence within an envelope that

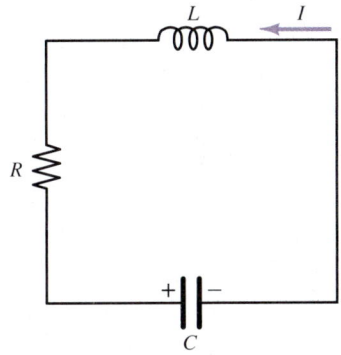

▲ **FIGURE 32–15** A basic *RLC* circuit.

TABLE 32–3 • Analogy Between *RLC* Circuits and Damped Harmonic Motion

	Damped Harmonic Motion Parameter	RLC Circuit Parameter
Variable	x	Q
Coefficient of variable	k	$1/C$
Coefficient of $\frac{d}{dt}$ (variable)	b	R
Coefficient of $\frac{d^2}{dt^2}$ (variable)	m	L

falls exponentially with time. Current is the time derivative of charge, and because the derivatives of sines, cosines, and exponentials are cosines, sines, and exponentials, the current will also be harmonic in time within an exponentially decaying envelope.

We now flesh this out in a quantitative way. We can follow the techniques of Section 13–7 to find a solution to the loop rule equation for the *RLC* circuit, Eqs. (32–27) or (32–28). Using our solution to the damped oscillator, Eq. (13–46), along with the equivalences in Table 32–3, we find

$$Q = Q_0 e^{-\alpha t} \cos(\omega' t + \phi). \qquad (32\text{–}30)$$

The constants α and ω' are determined either by substitution back into the original loop rule equation, Eq. (32–28), or simply by using the correspondences in Table 32–3 together with the results of Section 13–7. They are

$$\alpha = \frac{R}{2L} \qquad (32\text{–}31)$$

and

$$\omega'^2 = \frac{1}{LC} - \frac{R^2}{4L^2} = \frac{1}{LC} - \alpha^2 = \omega^2 - \alpha^2. \qquad (32\text{–}32)$$

The quantity α determines the rate of exponential damping and has dimensions of inverse time. The constants Q_0 and ϕ are determined from the initial conditions.

The exponential damping constant α depends only on L and R, as in *RL* circuits. This factor shows that the resistive element is the crucial element in damping; when there is no resistance, there is no damping. (Recall that pure *LC* circuits oscillate without damping.) The damping factor $e^{-\alpha t}$ forms a decreasing envelope for the harmonic behavior within the envelope. Note how the angular frequency ω' differs from the angular frequency $\omega = 1/\sqrt{LC}$ of the undamped circuit (an *LC* circuit). If the damping constant α is *small* compared to ω, then ω' is only slightly less than ω. In Fig. 32–16, we plot the behavior of the capacitor charge for a circuit in which $L = 1$ H, $C = 1$ F, and $R = 0.3\ \Omega$, and compare it to the previous case of $R = 0$. The period in the damped case is only slightly larger than the period for the undamped case, and this difference is not very apparent in the figure.

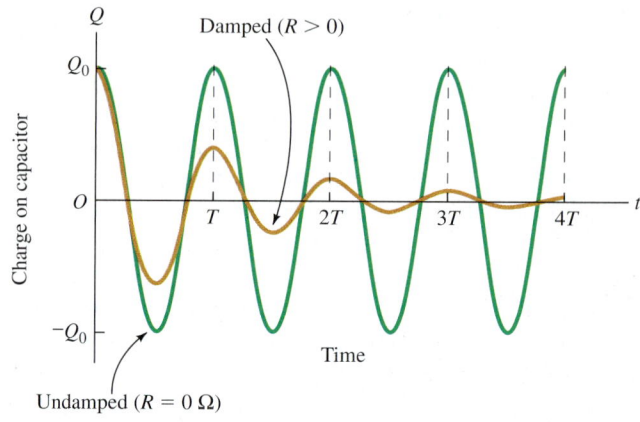

▶ **FIGURE 32–16** Comparison of an *RLC* circuit with and without damping: the charge on the capacitor versus time. The damped case has a very slightly larger period, so close to that of the undamped case that it cannot be seen in the figure. T is the period of the undamped oscillator.

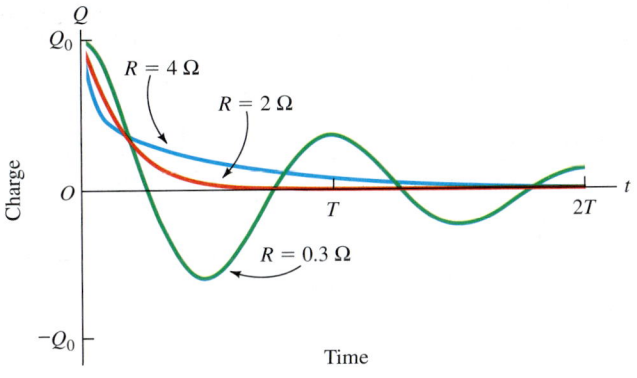

► FIGURE 32–17 The *RLC* circuit of Fig. 32–15 for various values of R. The critical value R_c is $R = 2\,\Omega$. There is no oscillatory behavior, only damping, when $R > R_c$.

Figure 32–17 shows what happens to the charge if R is increased to $2\,\Omega$ and to $4\,\Omega$. What is the explanation for this behavior? Equation (32–32) shows that when R is increased to a critical value R_c, ω'^2 decreases to zero. For $R = R_c$, we have

$$0 = \frac{1}{LC} - \frac{R_c^2}{4L^2},$$

which has the solution

$$R_c = 2\sqrt{\frac{L}{C}}. \qquad (32\text{–}33)$$

When ω'^2 is zero, there is no more oscillation; as in Chapter 13, we refer to this case as *critical damping*. Judicious use of a resistance that will lead to critical damping will *eliminate* current oscillations in situations where it is not wanted. The value $R = 2\,\Omega$ in Fig. 32–17 represents this case: When $L = 1\,\text{H}$ and $C = 1\,\text{F}$, we have $R_c = 2\,\Omega$. For values of R that are larger than the critical value, there is *overdamping*. This kind of motion has its mechanical analogue in the motion of a mass at the end of a spring when the mass is moving in a jar of thick molasses. There is no oscillation, just a slow movement directly to the equilibrium position. An example of this behavior is illustrated in Fig. 32–17, where we have chosen $R = 4\,\Omega$ (a value greater than R_c).

EXAMPLE 32–10 A switch is used to introduce a resistor of $0.052\,\Omega$ as a series element into an *LC* circuit that is undergoing oscillatory behavior (Fig. 32–18). The values of the inductance and the capacitance are $75\,\text{mH}$ and $16\,\mu\text{F}$, respectively. How much time passes between $t = 0$, when the switch is thrown, and the moment when the amplitude of the oscillations has decreased to one half of its value prior to $t = 0$? How many oscillations does the circuit undergo during this time?

Strategy This problem is a straightforward application of our expression for the capacitor charge Q, Eq. (32–30). The envelope factor is $\exp(-\alpha t)$, and at the time $t_{1/2}$ when this factor is one half, we say by definition that the amplitude of the oscillations will have been cut in half. We know all the factors that go into α, so the relation $\exp(-\alpha t_{1/2}) = \frac{1}{2}$ is a relation that we can solve for $t_{1/2}$. We then see how many periods $T = 2\pi/\omega'$ will fit into $t_{1/2}$, and this is the answer to the second part of the question.

It is worth noting that if $\omega \gg \alpha$, the envelope that characterizes the decay of the oscillations changes slowly compared to the oscillation time. Many oscillations take place during the time that the envelope decreases only a little, and it is meaningful to speak of the decay of the amplitude itself, according to

$$\text{amplitude} = A_0 \exp(-\alpha t).$$

We can verify numerically whether this situation holds or not.

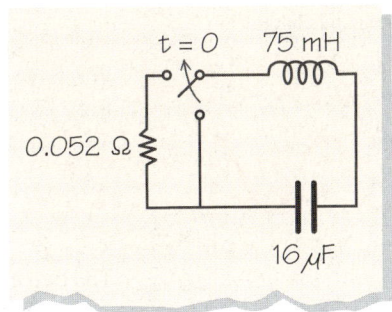

▲ FIGURE 32–18

Working It Out Equation (32–31) is used to determine the damping factor α:

$$\alpha = \frac{R}{2L} = \frac{(0.052\,\Omega)}{2(75 \times 10^{-3}\,\text{H})} = 0.35\,\text{s}^{-1}.$$

By comparison, the angular frequency of the oscillations of the (undamped) *LC* circuit is

$$\omega = \frac{1}{\sqrt{LC}} = \frac{1}{\sqrt{(75 \times 10^{-3}\,\text{H})(16 \times 10^{-6}\,\text{F})}}$$

$$= 9.1 \times 10^2\,\text{rad/s}.$$

(continues on next page)

Indeed $\omega \gg \alpha$, and the concept of the decay of the amplitude is reasonable.

For $t_{1/2}$, we have $\exp(-\alpha t_{1/2}) = \frac{1}{2}$, or taking the natural logarithm of both sides,

$$-\alpha t_{1/2} = \ln(1/2) = -\ln 2 = -0.69.$$

Thus

$$t_{1/2} = \frac{0.69}{\alpha} = \frac{0.69}{0.35 \text{ s}^{-1}} = 2.0 \text{ s}.$$

For the second part of the question, with $\omega \gg \alpha$, $\omega \cong \omega'$, and we can take $T = 2\pi/\omega$ as the period of the oscillator. Then during the time $t_{1/2}$, the circuit undergoes

$$\frac{t_{1/2}}{T} = \frac{t_{1/2}}{2\pi/\omega} = \frac{\omega t_{1/2}}{2\pi}$$

$$= \frac{(9.1 \times 10^2 \text{ rad/s})(2.0 \text{ s})}{2\pi} \cong 290 \text{ oscillations}.$$

What Do You Think? The original resistor is replaced by one that is 10 times larger. What effect does this have on the number of oscillations?

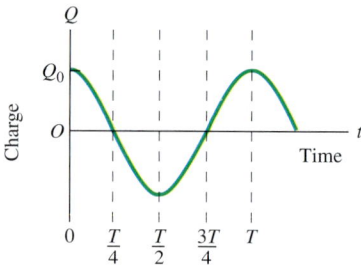

32–7 Energy in *LC* and *RLC* Circuits

Energy is a useful concept in *RLC* circuits, just as it is for the harmonic oscillator. Let's reconsider instances without and with damping.

No Resistance: We set $R = 0$, i.e. there is no damping. We take initial conditions such that the charge on the capacitor is

$$Q = Q_0 \cos \omega t. \quad (32\text{–}34)$$

The current in the circuit is then

$$I = \frac{dQ}{dt} = -\omega Q_0 \sin \omega t. \quad (32\text{–}35)$$

One full period of both of these functions is plotted in Fig. 32–19.

The energy contained in a capacitor is given by Eq. (26–8), $U_C = Q^2/2C$, or

$$U_C = \frac{Q_0^2}{2C} \cos^2(\omega t). \quad (32\text{–}36)$$

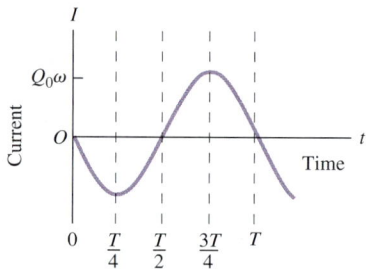

▲ **FIGURE 32–19** The capacitor charge and the current in a circuit that contains only inductance and capacitance are harmonic functions of time. Note the relative positions of the maxima and minima for current and charge.

Equation (32–14) gives the magnetic energy in an inductor, $U_L = \frac{1}{2}LI^2$:

$$U_L = \frac{1}{2}L\omega^2 Q_0^2 \sin^2(\omega t) = \frac{Q_0^2}{2C} \sin^2(\omega t). \quad (32\text{–}37)$$

We have used $\omega = 1/\sqrt{LC}$. The (positive) functions U_C and U_L are plotted in Fig. 32–20. The energy of one rises to a maximum as the other falls to zero. But the *total* energy in the inductor and capacitor is *constant*:

$$U = U_C + U_L = \frac{Q_0^2}{2C}[\cos^2(\omega t) + \sin^2(\omega t)] = \frac{Q_0^2}{2C}. \quad (32\text{–}38)$$

The two circuit elements swap the constant total energy back and forth harmonically, just as in the mechanical oscillator, for which the constant total energy is made up of a back-and-forth exchange of the potential energy of the spring and the kinetic energy of the attached mass. In both the mechanical and circuit case, this is an expression of the conservation of energy. In Figs. 32–21a to 32–21e, we have drawn the progression in a series of snapshots over a full period T, starting with no current and a fully charged capacitor at $t = 0$. Without a resistor to dissipate energy, the oscillation will continue forever.

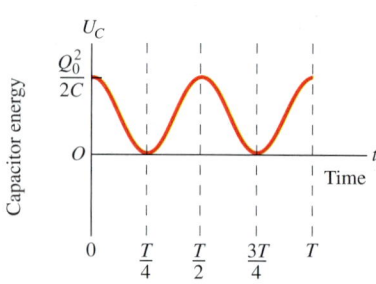

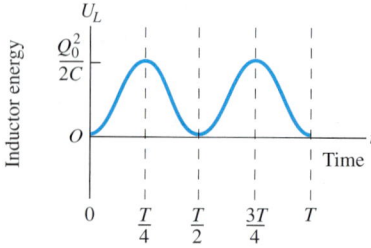

▲ **FIGURE 32–20** The energies of capacitor and inductor for charge and current of the circuit in Fig. 32–15, but with $R = 0$. These oscillate out of phase.

Resistance Is Introduced: We can easily understand the role of resistance in terms of energy. The power P_R dissipated in the resistor is voltage times current:

$$P_R = IV_R = I^2 R, \quad (32\text{–}39)$$

where we have used Ohm's law, $V_R = IR$. This power is proportional to the current squared and is always positive. Energy is *always* lost to Joule heating in a resistor, regardless of the sign of the current. This is the origin of the exponential damping in *RLC* circuits. The power loss in resistors should be contrasted to the equivalent expressions for inductors ($P_L = IV_L$) or capacitors ($P_C = IV_C$). In each case, the rate of energy expenditure can be positive or negative, according to the situation [see Eq. (32–11)]. Unlike the resistor, these elements sometimes take energy from the other circuit elements and sometimes give it back.

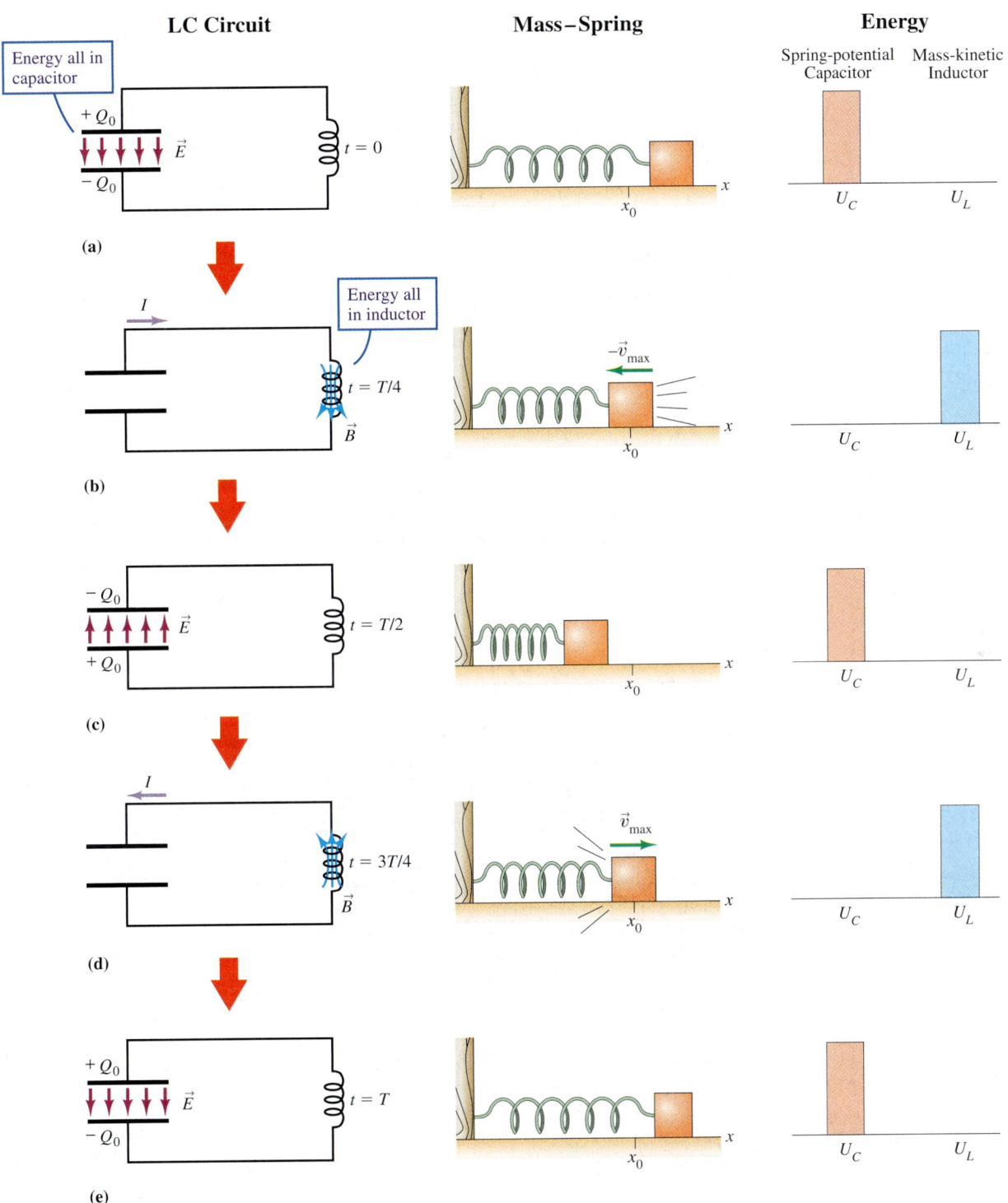

▲ **FIGURE 32–21** (a)–(e) A time sequence showing how the electric and magnetic fields, and therefore the energies, of Fig. 32–20 are realized within the capacitor and inductor. We also show the analogous mechanical system at corresponding times, as well as the balance of energy in the capacitor and inductor for those times.

Summary

An inductor is a circuit element that behaves as a current-carrying loop, or solenoid. It has an inductance L defined by the ratio between the magnetic flux and the current that passes through it:

$$\Phi_B = LI. \tag{32-1}$$

By Faraday's law, the emf induced in this circuit element is

$$\mathcal{E} = -\frac{d\Phi_B}{dt} = -L\frac{dI}{dt}. \tag{32-2}$$

Faraday's law also shows that when there are adjacent loops in a circuit (or pair of circuits), the changing current in one loop induces an emf in the adjacent loop. In this case, the geometrical factor is the mutual inductance M, which measures both the emf induced in loop 1 due to the current in loop 2 and the emf induced in loop 2 due to the change in current in loop 1,

$$\mathcal{E}_{21} = -M\frac{dI_1}{dt}. \tag{32-5}$$

Inductance is measured in henries (H) in the SI. The emf in an inductor is one more term to add to the loop rule.

A simple, calculable inductance is that of an ideal solenoid:

$$\text{for an ideal solenoid: } L = \mu_0 A \ell n^2. \tag{32-8}$$

Here, A is the area, ℓ is the length of the solenoid, and n is the number of turns per unit length.

The energy carried in an inductor is given by

$$U_L = \frac{1}{2}LI^2. \tag{32-13}$$

Just as the energy of a capacitor is carried by the electric field in the capacitor, the energy of an inductor is in the magnetic field. The energy density, or energy per unit volume, carried by a magnetic field is found by comparing the known field within a solenoid with the energy carried by the solenoid, and is given by

$$u_B = \frac{1}{2}\frac{B^2}{\mu_0}. \tag{32-16}$$

The combination of inductance, capacitance, and resistance in circuits with and without batteries leads to interesting time dependence for currents and charges. The current in the inductor cannot change instantaneously. When an inductor is placed in a circuit with a battery and a resistor, we have an RL circuit, and the loop rule produces an equation for the current characterized by transient exponential behavior, with time constant L/R. When a capacitor is added to the circuit, the loop rule produces an equation for the charge on the capacitor whose solution is a capacitor charge that varies harmonically when there is no resistance (we have an LC circuit) and whose capacitor charge behaves as damped harmonic motion when there is resistance (an RLC circuit). The current in these two cases also exhibits harmonic oscillations or damped harmonic oscillations, respectively. These phenomena can also be viewed in terms of energy, with the harmonic oscillations of the LC circuit characterized by a continual exchange of energy between the capacitor and the inductor, and the resistance providing a mechanism of energy loss. The angular frequency of the free oscillations in an LC circuit is

$$\omega = \frac{1}{\sqrt{LC}}. \tag{32-23}$$

When resistance is added, the capacitor charge, for example, has the time dependence

$$Q = Q_0 e^{-\alpha t} \cos(\omega' t + \phi). \tag{32-30}$$

The exponential damping factor is governed by the size of the resistance,

$$\alpha = \frac{R}{2L}, \tag{32-31}$$

while the frequency is shifted to the value

$$\omega'^2 = \omega^2 - \alpha^2. \tag{32-32}$$

Understanding the Concepts

1. Two electric circuits are placed near one another. Each circuit has self-inductance. Must there be a mutual inductance?
2. Does it take more work to cause current to flow through a coil of wire than through the same wire when it is straight?
3. Consider two circular coils that are in a variety of configurations (Fig. 32–22). Assuming that the separation between the coils is roughly the same for the various configurations, can you order the mutual inductances from largest to smallest?

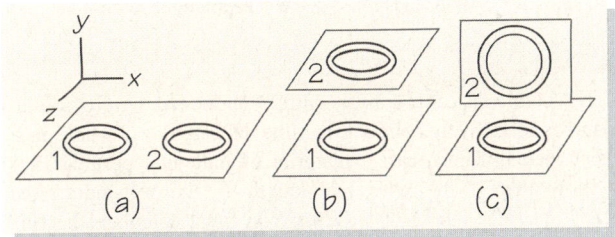

▲ **FIGURE 32–22** Question 3.

4. Why might Faraday's law cause the lights in a house to dim when an electrical appliance that uses a lot of energy, such as an electric clothes dryer, is turned on?
5. Is it possible to calculate the mutual inductance between a straight wire and a wire loop?
6. If the energy in a given region is equally split between energy in the electric and the magnetic fields, what does this tell you about the relative magnitude of the magnetic and electric fields?
7. The time for the current amplitude to drop to $\frac{1}{2}$ its initial value in Example 32–10 is (a) larger than, (b) the same as, or (c) smaller than the time for the charge amplitude to be one half its initial value.
8. Why do you sometimes see a spark at a light switch when the switch is turned off? Is there a spark when the switch is turned on? Why or why not?
9. Describe how you could measure inductance with a battery of known emf, a known resistor, a voltmeter, and a timer.
10. A lightbulb is placed in series with a resistor and in parallel with a coil of large inductance and negligible resistance. When a switch that connects a battery to this circuit is closed, the lightbulb flashes before glowing dimly. When the switch is opened, the bulb flashes again before going out. Explain.
11. In the oscillations of an *LC* circuit, the energy is transferred from the electric field in the capacitor to the magnetic field in the inductor. How does the energy get from one place to the other?
12. Given your knowledge of the largest and smallest practical sizes of capacitors and inductors, what would you estimate is the electronic oscillator with the smallest frequency possible? The largest?
13. Consider Example 32–9. What determines the value of the maximum charge on the capacitor in that example?
14. Given a certain length of wire, you are asked to wind it in the form of a solenoid in a way that maximizes the self-inductance. You may change the radius of the core and the number of turns per unit length, but must take into account that the wire has a certain fixed thickness. What should you do?
15. A solenoid has magnetic flux outside as well as inside, because magnetic field lines must close on themselves. Does this mean that there is magnetic energy outside the solenoid as well as inside it?
16. In Section 32–3, the magnetic energy calculation for a solenoid used the inductance of a portion of the solenoid, and that was translated into the energy density expression $B^2/2\mu_0$. The magnetic field must come around the outside of the solenoid because all magnetic field lines are closed. Why then does the above calculation give the correct answer for the energy density?
17. When $R = 0$, the time constant for the *RL* circuit is infinite. Physically, why?
18. If a diamagnetic material is placed inside the solenoidal coil, the self-inductance of the coil will (a) increase, (b) decrease, (c) stay the same.
19. We have made an analogy between a damped harmonic oscillator and an *RLC* circuit. What mechanical quantities are analogous to the energies $LI^2/2$ and $Q^2/2C$ of the *RLC* circuit?
20. How would you go about finding a generalization of Eq. (32–13) when two circuits with different currents are placed in such close proximity that their mutual inductance plays a role?
21. The magnetic energy density $B^2/2\mu_0$ has the dimensions of pressure and may be viewed as a magnetic pressure. Use this interpretation to justify the attraction/repulsion of two parallel wires that carry currents in the same/opposite directions.
22. If you have two parallel wires carrying currents, then the sign of the force (attraction or repulsion) depends on whether the currents travel in the same or opposite directions. Can you interpret this in terms of the magnetic energy density in the region between the wires?

Problems

32–1 Inductance and Inductors

1. (I) A wire loop has an inductance of 2 mH when a current of 30 mA passes through the circuit. What is the value of the magnetic flux that passes through the loop?
2. (I) What is the self-inductance per unit volume of a solenoid?
3. (I) Calculate the mutual inductance of a solenoid 25 cm long of radius 1.8 cm with 600 turns, and a single loop of radius 3.0 cm centered on the solenoid, with its area perpendicular to the axis of the solenoid.
4. (I) A current that changes at the rate of 1.25 A/s passes through a solenoid; an induced emf of 520 mV is the result. The length and diameter of the solenoid are 12 cm and 0.50 cm, respectively. What is the number of turns?
5. (I) The emf induced in an isolated circuit when the current in the circuit is changing by 10 A/s is 0.3 V. What is the self-inductance?
6. (II) An electrical engineer needs an inductor capable of producing an emf of 150 mV. The current source available produces current of the form $I = I_0 \cos(\omega t)$, with $I_0 = 0.60$ A and $\omega = 2.7 \times 10^2$ rad/s. What size inductor should be used?
7. (II) A current of 1 A flows through a circuit placed in isolation. A magnetic flux of 0.010 T·m² passes through the circuit area. When this circuit is placed near another circuit with a current flow of 2 A, the magnetic flux through the first circuit increases to 0.012 T·m². (a) What is the mutual inductance of the two circuits? (b) How much magnetic flux passes through the second circuit, whose self-inductance is 1 mH?

8. (II) What is the self-inductance of the single inductor that is equivalent to two inductors of values L_1 and L_2, respectively, placed in series? Neglect the mutual inductance.

9. (II) A solenoid of length L consists of two coils tightly placed on top of each other. One coil has N_1 turns, the other N_2, and the area of the coils is A. Calculate the self-inductances of the coils (a) if only one of the two coils is used; (b) if the two coils are connected in series with their windings going in the same direction; (c) if the two coils are connected in series with their windings going in the opposite direction. (d) Calculate the mutual inductance of the two coils.

10. (II) Consider two inductors with inductances L_1 and L_2, respectively, connected in parallel. What is the value of the single equivalent inductance that could replace the two inductances, assuming that the mutual inductance can be neglected?

11. (II) A solenoid of length L and area A contains two windings—one tightly placed on top of the other—with N_1 and N_2 turns, respectively. What happens if the two windings are connected in parallel and the composite coil is included in a circuit with a variable current?

12. (II) Equation (32–8) was derived for an ideal cylindrical solenoid. Show that this result holds also for a solenoid of any shape cross section, provided that the length is large compared to any cross-sectional measure.

13. (II) Consider the cylindrical solenoid and ring illustrated in Fig. 32–23. The solenoid, of diameter $d_1 = 2.0$ cm, has length $\ell = 20$ cm and 120 turns of wire. The ring of wire inside, with diameter $d_2 = 1.5$ cm and area perpendicular to the solenoid's axis, is connected by two wires to a single resistor of resistance $R = 33\ \Omega$. The current I_1 is in the form of a pulse that starts to rise linearly at $t = 0$ s. The current reaches a maximum of 30 A at $t = 0.30$ s, then starts to descend linearly; when the current reaches 0 A at $t = 0.60$ s, it ceases to flow. Find the current I_2 induced in the ring as a function of time.

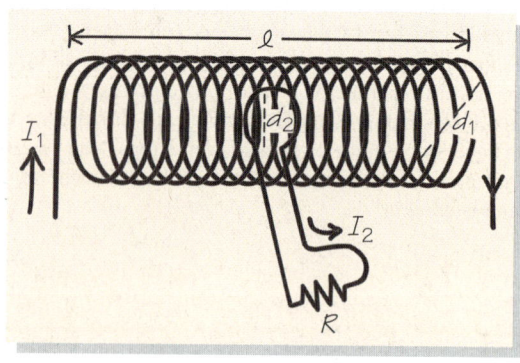

▲ FIGURE 32–23 Problem 13.

14. (II) The ring contained within the cylindrical solenoid in Fig. 32–6a is replaced by a second cylindrical solenoid, of length ℓ_2, radius R_2, and turn density n_2. Calculate the mutual inductance of this system.

15. (II) The current in an inductor has the periodic triangular form plotted in Fig. 32–24, with an amplitude of 0.50 A and a period of $T = 0.45$ s. What is the voltage across the inductor as a function of time if $L = 2.3 \times 10^{-3}$ H? Express your answer algebraically or plot it.

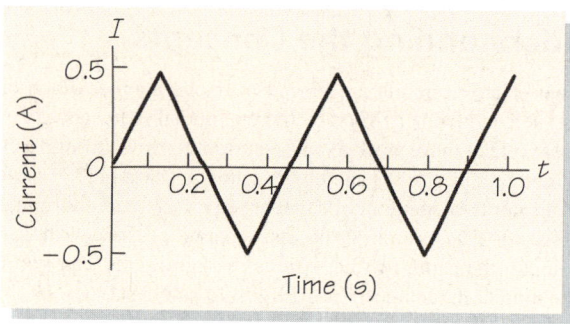

▲ FIGURE 32–24 Problem 15.

16. (II) A coaxial cable has a central conducting wire of radius r_0 surrounded by a conducting tube of radius r_1. The space in between is filled with a material of magnetic permeability μ. Show that if the wire has length ℓ, the self-inductance is $L = (\mu\ell/2\pi)\ln(r_1/r_0)$. [Hint: You must calculate the flux in the region between the cylinders.]

17. (II) Consider two identical solenoids placed end to end, with the windings going in the same direction. Prove that the total self-inductance of the combined system is $2(L + M)$, where L is the self-inductance of either solenoid and M is the mutual inductance.

18. (II) A torus of rectangular cross section with width w, height h, and inner radius R is wound with N turns of wire (Fig. 32–25). What is the self-inductance of the torus? Use the approximation $\ln(1 + x) = x$, valid for $x \ll 1$, to discuss the case where $R \gg w$ and its relation to the self-inductance of a solenoid.

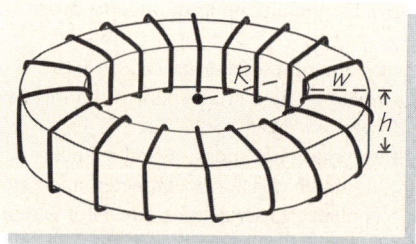

▲ FIGURE 32–25 Problem 18.

19. (II) Consider a torus of square cross section. The radius of the torus (distance from the symmetry axis to the center of the square) is 35.0 cm; the sides of the square are 5.00 cm. The torus is wound with 1650 turns of wire. (a) What is the self-inductance of the torus? (b) What is the self-inductance if the core of the torus is made of soft iron, with $\mu = 4200\mu_0$?

20. (II) Consider a toroidal coil wound around an empty core whose self-inductance is 15 mH. The current in the coil changes uniformly by 120 mA in 0.50 s. (a) What is the induced emf? (b) If the hollow center of the torus is filled with an iron core, with $\mu = 3400\mu_0$, what is the induced emf?

21. (II) Calculate the inductance of an elongated rectangular circuit, such as a length of a two-wire ribbon cable (Fig. 32–26). The length of the circuit is L and its width is $a \ll L$. If the radius of the wire is much less than a, can it (the radius) be neglected?

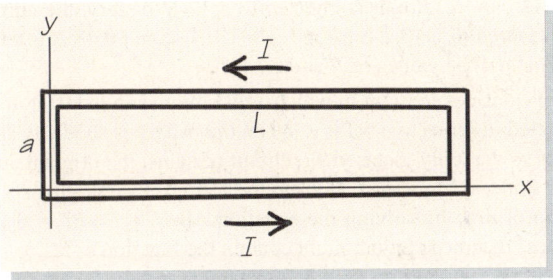

FIGURE 32–26 Problem 21.

22. (II) Figure 32–27 shows a straight wire that carries a current I and a square loop of wire with one side oriented parallel to the straight wire a distance d away. The square has sides of length a. Calculate the mutual inductance of this system. [*Hint*: The magnetic field due to the straight wire through a slice of the square of width dx parallel to the wire is constant, so the flux through this slice is easily calculable. Integrate to find the total flux through the square.]

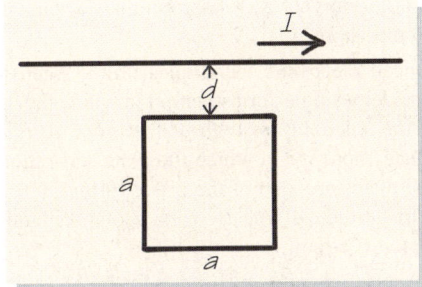

FIGURE 32–27 Problem 22.

23. (III) Calculate the mutual inductance of the two elongated rectangular circuits shown in Fig. 32–28. Assume that $L \gg a, b$.

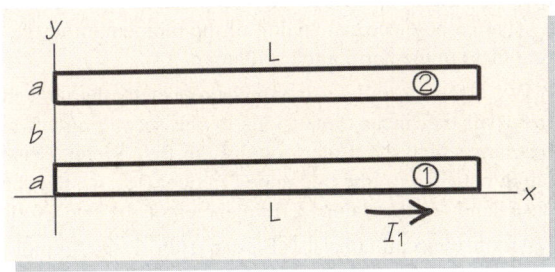

FIGURE 32–28 Problem 23.

32–2 Energy in Inductors

24. (I) A cylindrical solenoid of radius 0.75 cm is wound with a turn density of 140 turns/m. It carries a current of 0.45 A. How much energy is stored per meter length of the solenoid?

25. (I) Consider an inductor with $L = 16$ H and an internal resistance of 0.10 Ω. We wish to use this inductor to store 0.10 MJ of energy. What is the rate at which energy is lost to Joule heating in this system? It is not practical to store large amounts of energy in large inductors unless the wire is superconducting.

26. (I) An inductor with $L = 1.6$ mH has an internal resistance small enough to be ignored. How much work would a battery have to do to increase the current through the inductor from 88 mA to 98 mA?

27. (I) A capacitor with $C = 0.020$ μF has a charge of 15 μC. What is the equivalent steady current that should be carried by an inductor of $L = 20$ μH if the inductor is to store the same amount of energy?

28. (I) A doorbell circuit contains a solenoid with 600 turns of wire, a cross section of 6.0 cm², and a length of 12 cm. When the doorbell button is pushed, 100 mA passes through the circuit. How much energy is contained within the solenoid at this time?

29. (II) A current with time dependence $I = I_0 e^{-\alpha t}$ passes through an inductor with $L = 2$ mH; $I_0 = 4.0$ A and $\alpha = 0.02$ s^{-1}. Compute the power expended in the inductor as a function of time.

30. (II) The voltage across an inductor with $L = 3.0$ mH is fixed at 6.0 V. The current is increased (a) from 0.00 A to 0.25 A, (b) from 0.25 A to 0.35 A, and (c) from 0.35 A to 0.40 A. What average power must be supplied from an external source in each step?

31. (II) Consider an inductor with $L = 1$ H and a capacitor with $C = 1$ F. (a) Compare the energy contained in the inductor when a current of 10 A flows through it with the energy in the capacitor if the charge is the amount of charge contained in the 10-A current, flowing for 1 s. (b) Repeat part (a) for a current of 1 mA.

32. (II) An electrical engineer constructs a cylindrical solenoid of area 8 cm² and length 25 cm from 150 m of thin wire. The wire will handle a maximum current of 50 mA. (a) What is the inductance of the solenoid? (b) How much energy can the inductor store?

33. (II) The inductance of a small superconducting solenoid is 8 H. The current is gradually increased from 0 A to 40 A. (a) How much energy is stored in the solenoid? (b) When the current reaches 40 A, the solenoid "quenches"; i.e. the wire of the solenoid loses its superconductive property because of the large magnetic field. The current decreases to zero rapidly, and the magnetic energy is dissipated in the liquid helium coolant. Given that the latent heat of vaporization of helium is 2.7×10^3 J/L, how much of the liquid helium coolant is evaporated during the quench?

32–3 Energy in Magnetic Fields

34. (I) The magnetic field in interstellar space has an approximate magnitude of 10^{-10} T. How much magnetic field energy does this contribute to the spherical region around the Sun of radius matching the mean radius of Neptune's orbit?

35. (I) The two circular pole pieces of a magnet are 63 cm in diameter and 21 cm apart. The magnetic field between them is 0.10 T. What is the magnetic energy stored in the field?

36. (I) An ideal cylindrical solenoid carrying a current of 115 mA has a winding density of 15 turns/cm. If the core is filled with iron, $\chi_m = 5500$, what is the energy density contained in the magnetic field within?

37. (II) A straight wire carries a current $I = 20$ A. Find the energy density in the surrounding magnetic field as a function of the distance r from the wire. At what distance from the wire does the energy density equal that of a parallel-plate capacitor with a charge of 10^{-7} C and a capacitance of 6.3×10^{-9} F, if the separation between the plates is 1.5 mm?

38. (II) (a) What is the energy density of the magnetic field outside a straight wire of radius a that carries a current I? (b) What is the total energy per unit length, due to that magnetic field, that is contained in a cylinder of radius R ($R > a$) centered about the wire?

914 | Inductance and Circuit Oscillations

39. (II) Consider a torus of radius R, wound with n turns per unit length of a wire that carries a current I. The cross section of the torus forms a square with sides of length b; $b \ll R$. We know that the magnetic field inside the torus has the nearly constant value $B = \mu_0 n I$. Use this result and the two expressions related to the magnetic energy [Eqs. (32–13) and (32–15)] to show that, for this torus, $L = 2\pi\mu_0 n^2 R b^2$.

40. (II) Consider an iron ring of cross-sectional area A. The upper half of it has a certain number of turns of wire wrapped around it. The wire is connected to a battery so that a current I flows through the wire. The ring is cut so that there is a small gap of width y. Within the gap, the magnetic field has magnitude B. (a) What is the force exerted by the upper half of the ring on the lower half? (b) Suppose the gap is filled with some material of large (positive) susceptibility. Will the force increase or decrease?

41. (II) Consider a hollow solenoid inside of which there is a magnetic field B. The solenoid is vertical, and a small cylinder of iron that just fits into the solenoid without touching the sides is dropped into the empty space. Will there be a force on the small cylinder? If so, in what direction will it point? How large will it be?

42. (II) (a) What is the magnetic field energy density inside a straight wire of radius a that carries current I uniformly over its area? (b) What is the total magnetic field energy per unit length inside the wire?

43. (II) A coaxial cable consists of a wire 0.15 cm in diameter with a return path for the current in the shape of a very thin cylindrical conductor of diameter 0.80 cm. A current of 0.25 A flows through the cable. Calculate the magnetic energy per unit length of cable within the inner wire. [*Hint*: It will help to do Problem 42 first.]

32–4 Time Dependence in RL Circuits

44. (I) You wish to make a circuit in which a resistor and an inductor are connected in series to a battery such that when the switch is closed, the current builds up to within 18% of its steady-state value in 5×10^{-4} s. You have a series of inductors with inductances ranging from 0.010 H to 0.10 H. You must acquire a set of resistors with what range of resistance?

45. (I) Show that the time constant L/R that characterizes RL circuits has dimensions of time.

46. (II) Consider the RL circuit of Fig. 32–10; the switch is closed at time $t = 0$. For the circuit elements, $\mathscr{E} = 6$ V, $R = 3.3$ kΩ, and $L = 2.5$ mH. Using Eq. (32–19), find how much charge flows in the circuit during the first (a) 1 μs, (b) 1 ms, (c) 1 s.

47. (II) Show by direct substitution that Eq. (32–19) is a solution of Eq. (32–18).

48. (II) Consider the RL circuit shown in Fig. 32–29. The switch is opened at time $t = 0$ after it had been closed for a long time. How long will it take for the current in the inductance to drop to 25% of its initial value? How long will it take for the energy in the inductance to drop to 25% of its initial value?

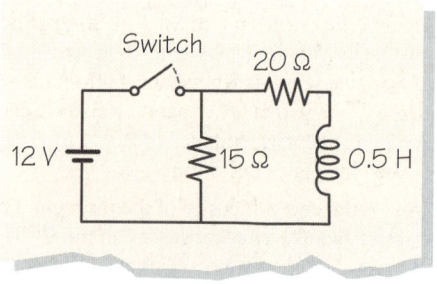

▲ FIGURE 32–29 Problem 48.

49. (II) In an RL circuit connected to a 12-V battery, the current is measured to be 0.2 A after 1.2×10^{-4} s, and 0.48 A after 10 s. What are the values of R and L?

50. (II) Consider a circuit in which a resistor and an inductor are connected in series to a battery. When the battery is suddenly shorted out so that only a closed RL circuit remains, the original current, V/R, decays to zero. Calculate the form of the current as a function of time by solving the equation $L(dI/dt) + RI = 0$ with a guess that the solution might contain the function $e^{-\alpha t}$.

51. (II) Consider the situation discussed in Problem 50. Calculate the total energy dissipated in the resistor from the time when the switch that shorts out the battery is thrown to the time $t = \infty$. Show that this is the energy stored in the inductor just before the switch is closed.

32–5 Oscillations in LC Circuits

52. (I) An electric oscillator consists of a parallel-plate capacitor and a long, cylindrical solenoid. If the resonant frequency of the oscillator is ω_0, what is the frequency of a similar oscillator in which both the capacitance and inductance are reduced by a factor of 12?

53. (I) You have an inductor with an inductance of 40 mH. Using it, you want to make a circuit with oscillations of frequency 20 Hz. What capacitor do you ask your roommate to pick up at the corner electronics store?

54. (II) Advanced electronic techniques utilize microscopic structures. Consider a single-turn solenoid in which the radius and the length of the solenoid are both of the order 10 microns, and a parallel plate capacitor in which the plate separation and the radius of the plates are also of the order 10 microns. *Estimate* the order of magnitude of the frequency of oscillation of such a microscopic LC circuit.

55. (II) Design an LC circuit—give values for C and L—that has an angular frequency of 4.32×10^4 rad/s and a stored energy of 0.30 mJ. The maximum voltage drop across the capacitor must be 20.0 V.

56. (II) Suppose that at time $t = 0$ the current in an ideal LC circuit is zero but the charge on the capacitor is $Q(0)$. (a) What is the energy in terms of $Q(0)$, L and C? (b) If the charge on the capacitor is zero at $t = 0$, and the current has the value $I(0)$, what is the energy? (c) Starting at $t = 0$, when is the first time that the energy is equally stored in the capacitor and in the inductor? (d) What is the electrical analog of the momentum of the oscillating mass in the harmonic oscillator?

57. (II) Two electric oscillators are made of exactly the same materials, but all the linear dimensions of the second circuit are ten times larger than the dimensions of the first circuit. Obtain the relation between (a) the undamped frequencies, (b) the damping factors, and (c) the damped frequencies of the two oscillators.

58. (II) An open circuit consists of a capacitor C and an inductor L connected in series. A charge q is placed on the capacitor, and the circuit is closed at time $t = 0$ by means of a switch. Find the maximum value of the current, as well as the times for which this maximum value occurs.

32–6 Damped Oscillations in RLC Circuits

59. (I) An RLC circuit is composed of a resistor $R = 0.883$ Ω, an inductor $L = 1.75$ H, and a capacitor $C = 133$ pF, all arranged in series. What is the angular frequency of current oscillations in this circuit?

60. (I) An RLC circuit has $R = 85$ mΩ, $L = 0.60$ mH, and $C = 55$ μF. (a) Find the damping factor and angular frequency. (b) If the resistance is variable, what value of R will give critical damping?

61. (I) Consider a series *RLC* circuit for which the initial capacitor charge is Q_0. If R is chosen such that there is critical damping, what is the instantaneous power consumption in the resistor? [*Hint*: Try the formula in Eq. (32–30).]

62. (II) Consider an *RLC* circuit at critical damping, with $L = 68$ mH. What is the value of R if the current decays by 15 percent in 8.0 ms?

63. (II) Show that Eqs. (32–30) through (32–32) solve Eq. (32–27).

64. (II) Suppose that the values of *R*, *L*, and *C* in a series *RLC* circuit are such that $\omega'^2 < 0$. Assuming that the solution for the charge on the capacitor takes the form $Q = Q_1 \exp(-\alpha_1 t) + Q_2 \exp(-\alpha_2 t)$, find the values of α_1 and α_2.

65. (II) Consider the basic *RLC* circuit. By making an appropriate approximation of Eq. (32–32), show that when α is small compared to $\omega = 1/\sqrt{LC}$ the modified angular frequency ω' of the damped *RLC* circuit is $\omega' \simeq \omega - R^2\sqrt{C/L}/8L$. Find a similar relation for the periods of the undamped and slightly damped cases.

32–7 Energy in *LC* and *RLC* Circuits

66. (II) Calculate the energy in an *LC* circuit, assuming that the initial conditions are such that the charge on the capacitor is $Q = Q_0 \cos(\omega t + \delta)$. Show that the energy is constant.

67. (II) A circuit consists of a capacitor of capacitance $C = 20$ nF connected in series with an inductor of inductance $L = 2 \times 10^{-5}$ H. If a charge of 30 nC is put on the capacitor, there is an oscillation in the circuit. (a) What is the maximum current that moves through this circuit? (b) Find the maximum energy within the inductor. (c) What is the ratio of the maximum energy in the inductor to the maximum energy in the capacitor?

68. (II) An *LC* circuit consists of a 15-mH inductor and a 120-μF capacitor. If the maximum energy stored in the circuit is 3.0×10^{-4} J, what are the maximum charge on the capacitor and the maximum current in the circuit? What are the minimum values?

69. (III) The 3-mF capacitor of an *RLC* circuit is initially charged to 30 μC. The 1.5-mH inductor has a very small resistance. At a particular instant, after 100 oscillations, the current through the inductor is zero while the capacitor is still charged to 5 μC. (a) What is the resistance of the circuit? (b) What are the energies of the circuit before and after the 100 oscillations? (c) Why are the two values of the energy in part (b) different? Where has the energy gone?

70. (III) Consider an *RLC* circuit. The energy is given by $E = LI^2/2 + Q^2/(2C)$. Show that the rate of change of this energy is equal to the power loss in the resistor (the ohmic heating power).

General Problems

71. (II) By considering the definition of inductance, show that if the voltage *V* across an inductor changes with time, the total current passing through the inductor in that time is given by
$$I = \frac{1}{L} \int V \, dt.$$

72. (II) Suppose that a square wave of voltage, as plotted in Fig. 32–30, is applied across an inductor with $L = 0.005$ H. Use the result of Problem 71 to plot the current as a function of time.

73. (II) The switch in the circuit shown in Fig. 32–31 has been closed for a long time. (a) What is the current in each leg of the circuit? (b) When the switch is opened, the current in the inductor drops by a factor of 2 in 8 μs. What is the value of the inductance? (c) What is the current passing in each leg at 12 μs?

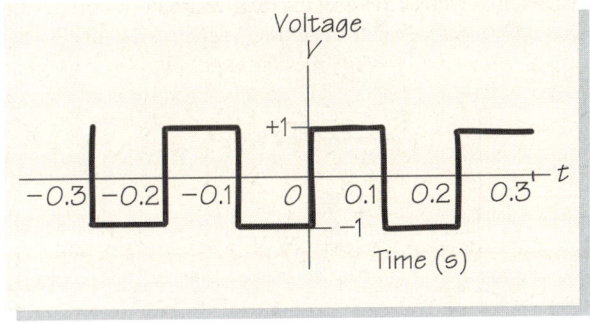

▲ **FIGURE 32–30** Problem 72.

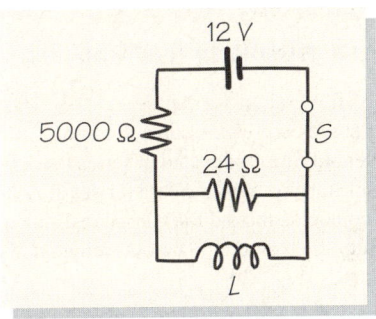

▲ **FIGURE 32–31** Problem 73.

74. (II) As a way of preventing arc formation between the terminals of a switch, a capacitor is connected to the two terminals (Fig. 32–32). What is the minimum capacitance of the capacitor if no voltage larger than 200 V is to be allowed in the circuit? [*Hint*: Assume that very little power can be dissipated during the time that the capacitor charges.]

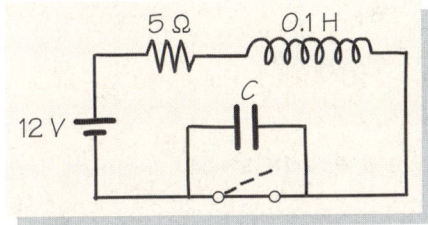

▲ **FIGURE 32–32** Problem 74.

75. (II) A coaxial cable has an inner, solid wire of radius r_1 and an outer, hollow wire of radius r_2. A current *I* flows through the inner wire and returns through the outer wire. Assuming that the cable is infinitely long, find the magnetic field energy per unit length. Include any field energy inside the inner wire and outside the outer wire.

76. (II) Molybdenum is paramagnetic, with a magnetic susceptibility of 1.2×10^{-4} at 300K, which is about one half its value at 20K. Suppose that the self-inductance of a solenoid filled with molybdenum is $L = 0.35$ mH at 300K. What is the fractional change in self-inductance between 300K and 20K?

77. (II) Consider a cavity uniformly filled with oscillating electric and magnetic fields. (a) Show that the ratio of the amplitude of these fields, E_0 and B_0, respectively, has the dimensions of $[\text{velocity}]^{-1}$. (b) For what value of this ratio is the magnetic energy density equal to the electric energy density?

78. (II) What are the currents in the three resistors of Fig. 32–33 immediately after the switch is closed? After a long time?

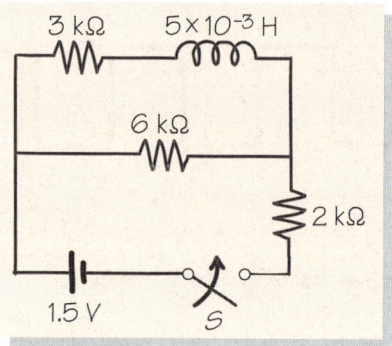

▲ FIGURE 32–33 Problem 78.

79. (II) Two solenoids are wound on a common soft iron core (Fig. 32–34). Solenoid S_1 is connected in series to a battery and a variable resistor. Starting with the resistor set at A (low resistance), the sliding contact is moved to B (large resistance) and back to A again. Sketch the voltage V across solenoid S_2 while this is happening.

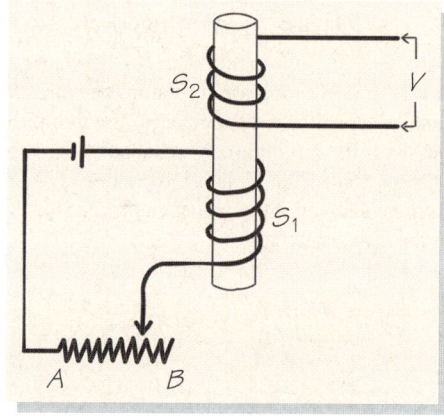

▲ FIGURE 32–34 Problem 79.

80. (II) The two identical coils in the circuit of Fig. 32–35 are placed close to each other, and their mutual inductance is 0.7 mH. Suppose that the switch has been closed for a long time and is then opened at $t = 0$. Calculate the current in the circuit at $t = 18$ ms.

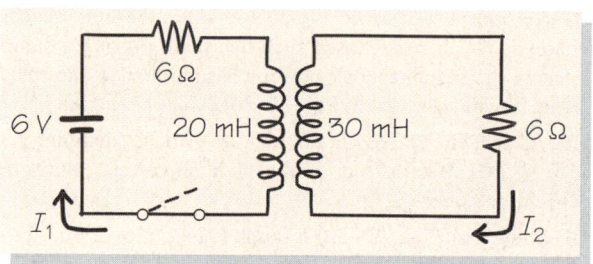

▲ FIGURE 32–35 Problem 80.

81. (II) A ferromagnetic torus is part of a device to be used in a region where the magnetic permeability has the constant value $\mu = 2500\mu_0$. The torus has a circular cross section of 4 cm². Over its total length of approximately 35 cm, the torus is wrapped with 220 turns of wire. Immediately surrounding this winding is a secondary winding of 40 turns of (insulated) wire. What is the mutual inductance of the two windings? What is the role of the iron core, if any, in determining this mutual inductance?

82. (II) A torus of inner radius r_i and outer radius r_0 has a square cross section (Fig. 32–36). It is wound with N turns of wire that carries a current I. (a) Use Ampère's law to find the magnetic field inside the torus. (b) Calculate the magnetic energy density within the torus. (c) Integrate the magnetic energy density to find the total magnetic energy within the torus. (d) Use the formula $U_L = \frac{1}{2}LI^2$ to compute the self-inductance of this torus.

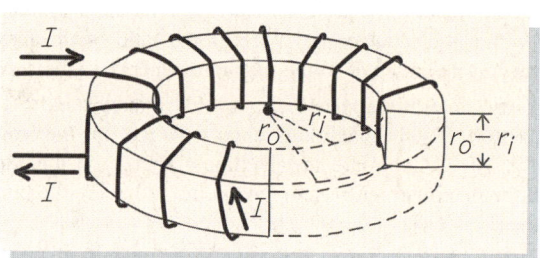

▲ FIGURE 32–36 Problem 82.

83. (II) An LC circuit oscillates with an angular frequency of 1.2×10^6 rad/s. When a second capacitor is inserted in series with the original one, the angular frequency becomes 1.6×10^6 rad/s. If the capacitors are replaced by a resistor of $0.02 \, \Omega$, the current drops to 1/2 of its initial value in 3.5 ms. What are the values of the two capacitors and of the inductance L?

84. (III) Consider two adjacent circuits as shown in Fig. 32–2. Show that the total energy is given by $U = \frac{1}{2}L_1I_1^2 + \frac{1}{2}L_2I_2^2 + MI_1I_2$ and that $M \le (L_1L_2)^{1/2}$.

Chapter 33

Alternating Currents

In Chapter 30 we learned how a changing magnetic flux induces an emf—Faraday's law. In particular, when a coil rotates in the presence of a magnet, an emf is induced in the coil that varies sinusoidally with time. The induced emf produces an alternating current (AC), which is a source of AC power. AC generators use induction to convert the mechanical energy of falling water or the pressure of hot steam into electric currents that vary with time. Such generators are the starting point for the delivery of electric power to home and industry. Alternating current circuits are at the heart of most household equipment, and alternating current flows every time you switch on a light. The ability to vary the maximum voltage of the harmonically oscillating emf that AC power provides is an important element in the delivery of electric power, and learning how to do this is one of our aims here. AC sources of emf in circuits that include resistors, inductors, and capacitors provide currents and voltages with new types of time-dependent behavior. In particular, such circuits exhibit the same kinds of resonance phenomena that we saw in mechanics and are the basis for devices like radio tuners.

◀ This bank of transformers at an electricity substation forms a crucial part of the electricity distribution system. Transformers use Faraday's law to raise or lower the amplitude of time-varying voltages, allowing electrical energy to be transported efficiently at high voltages and consumed at safer low voltages.

33–1 Transformers

An alternating current is characterized by harmonic (sine and cosine) time dependence, as are the other variables of the circuit, such as voltages[†]. The possibility of being able to vary the maximum AC voltage (the *voltage amplitude*) is of interest because high or

---

[†] "AC" stands for any kind of current or voltage that varies harmonically in time.

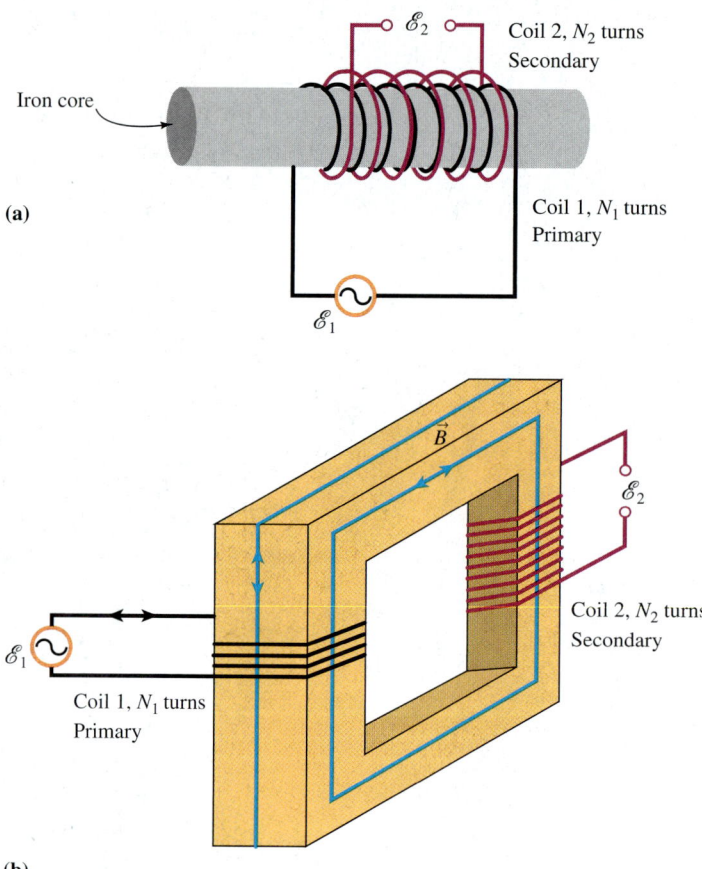

▲ **FIGURE 33–1** Two methods for creating fully linked coils: (a) one coil tightly wound over the other; (b) two coils wrapped around a common core of ferromagnetic material, which has the property of keeping the magnetic field lines within it.

low voltages are useful in differing circumstances. For example, it is more economical to transport electric energy at high voltage (see the "Think About This" box on p. 920). High voltages, however, are dangerous, as well as inefficient in small appliances.

Let's suppose that an AC generator produces an emf of the general form

$$\mathcal{E} = V_0 \sin \omega t \tag{33-1}$$

[see Eq. (30–14)]. The factor V_0 is the voltage amplitude of the source of emf, and this is the quantity we want to vary. In this section we describe a device that can take an AC emf as input and produce another AC emf with a *different* voltage amplitude. This device is called a **transformer**, and it is constructed using the principle of mutual inductance.

Consider two fully linked ideal solenoids. We can arrange this in either of the two ways shown in Fig. 33–1. What we mean by "fully linked" is that the magnetic flux through one turn is equal to the magnetic flux through one turn in the other. The solenoids have a different total number of turns: N_1 and N_2, respectively. Across the first coil—the *primary coil*—there is an AC emf $\mathcal{E}$ with an amplitude V_0, as in Eq. (33–1):

$$\mathcal{E} = V_0 \sin \omega t. \tag{33-2}$$

This emf is changing with time, and you can see from Fig. 33–2 that the voltage drop $V_1(t) = V_{10} \sin \omega_1 t$ across the full primary coil has exactly the same form as Eq. (33–2). The source of this potential is Faraday's law, and this law tells us that the voltage drop across *each turn* is proportional to the rate of change of the flux through the primary coil. It is irrelevant for our purposes that this proportionality has something to do with the self-inductance of the coil; the only thing that matters is that the voltage per turn, V_1/N_1, across the primary coil is proportional to this change in flux. Now, since the coils are fully linked, we see that the voltage across *each turn* of the secondary coil is proportional in exactly the same way to the rate of change of the magnetic flux.

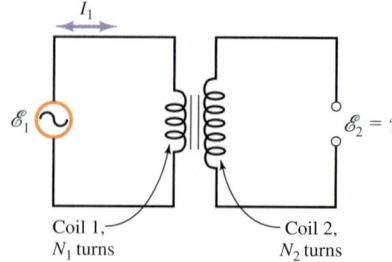

▲ **FIGURE 33–2** Schematic circuit-diagram symbol for two fully linked coils. When one coil is in the same circuit with an AC source of emf, this combination acts as a transformer.

The voltage drop across *each turn* of the secondary coil is thus identical to the voltage drop across each turn of the primary. We learn two things. First, the time dependence of the voltage across the secondary coil is exactly the same as the time dependence of the voltage across the primary coil. Thus the voltage across the secondary coil is $V_2(t) = V_{20} \sin \omega_2 t$, and the time dependence of the voltage across both coils is identical:

$$\omega_1 = \omega_2 = \omega. \tag{33-3}$$

The second thing we learn is a relation between the voltage *amplitudes* in the two coils. We have argued that

$$V_1(t)/N_1 = V_2(t)/N_2, \tag{33-4}$$

RATIO OF VOLTAGES IN A TRANSFORMER

and when we cancel the identical time-dependent factors $\sin \omega t$, we are left with $V_{10}/N_1 = V_{20}/N_2$ or, rearranging,

$$\frac{V_{20}}{V_{10}} = \frac{N_2}{N_1}. \tag{33-5}$$

The two-solenoid device we have described is a transformer, a tool for manipulating voltage amplitudes. Transformers find use throughout the power distribution network, and you have no doubt frequently used them for your electronic tools, which may require voltage amplitudes a good deal less than the 120-V or 220-V voltages supplied by a wall plug.

We see from Eq. (33–5) that when the number of turns on the secondary coil is greater than the number of turns on the primary coil ($N_2 > N_1$), the transformer is a *step-up transformer*, and the voltage amplitude in the secondary coil is greater than that in the primary coil. When $N_2 < N_1$, the transformer is a *step-down transformer*; the voltage amplitude in the secondary coil is smaller than that in the primary coil. Note that the terms primary and secondary do not imply any fundamental distinction between the two coils, and either coil could be primary or secondary according to the application.

CONCEPTUAL EXAMPLE 33–1 Assuming that resistance has been reduced to a minimum, how could you use the power associated with an inductor to find an expression for the ratio of currents in the primary and secondary coils of a transformer?

Answer The meaning of the resistance being reduced to a minimum is that there is minimal energy loss to Joule heating—we can ideally assume this to be zero—and any energy flow into or out of the primary coil will be matched by an energy flow out of the secondary coil. The power, or rate of energy flow, in an inductor is given by the product IV. Thus the equality $I_1 V_1 = I_2 V_2$ can be used to find the ratio of currents in terms of the ratio of emfs, and Eq. (33–4) gives us the latter quantity in terms of the number of turns in each solenoid when the solenoids are fully linked.

Conceptual Example 33–1 shows us how we can find the ratio of currents in the two coils in terms of their numbers of turns: Assuming insignificant losses to Joule heating, conservation of energy gives us $I_1 V_1 = I_2 V_2$. Rearrangement gives $I_1/I_2 = V_2/V_1$, and then use of Eq. (33–4) implies

$$\frac{I_1}{I_2} = \frac{N_2}{N_1}. \tag{33-6}$$

RATIO OF CURRENTS IN A TRANSFORMER

In other words, the current amplitude in the secondary coil of a step-up transformer ($N_2 > N_1$) is decreased, while it is increased in the secondary coil of a step-down transformer ($N_2 < N_1$). We should add here that when there is resistance in the secondary coil or in the circuit of which the secondary coil is a part—we say that there is a *load* on the secondary coil—then things are more complicated, with the inductances playing an explicit role.

EXAMPLE 33–2 A step-down transformer has 5,000 turns in the primary coil, which handles an AC current with voltage amplitude $V_{10} = 20{,}000$ V, and 220 turns in the secondary coil. If the current amplitude desired in the secondary coil is 100 A, what is the maximum power that must be delivered by the primary coil to the secondary coil? Assume negligible resistance.

Setting It Up Imagine a transformer similar to those shown in Fig. 33–1. We know the number of turns of wire, N_1 and N_2 respectively, in both the primary and secondary coils.

Strategy We must find the power P_2 delivered to the secondary coil. Maximum power is related to the unknown voltage amplitude V_{20} and to the known current amplitude I_{20} by the relation $P_{2,\max} = I_{20}V_{20}$. (This is actually true only when the load is zero, i.e. when there is no resistance. We'll assume this to be the case here.) We can find V_{20} from the known V_{10} and the known numbers of turns using Eq. (33–5).

Working It Out From Eq. (33–5),

$$V_{20} = \frac{N_2}{N_1} V_{10} = \frac{220}{5{,}000}(20{,}000 \text{ V}) = 880 \text{ V}.$$

With the maximum current I_{20} carried by the secondary coil at 100 A, the maximum power that it carries is $P_{2,\max} = (100 \text{ A})(880 \text{ V}) = 88$ kW. With negligible resistance, this is equal to the maximum power carried in the primary coil.

What Do You Think? Must the currents in the primary and secondary coils be wound in the same direction for a transformer to work? Answers to **What Do You Think?** questions are given in the back of the book.

THINK ABOUT THIS...
WHY IS ELECTRIC POWER TRANSMITTED ALONG HIGH-VOLTAGE LINES?

The long-distance transmission lines that bring power from generating plants to users carry AC with a voltage amplitude as high as $V = 1.0$ MV. (Compare this to the 160-V amplitude—rms 110 V—for typical appliance use.) Why such high voltages? If we label the rate of energy delivery as P, then the current that runs through these lines can be characterized according to $P = IV$. (We are ignoring the "alternating" aspect here—this doesn't have much effect on our argument.) On the way, some of this power will be dissipated as Joule heating. This depends on the resistance, R, of the transmission line—the power dissipated, or lost, in the line will be $P_{\text{lost}} = I^2 R = P^2 R / V^2$. Now a measure of the efficiency of transmission is the ratio of the power delivered to the power lost in Joule heating—one would want this ratio to be as large as possible. Using our expressions for P and P_{lost},

$$\frac{P}{P_{\text{lost}}} = \frac{V^2}{PR}. \qquad (33\text{–}7)$$

This ratio increases rapidly as V increases, and that is why it is best to deliver the power at as high a voltage as possible. For example, a transmission line delivering power $P = 1.0$ MW may have a total resistance of 10 Ω. If the power is delivered at 110 V, the ratio in Eq. (33–7) would be intolerably low: $(110 \text{ V})^2 / (1.0 \text{ MW})(10 \text{ }\Omega) = 1.2 \times 10^{-3}$; 99.9 percent of the electrical power that set off from the generating plant would be lost through Joule heating in the wires! If the power is delivered at 500,000 V, the ratio is $(500{,}000 \text{ V})^2 / (1.0 \text{ MW})(10 \text{ }\Omega) = 2.5 \times 10^4$, and nearly all of the electric energy produced is delivered to the end of the line. Then transformers can reduce the voltage amplitude and make the energy safe to use.

In addition to their ubiquitous use in power transmission and distribution, and their role in allowing electronic instruments to take their power from a wall socket, transformers are crucial to the operation of every automobile. The spark that is fired by a spark plug and ignites gasoline in an automobile cylinder requires a very high potential difference between the elements of a spark plug. However, the voltage differences supplied by the automobile battery or alternator are much more modest—the battery maintains 12 V DC, for example. The spark coil is the device that makes the large potential. The coil contains a secondary coil with a large number of windings. When the current in a primary coil is interrupted suddenly, by either mechanical or electronic means, the large change in flux induces a potential difference across the secondary coil that is large enough to cause a spark to form. Timing devices ensure that the potential is set up at the right time for the engine operation.

33–2 Single Elements in AC Circuits

We now turn to the question of how AC is used in circuits. We start with an examination of the effects of placing an AC source of emf, for which

$$\mathcal{E} = V_0 \sin \omega t,$$

in a circuit containing only a single resistor, capacitor, or inductor.

Resistive Circuit

We begin with the resistive circuit shown in Fig. 33–3a. The loop rule for the potential change around the circuit is

$$V_0 \sin \omega t - IR = 0. \tag{33-8}$$

The voltage across the resistor is then $V_R = IR = V_0 \sin(\omega t)$, and the current I through the resistor is

$$I = \frac{V_R}{R} = \frac{V_0 \sin(\omega t)}{R}. \tag{33-9}$$

The current through the resistor and the voltage across the resistor have the same sinusoidal time dependence (Fig. 33–3b). With peaks and valleys that occur at the same time, we say that the current and the voltage are *in phase*. The current amplitude is $I_{max} = V_0/R$.

Capacitive Circuit

We now place the AC emf across a pure capacitance in the circuit, so that the voltage across the capacitor is $V_C = V_0 \sin(\omega t)$ (Fig. 33–4a). To find the current in the circuit, we apply the loop rule:

$$V_0 \sin(\omega t) - \frac{Q}{C} = 0. \tag{33-10}$$

This expression gives us the charge Q on the capacitor,

$$Q = CV_0 \sin \omega t. \tag{33-11}$$

The current is then the time derivative of the charge:

$$I = \frac{dQ}{dt} = \omega C V_0 \cos(\omega t). \tag{33-12}$$

With the aid of the identity $\sin[\theta + (\pi/2)] = \cos \theta$, Eq. (33–12) becomes

$$I = \omega C V_0 \sin\left(\omega t + \frac{\pi}{2}\right). \tag{33-13}$$

The current amplitude in this circuit is

$$I_{max} = \omega C V_0. \tag{33-14}$$

We can compare this equation to a similar equation for the resistive circuit, for which the corresponding relation was $I_{max} = V_0/R$. The *effective resistance* for a capacitive circuit is called the **capacitive reactance**, X_C, defined by

$$X_C \equiv \frac{1}{\omega C}. \tag{33-15}$$

Equation (33–14) now takes the form

$$I_{max} = \frac{V_0}{X_C}. \tag{33-16}$$

The capacitive reactance has units of ohms.

For this circuit, the effective resistance to current flow *decreases* at higher frequencies. This is physically reasonable because a high frequency means that the voltage and current are changing rapidly, and for very rapid oscillations the current must be large enough to allow the charge on the capacitor to build rapidly and thereby oppose the driving emf. In other words, the current grows with frequency. On the other hand, a low frequency means an almost constant current, and no constant current can pass through the wires leading to a capacitor. Once enough charge has been deposited on the plates, there is a large enough potential difference to stop a *steady* current. To repeat this in another way, keep in mind that while current never actually flows across the capacitor, current can flow in the wires leading to and from the capacitor plates as the charge builds up or

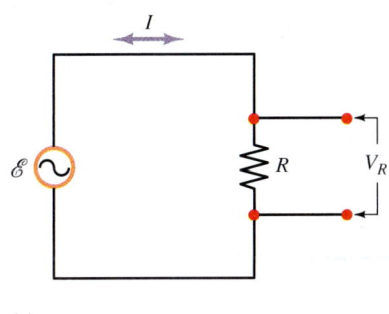

(a)

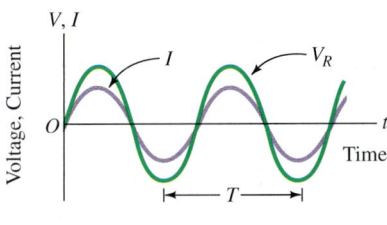

(b)

▲ **FIGURE 33–3** (a) A resistor connected in series with an AC source of emf. (b) The voltage across the resistor and the current through it are in phase. T is the period of the oscillation.

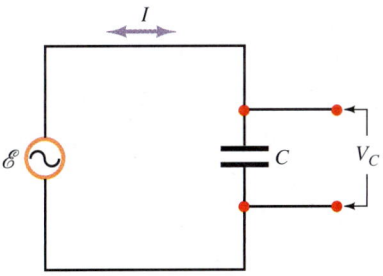

(a)

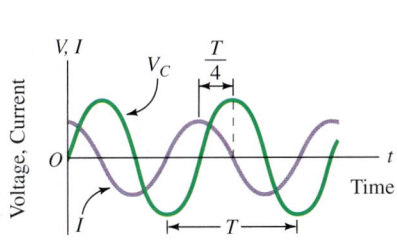

(b)

▲ **FIGURE 33–4** (a) A capacitor connected in series with an AC source of emf. (b) The current in the circuit leads the voltage across the capacitor by 90°.

EXAMPLE 33–3 The circuit shown in Fig. 33–4a has an emf given by $\mathcal{E} = V_0 \sin \omega t$, with $V_0 = 6.0$ V, and a capacitance $C = 1.0 \, \mu F$. (a) What are the peak currents for frequencies of 60.0 Hz and 6.00 MHz? (b) What are the currents I and voltages V_C at time 2.0 ms for the 60-Hz frequency?

Strategy (a) The unknown peak current is found from Eq. (33–14) and depends on the angular frequency. The angular frequencies ω for the two cases are found from $\omega = 2\pi f$. (b) Once we have the current amplitude, Eq. (33–13) gives us the full time dependence of the current, and this can be evaluated at any specific time, as can the given voltage, $V_0 \sin \omega t$.

Working It Out We begin with the evaluation of the angular frequencies, $\omega = 2\pi(60.0 \text{ Hz}) = 2\pi(60.0 \text{ s}^{-1}) = 377$ rad/s and $2\pi(6.00 \times 10^6 \text{ s}^{-1}) = 3.77 \times 10^7$ rad/s, respectively.

(a) From Eq. (33–14),

for 60 Hz: $I_{max} = (377 \text{ rad/s})(1.0 \times 10^{-6} \text{ F})(6 \text{ V}) = 2.3$ mA;

for 6 MHz: $I_{max} = (3.77 \times 10^7 \text{ rad/s})(1.0 \times 10^{-6} \text{ F})(6 \text{ V})$

$= 230$ A.

The maximum current is proportional to the frequency.

(b) We evaluate Eqs. (33–13) and the input voltage $V_0 \sin \omega t$ at 2.0 ms. For $f = 60$ Hz, we have $I_0 = 2.3$ mA, $\phi = \pi/2$ rad, $V_0 = 6.0$ V, and $\omega = 377$ rad/s. Thus

$$I(t = 2.0 \text{ s}) = (2.3 \text{ mA}) \sin\left[(377 \text{ rad/s})(0.0020 \text{ s}) + \frac{\pi}{2}\right]$$

$$= 1.7 \text{ mA},$$

and

$$V_C(t = 2.0 \text{ s}) = (6.0 \text{ V}) \sin[(377 \text{ rad/s})(0.0020 \text{ s})]$$

$$= 4.1 \text{ V}.$$

By $t = 2.0$ ms, the current is coming down from its peak toward zero, while the voltage is rising toward its peak. At $t = 2.0$ ms, they are both about 70 percent of their peak values.

What Do You Think? Why does the higher frequency make such a big difference in the maximum current?

Inductive Circuit

Repeating the previous procedure for the circuit in Fig. 33–5a, which contains a single inductor, we first note that the voltage drop across the inductor is the emf $V_L = V_0 \sin \omega t$. For the current, we apply the loop rule to the potentials around the circuit:

$$V_0 \sin \omega t - L\frac{dI}{dt} = 0. \tag{33–17}$$

This gives us dI/dt in terms of known quantities; to find I itself, we integrate over time:

$$I(t) = \int^t \frac{dI}{dt'} dt' = \frac{V_0}{L} \int^t \sin(\omega t') \, dt' = -\frac{V_0}{\omega L} \cos \omega t + \text{a constant.}$$

The constant must equal zero, because there is no constant emf to drive a constant current term. We now use the trigonometric identity $\cos \theta = -\sin[\theta - (\pi/2)]$ (Appendix IV–4) to rewrite our equation as

$$I = \frac{V_0}{\omega L} \sin\left(\omega t - \frac{\pi}{2}\right). \tag{33–18}$$

The maximum current through the inductor, or current amplitude, is

$$I_{max} = \frac{V_0}{\omega L}. \tag{33–19}$$

If we compare our expressions for I_{max} with the similar one from the purely resistive circuit, $I_{max} = V_0/R$, we see that the effective resistance for an inductive circuit is ωL. We call this the **inductive reactance**, defined by

$$X_L \equiv \omega L. \tag{33–20}$$

The inductive reactance has units of ohms.

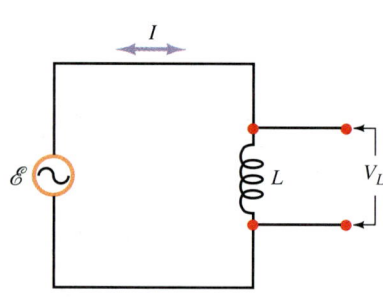

(a)

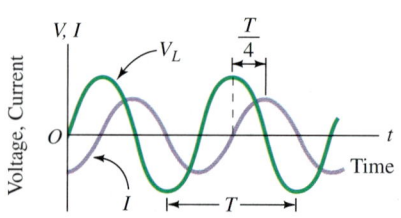

(b)

▲ **FIGURE 33–5** (a) An inductor connected in series with an AC source of emf. (b) The current in the circuit lags the voltage across the inductor by 90°.

For this circuit the effective resistance to current flow *increases* at higher frequencies. This is physically reasonable because inductors react to oppose any *change* in the current flow through them. The more rapidly the current is changing, the larger the opposing induced emf. On the other hand, the inductor is transparent to the unchanging current that corresponds to the limit $\omega = 0$.

We plot the current and voltage of the inductor versus time in Fig. 33–5b. As for the capacitive circuit, one sinusoidal curve is displaced from the other by a quarter cycle, although the role of the current and voltage curves is reversed in the two cases. This time the current in the circuit *lags* the voltage across the inductor.

EXAMPLE 33–4 Use the parameters of Example 33–3, but replace the capacitor with an inductor of inductance $L = 1.00$ mH. Calculate the inductive reactances.

Strategy We simply evaluate Eq. (33–20) to determine the inductive reactances. We have evaluated the angular frequencies in Example 33–3.

Working It Out The inductive reactances are

for 60.0 Hz:
$$X_L = \omega L = (377 \text{ rad/s})(1.00 \times 10^{-3} \text{ H}) = 0.377 \text{ }\Omega;$$

for 6.00 MHz:
$$X_L = \omega L = (3.77 \times 10^7 \text{ rad/s})(1.00 \times 10^{-3} \text{ H}) = 3.77 \times 10^4 \text{ }\Omega.$$

What Do You Think? A capacitor acts as an open switch when there is no time dependence in the current. Under what conditions does the inductor act as an open switch?

*Some Mathematical Devices

Two techniques simplify the treatment of circuits with time dependence. We treat them each here in abbreviated and optional form—in particular, we refer to time dependence with a single frequency. Further study in electrical engineering will involve one or the other of these techniques in more detail than we can supply here. The first involves **phasors**, which make it easier to follow phases by graphically summarizing phase and amplitude relations in AC circuits. The second, **complex analysis**, is a powerful tool that working engineers use a great deal. It simplifies all aspects of circuit analysis problems with time dependence.

Phasors: Any quantity in an AC circuit (or in any other problem) that has harmonic time dependence can be associated with a rotating vector known as a phasor. Thus for the function

$$f(t) = f_0 \sin(\omega t + \phi), \quad (33\text{–}21)$$

we define the phasor as follows: The phasor lies in the xy-plane with its tail fixed at the origin, and its length is the function amplitude, f_0. The time dependence is described by a counterclockwise rotation of the phasor with angular speed ω such that the function $f(t)$ itself is the instantaneous projection of the phasor on the y-axis (Fig. 33–6). For an alternating current, the length of the phasor represents the amplitude of the voltage or current, with its angular speed given by the angular frequency of the alternating current. For example, the function $V(t) = V_0 \sin \omega t$ has a phasor that starts at $t = 0$ aligned with the positive x-axis. As time increases and the phasor rotates counterclockwise, the y-component of the phasor increases until it reaches a maximum when $\omega t = \pi/2$. The y-component then decreases as the phasor moves into the second quadrant.

By comparing the phasors of different harmonic functions that appear in an AC problem, we learn about the relative phases of these functions; that is, which function leads or which function lags. To see how this works, let's apply it to the purely inductive AC circuit of Fig. 33–5. The input voltage—and hence the voltage V_L across the inductor—has the form $V_0 \sin \omega t$, whereas the current I_L takes the form $I_{\max} \sin(\omega t - \pi/2)$ according to Eq. (33–18). The phasors for these two quantities (Fig. 33–7a) rotate in the counterclockwise sense as time advances; the phasor for the voltage is always ahead of the phasor for the current. In this diagram, the idea that the voltage *leads* the current for the inductive circuit is easy to visualize and understand. We can easily perform the same exercise for the purely capacitive and purely resistive circuits, whose phasor diagrams are shown in Fig. 33–7b and 33–7c, respectively. Again, we see the phase relationship

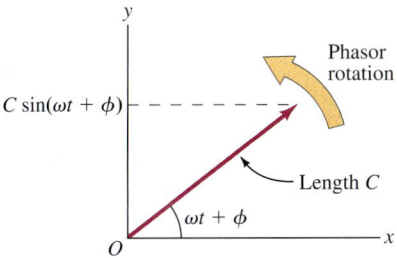

▲ **FIGURE 33–6** The projection of the phasor on the y-axis gives the value of the associated harmonic function.

924 | Alternating Currents

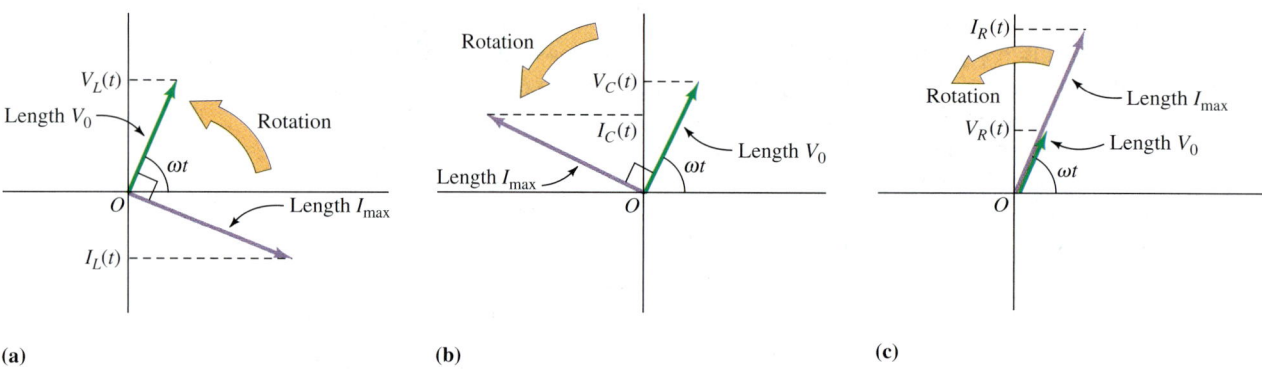

▲ **FIGURE 33–7** Phasors associated with the voltage across and current through (a) the inductor; (b) the capacitor; and (c) the resistor of an *RLC* circuit.

clearly in these phasor diagrams: The voltage across the capacitor in the purely capacitive AC circuit lags the current, and the corresponding current and voltage are in phase in the purely resistive circuit.

When elements are combined in certain AC circuits, the phasors combine as vectors, and this can simplify the analysis of these circuits. This is one way of treating the series *RLC* circuit to be discussed in Section 33–3. Here we content ourselves with a brief example.

EXAMPLE 33–5 Consider an AC circuit with a driving emf $V_0 \sin \omega t$, an inductor of inductance L, and a resistor of resistance R in series. Does the driving voltage lead or lag the current, and by what amount?

Setting It Up Our circuit diagram (Fig. 33–8a) is accompanied by a phasor diagram (Fig. 33–8b) that specifies the voltages across the elements. In accordance with the discussion above, the projection on the y-axis of the respective phasors $\vec{V}_L$, $\vec{V}_R$, and $\vec{I}$ represents the real values of the corresponding physical quantities. (Note that here the overarrow refers to the fact that the phasor is a rotating vector, not to any new "vector" properties of potential or current.) We also note here what is being asked: When we say that the driving voltage leads or lags the current, we want to know whether its phasor is ahead of or behind the current phasor in its counterclockwise rotation, and when we ask "by how much," we are asking for the value of the angle between these two phasors.

Strategy Figure 33–8b gives only the phasors $\vec{V}_L$, $\vec{V}_R$, and $\vec{I}$, with $\vec{V}_L$ leading the current by 90°. Since $\vec{V}_R$ and $\vec{I}$ are in phase (Ohm's law), these phasors are drawn together. To find the phasor $\vec{\mathcal{E}}$ for the driving emf, we note that at any given time its projection on the y-axis must match the sum of the projections of the phasors for the voltage across the inductor and the resistor. This is guaranteed for vectors if we take $\vec{\mathcal{E}} = \vec{V}_L + \vec{V}_R$. Thus graphical vector addition will give us the phasor for the driving emf; simple geometry will give us its angular location relative to the current.

Working It Out Figure 33–8c shows the required addition of the phasors. We see from the figure that the driving emf leads the current and that the angle between them is determined by

$$\tan \phi = \frac{V_L}{V_R} = \frac{IX_L}{IR} = \frac{X_L}{R}.$$

The angle ϕ is less than 90°, as we can see most easily from Fig. 33–8c.

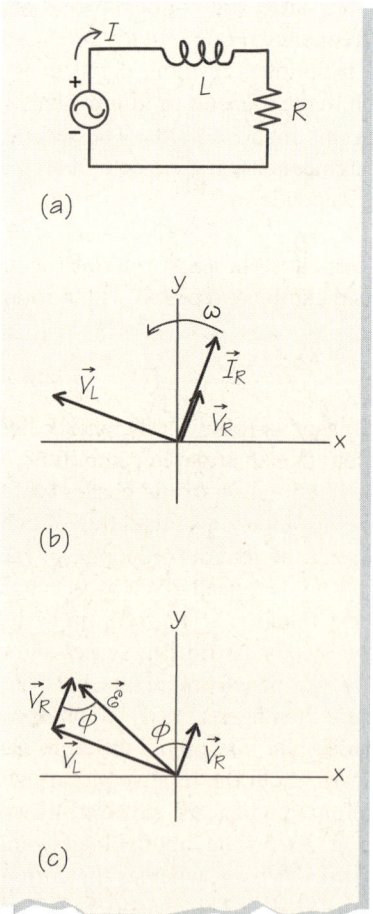

▲ **FIGURE 33–8** (a) A series *RL* circuit. (b) Phasor diagram of the current, voltage across the resistor, and voltage across the inductor. (c) The phasor for the driving emf is the sum of the phasors for voltage across the elements *R* and *L*.

Complex Analysis: Complex analysis is a powerful technique that is the working engineer's basic tool for the analysis of circuits with time dependence. Complex analysis depends on two ideas. The first is that a *complex variable* z can represent two physical quantities through the two variables x and y. This is accomplished through the use of $i \equiv \sqrt{-1}$; z is then given the form $z = x + iy$. x and y are extracted from z by taking the real and imaginary parts of z—just the coefficients of i^0 and i^1, respectively. We say that $x = \text{Re}(z)$ and $y = \text{Im}(z)$. (Higher powers of i, which may appear through algebraic manipulation, reduce to i^0 or i^1; for example, $i^2 = -1 = -i^0$.) In Fig. 33–9, we locate z on the xy-plane. You can see that if we think of z as a vector, then it is just the phasor that we spoke of above. But the general use of complex variables is more algebraic than pictorial.

The second idea of complex analysis is one of the most remarkable relations in mathematics:

$$e^{i\theta} = \cos\theta + i\sin\theta.$$

This result follows from the series expansions of the sine, the cosine, and the exponential. Using this identity, we can write an input voltage like that of Eq. (33–1) as

$$V_0 \sin\omega t = \text{Im}(V_0 e^{i\omega t}),$$

where $\text{Im}(z)$ is the imaginary part of z, that is, the coefficient of i in the expression for z. We then represent all oscillating functions—currents, or potentials across individual elements—as exponentials rather than as sines or cosines, including possible phases. At the end, we take the imaginary part of the complex quantity that we have calculated. Using the relation $e^{i\theta} = \cos\theta + i\sin\theta$, one additional representation of z is useful in the procedure, namely

$$z = x + iy = \rho e^{i\theta},$$

where $\rho = \sqrt{x^2 + y^2}$ and $\tan\theta = y/x$. We refer to ρ as the magnitude of z and to θ as the phase of z. Figure 33–9 labels these alternative variables and makes clear the relation between (x, y) and (ρ, θ).

What is the advantage of this procedure? The answer is simple: Upon differentiation, exponentials remain exponentials. Therefore, the differential equations that describe the circuit behavior contain overall powers of exponentials that ultimately cancel. Accordingly, these differential equations reduce easily to algebraic equations, which are simpler to deal with. Problems 91 to 95 illustrate complex analysis. We present a simple example here.

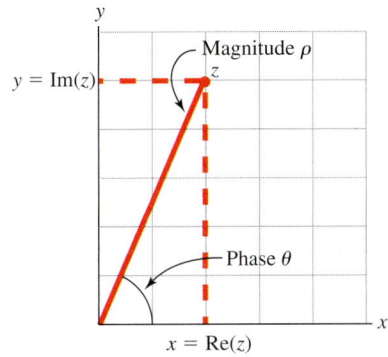

▲ **FIGURE 33–9** A complex number z is characterized either by its real and imaginary parts, x and y, respectively, or by a magnitude and phase, ρ and θ, respectively.

EXAMPLE 33–6 Consider the AC circuit treated in Example 33–5: a driving emf $\mathcal{E} = V_0 \sin\omega t$ and an inductor and resistor in series. Find the current amplitude in the circuit as a function of V_0, R, L, and ω.

Strategy Let us use the subscript "C" to indicate complex quantities, and employ the rule that any physical quantity A is found from the imaginary part of its complex counterpart A_C: $A = \text{Im}(A_C)$. We therefore write the driving emf as $\mathcal{E}_C = V_0 \exp(i\omega t)$, as well as $I_C = I_0 \exp(i\omega t)$ for the current. We will want to use these expressions in the loop rule. We should find that after derivatives are taken, the loop rule becomes a very simple algebraic expression for I_0, the desired current amplitude.

Working It Out Referring to Fig. 33–8a, the loop rule reads $\mathcal{E}_C - L(dI_C/dt) - I_C R = 0$. With the explicit complex forms for $\mathcal{E}_C$ and I_C above, and with $\dfrac{d}{dt} e^{i\omega t} = i\omega e^{i\omega t}$, we have

$$V_0 \exp(i\omega t) - i\omega L I_0 \exp(i\omega t) - I_0 \exp(i\omega t) = 0.$$

The factor $\exp(i\omega t)$ cancels out, and we are left with an algebraic equation for I_0 whose solution is

$$I_0 = \frac{V_0}{R + i\omega L}, \quad \text{or} \quad I_C = \frac{V_0}{R + i\omega L} e^{i\omega t}.$$

We are asked to find the amplitude of the current; that is, a real quantity, the *magnitude* of I_C. But what is this magnitude? For that we write the combination $R + i\omega L$ that appears in the denominator of our expression for I_C as a magnitude and a phase:

$$R + i\omega L = \sqrt{R^2 + (\omega L)^2} \exp(i\phi),$$

so that

$$I_C = \frac{V_0 \exp(i\omega t - i\phi)}{\sqrt{R^2 + (\omega L)^2}}.$$

The magnitude of this quantity is the coefficient of the exponential, and that is the required relation. We have not written the phase explicitly, because that is not called for in the question. But we could have, and it would have given us the phase that was found in Example 33–5. The square root that appears in our answer is called the impedance—it plays the role of a resistance and will be discussed in more detail below.

The main thing to note here is the simplicity of the algebraic steps.

33–3 AC in Series RLC Circuits

In mechanics, a *driven harmonic oscillator* is a device in which a harmonic external force acts on, for example, a mass fixed to a spring. If the driving force has been acting for some time, the mass will move with the *driving frequency*—the angular frequency ω of the force—even though the mass is attached to a spring and undergoes some damping. As we saw in Section 13–8, this system illustrates the important physical phenomenon of *resonance*, characterized by a large amplitude motion when the driving frequency is near the natural frequency[†] ω_0, the frequency with which the mass would move if there were no driving force.

In Chapter 32, we noted the similarities between the damped mass-spring system and series RLC circuits without a driving term. If we add an AC source of emf to a series RLC circuit, the analogy is extended to a damped mass-spring system with a harmonic driving force. Using this analogy, we can expect such a circuit to exhibit *resonant behavior*.

Let us treat our circuit—the RLC circuit with an AC source of emf—more explicitly (Fig. 33–10). Applying the loop rule for the potential changes around the circuit gives

$$V_0 \sin \omega t - L\frac{dI}{dt} - \frac{Q}{C} - IR = 0. \tag{33-22}$$

Because $I = dQ/dt$, we can reexpress this result in terms of the single variable Q, the charge on the capacitor:

$$V_0 \sin \omega t - L\frac{d^2Q}{dt^2} - R\frac{dQ}{dt} - \frac{Q}{C} = 0. \tag{33-23}$$

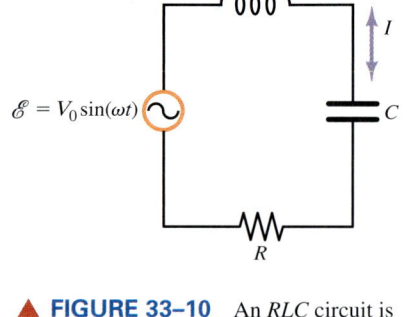

▲ **FIGURE 33–10** An RLC circuit is driven by an AC emf.

The unknown quantity in this differential equation is Q, and the solution will give us Q as a function of time. Once we solve for Q, differentiation with respect to time will give the current. We can then find the voltage drops across the various circuit elements.

The analogy to the mechanical system is evident in the comparison between Eq. (33–23) and Eq. (13–52), which expresses Newton's second law for a driven harmonic oscillator with damping (Fig. 33–11). The equation for the mechanical system is rewritten here in the form

$$F_0 \sin \omega t - m\frac{d^2x}{dt^2} - b\frac{dx}{dt} - kx = 0, \tag{33-24}$$

where the first term is the driving force, with amplitude F_0. The structure of Eqs. (33–24) and (33–23) are identical, and once we recall the solution for the mechanical system, we can immediately make the substitutions summarized in Table 33–1 to find the solution for the driven RLC circuit. Now, as we described in Section 13–8, Eq. (33–24) for the

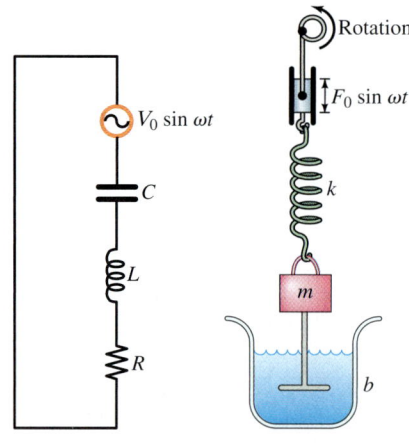

▲ **FIGURE 33–11** The RLC circuit and a damped mass-spring system are analogous. The analogies between the different elements of the two systems are understood in the vertical ordering of the elements.

TABLE 33–1 • Analogy Between Driven RLC Circuits and Driven Spring Motion

	Circuit	Mass–Spring
Variable	Charge Q	Position x
Coefficient of variable	$\dfrac{1}{C}$	k
Coefficient of $\dfrac{d(\text{variable})}{dt}$	R	b
Coefficient of $\dfrac{d^2(\text{variable})}{dt^2}$	L	m
Driving term	$V_0 \sin(\omega t)$	$F_0 \sin(\omega t)$
Natural frequency	$\dfrac{1}{\sqrt{LC}}$	$\sqrt{\dfrac{k}{m}}$

[†]When no confusion is possible, we use the term "frequency" rather than "angular frequency" for ω.

position $x(t)$ has a solution in which the position of the mass oscillates with the angular frequency of the driving force. This solution is given by Eq. (13–53); it is convenient to shift the phase of that solution and write it as

$$x = -A\cos(\omega t - \phi). \tag{33-25}$$

Not only is the frequency determined, but *the amplitude A and phase ϕ are also determined*. We find their values by direct substitution into the differential equation, with the results

$$A = \frac{F_0}{\sqrt{m^2(\omega^2 - \omega_0^2)^2 + b^2\omega^2}} \tag{33-26}$$

and

$$\tan\phi = \frac{1}{b}\left(\omega m - \frac{k}{\omega}\right). \tag{33-27}$$

Here, ω_0 is the natural frequency of the oscillator, given by $\omega_0 = \sqrt{k/m}$.

We also recall that for the mechanical oscillator, the force and the position are both harmonic with the same frequency but out of phase. For example, the function $\sin\omega t$—proportional to the force—rises from zero at $t = 0$, whereas the function $\cos(\omega t - \phi)$—proportional to the position—rises from zero when the argument of the cosine is $-90° = -\pi/2$ rad, that is, when $t = (\phi - \pi/2)/\omega$.

Now we can move on to the *RLC* circuit with an AC source of emf. The solution for the charge on the capacitor in the loop equation for this circuit [Eq. (33–23)] is found by making the formal substitutions of Table 33–1, which relates the parameters of the harmonic oscillator to the circuit parameters. The solution for the charge is

$$Q = -Q_{\max}\cos(\omega t - \phi) = -\frac{V_0}{\sqrt{L^2(\omega^2 - \omega_0^2)^2 + R^2\omega^2}}\cos(\omega t - \phi), \tag{33-28}$$

where $Q_{\max}$ is defined in comparing the second and third terms, and

$$\tan\phi = \frac{1}{R}\left(\omega L - \frac{1}{\omega C}\right). \tag{33-29}$$

The current is found from evaluation of dQ/dt:

$$I = \frac{dQ}{dt} = I_{\max}\sin(\omega t - \phi) = \frac{\omega V_0}{\sqrt{L^2(\omega^2 - \omega_0^2)^2 + R^2\omega^2}}\sin(\omega t - \phi). \tag{33-30}$$

Again $I_{\max}$ is directly defined here. Note that $I_{\max} = Q_{\max}\omega$.

As we would expect, all the results given here reduce to the cases we treated in Section 33–2, in which only one circuit element—the resistor, the capacitor, or the inductor—is present at a time. We need only replace the values of L, R, or $1/C$ by zero, as appropriate (keeping in mind that $\omega_0 = 1/\sqrt{LC}$).

When we dealt in Chapter 32 with *RLC* circuits without AC sources of emf, neither the amplitude nor the phase were determined; rather, they were a matter of initial conditions. What is different here? A clue is provided by the equation satisfied by the charge, Eq. (33–23). If the AC source is not present, then the equation is homogeneous in Q, meaning that if there is a solution for some charge, then a multiple of that charge is also a solution. The overall multiplicative constant would cancel from the equation. But once we add the AC emf, Eq. (33–23) is no longer homogeneous in Q, and the overall scale of Q is determined. The same reasoning applies to the phase, where without the AC source, any phase is possible. With the AC emf present, that emf "sets the clock" with its definite $\sin\omega t$ form that vanishes at $t = 0$, and the phase of the solution is no longer arbitrary. These remarks about amplitude and phase actually apply to long-time solutions—transient solutions do have properties set by initial conditions, but these solutions die away with time.

Impedance

We have already defined the reactances $X_L = \omega L$ and $X_C = 1/\omega C$ [Eqs. (33–15) and (33–20)]. They enter immediately into the phase [Eq. (33–29)] that appears in the driven RLC circuit according to

$$\tan \phi = \frac{1}{R}(X_L - X_C). \tag{33-31}$$

The reactances play the role of an effective resistance, dependent on frequency, for the single element circuits. As we shall see, the effective resistance of the *RLC* circuit is the **impedance** Z defined by

$$Z \equiv \sqrt{\left(\omega L - \frac{1}{\omega C}\right)^2 + R^2} = \sqrt{(X_L - X_C)^2 + R^2}. \tag{33-32}$$

The impedance has units of ohms.

In contrast to resistance, impedance depends on the frequency; we can understand this on physical grounds. Inductance opposes a change in current, and larger values of angular frequency mean more rapid changes in the current. However, an inductor is transparent to a static potential, corresponding to $\omega \to 0$. These properties are reflected in the frequency dependence of $X_L = \omega L$. A capacitor has just the opposite properties: No constant current can pass through a capacitor, but the capacitor has little effect when the current changes so rapidly that little charge can accumulate. These properties are reflected in the frequency dependence of $X_C = 1/\omega C$. The combination of the effects of both the inductance and the capacitance implies that the impedance is high in the limit of both large and small values of the frequency ω.

To see in more concrete terms how the impedance plays the role of a resistance, let's express the current in terms of it. It is a matter of a little algebra (see Problem 38) to show that, in terms of these quantities, the amplitude in Eq. (33–28) becomes $Q_{max} = V_0/\omega Z$ and hence, with $I_{max} = \omega Q_{max}$, we have $I_{max} = V_0/Z$. In other words,

$$I = I_{max}\sin(\omega t - \phi) = \frac{V_0 \sin(\omega t - \phi)}{Z}. \tag{33-33}$$

The current takes the form of an AC emf divided by the impedance. This equation is analogous to the DC equation $I = V/R$. *Impedance thus plays the role of resistance in an AC circuit.* Note that $I_{max} = V_0/Z$, but because of the phase, $I(t)$ is not generally equal to $V(t)/Z$.

EXAMPLE 33–7 The series *RLC* circuit in Fig. 33–10 is driven with an AC source of emf of the form $\mathcal{E} = V_0 \sin \omega t$, where V_0 is exactly 110 V and the frequency f is exactly 60 Hz. If $R = 20.0\ \Omega$, $L = 5.00 \times 10^{-2}$ H, and $C = 50.0\ \mu$F, find the potential drops across the inductor at $t = 0$ and at time t_1, the first time after $t = 0$ that $\mathcal{E}$ reaches a maximum.

Strategy We use $V_L = -L\, dI/dt$ to find the potential drop across the inductor, and utilize Eq. (33–33) for the current I in this circuit:

$$V_L = -L\frac{d}{dt}\left[\frac{V_0}{Z}\sin(\omega t - \phi)\right] = -\frac{LV_0\omega}{Z}\cos(\omega t - \phi).$$

This can easily be evaluated at $t = 0$. For the next part of the problem, we note that the first time after $t = 0$ that the source emf reaches a maximum is time t_1, determined by $\omega t_1 = \pi/2$, or

$$t_1 = \frac{\pi}{2\omega} = \frac{\pi}{4\pi f} = \frac{1}{4f}.$$

Again, V_L can be evaluated at this time.

Working It Out At $t = 0$, V_L takes the form

$$V_L = -\frac{LV_0\omega}{Z}\cos(-\phi) = -\frac{LV_0\omega}{Z}\cos\phi,$$

while at $t = t_1$ we have

$$V_L = -\frac{LV_0\omega}{Z}\cos\left(\frac{\omega}{4f} - \phi\right) = -\frac{LV_0\omega}{Z}\cos\left(\frac{\pi}{2} - \phi\right)$$

$$= -\frac{LV_0\omega}{Z}\sin\phi.$$

To evaluate these results numerically, note that $\omega = 2\pi(60\text{ Hz}) = 377$ rad/s. We insert the values of ω, R, L, and C to obtain $X_L = \omega L = 18.9\ \Omega$, $X_C = 1/\omega C = 53\ \Omega$, $Z = \sqrt{(X_L - X_C)^2 + R^2} = 39.6\ \Omega$, and $\tan \phi = -1.71$ [or $\phi = -1.04$ rad $(-59.7°)$]. Thus at $t = 0$,

$$V_L = -\frac{LV_0\omega}{Z}\cos\phi = -26.4\text{ V}.$$

At $t = t_1$,

$$V_L = -\frac{LV_0\omega}{Z}\sin\phi = 45.2\text{ V}.$$

What Do You Think? At what time after t_1 will the voltage across the inductor first reach a negative maximum?

Resonance in Driven *RLC* Circuits

The current amplitude contained in Eq. (33–30) exhibits the phenomenon of resonance—as do the amplitude of the charge on a capacitor and the voltage amplitudes across any of the elements of a driven *RLC* circuit. The amplitudes are inversely proportional to the impedance. When Z is a minimum, these amplitudes reach a maximum—we say they are *peaked*, or *resonate*. This occurs when the driving frequency ω is at the undamped natural frequency ω_0, which as we know from Section 32–5 is given by $\omega_0^2 = 1/LC$ (Fig. 33–12). (This can be seen by setting the derivative of $I_{\max}$ with respect to ω equal to zero.) The amount and sharpness of the peaking depends on the damping factor, which is the resistance R. Without R, the amplitudes would be infinite when the driving frequency ω equals ω_0, something you can easily see from the form of the impedance, Eq. (33–32). However, there is always some resistance in real circuits. Figure 13–25 illustrates the resonance peak for several values of the damping parameter b of a mechanical oscillator. Similar plots apply to *RLC* circuits. The smaller the damping—the smaller the resistance—the larger and sharper the resonance peaks. The maximum in this peak occurs when $X_L = X_C$,

$$\omega L = \frac{1}{\omega C}.$$

This condition implies that

$$\omega^2 = \omega_0^2 = \frac{1}{LC}. \quad (33\text{–}34)$$

At this value of the angular frequency, $Z = R$. The current amplitude at this value of the driving frequency is given simply by

$$\text{at resonance: } I_{\max} = \frac{V_0}{R}. \quad (33\text{–}35)$$

At resonance both *the maximum value of current amplitude and the sharpness of the resonance peak* increase as R decreases. It is useful to note one last feature: At resonance, the tangent of the phase [Eq. (33–29)], and hence the phase itself, is zero.

Resonance phenomena have many applications in circuits. The most familiar concern radio and television reception (Fig. 33–13). A receiver can be tuned by changing either an inductance or a capacitance, thus changing the resonant frequency. The receiver then preferentially picks up broadcast signals at that frequency.

33–4 Power in AC Circuits

Energy flows in an *RLC* circuit. The capacitor and inductor store this energy for some parts of the cycle and release it for others, but no energy is lost (dissipated) in these elements. Energy is, however, lost in the resistance of the circuit. The power dissipated in the circuit is accordingly

$$P = I^2 R = \frac{[V_0 \sin(\omega t - \phi)]^2}{Z^2} R. \quad (33\text{–}36)$$

This power is always positive, oscillating between zero and a maximum of $V_0^2 R/Z^2$. For engineering purposes, it is often sufficient to know the *average* power dissipated over time. Let's indicate time-averaged quantities with angle brackets, $\langle \; \rangle$. The average of a sine (or cosine) squared over one cycle is one-half:

$$\langle \sin^2(\omega t - \phi) \rangle = \frac{1}{2},$$

so

$$\langle P \rangle = \frac{V_0^2 R}{Z^2} \langle \sin^2(\omega t - \phi) \rangle = \frac{1}{2} \frac{V_0^2 R}{Z^2}. \quad (33\text{–}37)$$

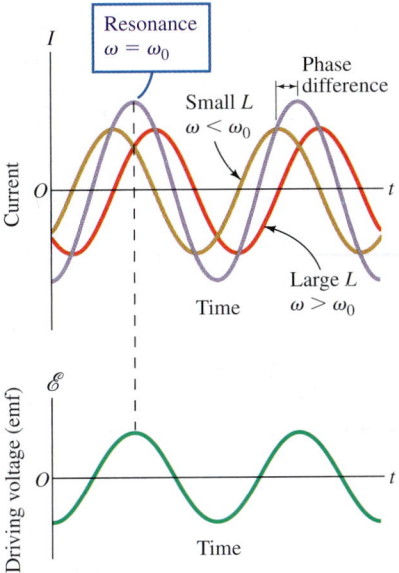

▲ **FIGURE 33–12** Given the driving voltage shown, the current through an *RLC* circuit varies harmonically with the frequency of the driving voltage but differs from it in amplitude and phase in a way that depends on whether or not the driving frequency ω equals the undamped natural frequency ω_0. At resonance ($\omega = \omega_0$), the current amplitude is a maximum and the phase is that of the driving voltage. In the plot of current versus time, we have controlled the natural frequency by varying L.

(a)

(b)

▲ **FIGURE 33–13** Receiver tuners can be constructed by (a) rotating the overlapping areas of the plates to change the capacitance or by (b) moving an iron core in and out of a solenoid to vary the inductance.

If we substitute Eq. (33–32) for the impedance, we find an explicit form for the average power:

$$\langle P \rangle = \frac{1}{2} \frac{V_0^2 R}{[\omega L - (1/\omega C)]^2 + R^2} = \frac{1}{2} \frac{V_0^2 R \omega^2}{L^2(\omega^2 - \omega_0^2)^2 + \omega^2 R^2}. \quad (33\text{–}38)$$

The resonant behavior of AC circuits is evident in this result for the power dissipated—the denominator is smallest, and so the average dissipated power is largest, when the driving angular frequency ω is near the "natural" angular frequency $\omega_0 = \sqrt{1/LC}$. As the driving frequency ω increases through ω_0, the power dissipated has the typical peaked behavior of resonance. (We'll discuss this in more detail below around Fig. 33–14.) At resonance, the expression for power simplifies to

$$\langle P \rangle_{\text{res}} = \frac{1}{2} \frac{V_0^2}{R}. \quad (33\text{–}39)$$

The power displays the same resonant behavior as does the current.

Equation (33–33) shows that the current oscillates with time. Just as it is useful to characterize the power with average values, we can also work with an *rms (root mean square)* value for the current (and other harmonically varying quantities in AC). The rms value, x_{rms}, of any quantity x is defined as the square root of the time average of the square of that quantity:

$$x_{\text{rms}} \equiv \sqrt{\langle x^2 \rangle}.$$

In particular, if x varies harmonically—if $x = x_0 \cos(\omega t - \phi)$—we can use the fact that the time average of the cosine squared is one-half to show that

$$x_{\text{rms}} = \frac{x_0}{\sqrt{2}}. \quad (33\text{–}40)$$

When we apply this concept to the AC current, we see from Eq. (33–33) that

$$I_{\text{rms}} = \frac{V_0}{(\sqrt{2})Z} = \sqrt{\frac{V_0^2 \omega^2/2}{L^2(\omega^2 - \omega_0^2)^2 + \omega^2 R^2}}. \quad (33\text{–}41)$$

From Eq. (33–38), we can see that I_{rms} and the average dissipated power, $\langle P \rangle$, obey the same power–current relation that DC quantities obey, namely

$$\langle P \rangle = I_{\text{rms}}^2 R. \quad (33\text{–}42)$$

Figure 33–14 plots I_{rms}^2 as a function of the driving angular frequency ω for three values of resistance. The sharpness in the peak of the average power (or of I_{rms}^2) versus ω is characterized by the *width* of the peak or, more precisely, the *total width at half-maximum* $\Delta\omega$. This is commonly called the **bandwidth** in the context of AC. To calculate the bandwidth, we find the angular frequencies at which the power drops to half the peak value and take the difference between these angular frequencies. This calculation shows that, for small values of resistance, the bandwidth is given by

$$\Delta\omega = \frac{R}{L}. \quad (33\text{–}43)$$

The smaller the resistance and the larger the inductance, the smaller the bandwidth. We can understand the importance of a small bandwidth by thinking about a radio or television receiver whose tuning circuit depends on the resonance phenomenon. If the resonance is sharp, the receiver will more effectively pick out only the desired frequency over others nearby (Fig. 33–14a). Conversely, if the resonance is broad, the circuit will respond to frequencies in the AC signal far from the desired frequency (Fig. 33–14b, c).

Another measure of the sharpness is the ratio $\omega_0/\Delta\omega$. This quantity is defined as the *quality factor*, or *Q-factor*,

$$Q \equiv \frac{\omega_0}{\Delta\omega} = \frac{\omega_0 L}{R}. \quad (33\text{–}44)$$

This factor is often used by electrical engineers to represent the sharpness of a resonant circuit.

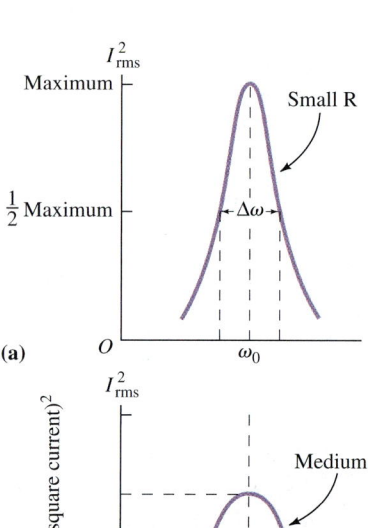

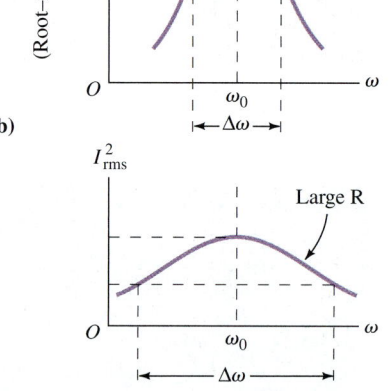

▲ **FIGURE 33–14** The rms current squared in a series *RLC* circuit with an AC source of emf such as the circuit of Fig. 33–10. (a) There is a resonance phenomenon when the driving frequency ω matches the natural frequency ω_0 of the circuit. The bandwidth $\Delta\omega$ measures the width of the rms current squared. (b, c) As R increases, the distribution of the current as a function of ω broadens.

EXAMPLE 33–8 Two FM radio stations broadcast at the same strength from the same nearby distance, one at a frequency of 91.3 MHz and the other at 91.1 MHz. You prefer the former broadcast over the latter, and you want to construct a simple series *RLC* circuit to act as a receiver that is unique to your favorite station. Given an inductor with an inductance L of exactly $1\,\mu\text{H}$ and adjustable resistance and capacitance, what values should you choose for R and C in order to limit the power received from the unwanted station to 1 percent of the power received from the desired station?

Strategy There are two requirements for the R and C values in our circuit. First, it must be resonant at $\omega_0 = 2\pi f = 2\pi(91.3\text{ MHz}) = 5.74 \times 10^8\text{ s}^{-1}$, and second, the resonant peak must be sharp enough to limit the power from the station broadcasting at $\omega_1 = 2\pi(91.1\text{ MHz}) = 5.72 \times 10^8\text{ s}^{-1}$. The resonant frequency is determined from L and C alone; since the value of L is known, we use this to find C. The sharpness requirement determines R. The power delivered by the signal at resonance is given by Eq. (33–39), whereas the power delivered off resonance is given by Eq. (33–38).

Working It Out We use $\omega_0^2 = 1/LC$ to find C:

$$C = \frac{1}{\omega_0^2 L} = \frac{1}{(5.74 \times 10^8\text{ s}^{-1})^2 (1\,\mu\text{H})} = 3.04 \times 10^{-12}\text{ F}.$$

For R, we look at the power delivered by the two stations. With signals arriving with the same strength, it is appropriate to use the same value of V_0. Thus

$$\frac{\langle P \rangle_{\omega_1}}{\langle P \rangle_{\text{res}}} = 0.01 = \left[\frac{1}{2}\left(\frac{V_0^2}{R}\right)\right]^{-1} \frac{1}{2} \frac{V_0^2 R \omega_1^2}{L^2(\omega_1^2 - \omega_0^2)^2 + \omega_1^2 R^2}$$

$$= \frac{R^2 \omega_1^2}{L^2(\omega_1^2 - \omega_0^2)^2 + \omega_1^2 R^2}.$$

This gives

$$[L^2(\omega_1^2 - \omega_0^2)^2 + \omega_1^2 R^2](0.01) = R^2 \omega_1^2.$$

To a good approximation, we can ignore $0.01\omega_1^2 R^2$ on the left compared to $\omega_1^2 R^2$ on the right. We then take the square root of both sides:

$$L(\omega_0^2 - \omega_1^2)(0.1) = L(\omega_0 - \omega_1)(\omega_0 + \omega_1)(0.1) = R\omega_1.$$

The factor $(\omega_0 + \omega_1)$ is, to a good approximation, equal to $2\omega_1$, so we have

$$2\omega_1 L(\omega_0 - \omega_1)(0.1) \cong R\omega_1.$$

Thus, canceling the factor ω_1,

$$R = 2L(\omega_0 - \omega_1)(0.1)$$
$$= 2(10^{-6}\text{ H})[(5.74 \times 10^8\text{ s}^{-1}) - (5.72 \times 10^8\text{ s}^{-1})](0.1)$$
$$= 0.4\,\Omega.$$

You can check with this value of R that our approximations were justified.

What Do You Think? Suppose instead of a factor 0.01 (1%), you wanted a "discrimination requirement" 100 times better, i.e. the power from the unwanted station is 0.0001 of that of the desired station. What value would you now choose for R?

The Power Factor

The power in AC circuits is commonly given in a form other than that given in Eq. (33–38). We find this form with the aid of the trigonometric identity

$$\cos^2 \phi = \frac{1}{\tan^2 \phi + 1}.$$

If we now use Eq. (33–31) for $\tan \phi$, we find that

$$\cos^2 \phi = \frac{1}{[(1/R)(X_L - X_C)]^2 + 1} = \frac{R^2}{(X_L - X_C)^2 + R^2} = \frac{R^2}{Z^2};$$

$$\cos \phi = \frac{R}{Z}. \tag{33–45}$$

Then, using Eqs. (33–41) and (33–43), we see that Eq. (33–42) becomes

$$\langle P \rangle = I_{\text{rms}}^2 R = I_{\text{rms}}^2 Z \cos \phi. \tag{33–46}$$

The term $\cos \phi$ in Eq. (33–46) is called the *power factor*. For a circuit without resistance, it is zero, whereas for a pure resistance it is a maximum, with a value of one.

33–5 Some Applications

Most electronic circuits in use today involve elements beyond those we have studied here. For example, transistors perform amplifying functions, while diodes have resistance that depends on the direction of current flow. Modern circuits are typically constructed in integrated form with many thousands of elements included together from the start, and perform many complicated functions. Nevertheless, we can understand a good deal more with a small addition to the elements we have in place.

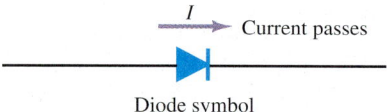

▲ **FIGURE 33–15** The symbol for a diode in a circuit diagram. Current can flow only in the direction shown.

Diodes and Rectifiers

While many sources of electric power supply AC voltage, many applications of that power require DC voltage. For example, rechargeable batteries require a DC current for charging but are charged with a device that plugs into household AC. A process called *rectification* transforms AC voltage to DC voltage using a diode. A **diode** is a semiconductor device with a high resistance to current that flows in one direction, but a low resistance to current that flows in the other direction—the direction of the arrow in the diode symbol (Fig. 33–15). In effect, the diode allows current flow only in one direction.

The diode can be used to construct a **rectifier**, a circuit element that changes AC into DC. Let's consider the circuit shown in Fig. 33–16a. The voltage across the load resistor can be negative or positive. When a diode is placed in the circuit, however, the negative voltages are blocked, leaving only positive voltages across the load resistor (Fig. 33–16b). Such a circuit is called a *half-wave rectifier*. This circuit can suffice as an approximation to a source of DC voltage although the voltage between points a and b, V_{ab}, is neither smooth nor constant.

We can improve the situation considerably with the circuit shown in Fig. 33–16c, called a *full-wave rectifier*. The diodes in this circuit are arranged such that *the voltage V_{ab} is always positive* even though the input voltage oscillates from positive to negative. When the emf produces positive voltage, positive current flows clockwise and passes through the path $cabd$ in the direction of the rectifier arrows. The voltage V_{ab} is positive. When the emf produces negative voltage, positive current flows counterclockwise and the path of the current is $dabc$. In this case also, the voltage V_{ab} is positive. Note that now the voltage V_{ab} is positive for all half-cycles, and the rms voltage is higher than it is for the half-wave rectifier. The use of *filters* (see below) allows the voltage peaks to be smoothed, producing a voltage that is much closer to being constant.

THINK ABOUT THIS...

HOW DOES A CAR PRODUCE THE ELECTRICITY IT NEEDS?

Although a car's battery is essential for starting, once the automobile engine is running, the engine itself produces the emf necessary to keep the engine running, to power the various "extras" such as lights or radio, and to recharge the battery. The mechanism involves an engine-driven belt that leads to the automobile's *alternator*. The belt runs over a pulley on whose shaft there is a wire coil rotating inside a magnetic field. We learned in Section 30–5 that such a device produces alternating current. But because AC current is not always suitable for the car's electrical system, and certainly not suitable for charging the battery, a rectifier circuit similar in principle to that in Fig. 33–16c is present to change the AC current to DC current.

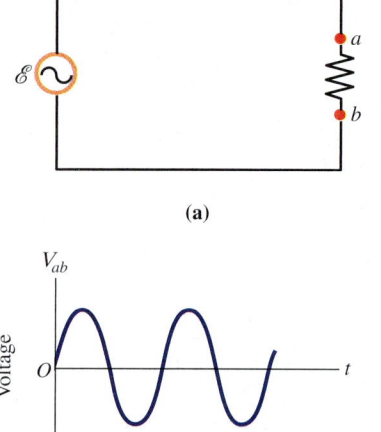
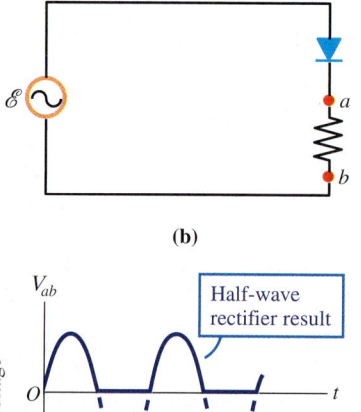
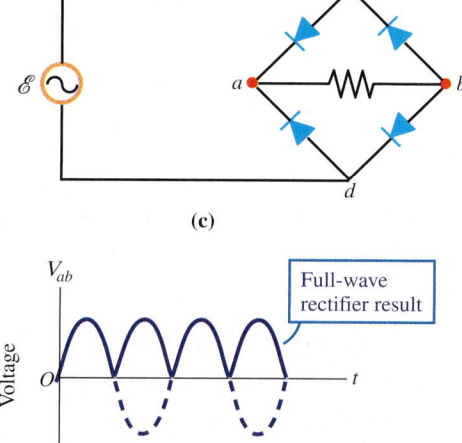

▲ **FIGURE 33–16** (a) The AC voltage across a resistor. (b) A half-wave rectifier is applied. (c) A full-wave rectifier is applied.

Filters

A **filter** is a device that takes an input signal from one part of a circuit that may be a mixture of AC and DC and passes only the AC or only the DC signal to a different part of the circuit. More generally, a filter will change the balance of high and low frequency components in a signal that may contain many different frequencies. Our discussion of the capacitive and inductive reactances shows that either a capacitor or an inductor can act as such a filter. Consider Fig. 33–17, in which the current I from the input is a mixture of DC and AC,

$$I = I_0 + I_1 \sin \omega t.$$

Here I_0 and I_1 are constants. A constant current cannot pass across a capacitor, whereas the impedance of the capacitor goes to zero if ω becomes large. Thus for the capacitor in Fig. 33–17a, only the AC part of the current passes to the other side. For the capacitor in Fig. 33–17b, AC passes through the capacitor to ground, and the DC passes through to the output side of the circuit. An inductor works in just the opposite way: DC passes through without impedance ($X_L \to 0$ as $\omega \to 0$, and $\omega = 0$ corresponds to DC), whereas the impedance is large for AC with large ω. Thus for the inductor in Fig. 33–17c, DC passes through to the output; for the inductor in Fig. 33–17d, AC passes through to the output and DC passes through the inductor to ground.

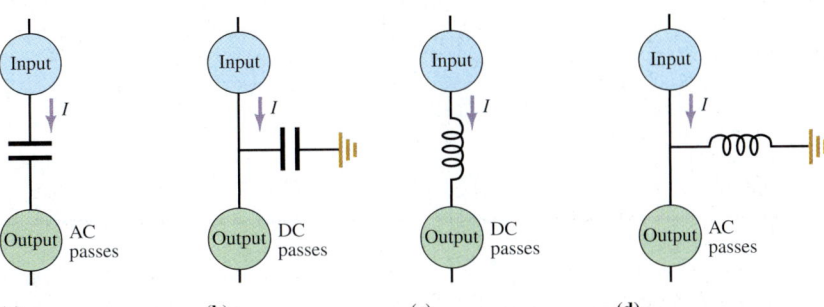

FIGURE 33–17 AC and DC filters formed from capacitors or inductors. With a capacitor: (a) AC current passes through to the output side. (b) DC current passes to the output side. With an inductor: (c) DC passes through to the output side. (d) AC passes through to the output side.

EXAMPLE 33–9 The generator in the circuit shown in Fig. 33–18 produces a combination of DC and AC in the form $\mathcal{E} = V_0 + V_1 \sin \omega t$, where $V_0 = 0.10$ V, and $V_1 = 0.25$ V. The values of the capacitance and resistance in the circuit have the values $C = 1.0$ μF and $R = 0.20$ Ω. What is the value of ω for which the voltage amplitude across the resistor is 50 percent of the value of the maximum voltage of the generator?

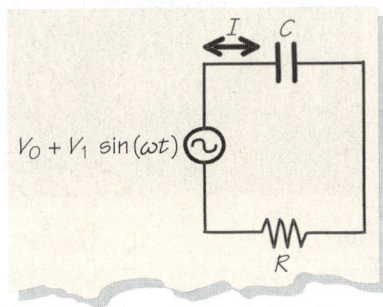

▲ **FIGURE 33–18**

Strategy We can apply the superposition principle by applying the loop rule that corresponds to the DC and AC terms separately, and then add the voltage drops. For the AC term, the current in the circuit is, from Eq. (33–33),

$$I_{AC} = \frac{V_1 \sin(\omega t - \phi)}{Z} = \frac{V_1}{\sqrt{(1/\omega C)^2 + R^2}} \sin(\omega t - \phi).$$

The voltage drop across the resistor from the AC is $I_{AC}R$ and therefore has the amplitude

$$\frac{V_1 R}{\sqrt{(1/\omega C)^2 + R^2}}.$$

For the DC term the capacitor acts as a perfect filter for the constant term in the input voltage because no constant current can pass. There is thus *no* voltage drop across the resistor associated with the V_0 term.

The maximum value of the input voltage is $V_0 + V_1$. We calculate the ratio of the voltage amplitude across the resistor to the maximum input voltage and set it equal to $\frac{1}{2}$. This equation determines ω.

Working It Out The equation for the ratio of potentials described above is

$$\frac{1}{V_0 + V_1} \frac{V_1 R}{\sqrt{(1/\omega C)^2 + R^2}} = 0.5.$$

We solve for ω:

$$\omega = \frac{1}{RC\sqrt{\dfrac{V_1^2}{(V_0 + V_1)^2 (0.50)^2} - 1}}$$

$$= \frac{1}{(0.20\ \Omega)(1.0 \times 10^{-6}\ \text{F})\sqrt{\dfrac{(0.25\ \text{V})^2}{(0.10\ \text{V} + 0.25\ \text{V})^2} \dfrac{1}{(0.50)^2} - 1}}$$

$$= 4.9 \times 10^6\ \text{s}^{-1}.$$

What Do You Think? At very high frequencies what becomes of the ratio of the voltage amplitude across the resistor to the maximum input voltage?

CONCEPTUAL EXAMPLE 33–10 Consider the RC circuit shown in Fig. 33–19a. The diode allows only the positive voltage to be passed through the circuit. Assume the values of R and C produce an RC time constant that is much longer than the period of the emf source voltage. What does the output voltage V_output across the resistor look like?

Answer During the rising voltage period of the source emf ($V = V_\text{max} \sin \omega t$), the capacitor will be charged to a full voltage V_max. As soon as the source voltage passes its peak and begins to decrease, the fully charged capacitor will start to discharge. Because the time constant is much greater than the period of the source voltage, the resulting voltage across the resistor will hardly decrease from its peak voltage V_max due to the capacitor discharging before the source voltage comes back to a peak again and recharges the capacitor. In effect, this circuit has rectified the AC voltage to one of full DC. We sketch the result in Fig. 33–19b. Application of a filter to the rectified voltage shown in Fig. 33–16b, for example, will further smooth out the peaks and valleys in the curve of voltage versus time.

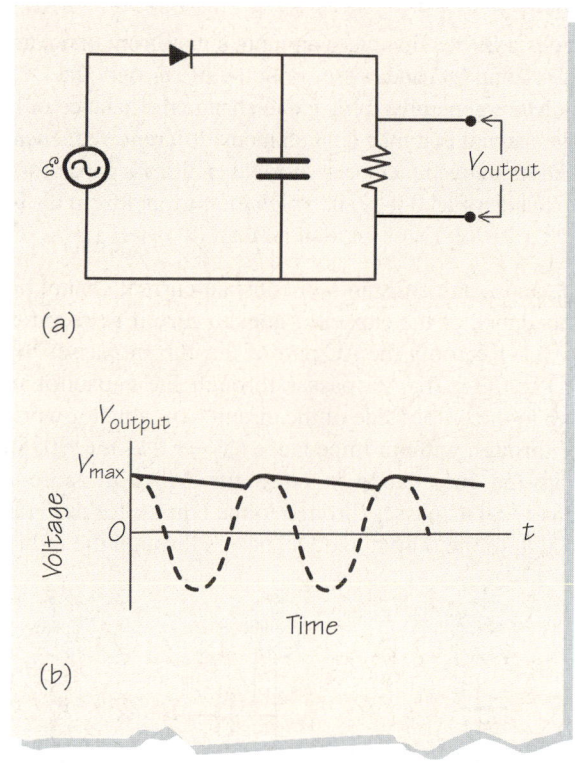

▶ **FIGURE 33–19** (a) An RC circuit acts as a filter for rectified AC voltage. Such a filter can produce a voltage that is nearly DC. (b) The slow decrease of the nearly constant-voltage segments is governed by the time constant of the RC circuit.

Figure 33–20 shows how filters can modify a signal that is a mix of many different frequencies. When you change the treble and bass balance on a music system you are applying filters. We can also add that when you sequentially apply a high- and low-pass filter, you are making a band-pass filter, one that allows a given band of frequencies to pass.

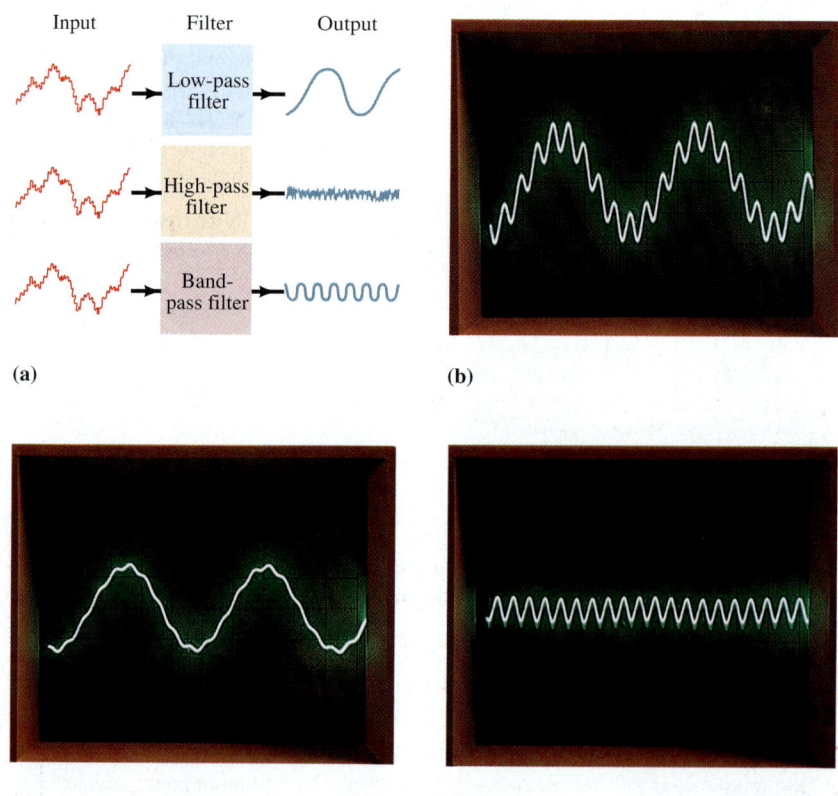

▶ **FIGURE 33–20** (a) Filters allow the low frequencies contained in an input signal to pass. (b) An input voltage containing both low and high frequencies. (c) The input voltage in part (b) has been sent through a low-pass filter, which eliminates the high frequencies. (d) The input voltage in part (b) has been sent through a high-pass filter, which eliminates the low frequencies.

*Impedance Matching

Another aspect of AC of practical importance concerns **impedance matching**, which refers, as in our discussion of filters, to the *connection* between different parts of a circuit. Figure 33–21a shows such a situation, in which some combination of circuit elements makes up circuit 1, connected at points a and b to circuit 2. The two circuits have impedances Z_1 and Z_2, respectively. We are not concerned here with the origin of currents in these circuits as much as we are with our ability to deliver power from circuit 1 to circuit 2. We therefore assume that the origin of these currents is within circuit 1 and break that circuit down as in Fig. 33–21b. The primary question is, if Z_1 is fixed, what are the requirements for Z_2 so that the power delivered to circuit 2 is a maximum? If, for example, a stereo amplifier is connected to a loudspeaker, what should the loudspeaker's impedance be in order that maximum power is delivered to it?

The answer is found by computing the average power $\langle P \rangle$ to circuit 2, which, from Eq. (33–42), is $I_{rms}^2 R_2$. The current in the loop of Fig. 33–21b is given by

$$I_{rms} = \frac{\mathscr{E}_{rms}}{Z_{total}}. \tag{33-47}$$

Here, $\mathscr{E}_{rms}$ is the rms value of the generator, whose maximum voltage, or amplitude, is V_0. If the generator produces a sinusoidal emf of the form of Eq. (33–1), then Eq. (33–40) shows that $\mathscr{E}_{rms} = V_0/\sqrt{2}$. The total impedance Z_{total} is found by separately adding the capacitive reactances, inductive reactances, and resistances, a result that follows from our knowledge of how series combinations of C, L, and R add (see Problem 44):

$$Z_{total} = \sqrt{[(X_{L_1} + X_{L_2}) - (X_{C_1} + X_{C_2})]^2 + (R_1 + R_2)^2}. \tag{33-48}$$

Thus the average power delivered to circuit 2 is

$$\langle P \rangle = \frac{\mathscr{E}_{rms}^2 R_2}{Z_{total}^2} = \frac{\mathscr{E}_{rms}^2 R_2}{[(X_{L_1} + X_{L_2}) - (X_{C_1} + X_{C_2})]^2 + (R_1 + R_2)^2}. \tag{33-49}$$

That there is a value of the parameters of Z_2 that maximizes this power is clear: If Z_2 is too small, the factor R_2 will also be small and $\langle P \rangle$ will be small; if Z_2 is too large, its parameters will dominate the denominator of Eq. (33–49), and $\langle P \rangle$ will again be small. An intermediate value of the parameters of Z_2 will give a maximum value of $\langle P \rangle$. Two independent parameters are involved here: the resistance R_2 and the total reactance term for circuit 2, $X_{L_2} - X_{C_2}$. Formally, we find the value of the parameters that maximize $\langle P \rangle$ by taking the derivative of $\langle P \rangle$ with respect to these quantities and setting it equal to zero. From this exercise the power is maximized when

$$R_2 = R_1 \quad \text{and} \quad X_{L_2} - X_{C_2} = -(X_{L_1} - X_{C_1}). \tag{33-50}$$

The second condition—that the reactance term of Z_2 is equal but opposite to that of Z_1—follows because it means that the reactance terms in the denominator of Eq. (33–49) cancel, thus maximizing $\langle P \rangle$ whatever the value of the resistances. The first condition—that the resistances be equal—is perhaps less intuitive but nevertheless follows directly from the requirement that the derivative of $\langle P \rangle$ is zero (see Problem 72). When the conditions of Eq. (33–50) are met, the impedances are said to be matched.

Impedance matching is desirable when you wish to deliver maximum power to one part of a circuit. It is worth noting that we do not always wish to deliver maximum power. An AC voltmeter, for example, should have an impedance *mismatch* because we want it to draw as little current as possible.

The subject of circuit analysis is highly developed. We have been able to do no more than describe its principles, and this chapter will not have taught you to fix, much less design, TVs or computers. But the principles we have described here apply to all AC circuits.

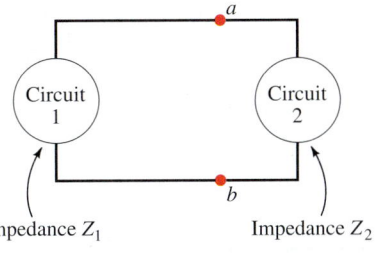

(a)

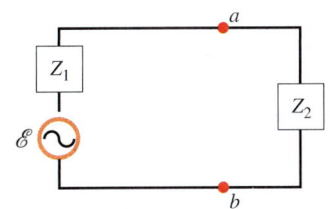

(b)

▲ **FIGURE 33–21** (a) Circuit diagram to illustrate impedance matching. (b) Circuit 1 of part (a) is broken down into a source of emf $\mathscr{E}$ and an impedance Z_1. Circuit 2 is assumed to include only an impedance Z_2.

Summary

The presence of AC sources of emf in circuits with resistors, inductors, and capacitors introduces a variety of new possibilities of phenomena and applications. Transformers allow us to vary the voltage amplitude of AC emfs, of importance in power transmission on large and small scales.

The relation between the voltages and the numbers of turns of the primary and secondary coils of a transformer is

$$\frac{V_2}{V_1} = \frac{N_2}{N_1}. \tag{33-4}$$

Conservation of energy implies that the currents carried by the respective coils are related inversely:

$$\frac{I_1}{I_2} = \frac{N_2}{N_1}. \tag{33-6}$$

A series *RLC* circuit with an AC source of emf of frequency ω behaves like a damped harmonic oscillator driven by a harmonically varying force. Solutions for currents, voltages, and charges in such circuits can be found by using the solutions already developed for the driven harmonic oscillator. For such circuits the impedance Z is a quantity that plays the role of a resistance. The impedance is frequency dependent:

$$Z \equiv \sqrt{\left(\omega L - \frac{1}{\omega C}\right)^2 + R^2} = \sqrt{(X_L - X_C)^2 + R^2}. \tag{33-32}$$

where X_C is the capacitive reactance and X_L is the inductive reactance. The current in the driven circuit is then

$$I = I_{\max} \sin(\omega t - \phi) = \frac{V_0 \sin(\omega t - \phi)}{Z}, \tag{33-33}$$

where $V_0 \sin \omega t$ is the driving emf. The phase ϕ is given by

$$\tan \phi = \frac{1}{R}\left(\omega L - \frac{1}{\omega C}\right). \tag{33-29}$$

Such circuits exhibit resonant behavior when the driving frequency is near the natural frequency $\omega_0 = \sqrt{1/LC}$. This type of behavior is most clearly seen in the power dissipated in driven *RLC* circuits. Averaged over time, the power lost is

$$\langle P \rangle = I_{\rm rms}^2 R = I_{\rm rms}^2 Z \cos \phi, \tag{33-46}$$

where $I_{\rm rms} Z = V_0/\sqrt{2}$ gives the rms current. Near resonance, the power dissipated is a maximum, and the width of the peak of average power versus driving frequency has a width at half-maximum of

$$\Delta \omega = \frac{R}{L}, \tag{33-43}$$

a result that holds as long as the resistance is not too large.

If we add diodes—devices that allow current to pass in one direction only—to our arsenal of circuit elements, we can construct a variety of electronic devices, including rectifiers, which produce a positive (or negative) emf from an AC source, and filters, which take a mixed AC and DC signal and pass predominantly either the constant (DC) part or the variable (AC) part of a given frequency. The term impedance matching refers to constraints that describe how different parts of a circuit can be connected together with minimal power loss.

Understanding the Concepts

1. Why is the material used to make the core of transformers so important?
2. What are some applications for step-up and step-down transformers?
3. Why does a capacitance act as a short circuit at high frequencies?
4. Without R, the current amplitude in a series *RLC* circuit would be infinite when the driving frequency ω equaled ω_0, but this possibility could never happen, because there is always some resistance in real circuits. How do you reconcile this statement with the existence of superconductors?
5. To find the rms current [Eq. (33–41)], we square the current, then take the time average, and then take the square root. Why do we not simply take the time average of the current?
6. Consider a lightbulb connected in series with an inductance L, a capacitor C, and an AC emf. The frequency ω of the power supply can be tuned. For what value of ω will the bulb be brightest?
7. In discussing AC current, we found it useful to describe its root mean square; why didn't we bother with this in our discussion of power?
8. In Example 33–9, we took the input emf to be a mixture of DC and AC, then treated the effect of the AC and DC parts separately and added the two parts. How is this procedure justified?
9. What is the distinction between which side of a transformer is the primary or secondary coil? Could it easily be switched around and operated the opposite way? What would change?
10. A capacitor and a lamp are connected in series with an AC generator of constant voltage but variable frequency (Fig. 33–22, see next page). Which of the following three statements is true? The lamp will (a) not light, because the capacitor is connected in series with the lamp; (b) burn brightest when the frequency is high; (c) burn with the same brightness for all frequencies.

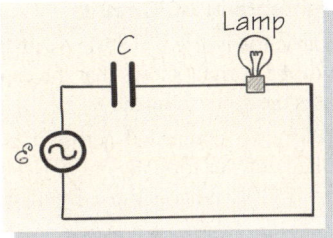

▲ FIGURE 33–22 Question 10.

11. A capacitor, lamp, and resistor are connected to an AC generator of constant voltage but variable frequency (Fig. 33–23). Which of the following statements is true? The lamp will (a) not burn, because the capacitor shorts out the lamp; (b) burn brightest when the frequency is low; (c) burn brightest when the frequency is high; (d) burn with the same brightness for all frequencies.

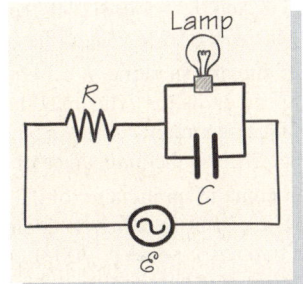

▲ FIGURE 33–23 Question 11.

12. A particular appliance or household circuit is rated for a maximum current. Why is that, and why must currents not exceed that maximum?
13. The purpose of a ballast in a fluorescent light is to start and control the lamp. A voltage at least three times as large as the steady state voltage is needed to start the lighting process, and current spikes must be controlled. Which of the three circuit elements (R, L, C) are you likely to find in a ballast? Explain what it might do.
14. The primary purpose of an electric heater is to produce heat. Why would such a heater require a 220-V socket rather than a 120-V socket?
15. In the complex analysis approach, two combinations of circuit elements have complex impedances Z_1 and Z_2, respectively. When these combinations are placed in series with one another, their (complex) total impedance is given by the addition $Z = Z_1 + Z_2$. For application to physical circuits, the actual impedance is the magnitude of the complex impedance. Will the magnitudes of the complex quantities Z_1 and Z_2 add to give the magnitude of the complex quantity Z?
16. The reactance X_C is infinite when the input voltage is DC. Does that mean that the impedance is not defined for this situation?
17. If a capacitor has a large impedance for DC and an inductor has a large impedance for AC, how can a series LC circuit pass any current?
18. If the frequency of an AC voltage across an inductance is doubled, the inductive reactance of the inductor is (a) increased by a factor of 4; (b) decreased by a factor of 4; (c) doubled, (d) none of these.
19. Television antenna wires normally have negligible resistance and an impedance of 75 Ω. Why is it important to use antenna wires with the same impedance throughout?
20. Some appliances that operate off a 120-V line and draw in excess of 20 A of current have different plugs than regular 120-V household devices have. Why are these plugs different? What might happen if they were not different?
21. Three lightbulbs, a capacitor, and an inductor are connected to an AC power supply as shown in Fig. 33–24. Under what circumstances do you expect bulbs 1 and 2 to be very bright, while 3 is quite dim?

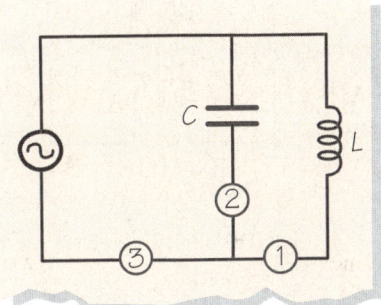

▲ FIGURE 33–24 Question 21.

22. Do all CRT-based television sets have transformers? How might transformers be used in a television set?
23. If electricity is transported in power transmission lines at 200 kV to 500 kV, does the power have to be generated at these voltages? Why or why not?
24. In a capacitor, there is a gap between the two plates. Why then is there any current at all flowing in a circuit with a capacitor? Is this current carried by sparks?

Problems

33–1 Transformers

1. (I) A transformer has 100 turns in the primary coil and 1,500 turns in the secondary coil. If the amplitude of the AC voltage in the primary coil is 600 V, what is the voltage amplitude in the secondary coil?
2. (I) Suppose that electric power costs 18 cents/kWH. Consider a transmission line that delivers 500 kW of power and has a resistance of 150 Ω. Calculate the dollars lost annually due to the transmission line if the power is delivered at (a) 750,000 V and (b) 1,440 V.
3. (I) Many electrical devices, such as doorbells or buzzers, operate on 12 V AC. A small transformer used to produce this voltage has a primary coil of 550 turns and takes an input of 110 V AC. How many turns must the secondary coil have?
4. (I) The primary coil of a step-down transformer is connected to house current, 115 V at 60 Hz. If the secondary coil of the transformer delivers a current with an amplitude of 2.0 A at 24 V, what is the current drawn by the primary coil? Ignore losses in the transformer.
5. (I) A transformer whose output voltage can be varied is used to obtain AC power from a 120-V, 10-A supply. The secondary coil consists of 1200 turns of wire. The variable transformer works by connecting different numbers of turns of wire on the secondary coil. The secondary voltage can thereby be regulated. When all 1200 turns act as the secondary coil, the output voltage has an amplitude of 120 V. How many turns of wire should be used to obtain 45 V (Fig. 33–25, see next page)? How much current will flow in this case?

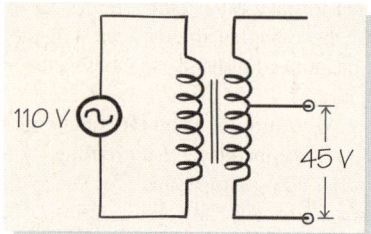

▲ FIGURE 33–25 Problem 5.

6. (I) A transformer has one coil with an inductance of 74 mH, area 35 cm², and length 20 cm. It is fully linked with another coil having the same area and length, but not the same number of turns. Their mutual inductance is 43.5 mH. How many turns does each coil have?

7. (II) Figure 33–26 shows an ideal transformer with 220 V on the primary coil supplying power to a resistor of resistance R. If the resistor dissipates 88 W, what is the current in the primary coil?

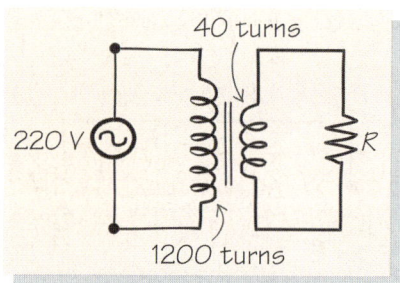

▲ FIGURE 33–26 Problem 7.

8. (II) A step-down transformer has a turn ratio (N_1/N_2) of 15:1. (a) If the primary coil is connected across a 220-V oscillating-voltage generator, what voltage appears across the secondary coil? (b) Assuming that there are no power losses in the transformer, what current would have to flow through the primary coil so that a 15-Ω resistor placed across the secondary coil draws all the power of the circuit? (c) What resistance connected across the 220-V voltage generator would draw the same total power?

9. (II) The transformer shown in Fig. 33–27 has two secondary windings; one supplies 220 V, the other, 11 V. The input voltage at the primary coil is 110 V. If the 220-V secondary coil has 1000 turns, how many turns does the 11-V secondary coil have?

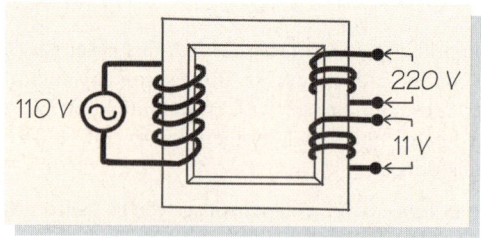

▲ FIGURE 33–27 Problem 9.

10. (II) Suppose that a transformer consists of two separate windings of wire on the same core. The core material has a magnetic permeability μ. How does the ratio of the emfs in the two coils depend on μ?

33–2 Single Elements in AC Circuits

11. (I) A 12-μF capacitor is used in series with an AC generator. Measurement of the current shows that the capacitive reactance is 1.0 Ω. What is the input frequency?

12. (I) A 25-Ω resistor is connected across a power supply that produces a voltage of the form $V_0 \sin(\omega t)$, where $f = \omega/2\pi = 60$ Hz and $V_0 = 130$ V. What is the current passing through the resistor?

13. (I) An alternating current of maximum value 2 A in a solenoid of self-inductance $L = 15$ mH induces an emf of maximum value 330 V. What is the angular frequency of the alternating current?

14. (I) An AC power supply with frequency 60 Hz is connected to a capacitor of capacitance $C = 40\,\mu$F. The maximum instantaneous current that passes through the circuit is 2.26 A. What is the maximum voltage?

15. (I) A current flowing through a circuit that contains only a capacitor and an AC power supply has the form $I_0 \cos[2\pi f t - (\pi/6)]$, where $I_0 = 2.45$ A and $f = 180$ Hz. If the maximum voltage supplied by the generator is 95 V, what is the capacitance?

16. (I) An AC power supply operating at a frequency of 220 Hz is connected across an inductor. The maximum voltage of the source is 4 V, and the maximum current is 65 mA. What is the inductive reactance? What is the inductance of the circuit?

17. (I) An AC circuit contains an inductor of 0.3 H and capacitor of 2 μC in series. The circuit is driven with an AC source of emf with an angular frequency range of 300–1,000 rad/s. What are the maximum values of the capacitive and inductive reactances?

18. (II) A current $I = I_0 \sin(\omega t - \pi/3)$ flows in a circuit for which $I_0 = 2.3$ A and $\omega = 2\pi(60\text{ Hz})$. (a) At what times does the peak current flow? (b) If the current flows through an inductance of 0.25 H, what is the peak voltage on the inductor? At what times does this peak voltage occur?

19. (II) The average of the square of the voltage in an inductive circuit (a circuit with no capacitors and no resistors) driven by an AC emf is $(30\text{ V})^2$, and the average of the square of the current is $(2\text{ A})^2$. What is the inductive reactance? If the inductance is 25 mH, what is the frequency of the alternating current?

20. (II) The voltage across an inductor takes the form $V(t) = (0.3\text{ V}) \sin[(400\text{ s}^{-1})t] + (0.3\text{ V}) \sin[(2{,}700\text{ s}^{-1})t]$. Determine the current through the inductor if $L = 40\,\mu$H.

21. (III) (a) Draw the phasor for the function $D\cos(\omega t + \phi)$ on the graph that contains the phasor for the function $C \sin(\omega t + \phi)$. Which phasor is more advanced in phase—that is, points in a direction corresponding to a larger angle, as measured from the $+x$-direction? (b) What is the angle between the two phasors on the plot you drew for part (a)? Which phasor leads? (c) Repeat the exercise for the function $f(t) = A\cos(\omega t) + B\sin(\omega t)$.

33–3 AC in Series RLC Circuits

22. (I) Consider an LC circuit driven by an AC source of emf. It differs from the one shown in Fig. 33–10 in that the inductor and the capacitor are in parallel, while there is no resistor. Given that the input voltage is $V_0 \sin \omega t$, determine the form of the current through the inductor without using Kirchhoff's rules.

23. (I) Consider a radio circuit with a fixed inductance of 14 μH. What is the value of the tunable capacitance for the reception of a 42-m radio wave?

24. (I) What is the range needed for a variable capacitor to be combined with a 12-mH coil so that a tuned circuit could be formed to cover the range of broadcast-band frequencies from 540 kHz to 1,600 kHz?

25. (I) The driving frequency in a driven RLC circuit is at the resonant frequency. The maximum current carried by the circuit is found to be insufficient for the desired application. By what factor should the resistance be changed to double the maximum current?

26. (I) An AC generator with a voltage amplitude of 50 V and a frequency of 750 Hz is built to drive a circuit meant to be resonant. The resistance of the circuit is 0.5 Ω, and the inductance is 5 mH. What must the value of the capacitance be?

27. (I) A series RLC circuit of frequency 60 Hz has a maximum current of 100 mA. What is the maximum charge on the capacitor? If the impedance is 40 Ω, what is the emf?

28. (I) FM radio stations operate at frequencies spaced at 0.2 MHz intervals from 88.1 to 107.9 MHz, but many radio tuners operate down to 87.9 MHz. However, TV channel 6 operates at 88.0 MHz, with the audio signal of TV channel 6 operating at 87.75 MHz. You have a RLC series tuning circuit with a 20 pF capacitor. What inductance do you need to receive the Channel 6 audio circuit?

29. (II) An inductance is connected in series with a resistor of 150 Ω. When these are placed in series across an rms 110 V AC line with $\omega = 60$ radians/s, the voltage across the resistance is 40 V. Calculate the value of L and its resistance.

30. (II) A series RLC circuit has parameters $R = 18\ \Omega$, $L = 30.0$ mH, and $C = 15.0\ \mu$F. Find the capacitive reactance, inductive reactance, and impedance for the frequencies (a) 60 Hz, (b) 500 Hz, and (c) 20,000 Hz.

31. (II) A series RLC circuit consists of a 1200.0-Hz AC emf with $V_0 = 80$ V; $R = 500\ \Omega$, $L = 92$ mH, and $C = 2\ \mu$F. Find X_C, X_L, Z, Q_{max}, ϕ, and I_{max}.

32. (II) You want to build an AC series circuit with the smallest possible impedance. You have a fixed frequency generator with an angular frequency of 1000 rad/s, and the following circuit elements available: two capacitors, of 1 μF and 100 μF; two inductors, of 10 mH and 25 mH, and two resistors, of $10^3\ \Omega$ and $3 \times 10^3\ \Omega$. You may use only one of each type element. What is the lowest circuit impedance and which values of R, L, and C would you choose to make this circuit?

33. (II) Find the voltages across the capacitor and inductor in the AC circuit of Problem 31 at $t = 0.10000$ s if the emf is switched on at $t = 0$ s. All circuit elements initially have no charge or current.

34. (II) An unknown impedance Z is investigated with an oscilloscope. It is connected in series with a 12-Ω resistor, and connected to a 60-Hz AC power supply. The horizontal deflection plates of the oscilloscope are connected to the known resistor so that the horizontal deflection of the electron beam is proportional to the potential drop on the resistor. The potential drop on the impedance Z is measured by the vertical displacement of the electron beam (Fig. 33–28). (a) Sketch the shape of the figure on the oscilloscope's screen if Z consists of a coil with inductance $L = 0.95$ mH and resistance $r = 650$ mΩ. (b) Repeat the calculation if Z is a 0.5-mF capacitor in series with a resistance $r = 650$ mΩ. (c) How can you tell if Z is capacitive or inductive?

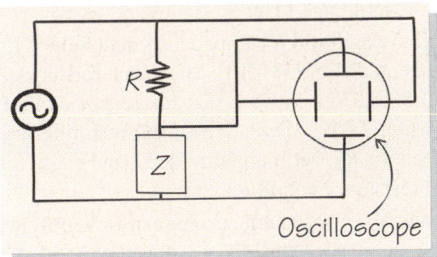

▲ FIGURE 33–28 Problem 34.

35. (II) Given that the maximum voltage in the circuit shown in Fig. 33–29 is 110 V and the frequency of oscillation is 60 Hz, calculate the maximum current and the maximum potential drops across the resistor, capacitor, and inductor.

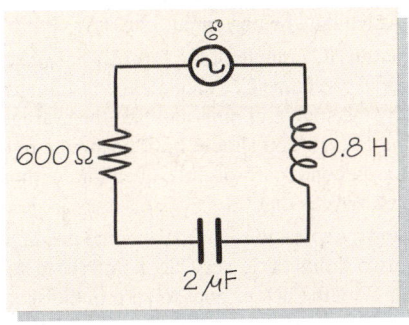

▲ FIGURE 33–29 Problems 35 and 36.

36. (II) What is the resonant angular frequency ω_0 of the circuit in Problem 35? Suppose that the voltage generator has a variable angular frequency ω. For what values of ω will the current have half the value it has at resonance?

37. (II) An AC circuit consists of a parallel-plate capacitor and a long, cylindrical solenoid. Suppose that all the dimensions of the apparatus, including the wire sizes, are scaled down by a factor of 2. (Note that the turn density doubles.) How would the resonant frequency of the circuit change? Assume that there are changes in resistance.

38. (II) Show that $Q(t)$ in Eq. (33–28) satisfies Eq. (33–23) by direct substitution. Determine the maximum charge Q_{max} on the capacitor in terms of the impedance.

39. (II) Sketch the current and voltage for the following AC series circuits: (a) a pure capacitive circuit, (b) a pure inductive circuit, (c) an RL circuit, (d) an RC circuit, and (e) an LC circuit.

40. (II) A resistor draws 3 A when connected to a 80-V, 60-Hz line. A capacitor of what capacitance, when connected in series with the resistor, will drop the current to 1.5 A? What are the voltage drops across the capacitor and the resistor?

41. (II) A 16-μF capacitor is connected in series with a coil whose resistance is 30 Ω and whose inductance can be varied. The circuit is connected across a 12-V, 60-Hz generator. What is the potential difference across the capacitor and across the inductor–resistor combination when the frequency is the resonant frequency?

42. (II) Suppose that the maximum voltages across the resistor, capacitor, and inductor of a series RLC circuit driven by an AC generator of frequency f are identical. If the resistor has a resistance R, find the values of C and L in terms of R and f.

43. (II) A series RLC circuit contains a 70-nF capacitor and a 0.2-Ω resistor. If the circuit is resonant at a frequency of 180,000 Hz, what is the inductance?

44. (II) Consider an RLC circuit in which two resistors, R_1 and R_2, are connected in series, as are two capacitors, C_1 and C_2, and two inductors, L_1 and L_2. Show that the resulting total impedance is of the form

$$Z_{total} = \sqrt{[(X_{L_1} + X_{L_2}) - (X_{C_1} + X_{C_2})]^2 + (R_1 + R_2)^2}.$$

45. (III) There is an AC source of emf in a single-loop circuit that produces a potential drop in the form $V(t) = V_0 \sin(\omega t)$, while the current in the circuit takes the form $I(t) = I_0 \sin(\omega t - \phi)$. Make a phasor diagram for the current and potential drop across each element if the circuit contains (a) a resistor and a capacitor, and (b) a resistor and an inductor.

33–4 Power in AC Circuits

46. (I) What is the average power dissipated in the resistor for the circuit in Problem 12?

47. (I) An AC power supply with a frequency of 75 Hz dissipates energy at a rate of 150 W in a 12-Ω resistor. If the current at time 0 s is 5.0 A, what is the current at time 0.04 s?

48. (I) Consider an AC voltage of the form $V_0 \sin(\omega t)$ connected to a capacitor of capacitance C. Calculate the instantaneous power VI delivered by the source of emf, and find the average power dissipated in the circuit. You should have been able to obtain the answer to the second part of this question without doing any calculations. Why is that?

49. (I) Write down expressions for the average power in an RLC circuit in the two limits (a) ω very large, and (b) ω very small. Can you explain why the power goes to zero in the second case?

50. (I) A portable electric heater operating on AC voltage of amplitude 110 V is rated at a power of 800 W. (a) What is the resistance of the heater? (b) Find the rms current. (c) Find the maximum current.

51. (I) What are the power factors for (a) pure capacitive circuits, (b) pure inductive circuits, and (c) pure resistive circuits?

52. (II) Show that, on average, no power is dissipated in a purely inductive circuit (a circuit with neither capacitors nor resistors).

53. (II) What are the power factors for (a) RL circuits, (b) RC circuits, and (c) LC circuits?

54. (II) A coil draws a current with a peak amplitude of 0.4 A from a peak 170 V source operating at a frequency $f = 60$ Hz. The power consumed is 18 W. What are the impedance, the resistance, and the inductance?

55. (II) Consider an RLC circuit, with $L = 170\ \mu\text{F}$ and $C = 24$ mH. (a) Calculate the resonant frequency. (b) How large should the resistance R be if a 1% shift in the frequency is to reduce the power received by a factor of 50?

56. (II) An AC source of emf operating at a frequency of 60 Hz produces an rms voltage of 78 V. Find the voltage amplitude. The source of emf is connected in series with an impedance of $Z = 20\ \Omega$. Find the rms current and the current amplitude.

57. (II) When a coil draws 200 W from a $V_{\text{rms}} = 110$-V, 60-Hz line, the power factor is 0.6. If the same coil with a capacitor added in series is to draw the same power from a $V_{\text{rms}} = 220$-V, 60-Hz line, what must the capacitance be? If the aim were to maintain the same power factor rather than the same rms power, how would your answer change?

58. (II) A machine shop uses 120 A from a 220-V, 60-Hz line. However, due to the primarily inductive load—motors—the voltage and current are out of phase by 40°, wasting a lot of heat in the cables. A large capacitor connected parallel to the machines can solve this problem (Fig. 33–30). (a) How large should the capacitance be? (b) What will be the current in the main cable with the capacitor attached? (c) What is the total power of the machines in the shop?

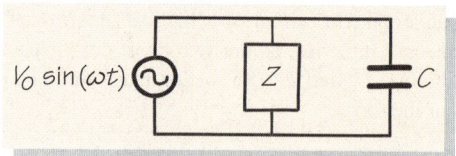

▲ **FIGURE 33–30** Problem 58.

59. (II) An electric motor consumes 5 kW of power at 220 V (voltage amplitude) with a power factor of 0.80. This motor is to be run at the end of a power transmission line with a total resistance of 2.5 Ω. What voltage and power must be supplied at the input end of the transmission line?

60. (II) An AC transmission line transfers energy to a device with a power factor of 0.85 at the rate of $\langle P \rangle = 4$ kW and a voltage of 220 V. If the transmission line has a resistance of 15 Ω, how much energy is lost to Joule heating in the transmission line?

61. (II) A 220-V generator has a current-carrying capacity of 80 A. What is the maximum rate at which energy can be taken from this generator by an impedance with a power factor of 0.55? for a power factor of 0.95?

62. (II) A 20-Ω resistor and a 4.5-μF capacitor are connected in series to a 110-V, 60-Hz power supply. What are the rms current, power, and power factor? How will these numbers change if an inductance of 0.035 H is connected in series with this circuit?

63. (II) An RLC circuit draws a peak 3.00 A from an rms 220 V AC power supply. The power consumed is 200 W. When the capacitor is short-circuited, the current falls to 2.20 A. Find the values of R, the capacitive reactance, and the inductive reactance, assuming that $X_L < X_C$.

64. (II) A series RLC circuit is known to have a resonance at 2.2 MHz. The half-power point is 3.0 KHz away from resonance. Given that the capacitance is 6.6 pF, what are L and R?

65. (II) House current, which has an rms voltage of 110 V and a frequency of 60 Hz, drives a resistor of a variable resistance set at $R = 50\ \Omega$, a capacitor of fixed capacitance $C = 20\ \mu\text{F}$, and an inductor of variable inductance, connected in series. (a) What is the power absorbed by the circuit if $L = 10$ mH? (b) What would the power drawn be if the resistance were halved without changing the setting of the inductance? (c) What is the maximum power drawn in part (b)?

66. (II) For a driven series RLC circuit, show that

$$\frac{R}{Z} = \frac{1}{\sqrt{1 + Q^2\left(\dfrac{\omega}{\omega_0} - \dfrac{\omega_0}{\omega}\right)^2}}.$$

67. (II) Plot R/Z in Problem 66 for values of ω/ω_0 from 0.4 to 2.5 and values of Q of 1, 10, and 100. Use a computer program and graphics output, if available.

68. (II) For a driven series RLC circuit, show that Q is related to $\Delta\omega$ by the relation

$$\frac{\Delta\omega}{\omega} = \frac{1}{Q}\frac{\omega_0}{\omega} \cong \frac{1}{Q}.$$

33–5 Some Applications

69. (II) Consider the circuit treated in Example 33–9 and drawn in Fig. 33–18. Take $C = 5$ nF and $R = 120\ \Omega$, but assume now that the input emf has the purely sinusoidal form $V_1 \sin(\omega t)$, where $V_1 = 0.20$ V. Calculate the potential across the capacitor for (a) $f = 100$ Hz, (b) $f = 10^5$ Hz, and (c) $f = 10$ MHz.

70. (II) Design a high-pass RC filter that will remove voltages with frequencies lower than 8 kHz.

71. (II) Design a high-pass RL filter for filtering out signals with frequencies lower than 8 kHz.

72. (II) The first condition for impedance matching is that the resistances are equal [Eq. (33–50)]. Show that this is true by starting with Eq. (33–49) in the case that the reactance terms are equal and opposite. Take a derivative of the resulting average power with respect to R_2, set it equal to zero, and show that this gives the equal resistance condition.

73. (II) A diode, through which current can flow only when the emf is positive, acts as a filter for an AC generator of angular frequency ω. The current has maximum magnitude I_0. Find its average and rms values.

74. (III) An *RC* filter circuit like that shown in Fig. 33–18 is called a *high-pass* filter circuit when the voltage output is taken across the resistor. Plot the ratio V_{out}/V_{in} as a function of frequency. Why does such a circuit block signals of low frequency but allow high-frequency signals to pass?

75. (III) An *RC* filter circuit like that shown in Fig. 33–18 is called a *low-pass* filter circuit when the voltage output is taken across the capacitor. Plot the ratio V_{out}/V_{in} as a function of frequency. Why does such a circuit block high-frequency signals but allow low-frequency signals to pass?

76. (III) Consider the *LC* filter of Fig. 33–31 with the emf $V_0 \sin(\omega t)$. Assume that $X_L \gg X_C$ (or $\omega \gg \omega_0$). (a) Show that $V_{out} = (X_C/X_L)V_0$. (b) Show that the circuit of Fig. 33–31 is generally effective in reducing the AC components, but not the DC components, of emf.

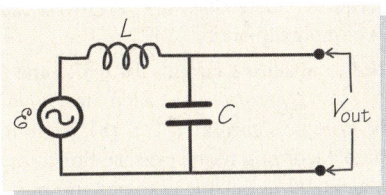

▲ **FIGURE 33–31** Problem 76.

77. (III) Given that the driving voltage of the *RLC* circuit shown in Fig. 33–32 is $V = V_0 \cos(\omega t)$, calculate the currents in the three elements. Is there a resonant frequency? [*Hint*: Write down the circuit equations, and substitute the trial solution $I = I_0 \cos(\omega t + \phi)$].

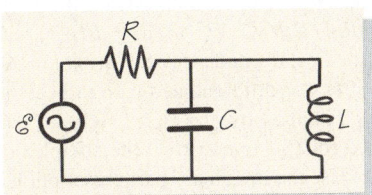

▲ **FIGURE 33–32** Problem 77.

78. (III) What is the relation between the resistances if the average power dissipated in them is the same at $\omega = 1/\sqrt{LC}$ (see Fig. 33–33)?

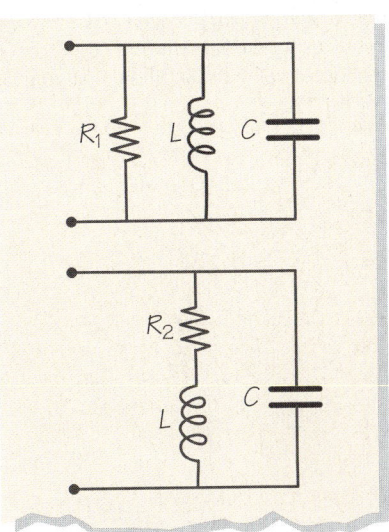

▲ **FIGURE 33–33** Problem 78.

General Problems

79. (II) Calculate (a) the maximum instantaneous voltage across each capacitor; and (b) the maximum instantaneous voltage across the inductor for the circuit shown in Fig. 33–34. Use the parameters specified in Problem 80.

80. (II) Consider the circuit shown in Fig. 33–34. The emf has an amplitude of $V_0 = 4$ V and a frequency of 600 Hz; $L = 70$ μH, $C_1 = 4$ μF, and $C_2 = 9$ μF. Find (a) the maximum current; (b) the resonant frequency.

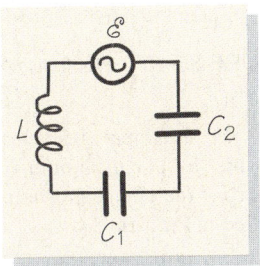

▲ **FIGURE 33–34** Problem 80.

81. (II) Show that Eq. (33–31) for the phase angle can be determined from

$$\tan\phi = \frac{(\omega^2 - \omega_0^2)L}{\omega R}.$$

82. (II) An amplifier with an equivalent impedance of 3,000 Ω is to be connected to an 8-Ω speaker through a transformer (Fig. 33–35). What should the turn ratio of the transformer be?

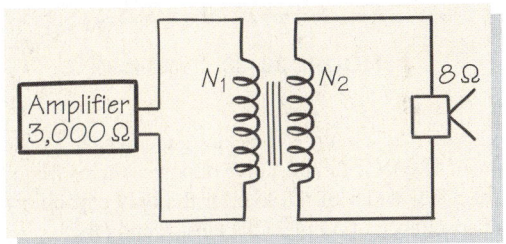

▲ **FIGURE 33–35** Problem 82.

83. (II) The impedance Z_1 in Fig. 33–36 can be regarded as a pure resistance $R_1 = 15$ Ω, whereas the impedance Z_2 is associated with a series resistance $R_2 = 8$ Ω and a capacitance $C = 2$ μF. If $f = 3000$ Hz and $V_0 = 3$ V, what is the power dissipated in Z_2?

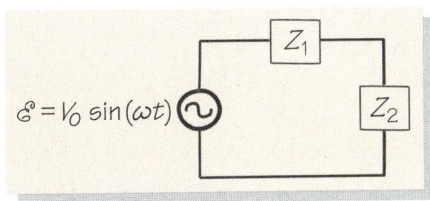

▲ **FIGURE 33–36** Problem 83.

84. (II) Consider the circuit shown in Fig. 33–37. The emf has an amplitude of $V_0 = 12$ V and a frequency of 400 Hz; $L = 10$ mH, $C_1 = 20$ μF, and $C_2 = 30$ μF. Find (a) the maximum current in each leg, and (b) the resonant frequency.

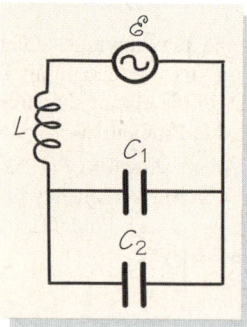

▲ **FIGURE 33–37** Problems 84 and 85.

85. (II) Calculate (a) the maximum instantaneous voltage across each capacitor; and (b) the maximum instantaneous voltage across the inductor for the circuit shown in Fig. 33–37. Use the parameters specified in Problem 84.

86. (II) Write down the two equations that specify the currents I_1 and I_2 in the two loops of the circuit shown in Fig. 33–38.

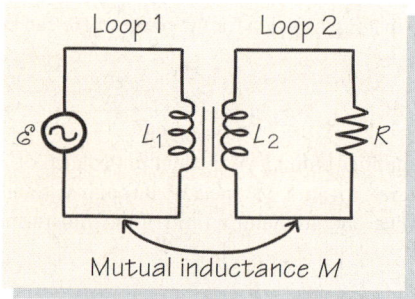

▲ **FIGURE 33–38** Problem 86.

87. (II) A series RLC circuit is to be designed to have a resonant frequency of 18 MHz, and the curve of power versus frequency f is to have a full width of 4.0 kHz. If the only capacitor available has a capacitance of 33 pF, what must R and L be?

88. (II) A resistor with $R = 62\ \Omega$ draws a current from a wall plug; a capacitor is connected in parallel with this resistor. The current source has an amplitude of 110 V and a frequency of 60 Hz, and the reactance of the capacitor is $3\ \Omega$ at this frequency. What is the current drawn by the parallel combination?

89. (II) A 15-μF capacitor connected in series with a resistor of variable resistance R is connected to a $V_{rms} = 110$-V, 60-Hz AC supply. Plot the variation of the rms current with R, and calculate the value of R for which the power delivered is maximum.

90. (II) An AC circuit supplies $V_{rms} = 220$ V at 60 Hz to a 10-Ω resistor, a 35-μF capacitor, and an inductor of variable self-inductance in the 30 mH to 300 mH range, all in series. The capacitor is rated to stand a maximum voltage of 1200 V. (a) What is the largest current possible that does no damage to the capacitor? (b) To what value can the self-inductance be safely set?

Problems with Complex Variables

In the following set of problems, we suppose there is an AC source of emf that produces a potential drop of the form $V_0 \sin(\omega t)$. In the complex variable technique, we work instead with the complex form $V_C(t) = V_0 \exp(i\omega t)$, with the instruction that the original potential drop is given by the imaginary part of this form, $\text{Im}[V_C(t)]$. All quantities with oscillating time dependence are given this treatment. Note: The subscript "C" here does not refer to a capacitor, but instead indicates a complex quantity.

91. (III) Consider a capacitive circuit, for which the loop rule takes the form $V_C(t) = Q_C(t)/C$. (a) Calculate $Q_C(t)$, and use it to calculate the complex current $I_C(t)$. (b) Show that the current $I(t)$ calculated according to the prescription at the head of these problems is identical to the one obtained in Eq. (33–12).

92. (III) Consider an inductive circuit, for which the loop rule takes the form $V_C(t) = L\, dI_C(t)/dt$. (a) Calculate $I_C(t)$ in two ways: (i) by direct integration of the equation; (ii) by noting that no matter how often the function $\exp(i\omega t)$ is differentiated with respect to time, the time dependence remains $\exp(i\omega t)$. (b) Show that the current $I(t)$ calculated according to the prescription at the head of these problems is the same as the one obtained in Eq. (33–18).

93. (III) In complex analysis, the loop rule for the series RLC circuit is $V_C(t) = L(dI_C(t)/dt) + Q_C(t)/C + RI_C(t)$. (a) Make use of the fact that the time dependence of $\exp(i\omega t)$ remains the same no matter how often it is differentiated with respect to time in order to calculate Q_{0C}, defined by $Q_C(t) = -iQ_{0C}e^{i\omega t}$. (b) Use the formula obtained for Q_{0C} to construct the complex quantities $Q_C(t)$ and $I_C(t)$. (c) Use the imaginary part prescription described at the head of these problems to calculate $Q(t)$ and $I(t)$.

94. (III) Consider the RLC circuit of Problem 93. Define the complex impedance Z_C to be $V_C(t)/I_C(t)$, and use the results of Problem 93 to show that Z_C is independent of time, with an absolute magnitude given by Eq. (33–32).

95. (III) Consider an RLC circuit without a driving emf, so that the loop rule reads $L(dI_C(t)/dt) + Q_C(t)/C + RI_C(t) = 0$. Solve this equation by using the substitution $Q_C(t) = Q_{0C}e^{i\omega t}$. Use the equation to obtain a value for ω. Show that your solution leads to the results of Eq. (32–32).

Chapter 34

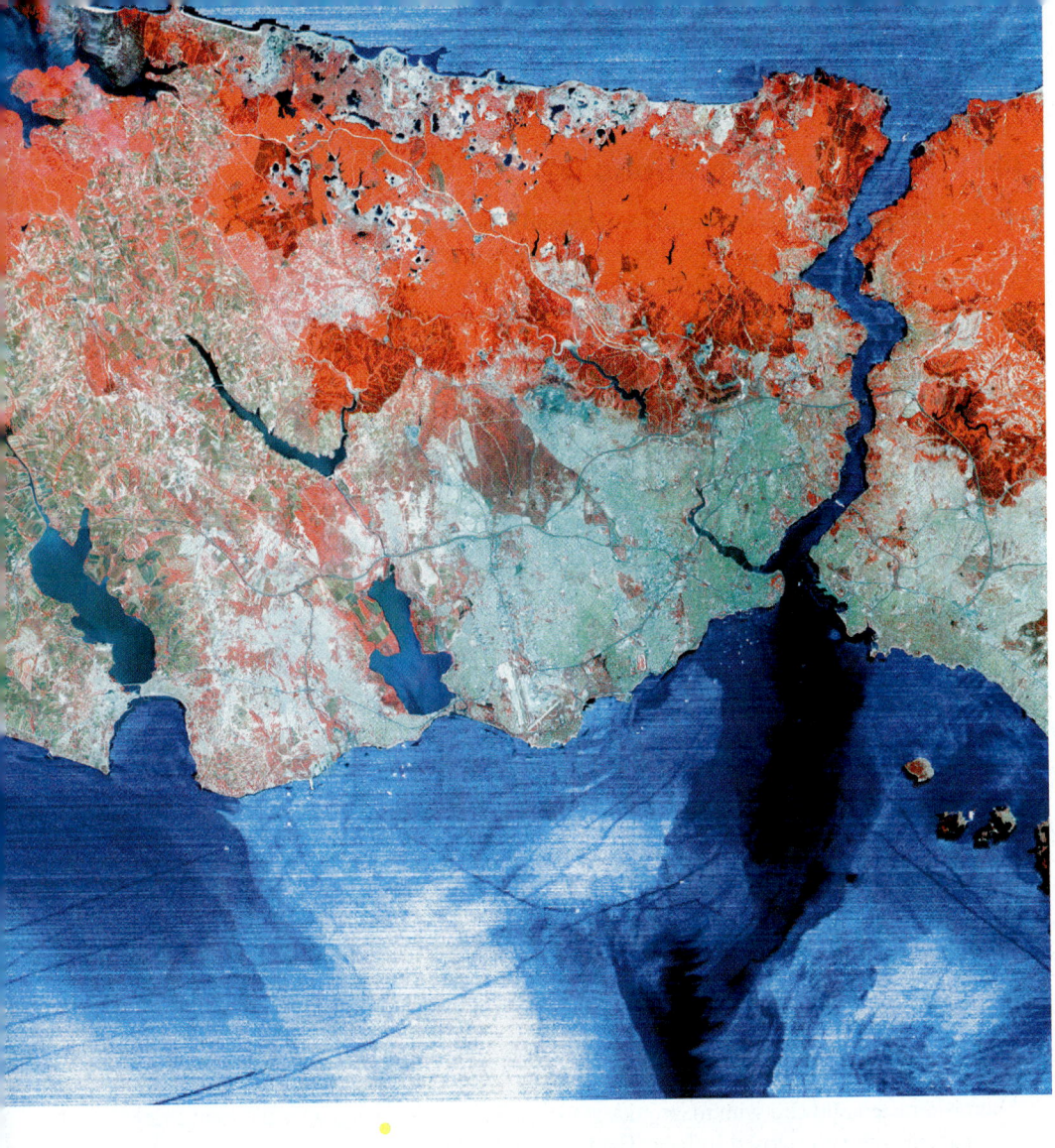

◀ Terra satellite image of the straits of Bosporus. This image is a combination of visible and infrared light images. Vegetation appears red while urban areas appear blue-green; the different water colors are measures of the water temperature.

Maxwell's Equations and Electromagnetic Waves

Faraday's law shows that electricity and magnetism are fundamentally connected; James Clerk Maxwell's introduction in 1864 of the displacement current enhances this connection. We have a complete, consistent set of fundamental laws for electricity and magnetism, collectively known as Maxwell's equations. The individual experiments that led to the discovery and completion of each of these equations—Gauss' law for electric fields, Gauss' law for magnetic fields, Ampère's law, and Faraday's law—never hinted at the rich implications of Maxwell's equations taken *together*. The most dramatic prediction of the full set of Maxwell's equations is the existence of electromagnetic waves that propagate through empty space at a predictable speed—the speed of light. Light itself is such a wave. It is the intimate coupling between electric and magnetic fields that leads to these waves, with, in effect, one field generating the other in a self-sustaining way. In this chapter, we discuss the nature and properties of these *electromagnetic waves*: the orientation and relationship of the electric and magnetic fields contained in them, the energy and momentum they carry, and polarization. While these waves are described in terms of the mechanical waves we met in Chapters 14 and 15, they differ in a profound way in that they do not require a medium in which to propagate. This observation contains the seeds of Einstein's special relativity. We'll also see here how quantum physical phenomena intrude into the properties of classical electromagnetic waves.

34-1 Maxwell's Equations

Let's first recall and then comment on **Maxwell's equations**, which fully describe the electric and magnetic fields in the presence of electric charges and currents. In this discussion we'll largely ignore the effects of matter, which although very well understood, would be a diversion from the important points to be described in this chapter.

I. Gauss' law for electric fields

$$\int_{\text{closed surface}} \vec{E} \cdot d\vec{A} = \frac{Q}{\varepsilon_0}. \tag{34-1}$$

II. Gauss' law for magnetic fields

$$\int_{\text{closed surface}} \vec{B} \cdot d\vec{A} = 0. \tag{34-2}$$

III. Generalized Ampère's law

$$\oint \vec{B} \cdot d\vec{s} = \mu_0 I + \mu_0 \varepsilon_0 \frac{d}{dt} \int_{\text{surface}} \vec{E} \cdot d\vec{A}. \tag{34-3}$$

IV. Faraday's law

$$\oint \vec{E} \cdot d\vec{s} = -\frac{d}{dt} \int_{\text{surface}} \vec{B} \cdot d\vec{A}. \tag{34-4}$$

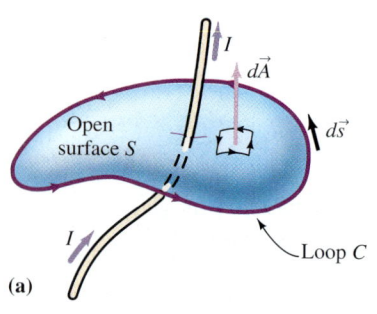

▲ **FIGURE 34-1** An infinitesimal surface element $d\vec{A}$ on the closed surface S.

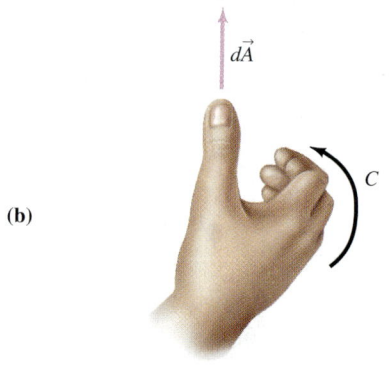

▲ **FIGURE 34-2** (a) A surface S bounded by the closed loop C. A current I passes through the surface. If the integration along C proceeds counterclockwise, (b) then the direction of the surface elements $d\vec{A}$ that make up S is given by a right-hand rule.

I. Gauss' law is equivalent to Coulomb's law in static situations and hence shows that the electric field due to a static charge varies inversely with the square of distance from the charge. It relates the electric flux through a closed surface (the surface can be imaginary) to the charge enclosed [see, for example, Eq. (23-7)] and states that charges are the sources of electric field. The surface element $d\vec{A}$ in Eq. (34-1) is normal to the surface S and is directed outward with magnitude dA (Fig. 34-1). The charge Q is the total charge contained within the closed surface. The factor ε_0 (the permittivity of free space) is associated with our choice of units. Gauss' law holds even for time-dependent electric fields.

II. Magnetic monopoles—which would be the magnetic analogues of electric charge—have never been discovered. Their nonexistence leads to Gauss' law for magnetic fields [Eq. (29-12)], which states that the flux of the magnetic field through a closed surface is zero, implying that magnetic field lines have no beginning or ending. This equation also holds even for time-dependent magnetic fields.

III. Ampère's law states that magnetic fields arise from currents or changing electric fluxes. The left-hand side of this equation is the expression for the integral of the magnetic field's tangential component along an arbitrary closed loop C (Fig. 34-2). The right-hand side has two contributions: One is the total current flowing through any surface S bounded by the closed loop C; the other is the rate of change of the electric field flux through such a surface, the displacement current contribution. As we described in Chapter 29 [see Eq. (29-31)], Maxwell was responsible for introducing the displacement current. As you can review in Chapter 29, he did this to fill a gap in the logic of Ampère's law. The presence of the parameter μ_0 (the permeability of free space) is a consequence of SI units.

IV. Faraday's law describes the induced electric field loop generated by a changing magnetic flux [see Eq. (30-2)] and shows that the electric field encircles any area in which the magnetic flux is changing. The left-hand side is the integral of the tangential component of the induced electric field around an arbitrary closed loop C. The right-hand side measures the rate of change of the magnetic flux through any surface S bounded by C, just as in Fig. 34-2. Equation (34-4), as well as Eq. (34-3), implies a sign convention given by a right-hand rule. The minus sign is very important: It represents the fact that the induced electric field, were it to act on charges, would give rise to an induced current that opposes the change in the magnetic flux (Lenz's law).

The four Maxwell equations display a degree of symmetry between electric and magnetic fields. This symmetry is not perfect, because magnetic monopoles apparently do not exist. Faraday's law contains no term like the $\mu_0 I$ term in Ampère's law, because there is no free magnetic charge to form a magnetic current. In a vacuum there are neither currents nor charges, and then the symmetry between $\vec{E}$ and $\vec{B}$ is complete. This is a situation in which the propagation of electromagnetic waves occurs.

In the presence of matter, Maxwell's equations are modified. Assuming that we don't have to deal with boundaries between different materials, then for most types of materials we can simply replace ε_0 by $\varepsilon = \kappa \varepsilon_0$, where κ is the dielectric constant. Except for ferromagnetic materials, μ is very close to μ_0, so that the additional rule that the permeability of the vacuum (μ_0) is to be replaced by the material's permeability (μ) does not much affect things.

34–2 Electromagnetic Waves

Equations (34–3) and (34–4) clearly show that time-dependent electric and magnetic fields influence each other; they are said to be *coupled*. As a consequence of this coupling, electric and magnetic fields can transport energy (and momentum) over much larger distances than might be suggested by the $1/r^2$ falloff of the electric field in Coulomb's law or the magnetic field in the Biot–Savart law. As we shall see, the coupled fields transport energy through traveling waves called **electromagnetic waves**. These waves are all around us in the form of radio, television, and cell phone signals, microwaves, visible light, and X rays, for example. In fact, electromagnetic waves can be produced with essentially an unlimited range of wavelength or frequency, and the examples listed above just correspond to different wavelength or frequency ranges.

The following argument conveys the physical mechanism by which the electric and magnetic fields couple and electromagnetic waves propagate. Consider a straight wire that is aligned with the *x*-axis and carries a time-varying current *I* (Fig. 34–3a). A magnetic field encircles the wire; if the current changes with time, so does the magnetic field. As the current increases, so does the magnetic field. As the magnetic field increases, so does the magnetic flux through an area A_1 in the *xz*-plane. According to Faraday's law, Eq. (34–4), a changing magnetic flux induces an emf around the boundary of this area. This emf is associated with the induced electric field shown in Fig. 34–3b. Lenz's law determines the orientation of the field. (Note that while we have drawn electric fields in this figure, all that Faraday's law really tells is that an induced electric field is present whose integral along the boundary is given. The drawn electric fields are simply present to help you follow this argument more closely. Later we'll see how to tighten up the argument.)

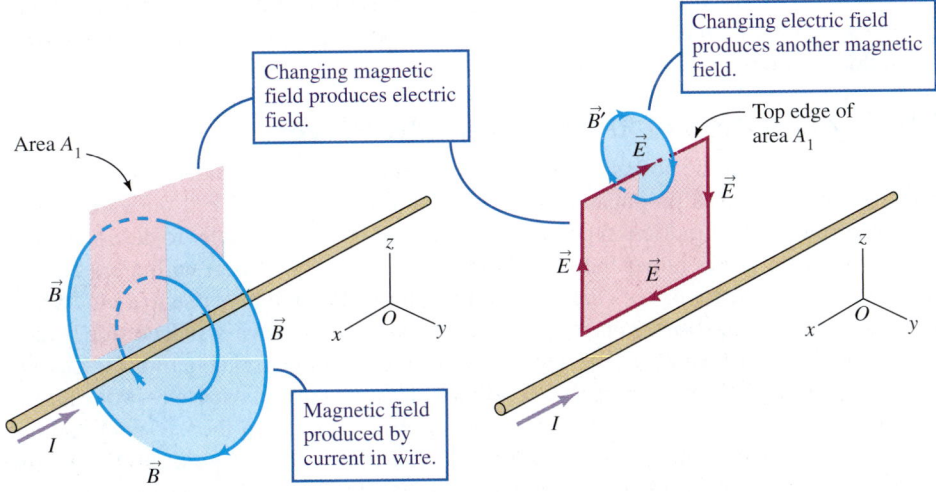

◀ FIGURE 34–3 (a) As we know from Ampère's law, a current-carrying wire aligned in the *x*-direction has a magnetic field that forms circles in the *yz*-plane. (b) If the current in the wire changes with time, the magnetic field it produces changes with time, inducing a changing electric field, which in turn induces a changing secondary magnetic field, and so forth.

Let's consider now the top edge of area A_1. Along that edge, the electric field has been induced in the $-x$-direction. This induced electric field is time varying because it is due to a changing magnetic field; at the instant shown in our example, the electric field is increasing. Now, according to the generalized Ampère's law, Eq. (34–3), we do not need flowing charges to induce a magnetic field. *A changing electric field also produces a magnetic field by giving rise to a displacement current* (Section 29–5). The displacement current in this case is along the direction of the changing electric field, which is the direction of the original current. The displacement current, however, is spread through space rather than localized at the wire like the original current. In particular, there is a displacement current away from the location of the original current. At this point, we can begin to see how the propagation works: The displacement current produces a secondary magnetic field $\vec{B}'$ at still larger values of z (Fig. 34–3b). The field $\vec{B}'$ is perpendicular to the xz-plane. Because the displacement current varies with time, $\vec{B}'$ is also changing. Therefore this secondary magnetic field produces an induced emf aligned in the x-direction at still larger values of z, and the process repeats itself to larger and larger z values.

We could have taken our plane area A_1 to intersect the cylinder around the wire at any angle relative to the z-axis, since our picture is symmetric about the x-axis. The same sequence of events results from the coupling however we rotate our figure about the x-axis. This means that when we use the words "larger value of z," we really mean "larger value of the perpendicular distance from the wire." As the current changes with time, so do the secondary fields $\vec{B}'$. We therefore have electric and magnetic fields that *propagate* on the surface of an ever growing cylinder whose axis lies along the wire.

CONCEPTUAL EXAMPLE 34–1 Suppose the current oscillates for a while and is then turned off. What happens to the fields?

Answer If you look at the discussion above, you can see that to learn about a field farther from the wire, we discuss the fields closer to the wire, not the current itself. The induced electric field, the secondary magnetic field whose source is the displacement current due to this induced electric field, and all their successors have acquired a life of their own. While they were initially created by the current in the wire, the fields have detached themselves from their original source and are propagating "alone." If the current is turned off at some time $t = T$, then from that moment on, no new fields are produced, but the fields produced earlier continue their cylindrical propagation outward.

What qualitative conclusions can we can draw from this discussion?

1. With the changing current *restricted to a line*, the fields propagate with cylindrical symmetry outward from the current line. The electric field is aligned parallel to the current, and the magnetic field is aligned perpendicular to both the electric field and to the direction of propagation. *These are general features of electromagnetic waves.*

2. The current must change in time if it is to give rise to propagating fields, as a steady current merely produces a static magnetic field. We can translate this into a statement about the charges whose flow gives rise to the current: *The charges that give rise to the propagating electric and magnetic fields must be accelerating.* Harmonically varying currents will give rise to harmonically varying electric and magnetic fields, as we will verify below.

The Propagation of Electromagnetic Waves

Let us now build on the points above to gain a more quantitative understanding of how electromagnetic waves propagate. In the situation above, we worked outward from a line of current to a cylindrical type of propagation. It is in fact simpler to describe plane electromagnetic waves rather than cylindrical ones. So we will start with a brief qualitative description of a *sheet of current*, a configuration that will give rise to a *plane electromagnetic wave*. A sheet of current can be formed by a set of wires placed side by side in the xy-plane, each carrying the same current in the x-direction (Fig. 34–4). Charges will move in the $+x$- and then in the $-x$-direction as the current oscillates. We can now go through the same argument as before. When the current is in the $-x$-direction, the magnetic field will lie in a plane parallel to the current sheet and point in the positive y-direction—it will be a superposition of cylindrical shapes, one from each wire, superimposed side by side, as in Fig. 34–5. In the limit of wires very close to each other, the

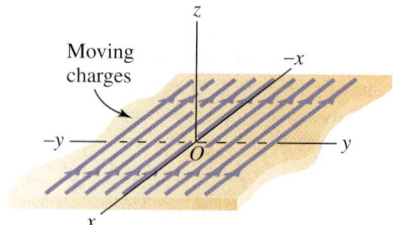

▲ **FIGURE 34–4** Current flows in a sheet along the $-x$-direction. It can be approximated by aligning wires side by side in the x-direction. If the current is oscillatory, charges move first in the $-x$-direction, then in the $+x$-direction.

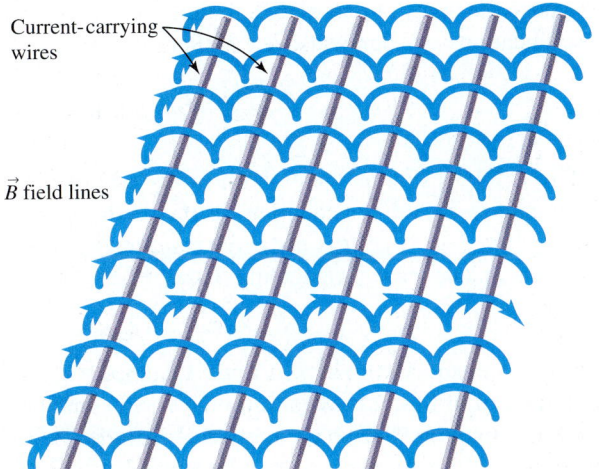

◀ **FIGURE 34–5** In the limit that the wires carrying the current are close to one another, the magnetic field of a series of closely spaced and parallel currents forms planes parallel to the plane of the currents.

field lines of the net magnetic field form a corrugated surface; taking the wires still closer to one another, the field forms two planes, one above the current sheet and the other the same distance away below the current sheet. The electric field will also lie in a plane parallel to the current sheet, and point in the direction of the current. These fields in turn lead to other fields, and if you follow the same argument that we made for the single wire, you will find that these fields will propagate in the z-direction (both $+$ and $-$), perpendicular to the sheet. These fields will depend on time in a way that mirrors the time dependence of the current. The electromagnetic fields form *plane waves*, which we recall from Chapters 14 and 15 refers to waves that advance along planar wave fronts—in this case, planes parallel to the xy-plane.

To understand quantitatively how we can get fields that behave like this, we assert that Maxwell's equations—in particular, Faraday's law and the generalized Ampère's law—are equivalent to a set of differential equations for the components of these fields. Not surprisingly, these are *coupled* differential equations—how the magnetic field changes determines how the electric field changes and vice versa. We derive these equations in the Appendix to this chapter on p. 967. Here, in order to get directly to plane waves propagating in the z-direction, we write them down for the special case of fields that are independent of x and y, but which do depend on z and, of course, on the time t. The components B_y and E_x in particular satisfy

$$-\frac{\partial B_y(z,t)}{\partial z} = \mu_0 \varepsilon_0 \frac{\partial E_x(z,t)}{\partial t} \tag{34-5}$$

and

$$-\frac{\partial B_y(z,t)}{\partial t} = \frac{\partial E_x(z,t)}{\partial z}. \tag{34-6}$$

Although a changing current was necessary to produce the initial changing fields, these equations are not directly dependent on that current—this is what we meant when we said that the fields take on "a life of their own." The field components B_y and E_x both depend on the value of z and on the time t. (Recall that partial derivatives appear whenever quantities such as fields depend on two or more variables. In taking a partial derivative with respect to one variable, the other variables are held fixed.)

Equations (34–5) and (34–6) lead directly to wave equations for the field components. These two equations couple the two fields. We can combine and simplify them by a straightforward procedure. The partial derivative of Eq. (34–5) with respect to time gives

$$-\frac{\partial^2 B_y(z,t)}{\partial t\, \partial z} = \mu_0 \varepsilon_0 \frac{\partial^2 E_x(z,t)}{\partial t^2}.$$

Similarly, the partial derivative of Eq. (34–6) with respect to z gives

$$-\frac{\partial^2 B_y(z,t)}{\partial z\, \partial t} = \frac{\partial^2 E_x(z,t)}{\partial z^2}.$$

Because the order of partial differentiation does not matter, the left-hand sides of these two equations are identical. We can therefore equate the right-hand sides:

$$\frac{\partial^2 E_x(z,t)}{\partial z^2} = \mu_0 \varepsilon_0 \frac{\partial^2 E_x(z,t)}{\partial t^2}. \tag{34-7}$$

There will be a similar equation for the y-component of the magnetic field, as we'll see in more detail later. The equation for E_x has the same form as an equation we have seen before [Eq. (14–10)]: *It is the wave equation!* One solution of this wave equation is a harmonic plane wave propagating in the $+z$-direction:

$$E_x = E_0 \cos(kz - \omega t + \phi), \tag{34-8}$$

where E_0 is an amplitude, k is a wave number, and ω is an angular frequency. Direct substitution of this expression into the wave equation, Eq. (34–7), will verify that it is indeed a solution. The phase angle ϕ is included because we shall want to look at how the phase of the magnetic field, which also has an oscillating solution, is related to that of the electric field. It is important to understand what this equation means: the electric field has an x-component (always aligned along the x-axis), and that component changes as one looks at different values of z and t.

The Propagation Speed of Electromagnetic Waves

We recall from our discussion in Chapter 14 of wave motion that Eq. (34–8) represents a wave of wavelength $\lambda = 2\pi/k$ and frequency $f = \omega/2\pi$. The propagation speed is $v = \lambda f = \omega/k$. This speed is found immediately from the wave equation itself, as comparison with the original form of the wave equation, Eq. (14–10), shows. We have

$$v^2 = \frac{1}{\mu_0 \varepsilon_0}. \tag{34-9}$$

When we use the numerical values for μ_0 and ε_0, we find

$$v^2 = \frac{1}{(1.257 \times 10^{-6}\,\text{T}\cdot\text{m/A})(8.854 \times 10^{-12}\,\text{C}^2/\text{N}\cdot\text{m}^2)} = 8.999 \times 10^{16}\,\text{m}^2/\text{s}^2$$
$$= (3.00 \times 10^8\,\text{m/s})^2.$$

Maxwell recognized that the magnitude of v is the *speed of light c!* One of the most commonplace of all the physical phenomena around us, light, was now explained in terms of the laws of electricity and magnetism, or more properly, electromagnetism. This surely represents one of the greatest discoveries of science, equal in every way to Newton's understanding of gravitation. (And there was more, much more, to come from Maxwell's equations, including special relativity and a great portion of today's technology.) Equation (34–9) tells us that the speed of light is determined by constants of the equations of electromagnetism:

$$c = \frac{1}{\sqrt{\mu_0 \varepsilon_0}}. \tag{34-10}$$

ELECTROMAGNETIC WAVE SPEED IN EMPTY SPACE

What distinguishes *visible* light? The wavelengths of electromagnetic waves are not restricted to any particular value or set of values. The kinds of experiments with optics that we'll be discussing later in the book show that what we call visible light corresponds to a limited range of wavelengths for which, through evolutionary adaptation, our eyes have become particularly good detectors. This range of wavelength is one in which the Sun emits radiation strongly and for which the waves pass easily through the atmosphere. Within this visible spectrum, we interpret different wavelengths as colors. The shortest wavelengths of the visible spectrum are violet; the longest wavelengths are red. The speed which Maxwell confirmed was the speed of visible light, because that was the phenomenon that to that point had been, quite literally, visible and whose speed had been measured. But Maxwell's discovery opens a door onto the full spectrum of electromagnetic radiation, a subject we'll discuss more fully later in this section.

Keep in mind the important relation $v = \lambda f$ between speed, wavelength, and frequency. If the speed of propagation is fixed, as it is in empty space, then wavelength and frequency have an inverse relation, with large wavelengths corresponding to small frequencies and vice versa.

The Relation Between *E* and *B* in an Electromagnetic Wave

To see how E and B for an electromagnetic wave are related, we can start with Eqs. (34–5) and (34–6) and show that B_y also obeys a wave equation similar to that of E_x; namely,

$$\frac{\partial^2 B_y(z,t)}{\partial z^2} = \mu_0 \varepsilon_0 \frac{\partial^2 B_y(z,t)}{\partial t^2}.$$

Like the *x*-component of the electric field, the *y*-component of the magnetic field forms a wave that propagates at speed c in the *z*-direction. However, because Eqs. (34–5) and (34–6) couple the fields, the waves of B_y do not propagate independently from those of E_x. If we have a wave solution for E_x, Eq. (34–8), then from Eq. (34–5),

$$\frac{\partial B_y}{\partial z} = \mu_0 \varepsilon_0 \frac{\partial E_x}{\partial t} = -\mu_0 \varepsilon_0 \frac{\partial}{\partial t}[E_0 \cos(kz - \omega t + \phi)] \qquad (34\text{–}11)$$
$$= -\mu_0 \varepsilon_0 \omega E_0 \sin(kz - \omega t + \phi).$$

Equation (34–6) becomes

$$\frac{\partial B_y}{\partial t} = -\frac{\partial E_x}{\partial z} = -\frac{\partial}{\partial z}[E_0 \cos(kz - \omega t + \phi)] = kE_0 \sin(kz - \omega t + \phi). \qquad (34\text{–}12)$$

From these two expressions for the derivatives of B_y, it is easy to check that the following equation has the correct spatial and time dependence:

$$B_y = B_0 \cos(kz - \omega t + \phi). \qquad (34\text{–}13)$$

Relations Between Amplitudes: The amplitude B_0 of the magnetic field wave is not independent of the amplitude E_0 of the electric field wave, as Example 34–2 shows.

EXAMPLE 34–2 Consider the electromagnetic traveling wave for which the electric and magnetic fields are given by Eqs. (34–8) and (34–13). Use the derivative relations we have found to show that the amplitudes are related by $E_0 = cB_0$.

Setting It Up Here we are proving a relation between the two coupled quantities E and B, so we must use the equations that couple them.

Strategy Equation (34–11) relates a derivative of B_y to a derivative of E_x. With E_x given by Eq. (34–8), and B_y given by Eq. (34–13), we can compute the partial derivative of B_y with respect to z using Eq. (34–11), and we can compute it directly. Comparing these two results will give us information about the magnitude of the amplitudes as well as the phases.

Working It Out We have from Eq. (34–11) with E_x given by Eq. (34–8)

$$\frac{\partial B_y}{\partial z} = -\mu_0 \varepsilon_0 \omega E_0 \sin(kz - \omega t + \phi).$$

With B_y given by Eq. (34–13), we can compute the partial derivative directly:

$$\frac{\partial B_y}{\partial z} = \frac{\partial}{\partial z}[B_0 \cos(kz - \omega t + \phi)] = -kB_0 \sin(kz - \omega t + \phi).$$

We equate these two results:

$$-kB_0 \sin(kz - \omega t + \phi) = -\mu_0 \varepsilon_0 \omega E_0 \sin(kz - \omega t + \phi).$$

The sine factor cancels, and we are left with

$$B_0 = \frac{\mu_0 \varepsilon_0 \omega}{k} E_0.$$

The factor $\omega/k = c$, whereas $\mu_0 \varepsilon_0 = 1/c^2$, so we are left with $B_0 = E_0/c$, the relation we needed to show.

What Do You Think? Suppose the wave in this example impinges on a metal sheet in which there are free electrons. Will it be the electric field or the magnetic field that predominantly determines the motion of the electrons? [*Hint:* Think about the size of B_0.] Answers to **What Do You Think?** questions are given in the back of the book.

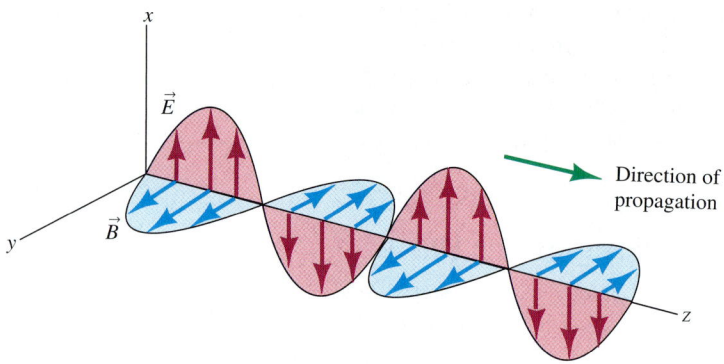

(a)

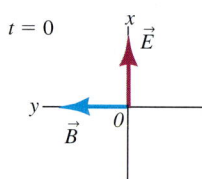
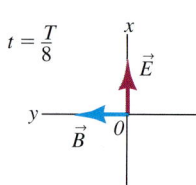
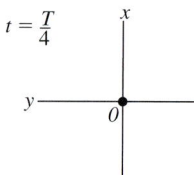
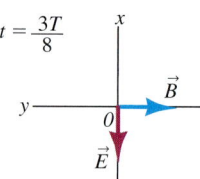
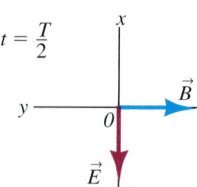

(b)

▲ **FIGURE 34–6** (a) A view at one particular time of the transverse electric and magnetic fields that propagate along the z-axis. (b) A view downward from the $+z$-direction of the electric and magnetic fields of an electromagnetic wave in an xy-plane at one particular z-value over time.

In the example above, we did not mention the currents that set up the original wave, and in fact the relation between the electric and magnetic field amplitudes in an electromagnetic wave is independent of those currents. We have, in general,

$$E = cB, \qquad (34\text{--}14)$$

RELATION OF FIELD AMPLITUDES

where E and B are the amplitudes of the fields in an electromagnetic wave.

The Transversality of Electromagnetic Waves: The fields described by Eqs. (34–8) and (34–13) and pictured in Fig. 34–6 form traveling waves that propagate in the z-direction. Even though the fields are oriented in the x- and y-directions, they do not depend on x or y—we are describing plane waves, as we stated earlier. The currents that we used to set up the waves in the first place guarantee that the waves will be plane waves, and by changing the originating current, we can form other configurations of waves. In particular, there is nothing special about the x- and y-directions. If we had set up our currents to run in the y- rather than the x-direction, we would have found another set of solutions, with $\vec{E}$ in the y-direction and $\vec{B}$ in the x-direction. The wave propagation would still have been in the z-direction. It is generally true that *the electric field and the magnetic field in an electromagnetic wave are perpendicular to each other*,

$$\vec{E} \cdot \vec{B} = 0. \qquad (34\text{--}15)$$

ORTHOGONALITY OF FIELDS

Moreover, an electromagnetic wave is *transverse* because the direction of the fields involved is perpendicular to the direction of wave propagation. *Neither the electric field nor the magnetic field has a component in the direction of propagation of the wave* (Fig. 34–6). We saw that both of these properties hold for the cylindrical wave emanating from the wire at the beginning of this discussion. We will limit ourselves to the plane and cylindrical wave discussions, but it is not difficult to see that these properties hold generally.

The Electric Field and Magnetic Field Are in Phase: The phases that appear in the harmonic expressions for B_y and E_x in Eqs. (34–13) and (34–8), respectively, are exactly the same. When the electric field is a maximum, the magnetic field is also a maximum; when one is zero, the other is zero, and so forth. The fields oscillate together as shown in Fig. 34–6. The fields are *in phase*.

Figure 34–6 illustrates each of the features of electromagnetic waves described above: The fields are in phase, transverse (perpendicular to the direction of propagation), and perpendicular to each other.

Electromagnetic Waves Are Real

When Maxwell introduced the displacement current and predicted electromagnetic waves in 1864, a number of the leading physicists of his time found these notions difficult to accept, and it was more than 20 years before experiment made all such resistance collapse. Experimental confirmation of the existence of electromagnetic waves was not possible when Maxwell proposed them because there was no technology to create AC currents of sufficiently high frequency and amplitude to provide detectable radiation. Heinrich Hertz devised the first direct test of Maxwell's waves in 1887. Hertz used the sparks that form when there is a large potential difference between the two points of a "spark gap" (Fig. 34–7a). The sparks have a rhythm associated with a back-and-forth motion of charge in the gap. To confirm that this oscillatory motion of charges produces electromagnetic waves, or *radiation*, Hertz took a wire bent into a circle with a (second) gap and placed it near the original spark gap (Fig. 34–7b). The electromagnetic wave that propagated in the space between the spark gap and the circular wire loop gave rise to sparks in the secondary gap, which thereby acted as a detecting antenna. Hertz also reflected waves from metallic surfaces, focused them with a concave metallic mirror, and found that they generally shared many of the properties of light that we shall study in Chapters 35 and 36.

The frequencies of the electromagnetic waves studied by Hertz are quite different from the frequencies of the waves that form visible light. The wave equation for electromagnetic waves admits solutions for *any* frequency, and the collection of all frequencies is known as the **electromagnetic spectrum**. In the century since Hertz's

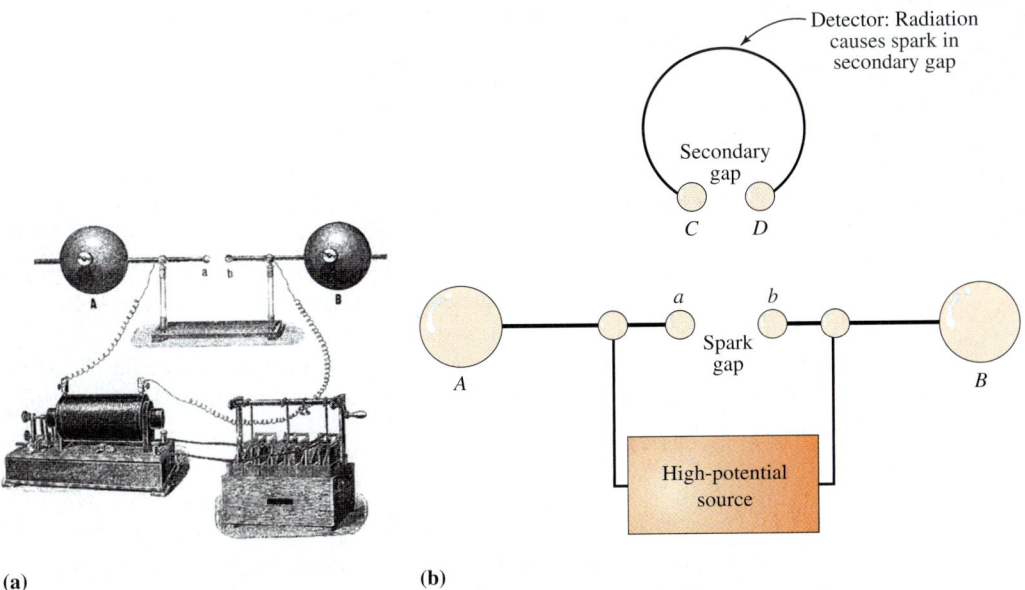

▲ **FIGURE 34–7** (a) Hertz's apparatus for the detection of electromagnetic radiation. (b) Schematic diagram of Hertz's apparatus. The radiation propagates from the region between the oscillating spark *ab* to the gap *CD*, which detects the radiation produced at gap *ab* by forming its own sparks.

952 | Maxwell's Equations and Electromagnetic Waves

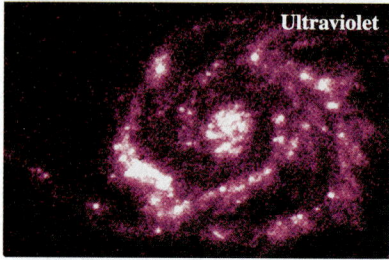

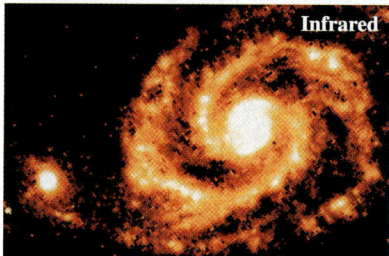
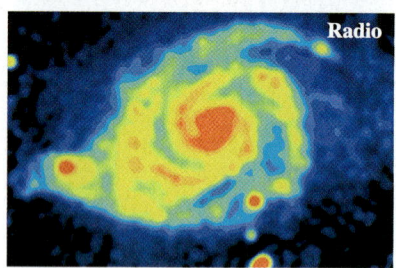

(a)

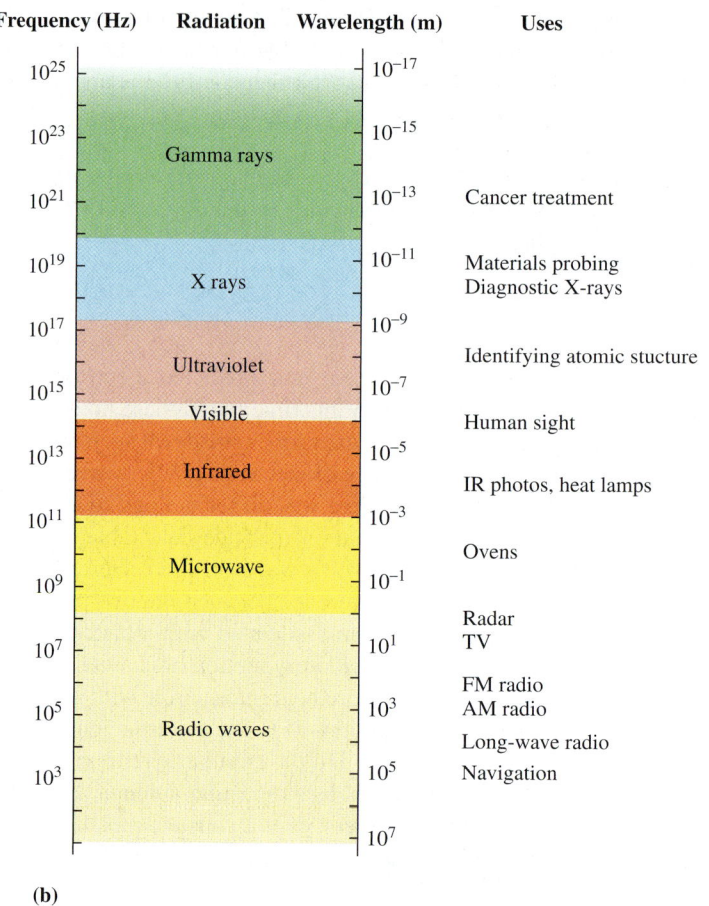

(b)

▲ **FIGURE 34–8** (a) These spectacular photos taken of the Whirlpool galaxy reveal different details, because they record radiation in different frequency ranges. (b) Different frequency regions of the full spectrum of electromagnetic radiation have specific names, as well as uses.

work, the electromagnetic spectrum has been explored across an enormous range of frequencies (Fig. 34–8). The radiation corresponding to a given part of the spectrum often has its own name—this includes visible light, ultraviolet radiation, infrared radiation, microwaves, radio waves, X rays, and gamma rays. The entire spectrum occurs in nature, but many of these names have been introduced in connection with technologies. You can say that we have been enormously successful at harnessing vast ranges of the spectrum.

Our discussion to this point has involved the formation of waves in empty space. Electromagnetic waves also propagate in matter. In transparent nonmetallic media, Maxwell's equations are modified only slightly. We mean by the term "transparent" that waves pass through—in other words, they propagate within the medium. Waves may or may not propagate depending on the wavelength and the medium. Ozone, a layer of which occurs high in the atmosphere, is transparent to visible light but much less so to ultraviolet light, which is composed of waves with wavelengths immediately below those of visible light. The speed of electromagnetic waves propagating through a transparent medium is reduced by a factor n according to

$$v = \sqrt{\frac{1}{\mu \varepsilon}} \equiv \frac{c}{n}. \tag{34–16}$$

The quantity n is the **index of refraction** of a given medium. In all except ferromagnetic materials (introduced in Chapter 31), μ is very close to μ_0 and $\varepsilon = \kappa \varepsilon_0$, where κ is the dielectric constant of the medium. Thus

$$v = \sqrt{\frac{1}{\mu_0 \varepsilon_0 \kappa}} = \frac{c}{\sqrt{\kappa}}; \tag{34–17}$$

in other words, the index of refraction $n = \sqrt{\kappa}$. It should be noted that the dielectric constant, and hence the wave speed, can depend on the frequency of the electromagnetic wave. When the speed of the wave depends on the frequency, the medium is said to be *dispersive*. The rainbow is an example of a phenomenon associated with dispersion.

THINK ABOUT THIS...
HOW DO RADIO WAVES TRAVEL AROUND THE GLOBE?

Driving at night in northern Minnesota, you may well hear a clear AM radio broadcast emanating from Florida. But the horizon seen from the antennas that produce these radio waves, which are electromagnetic radiation with wavelengths measured in tenths of kilometers, is substantially less than 100 km. The AM radio waves that you receive from so far away must travel through the atmosphere in what is effectively a curved path around Earth. The mechanism for this process lies in the *ionosphere*. This is a complex region of the atmosphere that extends from about 50 km to about 300 km from the ground. Short-wavelength radiation from the Sun and cosmic-ray particles that strike the upper atmosphere are energetic enough to separate some electrons from the atoms in this region, and this process of *ionization* leads to the presence of some *free electrons* in the ionosphere, forming a gas of electrons and ions known as a *plasma*. The structure of the ionosphere is that of a series of layers, with the most persistent ionization in the upper layers.

Radio waves that arrive at the ionosphere cause the free electrons to oscillate, and we have already pointed out that an accelerating charge produces radiation. This reradiation makes waves whose frequency matches the frequency of the oscillating charge. It sends back radio waves like those that have impinged on the plasma—we can refer to this as reflection. Through a process that involves the group behavior of large numbers of free electrons in the plasma, strong reflection occurs over a limited range of frequencies: Only radiation with a frequency below what is called the *plasma frequency* is reflected, or, in other words, radiation with wavelengths above a certain limit. The plasma is transparent to waves above that frequency. Now the plasma frequency varies with the density of free electrons n_e as $\sqrt{n_e}$, and for typical numbers in the ionosphere the plasma frequency is of the order of 10^7 Hz. This frequency is in the range of FM radio transmission but a factor of 100 above the frequency range of AM transmission. Thus AM radio signals are well reflected by the ionosphere, whereas FM signals pass through it. This allows AM signals to "bounce" their way to other locations through successive reflections from the bottom of the ionosphere and Earth's surface (Fig. 34–9).

Where, then, does the day–night difference come in? During the day, the Sun's radiation creates ionization. At night, when there is no ionizing radiation from the Sun, electrons recombine with their parent ions, and the ionosphere's composition changes. In particular, the lowest layer of the ionosphere, the layer from 50 km to 100 km high, essentially disappears as a plasma. The layer from which the radio waves reflect is now at a much higher altitude than during the day, and many fewer bounces are needed to reach a distant location. Since some energy is lost from the wave with each bounce, the distance the wave will travel at night is greatly increased over the distance it travels during the day.

The entire subject of the relation between the ionosphere and different communication bands (ranges of wavelength) is a rich one, and today's technologies make extensive use of knowledge about these effects.

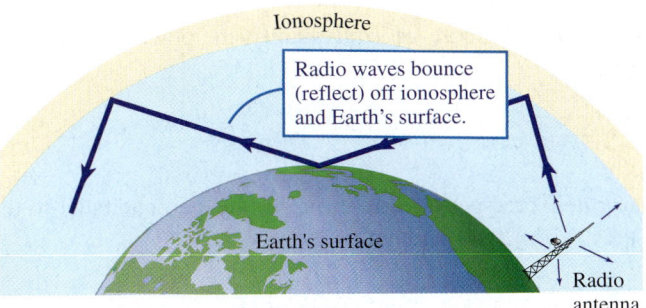

▲ FIGURE 34–9 For a certain range of frequencies the waves emitted by an antenna will be reflected back toward Earth by the ionosphere. That range includes AM radio waves.

34–3 Energy and Momentum Flow

The Energy of Electromagnetic Waves

We can get sunburned because electromagnetic waves carry energy. We can calculate the energy in electromagnetic waves from our earlier results, in particular Eq. (32–17), on the energy density in electric and magnetic fields:

$$u = \frac{1}{2}\left(\frac{B^2}{\mu_0} + \varepsilon_0 E^2\right) = \frac{\varepsilon_0}{2}\left(\frac{B^2}{\mu_0 \varepsilon_0} + E^2\right) = \frac{\varepsilon_0}{2}(c^2 B^2 + E^2). \quad (34\text{–}18)$$

Let's apply this result to an electromagnetic wave traveling in the z-direction. The fields are given by Eqs. (34–8) and (34–13):

$$E_y = E_0 \cos(kz - \omega t + \phi) \quad \text{and} \quad B_x = -B_0 \cos(kz - \omega t + \phi),$$

where $E_0 = cB_0$. For this wave, the energy density is

$$u = \frac{\varepsilon_0}{2}(c^2 B_0^2 + E_0^2) \cos^2(kz - \omega t + \phi). \quad (34\text{–}19)$$

In this expression, the two terms are the contributions of the magnetic and electric parts of the wave, respectively. Because $E_0 = cB_0$, *the energy contained in an electromagnetic wave is shared equally between the magnetic field and the electric field.* Equivalently, we could take the contribution of either the electric or the magnetic terms and multiply by 2 to find the total energy density in an electromagnetic wave:

$$u = \varepsilon_0 E^2 = \frac{1}{\mu_0} B^2. \quad (34\text{–}20)$$

For practical purposes, the oscillations in electromagnetic waves are so rapid that we can simply consider the average of the energy density over one period, which we write as $\langle u \rangle$. The average of the cosine-squared factor in Eq. (34–19) over one period is one-half, so that

$$\langle u \rangle = \frac{\varepsilon_0}{2} E_0^2 = \frac{1}{2\mu_0} B_0^2. \quad (34\text{–}21)$$

The Transport of Energy

The $\cos^2(kz - \omega t + \phi)$ time and space dependence of the energy density in Eq. (34–19) shows that the energy in an electromagnetic wave is itself *transported* as a wave; it travels at speed $v = \omega/k = c$ in the z-direction. The amount of energy dU_t transported across a surface of area A perpendicular to the transport direction in a time interval dt is the energy contained in the volume of area A times the distance $c\,dt$ (Fig. 34–10); that is, the energy density u times this volume,

$$dU_t = u(Ac\,dt).$$

Thus the rate of energy transport, or, equivalently, the power delivered by the electromagnetic wave, is

$$\frac{dU_t}{dt} = cuA.$$

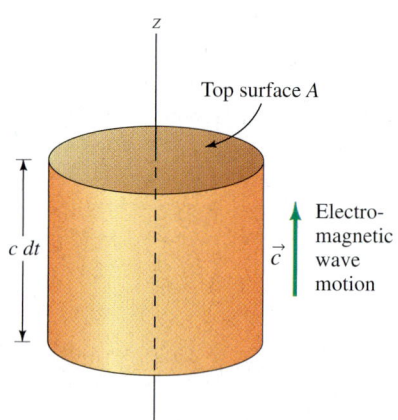

▲ **FIGURE 34–10** Electromagnetic energy contained in a volume $Ac\,dt$ is delivered in time dt to the area A.

Finally, the power delivered per unit area to a surface perpendicular to the direction of propagation—the *energy flux*—is given by

$$S = \frac{1}{A}\frac{dU_t}{dt} = cu. \quad (34\text{–}22)$$

This flux has a direction associated with it and is more properly described as a vector. The vector $\vec{S}$ that describes the energy flux is the **Poynting vector**, given by

$$\vec{S} = \frac{1}{\mu_0}\vec{E} \times \vec{B}. \qquad (34\text{–}23)$$

THE POYNTING VECTOR

Let's check that the magnitude and direction of the vector $\vec{S}$ are indeed correct. Because the fields $\vec{E}$ and $\vec{B}$ are at right angles to each other, we notice that the magnitude of $\vec{S}$ is just EB/μ_0. This can be rewritten in terms of the electromagnetic energy density, $u = \varepsilon_0 E^2$, as

$$S = \frac{1}{\mu_0}EB = \frac{\varepsilon_0}{\varepsilon_0\mu_0}E\left(\frac{E}{c}\right) = \varepsilon_0 c^2 \frac{E^2}{c} = c\varepsilon_0 E^2 = cu. \qquad (34\text{–}24)$$

This is the same magnitude we found in Eq. (34–22). As for the direction, recall that the vector product of two vectors is perpendicular to both of them. From Fig. 34–6, we see that the direction of the vector product $\vec{E} \times \vec{B}$ is the $+z$-direction, the direction of wave propagation. More generally, the transversality of the electromagnetic wave will always lead to a Poynting vector that lies along the direction of propagation.

The energy density and the magnitude of the Poynting vector each vary with time. The value of the magnitude of $\vec{S}$ time-averaged over one cycle of the electromagnetic wave is called the **intensity** I of the radiation. Equation (34–24) allows us to relate the intensity to the amplitude E_0 of the electric field in the wave:

$$I = \langle S \rangle = c\varepsilon_0 \langle E^2 \rangle = \frac{1}{2}c\varepsilon_0 E_0^2. \qquad (34\text{–}25)$$

Note from Eq. (34–21) that the intensity is also related to the average energy density in the wave, $I = c\langle u \rangle$.

EXAMPLE 34–3 A characteristic number for the rate per unit area at which solar energy is delivered to a spot on Earth's surface is 1000 W/m². (This energy consists mainly of electromagnetic radiation in the visible range of wavelengths.) Use this number to estimate the amplitude of the electric and magnetic fields in the waves that deliver this energy.

Strategy We use the fact that the power (energy per unit time) delivered by an electromagnetic wave per unit area is S, whose average value is simply related to the square of the amplitude of the electric field. This result is independent of the wavelength of the radiation. More particularly, the equation $I = \langle S \rangle = \frac{1}{2}c\varepsilon_0 E_0^2$ gives the electric field amplitude in terms of the power delivered per unit area—we can solve this for E_0. Once this is calculated, we can easily obtain B_0 using $B_0 = E_0/c$.

Working It Out The relation between I and E_0 gives

$$E_0 = \sqrt{\frac{2I}{c\varepsilon_0}} = \sqrt{\frac{2(1000\text{ W/m}^2)}{(3 \times 10^8\text{ m/s})(9 \times 10^{-12}\text{ C}^2/\text{N}\cdot\text{m}^2)}}$$
$$= 0.9 \times 10^3\text{ V/m}.$$

In turn,

$$B_0 = \frac{E_0}{c} = \frac{0.9 \times 10^3\text{ V/m}}{3 \times 10^8\text{ m/s}} = 0.3 \times 10^{-5}\text{ T}.$$

What Do You Think? An inventor seeking financing claims that he can block just the electric field in a light wave, leaving all the power in the magnetic field. Should you invest?

CONCEPTUAL EXAMPLE 34–4 Compare the electric field in sunlight to the electric field in the beam of a 0.1-W laser that covers an area of 1 cm². (The laser rating is the power in the beam.)

Answer The field is proportional to $\sqrt{I}$. For sunlight I is known to be 1000 W/m², and for the laser it is $(0.1\text{ W})/(1\text{ cm}^2) = 10^3\text{ W/m}^2$. The fields will be comparable.

Lasers with much higher power than the one in the example above can generate fields strong enough to rip atoms apart! On a less dramatic level, an interesting device makes direct use of the electric fields in light for manipulation of large molecules and nanostructures. In the *optical tweezers*, a laser beam is focused to a very sharp point; the rms electric field is largest at that point and falls off away from it. When the focal point is near a tiny piece of dielectric material that may be chemically attached to a large molecule such as DNA, a dipole moment is induced in the material and in the same way that bits of paper with induced dipole moments are attracted to larger values of nonuniform

956 | Maxwell's Equations and Electromagnetic Waves

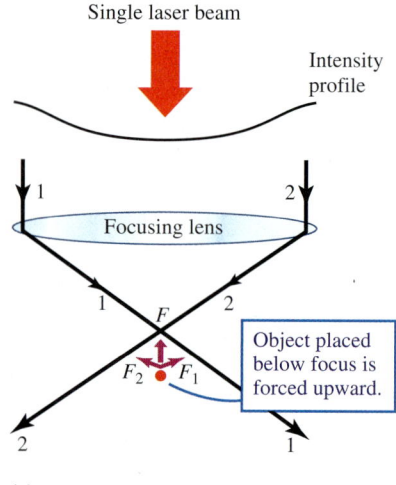

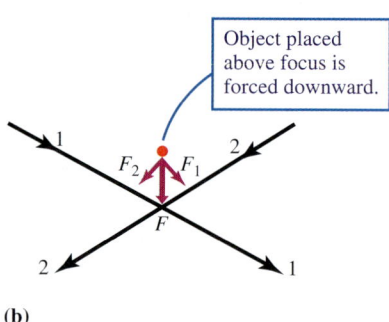

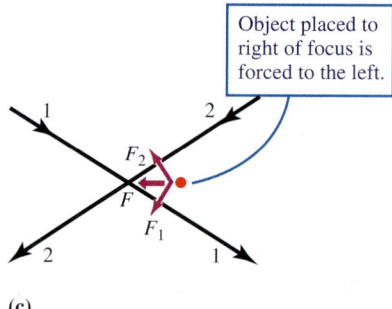

▲ **FIGURE 34–11** The focal point of an intense beam of light, a location where the beam is concentrated, acts to attract objects within which the electric field associated with the beam induces a dipole moment. When a laser provides the beam, the focal point can be so "tight" that the forces it exerts make a kind of optical tweezer that can manipulate nanoscale objects such as DNA molecules.

fields, it is attracted to the focal point of the nonuniform rms electric field of the laser beam. Figure 34–11 describes the action of the tweezers in schematic form. Once the system has been "gripped," the laser beam can be moved, moving the system with it.

Momentum in Electromagnetic Waves

An electromagnetic wave carries momentum as well as energy. To see this qualitatively, let's reconsider a plane wave that travels in the z-direction, with the electric and magnetic fields along the x- and y-directions, respectively. When such a wave impinges on a particle (you can think of an electron if you like, although we'll assume for simplicity that the charge is positive) with charge $+q$, the fields exert forces on the particle. Suppose that, at a given time, the oscillating electric field of the wave points in the $+x$-direction so there is a force qE in the $+x$-direction. The charge accelerates and moves with some velocity $\vec{v}$ in the $+x$-direction. If $\vec{E}$ points in the $+x$-direction, then $\vec{B}$ (which oscillates in phase with $\vec{E}$) must point in the $+y$-direction. The magnetic force $q\vec{v} \times \vec{B}$ on the charge acts in the $+z$-direction and pushes the charge in that direction. When the electric field later reverses sign, the electric force on the charge acts in the $-x$-direction, and the velocity then has a component in the $-x$-direction. The magnetic field has also reversed sign, but the magnetic force *continues to act in the $+z$-direction*. All the forces in the x- and y-directions average to zero, but the force in the z-direction is always positive, and there is a net force in the $+z$-direction. The charge has an increased momentum in the $+z$-direction; by momentum conservation, this momentum had to have been supplied by the electromagnetic wave.

By evaluating the amount of momentum the charged particle above picks up in a period of time dt, and recalling that the wave advances with speed c, so that a length $c\,dt$ of wave impinges on the particle in that time, we can show that the **momentum density** of an electromagnetic wave—the amount of momentum carried by the wave per unit volume—is $\vec{S}/c^2$. The magnitude of the momentum density is given by

$$\frac{S}{c^2} = \frac{u}{c}, \qquad (34\text{–}26)$$

and the direction is that of $\vec{S}$, along the direction of wave propagation.

Radiation Pressure: When electromagnetic waves are absorbed or reflected as they enter into matter, they transfer momentum to the material on which they impinge. The particle upon which the wave fell above is an example. The rate at which momentum is transferred per unit area is a force exerted per unit area; that is, a pressure: **radiation pressure**. When an electromagnetic wave is absorbed, which happens when light falls on a black surface, all the momentum carried by the wave is transferred to the surface. The amount of radiation-produced momentum that falls perpendicularly on a surface A in a time interval dt is given by the momentum density multiplied by the volume $A(c\,dt)$. Thus the momentum dp transferred is

$$dp = \left(\frac{S}{c^2}\right)(Ac\,dt) = \frac{S}{c}A\,dt.$$

The force per unit area (radiation pressure) is given by

$$\frac{F}{A} = \frac{1}{A}\frac{dp}{dt} = \frac{1}{A}\frac{S}{c}A = \frac{S}{c} = u, \qquad (34\text{–}27)$$

where we have used Eq. (34–22). This expresses the radiation pressure when radiation is totally absorbed. When the electromagnetic wave is reflected, which happens when it falls on a shiny, metallic surface, then the momentum of the wave is reversed upon reflection. Thus the momentum density transferred to the metallic surface is double the previous result, $2u/c$, and the radiation pressure is $2u$. (As we remarked above, a microscopic view of reflection is that the wave imparts momentum to a charged constituent, which then in turn reradiates and recoils in the process. This accounts for the factor of two.)

CONCEPTUAL EXAMPLE 34–5 Above we assumed that when light falls on a mirror and is reflected, the energy reflected is equal to the incident energy. Is this reasonable?

Answer Reflection is actually the reradiation of the wave by free electrons in the material. Metallic reflectors contain some free electrons, and that is why they make mirrors. (As for the shiny surface, that is another story, to be discussed in Chapter 35.) But the motion of these electrons is a local current, and there will be some small resistance, ohmic heating, and associated energy absorption. This means that the reflected energy will be slightly lower than the incident energy. This is typically a small effect.

EXAMPLE 34–6 Consider a 10^4-W searchlight that projects a cylindrical beam 0.6 m in diameter. What is the radiation pressure on a metallic mirror placed at right angles to the beam? Ignore the spreading of the beam.

Strategy The power delivered by the electromagnetic wave to a surface at right angles to the beam is given by

$$P = (\text{energy flux})(\text{area}) = SA = cuA,$$

where u is the energy density in the beam at the surface and A is the area of the beam. Given the area of the beam $A = \pi r^2$, we can calculate the energy density u and therefore the radiation pressure.

Working It Out The area is $A = \pi r^2 = \pi(0.3 \text{ m})^2 \cong 0.3 \text{ m}^2$. Thus

$$u = \frac{P}{Ac} = \frac{10^4 \text{ J/s}}{(0.3 \text{ m}^2)(3 \times 10^8 \text{ m/s})} \cong 10^{-4} \text{ J/m}^3.$$

In turn, the radiation pressure is

$$\frac{F}{A} = 2u \cong 2 \times 10^{-4} \text{ N/m}^2.$$

What Do You Think? Just how small is this pressure? What thickness of a water layer on the ground would exert this kind of pressure?

EXAMPLE 34–7 The intensity (average energy flux) of solar radiation that falls on Earth is 1.4×10^3 W/m². Compare the force exerted by solar radiation on a totally absorbing dust particle of diameter 10^{-6} m and mass density 3×10^3 kg/m³ with the gravitational force on the particle due to the Sun. The particle is located at a distance from the Sun equal to the Earth–Sun distance, $R = 1.5 \times 10^{11}$ m. The mass of the Sun is $M_{\text{Sun}} = 2 \times 10^{30}$ kg.

Strategy We are given the intensity $I = uc$, so that we have the data to find the radiation pressure, just u for an absorbing object like the dust particle. (We assume time averages throughout.) Since we can work out the area A of the dust particle, the radiation force F on the particle is pressure × area = $uA = IA/c$. Knowing the size of the particle and its density, as well as the mass of the Sun, and the distance from the Sun, Newton's law of gravitation determines the gravitational force on it, magnitude F_g.

Working It Out The area presented by the dust particle is $A = \pi(d/2)^2 = \pi(0.5 \times 10^{-6} \text{ m})^2 = 0.8 \times 10^{-12}$ m². This gives for the radiation force

$$F = uA = \frac{IA}{c} = \frac{(1.4 \times 10^3 \text{ W/m}^2)(0.8 \times 10^{-12} \text{ m}^2)}{3 \times 10^8 \text{ m/s}}$$

$$= 0.4 \times 10^{-17} \text{ N}.$$

As for the gravitational force on the particle, the mass of the dust particle is $m = \rho V$, and the force of gravity is $F_g = \dfrac{GmM_{\text{sun}}}{R^2}$. Numerically, the mass m of the dust particle is

$$m = \rho V = \frac{4}{3}\pi\left(\frac{d}{2}\right)^3 \rho = \frac{4}{3}\pi(0.5 \times 10^{-6} \text{ m})^3(3 \times 10^3 \text{ kg/m}^3)$$
$$= 1.6 \times 10^{-15} \text{ kg},$$

hence the gravitational force on it is

$$F_g = \frac{GmM_{\text{sun}}}{R^2}$$
$$= \frac{(6.67 \times 10^{-11} \text{ N} \cdot \text{m}^2/\text{kg}^2)(1.6 \times 10^{-15} \text{ kg})(2 \times 10^{30} \text{ kg})}{(1.5 \times 10^{11} \text{ m})^2}$$
$$= 0.9 \times 10^{-17} \text{ N}.$$

We see that the two forces are comparable, so the radiation pressure can keep the dust particle at its current location. It is therefore no coincidence that this kind of dust grain is typical of those found in interplanetary space.

34–4 Dipole Radiation

Accelerating charges produce electromagnetic waves. For waves with wavelengths much longer than light (radio, TV, cell phones, etc.), we refer to the systems in which accelerating charges initiate electromagnetic waves as *broadcasting antennas*; we refer to the systems in which we detect the response of charges to the fields of an electromagnetic wave as *receiving antennas*. Here, we shall describe one of the simplest systems that can act as an antenna, the *dipole antenna*. Radiation emitted with the characteristic pattern of this antenna is called **dipole radiation**.

A dipole antenna is formed by charges that move back and forth in harmonic motion along a line. The charge configuration within the antenna is that of a set of electric dipoles, with one or both of the two charges making up each dipole oscillating (Fig. 34–12). Such an antenna is easy to construct using an AC generator. When the dimensions of the antenna are small compared with the wavelength of the radiation, the current

958 | Maxwell's Equations and Electromagnetic Waves

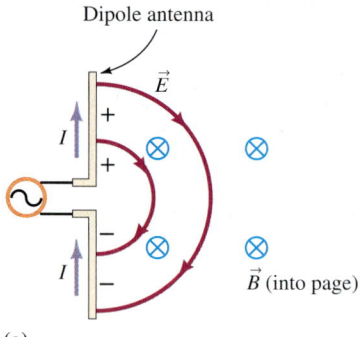

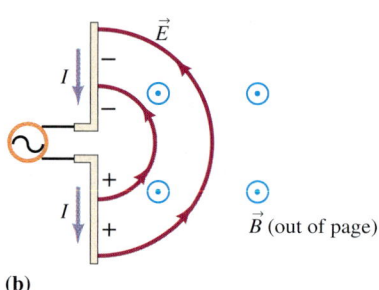

▲ **FIGURE 34–12** Pairs of equal but opposite charges move in simple harmonic motion along a line (vertical, here). These pairs of charges form a dipole antenna. In (a) the current is moving in one direction and in (b) it is moving in the opposite direction. As a result in (a) and (b) the two sides of the antenna are oppositely charged, and the field directions are reversed.

throughout the antenna is in phase, and the resulting electric and magnetic fields are oriented as shown in Fig. 34–12. They form an outgoing electromagnetic wave whose frequency is that of the oscillating charges.

How the Intensity of Radiation from an Antenna Decreases with Distance

Because electromagnetic waves are so often used for communication, it is clear that one of the most important characteristics of the electromagnetic waves radiated by an antenna is the rate at which the intensity decreases with increasing distance from the antenna. To understand this feature, we can consider any radiator—it is not important that the antenna be a dipole antenna. Let us consider an antenna that radiates electromagnetic waves symmetrically in all directions, such as the Sun or a lightbulb. From a distance, the electromagnetic waves emitted by the Sun appear to come from a point source, and we can use this fact to study the magnitudes of the electric and magnetic field strengths. As we learned in Section 34–3, the energy flux (the rate of flow of energy per unit area) is given by $S = cu = c\varepsilon_0 E^2$. The total energy flow per unit time (the power) across any surface is

$$P = \int_{\text{surface}} \vec{S} \cdot d\vec{A}.$$

If the magnitude of the electric field is independent of direction, as we would expect for a point source, then the total rate of energy flow across a sphere of radius R centered on the source is

$$P = c\varepsilon_0 E^2 (4\pi R^2). \tag{34-28}$$

But all the radiation emitted must eventually pass through any sphere that surrounds the source, whatever its radius, so P does not depend on R. From our expression for P, we see that this is possible only if the electric field decreases as $1/R$. The magnetic field must similarly fall off as $1/R$, because the magnetic and electric fields only differ by a factor of c in an electromagnetic wave. Contrast this result with the typical $1/R^2$ behavior of static electric fields (Chapter 22).

We can express the result of Eq. (34–28) in terms of intensity. The quantity $c\varepsilon_0 E^2$ is the magnitude of the Poynting vector, and its average value, which is defined as the intensity I, is one-half this value [Eq. (34–25)]. Thus

$$P = 2I(4\pi R^2). \tag{34-29}$$

Because P is independent of R, the intensity of the electromagnetic wave from a point source decreases as $1/R^2$. The next two examples illustrate this important property.

CONCEPTUAL EXAMPLE 34–8 In Example 34–7 we studied the balance of radiation pressure from sunlight with the gravitational force from the Sun for a dust particle as distant from the Sun as is Earth. We found that these forces balanced, suggesting that this is a mechanism for holding these particles at their current location. How would the argument change if the dust particle were as far away from the Sun as Jupiter?

Answer There is no change; the forces would still balance. Both the intensity of the Sun's radiation and the gravitational force due to the Sun have inverse square dependence on the distance from the Sun. If they balance at Earth's orbit, they will balance at Jupiter's.

EXAMPLE 34–9 A 100-W lightbulb emits electromagnetic radiation equally in all directions. Assume that 10 percent of the 100 W is converted into radiation in the visible spectrum. What is the intensity of the visible radiation 1.5 m from the bulb?

Setting It Up We are given the total power P and the power P_0 that appears as visible light, as well as the radius R of the sphere over whose surface the power is distributed. The desired intensity is denoted as I.

Strategy We may then use the relation $P = 2I(4\pi R^2)$ to calculate I as a function of R.

Working It Out We have $P_0 = 10\%$ of 100 W, so that $P_0 = 10$ W.

This gives

$$I = \frac{P_0}{8\pi R^2} = \frac{10 \text{ W}}{8\pi (1.5 \text{ m})^2} = 0.2 \text{ W/m}^2.$$

Compare this value to the 1400 W/m² in sunlight incident at the top of Earth's atmosphere, or to the 1000 W/m² of solar energy that reaches Earth's surface. About half this solar energy is in light in the visible part of the spectrum, whereas the lightbulb emits most of its energy in the infrared region of the spectrum.

What Do You Think? Does the fact that a lightbulb radiates equally in all directions mean that it is not an antenna?

In our discussion of static electric fields due to a point source, we used symmetry considerations to argue that the *electric field vector points in a radial direction*. This cannot be the case for electromagnetic waves emitted from a point source, however, because we showed that the fields for such waves are transverse to the direction in which they travel. When the waves come from a point source, they travel outward radially. So given the symmetry, how does the electric (or magnetic) field "know" in what direction to point for a point source? The resolution to what appears to be a paradox is simple: *There are no truly pointlike sources of electromagnetic radiation*. The Sun is not a point; it radiates because charges within it move and accelerate. As far as this discussion is concerned, we can think of the Sun as a large collection of dipole antennas with random orientation.

The Angular Pattern of Dipole Radiation

The variation of the intensity of electromagnetic radiation with the angle of observation is an important property of radiation from an antenna. Such patterns are an important element of the design of real antennas. No broadcaster wants to use expensive electric power to send a signal where nobody lives. In our simple dipole antenna (Fig. 34–12), charges execute simple harmonic motion along the antenna direction (we shall call this the z-axis). The motion of the charge determines a preferred direction—along the z-axis. An observer looking along the z-axis would see no motion. An observer looking along a line perpendicular to the z-axis would see the full range of motion of the charges. An observer at an angle θ to the z-axis would see the charges move harmonically with an amplitude reduced from the full amplitude by a factor $\sin \theta$. The electric field that the observer sees is thus proportional to $\sin \theta$. Because the intensity is proportional to the square of the electric field in the wave, the intensity of the radiation emitted by a dipole antenna along the direction of θ is proportional to $\sin^2 \theta$:

$$S \propto \frac{\sin^2 \theta}{R^2}. \qquad (34\text{–}30)$$

Here we have also included the $1/R^2$ factor that describes how the intensity varies with the distance R from the antenna. This intensity pattern describes the **angular distribution** of the power emitted by charges oscillating along a line (Fig. 34–13): no signal along the direction of the antenna and a maximum signal perpendicular to it.

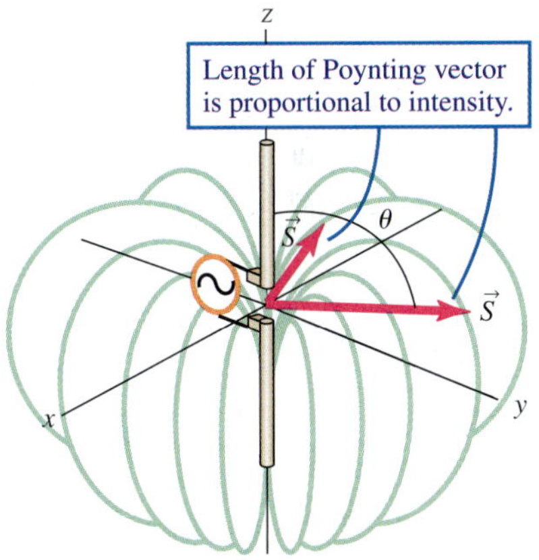

▲ **FIGURE 34–13** The intensity distribution S for a radiating dipole antenna. The curve illustrates the relative amount of power emitted as a function of the angle θ. The Poynting vectors for two different positions are drawn.

34–5 Polarization

A little experimentation with *polarizing* sunglasses at the seashore shows that a change in the orientation of the glasses' axis results in a change of the intensity of the light transmitted. This occurs because the sunglasses are made of a material that is sensitive to the direction of the electric field. As we shall see, light reflected from water or sand is **polarized**, meaning that its electric field is oriented in a particular way; the glasses "detect" the polarization of the electromagnetic wave (light). These sunglasses also have the effect of passing light whose electric field is aligned in a certain direction, and if two pairs of glasses are aligned in a "crosswise" fashion, no light passes the pair (Fig. 34–14).

If polarized light contains its electric field oriented in a particular direction, what is unpolarized light? Unpolarized light consists of a mixture of light waves with their electric fields aligned in different directions—always perpendicular to the propagation direction. The alignment must be such that *no direction of $\vec{E}$ is preferred*. In sampling a beam of unpolarized light, one would be just as likely to find the electric field aligned in one transverse direction as in another.

Let us reconsider a charge that oscillates along the z-axis, as in Fig. 34–13. We found that, if we look along the x-direction, we would detect an electromagnetic plane wave that propagates along the x-direction, with an electric field aligned along the z-direction: $\vec{E} = E_z \hat{k}$, with $E_z = E_0 \cos(kx - \omega t)$. (Setting the phase $\phi = 0$ won't

▲ **FIGURE 34–14** The extent to which polarizing materials pass light or other electromagnetic waves depends on how an internal "axis" that they possess is oriented relative to the orientation of the electric field vector of the light. Little light passes through the region where the glasses' axes are crossed—light passing through one pair cannot pass through the second pair. But when the axes are aligned, the same amount of light passes as for a single lens.

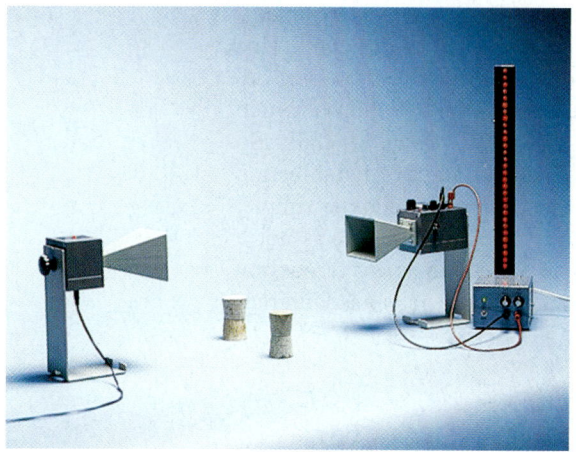
(a)

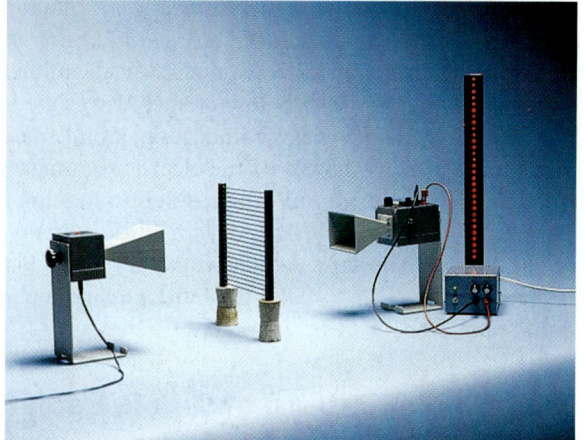
(b)

▶ FIGURE 34–15 (a) A receiver and detector for determining the polarization of microwave radiation. The red lights indicate the presence of a signal. (b) A horizontal grid is placed between them, oriented so that the radiation passes. (c) The grid is now oriented vertically so that the radiation cannot pass. The grid's orientation reveals the polarization of the radiation, which is vertical.

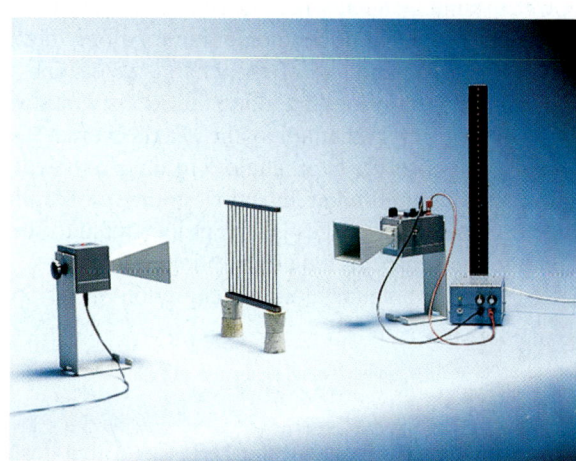
(c)

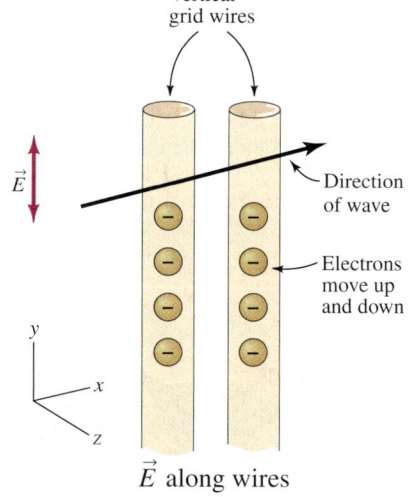
(a) $\vec{E}$ along wires

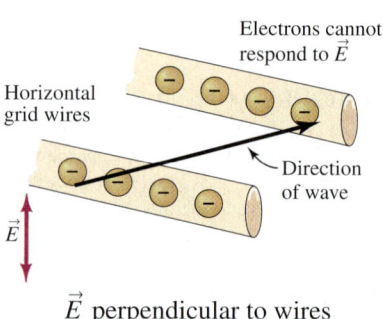
(b) $\vec{E}$ perpendicular to wires

change anything here.) We say that the light, or indeed electromagnetic radiation of any wavelength, is **linearly polarized** along the direction of the electric field vector—in this case, the light propagating along the x-direction is polarized along the z-direction. Suppose that a dipole antenna emits radiation with a wavelength in the centimeter range. The polarization can be detected as follows: A current is induced in a receiving antenna, and the rms current detected can be measured (Fig. 34–15). Place a metal grid (such as an oven rack) between the transmitter and the receiver. The diameter of the wires in the grid should be much less than the wavelength of 1 cm, and the grid spacing should be on the order of 1 cm or less. Then, the intensity of the radiation at the receiver depends on the orientation of the metal grid, and we say that the grid acts as an **analyzer** for the polarization.

Here is why the grid acts as an analyzer. The electrons in the grid wires are accelerated by the electric field of the wave along the field direction. When the wires in the grid are parallel to the electric field, the electrons in the grid wires can move in response to the field (Fig. 34–16a). Because they are set into motion, they *absorb large amounts*

◀ FIGURE 34–16 (a) When the grid wires are oriented in the direction of the electric field of an incoming wave, the electrons of the grid wires can respond and absorb energy from the wave. The transmitted wave is reduced in amplitude. (b) When the wires are perpendicular to the electric field of the wave, the electrons of the grid wires are constrained and cannot respond. Little energy is absorbed, and the wave passes through with little attenuation.

of energy from the field. This energy is lost in ohmic heating. The electric field of the radiation that passes through is reduced in magnitude because energy has been removed from the incident wave. In effect, the grid is opaque to the polarized radiation when it is oriented along the electric field vector. When the wires in the grid are perpendicular to the z-direction (Fig. 34–16b), the electrons in the metal are accelerated across the diameter of the wire. But, because the diameter is small, the electrons in the grid wires cannot respond fully and cannot absorb large amounts of energy from the incident wave. The energy remains in the transmitted wave. The grid acts as if it were transparent when it is oriented perpendicular to the polarization direction of the wave.

Certain materials, such as Polaroid, are analyzers for visible light. They are made of long molecules aligned parallel to each other. Electrons can easily move along the molecules but not across them and, because the molecular spacings are appropriate to the wavelengths of visible light, these materials behave like the microwave grid does.

A microwave grid or a piece of Polaroid is not simply an analyzer; it is also a **polarizer**: The microwave radiation that passes through the grid becomes polarized perpendicular to the grid wires. This is easily understood. Suppose that unpolarized microwave radiation approaches the grid. *Unpolarized radiation* is radiation that consists of a mixture of waves whose electric field vectors are as likely to point in any one direction as in another, as long as the direction is perpendicular to the direction of wave propagation. As we have seen, only those waves with the electric field oriented perpendicular to the grid can pass through, whereas the waves with the electric field parallel to the grid are absorbed. Thus the radiation that passes through the grid has become polarized perpendicular to the direction of the grid.

Malus's Law

When unpolarized radiation moving in the z-direction falls on a polarizer whose polarizing axis (the axis perpendicular to the "grid wires" within the polarizing material) makes an angle θ with the x-axis, for example, then only the component of any electric field along the polarizing axis will pass through. What emerges is radiation that is linearly polarized along a line that makes an angle θ with the x-axis. We take the magnitude of the electric field that has passed through the polarizer to be E_0. The corresponding intensity is then

$$I_0 = \langle S \rangle = (\text{a constant})E_0^2. \tag{34–31}$$

Let's now place a second polarizer so that its axis lies along the x-axis (Fig. 34–17). The amplitude for the electric field in the wave incident on the polarizer is

$$\vec{E}_0 = (E_0 \cos\theta)\hat{i} + (E_0 \sin\theta)\hat{j}. \tag{34–32}$$

Only the component that is parallel to the axis of the second polarizer—the x-axis—passes through. Thus the field behind the second polarizer (which acts here as an analyzer) is given by $E_0 \cos\theta \,\hat{i}$. The intensity of the transmitted light is therefore

$$I = (\text{a constant})(E_0 \cos\theta)^2, \tag{34–33}$$

and the intensity of the light is reduced:

$$I = I_0 \cos^2\theta. \tag{34–34}$$

Equation (34–34) is known as **Malus's law**. In particular, when the axes of the polarizer and analyzer are perpendicular to each other ($\theta = \pi/2$), radiation is not transmitted. This case is illustrated well for the sunglasses of Fig. 34–14.

One of the important consequences of Malus's law is that when unpolarized light passes through a polarizer, it has *half* its original intensity (see Example 34–10).

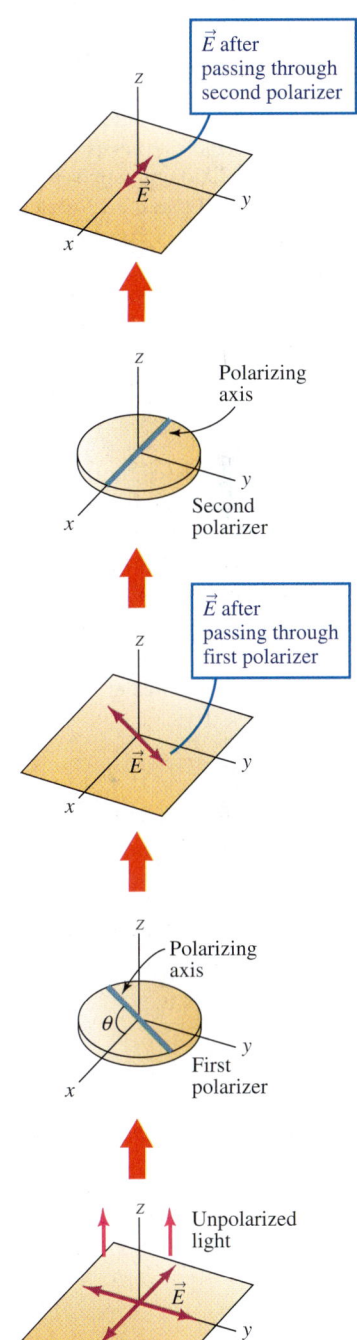

▲ **FIGURE 34–17** An unpolarized beam passes first through a polarizer whose axis makes an angle θ with the x-axis, and the beam is then polarized linearly in this direction. A second polarizer aligned with the x-axis allows only the component of the electric field aligned along the x-axis to pass.

EXAMPLE 34–10 Light passes through the glass plate of a transparency projector and emerges unpolarized with intensity I_0. (a) A Polaroid sheet is placed on the glass plate with its polarizing axis aligned with the 12-o'clock position. What are the polarization and intensity of the emerging light? (b) A second Polaroid sheet, with its polarizing axis along the 2-o'clock position, is placed over the first. Again find the polarization and intensity of the emerging light.

Setting It Up We must clarify what is meant by *unpolarized* light to understand the intensity of the emerging light in part (a). We set up the solution by supposing that the light wave propagates in the z-direction and that the 12-o'clock position is aligned along the $+y$-axis, as sketched in Fig. 34–18.

Strategy (a) The first polarizing sheet passes light with its polarization in the y-direction, so the emerging light is polarized in the y-direction. To find its intensity, we recall that unpolarized light is a mixture of waves with the electric field equally likely to lie along any (transverse) direction. If the projection of the incoming electric field on the y-axis for some particular wave in the mixture is $E \cos \theta$, then the intensity passed for that wave is $I = I_0 \cos^2 \theta$. Note that I_0 will be the same for all the incoming waves—that is what we mean by saying that no particular orientation is preferred. We must *average* this intensity over all θ, and that will give us the intensity of the light that has passed the first polarizer.

(b) The light emerging polarized along the 12-o'clock axis will be reduced in intensity according to Malus's law. The angle between the 12-o'clock direction and the 2-o'clock direction is $\pi/3$ radians.

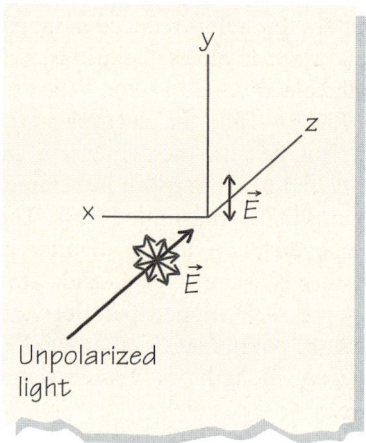

▲ **FIGURE 34–18**

Working It Out The average value of $\cos^2 \theta$ is $1/2$, giving the following answers for part (a): The polarizer is aligned with the x-direction, so that the light will be polarized in the x-direction, and the intensity of the light will be $I = I_0 \langle \cos^2 \theta \rangle = \frac{1}{2} I_0$.
(b) From Malus's law we get the intensity

$$I_1 = I \cos^2(\pi/3) = \frac{I_0}{2} \frac{1}{4} = \frac{I_0}{8}.$$

What Do You Think? Both Polaroid sheets will experience some heating due to absorption of light. In what ratio do you expect the two sheets to acquire thermal energy?

How to Produce Polarized Radiation

We have already described two ways to produce polarized radiation: by accelerating charges in an oriented dipole antenna and by passing unpolarized radiation through a polarizer. Two more ways are important in many situations.

Polarization by Scattering: If you look at the beam of an automobile headlight from the side in a rainstorm or a snowstorm, it is quite visible, because there is a pronounced scattering of the light. The light beam is much less visible from the side on a dry night. Nevertheless, even in the absence of water droplets or dust particles, light and other forms of electromagnetic radiation are scattered by air molecules. The scattering mechanism is quite different from the mechanism that operates when the light scatters from droplets, because the size of the droplets is much larger than the wavelength of the light. It can be described as follows: The oscillating electric field $\vec{E}$ of the incoming radiation sets in motion the electrons in the air molecules. The electrons act like oscillators subject to an external harmonic force and oscillate with the frequency of the incoming field. The electrons move in a plane perpendicular to the incident radiation and, if the incident wave is unpolarized, then there is no preferred direction to the electron motion as long as it occurs in the plane. An observer looking at an electron from a direction close to that of the incident radiation will see a radiated field that is unpolarized because there is no preferred direction. In contrast, an observer looking at the electron from a direction perpendicular to the direction of the incident radiation will see the electron moving in just one direction (and will not see the component of the motion toward or away from him or her). This observer thus sees 100 percent linearly polarized light (Fig. 34–19). The polarization is partial for angles between these directions. If you live where the atmosphere is clear, you can easily observe this by holding a piece of Polaroid and looking 90° away from (*but not at*) the Sun. The light intensity will change when the Polaroid is rotated, showing that the light scattered by the air molecules is polarized.

34-5 Polarization

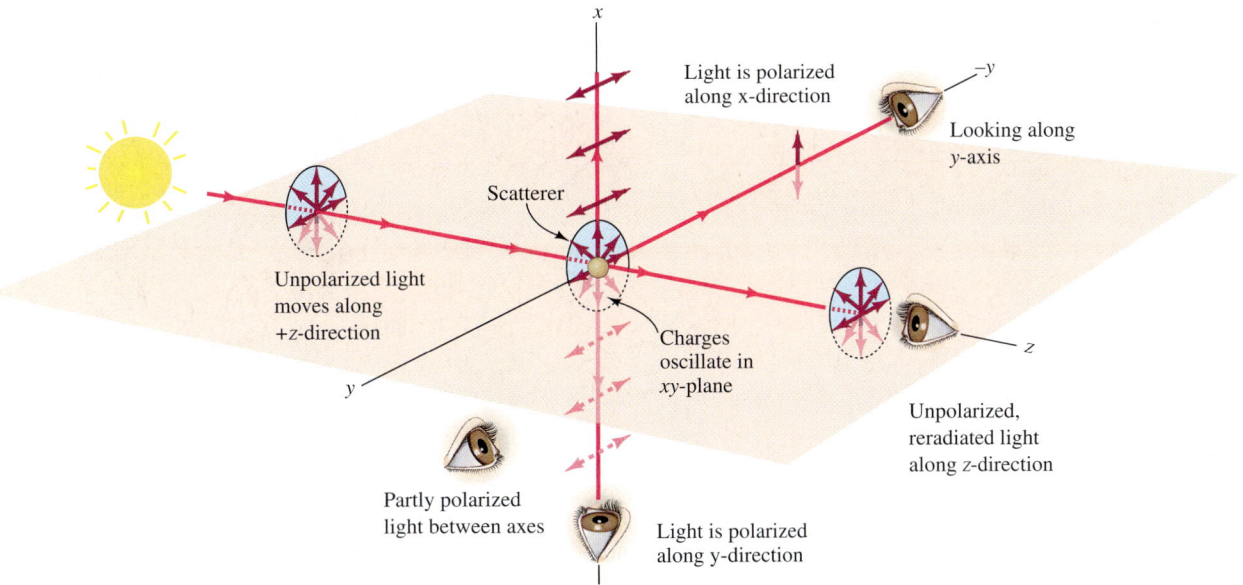

▲ **FIGURE 34–19** The polarization of radiation by scattering. An electromagnetic wave can propagate through a material because the wave's electric and magnetic fields cause electrons in the material to oscillate at the radiation frequency; these electrons in turn radiate new waves of the same frequency. The electric field of these new waves is aligned with the electrons' motion. Here, unpolarized radiation is perpendicularly incident on the xy-plane in a gas; the electric fields of the radiation lie in that plane but are otherwise unrestricted. An observer along the z-axis sees the full range of motion of the electrons in that plane and hence sees unpolarized light. An observer at 90° to the original wave direction can see a side view of the plane from which the light is radiated and hence sees light fully polarized; the polarization direction is parallel to the plane's edge. At intermediate angles, the reradiated light is partly polarized.

Polarization by Reflection: When unpolarized radiation is reflected from a surface such as glass, the reflected light is partly polarized (Fig. 34–20a). When the angle of incidence is just right, the reflected light is fully polarized (Fig. 34–20b). This is for much the same reason that scattered light is polarized (Fig. 34–19). Unpolarized light incident at an angle θ_i (*the angle of incidence*) impinges on a surface. In general, we may decompose the electric field of the incident wave into two components—each perpendicular to the direction of propagation. As Fig. 34–21 illustrates, one of these directions, the z-direction, is perpendicular to the surface of the page and parallel to the reflecting surface; we label the other the a-direction. When the wave arrives at the surface, its electric field accelerates electrons. These accelerated charges reradiate and give rise to both the transmitted and the reflected wave.

◀ **FIGURE 34–20** Radiation is polarized by reflection. (a) Here we see a shop window with oblique reflections, which are partially polarized. (b) The same scene, but with the camera lens fitted with a polarizing filter. The reflected light passing through the filter is greatly reduced.

(a)   (b)

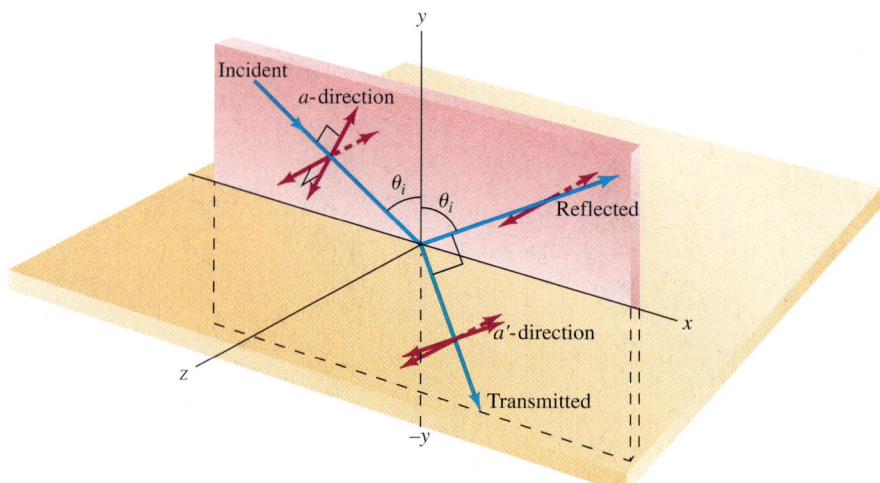

▶ **FIGURE 34–21** There is an angle of incidence θ_i for which the reflected wave is fully polarized. The electric field components are perpendicular to the rays, both in the marked plane and out of it.

Let's first discuss the radiation caused by the component of $\vec{E}$ in the z-direction, which is perpendicular to the plane of the paper. The electrons accelerated by that component of the incoming electric field move at right angles to the direction of the reflected wave. An observer looking back along the line of the reflected wave sees the full motion of these electrons. Thus there is strong reflection of this part of the incident wave. Next, let's consider the radiation induced by the component of the electric field of the incident wave in the a-direction. The electrons that absorb this incident radiation move parallel to the a-direction. An observer who looks along the line of the reflected wave sees a foreshortened motion of the electrons and thus only a limited amount of reflected radiation. Thus there is a preferential polarization direction for the reflected light. In the special case that the direction of the reflected wave is along the a'-direction (perpendicular to the direction of the transmitted radiation), there is no reflected radiation polarized along the a'-direction because the motion of the absorbing and reradiating electrons along the a'-direction cannot be seen. *The reflected radiation is plane-polarized with an electric field in the z-direction, parallel to the plane of the reflecting surface.* For this special angle, the reflected and transmitted waves must be at 90° to one another.

The angle of incidence for which the reflected and transmitted (or refracted) rays are perpendicular to one another are easily found once the rules for these rays (Snell's law) are developed, which we will do in Chapter 35. We give the result here: When the *angle of incidence* is the angle θ_B, known as **Brewster's angle**, the reflected ray is linearly polarized. This angle, which is the incident angle in Fig. 34–21, is given by

$$\tan \theta_B = \sqrt{\frac{\varepsilon}{\varepsilon_0}} = n. \tag{34-35}$$

As we have already noted, the effect is present but less dramatic for other angles. An analyzer whose polarizing axis is oriented in a direction perpendicular to the z-direction (the direction of polarization of the reflected wave) will absorb most of the reflected radiation. Thus Polaroid sunglasses, worn to cut down the glare of reflected light from roads, beaches, car hoods, and other horizontal surfaces, must have their polarizing axis aligned in a vertical direction.

*34–6 Electromagnetic Radiation as Particles

In one of the most astonishing discoveries of the early part of the twentieth century, we learned that *electromagnetic radiation consists of particles.* The research of Max Planck, of Albert Einstein, and of Arthur Compton established that what we call an electromagnetic wave consists of a large number of individual particles called **photons**. These particles are indivisible: It is not possible to have 0.3 photons, for example. For radiation characterized by a frequency $f = \omega/2\pi$, the energy carried by a single photon is

$$E = hf, \tag{34-36}$$

where $h = 6.63 \times 10^{-34}$ J·s is Planck's constant. A photon also carries momentum, given by

$$p = \frac{E}{c} = \frac{hf}{c} = \frac{h}{\lambda}. \qquad (34\text{--}37)$$

The particle nature of electromagnetic radiation was established through Compton's experiments on the scattering of radiation by free electrons in carbon. Photons scattered through a given angle have energy that can be calculated by treating each photon as a relativistic billiard ball that collides elastically with an electron at rest. The momentum of the outgoing photon depends on the collision angle. Equation (34–37) then implies that the frequency of the scattered radiation also depends on the collision angle in a way that can easily be calculated.

That h is small explains why we think of light as a continuous phenomenon rather than a series of individual photons. Someone standing under Niagara Falls wouldn't feel as if he or she is being bombarded by droplets of water! We now have instruments that can routinely detect individual photons. Had the evolutionary history of the human eye been somewhat different—so that the eye could easily respond to a single photon—the notion of radiation as consisting of particles would have been obvious to everyone.

EXAMPLE 34–11 At what rate does a 60-W lightbulb emit photons? For simplicity, assume that the light is emitted with a single wavelength of 590 nm.

Strategy We are given the wavelength, and therefore the frequency of the light, from which we can calculate the energy per photon, namely $E = hf = h(c/\lambda)$. The given wattage P of the bulb is the energy emitted per second, and P is the number of photons per second N times the energy per photon, $P = NE$. We can then solve for N.

Working It Out We have

$$E = h\frac{c}{\lambda} = \frac{(6.6 \times 10^{-34} \text{ J}\cdot\text{s})(3 \times 10^8 \text{ m/s})}{(590 \times 10^{-9} \text{ m})} = 3.4 \times 10^{-19} \text{ J}.$$

From this we learn the number of photons per second,

$$N = \frac{P}{E} = \frac{60 \text{ W}}{3.4 \times 10^{-19} \text{ J}} \cong 1.8 \times 10^{20} \text{ photons/s}.$$

This is a very large number, one that would not permit you to sense the presence of individual photons.

What Do You Think? Do you expect photons to carry momentum? What do you think is the momentum of a photon with frequency f?

THINK ABOUT THIS...
WHEN PHOTONS GET REFLECTED BY A MIRROR, DO THEY DROP IN FREQUENCY?

We argued earlier that when electromagnetic waves are reflected by a mirror, a little of the energy that sets the electrons in motion is absorbed by ohmic heating, so that a little of the incident energy is absorbed, and the energy reflected is a little smaller than the incident energy. In terms of photons, which for monochromatic incident waves each have the same energy hf, a loss of energy would imply a drop in the frequency, and therefore an increase in the associated wavelength ($f = c/\lambda$). Since color of radiation is associated with frequency, does this mean that the reflected light is somewhat redder than the incident light? This is certainly not what is observed. A charge that is set into oscillation by an external field will reradiate with the same frequency, which is what we *do* observe. The loss in energy is not due to a loss of energy for individual photons, but rather in the reflection of *fewer* photons. Some of the photons emitted by individual oscillators are absorbed in the medium and their energy goes into kinetic energy of the atoms that make up the medium—causing ohmic heating.

Summary

Maxwell's equations—which comprise Gauss' laws for electric and magnetic fields, the generalized Ampère's law, and Faraday's law [Eqs. (34–1) to (34–4)]—imply that it is possible to have propagating electric and magnetic fields even in the absence of currents and charges. In the absence of free charges, the electric and magnetic fields obey the wave equation, which has the generic form, here written for the x-component of the electric field,

$$\frac{\partial^2 E_x(z, t)}{\partial z^2} = \mu_0 \varepsilon_0 \frac{\partial^2 E_x(z, t)}{\partial t^2}. \qquad (34\text{--}7)$$

In the case that $E_z = 0$ and $B_z = 0$, the waves propagate along the z-direction. Whatever the direction, the speed of propagation is given by

$$v^2 = \frac{1}{\mu_0 \varepsilon_0}. \tag{34-9}$$

This speed is the speed of light, $v = c \cong 3 \times 10^8$ m/s. In material media characterized by the dielectric constant κ, the speed of propagation is $c/\sqrt{\kappa} = c/n$, where $n = \sqrt{\kappa}$ is the index of refraction. There are solutions of the wave equation (electromagnetic waves) in which the fields have the harmonic form

$$E_x = E_0 \cos(kz - \omega t + \phi), \tag{34-8}$$

and

$$B_y = B_0 \cos(kz - \omega t + \phi). \tag{34-13}$$

These waves propagate in the z-direction. More generally, electric and magnetic fields of waves that propagate in a given direction are transverse to that direction. The electric and magnetic field amplitudes are related by

$$E = cB, \tag{34-14}$$

and the fields are perpendicular to each other:

$$\vec{E} \cdot \vec{B} = 0. \tag{34-15}$$

Electromagnetic waves carry energy with energy density

$$u = \frac{\varepsilon_0}{2}(c^2 B^2 + E^2). \tag{34-18}$$

This energy is carried in equal amounts by the electric and magnetic fields. Electromagnetic waves also carry momentum, with momentum density $\vec{S}/c^2$, where $\vec{S}$ is the Poynting vector, given by

$$\vec{S} = \frac{1}{\mu_0} \vec{E} \times \vec{B}. \tag{34-23}$$

Thus radiation can transfer momentum; when a material absorbs radiation, there is a radiation pressure on the material, given by

$$\frac{S}{c} = u. \tag{34-27}$$

Charged particles radiate when they are accelerated. For a charge q undergoing an acceleration along the z-direction, the energy flux is proportional to

$$S \propto \frac{\sin^2 \theta}{R^2}, \tag{34-30}$$

where θ is the angle with the z-axis and R is the distance from the charge. Radiation with a $\sin^2 \theta$ angular dependence is called dipole radiation.

The polarization of an electromagnetic wave is the direction of the transverse electric field vector. It can be measured because polarizers transmit electromagnetic waves only along a particular polarization axis. Polarizers may be used to detect as well as to polarize electromagnetic waves. If a second polarizer is placed with its axis making an angle θ with the first one, then the electric field E of the transmitted wave is reduced in magnitude from the electric field E_0 of the incident wave according to $E = E_0 \cos \theta$. Thus the intensity I (the average of the energy flux) of the transmitted light is reduced from the incident intensity I_0 according to Malus's law:

$$I = I_0 \cos^2 \theta. \tag{34-34}$$

Waves can be polarized by reflection. If light falls on a medium of dielectric constant κ at an angle θ_B (Brewster's angle), for which

$$\tan \theta_B = n, \tag{34-35}$$

(n is the material-dependent index of refraction), then the reflected light is polarized in a direction perpendicular to both the incoming direction and the reflected direction of the wave. Light can also be polarized by scattering.

Appendix Getting Maxwell's Equations in Differential Form

Starting from a set of accelerating charges and Maxwell's equations, let's derive the equations that lead us directly to electromagnetic waves. The particular set of charges that we use form currents in the xy-plane, oscillating back and forth in the x-direction, as in Fig. 34–4. As in the qualitative discussion of Section 34–2, we know that the moving charges will give rise to changing electric and magnetic fields. We concentrate on time-dependent fields that vary with z but not with x and y. This implies that for a given z, the fields are the same out to infinity in the x- and y-directions. This cannot strictly be true in a physical situation; thus, we shall keep in the back of our minds that somewhere, for large enough values of x and y, the fields actually taper off to zero.

Let's draw an imaginary loop C in the yz-plane (at $x = 0$) that goes from $y = b$ to $y = -b$ at some value of z and returns from $y = -b$ to $y = b$ at $z + dz$ (Fig. 34A–1). We are going to apply the generalized Ampère's law to the loop. Sides at $y = \pm b$, going from z to $z + dz$, are very short. We shall ignore the contribution from the short sides because we can make these sides infinitesimally short. Moreover, our qualitative argument in Section 34–2 gives us no reason to believe that there is a field B_z. (This can be verified with the help of Gauss' law.) Application of the generalized Ampère's law, Eq. (34–3), now becomes easy. All we need to calculate for the line integral in Ampère's law are the contributions from the long (horizontal) sides of the loop. We have

$$B_y(z + dz, t)(2b) - B_y(z, t)(2b) = \mu_0 \varepsilon_0 \frac{d}{dt} \int_{\text{loop area}} \vec{E} \cdot d\vec{A}. \tag{34–A1}$$

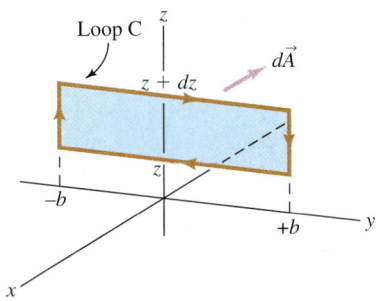

▲ **FIGURE 34A–1** A loop used to derive a relation between $\partial B_y/\partial z$ and $\partial E_x/\partial t$ using Ampère's law.

From the definition of a derivative, the difference $B_y(z + dz, t) - B_y(z, t)$ is the rate of change of B_y with respect to z times dz, so

$$2b[B_y(z + dz, t) - B_y(z, t)] = 2b\left(\frac{\partial B_y}{\partial z} dz\right). \tag{34–A2}$$

The partial derivative appears because we keep t constant in $B_y(z, t)$.

Now let's consider the right-hand side of Eq. (34–A1). In using Ampère's law, a right-hand rule dictates that for loop C in the direction shown in Fig. 34A–1, the surface element $d\vec{A}$ is oriented in the $-x$-direction, so $\vec{E} \cdot d\vec{A} = -E_x\, dA$. In addition, the area $A = 2b\, dz$ is infinitesimally small, so we can assume that E_x does not vary over the surface and we can remove it from the integral. Finally, the time derivative on the right-hand side of Eq. (34–18) acts only on E_x, because the surface is itself fixed. Thus

$$\mu_0 \varepsilon_0 \frac{d}{dt} \int_{\text{loop area}} \vec{E} \cdot d\vec{A} = -\mu_0 \varepsilon_0 \frac{\partial}{\partial t} E_x \int_{\text{loop area}} dA = -\mu_0 \varepsilon_0 \frac{\partial E_x}{\partial t} A$$

$$= -\mu_0 \varepsilon_0 \frac{\partial E_x}{\partial t}(2b)\, dz. \tag{34–A3}$$

We have used a partial derivative because z is a second variable that is held fixed. We now equate the two right-hand sides of Eqs. (34–A2) and (34–A3):

$$2b\left(\frac{\partial B_y}{\partial z} dz\right) = -\mu_0 \varepsilon_0 \frac{\partial E_x}{\partial t} 2b\, dz;$$

that is,

$$\frac{\partial B_y}{\partial z} = -\mu_0 \varepsilon_0 \frac{\partial E_x}{\partial t},$$

which is Eq. (34–5).

We next make use of Faraday's law, Eq. (34–4), the fourth of Maxwell's equations. We apply it to a loop C' that goes from $x = a$ to $x = -a$ at some value of z and returns from $x = -a$ to $x = a$ at $z + dz$ (Fig. 34A–2). Then, a nearly identical derivation to the one that led us to Eq. (34–5) leads us to Eq. (34–6),

$$\frac{\partial B_y}{\partial t} = -\frac{\partial E_x}{\partial z}.$$

Equations (34–5) and (34–6) are the ones we use in Section 34–2 to find the wave equation for electromagnetic waves.

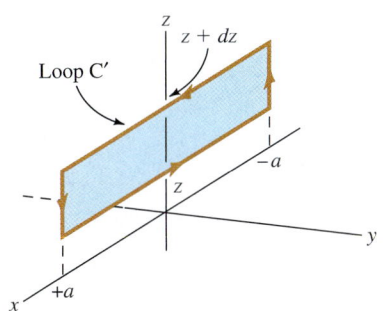

▲ **FIGURE 34A–2** A loop used to derive a relation between $\partial B_y/\partial t$ and $\partial E_x/\partial z$ using Faraday's law.

Understanding the Concepts

1. Is polarization of light a general phenomenon for all waves? Can sound waves be polarized?
2. Stable charged particles that move uniformly produce no electromagnetic waves. How does the conservation of energy suggest that this must be true?
3. Short-wave radio signals have wavelengths of several tens of meters. Such waves are particularly well reflected by Earth's ionosphere (an upper layer of the atmosphere that contains many free charges). Why would Earth's ionosphere reflect, rather than absorb, these waves?
4. Two identical wires with identical lightbulbs connected to the middle of each wire are placed in an electromagnetic plane wave moving out of the plane of the page. The two wires make an angle of 30 degrees with each other. Bulb B1 is totally dark, bulb B2 is not. Can bulb B1 be made to shine brightly, and if so, how?
5. The production and detection of polarization depends on the electric field vector. Can there be, in principle, a polarization associated with the magnetic field vector?
6. Can there be standing electromagnetic waves as well as traveling ones? Recall that mechanical standing waves on a string are possible when certain boundary conditions are satisfied, such as the ends of the string being fixed. How can we control the values of electric or magnetic fields on fixed boundaries?
7. You have a series of simple polarizers that you can orient as you choose in succession along a beam. Can you arrange them so that if the beam were polarized to start with it would be unpolarized when it has passed through your succession of polarizers?
8. Would the presence of magnetic monopoles analogous to electric charges change the nature of electromagnetic waves in free space?
9. A *solar sail* is a large surface on which the radiation pressure of the Sun's radiation can act and thereby push the sail along. Solar sails have been proposed for spaceships to travel throughout our solar system. What properties must such a sail have, and what difficulties do you see in the proposal?
10. Rockets are propelled forward when mass is ejected backward from them. Could a source of light (or other electromagnetic radiation) be used in place of the mass?
11. How can you tell whether or not light is linearly polarized?
12. Incident light is linearly polarized along the x-axis. We would like to rotate the direction of polarization so that it lies along the y-axis. Can this be done with one polarizer? Can it be done with two? What is the minimum reduction in intensity when two polarizers are used? Can there be even less intensity reduction with three polarizers?
13. Consider a metal rod with a lightbulb connecting the two halves (Fig. 34–22). The lightbulb glows when the rod is in the vicinity of a radio station. What happens when the rod is rotated in a plane perpendicular to the line between the rod and the station?

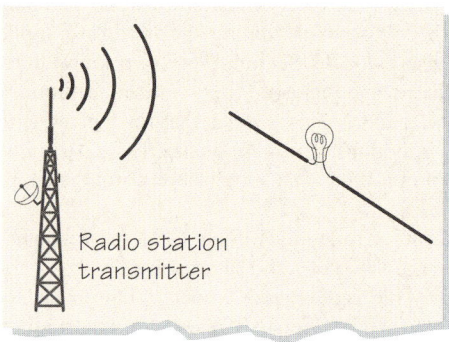

▲ **FIGURE 34–22** Question 13.

14. In Hertz's test of the existence of electromagnetic waves, sparks appear in a secondary gap as the result of an AC current in a primary circuit. Hertz interpreted these sparks as due to the effect of electromagnetic waves and not to the effects of Faraday induction. What sort of checks did Hertz need to make in order to rule out Faraday induction?
15. A 100-W lightbulb hangs from the ceiling of a windowless and thermally isolated room with the door closed. How much of the 100 W goes into heating the room?
16. Can you use the example and the arguments given in Section 34–2 to prove in a more general way than was done there that electromagnetic waves are transverse? Are there any pitfalls?
17. We showed that electromagnetic radiation carries momentum by thinking about its effect on a free charge. Consider its effect on an electric dipole to see whether it might carry angular momentum as well. Start by orienting the dipole with its axis along the direction of the electric field vector of the electromagnetic wave.
18. Consider the electromagnetic wave shown in Fig. 34–6(a). Suppose the green arrow showing the direction of the propagation of the wave were missing. Could you figure out how to replace it?
19. Consider the solar sail described in Question 9. Is it better to make a solar sail reflective (shiny) or absorbing (black)?
20. In the subsection on momentum in electromagnetic waves, we mentioned that an electromagnetic wave will accelerate a charged particle, giving it momentum at the same time. Does this mean the wave loses energy and momentum? In other words, if we sit behind a receiving antenna, do we pick up less radiation because of the presence of the antenna?
21. When electromagnetic radiation interacts with matter—for example, when light propagates in a crystal—it is always the electric and not the magnetic field that determines the behavior. Can you explain this, recalling the role of the two fields in the Lorentz equation?
22. How is it possible to clearly hear a 50,000-watt radio station that is 375 miles away, but not possible to clearly hear a local 5000-watt radio station that is 5 miles away?

Problems

34–1 Maxwell's Equations

1. (I) Verify the consistency of the dimensions of both sides of each of the four Maxwell equations.
2. (II) Gauss' laws for electric fields and for magnetic fields differ due to the lack of magnetic charges. Assume that magnetic monopoles (magnetic charges) exist; denote them by the symbol M. Rewrite Gauss' law for magnetic fields, and give the SI units of M.
3. (II) Ampère's and Faraday's laws differ due to the lack of a currentlike term in Faraday's law. Assume that magnetic monopoles exist (call them M), and rewrite Faraday's law. Discuss the physical significance of any new terms added.
4. (II) A region is bounded by an imaginary closed surface. Cut the region into two subregions with an arbitrary surface. Show that if Maxwell's first and second equations (which involve the surface

34–2 Electromagnetic Waves

5. (I) If the electric field for a plane wave is given by $E_x = 0$, $E_y = E_0 \cos(kz + \omega t)$, what are $\vec{B}$ and the direction of propagation of the wave?

6. (I) Use dimensional analysis to show that $1/\sqrt{\mu_0 \varepsilon_0}$ has the dimensions of speed, $[LT^{-1}]$.

7. (I) An FM radio station announcer identifies the station as "Q94;" the number 94 stands for the frequency in some units. What is the wavelength and frequency of the waves emitted by the radio station?

8. (I) What is the relation between the amplitudes of the electric and magnetic fields in an electromagnetic wave propagating in a medium whose dielectric constant is κ? Assume the magnetic permeability of the medium is that of the vacuum.

9. (I) A superposition of electromagnetic waves traveling in the $+z$-direction and electromagnetic waves traveling in the $-z$-direction gives rise to standing waves. Check that a standing wave whose x-component of electric field has the form $E_0 \sin(kz) \cos(\omega t)$ satisfies the wave equation [Eq. (34–7)].

10. (II) Find the approximate wavelength, wave number, frequency, and angular frequency for electromagnetic waves associated with (a) your favorite AM station; (b) your favorite FM station; (c) a microwave oven; (d) yellow light; (e) X rays.

11. (II) Use Gauss' law to show that electromagnetic waves must be transverse. [*Hint*: Choose as your Gaussian surface a pill-box, with one of the plane surfaces chosen such that $\vec{E}$ or $\vec{B}$ vanishes on it.]

12. (II) Starting from Eqs. (34–5) and (34–6), derive a wave equation for the y-component of the magnetic field. What is the speed of the resulting wave?

13. (II) Write the counterparts of Eqs. (34–5) and (34–6) for electromagnetic fields B_y and E_z that lie in the yz-plane and propagate in the x-direction. [*Hint*: Start with Figs. 34A–1 and 34A–2. Then relabel the axes according to $x \to y \to z \to x$.]

14. (II) A plane harmonic wave of electromagnetic radiation with wavelength λ is propagating in the $-x$-direction. The z-component of the electric field has magnitude E_0, and there is no y-component. (a) Write an expression for the electric field. (b) Use this expression and the result of Problem 13 to calculate the magnetic field. What vector components will this field have?

15. (II) A plane wave propagates along the direction in the xy-plane that makes an angle θ with the x-axis. Show that the electric field is given by $\vec{E}_0 \cos(kx \cos\theta + ky \sin\theta - \omega t + \phi)$. What directions can $\vec{E}_0$ have?

16. (II) A plane wave of wavelength 17 m propagates in the z-direction. The electric field points in the y-direction and has an amplitude of 0.16 V/m. Write an expression for the magnetic field, including its amplitude in SI units. Assume that the electric field is at its maximum at $z = 0$, $t = 0$.

17. (II) An electromagnetic wave of wavelength 600 nm propagates in the z-direction. The magnetic field points in the y-direction, and has a magnitude of 10^{-8} T. Write an expression for the electric field, including numerical values and units. Assume that the magnetic field is maximum at $z = 0$ m, $t = 0$ s.

18. (II) An electromagnetic traveling wave is generated at the left-hand end of a tube oriented in the z-direction; the wave travels in the $+z$-direction. At the ends of the tube, $z = 0$ and $z = L$, are highly reflective mirrors. The electric field of the incident wave is $\vec{E} = E_1 \cos(kz - \omega t)\hat{i}$, and the electric field of the wave reflected at $z = L$ is $\vec{E} = E_1 \cos(kz + \omega t + \phi)\hat{i}$. Show that the net electric field forms a standing wave and, by computing B_y, that the associated magnetic field B_y also has the form of a standing wave.

19. (II) Consider the standing electromagnetic wave in Problem 9. If the standing wave is confined to a region lying between $z = 0$ and $z = L$ by two metallic plates, what is the relation between the allowed wavelengths of the radiation and L? (Recall from Chapter 22 that the electric field along a conducting surface must vanish on that surface.)

20. (III) A pulse of electromagnetic radiation travels in the $-z$-direction. The electric field is oriented in the x-direction and is given by $\vec{E} = E_0 \exp(-(z + ct)^2/a^2)\hat{i}$. What is the orientation of the magnetic field? Make a guess of the space–time dependence of the magnetic pulse, and use Eqs. (34–5) and (34–6) to find a form for $\vec{B}$ that satisfies Maxwell's equations.

34–3 Energy and Momentum Flow

21. (I) The intensity of an electromagnetic wave is 6×10^6 W/m^2. What is the amplitude of the magnetic field in this wave?

22. (I) A radio station emits a signal with a power of 18 kW. What are the values of the electric field and magnetic field at distances of 3.5 km and 10.5 km? Assume that the signal far from the antenna is transmitted with equal intensity in all directions. (Real radio stations cannot afford to transmit their energy in this way, and their antennas distribute energy with a high degree of directionality.)

23. (I) The electric field for a given electromagnetic wave has a peak value of 140 mV/m. What is the intensity of the wave?

24. (I) A laser emits a beam with an intensity of 0.40×10^{13} W/m^2 across an area of 1.5 mm^2. What force would the laser beam exert on a black (perfectly absorbing) object?

25. (I) A harmonic plane wave of wavelength 0.45 μm and an electric field amplitude of 3 V/m impinges on a totally reflecting surface of area 200 cm^2. What is the radiation pressure exerted by the wave?

26. (I) A plane electromagnetic wave with maximum electric field amplitude of 120 V/m is incident on a perfectly absorbing surface perpendicular to the direction of propagation. What is the rate of energy absorption per unit area of the surface?

27. (I) The rate at which the Sun emits energy in the form of radiation is 3.8×10^{26} W. (a) Calculate the magnitude of the Poynting vector at a distance of 1.5×10^{11} m from the Sun. (b) What is the radiation pressure exerted on a totally absorbing surface perpendicular to the direction of the radiation?

28. (II) (a) Sketch on the same graph $\sin x$ and $\cos x$ as a function of x. (b) On a separate graph sketch $\sin^2 x$ and $\cos^2 x$. Observe that the periodic functions $\sin^2 x$ and $\cos^2 x$ are identical to one another, except that one is displaced from the other by an interval $\pi/2$. (c) Use your sketch from part (b) to show that the area under the $\sin^2 x$ curve in the interval $0 \le x \le 2\pi$ is the same as the area under the $\cos^2 x$ curve in the same interval. (d) Given the fact that $\sin^2 x + \cos^2 x = 1$, use the results obtained in parts (a)–(c) to show that the averages $\langle \sin^2 x \rangle$ and $\langle \cos^2 x \rangle$ are equal to each other and thus equal to $1/2$.

29. (II) The magnetic field for a given electromagnetic wave has an rms value of 7×10^{-9} T. What is the intensity of the wave? How much energy is transported per minute through a 0.1-m^2 area?

30. (II) A typical lecture-demonstration laser of power 0.75 mW has a beam of diameter 0.90 mm. (a) What are the peak values of the electric and magnetic fields? (b) Suppose—as is in fact possible—that the beam is focused to a circular area with diameter of one wavelength. What is the peak value of the electric field, given that $\lambda = 650$ nm?

31. (II) Assume that a 75-W lightbulb emits light equally in all directions. What are the peak and rms values of the electric and magnetic fields at a distance of 0.50 m?

32. (II) A 75-W lightbulb radiates uniformly in all directions, and 9 percent of this energy is emitted as electromagnetic radiation in the visible light range. What is the electromagnetic energy density of visible light at a distance of 130 cm from the bulb? What are the rms values of the corresponding electric and magnetic fields there?

33. (II) The total electromagnetic power emitted by the Sun is 3.8×10^{26} W. What is the radiation pressure exerted on a totally reflecting surface a distance $r = 1.0 \times 10^{10}$ m from the Sun?

34. (II) What are the dimensions and SI units for the Poynting vector? Reduce your answer to the dimensions and units of mass, length, and time, then reexpress it in terms of watts and meters.

35. (II) Solar energy delivered to a horizontal surface in Washington, D.C., averaged over a full year is 160 W/m². Assuming that this radiation is fully absorbed on a particular square meter of ground, what is the approximate total momentum delivered to this area in 1 y? Compare this number to an estimate of the momentum absorbed by a baseball catcher in catching a single pitch.

36. (II) The radiation pressure of a beam of electromagnetic radiation is equal to atmospheric pressure. Calculate the intensity, energy density, and rms electric and magnetic fields of this beam. Assume that the beam is totally absorbed.

37. (II) Sunlight exerts an average radiation pressure of 5×10^{-6} N/m². Consider a rectangular mirror of dimensions 20 cm × 500 cm attached to a vertical wire, so that it is oriented perpendicular to the sunlight. How should you attach it to get the maximum torque exerted by the sunlight? What is the torque?

38. (II) What is the radiation pressure on the walls of a microwave oven in which the rms electric field is 500 V/m? Assume that the waves are traveling waves of the sort we have studied in this chapter, and if it is necessary to know the frequency, estimate the appropriate value.

39. (II) Suppose that you want to use the radiation pressure from a beam of light to suspend a piece of paper in a horizontal position; the paper has an area of 50 cm² and a mass of 0.20 g (Fig. 34–23). Assume that there is no problem with balance, that the paper is dark and absorbs the beam fully, and that the entire beam can be used to hold the paper against the pull of gravity. How many watts must the light produce? Given your answer, what do you suppose would happen to the paper?

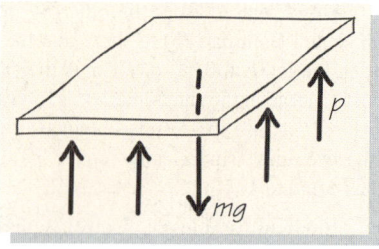

▲ FIGURE 34–23 Problem 39.

40. (II) Tiny flakes of mica are kept aloft by a beam of light projected vertically upward. If the mass of a typical flake is 5.4×10^{-9} kg, and if on the average the area presented to the beam by a flake is 0.06 mm², what is the intensity of the beam? Assume that all of the light is reflected.

41. (II) A light beam with a given Poynting vector falls on a flat, fully reflecting surface at an angle of incidence θ (with respect to the vertical) (Fig. 34–24). What is the momentum transferred to the surface per unit area?

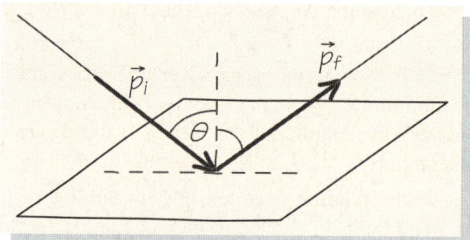

▲ FIGURE 34–24 Problem 41.

42. (II) A laser delivers 1.8×10^2 J of energy in a pulse that lasts 5×10^{-9} s. What are the peak electric and magnetic fields for a laser beam of diameter 0.5 mm?

43. (II) Consider an electromagnetic wave propagating in the negative x-direction with frequency of 6×10^9 Hz. The wave exerts a pressure of 10^{-4} N/m². Write down an expression for the electric and magnetic fields. You may choose your axes such that the electric field is polarized along one of the axes. Specify your choice.

44. (II) Consider a standing electromagnetic wave for which the electric field is $\vec{E} = \hat{j} E_0 \sin kx \cos \omega t$. (a) What is the wavelength of the electromagnetic wave? (b) What is the magnetic field? (c) What is the value of the Poynting vector?

45. (III) The short side of a thin, stiff rectangle 3.0 cm × 1.0 cm is attached to a vertical axis. Half of each side is painted black and is fully light absorbent; the other half is a shiny, reflecting metal (Fig. 34–25). The back of each half is different from the front. There is no friction at the axis. The apparatus is bathed in a well-collimated (nonspreading) beam of light whose Poynting vector has magnitude 0.5 kg/s³ and travels perpendicular to the vertical axis. Is there a net torque on the rectangle's surface? If so, what is its average value due to the light over a full, uniform rotation of the rectangle about the axis?

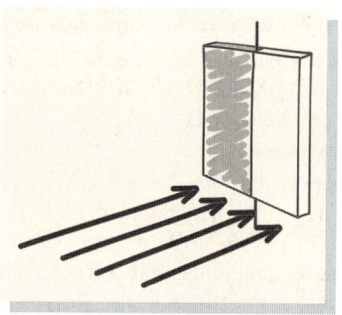

▲ FIGURE 34–25 Problem 45.

46. (III) The total power of a broadcasting dipole antenna is 20 MW. Calculate the intensity of its radiation at a distance of 1000 m, in the direction of the intensity maximum. Compare this to the intensity that would have been obtained if the intensity were distributed uniformly in every direction.

34–4 Dipole Radiation

47. (I) Suppose that a vertical tower 120 m tall acts as a dipole antenna, with currents running back and forth along the tower to generate electromagnetic waves in a dipole pattern. If the wavelength of each electromagnetic wave is the height of the tower, what is the period of the current oscillation in the tower?

48. (I) A charge moves harmonically along an 8-m length in the z-direction, emitting dipole radiation. Two observers detect this radiation. Observer A is at a position that is 10 km from the charge and at an angle of 25° with respect to the z-axis, while observer B is at a position that is also 10 km from the charge, but at an angle of 58° with respect to the z-axis. What is the ratio of the intensity detected by the two observers?

49. (II) A broadcasting dipole antenna is oriented along the y-axis. For the geometry shown in Fig. 34–26, give the following information for a point P far away along the z-axis: (a) the direction of the electric field; (b) the direction of the magnetic field; (c) the direction of the Poynting vector. (d) Repeat parts (a)–(c) for the electromagnetic wave one-half cycle later.

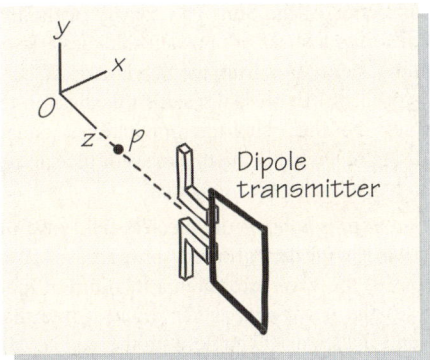

▲ **FIGURE 34–26** Problem 49.

50. (III) A cross-shaped antenna lies in the xy-plane, centered at the origin (Fig. 34–27). The charges oscillate with the same frequency within each arm of the cross. Find the Poynting vector along the z-axis as a function of z if charges moving in the $+x$-direction in the x-arm pass the origin at the same moment that (a) charges moving in the $+y$-direction in the y-arm pass the origin; (b) charges moving in the $-y$-direction in the y-arm pass the origin.

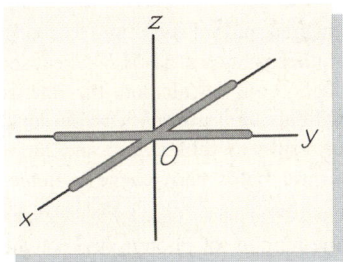

▲ **FIGURE 34–27** Problem 50.

34–5 Polarization

51. (I) At what angle should the axes of two ideal Polaroid sheets be placed to reduce the intensity of a given source of unpolarized light to (a) 7/10; (b) 3/10; (c) 3/20; (d) 1/20?

52. (I) The axes of four ideal Polaroid sheets are stacked, each at 28° with respect to the previous one. What fraction of initially unpolarized light passes through all four sheets?

53. (I) The Moon reflects light off a still pond at night. At what angle above the horizon is the polarization a maximum? The index of refraction of water is 1.33.

54. (I) Polarized light of intensity 1.0×10^6 W/m² is incident on a Polaroid sheet placed perpendicular to the light beam with the polarizing axis of the sheet at an angle of 40° to the polarization vector of the light. What is the intensity of the beam after passing through the polarized sheet?

55. (I) The beam of Problem 54, after passing through the Polaroid sheet described in that problem, then passes through another Polaroid sheet, this one with its polarizing axis at an angle of 80° to the original polarization vector. What is the final intensity of the beam?

56. (II) A beam of light propagating in the z-direction is polarized in the y-direction. Two superposed Polaroid sheets are placed perpendicular to the beam. The polarization axis of one makes a 33° angle with respect to the y-direction, and the axis of the other makes a 51° angle with respect to the axis of the first sheet (Fig. 34–28). What is the intensity of the transmitted beam?

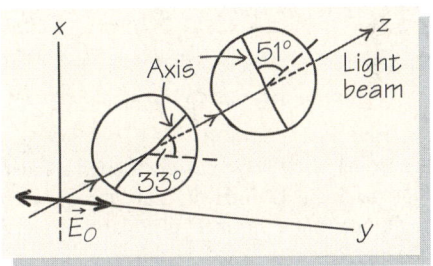

▲ **FIGURE 34–28** Problem 56.

57. (II) What fraction of initially unpolarized light passes through two Polaroid sheets placed at right angles to each other? What happens if a third sheet is placed between the two sheets, with its axis at an angle of 45° to the two?

58. (II) If light of intensity I_0 moving in the z-direction is polarized linearly in the x-direction, it will not pass through a piece of Polaroid that passes light polarized in the y-direction. Figure 34–29 shows a way in which this light can pass the y-direction analyzer if a second analyzer is used. If the lower analyzer makes an angle of θ with respect to the x-direction, what is the intensity of the light that passes the upper analyzer?

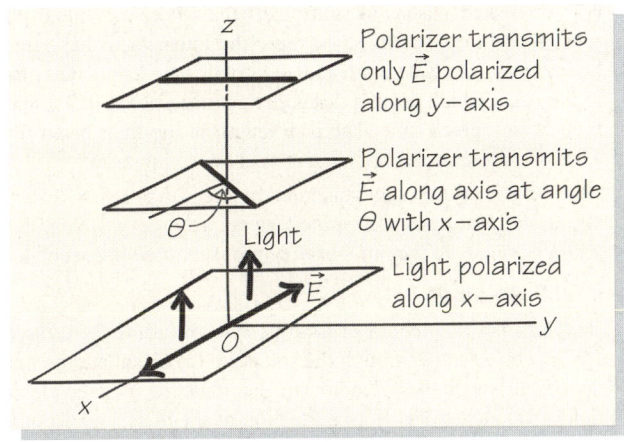

▲ **FIGURE 34–29** Problem 58.

59. (II) Unpolarized light of intensity I_0 passes through two pieces of Polaroid successively. (a) What is the intensity of the light after it passes through the first piece of Polaroid? (b) The second piece is rotated so that the intensity of the transmitted light goes to zero. What angle does the polarizing axis of the second piece make with that of the first piece? (c) A third piece of Polaroid is inserted between the two pieces in place. Calculate the intensity as a function of the angle θ that the axis of the third piece makes with the axis of the first piece (Fig. 34–30). (d) Show that the intensity of the transmitted light is no longer zero unless the axis of the third piece is parallel to that of either of the other pieces.

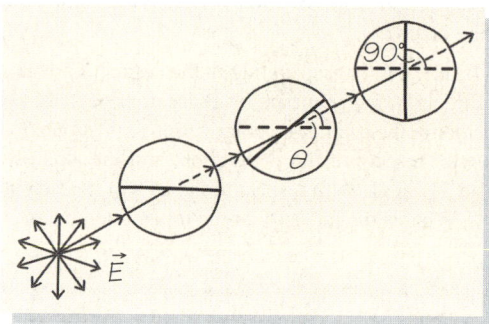

▲ **FIGURE 34–30** Problem 59.

60. (II) An electromagnetic wave passes through a sheet of Polaroid, and 3/4 of the incident intensity gets through. Through what angle should one rotate the sheet of Polaroid to let all the radiation get through?

61. (III) We have *circular polarization* when the electric field of a wave propagating in the z-direction takes the form of the superposition of a wave linearly polarized with the electric field along the x-direction and another wave linearly polarized with the electric field along the y-direction. The phases of these waves must differ by $\pi/2$, and this gives a net electric field of the form $\vec{E} = [E_0 \cos(kz - \omega t)]\hat{i} + [E_0 \sin(kz - \omega t)]\hat{j}$. Show that this electric field at any given z-value forms a vector that rotates uniformly in the xy-plane. Why is this superposition of solutions a physically acceptable wave?

*34–6 Electromagnetic Radiation as Particles

62. (II) Calculate the number of photons emitted by an FM radio station that broadcasts at a frequency of 4.5×10^8 Hz that are required to equal the energy contained in one photon of visible light at a wavelength of 450 nm.

63. (II) A scientist wishes to study the behavior of individual photons. To do that, she must decrease the intensity of her 1-mm² laser beam—the laser emits radiation with wavelength 630 nm—to a level at which there is no more than one photon in her apparatus at any given time. The path length of the light beam from source to detector is 2 m. What should be the intensity?

64. (II) The power currently generated by the Sun is 3.8×10^{26} W. Assuming that it is all emitted at an average wavelength of 550 nm, calculate the number of photons emitted per second.

General Problems

65. (II) Consider a solenoid of n turns/m with radius R. A current $I = I_0 \cos \omega t$ goes through the solenoid. (a) Calculate the magnetic field inside the solenoid. (b) Calculate the induced electric field inside the solenoid as a function of the distance r from the axis. (c) Calculate the Poynting vector, $\vec{S}$. In particular, find its direction at different times during one cycle.

66. (II) The electric and magnetic fields of an electromagnetic wave act on a charge q. With what speed must the charge move so that the magnetic force on the charge is, at most, 30 percent of the electric force? If the electromagnetic wave is traveling in the z-direction and the electric field has only an x-component, what is the direction (or directions) of motion of q so that the magnitude of the magnetic force is greatest?

67. (II) Consider the solar sail described in Question 9. A solar sail can be aligned with its area perpendicular to a radial line from the Sun so that the sail is pushed straight outward. Show that in this configuration the force on the sail always has the same sign and is proportional to $1/r^2$, where r is the distance from the sail to the Sun. (Assume that only the radiation pressure and the gravitational force due to the Sun act on the sail.) This economical method of propulsion has been proposed for travel to the far reaches of the solar system when transit time is not an important factor.

68. (II) A solar sail (see Question 9) is to be designed such that, when it is aligned perpendicular to the Sun's rays and is 1.5×10^{11} m from the Sun, the radiation pressure on it, P, just cancels the gravitational attraction of the Sun. The density of the material of the sail, which forms a sheet of constant thickness, is ρ. (a) Find P, given that the energy flux from the Sun is 1.4 kW/m² at the radius of Earth's orbit. (b) Express the sail's thickness in terms of ρ, P, the mass of the Sun, and the gravitational constant. If ρ is 2.0×10^3 kg/m³, what is the thickness of the sail material? Your result is independent of the sail's area.

69. (II) Find an expression for the electric field of a plane electromagnetic wave with the following properties: (a) the frequency is 10^{14} Hz; (b) the wave travels in a medium of index of refraction 1.4; (c) the wave propagates along a line that lies in the xy-plane and makes a 30°-angle with the x-axis; (d) the wave is polarized along the z-axis; (e) the average value of the Poynting vector is 500 W/m².

70. (II) A swimming pool has underwater lights. What is Brewster's angle for reflection off the upper surface of water? The index of refraction of water is 1.33.

71. (II) The amount of solar energy reaching your body when you sunbathe on an ocean beach in summer is about 800 W/m². Assume that your body absorbs 40 percent of this incident radiation and that your exposed body area is 0.5 m². How much solar energy do you absorb in 1 h? Estimate how much perspiration must evaporate to dissipate this energy (see Chapter 17).

72. (II) A high-powered, pulsed laser used to confine plasma for nuclear fusion studies is rated at 15 MW. The laser beam is focused on an area of 0.60 mm². Calculate the intensity, peak electric and magnetic fields, and average energy density in this beam. Compare your results to Tables 22–1 and 28–1, which list some values for electric fields and magnetic fields, respectively, in other contexts.

73. (II) What is the number of photons/m³ contained in a beam of electromagnetic radiation in a plane wave with a wavelength of 2 cm and an electric field amplitude 10 V/m?

74. (II) The solar energy flux at a distance $R_0 = 1.5 \times 10^{11}$ m from the Sun (the radius of Earth's orbit) is 1400 W/m². (a) What is the total energy flow from the Sun in watts? (b) Use your result to calculate the rate at which photons are emitted. Assume an average wavelength of 600 nm. (c) Using the result of part (b), find the number of photons/s that strike a 1 mm × 1 mm surface at a distance R_0. The surface is oriented perpendicular to the Sun.

75. (II) A laser emits N photons of frequency f. The beam strikes a mirror that is moving with speed v in the direction of propagation of the laser beam. Assuming that the kinetic energy of the mirror is much larger than that of the beam, use energy conservation and momentum conservation to find the frequency of the reflected beam. Treat the photon as a particle of energy hf and momentum hf/c.

76. (II) A wire is bent into a loop. The two ends are attached to the terminals of a battery and a current flows through the wire. Sketch the direction of the Poynting vector field near the wire.

77. (III) Many people believe solar sails are a useful source of propagation in space. What acceleration would a 150-kg space probe have if it had a 100 m^2 sail perpendicular to the line to the Sun and starts at Jupiter's orbit? The intensity of sunlight at Earth is 1400 W/m^2; use this fact to find the corresponding quantity at the radius of Jupiter's orbit, all the way to Saturn's orbit. How long would it take this probe to reach the orbit of Saturn?

78. (III) Consider a current I that flows through a cylindrical wire of length L, radius b, and resistance R (Fig. 34–31). The current flows uniformly across the cross section of the wire. Calculate the electric fields inside and on the surface of the wire. The current in the wire gives rise to a magnetic field, which you can calculate. Use these fields to find the direction and magnitude of the Poynting vector on the surface of the wire. Show that the rate of energy flow into the wire through its surface is IR^2, the power dissipated in ohmic heating.

79. (III) Consider Eq. (34–3), in which the current I passes through the surface dA, so that $I = \int_{\text{surface}} \vec{J} \cdot d\vec{A}$. If the wave propagates in a medium in which there is some conductivity, we may write the right hand side of Eq. (34–3) as

$$\mu_0 \varepsilon_0 \frac{d}{dt} \int \vec{E} \cdot d\vec{A} + \frac{\mu_0}{\rho} \int \vec{E} \cdot d\vec{A}.$$

This means that in Eq. (34–5) we should make the replacement $\frac{\partial}{\partial t} E_x \to \frac{\partial}{\partial t} E_x + \frac{1}{\varepsilon_0 \rho} E_x$. How does this affect the wave equation for E_x?

80. (III) Consider a plane electromagnetic wave of frequency f that propagates in the z-direction in a cubic box whose sides are length L, with L much larger than the wavelength. The electric field of the radiation has the form $\vec{E} = E_0 \sin(kz - \omega t)\hat{i}$. Alternatively, we can say that the radiation consists of N photons, each propagating in the z-direction with energy hf, where h is Planck's constant. Use two alternative expressions for the energy of the radiation to express E_0 in terms of h, f, N, and L.

81. (III) Consider a capacitor that consists of two circular metal plates of radius R a distance d apart (Fig. 34–32). R is so much larger than d that all fringe fields can be neglected. If the charge on the plates, Q, changes with time, then according to Ampère's law a magnetic field will be induced in the region between the plates. (a) What is the induced magnetic field? (b) Using the induced magnetic field and a calculation of the electric field between the plates, find the Poynting vector. (c) Show that with this Poynting vector, the net energy flow into the capacitor is the rate of change of the capacitor energy $Q^2/2C$.

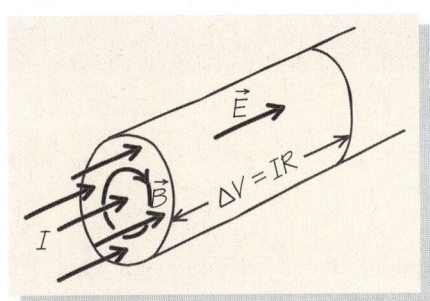

▲ **FIGURE 34–31** Problem 78.

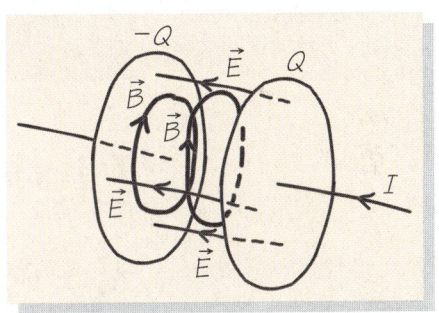

▲ **FIGURE 34–32** Problem 81.

Chapter 35

▶ This spectacular shuttle launch produced a very special and memorable effect. The launch took place just after sunset, and the lower part of the plume was in shadow while the upper part still received the rays of the Sun. The launch also took place very close to a full Moon, so that the Sun, Moon and Earth were nearly aligned. The upper part of the plume cast a shadow, visible in the form of an "anticrepuscular ray," a shadow that is approximately aligned along the direction to the Moon.

Light

You may think of light as something that travels in straight lines. You may have come to this conclusion from observing the rays that appear when light penetrates a forest on an early morning. This property of light strongly suggests that light is composed of particles emitted by a source, and Isaac Newton, whose earliest work was on optics, supported that view. The phenomena he considered in coming to a conclusion include many with which we have some everyday experience: the reflection of light from mirrors, refraction as light passes through glass lenses or water, and observations of rainbows and the prismatic separation of colors. The particle model provided such a good explanation of these observations that it is surprising that the idea that light consists of waves could have taken root in Newton's day. Yet Robert Hooke's idea that light is some type of oscillatory activity in an unidentified medium led Christian Huygens to propose a wave theory of light in 1687. In this chapter, we shall show that the wave theory of light can explain almost everything that the particle theory can, as well as the interference and diffraction phenomena that cannot be explained by a particle picture.

By the early nineteenth century, it had become apparent that certain observations could not be explained by the particle theory; an explanation of these observations demanded that light behave like a wave. For example, when we look very closely, light does not cast *sharp* shadows, and so to some extent light bends around corners. (Newton did not have the equipment to make this observation, and, in fact, he argued against the wave theory on the basis that light does *not* appear to bend around corners!) Under controlled conditions, we can also see that beams of light interfere with each other in just the same way as the waves discussed in Chapter 15 interfere. Definitive experiments by

Thomas Young in 1801 on the wave aspects of light eventually established the preeminence of the wave theory (Fig. 35–1). The phenomena associated with the wave aspects of light are the subject of Chapters 37 and 38. The prediction from Maxwell's equations that light is an electromagnetic wave would seem to have settled the question of whether light is a particle or a wave once and for all. In the twentieth century, however, we had to revise our view once more, as new experimental evidence suggested that some aspects of light can be explained only if light sometimes behaves as particles. Today we are not forced to choose between a particle picture and a wave picture of light; a quantum mechanical explanation encompasses them both.

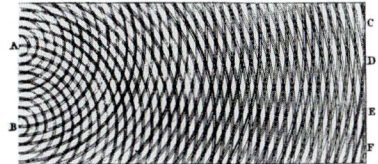

▲ **FIGURE 35–1** Young's view of the wave nature of light. In this sketch published in 1807 from his lectures, points A and B represent pinholes; points where the waves reinforce each other on the screen are marked.

35–1 The Speed of Light

Light travels so rapidly that it requires a good deal of ingenuity to show that its transmission is not instantaneous. Although Galileo had thought that the speed of light might be finite, he failed to find a time delay in the passage of light from one mountaintop to another. Ole Roemer observed the eclipses of the moons of Jupiter in 1675 and found that the timing of these phenomena could be explained if light traveled with a large, but finite, speed. The solar system data available to Roemer at the time gave a value of 2×10^8 m/s for the speed of light—certainly a result of the correct order of magnitude. The first terrestrial measurements were made in 1849 by Hippolyte Fizeau, who used the device shown in Fig. 35–2. A light source is placed behind a toothed wheel that can be rotated at high speeds. The light passes through an inclined glass plate and then between two teeth of the rotating wheel. It then travels to a mirror and is reflected straight back. If the speed of light were infinite, light would be reflected back through the gap before the wheel had moved at all (Fig. 35–2b). Light traveling at a finite speed would also pass

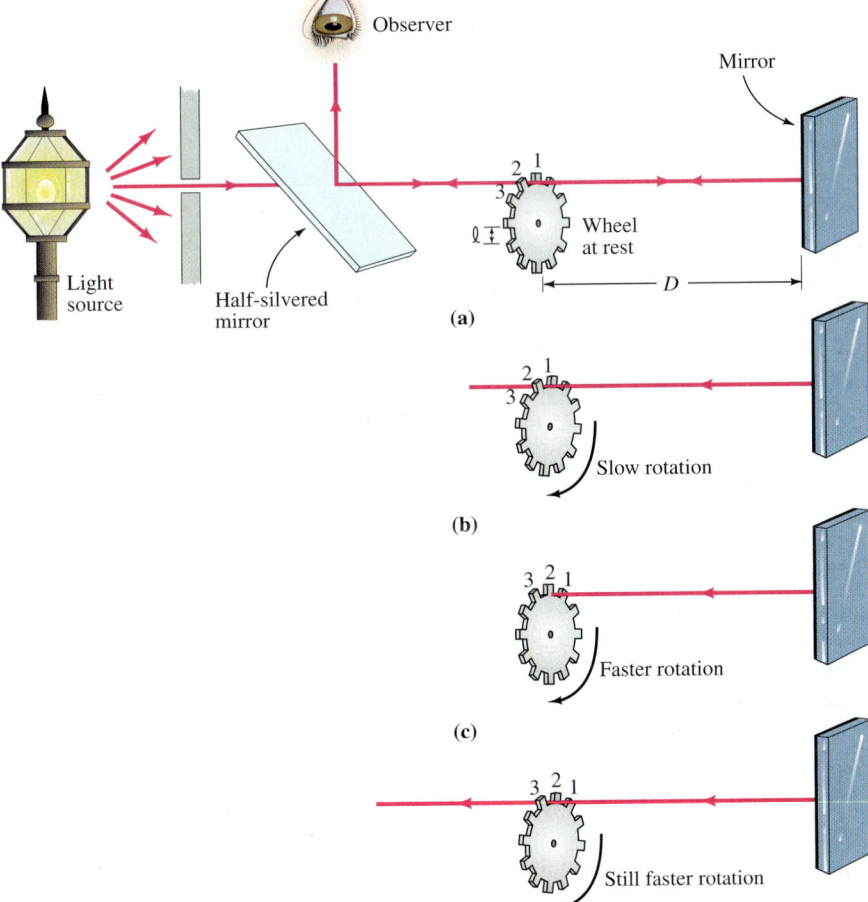

◀ **FIGURE 35–2** Fizeau's method for measuring the speed of light: (a) a sketch of the apparatus he used. (b) Incident and reflected light pass through the same gap when the wheel rotates slowly. (c) As the rotation of the wheel increases, it will become fast enough so that the reflected light fails to pass through the original gap, and instead strikes the cog between. (d) When the rotation speed is still higher, the reflected light passes through the next gap by traveling the distance $2D$ in the time the wheel rotates to the next gap.

through the same gap if the rotational speed of the wheel were small. But if the wheel is rotating fast enough and the speed of light has a finite value c, then, at a certain value of angular velocity, the wheel moves enough during the time the light travels to the mirror and is reflected back that the light will strike cog 2 (Fig. 35–2c), and no light reaches the observer in this case. As the wheel rotates even faster, light once again reaches the observer, but this time it passes through the *next* gap in the wheel, the one between cogs 2 and 3 (Fig. 35–2d). If the edge of the wheel is moving at speed v when Fig. 35–2d first applies, then we can equate the time $2D/c$ for light to make a round trip from the wheel to the mirror (D is the distance from wheel to mirror) and back to the time ℓ/v (ℓ is the cog spacing) for the wheel to move a distance of one gap:

$$\frac{\ell}{v} = \frac{2D}{c}.$$

If D is much larger than ℓ, then a value of v much smaller than c would suffice to measure c accurately. In Fizeau's 1849 experiment, the light traveled almost 20 km, while a wheel with more than 700 cogs rotated tens of times per second. Fizeau reported a result of about 3.1×10^8 m/s, a few percent different from the correct value.

Better measurements improved the value of c. Ultimately, because a measurement of c is the ratio of a distance to a time, the best measurement of c would be limited by our ability to measure time (today, to one part in 10^{13}) and distance (today, to four parts in 10^9). Because distance is harder to measure accurately than time, today we pass this difficulty by *defining* the speed of light in a vacuum to be $c = 299{,}792{,}458$ m/s and use time along with the definition of c to measure distances. In this way of doing things, the meter is no longer defined but is rather measured: One meter is $1/299{,}792{,}458$ times the distance traveled by light in 1 s (see Section 1–2). For practical purposes, you can use $c = 3.00 \times 10^8$ m/s.

The Index of Refraction

Fizeau also found that the speed of light in transparent materials such as water or glass is *less* than the speed of light in empty space. We reserve the symbol c for the speed of light in empty space, and express the speed of light in a material as

$$v_m = \frac{c}{n}, \tag{35-1}$$

where n is the *index of refraction* of the material, a quantity introduced in Chapter 34. Table 35–1 lists indices of refraction for a variety of materials.

The speed of light in materials also generally varies with wavelength, or, put another way, *the index of refraction is a function of wavelength*. For example, violet light, which has a shorter wavelength than red light, travels more slowly in glass than does red light. We shall see that this property explains the separation of white light into the colors of a rainbow by a prism or by water droplets in the atmosphere.

We saw in Chapter 34 [Eqs. (34–16) and (34–17)] that the index of refraction for a material with a dielectric constant κ is

$$n = \sqrt{\kappa}. \tag{35-2}$$

(We have assumed that the relative magnetic permeability $\mu_m/\mu_0 \cong 1$, which is a good approximation for substances that are transparent to light.) The variation of n with wavelength occurs because the dielectric constant generally varies with wavelength to some extent. We must therefore use κ at the appropriate wavelength rather than its static value in Eq. (35–2). Frequency f and wavelength λ are related by $f\lambda = v$, so we find from Eq. (35–1) that in a medium of index of refraction n,

$$f\lambda = \frac{c}{n}. \tag{35-3}$$

Equation (35–3) shows that the product of f and λ is inversely proportional to n, and you might think that both f and λ could change as light passes from one medium to another. But when the medium changes *the frequency does not change*. This is easy to understand: Consider two observers on either side of an air–glass interface. Each wave front that passes one observer must pass the other—otherwise wave fronts would pile

TABLE 35–1 • Indices of Refraction for Various Substances ($\lambda = 600$ nm)

Material	Index of Refraction, n
Air (1 atm, 0°C)	1.00029
Carbon dioxide (1 atm, 0°C)	1.00045
Ice	1.31
Water (20°C)	1.33
Ethyl alcohol	1.36
Castor oil	1.48
Benzene	1.50
Fused quartz	1.46
Glass (crown)	1.52
Glass (flint)	1.66
Diamond	2.42

up or disappear, neither of which happens. As a consequence, *it is the wavelength of light that changes with the index of refraction* in such a way that $c/f = n\lambda$ is constant [see Eq. (35–3)]. Thus, when light passes between media 1 and 2,

$$n_1\lambda_1 = n_2\lambda_2. \tag{35–4}$$

35–2 When Can Light Waves Be Treated as Rays?

A particle model of light accounts for many of the apparent features of light propagation, including the facts that light seems to travel in a straight line and casts a sharp shadow. But a wave model can also explain these features, as well as how they break down when we take a closer look. These issues are of some importance because one of the goals of this chapter is to establish the validity of a wave picture of light.

In Chapter 34, we discussed electromagnetic waves that propagate along the $+z$-axis. It will be helpful to review the properties of those waves here. The space dependence and time dependence of the electric or magnetic fields are described by a function such as $\cos(kz - \omega t)$. This function has a series of crests and troughs, with, for example, crests occurring at $kz - \omega t = 0$, or

$$z = \frac{\omega t}{k} = ct.$$

Thus the crest propagates at speed c. We referred to this as a plane wave because all points in the xy-plane defined by a fixed value of z have the same fields, whatever the x- or y-value (Fig. 35–3a). This allows us to make the idea of the wave fronts in electromagnetic waves more explicit: The planes for which the argument $kz - \omega t$ is constant represent the fronts. Figure 35–3b shows a sequence of wave fronts transverse to the direction of the electromagnetic (light) wave. It is useful to think of these fronts as representing a particular set of field values along the wave. For example, the wave fronts could represent the planes on which the electric field is a maximum or the points where the field is zero. (Figure 35–3b arbitrarily sets the wave fronts at the points where the electric field is maximal; since the magnetic and electric fields are in phase, these are also the points where the magnetic field is maximal.) In empty space the sequence of wave fronts moves at speed c along the original direction of propagation. Light is observed to move in a straight line, and the description above is in fact in accord with the observation. Christian Huygens, who was one of the earliest proponents of the wave theory of light, was able to strengthen this agreement between the wave picture and observation with a special approach to the process.

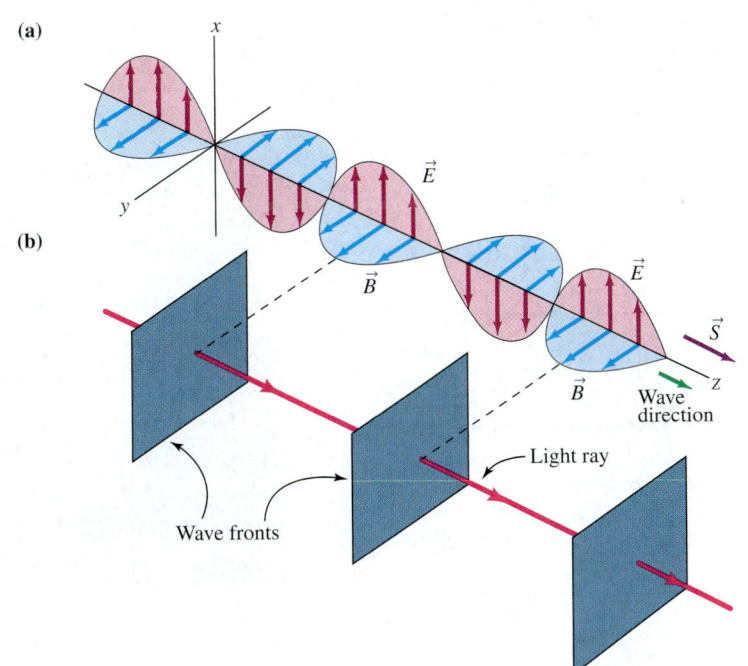

◀ **FIGURE 35–3** (a) The electric and magnetic fields of an electromagnetic wave propagating in the direction of the Poynting vector $\vec{S}$ (see Chapter 34). (b) Wave fronts are chosen arbitrarily at points where the electric and magnetic fields are maximal; the fronts could be chosen at the fields' zero points instead.

978 | Light

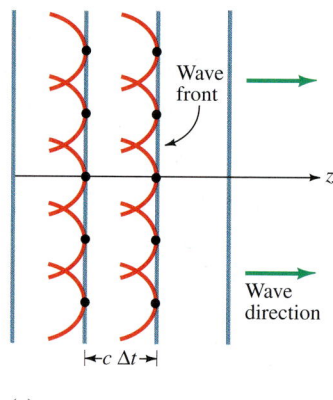

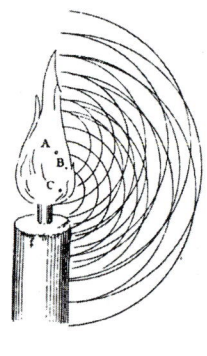

▲ **FIGURE 35–4** (a) Huygens' construction of wave fronts. Wavelets emitted at each point along the wave front add to a new wave front and produce plane waves. (b) Huygens' illustration of wavelets, here for a candle flame, from his book *Traité de la Lumière* (1885).

Huygens' Principle

Consider, as Huygens did in a publication dating from 1690, a wave front perpendicular to the $+z$-direction (Fig. 35–4a). Huygens located the wave front after a time interval Δt by viewing *every* point on the original wave front as a source of light emitting a spherical pulse (or *wavelet*) of radiation (Fig. 35–4b). The radius of the wavelet sphere in empty space is $c\,\Delta t$, the distance the light travels in time Δt. (In a medium in which the speed of light is c/n, the radius of the sphere is reduced by a factor of n to $(c/n)\,\Delta t$.) In the limit that the separation between all the emission points is small, the envelope of all these tiny spheres, taken in the direction of propagation of the initial wave front, is the new wave front. As long as no matter boundaries are encountered, the wave fronts generated in this way remain planes parallel to the *xy*-plane, and the straight-line propagation of wave fronts is assured. This treatment of waves, in which each point on an advancing wave front emits wavelets, and the wavelets add to the future configuration of the entire wave, is called **Huygens' principle**. Huygens' principle can be justified from a detailed study of the behavior of waves in Maxwell's equations, although we shall not do so here.

We know from our study of mechanical waves in Chapter 15 that when a wave front such as that formed by water waves approaches a slot, or opening, in a barrier, the wave spreads as it passes the edge of the opening, a phenomenon known as *diffraction*. Huygens' principle can be applied to this situation (Fig. 35–5). When the wave front arrives at the wall, only the part of the wave at the opening can continue to propagate beyond the wall. This part of the wave front generates waves that travel past the slot, with the additional feature that the spherical wavelets emitted near the edges of the slot have no neighboring wavelets to maintain a parallel wave front, and *a wave that spreads away from the slot edges is generated past the slot*. In effect, the light bends around the corner of the opening. Huygens' principle suggests that the spreading is significant (in terms of the fraction of energy in the bent waves) only if the wavelength is about the same as, or larger than, the size of the slot. If the slot width is much larger than the wavelength, only a small fraction of the energy goes into the bent waves, and it is adequate to view the entire slot as a source of a plane wave front. Light has wavelengths around 5×10^{-7} m; therefore the slot must not be too much larger than this size for the effect to be significant. Diffraction had in fact been observed in Newton's day, and it presented a difficulty, although for some reason not a mortal one, for the particle picture. We'll study this phenomenon as it applies to light in Chapter 38.

If the slot is large compared with the wavelength of light, then diffraction can be ignored. The wall casts a shadow, but to a good approximation that is its only effect. We have a **beam** of light, and we can talk about **rays** of light. It is important to remember that when we draw a single ray, this merely represents the *direction* of the motion of wave fronts; since wave fronts have by definition an area, a more accurate graphical description is a *bundle* of parallel rays, each perpendicular to the plane of the wave front.

▶ **FIGURE 35–5** (a) Huygens' construction of wave fronts that approach and pass through an open slot in a wall. Past the slot, the wave fronts bend around the slot edges. (b) A concrete realization of the effect, in a ripple tank, with an opening small enough to act as a single point source.

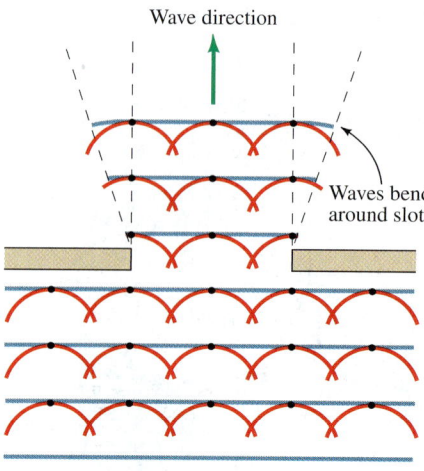

(a)   (b)

Light entering a darkened room through a pinhole or the beam of a searchlight or laser provides vivid images of the propagation of light in the form of rays. We normally do not *see* rays, but they can be made visible by, for example, putting dust particles in the path of the ray (Fig. 35–6). A small part of the light is scattered to the side and is visible from there. The description of light based on the straight-line propagation of rays is called **geometric optics**. This description is perhaps more intuitively referred to as particle-like, because straight-line propagation in free space is a particle characteristic.

In the next section we introduce the two principal laws that allow a complete treatment of geometric optics, laws that describe the reflection and refraction of light rays. In Section 35–4 we describe how both of these laws can be derived directly from Huygens' principle. Even though Huygens' principle is based firmly on the idea that light is a wave phenomenon, to the extent that it explains the behavior of geometric optics, it describes something that more clearly resembles a particle phenomenon. And, as we have seen in our discussion of the passage of light through slots, Huygens' principle will also account for the characteristically wavelike features of light. As we proceed, keep in mind that our treatment of wave phenomena is reserved for Chapters 37 and 38. Here we concentrate on geometric optics.

▲ **FIGURE 35–6** A beam of light generated by a laser. The beam is visible because particles in the air have scattered the light. Lasers have many uses—this beam is used to make a reference for astronomical instruments that can adjust for atmospheric turbulence. It supplies a view of the instantaneous turbulence, which can then be corrected for.

35–3 Reflection and Refraction

Reflection

A light ray **reflects**—it "bounces back"—when it strikes a smooth surface such as that formed by a mirror. (This, of course, oversimplifies much of what happens to reflected light. Virtually everything we see comes from light that has reflected from objects. But these reflections are for the most part quite complicated, because of both the nature of the material that is doing the reflecting and the surface structure of that material. Here you should be thinking of a mirror, something whose surface is essentially invisible but that reflects light rays in the ways we'll describe now.) The *incident ray* makes an angle θ with a line normal to the surface at the point of reflection (Fig. 35–7). The *reflected ray* lies in the plane formed by the incident ray and the normal. The angle θ' that the reflected ray makes with the normal obeys the equation known as the **law of reflection**:

$$\theta' = \theta. \tag{35–5}$$

LAW OF REFLECTION

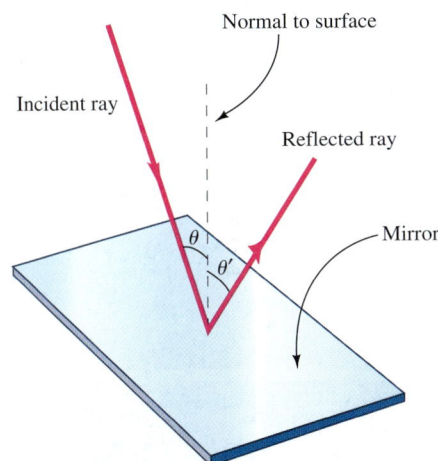

▲ **FIGURE 35–7** The angle of incidence θ equals the angle of reflection θ'.

The consequences of this law are shown in Fig. 35–8a for the reflection of a set of parallel incident rays (a *bundle* of rays) from a flat mirror and in Fig. 35–8b for that from a smooth, curved surface. For the curved surface, the angles of incidence and reflection are indeed equal, but the direction of the normal to the surface varies from point to point, and the reflected rays radiate in various directions.

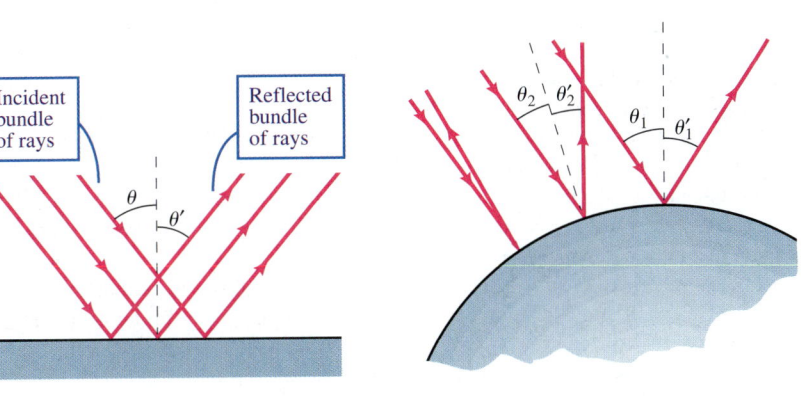

◀ **FIGURE 35–8** Reflection of rays from (a) flat and (b) curved surfaces.

CONCEPTUAL EXAMPLE 35–1 Show that two flat mirrors placed at 90° with respect to each other form a perfect reflector if the incident ray is in the plane perpendicular to the two mirrors. By a perfect reflector we mean that after two successive reflections the outgoing ray will travel back parallel to the line of incidence for an incoming ray at any incident angle.

Answer By drawing the two mirrors perpendicular to the page, as in the sketch of Fig. 35–9, the incident ray is in the plane of the page. Then we need only note that angle ACB—the angle between two normals to surfaces perpendicular to each other—is 90°. We can then establish that the sum of the remaining two interior angles of triangle ACB add to $180° - 90° = 90°$, that is, $\theta + \phi = 90°$. That means the angle of the outgoing ray with respect to the incoming ray, which the diagram shows is $2\theta + 2\phi$, is 180°. The incoming and outgoing rays are parallel.

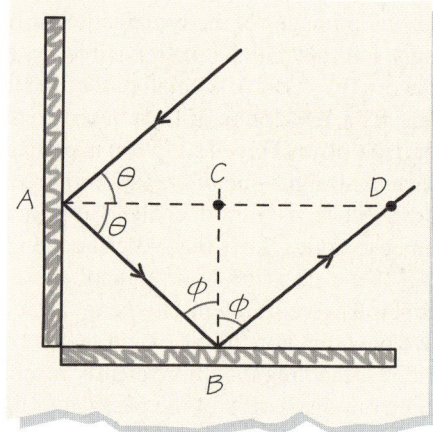

▲ **FIGURE 35–9** The incoming ray that strikes point A eventually emerges along BD, parallel to the incoming ray.

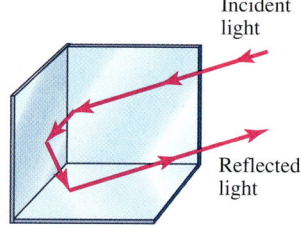

▲ **FIGURE 35–10** A corner reflector. A laser beam that has been sent into the device produces a reflected beam that is parallel to the incoming beam.

Corner Reflectors: Example 35–1 illustrates the principle behind the construction of *corner reflectors* (Fig. 35–10). In a corner reflector, *three* mirrors are placed together at mutual 90° angles, like the walls of an interior corner of a room, or an array of such corners is placed adjacent to one another. Geometry of the kind used in Example 35–1 shows that a ray incident at *any* angle reflects out from a corner reflector along a ray parallel to the incident ray. Corner reflectors are used on some highway signs so that the reflection of light from a vehicle's headlight automatically heads back to the vehicle, making the presence of the reflector very evident to the driver. Other applications stem from the fact that it is possible to measure time delays with great accuracy; a pulse from a laser directed at a corner reflector comes back, and the time delay gives a measurement of distance. In this way, for example, slight movements across fault lines on Earth's surface can be accurately surveyed.

THINK ABOUT THIS...
WHAT HAVE WE LEARNED FROM CORNER REFLECTORS ON THE MOON?

Starting with the *Apollo 11* lunar landing in 1969, several corner reflectors have been placed on the Moon (Fig. 35–11). Flashing laser beams sent through terrestrial telescopes and aimed at the reflectors allow us to capture a fraction (actually only one in about 50 of the very brief flashes ever sends back as much as a single photon—the minimum possible amount—to the detectors on Earth) of the sent beam and measure the time elapsed between departure and arrival. Precision atomic clocks that time the return of the light allow us to measure the distance between Earth and the Moon with an accuracy of less than 1 cm! We have found with these measurements that the Moon recedes from Earth at the rate of 3.8 cm/year. This increase in separation agrees with predictions that start with the fact that the tides dissipate energy as water moves along the bottom of the sea and on the beaches. This has the effect of slowing Earth's rotation about its axis. But the angular momentum of the Earth–Moon system is conserved, so that the angular momentum associated with the Moon's orbiting the Earth must increase to compensate, and for this to happen, the radius of the Moon's orbit increases—in other words, the Earth–Moon separation increases. The measurement of the rate gives us confidence that we understand the process. We have learned other things from the Earth-Moon distance measurements: (1) There is a constant change in Earth's shape, resulting from the fact that land masses are slowly recovering from the compression caused by the weight of glaciers in the last Ice Age. More local variations in the height of tectonic plates, such as the changing difference between the two sides of the San Andreas Fault in California, can also be followed in time. (2) With the distance data in hand, it is possible to retrodict the Moon's orbit to very high precision, allowing an accurate determination of solar eclipses as far back as 1400 B.C.E., something that is useful for historical studies. (3) The data allows us to look in more detail at the Moon's rotation and to show that the Moon has a liquid core.

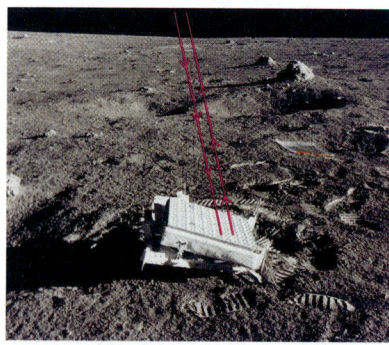

▲ **FIGURE 35–11** This array of corner reflectors was left on the Moon by the *Apollo 14* astronauts. By shining a laser beam from Earth at the reflector and looking for the return light on Earth, the distance to the Moon can be accurately measured. The red lines indicate the directions of the incoming and outgoing laser beams.

35–3 Reflection and Refraction | 981

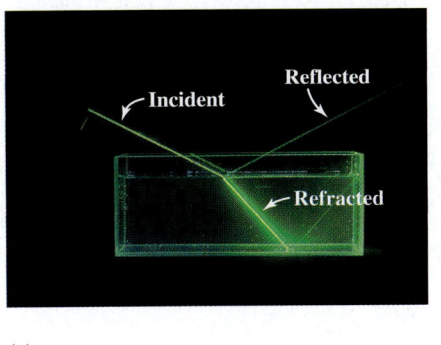

(a)

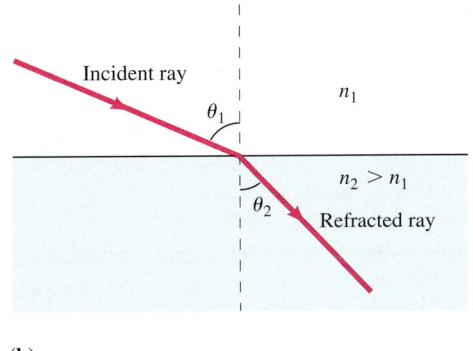

(b)

◀ **FIGURE 35–12** (a) A beam of light is refracted as it enters a tank of water. (b) Refraction from a medium with index of refraction n_1 into a medium with index of refraction n_2. In this case $n_2 > n_1$, and the refracted ray is bent toward the normal to the boundary surface. If n_2 had been less than n_1, the refracted ray would have bent away from the normal.

Refraction

We are all aware of the fact that some media are transparent and that light can pass from one transparent medium into another; air and water form one familiar pair, as do air and glass. We have referred to this process as *transmission*. When the light forming a ray undergoes transmission, the incident ray does not continue along a single straight line but instead changes direction at the boundary between the media. The ray is said to undergo **refraction** (Fig. 35–12a).

Let the index of refraction of the medium with the incident ray be n_1 and that of the medium with the *refracted ray* be n_2. The angles that the incident and refracted rays make with the line normal to the boundary between the media are θ_1 and θ_2, respectively (Fig. 35–12b). Then one finds that

$$n_1 \sin \theta_1 = n_2 \sin \theta_2. \tag{35-6}$$

SNELL'S LAW

This is **Snell's law**, discovered by Willebrord Snell in 1621. The index of refraction of air is very close to unity, so the angle of the refracted ray θ_2 at the interface for light that passes from air into a medium with index of refraction n is given by

$$\sin \theta_1 = n \sin \theta_2. \tag{35-7}$$

Because n is generally larger than one, it follows that $\theta_2 < \theta_1$; that is, *the light is bent toward the normal to the boundary surface*. Equation (35–6) also shows that when light enters a medium with a lower index of refraction, such as when a ray of light travels from water to air, the ray is bent farther away from the line normal to the boundary. This makes a ray coming from an underwater object move closer to the horizontal upon reaching the air, so that an eye receiving this ray and mentally tracing it back will "place" the object at a shallower position (Fig. 35–13).

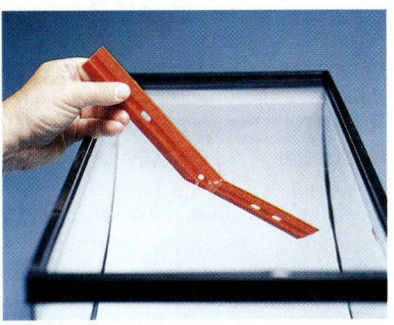

▲ **FIGURE 35–13** Water has a higher index of refraction than air, so the immersed part of this ruler appears in a position that is shallower than its actual position.

EXAMPLE 35–2 Consider a horizontal ray of light approaching a prism, a piece of glass shaped as a rod whose cross section makes a triangle. In this example, the triangle is an equilateral triangle, and the ray is perpendicular to the axis of the rod. The ray is refracted once as it enters the glass and then a second time as it exits to the air. What is the total deflection of the ray, given that the index of refraction of the prism glass is 1.50?

Setting It Up The essential part of this problem is the sketch (Fig. 35–14), because the tools necessary for this problem are those of simple geometry. The sketch defines a series of angles that must be calculated to find the angle of the final ray.

▶ **FIGURE 35–14** Geometric construction of the path followed by a ray incident on a prism.

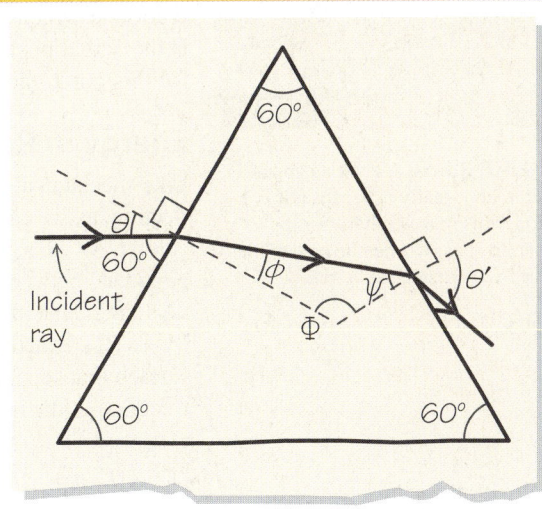

(continues on next page)

Strategy Simple geometry can be used to obtain the angle that the incident ray, drawn parallel to the base, makes with the line perpendicular to the face of the prism, namely $\theta = 30°$. Snell's law then gives us the angle ϕ that the beam inside the prism makes with the perpendicular to the first face. We need to know the angle ψ at which the interior ray meets the second face, and then another application of Snell's law will give us the final angle θ'. To find ψ, note that from the figure, the interior angle Φ is an interior angle of a four-sided figure (the two dashed lines and the lines that rise to the apex) for which the other three interior angles are 90°, 60°, and 90°. But the sum of the interior angles of a four-sided figure is 360°, so $\Phi = 120°$. Then we find ψ by using $\phi + \psi + \Phi = 180°$.

Working It Out To find ϕ, we use Eq. (35–7):

$$\sin \theta = n \sin \phi.$$

When $n = 1.50$ and $\theta = 30°$, this expression gives $\phi = 19.5°$.

With $\Phi = 120°$, the angle ψ that our once-refracted ray makes with the line normal to the second surface is given by $\phi + \psi + 120° = 180°$, or $\psi = 60° - \phi = 60° - 19.5° = 40.5°$. The angle θ' that the second refracted ray makes with the line normal to the second surface is then given by a second application of Snell's law,

$$\sin \theta' = n \sin \psi = 1.50 \sin(40.5°) = 0.97.$$

Thus, $\theta' = \arcsin(0.97) = 77°$, and we see from Fig. 35–14 that the angle the outgoing light ray makes with the base of the prism is $\theta' - \theta = 77° - 30° = 47°$.

What Do You Think? If we want to make the exit angle θ' smaller, how do we change the angle θ? *Answers to What Do You Think? questions are given in the back of the book.*

CONCEPTUAL EXAMPLE 35–3 The study of refraction originated with the discovery by Isaac Newton that white light falling on a prism is split into the colors of the rainbow, and his interpretation of this as white light being composed of all the colors at once. Some people argued that another interpretation was possible: that sunlight, falling on the prism, generated the emission of a variety of colors. Can you propose one or two experiments, using a second, identical prism, to argue that it is indeed white light that is composed of the rainbow colors?

Answer (a) Put the second prism upside down so that the face of the first prism from which the spread of colors emerges is parallel to the face of the second prism that the colored rays will enter (Fig. 35–15). This reverses the refraction process of the first prism. What emerges from the far side of the two prisms is white light, illustrating the reconstitution of the white light through direct combination of its colorful components. The only way this is possible is that the white light is simply constituted of light of the rainbow colors, and that in adding the second prism we have put those colors back together again. Note that in the figure we have drawn a single ray entering the prism, whereas on the emerging side of the double prism we have parallel rays. But we noted before that rays represent only a direction, with the actual incoming light more properly represented as a bundle of rays. The correct representation of our situation is that there is a bundle of incoming white light rays and a bundle of outgoing white light rays. (b) Again insert the second prism, but this time place a mask on the surface of the second prism so that only light of one color enters (say green). The result of this is that only green will emerge on the far side of the prism. This can be done with each of the colors. Again, the constituent colors of white light are nothing more than that—constituents—and the only role of the first prism is to separate them out.

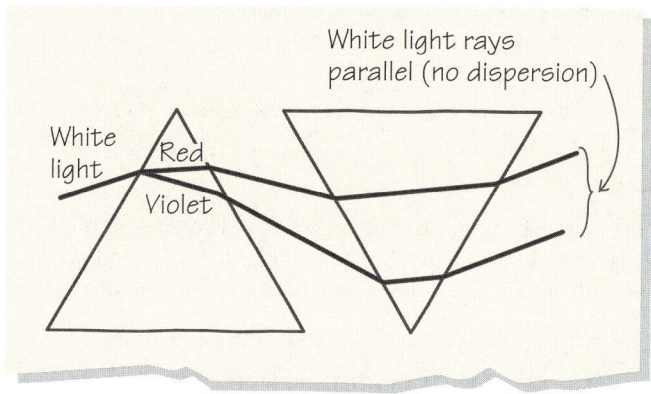

▲ **FIGURE 35–15** Arrangement of two prisms that shows that the colors are the components of white light.

▲ **FIGURE 35–16** The fish appears less deep than it really is because of refraction. This makes it hard for the fisherman to spear it when he or she aims the spear from outside the water.

Refraction is responsible for some curious optical effects, as Fig 35–13 shows. You may be familiar with the problem of spearing a fish (Fig. 35–16). The source of the light ray from the fish is seen to originate from a point that lies on a straight line along the direction at which the ray enters the observer's eye. Thus the observer thinks the image I is the actual position of the fish, and if the observer is a spear fisherman who is aiming for this point, the fish will not be touched.

Energy in Reflection and Refraction

Like mechanical waves, transmission (with refraction) is generally accompanied by reflection, as we can see in Fig. 35–12a where a reflected ray is visible as well as a refracted one. As we saw in Chapter 34, the incident ray carries electromagnetic energy. At the boundary between media, this energy is apportioned among the reflected and refracted rays such that the total energy is conserved. Here we quote a result without proof: Maxwell's equations can be used to show that when light is perpendicularly incident on a surface that separates a medium of index of refraction n_1 from a medium of index of refraction n_2, the intensity of the reflected light, I_r, is related to the incident intensity, I_0, by

$$\frac{I_r}{I_0} = \frac{(n_2 - n_1)^2}{(n_2 + n_1)^2}. \tag{35–8}$$

For light perpendicularly incident—the angle of incidence is zero—from air ($n = 1.0$) into glass ($n = 1.5$), only 4 percent of the incident light is reflected. Note the squares in this equation. The ratio is the same for light coming from air and moving into glass as it is for light coming from glass and passing into air. More generally the intensity of the reflected (and refracted) light varies with the angle of incidence, with the intensity of the reflected ray a minimum at perpendicular incidence.

CONCEPTUAL EXAMPLE 35–4 What is the intensity of the transmitted light that falls in a perpendicular direction on a thin sheet of glass ($n = 1.5$) and then passes out of the far side of the sheet?

Answer The index of refraction of air is approximately unity, so from Eq. (35–8) the fraction of energy reflected at the first surface is $f_r = (1.5 - 1.0)^2/(1.5 + 1.0)^2 = 0.04$. By energy conservation, the fraction that penetrates is $1 - 0.04 = 0.96$. This intensity of light falls on the second interface, and here again 4 percent is reflected, and therefore the total fraction of light transmitted is $(0.96)^2 = 0.92$. Some of the light reflected at the second interface will be re-reflected at the first interface and head back out the far side, but there we are dealing with fractions like 4 percent of 4 percent, which we can neglect. We should add that when monochromatic light is involved, wavelike phenomena can make the fraction reflected vary dramatically with wavelength or glass thickness (see Chapters 37 and 38).

Total Internal Reflection

For some incident angles, *all* the incident energy is contained in the reflected ray. This situation, known as **total internal reflection**, can occur only when light travels from a medium with a larger index of refraction toward a medium with a smaller index of refraction, such as when light passes from water toward air. Simple geometry explains this phenomenon.

Let's consider a light-ray incident from a medium with an index of refraction n_1 to a medium with an index of refraction n_2; this time $n_1 > n_2$. Snell's law, Eq. (35–6), may be written in the form $\sin\theta_2 = (n_1/n_2)\sin\theta_1$. As θ_1 increases, θ_2 reaches 90° before θ_1 does, because the factor n_1/n_2 is larger than unity. Figure 35–17 shows the reflected and refracted rays for various values of θ_1. When $\theta_2 = 90°$, the ray in medium 2 skims along the interface of the two media. This occurs when θ_1 reaches a critical angle θ_c such that $(n_1/n_2)\sin\theta_c = \sin 90° = 1$, or

$$\sin\theta_c = \frac{n_2}{n_1}. \tag{35–9}$$

When θ_1 exceeds θ_c, there is no angle θ_2 that can satisfy Snell's law. The electromagnetic energy carried by the incident ray must go somewhere, and the ray is reflected. There is no diminution of the intensity of the reflected ray, and the reflection is total.

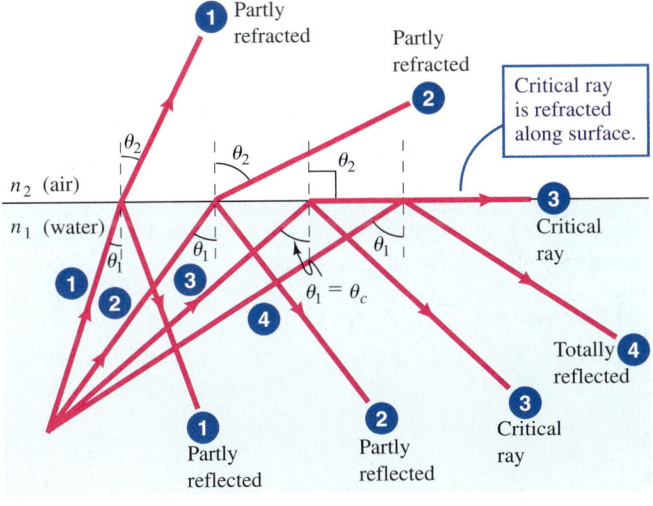

(a)

(b)

▲ **FIGURE 35–17** (a) Various rays traveling from a medium with a larger index of refraction (water) to a medium with a smaller index of refraction (air). When the incident angle is θ_c, there is total internal reflection. (b) Refraction and total internal reflection off the air–water interface in a water tank.

CONCEPTUAL EXAMPLE 35–5 Suppose a ray of white light originates inside the tank in Fig. 35–17 and that the ray is at an angle larger than the critical angle for all the colors in white light. The ray is then rotated so that the incident angle decreases. What color light emerges first into the air?

Answer The observation of the decomposition of white light by a glass prism shows that the index of refraction for blue light is larger than that for red light: $n(B) > n(R)$. Taking the index of refraction of air as unity, Eq. (35–9) reads $\sin \theta_c = 1/n$ and thus $\theta_c(B) < \theta_c(R)$. This means that *red* light emerges first as the white-light source rotates upward to make a smaller and smaller angle of incidence in Fig. 35–17.

EXAMPLE 35–6 A fisherman knows of a fishing spot whose horizontal distance from a river bank is a distance $R = 1.5$ m from the bank. Thinking to sneak up on a fish which might be at any water depth, he stealthily approaches the bank with his eyes at ground (and water) level. Describe what he might see.

Setting It Up We draw in Fig. 35–18 an illustration. We have drawn a particular ray, with angle of incidence θ, that meets the surface very close to where the eye is located.

Strategy and Working It Out If we model the fish as a point, then for a given depth D, there is only a single ray from the fish that reaches the fisherman's eyes. When D is large (the fish is deep) and the angle of incidence is smaller than the critical angle, then when the ray leaves the water it continues upward, although more parallel to the surface than the incident ray. The only such ray able to reach the eye at the edge must then strike the surface very close to the edge. The fisherman extrapolates this ray back and mentally places the fish closer to the surface, just as for the spear fisherman of Fig. 35–16.

We can now imagine that D decreases, until at a critical depth D_c the incident angle in the figure is the critical angle θ_c. The refracted ray comes off parallel to the surface, so the fisherman sees the fish right at the surface. The corresponding ray that comes from any fish closer to the surface than D_c does not make it out of the water, because the incident angle is greater than the critical angle. But a ray from that fish that is incident at the critical angle, and which intersects the surface at a point in between the edge and the horizontal location of the fish, will propagate in air parallel to the surface and ends at the eye. The fisherman sees *all* the fish from just below the surface to a

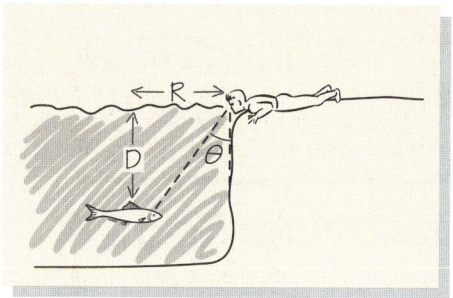

▲ **FIGURE 35–18**

depth of D_c at once, squashed into a flat pancake at the surface and hardly recognizable. Hanging in the region from $D = 0$ to $D = D_c$ is a good way for the fish to hide from this particular fisherman!

What is D_c? We can easily find it from geometry. The figure shows that generally $\sin \theta = R/\sqrt{D^2 + R^2}$. With $\sin \theta_c = 1/n_\text{water}$, we then have

$$\frac{R}{\sqrt{D_c^2 + R^2}} = \frac{1}{n_\text{water}} \quad \text{or} \quad \frac{\sqrt{D_c^2 + R^2}}{R} = n_\text{water}.$$

We solve this for D_c:

$$D_c^2 = R^2(n_\text{water}^2 - 1) = (1.5 \text{ m})^2(1.33^2 - 1) = 1.7 \text{ m}^2;$$

$$D_c = 1.3 \text{ m}.$$

What Do You Think? Can a fish see the fisherman sneaking up to the bank?

THINK ABOUT THIS...
HOW DOES LIGHT PROPAGATE WITHIN OPTICAL FIBERS?

Fiber optics represents one of the most important technological applications of total internal reflection. The principle behind this technique of conducting light from one place to another is straightforward: A transparent quartz fiber (typically of diameter 50 μm—the thickness of a human hair) will serve as a conductor of light if any ray inside the fiber undergoes total internal reflection upon striking the side of the fiber (Fig. 35–19a). Figure 35–19b shows a ray in air ($n = 1$) entering a cylinder of diameter D at an angle θ_i with the axis of the cylinder. If n_f is the index of refraction of the fiber, then the angle that the ray makes with the axis inside the fiber is θ_f, where $\sin \theta_f = \sin \theta_i / n_f$. This ray will strike the wall of the cylinder at an angle $(90° - \theta_f)$ with the normal to the wall. There will be total internal reflection if $n_f \sin(90° - \theta_f) > 1$; that is, if $n_f \cos \theta_f > 1$. We have

$$n_f \cos \theta_f = n_f \sqrt{1 - \sin^2 \theta_f}$$
$$= n_f \sqrt{1 - \frac{\sin^2 \theta_i}{n_f^2}}$$
$$= \sqrt{n_f^2 - \sin^2 \theta_i} > 1.$$

Because $\sin^2 \theta_i \leq 1$, we have

$$\sqrt{n_f^2 - \sin^2 \theta_i} \geq \sqrt{n_f^2 - 1}.$$

Thus we automatically satisfy the condition for total internal reflection, $n_f \cos \theta_f > 1$, if

$$\sqrt{n_f^2 - 1} > 1. \qquad (35\text{–}10)$$

Because the largest value of $\sin \theta_i$ is 1 (the light first enters the cylinder from the end), Eq. (35–10) is a condition for internal reflection for *all* of the light that enters the fiber. Equation (35–10) is satisfied for any material with $n_f > \sqrt{2}$. A typical fiber has an index of refraction of 1.62, which is larger than the critical value. Note that once a ray is in the fiber, it remains inside *even if the fiber curves*, at least if the bend is not too sharp. Information is carried when the light is pulsed in an order that contains the information, and in this way the light carries the information of a telephone call or an exchange between computers. There is a considerable advantage to optical fibers: Compared to the density of information that can be carried in an old-fashioned coaxial cable, a fiber can carry a much greater density of information. That translates into a capacity to carry many telephone calls at once, or to carry the enormous amounts of information exchanged by modern computers. This is in addition to the fact that for the same weight or diameter of an ordinary cable, one can put together a great number of fibers to carry information in parallel.

The situation outlined above is only an ideal. The internal reflection is somewhat less than total if there are impurities such as moisture, dust, or oil on the surface, because electromagnetic energy can leak across the thin "barrier" formed by the air layer between the fiber and the impurity. In long-distance transmission, light may be reflected many times, and it is therefore important to have no leakage of light. This problem is controlled largely by *cladding*—coating each fiber with a transparent covering whose index of refraction is lower than that of the fiber. In addition, the light intensity generally decreases as the ray propagates in a medium because the medium is not perfectly transparent. This effect is reduced by making the fiber from fused quartz, a highly transparent material, and purifying it to remove all traces of water. For the trans-Atlantic cable TAT-8, which can carry 40,000 conversations over two pairs of glass fibers simultaneously, it is necessary to boost the signal only every 50 km with a repeater station. This is much less expensive than systems of metal wire, which require boosting every kilometer.

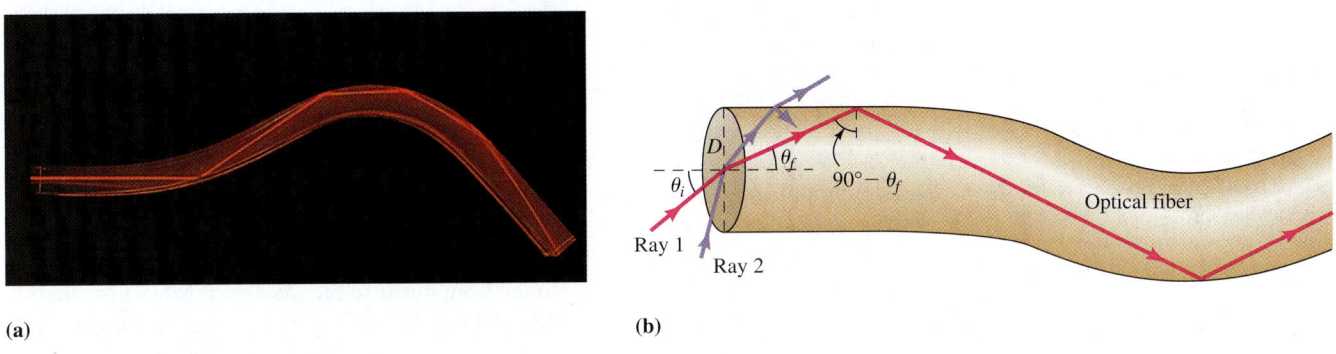

(a) (b)

▲ **FIGURE 35–19** (a) Total internal reflection in an optical fiber. (b) Detailed construction of ray angles in a curved fiber.

*35–4 Fermat's Principle

Fermat's principle provides us with an interesting way to understand reflection and refraction. In order to set the stage for Fermat's principle, let us first look at how Huygens' principle leads to the law of reflection [Eq. (35–5)]. Figure 35–20a shows a sequence of wave fronts as they approach a mirror. In Fig. 35–20b, point C_2 is the center of a reflected spherical wave, one of many along the mirror. An outgoing (reflected) wave front—here, the line tangential to point D_2 of the semicircle centered on point C_2—forms. The distance the wave travels in time Δt is the same for incoming and outgoing waves, so a simple geometrical argument yields the result described by Eq. (35–5); namely, that the angle of reflection equals the angle of incidence. Figure 35–20c shows a later part of the sequence.

Snell's law may also be obtained by an application of Huygens' principle. The bending of the wave front is associated with the slowing down of the light waves in the medium. The bending can be visualized by analogy with the direction change of a band whose members march at an angle toward a sidewalk. Each row is like a wave front. The band members are given orders that each one must walk more slowly, without changing the distance between marchers in each row, as soon as that marcher steps on the sidewalk.

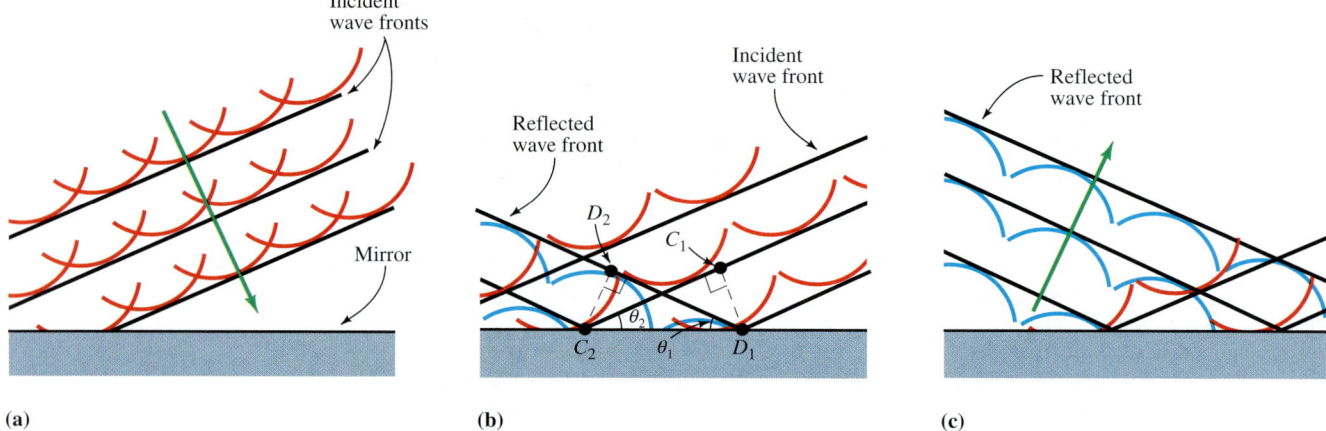

▲ FIGURE 35–20 (a) Incident wave fronts approaching a plane mirror. (b) The wave fronts reflect. (Note that for clarity, fewer wavelets are shown.) The reflected wave fronts are generated by Huygens' construction. The relation $C_2D_2 = C_1D_1$ leads by geometrical reasoning to equal angles of incidence and reflection. (c) Most of the wave fronts have been reflected.

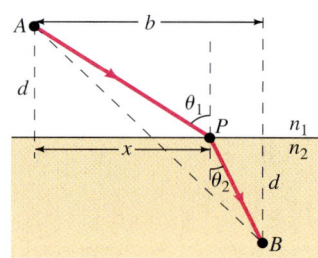

▲ FIGURE 35–21 A marching band forms rows that change direction when the speed of the marchers change across a row. Here the rows bend sharply at a boundary where the marchers begin to march more slowly, holding the spacing between the marchers within each row constant.

The result is that the front formed by the row of band members bends (Fig. 35–21). Rather than going through the derivation of Snell's law from the Huygens construction, which leads to a picture like that of the marching band, let's demonstrate it from the principle enunciated by Pierre de Fermat in 1657. **Fermat's principle** states that

> **The path of a ray of light between two points is the path that minimizes the travel time.**

To derive Snell's law from Fermat's principle, let's consider a point A in medium 1 with index of refraction n_1 and a point B in medium 2 with index of refraction n_2 (Fig. 35–22a). We want to find the path between points A and B that takes a ray of light the least amount of time to travel. We choose A to be a distance d above the boundary and B a distance d below the boundary; we choose the horizontal distance between A and B to be b. The straight line connecting A to B crosses the boundary at a distance $b/2$ from the normal dropped from A onto the boundary, but because the indices of refraction are different, this will not be the ray's path; instead the ray's path crosses the boundary at some point P that we want to find. Figure 35–22a shows that the distance from A to the intersection point P is $\sqrt{d^2 + x^2}$, and the distance from the intersection point P to B is $\sqrt{d^2 + (b-x)^2}$. The time for the ray to travel a distance D in a medium of index of refraction n is given by $t = D/v = D/(c/n) = nD/c$. Thus the total travel time is

$$t_{AB} = t_{AP} + t_{PB} = \frac{n_1\sqrt{d^2 + x^2} + n_2\sqrt{d^2 + (b-x)^2}}{c}. \quad (35\text{–}11)$$

Figure 35–22b is a graph of t_{AB} as a function of x. The minimum travel time is obtained by finding the place at which the slope of t_{AB} as a function of x is flat; that is, the value of x at which

$$\frac{dt_{AB}}{dx} = 0.$$

This condition implies that

$$\frac{dt_{AB}}{dx} = \left(\frac{1}{c}\right)\left[\frac{n_1 x}{\sqrt{d^2 + x^2}} - \frac{n_2(b-x)}{\sqrt{d^2 + (b-x)^2}}\right] = 0. \quad (35\text{–}12)$$

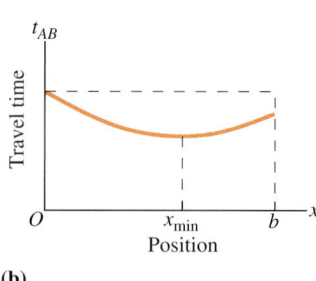

▲ FIGURE 35–22 (a) Geometry for proving Snell's law by Fermat's principle. (b) The travel time t_{AB} for the ray as a function of x.

Now observe from Fig. 35–22a that

$$\frac{x}{\sqrt{d^2 + x^2}} = \sin\theta_1 \quad (35\text{–}13\text{a})$$

and that

$$\frac{b-x}{\sqrt{d^2+(b-x)^2}} = \sin\theta_2, \quad (35\text{-}13b)$$

where θ_1 and θ_2 are the angles the two rays make with respect to the normals in their respective media. Thus Eq. (35-12) may be rewritten as

$$n_1 \sin\theta_1 = n_2 \sin\theta_2, \quad (35\text{-}14)$$

which is just Snell's law.

Both the straight-line propagation of light in a single medium and the law of reflection can also be derived from Fermat's principle. Fermat's principle in fact follows directly from Maxwell's equations, although we do not perform the derivation here. As Example 35-7 shows, principles such as Fermat's principle (more generally termed minimum principles) can apply in surprising circumstances.

EXAMPLE 35-7 A girl located at point B in Fig. 35-23a spots a ball at point A. Point A is in tall grass, where the girl can run at 1.1 m/s, and point B is in short grass, where the girl can run at 2.2 m/s. The whole area is flat. At what point should she cross the boundary between the grasses so that she retrieves the ball as quickly as possible?

Strategy Fig. 35-23a shows that the problem is conceptually identical to the problem of deriving Snell's law from Fermat's principle. We want an expression for the total travel time, consisting of the time in the short grass where the speed is 2.2 m/s and the time in the tall grass, where the speed is 1.1 m/s. The distances in the two regions can be calculated in terms of the unknown crossing point, whose position is at x.

For a given (to be found) value of x, the distance to be covered in the short grass is given by $L_\text{short} = \sqrt{d^2+(d-x)^2}$. If the speed of travel in the short grass is v_short, then the time spent in the short grass is $t_\text{short} = \dfrac{\sqrt{d^2+(d-x)^2}}{v_\text{short}}$. The distance traveled in the tall grass is $L_\text{tall} = \sqrt{d^2+x^2}$, hence the time in the tall grass is $t_\text{tall} = \dfrac{\sqrt{d^2+x^2}}{v_\text{tall}}$. Our task then is to consider the total time as a function of x and find the value of x for which this sum is a minimum.

Working It Out The expression for the total time involves $d = 5.0$ m, $v_\text{short} = 2.2$ m/s, and $v_\text{tall} = 1.1$ m/s. We thus want to minimize

$$t(x) = \frac{\sqrt{25+x^2}}{1.1} + \frac{\sqrt{50-10x+x^2}}{2.2}.$$

We plot $t(x)$ as a function of x in Fig. 35-23b. There is indeed a minimum, at about $x = 1.5$ m. Compare this to the 2.5-m value x would take if the child were to run in a straight line. It is best to run a little farther in the short grass.

Rather than plotting, we could also find the desired value of x by solving for x in the algebraic equation $\dfrac{dt(x)}{dx} = 0$. But this is a rather messy equation and needs more numerical work than the plot of Fig. 35-23b.

The path followed looks like the path a light ray would take in traveling from a medium with a larger value for the speed of light to a medium with a smaller value for the speed of light.

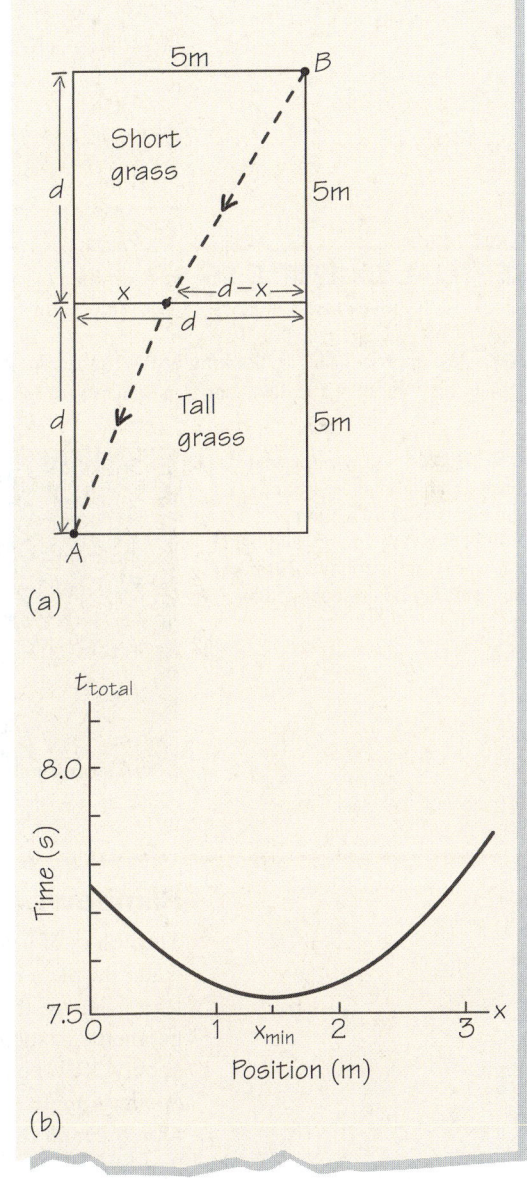

▲ **FIGURE 35-23** (a) Bird's-eye view of girl's path. (b) The travel time as a function of where the child crosses the border between the tall and short grass.

35–5 Dispersion

In this section, we explore another property of the index of refraction, a property with some truly spectacular consequences: In general, the index of refraction depends on the wavelength (or color) of the light being transmitted. Table 35–2 shows how n varies with wavelength for glass, for wavelengths near and including the wavelengths of visible light. Different wavelengths are refracted to different degrees. In this way, white light (a mixture of different wavelengths) can be separated into its constituent colors of the rainbow (Fig. 35–24). As we mentioned in Section 34–2, the dependence of refraction on the wavelength of light is called *dispersion*.

TABLE 35–2 • Index of Refraction of Glass as a Function of Wavelength

Wavelength in Air (nm)	$\omega = (2\pi c/\lambda)$ rad/s in units of 10^{15}	n	Color
361	5.22	1.539	Near ultraviolet
434	4.35	1.528	Blue
486	3.87	1.523	Blue-green
589	3.19	1.517	Yellow
656	2.86	1.514	Orange
768	2.45	1.511	Near infrared
1200	1.58	1.505	Infrared

CONCEPTUAL EXAMPLE 35–8 Consider the setup in Fig. 35–24. Which moves faster in glass, red or violet light?

Answer We approach this by looking at the figure. We see that violet light is bent more. (The fact that it is bent twice because it passes through two surfaces does not change this conclusion.) By Snell's law this implies that n for violet light is larger than for red light, and since the speed of light is c/n, this implies violet light moves more slowly than red light. We confirm this by looking at Table 35–2, which shows that n is indeed larger for violet light.

▶ **FIGURE 35–24** (a) Bright white light is formed into a beam by a slit before being dispersed by a prism. (b) White light enters the prism, and light of different wavelengths follows different paths. The result is a beam separated by color.

(a)

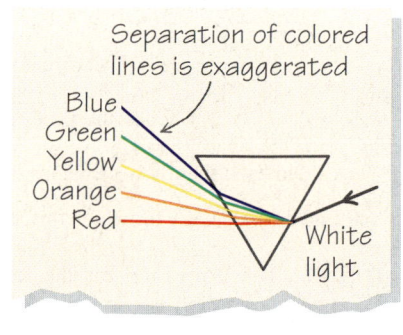
(b)

Rainbows and the Blue Sky

The colors of a rainbow result from dispersion in the reflection of light from individual water droplets in the air. When sunlight shines on a raindrop, light is reflected once before it leaves the drop. Many paths are possible; two are shown in Fig. 35–25a. The geometry is such that no ray can emerge after one reflection at an angle *steeper* than about 42°. When the Sun is behind the viewer, only drops that lie within a cone with an opening angle of about 42° reflect sunlight back to the observer's eye (Fig. 35–25b); moreover, *all* the drops in this cone reflect light to the observer. (We shall refer to a disk that fits into the cone because the depth of the cone is irrelevant.) One other feature of the disk is that light is reflected most strongly from raindrops at the edge, around 42°.

Dispersion has played no role in our discussion so far. The effect of dispersion is to make the angle of the outer radius of the disk slightly different for different colors. As Fig. 35–25c indicates, the disk for red light is larger than the disk for blue light. Because

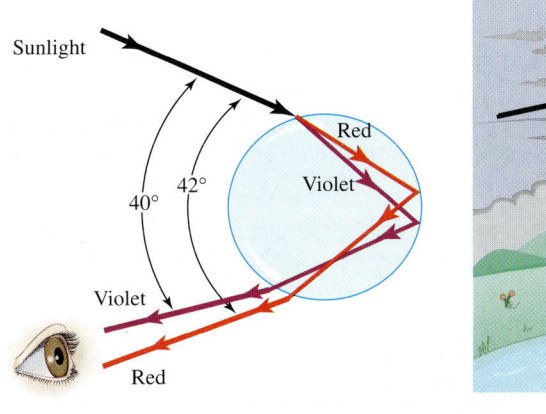
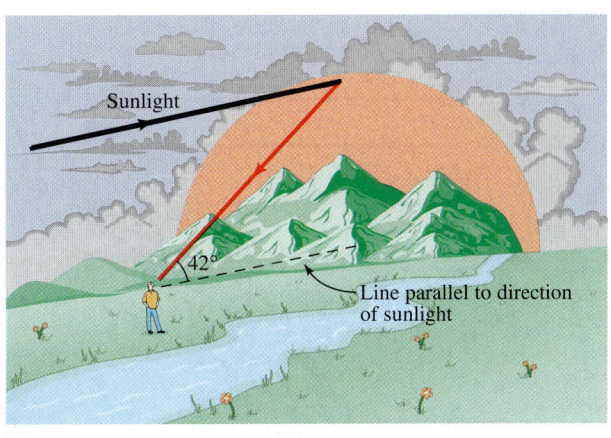

▲ **FIGURE 35–25** (a) When sunlight enters a raindrop and exits after one reflection, no light exits at an angle steeper than about 42° from the incident ray. (b) As a result, light comes back to an observer with the Sun behind him from all the raindrops that lie within a cone with an angle of about 42°, as seen by the observer. (c) Sunlight is a variety of colors. Due to dispersion, the disks that fit into the cones are of slightly different sizes for different colors, with red forming the largest cone and violet the smallest. This figure exaggerates the effect.

the intensity of the light in the disk is strongest at the edges, we see a red ring outside a blue ring (with other colors placed accordingly). All the disks overlap inside the rainbow, giving white light. A *secondary rainbow* can be produced when there are two internal reflections within the raindrops (Fig. 35–26). The order of the colored disks produced by the raindrops will now be reversed, with red light at the bottom and blue light at the top of the secondary rainbow. Figure 35–27a illustrates how an observer sees rainbows and how the pattern of dispersion leads to the color inversion of a secondary rainbow compared with a primary rainbow. The light is brightest inside (below) the primary rainbow and outside (above) the secondary rainbow because the disks of each of the colors overlap in these regions; it is relatively darker between the two rainbows (Fig. 35–27b).

Dispersion in the more general sense of phenomena that depend on frequency (or wavelength) is quite common; for example, the scattering of light by matter has such dependence. In 1872 Lord Rayleigh showed that the fraction of incident light scattered by air molecules varies as f^4 for light in the visible range. This explains the color of the setting Sun. As the Sun sets, the rays of light pass through more and more atmosphere to reach our eyes, and an increasing number of the high-frequency (low-wavelength) components are scattered away, leaving the lower frequency in the direct rays from the Sun. The Sun's color changes from white to yellow to orange and finally to red, as it moves closer to the horizon, as the higher frequencies are scattered away from the observer.

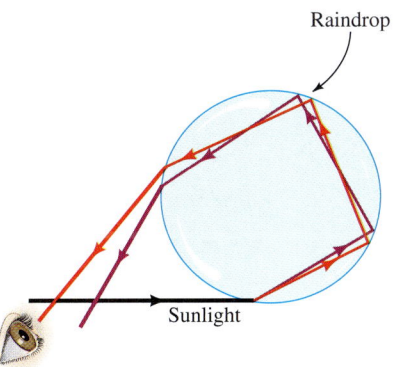

▲ **FIGURE 35–26** The light that reaches the eye from a secondary rainbow has undergone two internal reflections in a set of raindrops. Light of shorter wavelengths (violet light) emerges at a steeper angle than light of longer wavelengths (red light) does, in contrast to the light that undergoes only one internal reflection, which forms the primary rainbow, and the colors of the secondary rainbow are inverted.

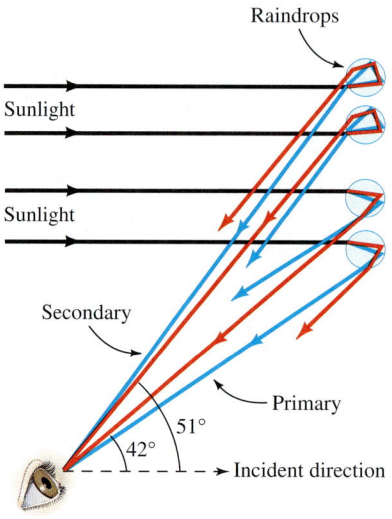 

▶ **FIGURE 35–27** (a) When the eye sees a region of sky that contains raindrops illuminated by sunlight, the light reflected from individual drops forms a primary rainbow (one internal reflection within each drop) and a secondary rainbow (two internal reflections). (b) The brighter, primary rainbow is on the bottom. The order of colors is reversed in the two rainbows due to the extra reflection, which produces the fainter secondary rainbow. The disks overlap, so it is relatively brighter below the primary rainbow and above the secondary rainbow, but darker between them.

THINK ABOUT THIS...
WHY IS THE SKY BLUE WHILE CLOUDS ARE WHITE?

The frequency dependence of the amount of light scattered by the atmosphere is responsible for the fact that the sky looks blue. Blue light has a higher frequency than red, and because there is more scattering in the higher frequencies, it is the blue component of sunlight that is preferentially scattered into our eyes by the atmosphere. Above the atmosphere, where there are no molecules to scatter the light, astronauts see a black sky.

If the sky is blue, why are clouds white? The answer to this question has to do with the size of the objects doing the scattering. The f^4 law (equivalently what we could call the $1/\lambda^4$ law) applies only to the scattering of light by objects much smaller than the wavelength of the light. Thus it applies to the scattering of light by individual molecules in the air. However, the water droplets that make up a cloud are larger than the wavelength of light impinging on them. For such objects, a set of molecules lying within an area whose linear size is λ will act *together* as a "single reflector" to reflect the light. The amplitude of the reflected wave is proportional to the effective area of the "single reflector," and this is proportional to λ^2. The intensity of the reflected light is in turn proportional to the square of the amplitude, that is, to λ^4. This just cancels the $1/\lambda^4$ dependence of the individual contributions of the molecules, and consequently there is no preferential wavelength in scattering by droplets. All wavelengths are scattered equally, so that white light scattered by a cloud stays white, and that is how we see the cloud.

The Atomic Theory of Dispersion

Dispersion occurs because of the atomic structure of dielectric media. The atoms that make up the medium contain electrons bound to their respective nuclei (in opposition to conductors, in which some of the electrons are in effect free to roam). The size of the bound system has linear dimensions of about 0.1 nm. Let's think of each atom as a single electron (of mass m) that oscillates about a positive ion as though the electron were bound to that ion by a spring. If the spring constant is k, then the angular frequency of oscillation (the natural frequency) is ω_0, given by $\omega_0^2 = k/m$. If no other forces act, the motion of the electron along the z-axis is of the form

$$z = A \cos \omega_0 t, \tag{35-15}$$

where A is the amplitude of motion. While this may not immediately remind you of an atom, you can see that it is relevant if you recall that an object in a circular orbit has a motion whose projection along an axis in the orbit plane is simple harmonic motion (see Chapter 13).

Now suppose that a plane electromagnetic wave oscillating with angular frequency ω is incident on an atom, with the electric field oriented in the z-direction. The electric force on an electron in the atom is then oscillatory with frequency ω. The situation is that of a driven harmonic oscillator. The motion of the electron is oscillatory with the driving frequency ω. The amplitude exhibits resonance, becoming large when $\omega_0 \cong \omega$. Thus, based on our knowledge of the driven harmonic oscillator, the motion is of the form

$$z \propto \frac{1}{\omega_0^2 - \omega^2} \cos \omega t. \tag{35-16}$$

We assumed here that ω_0 is not exactly at ω; in fact, for materials such as water and glass, ω_0 is on the order of 5 to 6 times larger than the characteristic angular frequencies of visible light. As you may recall from our work on the driven harmonic oscillator, there is no actual singularity when $\omega = \omega_0$, because there is an additional additive term in the denominator of Eq. (35-16) associated with damping. This term can be ignored unless ω_0 and ω are very close.

An accelerating charge (the electron) *radiates* electromagnetic energy, and the intensity of the radiation is proportional to the average of the acceleration squared. (While we are not going to show this in detail, it is implicit in our original discussion of electromagnetic waves in Chapter 34.) From Eq. (35-16), the average acceleration is proportional to

$$\frac{d^2z}{dt^2} \propto \frac{\omega^2}{\omega_0^2 - \omega^2} \cos \omega t \cong \frac{\omega^2}{\omega_0^2} \cos \omega t.$$

In the last step, we have used the fact that $\omega_0^2 \gg \omega^2$ for visible light in these materials. The intensity I of the radiation is proportional to the acceleration squared. Thus I varies as ω^4 or, equivalently, as f^4. (The average of $\cos^2 \omega t$ over many periods is just $1/2$.) The wavelength λ is related to ω by $\lambda = c/f = 2\pi c/\omega$, so *the intensity of the radiation emitted by a charge set in oscillation by an external electric field is proportional to $1/\lambda^4$, where λ is the wavelength of the oscillating field*. This fact, first obtained by Lord Rayleigh, explains the blue sky: blue light, which is a high-frequency component of visible light, is scattered preferentially from atmospheric molecules—they reradiate this light equally in all directions (see the "Think About This" box on p. 990).

This discussion shows how the original electric field is modified by the addition of a radiated field as a result of the electron's motion; in fact, it is such a modification that is described by a dielectric constant κ, and hence an index of refraction n given by Eq. (35-2), $\kappa = n^2$. We accordingly expect n to depend on the frequency; that is, to exhibit dispersion. A more advanced treatment of how the external field is modified in the presence of atomic electrons shows that

$$\frac{1}{n^2} = 1 - \frac{C}{\omega_0^2 - \omega^2}, \tag{35-17}$$

where C is a constant that is proportional to the density of atoms. This equation describes how the index of refraction varies with frequency. Because $\omega_0 \gg \omega$ for visible light, *the index of refraction increases as the frequency of the light increases*. This is certainly in accord with Table 35-2. In fact, atoms and molecules have many resonant frequencies, so a more accurate version of Eq. (35-17) must contain several terms of the form $C_k/(\omega_{0k}^2 - \omega^2)$ added together.

Summary

The wave theory of light allows us to understand almost all the classical properties of light, in particular straight-line propagation in open space, reflection, and refraction. Light waves propagate with speed $c = 3.00 \times 10^8$ m/s in a vacuum; in transparent media, the speed of propagation is c/n, where n is the index of refraction of the medium. In general, the index of refraction depends on the wavelength of the light.

The propagation of light can be described either in terms of wave fronts, which form an envelope of spherical wavelets built upon earlier wavelets (Huygens' principle), or in terms of rays, which are lines perpendicular to the wave fronts. Light rays travel in straight lines unless they meet boundaries. Upon reflection from a surface, the angle θ that the incident ray makes with the normal to the surface is equal to the angle θ' that the reflected ray makes with the surface (the law of reflection):

$$\theta' = \theta. \tag{35-5}$$

In the passage from a medium of index of refraction n_1 to a medium of index of refraction n_2, the incident angle θ_1 and the refracted angle θ_2 are related by Snell's law of refraction:

$$n_1 \sin \theta_1 = n_2 \sin \theta_2. \tag{35-6}$$

These results can be established using the geometry of wave fronts. They can also be derived with the help of Fermat's principle, which states that the path taken by a light ray between two points is the path that takes the shortest time. One consequence of Snell's law is that total internal reflection occurs when light moving in a medium with index of refraction n_1 strikes a boundary of a medium with index of refraction n_2, where $n_1 > n_2$. This holds true provided that the angle of incidence is larger than a critical angle θ_c, given by

$$\sin \theta_c = \frac{n_2}{n_1}. \tag{35-9}$$

The dependence of the index of refraction on wavelength is called dispersion. Dispersion causes the different wavelengths in a beam of white light to refract through different angles. The colors of the rainbow and the blue sky are naturally occurring dispersion phenomena. Dispersion can be understood in terms of the atomic theory of matter.

Understanding the Concepts

1. If light travels only in straight lines, how does a light burning in one room give light in another room?
2. When light reflects from a surface, there is a change in (a) frequency, (b) wavelength, (c) speed of light, (d) all of these, (e) none of these.
3. How difficult would it be to reflect light back to Earth from the Moon by using two perpendicular plane mirrors? Why does it help if there are three mutually perpendicular mirrors?
4. If fish could think, they might realize that the relative indices of refraction of water and air allow them to outwit fishermen. Why?
5. A person swimming underwater sees a lifeguard who is standing in the shallow part of the pool; the water comes up to the lifeguard's waist. In what way does the swimmer see the lifeguard's upper body distorted?
6. If you lie at the bottom of a pool and look up, it appears that you are lying at the bottom of a conical hole. Why is that? Estimate the angle that apparent walls make with the vertical.
7. A fisherman standing up to his waist in a lake appears, to an observer outside the lake, to have shorter-than-normal legs. How will a fish in a horizontal position near the bottom of the lake appear to the observer?
8. A plane wave of radiation has an electric field of the form $\vec{E}_0 \cos(kz - \omega t)$ when it propagates in empty space. How do k and ω change when the plane wave enters a medium with index of refraction n?
9. When a beam of light of frequency f passes from air into glass (whose index of refraction is 1.5), which of the following happens: (a) f increases by 1.5; (b) f decreases by 1.5; (c) f becomes zero; (d) f is unchanged; (e) f decreases very slightly.
10. As the Sun sets, its color changes from white to yellow to orange and finally to red. As the lowest part of the Sun sinks below the horizon, the Sun appears squashed, more egg-shaped than circular. Why?
11. A coin lies at the bottom of a pool of water. Starting from a point immediately above the coin, you observe the coin from the level of the surface. You then move your head horizontally away from the coin across the surface of the water. Is there a horizontal distance at which the coin is no longer visible?
12. Light from the sky refracts near the surface of hot sand, giving the impression that there is a bright surface that could be interpreted as water: a mirage (Fig. 35–28). The air near the surface of hot sand is hotter than the surrounding air. Does light travel faster or slower in hot air than in cold air?

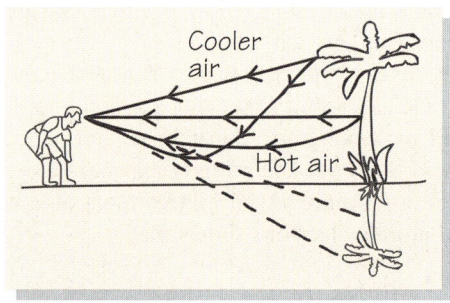

▲ **FIGURE 35–28** Question 12.

13. Mirages can occur when a layer of cold air lies closer to the surface. How would such an air layer affect the appearance of distant houses?
14. Consider Fizeau's experiment modified so that the light between the rotating wheel and the reflecting mirror goes through a pipe filled with water. Should the wheel be speeded up or slowed down in the repetition of the experiment described in the text? What would the result be if light traveled faster in water than in air?
15. Why does the sky look black rather than blue, as it does from Earth, to astronauts in orbit?
16. What is the index of refraction of a vacuum?
17. For a moment, you are lying in the middle of a circular swimming pool—at the bottom of the pool—which is filled to a depth of 1 m with water and is surrounded by trees. A 2-m-tall lifeguard is standing in the water about 3 m from you. What do things look like as you scan in all directions?

18. Laser light directed into the end of a glass rod comes out the other end with almost the original intensity. If another glass rod touches the side of the first rod, making a 30° angle with the lengthwise direction of the first rod, nothing happens. But if the point of contact is lubricated with glycerin, some of the original light beam is "stolen" by the second rod. Explain what happens.

19. White light is incident onto a pane of glass. Is there a dispersion of colors in the reflected light?

20. Stick a pin into the underside of a cylinder of cork, then float the cork in water. Even if you do not stick the pin in very far, you may not be able to see it from outside the water. Why not?

21. Why isn't the Moon red when it sets?

Problems

35–1 The Speed of Light

1. (I) What are the speeds of light in ice, ethyl alcohol, benzene, and diamond?

2. (I) The nearest star to our solar system (aside from the Sun) is Alpha Centauri, some 4.2 ly from Earth. How far is this in meters?

3. (I) A light wave of red light ($\lambda = 650$ nm) passes from air into water, where the index of refraction is 1.32. What are the wavelength and frequency of the light in water?

4. (I) Light of frequency 5.6×10^{14} Hz impinges on glass, $n = 1.45$. What are the wavelengths of this light in a vacuum and in glass? What is the index of refraction of a material within which the wavelength of yellow light is one-half its value in a vacuum?

5. (II) Suppose that you have a version of Fizeau's apparatus in which the round-trip distance for the light beam is $2D = 1,000$ m. The width of the opening between the teeth on the cogged wheel is 0.70 mm, and the center-to-center distance between these gaps is 1.5 mm. The wheel has a radius of 15.0 cm. What would the minimum rotational speed be, in revolutions per minute, so that light entering through the center of one gap would come out through the center of the next gap? Is such an apparatus realizable?

6. (II) Figure 35–29 shows an exaggerated view of the eclipsing of Io, the innermost moon of Jupiter, as seen from two different points on Earth's orbit around the Sun. If Earth were stationary at a point nearest Jupiter, N, a particular eclipse would begin at a precise time. When Earth is at point F, the eclipse starts somewhat later than expected because the light has to travel the additional distance of a diameter of the Earth–Sun orbit. The mean distance from Earth to the Sun is 1.50×10^{11} m. How much later will the eclipse be seen at point F compared with point N?

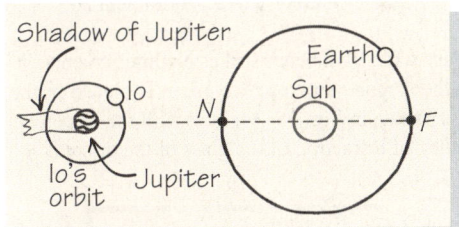

▲ FIGURE 35–29 Problem 6.

7. (II) Personal computers can perform as many as 5×10^7 steps every second. This means that some leads connecting different parts of the computer may carry this many pulses per second. If you assume that the pulses travel at the speed of light, what is the distance between pulses? Does this result have implications for the design of these machines?

8. (II) Telephone connections between Europe and North America can be carried by cable or by the use of a geosynchronous communication satellite. Estimate the time it takes for a signal to travel 10,000 km via cable, assuming the speed is close to the speed of light. How does this compare to the time required for the same signal to travel via satellite, 40,600 km from the center of Earth?

9. (II) The speed of light in a vacuum is defined to be 299,792,458 m/s. A lunar-ranging experiment measures the time for a light pulse to reach the Moon and reflect back to Earth. Such experiments allow us to determine the distance between the Moon and Earth, which is approximately 3.84×10^8 m, to an accuracy of 15 cm. What is the smallest time interval that can be measured by the clock used to determine the time it takes for light to go to the lunar reflector and back?

10. (II) Galileo attempted to measure the speed of light with the help of lights and a clock on two adjacent mountains. In essence, a shutter over a light was opened on the first mountain, an observer on the second mountain saw that signal and returned a second signal, and the experimenter on the first mountain looked for a delay between the time the shutter was opened and the time the signal was returned. Use your knowledge of human reaction time to estimate the time measured by the first experimenter for the total round trip. How long would it actually take for light to travel back and forth between two mountaintops separated by 4 km? Your answers explain why Galileo's attempt did not work.

11. (II) Imagine an experiment similar to Fizeau's, with a cogged wheel of diameter 20 cm. A laser beam shines through one opening, travels 1,500 m, and is reflected back. Given that the fastest rotation rate of the wheel is 1.2×10^5 rev/min, what should be the separation between adjacent cogs on the rim of the wheel?

35–3 Reflection and Refraction

12. (I) A fixed projector emits a narrow beam of light onto a plane mirror. At what angle with respect to the beam should you place the mirror in order to turn the beam by 75°?

13. (I) The critical angle for a particular material (used in air) is observed to be 38°. What is the material's index of refraction?

14. (I) A horizontal beam of light is reflected from a plane mirror that revolves about a vertical axis at a rate of 30 rev/min. The reflected beam sweeps across a screen that, at the point nearest the mirror, is 20 m away (Fig. 35–30). With what speed does the spot of light move across the screen at the point nearest the mirror?

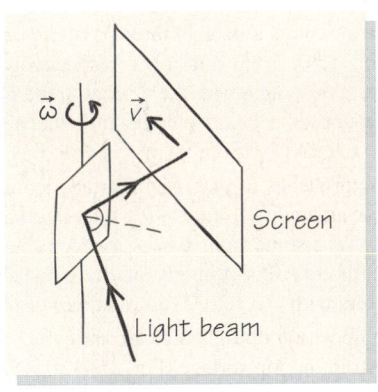

▲ FIGURE 35–30 Problem 14.

15. (I) An intense light beam is incident at 45° to the surface of a clear lake. If the lake is 500 m deep and has a flat bottom, how far does the light beam travel before it hits bottom?

16. (I) A beam of light is sent from medium 1, index of refraction n_1, into a medium 2, index of refraction n_2; here, $n_2 > n_1$. We know that a refracted ray is bent toward the perpendicular to the boundary. Are there any incident angles for which the angle of the refracted ray is 90°? If not, what is the largest possible angle of refraction? Give a numerical value to your answer in the case of water-to-glass, $n_1 = 1.33$ and $n_2 = 1.50$.

17. (I) A burglar stands in front of a department store window and directs his flashlight into the store. What fraction of the light is reflected at the window's surface, assuming that the index of refraction of the glass is 1.43? Ignore all reflections except for that at the outside interface between the glass and air.

18. (I) A swimmer is at the bottom of a large, shallow swimming pool. Through what angle must she move her eyes so that her direct gaze swings across the whole sky? Water's index of refraction is 1.33.

19. (I) The index of refraction of air is $1 + (2.93 \times 10^{-4})$. Assume that the atmosphere may be treated as a uniform medium of thickness 8.3 km, which covers Earth's surface; further, suppose a ray of light hits the top of the atmosphere parallel to the top of the atmosphere—grazing incidence. What is the angle that the refracted ray makes with the *horizontal*?

20. (II) What is the critical angle for total internal reflection in a glass (used in air) for which $n = 1.46$? Is it possible to use a 45°–45°–90° triangular prism of crown glass (see Table 35–1) to make a perfect reflector of light (Fig. 35–31)?

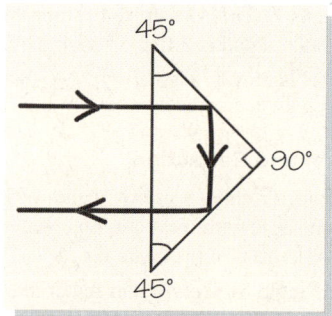

▲ **FIGURE 35–31** Problem 20.

21. (II) A thick glass plate ($n = 1.53$) lies on the bottom of a tank of water ($n = 1.33$). A light ray enters the water from air, making an angle of 72° with the normal to the surface. What angle does the ray make with the normal when the ray is in the water? What angle does it make with the normal when it is in the glass?

22. (II) Light in air enters a stack of three parallel plates with indices of refraction 1.50, 1.55, and 1.60, respectively. The incident beam makes a 60° angle with the normal to the plate surface. At what angle does the beam emerge into the air after passing through the stack?

23. (II) Light approaches a glass–air interface from the glass side ($n = 1.6$) at an angle θ_i. If $\theta_i > \theta_c$, total internal reflection occurs; if $\theta_i < \theta_c$, some light passes through the interface. Is it possible to modify this property of the interface by adding a stack of layers with carefully chosen indices of refraction?

24. (II) The composition of a glass block varies as a function of the distance x from the top surface (Fig. 35–32). As a consequence, the index of refraction increases as a function of x according to $n(x) = 1.54 - (0.18 \text{ cm}^2)/(x + 1 \text{ cm})^2$, with x in centimeters. A beam of light strikes the surface at an angle of incidence

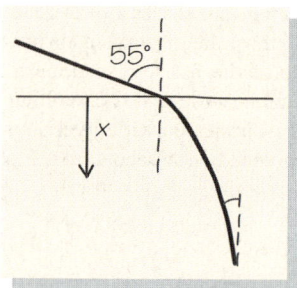

▲ **FIGURE 35–32** Problem 24.

of 55° from the vertical. What will be the direction of the beam deep inside the block?

25. (II) A glass sphere ($n = 1.6$) is centered at the origin of a coordinate system, with its equatorial plane defining the xy-plane. A beam of light enters the glass sphere at a latitude of 40°, parallel to the x-axis in the xz-plane. Make a careful drawing to determine the angle at which the beam will strike the back of the sphere. Will there be total internal reflection?

26. (II) White light is refracted by the triangular prism shown in Fig. 35–33. A beam of light enters the prism along a path parallel to the prism base. The light is observed on a screen that is located 10 m from the prism and is perpendicular to the emerging rays. How far apart on the screen are the spots of blue light ($n = 1.528$) and red light ($n = 1.514$)?

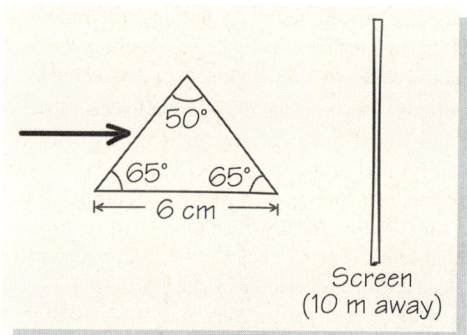

▲ **FIGURE 35–33** Problem 26.

27. (II) A very wide light beam strikes a white screen at 90° to the surface of the screen. An isosceles prism is placed in the way of the beam, as shown in Fig. 35–34. How will the screen be illuminated if the index of refraction of the glass of the prism is $n = 1.5$?

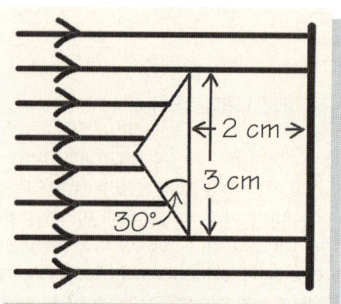

▲ **FIGURE 35–34** Problem 27.

28. (II) A lifeguard whose eyes are 1.78 m from his feet stands in water 90 cm deep. From a vantage point at the bottom of the pool, a swimmer sees the lifeguard's head to be along a line at a 46° angle to the vertical. How far is the swimmer's eye from the lifeguard's feet? (For water, $n = 1.33$.)

29. (II) Suppose that you look at an aquarium with your eyes at the level of the water surface (Fig. 35–35). A duck swims on the surface of the water. When you look at the duck from the front, everything seems normal. However, when you look at the duck at an angle to the glass surface, the duck seems to be split in half, with the feet paddling ahead of the upper body. Explain this phenomenon. Suppose that both the duck and your eyes are at a distance of 1 m from the glass, and the line connecting them forms a 30° angle with the glass. Calculate the difference between the directions of the line of sight of the upper and lower halves of the duck.

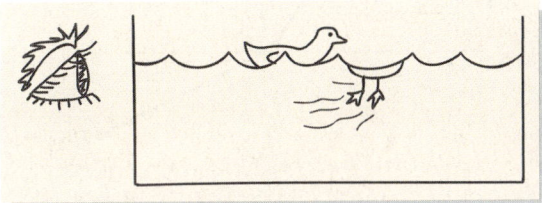

▲ **FIGURE 35–35** Problem 29.

30. (II) A transparent tank of water, of index of refraction n_1 and thickness t_1, is placed on top of a slab of glass of index of refraction n_2 and thickness t_2. A laser beam strikes the upper surface of the tank at an angle θ. At what horizontal distance from the point of entry will the beam emerge from the bottom surface of the glass? What is the answer if the tank and the slab of glass are interchanged?

31. (II) A beam of light from a flashlight is reflected by a mirror which is placed under a sheet of glass (index of refraction 1.6) that is 5 cm thick. The beam makes an angle of 60° with the vertical. If the light source is 15 cm above the surface of the glass, how far below the surface of the glass will you see a source of light matching the original? You are locating the *image* of the light source. [*Hint*: In this exercise in trigonometry, use $\tan \theta = \sin \theta / \cos \theta$; $\cos \theta = \sqrt{1 - \sin^2 \theta}$.]

32. (II) A narrow beam of light is incident at a 30° angle from the normal onto a glass pane 6 mm thick. Describe the position of the exit beam of light. What is its direction? Is it displaced from the incident beam? If so, by how much? (For the glass, $n = 1.60$.)

33. (II) Light is incident on an equilateral triangular prism ($n = 1.55$) at a 35° angle from the normal to one of the faces (Fig. 35–36). What is the exit angle?

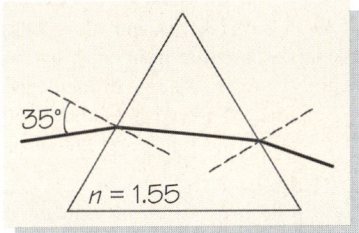

▲ **FIGURE 35–36** Problem 33.

34. (II) Consider a solid glass rod of length 75 cm and diameter 1.5 cm, with index of refraction 1.46. The ends of the rod are perpendicular to the lengthwise direction. (a) Light enters the center of the end of the rod from air. What is the maximum angle of incidence for which the light is totally reflected inside the rod? (b) Repeat part (a) for a similar rod totally immersed in water ($n = 1.33$).

35. (II) A ray of light impinges at a 60° angle of incidence on a glass pane of thickness 5 mm and index of refraction 1.54. The light is reflected by a mirror that touches the back of the pane (Fig.

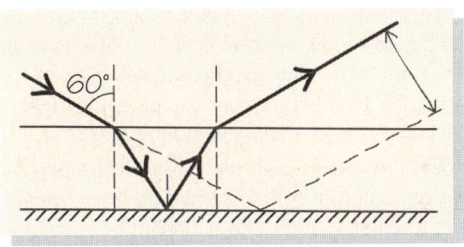

▲ **FIGURE 35–37** Problem 35.

35–37). By how much is the beam displaced compared with the return path it would have if the pane were absent?

36. (II) When a light beam is reflected by a conventional mirror, part of the light is reflected by the front surface of the glass pane, and part by the silvered back surface. What is the distance between the two reflected beams if the mirror is 2.0 mm thick, if it is made of glass with $n = 1.45$, and if the angle of incidence is 70°? (To avoid this double reflection, many optical instruments use mirrors with their front surfaces silvered.)

37. (II) At noon, a 2.0-m-long vertical stick casts a shadow 1.0 m long. If the same stick is placed in a flat-bottomed pool of water half the height of the stick (still at noon), how long is the shadow on the floor of the pool? (For water, $n = 1.33$.)

38. (II) You have three transparent liquids labeled 1, 2, and 3 that do not mix. When light is sent from liquid i to liquid j, there is an angle of incidence θ_i and an angle of refraction θ_j. Two separate experiments show the following: $1 \rightarrow 2$, $\theta_i = 22°$ and $\theta_j = 29°$; $2 \rightarrow 3$, $\theta_i = 35°$ and $\theta_j = 53°$. Find the ratios of the indices of refraction for each pair of liquids.

39. (II) Consider light that is perpendicularly incident on a triangular prism of the kind shown in Fig. 35–38. The index of refraction of the prism material is $n_1 = 1.814$. Suppose that the two reflecting sides are coated with a thin, uniform layer of a dielectric with index of refraction $n_2 = 1.380$. Will the glass–dielectric interface be totally reflecting? How large can n_2 be so that the interface is still totally reflecting?

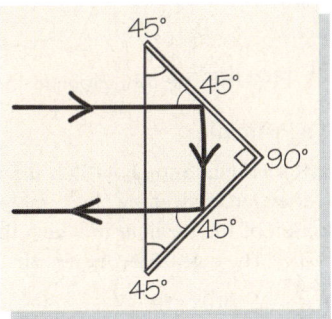

▲ **FIGURE 35–38** Problem 39.

40. (II) A sodium lamp emitting light with $\lambda = 589$ nm is placed at the bottom of a pool 4 m deep. (a) Seen from the edge of the pool, where will the light appear? (b) The lamp is taken out of the pool and the light now shines into the pool. What will be the frequency, wavelength, and speed of the light as it appears to a swimmer under the surface of the pool? ($n_{\text{water}} = 1.33$)

41. (II) A prism made of glass with index of refraction n and whose cross section is an equilateral triangle deflects light of wavelength 550 nm. As the beam is moved to make different angles with respect to the prism, the minimum angle through which the beam is deflected is 35°. What is the value of n?

42. (II) A prism has a cross section in the shape of an isosceles triangle with a base-to-height ratio of 1/2.5. A beam of light is incident upon the left side, parallel to the base. At what angle relative to the base will the beam leave the right side of the prism, which is made of glass with $n = 1.58$?

43. (II) Suppose you have eight perfectly clean microscope slides, index of refraction $n = 1.5$, stacked on top of each other. Estimate the fraction of light that is transmitted. There are tiny air gaps between the slides; ignore multiple reflections.

44. (II) Zeno and his friends are discussing light incident perpendicularly on a glass pane in air. The pane has two surfaces at which reflection can take place—we'll call them A and B, with A being the first surface encountered by the light. The discussion centers around the fact that some light reflects back from A directly, but more light comes back to the A side of the pane after having undergone reflection at surface B, then from double reflection at surface B, having bounced once from A, and so on. They reason that since each bounce contributes something to the A side, an infinite amount will end up there! But of course this is just Zeno's paradox, and you know that this is really a matter of a limit. What is the total intensity of the light reflected back to the A side, including an arbitrary number of bounces at the boundaries? Assume an incident intensity I_0. Take the index of refraction of air to be 1, that of the glass n, and assume that there is no absorption within the glass.

45. (III) Use Huygens' construction to prove Snell's law by working out the geometrical details in Fig. 35–39.

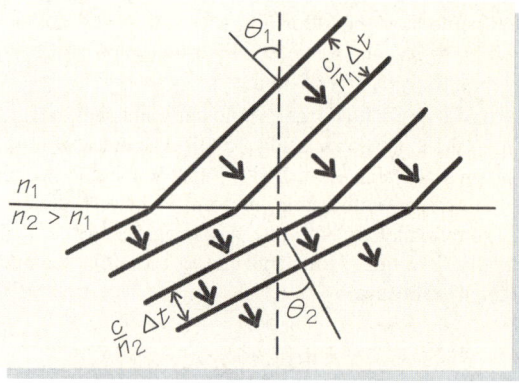

▲ **FIGURE 35–39** Problem 45.

*35–4 Fermat's Principle

46. (II) Use Fermat's principle to show that the critical angle for total internal reflection is given by $\sin\theta_c = 1/n$, where n is the index of refraction of the medium in which the light ray originates (Fig. 35–40). The outside medium is air.

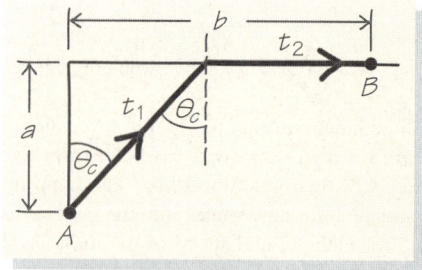

▲ **FIGURE 35–40** Problem 46.

47. (II) Show that the law of reflection follows from Fermat's principle.

48. (II) By using Fermat's principle, show that if two media have exactly the same index of refraction, then a beam of light travels in a straight line when it crosses the boundary between them.

49. (II) Use Fermat's principle to show that a beam of light that enters a plate of glass of uniform thickness emerges parallel to its initial direction (Fig. 35–41).

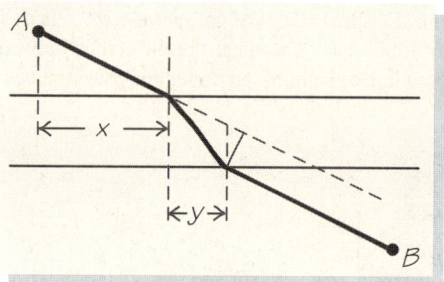

▲ **FIGURE 35–41** Problem 49.

50. (III) Calculate the parallel displacement of a beam of light that strikes a vertical slab of glass, with index of refraction n and thickness D, at an angle ϕ with the horizontal (use Fermat's principle). (In Problem 49, we showed that a ray of light that passes through a slab of glass emerges parallel to its initial direction. The ray is, however, displaced from its original line, and that displacement is what we want here.)

35–5 Dispersion

51. (I) At 0°C, the index of refraction in water of light of wavelength 397 nm (violet) is 1.3444, whereas it is 1.3319 for a wavelength of 656 nm (red). What is the difference in angles of refraction for rays refracting from water near the freezing point into air for these two wavelengths? The angle of incidence is exactly 30° in each case. Take $n = 1$ for air and ignore dispersion in this medium.

52. (I) By what percent does the speed of red light in a type of glass ($\lambda = 656$ nm, $n = 1.522$) exceed that of blue light in the same glass ($\lambda = 486$ nm, $n = 1.545$)?

53. (II) A beam of white light, whose frequencies are mixed with equal intensity, passes within a piece of glass and impinges on a boundary to the air at an angle of incidence θ. The index of refraction of the glass increases with increasing angular frequency according to the formula $n^2 = 1 + [C/(\omega_0^2 - \omega^2 - C)]$, where $C = 529 \times 10^{30}$ rad^2/s^2 and $\omega_0^2 = 685 \times 10^{30}$ rad^2/s^2. (a) What is the largest angular frequency that passes through the glass into the air? (b) At what angle of incidence should the light approach the boundary if we wish to allow only frequencies of $\omega = 3.2 \times 10^{15}$ rad/s (red light) and below to pass through to the air?

54. (II) Use the data in Problem 53 to calculate the critical angles for total internal reflection for five values of wavelengths in the range 430 nm to 770 nm. Plot your results.

55. (III) We wish to select a glass to construct a prism that can separate the yellow ($\lambda = 590$ nm) component of light from the blue-green ($\lambda = 490$ nm) component. The prism is to be a bar with the cross section of an equilateral triangle. If a ray of white light arrives parallel to the base of the prism, it must leave the prism with the two colors separated by at least 2°. What must the difference in indices of refraction be for the two colors? [*Hint*: Because the difference of angles is small, so is the difference of indices of refraction. Keep only leading terms in differences of angle and of index of refraction.]

General Problems

56. (I) Light of wavelength 450 nm enters a piece of glass with index of refraction 1.50. What are the wavelength, frequency, and speed of that light in the glass?

57. (II) A pin is partly inserted perpendicularly into the flat surface of a cork with a 1.5-cm radius (Fig. 35–42). The cork, with the pin on the underside, is set afloat in a pool. A length of 1.2 cm of cork is under the water surface. Because of the effects of refraction, much of the pin is hidden from view from above the surface. What length of pin can be hidden in this way?

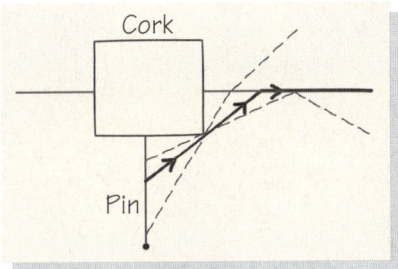

▲ **FIGURE 35–42** Problem 57.

58. (II) A beam of light is incident at an angle of 30° to the vertical on a horizontal glass plate of thickness 2.0 cm. The index of refraction of the glass is $n = 1.52$. The beam emerges on the other side. What is the perpendicular distance between the straight-line extrapolation of the incident ray and the ray refracted by the glass plate?

59. (III) Show that if an incident ray of white light that is parallel to the base of a prism in the shape of an isosceles triangle (apex angle 2ϕ) is separated into two components that exit the prism with an angular separation $\Delta\theta \ll 1$, then the difference in the indices of refraction for the two colors is proportional to $\Delta\theta$. Find the equation that expresses the relation between the differences in the indices of refraction and in $\Delta\theta$. [*Hint*: Consider the angle of emergence for a given n, and then find Δn as a function of $\Delta\theta$.]

60. (III) Sound can refract like light. Suppose that a submarine lies flat 180 m below the water surface, and that there are three thermal layers of water (each 60 m deep) of different temperatures (Fig. 35–43). The speed of sound in water depends on temperature. In the bottom layer, the speed is 1.16 times that in the top layer; in the middle layer, the speed is 1.05 times that in the top layer. A detection device at surface level determines that sound from the submarine arrives at the surface at a 36° angle with the horizontal. What is the horizontal distance between the detector and the submarine?

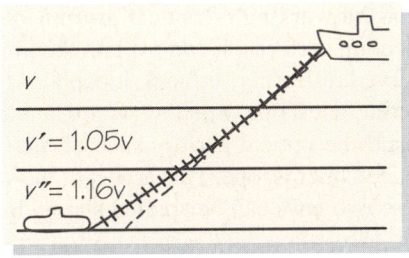

▲ **FIGURE 35–43** Problem 60.

61. (III) A ray of light is incident at an angle of incidence θ_i on one surface of a prism whose cross section is an isosceles triangle (apex angle 2ϕ). The light exits the prism at a total deflection angle θ (Fig. 35–44). The prism has index of refraction n and is in a vacuum, which has an index of refraction of exactly 1. For what angle θ_i is the angle of deflection θ a minimum?

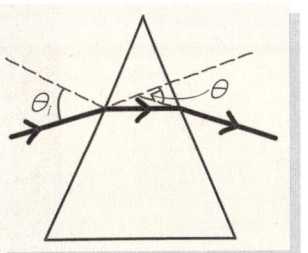

▲ **FIGURE 35–44** Problem 61.

62. (III) A ray of light incident from air onto a glass pane is partly reflected and partly refracted at the two surfaces of the pane (Fig. 35–45). The glass has an index of refraction n and a thickness d. Express in terms of n, d, and θ_i the displacement d' of the ray drawn, which enters the glass, reflects off the back surface, and exits.

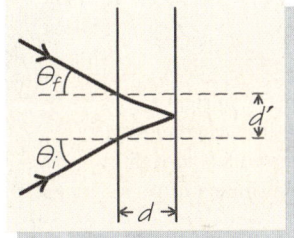

▲ **FIGURE 35–45** Problem 62.

63. (III) The first successful measurement of the speed of light, made by Ole Roemer in 1675, was based on the following method. The mean orbital period of Io, a moon of Jupiter, is 42.5 h; however, that period is measured to be about 15 s less than this value when Earth in its orbit is approaching Jupiter, and about 15 s more when Earth is receding from Jupiter. (a) Given that Earth's orbital speed around the Sun is about 30 km/s, and that Earth is on a part of its orbit when it is moving toward Jupiter, how much closer will Earth have moved toward Jupiter during one orbit of Io? (b) Use the information given to estimate the speed of light.

Chapter 36

▶ Two mirrors placed at right angles gives us three images of a candle. Are all three images reversed left-to-right, as a single image in a single mirror would be?

Mirrors and Lenses and Their Uses

Instruments that can explore previously inaccessible domains often open new doors to understanding nature. For example, astronomy owes its progress to the invention of the telescope, and modern biology could not have been created without the microscope. In this chapter, we shall discuss the ideas that govern the construction of optical instruments such as these. The law of reflection and Snell's law, both introduced in Chapter 35, provide the foundation for the working of optical instruments. When the law of reflection is applied to flat reflecting surfaces, it can explain the images that we see in mirrors, and when applied to curved reflecting surfaces, it explains the functioning of rearview mirrors and the reflecting telescope. Snell's law applied to curved refracting surfaces can help us understand the optical performance of the eye, camera, magnifying glass, refracting telescope, and microscope. This aspect of the study of light is called geometric optics because these two laws can be applied simply by tracing the geometrical paths of light rays.

36–1 Images and Mirrors

The simplest reflecting surface is a flat (or plane) mirror. When you look into a mirror, you see an image of yourself. What is an image, and how is it formed? Let's begin with rays going directly from a point source S—think of a small lightbulb—to a person's eye

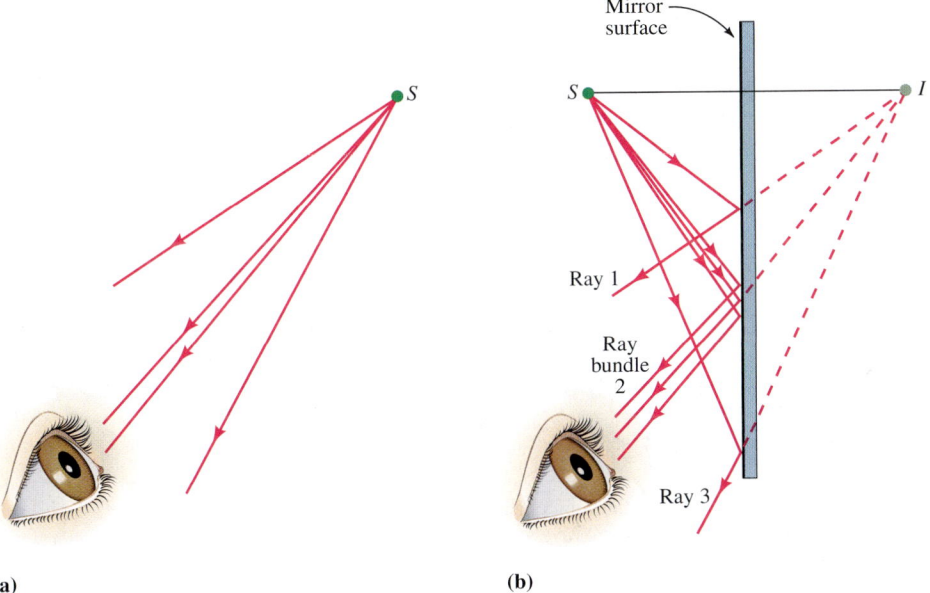

▲ **FIGURE 36–1** Rays leaving source point S (a) go to an eye and (b) reflect from a plane mirror before going to the eye. A bundle of such rays enters the eye, apparently from point I.

(Fig. 36–1a), together with the light from the source reflecting from a plane mirror according to the law that the angles of incidence and reflection are equal (Fig. 36–1b). The rays we draw follow the direction of motion of the wave front that emanates from the light source, and we could, in fact, draw an infinite number of such rays as close to one another as we like. As we described in Chapter 35, rays that are near one another form *bundles*, visible in Fig. 36–1b.

When we look toward the mirror, we see an *image* of the light source. What exactly do we mean by this? The simple geometry in Fig. 36–1b allows us to see that *all the reflected rays from S trace back to the same point I.* To see this, look at rays 1 and 3 in Fig. 36–2, which shows in more detail the situation in Fig. 36–1. We have indicated the equal angles of incidence and reflection θ_1 and θ_3 for these rays, respectively, as well as the angles α_1 and α_3. The angle formed by BP_1I is then equal to α_1. If point B is formed by dropping a perpendicular line to the mirror from point S and if point I lies along the continuation of this line, triangles BIP_1 and BSP_1 are similar triangles. By the same method, so are triangles BIP_3 and BSP_3. Because both rays 1 and 3 emanate from the same point S, the distance BS forms the base of both triangles to the left of the mirror (the *object side*), and the distance BI forms the base of both triangles to the right of the mirror (the *image side*). The (imaginary) continuations of rays 1 and 3 to the image side meet at point I, as would the continuation of *any* reflected ray.

We have calculated the location of point I. Because BIP_1 and BSP_1 are similar triangles, the distances BS and BI are equal. How does the eye/brain "know" where to put I? Two eyes (or one eye that moves a little) sense a bundle of rays rather than a single ray. The eye/brain can measure their degree of divergence and is capable of extrapolating this diverging bundle back to point I.

The Image of an Extended Object

Suppose now that our light source is extended rather than being a point source—think this time of a candle flame or, perhaps better yet, think of an object that is reflecting light and acting as a source, such as a person. Two different points on the source form two different image points in the mirror. Moreover, the second image point is as close to the first image point as the second source point is to the first source point. Indeed, a set of nearby source points forms nearby image points (Fig. 36–3). The entire **object**, or *source* (we use the terms interchangeably), forms a set of matching image points, which together constitute an **image**. *An image is a set of contiguous points to which reflected rays lead when the rays are extrapolated back in straight lines.* Figure 36–3 illustrates

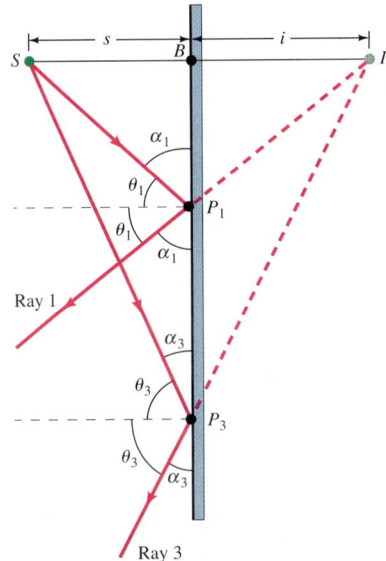

▲ **FIGURE 36–2** All the reflected rays from point S trace back to point I. The geometry implies that the perpendicular distance s from the mirror to point S equals the distance i from the mirror to point I.

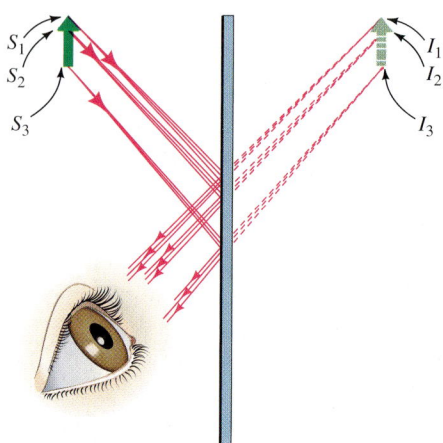

▲ **FIGURE 36–3** When the source is an extended one, there is an image point for every source point. This means that a bundle of rays will enter your eye from every image point no matter what your position before the mirror. Geometry can be used to locate the image $I_1 I_2 I_3$ of source $S_1 S_2 S_3$.

that the source and the image formed by a plane mirror have the same size. The idea of the size of the image will be of some importance throughout this chapter. We can also see that the image is as far behind the mirror as the object is in front.

No light rays actually emanate from the image formed by a plane mirror; thus, we call it a **virtual** image. Light rays actually do pass through a **real** image on their way to your eye, and we shall see how this can occur when we discuss mirrors with curved surfaces. We could summarize the difference between real and virtual images by saying that if you were to place a screen at the location of a virtual image, no light would strike the screen, whereas a screen placed at the location of a real image would reveal the presence of the image.

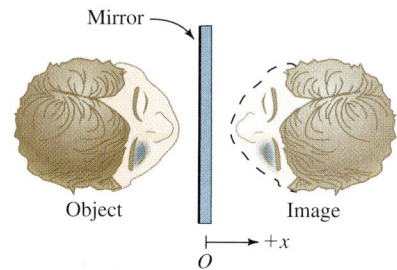

▲ **FIGURE 36–4** An image is reversed front to back. This means that if your right eye is black, your image has a black left eye.

There is one striking peculiarity of plane mirrors. If your right eye is blackened, your image, viewed as if you were meeting yourself on the street, has a left eye that is blackened. From Fig. 36–4, however, we can see that the *actual* reversal is a front-to-back reversal (the nose of the object points in the $+x$ direction in the figure, whereas the nose of the image points in the $-x$ direction), and this is what lies behind the left-to-right reversal of the image.

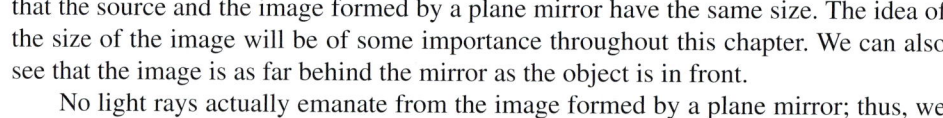

CONCEPTUAL EXAMPLE 36–1
How are the multiple reflections shown in Fig. 36–5 formed?

Answer The image of some source that is made by a mirror comes from the reflections of the rays emitted by the source. These reflected rays form a set of diverging rays, which result in the image. But the reflected rays can approach and reflect from a second mirror, and we have no way to tell whether the diverging set of rays incident on the second mirror come from an actual source or from the reflection from the first mirror of the source. In other words, *an image can act as a source for a second image*. It makes no difference whether an image is virtual or real—the only thing that matters is that the rays are diverging as they approach and then reflect from the second mirror. In the case of Fig. 36–5, the successive images must be due to a second mirror in front of the chess pieces, not visible in the photo. You might note that while the image produced by one reflecting surface is reversed left-to-right, the image of the original source produced by successive reflection from two (or more generally an even number of) mirrors is not.

▲ **FIGURE 36–5** Multiple reflections can be obtained with two plane mirrors. Where is the second mirror in this case?

EXAMPLE 36–2 A horizontal ray of light is incident at angle θ on a vertically suspended plane mirror. If the mirror is rotated about a vertical axis through an angle α, by what angle ϕ is the reflected ray rotated?

Setting It Up Figure 36–6 is an overhead view of the situation, including the initial and final normals to the mirror, N and N', respectively, as well as the reflected ray, labeled I.

Strategy A rotation of the mirror through an angle α means that the new angle of incidence is $\theta + \alpha$, which is also the new angle of reflection from N'. It is then a matter of geometry to find the required angle ϕ.

Working It Out From the figure, the angle between the incident ray and the new reflected ray I' increases from 2θ to $2(\theta + \alpha)$. Because the incident ray has not moved, the reflected ray is rotated by $\phi = 2\alpha$.

What Do You Think? Suppose you want to measure by how much an object has rotated and that the rotation angle is very small. How might you go about it? Answers to **What Do You Think?** questions are given in the back of the book.

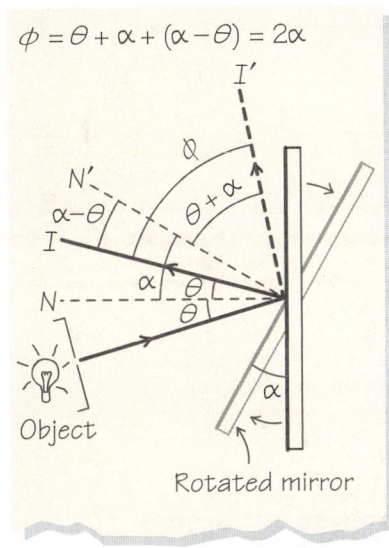

▲ **FIGURE 36–6** Overhead view of a horizontal ray of light incident on a mirror that rotates about a vertical axis.

36–2 Spherical Mirrors

Figure 36–3 shows that plane mirrors produce images that are the same size as the object. We can construct mirrors that produce images of altered sizes by using curved surfaces. In this section, we shall study mirrors whose surfaces form a segment of a sphere. We look at both *concave* (Fig. 36–7a) and *convex* (Fig. 36–7b) mirrors: For concave mirrors the source (object) is on the same side of the surface as the sphere's center, while for convex mirrors the source is on the other side. There are some particularly simple and significant rays to follow in these situations, and the ray-tracing techniques of Section 36–1 will be useful here. We simplify things by studying only objects on or near the *axis*, the line perpendicular to the center point of the mirror, as shown in the figure. The tips of the arrows shown in Fig. 36–7 are a distance h from the axis, and if h is small compared to the radii of curvature of the mirrors, we say that the object is *near the axis*. We consider only rays that are so close to being parallel to the axis that we can use small-angle approximations in studying their reflections. Such rays are said to be **paraxial**.

The Concave Mirror

We will first consider rays from a very distant point source (to the left of the figure) on the axis CB of the concave mirror in Fig. 36–8a. When a source is so far away that *all the rays from it arrive practically parallel to each other*, we say that the source is *at infinity*. Point C indicates the position of the center of the sphere (of radius R) of which the mirror is a segment. The position of C (called the *center of curvature*) is therefore a distance R from the mirror surface, and all lines from point C to the mirror are perpendicular to the mirror.

Location of the Focal Point: Look at ray 1, which is reflected at point A in the direction AF in Fig. 36–8a. Angle θ is the angle of incidence and the angle of reflection (line CA is perpendicular to the mirror). Note that triangle ACF is isosceles with a base of length R. Thus, by dropping a perpendicular from point F to the base of the triangle AC, we see that the distance CF is $CF = (R/2)/\cos\theta$. For small θ, $\cos\theta \cong 1$; hence $CF = R/2$, or $BF = R - CF = R/2$, independent of θ.† All the parallel rays near the axis reflect through point F, a distance $R/2$ from the mirror (Fig. 36–8b, c). This is the position at which we will see an image point of a very distant source point. Is the image we see at point F a real or virtual image? Unlike the image points produced by a plane mirror, point F is a real image point because rays actually cross there.

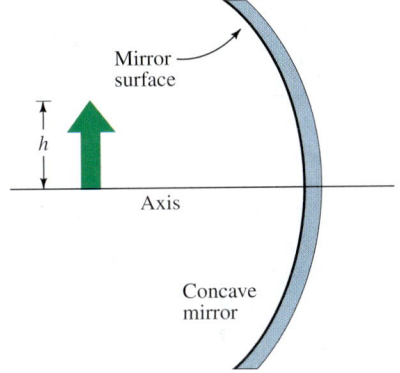

(a)

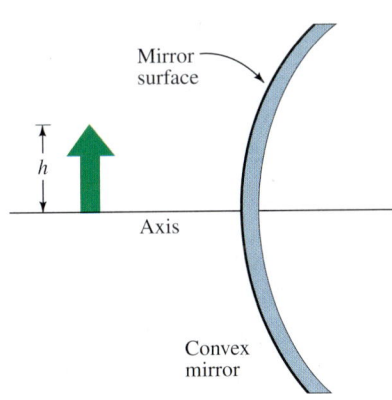

(b)

▲ **FIGURE 36–7** (a) Concave and (b) convex spherical mirrors. The object shown, an arrow, acts as an extended source.

---
†This result is accurate to 1 percent for angles θ less than about 10°.

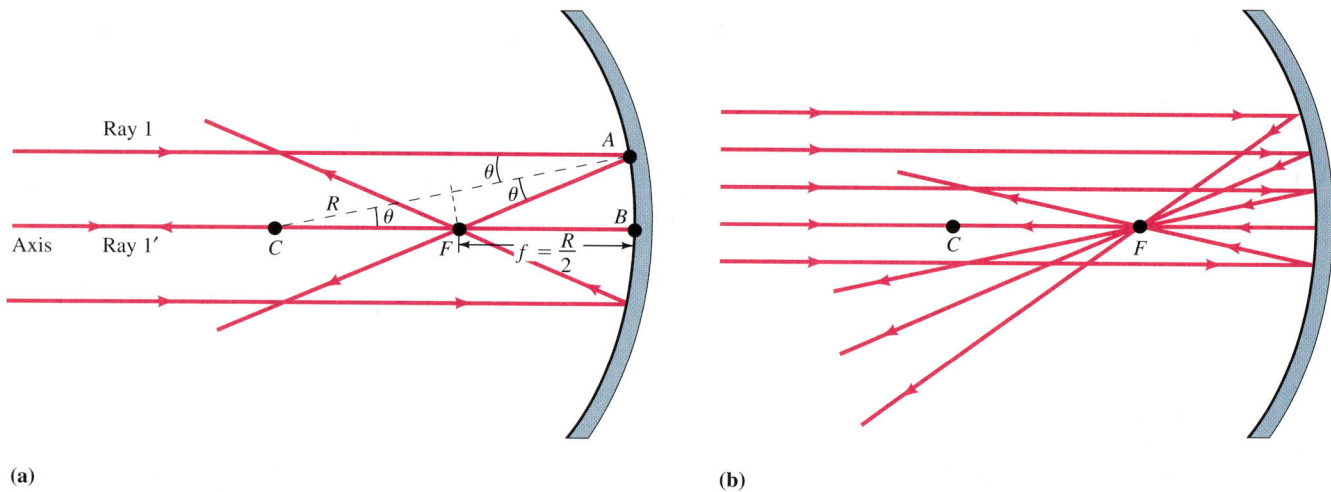

(a)   (b)

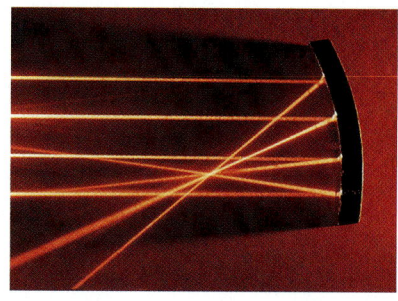

(c)

▲ FIGURE 36–8 (a) Rays emitted by an object at infinity are all parallel to the axis. Ray 1 is reflected by the concave mirror surface and passes through the focal point, F. (b) To a good approximation, F is independent of θ; that is, any incoming ray parallel to the axis is reflected through it. Thus all rays from infinity cross the axis at F, which is therefore an image point. (c) A demonstration of the construction of part (b).

The paraxial rays from a source point at infinity are brought together to form an image at the point F; we say they are *focused* at point F, which is known as the **focal point**, or *focus*, and we say that its distance f from the mirror is the **focal length**. These terms can be applied to any optical system that produces images—including plane mirrors, whose focal length is infinite. We have shown that for concave mirrors

$$f = R/2. \tag{36–1}$$

CONCEPTUAL EXAMPLE 36–3 How could you use a concave mirror to make a flashlight?

Answer The purpose of a flashlight is to send out a parallel beam—a beam that in effect goes to a point far away. Since the path of any ray can be followed in reverse, this can be done by reversing the picture in Fig. 36–8a. We place a source of rays (a lightbulb) at the focal point, and the result will be a set of parallel rays leaving the mirror.

The Image of an Extended Object: Let's take an extended object—as, for example, the arrow in Fig. 36–7—that is small compared to the radius of curvature of the mirror and close enough to the axis so that the rays are paraxial. In Fig. 36–9a, we label two points on the object, which is *upright*, with the letters S and S'. If we follow bundles of light rays coming from a given spot on the object, we see that after reflection the rays pass through a corresponding spot in space, thus forming image points I and I' of the two points, and indeed an image of the entire object. We want to determine the position and size of the image. To do so, we use the principal-ray technique described in detail in the Problem-Solving Techniques, p.1004. In Fig. 36–9b, we draw the principal rays to find point I of the image that corresponds to point S of the source, and in Fig. 36–9c, we draw the principal rays to find the image point I'' that corresponds to source point S''. These sets of rays are

- ray 1, which approaches the mirror parallel to the axis and is reflected through the focus F;
- ray 2, which passes through the focus and reflects off the mirror parallel to the axis;
- ray 3, which passes through the center of curvature of the mirror and is reflected back in the direction from which it came;
- ray 4, which goes to the center of the mirror surface and whose reflection makes the same angle with the axis as the incident angle.

The rays leaving source point S do indeed cross at the image point I, and those from S'' cross at I''. All other points of the source have image points that can be constructed in

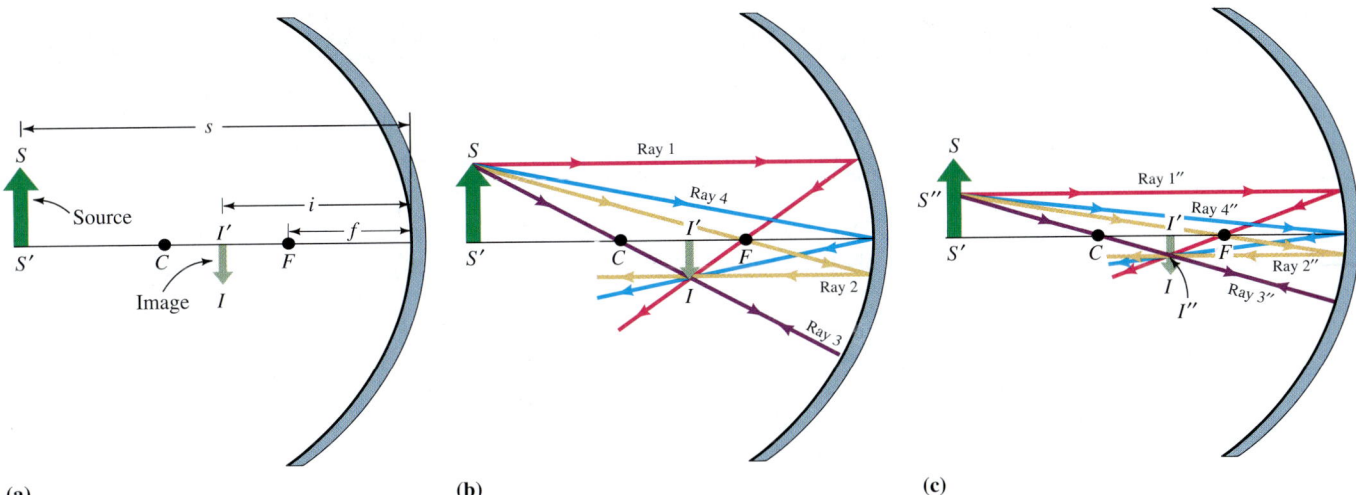

▲ **FIGURE 36–9** (a) An extended object a distance s from a mirror forms an image at a distance i from the mirror. The object is outside the focal length of the mirror. (b) Ray tracing for the concave mirror, with the principal rays for source point S. (c) Principal rays for source point S''. By repeating this exercise, we can build up the entire image. The principal rays are a guide; *any* ray from S that reflects will cross the image point I. (d) The object has produced an inverted and reduced image in a concave mirror.

the same way, and the image is thus constructed as in Figs. 36–9b and 36–9c. Note that both image point I' and source point S' lie along the optical axis. Our construction shows (and it is generally true) that a vertical source gives a vertical image, so that we can compute the location of just one image point rather than many. For example, the entire image can be constructed if we find only image point I of the top of the source (that is, of source point S), and, in fact, any two of the four principal rays are sufficient to determine point I. For example, rays 1 and 2 are sufficient for locating point I in Fig. 36–9b.

In this discussion, the source is *outside* the focal point, i.e., further away from the mirror than F. The image in Fig. 36–9a, located by the procedure in Figs. 36–9b and 36–9c, is *real* (real light rays pass through points I and I' and those in between), in contrast to the virtual image produced by plane mirrors. The image of the object in Fig. 36–9d is *inverted* (*upside down*) and *reduced in size*.

Let's now consider the situation depicted in Fig. 36–10, with the source *inside* the focal point of the mirror. In this case rays 2 and 3 only *behave* as though they pass through points F and C, respectively. The reflected rays from source point S do not actually cross but are aligned as though they come from behind the mirror at image point I. In other words, the image is virtual. It is also *upright* and enlarged. These are useful features of the concave mirrors used for shaving or applying makeup, or in dentistry.

(d)

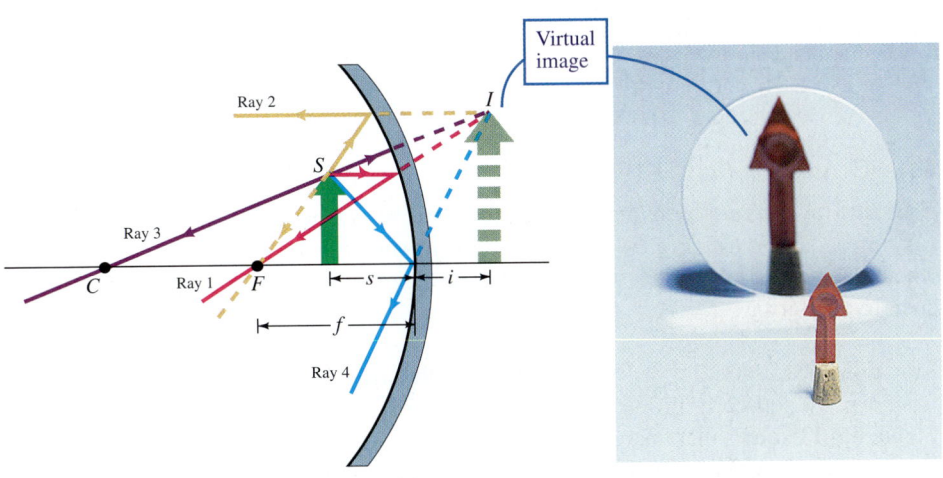

◀ **FIGURE 36–10** (a) Ray tracing with the principal rays for a concave mirror, for a source closer to the mirror than that of Figure 36–9b. The image becomes virtual when the source moves inside focal point F. (b) An object is placed within the focal length of a concave mirror. This time the image it produces is upright and magnified.

Problem-Solving Techniques

Principal Rays

Given a source (or object) and an optical system made of mirrors and lenses, one generally wants to find the size and location of the image. Because all rays cross at the image point of a source point (or behave as though they do in the case of virtual images), we need only find the crossing points of a few rays from any point on the object to be able to find both the position and size of the image. We refer to these rays as *principal rays*. Even if the optical system is such that a ray does not actually exist—for example, there may be a hole at the center of a mirror—we can pretend that it does and draw a ray through it. That is because the image is actually formed by rays coming from all parts of the mirror, and if part of the mirror is missing, the image is still formed in the same way (although it may be less intense). In other words, we simply use the principal rays as a tool to learn where the rest of the rays go.

In this box, we describe the principal rays for a convex mirror (Fig. 36B–1) and a converging lens (Fig. 36B–2; see Section 36–4), but the method applies to concave mirrors, diverging lenses, and single refracting surfaces. Although we have not yet introduced all of these cases, we will eventually come to them—they are all illustrated in this chapter.

We count four principal rays, numbered 1 through 4, from a given source point S. You can follow each ray in Fig. 36B–1 for the example of a convex mirror:

1. Rays that enter the system parallel to the optical axis. By definition, these paraxial rays are reflected or refracted to the focal point F.

2. Rays that pass through (or for a virtual image are aligned so that if they are extrapolated, they would pass through) the focal point as they enter the system. These rays are just reversed versions of type 1 rays and thus after reflection or refraction leave the system parallel to the axis.

3. Rays that pass through (or are aligned as though they pass through) the center of curvature C of the sphere from which a mirror or refracting surface is formed. These rays are perpendicularly incident on the surface and will be reflected or refracted back along the line of arrival. (As we will see later in the chapter, there is no useful analogous ray when the optical system is a thin lens.)

4. Rays that strike the center of the mirror surface. The reflected rays make the same angle with the axis as do the incident rays (except for sign). (As we will see later in the chapter for the case of thin lenses, the ray drawn directly to the center of the lens passes through it in a straight line. Also, there is no useful analogous ray for a single refracting surface.)

By drawing these principal rays from any given point S on a source, we find where the reflected or refracted rays cross (or for a virtual image appear to cross) and learn the location of the image point I of source point S. When an optical system has more than one reflecting or refracting surface (an "element"), we can apply the simple rule that the image formed by one element serves as a new object for the next element. In that case, *the principal rays must be redrawn for the new object* as they apply to the next optical element to locate the next image.

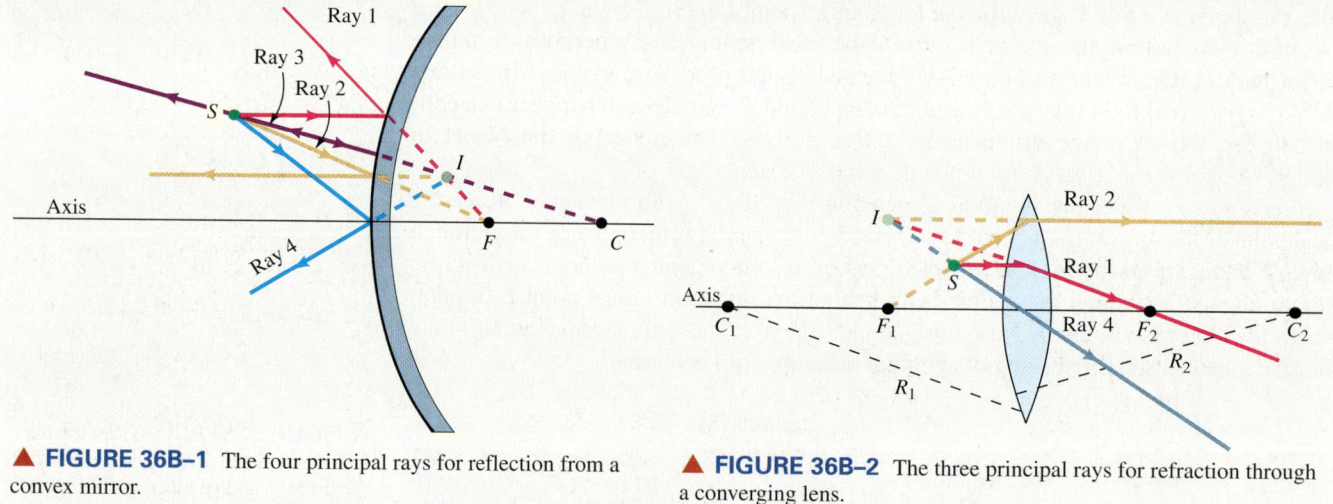

▲ FIGURE 36B–1 The four principal rays for reflection from a convex mirror.

▲ FIGURE 36B–2 The three principal rays for refraction through a converging lens.

The Convex Mirror

The same ray-tracing techniques we used for concave mirrors allow us to understand convex mirrors. Point C in Fig. 36–11 is the center of curvature of the sphere (of radius R) of which the convex mirror is a segment. All lines from point C to the mirror are perpendicular to the mirror.

Location of the Focal Point: We start by finding the focal point, the spot where rays from a point source at infinity (that is, a set of rays parallel to the axis) are focused. Figure 36–11 shows that the reflected rays diverge, so *the image is virtual*, with the rays appearing to originate at a common point F behind the mirror. By following the same trigonometric reasoning we used for the concave mirror and by using Fig. 36–11, we

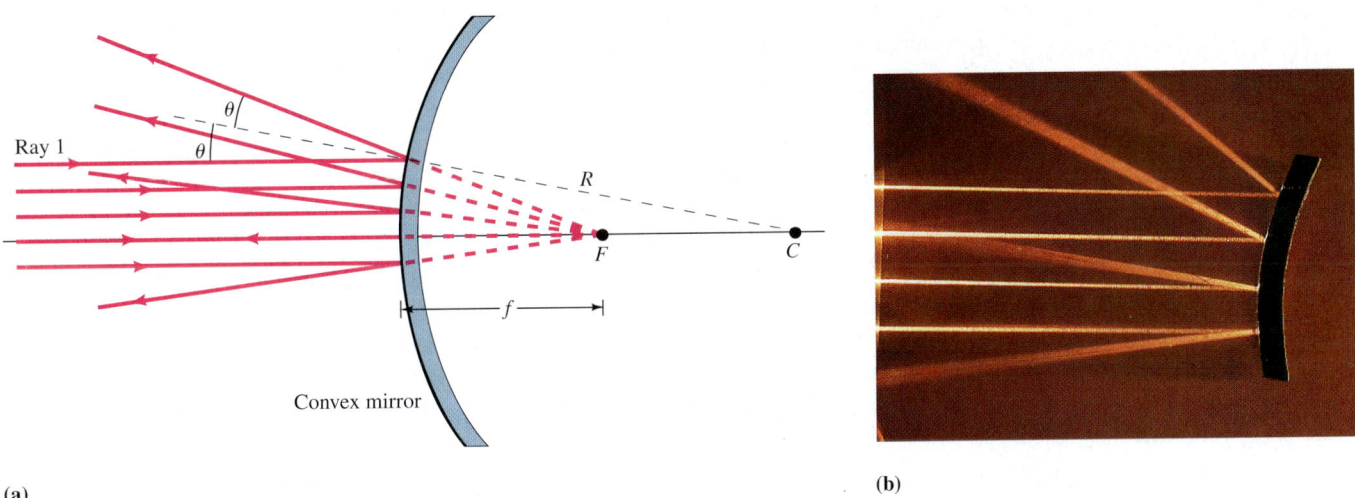

(a)　　　　　　　　　　　　　　　　　　　(b)

▲ **FIGURE 36–11** (a) When a spherical mirror is convex, the focal point lies behind the mirror, as ray tracing shows. The reflected rays diverge, and their extensions all lead back to the focal point. (b) Parallel rays of light reflected by a convex mirror.

can see that the distance f is again given by Eq. (36–1), $f = R/2$ (see Problem 14). Note that the focal point of a convex mirror is on the side opposite the object, unlike the case of the concave mirror, for which the focal point is on the same side as the object.

The Image of an Extended Object: We trace the four principal rays from source point S of the extended object shown in Fig. 36–12: ray 1—parallel to the optical axis, and whose reflection extends back along the line from the mirror to point F; ray 2—drawn as though it would pass through F, and whose reflection is parallel to the axis; ray 3—drawn as though it would pass through C, and whose reflection returns along the line of incidence; and ray 4—striking the center of the mirror surface, and whose reflection makes the same angle with the axis as the incident angle. A careful drawing shows that the reflected rays diverge from each other, but all four (indeed, *all* rays from S) would originate at point I if they were traced back through the mirror. *Point I is the virtual image of point S.*

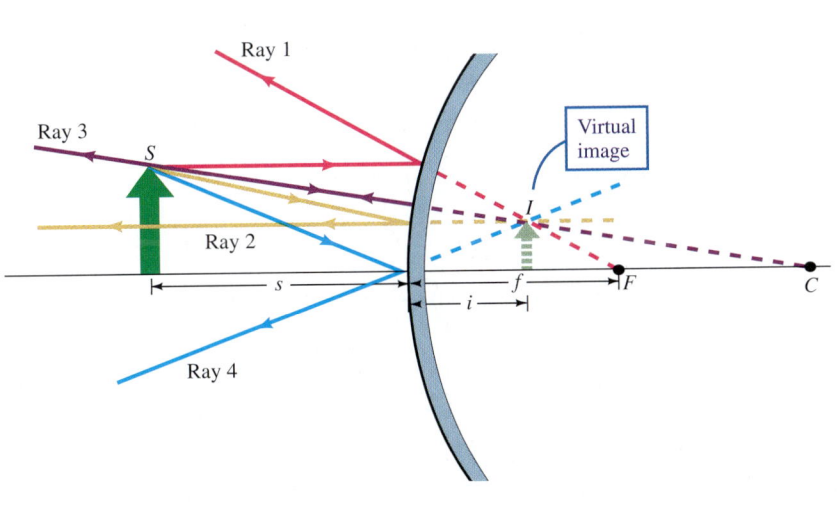

(a)　　　　　　　　　　　　　　　　　　　(b)

▲ **FIGURE 36–12** (a) Ray tracing describes the formation of a virtual image by a convex spherical mirror. (b) The image produced by a convex mirror is upright, reduced, and behind the mirror.

We can similarly find the virtual image of the entire source, which will always be upright and smaller than the source, whatever the position of the object. When the source moves farther away, the image becomes smaller and remains upright, but there is no transition from virtual to real image, as there is in the concave case. These properties make convex spherical mirrors, which with a smaller image encompass a wider range of view, useful for vehicle rearview mirrors.

The Relation Between Source Distance and Image Distance

In Figs. 36–9a, 36–10, and 36–12 we have indicated the distance s from the mirror to the source, the distance i to the image, and the focal length $f = R/2$ (with R the radius of curvature). Using geometrical arguments, we can find a relation between these three quantities. In addition to its direct usefulness, we will also be able to use the relation to help us find image height and whether or not the image is inverted. The relation is

$$\frac{1}{s} + \frac{1}{i} = \frac{1}{f}. \tag{36–2}$$

SOURCE–IMAGE–FOCAL-LENGTH RELATION

(The details of the argument for a concave spherical mirror are contained in the optional subsection that follows.) With an appropriate set of conventions about signs, *the same relation holds for the convex mirror*. Equation (36–2) is immediately understood in two limits. When the object is far away ($s \to \infty$), then $1/s \to 0$ and $i = f$ (which is the definition of f). When the object is at the focus, $s = f$, then $1/i = 0$: The image is very far away.

Equation (36–2) contains information on the sign of the image point i. We'll discuss this information here, but point out that later the information on the sign will be summarized in a table. To start, note that if the object is between the concave mirror and the focus, as in Fig. 36–10, then s is smaller than f, and Eq. (36–2) implies that *i must be negative*. We associate a negative i with the image on the far side of the mirror—that is, with a virtual image. Likewise, i is positive when the image is real. Equation (36–2) may be applied to a convex mirror if we follow the convention that *the focal length f is negative when the focus is on the "virtual image" side of the mirror*. This is equivalent to saying that if the mirror's center of curvature is on the back (nonreflecting) side of the mirror, f is negative. In the application of Eq. (36–2) to a convex mirror, s is always positive and f is always negative. If s is positive (as it is when our optical system consists only of the convex mirror), then i will be negative (the image is virtual). Furthermore, the image in this case must be between the mirror and the focal point. That is because $1/s = 1/f - 1/i = -1/|f| + 1/|i|$ is positive, so $1/|i| > 1/|f|$, and hence $|i| < |f|$.

The rules for the sign of the object distance s will be discussed in Section 36–3.

CONCEPTUAL EXAMPLE 36–4 You are given a concave mirror and cannot measure its radius of curvature directly. How could you determine this radius with optical techniques?

Answer If we can determine the focal point of the mirror, then its radius of curvature is $R = 2f$, according to Eq. (36–1). As Eq. (36–2) confirms, this is done with a source at infinity, $s \to \infty$, so that the image is at the focal point, $i = f$. Sunlight provides such a source, and if we allow it to shine on the concave mirror, we need only to measure the distance from the mirror to the "hot spot"—the image location. Then $R = 2f = 2i$.

---

*How to Obtain Eq. (36–2)

To arrive at Eq. (36–2), we consider two points on an optical axis—the light source (or object), S, and its image, I—and a concave spherical surface. We can see from Fig. 36–13 and from the fact that the sum of the internal angles (in radians) of a triangle is π, that the following relationships hold:

$$\gamma = \beta + \alpha; \tag{36–3}$$

$$\delta = \gamma + \alpha = \gamma + (\gamma - \beta) = 2\gamma - \beta. \tag{36–4}$$

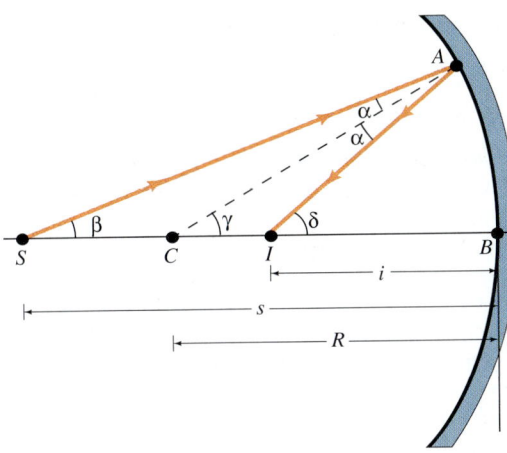

▼ **FIGURE 36–13** Geometric construction for deriving Eq. (36–2) for a spherical mirror.

In arriving at Eq. (36–4), we have used Eq. (36–3) as an intermediate step to eliminate α. The distances of Fig. 36–13 are related to the angles by the exact relation $AB = R\gamma$ and by the *approximate* (small-angle) equations $AB = i\delta = s\beta$. These relations allow Eq. (36–4) to be rewritten as

$$\frac{AB}{i} = \frac{2AB}{R} - \frac{AB}{s}.$$

We divide out the common factor AB and use the focal length $f = R/2$ for a spherical surface. We immediately obtain Eq. (36–2).

Magnification

Our geometric constructions show that an image may not be the same size as its source. Consider the convex mirror in Fig. 36–14. Ray 4 to the center point A of the mirror is useful because all the angles marked θ are the same, so triangles $AS'S$ and $AI'I$ are similar triangles. Thus, the magnitude of the **magnification** M defined as the ratio of the heights of the source and image, is

$$|M| \equiv \frac{|II'|}{|SS'|} = \frac{|i|}{|s|}. \qquad (36\text{--}5)$$

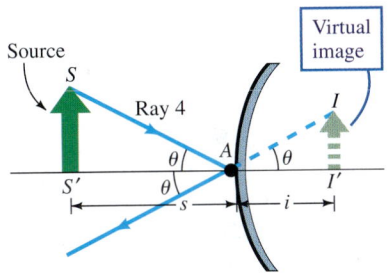

▲ **FIGURE 36–14** Geometry for the calculation of magnification.

We also specify whether the image is upright or inverted by writing

$$M = -\frac{i}{s}. \qquad (36\text{--}6)$$

MAGNIFICATION

If M is negative, the image is inverted; if M is positive, the image is upright. We can verify that this form works in the following explicit cases:

1. When the mirror is concave and the source is outside the focal point, the image is real (i is positive). By Eq. (36–6), M is then negative and the image should be inverted, as it is in Fig. 36–9a.

2. When the mirror is concave and the source is inside the focal point, the image is virtual (i is negative). By Eq. (36–6), M is then positive and the image should be upright, as it is in Fig. 36–10.

3. When the mirror is convex, the image is virtual (i is negative). By Eq. (36–6), M is then positive and the image should be upright, as it is in Fig. 36–12.

Equation (36–2) can be rewritten as

$$\frac{1}{i} = \frac{1}{f} - \frac{1}{s} = \frac{s-f}{fs}.$$

We can thereby find M in a form in which the image distance does not appear:

$$M = -\frac{i}{s} = -\frac{fs/(s-f)}{s} = \frac{f}{f-s}. \quad (36\text{-}7)$$

This equation can be applied to both concave and convex mirrors if we recall that f is negative for convex mirrors. In the convex case, $f - s$ is always negative, so M is always positive; the image is always upright. Also, $|f - s| = |f| + |s|$ is always larger than $|f|$ for convex mirrors, so the image is always reduced in size.

EXAMPLE 36–5 A convex spherical mirror of radius of curvature R of magnitude 20.0 cm produces an upright image precisely one-quarter the size of the candle that is the object. What is the separation distance between the object and its image?

Setting It Up We draw the mirror in Fig. 36–15, indicating the center of curvature C and the focal point F. We know the position of neither source nor image.

Strategy We find f from Eq. (36–1), $f = R/2$. We then rearrange Eq. (36–7) to find s in terms of the known quantities f and M. We can then use Eq. (36–6) to find i.

Working It Out First we have numerically $f = R/2 = -10.0$ cm. (The negative sign indicates that the mirror is convex.) The appropriately rearranged Eq. (36–7) is

$$s = f\left(1 - \frac{1}{M}\right) = (-10.0 \text{ cm})\left(1 - \frac{1}{\frac{1}{4}}\right)$$
$$= (-10.0 \text{ cm})(-3) = 30.0 \text{ cm}.$$

(We have taken $M = \frac{1}{4}$, positive because the image is upright.) Next we rearrange Eq. (36–6) to determine i:

$$i = -sM = -(30.0 \text{ cm})\left(\frac{1}{4}\right) = -7.50 \text{ cm}.$$

The minus sign is consistent with our knowledge that the image of a convex mirror is virtual (on the far side of the mirror). Finally, the distance between object and image will be $|s| + |i| = 30.0$ cm + 7.50 cm = 37.5 cm.

Alternative Solution The ray-tracing construction in Fig. 36–15 confirms these conclusions qualitatively; to confirm the actual numbers, the ray-tracing technique requires a very accurate drawing.

What Do You Think? What happens to the image when the candle is slowly moved farther away from the mirror?

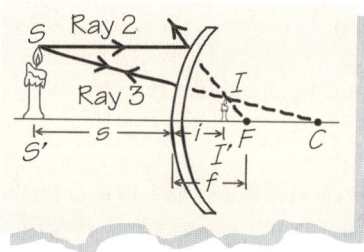

▲ **FIGURE 36–15** Ray tracing to find the image.

CONCEPTUAL EXAMPLE 36–6 What are the focal length f and magnification M of a plane mirror?

Answer A plane mirror is a special case of a concave or convex mirror in which $R \to \infty$. Since $f = R/2$, if $R \to \infty$, then $f \to \infty$ as well. We have seen that by using the relation between s, i, and f, we can determine the magnification in terms of any two of these. Here we use Eq. (36–7), which gives M in terms of f and s:

$$M = \frac{f}{f-s} \to \frac{f}{f} = 1.$$

We have used the fact that $f \gg s$, and the answer is no magnification (image size = object size) for any value of s. Note that M is positive, indicating an upright image—every morning, when you look in the bathroom mirror, you verify this!

THINK ABOUT THIS...
ARE RAY-TRACING TECHNIQUES REALLY USEFUL?

In this chapter our primary tools for analyzing optical systems are algebraic equations, not ray-tracing techniques. But ray tracing is the basis for the design of real optical systems, especially the most sophisticated. We may want an optical system to produce a very sharp image over a very limited range of source distances; for example, the lenses used in orbital satellites to image Earth's surface will never have to make an image of a very close source. Or, as in many cameras, we may want to sacrifice a sharp image in order for an optical system to operate in dim light. Real systems may have nonspherical mirrors, or thick, multielement lenses in which the elements move relative to one another, as in zoom lenses. To attain the desired optical properties, designers of such systems use computer programs capable of tracing large numbers of rays in a system design, which allows them to preview the quality and placement of the image and to test modifications in the design.

The relation between source distance, image distance, and focal length [Eq. (36–2)], the expression for magnification [Eq. (36–6)], and the ray-tracing techniques are applicable to lenses as well as to mirrors. For the lenses that we will study in this chapter, these three elements provide all the information we need.

We conclude this section with a comment on signs. Table 36–1 gives the signs of all the quantities necessary for mirrors, refracting surfaces, and lenses. However, in our opinion it is not necessary to keep track of the signs of the various quantities we have discussed. Develop your ray-tracing techniques, and you will be able to rederive the signs on your own. In this way a ray-tracing diagram is perhaps as useful as a free-body diagram.

TABLE 36–1 • Sign Conventions for Mirrors, Refracting Surfaces, and Lenses

In applying the information in this table, we must distinguish two "sides" to a reflecting or refracting surface:

Side A, the side from which light originates, and

Side B, the side to which light passes.

For mirrors, side B is identical to side A; for refracting surfaces and lenses, the two sides are opposite. Only the sign of the source position is determined by side A. All other quantities are determined by reference to side B.

Determined by Side A

Source distance s	Positive if object is on side A (real object)
	Negative if object is on side opposite to side A (virtual object)

Determined by Side B

Image distance i	Positive if image is on side B (real image)
	Negative if image is on side opposite to side B (virtual image)
Curvature R	Positive if center of curvature is on side B
	Negative if center of curvature is on side opposite to side B
Focal point	Positive if on side B
	Negative if on side opposite to side B

36–3 Refraction at Spherical Surfaces

Mirrors change the direction of rays of light and create real or virtual images of objects. Lenses do the same, using pieces of transparent material to refract light. With Snell's law of refraction, we can use rays to determine the behavior of lenses in the same way that we handled mirrors using the law of reflection. The most basic type of lens typically has two curved refracting surfaces, and it is best to approach this case by first thinking about refraction through a single curved surface, which is the subject of this section. By repeatedly applying the rules we develop for a single boundary, we shall be able to understand the passage of light through lenses. We study, in particular, surfaces that are spherical sections without too much curvature, and we consider only paraxial rays. This simplifies the calculations and leads to the right qualitative conclusions.

Consider, then, light that crosses the boundary between one medium with index of refraction n_1 and another medium with index of refraction n_2 (Fig. 35–9). The angles of incidence and refraction satisfy Snell's law, Eq. (35–6):

$$n_1 \sin \theta_1 = n_2 \sin \theta_2.$$

Here we apply this law to a boundary that is not flat but rather forms a segment of a sphere of radius of curvature R. Let's take a convex surface, one whose center of curvature—point C in Fig. 36–16—is in the region to which light passes. Although we choose $n_1 < n_2$, so that the light that passes from medium 1 to medium 2 bends toward the perpendicular to the surface, the results will be more general.

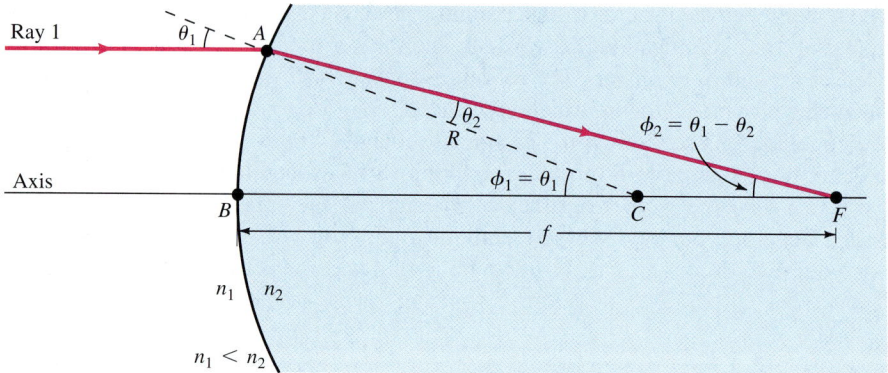

FIGURE 36–16 Ray tracing of a ray that enters a medium whose index of refraction is different from that of the medium from which the ray came requires us to use Snell's law of refraction. Here we see refraction at a convex spherical surface.

The Focal Point of a Single Refracting Surface

As for a spherical mirror, a single refracting surface has a focal point F that we find by tracing rays that come from a very distant source, parallel to the axis. For the convex surface in Fig. 36–16, ray 1 bends toward the axis and crosses it at a point F. This point will be a focal point if all the incident rays that are parallel to the axis cross at F. Here we show that this is the case for paraxial rays, where the angle of incidence θ_1 and that of refraction θ_2 are both small, so the relation $\sin\theta \cong \theta$ is a good approximation. In that case Snell's law becomes

$$n_1\theta_1 \cong n_2\theta_2. \tag{36-8}$$

Simple geometry shows that in Fig. 36–16 $\phi_1 = \theta_1$, and therefore $\phi_2 = \theta_1 - \theta_2$. For small angles, the relation between BF and the arc length AB is given by

$$BF(\theta_1 - \theta_2) \cong AB.$$

Because $AB = R\theta_1$, this result, along with Eq. (36–8), implies that

$$BF \cong \frac{R\theta_1}{\theta_1 - \theta_2} \cong \frac{Rn_2}{n_2 - n_1}.$$

This distance is independent of θ_1 for small angles, so *all* parallel rays near the axis pass through point F, and F is the image of a point source at infinity. The focal length f is the distance BF:

$$f = \left(\frac{n_2}{n_2 - n_1}\right)R. \tag{36-9}$$

The focal point for a single refracting surface is farther from the surface than the center of curvature, as in Fig. 36–16 or by noting that in Eq. (36–9), $f > R$ if $n_2 > n_1$. Although we have derived Eq. (36–9) for a convex surface, we can derive it for a concave surface just as easily (Fig. 36–17). The center of curvature C of the concave surface is on the side from which the light is incident. We find exactly the same formula—except that the focal point is to the left of the surface, on the same side as C. We see in Fig. 36–17 that an image at the focal point F for such a surface is virtual.

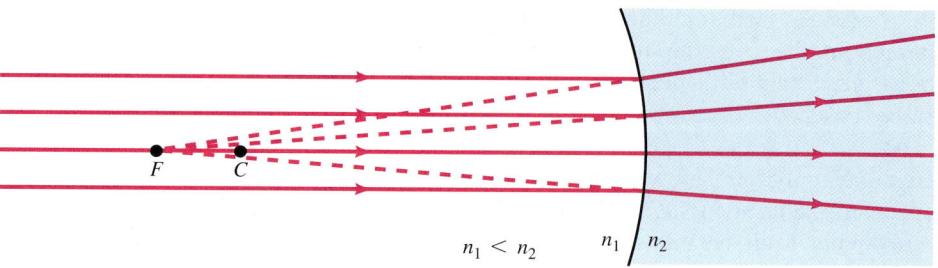

▲ **FIGURE 36–17** Refraction at a concave spherical surface.

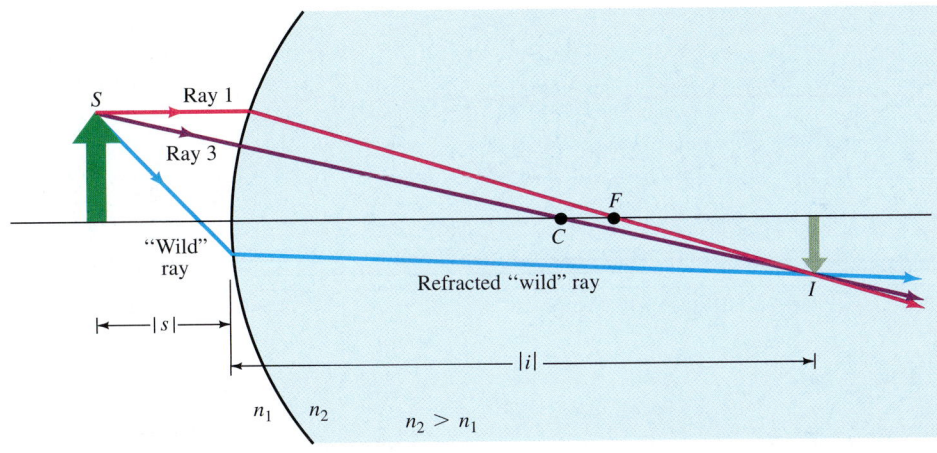

◀ **FIGURE 36–18** Ray tracing shows how a real image is formed by a convex spherical refracting surface.

The Image of an Extended Object

Convex Surface: Consider next a vertical object that stands erect on the optical axis. We already know enough about the principal rays to proceed with ray tracing. For a single refracting surface, only two of the four principal rays from any source point, such as point S, are useful. In Fig. 36–18, ray 1 is incident parallel to the axis and refracts such that it crosses the axis at F; ray 3 forms the straight line through C. Ray 3 is perpendicular to the surface and so passes into the medium without deflection. The two refracted rays meet at point I. Again, these rays are only two of an unlimited number of rays that leave point S and pass through I. For example, we have drawn a "wild" ray in Fig. 36–18. We will not carry out the detailed geometry that shows that the wild ray passes through I, but this is the case. By drawing the principal rays for any point on the object, we can reconstruct the entire image, which, given the distance s, is *real* and *inverted*.

Concave Surface: In Fig. 36–19, we offer three possibilities that depend on whether the source distance s is larger than the focal length of the lens f, smaller than the radius of curvature R, or smaller than f but larger than R. In each case, we use the same two principal rays from point S that we did for the convex case to locate the image of S at point I. In Figs. 36–19b and 36–19c, ray 3 may not actually pass through the curved surface. This is not a problem, because the principal rays are just tools for determining where all the rays that do pass through the surface cross. In each case, the image is upright and virtual (the diverging rays appear to come from point I).

The Relation Between Source Distance and Image Distance

The relationship between the positions of a source and an image for a single (concave) refracting surface is analogous to Eq. (36–2), which holds for a mirror. It takes the form

$$\text{for a refracting surface:} \quad \frac{n_1}{s} + \frac{n_2}{i} = \frac{n_2 - n_1}{R}. \tag{36–10}$$

The derivation of this equation is like that of Eq. (36–2). We will not derive Eq. (36–10) but rather leave the derivation to Problem 28. In the derivation it is easiest to assume that the surface is convex; that is, that the center of curvature of the surface is on the side of the surface to which the light passes. Let's suppose that this corresponds to a positive value of R, as for the concave mirror. In addition, s is positive from the start. If we then find i from Eq. (36–10), it could be either positive or negative. When i is positive, it is on the side of the surface to which light passes and the image is real, meaning that light passes through it. When i is negative, it is on the side from which light is emitted and the image is virtual, meaning that light only *appears* to radiate from it when it is observed from medium 2. At the start of the discussion, we assumed that $n_2 > n_1$, but the result does not depend on this (see Problem 29), although a geometrical drawing does.

We can also repeat this exercise with a concave spherical surface between the two media (see Problem 30). In this case, the center of the spherical surface is on the side of the light source, and the important result is that *Eq. (36–10) continues to hold, but with a*

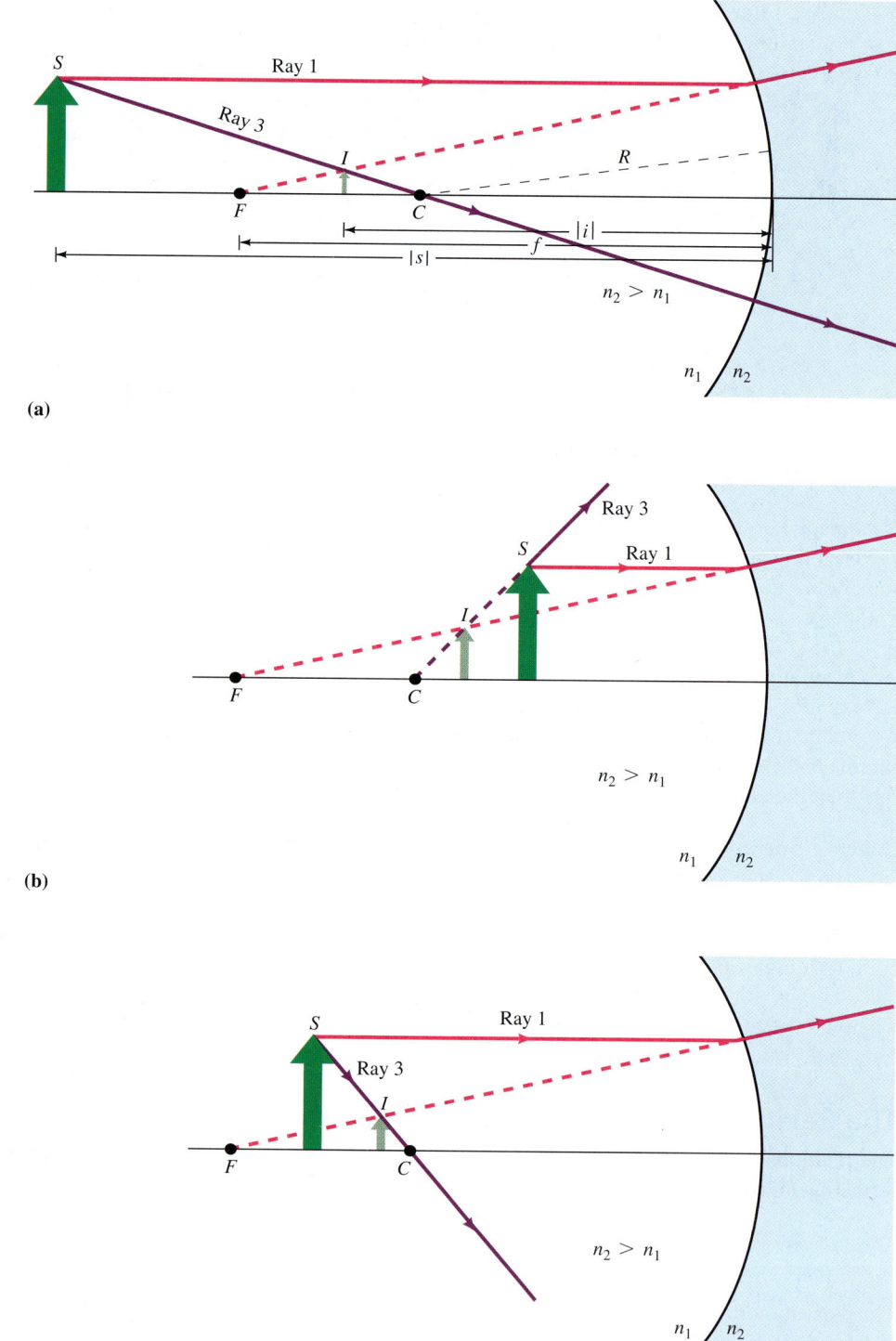

▶ **FIGURE 36–19** Ray tracing with principal rays for image formation for a concave spherical refracting surface: situations in which (a) $s > f$; (b) $R > s$; (c) $f > s > R$.

negative value of R. Summarizing, when the image is on the side to which light goes, i is positive and the image is real, while R is positive when it is on that side. When the image is on the side from which light radiates, i is negative and the image is virtual, while R is negative when it is on that side. Table 36–1 contains a summary of the various signs.

The Sign of the Object Distance

In order to be able to apply the results of this section to material with two surfaces—lenses—it will be necessary to understand one more aspect of signs. We have established that a negative image distance and radius of curvature have meaning. Is it possible for s

to take on a negative value? A positive s corresponds to a real object on the side from which the light radiates. For positive s, the light rays diverge from the object as they approach the boundary surface. Negative values of s correspond to the rays that *converge* as they approach the boundary, so their extrapolation would be on the side of the boundary to which the light passes. This cannot occur for a real object. But it is possible if the image produced by one surface—boundary 1, say—acts as the source object for a second surface: boundary 2 (Fig. 36–20a). We can break up the problem and find first the image point I_1 produced by boundary 1 (Fig. 36–20b). In actuality, the light never forms the image point I_1 because boundary 2 intervenes. However, the image I_1 becomes the virtual object S_2 for the light refracted at boundary 2 (Fig. 36–20c). According to our convention, the source distance s_2 is negative because the rays converge toward boundary 2. The light comes from the left side of boundary 2, but the virtual object is on the right side. Equation (36–10) holds for the refraction at boundary 2, with a negative object distance s_2. The actual paths of the light rays through both boundaries are shown in Fig. 36–20d.

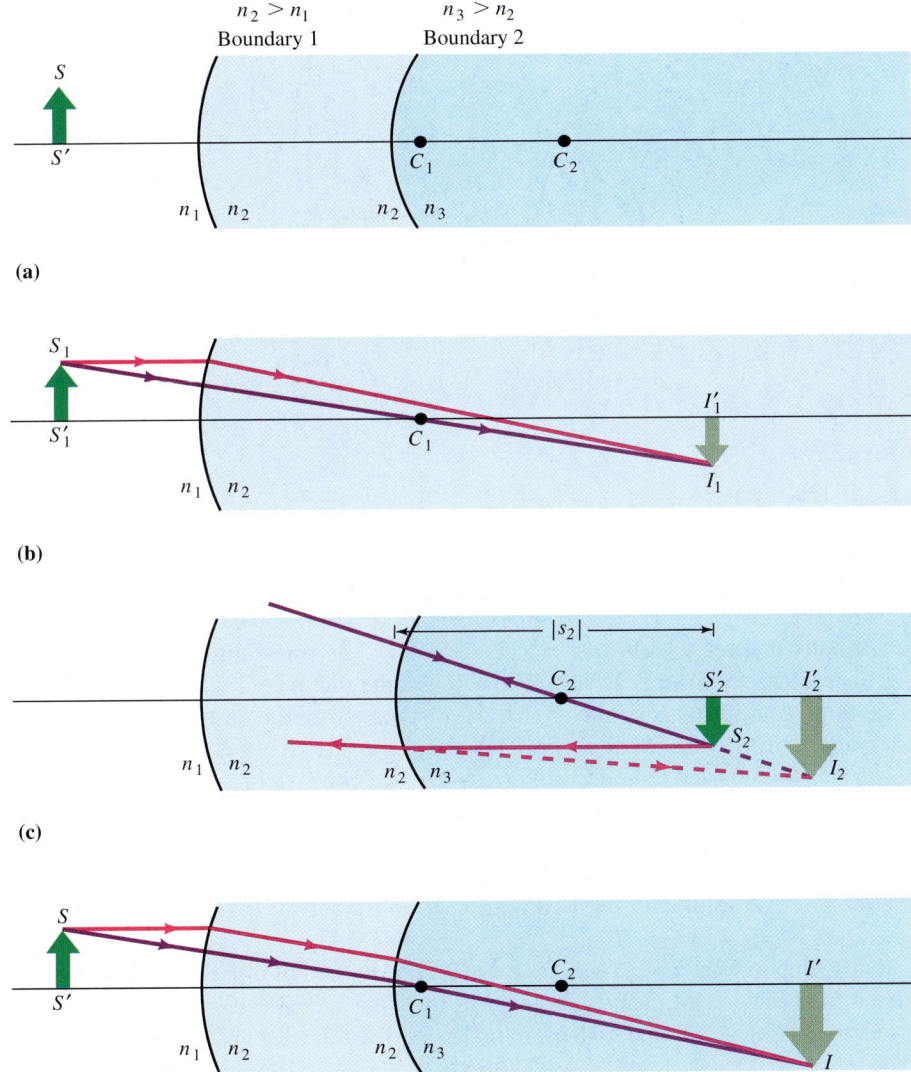

▲ **FIGURE 36–20** (a) Image construction when two refracting surfaces are involved. To simplify, (b) we can split the problem up by first finding the image point I_1 from boundary 1. (c) The resulting image serves as a virtual object; we use virtual source point S_2 for the interaction with boundary 2 to find the final image, at image point I. (d) The actual ray path.

EXAMPLE 36–7 Consider a cylinder of glass 50 cm long, with $n = 1.6$, in air (Fig. 36–21). Surface 1 has a radius of curvature $R_1 = 0.20$ m; surface 2 has a radius of curvature $R_2 = 0.40$ m. Both surfaces "bulge out." A small object (a leaf) is placed perpendicular to the optical axis at a distance of 120 cm from surface 1. (a) Find the location of the object's image due to refraction at surface 1. (b) Let this image be the source object for surface 2, and find the location of *its* image as light passes through surface 2.

Strategy Equation (36–10) applied successively to the two surfaces leads to the location of the final images. We first find the image formed by surface 1, and this image then becomes the object for surface 2.

Working It Out (a) We calculate the distance i_1 of the image point I_1 from surface 1. We have $n_2 = 1.6$ and $n_1 = 1.0$ (air). The center of curvature is on the side to which light passes, so R_1 is positive. Finally, $s = +1.20$ m. We then have

$$\frac{n_1}{s} + \frac{n_2}{i} = \frac{n_2 - n_1}{R}, \text{ or } \frac{1.0}{1.20 \text{ m}} + \frac{1.6}{i_1} = \frac{1.6 - 1.0}{0.20 \text{ m}};$$
$$i_1 = +0.74 \text{ m}.$$

The image is real and located (not shown) 74 cm to the right of surface 1.

(b) Because the surfaces are separated by 50 cm, this new object S_2 (which is the image I_1 for surface 1) is 24 cm to the right of surface 2. The object is on the side to which light passes, so its distance from surface 2 is negative: $s_2 = -0.24$ m. For this second step, $n_1 = 1.6$, $n_2 = 1.0$, and $R_2 = -0.4$ m (surface 2 is concave, so its center of curvature is on the side from which light comes). Thus Eq. (36–10) now gives

$$\frac{1.6}{-0.24 \text{ m}} + \frac{1}{i_2} = \frac{1.0 - 1.6}{-0.4 \text{ m}};$$
$$i_2 = +0.12 \text{ m}.$$

This is positive, so the second image is real, or to the right of surface 2.

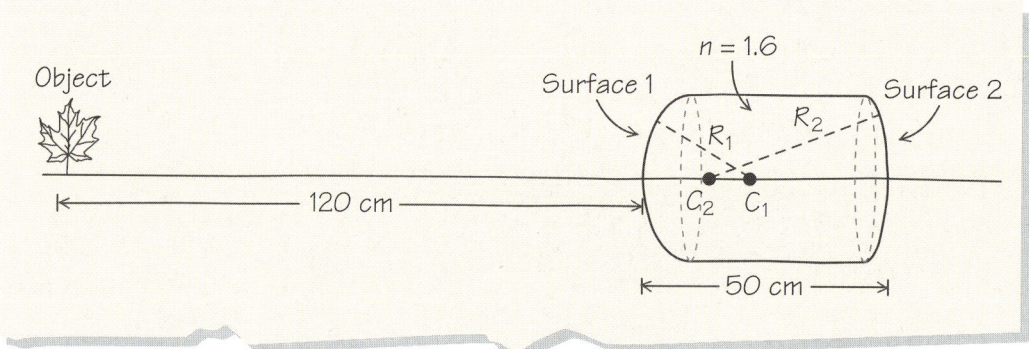

▲ **FIGURE 36–21**

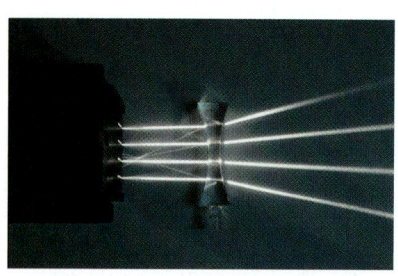

(a)

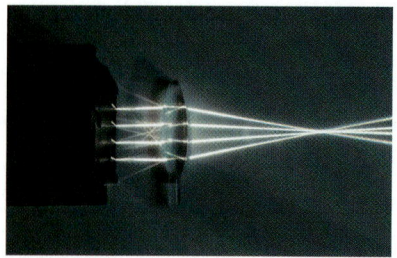

(b)

▲ **FIGURE 36–22** (a) Light passing through a diverging lens and (b) light through a converging lens.

36–4 Thin Lenses

A real lens can consist of a complicated combination of individual elements. We can consider a basic element of a more complicated lens to be a single element made of a transparent material of refractive index n embedded in a material of refractive index n_1, normally air, for which $n_1 = 1$ (Fig. 36–22). It is this simpler element that we refer to as a **lens** in this discussion. We shall assume that $n > 1$ and $n_1 = 1$. We shall also assume that our lens is thin, so that the distance from the object and the image to each surface of the lens is the same. This simplifies the treatment considerably. The two surface boundaries (1 and 2 in Fig. 36–23) are concave or convex spherical segments (or planar surfaces) with respective radii of curvature R_1 and R_2. Whether these radii are positive or negative depends on whether the center of curvature is on the side to which light passes (positive R) or the side from which light radiates (negative R). For example, with light coming in from the left in Fig. 36–23a, R_1 is positive and R_2 is negative.

Let's suppose that a real object is a distance s_1 to the left of a thin lens. We can locate the final image and identify its features by using Eq. (36–10) twice in succession for image-making at a single surface, much as we did in Example 36–7. The image produced by the first surface serves as the object for the second surface. We do not have to worry about whether the various objects and images are real or virtual, upright or inverted, because the equation will automatically handle these questions. At surface 1, we have

$$\frac{1}{s_1} + \frac{n}{i_1} = \frac{n-1}{R_1}, \tag{36–11a}$$

which we rewrite as

$$\frac{1}{i_1} = \frac{n-1}{nR_1} - \frac{1}{ns_1}. \qquad (36\text{–}11b)$$

Now, the image point I_1 produced by surface 1 serves as an object point S_2 for surface 2, producing a final image point at I_2. What is the sign of i_1? If i_1 is positive, the image is on the right of surface 1 and hence on the right of surface 2. This corresponds to an object distance s_2 for surface 2 that is negative. Similarly, if i_1 is negative, the image is to the left of both surfaces, corresponding to a positive object distance s_2 for surface 2. We must then reverse the sign of i_1 when we use it as the source distance s_2 for surface 2. Finally, note that in applying Eq. (36–10) a second time, $n_1 = n$ and $n_2 = 1$. Thus

$$\frac{n}{s_2} + \frac{1}{i_2} = \frac{1-n}{R_2};$$

$$-\frac{n}{i_1} + \frac{1}{i_2} = \frac{1-n}{R_2}.$$

When we substitute Eq. (36–11b), we find that

$$-n\left(\frac{n-1}{nR_1} - \frac{1}{ns_1}\right) + \frac{1}{i_2} = \frac{1-n}{R_2}.$$

If we now write $s_1 = s$ for the original object and $i_2 = i$ for the final image, we find (upon rearrangement)

$$\text{for a thin lens in air: } \frac{1}{s} + \frac{1}{i} = (n-1)\left(\frac{1}{R_1} - \frac{1}{R_2}\right). \qquad (36\text{–}12)$$

LENS-MAKER'S EQUATION

Equation (36–12), which applies *only* to thin lenses in air, is the *lens-maker's equation*. By Eq. (36–12), the image can be positive or negative; that is, real or virtual. The signs are summarized in Table 36–1, and ray tracing will alternatively allow you to understand the image—large or small, upright or inverted, real or virtual.

Equation (36–12) can be used to find the focal point of a lens. By definition, the image is at the focal point when $s \to \infty$. Therefore in this limit

$$\frac{1}{f} = (n-1)\left(\frac{1}{R_1} - \frac{1}{R_2}\right). \qquad (36\text{–}13)$$

If we substitute this result into Eq. (36–12), we get Eq. (36–2), which we originally derived for mirrors—Eq. (36–2) is a general one that holds for most of the optical systems we study. The sign of f is determined by the signs of the radii of curvature, but we can say that f is positive if the image of a point source at infinity is on the side to which light passes (real image); f is negative if the image of the source at infinity is on the side from which light radiates (virtual image).

We now turn to some of the rays that help us understand the image. Ray 1, which comes in parallel to the optical axis of the lens, crosses (or behaves as though it crosses) the axis at f. This ray is drawn in Fig. 36–24a. Note that there is a symmetry in Eq. (36–12). When light arrives from the right of the lens rather than the left, R_1 and R_2 reverse their signs, and light from infinity coming from the right is focused the same distance from the lens as the first focal point, but on the opposite side. In turn, if light radiates (or behaves as though it does) from one of the two symmetric focal points of the lens (ray 2), the light emerges as a set of parallel rays.

Ray 4, which is drawn in Fig. 36–24a to the center of the lens and behaves as though it passes straight through, is a last useful ray. (Remember, principal ray 3 is not applicable to lenses.) This ray is shown in more detail at the enlarged section (Fig. 36–24b).

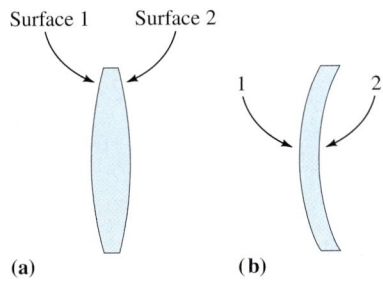
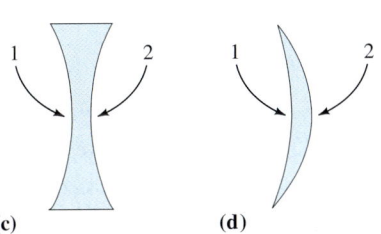
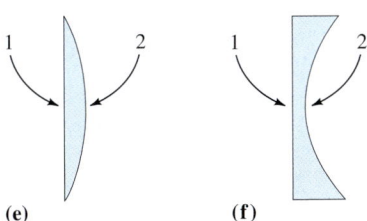

▲ **FIGURE 36–23** Six types of simple thin lenses with surfaces of different radii of curvature: (a) $R_1 > 0$, $R_2 < 0$; (b) $R_1 > 0$, $R_2 > 0$; (c) $R_1 < 0$, $R_2 > 0$; (d) $R_1 < 0$, $R_2 < 0$; (e) $R_1 = \infty$, $R_2 < 0$; (f) $R_1 = \infty$, $R_2 > 0$.

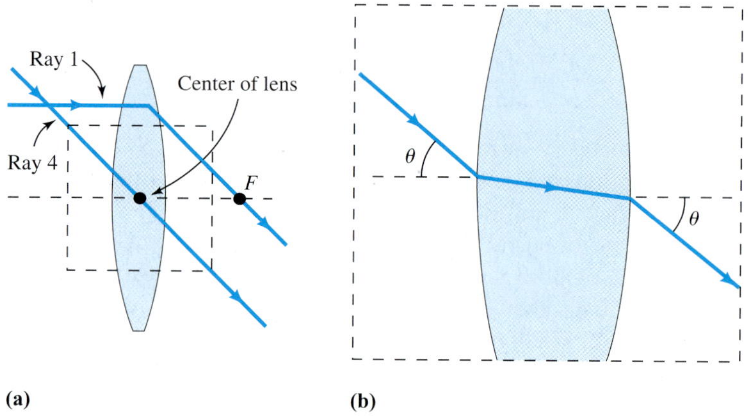

▶ **FIGURE 36–24** (a) The ray to the center of a lens passes through without changing angle, because at its axis the lens is like a pane of glass. (b) An enlarged view of the same lens. If the lens is thin, the displacement of the ray is small.

Since the two lens surfaces are in the middle of the lens, the ray behaves to a good approximation like a ray that passes through a thin pane of glass. There is a *small* displacement of the ray, but it drops to zero as the lens becomes thinner.

These three principal rays can be used to find an image. For example, Fig. 36–25 shows a lens that collects light from an object, and with the principal rays, we can easily find the image point I of object point S. The construction works for any point on the object. The image is real and inverted in this case. In general, if a lens causes rays that pass through it to come together, it is called a *converging lens*, and if it causes rays that pass through it to spread out, it is a *diverging lens*. Converging lenses have positive focal lengths, whereas diverging lenses have negative focal lengths. Some simple ray tracing will show that a lens like that of Fig. 36–23a is a converging lens, and one like that of Fig. 36–23c is a diverging lens.

Magnification

A thin lens produces a perfect image, to the extent that the small-angle approximation is valid. Thus we can find the magnification by direct use of similar triangles. In Fig. 36–25, the magnification of the image has magnitude

$$M = \frac{II'}{SS'}.$$

From the geometry of the similar triangles $SS'P$ and $II'P$, we see that the magnitude of the magnification is $|M| = |i|/|s|$. Just as for mirrors, a systematic look at signs shows that we can decide with a single sign whether the image is upright or inverted:

$$M = -\frac{i}{s}.$$

This is Eq. (36–6)—the same form we found for mirrors. If M is positive, the image is upright; if it is negative, the image is inverted. From Eq. (36–2), we have the alternate form

$$M = \frac{f}{f - s},$$

which is Eq. (36–7), also applicable to mirrors.

▶ **FIGURE 36–25** Ray tracing shows how a real image is formed with one type of thin lens. Point P marks the center of the lens.

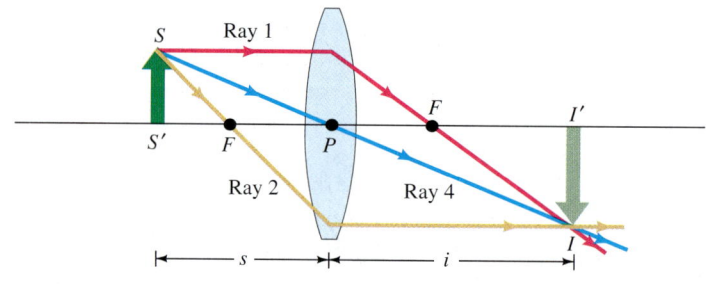

EXAMPLE 36–8 A converging lens like that shown in Fig. 36–23a has surfaces with radii of curvature $R_1 = 80$ cm and $R_2 = 36$ cm. An emerald that is 2.0 cm tall is placed 15 cm to the left of the lens, for which $n = 1.63$. Where will the image be located, and what will its size be?

Setting It Up We have sketched Fig. 36–26 to include three rays. Such a diagram is useful if one wants to use ray tracing to verify results.

Strategy We first calculate the focal length from Eq. (36–13). Then we use Eq. (36–2) to determine the image distance i. At that point we will have both s and i and can find M.

Working It Out The radius of curvature of the first surface is positive, $R_1 = 80$ cm, whereas the second surface has negative curvature, $R_2 = -36$ cm. Thus Eq. (36–13) gives

$$\frac{1}{f} = (n - 1)\left(\frac{1}{R_1} - \frac{1}{R_2}\right)$$

$$= (1.63 - 1)\left(\frac{1}{80 \text{ cm}} - \frac{1}{-36 \text{ cm}}\right) = 0.025 \text{ cm}^{-1}.$$

The object distance is positive, $s = 15$ cm, so Eq. (36–2) gives

$$\frac{1}{i} = \frac{1}{f} - \frac{1}{s} = 0.025 \text{ cm}^{-1} - \frac{1}{15 \text{ cm}} = -0.041 \text{ cm}^{-1}.$$

Thus $i = -24$ cm. The minus sign indicates that the image is virtual and on the same side as the light source. The magnification is given by

$$M = -\frac{i}{s} = -\frac{-24 \text{ cm}}{15 \text{ cm}} = 1.6.$$

The positive value indicates that the image is upright.

The rays in Fig. 36–26 confirm the qualitative aspects of our results.

What Do You Think? Suppose the lens were replaced by one with a smaller value of n. How would the magnification change?

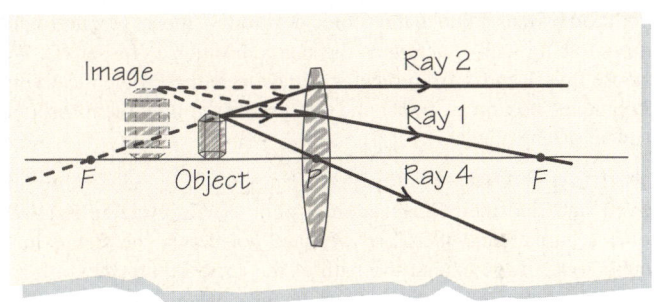

▲ **FIGURE 36–26** The horizontal and vertical scales are different here. When the object lies inside the focal point of the lens, ray tracing shows that the image formed is virtual.

We saw in Section 36–3 how the image produced by refraction at one surface acts as an object for the second surface. This principle extends to combinations of two or more lenses and lies at the heart of the design of more complicated lenses or optical instruments. Figure 36–27 gives an image construction for two thin converging lenses. The object SS' lies inside the focal length of lens 1 and thus gives rise to a virtual, enlarged image $I_1 I_1'$. That image serves as an object $S_2 S_2'$ for lens 2. Ray tracing uses the parallel ray $I_1 A_2 F_2 I$ and $I_1 P_2 I$ to determine the position of the real image, but the particular rays chosen really follow paths like $SA_1 F_1 AI$ and $SP_1 BI$. This example shows that it is possible to obtain a magnified *real* image with two converging lenses in conditions where it is not possible with one lens.

In Example 36–9, the object for the second lens is a negative distance from the lens.

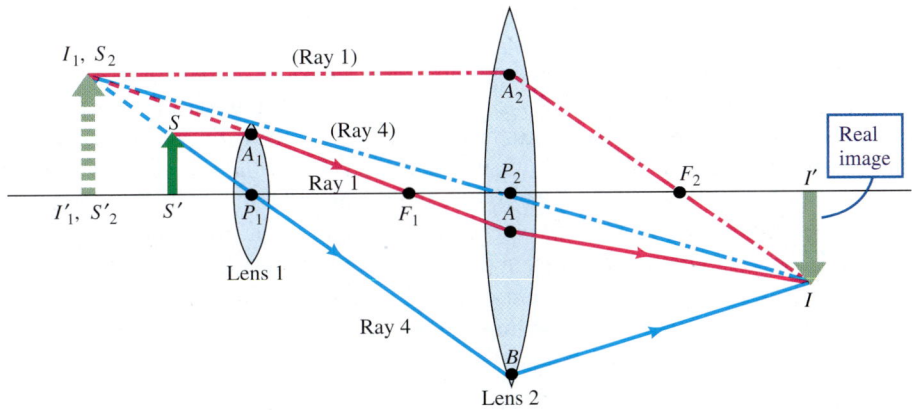

▲ **FIGURE 36–27** Ray tracing shows how two converging lenses produce a real magnified image.

EXAMPLE 36–9 Consider the two-lens arrangement shown in Fig. 36–28, with the source just outside the focal point of lens 1. The focal length of each lens is known. You place your eye at the position shown and look for the image. Use ray-tracing techniques to describe qualitatively that image.

Setting It Up If these were not already included in the figure, the starting point would be to draw in the focal points F_1 and F_2, the source distance s_1, and the lens separation distance L. The figure should be to scale. Note that with the eye placed as shown, the "image" in question is the one that is produced by lens 2 using the image produced by lens 1 as a source.

Strategy We must use ray tracing to find the image position i_1 produced by lens 1 due to the source position s_1, then a second round of ray tracing with i_1 acting as the source position s_2 for lens 2. We choose rays 1 and 4 from our list of rays (see the Problem-Solving Techniques box on p. 1004) and successively follow them through the lens arrangement.

Working It Out We trace rays 1 and 4 in Fig. 36–28 through lens 1 and find their intersection at point i_1. Ray tracing requires some accuracy, and therefore we have not drawn the figure in a rough style. If we extend the path of the rays, we see they are refracted by lens 2 and pass into the eye. Extensions of the rays that enter the eye give an image i_2 that is behind the lens and very far away. That is because we have placed the relative position of the two lenses so that the image I_1 formed by lens 1 is very close to the focal point F_2 of lens 2, so that the image I_2 formed by lens 2 is very distant. The image in this case will be very large. (Note that to keep the exercise simple, we have not actually traced the principle rays for lens 2 formed by I_1.)

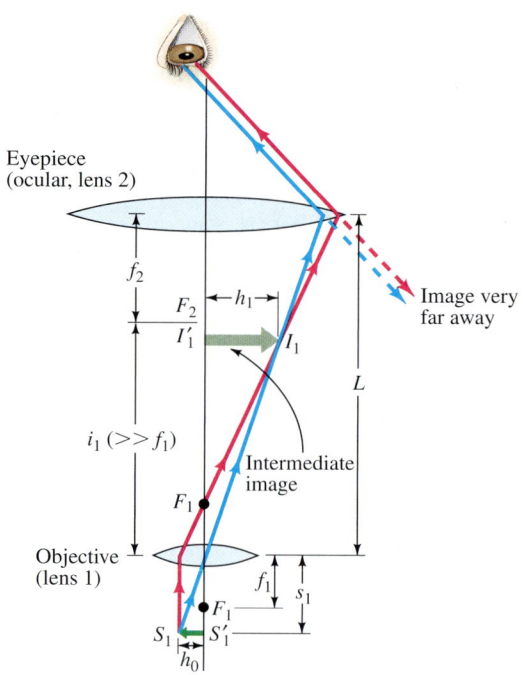

▲ **FIGURE 36–28** Ray tracing for the arrangement of the example.

This arrangement describes a **compound microscope**. The lens near the original source is called the *objective* lens, and the lens near the eye is called the *ocular* or *eyepiece*.

What Do You Think? Should lens 2 be moved up or down to obtain a taller final image?

36–5 Optical Instruments

The example of the compound microscope above shows how an arrangement of thin lenses can work together to produce some desired optical goal. As we implied, real camera lenses are in fact combinations, often rather complicated ones, of thin (or thick) simple lenses. There are many other examples of combinations that fulfill precise needs, and we look at some in this section. We start with the instrument that takes the light reflecting off the ink on this page and transforms it into a form that your brain can use to reconstruct the image.

The Eye

The typical vertebrate eye—the basic structure of which is shown in Fig. 36–29—is a remarkable optical instrument. Light enters the eye through the *pupil*, the size of which can be changed by contraction or expansion of a membrane called the *iris* according to the intensity of the incident light. The light then passes through a convergent *crystalline lens* into a chamber filled with the *vitreous humor*, a fluid with index of refraction near that of water. The light is focused onto the back of the eye, the *retina*, which is covered with sensitive receptor cells. The stimulation of these cells by light produces a message that is sent to the brain along the *optic nerve*, and the brain reconstructs the image.

When a normal eye is relaxed, objects at infinity form an image precisely on the retina, a distance of about 1.7 cm from the lens. When objects are brought closer, the lens is compressed by surrounding muscles and becomes more convergent. The focal length is reduced, and the image continues to be focused on the retina. There is a limit to how much the muscles can compress the lens—this limit is the lens's power of *accommodation*. Objects closer than the *near point*, about 25 cm from the lens (or less for younger people), appear blurred. The near point tends to increase with age because the lens becomes unable to compress as far as it once did, and the image of a near object

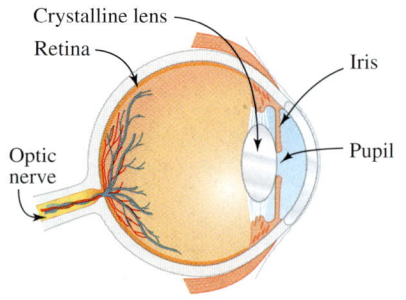

▲ **FIGURE 36–29** Schematic diagram of the human eye and some of its important features.

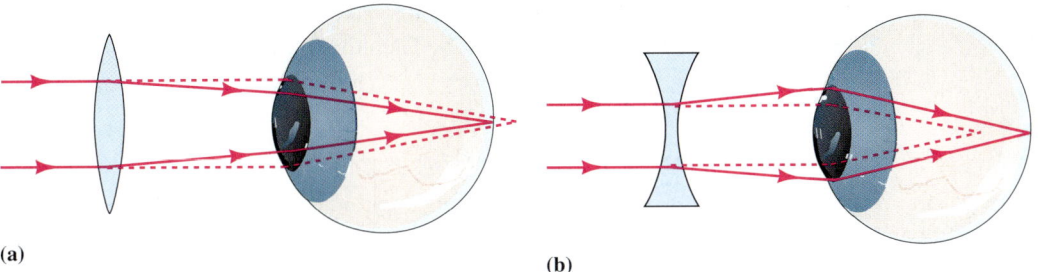

▲ **FIGURE 36–30** The dashed lines indicate the paths rays would take if no correcting lens were present. The solid lines mark the path of rays when a correcting lens is included. (a) A converging lens causes rays from an object, in this case at infinity, to focus closer to the lens of the eye. Such a lens corrects farsightedness by allowing the near point to be moved closer to the eye. (b) A diverging lens causes rays from an object, in this case at infinity, to focus farther from the lens of the eye. Such a lens corrects nearsightedness.

is beyond the location of the retina. Converging lenses correct this problem (Fig. 36–30a). In cases of nearsightedness, the image of an object at infinity is in front of the retina. A diverging lens will provide the necessary correction (Fig. 36–30b).

The Camera

With one important difference, the camera is optically equivalent to an eye. There is a converging lens in front, and the *film* (or receiving surface in a digital camera), which plays the role of the retina, is in back. There is an *aperture*, an opening equivalent to the pupil, and a *shutter*, which provides an approximation to an instantaneous image to avoid blurring of the picture due to motion. The difference between the simple camera and the eye is that the focal length of the lens changes in the eye, whereas the focal length is fixed in a simple camera. Instead, the camera lens moves in and out (changing the image distance) to enable objects of different source distances to produce a focused image on the film.

Angular Magnification

For optical instruments used for observing the world closely, *angular magnification* is a critical concept, and we shall discuss it before we cover some other instruments.

From Eq. (36–7), we see that the magnification of a lens or mirror is infinite when $s = f$. This is less important than it might appear to be because the image distance i also becomes infinite in that case. More important than the actual size of the image is the angle the image takes up in our field of vision. Given the limits of our own vision, *it is this angular coverage that determines how much detail we can see in an observed source.*

Imagine that you are a distance d from some object of height h (Fig. 36–31). For a source that does not cover an enormous part of your vision, the angular size θ_s of the source is

$$\theta_s \cong \frac{h}{d}. \tag{36–14}$$

For normal, unaided vision, this angular size can be maximized when the object is brought to the near point of vision, around $d = d_{\min} = 25$ cm, and it is $\theta_s \cong h/(25 \text{ cm})$ that is used as a reference for the angular magnification. Suppose now

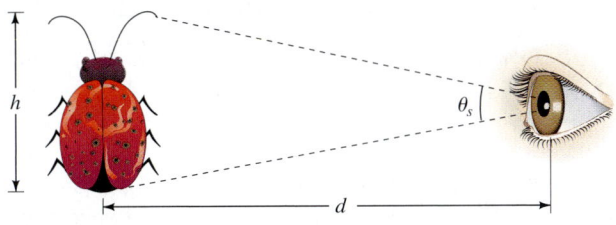

◀ **FIGURE 36–31** The angular size of an object, θ_s, is the relevant quantity for our ability to see detail in the object.

that we use an optical system to observe our source and that the image of the source as seen through the system has an angular size θ_i. Then the **angular magnification** of the system is

$$M_\theta \equiv \frac{\theta_i}{\theta_s}. \tag{36-15}$$

ANGULAR MAGNIFICATION

We do not bother with signs here and keep track only of the magnitudes of the angular sizes. If we know the angular magnification of two elements that are superposed in an optical system, then the net angular magnification is the product of the angular magnifications of each element.

The Simple Magnifier

A converging lens has a positive focal length. By Eq. (36-2), $\frac{1}{i} = \frac{1}{f} - \frac{1}{s}$. For a real object, i passes from positive (real image) to negative (virtual image) as the object moves toward the lens through the point $s = f$. At this point, i shifts to $-\infty$. A *simple magnifier* is a converging lens with the object placed near $s = f$ (Fig. 36-32). If the object size is h, the image size is, by definition, $h_i = Mh$, where $M = i/s$ is the magnitude of the magnification. The image size is infinite if i is infinite, but the *angular size of the image is finite*. When $s = f$, we have for the angular size

$$\theta_i = \frac{h_i}{i} = \frac{Mh}{i} = \frac{h}{s}\bigg|_{s=f} = \frac{h}{f}. \tag{36-16}$$

Note that we have no trouble seeing an image at infinity. At the near point, $d_{\min} = 25$ cm, the angular size of our object is $\theta_{\text{object}} = h/d_{\min}$. Thus the angular magnification of the magnifier is

$$M_\theta = \frac{\theta_i}{\theta_{\text{object}}} = \frac{h/f}{h/d_{\min}} = \frac{d_{\min}}{f}. \tag{36-17}$$

If we choose a converging lens with a focal length of 2 cm, we get an angular magnification of $(25\ \text{cm})/(2\ \text{cm}) = 12.5$.

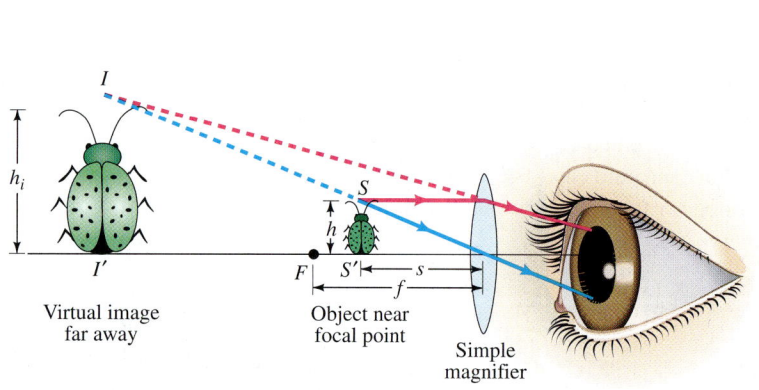

(a)

(b)

▲ **FIGURE 36-32** (a) The simple magnifier is a converging lens with an object placed near the focal point. The image is virtual and far away. (b) A simple magnifier in use.

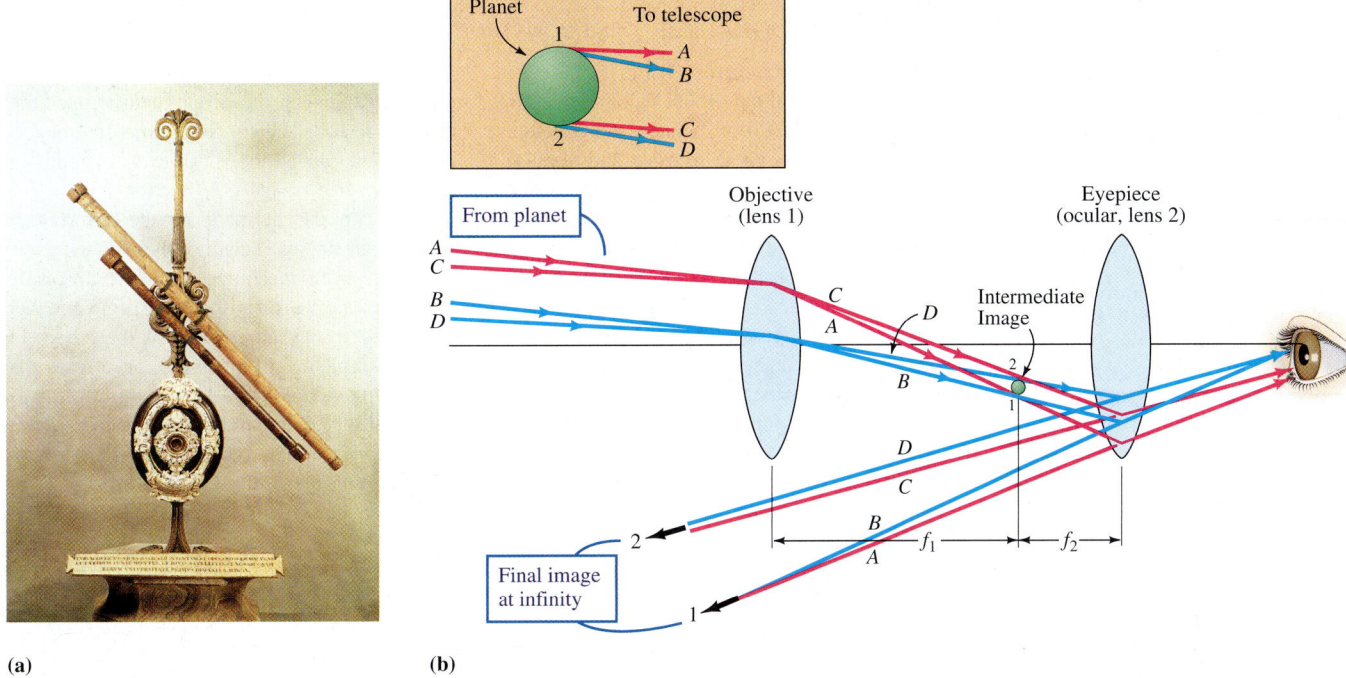

▲ FIGURE 36–33 (a) Galileo's refracting telescope, used for viewing distant objects. (b) Schematic diagram of a refracting telescope.

The Telescope

The **telescope** magnifies very distant objects. It was invented in Holland at the beginning of the seventeenth century and made an impact on astronomy soon thereafter. Galileo built his own telescope in 1609 (Fig. 36–33a).

The *refracting telescope* (a telescope with only refracting elements) is a system designed around the fact that the original object is very distant—in effect at infinity (Fig. 36–33b). The first lens, the *objective*, creates an intermediate image very close to the focal point of that lens. If that point coincides with the focal point of the eyepiece, then the eyepiece acts as a simple magnifier. The final image is magnified. Let's calculate the angular magnification for an object that has angular size θ_s. (The Moon, for example, has angular size of $1/2°$. With the naked eye, we can distinguish stars separated by about $1'$ of arc [$1/60$ of $1°$].) If the original object has size h_0, the objective produces an image of size $h_1 = Mh_0 = ih_0/s = i\theta_s = f_1\theta_s$. The final image then has an angular size given by Eq. (36–16) with $h \to h_1$ and $f \to f_2$, namely, $\theta_i = h_1/f_2 = \theta_s f_1/f_2$. In turn, the angular magnification is

$$M_\theta = \frac{\theta_i}{\theta_s} = \frac{f_1}{f_2}.$$

This quantity will be large for large f_1: The objective lens of a telescope should have as long a focal length f_1 as is practical, and that is why a refracting telescope is long.

The study of distant galaxies depends on an examination of the spectrum of the light they emit and of their energy output. The incident light from very distant objects is rather low in intensity, and more light is needed at the eyepiece in order to study spectra. To be most efficient at collecting light, the diameter of the optical system must be large. Large lenses are more difficult to construct than large mirrors, so most large telescopes are *reflecting telescopes* rather than refracting telescopes. In a reflecting telescope, a mirror replaces the objective for the purpose of creating an intermediate image, which is then magnified by the eyepiece (Fig. 36–34). Another advantage of such a telescope is that it has no chromatic aberration (see Section 36–6).

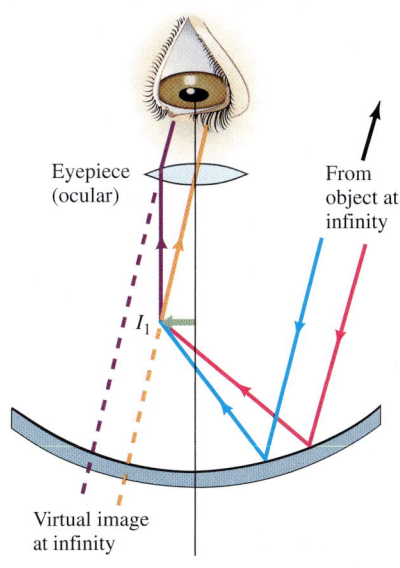

▲ FIGURE 36–34 Schematic diagram of a reflecting telescope.

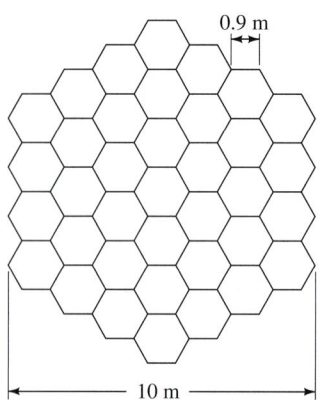

▲ FIGURE 36–35 A large collecting mirror is made by assembling hexagonal elements.

THINK ABOUT THIS...
HOW ARE MODERN TELESCOPES CONSTRUCTED?

Useful single large mirrors like that of the 5-m-diameter mirror in the Mount Palomar telescope are hard to produce, both because it is difficult to cast a single large piece of glass and because a large mirror will distort due to its own weight. Large, modern telescopes may be made from an array—a mosaic—of spherical mirrors, arranged on a frame that positions the mirrors across a paraboloidal surface. As any one mirror of the array is relatively small, its production is less problematic. In one type of design, the elements of the array are hexagons that fit together neatly (Fig. 36–35). Computer-controlled positioning devices are then used to mount the hexagons onto the main support frame. Each hexagonal mirror can be oriented by computer command; as the elevation of the telescope is changed or as thermal expansion due to ambient temperature changes distorts the support frame, the individual mirrors can be repositioned to maintain an aberration-free image (see Section 36–6). Corrections can also be made for the distortion of light by our fluctuating atmosphere by the use of *adaptive optics*, a technique by which the distortion is first measured with a laser beam and then accounted for by adjustment of the array elements. Recall from our discussion of rays that one can make holes in a lens or a mirror and it would still work to gather light and focus it at a detector. Indeed, in other versions of today's telescopes the elements of the array that make up the device may be widely separated.

▲ FIGURE 36–36 This poor image is the result of spherical aberration, a type of monochromatic aberration.

*36–6 Aberration

An accurate calculation would show that all rays that arrive at a spherical mirror or refracting surface from infinity cross in a small but finite region rather than at a single point. This is but one example of **aberration** (Fig. 36–36). Aberration should be distinguished from *distortion*, in which an image is not identical in form to the object, as in a fun-house mirror. For scientific purposes, the fun-house image is not necessarily a bad one, because every ray from the object is correctly positioned at a precise location in the image. Aberration concerns what we might call the quality of an image, not its geometric form.

We can distinguish two important types of aberrations in geometric optics. *Monochromatic aberrations* describe the fact that, in real optical systems, the rays from a given point on an object are not focused on a single image point (Fig. 36–37a). The correction for this type of aberration depends on the application. An optical system that collects images only from distant objects will have no aberration when a parabolic surface is used (Fig. 36–37b). Although such surfaces are difficult to construct from glass, a pool of mercury spinning about a vertical axis forms a parabolic surface, and such surfaces are employed in some modern telescopes. These telescopes only point up! Mosaic telescopes can also avoid monochromatic aberration. Alternatively, this type of aberration is minimized when the spherical section of the lens surface or mirror is small, although then the system collects less light.

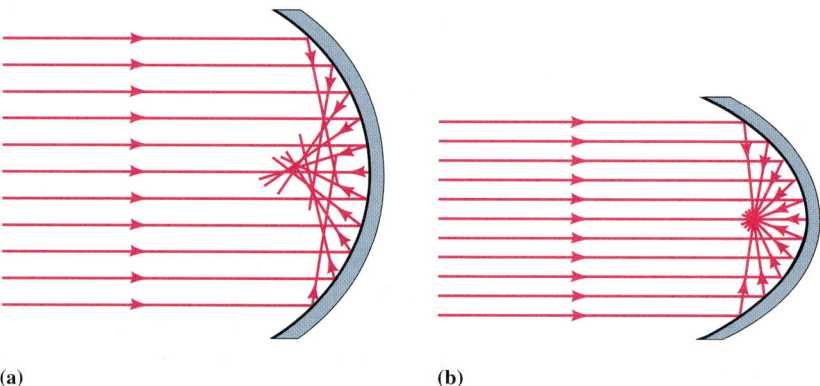

▲ FIGURE 36–37 (a) In monochromatic aberrations, all the rays from infinity do not pass through the same point for a spherical mirror, so the focus is not sharp. (b) This type of aberration is eliminated by use of a parabolic mirror.

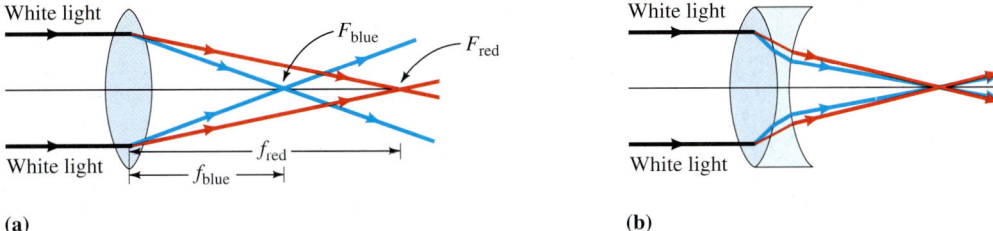

▲ **FIGURE 36–38** (a) In chromatic aberrations, the focal point of a converging lens may be different for different wavelengths. Here we show rays only for red light and blue light. (b) This type of aberration is eliminated by combining the lens with another lens with a different dispersion.

Chromatic aberrations appear in refracting systems but not in mirrors. We treated dispersion in Chapter 35, in which the index of refraction of a material depends on the wavelength of the light. The optical path of a ray at one wavelength will differ from that of a ray at another wavelength (Fig. 36–38a). If a given point on an object is the source of a mixture of wavelengths (as is true of white light), then the image of this point is spread out according to the wavelength. A simple correction for chromatic aberration is to use filters that allow only a narrow band of wavelengths to pass. More commonly, several lenses are superposed (Fig. 36–38b). Different elements are designed to have canceling dispersion to minimize the net dispersion.

A good camera lens may consist of a dozen elements of different types of glass, with complicated geometric relations, which, in the case of zoom lenses, are also variable. Unfortunately, even with all the corrections such lenses provide, the wave nature of light provides a fundamental and unavoidable limitation on the ability of optical systems to produce sharp images (see Chapters 37 and 38).

Summary

Geometric optics is based on two basic laws of the behavior of light rays: In reflection, the angle of incidence on a reflecting surface is equal to the angle of reflection. In refraction, Snell's law, $n_1 \sin \theta_1 = n_2 \sin \theta_2$, holds, where θ_1 and θ_2 are the angles of incidence and refraction, respectively, for a ray incident from medium 1 on medium 2. Ray tracing is a technique that allows us to locate the image of a given source.

A spherical reflecting or refracting surface forms an image of an object that is a distance s from the surface. Bundles of light rays pass through an image point (for a real image) or are directed as though they all come from such a point (for a virtual image). The image point is a distance i from the surface. One limit of such a surface is the plane mirror, for which the distance of the virtual image from the mirror is given by $i = -s$.

Parallel rays falling on a spherical reflecting or refracting surface approximately converge to the focal point, a distance f from the element. For a mirror, $f = R/2$ [Eq. (36–1)], where R is the radius of curvature of the spherical section. The distance of the object and the distance of the image from the surface, and the focal length for a spherical mirror are related by

$$\frac{1}{s} + \frac{1}{i} = \frac{1}{f}. \tag{36–2}$$

Equation (36–2) applies to both convex and concave mirrors if proper account is taken of the signs of s, i, and R. The image size is magnified by a factor M, the magnification, times the object size, where

$$M = -\frac{i}{s} = \frac{f}{f - s}. \tag{36–6, 36–7}$$

For a positive M, the image is upright; for a negative M, the image is inverted.

For a spherical boundary between a medium of refractive index n_1 and a medium of refractive index n_2, with light incident from medium 1, Eq. (36–2) is replaced by

$$\frac{n_1}{s} + \frac{n_2}{i} = \frac{n_2 - n_1}{R}. \tag{36–10}$$

The same formula applies to both convex and concave spherical surfaces if proper account is taken of the signs of s, i, and R.

Thin lenses are understood by thinking of the image due to refraction at the surface nearest the object as the object for refraction at the second lens surface. For thin lenses in air, Eq. (36–2) again applies, with

$$\frac{1}{f} = (n-1)\left(\frac{1}{R_1} - \frac{1}{R_2}\right). \tag{36-13}$$

Moreover, Eqs. (36–6) and (36–7) apply also to thin lenses.

Thin lenses can be used singly or in combination to make up optical instruments, including magnifiers, eyeglasses, cameras, microscopes, and telescopes. Angular magnification, which measures the ratio of the angular size of an object as seen through the instrument to the object's angular size as the naked eye sees it, is a fundamental consideration for those instruments whose explicit purpose is to magnify, as is the quality of the image they produce.

Understanding the Concepts

1. Why is "AMBULANCE" written on the front of an ambulance?
2. Consider a large room with walls covered with mirrors; at the center of the room is a candelabra with burning candles. Is the room brighter than a comparable room with black drapes in place of the mirrors?
3. The image of a distant candle is projected by a converging lens on a screen placed at the focal length of the lens. A piece of paper is taped over the lower half of the lens. Will only half of the image be seen?
4. Draw a right-handed coordinate system and its image in a plane mirror. Is the image a right-handed or a left-handed coordinate system? (In a right-handed system, the vector product $\hat{i} \times \hat{j}$ points along $\hat{k}$.)
5. A physicist stands in front of a mirror. The floor is the xy-plane at $z = 0$; the mirror is the xz-plane. He has learned that reflections involve the change $(x, y, z) \rightarrow (-x, -y, -z)$. Will he be surprised when he looks in the mirror? What is the rule that gives him the correct description of what he sees?
6. Would a dental mirror, the small mirror a dentist uses to examine your teeth, be concave, convex, plane, or sometimes one or another?
7. The sideview mirrors of some cars are labeled "Objects seen in this mirror may be closer than they appear." Is the mirror plane, convex, or concave?
8. For each of the simple lenses shown in Fig. 36–23, Eq. (36–6)—for magnification—shows that the size of the image of a ball placed at the focal point is infinite. Can you see by ray tracing why this must be?
9. Figure 36–39 shows the reflection made by a spherical surface. Parts of all four walls of the room are visible in the image. Why?
10. To form a taller image in Example 36–5, the object in Fig. 36–15 should be moved (a) to the left, (b) to the right, (c) can't be done.
11. Does the focal length of a lens change when the lens is in water?
12. When a magnifying glass is lined up perpendicular to the line between it and the Sun, a hot spot forms on the side of the lens away from the Sun. What is the relation between the distance of this hot spot from the lens and the focal length of the glass? Why does the spot become hot?
13. A camera works by forming a real image on a film plate. Can a camera take a picture of a virtual image?
14. In William Golding's novel *The Lord of the Flies* (1954), some boys rediscover fire with the aid of the Sun shining through the eyeglasses of Piggy, a nearsighted boy. Has Golding made a mistake?

▲ FIGURE 36–39 Question 9.

15. Are any principal rays useful for a point *on* the axis of an optical system?
16. You recently noticed that someone who used to wear thick glasses now wears glasses that are much thinner. Assuming that their eyesight has not improved, why might this be?
17. When you have an eye exam, even one as simple as reading an eye chart, the examiner may dilate (open) the pupil by putting drops in your eye. Why is that useful?
18. The n-dependence of the bending of light allows for the making of a flat eyeglass lens with material in which the index of refraction varies with position. Sketch the profile of the index of refraction for a converging lens constructed in this way.
19. Legend has it that Archimedes, acting as an advisor to the ruler of Syracuse, devised an optical system made of shields that could concentrate sunlight sufficiently well to set enemy boats on fire from a distance. How plausible is this legend?
20. Can the image ever be smaller than the object for a converging lens?

Problems

36–1 Images and Mirrors

1. (I) Consider two mirrors at right angles to each other (Fig. 36–40). How many virtual images will a pointlike light source have?

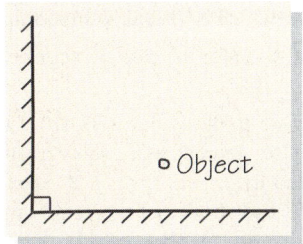

▲ FIGURE 36–40   Problem 1.

2. (I) Consider two parallel mirrors that face each other, placed along the x-axis at $x = a$ and $x = -a$. Assume that a point source of light is placed at $x = x_0$ between the mirrors. What are the locations of the four images of the point source with the smallest values of image distance i?

3. (I) Two mirrors, each 2.0 m wide, are placed facing each other and parallel to each other and are separated by 10 cm. A ray of light enters the gap between them, grazes the edge of one mirror and strikes the other mirror at an angle of 30° with respect to the normal to the mirrors. At each reflection, the intensity of the light beam is attenuated by 5 percent. By how much is the beam attenuated when it finally leaves the space between the mirrors?

4. (II) A mirror is exactly half your height, and the top of the mirror is aligned with the top of your head. (a) If your eyes were at the top of your head, how close would you need to be to the mirror in order to be able to see your feet? (b) If your height is 158 cm and your eyes are 12 cm below the top of your head, what would have to be done with the mirror so that you could see both the top of your head and your feet?

5. (III) A kaleidoscope contains three plane mirrors forming a prism with an equilateral triangular base of side a (Fig. 36–41). Consider a small object placed on the axis of the kaleidoscope. Construct the position of the images formed by single as well as double reflection. How far are these images from the axis?

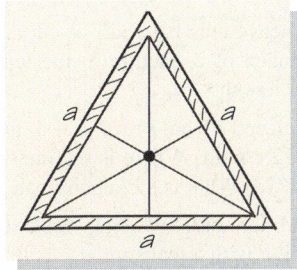

▲ FIGURE 36–41   Problem 5.

6. (III) Suppose that two plane mirrors meet at an angle of 60° (Fig. 36–42). An object is placed between the mirrors on the line that bisects this angle. Use graphical methods or trigonometric methods to locate all the images.

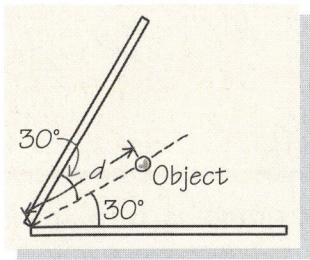

▲ FIGURE 36–42   Problem 6.

36–2 Spherical Mirrors

7. (I) A dime 60 cm away from and on the optical axis of a concave spherical mirror produces an image 20 cm away from the mirror. If the dime is moved on the axis to 35 cm from the mirror, where will the image move? How large is the radius of the sphere of which the mirror is a section? Draw the system for the second case described.

8. (I) A paper clip is placed on the axis 28 cm away from a convex mirror, part of a sphere of radius 44 cm. Where will the image be located, and what is the magnification? Make a sketch, including rays.

9. (I) An object of height 2.0 cm is placed 20 cm from a concave mirror. The real image is found to be 8.0 cm from the mirror. On which side of the mirror is the image and how tall is it? Is it inverted?

10. (I) A concave mirror has a radius of curvature of 176 cm. What is the size of the image of an object 6.00 cm tall that is placed 133 cm from the mirror?

11. (I) A concave mirror is cut from a spherical surface of radius of curvature 2.0 m. A pencil 10 cm long is placed perpendicular to the axis of the mirror at a distance of 80 cm from the mirror. Where is the image and how large is it?

12. (II) Consider an object in front of a convex mirror. A plane mirror is placed along the axis between the object and the convex mirror, and moved along till the images of the object in the two mirrors coincide. If the distance of the object to the plane mirror is x and the distance of the plane mirror to the convex mirror is y, what is the focal length of the convex mirror?

13. (II) A bird flies from far away toward a concave mirror, with constant speed v. What is the velocity of the image as a function of the distance of the bird from the mirror? When does the bird meet its image? [*Hint*: Use the relation between source distance and image distance and differentiate with respect to time.]

14. (II) By using the same reasoning that we used in the text for the case of the concave mirror, show that the reflection of ray 1 in Fig. 36–11 appears to originate at point F, independent of the angle θ. Your argument shows that, at point F, there is an image of a source point at infinity.

15. (II) Use ray tracing for parallel rays far from, as well as near to, the optical axis to show that parabolic mirrors more accurately focus parallel rays than do spherical mirrors.

36–3 Refraction at Spherical Surfaces

16. (I) A sphere of glass ($n = 1.50$) of radius 3.0 cm is immersed in water ($n = 1.33$). A small flower (at point B) is 1.5 cm outside the sphere (Fig. 36–43, see next page). What are the location and nature (real or virtual) of the flower's image made by refraction at the first surface?

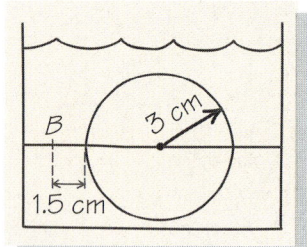

▲ **FIGURE 36–43** Problem 16.

17. (I) An object is placed 15 cm in air from the convex surface (radius of curvature 10 cm) of a very thick piece of glass ($n = 1.5$). Where is the image?

18. (I) The single refracting surface of a piece of glass in air has a radius of curvature $R = 8.5$ cm. A ray parallel to the axis of the curved piece of glass is bent toward the axis inside the piece of glass and crosses that axis at a point 13 cm into the glass. What is the index of refraction of the glass?

19. (I) A fish is located at a distance of 40 cm from the glass pane of an aquarium. How far from the glass does the fish appear to be located to an observer looking from the outside? (Use $n_{water} = 1.33$.)

20. (I) A small fish is cast into the center of a glass sphere of radius $R = 5$ cm and $n = 1.5$. Where will an observer see the fish? Where will the observer see a decorative background pattern painted on the back side of the sphere?

21. (II) A glass rod of refractive index $n = 1.6$ and diameter 1.6 cm has a hemispherical cap (Fig. 36–44). There is a fault in the glass 2.3 cm from the end. Can you see this fault if you look at the rod through the spherical cap? From about how far away should you look?

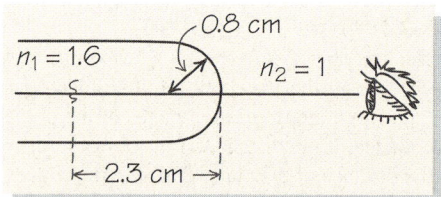

▲ **FIGURE 36–44** Problem 21.

22. (II) By applying Eq. (36–10), show that, if light is incident on a convex refracting surface with $n_2 > n_1$ (see Fig. 36–18), there is a critical distance s_c such that the image of an object closer than s_c will be virtual. Find s_c, and show by ray tracing that the virtual image when $s < s_c$ is upright and magnified.

23. (II) Consider the situation described in Problem 22. Use ray tracing to find the image when $s = s_c$, and when $s = 3s_c$.

24. (II) Consider a convex spherical boundary between two media with an upright object whose extreme point is at S, as in Fig. 36–18. Suppose that $n_2 < n_1$ rather than $n_2 > n_1$. Find the nature of the image (inverted or upright, virtual or real, reduced or magnified) by tracing rays from S. Is there a critical distance at which the nature of the image changes, as in Problem 22?

25. (II) A convex spherical boundary produces an image whose distance from the boundary surface is governed by Eq. (36–10). Suppose that $n_2 > n_1$. (a) Show that when an object is very far from the surface, the image is a distance $i = n_2 R/(n_2 - n_1)$ from the surface, and that the image is inverted, reduced, and real. (b) What is the distance s at which the image distance becomes infinite? (c) What is the position of the image for s just less than the critical value found in part (b)? Is it real? (d) As s continues to decrease, what happens to the position of the image?

26. (II) Consider a concave surface of radius of curvature R that separates two media with indices of refraction n_1 and n_2, where $n_2 > n_1$ (see Fig. 36–19). Find the distance s of an object for which the image, which is virtual, is superimposed on the object.

27. (II) Derive Eq. (36–9) for the case of a concave surface shown in Fig. 36–17.

28. (II) Use Fig. 36–45 to derive Eq. (36–10). Let $n_1 < n_2$ and use small angles. Use geometry to show that $\theta_2 = \beta - \alpha$ and $\theta_1 = \beta + \gamma$, and $n_1(\beta + \gamma) = n_2(\beta - \alpha)$. Use the exact relation $AB = R\beta$ and show that the small-angle approximations give $AB = s\gamma = i\alpha$ to finally obtain Eq. (36–10).

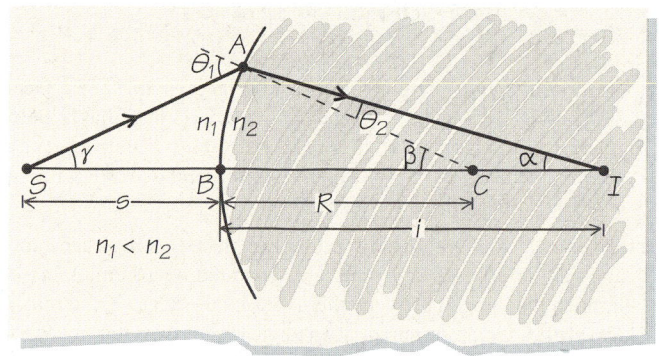

▲ **FIGURE 36–45** Problem 28.

29. (III) In deriving Eq. (36–10) in the previous problem, we took $n_1 < n_2$ in Fig. 36–45. Make a new drawing appropriate to $n_2 < n_1$ for a convex surface. Apply the same kind of reasoning, using small angles to show that the same algebraic formula applies whatever the relative sizes of n_2 and n_1.

30. (III) Show that Eq. (36–10) holds for a concave refracting surface with $n_2 > n_1$ by drawing a figure analogous to Fig. 36–45 (Problem 28) and by making small-angle assumptions.

36–4 Thin Lenses

31. (I) The image of an object placed 24 cm away from a thin lens forms at a distance of 51 cm on the other side of the lens. (a) What is the focal length? (b) What type of lens is it? (c) Is the image real? upright? (d) What is the magnification?

32. (I) A double concave lens has radii of curvature of 6.0 cm and 7.5 cm. If the index of refraction of the lens material is 1.56, what is the focal length?

33. (I) An apple is placed 15 cm in front of a diverging lens with a focal length of 22 cm. (a) Where is the image? (b) Is the image real? (c) upright? (d) What is the magnification?

34. (I) Find the condition under which a single thin lens produces a real image, starting with a real source.

35. (II) We want to form an image of an insect magnified twofold by using a converging lens with a focal length of 25 cm. (a) Where should the object be placed for the image to be real? (b) Repeat part (a) for a virtual image.

36. (II) An object 4.5 cm high is placed on one side of a thin converging lens of focal length 43 cm. What are the location, size, and orientation of the image when the object is (a) 86 cm from

the lens, (b) 50 cm from the lens, (c) 40 cm from the lens, (d) 15 cm from the lens?

37. (II) The two surfaces of a thin lens have radii of the same sign and magnitude. Show by ray tracing that the focal length of this lens is infinite. Is the image produced by this lens real or virtual?

38. (II) A thin converging lens forms an image of a distant mountain at a distance of 38 cm from the lens. (a) What is the focal length of the lens? (b) A pine cone is placed 75 cm from the lens. Describe the resulting image: its magnification and distance from the lens, and whether it is real or virtual, upright or inverted. (c) The lens glass has an index of refraction of 1.55. The lens is immersed in water, index of refraction 1.33. What is its focal length in water?

39. (II) Consider the thin lenses shown in Figs. 36–23a through 36–23d. Suppose that in each case the magnitudes of the radii of curvature are $R_1 = 25$ cm and $R_2 = 60$ cm, and that $n = 1.55$. (a) Find the focal lengths for each of the four lenses, and use the sign of the focal lengths to obtain the locations of the image of a source 10 m from each lens. (b) In each case, is the image upright or inverted, real or virtual? (c) Calculate the magnification M from Eq. (36–6) and check that it is consistent with your results in part (b).

40. (II) An object is placed 25 cm to the right of each of the lenses of Problem 39. For each case, locate the image, state whether it is upright or inverted and real or virtual, and give the magnification.

41. (II) Repeat Problem 40 for an object placed 65 cm to the right of each of the lenses.

42. (II) Consider an object on the left side of a thin lens with $R_1 > 0$ and $R_2 < 0$. The image will be on the right of the lens. Let X_0 be the distance of the object from the focal point on the left and X_1 be the distance of the image from the focal point on the right. Show that $X_0 X_1 = f^2$. This was Newton's original formulation of the lens equation, given in his *Opticks* in 1704.

43. (II) Show that the thin lens equation [Eq. (36–12)] follows from Fermat's Principle, according to which all light rays leaving an object and traveling to the image do so in minimum times. [*Hint*: You may need the following geometric result. Consider a line that is drawn perpendicular to a particular radius at a distance $R - x$ from the center. Then the distance from the radius to the point where the line touches the circle, y, is given by the relation $y^2 = x(2R - x)$ which for small x reduces to $y^2 = 2Rx$.]

44. (II) Two thin lenses of focal length f_1 and f_2, respectively, are aligned along the same axis and placed very close together. Show that the focal length f of the combination is given by

$$\frac{1}{f} = \frac{1}{f_1} + \frac{1}{f_2}.$$

36–5 Optical Instruments

45. (I) The eyes of an elderly person have near points of 70 cm. What must the focal length of corrective lenses be in order for this person to read a book at a distance of 30 cm?

46. (I) A nearsighted person has near and far points of 12 cm and 41 cm, respectively. (The *far point* is the farthest point at which a person can see clearly.) (a) Determine the lens required for this person to be able to see clearly at infinity. (b) What does the lens correction of part (a) do to the near point? Can the person still easily read a book?

47. (I) You have a thin lens with $f = 9$ cm. If you want to see an insect magnified by a factor of 3, how close should you hold the glass to the insect? (Let the image be virtual.)

48. (I) What is the magnification of a telescope that has an objective lens with a focal length of 80 cm and an eyepiece with a focal length of 1.7 cm?

49. (I) You are trying to construct a compound microscope given two lenses with focal lengths $f_1 = 1.0$ cm and $f_2 = 4.0$ cm. How far apart should you place the lenses in order to obtain an angular magnification of 60?

50. (II) The two lenses of a telescope with magnification of 120× are separated by 70 cm. What are the focal lengths of the lenses?

51. (II) Calculate the angular magnification of the reflecting telescope shown in Fig. 36–33b.

52. (II) Galileo's original telescope had a convex objective and a concave eyepiece. The focal points of the two lenses coincided, as shown in Fig. 36–46. What is the angular magnification for a distant (but not infinitely far) object, and what is it for an infinitely far object?

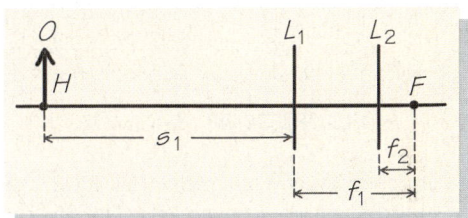

▲ FIGURE 36–46 Problem 52.

*36–6 Aberration

53. (II) Consider a spherical mirror without making the paraxial approximation (Fig. 36–47). Show that when a ray parallel to the axis makes an angle θ with the radius R at the point of contact, then f, here the distance at which the ray crosses the axis, is given by

$$f = R\left(1 - \frac{1}{2\cos\theta}\right).$$

Show that for small angles, this formula reduces to $f = R/2$. Note that F is not the focal point here (there is no sharp focus), but only the point at which some particular ray crosses the axis.

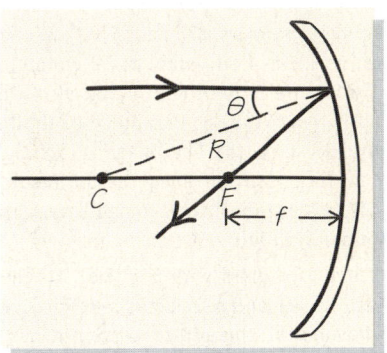

▲ FIGURE 36–47 Problem 53.

54. (II) Use the result of Problem 53 to calculate the spread in values of f for a spherical mirror of radius 0.18 m and arc length 46 cm.

55. (II) The index of refraction of optical glass used for a thin lens with $R_1 = +20.00$ cm and $R_2 = +28.75$ cm is $n = 1.48523$ for light of wavelength $\lambda = 587.6$ nm and $n = 1.48135$ for light of wavelength $\lambda = 768.2$ nm. What is the difference in the focal length for these two wavelengths?

Mirrors and Lenses and Their Uses

General Problems

56. (II) Consider a circular concave mirror of focal length f and diameter d, where $f \gg d$. This mirror's optical axis is aligned with the Sun. What is the area of the spot that contains the reflected rays as a function of the distance L from the mirror if $L < f$ (Fig. 36–48)? Sunlight has an intensity I as it arrives at the mirror. Find the intensity of the reflected rays as a function of L. Treat the Sun as a point source.

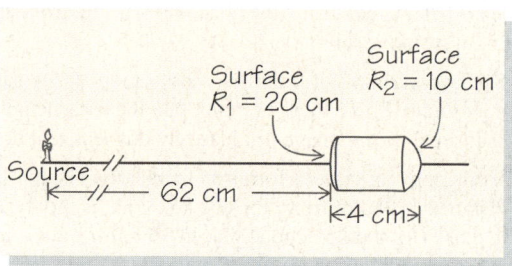

▲ **FIGURE 36–49** Problem 60.

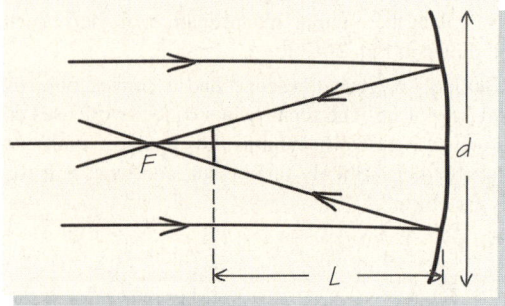

▲ **FIGURE 36–48** Problem 56.

57. (II) You are given a converging lens (Fig. 36–23a) with equal radii of curvature and a diverging lens (Fig. 36–23c) with the same radii of curvature as those of the converging lens. The lenses are made of material with $n = 1.50$, and the radii of curvature are all 35 cm. They are placed at opposite ends of a tube 15 cm long, and the nearer lens is 10 cm from an object. What is the location of the image that results from the two refractions? Does it make a difference whether the converging or the diverging lens is closer to the object?

58. (II) Consider a 45-cm-long cylinder of glass in air, with $n = 1.6$, like the cylinder shown in Fig. 36–21. The two ends are shaped into sections of spheres; each has radius 18 cm. A small object is placed perpendicular to the optical axis at a distance of 12 cm from one of the spherical surfaces. (a) Find the location of the object's image due to refraction at surface 1. (b) Let this image be the object for surface 2, and find the location of *its* image as light passes through surface 2. (c) Use ray tracing to determine if the final image is upright or inverted.

59. (II) Two concave mirrors M_1 and M_2 face each other. They have respective radii of curvature of 32 cm and 14 cm and are separated by 50 cm. A lightbulb is placed on the optical axis 7 cm from M_1. (a) Where is the image of the bulb formed by M_1? Draw the system. (b) The image of the lightbulb formed by M_1 can in turn form an image as the result of reflections from M_2. Construct this second image by ray tracing, starting from the source lightbulb.

60. (II) A lens, made of glass with $n = 1.5$, has the configuration shown in Fig. 36–49, and a candle is placed 62 cm from surface 1. The lens cannot be thought of as thin; it has a thickness of 4 cm. (a) Where is the image made by surface 1? Is it inverted? What is the magnification? (b) By using the image made by surface 1 as an object for surface 2, find the final image's location relative to the candle as well as the magnification of the image. Is it inverted or upright?

61. (II) A thin lens with focal length f_1 is placed a distance d in front of a concave mirror with focal length f_2. What is the focal length of the combination?

62. (II) Consider the sphere of glass with $n = 2$ in Fig. 36–50. Any incoming ray is parallel to an axis through the middle of the sphere and will be refracted, striking the rear surface of the sphere at the axis. Demonstrate that this holds true for paraxial rays. If the back surface is painted with a reflecting material, symmetry shows that the ray will come back out in the opposite direction. Tiny spheres of this type are used for highway reflectors.

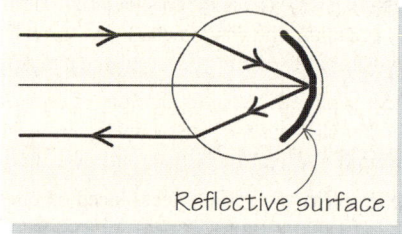

▲ **FIGURE 36–50** Problem 62.

63. (III) Rays of light strike a spherical glass surface parallel to the optical axis (Fig. 36–51). The incoming ray makes an angle θ with the normal to the surface. Show that the rays will cross the optical axis at a distance $d = R/\left(\sqrt{n^2 - \sin^2\theta} - \cos\theta\right)$ beyond the center of the sphere, where n is the index of refraction of the glass. To what does this expression reduce for small angles?

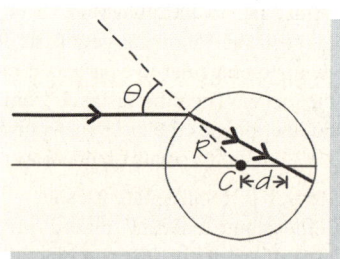

▲ **FIGURE 36–51** Problem 63.

64. (III) The index of refraction of a particular type of glass varies from 1.615 (for blue light) to 1.596 (for red light). Use the result of Problem 63 to calculate the color spread on the axis for light that strikes a hemispherical cap at the end of a glass rod, at an angle $\theta = 0.8$ rad. Take the radius of curvature of the sphere to be $R = 1.50$ cm. What is the spread for paraxial rays?

65. (III) An optical system contains a thin lens, $n = 1.4$, with positive curvature of radius $R_1 = 25$ cm for surface 1 and negative curvature of radius $R_2 = -25$ cm for surface 2. This lens collects light from the right side. Where to the left of the lens should a flat plate of thickness t of the same glass be placed, and how thick should it be, if you want the light that radiates from a distant object to be focused on a screen 35 cm to the left of the lens?

◀ The brilliant colors of the peacock's feathers are due not to pigmentation, but to interference of the light reflected from them.

Interference

In Chapters 35 and 36, we emphasized the geometrical properties of light. We discussed reflection and refraction by treating light in terms of rays, but did not address the fact that light is a wave phenomenon. However, if we look more carefully at the behavior of light when obstacles or holes have dimensions comparable to the wavelength of the light, geometric (or ray) optics is inadequate, and the wave nature of light becomes important. Geometric optics cannot explain the colors observed in oil slicks or soap bubbles, and if we look closely at shadows, we find that they are not completely sharp, in contradiction to the predictions of geometric optics. These phenomena are due to interference and diffraction, the subjects of this chapter and Chapter 38, subjects already discussed in terms of mechanical waves in Chapter 15. *Physical optics*, which takes into account the wave nature of light, explains a wider range of observations than does geometric optics.

37–1 Young's Double-Slit Experiment

When two or more harmonic waves superpose, they interfere—whether they are water waves, waves on a string, sound waves, or light waves. We saw how this worked for waves in mechanical systems in Chapter 15. Here we shall see how interference between two light waves occurs because *the electric (or magnetic) fields of the two*

1030 | Interference

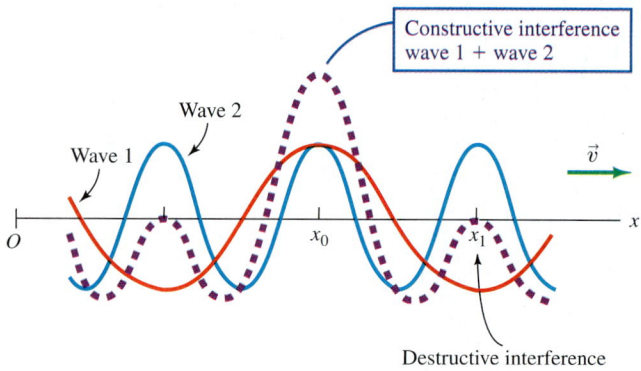
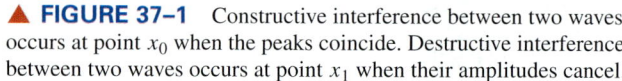

▲ **FIGURE 37–1** Constructive interference between two waves occurs at point x_0 when the peaks coincide. Destructive interference between two waves occurs at point x_1 when their amplitudes cancel.

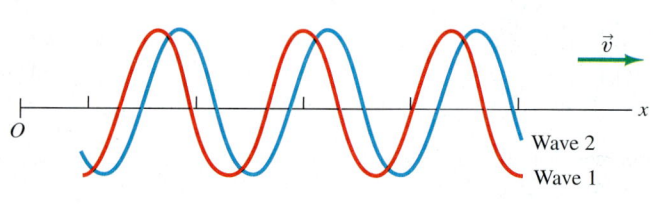

▲ **FIGURE 37–2** Two coherent waves have a constant phase difference.

waves add vectorially. Consider the superposition of two light waves from different sources at particular points in space that are propagating along the *x*-axis at a given time (Fig. 37–1). Where the two waves add to produce a wave with a larger amplitude (we refer here to, say, the electric field amplitude), we say that the waves interfere constructively, as, for example, at point x_0 in Fig. 37–1. The two waves interfere destructively where they cancel each other, as they do, for example, at point x_1. Any degree of interference between maximally constructive or maximally destructive is possible.

Light waves from two sources can produce an interference pattern in space when there is a definite relation between the respective wavelengths and phases *at their respective sources.* In other words, the waves must be *coherent.* The two light waves shown in Fig. 37–2 each have the same wavelength and a constant phase difference. We can also call monochromatic waves from a single source coherent when those waves form a long train of purely harmonic form. The "coherent" single source becomes two coherent sources that can interfere when the light is split; the interference occurs when the optics are arranged so that the two parts recombine later. A laser emits very long wave trains like these, and it is easy to demonstrate the interference pattern produced by laser light in the classroom (see Chapter 38). If the waves emitted at one or both sources consist of a mixture of waves of different wavelengths and phases, then there is no interference pattern. For example, an incandescent lightbulb produces light from many independent atomic sources at different times and places within the filament—this light is *incoherent.*

Thomas Young observed interference phenomena between two sources of light at the very beginning of the 19th century—he was the first to see such effects in light. One can produce coherent light at two sources by illuminating a single aperture S (a slit or a hole) with a source of monochromatic light. If the source of this light is a lightbulb, the light comes in series of bursts from individual atomic sources, a quantum phenomenon. (The monochromaticity can be taken care of with a prism.) These bursts last for a time on the order of $\tau \cong 10^{-8}$ s, and the length of the resulting individual wave trains is therefore $c\tau$, or several meters. The aperture S must be so small that only one wave train enters at a time. The single wave train of light that passes through illuminates two other apertures, S_1 and S_2 (Fig. 37–3). We can suppose that the size of the two apertures is the same, so that the amplitudes of the waves that come from them are identical. *These two apertures are two sources of coherent light.* If S_1 and S_2 are equidistant from S, the light from S travels the same distance to reach S_1 and S_2, and the light is in phase as it passes through the two apertures (Fig. 37–3). If S_1 and S_2 are not equidistant from S, the light waves that pass through them are still coherent because they have a definite, time-independent phase difference.

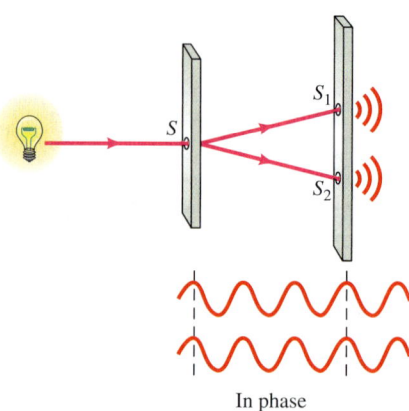

▲ **FIGURE 37–3** The light waves that pass through slits S_1 and S_2 are coherent. These waves are in phase if the light travels the same distance from S to S_1 as the distance from S to S_2. Even when the path lengths are different, there is a definite constant phase difference.

CONCEPTUAL EXAMPLE 37–1 Will there be interference between the (monochromatic) light passing through the slits of S_1 and S_2 if the light exiting the two slits is exactly 180° out of phase?

Answer We rearrange the slits of Fig. 37–3 as shown in Fig. 37–4 to make the light exiting the two slits to be π rad out of phase. The two sources of light have a strict coherence. At any point equidistant from the two slits (point A in Fig. 37–4, for example), the sources of light will remain out of phase and will interfere destructively. On the other hand, the two waves will be in phase at a point such as B placed so that its distance from the two slits differs by $\lambda(n + 1/2)$, where $n = 0, 1, 2, \ldots$ is an integer. At these points there will be constructive interference (Fig. 37–4). Similar remarks can be made if the phase difference for the light passing through the two slits has *any* value, as long as it is a definite value and not something that is constantly changing.

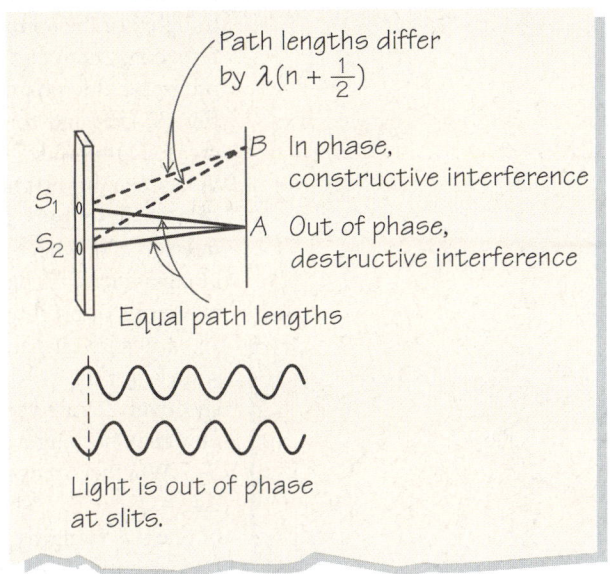

▶ **FIGURE 37–4** The light sources at S_1 and S_2 are out of phase. Whether they are out of phase or in phase at points A and B depends on the difference in path lengths to these points.

The Two-Source Interference Pattern

Let's review the spatial interference pattern produced when light from two sources of coherent waves interfere, a subject we developed in Chapter 15. We suppose that we have two sources, S_1 and S_2 (vertical slits), that emit monochromatic light in phase (Fig. 37–5a). Figure 37–5b is a view from above. Waves of the same frequency and phase emanate from S_1 and S_2 in the form of spreading cylinders from the sources, and these

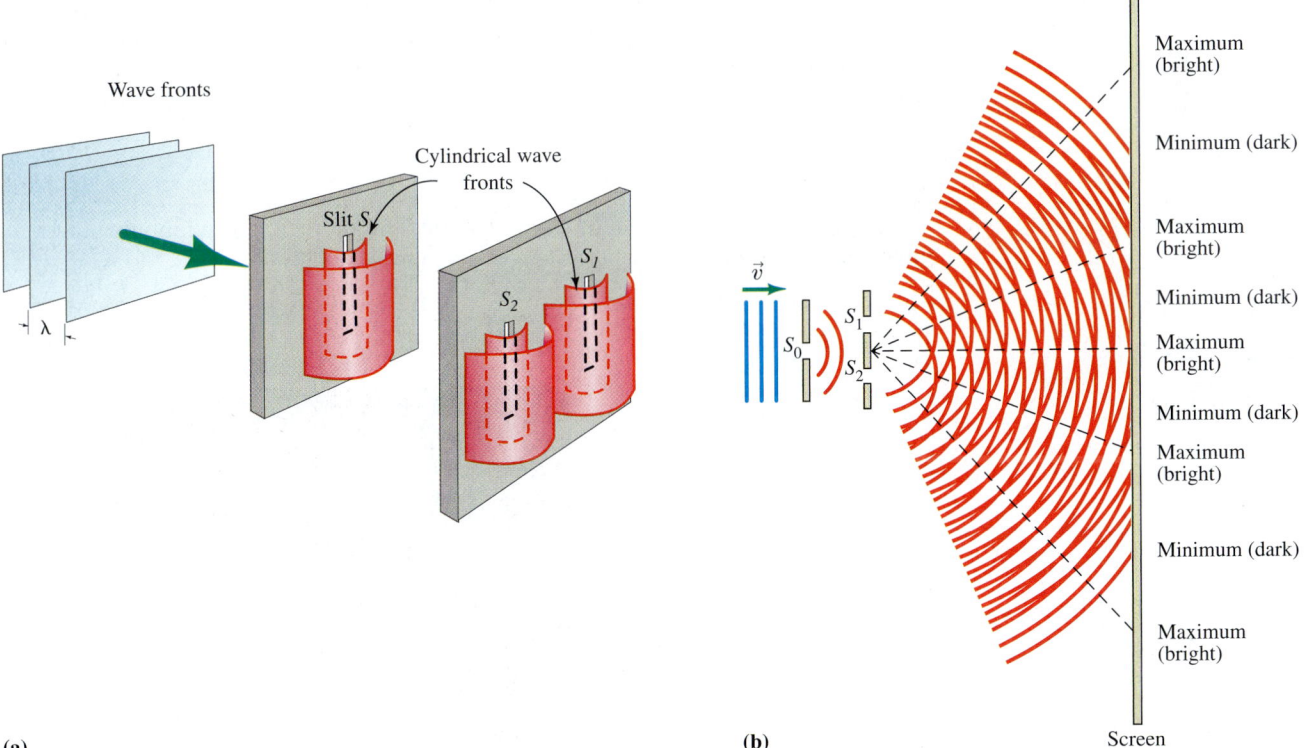

▲ **FIGURE 37–5** (a) Light coming from a single source S is split when it passes through the two slits S_1 and S_2, which act as two coherent sources, producing coherent cylindrical waves. (b) The view from above. Constructive interference occurs everywhere along the directions where the concentric circles, representing the crests of the spreading waves, overlap, because the waves are in phase along these directions. Alternating bright and dark places will be observed on a distant screen.

are seen in cross-section as circles in the figure. The circles represent the crests (or troughs) of the spreading waves. Where the crests (or troughs) overlap, the waves interfere constructively. The pattern of these overlap points is apparent: As the waves progress, the positions of these points advance (indicated by the dashed lines in Fig. 37–5b) and form lines. There is constructive interference—wave motion with increased amplitude—*all along* these lines; therefore, the places where the lines intersect the screen are bright. There is partial constructive interference in the regions on either side of these lines, and destructive interference in between the regions of constructive interference, where the screen is dark. (Again, the interference is maximally destructive along a line, with partial destructive interference in a region around that line.) The result is a series of bright and dark areas on the screen.

Let's investigate this double-slit configuration more closely. Consider the geometry shown in Fig. 37–6. Along ray 1 and ray 2, the waves travel distances L_1 and L_2, respectively, to arrive at point P on the screen. Because the rays travel different distances, they may no longer be in phase at P, although they were in phase at the sources S_1 and S_2. Whether they are in phase or not depends on the *path-length difference* $\Delta L = L_2 - L_1$. The waves arrive in phase if ΔL is zero or if ΔL is an integral multiple of one wavelength ($\Delta L = \lambda n$), but will be 180° out of phase if this difference is a half-integer multiple of one wavelength $[\Delta L = \lambda(n + 1/2)]$. The interference is constructive where the waves are in phase and destructive when the waves are 180° out of phase:

for constructive interference: $\Delta L = n\lambda, \quad n = 0, \pm 1, \pm 2, \ldots;$ (37–1a)

for destructive interference: $\Delta L = \left(n + \dfrac{1}{2}\right)\lambda, \quad n = 0, \pm 1, \pm 2, \ldots.$ (37–1b)

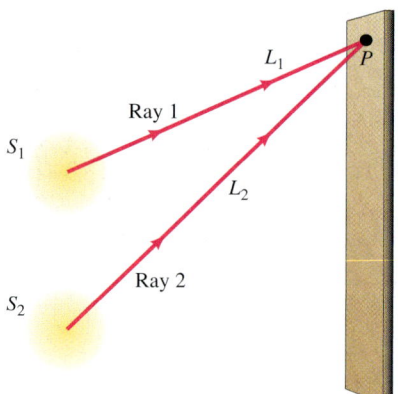

▲ **FIGURE 37–6** Light waves, indicated by rays 1 and 2, may not be in phase at point P despite being in phase at their sources, S_1 and S_2.

The resulting series of bright and dark lines on the screen is indicated in Fig. 37–5 and Fig. 37–7.

The geometry shown in Fig. 37–8 determines conditions for constructive and destructive interference. We assume that the distance R to the screen is much greater than the distance d between the two slits. A very distant screen means that the rays from the slits to a given point on the screen become parallel, so we obtain a good approximation to the path difference by using only the single angle θ made by the line from the center point between the slits to P. From Fig. 37–8, we see that the angle formed by $S_2 S_1 K$ is also θ. Thus

$$\Delta L = d \sin \theta. \qquad (37\text{–}2)$$

According to Eqs. (37–1a) and (37–1b), maxima (bright regions) and minima (dark regions) thus occur on the screen for angles given by

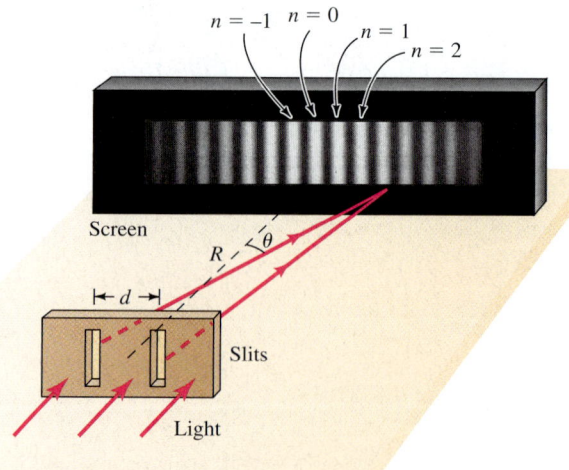

▲ **FIGURE 37–7** The interference pattern produced by double vertical slits is a series of alternating bright and dark vertical lines on a screen. The falloff of the intensity toward the edge of the figure is a single-slit effect (see Chapter 38).

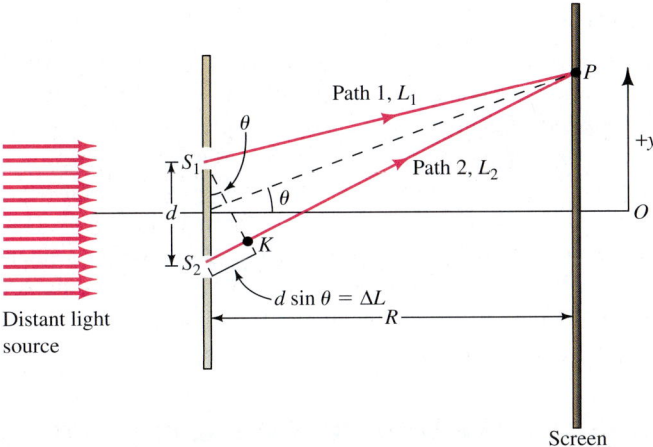

▲ FIGURE 37–8 The geometry used to find the interference-pattern conditions for the light that reaches point P. The path-length difference $\Delta L = d \sin \theta$.

for constructive interference: $\sin \theta = n \dfrac{\lambda}{d}, \quad n = 0, \pm 1, \pm 2, \ldots;$ (37–3a)

for destructive interference: $\sin \theta = \left(n + \dfrac{1}{2} \right) \dfrac{\lambda}{d}, \quad n = 0, \pm 1, \pm 2, \ldots.$ (37–3b)

INTERFERENCE CONDITIONS FOR TWO SOURCES

The point that is aligned with the sources ($\theta = 0$) is a maximum ($n = 0$). Alternating minima and maxima lie on either side of this centerline. The value of n that labels the maxima is known as the *order*. The central maximum is the zeroth order, and the maxima on either side of the central maximum are first-order maxima ($n = \pm 1$). If θ is small, so that $\sin \theta \cong \theta$, the maxima and minima are equally spaced in θ. This experiment shows conclusively that light behaves as a wave. Geometric optics cannot explain the result shown in Fig. 37–7.

EXAMPLE 37–2 In a double-slit experiment, y is the distance along the screen from the center maximum. Find the positions of the maxima as a function of y. If the source-to-screen distance $R = 3.0$ m, the source separation $d = 0.20$ mm, and light comes from a helium–neon laser ($\lambda = 633$ nm) far away, determine y for the ninth-order maximum.

Setting It Up Fig. 37–8 illustrates the experimental schematic and shows the path difference $\Delta L = L_2 - L_1$.

Strategy Geometry tells us that $\Delta L = d \sin \theta$, and then the values of θ for which maxima occur is given by Eq. (37–3a). After determination of the angles, we use $y = R \tan \theta$ to find the desired y-values. If $R \gg y$, $\tan \theta \cong \sin \theta$, and we can directly insert the value of $\sin \theta$ from Eq. (37–3a) for the maxima.

Working It Out The maxima are located at

$$y = R \tan \theta \cong R \sin \theta = R \dfrac{n\lambda}{d}. \quad (37\text{–}4)$$

For the ninth order, we have $n = 9$, and $y = 9\lambda R/d$. If we substitute the given values for λ, R, and d, we get for the ninth-order maximum

$$y = \dfrac{9(633 \times 10^{-9}\text{ m})(3.0\text{ m})}{0.20 \times 10^{-3}\text{ m}} = 8.5 \text{ cm}.$$

The distance between each of the maxima is therefore about 1 cm.

What Do You Think? Could such measurements be used to determine the wavelength λ? *Answers to **What Do You Think?** questions are given in the back of the book.*

CONCEPTUAL EXAMPLE 37–3 Show that the observed pattern of lines on the screen in the double-slit experiment spreads out as the wavelength increases and/or the separation between the sources decreases.

Answer Equation (37–4) shows that the distance between adjacent maxima on the screen equals $\lambda R/d$. With R a constant, the separation Δy is proportional to λ/d. The pattern is most evident—it spreads over large angles—when the wavelength is comparable to the source separation. Thus to see the interference pattern in a water ripple tank where the wavelength is measured in centimeters, the slit separations should be on the order of centimeters. This is a feature characteristic of all wave phenomena.

Waves or Particles?

Young's observations seemed to resolve once and for all the controversy about whether light was a particle or a wave phenomenon, particularly since Maxwell's equations successfully predicted electromagnetic waves and their properties. As we have already described in Chapter 34, the question was reopened toward the end of the 19th and through the first part of the 20th century. In certain circumstances far beyond the physical realms that we study in this chapter, a particle aspect of light reveals itself. Light is neither a pure wave nor a pure particle phenomenon, but is rather a phenomenon that displays behavior characteristic of waves and of particles, depending on the circumstances. The fact that light has this double nature is completely nonclassical, explicable only in terms of quantum mechanics.

37–2 Intensity in the Double-Slit Experiment

The previous discussion relied on geometrical arguments to determine the angles for which maxima and minima can be obtained. We now turn our attention to the *intensity* of the light that reaches the screen. In this section, we are going to assume that the two sources are both slits that are very narrow. The case where even a single slit has a finite width is by itself interesting, and we'll reserve that case for Chapter 38.

The intensity (or brightness, for light) measures the energy delivered by a wave per unit time per unit area. The energy in a given mechanical wave or superposition of waves is proportional to the displacement squared. For light, the quantity that plays the role of displacement is the electric (or magnetic) field. The intensity of a light wave (the energy delivered by the wave per unit time per unit area) is the time average of the Poynting vector (see Chapter 34), which is proportional to the product of electric and magnetic field vectors in the wave. Because the magnetic field is itself proportional to the electric field in an electromagnetic wave, the intensity is proportional to the electric field squared ($I \propto E^2$). In order to find the intensity of a collection of waves, we add the electric fields of all the waves and square the sum of the net field. For example, with two sources of equal intensity I_0, the maximum electric field is twice the electric field E_0 from each source, so

$$\frac{I_{\max}}{I_0} = \frac{(E_0 + E_0)^2}{E_0^2} = 4,$$

where $I_{\max}$ is the maximum intensity. The maximum intensity $I_{\max} = 4I_0$. Similarly, the minimum intensity occurs when the electric fields exactly cancel, and $I_{\min} = 0$.

The simple argument just given, which is based on energy, is so useful that we shall develop it further. The intensity at any point P on the screen in Fig. 37–8 is proportional to the net Poynting vector, which is in turn proportional to the square of the net electric field. The net instantaneous electric field $\vec{E}_{\text{net}}$ at P is the sum of the instantaneous electric fields of the light waves emitted at the two sources: $\vec{E}_{\text{net}} = \vec{E}_1 + \vec{E}_2$. The net Poynting vector therefore has magnitude $S = (\vec{E}_1 + \vec{E}_2)^2 = E_1^2 + E_2^2 + 2\vec{E}_1 \cdot \vec{E}_2$. But light waves oscillate rapidly, and thus it is the *time average* of the Poynting vector (that is, the intensity, I) at P that is of interest.

It is the coherence of the two light waves that is important for the time-averaged intensity value. If we denote time averages with triangle brackets, then

$$I_{\text{net}} \propto \langle E_1^2 \rangle + 2\langle \vec{E}_1 \cdot \vec{E}_2 \rangle + \langle E_2^2 \rangle. \tag{37–5}$$

For incoherent light, there is no correlation—no definite phase relation—between the electric fields from the two sources. One moment the sources have one relative phase, the next moment the relative phase is different, and *the term $\langle \vec{E}_1 \cdot \vec{E}_2 \rangle$ is zero*. Thus

$$I_{\text{incoh}} = I_1 + I_2. \tag{37–6}$$

For coherent waves, $\langle \vec{E}_1 \cdot \vec{E}_2 \rangle$ in Eq. (37–5) is not zero. If, at a given time, there is constructive interference at point P, where $\vec{E}_1 = \vec{E}_2$, the constructive interference will persist because the waves are coherent. Similarly, if there is destructive interference at a given time, where $\vec{E}_1 = -\vec{E}_2$, it also persists through later times. For destructive interference, $\langle \vec{E}_1 \cdot \vec{E}_2 \rangle \propto -I_1$, and Eq. (37–5) gives $I_{net} = I_1 - 2I_1 + I_1 = 0$.

Suppose then that the electric fields of the light waves from our coherent sources S_1 and S_2 at a single point P in space are identically oriented and have magnitudes

$$E_1 = E_0 \sin \omega t, \quad (37\text{–}7a)$$

$$E_2 = E_0 \sin(\omega t + \phi). \quad (37\text{–}7b)$$

The phase difference ϕ for E_2 results from the path-length difference between the waves. If $\phi = 2\pi n$, where n is an integer, the fields are identical, and there is constructive interference. This phase difference of $2\pi n$ corresponds to a path-length difference of $\Delta L = n\lambda$. The ratio of ϕ to $2\pi n$ is the same as ΔL to $n\lambda$, so we have

$$\frac{\phi}{2\pi n} = \frac{\Delta L}{n\lambda};$$

$$\frac{\phi}{2\pi} = \frac{\Delta L}{\lambda}. \quad (37\text{–}8)$$

For the distant-screen geometry of Fig. 37–8, we can use $\Delta L = d \sin \theta$ [Eq. (37–2)] to transform Eq. (37–8) to

$$\phi = 2\pi \frac{\Delta L}{\lambda} = \frac{2\pi}{\lambda} d \sin \theta. \quad (37\text{–}9)$$

Now the net electric field at P has magnitude

$$E_{net} = E_1 + E_2 = E_0[\sin \omega t + \sin(\omega t + \phi)].$$

If we apply the equation $\sin \theta_1 + \sin \theta_2 = 2 \cos[(\theta_1 - \theta_2)/2] \sin[(\theta_1 + \theta_2)/2]$ (see Appendix IV–4), with $\theta_1 = \omega t$ and $\theta_2 = \omega t + \phi$, we find

$$E_{net} = 2E_0 \cos\left(\frac{\phi}{2}\right) \sin\left[\omega t + \left(\frac{\phi}{2}\right)\right]. \quad (37\text{–}10)$$

The Poynting vectors $\vec{S}_1$ and $\vec{S}_2$ of the light from the individual sources have magnitudes

$$S_1 \propto E_1^2 = E_0^2 \sin^2(\omega t) \quad \text{and} \quad S_2 \propto E_2^2 = E_0^2 \sin^2(\omega t + \phi), \quad (37\text{–}11)$$

respectively, whereas the net Poynting vector at P has magnitude

$$S_{net} \propto E_{net}^2 = 4E_0^2 \cos^2(\phi/2) \sin^2\left[\omega t + \left(\frac{\phi}{2}\right)\right]. \quad (37\text{–}12)$$

To find the intensities (the time averages of the Poynting vectors), we need know only that the time average of $\sin^2(at + b) = \frac{1}{2}$. If we write the individual intensities as $I_0 \propto E_0^2/2$, then the net intensity from the two sources in terms of I_0 is

$$I_{net} = 4I_0 \cos^2\left(\frac{\phi}{2}\right). \quad (37\text{–}13)$$

When the phase ϕ in Eq. (37–7b) is related to the path-length difference by Eq. (37–8), then Eq. (37–13) for the intensity on the distant screen becomes

$$I_{net} = 4I_0 \cos^2\left(\frac{\pi d}{\lambda} \sin \theta\right). \quad (37\text{–}14)$$

This is the expression for the intensity in Young's classic double-slit experiment.

The maxima and minima occur at the angles specified by Eq. (37–3). Figure 37–9 is a plot of the intensity at the screen as a function of $\sin \theta$. This figure also serves as a plot of the intensity as a function of the distance y from the center maximum along the screen: For small θ, $y \cong R \sin \theta$, where R is the distance from the screen. If the apertures are narrow vertical slits, then the bright maxima on the screen are vertical lines called *fringes*.

▶ **FIGURE 37–9** The net intensity of light from the double slit as a function of the distance from the center point on the screen ($y \cong \sin \theta$). Compare the results for coherence and incoherence. The same amount of light energy reaches the screen in both cases, but in the coherent case, it occurs in peaks and valleys. We have assumed that each slit is infinitely narrow; that is why the falloff visible at the edges of Fig. 37–7 is not visible here.

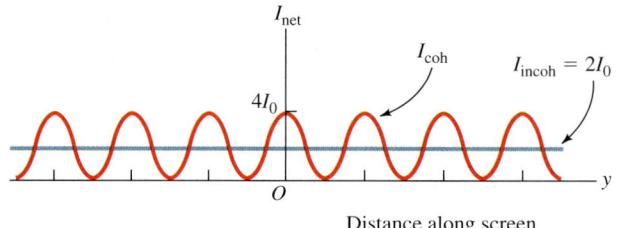

We have also plotted in Fig. 37–9 the intensity $2I_0$ that would be present on the screen if the light sources were incoherent, represented by the blue line in the plot. The result for incoherence is constant and shows no interfering maxima and minima. However, *averaged over the entire screen*, the energy reaching the screen is exactly the same in the two cases, as required by the conservation of energy. To average the energy that reaches the screen in the case for coherence, we need only use the fact that the average of the cosine-squared factor in Eq. (37–14) is $\frac{1}{2}$, and $4I_0 \times \frac{1}{2} = 2I_0$. The energy emitted at each source is the same whether the light from these sources is coherent or incoherent, and the total energy arriving at the screen must also be the same as that emitted. The energy is spread evenly over the screen when the sources are incoherent, whereas it is distributed in peaks and valleys when the sources are coherent.

EXAMPLE 37–4 Suppose you live at point H, 20 km from a vertical radio dipole antenna that broadcasts at a frequency of 1100 kHz from point B. How well your radio picks up the signal is a direct function of the intensity of the signal. A second antenna is constructed at point A, located $d = 100$ m from the first, so that AB makes an angle of 15° (0.26 rad) with AH. The two antennas are fed by the same source, and the two signals are in phase at the sources—they have the same wavelength and are coherent. Find the new intensity at your radio in terms of the old intensity. Is your signal improved?

Setting It Up The diagram in Fig. 37–10, which is a vertical view, helps us determine the path difference to the house from the two sources.

Strategy The situation is like the double-slit situation for light because there are two sources of coherent radiation, with the difference that in this case the relevant wavelengths are much longer. Because the distance from the transmitting antennas is much greater than their separation distance, the geometrical approximations we used in discussing the double-slit experiment apply. These approximations tell us that the difference in distances between the house and the two antennas is $R_A - R_B = d \cos \theta$ (Fig. 37–10). To find the net electric field at the house, we add the fields from each source, using the same methods that we did for light. Note that the electric fields, which are parallel to the antennas, are vertical (perpendicular to the page).

Working It Out Suppose that the first antenna (at point B) broadcasts a signal whose z-component at point H is $E_B = E_0 \cos(\omega t)$ and that the original intensity is $I_0 = CE_0^2$, where C is some constant. The second antenna's electric field takes the same form except that there is a phase difference due to the fact that the antenna is a distance $R_A - R_B$ farther away:

$$E_A = E_0 \cos(\omega t + \phi),$$

where $\phi = 2\pi(R_A - R_B)/\lambda = (2\pi d \cos \theta)/\lambda$. The net field at point H is then

$$E_{net} = E_A + E_B = E_0[\cos \omega t + \cos(\omega t + \phi)].$$

This sum is similar to the one we needed in order to find Eq. (37–10), $E_{net} = 2E_0 \cos(\phi/2) \times \sin[\omega t + (\phi/2)]$. The net intensity is given similarly in Eq. (37–13),

$$I_{net} = CE_{net}^2 = 4CE_0^2 \cos^2\left(\frac{\phi}{2}\right) = 4I_0 \cos^2\left(\frac{\phi}{2}\right).$$

It remains to calculate the factor $\cos^2(\phi/2)$ and see by what factor the intensity is changed. We require that

$$\cos\left(\frac{\phi}{2}\right) = \cos\left(\frac{2\pi d \cos \theta}{2\lambda}\right) = \cos\left(\frac{\pi d \cos \theta}{\lambda}\right).$$

We have $d = 100$ m and $\theta = 0.26$ rad. The wavelength λ comes from the frequency $f = 1100$ kHz $= 1.1 \times 10^6$ Hz. We get $\lambda = c/f = (3.0 \times 10^8 \text{ m/s})/(1.1 \times 10^6 \text{ s}^{-1}) = 270$ m. Thus

$$\cos\left(\frac{\phi}{2}\right) = \cos\left[\frac{\pi(100 \text{ m})(\cos 0.26 \text{ rad})}{270 \text{ m}}\right] = \cos(1.12 \text{ rad}) = 0.43.$$

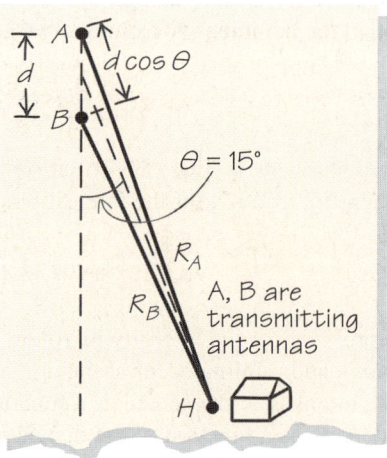

▲ **FIGURE 37–10** R_A and R_B are the distances between your home and two antennas at points A and B, respectively.

The net intensity $I_{net} = 4I_0 \cos^2(\phi/2)$ is a factor $4[\cos^2(\phi/2)] = 4(0.44)^2 = 0.77$ times the original intensity I_0. The signal you receive at your radio at point H has actually become weaker because there is partial destructive interference between the signals of the two antennas at that point.

What Do You Think? Suppose that the length of the cables from the radio station to the two transmitting antennas were each changed by different amounts. Will the signals still interfere at your house?

37–3 Interference from Reflection

A ray in air, say, may be partly reflected from a surface and partly transmitted through the surface into a medium such as glass. The transmitted ray may then be subsequently reflected from a second surface and reemerge into the air. If the original wave train is long compared with the distance traversed in the glass, the two reflected rays are coherent and can interfere.

Interference Fringes from the Space Between Two Glass Plates

Let's look at two glass plates, each with one very flat surface. They have been placed together with a spacer at one side, so that the flat surfaces touch along one side and are separated by a small distance d at the other (Fig. 37–11). When these plates are illuminated from above with monochromatic light, a series of alternating light and dark bands is seen, starting with a black band along the side where the plates touch.

To study this phenomenon further, let the plates have indices of refraction n, the space between the plates be air ($n = 1$), and the wavelength of the light be λ. We can think of the light as consisting of rays perpendicularly incident on the plates. These rays are *perpendicular to the wave fronts*; when we speak of rays interfering, we are referring to the interference of the waves that form those fronts. In this situation, the rays can reflect and refract to form new rays in various ways, but one possibility leads to the observed interference here: An incident ray in Fig. 37–11 passes almost vertically through the top piece of glass and is partly reflected (ray 1) and partly transmitted (ray 2) at point P_1 on the bottom surface of the top plate. Ray 2 then continues through the air gap and is reflected at point P_2 at the top of the bottom plate. To a good approximation, ray 2 travels a distance $2P_1P_2$ in air farther than ray 1, because the incident ray is almost vertical. Rays 1 and 2, exiting at the top, will interfere, resulting in the series of bright and dark bands corresponding to different values of $2P_1P_2$.

Now, if the points P_1 and P_2 move systematically closer to the side where the plates touch (point C), the path-length difference distance $2P_1P_2$ decreases, and we expect to see a series of bright and dark regions corresponding to constructive and destructive interference. What happens when the plates touch at point C? Observation reveals that there is a dark line at the edge where the plates touch, indicating destructive interference. But at point C each of the rays travel exactly the same distance, so the destructive interference cannot come from any path-length difference. The only explanation is that *destructive interference occurs because one of the rays undergoes a 180° phase change during reflection*. The light undergoes a phase change of 180° when it is reflected at P_2, whereas the light undergoes no such change when it is reflected at P_1. Let us consider the origin of this fact.

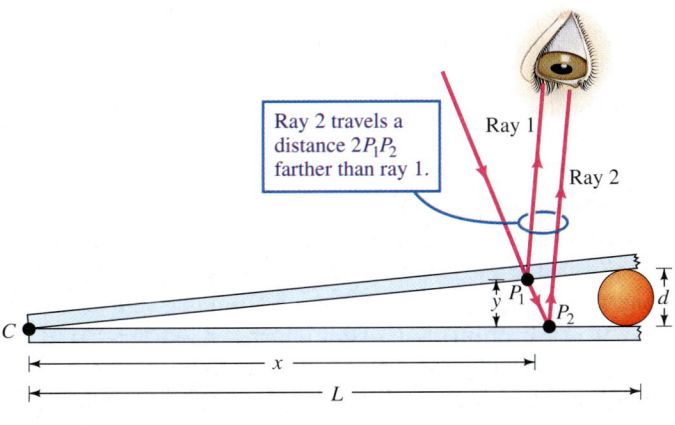

◀ **FIGURE 37–11** Two pieces of glass are placed together with a spacer (exaggerated in size) at one edge. The adjacent surfaces are flat on the scale of the wavelength of light.

1038 | Interference

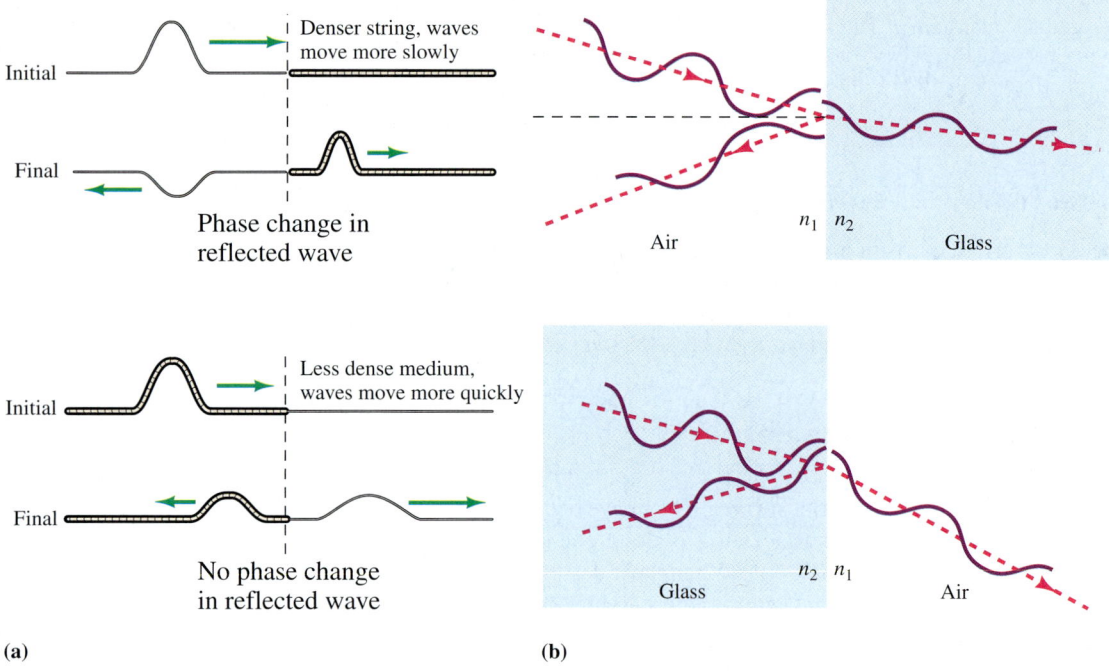

◀ FIGURE 37–12 (a) The phase change of light upon reflection is similar to the inversion that occurs when a pulse that moves along a string meets a denser string. No phase change occurs when the string encounters a less dense string. (b) For light, a phase change of 180° occurs when the second medium has a higher index of refraction. When the second medium at the reflection boundary has a lower index of refraction, there is no phase change in the reflection.

In Section 15–5, we discussed a corresponding phenomenon in the reflection of one-dimensional waves on strings. When two strings of different densities are connected (Fig. 37–12), the connection point forms a boundary at which there may be reflection and transmission. We showed that if the density on the side from which the wave comes is less than the density on the far side of the boundary, then the reflected wave is inverted, corresponding to a phase shift of 180° if the wave is harmonic. Similarly, if the pulse comes from a string of larger density than the string on the far side of the connection point, the pulse is not inverted on reflection.

Electromagnetic waves behave similarly: Maxwell's equations determine the form of the electric and magnetic fields of an electromagnetic wave at a boundary between two dielectric media. The electric field changes sign or does not change sign according to whether the wave speed ($v = c/n$) in the medium on the far side of the boundary is greater than or less than the wave speed in the medium from which the wave comes. (There is no such change of sign for the magnetic field of the wave.) When the electric field changes sign, the result is a 180° phase change in the reflected electromagnetic wave. The result for the phase change can therefore be stated as:

> **The phase of an electromagnetic wave that moves from a medium of index of refraction n_1 toward a medium with index of refraction n_2 will change by 180° upon reflection when $n_2 > n_1$ and will not change when $n_2 < n_1$.**

This information allows us to understand the interference pattern for the plates. The phase change of 180° occurs only for the reflection at P_2 because ray 2 goes from air ($n_1 = 1$) to glass ($n_2 > 1$). Because a phase shift of π rad (180°) corresponds to a shift of one-half wavelength, the condition for constructive interference becomes

$$\text{for constructive interference:} \quad \Delta L = 2P_1P_2 = \left(m + \frac{1}{2}\right)\lambda, \quad m = 0, \pm 1, \pm 2, \ldots . \tag{37-15}$$

The alternate bright and dark bands correspond to plate separation distances for which a monochromatic wave is in alternate constructive and destructive interference. In particular, the touching edge, where $\Delta L = 0$, has destructive interference; there is a dark band along that edge. This is consistent with a much simpler model of what happens at that edge: We have a film of air of zero thickness—that is, no film at all—and it is as though there is no boundary, hence no reflection, hence a dark region. The consistency of these two pictures gives us confidence that our discussion of the phase change is correct.

EXAMPLE 37–5 Two flat glass plates of length $L = 10$ cm touch at one end but are separated by a wire of diameter $d = 0.01$ mm at the other end (Fig. 37–11). Light shines almost perpendicularly on the glass and is reflected into the eye as shown. What is the distance x between the observed maxima if the incident (violet) light has $\lambda = 420$ nm?

Strategy We must here take the conditions for constructive interference that we previously described and translate those conditions into a band separation. We utilize Eq. (37–15) for the constructive interference condition, and by noting that $2y \cong 2P_1P_2$, where y is the spacing between the plates, we have

$$2y = \left(m + \frac{1}{2}\right)\lambda, \quad m = 0, \pm 1, \pm 2, \ldots.$$

To translate this condition into a condition for the x-position of the bands we use a geometrical relation between similar triangles in Fig. 37–11: $x/y = L/d$ or $x = yL/d$.

Working It Out We insert the y-values for constructive interference into our equation for x and find the x-positions where there is constructive interference:

$$x = \frac{L}{d}y = \frac{L}{d}\frac{1}{2}\lambda\left(m + \frac{1}{2}\right).$$

The difference in x from one maximum to another corresponds to a shift in m of 1, so

$$\Delta x = \frac{L}{d}\frac{\lambda}{2}\left\{\left(m + \frac{1}{2}\right) - \left[(m-1) + \frac{1}{2}\right]\right\} = \frac{L}{d}\frac{\lambda}{2}$$
$$= \frac{(10 \times 10^{-2}\text{ m})(420 \times 10^{-9}\text{ m})}{(0.01 \times 10^{-3}\text{ m})2} \cong 2 \text{ mm}.$$

For glass plates 10 cm long, there will be about 50 bands of constructive interference.

What Do You Think? Would the number of bands of constructive interference increase or decrease if red light is used instead of blue light?

Newton's Rings

Newton's rings form a variant on Example 37–5; they are named after Isaac Newton, who studied them in the 1700s along with his contemporaries Robert Hooke and Robert Boyle. Here, a curved piece of glass is placed on a flat piece of glass and illuminated from above with white light (Fig. 37–13a). Observation from above reveals rings of color (Fig. 37–13b). If monochromatic rather than white light shines down on the glass, then a series of bright and dark concentric rings appear (Fig. 37–13c). Most significantly, the region around point *C*—the place where the glass pieces touch—is dark (Fig. 37–13c). From our earlier discussion, we know why this occurs: The region around *C* can be thought of as a region without a reflecting boundary at all or the result of a phase change where ray 2 reflects from the second glass surface with no phase shift due to different path lengths.

How do we explain the colors we observe when white light such as sunlight, which consists of all wavelengths, is incident on the glass? Refer to Fig. 37–14. For a particular radial distance from point *C*, there may be only one wavelength in the visible range, say, for the color blue, for which there is *destructive* interference. At that radius, we see the color of sunlight with blue subtracted. If we look slightly farther away from *C*, where the distance P_1P_2 is greater, there will be destructive interference for a slightly larger wavelength, say, for green, and we see sunlight with green subtracted. These colors are not as vivid as those of the rainbow because they remain mixtures of different frequencies with some frequencies subtracted.

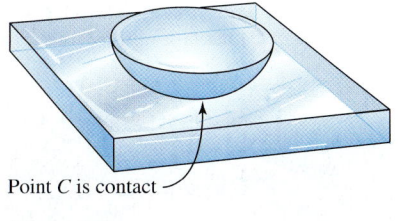

Point *C* is contact

(a)

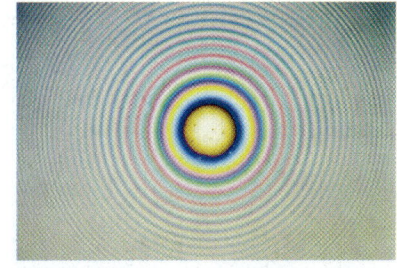

(b)

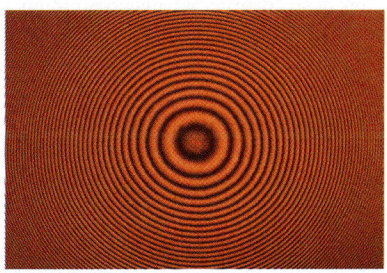

(c)

▲ **FIGURE 37–13** (a) A piece of glass whose bottom surface is curved rests on a flat glass surface. (b) The system is illuminated with white light from above, and a vertical view reveals colored rings called Newton's rings. (c) Concentric rings of alternating bright and dark appear when Newton's rings form from monochromatic light.

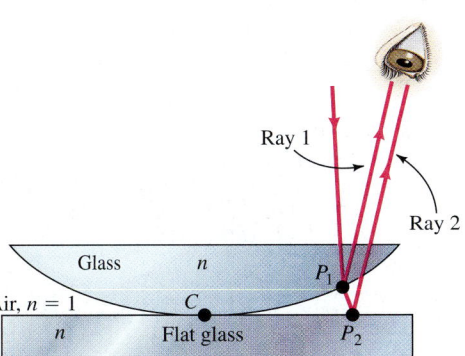

▲ **FIGURE 37–14** The geometry used to obtain the conditions for constructive interference in Newton's rings. The two rays moving toward the eye interfere after they reflect from different surfaces.

1040 | Interference

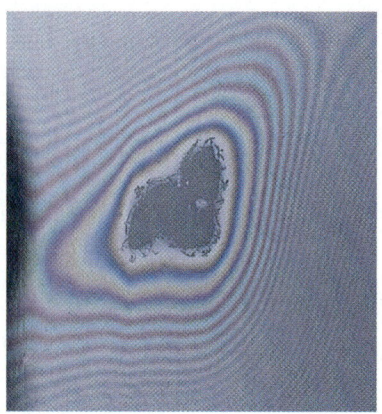

▲ **FIGURE 37–15** Fringes that occur when two nonflat pieces of glass are placed on top of one another. If each surface were perfectly flat and parallel, there would be no interference pattern. If the surfaces were flat but not parallel, the fringes would form straight lines. This is a good test of flatness for mirrors.

▲ **FIGURE 37–16** The colors of films of soapy water are due to thin-film interference of light.

THINK ABOUT THIS...
HOW WOULD YOU CHECK THE FLATNESS OF A GLASS SURFACE?

The interference phenomena we have described in this section give us a practical method for determining how flat a given glass surface is (Fig. 37–15). If the surface to be tested is placed on top of an *optical flat* (a glass with a surface known to be flat to a fraction of a wavelength of visible light), then if the tested surface is indeed flat, no regions of constructive interference will appear. In practice, a tiny spacer may be placed between the piece to be measured and the optical flat to open a wedge-shaped space between them. If the two pieces are so nearly flat and parallel that, without the spacer, *no* interference fringes are observed, the pieces may bond together and be virtually impossible to pull apart! In effect, the two pieces become one. (Weeks of grinding and much money will have been lost if this happens).

Mirrors are commercially available that are flat to better than 5 percent of one wavelength, or about 25 nm. They are used in lasers as well as in *interferometers*—devices that use the interference of light to measure distance down to a fraction of one wavelength (see Section 37–4).

Thin-Film Interference

The colors seen in soap bubbles and oil slicks are a manifestation of *thin-film interference*, which is another example of interference from reflection (Fig. 37–16). The interference occurs between the light reflected from the two surfaces of the thin film that forms the bubble. Consider light ray 1 that is incident on the thin film in Fig. 37–17. Part of the light is reflected at boundary I and forms ray 2. Part of ray 1 is refracted at boundary I and then reflected at boundary II. This light wave is partly refracted again at boundary I before forming ray 3.

Because rays 2 and 3 both originate from ray 1 at point P_1, the conditions for constructive or destructive interference depend on the path-length difference $\Delta L = P_1 P_2 P_3 - P_1 P_4$ as well as on any phase changes that may occur during the reflection. The rule discussed earlier for phase changes upon reflection indicates that ray 1 undergoes a phase change upon reflection at surface I but not at II. Thin films such as soap bubbles have varying thicknesses, so different wavelengths will interfere destructively on different parts of the bubble, and the colors that appear in the reflected light represent the original light minus the wavelength that interferes destructively. There is also an enhancement of those colors for which there is constructive interference. For an oil film floating on water, the oil may have an index of refraction between that of air and water, in which case there is a 180° phase change at *both* the air-oil surface and the oil-water surface. The phase changes due to reflection then cancel each other in the interference from light reflected from these two surfaces, and any phase difference is due only to a difference in the path length.

A feature in thin-film interference that does not arise in the case of Newton's rings is that the additional path length lies *within* the thin film of material. The phase change is then computed according to Eq. (37–9), but *the wavelength that appears is the wavelength within the material*, $\lambda_{\text{film}} = \lambda/n$, where n is the index of refraction of the material.

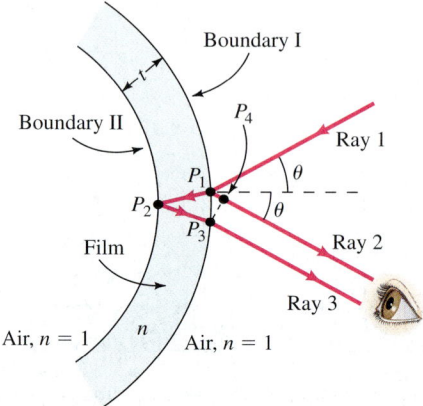

▲ **FIGURE 37–17** Geometry for thin-film interference. A light ray reflects from both the front and back boundaries of the film, and the reflected waves interfere.

EXAMPLE 37–6 The soap bubble shown in cross section in Fig. 37–17 has thickness t and index of refraction n. Light of wavelength λ in air falls vertically on the bubble and is reflected back. (a) Express the condition for constructive interference for the reflected light. (b) If $t = 400$ nm and $n = 1.3$, what color or colors will interfere constructively in the reflected light?

Strategy (a) There is constructive interference when $\phi = 2\pi m$, where m is an integer. To find the value of ϕ, note that when the incident light in Fig. 37–17 is vertical, distance $P_1 P_4 = 0$. Light reflects at P_1 and P_2, undergoing a 180° phase change only at P_1, where the index of refraction of the medium on the far side—the soap film—is greater than that of air. The path-length difference is $\Delta L = 2t$, and the phase difference $\phi_{\Delta L}$ between the two reflected waves due to ΔL is found from Eq. (37–8) in terms of the wavelength λ_n in the soap film:

$$\frac{\phi_{\Delta L}}{2\pi} = \frac{2t}{\lambda_n} = \frac{2tn}{\lambda}.$$

The overall phase difference between the two reflected waves is then $\phi = \pi + \phi_{\Delta L}$, where π appears as an additive factor because of the 180° phase change at P_1.

Working It Out We have

$$\phi = \pi + \phi_{\Delta L} = \pi + \frac{4\pi t n}{\lambda} = 2\pi m.$$

Thus

for constructive interference: $4nt = (2m - 1)\lambda, \quad m = 1, 2, 3, \ldots$.
(37–16)

Strategy (b) We need only solve the condition for constructive interference [Eq. (37–16)] for λ.

Working It Out Given $t = 400$ nm and $n = 1.3$, we have

$$\lambda = \frac{4nt}{2m - 1} = \frac{4(1.3)(400 \text{ nm})}{2m - 1} = \frac{2100 \text{ nm}}{2m - 1}.$$

For $m = 1$ through $m = 4$, these values of λ are

$$\lambda \cong 2100 \text{ nm}, 700 \text{ nm}, 420 \text{ nm}, 300 \text{ nm}.$$

Only the wavelengths 700 nm (red) and 420 nm (violet) are in the visible spectrum and these are the colors that interfere constructively in the reflected light.

What Do You Think? Estimate the wavelength of the color, if any, that is missing between the maxima at 700 nm and 420 nm.

EXAMPLE 37–7 A thin film of water, $n_2 = 1.33$, floats on cinnamon oil, which is denser than water and has index of refraction $n_3 = 1.65$. White light reflected at 45° has a maximum intensity for a wavelength around 600 nm. What is the minimum possible thickness of the film?

Setting It Up Figure 37–18 illustrates the paths of the two reflected rays that arrive in parallel at a distant point.

Strategy Both reflected rays undergo a phase shift of 180°, so that no net phase difference is associated with reflection. There is still a phase shift associated with different path lengths for the two rays—the phase shift ϕ_1 from the additional path $P_1 P_4$ of ray 1,

$$\frac{\phi_1}{2\pi} = \frac{P_1 P_4}{\lambda};$$

and the phase shift ϕ_2 for the additional path $P_1 P_2 + P_2 P_3$ of ray 2,

$$\frac{\phi_2}{2\pi} = \frac{P_1 P_2 + P_2 P_3}{\lambda_{H_2O}} = \frac{(P_1 P_2 + P_2 P_3)n_2}{\lambda}.$$

Note that the relevant wavelength for ray 1 is λ, the wavelength in air $(n_1 = 1)$, whereas the relevant wavelength for ray 2 is $\lambda_{H_2O} = \lambda/n_2$. It is the difference $\phi_2 - \phi_1$ that enters into the net phase difference between the two waves. The geometry necessary to calculate the path lengths is shown in Fig. 37–18. Once we know the phase difference, the usual conditions relate it to constructive and destructive interference; we will want to solve the condition for a first maximum to find the thickness t.

Working It Out The net phase difference is

$$\phi = \phi_2 - \phi_1 = (P_1 P_2 + P_2 P_3)\frac{2\pi n_2}{\lambda} - (P_1 P_4)\frac{2\pi}{\lambda}.$$

The incident ray enters at an angle θ perpendicular to the surface and is either reflected at angle θ or refracted into the water at angle θ'. We note that $P_1 P_2 = P_2 P_3 = t/(\cos \theta')$. Also, $P_1 P_4 = P_1 P_3 \cos(90° - \theta) = (2t \tan \theta') \cos(90° - \theta)$. The phase difference is then

$$\phi = \left(\frac{2t}{\cos \theta'}\right)\frac{2\pi n_2}{\lambda} - (2t \tan \theta') \cos(90° - \theta)\frac{2\pi}{\lambda}.$$

We solve the equation for t, the film thickness:

$$t = \frac{\lambda \phi}{4\pi} \frac{1}{(n_2/\cos \theta') - \tan \theta' \cos(90° - \theta)}.$$

From Snell's law, $\sin \theta' = (\sin \theta)/n_2$, and with $\theta = 45°$, $\theta' = 32°$. The problem states that there is a maximum in the intensity (that is, constructive interference) for $\lambda = 600$ nm. For constructive interference, $\phi = m(2\pi), m = 1, 2, 3, \ldots$. (Is $m = 0$ allowed here?) Thus

$$t = \frac{\lambda m(2\pi)}{4\pi} \frac{1}{(n_2/\cos \theta') - \tan \theta' \cos(90° - \theta)}$$

$$= \frac{600 \text{ nm}}{2} m \frac{1}{(1.33/\cos 32°) - \tan 32° \cos(90° - 45°)}$$

$$= (266 \text{ nm})m.$$

The minimum film thickness, for $m = 1$, is 266 nm.

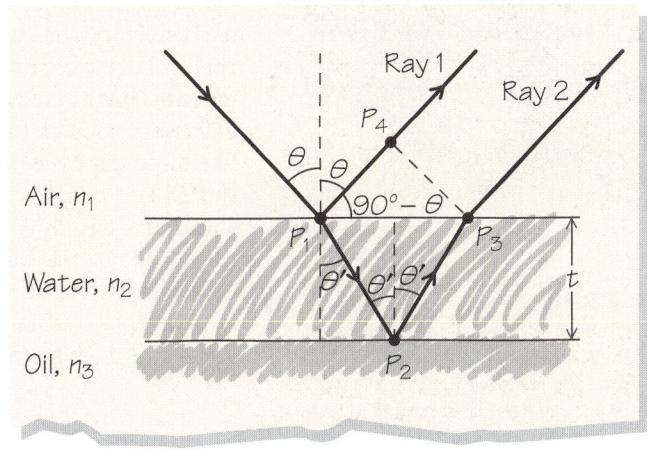

▲ **FIGURE 37–18**

1042 | Interference

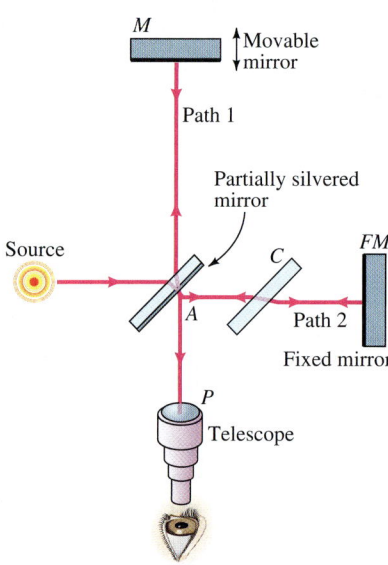

▲ **FIGURE 37–19** Thin coatings of materials can serve to reduce reflection by taking advantage of destructive interference. MgF_2 is often used as a coating. Here the incident and reflected rays are essentially vertical.

(a)

(b)

▲ **FIGURE 37–20** (a) Schematic diagram of a Michelson interferometer. Light is split by the partially silvered mirror. The resulting light travels two different paths before it returns to and interferes at point A and is subsequently observed through the telescope. (b) A modern Michelson interferometer.

THINK ABOUT THIS...
WHY DO SOME EYEGLASSES REFLECT LIGHT WHILE OTHERS DON'T?

The lenses of many eyeglasses, cameras, and other optical devices are coated with a thin layer of material whose purpose is to reduce the intensity of the reflected light through destructive interference. Many optical devices have multiple lenses, and if each lens typically reflects 4 percent of the energy of incident light, there may be an intolerable loss of light that is meant to be transmitted.

If light is reflected directly back from both surfaces in Fig. 37–19, and if the coating material has an index of refraction between that of air and the glass beneath it, there is a phase change in each reflection. Destructive interference will occur when the difference $2t$ in optical path lengths is a half-integer multiple of the wavelength λ_n within the coating and t is the thickness of the coating; in other words,

$$2t = \left(m + \frac{1}{2}\right)\lambda_n = \left(m + \frac{1}{2}\right)\frac{\lambda}{n}. \quad \text{(AB–1)}$$

The coatings are applied as thinly as possible (that is, for $m = 0$). For $m = 0$ the coating thickness must be

$$t = \frac{\lambda}{4n}, \quad \text{(AB–2)}$$

what is called a *quarter-wave* thickness. Of course, there can be destructive interference for only one wavelength in the visible range of light. A 100-nm thickness of MgF_2, a material commonly used to coat glass lenses, reduces reflectivity at 550 nm, in the middle of the visible range. The reflected light consists of red and blue light, which explains why such lenses have a purple cast to them. Antireflective coatings also increase the intensity of the transmitted light, and for this reason they are used to coat solar cells (components of solar batteries).

Sometimes we want to *increase* the intensity of reflected light. Lasers require mirrors that reflect light strongly at their operating wavelengths. For this purpose, a single coating must have thickness $t = \lambda/2n$, so that there will be constructive interference between the two reflected light waves. Multiple coatings are even more effective in increasing the reflected intensity. It is possible in this way to reflect more than 99 percent of the energy.

*37–4 Interferometers

Optical interferometers are devices that utilize the interference between light waves to measure quantities such as wavelength, small path-length differences, wave speeds, and indices of refraction, all to high precision. Figure 37–20a is a schematic diagram of one type of optical interferometer called the **Michelson interferometer** (Fig. 37–20b). In this device, developed by Albert Michelson in the 1880s, a light source is split by a beam splitter — for example, a partially silvered mirror—into two coherent waves that may travel different distances or through different media before they rejoin and interfere.

Monochromatic light from the source in Fig. 37–20a is split at the mirror at point A. The two beams then travel along paths 1 and 2 before they rejoin at A. The recombined beam is formed from the superposition (and therefore interference) of the two beams that arrive at A. The element C (a *compensator*) is added to make sure that the two light waves travel through the same amount of glass. If the path lengths are exactly the same, the two light waves will constructively interfere. With mirrors M and FM precisely perpendicular, the combined beam undergoes constructive interference and is bright. But if the path lengths are not precisely the same because the mirrors are not quite perpendicular to one another, the interference will produce alternating dark and bright lines, much like those discussed in Example 37–3, between two flat glass plates. The fringes will shift if the (screw-mounted) movable mirror is moved slightly. A movement of the movable mirror of only $\lambda/2$ will cause a shift from one fringe maximum to the adjacent one.

Unknown wavelengths of light can be determined by accurately measuring the movement of the mirror and counting the number of maxima that pass across the telescope eyepiece as the mirror moves. If ΔL is the mirror movement distance and N is the number of maxima that pass across the eyepiece for this movement, then

$$N = \frac{\Delta L}{\lambda/2} = \frac{2\Delta L}{\lambda},$$

or solving for the unknown wavelength,

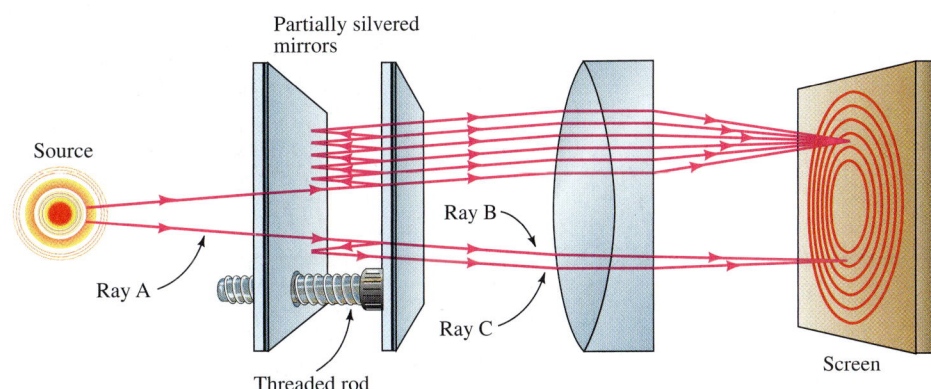

(a)　　　　　　　　　　(b)

$$\lambda = \frac{2\Delta L}{N}. \qquad (37\text{-}17)$$

If the wavelength is known, the same technique can be used to measure very small distances, in this case, ΔL.

The Fabry–Perot Interferometer

The most widely used interferometer is the *Fabry–Perot interferometer* (Fig. 37–21a), invented by Charles Fabry and Alfred Perot and illustrated schematically in Fig. 37–21b. It contains two end plates (partially silvered mirrors) that are precisely parallel and flat and are connected by a rod that allows the distance between the plates to be changed smoothly by a screw thread. An incident laser beam (ray A) is partly transmitted (ray B) and partly reflected; the reflected ray is in turn partly reflected and partly transmitted to make ray C, which interferes with ray B. The major improvement incorporated into the Fabry–Perot interferometer is that the plates are silvered such that multiple reflections are possible, and the interference is between the multiple rays formed by these multiple reflections. The multiple reflections reinforce the regions of constructive interference, making the maxima stronger. The maxima become easier to locate and the pattern becomes more distinct. It then becomes easier to tell when these maxima have been shifted, so distances can be measured with more precision. If we compare Figs. 37–22a and 37–22b—interference patterns produced by a Michelson and by a Fabry–Perot interferometer, respectively—the greater sharpness of the Fabry–Perot pattern is evident.

To measure distances with a Fabry–Perot interferometer, we begin with the plates at known positions and then count the number of interference maxima changes (fringes) as the plate separation varies by the desired distance. Because the location of maxima can be determined to great accuracy, the distance change can be measured to within an error of only a fraction of a wavelength of the laser light. The fringe counting is done automatically by electronic sensors. The number of fringes involved in distances of about 1 m is on the order of the number of wavelengths of visible light contained in 1 m, around 50 million.

▲ **FIGURE 37–21** (a) A Fabry–Perot interferometer. (b) Schematic diagram of a Fabry–Perot interferometer. Some of the light is transmitted and some reflected at the partially silvered mirrors. Rays B and C may constructively or destructively interfere, depending on the extra path length of ray C. Multiple reflection is an important part of the operation of the Fabry–Perot interferometer and serves to make the maxima sharper and more easily locatable.

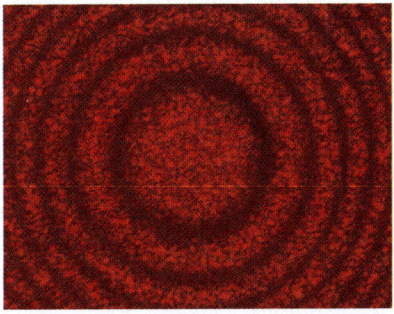
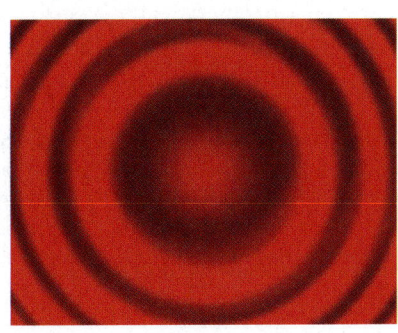

(a)　　　　　　　　　　(b)

◀ **FIGURE 37–22** The interference patterns produced by (a) a Michelson interferometer and (b) a Fabry–Perot interferometer. The Fabry–Perot fringes are noticeably sharper.

Summary

The wave theory of light explains experimental phenomena such as interference and diffraction that geometric optics cannot explain. Thomas Young was the first to substantiate the wave nature of light through his double-slit experiment, which produces interference maxima and minima. If λ is the wavelength and d is the distance between the (narrow) slits, there will be maxima and minima at angles θ on a distant screen given by

$$\text{for constructive interference:} \quad \sin\theta = n\frac{\lambda}{d}, \quad n = 0, \pm 1, \pm 2, \ldots; \quad (37\text{–}3\text{a})$$

$$\text{for destructive interference:} \quad \sin\theta = \left(n + \frac{1}{2}\right)\frac{\lambda}{d}, \quad n = 0, \pm 1, \pm 2, \ldots. \quad (37\text{–}3\text{b})$$

The intensity pattern for the double slit is

$$I_{\text{net}} = 4I_0 \cos^2\left(\frac{\pi d}{\lambda}\sin\theta\right). \quad (37\text{–}14)$$

where I_0 is the maximum intensity for a single slit and I_{net} is the total observed intensity.

We have looked at a variety of common situations where the geometry and other properties of reflecting surfaces results in characteristic interference patterns. Newton's rings appear when light shines vertically down upon a curved glass that rests on a flat piece of glass. The light reflecting from two surfaces interferes. For monochromatic light, alternating rings of bright and dark are observed. For sunlight, colored rings are observed. In both cases, the center is dark because the phase of an electromagnetic wave will change by 180° upon reflection when the wave moves from one medium to another of higher index of refraction. The condition for constructive interference in Newton's rings is

$$\Delta L = \left(m + \frac{1}{2}\right)\lambda, \quad m = 0, \pm 1, \pm 2, \ldots, \quad (37\text{–}15)$$

where ΔL is the difference in path length of the interfering rays.

Thin-film interference is responsible for the colors observed in bubbles and oil slicks. For light falling perpendicularly from air onto the surface of the film of thickness t, the condition for constructive interference is

$$4nt = (2m - 1)\lambda, \quad m = 1, 2, 3, \ldots, \quad (37\text{–}16)$$

where n is the index of refraction of the film. Light reflected from the two surfaces of the thin film interferes destructively for some wavelengths but not for others. If the incident light is white, the wavelength undergoing destructive interference is subtracted from the reflected light. This phenomenon is used in the application of thin coatings to lenses and other optical surfaces in order to reduce the intensity of reflected light.

Optical interferometers utilize the interference of light waves to measure distances with great precision. Two interferometers with practical applications are the Michelson interferometer and the Fabry–Perot interferometer.

Understanding the Concepts

1. Why is a laser a more practical source of coherent light than a bright lightbulb and a series of slits?
2. In Fig. 37–5b, the intersections represent places where constructive interference occurs on the crests of the advancing wave fronts. The troughs also reinforce each other between the circles. Do such reinforced troughs also represent a region of constructive interference? In the regions between the crests and troughs, where each wave is zero, the waves add to zero. Is this a region of constructive interference?
3. In Fig. 37–5b, the intersections represent sites where constructive interference occurs. Will these bright spots be seen in air, or is a screen required for us to see the alternating bright and dark spots?
4. We refer to lines, or fringes, when we use two elongated slits for interference. If we used two holes rather than two slits, would we still observe lines or would we observe something else?
5. Why does a single thin coating produce destructive interference in the reflection of light for only one wavelength in the visible region of light?
6. When we look up, stars twinkle and planets do not, while astronauts in orbit do not see any twinkle at all. Can you explain this?
7. In discussing interference from thin films, we have mentioned viewing the reflected light from afar. Why? Is there no interference if the light is viewed from close to the film?
8. Why don't we study (or see) interference effects with thick films?
9. There is no light intensity and therefore no energy at an interference minimum from two sources of coherent light. Yet each of the two interfering sources alone would produce energy at that point. Why does this situation not violate the conservation of energy?
10. When there is a minimum in the intensity of light reflected from thin films, there is less energy in the reflected beam. Does this mean that energy conservation is violated? If not, what happens to the energy that would otherwise have been in the reflected light? It may be helpful to think about what the light that passes all the way through a Newton's rings apparatus must look like.
11. Is it possible to use two beams of light that travel in the z-direction—one polarized with its electric field vector aligned

along the x-direction, and the other polarized with its electric field vector aligned along the y-direction—to make an interference pattern?

12. We estimated in Example 37–5 that almost 50 bands of constructive interference can occur along the glass plates. Can all these bands be observed? Will the color still appear blue?

13. The discussion of antireflective coatings suggested that the index of refraction of the coating material should be in between that of air and that of the glass that it coats. Why? (See Problem 52.)

14. In our discussion of Newton's rings, we did not consider light that reflects from the top surface of the curved piece of glass, nor that from the bottom surface of the bottom plate. Why not?

15. In Example 37–4, a second antenna has been constructed whose signal is as strong as that of the first, yet the signal you receive weakens. How is this consistent with the conservation of energy?

16. For small angles, the maxima for the double slit interference pattern (see Fig. 37–7) are equally spaced. Is this true for larger angles, ones for which the small-angle approximations are no longer useful?

17. How is it possible to use interferometry to measure a small fraction of a wavelength when the distance between maxima represents a full-wavelength difference?

18. When you look into a good-quality camera lens, you will see a color tint (generally purple). What is the origin of the observed color?

19. It might appear that a given antireflective coating will work for any surface to which it is applied, provided that the thickness of the coating is given by Eq. (AB–2) on p. 1042; however, this is not true. Why not? You can go back to the discussion of the energy in reflection (in Chapter 36) to find out why the index of refraction of the surface to be coated must be considered.

20. Why is it important that each slit be as narrow as possible in the double-slit experiment?

21. When you look out of a window with mini-blinds, you don't see interference effects. Why?

22. Antireflective coatings are always applied to the front surface—the side from which light comes—of an optical element, never to the back surface. Why?

Problems

37–1 Young's Double-Slit Experiment

1. (I) A coherent source of monochromatic light of unknown wavelength shines on double slits separated by 0.20 mm. Bright spots separated by 0.70 cm appear on a screen 3.0 m away. What is the wavelength of the light?

2. (I) Red light ($\lambda = 630$ nm) shines on a double slit with slit separation $d = 0.25$ mm. How far away from the central axis will the first minimum be on a screen 2.0 m from the double slit?

3. (I) A double-slit interference experiment is done in a ripple tank. The slits are 3.5 cm apart, and a viewing screen is 0.8 m from the slits. The wave speed of the ripples in water is 0.12 m/s, and the frequency of the vibrator producing the ripples is 12 Hz. How far from the centerline of the screen will the first maximum be found?

4. (I) The source for a double-slit experiment has a wavelength of 525 nm. The slits are a distance of 120 μm apart, and a screen is 40 cm away from the wall that contains the slits. How far from the center will the third maximum occur?

5. (I) Light of wavelength 590 nm falls on a wall with two slits 0.12 mm apart. A photographic plate is placed at a distance R from the wall. The $n = 3$ maximum appears 18 cm from the central maximum on the photographic plate. How far is the plate from the wall?

6. (I) Two small speakers are 65 cm apart. They broadcast a signal of frequency $f = 380$ Hz; the signals are in phase. A sensitive microphone is placed 2.5 m from the midpoint of the two speakers, on the line perpendicular to the line joining the two speakers. Is the intensity of sound a minimum or a maximum there? How far would the microphone have to be moved along an arc of radius 2.5 m centered on the midpoint in order to pick up a signal of maximum intensity?

7. (II) A double-slit experiment produces fringes on a distant screen. How does the linear separation between the bright maxima on the screen change when (a) the wavelength of the light doubles? (b) The separation between the slits doubles? (c) The distance between the slits and the screen doubles? (d) The intensity of the light doubles?

8. (II) A laser emitting light with $\lambda = 595$ nm shines on a double slit with a separation of 0.15 mm and produces interference fringes. If the maxima are separated by 0.50 cm, how far away is the screen on which the fringes are observed?

9. (II) In a double-slit experiment to determine an unknown wavelength of light, the measured total distance between 16 maxima (8 on each side of the central maximum) is 16.8 cm. The screen is located 3.45 m from the double slits, whose centers are 0.21 mm apart. What is the wavelength?

10. (II) Suppose that a double slit illuminates a distant screen. The light from sources S_1 and S_2 has come from a single monochromatic source S that is one-half wavelength closer to S_1 than to S_2. Use the geometry of Fig. 37–8 for the relation between the screen and the double slit to express the locations of maxima and minima. Is the point at $\theta = 0$ a maximum, a minimum, or neither?

11. (II) A double slit with variable separation d is superimposed on a single slit at right angles to the double slit, leading to a two-point source (Fig. 37–23). Then the double slit is rotated relative to its original direction by an angle ϕ. If light of wavelength λ shines through the system, determine the position of the interference maxima on a screen a distance R away. Do the fringes move inward or outward as ϕ increases?

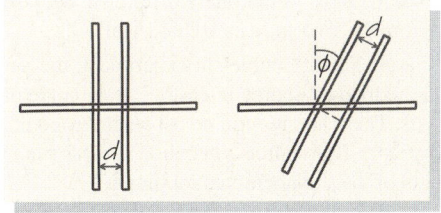

▲ **FIGURE 37–23** Problem 11.

12. (II) Two microwave sources are 18 cm apart. They radiate coherently with a frequency of 3.7×10^{10} Hz but with a phase difference α between the two sources. A microwave detector is moved along a line 1.6 m away from the sources. How far from the center ($\theta = 0$) will the first maximum occur, as α varies from 0 to 2π?

13. (II) Two sources radiate at almost identical frequencies f and $f + \Delta f$. How fast will the interference fringes on a screen a distance R away move, assuming that the sources are a distance d apart and that the velocity of wave propagation is v? Evaluate the result for two sources in a ripple tank ($f = 10$ Hz, $\Delta f = 10^{-6}f$, $R = 1$ m, $d = 5$ cm, $v = 0.15$ m/s). Do the same for an optical double-slit experiment ($f = 4.7 \times 10^{14}$ Hz, $\Delta f = 10^{-6}f$, $R = 1$ m, $d = 0.25$ mm, $v = 3.0 \times 10^8$ m/s). What can you conclude about the degree of coherence in the two cases?

14. (II) Light of two different wavelengths, λ_1 and λ_2, is incident on a double slit. On a distant screen, the twentieth maximum of λ_1 overlies the nineteenth minimum of λ_2. Show that the relative difference $(\lambda_1 - \lambda_2)/\lambda_1$ is small, and find a numerical value for this ratio.

15. (II) Consider two narrow slits illuminated from behind; the light impinges on a screen that is close rather than far (Fig. 37–24). Assume that the wavelength of the light λ is comparable to the separation d between the slits and to the screen distance R. The angle θ is measured from the point midway between the slits. (a) Show that the point corresponding to $\theta = 0$ continues to be a maximum of the pattern of light that reaches the screen. (b) At what angle θ is there a first maximum? (c) Show that your result for part (b) reduces to the distant-screen result for $R \gg d$ (and $R \gg \lambda$).

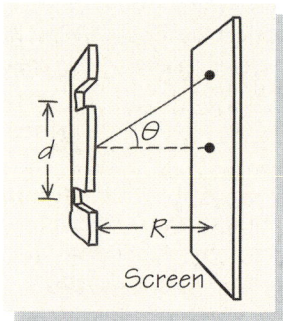

▲ **FIGURE 37–24** Problem 15.

37–2 Intensity in the Double-Slit Experiment

16. (I) Light of wavelength 500 nm shines on two slits separated by 0.3 mm. Find the intensity ratio I/I_0 at positions 0.6 mm and -0.5 mm from the central maximum on a screen 1 m from the slits.

17. (I) Two coherent light sources of the same wavelength, each with intensity of 1.0×10^3 W/m², interfere at a point at which the phase difference is 60°. What is the net intensity of light at this point?

18. (I) The net intensity from two coherent sources of equal strength is exactly twice the intensity due to either source at a certain point in space. What is the phase difference between the waves arriving from the two sources at that point?

19. (I) In a classic Young double-slit experiment, the net intensity is 1/4 of the individual source intensities at an angle of 27° off the central axis. There are no minima between this point and the central maximum. The source separation is 485 nm. What is the wavelength of the (monochromatic) light?

20. (I) The intensity at the central maximum of a double-slit diffraction pattern is I_{max}. If the wavelength of the light is 490 nm and the nearest location of maximum intensity is 1.7×10^{-3} rad from the central axis, what is the separation of the slits?

21. (II) Use the result of Young's classic double-slit experiment to find the average intensity on the screen by integrating the intensity over the surface of the screen. This result should be twice the average intensity from one slit alone and shows that energy conservation holds even when there is interference.

22. (II) Consider the double-slit arrangement shown in Fig. 37–25. The center of the screen C is a point of constructive interference. A container of thickness w, holding a liquid of refractive index n, is placed in the path of the ray from slit S_2 to C. Plot, qualitatively, the intensity of light at C as a function of w, assuming that the separation between the slits is d, and that the screen is a distance L from each slit. Ignore any intensity loss due to absorption in the fluid or at the fluid boundaries.

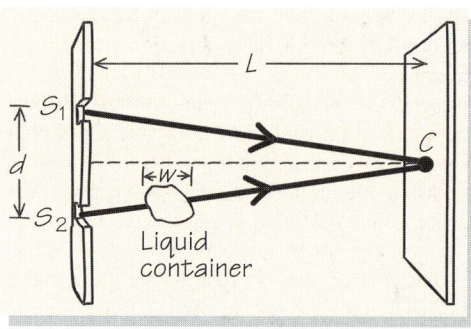

▲ **FIGURE 37–25** Problem 22.

23. (II) The *angular width* of a maximum of the intensity pattern due to double-slit interference is defined to be the angular separation $\Delta\theta$ of the points where the intensity is half its maximum value. Express the width of the central maximum in terms of the wavelength and the slit separation. Are the widths of all the maxima the same?

24. (II) A He–Ne laser, $\lambda = 633$ nm, shines on double slits separated by 0.35 mm. At what minimum angle θ is the intensity 50 percent of the maximum? If the screen is located 1.8 m away, what is the distance between the two angles on either side of the maximum for which this intensity occurs? This distance is the *full width at half maximum* of the central peak.

25. (II) Two point sources of radio waves, 12 m apart, radiate in phase with a frequency of 3.8×10^7 Hz. (a) If the average intensity of each single source is 5.0×10^{-4} W/m² at a certain distance, what is the direction in which the combined intensity is maximized? (b) What is the magnitude of the maximum intensity? (c) At what angle will the intensity have fallen to half its maximum value?

26. (II) Suppose that the two slits in a double-slit experiment are not exactly the same size, so the electric field from one of the slits at a particular point P on the screen is $E_1 \sin(\omega t)$, whereas the other one is $E_2 \sin(\omega t + \phi)$, where the phase ϕ is due to the path-length difference [compare Eq. (37–7a, b)]. Show that the intensity at P is given by

$$\langle I_{net} \rangle = \langle I_1 \rangle + \langle I_2 \rangle + 2\sqrt{\langle I_1 \rangle \langle I_2 \rangle} \cos\phi,$$

where $\langle I_1 \rangle$ and $\langle I_2 \rangle$ are the intensities due to the light from the individual slits. [*Hint*: You will need $\langle \sin^2(\omega t) \rangle = \frac{1}{2}$; $\langle \sin(\omega t) \cos(\omega t) \rangle = 0$.]

27. (II) Point sources S and S' radiate with the same intensity and the same frequency, corresponding to a wavelength of 0.020 m (Fig. 37–26). They are 45° out of phase and 2.5 m apart. Plot the intensity as a function of distance along the *x*-axis for values of *x* much larger than the source separation.

▲ **FIGURE 37–26** Problem 27.

28. (III) Find an expression for net intensity for the situation in Problem 27 in terms of x, the wavelength (λ), and the source separation (d). Take into account that the electric field of the electromagnetic wave decreases as the inverse power of the distance between the source and the receiver, and that the distance between source S and any point on the axis is different from the distance between source S' and that point. [*Hint*: Use the results of Problem 26.]

37–3 Interference from Reflection

29. (I) Consider the two glass plates of Example 37–5 and the configuration of Fig. 37–11. The plates are 25 cm long. When light of wavelength 656 nm from hydrogen shines down perpendicularly to the glass, 102 interference fringes appear. How thick is the wire that separates the two glass plates at one end?

30. (I) Two rectangular pieces of glass are laid on top of one another on a plane surface. A thin strip of paper is inserted between them at one end, so that a wedge of air is formed. The plates are illuminated by perpendicularly incident light of wavelength 615 nm, and 17 interference fringes per centimeter-length of wedge appear. What is the angle of the wedge?

31. (I) For what thicknesses of a soap bubble ($n = 1.3$) that are less than 500 nm thick will blue light ($\lambda = 420$ nm) interfere constructively?

32. (I) A thin, uniform film of oil is spread on a glass plate. The oil has an index of refraction between that of air and of the glass. Write the condition for constructive interference for light of wavelength λ in air perpendicularly incident from air and reflecting back into the air from the air–oil–glass interface.

33. (II) You are an inventor who has never heard of tinting, and you want to use a plastic film of index of refraction 1.32 to place on the outside of glass to increase reflection, thereby keeping the intensity of the transmitted light down. What is the minimum thickness of the film that will accomplish this for light of average wavelength 550 nm? As this might be too thin to work with, suggest two other thicknesses that would work.

34. (II) A gardener accidentally spills oil (index of refraction 1.38) into a swimming pool. The oil spreads over the surface in a uniform way. (a) As you look directly down at the surface, you see light whose color suggests that the wavelength 520 nm is not being reflected. What is the minimum thickness of the oil film? (b) What will be the primary transmitted light frequency observed by someone swimming under water in the pool?

35. (II) A curved piece of glass in the form of the cap of a sphere of radius R is placed on a plane surface of glass. What will be the radius of the first dark Newton ring for light of wavelength λ? [*Hint*: It follows from plane geometry that, for a spherical cap, the radius of the circle bounding the cap, r, is related to the radius of the sphere R and h, the maximum thickness of the cap, by the relation $r^2 = h(2R - h)$, which can be approximated by $2Rh$ for $h \ll R$.]

36. (II) When two flat glass plates are placed on top of one another and a slip of paper is inserted between them at one edge, a thin wedge filled with air is produced between them. Interference bands form in reflection when monochromatic light falls vertically on the plates. Is the first band near the edge where the plates are in contact light or dark? Why?

37. (II) A wedge of air is formed between two glass plates that are 8.0 cm long, with one edge touching, and the other separated by a wire that is 1.0 mm in diameter. How far apart are the maxima for the reflected light with $\lambda = 655$ nm?

38. (II) Two plane glass plates touch at one end and are separated at the other by a wire, so that they form a wedge of air. When light with wavelength $\lambda = 590$ nm is reflected by the combination, 110 dark stripes appear. What can you say about the diameter of the wire?

39. (II) In a standard Newton's rings experiment, there is a dark spot where the convex surface touches the flat plate. Light of wavelength 500 nm is perpendicularly incident on the system. The convex lens is pulled slowly away from the flat plate until the minimum of the convex lens is 0.25 mm from the flat plate. A series of maxima and minima will appear at the center as the lens moves. How many maxima pass? Do the rings appear to move in to the center or away from the center?

40. (II) Consider the Newton's rings apparatus of Problem 39, with the minimum of the convex lens 0.85 mm from the flat plate. Water is poured into the space between the plates. Do the rings appear to move in to the center or away from the center, and how many maxima pass?

41. (II) A lens whose curved surface is part of a sphere of radius 5.0 m is placed over a flat glass plate and Newton's rings are observed. Determine the diameter of the fourth and seventh dark fringe for a wavelength of 520 nm.

42. (II) Light with a wavelength of 560 nm gives rise to a system of Newton's rings formed with a convex lens resting on a plane surface. The twentieth bright ring is at a radial distance of 0.98 cm. What is the thickness of the air film there, and what is the radius of curvature of the lens surface?

43. (II) The radius of curvature of a convex surface used for a Newton's rings apparatus is R (Fig. 37–27). Find the position x, measured from the point where the convex surface touches the flat surface, of the nth dark ring for light of wavelength λ perpendicularly incident from above.

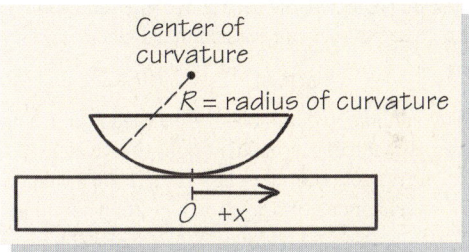

▲ **FIGURE 37–27** Problem 43.

44. (II) Constructive interference occurs when a soap bubble reflects light of wavelength 460 nm. What is the minimum thickness of the bubble if its index of refraction is 1.35?

45. (II) What minimum thickness of antireflective coating of MgF_2 is required to minimize the reflection of red light at 650 nm (the wavelength in air)? For MgF_2, $n = 1.38$.

46. (II) For light emitted by a He–Ne laser, $\lambda = 633$ nm, what *nonzero* minimum thickness of MgF_2 coating allows *maximum* reflectivity?

47. (II) The reflected light from an oil film floating on water shows constructive interference for light of wavelengths 434 nm and 682 nm incident along the normal. The index of refraction of the oil is $n = 1.51$. What is the film's minimum possible thickness?

48. (II) The reflected light from an oil film floating on water shows destructive interference for light of wavelengths 550 nm, 610 nm, and 685 nm incident along the normal. The index of refraction of the oil is $n = 1.40$. What is the film's minimum possible thickness?

49. (II) White light reflected at perpendicular incidence from a uniform soap film has an interference maximum at 666 nm and a minimum at 555 nm with no minima between 666 nm and 555 nm. If $n = 1.34$ for the film, what is the film thickness?

50. (II) A thin layer of CaF$_2$ ($n = 1.41$) is deposited onto glass, with $n = 1.52$. The layer is viewed in reflected light at 45° using a white light source (Fig. 37–28). For what layer thicknesses will light of 640 nm show constructive interference? Are there possible thicknesses for which light of 480 nm interferes constructively as well?

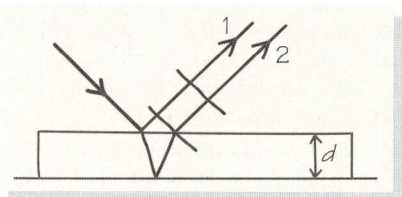

▲ **FIGURE 37–28** Problem 50.

51. (II) Light is perpendicularly incident on an oil film with $n = 1.2$, suspended in air. (a) If green light ($\lambda = 550$ nm) is reflected back most strongly, what is the minimum thickness of the film? (b) If n were increased, would the maximally reflected light have a longer or shorter λ? (c) If the film were suspended on the interface between water ($n = 1.33$) and air, what would be seen?

52. (II) We mentioned in Chapter 36 that when light is perpendicularly incident from a medium of index of refraction n_1 and refracts into a medium of index of refraction n_2, then the intensity of the reflected light, I_r, is related to the incident intensity, I_0, by $I_r/I_0 = (n_2 - n_1)^2/(n_2 + n_1)^2$. In order for a coating to eliminate reflections, it is not enough that the light reflecting from the two surfaces differs in phase by 180°; the interference is totally destructive only if the amplitudes are equal. Show that, when multiple reflections at interfaces are neglected, the destructive interference in light reflecting from coated glass in air is maximized when

$$(n_{\text{air}}/n_{\text{coat}}) = (n_{\text{coat}}/n_{\text{glass}}).$$

53. (II) You would like to eliminate the reflected light from a flat glass pane for perpendicularly incident light of wavelength 600 nm. If the index of refraction of the glass is 1.55 and you have a coating material with an index of refraction of 1.25, what minimum thickness of coating material will have the desired effect?

54. (II) Blue light, $\lambda = 485$ nm, is perpendicularly incident on a vertical soap film held in a plane. A sequence of bright horizontal bands appears in the reflected light (Fig. 37–16). Note the film is illuminated with white, not monochromatic, light. What is the rate of change with height of the thickness of the soap film if the horizontal bright bands are 0.6 cm apart? For the soap solution, $n = 1.36$.

55. (II) The material to be used for an antireflective coating has index of refraction of 1.25. How thick should the coating be to give the best result for $\lambda = 550$ nm and an angle of incidence of 30° with the normal?

*37–4 Interferometers

56. (I) Laser light with $\lambda = 633$ nm enters a Michelson interferometer. How many fringes will pass through the field of view if one of the mirrors is moved 0.10 mm?

57. (I) If the mirror of one arm of a Michelson interferometer is moved along the arm by 0.31 mm, 980 fringes traverse the field of view. What is the wavelength of the light used?

58. (II) A very sharp wedge of glass of index of refraction 1.55 is introduced perpendicularly in the path of one of the interfering beams of a Michelson interferometer illuminated by a narrow beam of light of wavelength 426 nm. This causes 750 dark fringes to sweep across the field of view. Calculate the thickness of the glass wedge at the point where the beam passes through it.

59. (II) Identical glass tubes, each 10 cm long, are placed in the two optical paths of a Michelson interferometer. Both of them are evacuated, then one is slowly filled with a gas until it reaches atmospheric pressure. During the filling time the pattern has moved by 90 fringes. What is the index of refraction of the gas?

60. (II) A thin glass plate of index of refraction 1.58 is introduced into one of the beams of a Michelson interferometer. This causes a displacement of 25 fringes for light with $\lambda = 589$ nm. What is the thickness of the plate?

61. (II) A scientist wants to measure the wavelength of yellow light ($\lambda \cong 590$ nm) to a precision of 0.1 percent. The minimum motion of the movable mirror is 0.03 mm. What is the minimum number of fringe shifts that must be counted?

62. (II) A laser with light of wavelength 582.5 nm is used to calibrate a Fabry–Perot interferometer. As the screw controlling the position of one end plate rotates exactly 100 turns, 714 fringes are counted. Calculate the wavelength of another light source that shifts only 593 fringes for 100 turns of the screw.

General Problems

63. (II) Two ordinary lightbulbs S_1 and S_2 are 1 m apart, each emitting light waves with intensity I, mainly at a wavelength of 550 nm. What is the pattern of intensity on a screen 100 m away?

64. (II) AM radio waves with a wavelength of 480 m travel 13 km to your home. Halfway between the transmitting tower and your home, but off to the side, is a building that reflects radio waves; the reflected wave has no phase shift due to the reflection. How far off the direct line is the building if destructive interference occurs between the direct waves and the reflected waves (Fig. 37–29)?

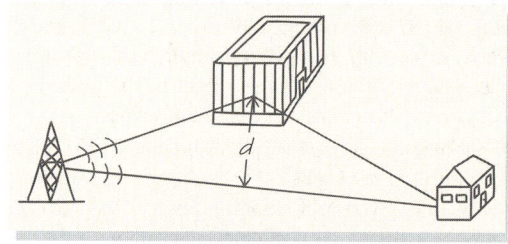

▲ **FIGURE 37–29** Problem 64.

65. (II) Two point sources of identical strength radiate in phase with the same frequency f. They are separated by a distance L. What is the energy density (discussed in Section 34–3) as a function of distance from one of the sources along the line that connects the sources in the central region, where the variation of the amplitudes with distance can be neglected (Fig. 37–30)?

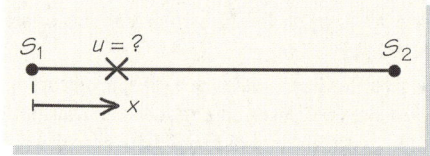

▲ **FIGURE 37–30** Problem 65.

66. (II) You want to hear a radio station that broadcasts at 97.9 MHz. You live on a direct line between the antenna and a large building that acts as a mirror for the radio waves broadcast by the station; you are exactly 100 m from the building (Fig. 37–31, see next page). Calculate the intensity of the signal you receive in terms of the intensity you would receive if the building were not present. Assume that the reflection from the building is total, with no phase shift, and ignore any decrease of the signal with distance.

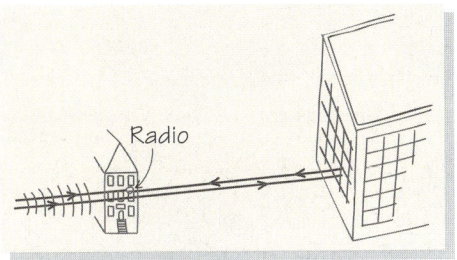

▲ **FIGURE 37–31** Problem 66.

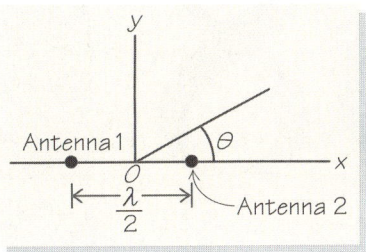

▲ **FIGURE 37–34** Problem 70.

67. (II) Coherent microwave radiation reflects from two identical obstacles (Fig. 37–32). Each obstacle, of size a, is much smaller than the wavelength of the radiation, λ, and much smaller than the obstacle separation d. Find an expression for the angles θ', defined in the figure, for which there are maxima on the distant screen.

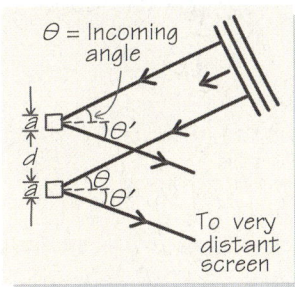

▲ **FIGURE 37–32** Problem 67.

68. (II) A radio wave undergoes a phase shift of 180° when it reflects from the calm surface of the ocean. A ship in a calm port nearing a shore station receives a 230-MHz signal from the station's antenna. This antenna is located 15 m above the sea surface, as is the ship's receiving antenna. The direct and reflected signals interfere, and a succession of maxima and minima are heard in the interfering signal at the ship (Fig. 37–33). How far is the ship from the station the first time the signal passes through a minimum? How fast is the ship moving if the time between this first minimum and the next one is 170 s?

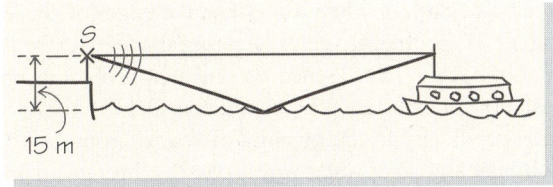

▲ **FIGURE 37–33** Problem 68.

69. (II) Sources S_1 and S_2 illuminate a distant screen; the distance to the screen is much larger than the separation between the sources. Each source emits light rays that are in phase at the sources, but the intensity I_1 of the light from S_1 is twice the intensity I_2 of the light from S_2. Give the ratio of the maximum to the minimum intensity of the light observed on the screen. (See Problem 26.)

70. (II) Figure 37–34 is an overhead view of two dipole radio antennas—vertical towers separated along the x-axis by a distance $\lambda/2$. This arrangement allows the radio signal to be beamed with greater intensities in some directions than in others, whereas either antenna alone would radiate its signal with the same intensity I_0 for any angle θ. (a) Find the intensity radiated by the antenna pair very far from the antennas as a function of θ, assuming that the signals of the two antennas are in phase. Describe the signal for all values of θ, from 0° to 360°. (b) The signal in the antennas is now 180° out of phase. How, if at all, does the distant intensity pattern change?

71. (II) Consider the pair of dipole antennas in Problem 70. Suppose that the antennas are separated by one-quarter wavelength and that the signal in antenna 1 lags the signal in antenna 2 by 90° (one-quarter cycle). Show that the signal has a maximum in one direction, not two.

72. (II) An interference experiment uses a triplet slit, with slit separations d. What are the positions of the maxima on a screen a distance R away, if light of wavelength λ shines through the slits. What are the intensities of the maxima in terms of the intensity that each slit would produce alone?

73. (II) Four identical loudspeakers are placed at the corners of a square with sides of length $\lambda/\sqrt{2}$. The loudspeakers emit sound coherently with wavelength λ. A listener is situated very far from the square along one of the diagonals. If the intensity with just one loudspeaker on is I_0, what are the intensities when two, three, and four speakers are on? In a table, list all combinations and the resultant intensities.

74. (II) Light of wavelength λ is incident at a normal angle θ_1 to a double slit with slit separation d. Show that if you observe the exit light at a normal angle of θ_2, interference peaks will be seen when $m\lambda/d = \sin\theta_1 + \sin\theta_2$, where m is an integer.

75. (III) Light of wavelength λ is incident on a slab of glass of thickness h and index of refraction n resting on a mirror. The ray makes an angle θ with the vertical. The light reflected at the top of the glass surface and that reflected at the mirror will interfere. For what angles will the interference be totally constructive? [For angles even slightly away from grazing incidence (see Problem 76), light undergoes no phase shift upon reflection by a conductor such as a mirror.]

76. (III) *Lloyd's mirror* is a mirror that reflects light at large angles of incidence from a point source to a screen (Fig. 37–35). As the angle of incidence nears 90° (a grazing angle), the source and image become close. (a) Will the direct light and reflected light interfere constructively or destructively in this limit? (b) Suppose that the (monochromatic) source is 20 cm from the center of the mirror and 0.50 mm above the plane of the mirror and that the screen is distant. What will the angular separation be between successive maxima of the interference pattern, where the angle is measured from the source point to the screen?

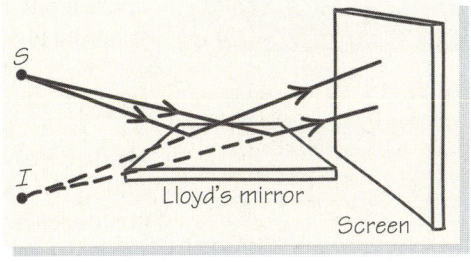

▲ **FIGURE 37–35** Problem 76.

Chapter 38

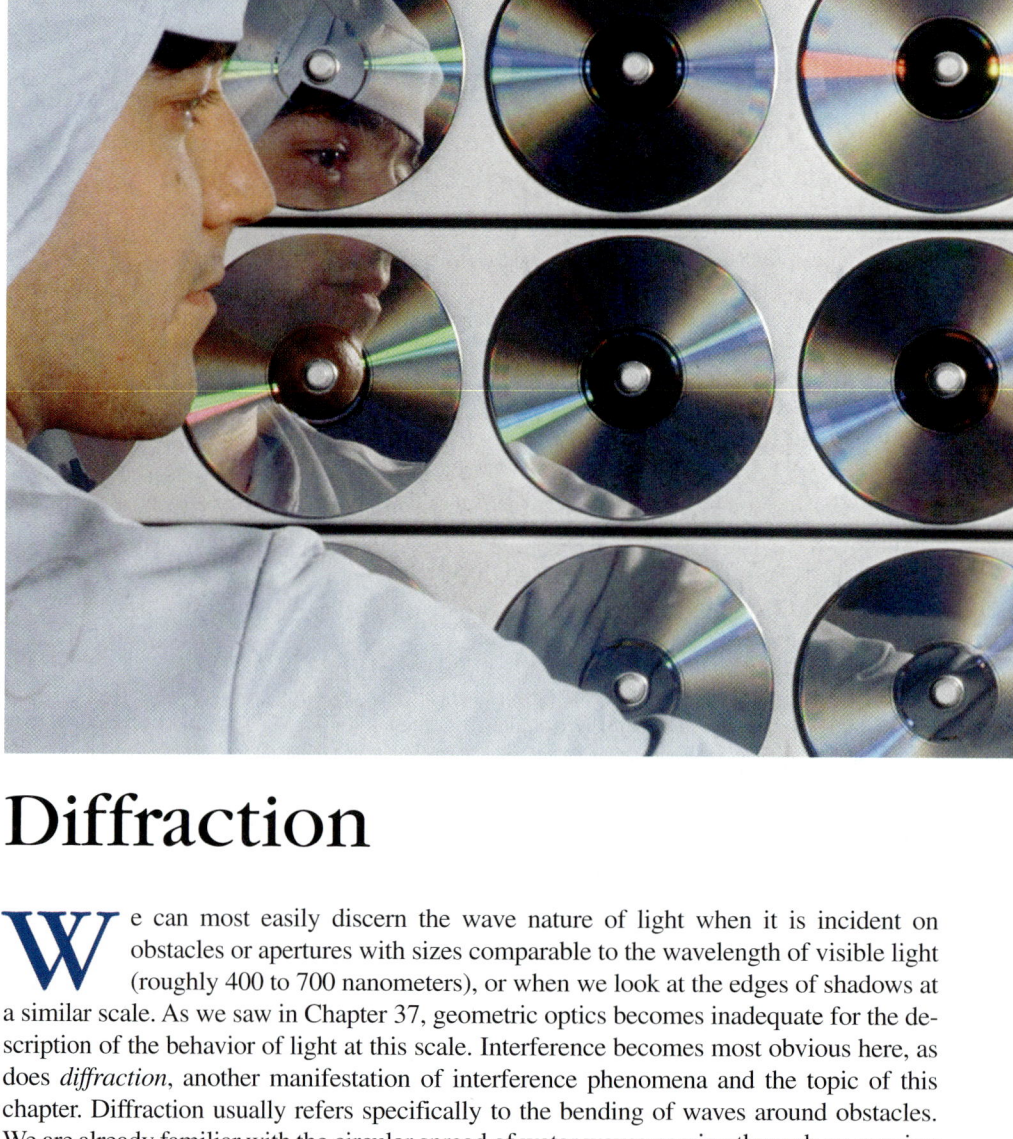

▶ The regular grooves in these compact disks act as diffraction gratings, which here exhibit maxima in reflection at angles that are different for light waves of different colors.

Diffraction

We can most easily discern the wave nature of light when it is incident on obstacles or apertures with sizes comparable to the wavelength of visible light (roughly 400 to 700 nanometers), or when we look at the edges of shadows at a similar scale. As we saw in Chapter 37, geometric optics becomes inadequate for the description of the behavior of light at this scale. Interference becomes most obvious here, as does *diffraction*, another manifestation of interference phenomena and the topic of this chapter. Diffraction usually refers specifically to the bending of waves around obstacles. We are already familiar with the circular spread of water waves passing through an opening narrower than their wavelength from Chapter 15, and we shall see that for light a single, narrow slit behaves similarly, producing characteristic patterns. This phenomenon is a diffractive effect. So is the pattern of maxima and minima that spreads across a screen in Young's double-slit experiment. The term diffraction also refers to interference between waves that emanate from a large number, or even a continuous set, of sources. Diffraction gratings consist of many slits or sources of coherent light, and such gratings have important applications in the study of atomic systems and crystalline materials. Holography is a spectacular application of the diffraction phenomenon with some interesting consequences.

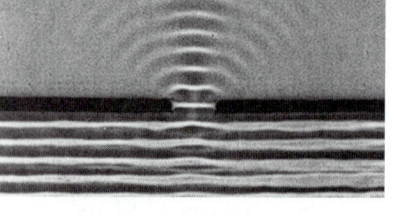

▲ **FIGURE 38–1** The way water waves bend around obstacles and through openings in obstacles is well known. Here, parallel water waves pass through a hole, producing circular wave fronts on the far side.

38–1 The Diffraction of Light

By the 1820s, Young's double-slit experiment had reinforced the wavelike interference effects associated with light, and serious attempts were under way to understand the consequences of these light waves. The bending of light waves around obstacles or at the edges of apertures (Fig. 38–1) is not part of our everyday experience. This is because *interference effects require coherent wave sources* and because *diffraction effects are typically most significant when the sizes of the apertures or obstacles involved are comparable to the wavelength.* For light, the coherence condition is not realized in most situations, and wavelengths—several hundred nanometers—are tiny compared to the sizes of familiar objects. Nevertheless, these effects can be very important for the kind of high-precision instruments that today's technology requires.

We can observe diffraction effects with a source of coherent light, an intermediate object in the form of an obstacle or a wall with holes, and a viewing screen (Fig. 38–2). For example, the shadow of a razor blade in Fig. 38–3 shows a diffraction pattern at its edges. Around 1810 the French physicist Augustin Fresnel systematically investigated the interference patterns from apertures, edges, and small obstacles, and these diffraction patterns form what is now known as *Fresnel diffraction*. Fresnel treated various interference phenomena using the Huygens construction (see Section 35–2). He showed that even a single aperture creates its own diffraction pattern, because waves passing through different parts of the aperture interfere with each other. Similarly, even a single obstacle creates a diffraction pattern because parts of the original plane wave have been blocked by the obstacle and no longer participate in the Huygens regeneration of the wave. By 1821, Fresnel's research had progressed to the point where he was able to use a primitive version of an interferometer to make the first quantitative measurement of the wavelength of light.

If both the source and the screen are far from the intermediate aperture or object that forms the pattern, the mathematics is considerably simplified (Fig. 38–4). This special case is known as *Fraunhofer diffraction*, after the Bavarian physicist Joseph von Fraunhofer, a contemporary of Fresnel. The Fraunhofer limit is easy to treat because the lines that originate at each position of the intermediate object and that reach a given point on the screen are nearly parallel; approximating them as parallel simplifies the calculation of path-length differences and phase differences. We used the geometry appropriate to the Fraunhofer case in Chapter 37 when we studied double-slit interference patterns. Starting with Section 38–2, we shall restrict ourselves to this case.

The types of things seen by Fresnel and Fraunhofer are visible to the naked eye with a little ingenuity. For example, try looking at a distant mercury-vapor street lamp through the narrowest possible slit you can make between your fingers. As you look through the slit, what do you see?

Before we proceed to more detail, let's look at one rather spectacular demonstration of diffraction effects observed by Arago in 1818. Suppose that we place a perfectly round obstacle in the path of a point source of coherent light, as in Fig. 38–5a. If every point on the rim of the disk is equidistant from the source, the light falling on the rim is perfectly in phase. According to the Huygens construction, we can think of each of these points as a new source, and they are all in phase. All the rim points are equidistant from the point P on a screen that lies on the symmetry axis between the disk and the source. Because the light reemitted in the Huygens construction from all points on the rim arrives at P in phase, there is constructive interference at P, and hence a bright spot appears on the screen at the center

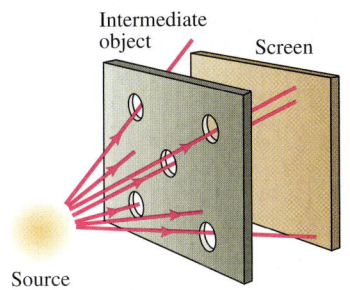

▲ **FIGURE 38–2** Diffraction effects can be observed by placing an intermediate object in the path of light that passes from a source to a viewing screen.

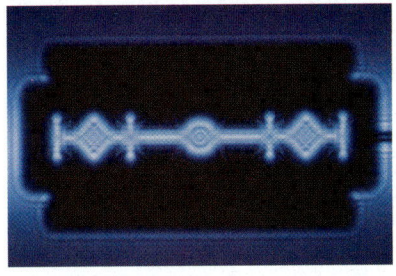

▲ **FIGURE 38–3** The diffraction of light around a razor blade.

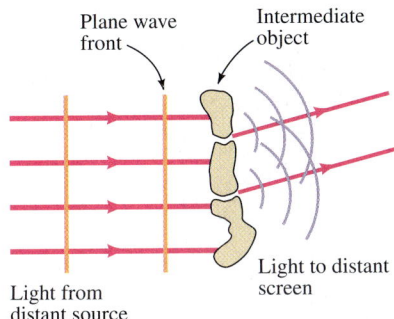

▲ **FIGURE 38–4** Parallel rays from a distant source are diffracted by an intermediate object and then viewed on a distant screen. When the screen is distant, we can treat the outgoing rays as parallel. This is the case of Fraunhofer diffraction.

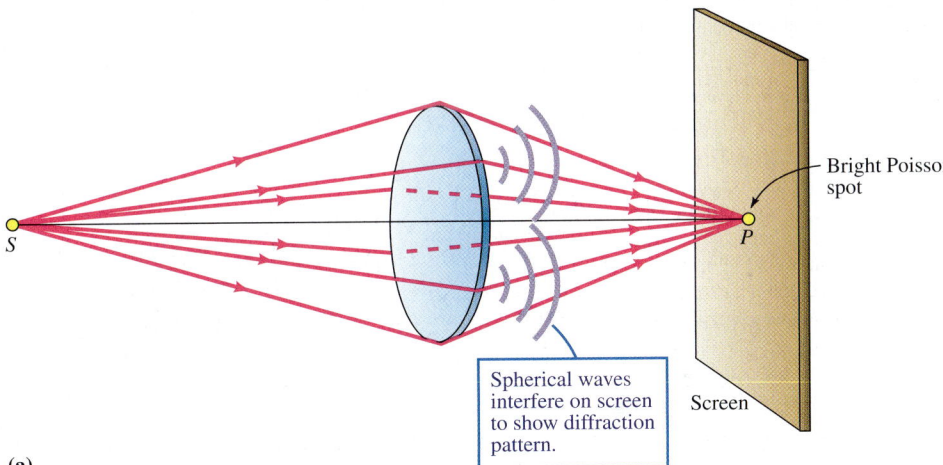

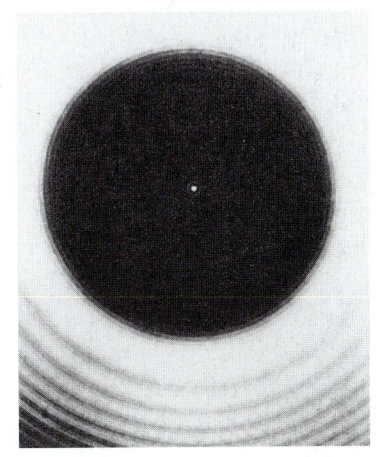

▲ **FIGURE 38–5** (a) A round, opaque object in the path of a point source of coherent light. Diffraction causes a bright spot (the Poisson spot) to be seen along the optical axis at point P. (b) The Poisson spot of a penny.

of the shadow, a central maximum known as the *Poisson spot*. This diffractive effect is certainly inconsistent with geometric optics! Even an ordinary penny can be used as the object; the Poisson spot at the center of its shadow in Fig. 38–5b is clearly visible.

38–2 Diffraction Gratings

A simple way to generalize the double-slit interference experiment that we studied in Chapter 37 is to increase the number of narrow slits. *If the slits are regularly spaced*, then a characteristic interference pattern is obtained; a screen with such an arrangement is called a **diffraction grating**. Diffraction gratings are important for two reasons: First, the multiple slits allow more light through than do two slits, thus increasing the intensity; second, the interference maxima are much sharper than they are for two slits, allowing the wavelength of the light to be measured more precisely.

A diffraction grating can take many forms. We simply require an array of obstacles to serve as pointlike sources for the reradiation of spherical wavelets. For example, when light passes through a glass plate scratched with rulings, the scratches act as sources for the regeneration of spherical wavelets. This type of grating is called a *transmission grating*. When the scratches are made on metal plates, they act as regular point sources of reflected rather than transmitted light; this grating is called a *reflection grating*. Even the marks on an ordinary ruler can be a reflection grating for laser light. What counts is that the light be scattered from regularly spaced centers.

THINK ABOUT THIS...
WHAT IS THE SIGNIFICANCE OF DIFFRACTION GRATINGS?

We stated above, and we'll show below in ample detail, that light that passes through a diffraction grating exhibits very sharp interference maxima and that this enables the measurement of light wavelengths to a very high precision. The ability to produce such gratings in a reliable way in the last half of the nineteenth century led to the discovery that light is emitted by excited atoms in discrete frequencies, and this discovery laid a good part of the foundations for quantum mechanics and, eventually, to an understanding of the structure of the atom. The unique set of wavelengths (or frequencies) produced by excited states of each type of atom, ion, or molecule is called its *spectrum*. Spectroscopes—optical devices used to measure so-called spectral lines, which are the intensity maxima corresponding to a particular wavelength—are typically based on diffraction gratings. Figure 38–6a is a schematic diagram of a spectroscope, and Fig. 38–6b shows the result of using a spectroscope to observe the light from a distant sodium-type street lamp. Diffraction gratings are still widely used in science and technology. Today we have a good understanding of the spectra of atoms and molecules. Knowing these wavelengths, which are completely characteristic of atoms and molecules of particular species, we can use a spectroscope as a very sensitive tool to recognize the presence of particular atoms or molecules in a distant star, or in a complex material. Observation of a given set of spectral lines is conclusive evidence of the presence of the emitting atom or molecule.

▶ **FIGURE 38–6**
(a) Spectroscopes that capture an entire spectrum spread over a large sheet of film, as in this schematic diagram, are sometimes called spectrographs. The spectrograph depicted here employs a grating that is located along the circle. Light coming from a point on that circle is reflected back to the film. (b) The characteristic light spectrum produced by a sodium-type street lamp, as observed through a spectroscope. The yellow lines are so bright that they mask other lines that are present.

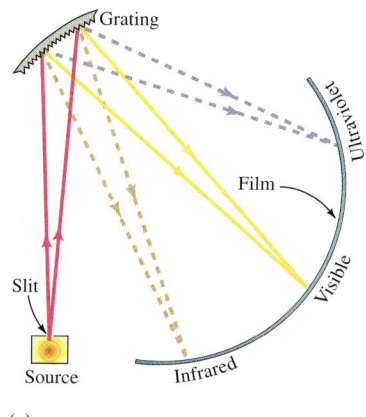

Energy Conservation and Intensity

Let's turn now to the analysis of the pattern made by a diffraction grating on a distant screen—the Fraunhofer diffraction pattern. Figure 38–7 shows a few of the slits of a grating with N slits, separated from each other by a distance d; a monochromatic plane wave of wavelength λ approaches from a distant source. The plane wave arrives with the same phase at each slit, and spherical waves are emitted in phase at each slit. Consider wave propagation along the lines labeled W, with lines $W1$, $W2$, $W3$, and so on oriented at an angle θ to the original wave-propagation direction. We allow the screen to be so distant that these lines are approximately parallel even though they all point toward a particular spot on the screen (or a lens focuses them on that spot).

If the wavelets are in phase along the front AA', defined by θ, then the light that eventually reaches the distant screen at this angle will also be in phase, and there will be constructive interference—a maximum. The condition under which the waves along AA' are in phase is that each path length differs from any other by integral multiples of the wavelength λ. The path-length difference between adjacent waves is $d \sin \theta$ (Fig. 38–7). Thus there are *principal maxima*, where the light from all the slits interferes constructively, at angles such that

$$d \sin \theta = m\lambda, \quad \text{where } m = 0, \pm 1, \pm 2, \ldots. \quad (38\text{–}1)$$

PRINCIPAL MAXIMA FOR A GRATING

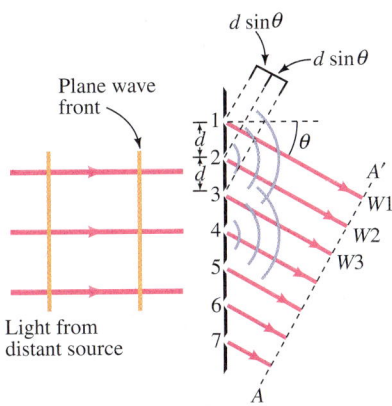

▲ **FIGURE 38–7** The geometry of a diffraction grating. Light passing through individual slits spreads in all directions. The interference of the light at a distant screen, along the lines indicated by $W1$, $W2$, $W3$, and so on, depends on the path-length differences $d \sin \theta$ between adjacent slits. The distant screen is not shown.

As is the case for the two-source pattern, the integer m specifies the *order* of the principal maxima. In this result, we see a universal diffraction phenomenon: *The pattern spreads in angle as the ratio λ/d increases.* Equation (38–1) is the same as Eq. (37–3a), which determines the maxima of the double-slit pattern. The intensity pattern on the screen, however, takes quite a different form.

Before performing the mathematical analysis for the intensity pattern, let's first do a qualitative analysis of intensity. If there were no interference whatsoever, the average intensity over the entire screen due to N slits would be NI_0, where I_0 is the average intensity for just one slit. Energy must be conserved whether or not there is interference, so the average intensity over the entire screen must be NI_0 even with interference. If the light intensity is zero in regions of destructive interference, there must be a higher light intensity in regions of constructive interference. What is the intensity at the principal maxima, where there is constructive interference? At such points, the electric or magnetic fields of the waves from all the slits add to N times the field from one slit. Because the intensity is proportional to the *square* of the fields, the intensity at any maximum is N^2 times the intensity I_0 due to one slit:

$$I_{\max} = N^2 I_0. \quad (38\text{–}2)$$

This result agrees with the calculation of the double-slit pattern, where we found that the heights of the maxima are $4I_0 = (2^2)I_0$. If the maximum intensity increases so markedly for N slits, then, in order to conserve energy, the space over which the maximum occurs must be much smaller. Suppose that the principal maxima have a *width* of $\Delta\theta$. (Qualitatively, you can think of the width as some reasonable measure of the spread in angle for which there is some substantial intensity. We make this quantitative by defining the width as the angular spread of the peak where the intensity is half its maximum height of $I_{\max}$.) To find $\Delta\theta$, we equate the intensity in a single maximum with the averaged intensity between maxima. Thus

$$I_{\max} \Delta\theta \cong NI_0 \times (\text{angular separation of successive maxima})$$
$$\cong NI_0(\lambda/d).$$

We solve this equation for the width, and when we use Eq. (38–2), we find

$$\Delta\theta = \frac{NI_0(\lambda/d)}{I_{\max}} = \frac{NI_0(\lambda/d)}{N^2 I_0} = \frac{1}{N}\frac{\lambda}{d}. \quad (38\text{–}3)$$

Intensity Pattern

Let us turn now to a more precise treatment of the pattern. We can sum the electric fields at the screen associated with each of the slits in the N-slit pattern and by squaring the time average of the *net* field we find the intensity. Although we shall not go through the details here, the result for the intensity is

$$\text{for multiple slits: } I = I_0 \left[\frac{\sin(N\beta)}{\sin \beta} \right]^2. \quad (38\text{-}4a)$$

The quantity 2β is the slit-to-slit phase difference. It is given by the product of the path length difference $d \sin \theta$ (see Fig. 38–7) and the wave number $2\pi/\lambda$:

$$2\beta = \frac{2\pi d \sin \theta}{\lambda}. \quad (38\text{-}4b)$$

We can show that this result agrees with the earlier qualitative argument based on energy. Using Eq. (38–1), we see that $\sin \beta$ is zero at principal maxima. As $\sin \beta \rightarrow 0$, l'Hôpital's rule tells us that the ratio $\sin(N\beta)/\sin \beta$ approaches N. In this limit, Eq. (38–4a) agrees with Eq. (38–2).

Equation (38–4a) is most easily understood by plotting it, as in Fig. 38–8 for $N = 2$, 4, and 10: The ratio I/I_0 is plotted against β. The figure shows the principal maximum at $\beta = 0$ (corresponding to $\theta = 0$), the central spot on the screen, as well as the first principal maxima to the sides, corresponding to $m = \pm 1$ in Eq. (38–1). The widths of the principal maxima do indeed decrease as $1/N$. (Note that there are also $N - 2$ small secondary maxima between each pair of principal maxima. They have intensities on the order of I_0 itself. For diffraction gratings in ordinary use, N is in the thousands, so the secondary maxima can safely be ignored. They occur because there is the possibility that the light from two or more of the non-neighboring slits may be in phase at a given angle, even though the light from *all* the slits is not in phase.)

We can check our general result for N slits by setting $N = 2$ and comparing with our result for the double slit. If we let $N = 2$ in Eq. (38–4) with $\sin(2\beta) = 2 \sin \beta \cos \beta$, we find

$$I = I_0 \left[\frac{\sin(2\beta)}{\sin \beta} \right]^2 = I_0 \frac{(2 \sin \beta \cos \beta)^2}{(\sin \beta)^2} = 4 I_0 \cos^2 \beta.$$

This is indeed the result for two slits, given in Eq. (37–13).

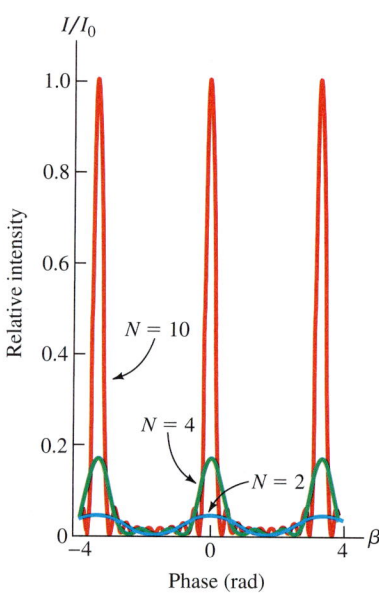

▲ **FIGURE 38–8** The ratio I/I_0 plotted against the phases $\beta = (\pi d \sin \theta)/\lambda$ for $N = 2$, 4, and 10 slits. The intensity pattern changes dramatically with N. The widths of the principal maxima decrease as $1/N$.

Resolution of Diffraction Gratings

Angular Dispersion: Diffraction gratings were formerly used to identify the characteristic wavelengths of elements. Now that these wavelengths are known, gratings are used primarily to *identify* elements, ions, and compounds through the characteristic light they emit or as a tool for understanding the structure of molecules. Because d, the distance between slits, is usually known, the angular location of a principal maximum ($m \neq 0$) gives λ according to Eq. (38–1). In this case, an important limitation is the ability to separate the spectral lines of nearly equal wavelengths λ_1 and λ_2. Two quantities determine the effectiveness of spectroscopic instruments. One is the **angular dispersion**, defined as $\Delta \theta / \Delta \lambda$, which measures the difference $\Delta \theta$ in the angles of the principal maxima of a given order due to two nearly equal wavelengths that differ by $\Delta \lambda$. Larger values of angular dispersion are desirable when one wants to distinguish two lines. We find the angular dispersion by considering two nearly identical wavelengths, λ and $\lambda + \Delta\lambda$, with maxima at the angles θ and $\theta + \Delta\theta$, respectively. Equation (38–1) gives us the difference in the angular position of their maxima

$$\sin(\theta + \Delta\theta) - \sin\theta = \frac{m}{d}[(\lambda + \Delta\lambda) - \lambda] = \frac{m(\Delta\lambda)}{d}.$$

If $\Delta\lambda$ is small, then so is $\Delta\theta$ and the difference $\sin(\theta + \Delta\theta) - \sin\theta$. We can approximate $\sin(\theta + \Delta\theta) = \sin\theta\cos(\Delta\theta) + \sin(\Delta\theta)\cos\theta \cong \sin\theta + \Delta\theta\cos\theta$, and hence $\sin(\theta + \Delta\theta) - \sin\theta \cong \Delta\theta\cos\theta$. The relation above can then be rearranged to read

$$\frac{\Delta\theta}{\Delta\lambda} = \frac{m}{d\cos\theta}. \quad (38\text{–}5)$$

The angular dispersion increases for higher orders, is inversely proportional to the distance between slits, and increases away from the central maximum.

Resolution: The angular dispersion alone does not tell us whether we can visually separate two similar wavelengths. This aspect of the effectiveness of a grating is characterized by the **resolving power** of the grating, defined by

$$R \equiv \frac{\lambda}{\Delta\lambda}. \quad (38\text{–}6)$$

Here, $\Delta\lambda$ is the smallest wavelength difference that can be observed with the grating. (The maxima of two wavelengths that are too close together lie so close to one another that they cannot be distinguished.) The larger the value of R, the better the grating can distinguish the relative wavelength difference of two closely spaced lines. Two closely spaced lines can be separated if the peaks are sharp. Detailed analysis of the peak widths of an N-slit system allows us to see that the resolving power of a grating is given by

$$R = mN. \quad (38\text{–}7)$$

The resolving power improves as the number of slits N increases and is better for larger orders; that is, for larger integers m. The improvement in the resolving power as N gets large is quite visible in Fig. 38–8.

EXAMPLE 38–1 Heated sodium is an easily available source of light. It emits light of a characteristic yellow–orange color with two intense wavelengths of 589.00 nm and 589.60 nm, called a *doublet*. (a) How many slits are required in a grating that resolves the doublet at the first-order maxima? (b) If the screen is 4.000 m from a grating with exactly 2000 slits/cm, what are the screen positions of the two principal maxima of first order? Assume that the screen is far enough away so that the conditions of Fraunhofer diffraction apply.

Strategy The solution to this problem lies in our discussion of the resolving power and the position of principal maxima. For part (a), the minimum resolving power needed to resolve two lines whose wavelengths differ by $\Delta\lambda$ is given by Eq. (38–6), $R = \lambda/\Delta\lambda$. Given R, Eq. (38–7) then allows us to find N. For part (b), the angular positions of the first-order principal maxima are given by Eq. (38–1) with $m = 1$.

Working It Out (a) We have $\Delta\lambda = 589.60$ nm $-$ 589.00 nm $=$ 0.60 nm, and hence the resolving power R is

$$R = \frac{\lambda}{\Delta\lambda} = \frac{589.0 \text{ nm}}{0.60 \text{ nm}} = 9.8 \times 10^2.$$

Equation (38–7), with $m = 1$, gives $N = R = 9.8 \times 10^2$. About 1000 slits are required.

(b) To use Eq. (38–1), we need the slit separation $d = 1/(2000 \text{ slits/cm}) = 5.000 \times 10^{-6}$ m. Then Eq. (38–1), with $m = 1$, gives

$$\text{for } \lambda_1: \sin\theta_1 = \frac{\lambda_1}{d} = \frac{589.0 \text{ nm}}{5.000 \times 10^{-6} \text{ m}} = 0.1178;$$

$$\text{for } \lambda_2: \sin\theta_2 = \frac{\lambda_2}{d} = \frac{589.6 \text{ nm}}{5.000 \times 10^{-6} \text{ m}} = 0.1179.$$

The respective angles are $\theta_1 = 0.1181$ rad and $\theta_2 = 0.1182$ rad, and the distance from the centerline of the screen is $y = L\tan\theta$. These respective distances are

$$y_1 = (4.000 \text{ m})\tan(0.1181 \text{ rad}) = 0.4745 \text{ m};$$
$$y_2 = (4.000 \text{ m})\tan(0.1182 \text{ rad}) = 0.4750 \text{ m}.$$

The images are separated by only 0.5 mm, but this resolution is sufficient to distinguish the two spectral lines.

What Do You Think? If the order $m = 1$ is not sufficient to separate the two sodium lines, would it help to look at higher orders? Answers to **What Do You Think?** questions are given in the back of the book.

CONCEPTUAL EXAMPLE 38–2 Take a compact disc (CD) and hold it below a lightbulb in a horizontal position such that when you look down on the CD, the reflection of the lightbulb overlaps the hole in the middle. As you tilt the CD away from the horizontal, you will see a swath of brilliant colors coming in from the rim toward the center. Why is this? Which colors come first? Will the colors repeat? Finally, could you estimate the spacing between the grooves on the CD from your experiment?

Answer The CD, which contains very finely spaced grooves, acts as a reflection grating. You start by looking straight down on the disc, that is, with an angle of reflection θ of zero. As you tilt it, you increase the reflection angle, and with it the groove-to-groove phase difference β [see Eq. (38–4b)]. This phase difference depends inversely on the wavelength λ so that it is larger for blue light than for red light, and the blue light will be seen first, with a sequence of the colors of the rainbow all the way to red light. Further tilt will bring in higher orders (secondary maxima), so there will be a repeat of the sweep of colors. Finally, β also depends on the groove spacing d, so that if you know the wavelength of, say, blue light and can estimate the tilt angle, you can measure d. You can certainly try this out for yourself (Fig. 38–9).

What Do You Think? What would you expect to see if you did the same experiment in the light of a helium-neon laser ($\lambda = 633$ nm) in an otherwise dark room?

▲ **FIGURE 38–9** The grooves on a CD can act as a diffraction grating, here separating the colors of a light bulb.

38–3 Single-Slit Diffraction

Coherent light passing through even a *single* slit produces a diffraction pattern. In the language of Huygens' principle, this pattern forms because wavelets that have regenerated at different places across the single slit interfere with each other. If the width a of the single slit is comparable to or smaller than the wavelength of coherent light that passes through it, the diffraction pattern is quite evident. We can explain it by treating the single slit as an infinitely large number of infinitesimal sources of wavelets.

Let's consider a plane light wave of wavelength λ that moves toward a wall with a narrow rectangular slit of width a ($a > \lambda$). A coherent wave arrives at the slit and regenerates spherical waves at each point across it. In Fig. 38–10a, we examine the light from the slit that continues along the initial direction toward the center point P of a distant screen. If the screen is far enough, we can assume that to a good approximation the waves from all parts of the slit travel the same distance, so that the regenerated wavelets are all in phase in this direction, and the central point of the screen is bright.

Along the direction that leads to point P' on the screen (Fig. 38–10b), with the angle given by $\sin\theta = \lambda/a$, there will be destructive interference under certain conditions. The path lengths from the tilted line s_1s_2 to point P' are all the same because the screen is distant, so we must consider only the phase relations of the light waves at line s_1s_2. The wave emitted at the top of the slit has traveled a distance λ to point s_2, and the wave emitted from the midpoint of the slit has traveled a distance $\lambda/2$ to point s_3. Thus, along the line s_1s_2, the wave emitted at the top is *out of phase* with the wave emitted at the center of the slit, and the waves will have this same phase relation at point P'. Similarly, the wave emitted just below the top of the slit is out of phase with the wave emitted from just below the center. We can follow the points *in pairs* along the slit. For every point in the top half of the slit, there is a point in the bottom half, and the waves from the two points are precisely out of phase with each other. The result is destructive interference—a minimum (or dark spot)—on the screen at P' at the angle given by $\sin\theta = \lambda/a$.

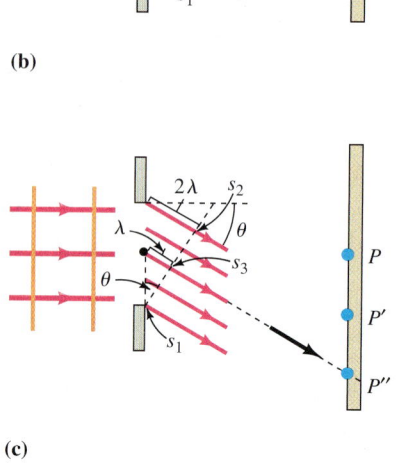

◀ **FIGURE 38–10** Monochromatic plane waves of light enter a narrow slit of width a (exaggerated here). We show three positions (P, P', and P'') on a distant screen. (a) We look along the incident direction at P. (b) We look along angle θ at P'. If the light from positions in the top half of the slit are out of phase with corresponding positions in the bottom half, destructive interference occurs at P'. (c) We look along a different angle θ at P''. Destructive interference can occur also at P'', as demonstrated by breaking the slit into halves and treating each half as we did for the direction toward point P'.

Along the direction given by $\sin\theta = 2\lambda/a$ (Fig. 38–10c), the wavelet emitted from the top of the slit travels a distance 2λ farther than the wavelet emitted from the bottom, and a distance λ farther than the wavelet emitted from the center point. We can think of the slit of width a as separated into two slits of width $a/2$. Along the direction chosen, we are in a situation similar to that in Fig. 38–10b. There is destructive interference at point P'' due to net destructive interference from both the top half and the bottom half of the slit. For example, for every wavelet emitted from a particular point in the top half of the slit, we can find a second wavelet emitted from another point in the top half that destructively interferes with the first wavelet, because their path-length difference is $\lambda/2$. The same happens for the bottom half. We will have another intensity minimum (dark spot) on the screen for angle $\sin\theta = 2\lambda/a$.

If we continue our analysis, we find that every time there is an additional path-length difference of λ between the top and bottom of the slit, destructive interference and a screen minimum result. Thus we have

$$\sin\theta = \frac{m\lambda}{a}, \qquad \text{where } m = \pm 1, \pm 2, \pm 3, \dots. \qquad (38\text{-}8)$$

MINIMA FOR A SINGLE SLIT

The value $m = 0$ is not part of this sequence of minima: For $m = 0$, $\sin\theta = m\lambda/a = 0$, and we have seen that this central point P must always be a maximum.

The interference pattern (Fig. 38–11) has the typical behavior of diffractive phenomena: Larger values of a/λ give a smaller angular spread in the interference pattern, and smaller values of a/λ give a larger angular spread. In the limit that $a \gg \lambda$, the spread decreases so much that there is only a bright central spot on the screen. Do not confuse the angular spread on the screen with a projection of a slit width in geometric optics. The screen is very distant, and if geometric optics were to hold, the projection of the slit on the screen due to direct incident light would be a line with precisely the width of the slit. *But geometric optics does not hold; the smaller the slit width compared to the wavelength, the larger the angular spread of the interference pattern.*

Approximately halfway between the successive minima, we have conditions for constructive interference and an intensity maximum (a bright region). In other words, in addition to the constructive interference at $\theta = 0$ (the central maximum), we also have

for constructive interference: $\sin\theta \cong \dfrac{(m + \frac{1}{2})\lambda}{a}, \qquad \text{where } m = \pm 1, \pm 2, \dots. \qquad (38\text{-}9)$

We'll work out more details of the intensity pattern in the two subsections to follow.

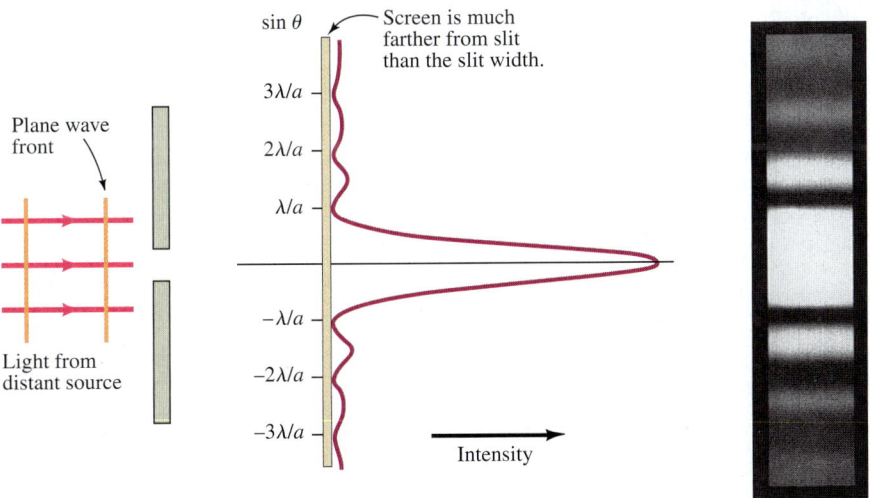

▲ FIGURE 38–11 The interference pattern of single-slit diffraction and the relative intensities of such a pattern. Most of the light energy is in the bright central peak. The central peak is twice as wide as the secondary peaks.

EXAMPLE 38–3 Helium–neon laser light of wavelength 633 nm passes through a single slit of width 0.10 mm. The diffraction pattern is observed on a screen 3 m away. What is the distance between the two minima on either side of the central maximum?

Strategy The conditions for the single-slit pattern apply here; in particular, the slit width a is some 160 times larger than the wavelength. We can use Eq. (38–8) to find the angular positions of the minima that correspond to $m = 1$ and -1, and with the angular positions known geometry gives the positions on the screen.

Working It Out The minima for $m = \pm 1$ occur at angles

$$\sin\theta = \frac{m\lambda}{a} = (\pm 1)\frac{633 \times 10^{-9}\text{ m}}{0.10 \times 10^{-3}\text{ m}} = \pm 0.0063;$$

$$\theta = \pm 0.36°.$$

At a distance of 3 m away, these minima will be $(3\text{ m})\tan 0.36° = 1.9$ cm from the central line ($\theta = 0$). The total distance between them is thus $2(1.9\text{ cm}) = 3.8$ cm.

What Do You Think? If coherent blue light is used instead of the red light of the helium-neon laser, would the distance between minima increase or decrease?

The Intensity Pattern of Single-Slit Diffraction

As we shall show in the next subsection, the intensity pattern on a distant screen due to a single slit of width a takes the form

$$\text{for single slits: } I = I_{max}\frac{\sin^2\alpha}{\alpha^2}, \quad (38\text{–}10)$$

where

$$\alpha = \frac{\pi a \sin\theta}{\lambda}. \quad (38\text{–}11)$$

As we learned at the start of this section, the angle α, which is measured in radians, is simply the phase difference between the top and the middle of the slit.

As is confirmed by Fig. 38–11, there is a central maximum at $\theta = 0$: At $\theta = 0$, Eq. (38–11) shows that $\alpha = 0$. Because the limit of $(\sin\alpha)/\alpha \to 1$ as $\alpha \to 0$, the intensity at this point on the screen is just I_{max}. The intensity drops off rapidly as α increases. The intensity is zero for angles given by Eq. (38–8), and this occurs when α is an integral multiple of π.

$$\text{for minima: } \alpha = n\pi = \frac{\pi a \sin\theta}{\lambda}, \quad \text{where } n = \pm 1, \pm 2, \pm 3,\ldots. \quad (38\text{–}12)$$

The intensities at the secondary maxima are estimated in Example 38–4.

By looking at Fig. 38–12, for which we have chosen $a = 4\lambda$, we can identify two important features. First, the secondary maxima are not very strong—most of the light is contained in the wide central maximum at $\theta = 0$. This contrasts with the pattern made by a grating, for which there is no difference in the strength of the successive maxima. Second, the central maximum is *twice* as broad as the secondary maxima are—a feature that distinguishes the single-slit pattern from the double- or multiple-slit patterns.

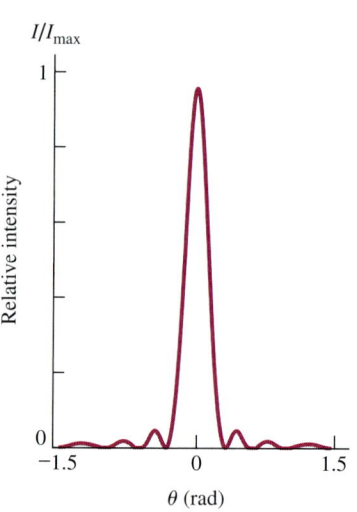

▲ **FIGURE 38–12** The intensity ratio I/I_{max} as a function of angle θ [from Eqs. (38–10) and (38–11)] for the slit width $a = 4\lambda$.

EXAMPLE 38–4 Estimate the ratios of the intensities of the first and second maxima to the intensity of the central maximum for a single slit.

Strategy We can use our approximate relation, Eq. (38–9), to locate the maxima. This relation places the maxima halfway between the minima, and in terms of α this means

$$\text{for maxima: } \alpha \cong (n + \tfrac{1}{2})\pi, \quad n = \pm 1, \pm 2, \ldots.$$

The intensity need only be evaluated at these locations.

Working It Out Denote the intensity at the secondary maxima by I_n. Then Eq. (38–10) gives

$$\frac{I_n}{I_{max}} = \left\{\frac{\sin\left[\left(n + \tfrac{1}{2}\right)\pi\right]}{\left(n + \tfrac{1}{2}\right)\pi}\right\}^2 = \frac{1}{\left(n + \tfrac{1}{2}\right)^2\pi^2}.$$

[We have used $\sin^2(\pi/2) = \sin^2(3\pi/2) = \ldots = 1$.] With $n = 1$ and $n = 2$, we get

$$\frac{I_1}{I_0} = 0.045 \quad \text{and} \quad \frac{I_2}{I_0} = 0.016.$$

The intensities of the secondary maxima fall off rapidly, with most of the intensity in the central bright maximum.

What Do You Think? We approximated a midway location for the maxima; how can we find an exact expression for the location of the maxima in single-slit diffraction?

*Deriving Single-Slit Intensity Values

Let's imagine breaking up a single slit of width a into N strips of width d (Fig. 38–13), so that $Nd = a$. Each strip acts as a separate slit. In the limit that the number of strips $N \to \infty$, d must approach zero in order to keep a constant. Because each strip is infinitely narrow, we can treat these strips as the thin slits of a grating. We can then use the diffraction grating result of Eq. (38–4) directly to find that the intensity at the screen at angle θ is

$$I = \lim_{N \to \infty} I_0 \left[\frac{\sin(N\beta)}{\sin \beta} \right]^2.$$

This expression includes a slit width $d = a/N$ that approaches zero as $N \to \infty$ through the quantity β. According to Eq. (38–4), $\beta = [\pi d \sin \theta]/\lambda = [\pi(a/N) \sin \theta]/\lambda = \alpha/N$, where the definition of α is contained in Eq. (38–11). Thus the intensity takes the form

$$I = \lim_{N \to \infty} I_0 \left[\frac{\sin \alpha}{\sin(\alpha/N)} \right]^2.$$

In the limit of large N, the factor $\sin(\alpha/N) \cong \alpha/N$, and

$$I = N^2 I_0 \frac{\sin^2 \alpha}{\alpha^2}.$$

The factor I_0 is the intensity due to one of the subslits of width d. We need only interpret the factor $N^2 I_0$ as the maximum possible intensity I_{max} of the single slit of width a to get Eq. (38–10).

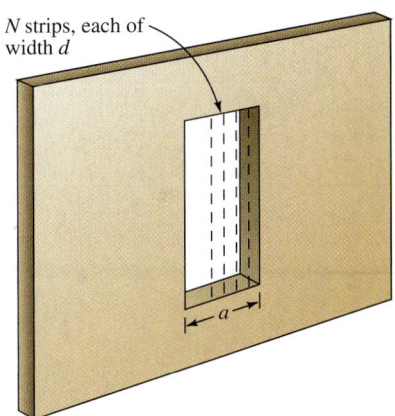

▲ **FIGURE 38–13** We can understand the intensity pattern of a single slit by supposing that the slit is composed of a large number of strips N and then using the result we derived for diffraction gratings.

38–4 Resolution of Optical Instruments

We have already seen that the ability of a grating to resolve closely spaced lines is limited by the width of the principal maxima. Similarly, the fact that there is some spreading of light in a single aperture due to diffraction *intrinsically* limits the capacity of optical instruments to resolve objects. The resolution of instruments such as telescopes or microscopes is also limited by lens aberration (see Chapter 36), but better lens design decreases aberration, and there is no theoretical limit to such improvement. The limitation due to diffraction, however, is set by the aperture of the instrument and the wavelength of light, and this limit is intrinsic to the wave nature of light.

Most optical instruments rely on circular lenses or mirrors, so we shall concentrate on circular apertures. Sir George Airy worked out the diffraction pattern from a distant point source that passes through a circular aperture (Fig. 38–14) in the 1830s. The bright central area—containing some 85 percent of the light intensity—is called an *Airy disk*. The rings outside the central area are the minima and secondary maxima of the diffraction pattern. The position of the first minimum occurs at an angle from the central axis given by

$$\theta_{min} = 1.22 \frac{\lambda}{D}, \qquad (38\text{–}13)$$

where D is the diameter of the aperture. This result can be compared to Eq. (38–8), which gives the angle of the first minimum for a slit of width $a \gg \lambda$ as $\theta_{min} = \lambda/a$. (Recall that $\sin \theta \cong \theta$ for small θ.) The factor 1.22 arises because the "width" of a circular aperture varies (in effect, $a \cong D/1.22$). But it is enough to use a factor 1 instead of the precise factor 1.22 to understand the effects of diffraction. Thus we give an approximate θ_{min} as

$$\theta_{min} \cong \frac{\lambda}{D}. \qquad (38\text{–}14)$$

MINIMUM OBSERVABLE ANGULAR SEPARATION

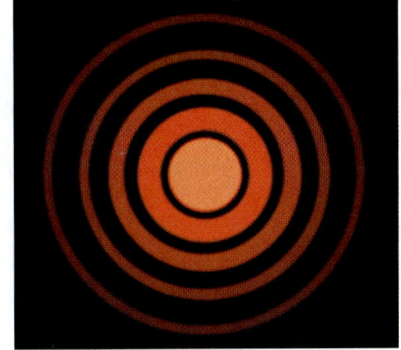

(a)

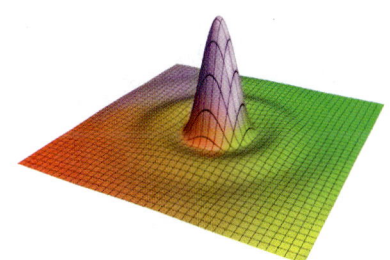

(b)

▲ **FIGURE 38–14** (a) Diffraction pattern of light from a distant source after the light passes through a circular aperture. Some 85 percent of the light intensity is contained in the bright central maximum, called the Airy disk. (b) A three-dimensional representation of the light intensity on a distant screen from a circular aperture.

How does the presence of a diffraction pattern limit our ability to make images? In many applications high resolution is a necessity; for example, we may want to observe two closely spaced objects, such as a double star, or we may need to see detail in an X

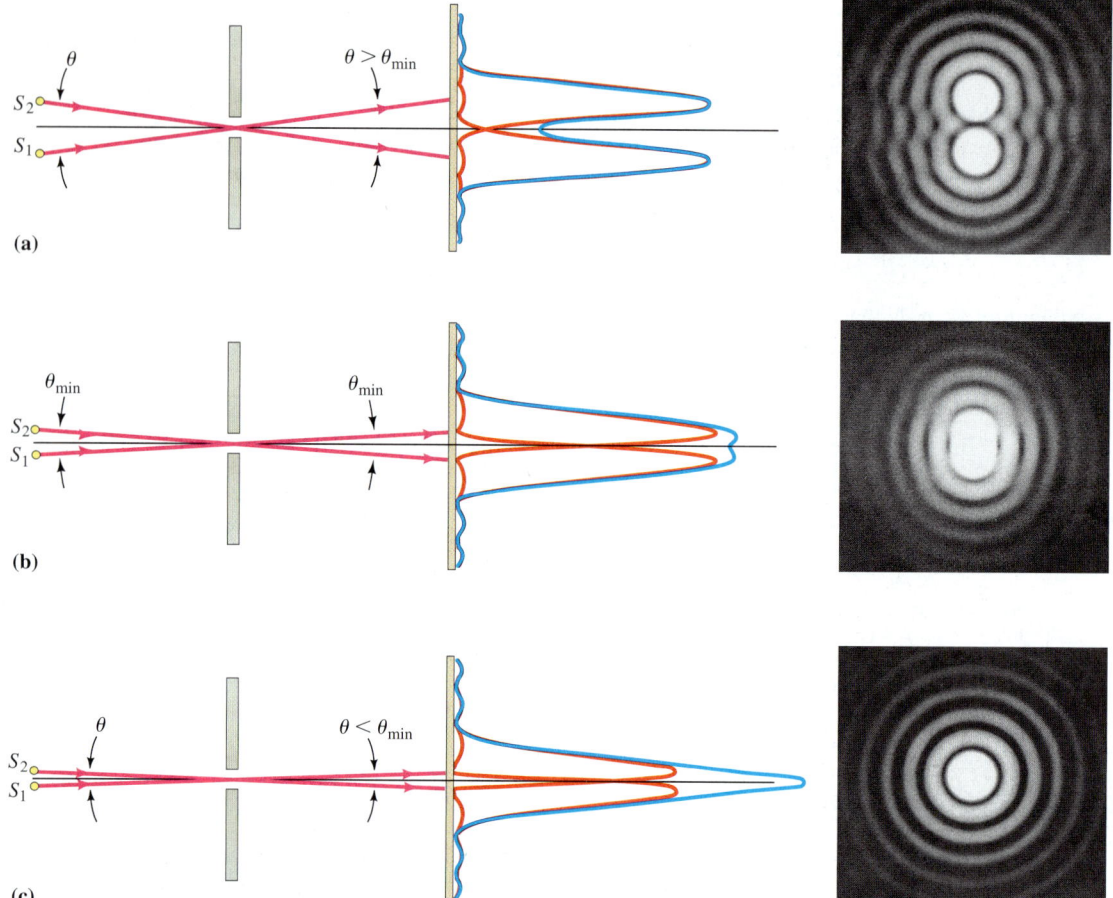

▲ FIGURE 38–15 Light from two sources passes through an aperture. (a) The objects are easily resolved. (b) The objects are just barely resolved. The angular separation is θ_{min}. (c) The objects are not resolved. The blue lines in each represent the sum of the two intensities.

ray taken of suspected stress fractures in piping for a power plant. The presence of the Airy disk means that even a very distant star does not produce a pointlike image but rather a disklike image with an angular spread described by Eq. (38–14). The image in a telescope of two stars that are so close together that their Airy disks overlap cannot be recognized as an image of two stars. In Fig. 38–15a, two objects have a sufficient angular separation θ that they are easily resolved. In Fig. 38–15b, the objects are just barely resolved. When they are even closer, as in Fig. 38–15c, the central diffraction peaks overlap too much to be resolved. It has become customary to describe the limiting case—shown in Fig. 38–15b—as the *Rayleigh criterion*:

> **Two point sources are just resolved if the peak of the diffraction image of the first source overlies the first minimum of the diffraction image of the second source.**

The Rayleigh criterion is satisfied when the angular separation of the objects is just θ_{min}, defined by Eq. (38–13). Equation (38–14) provides an approximate alternative. Remember, the quality of the optical instruments is not the issue here; the limitation on the resolution of images is due to light's intrinsic wave character and can be improved only by making the instrument aperture larger.

The angular separation of an image formed by an optical system is also the angular separation of the objects. We can show this using the methods we used when we studied optical instruments in Chapter 36. These angular separations are the same because the principal rays through the center of a lens are undeviated. These rays are shown in Fig. 38–16 for two objects at P_1 and P_2 respectively, with the objects a distance L from a lens of diameter D. When this angle is less than that specified by the Rayleigh criterion, the objects cannot be separated. This translates into a minimum spatial separation S_{min} of the two objects through the relation $S_{min} = L\theta_{min}$:

$$S_{min} = L\theta_{min} \cong \frac{L\lambda}{D}. \qquad (38\text{–}15)$$

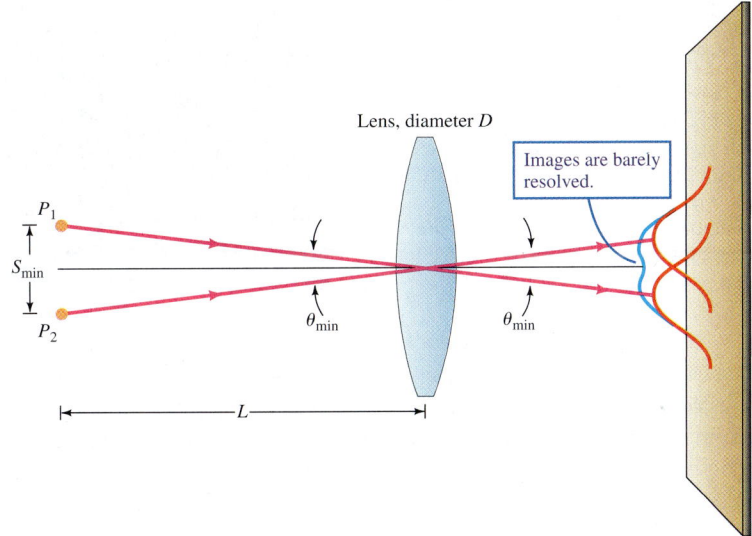

► **FIGURE 38–16** The minimum separation distance that is resolvable for a lens depends on θ_{min}.

EXAMPLE 38–5 Estimate the minimum separation between two objects such that the human eye can still perceive them as separate (the *minimum visible object separation*) if the objects are (a) at the near-point distance (25 cm) and (b) at a distance of 5 m. Take the pupil diameter to be 2.5 mm.

Setting It Up We can apply Eq. (38–15) if we can estimate the eye's aperture as well as a wavelength. You can look in a mirror and estimate the pupil diameter to be 2.5 mm, and for the wavelength, we choose a median range of visible wavelengths, about 550 nm.

Strategy This is straightforward application of Eq. (38–15).

Working It Out (a) For $L = 25$ cm (the near point), we have

$$S_{min} = \frac{L\lambda}{D} = \frac{(0.25 \text{ m})(550 \times 10^{-9} \text{ m})}{0.0025 \text{ m}} = 0.055 \text{ mm}.$$

This is about the diameter of a thin thread or a human hair. It is also, roughly, the separation between the cells of the retina. In other words, the cells that receive light and send that message to the brain are no closer than the minimum separation that could ever be resolved, an admirable example of the economy of biological systems.

(b) At 5 m, the minimum separation distance becomes

$$S_{min} = \frac{L\lambda}{D} = \frac{(5.0 \text{ m})(550 \times 10^{-9} \text{ m})}{0.0025 \text{ m}} = 1.1 \text{ mm}.$$

Thus, from a few meters away, we have an *intrinsic* inability to distinguish the millimeter markings on a meter stick.

What Do You Think? What is the separation distance that can be resolved when the two separated objects are at infinity?

The Resolution of Telescopes

Depending on the case, diffraction effects can limit the effectiveness of telescopes. For the visible-light region of the electromagnetic spectrum, one of the world's largest telescopes is the 200-in-diameter Hale telescope on Mt. Palomar in southern California. Its diffraction limit alone implies an angular resolution of about 0.03 seconds of arc. Its real resolution is some 300 times worse than this due to the effects of atmospheric turbulence and aberration. The New Technology Telescope in Chile, another visible-light telescope, has resolved 0.36 seconds of arc, much closer to the diffraction limit. However, as the next example shows, the Hubble Space Telescope achieves a resolution that is very close to the diffraction limit.

EXAMPLE 38–6 The primary mirror on the Optical Telescope Assembly on the Hubble Space Telescope (HST), which orbits 600 km above Earth, has a diameter of 2.4 m. (a) Calculate the minimum angular separation that it might resolve for visible light (about 550 nm). (b) Assume that the telescope is viewing Earth's surface. What is the separation of the most closely spaced objects that it might resolve? Ignore all atmospheric effects.

Strategy (a) This problem is a direct application of the resolution angle of Eq. (38–14). For (b) we simply use $S_{min} = L\theta_{min}$, where $L = 600$ km and S_{min} is the separation distance.

Working It Out (a) Equation (38–14) gives

$$\theta_{min} \cong \frac{\lambda}{D} = \frac{(550 \times 10^{-9} \text{ m})}{2.4 \text{ m}} = 2.3 \times 10^{-7} \text{ rad} = 0.047'' \text{ of arc}.$$

The actual resolution of the instrument is 0.1 seconds of arc, to a good approximation the diffraction limit. The HST can do so much better than comparable Earth-based telescopes because the HST avoids all atmospheric turbulence.

(b) $S_{min} = L\theta_{min} = (600 \text{ km})(2.3 \times 10^{-7} \text{ rad}) = 0.14 \text{ m} = 14 \text{ cm}.$

The HST would not avoid atmospheric turbulence if it looked down, so 14 cm is better than the actual resolution that would be obtainable with this instrument if we were to point it toward the surface. Earth-observation satellites such as Landsat and Spot have resolution capabilities of a few meters, and classified spy satellites are reported to have resolutions of less than 1 m.

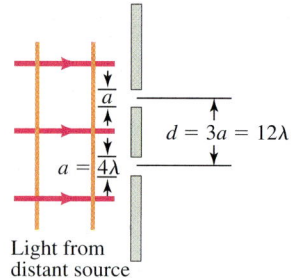

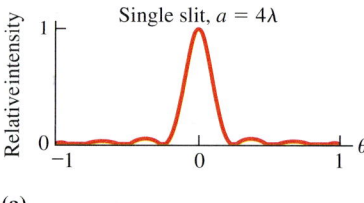

(a)

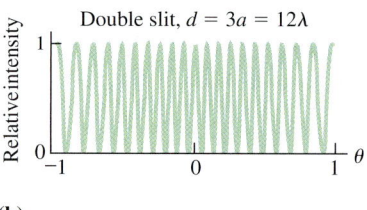

(b)

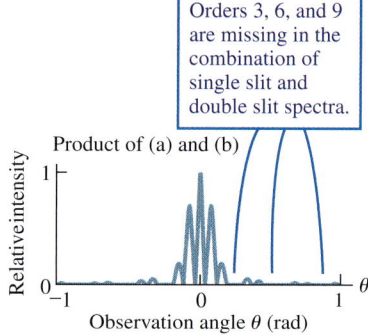

(c)

▲ **FIGURE 38–18** Intensity patterns as a function of observation angle θ for diffraction from multiple slits must include the effects of single-slit diffraction. For $a = 4\lambda$, (a) the intensity pattern for a single slit, (b) a double slit ($d = 12\lambda$), and (c) their product, which is the observed pattern. Missing orders occur; in this case, $d = 3a$, and the 3rd, 6th, 9th, ... orders are missing from the overall pattern.

▲ **FIGURE 38–19** Diffraction pattern for multiple slits where $d = 10a$. Note the missing orders.

▶ **FIGURE 38–17** The complex of telescopes on the Mauna Kea volcano in Hawaii is one of the great observatories of the world. The twin Keck telescopes, visible at the center, can act as a single telescope whose mirror is separated into two pieces.

An optical device need not consist of a single aperture. The mirrors of many of the largest modern reflecting telescopes are separated into several pieces (Fig. 38–17), like a mosaic, in part because a single large piece of glass is difficult to form and handle. Such instruments diminish diffractive effects because the minimum angle that can be resolved is determined by the interference from the most widely separated pieces of the apparatus. If this maximum separation is D, then the minimum resolvable angle between objects is λ/D. (Note that D refers to a *transverse* separation; the distance between an eyepiece and an objective does not enter into diffractive effects.) An array of small electronically connected optical telescopes on the Moon spread over a region with a diameter of 10 km would have a resolution 100,000 times better than that of the best telescope on Earth. Were it not for Earth's atmosphere, such an instrument could pick up a newspaper headline on Earth!

38–5 Slit Width and Grating Patterns

Our discussions of double- and multiple-slit diffraction patterns have treated each slit as a point source of light that emits a single spherical wave. The angular spread of the diffraction pattern depends on the parameter d/λ, where d is the slit separation. We have now seen that single slits of finite width have their own diffraction pattern. The angular spread of this pattern depends on a/λ, where a is the slit width. What effect does a finite slit width have on the multiple-slit pattern? For Fraunhofer diffraction, the overall intensity distribution is the product of the two intensity patterns. The pattern I_{mult}, corresponding to the double or multiple slit, is multiplied by the pattern I_{single}, corresponding to the single slit. The multiple-slit intensity is given by Eq. (38–4), whereas the single-slit intensity is given by Eq. (38–10). Thus their product is

$$I = I_{\text{multi}} I_{\text{single}} = I_{\text{max}} \left[\frac{\sin(N\beta)}{\sin \beta} \right]^2 \left(\frac{\sin \alpha}{\alpha} \right)^2, \quad (38\text{–}16)$$

where we recall that

$$\beta = \frac{\pi d \sin \theta}{\lambda} \quad \text{and} \quad \alpha = \frac{\pi a \sin \theta}{\lambda}.$$

In the equation for the intensity, we have combined the maximum intensity factors into a single maximum intensity I_{max}.

The fact that the intensity patterns are multiplied means that the broader pattern (usually due to the single slit) acts as an envelope for the narrower pattern. For example, suppose that $d = 3a$ for a double slit ($N = 2$). In this case, the individual slit pattern is much broader than the multiple-slit pattern. At the same time, let $a = 4\lambda$, so that the single-slit pattern is easily distinguishable. Figure 38–18 shows the single-slit pattern, double-slit pattern, and combined pattern. Note that certain maxima of the double-slit pattern are absent from the combined pattern because they fall where the minima of the single-slit diffraction pattern occur. These missing maxima are called *missing orders*. The locations of missing orders are independent of λ, as Problem 43 illustrates. A measurement of the pattern described here, but with $d = 10a$, is shown in Fig. 38–19.

*38–6 X-Ray Diffraction

We have been emphasizing the use of gratings as a tool for the exploration of diffracted light. Light is just one form of electromagnetic radiation, and other electromagnetic waves can also be diffracted. Let's take a look at the diffraction of X rays, an important tool in the understanding of crystalline solids. Diffraction gratings work because the apertures or obstacles serve as rescatterers. A powerful constructive interference occurs among rescattered light from *all* the apertures or obstacles because the sources of wavelets are in a *regular* pattern. A crystalline solid contains a regular array of obstacles and therefore forms a natural diffraction grating. In a crystalline solid the array is spread over three dimensions rather than two, but the effect works in just the same way. If the electromagnetic waves can penetrate the material, each atom in the array can serve as a rescatterer, and the diffraction pattern will be characteristic of the particular way in which the rescatterers are arranged.

In 1895, Wilhelm Roentgen discovered that radiation was produced when he bombarded metal with high-energy *cathode rays* (now called electrons). This radiation was unlike any seen previously, and Roentgen called it *X rays*. Shortly thereafter, he produced the first X-ray picture, a human hand (Fig. 38–20). We know now that X rays are just electromagnetic radiation with wavelengths in the range of about 0.01 nm to 10 nm. This radiation is produced when atomic electrons change states within atoms, or when electrons are accelerated (or decelerated).

In the early 1900s, it was suspected that X rays might be some form of electromagnetic radiation. A diffraction experiment reported in 1899 vaguely suggested that X rays might have wavelengths of about 0.1 nm, much smaller than those of visible light. At the same time, some scientists suspected that solids might be made of atoms arranged in regular arrays. In 1912, Max von Laue had the idea of scattering X rays from solids. If X rays had about the same wavelength as the distance between the arrays of atoms (about 0.1 nm), then diffraction effects would be significant. Von Laue was interested in both a tool for the precise measurement of the wavelengths of X rays and a tool for the exploration of crystals. He convinced two of his colleagues, Friedrich and Knipping, to perform an experiment, and the observation of X-ray diffraction soon followed. Von Laue's idea was a crucial step in the measurement of X-ray spectra and led to a revolution in our ability to study the nature of solids and the molecules that compose them. The precise knowledge that table salt, NaCl, has the three-dimensional structure shown in Fig. 38–21 is a consequence of X-ray diffraction experiments; virtually all of our knowledge of crystalline structure comes from such experiments. It was the use of X-ray diffraction on a crystallized form of DNA that led to the discovery of that molecule's double-helical structure (Fig. 38–22).

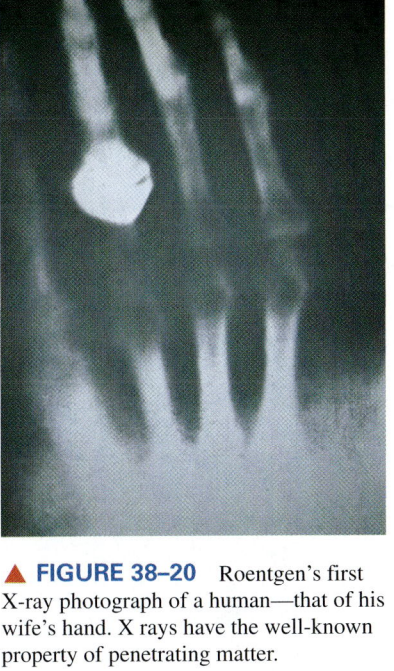

▲ **FIGURE 38–20** Roentgen's first X-ray photograph of a human—that of his wife's hand. X rays have the well-known property of penetrating matter.

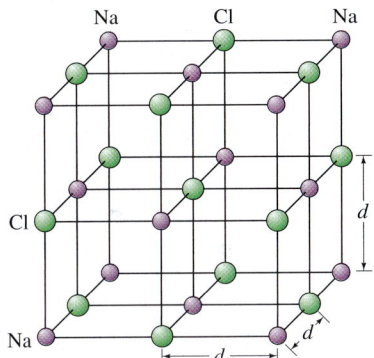

▲ **FIGURE 38–21** Crystals have three-dimensional structure with their atoms in regular arrays. This diagram shows one of the simplest, NaCl (table salt), which has a cubic structure.

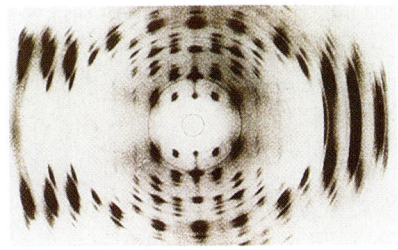

(a)

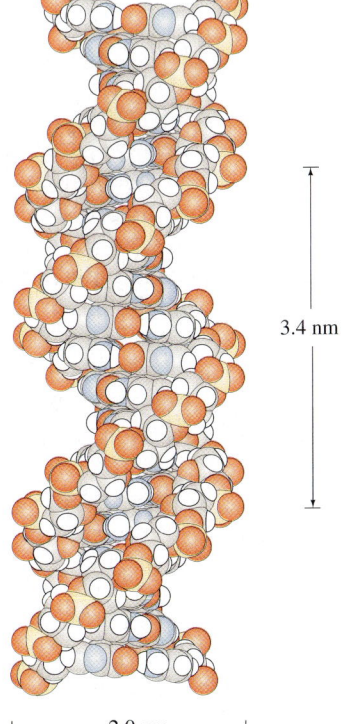

(b)

▲ **FIGURE 38–22** (a) Analysis of thousands of diffraction patterns produced by crystals of the large biological molecule deoxyribonucleic acid (DNA) showed that (b) the molecule has the shape of a double helix.

1064 | Diffraction

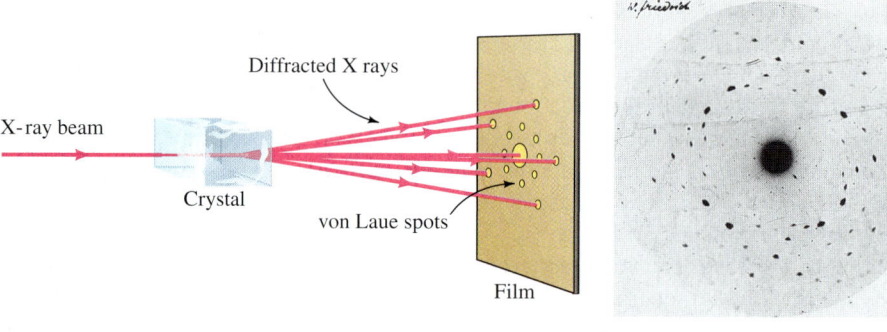

▶ **FIGURE 38–23** (a) Schematic diagram of the von Laue experiment for the diffraction of X rays. (b) Von Laue spots in one of the first X-ray diffraction patterns. The large spot is undiffracted radiation.

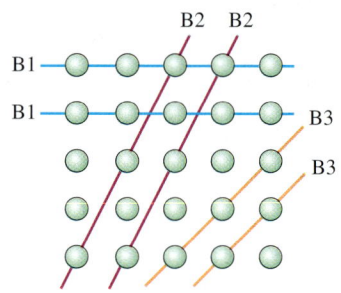

▲ **FIGURE 38–24** Many parallel Bragg planes can be drawn in a three-dimensional crystal.

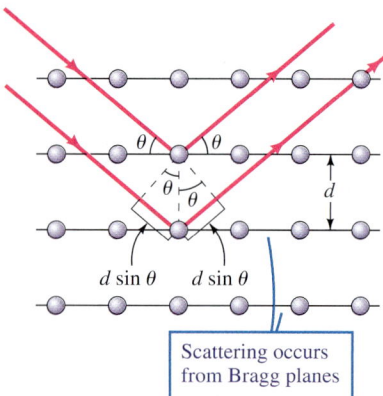

▲ **FIGURE 38–25** The geometry of X-ray diffraction between adjacent Bragg planes.

Because the rescattering centers (the atoms of a solid) are pointlike and three-dimensional rather than slitlike, the diffraction pattern of a crystalline solid consists of a regular array of spots rather than lines. The von Laue experiment on a crystalline solid, shown schematically in Fig. 38–23a, leads to a set of spots like those shown in Fig. 38–23b. Von Laue's idea was clarified almost immediately by W. L. Bragg in 1912, who proposed a simple and systematic way of showing just how the positions of the spots would be determined by the solid's crystalline structure. Bragg pointed out that in any crystal, many sets of parallel planes (called *Bragg planes*) can be drawn that pass through the positions of the atoms, and that the planes of a set are separated by characteristic distances (*Bragg spacings*). Figure 38–24 shows where some of these planes cut a two-dimensional cross section of a cubic lattice similar to that formed by NaCl. We can think of each family of parallel planes as a slit-type diffraction grating for the X rays. Figure 38–25 shows two rays scattered from two parallel planes within a crystal. These rays scatter from a given plane for which the reflection angle equals the incident angle because, at that reflection angle, the wavelets re-emitted by each atom within that plane add constructively. Now consider the interference between the scattered waves of *different* planes, which occurs because the X rays penetrate the crystal. If the separation between the planes is d, then, from the geometry of Fig. 38–25, the difference in path lengths for the two lines is $2d \sin \theta$. Note that angle θ is measured from the plane surface rather than from the normal to the plane. Constructive interference for scattering from these two adjacent planes occurs when this path-length difference is an integer multiple of the wavelength. This relation is known as **Bragg's law**, or the **Bragg condition**:

$$2d \sin \theta = n\lambda, \quad \text{where } n = 1, 2, 3, \ldots. \quad (38\text{--}17)$$

BRAGG'S LAW

Because the planes are equally spaced, the waves that scatter from the atoms in the entire set of planes in the direction specified by the Bragg condition *all* add constructively and, as in the case of a diffraction grating, the maximum is large and narrow.

This discussion, while laying out the principles behind the X-ray spectrometer (Fig. 38–26), is inadequate to explain the *intensities* of the spots. We may state generally that, if a particular family of planes contains more atoms than another does, the maxima those planes give are more intense, and intensity information is very important in determining the crystalline structure. More advanced mathematical methods are necessary for a complete model of the diffraction pattern, which is quite complicated because of the many possible planes and orders of scattering.

EXAMPLE 38–7 Figure 38–26 shows an X-ray tube, which produces a continuous distribution of wavelengths in the X-ray range. If these wavelengths are scattered from a particular set of parallel planes of rock salt (NaCl) with a spacing $d = 0.282$ nm, what wavelengths will appear in the first and second orders at 25°?

Strategy To use Bragg's law, either the wavelength or the atomic-plane spacing must be known. In this case, the plane spacing is known, and Bragg's law, Eq. (38–17), can be used to identify unknown wavelengths.

Working It Out We use Eq. (38–17) to determine the wavelengths:

$$\lambda = \frac{2d \sin \theta}{n} = \frac{2(0.282 \text{ nm})(\sin 25°)}{n} = \frac{0.238 \text{ nm}}{n}.$$

The wavelengths at 25° are 0.238 nm and 0.119 nm for the first ($n = 1$) and second ($n = 2$) orders, respectively.

What Do You Think? If $\theta = 25°$, what is the actual overall deflection from the original beam?

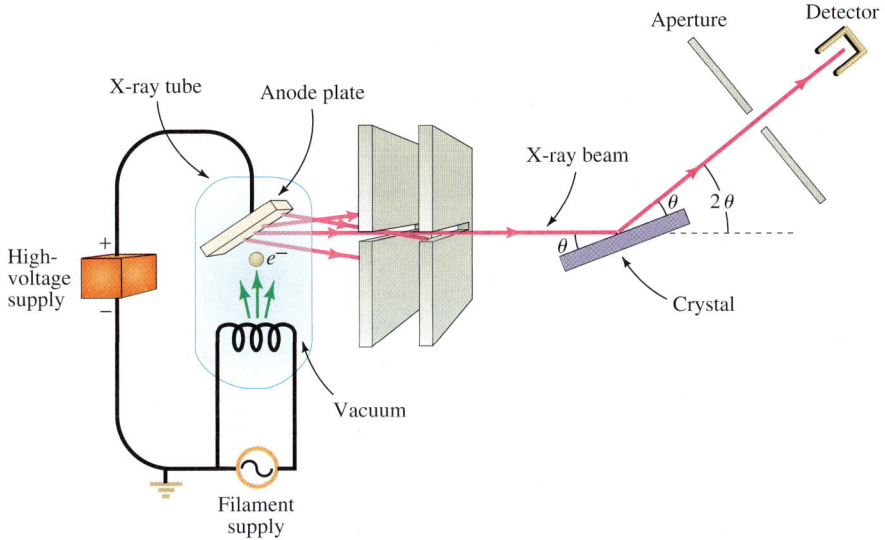

FIGURE 38–26 Schematic diagram of an X-ray spectrometer used to study properties of crystals. Electrons bombard the anode plate, producing X rays that are collimated before being scattered by the crystal. A movable detector records X-ray intensity as a function of θ to determine where constructive interference occurs.

*38–7 Holography

In 1947, Dennis Gabor proposed that interference effects between light emitted by an object (a source) and a second coherent beam can be recorded on film, which effectively becomes a diffraction grating. When light is passed through this diffraction grating, it is diffracted and forms a fully three-dimensional image of the object, an image that can be viewed from different positions and angles, just like the original object. This process is **holography**, and the film on which the interference pattern is stored is a *hologram*.

In order to understand the principles, let's start with a distant point source that sends plane waves directly toward a piece of film (Fig. 38–27). At the same time, we send a second beam—*coherent with the light from the source*—toward the film from an angle θ_r. This second beam is known as the *reference beam*. The reference beam interferes with the light from the source. Suppose that the wave in the reference beam interferes constructively with the source wave at point P_1. There will also be constructive interference at P_2, a distance d from P_1, if the wave path along line ℓ_2 differs from the path along line ℓ_1 by λ (or an integer m times λ; we consider only the case $m = 1$). When there is constructive interference at both P_1 and P_2, the relation between θ_r, the wavelength of the light, λ, and the separation d is

$$d \sin \theta_r = \lambda. \tag{38–18}$$

On the slice of film shown in Fig. 38–27, constructive interference will occur at a series of equally spaced points, which will be recorded as dark spots on the film. (Recall that the film makes a negative.) These points are parts of continuous lines into or out of the page. The full interference pattern on the film thus consists of a set of curving lines that represent all the places where there is constructive interference. The film can record with shadings of gray places where the interference is not totally destructive.

Let's now turn to the question of how the image is viewed (or *reconstructed*). Suppose that we project a beam just like the reference beam, and at the same angle, onto the back of the film (Fig. 38–28). The dark areas on the film act as obstacles that rescatter the light. The direction indicated, *that of the original light from the source*, is a direction for which the diffracted light is a maximum as the geometry in Fig. 38–28 shows. Thus a viewer placed at point E will see light as though it comes from a distant point I, which we may think of as an image of the original source. Note that there is no requirement that the beam that produces the image be identical to the original reference beam, as long as it is coherent across the film. If the angle of the new beam is different, the only effect is to shift the angle of the viewed image.

Suppose now that the point source is closer to the film when the image is made. In this case, the spots of constructive interference will not be spaced equally across the screen (Fig. 38–29). At region A_1, the situation we have described is reproduced, but at

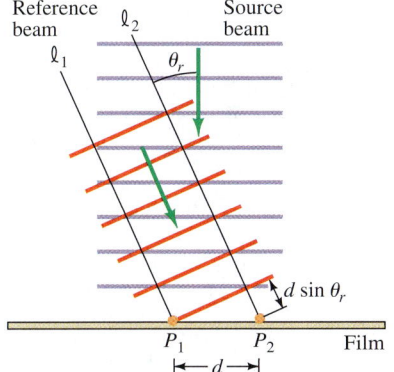

▲ **FIGURE 38–27** Coherent light from a distant source beam and from a reference beam interfere on film, producing a hologram.

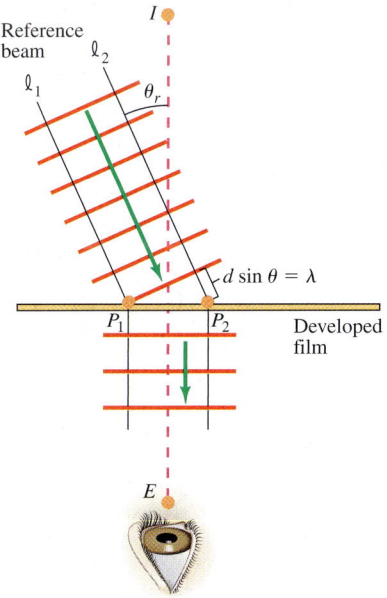

▲ **FIGURE 38–28** When a reference beam shines on holographic film, the image of the original object is reconstructed.

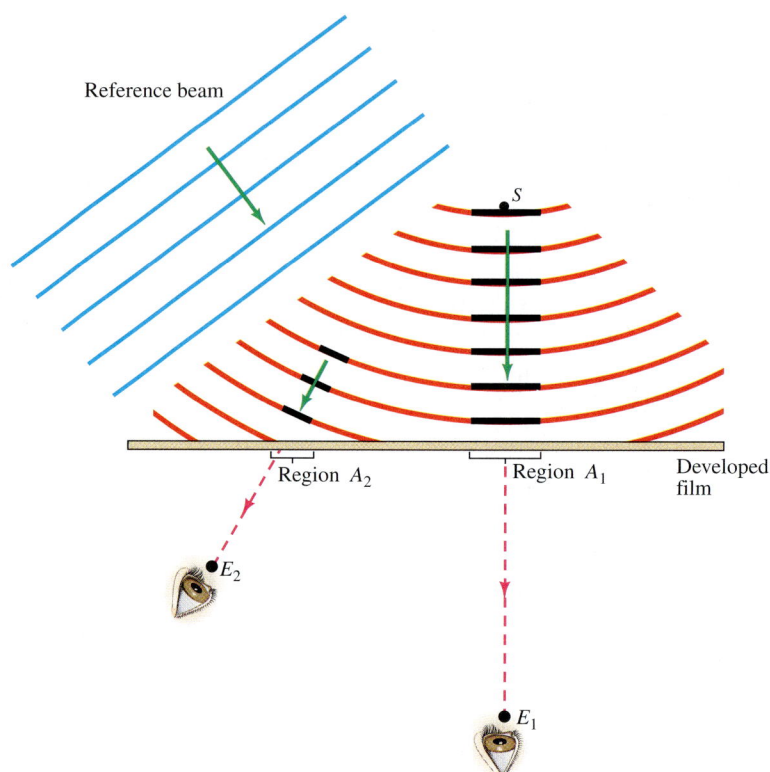

▶ FIGURE 38–29 Holograms produce a three-dimensional image that allows the original object to be seen from different positions. A different image will be observed in each place.

region A_2, the points where there is constructive interference between the beams are different. When the exposed film is illuminated by a reference beam, a viewer at E_1 will see a plane wave along the direction from E_1 back to region A_1; that is, the observer will be looking back at the source from one angle. However, when the viewer is at E_2, the maximum of the diffraction pattern will indicate an image back along the direction from E_2 to A_2. *The viewer will be looking at the source from another angle.* There is a true three-dimensional image, which can be viewed from different angles.

When the object is more complicated than a point source, light arrives at any given point on the film from many points of the object. The interference pattern that this light makes with the reference beam is far more complicated and irregular, but it is nevertheless unique to the object. Once this pattern is recorded, it serves as a diffraction grating for light from a reference beam to make a unique pattern that reproduces the light emitted by the original object, and from many angles. Although we have treated the film as a transmission grating, if the interference pattern can be recorded as scratches on a shiny surface, then it will act as a reflection grating, such as that in a hologram on a credit card.

How is the reference beam made coherent with the light from the source? We can take a laser beam and split it in two. One part illuminates the object while the other is routed to serve as a reference (Fig. 38–30). Figure 38–31 shows one of the many uses of holograms.

▶ FIGURE 38–30 (a) Schematic diagram of the formation of a hologram. The reference beam and the light reflected from the object must be coherent so that they can interfere to make the hologram. (b) The holographic image is formed when the hologram acts as a diffraction grating.

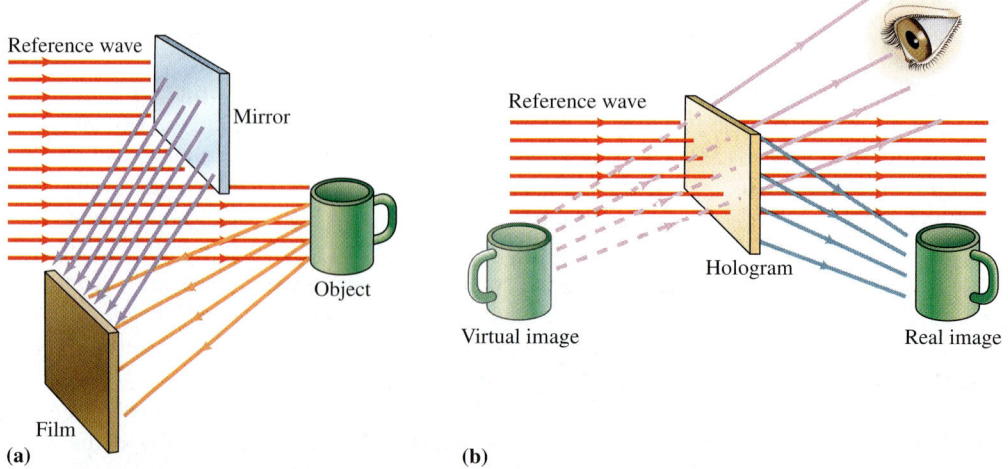

► **FIGURE 38–31** You can observe a hologram like those visible on these visa documents by looking at a credit card. Holograms such as these are difficult to counterfeit.

THINK ABOUT THIS...
WHAT ARE THE USES OF HOLOGRAPHY?

The uses of holograms go beyond the simple beauty of the image. Holography has the potential to provide an extremely compact system of information storage. Because the light from every point on a printed page reaches every point of a hologram, every region of the film larger than several wavelengths across can reproduce the entire page, albeit with less detail. Moreover, successive holograms of successive pages can be made within thick photographic emulsions. If the exposure of each page is made with a reference beam oriented at a slightly different angle, then, by illuminating the resulting hologram with a light beam of that particular angle, only the corresponding page appears from a particular vantage point. For example, all the paintings in a museum could be so recorded with great accuracy in a very small space indeed.

Another important use of holography involves making two holograms of the same object on the same film at successive times. If the object has moved slightly between the moments when the holograms are made, then the two images interfere with each other like the light from two surfaces of a soap film does. Figure 38–32 shows an interference hologram of a violin in motion under the influence of a vibrating string. This interference pattern, not to be confused with the interference that makes the hologram itself, reveals detail about the motion not otherwise visible. Similarly, density variations in air are visible as the interference between two successive images of the air made on the same hologram. In this way, the mechanisms by which a candle heats the air above it or an airplane produces shock waves can be studied.

▲ **FIGURE 38–32** Interference and diffraction were crucial in producing this image of a violin. The lines indicate the motion of the vibrating instrument and are associated with the interference of two holographic images.

Summary

Diffraction is a manifestation of interference among waves. Examples include the pattern produced by screens with evenly spaced multiple slits (diffraction gratings) and the patterns made by light that passes through single apertures or around obstacles.

If d is the distance between slits in a diffraction grating and θ is the angle of observation from the direction of incident light, principal maxima are observed for the condition

$$d \sin \theta = m\lambda, \quad \text{where } m = 0, \pm 1, \pm 2, \ldots. \tag{38–1}$$

Here, m is the order of the principal maxima. If the average intensity reaching the screen from any one slit is I_0, then the intensity of the light from the grating at the principal maxima is $N^2 I_0$, where N is the number of slits. In addition, the width of the principal maxima depends on N as $1/N$, so the diffraction peaks become sharper as N increases. This dependence on the parameters of the grating is contained in the expression for the intensity pattern:

$$I = I_0 \left[\frac{\sin(N\beta)}{\sin \beta} \right]^2, \tag{38–4a}$$

where

$$\beta = \frac{\pi d \sin \theta}{\lambda}. \tag{38–4b}$$

Angular dispersion represents the change in observation angle θ as a function of a change in wavelength and is given by

$$\frac{\Delta \theta}{\Delta \lambda} = \frac{m}{d \cos \theta}. \tag{38–5}$$

The resolving power, R, is the ability of a grating to separate closely spaced lines:

$$R = \frac{\lambda}{\Delta \lambda} = mN. \tag{38–6, 38–7}$$

A single slit produces a diffraction pattern that can be derived by considering the slit to be composed of a large number of very thin slits. The criterion for destructive interference is

$$\sin\theta = \frac{m\lambda}{a}, \quad \text{where } m = \pm 1, \pm 2, \pm 3, \ldots \qquad (38\text{–}8)$$

and a is the width of the slit. The intensity pattern of a single slit is

$$I = I_{max}\frac{\sin^2\alpha}{\alpha^2}, \qquad (38\text{–}10)$$

where the angle α is given by

$$\alpha = \frac{\pi a \sin\theta}{\lambda}. \qquad (38\text{–}12)$$

The minima are given in terms of α by

$$\alpha = n\pi, \quad \text{where } n = \pm 1, \pm 2, \pm 3, \ldots. \qquad (38\text{–}12)$$

Most of the light from the single slit is contained in the central peak; the secondary peaks are much less intense. The narrower the slit, the broader the diffraction pattern.

The Rayleigh criterion specifies that two point sources are just resolved if the peak of the diffraction image of the first source falls on the first minimum of the diffraction image of the second source. The minimum separation angle of two closely spaced sources obtained by a circular aperture of diameter D is approximated by

$$\theta_{min} \cong \frac{\lambda}{D}. \qquad (38\text{–}14)$$

The minimum separation S_{min} of two closely spaced objects a distance L from a lens of diameter D is given by

$$S_{min} \cong \frac{L\lambda}{D}. \qquad (38\text{–}15)$$

The practical limitation of Earth-based telescopes is due to air turbulence and not to diffraction limits.

X rays are diffracted by the atom centers of regularly spaced Bragg planes—planes formed by the regular array of atoms in a crystal. The technique is important for studies of crystalline structure. Bragg's law gives the observation angles θ (as measured from a plane surface in a crystal lattice) for which constructive interference is obtained from planes of spacing d:

$$2d\sin\theta = n\lambda, \quad \text{where } n = 1, 2, 3, \ldots. \qquad (38\text{–}17)$$

Holography represents a special process by which three-dimensional images are captured. A hologram is a kind of diffraction grating formed on film by scattering coherent light from an object; the three-dimensional image of the object can be reconstructed from the film when an appropriate light beam passes through it.

Understanding the Concepts

1. Describe what happens to the pattern you observe for single-slit diffraction as the width of the slit is slowly reduced.
2. Would the diffraction of water waves around the timbers of a pier be reduced by decreasing the diameter of the support poles? By increasing their diameter?
3. There are tentative plans to build telescopes for waves of various wavelengths, including visible light, on the Moon. What would the advantages of such facilities be?
4. Discuss how a Poisson spot might be obtained from a bowling ball. Would you want the source and screen to be close to or far away from the bowling ball? Explain.
5. A CD is "burned" with a laser—information is inscribed with the beam on the disk. If you had to maximize the information the CD contained, should you use a short or a long wavelength laser?
6. Is it possible to obtain better resolution with a microscope with blue light than with red light? Why or why not?
7. Two waves are linearly polarized. The electric field of one wave is aligned with the x-axis and the other is aligned with the y-axis. In the absence of matter that might change the polarization, can these waves interfere with each other?
8. In a demonstration of diffraction peaks that involves the reflection of laser light from an ordinary ruler, does the light have to be at a glancing angle?
9. The spreading of light due to diffraction in an optical instrument is greater when the instrument uses red as opposed to blue light. Why?
10. A hologram contains information about an entire object, even in just a small portion of the film. Would you expect the image made by a small portion of the hologram to be as sharp as the image made by the entire hologram?
11. For diffraction by two slits whose centers are a distance a apart, the condition for maximum intensity is $n\lambda = a\sin\theta$. When the center between the slits is removed, so that we now have a single slit of width b, the same formula with a replaced by b describes minima. What is the role of the centerpiece in producing opposite effects in the apparently identical formulae?
12. What are the differences between the interference patterns formed on a distant screen by coherent light that passes through a diffraction grating with thousands of rulings at a particular spacing and a double slit separated by the same spacing?

13. A lightbulb emits light with a spectrum characteristic of blackbody radiation (see Chapter 17). What pattern will this light produce when it is observed through a grating?
14. You are standing in the ocean, and a wave passes around you. Is this an example of diffraction?
15. How do the X rays used in X-ray diffraction "know" that there is a given set of planes of atoms for which a diffraction pattern appears?
16. Does the fact that light bends around corners mean that, with a sensitive camera, you could read a newspaper from around a corner? (This is a serious question: Try to estimate the amount of bending that the smallest obstacle would give for light, and how much information that light could contain.)
17. In the so-called 3-D movies introduced in the 1950s, a three-dimensional effect is achieved when different images are sent to each of your two eyes. How could you tell that they are not holographic images?
18. When monochromatic light falls with normal incidence on a screen with a circular hole, and we look at the transmitted light falling on a second parallel screen, there is a diffraction pattern on the second screen. At a given distance between the two screens the intensity at the center of the diffraction pattern is zero. As the second screen is moved away, the center of the pattern brightens, darkens again and then brightens, remaining the brightest part of the pattern at all greater distances. Explain these observations.

Problems

38–2 Diffraction Gratings

1. (I) Laser light is diffracted from a grating with 400 lines/cm. The central peak and the fourth peak are 10.34 cm apart on a screen 1.44 m away. The screen is perpendicular to the ray that makes the central peak. What is the wavelength of the light?

2. (I) A grating has a line density of 1200/cm, and a screen perpendicular to the ray that makes the central peak of the diffraction pattern is 1.5 m from the grating. If light of two wavelengths, 620 nm and 635 nm, passes through the grating, what is the separation on the screen between the second-order maxima for the two wavelengths?

3. (I) A student finds a diffraction grating but does not know the spacing of the ruled lines. She shines light from a laser with $\lambda = 680$ nm through the grating and examines the maxima on a screen 265 cm away. If the distance between the tenth maxima on either side of the central peak is 14.3 cm, what is the rule spacing of the grating?

4. (I) A spectrum is formed with a diffraction grating, which has 4500 lines per cm. Sodium light has two nearby lines with $\lambda = 589.0$ nm and $\lambda = 589.6$ nm. What is the resolution in the second-order spectrum for this grating? What is the angular separation for the sodium lines in the second-order spectrum?

5. (II) Mercury has visible optical spectral lines of wavelengths 422.729 and 422.787 nm. You have an apparatus that allows the light to fall on only 0.800 cm of a transmission diffraction grating. How many lines per centimeter does your diffraction grating need in order to resolve the third-order maxima of the two mercury lines?

6. (II) A grating with 2.0×10^4 rulings spaced uniformly over 3.0 cm is illuminated at normal incidence by light of wavelength 530 nm. (a) What is the dispersion of the grating in the second order? (b) What is the smallest wavelength interval that can be resolved in the second order near $\lambda = 530$ nm?

7. (II) What is the resolving power of a 3.0-cm-wide diffraction grating with 5000 lines/cm, for the first three orders? If light consisting of a series of discrete wavelengths around 420 nm is incident on the grating, what is the minimum wavelength separation that can be resolved in these three orders?

8. (II) Estimate the line spacing between two closely spaced lines near 530 nm if they are barely resolved in the third order by a grating with $N = 11,000$.

9. (II) A grating is to be inscribed on a 4.00-cm-wide glass plate so as to resolve two spectral lines with wavelengths 618.32 nm and 618.34 nm, respectively, in the first order. What is the minimum number of lines that must be ruled on the plate? What is the dispersion of the grating with this number of lines?

10. (II) Argon has wavelengths of 454.6 and 457.9 nm. (a) What resolving power does a transmission diffraction grating need to resolve the two wavelengths? (b) If the diffraction grating has 450 lines/mm, how wide does the grating have to be in order to resolve the second-order maxima?

11. (II) Parallel light falls at an angle of 30° from the vertical on a horizontal reflection grating with 4000 lines per cm. At what angle will you see the first-order spectrum of sodium light ($\lambda = 589$ nm) reflected from the grating?

12. (II) A grating is made of five similar, uniformly spaced, narrow slits. For light of wavelength $\lambda = 633$ nm perpendicularly incident on the slits, the angular position of the first principal order is 0.18° to the normal. What is the slit separation? What is the angular position of the first principal order when the first and fifth slits are covered? When the second and fourth slits are covered?

13. (II) The resolving power of a certain grating for the first-order spectrum is 10^4. If the grating is 2 cm long, what angle separates the first- and second-order images for light with $\lambda = 580$ nm at normal incidence?

14. (II) White light shines on a diffraction grating with 3600 lines/cm. The diffracted light is observed on a screen 0.90 m away. Find the second- and third-order positions for blue light (440 nm), green light (560 nm), and red light (720 nm). Sketch a view of the screen.

15. (II) Visible light extends from wavelengths of 430 nm to 680 nm. If blackbody radiation, which contains all these wavelengths, is incident on a 5.0-cm-wide grating with 2500 slits/cm, what range of angles is covered for these wavelengths in the first-order maximum? In the second-order maximum?

16. (II) An atomic source emits two strong spectral lines, a red one of wavelength 615 nm and a blue one of wavelength 475 nm. The light falls on a diffraction grating with 5000 lines/cm that is 1.2 cm across, and passes to a screen 2.00 m away. On the screen, how far from the central maximum are the second-order maxima ($m = 2$) of the spectral lines? What is the width of these maxima?

17. (II) Light of wavelength λ is incident at an angle α to the normal of a transmission grating with spacing d between each slit (Fig. 38–33). At what angles β to the normal will diffraction maxima be located?

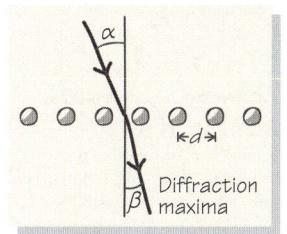

▲ FIGURE 38–33 Problem 17.

38–3 Single-Slit Diffraction

18. (I) Light of wavelength $\lambda = 535.0$ nm falls on a slit of width $a = 0.35$ mm. At what angle θ from the normal to the wall in which the slit is cut does the second dark fringe occur?

19. (I) A single slit of width 2.8×10^{-5} m diffracts light of wavelength 495 nm to a screen. The distance between the minima on either side of the central maximum is 1.8 cm. How far away is the screen?

20. (I) A single slit diffracts laser light of wavelength 635 nm onto a screen 2.5 m away. The distance between the first-order maxima on either side of the central peak is 6.0 mm. How wide is the slit?

21. (I) Microwaves of wavelength 2.6 cm are perpendicularly incident on a single slit of width 4.2 cm, and then onto a screen. (a) At what angles are maxima and minima found on the screen? (b) If the screen is 0.8 m away from the slit, what is the distance of the secondary maximum from the central maximum?

22. (II) Plane waves of light of wavelength 610 nm are incident on a single slit of width 75 μm. A lens focuses the plane waves on a screen 120 cm away (Fig. 38–34). (a) What is the width of the central maximum on the screen? (b) What is the intensity ratio between the central maximum and the first-order maximum?

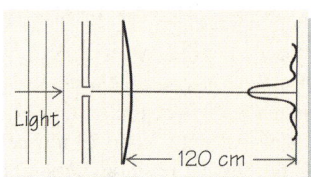

▲ **FIGURE 38–34** Problem 22.

23. (II) Blue light ($\lambda = 470$ nm) passes through a slit 10 μm wide. What is the ratio between the maximum intensity of the central peak and the maximum intensity of the next adjacent peak? At what angle θ from the horizontal will the intensity of the central peak be half its maximum value? Would θ increase or decrease if red light ($\lambda = 670$ nm) were used instead?

24. (II) A diffraction pattern is formed by an adjustable slit. If the width of the slit is doubled, how do the following quantities change? (a) The distance of the first minima on the two sides of the central maximum; (b) the intensity at the central maximum; (c) the total power reaching the screen.

25. (II) A single slit produces a diffraction pattern on a distant screen. Show that the separation distance between the two minima on either side of the central maximum is twice as large as the separation distance between all the other neighboring minima. Compare your result to the corresponding case for a double-slit pattern with very narrow slits.

26. (II) When light of wavelength 540 nm passes through a single slit of unknown width, the diffraction pattern displays a second maximum where the first minimum of light of an unknown wavelength had been observed to fall (Fig. 38–35). What is the unknown wavelength?

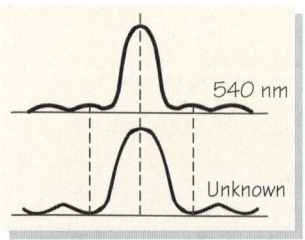

▲ **FIGURE 38–35** Problem 26.

27. (II) The width of the central peak of a single-slit diffraction pattern can be characterized by the distance of the first-order minima on both sides of the maximum or the full width at half maximum, the latter defined by the points where the intensity decreases to 50 percent (Fig. 38–36). Compare the values of these widths.

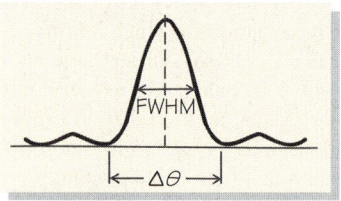

▲ **FIGURE 38–36** Problem 27.

28. (II) Monochromatic light falls with normal incidence on a screen with a circular hole of diameter 2.2 mm. When a second screen is placed 0.60 m away, the intensity at the center of the diffraction pattern is zero (first minimum). As this screen is moved away, the center of the pattern brightens, darkens again and then brightens, remaining the brightest part of the pattern at all greater distances. What is the wavelength of the light? What was the distance between the screens at the second minimum?

29. (II) Light of wavelength λ arrives at a single slit of width a; the plane wave fronts arrive at the slit at an angle θ_i (Fig. 38–37). Find the angles θ for which minima appear on a very distant screen. Is there a "central maximum" in the direction defined by the incoming wave; that is, at $\theta = \theta_i$?

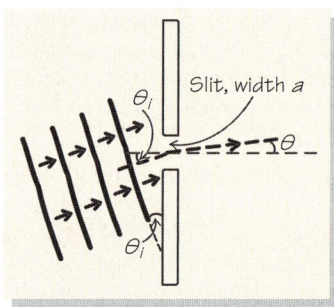

▲ **FIGURE 38–37** Problem 29.

30. (II) Suppose that light falls on a single slit at an angle ϕ with the normal to the wall that contains the slit (Fig. 38–38). Show that Eq. (38–10) still holds, but $\sin \theta$ must be replaced by $(\sin \theta + \sin \phi)$ in the expression for α [Eq. (38–11)].

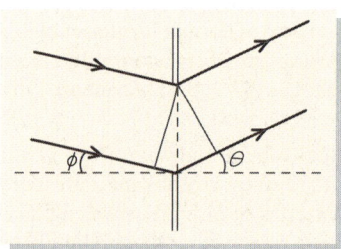

▲ **FIGURE 38–38** Problem 30.

31. (III) When we determined the position of the minima of the Fraunhofer diffraction pattern for a single slit, we argued that the maxima are located midway between the minima. To look at the accuracy of this assumption, (a) show that the maxima of the intensity pattern $(\sin^2 \alpha)/\alpha^2$ are determined by the solutions of the transcendental equation $\alpha = \tan \alpha$. (b) Compare a numerical solution of this equation for the first and second maxima with

angles that are midway between the first and second, and second and third, minima (you will need a calculator for the trigonometric values). (c) By plotting the intersection points of $y = \tan \alpha$ and $y = \alpha$, show that the approximation improves as the order of the maximum increases.

38–4 Resolution of Optical Instruments

32. (I) A plane wave of microwave radiation, $\lambda = 2.3$ cm, passes through a circular aperture of diameter 4.7 cm. What is the angular position of the first minimum of the resulting Fraunhofer diffraction pattern?

33. (I) Astronauts leave two lunar rovers 5.00 km apart on the Moon. An Earth-based telescope of what minimum diameter is required to resolve laser beams ($\lambda = 650$ nm) emitted by the rovers toward the telescope? The rovers are 3.0 m long. A telescope of what diameter is required for the rovers themselves to be detected? Ignore air turbulence. (The Earth–Moon distance is 3.83×10^8 m.)

34. (I) An amateur astronomer uses a reflecting telescope of diameter 20 cm and focal length 200 cm to observe light of $\lambda \cong 600$ nm from a star. (a) What minimum angular resolution can the astronomer obtain? (b) What is the diameter of the Airy disk? (c) What is the minimum separation distance of two objects on the Moon that the telescope can resolve?

35. (II) An astronaut in a satellite can barely resolve two point sources on Earth 220 km below. What is the separation distance between the sources, assuming ideal conditions, $\lambda = 480$ nm, and a pupil diameter of 1.5 mm?

36. (II) The two stars of a binary star system are just resolvable when observed by a telescope with a resolution of 18 seconds of arc and are 125 ly from Earth. Estimate their separation.

37. (II) A spy satellite is announced to be capable of distinguishing detail 10 inches across. If the satellite orbits at a height of 220 mi, what must be the minimum size of the lens aperture, assuming maximum sensitivity at $\lambda = 525$ nm? Would it be better if the film (or other sensor) were sensitive to shorter or to longer wavelengths?

38. (II) What must be the lateral separation between two objects located 1.0 km from a camera that must resolve them? The camera lens's aperture is 5.0 mm in diameter, and the film is sensitive to light of wavelength 550 nm.

39. (II) The lens of a 35-mm camera is set in such a way that the image of a very distant object is ideally sharp on the film. The focal length of the objective is 50 mm. At what setting of the camera's aperture will the sharpest possible image of an object 5.0 m away form? [*Hint*: The image will be blurred both because it is off focus and because of diffraction. Find the aperture at which these two sources give equal contributions.]

40. (II) Use the Rayleigh criterion and make assumptions to estimate the distance at which the human eye should be able to resolve the headlines in a newspaper. Carry out an experiment to see how good your estimate is!

41. (II) The SR-71 Blackbird reconnaissance airplane could fly at over 70,000 ft. If the pilot's pupil has a diameter of 1.5 mm on a bright day, what is the distance between two objects on Earth that the pilot could just resolve from 60,000 ft? Take the wavelength of light to be 520 nm.

42. (II) The headlights of a car are 1.8 m apart. At night, the pupils of an oncoming driver have expanded to 4.2 mm. How close must the two cars approach before the headlights can be resolved? Take the wavelength of the light to be 510 nm.

38–5 Slit Width and Grating Patterns

43. (II) Calculate the lowest missing order of a double-slit interference pattern if the separation of the two slits is three times their individual widths, $d = 3a$.

44. (II) The separation distance between two narrow slits is ten times the width of either slit. What is the intensity of the tenth interference maximum, taking the center as the first, when monochromatic light passes through the two slits and falls on a distant screen?

45. (II) Light of wavelength 690 nm from a ruby laser impinges on two slits 1.1 mm apart. Each slit is 0.20 mm wide. Find the intensity ratio I/I_0 on a screen 3.0 m away at the following distances from the central maximum: 0.050 mm, 0.50 mm, 1.5 mm, and 3.0 mm.

46. (II) Figure 38–39 shows the intensity as a function of diffraction angle (in radians) for a double slit with light of wavelength 550 nm. Estimate the separation of the slits as well as their widths.

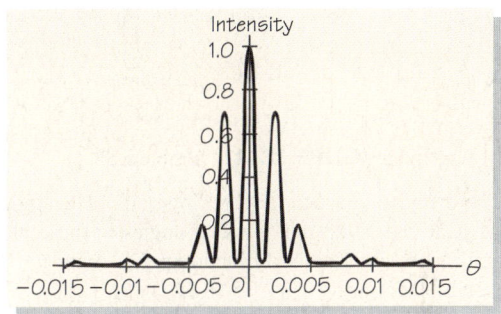

▲ **FIGURE 38–39** Problem 46.

47. (II) Figure 38–40 shows the intensity as a function of diffraction angle (in radians) for a multiple slit with light of wavelength 600 nm. Compare the pattern with the double-slit pattern shown in Fig. 38–39 of the previous problem and determine the number of slits, their separation, and their width. [*Hint*: Look at Fig. 38–8.]

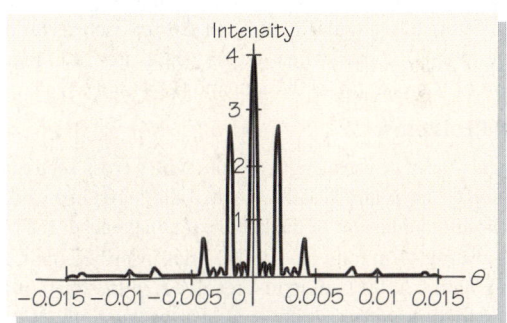

▲ **FIGURE 38–40** Problem 47.

48. (II) Light of wavelength 600 nm is perpendicularly incident on a diffraction grating. Two adjacent maxima occur at $\sin \theta = 0.30$ and $\sin \theta = 0.36$, respectively. The fourth order is missing. (a) What is the separation distance between adjacent slits? (b) What is the smallest possible individual slit width? (c) Name all orders that appear on the screen, consistent with the answers to parts (a) and (b).

49. (II) The centers of a double slit are separated by 1.2 mm; each slit is 0.4 mm wide. Are there missing orders? If so, at what angles are they missing on a distant screen if $\lambda = 589$ nm?

50. (II) The slit widths of a grating with 900 slits/cm are one-third the slit spacing. What is the ratio of the intensities of the second-order and first-order principal maxima of the grating?

51. (II) Light of wavelength 625 nm is perpendicularly incident on a screen in which double slits of width $a = 0.25$ mm have been cut. The slits are a distance $d = 0.30$ mm apart. Find the first angle away from the central axis for which the intensity on a distant screen is exactly one-half the maximum intensity.

1072 Diffraction

52. (II) A grating consists of slits of width a whose centers are separated by a distance d. Sketch the diffraction pattern for (a) $d \gg a$ and (b) $d - a \ll a$ (the slits are wide compared to the strips between them).

*38–6 X-Ray Diffraction

53. (I) X rays of wavelength 0.14 nm are aimed at an unknown crystal in a diffractometer. A first-order peak occurs at 38.2° (Fig. 38–41). What is the corresponding Bragg-plane spacing for the crystal?

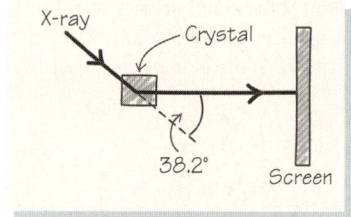

▲ **FIGURE 38–41** Problem 53.

54. (II) The distance between neighboring pairs of Bragg planes in calcite ($CaCO_3$) is 0.3 nm. At what angles to these planes will the first- and second-order diffraction peaks occur for X rays of wavelength 0.55 nm?

55. (II) Consider a crystal consisting of identical cubes with atoms at the vertices. The spacing between adjacent atoms is 0.28 nm. X rays of wavelength 0.14 nm scatter elastically from a set of planes parallel to the face of the cubes. At what angles will first-order Bragg diffraction be observed?

56. (II) Mica has a set of Bragg planes with spacing of 1.0 nm, whereas a set of planes in rock salt has a spacing of 0.28 nm. For an X ray of wavelength 0.10 nm, which material produces a diffraction pattern with the greater angular separation? What is the difference in angular separation $\Delta\theta$ for each material for the Bragg planes above when the crystals are illuminated with X rays of wavelengths 0.096 nm and 0.104 nm?

General Problems

57. (I) A ruby laser of wavelength 690 nm with a cross-sectional area of 1.0×10^{-3} m^2 is aimed at the Moon, 3.84×10^8 m away. Estimate the minimum diameter of the light beam that reaches the Moon.

58. (I) A grating 4 cm long has 16,000 lines inscribed on it. A line of wavelength 623.000 nm is just resolved, in the third order, from a second line with a slightly longer wavelength. What is the wavelength of the second line?

59. (II) Radar is used to study the shapes of airplanes from as far away as 100 km. (a) Assuming that the distance scale determining a plane's shape (the size of the curves that distinguish one plane from another) is 1 m, what angular resolution is needed in the radar system? (b) Estimate the wavelength of the radar waves if the reflected radar signals are gathered in a dish of diameter 2.5 m.

60. (II) Deep-ocean waves move in linear fronts directly toward a harbor opening of width 50 m (Fig. 38–42). For what wavelength will there be a minimum within the harbor at an angle of 50° from the axial line of the opening?

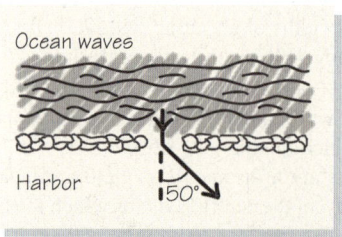

▲ **FIGURE 38–42** Problem 60.

61. (II) By varying the spacing between two vertical dipole antennas as well as the phase of the signal generated by each antenna, the antennas can give signals that are stronger in some directions than in others (see Chapter 37). Suppose that N antennas are lined up along the x-axis (Fig. 38–43). The total distance between the first and last antennas is λ, so the spacing between the antennas is $\lambda/(N - 1)$. Any one antenna would radiate its signal with the same intensity I_0 for any angle θ. (a) Find the intensity radiated very far from the array by the system of antennas as a function of angle θ in terms of I_0, assuming that the signals of all the antennas are in phase. (b) Describe the signal for all values of θ from 0° to 360°.

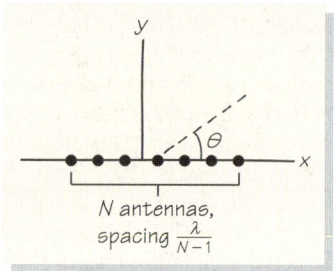

▲ **FIGURE 38–43** Problem 61.

62. (II) A grating with alternate perfectly clear and perfectly opaque spaces gives a spectrum in which all the even orders are missing. What is the ratio of the width of the clear and opaque spaces?

63. (II) A researcher embeds a single slit of width 5×10^{-4} m in a piece of glass of index of refraction n. He then shines laser light of wavelength 560 nm through one surface of the glass, which is diffracted through the slit and passes through another surface of the glass and on to a distant screen. The screen in which the slit is cut, the second surface of the glass, and the distant screen are all parallel to one another and perpendicular to the incoming laser beam. He hopes to measure the index of refraction n of the glass by seeing how the position of the first diffraction minimum differs from the position of the same minimum when the slit is in air. Can the researcher measure n this way?

64. (II) In Chapter 40, we shall see that an electron behaves like a wave whose wavelength, λ, is related to its momentum p by $\lambda = h/p$. Here, h is Planck's constant, $h = 6.63 \times 10^{-34}$ J·s. Electrons used in an electron microscope can be diffracted, and electron microscopes have a diffraction limit. If the energy of the electrons used in an electron microscope is 8.0 keV and if the aperture through which the electrons are channeled has a diameter of 0.06 mm, what, approximately, is the smallest angular separation the microscope can distinguish in an object?

65. (III) *Babinet's principle* is useful for the treatment of the diffraction of light by obstacles. It states that if light is incident on an opaque screen in which a hole (of any shape) is cut, then the diffraction pattern produced is the same (except at $\theta = 0$) as that obtained if the screen were removed and the hole were replaced by an obstacle. Use Babinet's principle to estimate the size of an opaque obstruction on a glass slide if a narrow laser beam (with $\lambda = 633$ nm) perpendicularly incident on the slide spreads to a spot of diameter 0.70 cm on a screen 2.5 m from the slide.

66. (III) What diffraction pattern is produced on a distant screen when light of wavelength λ is perpendicularly incident on a plane that contains N very thin hairs, each spaced a distance d apart from the next hair. [*Hint*: See Problem 65.]

67. (III) Electromagnetic radiation of frequency 1.25×10^{23} Hz is scattered by a nucleus of radius 3.2×10^{-15} m. The nucleus is totally radiation-absorbent and thus is a perfect obstacle. At what angle will the first diffraction minimum lie? [*Hint*: Use Babinet's principle (Problem 65).]

APPENDIX I

The Système Internationale (SI) of Units

I–1 SOME SI BASE UNITS

Physical Quantity	Name of Unit	Symbol
length	meter	m
mass	kilogram	kg
time	second	s
electric current	ampere	A
thermodynamic temperature	kelvin	K
amount of substance	mole	mol

I–2 SOME SI DERIVED UNITS

Physical Quantity	Name of Unit	Symbol	SI Unit
frequency	hertz	Hz	s^{-1}
energy	joule	J	$kg \cdot m^2/s^2$
force	newton	N	$kg \cdot m/s^2$
pressure	pascal	Pa	$kg/m \cdot s^2$
power	watt	W	$kg \cdot m^2/s^3$
electric charge	coulomb	C	$A \cdot s$
electric potential	volt	V	$kg \cdot m^2/A \cdot s^3$
electric resistance	ohm	Ω	$kg \cdot m^2/A^2 \cdot s^3$
capacitance	farad	F	$A^2 \cdot s^4/kg \cdot m^2$
inductance	henry	H	$kg \cdot m^2/A^2 \cdot s^2$
magnetic flux	weber	Wb	$kg \cdot m^2/A \cdot s^2$
magnetic flux density	tesla	T	$kg/A \cdot s^2$

I–3 SI UNITS OF SOME OTHER PHYSICAL QUANTITIES

Physical Quantity	SI Unit
speed	m/s
acceleration	m/s^2
angular speed	rad/s
angular acceleration	rad/s^2
torque	$kg \cdot m^2/s^2$, or $N \cdot m$
heat flow	J, or $kg \cdot m^2/s^2$, or $N \cdot m$
entropy	J/K, or $kg \cdot m^2/K \cdot s^2$, or $N \cdot m/K$
thermal conductivity	$W/m \cdot K$

I–4 SOME CONVERSIONS OF NON-SI UNITS TO SI UNITS

Energy:
1 electron-volt (eV) = 1.6022×10^{-19} J
1 erg = 10^{-7} J
1 British thermal unit (BTU) = 1055 J
1 calorie (cal) = 4.185 J
1 kilowatt-hour (kWh) = 3.6×10^6 J

Mass:
1 gram (g) = 10^{-3} kg
1 atomic mass unit (u) = 931.5 MeV/c^2 = 1.661×10^{-27} kg
1 MeV/c^2 = 1.783×10^{-30} kg

Force:
1 dyne = 10^{-5} N
1 pound (lb or #) = 4.448 N

Length:
1 centimeter (cm) = 10^{-2} m
1 kilometer (km) = 10^3 m
1 fermi = 10^{-15} m
1 Angstrom (Å) = 10^{-10} m
1 inch (in or ″) = 0.0254 m
1 foot (ft) = 0.3048 m
1 mile (mi) = 1609.3 m
1 astronomical unit (AU) = 1.496×10^{11} m
1 light-year (ly) = 9.46×10^{15} m
1 parsec (ps) = 3.09×10^{16} m

Angle:
1 degree (°) = 1.745×10^{-2} rad
1 min (′) = 2.909×10^{-4} rad
1 second (″) = 4.848×10^{-6} rad

Volume:
1 liter (L) = 10^{-3} m^3

Power:
1 kilowatt (kW) = 10^3 W
1 horsepower (hp) = 745.7 W

Pressure:
1 bar = 10^5 Pa
1 atmosphere (atm) = 1.013×10^5 Pa
1 pound per square inch (lb/in^2) = 6.895×10^3 Pa

Time:
1 year (yr) = 3.156×10^7 s
1 day (d) = 8.640×10^4 s
1 hour (h) = 3600 s
1 minute (min) = 60 s

Speed:
1 mile per hour (mi/h) = 0.447 m/s

Magnetic field:
1 gauss = 10^{-4} T

APPENDIX II

Some Fundamental Physical Constants[†]

Constant	Symbol	Value	Error
speed of light in a vacuum	c	2.99792458×10^8 m/s	exact
gravitational constant	G	6.67259×10^{-11} m^3/kg·s^2	128
Avogadro's number	N_A	6.02214×10^{23} mol^{-1}	0.1
universal gas constant	R	8.31447 J/mol·K	8.4
Boltzmann's constant	k	1.38065×10^{-23} J/K	1.7
elementary charge	e	1.60218×10^{-19} C	0.004
permittivity of free space	ε_0	$8.85418781762 \times 10^{-12}$ C^2/N·m^2	exact
	$1/4\pi\varepsilon_0$	8.987552×10^9 kg·m^3·s^{-2}·C^{-2}	
permeability of free space	μ_0	$4\pi \times 10^{-7}$ T·m/A	exact
electron mass	m_e	9.10939×10^{-31} kg	0.1
proton mass	m_p	1.67262×10^{-27} kg	0.1
neutron mass	m_n	1.67493×10^{-27} kg	0.1
Planck's constant	h	6.62607×10^{-34} J·s	0.1
$h/2\pi$	$\hbar$	1.05457×10^{-34} J·s	0.1
		$= 6.58212 \times 10^{-22}$ MeV·s	0.1
	$\hbar c$	197.327 MeV·fm	0.3
electron charge-to-mass ratio	$-e/m_e$	-1.75882×10^{11} C/kg	0.1
proton-electron mass ratio	m_p/m_e	1836.15	0.15
molar volume of ideal gas at STP		22414.0 cm^3/mol	1.7
Bohr magneton	μ_B	9.27401×10^{-24} J/T	0.1
magnetic flux quantum	$\Phi_0 = h/2e$	2.06783×10^{-15} Wb	0.1
Bohr radius	a_0	0.529177×10^{-10} m	0.005
Rydberg constant	R_∞	1.09737×10^7 m^{-1}	0.00001

[†] P. J. Mohr and B. N. Taylor, "The 1998 CODATA Recommended Values of the Fundamental Physical Constants, Web Version 3.1," available at physics.nist.gov/constants (National Institute of Standards and Technology, Gaithersburg, MD 20899, 3 December 1999).

We have given values of the measured constants to six significant figures, even though they may be known to greater accuracy. The error, which expresses the uncertainty in the values of these constants, is in parts per million. Defined constants have no error, and we give their full definition; they are indicated by the notation "exact" in the error column.

APPENDIX III

Other Physical Quantities

III–1.1 SOME ASTRONOMICAL CONSTANTS

Constant	Symbol	Value
standard gravity at Earth's surface	g	9.80665 m/s^2
equatorial radius of Earth	R_e	6.378×10^6 m
mass of Earth	M_e	5.976×10^{24} kg
mass of Moon		7.350×10^{22} kg $= 0.0123 \, M_e$
mean radius of Moon's orbit around Earth		3.844×10^8 m
mass of Sun	$M_\odot$	1.989×10^{30} kg
radius of Sun	$R_\odot$	6.96×10^8 m
mean radius of Earth's orbit around Sun	AU	1.496×10^{11} m
period of Earth's orbit around Sun	yr	3.156×10^7 s
diameter of our galaxy		7.5×10^{20} m
mass of our galaxy		2.7×10^{41} kg $= (1.4 \times 10^{11}) \, M_\odot$
Hubble parameter	H	$2.5 \times 10^{-18} \text{ s}^{-1}$

III–1.2 PLANETARY DATA

Planet	Diameter (in km)	Relative[†]	Relative Mass[†]	Average Density (in g/cm³)	Period of Rotation	Surface Gravity[†] (in g)	Escape Speed (in km/s)	Semimajor Axis (AU)	Period of Solar Orbit	Average Orbital Speed (in km/s)
Mercury	4,800	0.38	0.05	5.4	58 d 15 h	0.38	4.3	0.387	87.96 d	47.8
Venus	12,100	0.95	0.82	5.2	243 d 4 h	0.90	10.3	0.723	224.7 d	35.0
Earth	12,750	1.00	1.00	5.5	23 h 56 min	1.00	11.2	1.000	365.26 d	29.8
Mars	6,800	0.53	0.11	3.9	24 h 37 min	0.38	5.0	1.524	687.0 d or 1.88 yr	24.1
Jupiter	142,800	11.21	317.8	1.3	9 h 50 min	2.53	59.5	5.20	11.86 yr	13.1
Saturn	120,660	9.45	95.2	0.7	10 h 39 min	1.07	35.5	9.58	29.46 yr	9.7
Uranus	51,000	4.00	14.5	1.3	17 h	0.91	21.3	19.20	84.01 yr	6.8
Neptune	49,500	3.88	17.1	1.6	16 h	1.14	23.5	30.05	164.79 yr	5.4
Pluto	2,390	0.18	0.002	0.32	6 d 9 h 17 min	0.05	1.1	39.24	247.68 yr	4.7

[†]Relative to Earth.

III–2 ENERGY SUPPLY AND DEMAND[†]

[†]From the *Physics Vade Mecum,* Ed. Herbert L. Anderson, American Institute of Physics (New York, 1981); and U.S. Congress, Office of Technology Assessment, *Changing by Degrees: Steps to Reduce Greenhouse Gases,* OTA-O-482 (Washington, D.C.: U.S. Government Printing Office, February 1991).

III–2.1 FUEL RESOURCES (1980, ESTIMATED)

Resource	U.S. Resources	World Resources
coal (recoverable)	5×10^{21} J	2×10^{22} J
oil (not including oil shales)	10^{21} J	10^{22} J
natural gas	2×10^{21} J	10^{22} J
hydroelectric	10^{22} J/yr (North America)	6×10^{22} J/yr

III–2.2 ANNUAL USAGE OF RESOURCE (2001, PERCENTAGE OF TOTAL)
Source: www.energy.gov

Resource	U.S. Usage (total = 1×10^{20} J)	World Usage (total = 4×10^{20} J)
coal	23	24
oil	40	39
natural gas	24	23
nuclear	8	6
hydroelectric	2	7
biomass	3	1

III–2.3 ENERGY CONTENT OF FUELS

Fuel	Energy Content (in J/kg)
bread	10×10^{6}
glucose ($C_6H_{12}O_6$)	16×10^{6}
white pine wood	20×10^{6}
methyl alcohol (CH_4O)	23×10^{6}
anthracite coal	31×10^{6}
domestic heating oil	45×10^{6}
propane (C_3H_8)	50×10^{6}
natural gas (96% CH_4)	51×10^{6}
fission of U^{235}	5.8×10^{11}
perfect mass-energy conversion	9×10^{16}

III–2.4 SOLAR ENERGY OUTPUT

total radiated power from the Sun	4×10^{26} W
power per unit area at the top of Earth's atmosphere	1.4 kW/m^2
average power per unit area delivered to an average horizontal surface in the United States in 1 yr	0.2 W/m^2

III–2.5 ENERGY CONSUMPTION IN TRANSPORTATION

Mode	Energy Consumption (J/passenger·km)
bicycle	5×10^{4}
foot travel	1.5×10^{5}
automobile	1.9×10^{5}
intercity bus	6×10^{5}
intercity train	9×10^{5}
747 jet airplane	2.3×10^{6}
snowmobile	6×10^{6}

III–2.6 ENERGY CONSUMPTION OF ELECTRICAL APPLIANCE (*See* http://www.ianr.unl.edu/pubs/consumered/heg94.htm)

Appliance	Power (in W)	Energy Use per Year (in kWh)
window air conditioner	3750	3750
clock	2	17
dishwasher	1200	363
window fan	200	170
hair dryer	750	38
iron	1000	144
microwave oven	1450	190
radio	71	86
refrigerator-freezer	615	1830
stove	12,200	1175
color television	200	440
vacuum cleaner	630	46
washing machine	512	107

APPENDIX IV

Mathematics

IV-1 SOME MATHEMATICAL CONSTANTS[†]

Constant	Value
π	3.14159
e (Euler's constant)	2.71828
$\sqrt{2}$	1.41421
$1/\sqrt{2}$	0.707107
$\ln(10)$	2.30259
$\ln(2)$	0.693147
1 rad	57.2958°
1°	0.0174533 rad

[†]To six significant figures.

IV-2 SOLUTION OF QUADRATIC EQUATIONS

Quadratic equation:
$$ax^2 + bx + c = 0$$

Two solutions:
$$x = \frac{-b \pm \sqrt{b^2 - 4ac}}{2a}$$

IV-3 BINOMIAL THEOREM

$$(x + y)^n = \sum_{k=0}^{n} \binom{n}{k} x^{n-k} y^k$$

where
$$\binom{n}{k} = \frac{n!}{(n-k)!k!}$$

The factorial $m! \equiv 1 \cdot 2 \cdot 3 \ldots \cdot m$; $0! \equiv 1$. Some particular cases of the binomial theorem:

(1) $(x \pm y)^2 = x^2 \pm 2xy + y^2$;
(2) $(x \pm y)^3 = x^3 \pm 3x^2 y + 3xy^2 \pm y^3$;
(3) $(x \pm y)^4 = x^4 \pm 4x^3 y + 6x^2 y^2 \pm 4xy^3 + y^4$.

IV-4 TRIGONOMETRY

1. For a right triangle with sides a, b, and c (the hypotenuse), where the angle opposite side a is θ_a (Figure A–1),

$$\text{sine of } \theta_a = \sin \theta_a = \frac{a}{c};$$

$$\text{cosine of } \theta_a = \cos \theta_a = \frac{b}{c};$$

$$\text{tangent of } \theta_a = \tan \theta_a = \frac{a}{b}.$$

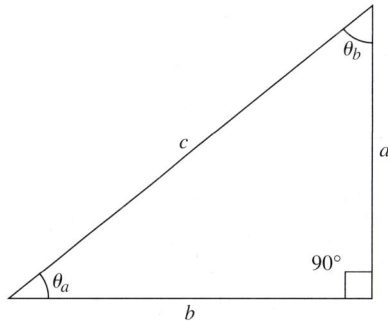

▲ FIGURE A–1

2. The cosine function is even, $\cos(-x) = \cos x$; the sine function is odd, $\sin(-x) = -\sin x$.

3. (1) $\tan \theta = \dfrac{\sin \theta}{\cos \theta}$

   (2) $\sec \theta = \dfrac{1}{\cos \theta}$

   (3) $\operatorname{cosec} \theta = \dfrac{1}{\sin \theta}$

   (4) $\cot \theta = \dfrac{1}{\tan \theta}$

4. (1) $\sin^2 \theta + \cos^2 \theta = 1$
   (2) $\sec^2 \theta - \tan^2 \theta = 1$
   (3) $\operatorname{cosec}^2 \theta - \cot^2 \theta = 1$

5. (1) $\sin(\theta_1 \pm \theta_2) = \sin \theta_1 \cos \theta_2 \pm \cos \theta_1 \sin \theta_2$

   (2) $\cos(\theta_1 \pm \theta_2) = \cos \theta_1 \cos \theta_2 \mp \sin \theta_1 \sin \theta_2$

   (3) $\sin \theta_1 \pm \sin \theta_2 = 2 \sin\left(\dfrac{\theta_1 \pm \theta_2}{2}\right) \cos\left(\dfrac{\theta_1 \mp \theta_2}{2}\right)$

   (4) $\cos \theta_1 + \cos \theta_2 = 2 \cos\left(\dfrac{\theta_1 + \theta_2}{2}\right) \cos\left(\dfrac{\theta_1 - \theta_2}{2}\right)$

   (5) $\cos \theta_1 - \cos \theta_2 = -2 \sin\left(\dfrac{\theta_1 + \theta_2}{2}\right) \sin\left(\dfrac{\theta_1 - \theta_2}{2}\right)$

   (6) $\tan(\theta_1 + \theta_2) = \dfrac{\tan \theta_1 + \tan \theta_2}{1 - (\tan \theta_1)(\tan \theta_2)}$

   (7) $\cos\left(\theta \pm \dfrac{\pi}{2}\right) = \mp \sin \theta$

   (8) $\sin\left(\theta \pm \dfrac{\pi}{2}\right) = \pm \cos \theta$

   (9) $\sin \theta_1 \sin \theta_2 = \dfrac{1}{2}[\cos(\theta_1 - \theta_2) - \cos(\theta_1 + \theta_2)]$

   (10) $\cos \theta_1 \cos \theta_2 = \dfrac{1}{2}[\cos(\theta_1 - \theta_2) + \cos(\theta_1 + \theta_2)]$

   (11) $\sin \theta_1 \cos \theta_2 = \dfrac{1}{2}[\sin(\theta_1 - \theta_2) + \sin(\theta_1 + \theta_2)]$

6. (1) $\sin(2\theta) = 2\sin\theta\cos\theta = \dfrac{2\tan\theta}{1+\tan^2\theta}$

   (2) $\cos(2\theta) = \cos^2\theta - \sin^2\theta = 2\cos^2\theta - 1 = 1 - 2\sin^2\theta$

   (3) $\tan(2\theta) = \dfrac{2\tan\theta}{1-\tan^2\theta}$

   (4) $\sin\left(\dfrac{\theta}{2}\right) = \pm\sqrt{\dfrac{1-\cos\theta}{2}}$

   (5) $\cos\left(\dfrac{\theta}{2}\right) = \pm\sqrt{\dfrac{1+\cos\theta}{2}}$

7. Expansions of trigonometric functions (θ in rad):

   (1) $\sin\theta = \theta - \dfrac{\theta^3}{3!} + \dfrac{\theta^5}{5!} - \dfrac{\theta^7}{7!} + \cdots \quad (\theta^2 < 1)$

   (2) $\cos\theta = 1 - \dfrac{\theta^2}{2!} + \dfrac{\theta^4}{4!} - \dfrac{\theta^6}{6!} + \cdots \quad (\theta^2 < 1)$

   (3) $\tan\theta = \theta + \dfrac{1}{3}\theta^3 + \dfrac{2}{15}\theta^5 + \dfrac{17}{315}\theta^7 + \cdots \quad \left(\theta^2 < \dfrac{\pi^2}{4}\right)$

IV-5 GEOMETRICAL FORMULAS

1. (circumference of a circle of radius r) = $2\pi r$

2. (area of a circle of radius r) = πr^2

3. (area of a sphere of radius r) = $4\pi r^2$

4. (volume of a sphere of radius r) = $\tfrac{4}{3}\pi r^3$

5. (area of a rectangle with sides of lengths L_1 and L_2) = $L_1 L_2$

6. For a right triangle with sides a, b, and c and angles θ_a and θ_b opposite the sides a and b, respectively (Fig. A–1):

   (1) $a^2 + b^2 = c^2$ (the Pythagorean theorem)

   (2) area = $\tfrac{1}{2}$(base)(height) = $\tfrac{1}{2}ab$

7. For a triangle with sides a, b, and c opposite the angles θ_a, θ_b, and θ_c, respectively (Figure A–2):

   (1) $\theta_a + \theta_b + \theta_c = 180° = \pi$ rad

   (2) $a^2 = b^2 + c^2 - 2bc\cos\theta_a$

   (3) $\dfrac{a}{\sin\theta_a} = \dfrac{b}{\sin\theta_b} = \dfrac{c}{\sin\theta_c}$

   (4) $a = b\cos\theta_c + c\cos\theta_b$

   (5) area = $\tfrac{1}{2}$(base)(height) = $\tfrac{1}{2}ab\sin\theta_c = \tfrac{1}{2}a^2\dfrac{\sin\theta_b\sin\theta_c}{\sin\theta_a}$

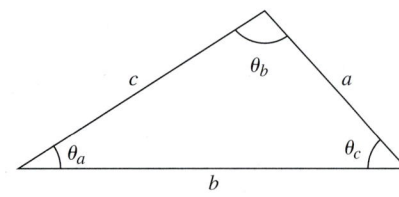

▲ **FIGURE A–2**

8. volume of a right cylinder of height h and radius r = $\pi r^2 h$

IV-6 SOME PROPERTIES OF ALGEBRAIC FUNCTIONS

1. General properties

   (1) $a^x a^y = a^{x+y}$

   (2) $a^0 = 1$

   (3) $(ab)^x = a^x b^x$

2. Properties of exponential of x, $\exp(x)$ or e^x:

   (1) $\exp(\ln x) = x$

   (2) $\exp(x_1)\exp(x_2) = \exp(x_1 + x_2)$

   (3) $\exp(0) = 1$

   (4) expansion: $e^x = 1 + x + \dfrac{x^2}{2!} + \dfrac{x^3}{3!} + \cdots$

3. Properties of the natural logarithm of x, $\ln(x)$:

   (1) $\ln(e^x) = x$

   (2) $\ln(x_1 x_2) = \ln(x_1) + \ln(x_2)$

   (3) $\ln(x_1/x_2) = \ln(x_1) - \ln(x_2)$

   (4) $\ln(1) = 0$

   (5) expansion:
   $$\ln(1+x) = x - \dfrac{x^2}{2} + \dfrac{x^3}{3} - \dfrac{x^4}{4} + \cdots \quad (x^2 < 1)$$

IV-7 DERIVATIVES

In the following, b and p are constants, and u and v are functions of x:

1. $\dfrac{db}{dx} = 0$

2. $\dfrac{d}{dx}(bu) = b\dfrac{du}{dx}$

3. $\dfrac{d}{dx}(u + v) = \dfrac{du}{dx} + \dfrac{dv}{dx}$

4. $\dfrac{d}{dx}(uv) = v\dfrac{du}{dx} + u\dfrac{dv}{dx}$

5. $\dfrac{dx^p}{dx} = px^{p-1}$

6. Chain rule: If u is a function of y, and y is in turn a function of x, then $\dfrac{du}{dx} = \dfrac{du}{dy}\dfrac{dy}{dx}$.

7. $\dfrac{d}{dx}(\sin x) = \cos x$

8. $\dfrac{d}{dx}(\cos x) = -\sin x$

9. $\dfrac{d}{dx}(\tan x) = \dfrac{1}{\cos^2 x}$

10. $\dfrac{d}{dx}(e^{bx}) = be^{bx}$

11. $\dfrac{d}{dx}\ln(x) = \dfrac{1}{x}$

IV–8 TAYLOR EXPANSION

If $f(x)$ is well behaved near point $x = x_0$,

$$f(x) = f(x_0) + \left.\frac{df}{dx}\right|_{x=x_0}(x - x_0) + \frac{1}{2!}\left.\frac{d^2f}{dx^2}\right|_{x=x_0}(x - x_0)^2 + \cdots$$

IV–9 INTEGRALS

In the following, b and p are constants, and u and v are functions of x:

1. $\int \frac{du}{dx}\,dx = u$

2. $\int_{x_1}^{x_2} \frac{du}{dx}\,dx = u(x_2) - u(x_1)$

3. $\int bu(x)\,dx = b\int u(x)\,dx$

4. $\int (u + v)\,dx = \int u\,dx + \int v\,dx$

5. $\int u\frac{dv}{dx}\,dx = uv - \int v\frac{du}{dx}\,dx$ (integration by parts)

6. If u is a function of y and y is in turn a function of x, then
$$\int u\,dy = \int u\frac{dy}{dx}\,dx$$

7. $\int x^p\,dx = \frac{x^{p+1}}{p+1}\quad(p \neq -1)$

8. $\int \frac{dx}{x} = \ln x$

9. $\int (\sin x)\,dx = -\cos x$

10. $\int (\cos x)\,dx = \sin x$

11. $\int e^{bx}\,dx = \frac{1}{b}e^{bx}$

12. $\int xe^{bx}\,dx = e^{bx}\left(\frac{x}{b} - \frac{1}{b^2}\right)$

13. Some definite integrals:

    (1) $\int_0^\infty x^n e^{-x}\,dx = n!$

    (2) $\int_0^\pi (\sin^2 x)\,dx = \int_0^\pi (\cos^2 x)\,dx = \frac{\pi}{2}$

    (3) $\int_0^\infty e^{-b^2x^2}\,dx = \frac{\sqrt{\pi}}{2b}\quad(b > 0)$

    (4) $\int_0^\infty xe^{-x^2}\,dx = \frac{1}{2}$

    (5) $\int_0^\infty x^2 e^{-x^2}\,dx = \frac{\sqrt{\pi}}{4}$

    (6) $\int_0^\infty \frac{b}{b^2 + x^2}\,dx = \begin{cases} \frac{\pi}{2} & (b > 0) \\ 0 & (b = 0) \\ -\frac{\pi}{2} & (b < 0) \end{cases}$

IV–10 SOME EXPANSIONS APPROPRIATE FOR $x^2 < 1$

1. The following expression is good for any n, positive or negative, integer or noninteger:
$$(1 + x)^n = 1 + nx + \frac{n(n-1)}{2!}x^2 + \frac{n(n-1)(n-2)}{3!}x^3 + \cdots$$

2. $\sin x = x - \frac{x^3}{3!} + \frac{x^5}{5!} + \cdots$

3. $\cos x = 1 - \frac{x^2}{2!} + \frac{x^4}{4!} + \cdots$

4. $\tan x = x + \frac{x^3}{3} + \frac{2}{15}x^5 + \cdots$

5. $e^{ax} = 1 + ax + \frac{(ax)^2}{2!} + \frac{ax^3}{3!} + \cdots$

IV–11 SOME MATHEMATICAL NOTATION

1. $=$ is equal to
2. $\cong$ is approximately equal to
3. $\propto$ is proportional to
4. $\equiv$ is defined to be
5. $\neq$ is unequal to
6. $>$ is greater than
7. $\geq$ is greater than or equal to
8. $<$ is less than
9. $\leq$ is less than or equal to
10. Δx the change in x
11. $|x|$ the absolute value of x
12. $O(N)$ on the order of the magnitude of N
13. $\pm$ plus or minus
14. $\mp$ minus or plus
15. $\langle x \rangle$ average of x
16. $\sum_{i=i_1}^{i_2} f_i$ the sum of all f_i over the integers i from a smallest integer i_1 to a largest integer i_2
17. $\ln(x)$ natural logarithm of x
18. $\log_{10}(x)$ logarithm to the base 10 of x
19. $\int$ integral
20. $\oint$ line integral around a loop

APPENDIX V

Periodic Table of the Elements

PERIODIC TABLE
Atomic Properties of the Elements

Adapted from:
U.S. DEPARTMENT OF COMMERCE
Technology Administration
National Institute of Standards and Technology
http://physics.nist.gov/PhysRefData/IonEnergy/periodic-table.pdf

For a description of the atomic data, visit physics.nist.gov/atomic

March 1999

Group	IA	IIA	IIIA	IVA	VA	VIA	VIIA	VIIIA			IB	IIB	IIIB	IVB	VB	VIB	VIIB	VIII
Period 1	**1 H** Hydrogen 1.00794																	**2 He** Helium 4.00260
Period 2	**3 Li** Lithium 6.941	**4 Be** Beryllium 9.01218											**5 B** Boron 10.811	**6 C** Carbon 12.0107	**7 N** Nitrogen 14.00674	**8 O** Oxygen 15.9994	**9 F** Fluorine 18.998403	**10 Ne** Neon 20.1797
Period 3	**11 Na** Sodium 22.98977	**12 Mg** Magnesium 24.3050											**13 Al** Aluminum 26.98154	**14 Si** Silicon 28.0855	**15 P** Phosphorus 30.97376	**16 S** Sulfur 32.066	**17 Cl** Chlorine 35.4527	**18 Ar** Argon 39.948
Period 4	**19 K** Potassium 39.0983	**20 Ca** Calcium 40.078	**21 Sc** Scandium 44.95591	**22 Ti** Titanium 47.867	**23 V** Vanadium 50.9415	**24 Cr** Chromium 51.9961	**25 Mn** Manganese 54.93805	**26 Fe** Iron 55.845	**27 Co** Cobalt 58.93320	**28 Ni** Nickel 58.6934	**29 Cu** Copper 63.546	**30 Zn** Zinc 65.39	**31 Ga** Gallium 69.723	**32 Ge** Germanium 72.61	**33 As** Arsenic 74.92160	**34 Se** Selenium 78.96	**35 Br** Bromine 79.904	**36 Kr** Krypton 83.80
Period 5	**37 Rb** Rubidium 85.4678	**38 Sr** Strontium 87.62	**39 Y** Yttrium 88.90585	**40 Zr** Zirconium 91.224	**41 Nb** Niobium 92.90638	**42 Mo** Molybdenum 95.94	**43 Tc** Technetium (98)	**44 Ru** Ruthenium 101.07	**45 Rh** Rhodium 102.90550	**46 Pd** Palladium 106.42	**47 Ag** Silver 107.8682	**48 Cd** Cadmium 112.411	**49 In** Indium 114.818	**50 Sn** Tin 118.710	**51 Sb** Antimony 121.760	**52 Te** Tellurium 127.60	**53 I** Iodine 126.90447	**54 Xe** Xenon 131.29
Period 6	**55 Cs** Cesium 132.90545	**56 Ba** Barium 137.327	57 *La	**72 Hf** Hafnium 178.49	**73 Ta** Tantalum 180.9479	**74 W** Tungsten 183.84	**75 Re** Rhenium 186.207	**76 Os** Osmium 190.23	**77 Ir** Iridium 192.217	**78 Pt** Platinum 195.078	**79 Au** Gold 196.96655	**80 Hg** Mercury 200.59	**81 Tl** Thallium 204.3833	**82 Pb** Lead 207.2	**83 Bi** Bismuth 208.98038	**84 Po** Polonium (209)	**85 At** Astatine (210)	**86 Rn** Radon (222)
Period 7	**87 Fr** Francium (223)	**88 Ra** Radium (226)	89 *Ac	**104 Rf** Rutherfordium (261)	**105 Db** Dubnium (262)	**106 Sg** Seaborgium (263)	**107 Bh** Bohrium (264)	**108 Hs** Hassium (265)	**109 Mt** Meitnerium (268)	**110 Uun** Ununnilium (269)	**111 Uuu** Unununium (272)	**112 Uub** Ununbium						

57 La Lanthanum 138.9055	**58 Ce** Cerium 140.116	**59 Pr** Praseodymium 140.90765	**60 Nd** Neodymium 144.24	**61 Pm** Promethium (145)	**62 Sm** Samarium 150.36	**63 Eu** Europium 151.964	**64 Gd** Gadolinium 157.25	**65 Tb** Terbium 158.92534	**66 Dy** Dysprosium 162.50	**67 Ho** Holmium 164.93032	**68 Er** Erbium 167.26	**69 Tm** Thulium 168.93421	**70 Yb** Ytterbium 173.04	**71 Lu** Lutetium 174.967
89 Ac Actinium (227)	**90 Th** Thorium 232.0381	**91 Pa** Protactinium 231.03588	**92 U** Uranium 238.0289	**93 Np** Neptunium (237)	**94 Pu** Plutonium (244)	**95 Am** Americium (243)	**96 Cm** Curium (247)	**97 Bk** Berkelium (247)	**98 Cf** Californium (251)	**99 Es** Einsteinium (252)	**100 Fm** Fermium (257)	**101 Md** Mendelevium (258)	**102 No** Nobelium (259)	**103 Lr** Lawrencium (262)

Metals ← → Nonmetals

☐ Solids
☐ Liquids
☐ Gases
☐ Artificially Prepared

Atomic Number — 58
Symbol — Ce
Name — Cerium
Atomic Weight[†] — 140.116

[†]Based upon ^{12}C. ( ) indicates the mass number of the most stable isotope. For a description and the most accurate values and uncertainties, see J. Phys. Chem. Ref. Data, **26** (5), 1239 (1997).

APPENDIX VI

Significant Dates in the Development of Physics

History can rarely be stated as a simple series of dates, and the history of science is no exception. Throughout the text we have alluded to important discoveries in physics. The list below is a personal choice and should be thought of as a guide. It oversimplifies some of the history, including stories that are covered more thoroughly in the text. Some of the dates are to be taken with a grain of salt, because discoveries are rarely made in a single identifiable moment. Our list includes some names (and discoveries) not mentioned in the text. Far more numerous are the names not listed, the names of those who built the experimental foundations, those who explored the false paths and cleared the way for those whose names we remember today, or those who verified the speculations that are now called laws.

Year	Scientist	Discovery
1583	Galileo	Pendulum motion
1600	Gilbert	Study of magnets
1602	Galileo	Early statement of Newton's first law
1602	Galileo	Laws of falling bodies
1609	Kepler	First two laws of planetary motion
1619	Kepler	Third law of planetary motion
1620	Snell	Law of refraction
1648	Pascal	Atmospheric pressure
1650	Grimaldi	Diffraction of light
1661	Boyle	Chemical elements
1669	Newton	Light dispersion in prisms
1678	Huygens	Wave propagation
1687	Newton	Laws of motion; universal gravitation
1760	Black	Calorimetry
1785	Coulomb	Coulomb's law
1789	Lavoisier	Conservation of mass
1798	Cavendish	Measurement of G
1800	Volta	Electric battery
1801	Young	Interference of light
1801	Dalton	Laws of chemical combination
1802	Charles; Gay-Lussac	Ideal gases
1807	Dalton	Atomic theory
1812	Fourier	Decomposition of waves
1815	Fraunhofer	Discrete spectral lines
1819	Fresnel	Wave picture of light
1820	Oersted	Magnetic fields from currents
1820	Biot; Savart	Law of magnetic field produced by current
1824	Carnot	Second law of thermodynamics
1827	Ohm	Ohm's law
1827	Ampère	Ampère's law
1831	Faraday; Henry	Magnetic induction
1842	Joule	Mechanical equivalent of heat
1847	Helmholtz	Conservation of energy
1849	Fizeau	Direct measurement of the speed of light
1865	Maxwell	The laws of electricity and magnetism; light waves
1877	Boltzmann; Gibbs	Statistical mechanics
1879	Stefan	Blackbody radiation
1885	Osmond	Crystalline structure of metals
1887	Hertz	Electromagnetic waves
1887	Michelson and Morley	Constancy of the speed of light
1896	Becquerel	Radioactivity
1897	Thomson	Charge-to-mass ratio of the electron

Year	Scientist(s)	Discovery
1900	Planck	Quanta in blackbody radiation
1903	Rutherford; Soddy	Isotopes
1905	Einstein	Special relativity; quanta in photoelectric effect
1908	Kammerlingh Onnes	Superfluidity
1911	Kammerlingh Onnes	Superconductivity
1911	Rutherford	Nuclear structure of atom
1911	Millikan	Quantization of charge
1912	von Laue	X-ray diffraction in crystals
1912	Bragg and Bragg	Analysis of crystal structure with x-rays
1913	Bohr	Atomic structure
1916	Einstein	General relativity
1922	Compton	Scattering of x-rays
1923	Hubble	Discovery of galaxies
1924	de Broglie	Wave nature of particles
1925	Pauli	Exclusion principle
1925	Heisenberg	Formulation of quantum mechanics
1925	Goudsmit and Uhlenbeck	Electron spin
1926	Davisson and Germer; Thomson	Diffraction of electrons by crystals
1926	Schrödinger	Alternate formulation of quantum mechanics
1926	Born	Probabilistic interpretation of quantum theory
1927	Lemaitre	Big bang universe introduced
1927	Heisenberg	Uncertainty relations
1929	Hubble	Hubble's law
1930	Dirac	Antiparticles
1932	Anderson	The positron
1932	Lawrence and Livingston	The cyclotron
1932	Chadwick	The neutron
1934	Yukawa	Nuclear forces and the pi meson
1942	Fermi	Nuclear chain reaction
1948	Feynman; Schwinger; Tomonaga	Electromagnetism as a quantum theory
1954	Townes	The maser
1956	Reines and Cowan	Neutrinos observed
1957	Lee and Yang	Nonconservation of parity
1957	Bardeen, Cooper, and Schrieffer	Theory of superconductivity
1962	Josephson	Josephson junction
1964	Gell-Mann; Zweig	Quarks
1964	Penzias and Wilson	Background radiation of the universe
1967	Bell and Hewish	Neutron stars discovered (pulsars)
1967–1970	Glashow; Salam; Weinberg	Unification of electromagnetic and weak forces
1981	Binnig and Rohrer	Scanning tunneling electron microscope
1983	Taylor and Hulse	Gravitational radiation
1983	CERN	$W^{\pm}$ and Z^0 particles discovered
1986	Bednorz and Müller	High-temperature superconductors
1990	Davis and Koshiba	Neutrino oscillations and inference of neutrino mass

Answers to "What Do You Think?" Questions

CHAPTER 21

1. The missing mass will be (number of electrons) × (mass of an electron) = $(6.9 \times 10^{11}$ electrons$)(9.11 \times 10^{-31}$ kg/electron$) \cong 10^{-18}$ kg. There is no instrument we can use to measure such a small mass, let alone a mass difference of that size.

2. (a) The charge left behind would have the same magnitude as the total electron charge, 1.1×10^6 C, because the coin was initially neutral. (b) If one of the 79 electrons in each gold atom were removed, the total charge would be only $1/79$ of the charge just found, or 1.4×10^4 C, which is a huge electric charge.

4. No, just the opposite: Charge opposite to that of the smaller piece would move closer to that smaller piece, because it is attracted, while charge the same as that of the smaller piece would move farther because it would be repelled. The net effect would be an even stronger attraction. This phenomenon, charge induction, is what lies behind the separation of the electroscope leaves in Conceptual Example 21–3.

5. Most objects have little or no net electrical charge, and there is no Coulomb law force between such objects. The Earth is practically neutral, so that there is little or no Coulomb force between us and Earth, as we are also electrically neutral. But Earth is extremely massive, and the gravitational force between us and Earth is adequate to hold us at the surface.

6. The electric force between the two cork balls would be unchanged. The lower ball's mass is irrelevant—it sits on a surface. The force of gravity on the upper ball would now be twice as large as above and would match precisely the Coulomb force acting on it. It would now be in equilibrium and would not move farther.

7. Charges q_1 and q_2 are identical. If we place q_3 precisely on the x-axis between the other two charges, q_3 will be equally attracted to the other two charges. If q_3 is displaced slightly toward either q_1 or q_2, its attraction to the nearer charge will be larger than its attraction to the farther one, because the Coulomb force increases as $1/r^2$ when r decreases. Thus q_3 will not return to equilibrium—the equilibrium is unstable.

9. When $L = 0$, the charge is at the middle of the ring, and, by symmetry, the net force should be zero, as in Conceptual Example 21–9. This is indeed a property of our result, and a second check of the calculation.

10. The right side of the rod is positively charged, so we expect on physical grounds that the repulsion on the point charge will become very strong. This is verified with a numerical evaluation of the force. We can also see it in the analytic expression: Our limit is $R \to L/2$, and the dominant term in the force in this limit is the term $1/(R - L/2)$, which is indeed positive. The force on the point charge is large and to the right.

CHAPTER 22

2. There would be no change in the electric field due to q. However, the force on a q' of opposite sign would be a force with the same magnitude but reversed direction. The force would be repulsive, not attractive.

3. Remembering that the electric field produced by a charge points away from the charge if the charge is positive and toward the charge if the charge is negative, we see that the fields from all three charges are to the right in the region between q_2 and q_3. Thus the net field also points to the right.

4. Very close to one of the charges the net field is to a good approximation the field of the close charge, and this does not depend on L in any way. It is also possible to show that the field depends on p alone at distances $r \gg L$, although this requires a detailed analysis.

6. Although the lines will initially bend toward $-q$, very far away they will be pointing radially outward from a net charge $+q$ ($+2q - q = +q$); at great distances we see a net charge of $+q$, and the field will be indistinguishable from the field of a point charge $+q$. The fact that 12 lines remain is consistent with our original choice of 24 lines for the line density (Fig. 22–13c).

7. To a point charge very close to the rod, the rod appears to have infinite length, and there is an infinite amount of charge in an infinite rod with a finite charge density. The summation over all the fields from charges in the rod, including the ones that are very distant from the point at which the field is measured, builds up a net field that decreases more slowly than the field of a finite charge distribution.

8. (b)

10. (c). The surface charge density would be larger because the area of the Moon is smaller than that of Earth, so the Moon's surface electric field would be correspondingly larger. If you calculate the (repulsive) Coulomb force between Earth and the Moon you would find about 10^{-5} N, a completely negligible effect.

11. The electron would not move, because it would be attracted to the positively charged plane.

12. (b) Our expression for y is proportional to E. This is the feature that makes the tube we have described here so useful.

CHAPTER 23

3. Gauss' law is now applied to a total charge of $q + 2q - 7q + 4q = 0$. There will be no *net* flux through the surface. Again we stress *net* flux, because there will certainly be flux going out of, and coming into, the surface, as could be seen by putting a cube around the two equal and opposite charges in Fig. 23–12b.

4. We would now use the principle of superposition of electric fields. Each face would get no contribution from the four charges at its edges, but would get an equal contribution from each of the four charges around the opposite face. Thus the flux through each face is $4 \times (q/24\varepsilon_0) = q/6\varepsilon_0$.

5. The symmetry is the same as in the example. If we take a Gaussian surface that is cylindrical with the same axis as the charged cylinder with a radius larger than that of the charged cylinder, exactly the same argument holds, and we obtain the same result. On the other hand, if we draw our Gaussian cylinder *inside* the charged cylinder, then the symmetries are exactly the same, but the enclosed charge is zero. This is what we would expect, as the electric field inside the cylinder is zero.

6. The Gaussian surfaces will again be concentric spheres. In each case we end up with the result $E = \dfrac{Q_{\text{enclosed}}}{4\pi\varepsilon_0 r^2}$. When r is larger than the radius of the outer shell, then the total enclosed charge is zero, so that $E = 0$. When r lies between the shells, then the total charge enclosed is Q and that is to be inserted in the above equation. When r is inside the inner shell, then no charge is enclosed and $E = 0$ again. There is a non-zero field only between the shells.

7. The motion is determined by the force, and the force on the charge is directed toward the center, with magnitude
$$F = -qE = -\frac{qQ}{4\pi\varepsilon_0}\frac{r}{R^3}$$
This is a "restoring" force proportional to the displacement from the origin. The motion is therefore simple harmonic motion, and the charge will oscillate about that point.

8. We know that at a great distance from the charge distribution the electric field lines from any localized charge, such as found on a finite plane, are spread in a spherically symmetric pattern, as if the charge were a point charge. This means that the field lines will ultimately bend and not point straight away from the charged surface. In a region close to the charged surface, where the distance to the surface is small compared with the distance to the nearest edge, we may make the approximation that the field lines are perpendicular to the charged surface. If we take for a Gaussian surface a very shallow squat cylinder half into the surface, the argument that led to Eq. (23–11) goes through as before, and the result applies here too.

CHAPTER 24

3. False. The sign of this result makes sense. The electric potential at point b is negative. The new charge is positive and will be attracted to the negative potential. Franklin would have to do positive work to bring the same charge back out to infinity. It is better to understand what is happening physically than to rely on your ability to avoid algebraic error.

4. The energy of the at-rest electron-proton system once the electron has been moved far away is zero. Therefore you must add positive energy to the system to make the separation.

5. If we could freeze the charges in position, then there would indeed be a dipole field. However, the two charges form a kind of planetary system, revolving around each other; from a given point P the angle θ that appears in Eq. (24–24) varies uniformly with time. Thus the average dipole potential involves the average of $\cos \theta$. But $\cos \theta$ varies between -1 and 1, and its average value is zero.

8. Yes. The electric field points from the positively charged plate to the negatively charged plate (see for example Fig. 22–19). We also know that the electric field points from the higher to the lower potential surface. This then implies that it is the lower potential surface—the left side plate—that carries the negative charge.

9. The solution is obtained by setting $x = 0$. It is $V = \dfrac{Q}{4\pi\varepsilon_0 R}$. The result follows very simply from superposition. The fact that all the charges whose potentials add up lie on a circle is irrelevant, because the potential is a *scalar* quantity that does not involve directions. The only thing that matters here is that the center is equidistant from all the elements of charge.

10. (d). This is evident from the fact that at a great distance the disk is indistinguishable from a point charge. If we want to take a mathematical limit, we cannot just drop the R^2 term in the square root, because we then get zero. We instead need to work out the term in parentheses by making use of the fact that

$$\frac{x}{\sqrt{x^2 + R^2}} - 1 = \frac{x}{x\sqrt{1 - (R^2/x^2)}} - 1 = \frac{1}{\sqrt{1 - R^2/x^2}} - 1,$$

then making an expansion of 1 divided by the square root for small R/x.

11. Writing the potential as

$$V = \int_{-\infty}^{\infty} \frac{\lambda\, dz}{4\pi\varepsilon_0 \sqrt{R^2 + z^2}}$$

is based on choosing the potential due to an element of charge as vanishing at infinity. As argued in the solution, this is not legitimate here. In fact, the integral written above is independent of R (easily seen by changing variables to $u = z/R$), and therefore gives a zero electric field, an incorrect result.

12. The relevant fact is that the potential is *constant*. This means that its derivatives are zero, consistent with zero electric field. The particular constant value of the potential ensures that the potential is *continuous*. If the potential were not continuous at $r = R$, then the motion of a point charge in that potential would not be physical; a jump in the potential energy would require a discontinuous jump in the kinetic energy, that is, in the velocity of a point charge.

CHAPTER 25

2. The potential difference doubles when Q doubles, but C is unchanged. C depends only on factors such as geometry, and is independent of the amount of charge on the capacitor.

4. The permittivity, ε_0, has units F/m and so must be multiplied by a length to give an acceptable capacitance. The only length in this problem is the radius of the sphere, so $C \propto \varepsilon_0 R$.

5. The energy stored in a battery is local and is held chemically, molecule by molecule. In a capacitor, each time we transfer another charge, it requires more work, because there is an increasing amount of charge already placed on the capacitor. We are fighting an uphill battle.

7. In regions outside the sphere itself, the sphere behaves as if all its charge is concentrated at the center. Thus it requires more work to bring charge from infinity and place it on the smaller conducting sphere because in effect you are closer to the center when you are charging a smaller sphere. Therefore the energy of the system also becomes large.

9. Consider the extreme case where we make this capacitance very small. From Eq. (25–17) we see that a small capacitor has a large effect when it is placed in series with other capacitors; indeed a capacitor with a very small capacitance C_{small} will dominate all the terms, giving as a good approximation $C_{eq} = C_{small}$. Thus decreasing the capacitance of the 2 μF capacitor will decrease the equivalent capacitance.

10. Yes, work was done because the electrical energy changed. The sign of the work done by the capacitor is positive, because the capacitor's energy decreases as the Teflon was inserted.

11. Bakelite has a larger value of the dielectric constant κ, so that for a given charge, the field within the Bakelite would be smaller, the voltage drop across the entire space would be smaller, and hence the capacitance larger.

12. The *change* in energy, which is what determines the force, depends only on how much *additional* length of plug is inserted. For example, the energy change is the same whether the plug is first inserted 1 mm in as it is when the plug is already 1 cm in and is then inserted an additional 1 mm.

CHAPTER 26

1. In a proton beam the charge carriers are positively charged, whereas in a wire the charge carriers are electrons, which have negative charge. Moreover, a wire is electrically neutral, because there is a background of positive ions that match the current-carrying electrons, whereas a proton beam in space is not neutral.

2. The question is, Is the rate at which charge goes into the annihilation region the same as the rate of charge that goes out? If we go to the right of the annihilation region, we only have a flux of antiprotons, and the entering rate of charge from the right is $N'e = Ne$. On the left of the annihilation region we have only a flux of incoming protons, and the rate at which charge enters is $N(-e)$. When we add these two terms together, we see that the net rate at which charge enters is zero, the same as the rate at which it leaves. Current conservation is a necessary consequence of charge conservation.

3. For a fixed current density, a quadrupling of the current implies a quadrupling of the area, and since the area is proportional to the radius squared, the radius of the wire will have to be doubled.

4. Table 26–2 shows that the resistivity of iron is about six times larger than that of copper. This means that for fixed E, the current density will be about six times smaller, and for a wire of the same cross section the current will also be a factor of six times smaller. Copper is used widely in the wires of circuits, and appliances, as it is a good conductor and has good mechanical properties.

5. A crucial requirement for a filament is that it should not melt! The other is that it is best to have a higher resistance, since it is the resistance that is responsible for the dissipation of energy, and this dissipation manifests itself in the heating of the coil, and ultimately in the radiation of the coil. Our result above tells us that at a temperature a bit below the copper melting point, the resistance per unit length of our platinum wire is 53/20 = 2.65 Ω/m. The coefficient α for platinum is the same, but ρ_0 is a factor of 6.2 larger, so that even at the same temperature, the resistance of the platinum filament is that much larger. In addition, the platinum filament can be heated to a higher temperature, increasing the resistance still further. For these reasons the platinum filament is to be preferred.

6. With this exchange, the resistances on both sides would be equal (10 Ω each). With two equal resistances in parallel, the currents would be divided equally. This can be calculated but is also evident from symmetry—two identical paths will carry the same current. This means that the current in each branch is 1.5 A, and the potential difference overall is the same on both sides, i.e. (1.5 A) $\times$ (10 Ω) = 15 V. The potential difference across the 4 Ω resistor is (4 Ω) $\times$ (1.5 A) = 6 V.

9. With $P = V^2/R$, we can immediately tell that a doubling of the voltage implies an increase of the power by a factor of 4.

10. For a fixed voltage, the power is given by V^2/R. Whatever the resistance R of the bulb is, the power will be halved in case (a), since the resistance is doubled. The power will be doubled in case (b), since the reciprocal of the resistance is doubled. If the purpose of the lighting is to generate heat to keep the paint from freezing, the first alternative may be inadequate and the second too expensive.

11. The current is given by $(P/R)^{1/2}$. The radiated power is proportional to the overall area of the (cylindrical) wire, hence proportional to d; the resistance is inversely proportional to the cross section of the wire, hence proportional to d^{-2}. Thus, for a given $I \propto (d/d^{-2})^{1/2} = d^{3/2}$. You should therefore reduce the diameter of the wire to decrease the current. That means that the power will be reduced, so the wire will not be quite as effective at heating its surroundings.

CHAPTER 27

2. If we start from the expressions for $\mathcal{E}$ and r, we see that in the formula for $\mathcal{E}$, R_1 must be taken to be zero to the accuracy that the resistances are measured. Similarly, in the numerator of the expression for r, the second term $I_1 R_1$ must also be taken to be zero to the accuracy under consideration. This then implies that $r = \mathcal{E}/I_1$. This can be calculated to two significant figures.

3. For a given battery, with characteristic emf and internal resistance, the expression $P = \mathcal{E} I$ shows that P is largest when I is largest. This is achieved in our circuit by reducing the external resistance as much as possible. Setting $R = 0$ gives $P_{max} = \mathcal{E}^2/r$.

4. As stated in "Setting It Up," the solution to the problem decides the current direction. If we had initially chosen the opposite current direction our algebraic solution would have been positive. Because the emf $\mathcal{E}_2$ is larger than the emf $\mathcal{E}_1$, we could have anticipated that the current will run counterclockwise.

5. If R_2 is large, there is so much resistance in the leg that contains it that the current I_2 should drop to zero. In effect, the segment containing R_2 would be eliminated from the circuit. In this limit our expressions give $I_1, I_3 \rightarrow (\mathscr{E}_1 + \mathscr{E}_2)/(R_1 + R_3)$, and $I_2 \rightarrow 0$, just as we expect.

6. As long as the loop equations are independent, the Kirchhoff rules will always give the correct answers. In this case it is particularly easy to see that this applies: All you have to do is flip the segment (d–battery–a) to the other side of the diamond. In that case the new loop is just the mirror image of the old loop.

7. (a) The resistance can never be negative, whatever the signs of the emf or the current direction. (b) The sign of the potential difference between the two plates shows that the potential increases when we go from plate 1 to plate 2, meaning that plate 2 is at a higher potential than plate 1. This is because positive charges have accumulated on plate 2.

9. Qualitatively the corrections to V are of the general form R/R_V, where R is the generic resistance in the circuit (e.g., combinations of R_1 and R_2 in this example). When the denominator is a factor 100 times larger, the error, instead of being in the 3% range, is in the 0.03% range.

10. A look at Fig. 27–20b shows that we are dealing with the very flat part of the curve on the extreme right. To answer this quantitatively we need to know something about exponential functions, or equivalently about natural logs. As we saw in the example, the time is determined by $e^{-t/\tau} = 1 - p$, where p is the desired percentage. In the flat part of the curve we are dealing with p very close to 1 (0.999, and now 0.9999), so that if we write $p = 1 - x$ (with x very small), we get $-t/\tau = \ln x$ or $t/\tau = \ln(1/x)$. If we look up $\ln(10) = 2.30$, we see that each factor of 10 in the percentage accuracy increases the time (in units of RC) by an additional 2.30 units. In this case we would increase 6.91 ms to 9.21 ms.

CHAPTER 28

2. When a rock is thrown, the gravity force perpendicular to its initial motion is constant, and it always points in the same direction. In the case of a magnetic field, the force changes direction as the velocity changes direction. Furthermore the velocity is constant in magnitude. The path of the electron is therefore quite different from the familiar motion of a thrown rock.

3. For a given charge (and the deuteron and the proton have equal charges), the radius of curvature is proportional to the momentum, the product mv. The deuteron mass is double the proton mass, so it would have the same momentum as the proton if the speed were halved.

4. Let us assume that the charged particle comes out of the opening hole traveling in the $+z$-direction. If the magnetic field is used to define the $+x$-direction, then the magnetic force points in the $+y$-direction, and the electric field must therefore point in the $-y$-direction. Suppose the particle is positive and has an excess velocity that is originally in the $+z$-direction. The magnetic force will now initially be in the $+y$-direction. The particle will acquire a small component of velocity in that direction and will start a circular motion (with angular frequency qB/m) in a clockwise direction as seen by somebody looking along the x-axis. The path will be similar to that of a point on the rim of a wheel whose center is moving with a uniform speed while the wheel is rotating in a direction opposite to what its rolling rotation would be like. A negative particle would have motion like a spot on the rim of a rolling wheel.

9. The wire and field form a plane, and the force is perpendicular to the plane. Whether it goes into or out of the plane depends on using the right-hand rule in accordance with Eq. (28–19).

11. We have already mentioned that bar magnets are influenced by magnetic fields in just the same way as are current loops. This suggests, correctly, that the compass needle, which is a small bar magnet, behaves much like the loop in this example, oscillating about the stable equilibrium position in the absence of damping.

12. The drift velocity does not depend on the width of the strip. The electric field will be of magnitude vB, so that it does not depend on the width of the strip. Therefore the potential $V = Ed$ will double if the width of the strip doubles.

CHAPTER 29

2. The magnitude would be unchanged. We didn't specify the direction of the current in the first place, so there is no real information on direction. But if the current is reversed, the direction of the magnetic field would be reversed. The direction of the field is determined by the right-hand rule: If the thumb of the right hand is in the direction of current, the fingers indicate the direction of the magnetic field.

3. The flux will be a maximum when $\cos \theta = 1$, so that $\theta = 0$. This corresponds to the loop perpendicular to the field.

4. With another row of wire, the number of turns per unit length n would double, and the magnetic field would also double.

6. We work directly from Eq. (29–21), which for $L \gg D$ becomes $B \cong \dfrac{\mu_0 I}{4\pi} \dfrac{L}{D(L/2)} = \dfrac{\mu_0 I}{2\pi D}$, which is the same result as given by Ampère's law for the infinitely long wire.

7. Yes. We curl the fingers of the right hand along the current, and the thumb indicates the direction of the field. This orientation rule is the same one that we found for the magnetic field of a solenoid; indeed, the loop is nothing more than a compressed solenoid.

8. An electric dipole; see Eq. (22–14).

9. Remember that the electric field between the plates depends only on the charge density and is independent of d as long as $R \gg d$. The displacement current and magnetic field between the plates will not change.

CHAPTER 30

2. (c) The induced current will always be in the direction to oppose the change in magnetic field through the loop. As the loop goes back into the region of field, the induced current will be opposite to what it was as the loop left the field. The magnetic field that the induced current creates will oppose what is now an increasing magnetic flux.

4. The motion of the loop toward the straight wire would produce the same change in magnetic flux through the loop as the motion of the straight wire toward the loop. The effect would be exactly the same.

5. (b) The induced emf is the same, but the resistance increases; hence the induced current will be less.

7. The Lorentz force is $q(\vec{v} \times \vec{B})$, and its direction on a positive charge is indeed radially outward, in agreement with the result we found using Faraday's law.

9. Because the kinetic energy doesn't change, the change in total energy (dissipated in Joule heating) comes from the change in gravitational potential energy.

10. We can imagine a large circle, radius R, well outside the pole faces but centered at the center of the faces. The magnetic flux through this circle is to a good approximation $\Phi_B = B\pi r_{\text{face}}^2$, where r_{face} is the radius of the magnet. The same symmetry argument gives the electric field lines forming circles centered at the center of the pole face. As in the example, the magnitude of the field is now determined by $E \times 2\pi R = \pi r_{\text{face}}^2 \alpha$, or $E = r_{\text{face}}^2 \alpha/(2R)$; the field falls off as $1/R$. If R is larger than the whole magnet then $E = 0$.

CHAPTER 31

1. (b) Far smaller. The susceptibility of copper is much less than 1, so the expression for B includes $(1 + \chi_m) \cong 1$ instead of the very large value of the χ_m for iron. This means that B inside will be practically unchanged from its vacuum value.

3. The expression for the magnetic moment is proportional to vr which can be rewritten in the form ωr^2. When the nucleus and the electron rotate about their center of mass, the angular velocity ω is the same for both of them. However, the center of mass is very close to the massive nucleus, so that to good approximation $r_N = (m_e/M_N)r_e$. Thus the contribution of the nucleus to the orbital magnetic moment is negligible.

4. The terrestrial environment, in particular terrestrial temperatures, are such that the thermal energy of the tiny magnets (of order kT) dominates the energy of their interaction with any magnetic field B. In particular, the difference in the energies between alignment and anti-alignment ($2m_BB$) is tiny compared with kT. Thermal effects destroy order, and in this case, with $kT \gg m_B B$, they dominate the ordering that the field attempts to impose. Only at ultra-low temperatures, say below 10K for realistic fields, would you expect the alignment to be almost perfect.

6. Our calculation was based on perfect alignment of all unpaired electrons. But the assumption of a single domain is not realistic for everyday situations. This does not mean that near-perfect alignment is impossible: A nearly perfect alignment in ferromagnetic materials occurs with the application of external fields of several Tesla.

CHAPTER 32

1. (d) Doubling the number of turns without changing anything else doubles the turn density, which appears squared in the expression for the inductance of a solenoid. This will increase the induced emf by the greatest amount.

3. (a) The area of solenoid 1 increases by a factor of 4 and so does the value of L_1. There is no change in M, which does not involve A_1. (b) There is no change in either L_1 or M. Even though R_2 increases, the overlap with R_1 does not if, as here, the areas started out with the same value, so that M does not change.

4. a. The permeability is larger, indeed much larger as the material in question is ferromagnetic.

6. We must change something in the inductor, and since the geometry is fixed, that leaves the number of turns, which increases the stored energy as N^2.

7. It is limits on the current that can be carried by wires that has the most effect on our ability to make large magnetic fields. The appropriate use of ferromagnetic cores does enable these fields to attain many Tesla.

8. This is incorrect and is a matter of understanding the limits. No matter how large L is, if it is fixed there will always be a time much greater than R/L for which V_L will drop to zero.

9. The voltage doesn't depend on the value of the inductor at all. It is solely determined by the original voltage across the capacitor that in turn depends on Q_0 and C. Even though V_L in the example has an L dependence, it will cancel out if we insert the value $\omega^2 = 1/LC$.

10. The decay constant α is ten times larger, while it remains much less than ω. The reasoning that allows us to approximate the period is the same, and since $t_{1/2}$ is ten times smaller, there are ten times fewer oscillations in that time span.

CHAPTER 33

2. No; the only conceivable effect would be on the direction in which the current is induced, and in AC that is obviously unimportant.

3. For the higher frequency, 6 MHz, the capacitive reactance is small and the circuit has a small resistance to current flow. This is a manifestation of a capacitor's transparence to rapidly changing currents.

4. The resistance to current flow in the inductive circuit increases dramatically for higher frequencies—the opposite behavior from that of the capacitive circuit. The inductor acts as an open switch in the limit of very high frequencies.

7. The time dependence of V_L is proportional to $-\cos(\omega t + \phi)$, and this takes its most negative value when $\cos(\omega t + \phi) = +1$, or $\omega t + \phi = 0$, $0, 2\pi, \ldots$, or $\omega t = -\phi, -\phi + 2\pi, \ldots$. If we recall that $\omega t_1 = \pi/2$, we see for $\omega t = -\phi \cong +60°$, $t < t_1$. Thus we take the next occurrence, $\omega t = -\phi + 2\pi$.

8. In the calculation of the example, R is proportional to the square root of the discrimination requirement, i.e. $(0.01)^{1/2} = 0.1$. Thus by choosing R a factor of 10 smaller we can improve the discrimination requirement by a factor of 100.

9. Examination of the equation for the ratio of potentials indicates that for $\omega \to \infty$, the left-hand side reduces to $V_1/(V_0 + V_1)$. For high frequencies, the circuit acts as though the capacitor is not present, and voltage V_1 is dropped across the resistor.

CHAPTER 34

2. Recall that the motion of the electron in the presence of fields is determined by the equation $\vec{F} = -e(\vec{E} + \vec{v} \times \vec{B})$. With the magnitude of B given by E/c, we see that the second term is of order v/c compared with the first one. Electrons in metals move with nonrelativistic velocities so that the first term is by far the more important one.

3. The electric and magnetic fields in a wave are intrinsically coupled through Maxwell's equations, and as Eq. (34–6), for example, indicates, the coupling is *local*, that is, at every point in space. This means that elimination of the electric field will also eliminate the magnetic field. Don't invest.

6. The pressure is the weight of a film of water of height h acting on an area of 1 m^2, namely $(10^3 \text{ kg/m}^3)hg$. If we set this equal to the radiation pressure and solve for h we find $h \cong 2.4 \times 10^{-8}$ m, a height only about 50 atoms deep! Radiation pressure is in most circumstances a small effect.

9. The light emitted by a lightbulb consists of the (incoherent) superposition of radiation by a huge number of oscillating charges. Each such oscillator emits dipole radiation, but because all these oscillators act incoherently, all the angular dependences average out and the distribution is spherically symmetric. Is this an "antenna"? It is as long as we agree that by this word we mean any system that radiates. If we insist upon *coherent* radiation, a lightbulb is not an antenna.

10. With the assumption that all of the power lost in the Polaroid sheets goes into heating, the ratio of acquisition of thermal energy is the ratio of intensity *lost* in the two sheets. In the first sheet, half the intensity is lost. In the second sheet, the intensity lost is the fraction $1 - \cos^2\theta = 3/4$. Thus the ratio is $(1/2)/(3/4) = 2/3$.

11. The classical answer (i.e. when there are *many* photons) is that the ratio of momentum to energy is $1/c$. This must also be valid for individual photons, and we therefore expect that the momentum of a single photon of frequency f will be hf/c.

CHAPTER 35

2. Figure 35–14 shows that if θ' is to be decreased, then ψ must decrease (Snell's law). This, however, means that ϕ will increase, and this can only happen if θ increases.

6. In this case we are talking about rays that go from air to water. Every such ray is bent to the vertical, meaning every such ray can enter the water and there is no analog to total internal reflection. The fish sees the entire upper hemisphere concentrated within a cone whose opening angle is the critical angle. That includes the fisherman's eyes at water's edge. Given that this is the case, the fisherman might be better off standing up—by being sufficiently above the surface, the fish can be visible to the fisherman at every depth.

CHAPTER 36

2. Attach a plane mirror to the object that rotates and then reflect a ray from the mirror onto a screen a distance L away during the rotation. The image on the screen will move across an arc length $2L\alpha$, and if L is large, this will be measurable. In effect, the distance magnifies the rotation.

5. It is easiest to see from rays 2 and 3 in Fig. 36–15 that the image will become smaller and move away from the mirror toward the focal point. Only ray 3 changes.

8. The magnification is less, as you can most easily see from the rays in Fig. 36–26 (the rays are refracted less).

9. Down. This is most easily seen if you used rays 2 and 3; ray 2 wouldn't change, but ray 3 would.

CHAPTER 37

2. Yes, if we measure y, R, and d, we can solve Eq. (37–4) for λ.

4. Changing the length of the feeding cables by different amounts will introduce a phase difference for the signals at the antennas, corresponding to the new path-length difference. However, the signals remain coherent, and there will be interference at your house. Whether this is destructive or constructive is a matter of the numerical value of the new phase difference.

5. We see from the solution that Δx is proportional to λ. Red light has a larger wavelength than blue, so that Δx will be larger for red light, and there will be fewer bands for red light than for blue.

6. While our list of wavelengths in the solution shows that the spacing in wavelengths between successive values for constructive interference is not equal, it will certainly be a good approximation to put a minimum (destructive interference) equidistant between 700 and 420 nm, namely $\lambda \cong 560$ nm. How would you find it exactly?

CHAPTER 38

1. Higher orders do indeed help, because as Eq. (38–7) shows the resolution increases with order, and it would be easier to resolve two closely spaced peaks.

2. The starting point will be the laser beam shining perpendicular to the CD. You could then either tilt the CD or move to the side so that you are observing the CD at an angle, and at an angle corresponding to the first- or higher-order maxima for the laser light, you will see a bright reflection.

3. The blue light has shorter wavelength, making the angles of the minima smaller, and the spacing between them on the screen is less than for red light.

4. In order to find the exact positions of the maxima, we must take the derivative of the intensity with respect to α (or θ) and set that equal to zero (see Problem 38–31).

5. If the two objects are arbitrarily far away—think of two stars—they may be arbitrarily far apart and still make an angle as small as that specified by the Rayleigh criterion. For very distant objects, it is in fact only the angular resolution—the minimum angular separation that can be resolved—that matters, and separation distance is not a quantity that is relevant here.

7. The overall deflection is $2\theta = 50°$.

Answers to Odd-Numbered Understanding the Concepts Questions

CHAPTER 21

1. The midpoint of the line joining the two charges.
3. The charges acquired by the balloon through rubbing attracts the opposite charges induced in the wall.
5. The force on each charge and its mass; if $m_1 = 5m_2$ then $d_1 = (1/5)d_2$.
7. The ones farther form the nucleus.
9. The "peanuts" are charged through rubbing and attract the opposite charges induced in your hand.
11. The equilibrium at the center of the ring is unstable.
13. Take the charged cork ball and touch one of the three uncharged ones, then simultaneously touch it with the two remaining balls.
15. The hand becomes charged after rubbing.
17. No.
19. Yes if each contains an equal number of electrons and protons, but that would probably not be the case.
21. The forces $\vec{F}_{21}$ and $\vec{F}_{23}$ both double, so $\vec{F}_1$ and $\vec{F}_2$ will both increase in magnitude and rotate counterclockwise.

CHAPTER 22

1. To prevent a dangerous charge buildup in the gasoline-carrying vehicle by channeling the excess charge to the ground.
3. For the same reasons for introducing the electric field; the gravitational field lines can only end at (not start from) matter.
5. $Q_1 : Q_2 = -5 : 1$.
7. Two locations, one at $-2 \text{ cm} < x < +4 \text{ cm}$ and the other at $x > 10 \text{ cm}$.
9. The electric field lines above the surface of Earth point down.
11. The charges on the comb cause the molecules in the paper to polarize, resulting in a net attraction.
13. The charge distribution is that of two dipoles touching at one end, resulting in a net dipole field.
15. The density of the field lines a distance r from the charge is proportional to $1/r^2$, which would be inconsistent with the electric field if it goes like $1/r^{2+\delta}$.
17. A sphere uniformly charged to some negative net charge would do.
19. Yes; the field is not zero since the two dipoles are not at the same location.
21. If $F_E > F_g$ it will accelerate upward, otherwise downward. The initial height is irrelevant.
23. Just find the field due to q_1 and neglect those due to q_1 and q_2.

CHAPTER 23

1. No.
3. The net charge enclosed by the surface is zero, but the electric field on the surface may not be.
5. For a spherical Gaussian surface of radius r centered at the location of a point charge, Φ would be $4\pi cr$, which depends on r of the Gaussian surface, rather than just the charge enclosed.
7. Zero.
9. Zero.
11. The electric field in between the two plates doubles while that elsewhere vanishes. The net flux over a Gaussian surface remains the same.
13. Only that it is invariant under a rotation about the axis of symmetry of the wire.
15. No net charge is present in the region.
17. The electric field is not uniform over the Gaussian cylinder for a charged line of finite length.
19. The charge density is independent of z.
21. No.

CHAPTER 24

1. 1 Joule.
3. Insert a uniformly charged plane within the sphere.
5. The electrostatic potential energy; the work that was done in assembling the charges together.
7. At $\theta = 0$.
9. It is always an equipotential except during the brief time interval when equilibrium is being established as charges redistribute themselves on its surface.
11. The person is charged and the hair (with like charges) repel each other; to prevent a dangerous current from running through the person.
13. No; $E_x = \Delta V/\Delta x$.
15. Yes.
17. The work done by the electrostatic force along any enclosed loop is zero.
19. Yes.
21. Yes.

CHAPTER 25

1. No.
3. If the field drops abruptly to zero outside the plates then the voltage drop around a closed path consisting of one segment leading from one plate to another and another segment that closes the path from outside would be non-zero.
5. The potential difference becomes infinitely large, as it takes an infinite amount of work to concentrate a charge on an infinitely thin wire.
7. The charge on one plate is Q and that on the other is $-Q$.
9. V decreases, C increases, and U decreases.
11. Parallel plates.
13. It reduces C.
15. Yes.
17. Not in classical physics.

CHAPTER 26

1. Yes.
3. The thinner one, as its resistance is greater.
5. The density of free electrons.
7. While water must first fill up the hose before streaming out, a piece of metal is already loaded with free electrons.
9. Yes, but it is a very small effect.
11. No.
13. Eq. (26–25) gives $\rho \approx 10^{-7} \, \Omega \cdot \text{m}$ with $\tau \approx 10^{-14}$ s and $n_e \approx 10^{29} \text{m}^{-3}$.
15. It melts as too much heat is generated to be entirely dissipated in time.
17. $R_{eq} = R_1 + R_2$.
19. Bulb 2 is brighter when the switch is open; bulb 1 is brighter when the switch is closed.
21. Increase.
23. To increase its length (and hence resistance) without taking up too much space.

CHAPTER 27

1. A large current can flow through the body as tap water is a good conductor.
3. The equation representing the loop rule is unchanged.
5. No.
7. $C \approx 1000 \, \mu\text{F}, R \approx 10 \, \Omega, RC \approx 0.01$ s.
9. If I reverses Q would become $-Q$.
11. Put the two emf's in series to drive the two lightbulbs in parallel.
13. I increases to $3I$.
15. R_2 and R_3 are in series (branch 1), as are R_5 and R_6 (branch 2). Combine these with R_4 (branch 3), with the three branches in parallel, to obtain a single equivalent resistance, R_{eq}, which forms a one-loop circuit along with R_1 and the emf. Impossible for Fig. 27–8(b).
17. The steady-state I_1 decreases.

CHAPTER 28

1. The positive charges (ions) have no collective velocity but the free electrons do.
3. By an electric field; $\vec{v}$ must change direction if the force is magnetic.
5. No.
7. No.
9. Incoming charged particles are concentrated in the polar region where the magnetic field is the strongest.
11. A single circular loop.
13. Repel; attract if one of the current is reversed.
15. Attract.
17. No; yes.
19. No.
21. (d).

CHAPTER 29

1. The needle will maintain an orientation that follows the circular path of the compass itself.
3. The direction of the magnetic force depends on that of the current flow.
5. No.
7. An extended wire is made of many segments, each producing a magnetic field of its own magnitude and direction.
9. It is much easier to measure force accurately.
11. Replace ε_0 with ε.

13. The field doubles for two loops with currents flowing in the same sense, and is zero if they flow in opposite sense.
15. The magnets attract when adjacent poles are opposite and repel when they are the same. The fields add up when unlike poles are adjacent and cancel when like poles are adjacent.

CHAPTER 30

1. No.
3. (e).
5. Yes; the magnetic field may increase in one part of the region and decrease in another part.
7. (a) No, (b) yes, (c) no, (d) no.
9. No.
11. Yes; to compensate for the thermal energy dissipated by the induced current in the metal plate.
13. No change.
15. The falling water gives up its gravitational potential energy and drives the rotary turbine blades in a magnetic field to generate induced current.
17. It falls with $a < g$ as it is above the tube and is entering it, and with $a = g$ when completely inside the tube.
19. An induced current in the ring antiparallel to that in the coil is induced; the ring will not jump.
21. An induced current in the ring antiparallel to that in the coil is induced; the size of the ring must allow the magnetic force on it to cancel with its weight.
23. The residual magnetic field in the iron frames of the moving car induced a current in the wire loop.

CHAPTER 31

1. The nucleus is much more massive than the electrons.
3. Make the core of a solenoid out of the material and check to see if the magnetic field increases or decreases due to the presence of the core.
5. Yes.
7. Yes; very difficult to measure due to ferromagnetism, try search for it at high temperature.
9. Magnetically hard material.
11. The alignment of magnetic moments causing paramagnetism becomes less effective as temperature increases.
13. Pulled into the region.
15. The orbital angular momentum of the electron does not change if $\vec{B}$ is perpendicular to its plane of motion, and changes if $\vec{B}$ is parallel to the plane.
17. 0.
19. The field of a bar magnet is non-uniform.

CHAPTER 32

1. Not necessarily.
3. (b), (a), (c).
5. Yes.
7. (b).
9. Construct an RL circuit and measure its time constant to find L.
11. Due to the current flow.
13. It depends on the total electromagnetic energy U in the circuit, or I_{max}, or V_{max} across the capacitor.

15. Yes.
17. There is no dissipation mechanism to lower the magnetic energy, so I does not drop.
19. $\frac{1}{2}mv^2$ and $\frac{1}{2}kx^2$.
21. The regions in between two wires carrying the same current has lower magnetic pressure than the region outside both wires, resulting in a net attraction between the wires.

CHAPTER 33

1. It must amplify and confine the magnetic field and have sufficient mechanical strength.
3. $X_C = 1/\omega C$ approaches zero as ω is very high.
5. The time average of the current is zero but that of the power is not.
7. The time average of the current is zero so one must use I_{rms}.
9. They can be reversed in principle; step-up becomes step-down, and vice versa.
11. (c).
13. An inductor.
15. No, unless they are of the same kind (say, both are capacitive).
17. The impedances are not infinite for any finite frequency.
19. To prevent signal loss due to abrupt changes in impedance.
21. When $\omega = (1/LC)^{1/2}$.
23. No; transformers can be used.

CHAPTER 34

1. No.
3. The ionosphere has an abundance of free charges, like a metal.
5. Yes.
7. No.
9. They must have very large area, lower density, and high reflectivity; it's difficult to sail into the solar wind.
11. Rotate a polarizer to see if you can get a near-zero transmission.
13. It goes out.
15. All of it.
17. The angular momentum is zero for a linearly polarized electromagnetic wave, and non-zero if it's circularly polarized.
19. Reflective.
21. $F_B \ll F_E$ since $v/c \ll 1$.

CHAPTER 35

1. Through reflection.
3. Extremely difficult; having three mutually perpendicular mirrors ensures that the incident beam is reflected straight back.
5. The bodyguard appears taller.
7. It appears to be shorter and closer to the surface of the water.
9. (d).
11. Not if you tilt your head toward the direction of the coin.
13. A distant house appears as two images, one of which inverted.

15. No air is present in space to scatter the sunlight.
17. As you scan downward from the vertical line you first see his upper body (which appears to be stretched), then you see the reflection of his lower body (upside-down), followed by a direct view of his lower body.
19. No.
21. Light reflected from the Moon has relatively little red component.

CHAPTER 36

1. The sign reads "AMBULANCE" for drivers who see a vehicle arrive from behind in the rear-view mirror.
3. No, but the image is dimmer.
5. $(x, y, z) \rightarrow (-x, -y, z)$.
7. Convex.
9. It is a convex mirror in which images are smaller than the objects, allowing a wider field of view.
11. Yes.
13. Yes.
15. Not very, except the one originated from the focal point.
17. It accentuates any imperfection of the eye by exposing the edge of the lens.
19. Not very plausible.

CHAPTER 37

1. There is no reduction in intensity due to the slits.
3. They are in principle visible when we look toward the double slits.
5. Adjacent wavelengths resulting in destructive interferences differ by a factor of 3 but $\lambda_{max} < 3\lambda_{min}$ for visible light.
7. Light rays are parallel when viewed from afar, simplifying the analysis.
9. The total energy over the entire space remains conserved.
11. No.
13. To minimize the reflection
$n_{coat} = (n_{air}n_{glass})^{1/2}$.
15. Energy is intensified in other locations.
17. A tiny change in λ can result in a measurable shift in the interference pattern.
19. Total destructive interference requires identical amplitudes of two waves.
21. The slit spacing is much greater than λ.

CHAPTER 38

1. The diffraction pattern spreads out and gets dimmer overall.
3. No image distortion due to the atmosphere.
5. Short wavelength.
7. No.
9. The spreading of the diffraction pattern is proportional to λ, which is longer for red light.
11. When in place it blocks all the light except the two beams at the edges of the single slit.
13. A continuous spectrum (rainbow bands).
15. It interacts with the atoms.
17. One sees "around" the object as one changes the viewing angle of a holographic image.

Answers to Odd-Numbered Problems

CHAPTER 21

1. 6.2×10^9 fewer electrons.
3. 4.82×10^4 C.
5. 7.9×10^{-12} per atom.
7. -2×10^{-10} C, 1.25×10^9 electrons; -1×10^{-10} C, 6.2×10^8 electrons; -1×10^{-10} C, 6.2×10^8 electrons.
9. 2.2×10^{-10}.
11. (a) 1.8×10^{51} electrons, (b) 3.5×10^{-39}.
13. (a) conserved, (b) not conserved, (c) conserved, (d) not conserved.
15. -8.5×10^{-24} C.
17. 46 N repulsion, 23 N attraction.
19. $q = 1.6 \times 10^{-19}$ C, 1 electron.
21. 3.5×10^{-10}, mass much larger.
23. (a) 3×10^9 esu in 1 C, (b) 4.8×10^{-10} esu.
25. $q_1/q = q_2/q = \frac{1}{2}$.
27. (a) 2.6×10^{-9} N toward the proton (centripetal), (b) 9.2×10^5 m/s, (c) 4.9×10^{14} Hz, (d) 8.6 N/m.
29. 1.4×10^{-8} C.
31. $q_1 = +1.1 \times 10^{-7}$ C, $q_2 = -3.2 \times 10^{-7}$ C, $q_3 = +5.3 \times 10^{-7}$ C.
33. 0.9 m.
35. $x = 44.4$.
37. zero.
39. $\vec{F}_+ = 39$ N $30°$ above the line joining the two "up" quarks, $\vec{F}_- = 40$ N toward the center of the line joining the two "up" quarks.
41. (a) $\Sigma \vec{F} = 0$, (b) unstable, (c) stable.
43. (a) $\sqrt{3}\, kq^2/2L^2$, $9.7°$ above the $-x$-axis, (b) $3.2kqQ/L^2$, $6.3°$ below the $-x$-axis.
45. $(2kq\lambda/x_0)\hat{i}$.
47. $kqQ/d(L+d)$, away from the rod.
49. Let the radius of each ring be R and the separation between them be $2L$. If $Qq(R^2 - 2L^2) < 0$ then the equilibrium is stable, otherwise unstable.
51. 1.4 N away from the center.
53. 3.0×10^{-7} C.
55. 0.52 N.
57. $3kq^2[3(2.25 + y_0^2)^{-3/2} - (0.25 + y_0^2)^{-3/2}]\hat{i}$, y_0 in cm.
59. 5.7 N/m^2; same.
61. 1.5×10^{-3} C.
63. $6.6°$.
65. (a) $v = (e^2/4\pi\varepsilon_0 mR)^{1/2}$, (b) $L = (e^2 mR/4\pi\varepsilon_0)^{1/2}$, (c) $v = e^2/4\pi\varepsilon_0 L$, (d) $R = 4\pi\varepsilon_0 L^2/me^2$, (e) $\tau = 32\pi^3\varepsilon_0^2 L^3/me^4$,

(f) $v = 2.2 \times 10^6$ m/s; $R = 5.3 \times 10^{-11}$ m; $\tau = 1.5 \times 10^{-16}$ s.
67. (a) $F_{net} = kq^2[\ell(\ell - 2x)/x^2(\ell - x)^2]$ away from the closer charge; $x = \ell/2$, (b) $F_{net} = -2kq^2\ell\,\Delta x/[(\ell/2)^2 - (\Delta x)^2]^2$, (c) $f = (1/2\pi)(32kq^2/\ell^3 m)^{1/2}$.
71. (a) $\vec{F} = kqQR \sum_{n=-\infty}^{\infty} \left\{ \frac{1}{[(na)^2 + R^2]^{3/2}} \right\} \hat{j}$,
(b) $\vec{F} = \frac{k\lambda Q}{R} \int_{-\infty}^{\infty} \frac{du}{(u^2+1)^{3/2}} \hat{j}$.

CHAPTER 22

1. $1.8 \times 10^7(-0.60\hat{i} + 0.80\hat{j})$ N/C.
3. $(1.14 \times 10^{11}$ N/C$)\hat{r}$; $-(1.82 \times 10^{-8}$ N$)\hat{r}$ (toward the nucleus).
5. 3.19×10^6 N/C, $78°$ above the $+x$-axis.
7. (a) $-(1/4\pi\varepsilon_0)(8q/\ell^2)\hat{i}$, (b) 0, (c) $(1/4\pi\varepsilon_0)\{2qd/[d^2 + (\ell/2)^2]^{3/2}\}$ away from the origin.
9. $\vec{E} = \frac{2\vec{p}}{4\pi\varepsilon_0 r^3} \left\{ \frac{1}{[1 + (L/2r)]^2[1 - (L/2r)]^2} \right\}$.
$\vec{E} = \frac{2\vec{p}}{4\pi\varepsilon_0 r^3}$.
11. Stable; $(1/2\pi)(Qq/\pi\varepsilon_0 a^3 m)^{1/2}$.
13. Tripled.
15.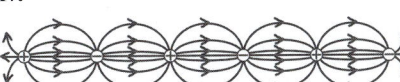
17.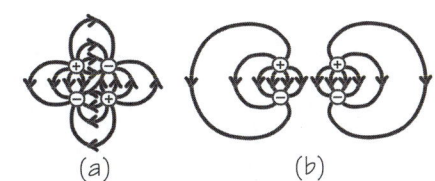
19.

(a) (b)

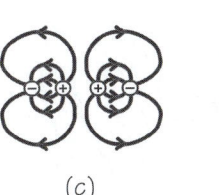
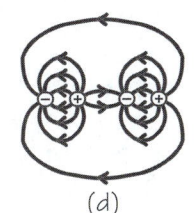

(c) (d)

21. 2.7×10^4 N/C perpendicular to and away from the line.
23.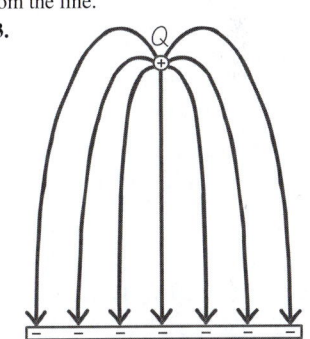
25.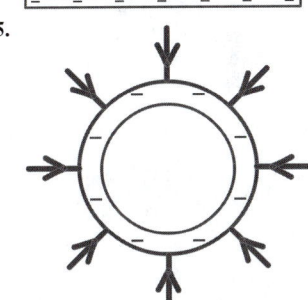
27.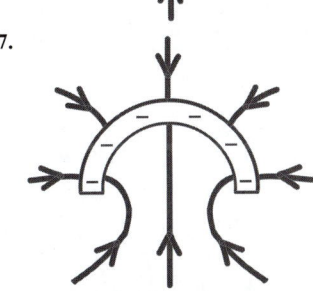
29. σ/ε_0 perpendicular to the plates and away from them; 0.
31. (a) $(z_0 Q/2\pi\varepsilon_0 R^2)[1/z_0 - (R^2 + z_0^2)^{-1/2}]\hat{k}$, (b) $(Q/4\pi\varepsilon_0 R^2)\hat{k}$, (c) $(Q/2\pi\varepsilon_0 R^2)\hat{k}$.
33. 7.3×10^4 N/C toward the point charge.
35. $Q/4\pi R^2 h$.
37. (a) $\frac{\sigma L z_0}{2\pi\varepsilon_0}\hat{k} \int_{-L}^{L} \frac{dx}{(x^2 + z_0^2)\sqrt{x^2 + L^2 + z_0^2}}$,
(b) $(\sigma/2\varepsilon_0)\hat{k}$, (c) $(\sigma/2\varepsilon_0)\hat{k}$.

O–1

39. 0.65 μC.
41. 1.7×10^{-7} C/m^2.
43. $(q\lambda/2\pi\varepsilon_0 m)^{1/2}$.
47. 3.4 cm.
49. 5.7 s^{-1}.
51. $-(1.41 \times 10^{-6}$ N·m$)\hat{k}$.
53. 2×10^6 N/C.
55. $-6p^2/4\pi\varepsilon_0 r^4$ (attraction).
57.

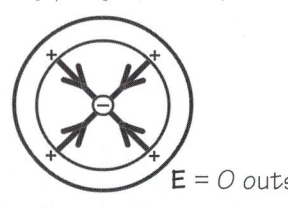

E = 0 outside

59.

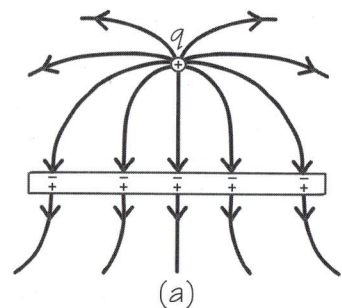

61. (a) $\{(\lambda R/2\pi\varepsilon_0)/[y^2 - (R/2)^2]\}\hat{j}$,
(b) $-\{(\lambda R/2\pi\varepsilon_0)/[x^2 + (R/2)^2]\}\hat{j}$.
63. $[(1 + 8.5t^2)\hat{i} + (1 - 14t^2)\hat{j}]$ m, with t in s.
65. (a) 7.2×10^{-14} N away from the plate,
(b) -3.2×10^{-13} J, (c) 4.4 m.
67.

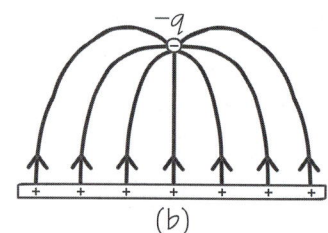

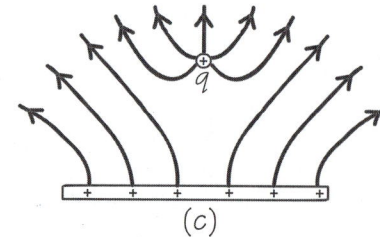

69. 19 kN; 0.
71. $\lambda_0 L^2/6$.

CHAPTER 23

1. (a) $\sigma\pi R^2/2\varepsilon_0$, (b) 0.866 $\sigma\pi R^2/2\varepsilon_0$.
3. $\lambda h/\varepsilon_0$.
5. $+6$ N·m^2/C.
7. $\pi E_0 R^2/3$.
9. $(q/\varepsilon_0)(1 + R^2/h^2)^{1/2}$.
13. (a) Zero, (b) 1.13×10^8 N·m^2/C.
15. (a) Net flux is zero.
17. (a) -3.54×10^{-9} C, (b) -3.54×10^{-9} C,
(c) -3.54×10^{-9} C.
19. 9.4×10^2 N·m^2/C out of the sides parallel to the xy- or yz-planes; 10.6×10^2 N·m^2/C out of the side perpendicular to the $-x$-axis;
8.2×10^2 N·m^2/C out of the side perpendicular to the $+x$-axis.
21. $\rho R^2/2\varepsilon_0 r$.
23. $E_{rod}/E_{pt\,chge} = 0.02 = 2\%$.
25. -2.3×10^{-4} C.
27. $(1.8 \times 10^4)\hat{r}$ N/C; $(1.8 \times 10^4)\hat{r}$ N/C.
29. $E = 0$ for $r < r_1$; $[\rho(r^2 - r_1^2)/2\varepsilon_0 r]\hat{r}$ for $r_1 < r < r_2$; $[\rho(r_2^2 - r_1^2)/2\varepsilon_0 r]\hat{r}$ for $r_2 < r$.
31. $E = 0$ for $r < R_1$;
$[Q(r^3 - R_1^3)/4\pi\varepsilon_0(R_2^3 - R_1^3)r^2]\hat{r}$ for $R_1 < r < R_2$; $(Q/4\pi\varepsilon_0 r^2)\hat{r}$ for $R_2 < r$.
33. (a) Point the x-axis from plate 1 (where $x = 0$) to plate 2 (where $x = a$), perpendicular to both plates: $\vec{E} = -[(\sigma_1 + \sigma_2)/2\varepsilon_0]\hat{i}$ $(x < 0)$;
$\vec{E} = [(\sigma_1 - \sigma_2)/2\varepsilon_0]\hat{i}$ $(0 < x < a)$;
$\vec{E} = [(\sigma_1 + \sigma_2)/2\varepsilon_0]\hat{i}$ $(x > a)$, (b) same; must also note that $E = 0$ inside the metallic plate.
35. 1st quadrant: E at $-\theta$; 2nd quadrant: E at $180° + \theta$; 3rd quadrant: E at $180° - \theta$; 4th quadrant: E at θ; $E = 3.3 \times 10^5$ N/C; $\theta = 31°$.
37. $\vec{E} = (-6.7 \times 10^8)\vec{r}$ N/C with r in m, where $r < 3$ cm; $\vec{E} = [(-1.8 \times 10^4)/r^2]\hat{r}$ N/C with r in m, where 3 cm $< r <$ 8 cm;
$\vec{E} = [(2.7 \times 10^4)/r^2]\hat{r}$ N/C, with r in m, where 8 cm $< r$.
39.

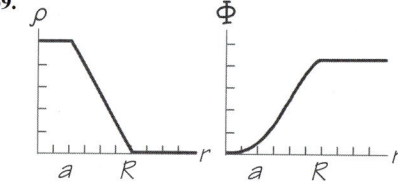

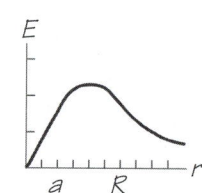

41. $r < R$: $\vec{E} = 0$; $R < r < 2R$:
$\vec{E} = (q/4\pi\varepsilon_0 r^2)\hat{r}$; $2R < r$: $\vec{E} = -(q/4\pi\varepsilon_0 r^2)\hat{r}$.
43. 2.7×10^{-5} C/m^2.
45. $\sigma_{sphere}/\sigma_{shell} = 1.96$.
47. $\sigma_{inner\,sphere} = Q/4\pi a^2$;
$\sigma_{shell,\,inside} = -Q/4\pi b^2$;
$\sigma_{shell,\,outside} = +Q/2\pi R^2$.
49.

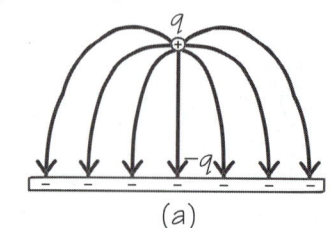

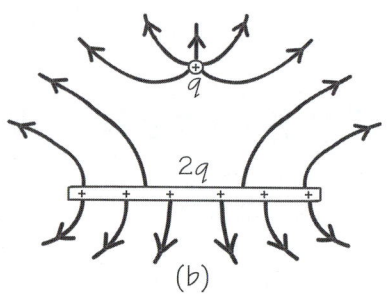

51. $\Phi_{x=0} = 0$; $\Phi_{x=a} = ba^4$;
$\Phi_{y=0} = \Phi_{y=a} = \Phi_{z=0} = \Phi_{z=a} = 0$;
$q = \varepsilon_0 ba^4$.
53. (a) $0.067Q/\varepsilon_0$, (b) $Q/4\pi\varepsilon_0 R^2$.
57. $0.433EL^2$.
59. $\rho(r) = \varepsilon_0 E/r$.
61. $+6000\varepsilon_0$ C/m^3 (constant).

CHAPTER 24

1. 4.6×10^{-14} J.
3. 9.5×10^{-2} J.
5. (a) 0, (b) -1.35 J, (c) $+1.35$ J.
7. 1.36×10^{-3} J; 0.
9. -4.8×10^{-2} J.
11. $r = \infty$.
13. 3.3×10^{-14} C.
15. $+4.2 \times 10^{-3}$ J.
17. -5.0×10^4 J.
19. (a) $+5.0 \times 10^3$ V, (b) -1.0×10^{-3} J.
21. $+2.4 \times 10^5$ V.
23. $+4.3 \times 10^5$ V.
25. (a) -3.7×10^5 V,
(b) $V(0, 0) = +1.6 \times 10^5$ V;
$\vec{E}(0, 0) = -(6.5 \times 10^7$ V/m$)\hat{j}$.
27. (a) $E_A = 0$; $E_B = (1.13 \times 10^6$ N/C·m$)x$, $x < 1$ cm; $E_C = 1.13 \times 10^4$ N/C, $x > 1$ cm,
(b) $V_B - V_A = -(5.65 \times 10^5$ V/m$^2)x^2$,
$x < 1$ cm;
$V_C = +5.65$ V $- (1.13 \times 10^4$ V/m$)x$,
$x > 1$ cm,
(c)

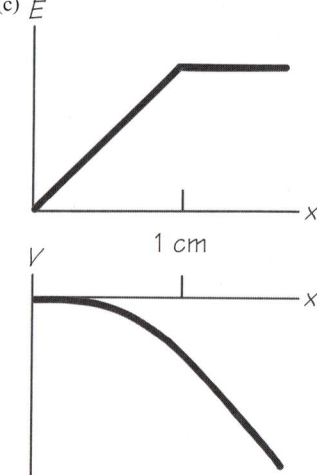

29. $V_{outside} = Q/4\pi\varepsilon_0 r$ $(r > R)$;
$(Q/8\pi\varepsilon_0 R)[3 - (r/R)^2]$ $(r < R)$.

31.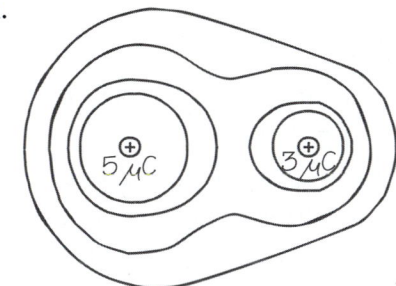

33.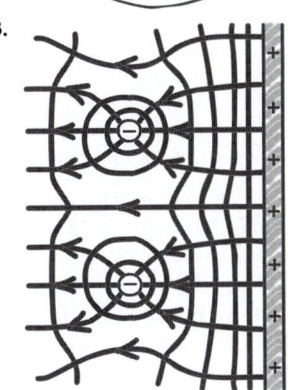

35.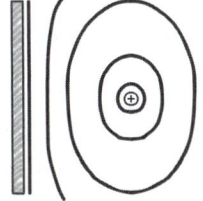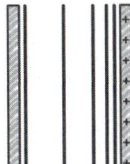

37.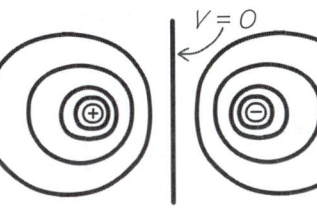

39. $(Q/4\pi\varepsilon_0 x^2)\hat{i}$.
41. $(6.68 \text{ V/m})\hat{i}$.
43. $-(p/4\pi\varepsilon_0 r^3)\hat{i}$.
45. $\vec{E}_{r<R} = (Q/4\pi\varepsilon_0)(6r/R^3)\hat{r}$; $\vec{E}_{r>R} = (Q/4\pi\varepsilon_0 r^2)\hat{r}$.
47. $-(2Qa_0^2/4\pi\varepsilon_0 Lr^3)(\sin\theta\,\hat{i} + 3\cos\theta\,\hat{j})$.
49. 1.0×10^{-10} C.
51. $-(2\lambda/4\pi\varepsilon_0)\ln(R)$ + (a constant).
53. 9.9×10^4 V.
55. (a) $\dfrac{q_0}{4\pi\varepsilon_0}\left(\dfrac{3}{x-x_0} - \dfrac{1}{x+x_0/2}\right)$,
(b) $\dfrac{q_0}{4\pi\varepsilon_0}\left(\dfrac{2}{x} + \dfrac{7}{2}\dfrac{x_0}{x^2} + \dfrac{11}{4}\dfrac{x_0^2}{x^3} + \dfrac{25}{8}\dfrac{x_0^3}{x^4} + \cdots\right)$,
(c) $q_{\text{net}} = 2q_0$, $p = 7q_0 x_0/2$, (d) $|x| > 12.1 x_0$.
57. $+4.2 \times 10^{-4}$ J.
59. $(q_1 r_2 - q_2 r_1)/(r_1 + r_2)$.
61. 3.7×10^{-4} C; 5.5 MeV; 8.8×10^{-13} J; 3.2×10^7 m/s.
63. $Q_1 = 3Q/4$; $Q_2 = Q/4$.
65. 4.4 km.
67. (a) $2.2\ \mu$C; $11\ \mu$C, (b) 4.95×10^7 V/m, radial; 9.90×10^6 V/m, radial.
69. (a) $+5.5 \times 10^6$ eV (8.8×10^{-13} J), (b) 3.3×10^7 m/s.
71. (a) 3.9×10^6 V, (b) 3.9 MeV, (c) 5.6×10^{-4} C.
73. $(Q/2\pi\varepsilon_0 R^2)\{[R^2 + (a+x)^2]^{1/2} + [R^2 + (a-x)^2]^{1/2} - 2a\}$.
75. $Qa/2\pi\varepsilon_0 x^2$.
77. $U = -(qQ/8\pi\varepsilon_0 R)(3 - r^2/R^2)$; $k = qQ/4\pi\varepsilon_0 R^3$.
79. 3.4×10^{-9} N toward the other Na$^+$; 1.2×10^{-18} J (7.4 eV).
81. $[Qx/4\pi\varepsilon_0(R^2 + x^2)^{3/2}]\hat{i}$.
83. $\Delta U = -7.68 \times 10^{-19}$ J; $\Delta K = +3.84 \times 10^{-19}$ J; $\Delta E = -3.84 \times 10^{-19}$ J.
85. $(1/4\pi\varepsilon_0)(q_1 + q_2)/r$ $(r_2 < r)$. $(1/4\pi\varepsilon_0)(q_2/r_2 + q_1/r)$ $(r_1 < r < r_2)$. $(1/4\pi\varepsilon_0)(q_1/r_1 + q_2/r_2)$ $(r < r_1)$.
87. (a) 9.9×10^{-3} kg, (b) $(0.023/\sin\theta) + 0.078(1 - \cos\theta)$.
89. $-\rho r^2/4\varepsilon_0 - (\rho R^2/2\varepsilon_0)[\ln(R/a) - 1/2]$ $(r < R)$; $-(\rho R^2/2\varepsilon_0)\ln(r/a)$ $(r > R)$.
91. $4\pi\rho^2 R^5/15\varepsilon_0$.

CHAPTER 25

1. (a) 44 pF, (b) 40 cm.
3. (a) 1 V, (b) 2.5 V, (c) 250 V.
5. 0.33 mm.
7. (a) $C \to 4\pi\varepsilon_0 r$, (b) $C \to 4\pi\varepsilon_0 r^2/(R - r) = \varepsilon_0 A/d$.
9. $(4.43 \times 10^{-7}$ m$) + (8.85 \times 10^{-10}$ m/s$)t$.
11. 6 kV.
13. 99 J.
15. 600 μF.
17. (a) 5.67×10^{-10} F, (b) 2.83×10^{-4} J; 2.83×10^{-2} J.
19. (a) $(\varepsilon_0 A V_0^2/2d_0^2)(d_1 - d_0)$, (b) $(\varepsilon_0 A V_0^2/2d_0^2)(d_1 - d_0)$, (c) $(\varepsilon_0 A V_0^2/2d_1 d_0)(d_0 - d_1)$, (d) energy has been stored in the battery.
21. $\lambda^2/8\pi^2\varepsilon_0 r^2$.
23. 1.1×10^{-2} J.
25. 1.7×10^{-7} C; $1.0 \times 10^{-5}/r^4$ J/m^3, with r in m; 7.0×10^{-4} J.
27. (a) 1.8 pF, (b) 2.3×10^{-7} J.
29. (a) 3.00×10^5 V/m, (b) 1.06×10^{-7} C, (c) 1.59×10^{-2} N, (d) 1.59×10^{-5} J.
31. $2.04\ \mu$F, $4.46\ \mu$F.
33. $1.93\ \mu$F.
35. 0.45 pF; 1.33 pF.
37. (a) $1.34\ \mu$F, (b) $Q_1 = 165\ \mu$C, $Q_2 = Q_5 = 237\ \mu$C, $Q_3 = Q_7 = 402\ \mu$C.
39. $1.6C$; C; $2C$.
41. (a) $5.0\ \mu$F, (b) $25\ \mu$C.
43. 370 V.
45. 2.1×10^{12} m^2.
47. (a) 10 nF, (b) 1.3, (c) a hyperbola.

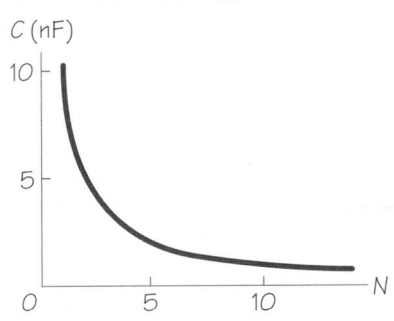

49. $C - C_0 = (\kappa - 1)4\pi\varepsilon_0 R$; $\sigma_{\text{ind}}/\sigma = (\kappa - 1)/\kappa$.
51. $q = 2.3 q_0$.
53. $\kappa\varepsilon_0 A/[(d + \kappa(D - d))]$.
55. air: 5.3×10^{-10} C; paper: $Q = 2.0 \times 10^{-9}$ C; neoprene: $Q = 3.6 \times 10^{-9}$ C; Bakelite: $Q = 2.6 \times 10^{-9}$ C; strontium titanate: $Q = 1.8 \times 10^{-7}$ C.
57. $(1 - \kappa)Q^2/2C$ (a decrease).
59. 0.32 nF; 0.90 nF.
63. $2.96\ \mu$F.
65. (a) $V_1 = C_2 V/(C_1 + C_2)$; $V_2 = C_1 V/(C_1 + C_2)$, (b) 17 V.
67. $Q \simeq 10^{-6}$ C.
69. (a) $\tfrac{1}{2}C_1 V^2/N$, (b) $\tfrac{1}{2}NC_1 V^2$, (c) $U_{\text{series}} = \tfrac{1}{2}Q^2 N/C_1$; $U_{\text{parallel}} = \tfrac{1}{2}Q^2/NC_1$.
71. $\tfrac{1}{2}(\kappa_0 + \kappa_1)(\varepsilon_0 L^2/d)$.
73. $V_A = 8.5 \times 10^2$ V; $V_B = 7.2 \times 10^2$ V; $V_C = 1.43 \times 10^3$ V.
75. 3.4 J; -2.2 J.
77. $(\kappa_1 - \kappa_0)[\varepsilon_0 L^2/D\ln(\kappa_1/\kappa_0)]$.

CHAPTER 26

1. 1.2×10^5 A/m^2; 0.46 C.
3. 1.0×10^4 s (2.8 h).
5. 5.8×10^{17} electrons.
7. 8.7×10^{-6} m/s; 2.0×10^{-5} m/s.
9. 4.5×10^{21} electrons; 4.2×10^9 electrons.
11. $\pi J_0 R^2/2$.
13. $v_+ = 1.7 \times 10^{-6}$ m/s, $v_- = -2.5 \times 10^{-6}$ m/s.
15. 3.1×10^{-15} A.
17. 2.8 A; 5.6×10^4 A/m^2; 5.8×10^{-6} m/s.
19. 8.5×10^{-5} m/s; 2.3×10^{-4} m/s.
21. $J_{\text{tube}} = I/2\pi Rd$ along the tube; $J_{\text{plate}} = I/2\pi rd$ radial.
23. $R_2 = 2R_1$.
25. (a) 1.2 Ω, (b) 0.15 cm.
27. $3.27 \times 10^{-2}\ \Omega$.
29. $9.6 \times 10^{-4}\ \Omega$; $3.0 \times 10^{-4}\ \Omega$.
31. 29 m.
33. -0.27.
35. $I_{25} = 19.6$ A; $I_{400} = 17.1$ A.
37. $r_{\text{Cu}}/r_{\text{Al}} = 1.74$.
39. 188 turns.
41. 2.0 Ω; 0.24 kg; 0.48 kg.
43. 33.4 kg.

O–3

45. 1.82×10^{-3} Ω; $r = 0.17$ cm; masses are the same.
47. Same brightness for bulbs 1 and 2; bulb 1 is 4 times as bright as bulbs 2 and 3.
49. 5 Ω; 24 V.
51. 90 Ω; 0.18 A; 2.8 W.
53. $R_{eq} = 100$ Ω.
55. 6 Ω.
57. $P_{24\,\Omega} = 11$ W; $P_{12\,\Omega} = 2.4$ W. $P_{8\,\Omega} = 14$ W; $P_{6\,\Omega} = 4.7$ W;
59. (a) 91.3 Ω, (b) 4.38 V, (c) 0.131 A.
61. 6.5×10^{-8} m.
63. $\rho = E/[nev_0(1 - r/R)]$.
65. 2.2 Ω.
67. 1¢.
69. (a) 0.18 A, (b) 35 mA.
71. $(PR)^{1/2}$.
73. 1.9×10^2 J.
75. 84 m.
77. 1.94×10^4 J (5.4 kWh).
79. Change the temperature.
81. 170 V; 43 Ω.
83. The power decreases by 3/4.
85. (a) $\rho L/[\pi r_0(r_0 + \alpha L)]$.
87. $I = [24(4\,\Omega + R_x)/(24\,\Omega + 5R_x)]$ A; $P_4 = [48(4\,\Omega + R_x)/(24\,\Omega + 5R_x)]^2$ W.
89. 0.9 kW; 35 s; 23 min.
91. (a) $R = (\rho_0 L/A)(1 + \alpha k t^2)$, (b) $VI_0/(1 + \alpha k t^2)$, (c) No.
93. $R_2/R_1 = 5.4$.
95. $dT/dt = k/[1 + \alpha(T - T_0)]$, where $k = V^2/mcR_0$; $I(t) = V/R_0\{1 + \alpha[T(t) - T_0]\}$.

CHAPTER 27

1. 0.15 Ω.
3. 0.0075 Ω.
5. 20 A.
7. 0.030 Ω; 300 W.
9. 11.8 V; 0.17 W; 8.7 V; 33 W.
11. $R_2 = R_1$; $R_2 \to \infty$.
13. $V_{gf} = 0$; $V_{ag} = -4.0$ V; $V_{ca} = 5.5$ V.
15. $I_4 = +0.5$ A; $I_5 = +1.0$ A; $I_6 = -1.0$ A.
17. (a) 7.5 Ω, (b) 7.0 Ω.
19. $I_1 = +0.45$ A; $I_2 = +0.38$ A; $I_3 = -0.068$ A.
21. $I_1 = 9.6$ mA; $I_2 = 5.7$ mA; $I_3 = 4.7$ mA.
23. $I_a = \mathcal{E}/[R + (r/N)]$; $I_b = \mathcal{E}/[r + (R/N)]$.
25. 0.38 V; 0.04 V; 0.004 V.
27. One independent junction; $I_1 = 3\mathcal{E}/7R$; $I_2 = I_3 = I_4 = \mathcal{E}/7R$.
29. (a) 60.5 mW; (b) 6.8 mW.
31. (a) 0.075 A, (b) 0.075 A.
33. 2.35 A; 2.52 A; no current.
35. $2V/5R$; $V/5R$.
37. 0.732 V; 0.737 V.
39. 0.010 Ω.
41. (a) 10 Ω, (b) 5×10^4 Ω, (c) 10^5 Ω.
43. 25 V.
47. $R = R_1 R_2/R_3$.

49. $R_x = (V/I)/(1 - V/IR_V)(R_V \gg R_x)$.
51. 20 μF.
53. 3.3 Ω.
57. $(\mathcal{E}/R_1)e^{-t/R_1 C} + \mathcal{E}/R_2$.
59. 300-Ω resistors in parallel with each other and in series with the 250-Ω resistor and the two capacitors.
61. 6.8×10^3 s (1.9 h).
63. 0.417 A; 0.500 A; 0.167 A.
65. Six 600-Ω resistors in parallel; six 16.7-Ω resistors in series, two series 150-Ω resistors in parallel with two more sets of two series 150-Ω resistors; three series 66.7-Ω resistors in parallel with three series 66.7-Ω resistors.
67. (a) 0.25 kA, (b) 12.7 V, (c) 1.1×10^6 J.
69. $I = \{-(R + \alpha) + [(R + \alpha)^2 + 4\beta\mathcal{E}]^{1/2}\}/2\beta$.

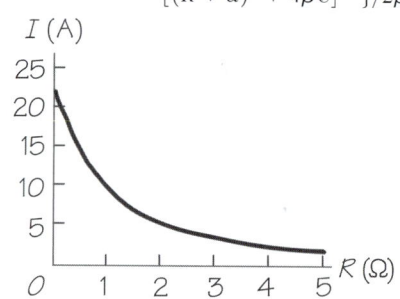

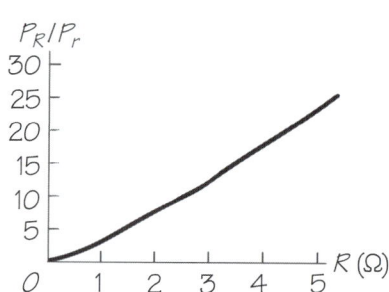

71. 667 W.
73. $I_{mixer} = 6.67$ A; $I_{vacuum} = 5.00$ A; $I_{chandelier} = 5.00$ A; 7 bulbs.
75. 0, $\mathcal{E}/2R$; $\mathcal{E}/R$.
79. (a) 21.4 μs, (b) 2.4 mC.
81. (a) V_0/d, (b) $\sigma \pi r^2 V_0/d$.

CHAPTER 28

1.

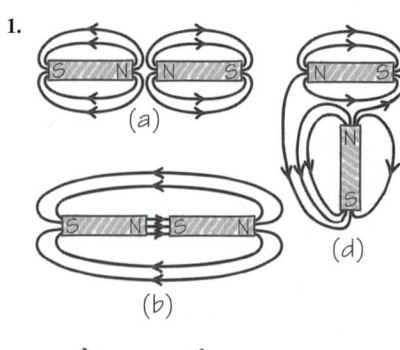

3. $-z$-direction.
5. $-(7.2 \times 10^{-3}\,\text{T})\hat{k}$.
7. 34 μT.
9. (a) $\theta = (eB\,\Delta t)/m$, (b) 0.15 T.
11. -4.2 mm (horizontal).
13. (a) 2.25×10^{-3} T, (b) 2.5×10^{-2} T.
15. 5.7×10^{-14} m; 1.4×10^{-6} N.
17. Magnetic field; 0.40 T; 1.20 T.
19. $R_e = 0.57$ m; $R_p = 10.4$ m; 1.04 km.
21. (a) 5.9×10^7 m/s, 3.4° from the original direction, (b) 1.2 cm.
23. (a) 24 J (15×10^{19} eV); (b) 16×10^{-8} kg·m/s; 48 J, (c) 8×10^{-8} kg·m/s; 24 J.
25. 6.8×10^{-5} m.
27. 500 eV.
29. The circular motion is parallel to the xz-plane with radius 5.7×10^{-5} m. The motion in the y-direction has a constant acceleration of -2.6×10^{14} m/s^2.
31. (a) 1.5×10^7 Hz, (b) 4.8×10^7 m/s tangential, (c) 1.9×10^{-12} J (1.2×10^7 eV), (d) 120, (e) 8.0×10^{-6} s.
33. 9.9×10^{-16} kg·m/s $< p < 45 \times 10^{-16}$ kg·m/s; 1.9 TeV $< E <$ 8.4 TeV; 4.2×10^{-3} m/s.
35. 3.54 T.
37. $F_B = 2.4 \times 10^{-16}$ N; $F_B/F_g = 2.6 \times 10^{13}$; $K = 5.2 \times 10^{-43}$ J (3.2×10^{-24} eV).
39. (a) $\vec{F}_B$ is to the left and $\vec{F}_E$ is to the right, (b) $F = q|E - vB|$, (c) $E = vB$.
41. $d\omega = qB/2m$.
43. 0.03 T.
45. 8.7×10^{-10} N perpendicular to the wire and to $\vec{B}$.
47. 5.0×10^{-2} N perpendicular to the wire and to $\vec{B}$.
49. $\Delta y = ILB/2k$.
51. 0.25 T.
53. $\mu = 0.11$ A·m^2 perpendicular to the loop; $\tau = 0.024$ N·m parallel to the loop and perpendicular to $\vec{B}$.
55. $IN\pi R^2 B(1 - \cos\theta)$.
57. -10^{-22} J $\leq U \leq +10^{-22}$ J.
59. (a) 5.0 mA, (b) $6 \times 10^{-5} \cos\theta_i$ J.
61. (a) 4×10^{10} m/s, (b) 10^{-23} N·m.
63. (a) $2IRB\,\hat{j}$, (c) $-2IRB\,\hat{j}$, (c) 0, (d) 0.
65. 0.43 rad/s.
67. 1.5×10^{27} carriers/m^3.
69. 0.11 mA.
71. 5.8×10^{11} N/C; 2.6×10^5 T.
73. 0.037 T.
75. $\Delta r/r = 0.05$ (decrease).
77. $e/m = 2V/B^2 R^2$.
79. (a) $L = NmvR$ perpendicular to the orbit, (b) $\frac{1}{2}NevR$, (c) $L/\mu = 2m/e$.
81. (b) horizontal direction, (c) $mg\hat{k} + qB(v_y\hat{i} + v_z\hat{j}) \times \hat{i} = m(dv_y/dt)\hat{j} + m(dv_z/dt)\hat{k}$; $\vec{v} = (mg/qB)\hat{j} + (mg/qB)(\sin\omega t\,\hat{k} - \cos\omega t\,\hat{j})$.

CHAPTER 29

1.
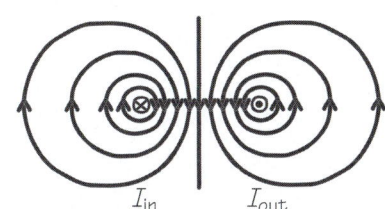
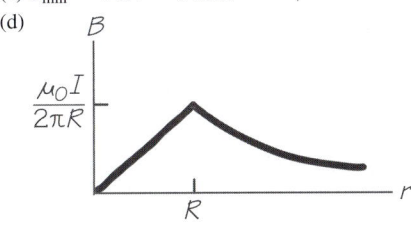

3. 1.7×10^3 N/m.
5. $0.25\ \mu$N/m attraction.
7. $[\mu_0] = [MLC^{-2}]$; $[\varepsilon_0] = [C^2M^{-1}L^{-3}T^2]$; 3.00×10^8 m/s.
9. (a) $r = R$, (b) $B_{max} = \mu_0 I/2\pi R$ at $r = R$, (c) $B_{min} = 0$ at $r = 0$ and $r = \infty$,
(d)

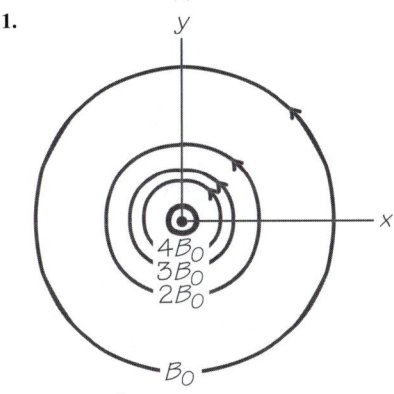

11.

13. 3.0×10^{-3} T.
15. (a) $\dfrac{\mu_0 I}{2\pi(x^2 + y^2)}(-y\hat{i} + x\hat{j})$,
(b) $\dfrac{\mu_0 I}{2\pi}\left\{-\left[\dfrac{y}{(x-a)^2 + y^2}\right.\right.$
$\left.+ \dfrac{y}{(x+a)^2 + y^2}\right]\hat{i}$
$+\left[\dfrac{x-a}{(x-a)^2 + y^2} + \dfrac{x+a}{(x+a)^2 + y^2}\right]\hat{j}\bigg\}$,
(c) $\dfrac{\mu_0 I}{2\pi}\left\{\left[\dfrac{-y}{(x-a)^2 + y^2} + \dfrac{y}{(x+a)^2 + y^2}\right]\hat{i}\right.$
$\left.+\left[\dfrac{x-a}{(x-a)^2 + y^2} - \dfrac{x+a}{(x+a)^2 + y^2}\right]\hat{j}\right\}$.
17. 6.7×10^{-4} T circular.
19. (a) away from the wire, (b) 5.9×10^{-3} m, (c) same energy.

21. $(\mu_0 Ia/2\pi) \ln[(a + d)/a]$.
25. 4.8×10^{-3} T.
27. 0.35 m.
29. 6.0×10^{-5} Wb.
31. 64 mA.
33. (a) 4.3×10^{-6} T, (b) 2.1×10^{-2} T.
35. $(\mu_0 NIL/2\pi) \ln[(R + L)/R]$.
37. $\sqrt{2}(\mu_0 I/2\pi x)$ in the yz-plane $45°$ below the y-axis.
39. $\sqrt{2}(\mu_0 I/4\pi H)$ in the xy-plane $45°$ from the $-x$-axis and the $-y$-axis.
41. (a) $-(2.2 \times 10^{-4}$ T/m$)\,dL\,\hat{j}$,
(b) $(0.56 \times 10^{-4}$ T/m$)\,dL\,\hat{k}$, (c) 0,
(d) $-(0.20 \times 10^{-4}$ T/m$)\,dL\,\hat{j}$.
43. 2.0×10^{-7} T along the axis.
45. $B = 2\mu_0 I(a^2 + b^2)^{1/2}/\pi ab$;
$B = 2\sqrt{2}\mu_0 I/\pi a$ $(a = b)$;
$B \approx 2\mu_0 I/\pi a$ $(b \gg a)$.
47. $(5.14 \times 10^{-7}$ T·m/A$)\,I/R$ out of the page.
49. 2.8×10^{-5} T into the page.
51. Circular path; $R = 2\pi mvd^3/q\mu\mu_0$.
53. 2.1×10^{-4} T into the page.
55. $\tfrac{1}{3}Q\omega R^2 \hat{k}$.
57. $(\mathscr{E}/R)e^{-t/RC}$.
59. 5.0×10^{-8} T·m.
61. $-CV_0 \omega \sin(\omega t)$.
63. $I = I_d = -4\pi R^2 \sigma_0/t_0$.
65. $(\mu_0 I/D)(2/\pi + 1)$ into the paper.
67. (a) $B = (\mu_0 I/2\pi R_1^2)r$ circular CCW, $r < R_1$, (b) $B = \mu_0 I/2\pi r$ circular CCW, $R_1 < r < R_2$, (c) $B = (\mu_0 I/2\pi r)(R_3^2 - r^2)/(R_3^2 - R_2^2)$ circular CCW, $R_2 < r < R_3$, (d) $B = 0$, $R_3 < r$.
69. (a) $-6.0 \times 10^{-5}\,\hat{j}$ N (attraction).
71. (a) $F_B/F_E = 4.3 \times 10^{-26}$,
(b) 2.1×10^8 electrons/cm, (c) 2.1×10^{-13}.
73. (a) $\dfrac{\mu_0 I}{2R}\left\{\dfrac{1}{[1 + (x/R)^2]^{3/2}}\right.$
$\left.+ \dfrac{1}{[2 - 2x/R + (x/R)^2]^{3/2}}\right\}$,
$B(0) = 0.677\mu_0 I/R$; $B(R/4) = 0.713\mu_0 I/R$; $B(R/2) = 0.716\mu_0 I/R$.
75. $NI = 336$ A·turns.

CHAPTER 30

1. Rotate the magnet around the coil; increase the speed of rotation.
3. 1.05×10^{-4} V.
5. $d\vec{B}/dt = (0.78$ T/s$)\hat{i}$.
7. $-BLv$ (clockwise) $(0 < t < L/v)$; $+2BLv$ (counterclockwise) $(L/v < t < 2L/v)$; $-BLv$ (clockwise) $(2L/v < t < 3L/v)$.
9. 18 V.
11. $-2B_0 v_0^2 t$ $(0 < t < L/v_0)$.
13. $-CDv^2 t/2\alpha(D + vt)$ clockwise.
15. $-CD^2 v/4\alpha(D + vt)$ clockwise.
17. $BA\omega \sin(\omega t)$.
19. 8×10^{-6} V.

21. 2.9×10^{-4} V.
23. 0.20 T.
25. 1.5×10^{-2} V, bottom end at the higher potential.
27. (a) 1.2×10^{-21} N, (b) 2×10^{-5} V.
29. $-2Bv(2Rvt - v^2 t^2)^{1/2}$ $(0 < t < 2R/v)$.
31.
(a)
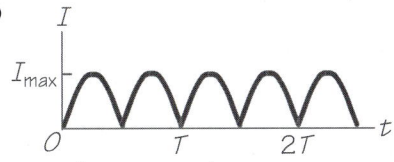

(b) $2NBA\omega/R\pi$, (c) $\pi I_{av} R/2NA\omega$.
33. 1.3×10^{-2} N; 0.46 W.
35. (a) 0.085 N, (b) 0.051 W.
37. $mgR/(BL)^2$.
39. (a) Speed increases until v_t is reached.
41. $(d\Phi_B/dt)^2/R$.
43. $aAB_0 e^{-at}$.
45. $E = +(\mu_0 n I_0 \omega r/2)\sin(\omega t)$, circular.
47. (a) 2.7×10^{-5} A, (b) in the same sense as the current in the solenoid.
49. -0.030 V (counterclockwise).
51. (a) 54 mA, (b) 4.4 mW.
53. 5.7 m/s.
55. Constant speed of E/B in the $-y$-direction; $B > 3 \times 10^{-5}$ T.
57. (a) -9.4×10^{-6} V (up), (b) $47\ \mu$A, (c) 9.7×10^{-10} J; external force.
59. (a) $(\pi r^2 \mu_0 N I \omega/2R)\sin(\omega t)$, (b) $90°$.
61. Counterclockwise; away from the wire.
63. 0.50 C.
65. $Q = (4.5 \times 10^{-9}$ C$)[1 - e^{-t/(3.0 \times 10^{-5}\,\text{s})}]$, with lower plate positive;
$I = (1.5 \times 10^{-4}$ A$)[1 - e^{-t/(3.0 \times 10^{-5}\,\text{s})}]$, clockwise.
67. $\mathscr{E} = (B\pi a^2 \omega/2)\sin(\omega t)$;
$P_{av} = B^2 \pi^2 a^4 \omega^2/8R$.
69. 2.4×10^{-3} V.
71. (a) 1.8×10^5 m/s, (b) 1.4×10^{-20} J, (d) -10%.

CHAPTER 31

1. 1.0×10^{-2} A·m^2 in the direction of the magnetic field.
3. 992.
5. 5 turns/m.
7. 4.6×10^2 A/m.
9. 0.014 A.
11. (a) $-(2.4 \times 10^{-5})\vec{B}_0$,
(b) $+(2.6 \times 10^{-4})\vec{B}_0$.
13. 0.064 A.
15. (a) 2.2×10^{25} electrons, (b) 2.2 A/m.
17. $\lambda \pi R^3 \omega$, along the axis of the ring.
19. $v = 4.1 \times 10^{10}$ m/s $> c$.
21. $M = 1.4 \times 10^6$ A/m; $\mu/\mu_0 = 3.6 \times 10^2$.
23. $(2.4 \times 10^{-3}$ T$)\chi_m$.
25. (a) 0.16 T, (b) 2.5×10^{-4} T.
27. 1.45×10^2 A/m; 0.88 T.
29. 289 A/m.

31. 6.3×10^{-4} T.
35. -2.2×10^{-4}.
37. $H = (1.6 \times 10^{-3} \text{ A})/r$ circular; $B = (2.0 \times 10^{-9} \text{ T} \cdot \text{m})/r$ circular; $\Delta B = (1.3 \times 10^{-12} \text{ T} \cdot \text{m})/r$.
39. 0.38 A/m; 4.7×10^{-7} T.
41. (a) With $\vec{B} = B\hat{k}$, $dm_x/dt = g_S m_y B$; $dm_y/dt = -g_S m_x B$; $dm_z/dt = 0$.
43. 4.2×10^6 Hz.
45. 1.96 T; 0.16 T.
49. $1.4 \text{ A} \cdot \text{m}^2/\text{cm}^3$.
51. $m = 8 \times 10^{22} \text{ A} \cdot \text{m}^2$; $M = 6 \times 10^2 \text{ A/m}$; 2×10^9 A.
53. $m_{\text{ring}}/m_{\text{disk}} = 2$.
55. (a) $C = (NmB/2\pi kT)/(e^{mB/kT} - e^{-mB/kT})$,
(b) $\langle \cos \theta \rangle = \dfrac{e^{mB/kT} + e^{-mB/kT}}{e^{mB/kT} - e^{-mB/kT}} - \dfrac{kT}{mB}$,
(c)

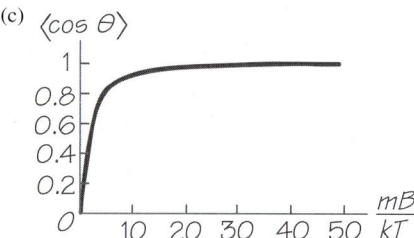

CHAPTER 32

1. 6.0×10^{-5} Wb.
3. 3.1 μH.
5. 30 mH.
7. (a) 1.0 mH, (b) 3×10^{-3} Wb.
9. (a) $\mu_0 A N_1^2/\ell$ or $\mu_0 A N_2^2/\ell$,
(b) $\mu_0 A (N_1 + N_2)^2/\ell$, (c) $\tfrac{1}{4}\mu_0 A (N_1 - N_2)^2/\ell$,
(d) $\mu_0 A N_1 N_2/\ell$.
13. $I = 0$ for $t < 0$; $I = -0.40$ μA for $0 < t < 0.30$ s; $I = +0.40$ μA for $0.30 \text{ s} < t < 0.60 \text{ s}$; $I = 0$ for $t > 0.60$ s.
15. $\mathcal{E} = -10$ mV for $0 < t < T/4$, $3T/4 < t < 5T/4, \ldots$; $\mathcal{E} = +10$ mV for $T/4 < t < 3T/4$, $5T/4 < t < 7T/4, \ldots$.
19. (a) 3.9×10^{-3} H, (b) 16 H.
21. $L = \dfrac{\mu_0 \ell}{\pi}\left[\tfrac{1}{2} + \ln\left(\tfrac{a}{r}\right)\right]$.
The radius of the wire cannot be neglected.
23. $(\mu_0 L/2\pi) \ln[(a+b)^2/b(2a+b)]$.
25. 1.2 kW.
27. 24 A.
29. $0.64 \, e^{-(0.04/\text{s})t}$ mW.
31. (a) Same, (b) same.
33. (a) 6.4×10^3 J, (b) 2.4 L.
35. 2.6×10^2 J.
37. 11 cm.
41. There is an upward force on the cylinder before it completely drops into the solenoid.
43. 1.6×10^{-9} J/m.
49. 25 Ω; 5.6 mH.
51. $V^2 L/2R$.
53. 1.6 mF.
55. 1.5 μF; 0.36 mH.
57. (a) $\omega_1/10$, (b) $\alpha_1/100$,
(c) $\omega_2'^2 = (\omega_1^2/100) - (\alpha_1/100)^2$.

59. $\omega' = 6.6 \times 10^4$ rad/s.
61. $2Q_0^2/(LC^3)^{1/2} e^{-2\alpha t}$.
65. $T + \tfrac{1}{4}\pi R^2 (C^3/L)^{1/2}$.
67. (a) 47 mA, (b) 2.2×10^{-8} J, (c) $U_{L\text{max}}/U_{C\text{max}} = 1$.
69. (a) 4.0×10^{-3} Ω, (b) 1.5×10^{-7} J at $t = 0$; 4.2×10^{-9} J at $t = 100$ oscillations, (c) Joule heating.
73. (a) $I_2 = 0$; $I_1 = I_\mathcal{E} = I_L = 2.4$ mA, (b) 0.28 mH, (c) $I_1 = I_\mathcal{E} = 0$; $I_L = I_2 = 0.86$ mA.
75. $(\mu_0 I^2/4\pi)[(1/4) + \ln(r_2/r_1)]$.
77. (b) c.
79.

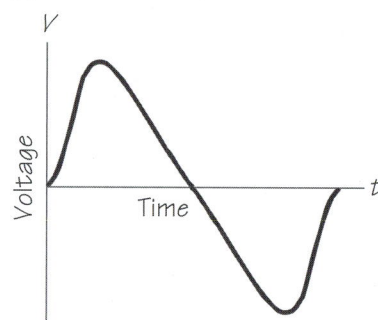

81. 3.2×10^{-2} H; the iron core increases B and M, and concentrates flux.
83. $L = 0.1$ mH; $C_1 = 3.9$ μF; $C_2 = 5.0$ μF.

CHAPTER 33

1. 9000 V.
3. 60 turns.
5. 450 turns; 27 A.
7. 0.40 A.
9. 50 turns.
11. 13 kHz.
13. 1.1×10^4 rad/s.
15. 22.8 μF.
17. $X_{C\text{max}} = 1.7 \times 10^3$ Ω; $X_{L\text{max}} = 300$ Ω.
19. 15 Ω; 95 Hz.
21. (a) $\vec{D}$ is more advanced in phase, (b) 90°; $\vec{D}$ leads, (c) $\tan \delta = A/B$.

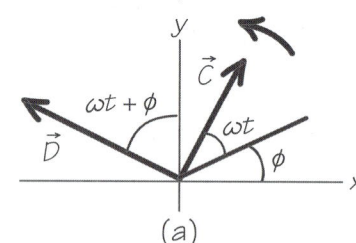

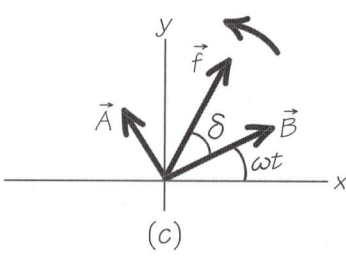

23. 35 pF.
25. $R_2 = \tfrac{1}{2} R_1$.

27. 2.7×10^{-4} C; 4.0 V.
29. 6.4 H.
31. $X_C = 66.3$ Ω; $X_L = 694$ Ω; $Z = 802$ Ω; $Q_{\text{max}} = 13$ μC; $\phi = +51.5°$; $I_{\text{max}} = 0.10$ A.
33. $V_C = -4.05$ V; $V_L = +43.1$ V.
35. $I_{\text{max}} = 92$ mA; $V_{R\text{max}} = 55.5$ V; $V_{C\text{max}} = 123$ V; $V_{L\text{max}} = 27.9$ V.
37. Double.
39.

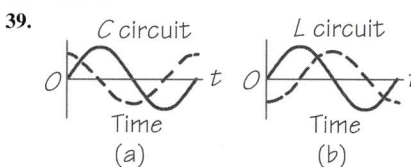

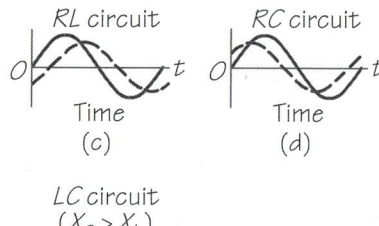

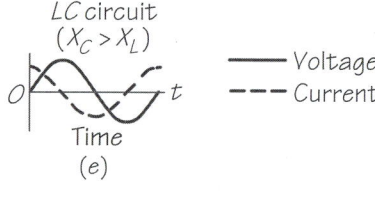

41. $V_{C\text{max}} = 66$ V; $V_{RL\text{max}} = 67$ V.
43. 11 μH.
45.

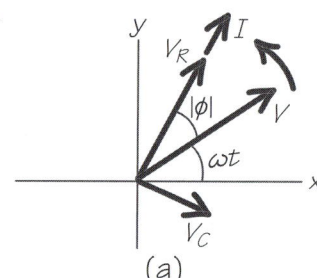

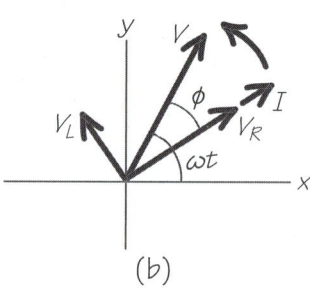

47. 5.0 A.
49. (a) $\langle P \rangle \to \tfrac{1}{2} V_0^2 R/\omega L \to 0$,
(b) $\langle P \rangle \to \tfrac{1}{2} V_0^2 R \omega C \to 0$; no current through the capacitor.
51. (a) 0, (b) 0, (c) 1.
53. (a) $R/(X_L^2 + R^2)^{1/2}$, (b) $R/(X_C^2 + R^2)^{1/2}$, (c) 0.
55. (a) 1.6 MHz; (b) 0.034 Ω.
57. 27 μF; 46 μF.
59. 345 V; 9.05 kW.
61. 9.7 kW; 16.7 kW.
63. $R = 22.2$ Ω; $X_L = 98$ Ω; $X_C = 168$ Ω.

65. (a) 31 W, (b) 17 W, (c) 34 W.
67.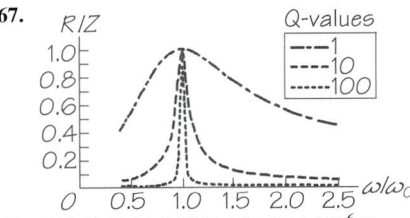
69. (a) 0.20 V, (b) 0.19 V, (c) 5.3×10^{-6} V.
71. $R/L = 5.0 \times 10^4$ s^{-1}.
73. $\langle I \rangle = I_0/\pi$; $I_{rms} = \frac{1}{2}I_0$.
75. At low frequency $Z \to X_C$; at high frequency $Z \to R$.
77. $I = (V_0/Z)\cos(\omega t + \phi)$;
$I_C = (V_0/X_C)[(R/Z)\sin(\omega t + \phi) - \sin(\omega t)]$;
$I_L = (V_0/X_L)[-(R/Z)\sin(\omega t + \phi) + \sin(\omega t)]$;
there is a resonant frequency.
79. (a) $V_{C1} = 5.14$ V; $V_{C2} = 2.20$ V, (b) 2.34 V.
83. 29 mW.
85. (a) 5.7 V, (b) 18 V.
87. 2.4 µH; 0.060 Ω.
89. 177 Ω.
91. (a) $Q_c(t) = CV_0 e^{i\omega t}$; $I_c(t) = i\omega CV_0 e^{i\omega t}$.
93. (a) $Q_{0C} = V_{0C}/\{\omega[i(L\omega - 1/\omega C) + R]\}$;
(b) $Q_c(t) = -i(V_0/\omega Z)e^{i(\omega t - \phi)}$;
$I_c(t) = (V_0/Z)e^{i(\omega t - \phi)}$,
(c) $Q = -(V_0/\omega Z)\cos(\omega t - \phi)$;
$I = (V_0/Z)\sin(\omega t - \phi)$.
95. $\omega = [(1/LC) - (R^2/4L^2)]^{1/2} + i(R/2L)$.

CHAPTER 34

3. $\oint \vec{E} \cdot d\vec{s} = \mu_0 \dfrac{d\vec{M}}{dt} - \dfrac{d}{dt}\iint_S \vec{B} \cdot d\vec{A}$.
5. $\vec{B} = (E_0/c)\cos(kz + \omega t)\hat{i}$, traveling in the $-z$-direction.
7. 94×10^6 Hz; 3.2 m.
13. $+\dfrac{\partial B_y}{\partial x} = \mu_0\varepsilon_0 \dfrac{\partial E_z}{\partial t}$ and $+\dfrac{\partial B_y}{\partial t} = \dfrac{\partial E_z}{\partial x}$.
15. Plane formed by the z-axis and the line $y = -(\tan\theta)x$.
17. $(3 \text{ V/m})\cos[(1.05 \times 10^7 \text{ m}^{-1})z - (3.15 \times 10^{15} \text{ rad/s})t]\hat{i}$.
19. $\lambda = 2L/n$, $n = 1, 2, 3, \ldots$.
21. 2.2×10^{-4} T.
23. 2.6×10^{-5} W/m^2.
25. 8.0×10^{-11} N/m^2.
27. (a) 1.3×10^3 W/m^2, (b) 4.5×10^{-6} N/m^2.
29. 1.2×10^{-2} W/m^2; 0.070 J.
31. $E_0 = 134$ V/m; $E_{rms} = 95$ V/m;
$B_0 = 4.47 \times 10^{-7}$ T; $B_{rms} = 3.16 \times 10^{-7}$ T.
33. 2.0×10^{-3} N/m^2.
35. 17 kg·m/s; 3 kg·m/s (about 1/5 of the solar result).
37. 2.5×10^{-5} N·m.
39. $P = 5.9 \times 10^5$ W; vaporize.
41. $\Delta p/A = 2(S/c)\cos^2\theta\,\Delta t$.
43. $\vec{E}(x,t) = (4.8 \times 10^3 \text{ V/m})\sin\{2\pi[(20 \text{ m}^{-1})x + (6 \times 10^9 \text{ Hz})t]\}\hat{k}$;
$\vec{B}(x,t) = (1.6 \times 10^{-5} \text{ T})\sin\{2\pi[(20 \text{ m}^{-1})x + (6 \times 10^9 \text{ Hz})t]\}\hat{j}$.
45. $\tau_{average} = 3.1 \times 10^{-16}$ N·m.
47. 4.0×10^{-7} s.
49. (a) $+y$-direction, (b) $-x$-direction, (c) $+z$-direction, (d) electric field in $-y$-direction; magnetic field in $+x$-direction; Poynting vector in $+z$-direction.
51. (a) no solution, (b) 39°, (c) 57°, (d) 72°.
53. 37°.
55. 3.5×10^5 W/m^2.
57. 0; 1/8.
59. (a) $\frac{1}{2}I_0$, (b) 90°, (c) $(1/8)I_0\sin^2(2\theta)$.
61. The superposition is acceptable since the sum of solutions to the Maxwell's equations is itself a solution.
63. 4.7×10^{-5} W/m^2.
65. (a) $\vec{B} = \mu_0 n I_0 \cos(\omega t)\hat{k}$ (along the axis),
(b) $\vec{E} = \frac{1}{2}\mu_0 n I_0 \omega r \sin(\omega t)$ (circular),
(c) $0 < t < \frac{1}{4}T$: $\vec{S}$ is in $+\hat{r}$;
$\frac{1}{4}T < t < \frac{1}{2}T$: $\vec{S}$ is in $-\hat{r}$;
$\frac{1}{2}T < t < \frac{3}{4}T$: $\vec{S}$ is in $+\hat{r}$;
$\frac{3}{4}T < t < T$: $\vec{S}$ is in $-\hat{r}$.
69. $\vec{E} = (5.19 \times 10^2 \text{ V/m})\cos[(2.54 \times 10^6 \text{ m}^{-1})x + (1.47 \times 10^6 \text{ m}^{-1})y - (2\pi \times 10^{14} \text{ s}^{-1})t]\hat{k}$.
71. 5.8×10^5 J; 0.25 kg.
73. 4.5×10^{13} photons/m^3.
75. $f' = [(c - v)/(c + v)]f$.
77. $a_J = 2.3 \times 10^{-7}$ m/s^2;
$a_S = 7.1 \times 10^{-8}$ m/s^2; 93 years.
79. $\dfrac{\partial^2 E_x}{\partial z^2} = \varepsilon_0\mu_0 \dfrac{\partial^2 E_x}{\partial t^2} + \dfrac{\mu_0}{\rho}\dfrac{\partial E_x}{\partial t}$.
81. (a) $\vec{B} = (\mu_0 r/2\pi R^2)\,dQ/dt$ circular, for $r < R$, (b) $\vec{S} = -(Qr/2\pi^2 R^4 \varepsilon_0)(dQ/dt)\hat{r}$.

CHAPTER 35

1. $v_{ice} = 2.29 \times 10^8$ m/s;
$v_{ethyl\ alcohol} = 2.21 \times 10^8$ m/s;
$v_{benzene} = 2.00 \times 10^8$ m/s;
$v_{diamond} = 1.24 \times 10^8$ m/s.
3. 492 nm; 4.62×10^{14} Hz.
5. 2.9×10^4 rev/min.
7. 6 m.
9. 1.0 ns.
11. 1.3 cm.
13. 1.62.
15. 590 m.
17. 0.031.
19. 1.4°.
21. $\theta_{water} = 45.7°$; $\theta_{glass} = 38.5°$.
23. Cannot be changed.
25. 7.4°; no total internal reflection.
27. Outside the "shadow" of the prism, there is direct illumination. For a distance of 0.57 cm from each edge, there will be no illumination. For the next distance of 0.77 cm − 0.57 cm = 0.20 cm, there will be illumination from one half. For the next distance of 0.73 cm to the center of the pattern, there will be illumination from both halves.
29. 8.3°.
31. 19 cm below the upper surface of the glass.
33. 74° from the normal.
35. 5.3 mm.
37. 0.86 m.
39. Prism will not be totally reflecting; $n_2 \leq 1.283$.
41. 1.5.
43. 52%.
51. 0.48°.
53. (a) $\omega \leq 12.5 \times 10^{15}$ rad/s, (b) 27.7°.
55. 0.009.
57. 1.31 cm.
59. $\Delta n = [(\cos\theta_4)(\cos\theta_2)/\sin(2\phi)]\,\Delta\theta$,
where $\cos\theta_2 = \cos\{\sin^{-1}[(\sin\phi)/n]\}$ and
$\cos\theta_4 = \cos\{\sin^{-1}[n\sin(2\phi - \{\sin^{-1}[(\sin\phi)/n]\})]\}$.
61. $\theta_i = \sin^{-1}(n\sin\phi)$.
63. (a) 4.6×10^6 km, (b) 3.1×10^8 m/s.

CHAPTER 36

1. 3 images.
3. 82.5% of the beam has been dissipated.
5. $0.577a, a$.
7. $i_2 = 26.3$ cm; $R = 30$ cm.

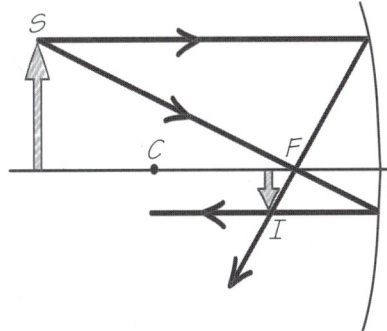

9. 0.80 cm tall, inverted, in front of the mirror.
11. $i = -400$ cm (behind the mirror); $+50$ cm tall.
13. $-f^2v/(s - f)^2$; $s = 2f$.
17. $i = -90$ cm (in air in front of glass).
19. $i = -30$ cm (behind the glass).
21. $i = +18.4$ cm; 43 cm from the surface.
23.
25. (b) $s = n_1 R/(n_2 - n_1)$, (c) image is very far in front of the boundary, (d) image approaches the boundary.
31. (a) $+16$ cm, (b) converging, (c) inverted, (d) -2.1.
33. (a) 8.9 cm in front of the lens, (b) no, (c) yes, (d) $+0.60$.
35. (a) $+38$ cm, (b) $+13$ cm.

37. Virtual image.

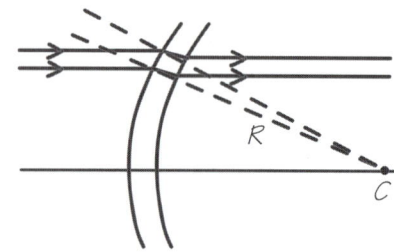

39. Lens a: (a) $f_a = 32.1$ cm, 33 cm beyond the lens, (b) inverted and real, (c) $M-$; Lens b: (a) $f_b = +77.9$ cm, 84 cm beyond the lens, (b) inverted and real, (c) $M-$; Lens c: (a) $f_c = -32.1$ cm, 31 cm in front of the lens, (b) upright and virtual, (c) $M+$; Lens d: (a) $f_d = -77.9$ cm, 73 cm in front of the lens, (b) upright and virtual, (c) $M+$.
41. Lens a: image is 63.4 cm to the left of the lens, inverted, real, with $M = -0.98$; Lens b: image is 393 cm to the right of the lens, upright, virtual, with $M = +6.0$; Lens c: image is 21.5 cm to the right of the lens, upright, virtual, with $M = +0.33$; Lens d: image is 35.4 cm to the right of the lens, upright, virtual, with $M = +0.55$.
45. $+53$ cm.
47. $+6$ cm.
49. 9.6 cm.
51. f_1/f_2.
55. 1.09 cm.
57. Near *positive* lens: image is 15.8 cm in front of the negative lens, or 0.8 cm from the positive lens on the object side; near *negative* lens: image is 65.4 cm in front of the positive lens, or 50.4 cm from the negative lens on the object side; the order of the lenses is important.
59. (a) First image is 12.4 cm behind M_1, (b) second image is 7.9 cm in front of M_2.

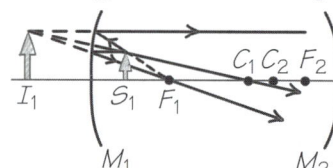

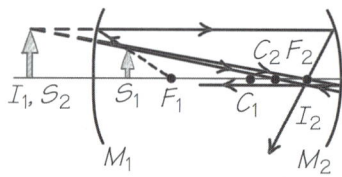

61. $f = [f_1(d^2 - f_1 d - 2f_2 d + f_1 f_2)] / (d^2 - 2f_1 d - 2f_2 d + 2f_1 f_2 + f_1^2)$.

63. $d \to R/(n-1)$.
65. The plate may be placed anywhere between the lens and the screen, with thickness 13.1 cm.

CHAPTER 37

1. 4.7×10^2 nm.
3. 0.24 m.
5. 12.2 m.
7. (a) Doubles, (b) reduces by $\frac{1}{2}$, (c) doubles, (d) no change.
9. 6.4×10^2 nm.
11. Inward.
13. $v_{\text{fringe, ripple}} = 3 \times 10^{-6}$ m/s; $v_{\text{fringe, optical}} = 1.2 \times 10^6$ m/s.
15. (b) $\theta = \tan^{-1}\{\lambda[1/(d^2 - \lambda^2) + 1/4R^2]^{1/2}\}$.
17. 3.0×10^3 W/m^2.
19. 151°; 525 nm.
21. $I_{\text{av}} = 2I_0$.
23. $\Delta\theta = 2\sin^{-1}(\lambda/4d)$.
25. (a) along the perpendicular bisector, (b) 2.0×10^{-3} W/m^2, (c) 9.5°.
27.

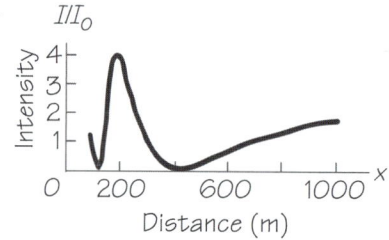

29. 33.6 μm.
31. 81 nm, 242 nm, 404 nm.
33. $t_{\min} = 0.208$ μm; 0.416 μm and 0.625 μm would also work.
35. $r = (\lambda R)^{1/2}$.
37. 26 μm.
39. 1000 maxima pass; the rings move in to the center.
41. 6.4 mm; 8.5 mm.
43. $x = (n\lambda R)^{1/2}$.
45. 118 nm.
47. 790 nm.
49. 621 nm.
51. (a) 115 nm, (b) longer, (c) no reflection of green light.
53. 120 nm.
55. 120 nm.
57. 633 nm.
59. 1.00020.
61. 100 fringes must be counted to 0.1 of a fringe.
63. Uniform.
65. $u \propto \sin^2[2\pi f(x - \frac{1}{2}L)/c]$.

67. $\sin\theta' = \sin\theta - (m\lambda/d)$, $m = 0, \pm 1, \pm 2, \ldots$
69. 34.0.
73. Two: 0 or $4I_0$; three: I_0; four: 0.
75. $\theta = \sin^{-1}\{n^2 - [\lambda(m - \frac{1}{2})/2h]^2\}^{1/2}$, $m = 1, 2, 3, \ldots$.

CHAPTER 38

1. 450 nm.
3. 39.7 lines/cm.
5. 3.0×10^3 lines/cm.
7. 15,000, 30,000, 45,000; 0.028 nm, 0.014 nm; 0.0093 nm.
9. 3.1×10^4 lines; 8.83×10^{-4} rad/nm.
11. 76°.
13. 18.6°.
15. First order: 6.17° to 9.79°; second order: 12.42° to 19.88°.
17. $\sin\beta = \sin\alpha - m\lambda/d$, $m = 0, \pm 1, \pm 2, \ldots$.
19. 0.51 m.
21. (a) Maxima: $-68°, 0°, 68°$; minima: $-38°, 38°$, (b) 2.0 m.
23. $I_0/I_1 = 22.2$; 1.19°; increase θ.
27. $\Delta\theta_h/\Delta\theta_1 = 0.443$.
29. $\sin\theta = \sin\theta_i - m\lambda/a$, $m = \pm 1, \pm 2, \ldots$; there is a "central maximum."
31. (b) $\alpha_{\max 1} = 257.43°$, midway between 1st and 2nd minima $= 270°$; $\alpha_{\max 2} = 442.61°$, while midway between 2nd and 3rd minima $= 450°$.
33. 5.0 cm; 83 m.
35. 70 m.
37. 0.73 m; shorter wavelengths.
39. 2.4 mm.
41. 21 ft.
43. Third order.
45. I/I_0: 3.97, 1.79, 2.43, 0.26.
47. $a = 0.1$ mm; $d = 0.3$ mm.
49. $n = 3m = 3, 6, 9, \ldots$; 0.084°, 0.169°, 0.253°, $\ldots$.
51. 0.0274°.
53. 0.21 nm.
55. 29.0°.
57. 15×10^3 m.
59. (a) 1.0×10^{-5} rad $= (5.7 \times 10^{-4})°$, (b) 25 μm.
61. (a) $I_0\{\sin[(N\pi\cos\theta)/(N-1)]/\sin[(\pi\cos\theta)/(N-1)]\}^2$, (b) slight decrease in I for θ close to 0° and 180° and then a buildup to the maxima at 90° and 270°.
63. No.
65. 0.45 mm.
67. 22°.

Photo Credits

FRONTMATTER Page ixL Adam Hart Davis/Photo Researchers, Inc. **Page ixR** © 2003 Peter Menzel/menzelphoto.com **Page xL** Ralph H Wetmore II/Getty Images Inc.—Stone AllStock **Page xR** Dr. Edwin Mueller/SPL Photo Researchers **Page xiL** AP/Wide World Photos **Page xiRT** Richard Megna/ Fundamental Photographs **Page xiRB** Taxi RM/Getty Images, Inc. **Page xiiLT** GSFC/MITI/ ERSDAC/ JAROS, and U.S./Japan ASTER Science Team/NASA Headquarters **Page xiiLB** Diane Hirsh/ Fundamental Photographs **Page xiiR** Richard Megna/Fundamental Photographs **Page xiiiL** Richard Megna/Fundamental Photographs **Page xiiiR** Michael Freeman **Page xviii** Tom Pantages **Page xix** Martin-Baker Aircraft Co Limited **Page xxix** Paul M Fishbane, Stephen G. Gasiorowicz, Stephen T. Thornton **CHAPTER 21** **CO-21** The Granger Collection, New York **21-1a** Tom Cogill/Stephen T. Thornton **21-2** Stephen T. Thornton **21-5** The British Library **21-7a** Sargent–Welch/ VWR International/ Courtesy of Central Scientific **21-8** LBNL/Photo Researchers, Inc. **CHAPTER 22** **CO-22** Adam Hart Davis/Photo Researchers, Inc. **22-2a** Harold M. Waage **22-11a** Harold M. Waage/Permission of the Princeton University Physics Department **22-12a** Harold M. Waage/ Permission of the Princeton University Physics Department **CHAPTER 23** **CO-23** © 2003 Peter Menzel/ menzelphoto.com **23-4a** Harold M. Waage/Permission of the Princeton University Physics Department **23-20** © Jim Krider/Arizona State University **CHAPTER 24** **CO-24** Ralph H Wetmore II/Getty Images Inc.-Stone Allstock **24-24** Harold M. Waage/ Princeton University **24-25** Adam Hart–Davis/Science Photo Library/Photo Researchers, Inc. **24-26** John Woodcock/ Dorling Kindersley Media Library **24-27** Lester V. Bergman and Associates, Inc. **24-29a** B. Loberg and H. Norden **24-30** Dr. Erwin Mueller/SPL/Photo Researchers, Inc. **24-34** National Institute of Standards and Technology (NIST). Geometric and Electronic Properties of Cs Structures on III–V (110) Surfaces: From 1–D and 2–D Insulators to 3–D Metals, L.J. Whitman, J.A. Stroscio, R.A. Dragoset, and R.J. Celotta, Phys. Rev. Le tt. 66, 1338 (1991). **24-35** Courtesy of International Business Machines Corporation. Unauthorized use not permitted. **CHAPTER 25** **CO-25** Randy Montoya/Sandia National Laboratories **25-4a** Harold M. Waage/ Princeton University **25-8a-b** Tom Cogill/ Stephen T. Thornton **CHAPTER 26** **CO-26** AP/ Wide World Photos **26-9** Tom Pantages **26-10b** The Image Works **26-21b** Dan McCoy/Rainbow **26-22** Dorling Kindersley Media Library **CHAPTER 27** **CO-27** AP/Wide World Photos **27-1a** NASA Headquarters **27-1b** Still Pictures/Peter Arnold, Inc. **27-1c** Loren Winters/Visuals Unlimited **27-13** Michael Dalton/Fundamental Photographs **27-14a** Tom Cogill/Stephen T. Thornton **27-15a** Tom Cogill/Stephen T. Thornton **27-20c** International Imaging Inc./Stephen T. Thornton **CHAPTER 28** **CO-28** David Miller, National Geophysical Data Center/NOAA Central Library Photo Collection **28-2a** Phil Degginger/ Color–Pic, Inc. **28-2b** Alfred Pasieka/ SPL/Photo Researchers, Inc. **28-2c** Tom Pantages **28-3a-b** International Imaging Inc./Stephen T. Thornton **28-12** Brookhaven National Laboratory/Science Photo/Photo Researchers, Inc. **28-13b** Lawrence Berkeley National Laboratory **28-16a** The Granger Collection, New York **CHAPTER 29** **CO-29** Princeton University **29-1a-b** Thomas J. Cutitta/ Interantional Imaging Inc./Stephen T. Thornton **29-2a, c** Tom Cogill/Stephen T. Thornton **29-09** Tom Cogill/Stephen T. Thornton **29-15b** Richard Megna/ Fundamental Photographs **29-17a** Richard Megna/Fundamental Photographs **29-25a** Richard Megna/Fundamental Photographs **CHAPTER 30** **CO-30** The Royal Institution **30-19a** IBM Research/Peter Arnold, Inc. **CHAPTER 31** **CO-31** Courtesy of American Superconductor Corporation (www.amsuper.com) **31-6** Dr. John Unguris **31-11** John Unguris/National Institute of Standards and Technology (NIST) **31-13** Richard Megna/ Fundamental Photographs **31-17a** Jack Plekan/Fundamental Photographs **31-17b** Berwin, Derek/Getty Images Inc.-Image Bank **CHAPTER 32** **CO-32** Richard Megna/Fundamental Photographs **32-2b** Sargent–Welch/VWR International **32-4** Michael A. Gallitelli/ Metroland Photo, Inc. **32-7** Tom Cogill/ Stephen T. Thornton **CHAPTER 33** **CO-33** Taxi (RM)/Getty Images, Inc. **33-13a-b** Tom Pantages **33-20b-d** Tom Cogill/Stephen T. Thornton **CHAPTER 34** **CO-34** GSFC/MITI/ERSDAC/ JAROS, and U.S./Japan ASTER Science Team/NASA Headquarters **34-7a** Courtesy of the Library of Congress **34-8a.1** Chandra X-Ray Center/NASA/CXC/UMd./ A.Wilson et al. **34-8a.2** Ultraviolet Imaging Telescope/ UIT Science Team/ NASA Headquarters **34-8a.3** Todd Boroson/NOAO/AURA/NSF. Copyright WIYN Consortium, Inc., all rights reserved. **34-8a.4** Infared Space Observatory/ESA/ISO, CAM, M. Sauvage et al./ **34-8a.5** ESA/ISO. CAM. M. Sauvage et al./Chandra X-Ray Center/ NASN/CXC/SAO **34-14** Stephen T. Thornton/Tom Cogill **34-15a-c** Stephen T. Thornton/Tom Cogill **34-20a-b** Leonard Lessin/Peter Arnold, Inc. **CHAPTER 35** **CO-35** Goddard Space Flight Center/ Image courtesy of NASA - National Aeronautics and Space Administration **35-1** The New York Public Library Photographic Services/Art Resource **35-4b** Courtesy of the Bancroft Library University of California, Berkeley **35-5b** Runk/Schoenberger/ Grant Heilman Photography, Inc. **35-6** Lawrence Livermore National/Photo Researchers, Inc. **35-11** NASA Headquarters **35-12a** Stephen T. Thornton/Tom Cogill **35-13** Stephen T. Thornton/Tom Cogill **35-17b** Thomas J. Cutitta/International Imaging/ Stephen T. Thornton **35-19a** Sargent-Welch/VWR International **35-24a** Grant Heilman Photography, Inc. **35-27b** Randy O' Rourke/Corbis/Stock Market **CHAPTER 36** **CO-36** Richard Megna/ Fundamental Photographs **36-5** Richard Megna/Fundamental Photographs **36-8c** Stephen T. Thornton/ Tom Cogill **36-9d** Stephen T. Thornton/ Tom Cogill

36–10b Stephen T. Thornton/ Tom Cogill **36–11b** Stephen T. Thornton/ Tom Cogill **36–12b** Stephen T. Thornton **36–22a-b** E. R. Degginger/Color-Pic, Inc. **36–32b** Jeff J. Daly/Fundamental Photographs **36–33a** Art Resource, N.Y. **36–36** Tom Pantages **36–39** M. C. Escher's, "Hand with Reflecting Sphere" © 2003 Cordon Art B. V.-Baarn-Holland. All rights reserved. **CHAPTER 37** **CO–37** Diane Hirsch/ Fundamental Photographs **37–7** Reproduced by permission from OHG. Copyright © 1962 by Springer-Verlag GmbH & Co KG. **37–13b** Richard Megna/Fundamental Photographs **37–13c** Tom Pantages **37–15** Color-Pic, Inc. **37–16** Richard Megna/Fundamental Photographs **37–20b** Courtesy of PASCO Scientific **37–21a** Sargent-Welch/VWR International **37–22a-b** Sargent-Welch/ VWR International **CHAPTER 38** **CO–38** Georg Fischer/ Bilderberg/Peter Arnold, Inc. **38–1** Pearson Education/PH College **38–3** Ken Kay/Fundamental Photographs **38–5a** Pearson Education/PH College **38–9** Don Farrall/Getty Images, Inc.-Photodisc. **38–11** Reproduced by permission. Copyright by Springer-Verlag GmbH & Co KG. **38–14** Tom Pantages **38–15a-c** Pearson Education/PH College **38–17** Richard J. Wainscoat **38–19** Reproduced by permission. Copyright by Springer-Verlag GmbH & Co KG. **38–20** Photo Researchers Inc. **38–22a** Omikron/ Photo Researchers, Inc. **38–23b** The Burndy Library, Dibner Institute for the History of Science and Technology, Cambridge, Massachusetts **38–31** Pascal Goetgheluck/SPL/Photo Researchers, Inc. **38–32** Michael Freeman

Index

Note: Italics indicate a definition or primary entry for multiple entries, where applicable.

Aberrations, 1022–23
  chromatic, *1023*
  monochromatic, *1022*
AC, *See* Alternating current (AC)
Accelerating charges, 636
  and electromagnetic waves, 957
Acceleration, units of, A–1
Accelerator, *703*
Acceptors, 755
Accommodation, 1018
Adaptive optics, *1022*
Airy disk, *1059*–60
Airy, Sir George, 1059
Algebraic functions, properties of, A–6
Allowed bands of energy levels, 753
Alternating current (AC), *739*, *917*–42
  AC circuits:
    power in, 929–31
    single elements in, 920–25
  applications, 931–35
  bandwidth, 930
  capacitive circuit, 921–22
  diodes, 932
  filters, 933–34
  impedance matching, 935
  inductive circuit, 922–23
  power factor, 931
  rectifiers, 932
  resistive circuit, 921
  in series *RLC* circuits, 926–29
  transformers, 917–20
Alternating-current (AC) generator, 863
Alternator, 932
Ammeters, *776*–77, 783
  analog, 776
  clip-on, 824
Ampere (A), 739, A–1
Ampère, André-Marie, 610, 739, 807, 820–21, 831
Ampère's law, 820–21, 823–24, *824*, 839, 886, 943, 944, 946, 967
  Maxwell's generalized form of, 839
  using to find the magnetic field, 825
  in solenoids, 829–31
Analog ammeters, 776
Analog voltmeter, 777
Analyzer, for polarization, 960–61
Angle, A–1
Angle of incidence, 963–64, 1023
Angular acceleration, units of, A–1
Angular dispersion, 1054–55
Angular distribution, of power, 959
Angular frequency, 926
  of precession, 891
Angular magnification, 1019–20, 1024
Angular speed, A–1
Antenna:
  broadcasting, 957

dipole, 957–59
  receiving, 957
Antiparticles, 628
Antireflective coatings, optical devices, 1042
Aperture, 1019
Arago, François, 807, 1051
Astronomical constants, A–3
Astronomical unit (AU), A–1
Atomic mass, 616
Atomic nuclei, 886
Atmosphere (unit), A–1
Atomic mass weight (u), A–1
Atoms:
  magnetic dipole moment of, 879–80
  as magnets, 878–81
Aurora, *803*
Average current, 739
Average magnetic moment, 881
Average speed of electrons in metals, 752
Avogadro's number, A–2
Axis, 1001

Babinet's principle, 1072
Band-pass filter, 934
Bandwidth, alternating current (AC), 930
Bar, A–1
Bar magnet, 838
  field lines for, 826–27
  levitated, 860
Bardeen, John, 755
Barrier tunneling, 706
Base units, SI, A–1
Batteries:
  capacitors vs., 719
  charging, 767
  conversion of chemical energy into an electromotive force (EMF), 767
  discharging, 767
  and electric power, 769–70
BCS theory, 755
Bednorz, J. George, 755
Betatron, 872
Binomial theorem, A–5
Biot, Jean-Baptiste, 833
Biot-Savart law, 832–36, *833*, 945
  using, 833–34
Bohr magneton, *880*, 888, A–2
Bohr, Niels, 880
Bohr radius, A–2
Boltzmann's constant, 690
Boyle, Robert, 1039
Bragg planes, *1064*
Bragg spacings, *1064*
Bragg, W. L., 1064
Bragg's law (Bragg condition), *1064*, 1068
Brewster's angle, *964*, 966
British thermal unit (BTU), A–1
Broadcasting antennas, 957

BTU (British thermal unit), A–1
Bubble chamber, *798*
Bulk magnetic behavior of materials, 876–77
Bundles of light rays, *979*, 999, 1023

Calculable mutual inductance, example of, 897
Calorie (unit) (cal), A–1
Camera, 1019
Capacitance, 714–23
  calculating, 716–18
Capacitive circuit, 921–22
Capacitive reactance, 921
Capacitors:
  batteries vs., 719
  charged, 718, 766
  discharge, 716
  electrolytic, 728
  energy in, 718–20
  equivalent, 621
  importance of, 715
  with large capacitance, construction of, 729
  multilayer ceramic, 728
  in parallel and in series circuits, 721–23
  parallel-plate, 696, 731
  voltage across, 726
Car battery, 932
Carbon film resistors, 757
Cartesian coordinates, determining the field in using, 695
Cathode rays, 1063
Cavendish, Henry, 610, 618, 675, 676
Center of curvature, *1001*
Centimeter (cm), A–1
Charge by induction, 613–14
Charge carriers, *740*
Charge density, 622
Charge distributions, electric potential of, 687–88, 696–700
Charge polarization, *613*
Charge quantization, 617
Charged by friction, *611*
Charged capacitors, 718, 766
Charged particles, 802
Charge-to-mass ratio of the electron, 800–801
Charging batteries, 767
Chemical energy, 719
Chromatic aberrations, *1023*
Circuit analysis, 767–68
Circuits, 767
Circular polarization, 972
Cladding, *985*
CMOS (Complementary Metal Oxide Semiconductor), 905
Coherent waves, 1030
Color-coded resistors, 745
Compensator, 1042
Complex analysis, 923, 925

Compound microscope, 1018
Compton, Arthur, 964–65
Concave mirror, 1001–4
  image of an extended object, 1002–3
  location of the focal point, 1001–2
Conducting surfaces, role of sharp points on, 702–3
Conduction band, 754
Conduction electrons, 752
Conductivity, *747*, *See also* Superconductivity
  and materials, 752–55
  and resistivity, 746–48
  values of, 747
Conductors, 611, 742
  and electric fields, 672–74
  electrons of, 672–73
  electrostatic fields near, 674
  field cancellation within, mechanism for, 673
  potentials and fields near, 700–703
Conservation of charge, 610, *616*, 619, 739
Conservation of energy, 769
Constant charge densities, 643–44
Constant current, 744
Constructive interference, 1031, 1032–33, 1044
Continuous distribution of charge, 622–23, 642–43
  linear charge density, 622–23
  surface charge density, 623
  volume charge density, 623
Converging lenses, *1016*
Conversion of units, A–1
Convex mirror, 1001, 1004–6
  location of the focal point, 1004–5
Convex, mirror, image of an extended object, 1005–6
Cooper, Leon, 755
Corner reflectors, *980*
Corona discharge, *703*
Cosmic rays, *703*, 803
Coulomb (C), 614, A–1
Coulomb, Charles, 610, 618, 676
Coulomb forces, 619
  between home appliances, 620
Coulomb's law, 609, *610*, 617–20, 627, 629, 653, 661, 832, 945
  correctness of, 675–77
  held over small and large distances, 676–77
Coupled differential equations, 947
Coupled fields, 945
Critical damping, 907
Critical fields, 886
Critical temperature, *752*, 755
Crossed fields, *800*
Crystalline lens, *1018*
Curie, Pierre, 882, 885
Curie temperature, 882

I–1

Curie's constant, 885
Curie's law, *731*, 885–86
Current conservation, 772
Current density, 741–42
   of moving charges, 742
   relation between field and, 747
Current(s), 738–65
   average, 739
   and the conservation of charge, 744–45
   current density, 741–42
   defined, *739*
   density, 744
   direction of, 740–41
   free, 875–76
   induced, 848
   instantaneous, *739*
   magnetic forces on, 804–7
   in materials, 742–45
   real, 875–76
   resistance, 745–52
   superconductors, 755
   values of, 740
   in wires, measuring, 824
Curvature, 1009
   center of, *1001*
   radius of, 799
Cyclotron, *798*
Cyclotron frequency, 798

Davy, Humphry, 847
Day (d), A–1
Degree, A–1
Density:
   charge, 622
   current, 741–42
   energy, 720, 900
   linear charge, 622–23
   momentum, 956
   number, *742*
   surface charge, 623
   turn, solenoids, 830
   volume charge, 623
Derivatives, A–1, A–6
Derived units, A–1
Destructive interference, 1032–33, 1039, 1042, 1044
Deuteron, 799
Diamagnetic materials, *875*, 877, 888
Diamagnetism, 877, 884
Dielectric breakdown, *703*, 726
Dielectric constant, *724*, 725, 732
Dielectric strength, 726, 732
Dielectrics, *723*–31
   consequences of the microscopic model of, 731
   defined, 723
   experimental evidence for the behavior of, 725–27
   Gauss' law and, 731
   microscopic description of, 729–31
   properties of materials, 724
Diffraction, 978, *1050*–72
   Fraunhofer diffraction, 1051
   Fresnel diffraction, *1051*
   holography, 1065–67
   resolution of optical instruments, 1059–62
   single-slit diffraction, 1056–59
   slit width and grating patterns, 1062–63
   X-ray diffraction, 1062–65
Diffraction gratings, *1052–56*, 1067
   angular dispersion, 1054–55
   energy conservation and intensity, 1053–54
   intensity pattern, 1054
   resolution of, 1054–56
   significance of, 1052
Diffraction of light, 1050–52
Diodes, *746*, 932
Dipole antenna, 957–59
Dipole radiation, *957–59*, *966*
   angular pattern of, 959
Direct current (DC), *739*
Direct-current (DC) circuits, 766–90
   circuits, 767
   internal resistance, 768–69
   Kirchhoff's junction rule, 772–76
   Kirchhoff's loop rule, 770–72
   measuring instruments, 776–79
   RC circuits, 779–82
Discharging batteries, 767
Dispersion, 988–91
   atomic theory of, 990–91
   rainbows and the blue sky, 988–90
Dispersive medium, 953
Displacement current, *839*, 946
Distortion, *1022*
Divergence of the vector field, 679
Diverging lenses, 1016
Domain walls, *882*
Domains, *888*
Donor electrons, 754
Doping, 754
Double-slit experiment, 1030–34, 1050
   intensity in, 1034–37
Doublet, *1055*
Drift velocity, *742*
Driven harmonic oscillator, *926*
Driven RLC circuits, resonance in, 929
Driven spring motion, analogy between driven RLC circuits and, 926
Driving frequency, *926*
Drude model, 750
Drude, Paul, 750, 751
Dufay, Charles, 610

Earnshaw's theorem, 627
Earth's magnetic field, 832
Eddy currents, 857–58
   and Joule heating, 857
Einstein, Albert, 791, 865, 943, 964
Electric charge(s), 609–32
   conservation of, 616–17
   continuous distributions of, 622–23
   Coulomb's law, 617–20
   electroscope, *615*
   forces involving multiple charges, 621–26
   by induction, 613–14
   and matter, 611
   motion of, in electric fields, 648–49
   as a property of matter, 609–17
   quantization of, 617
   significance of electric forces, 610
   spherically symmetric charge distribution, 626
   superposition, 637–38
   of two types, evidence of, 612–13
   units of, 614
Electric current, 706, *739*, *See also* Current
Electric dipole moments, 638, 808
   induced, 639
   permanent, 639
Electric dipoles, *638*
   and electric fields, 638
   energy of, in electric fields, 652–53
   in external electric fields, 650–52
Electric energy, 610, 755
Electric field lines, 639–43
   constant charge densities, 643–44
   continuous distributions of charge, 642–43
   defined, 639
   density in space of, 639
   drawing, 641
   from equipotentials, 692–93
   examples, 641
   properties of, 639–40
Electric fields, *633*–39
   and conductors, 672–74
   defined, 633
   determining from potentials, 694
   electric dipoles and, 638
   energy in, 720
   external:
      electric dipole in, 650–52
      energy of a dipole in, 652–53
   Gauss' law for, 944
   mixing of magnetic fields and, 801–2
   motion of a charge in, 648–49
   moving charged particles, deflection of, 649–50
   of a point charge, 635
   between two uniformly charged planes with opposite charge, 647–48
   usefulness of the field concept, 636
   using Gauss' law to determine, 668–72
   values of, 634
Electric flux, 662–64
Electric force, 610, 627
Electric motors, 864
   how they work, 809
Electric potential, 683–713
   of charge distributions, 687–88, 696–700
   defining due to a charge distribution, 686
   of a point charge, 686–87
   potentials and fields near conductors, 700–703
   and quantum engineering, 706–7
   in technology, 703–7
      field-ion microscope, 704–5
      quantum engineering, 706–7
      Van de Graaff accelerator, 703–4
      xerography, *705*
   units of, 688
Electric potential difference, 686–87
Electric potential energy, 684–86
   of a system of charges, 688–89
Electric power, 755–57, *769*
   and batteries, 769–70
Electric quadrupole, 660, 713
Electrical appliances, energy consumption of, A–4
Electrical energy, 719
Electrical resistance, *See* Resistance
Electrically neutral atom, *611*
Electricity, history of the study of, 609–10
Electrolytic capacitors, *728*–29
Electromagnetic energy, radiation of, 991
Electromagnetic radiation, as particles, 964–65
Electromagnetic spectrum, *951–52*
Electromagnetic waves, 943, *945–53*
   dipole radiation, 957–59
      angular pattern of, 959
   energy of, 954
   in-phase electric and magnetic fields, 951
   momentum in, 956–57
   propagation in matter, 952
   propagation of, 946–49
   radiation, 951–52
      intensity of, 958–59
   radiation pressure, 956
   relation between E and B in, 949–51
   relations between amplitudes, 949–50
   resistance to Maxwell's notion of, 951
   transport of energy, 954–56
   transversality of, 950
Electromagnetism, 661, 948
Electrometer, 628
Electromotive force (emf), *766*–70, 894
   AC sources of, in circuits, 917
   defined, 767–68
   induced, 848
   by mutual inductance, 894–95
   by self-inductance, 894
   sources of, 766
Electron capture, *616*
Electron charge-to-mass ratio, A–2
Electron mass, A–2
Electron-volt (eV), 690, A–1
Electroscope, *615*
Electrostatic fields near conductors, 674
Electrostatic unit, 629
Electrostatics, 611
Elementary charge, A–2
Elementary particles, 616
Emergency backup systems, and capacitors, 715
emf, *See* Electromotive force (emf)
Energy, A–2, *See also* Kinetic energy; Potential energy
   chemical, 719
   electric, 610, 719, 755
   electric potential, 684–89
   electromagnetic, radiation of, 991
   of electromagnetic waves, 954
   in inductors, 898–900
   in LC and RLC circuits, 908–9
   in magnetic fields, 900–901
   in motional emf, 858–60
   potential, 683, 688–89, 810
   in reflection and refraction, 982–83
   transport of, 954–56
Energy conservation, *See* Conservation of energy
Energy consumption, A–4

Energy content of fuels, A–4
Energy density, *720, 900*
Energy flux, 954, 958
Equilibrium, 767
Equipotential surfaces, 691–92
Equipotentials, *691–93*
   compared to contour lines on a topographic map, 692
   determining electric fields from, 694
   determining the field in Cartesian coordinates, 695
Equivalent capacitor, 721
Erg, A–1
Euler's constant, A–6
Excited electrons, 753
Experimental Researches in Electricity (Faraday), 848
Extended object, image of, 999
External electric fields:
   electric dipole in, 650–52
   energy of, 652–53
Eye, 1018–19
   accommodation, 1018
   basic structure of, 1018
   far-sightedness, 1018–19
   near point, 1018–19
   near-sightedness, 1019
Eyeglasses, antireflective coatings, 1042
Eyepiece, 1018

Fabry, Charles, 1043
Fabry-Perot interferometer, 1043–44
Farad (F), 716, A–1
Faraday cages, *676*
Faraday ice-pail experiment, *675*
Faraday, Michael, 610, 633, 639, 675, 716, 724, 791, 847, 848
Faraday's law, 847–72, 888, 893–94, 943, 944, 965
   discovery of, 847
   eddy currents, 857–58
   frame dependence of fields, 864–65
   generators, 863–64
   magnetic induction, 849–50
   motional emf, 855–56
   surface formed by the loop, 852–54
   time-varying magnetic fields, 861–63
Faraday's ring, 848
Far-sightedness, 1018–19
Fermat, Pierre de, 986
Fermat's principle, 985–87
Fermi, A–1
Fermilab (Fermi National Accelerator Laboratory), 815
Fermilab proton accelerator, 802
Ferromagnetic materials, 792, *875, 877, 881*, 888
Ferromagnetism, 877, 881–84
Fiber optics, *984*
Field amplitudes, relation of, 950
Field cancellation within conductors, mechanism for, 673
Field-ion microscopy, *704–5*
Film, 1019
Filters, 932, 933–34
First-order maxima, 1033
Fizeau, Hippolyte, 975–76
Flatness, test for, 1040
Flip coil, 872

Focal length, 1002
Focal point, 1002, 1009
Focus, 1002
Foot (ft), A–1
Forces, A–1
   and Lenz's law, 860–61
   in motional emf, 858–60
Frame dependence of fields, 864–65
Franklin, Benjamin, 609, 610, 617, 689
Fraunhofer diffraction, 1051, 1053, 1062
Fraunhofer, Joseph von, 1051
Free charge, *617*, 730
Free currents, 875–76
Free electrons, 953
Free space, permittivity of, *618*, 944, A–2
Free-electron model, *750*
   failure of, 752
   of resistivity, 750–52
Frequency, angular, *See* Angular frequency
Fresnel, Augustin, 1051
Fresnel diffraction, *1051*
Friction, charged by, 611
Fringe fields, 716
Fringes, *1035*, 1040
Fuel resources, A–4
Full-wave rectifier, 932
Fundamental physical constants, A–2

Gabor, Dennis, 1065
Galileo, Gallilei, 975
Galvanometers, *776, 809*–10, 848
Gas(es):
   image, 705
   magnetic properties of, 874
Gas constant, A–2
Gauss (G), 795, A–1
Gauss, Karl Friedrich, 661
Gauss' law, 661–82, 832
   correctness of, 675–77
   and Coulomb's law, 666–67
   defined, *662, 665*
   electric flux, 662–64
   electrostatic fields near conductors, 674
   essence of, 662
   field cancellation within conductors, mechanism for, 673
   general statement of, 666
   purpose of, 661–65
   testing with a null experiment, 675
   using to determine electric fields, 668–72
   using to find the field of a finite line of charge, 670–72
Gauss' law for electric and magnetic fields, 943
Gauss' law for magnetism, *826*, 826–28
   field lines of a bar magnet, 826–27
   and magnetic flux, 826–28
Gauss' laws, for electric and magnetic fields, 965
Gaussian surface, *663*
Geiger-Muller tube, 733
Gell-Mann, Murray, 617
Generalized Ampère's law, 944, 946, 947, 965, 967

Generators, 863–64
Geometric optics, *979*, 998, 1022, 1023, 1029, 1033
Geometrical formulas, A–6
Germanium, 611
   energy gap, 754
Gilbert, William, 791
Gram (g), A–1
Gravitation, 609
Gravitational constant, A–2
Gravitational field, 680
Gravitational forces, and electric charge, 610
Gray, Stephen, 610
Grounded objects, 612
Gyromagnetic ratio, *879*, 888

Half-wave rectifier, 932
Hall, Edwin H., 811
Hall effect, *810*–11
   technological uses, 811
Heat flow, units of, A–1
Heisenberg, Werner, 881
Helix, 799
Henry (H), 895, 910, A–1
Henry, Joseph, 895
Hertz, Heinrich, 951–52
Hertz (Hz), A–1
High-pass filter circuit, 941
High-voltage lines, transmission of electric power along, 920
Holes, *754*
Hologram, 1065
Holography, 1050, 1065–67
   uses of, 1067
Homopolar generator, 871
Hooke, Robert, 974, 1039
Hubble Space Telescope (HST), 1061
Huygens, Christian, 974, 977
Huygens' principle, *978*–79, 985, 991
Hybrid semiconductors, 754
Hysteresis, *883*–84, 888
Hysteresis loop, *883*

I^2R loss, 756
Image, *999*–1001
   of an extended object, 999, 1002–3
   inverted, 1003
   reconstructed, 1065
   upright, 1003
   virtual, 1000
Image distance, 1009
Imaging gas, *705*
Impedance, 925, 928
Impedance matching, 935
In phase current and voltage, 921
Inch (in), A–1
Incident rays, 979
Incoherent waves, 1030
Index of refraction, 952–53, 976–77
Induced charge, 613
Induced current, *848*
   and presence of a magnetic field, 862
Induced electric dipole moments, *639*
Induced electric field, 730
Induced emf, 848
Inductance, 893–98
   effects of magnetic materials on, 897–98
   finding, 896
   LC circuits, 910
   oscillations in, 903–5
   measuring, 903, 910

RL circuits, analogy between *RC* circuits and, 901
RLC circuits, analogy between damped harmonic motion and, 906
Inductive circuit, 922–23
Inductive reactance, 922–23
Inductors, 893, *896*, 910
   energy in, 898–900
Infinity, source at, 1001
Instantaneous current, *739*
Insulators, *611*, 626, 723, 742, 753
Integrals, A–7
Intensity, in the double-slit experiment, 1034–37
Intensity of radiation, 955
Interference, 1029–49
   constructive interference, 1032–33
   destructive interference, 1032–33
   fringes, 1035
   from the space between two glass plates, 1037–38
   interferometers, 1042–43
   Newton's rings, 1039–40
   from reflection, 1037–42
   thin-film interference, 1040–42, 1044
   two-source interference pattern, 1031–33
   waves vs. particles, 1034
   Young's double-slit experiment, 1029–34
   intensity in, 1034–37
Interference hologram, 1067
Interferometers, *1040*, 1042–43
   Fabry-Perot interferometer, 1043, 1044
   Michelson interferometer, 1042, 1044
Internal magnetic moment, 880
Internal resistance, 768–69
Intrinsic elemental semiconductors, 754
Intrinsic magnetic moments, 880, 881
Inverse-square law, 676
Inverted image, 1003
Ions, 611
Ion source, *703*
Ionization, 953
Ionosphere, 953
Iris, *1018*

Joule (J), A–1
Joule heating, 756, 886, 919, 920
   and eddy currents, 857
Joules per coulomb (J/C), 688

Keck telescope complex (Hawaii), 1062
Kelvin (K), A–1
Kilocalorie (kcal), A–1
Kilogram (kg), A–1
Kilometer (km), A–1
Kilowatt-hour (kWh), A–1
Kinetic energy, calculating the rate of change of, 796
Kinetic theory, 731
Kirchhoff, Gustav, 770
Kirchhoff's junction rule, *772*–76, 783
   solving for the behavior of multi-loop circuits, 773–76
Kirchhoff's loop rule, *770*–72, 783, 896, 905

I–3

Larmor precession, *887*
Law of reflection, 979, 992, 998, 1009
LC circuits, *903*, 910
  energy in, 908–9
  oscillations in, 903–5
Lenses, *1014*, *See also* Optical instruments
  converging, 1016
  diverging, 1016
  objective, 1018
  sign conventions for, 1009
Lens-maker's equation, 1015
Length, units of, A–1
Lenz, Heinrich Emil, 851
Lenz's law, *851*, 893, 898, 944, 945
  and the direction of induced current, 850–52
  and forces, 860–61
Levitated magnet, 860
Light, 974–97
  diffraction of, 1050–52
  dispersion, 988–91
  Fermat's principle, 985–87
  Huygens' principle, 978–79
  index of refraction, 976–77
  reflection, 979–81
  refraction, 981–82
  speed of, 943, 948, 966, 975–77
  total internal reflection, 983–85
Light switch, and current, 744
Light waves, treating as rays, 977–79
Lightning, *703*
Lightning rods, 703
Light-year (ly), A–1
Linear charge density, 622–23
Linked circuits, 894
Liquids, magnetic properties of, 874
Liter (L), A–1
Lloyd's mirror, 1049
Load resistance, *767*
Local charge conservation, 616
Local effect of collisions, 746
Lorentz force, 864
Lorentz force law, 796, 800, 802, 812, 855, 861
Lorentz, Hendrik A, 751, 796
Low-pass filter circuit, 941

Macroscopic quantities, connection between microscopic quantities and, 881
Magnetic bulk properties, 876–77
Magnetic charges, 792, 826
Magnetic dipole, 808
Magnetic dipole field, 836
Magnetic dipole moment, 808
Magnetic dipole moment of atoms, 879–80
  alignment of, bulk effects due to, 880
Magnetic dipole moment per unit volume, 874
Magnetic dipoles, 836–38
Magnetic domains, *882*
Magnetic field lines, 826
  properties of, 852
Magnetic fields, 781–819, 888, 945, A–1
  Ampère's law, 820–21, 823–24
  Biot-Savart law, 832–36
  charge-to-mass ratio of the electron, 800–801
  common, 795

constant, circular motion in, 796–97
energy and torque on loops, 809–10
energy in, 900–901
Gauss' law for, 944
Hall effect, 810–11
Lorentz force law, 796
magnetic force on an electric charge, 793–96
Maxwell displacement current, 838–40
mixing of electric fields and, 801–2
in outer space, 802–3
production and properties of, 820–46
solenoids, 828–32
with solenoids, 830
static, energy of a charged particle in, 796
of a straight wire, 821–23
using Gauss' law to find, 827–28
velocity selectors, 800
Magnetic flux, *826*
  and Gauss' law for magnetism, 826–27
Magnetic force law, *794*, 796
Magnetic forces, *792*
  on current loops, 807–10
  on currents, 804–7
  on finite wires with currents, 805–7
  on infinitesimal wires with currents, 804–5
  marked directional character of, 792
  vectors oriented perpendicular to the page, 796
Magnetic induction, 848, 854–55
Magnetic intensity, *875*
Magnetic levitation, 860
Magnetic materials, effects on inductance, 897–98
Magnetic monopoles, 792, 944
Magnetic properties:
  of bulk matter, 874
  of material, 873
Magnetic resonance imaging (MRI), *888*
Magnetic susceptibility, *875*, 888
Magnetically hard materials, 884
Magnetically levitated trains, 860
Magnetically soft materials, *884*
Magnetism, 791
  atoms as magnets, 878–81
  diamagnetism, 884
  ferromagnetism, 881–84
  Gauss' law for, 826–28
  history of the study of, 609–10
  hysteresis, 883–84
  and matter, 873–92
  nuclear magnetic resonance, 886–88
  paramagnetism, 884–86
  and superconductivity, 886
Magnetization, *874*
  direction of, 875
Magnetization curve, *883*
Magnetohydrodynamic (MHD) generator, 866
Magnification, 1007–9
  thin lenses, 1016–18
Malus's law, 961, 966
Mass, A–1
Materials, and conductivity, 752–55

Mathematics, A–5 to A–7
  algebraic functions, properties of, A–6
  binomial theorem, A–5
  constants, A–5
  derivatives, A–6
  expansions, A–7
  geometrical formulas, A–6
  integrals, A–7
  notation, A–7
  quadratic equations, A–5
  Taylor expansion, A–7
  trigonometry, A–5 to A–6
Matter, and electric charge, 611
Maxwell displacement current, 838–40
Maxwell, James Clerk, 610, 676, 791, 838–39, 943, 951
Maxwell's equations, *943*, 944–45, 948, 975, 982, 1038
  getting in differential form, 967
Mean free path, *752*
Measuring instruments, 776–79
  ammeters, *776*–77, 783
  analog measuring devices, constructing, 776–79
  voltmeters, *776*, 777, 783
Meissner effect, 886
Meter (m), A–1
Michelson, Albert, 1042
Michelson interferometer, 1042, 1043, 1044
Microamps (μA), 739
Microscopes, 998
  compound, 1018
  resolution of, 1059
Mile (mi), A–1
Mile per hour (mi/h), A–1
Milliamps (mA), 739
Millikan, Robert, 617, 648, 659
Minimum observable angular separation, 1059–60
Minimum visible object separation, 1061
Minute (min), A–1
Mirrors, 998–1028
  aberrations, 1022–23
  image of an extended object, 1011
  images and, 998–1001
  relation between source distance and image distance, 1011–12
  sign conventions for, 1009
  sign of the object distance, 1012
  single refracting surface, focal point of, 1010
  spherical mirrors, 1001–9
Missing orders, *1062*
Modern telescopes, construction of, 1022
Mole (mol), A–1
Momentum density, *956*
Monochromatic aberrations, *1022*
Monopoles, *826*
Moon, corner reflectors on, 980
Mosaic telescopes, 1022
Motional emf, 855–56, 861
  forces and energy in, 858–60
Mount Palomar telescope, 1022
Moving charged particles, deflection of, 649–50
Muller, K. Alex, 755
Multilayer ceramic capacitors, *728*
Multi-loop circuits, 783
  solving for the behavior of:

Multimeters, 776
Mutual inductance, 894–95, *See also* Inductance

Nanoamps (nA), 739
Near point, 1018–19
Near-sightedness, 1019
Neutrino, *616*
Neutron mass, A–2
Neutrons, *803*, 886–87
New Technology Telescope, 1061
Newton, Isaac, 791, 974, 1039
Newton's rings, 1039–40, 1044
Newton's second law, 648, 797
Newtons (N), A–1
Nonohmic materials, 746
Nonpolar molecules, *730*
North pole, 792
Northern lights, Earth's magnetic field and, 803
n-type semiconductors, 754
Nuclear magnetic resonance (NMR), 886–89
Null experiments, *675*
Number density, *742*

Object, 999
Objective lens, 1018
  refracting telescope, 1021
Ocular, 1018
Oersted, Hans Christian, 610, 791, 807, 820–21
Ohm, *745*, A–1
Ohmic heating, 756
Ohmic material, 745
Ohmmeters, 776, 783
Ohm's law, *745*, 746, 758
Oil drop experiment, 659
Onnes, H. Kammerlingh, 755
Optic nerve, 1018
Optical devices, antireflective coatings, 1042
Optical flat, *1040*
Optical instruments, 1018–22, 1024, *See also* Eye; Lenses
  angular magnification, 1019–20
  camera, 1019
  reflecting telescopes, *1021*
  resolution of, 1059–62
  simple magnifier, 1020–21
  telescope, 1021–22
Optical interferometers, *1042*–44
Optical tweezers, *955*–56
Optics:
  adaptive, 1022
  fiber, 984
  geometric, 979, 998, 1022, 1023, 1029, 1033
  physical, 1029
  ray, *See* Geometric optics
Orbital magnetic dipole moment, 879
Order, 1033
Orthogonality of fields, 950
Outward radial direction, 695
Overdamping, 907
Ozone, 952

Parallel connection, 722
Parallel-plate capacitors, *696*, 731
Paramagnetic materials, *875*, 877, 888–89
Paramagnetism, 877, 884–86
Paraxial rays, 1001, 1009
Parsec (pc), A–1

Pascal (Pa), A–1
Path-length difference, 1032
Pauli exclusion principle, *753*
Pauli, Wolfgang, 753
Peaked amplitudes, 929
Periodic table of elements, A–8
Permanent electric dipole moments, *639*
Permeability of a material, 875
Permeability of free space, 823, A–2
Permittivity of free space, *618*, 944, A–2
Perot, Alfred, 1043
Phasors, 923–24
Photons, *964–65*
 reflected by a mirror, 965
Photoreproduction, 705
Physical optics, *1029*
Physical quantities, A–3 to A–4
 SI units, A–1
Physics:
 quantum, 752, 886
 significant dates in development of, A–9 to A–10
Piezoelectricity, 731
Pioneer 10 spacecraft, 677
Planck, Max, 964
Planck's constant, 887, 965, A–2
Plane electromagnetic wave, 946
Plane mirrors, peculiarity of, 1000
Plane waves, *947*, 977
Planetary data, A–3
Planets, A–3
Plasma, 953
Plasma frequency, 953
Plexiglas, 725
Poisson spot, 1051–52
Polar molecules, *729*
Polar polymer, 789
Polarization, 959–64
 analyzer for, 960–61
 charge, 613
 circular, *972*
 linearly polarized light, 960
 Malus's law, 961
 polarizers, 961
 by reflection, 963–64
 by scattering, 962–63
 unpolarized radiation, 961
Polarized radiation, how to produce, 962–64
Polarizers, 961, 966
Polaroid, 961–63
Polycarbonate, 789
Positrons, *616*
Potential energy:
 advantages of using, 683
 of a magnetic dipole, 810
 of a system of charges, 688–89
Potentiometer circuit, 790
Pound (lb), A–1
Power, A–1
 in AC circuits, 929–31
 angular distribution of, 959
Power factor, 931
Poynting vector, 954–55, 958, 966
 time average of, 1034
Precessional motion, 887, 889
Pressure, A–1
Priestley, Joseph, 610, 617
Primary coil, *918*
Primary rainbow, 989
Principal maxima, *1053*
Principal rays, *1004*

Principle of superposition, 621, 627
Proton magnetometer, 891
Proton mass, A–2
Proton-electron mass ratio, A–2
Protons, 886–87
p-type semiconductors, 755
Pupil, eye, *1018*

Quadratic equations, solution of, A–5
Quality factor (Q-factor), 930
Quantum engineering, 707
 and electric potential, 706–7
Quantum mechanics, 747
 orbital angular momentum, 880
Quantum physics, *752*
 and magnetic flux, 886
Quantum-mechanical exclusion principle, 753, 881
Quarks, *617*, 627
Quarter-wave thickness, 1042

Radians, A–1
Radiation, 951–52
Radiation pressure, *956*
Radio waves, 953
"Radius of curvature," 799
Rainbows, 988–90
 secondary rainbow, 989–90
Ray optics, *See* Geometric optics:
Ray tracing, 1005, 1017, 1023
Rayleigh criterion, *1060*
Rayleigh, Lord, 989, 991
Ray-tracing techniques, 1001, 1008, 1009
RC circuits, 779–82
 energy in, 782
Real currents, 875–76
Receiving antennas, 957
Reconstructed image, 1065
Rectification, 932
Rectifiers, 932
Reference beam, *1065*
Reflected rays, 979
Reflecting telescopes, *1021*
Reflection, 979–81, 991
 corner reflectors, *980*
 energy in, 982–83
 interference from, 1037–42
Reflection grating, *1052*
Refracted rays, 981
Refracting surfaces, sign conventions for, 1009
Refracting telescope, 1021
Refraction, 981–82, 991, 1023
 energy in, 982–83
 index of, 976–77
 and optical effects, 982
 spherical mirrors, 1009–14
Resistance, *745*
 internal, 768–69
 in series and in parallel, 749–50
Resistive circuit, 921
Resistivity:
 and conductivity, 746–48
 free-electron model of, 750–52
 temperature dependence of, 748
Resistors, 745–46
 carbon film, 757
 color-coded, 745
 defined, 745
 in parallel, 740–50
 in series, 749
 shunt, 788
 wire-wound, 757

Resolution:
 of diffraction gratings, 1054–56
 of microscopes, 1059
 of optical instruments, 1059–62
 of telescopes, 1059, 1061–62
Resolving power, of diffraction grating, 1055
Resonance, 887, 926
 in driven *RLC* circuits, 929
Resonance peak, 929
Resonant behavior, 926
Resources, annual usage of, A–4
Retina, *1018*
rf (radio-frequency) range, 891
RL circuits:
 analogy between *RC* circuits and, 901
 time dependence in, 901–3
RLC circuits, 905
 analogy between damped harmonic motion and, 906
 damped oscillations in, 905–8
 energy in, 908–9
rms (root mean square), 930
Roemer, Ole, 975
Roentgen, Wilhelm, 1063
Rotating coil, 868
Rowland ring, *883*
Rydberg constant, A–2

Saturation phenomenon, 886
Savart, Félix, 833
Scalar field, 636
Scanning tunneling microscope, 706
Schrieffer, Robert, 755
Second (s), A–1
Second law of thermodynamics, 755
Self-inductance, 894, *See also* Inductance
Semiconductors, *611*, 742, 754
Series connection, 722–23
Series *RLC* circuits:
 AC in, 926–29
 impedance, 928
 resonance in driven *RLC* circuits, 929
Sheet of current, 946
Shielded rooms, *See* Faraday cages
Shunt resistor, 788
Shutter, 1019
SI (Système International), A–1
 base units, A–1
 conversion to non-SI units, A–1
 derived units, A–1
Silicon, energy gap, 754
Simple magnifier, 1020–21
Single refracting surface, focal point of, 1010
Single-loop circuit, 770
Single-slit diffraction, 1056–59
 deriving single-slit intensity value, 1059
 intensity pattern of, 1058
Smoke detectors, dependence on charges, 616
Snell, Willebrord, 981
Snell's law, 964, 981, 985–87, 992, 998, 1009, 1010, 1023
Solar energy output, A–4
Solar sail, 968
Solar wind, *803*
Solenoids, 828–32
 magnetic field with, 830
 toroidal solenoid, 831–32
 turn density of, 830

 using Ampère's law to find the magnetic field in, 829–31
Solids, magnetic properties of, 874
Sound synthesizers, 905
Source, 999
Source distance, 1009
Source-image-focal-length relation, 1006
Sources of electromotive force, *766*
South pole, 792
Spark coil, *920*
Spectral lines, *1052*
Spectrographs, *1052*
Spectroscopes, *1052*
Spectrum, *1052*
Speed, units of, A–1
Speed of light, 943, 948, 966, 975–77
Speed of light in a vacuum, A–2
Spherical mirrors, 1001–9
 concave mirror, 1001–4
 convex mirror, 1004–6
 magnification, 1007–9
 refraction, 1009–14
 relation between source distance and image distance, 1006
Spherically symmetric charge distribution, 626
Spin, 887
Spiral, 799
Split-ring commutator, *809*
St. Elmo's fire, *703*
Static magnetic fields, 796
Steady-state behavior, 767
Step-down transformers, 919
Step-up transformers, 919
Stoke's law, 659
Sun, energy output of, A–4
Superconductivity, 755
 and magnetism, 886
Superconductors, *611*, 742, 752, 755, 886, 889
Superposition, 637–38
Superposition principle, 621, 627
Surface charge density, 623
Symmetry, in drawing electric field lines, 641
Système International (SI), *See* SI (Système International)

TAT-8, 985
Taylor expansion, A–7
Technology:
 electric potential in, 703–7
 field-ion microscope, 704–5
 quantum engineering, 706–7
 Van de Graaff accelerator, 703–4
 xerography, 705
Teflon, 747
Telescopes, 1021–22
 modern telescopes, construction of, 1022
 mosaic telescopes, 1022
 resolution of, 1059, 1061–62
Temperature coefficient of resistivity, 748
Terminal voltage, 767
Tesla, Nikola, 795
Tesla (T), 794–95, 812, A–1
Test charge, 633
Theory of special relativity, 791, 865, 943
Thermal conductivity, SI unit, A–1
Thermodynamics, second law of, 755

I–5

Thin lenses, 1014–18, 1024
  lens-maker's equation, 1015
  magnification, 1016–18
Thin-film interference, 1040–42, 1044
Thomson, J. J., 648, 800–801
Thunder, 703
Time, units of, A–1
Time constant, 780, 901–2
Time rate of change, 850
Time-varying magnetic fields, 861–63
Tokamak, 820
Toroidal solenoid, 831–32
Torque, units of, A–1
Torque on a current loop, 808
Torus, *831*
Total energy flow per unit time, 958
Total internal reflection, *983*–85
Transformers, 895, 917–20
  ratio of currents in, 919
  ratio of voltages in, 919
  step-down transformers, 919
  step-up transformers, 919
Transistors, 755, 931
Transmission grating, 1052
Transmission of light, *981*
Transport of energy, 954–56
Transportation, energy consumption in, A–4
Transverse separation, 1062
Turn density, of solenoids, 830
Two-source interference pattern, 1031–33
Type I superconductors, 886
Type II superconductors, 886

Uniform charge distribution, 643
Units, A–1, *See also* SI (Système International)
  conversions, A–1
Universal gas constant, A–2
Unpolarized light, 959
Unpolarized radiation, 961
Upright image, 1003

Values of electric fields, 634
Van de Graaff accelerators (Van de Graaff generators), *703*–4
Van de Graaff, Robert, 703
Vector sum, 621
Velocity selectors, 800
Virtual image, 1000
Visible light, 948–49
Vitreous humor, *1018*
Volt (V), 688, A–1
Volta, Alessandro, 688
Voltage, *767*
Voltage amplitudes, 917, 919
Voltmeters, 776, *777*, 783
  analog voltmeter, 777
Volume, A–1
Volume charge density, 623
von Laue, Max, 1063–64

Water, as a polar dielectric, 731
Watt (W), A–1
Wave aspects of light, 975

Weakly bound electrons, 611
Weber (Wb), 826, A–1
Weber, Wilhelm Eduard, 826
Wheatstone bridge, 788
Wire-wound resistors, 757

X rays, 1063
Xerography, *705*
X-ray diffraction, 1062–65
X-ray spectrometer, 1064

Year (yr), A–1
Young, Thomas, 975, 1030, 1044
Young's double-slit experiment, 1029–34, 1050
  intensity in, 1034–37

Zener diode, 781
Zeroth-order maximum, 1033
Zoom lenses, 1008, 1023
Zweig, George, 617

TABLES IN THE TEXT

Table	Name of Table	Page
21–1	Mass and Charge of Atomic Constituents	614
22–1	Values of Electric Fields (N/C)	634
22–2	An Electric Dipole Rotating in a Uniform Electric Field	652
23–1	Experimental Measurements of Deviation from an Inverse-Square Force Law: Force $\propto 1/r^{2\pm\delta}$	676
24–1	Electric Fields and Potentials for Various Charge Configurations	701
25–1	Dielectric Properties of Materials	724
26–1	Values of Various Currents	740
26–2	Resistivities, Conductivities, and Temperature Coefficients	747
28–1	Some Magnetic Fields	795
31–1	Some Magnetic Susceptibilities	875
31–2	Magnetic Bulk Properties and Their Relations	876
31–3	Magnetic Forces on Materials Near a Large Electromagnet	877
32–1	Analogy Between RC and RL Circuits	901
32–2	Analogy Between LC Circuit and Mass on a Spring	904
32–3	Analogy Between RLC Circuits and Damped Harmonic Motion	906
33–1	Analogy Between Driven RLC Circuits and Driven Spring Motion	926
35–1	Indices of Refraction for Various Substances ($\lambda = 600$ nm)	976
35–2	Index of Refraction of Glass as a Function of Wavelength	988
36–1	Sign Conventions for Mirrors, Refracting Surfaces, and Lenses	1009

PROBLEM-SOLVING TECHNIQUES BOXES

Chapter	Subject	Page
21	Calculating electric forces	621
23	Using Gauss' Law	668
27	Multi-loop circuits	774
29	Static magnetic fields	835
36	Ray tracing	1004

SOME MATHEMATICAL CONSTANTS[†]

Constant	Value
π	3.14159
e (the "exponential")	2.71828
$\sqrt{2}$	1.41421
$1/\sqrt{2}$	0.707107
$\ln(10)$	2.30259
$\ln(2)$	0.693147
1 rad	57.2958°
1°	0.0174533 rad

[†]To six significant figures.

THE GREEK ALPHABET

Alpha	A	α	Nu	N	ν
Beta	B	β	Xi	Ξ	ξ
Gamma	Γ	γ	Omicron	O	o
Delta	Δ	δ	Pi	Π	π
Epsilon	E	ϵ	Rho	P	ρ
Zeta	Z	ζ	Sigma	Σ	σ
Eta	H	η	Tau	T	τ
Theta	Θ	θ	Upsilon	Y	υ
Iota	I	ι	Phi	Φ	ϕ
Kappa	K	κ	Chi	X	χ
Lambda	Λ	λ	Psi	Ψ	ψ
Mu	M	μ	Omega	Ω	ω

SOME USEFUL CONSTANTS

Constant	Value
Acceleration of gravity	$g = 9.81$ m/s^2
Density of air (STP)	1.29 kg/m^3
Specific heat of air at constant pressure	1.01×10^3 J/kg·K
Specific heat of air at constant volume	0.72×10^3 J/kg·K
Density of water (STP)	10^3 kg/m^3
Density of ice (STP)	0.917 kg/m^3
Speed of sound in air (STP)	331 m/s
Average range of audible frequencies	20 Hz–16,000 Hz
Index of refraction of water	1.33
Range of wavelengths for visible light	380 nm–750 nm
Resistivities of typical conductors	10^{-8}–10^{-6} Ω·m
Resistivities of typical insulators	10^9–10^{14} Ω·m